10th International Scientific Symposium on Electric Power Engineering (ELEKTROENERGETIKA 2019)

Stara Lesna, Slovakia
16 – 18 September 2019

Editors:

Michal Kolcun
Iraida Kolcunova
Juraj Kurimsky

ISBN: 978-1-5108-8871-5

Printed from e-media with permission by:

Curran Associates, Inc.
57 Morehouse Lane
Red Hook, NY 12571

Some format issues inherent in the e-media version may also appear in this print version.

Printed by Curran Associates, Inc. (2019)

For permission requests, please contact
Technical University of Kosice – Department of Electrical Power Engineering
at the address below.

Technical University of Kosice – Department of Electrical Power Engineering
Letna 9
042 00 Kosice
Slovak Republic

Phone: 421 55 602 1111

http://www.tuke.sk

Additional copies of this publication are available from:

Curran Associates, Inc.
57 Morehouse Lane
Red Hook, NY 12571 USA
Phone: 845-758-0400
Fax: 845-758-2633
Email: curran@proceedings.com
Web: www.proceedings.com

TECHNICAL UNIVERSITY OF KOŠICE

Faculty of Electrical Engineering and Informatics

Department of Electrical Power Engineering

Proceedings of the 10th International Scientific Symposium on Electrical Power Engineering

ELEKTROENERGETIKA 2019

September 16–18, 2019, Stará Lesná, Slovak Republic

TECHNICAL UNIVERSITY
OF KOŠICE

Faculty of Electrical Engineering and Informatics
Department of Electrical Power Engineering

Proceedings of the 10th International Scientific Symposium on Electrical Power Engineering

ELEKTROENERGETIKA 2019

September 16–18, 2019, Stará Lesná, Slovak Republic

Name:	Proceedings of the 10th International Scientific Symposium on Electrical Power Engineering ELEKTROENERGETIKA 2019
Publisher:	Technical University of Košice Faculty of Electrical Engineering and Informatics Department of Electrical Power Engineering
Published:	September 2019
Editors:	Michal Kolcun, Iraida Kolcunová, Juraj Kurimský
Edition:	First
Pages:	586
Circulation:	250
CDROM:	Technical University of Košice, 2019
Copyright :	Technical University of Košice, Slovak Republic © 2019

Manuscripts of papers submitted to Symposium Proceedings are peer reviewed by two reviewers. The review results are compiled according following procedure: the paper is accepted or rejected according match of recommendations of both reviewers. If there is conflict of review results, the paper are reviewed by third reviewer and his result is conclusive.

The Authors are responsible for contentual and lingual accuracy of their papers and the materials presented.

Publisher address

Technical University of Košice
Letná 9
Košice
Slovak Republic

ELEKTROENERGETIKA 2019 – General Informations

Symposium Guarantee

Dr. h.c. prof. Ing. Michal Kolcun, Ph.D., Technical University of Košice, SK

Members of International Scientific Committee

- Dauren S. Akhmetbaev, L.N.EA. National University, Astana, KAZ
- Juraj Altus, Žilinská univerzita, SK
- Ivan Artyukhov, Yuri Gagarin State Technical University of Saratov, RUS
- Anton Beláň, STU Bratislava , SK
- Július Bems, ČVUT Praha, CZ
- Ľubomír Beňa, Technical University of Košice, SK
- István Berta, BME Budapest, HU
- Kristina Berzina, Riga Technical University, LV
- Peter Birkner, Mainova AG, DE
- Nadezhda S. Buryanina, North-Eastern Federal University, Yakutsk, RUS
- Roman Cimbala, Technical University of Košice, SK
- Ivo Doležel, AV ER Praha, CZ
- Nikolay Djagarov, TU Varna, BG
- Emil Dvorský, ZČU Plzeň, CZ
- Anna Gawlak, Czestochowa University of Technology, PL
- Sergej Gerasimov, St. Peterburg, RUS
- Georgi Georgiev, 3G, BG
- Janusz K. Grabara, Czestochowa University of Technology, PL
- Valerij G. Goldstein, TU Samara, RUS
- Radomír Goňo, VŠB TU Ostrava, CZ
- Rainer Haller, Regensburg, DE
- Zdeněk Hradílek, VŠB - TU Ostrava, CZ
- Yuriy Ya. Chukreyev, RAS, Syktyvkar, RUS
- František Janíček, STU Bratislava, SK
- Péter Kádár, Óbuda University, Budapest, HU
- Rumen Kirov, TU Varna, BG
- Nikolay Kinsht, RAS, Vladivostok, Russia, RUS
- Jaroslav Knápek, ČVUT Praha, CZ
- Iraida Kolcunová, Technical University of Košice, SK
- Sebastian Kot, Czestochowa University of Technology, PL
- Hristo Krachunov, TU Varna, BG
- Tomáš Králik, ČVUT Praha, CZ
- Dušan Krokavec, Technical University of Košice, SK
- Juraj Kurimský, Technical University of Košice, SK
- Jaroslav Lelák, STU Bratislava, SK
- Tatiana Lomane, TU Riga, LV
- Anatolij Mahnitko, TU Riga, LV
- Tatjana Malinina, St. Petersburg State Polytech.University, RUS
- Lubomir Marciniak, Czestochowa University of Technology, PL
- Sergey Margulis, Kazan State Power Engineering University, RUS
- Zbyněk Martinek, ZČU Plzeň, CZ
- Marián Mešter, Východoslovenská distribučná, a.s., SK
- Alexander Mészáros, Technical University of Košice, SK
- Ján Michalík, Žilinská univerzita, SK
- György Morva, University of Budapest, HU
- Hans Michael Muhr, TU - Graz, AT
- Zdeněk Müller, ČVUT Praha, CZ
- Vladimir Nagay, NPI, RUS
- Alexander Nazaricev, PEIPK, St. Petersburg, RUS
- Negnevitsky Michael, University of Tasmania, Australia, AU
- Vladimir Nepomnyashchy, PEIPK, St. Petersburg, RUS
- Karel Nohač, ZČU Plzeň, CZ
- Maria Nowicka-Skowron, Czestochowa University of Technology, PL
- Irina Oleinikova, TU Riga, LV
- Andrey V. Pazderin, Ural Federal University, RUS
- Daniela Perduková, Technical University of Košice, SK
- Anton Prokhorov, Tomsk Polytechnic University, RUS
- Stanislav Rusek, VŠB - TU Ostrava, CZ
- Pavel Santarius, VŠB - TU Ostrava, CZ
- Antanas Sauhatas, TU Riga, LV
- Rasid Sitdikov, TSTU Tashkent, UZB
- Karel Sokanský, VŠB - TU Ostrava, CZ
- Viktor Starovojtenkov, Saint Petersburg, RUS
- Tynys B. Suleimenov, L.N.EA. National University, Astana, KAZ
- Konstantin Suslov, Irkutsk NR Technical University, RUS
- Alexey Tadzhibaev, PEIPK, St. Petersburg, RUS
- Vasily V. Titkov, St. Petersburg State Polytech. University, RUS
- Josef Tlustý, ČVUT Praha, CZ
- Petr Toman, VUT Brno, CZ
- Toomas Vaimann, TU Tallin, EE
- Zdeněk Vostracký, ZČU Plzeň, CZ
- Irena Wasiak, Polytechnika Lodz, PL
- Markus Zdrallek, Bergische Universität Wuppertal, DE
- Diana Zalostiba, Riga Technical University, LV
- Inga Zicmane, Riga Technical University, LV

Symposium Steering Committee

- Iraida Kolcunová, Technical University of Košice, SK
- Juraj Kurimský, Technical University of Košice, SK
- Jaroslav Džmura, Technical University of Košice, SK
- Jana Varnavčinová, Technical University of Košice, SK
- Dagmar Kramolišová, Technical University of Košice, SK

Reviewers for Proceedings of the 10th International Scientific Symposium on Electrical Power Engineering ELEKTROENERGETIKA 2019

Many experts have given generously of their time and expertise to review manuscripts for Proceedings International Scientific Symposium on Electrical Power Engineering ELEKTROENERGETIKA during years. We are extremely grateful to each of these reviewers for their commitment to the Proceedings. — Editors.

- Artyukhov Ivan
- Beláň Anton
- Bems Július
- Beňa Ľubomír
- Berzina Kristina
- Bucko Samuel
- Buryanina Nadezhda
- Cimbala Roman
- Čonka Zsolt
- Dolnik Bystrik
- Dvorský Emil
- Dzmura Jaroslav
- Gawlak Anna
- Georgi Georgiev
- Goňo Radomir
- Gyurov Valentin
- Chukreev Yuri
- Kanalik Martin
- Kinsht Nikolay
- Király Jozef
- Knapek Jaroslav
- Kolcun Michal
- Kolcunova Iraida
- Kot Sebastian
- Kralikova Ruzena
- Krokavec Dušan
- Kurimský Juraj
- Lelák Jaroslav
- Liptai Pavol
- Lomane Tatjana
- Malinina Tatjana
- Martinek Zbyněk
- Medved Dusan
- Mészáros Alexander
- Mešter Marián
- Müller Zdeněk
- Nagay Vladimir
- Negnevitsky Michael
- Noháč Karel
- Pavlík Marek
- Pazderin Andrey
- Perduková Daniela
- Petras Jaroslav
- Rusek Stanislav
- Rusnak Jozef
- Skurčák Ľuboš
- Suslov Konstantin
- Titkov Vasilii
- Toman Petr
- Vaimann Toomas
- Zbojovsky Jan
- Zicmane Inga

Table of Contents

Invited Lectures 15

Michael Negnevitsky
High Renewable Energy Penetration and Power System Security: New Challenges and Opportunities 16

Karel Máslo
Dynamic Models in the Context of the EU Commission Regulations 26

A: Generation of Electricity, Transmission, Distribution and Consumption of Electricity 31

Hormádka Aleš, Jiřičková Jana, Martínek Zbyněk, Sirový Martin
Operation Assessment of Cogeneration Power Plant by Thermodynamic Model 32

Bulatov Yuri, Kryukov Andrey, Suslov Konstantin, Shamarova Natalia
Ensuring Postemergency Modes Stability in Power Supply Systems Equipped with Distributed Generation Plants 38

Pazderin Andrey A., Pazderin Andrey V., Morozenko Nikita, Chernykh Ilya
Electric Losses Flow Distribution Method for Power Systems (utilities) 43

Kolosok Irina, Korkina Elena, Tikhonov Alexander, Mahnitko Anatolijs
Use of PMU Data for HVDC Link Modeling in the Power System State Estimation 48

Georgiev Georgi, Negnevitsky Michael, Kim Dmitriy
Probabilistic Approximation of Voltage Values for Power Distribution Network with Large Number of Nodes and Solar Generation 53

Georgiev Georgi, Negnevitsky Michael, Kim Dmitriy
Probabilistic Approach to State Estimation in Electricity Networks with Distributed Generation 58

Knápek Jaroslav, Vašíček Jiří, Bemš Július, Vávrová Kamila
Effectiveness of Biogas Power Plant Transformation into Biomethane Production 62

Pavlov Ľuboš, Skurčák Ľuboš, Gbelec Ján, Bojda Peter
Evaluation of Unbalance in the Transmission System of the Slovak Republic 67

Ilenin Stanislav, Varga Ladislav
Measurement the Ground Resistance of Masts of Overhead Power Lines in HV and EHV 71

Medveď Dušan, Gdovin Dominik
Solution of Electromagnetic Field Distribution Around the 22 kV Poles Embedded with Bird-barrier 75

Medveď Dušan, Žvanda Peter
Electromagnetic Field Distribution Modeling and Measuring of the 110 kV Substation 81

Kanalik Martin, Margitová Anastázia, Urbanský Jakub
Calculation of Power Transformer Losses due to Harmonic Current Flow 87

Velicka Richard, Bilik Petr, Krejci Petr, Velicka Jan
Behavior of Electricity Meters and Quality Analyzers Using Equipment to Optimize Subscriptions in Small Photovoltaic Power Plants 92

Velicka Richard, Bilik Petr, Koval Ludvik, Velicka Jan
Testing Behavior of Electricity Meters when Measuring Non-Standard Electricity Consumption and Deliveries 97

Velicka Richard, Santarius Pavel, Velicka Jan, Bilik Petr
Verification of Electrical Properties of Photovoltaic Panels 102

Horák Martin
Prediction of Household Consumption Based on the Model 107

Tesarova Miloslava, Kaspirek Martin
Evaluation of Short Interruptions of Supply Voltage in Distribution System 111

Belik Milan
Emergency Island Grids with Small Hydro Power Stations 116

Oboskalov Vladislav P., Mahnitko Anatolijs, Varfolomejeva Renata, Gubin Pavel Y., Gusev Sergey A., Valiev Rustam T.
Adequacy Analysis in Bulk Power Systems Based on the Analytical-Probabilistic Method 122

Oboskalov Vladimir, Menaem Abdel Amir, Valiev Rustam T., Mahnitko Anatolijs, Varfolomejeva Renata
Mathematical Methods for Probabilistic Estimation of Power Shortage in Concentrated Electric Power Systems **128**

Artyukhov Ivan, Stepanov Sergey, Zemtsov Artem
Transient Processes in the Microgrid Based on a Group of Autonomous Operating Synchronous Generators **133**

Nohacova Lucie, Zak Frantisek, Hulec Martin, Bělík Milan, Olkhovskiy Mikhail
The Influence of Load on the Transformers on the Size of their Own Losses **139**

Frolov Vladimir, Ivanov Dmitrij, Sivaev Alexander
Simulation of Arc Extinction in Discharge Chambers of Lightning Protection Equipment **143**

Papkov Boriss, Mahnitko Anatolijs, Zicmane Inga, Berzina Kristina, Oboskalov Vladislav, Gusev Sergey, Tavlincev Alexander

Analysis of the Quality Problem of Electric Power and the Management of Reliability of Power Supply **147**

Sowinski Janusz
Modeling of Electricity Demand in the Region under Uncertainty **153**

Gawlak Anna
Cost-efficiency of Constructing Medium Voltage Cable Lines **159**

Gawlak Anna, Malinowski Andrzej
Renewable Sources of Energt and Hybrid Electrical Wiring in Individual Houses **163**

Gyurov Valentin, Simeonova Neli, Proikov Mladen, Sotirov Sotir, Kirov Rumen
Study on Interconnections Between Electromagnetic Compatibility and Techno-Economic Characteristics of Industrial Enterprises from Chemical and Mining Industry in Republic of Bulgaria **168**

Gyurov Valentin, Simeonova Neli, Simeonov Stanislav, Sotirov Sotir
Study on Electromagnetic Compatibility of Powerful Industrial Loads in Different Working Regimes **172**

B: High Voltage Technique and Diagnostics in Power Engineering 177

Kinsht Nikolay, Petrunko Natalia
Multiple Partial Discharges and Set of Their Sources **178**

Khrennikov Alexander Yu., Aleksandrov Nikolay M.
Diagnostics of High Voltage Electrical Equipment in Power Engineering **183**

Pavlík Marek, Zbojovský Ján, Oliinyk Maksym, Ivančák Michal
Compare of Shielding Effectiveness for Building Materials with Different Air Gap **189**

Petráš Jaroslav, Džmura Jaroslav
Pollution Ash Powder Electric Parameter Measurement **193**

Petráš Jaroslav, Džmura Jaroslav
New Methods of De-Noising Partial Discharge Signal by Morphological Filters **197**

Prochazka Radek, Sefl Ondrej
Simulation Model of Partial Discharges in Internal Void **202**

Zbojovský Ján, Kurimský Juraj
Towards to Thermal Stability of Equivalent Series Resistance and Capacity of Flexible Sensor **207**

Zbojovský Ján, Pavlík Marek, Kolcunova Iraida
Physical Characteristics Influence of Shielding Coating on the Electromagnetic Field Penetration **211**

Dolník Bystrík, Molčan Patrik, Hájnik Samuel
The Changes of Electrical Properties of ZnO-based Varistors Initiated with Current Pulses **215**

Kolychev Alexander V., Titkov Vasilyi V., Tukeev Pavel D., Lubkov Alexander N.
Several Approaches to Development of Dynamic Predictive Model Thermal Modes of Underground Power Transmission Cable Lines **221**

Vostracký Zdeněk, Sedláček Jan, Mužík Václav
Coupled Tests Including Parameters TF and SLF of High Voltage SF6 Circuit Breakers **225**

Shevchenko Sergey, Danylchenko Dmytro
Method of Determination of the Current-Voltage Characteristics of Surge Arresters in the Area of Leakage Current **229**

Vysogorets Svetlana, Tadjibaev Alexey, Gorec Ilia
New Express-method for Liquid Dielectric Quality Valuation 233

Vysogorets Svetlana
Development of a Predictive Model for Estimating the Liquid Dielectric Resource 238

Džmura Jaroslav, Bernát Milan, Rudolf Ladislav
Computer-Aided Design in Technological Process of Applying Textile Fibres and Flocks Using the Force Effect of Electric Field 242

Cimbala Roman, Hrabčák Ondrej
Dielectric Properties of Segmented Sensor 247

Kolcunova Iraida, Zbojovský Ján, Kurimský Juraj
Measurement of Corona Discharges Using UV Camera 251

C: Power Systems Relaying, Faults Analysis 257

Chusovitin Pavel, Balakh Eduard, Dekhtiar Sergei, Valiev Rustam, Tashchilin Valeriy
Off-grid Distribution System Logical Protection for Resilience Enhancement 258

Lomane Tatjana, Berzina Kristina, Rubcovs Sergejs, Zicmane Inga
Estimation of Earth Fault Distance Protection Algorithms Effectiveness for High Voltage Cable Line 264

Buryanina Nadezhda, Korolyuk Yuriy, Koryakina Maya, Lesnykh Elena, Suslov Konstantin, Shamarova Natalia
Using a New Algorithm of Current Protection for Substations 269

Noháč Karel, Noháčová Lucie, Mužík Václav
Actual Possibilities of Open-Source Applications for Analysis of Power Systems 274

Khrennikov Alexander Yu., Aleksandrov Nikolay M.
Approaches and Experience in Testing of Digital Substation Primary Equipment and Relay Protection. 280

Fedotov Aleksander, Vagapov Georgii, Grackova Larisa
Monitoring of Overhead Power Lines and Detection of Damages 284

Grishkevich Andrey
Usage of Empirical Statistical Distributuions in Simulation Models to Estimate Structural Reliability Indices of a Complex Systems 289

Máslo Karel, Krištof Vladimír
Simlified Criterium for Transient Stability Assessment 293

Sarry Sergey, Nagay Vladimir, Kireev Pavel, Nagay Ivan, Trotsenko Igor
Intellectual Control System of Switching Devices of High-voltage Substations 297

Nagay Ivan, Nagay Vladimir, Kireev Pavel, Nagay Vladimir Jr., Sarry Sergey
Designing a Backup System of Relay Protection for Multi-end Overhead Lines 302

Nagay Vladimir, Nagay Vladimir Jr., Nagay Ivan, Kireev Pavel, Sarry Sergey, Trotsenko Igor
Recognition of longitudinal-transverse Asymmetric Faults in Electrical Networks 306

Cialla Björn, Bachratá Barbora
System Testing of Protection Devices 310

Český Lukáš, Janíček František, Skudřík Filip, Zuščak Jozef
Design and Testing of Voltage Instrument Transformers for Railway Application 315

D: Power System Control, Liberalisation of Electricity Market 319

Mukhlynin Nikita D., Pazderin Andrey V.
Electric Utility Costs Control by Means of Increasing Net Generation and Reducing Losses in the Distribution Grid 320

Malinina Tatiana, Novikova Olga
Comparative Evaluation of the Functioning and Development of the Energy Markets of Russia and Germany 325

Chukreyev Iurii, Chukreyev Mikhail
Problems of Ensuring the Balance Reliability of Perspective Development Schemes UES of Russia 329

Chukreyev Iurii, Chukreyev Mikhail
Methodical and Information Aspects of The Task of Balance Reliability and Their Impact on Standards Power Reserve UES of Russia 333

Chukreyev Iurii, Chukreyev Mikhail
Evaluation of Various Programming Languages in Relation to Solving the Problem of Determining Indicators of Balance Reliability **338**

Máslo Karel, Kasembe Andrew
Frequency Stability of Continetal Europe Synchronnous Area **342**

Grega Michal, Schnierer Tomáš
The Flow-Based Capacity Calculation Approach and Comparison with Currently Used Method **347**

Fedor Pavol, Perdukova Daniela, Olexa Richard
Modeling a Small Hydropower Plant **352**

Čonka Zsolt, Hocko Pavol
Assessment of the Primary Load-Frequency Control Operation by Considering the Effect of Photovoltaic Power Plants **357**

Shamarova Natalia, Suslov Konstantin, Gerasimov Dmitriy, Shushpanov Ilia, Altuhov Igor, Lombardi Pio, Komarnicki Przemyslaw
Stabilizing the Control of a Plant Material Drying Process in Off-Grid Power Systems **363**

Dvorsky Emil, Hejtmankova Pavla, Rakova Lenka
How to Replace Electric Power System Inertia **368**

Chladová Miloslava, Máslo Karel, Krištof Vladimír
Short Circuit Calculations and Dynamic Stability Assessment within Dispatch Control Environment, Operational Requirements and SW Reality **373**

Krokavec Dušan, Filasová Anna
Model-based Fault Diagnosis of Wind Turbines Built on Takagi-Sugeno Fuzzy Observers **377**

Prieložný Stanislav, Jedinák Martin, Gamboš Peter, Kret Miroslav, Grega Michal, Beláň Anton
The Real Test of Black-start Capacity and Island Operation in Electricity System of Slovakia on June 20th, 2018 ..... **383**

Tomin Nikita V., Kurbatsky Victor, Negnevitsky Michael
Development a Partially Observable Markov Decision Processes-based Intelligent Assistant for Power Grids using Monte Carlo Tree Search **389**

Lachmanová Michaela
Evaluation of Energy efficiency and RES Subsidies Metrics – Why is Important to do it? **394**

E: Smart Grids and Electromobility 399

Ziegler Andreas, Arndt Bernhard, Oeser David, Ackva Ansgar
A Universal, Modular and Bidirectional Active Balancing System for Lithium-Ion Cells **400**

Králik Tamáš, Vecka Jiří, Vašíček Jiří, Beneš Martin
District heating sector in the Czech Republic - Status, Challenges and Perspectives **406**

Glazunova Anna
Model Selection for Wind-Driven Asynchronous Machines in Electric Power System State Estimation Problems **411**

Kováč Zoltán, Sadloň Matej, Poničan Ján, Janíček František
Model of Smart Greenhouse for Future Grid **415**

Jirinec Jakub, Rot David, Senkyr Vojtech
Measurement of Indoor Environment Parameters in a Classroom Equipped with Heat Recovery Unit **421**

Jirinec Jakub, Rot David, Jirinec Stanislav
Optimization of the Energy Consumption of a Building with PV **425**

Dvořáček Lukáš
Shared Battery Storage, a Key Solution for the Construction of Charging Stations in the Urban Area **429**

Beranovský Jiří, Knápek Jaroslav
Economic Evaluation of Heating Systems for Energy Efficient Houses **434**

Sikora Tadeusz, Vantuch Tomáš, Fulneček Ján, Prokop Lukáš, Mišák Stanislav
Infrared Monitoring of Industrial Uniterruptible Power Supply Batteries **440**

Bilik Peter, Martinek Radek, Vanus Jan, Gorjani Ojan M., Velicka Richard, Koval Ludvik
Power Quality Analysis of Grid Connected Photovoltaic Inverter **444**

Moldrik Petr, Minarik Daniel, Chesalkin Artem
Research of Hydrogen Fuel Cells Based on PEM Technology **448**

Šimon Pavel, Holá Michaela
Small Renewable Energy Systems on Cars **453**

Zicmane Inga, Zamulko Anatolii, Berzina Kristina, Veremiichuk Yurii
The Use of Electromobiles to Align Daily Electrical Load Schedule of Power System **458**

F: Works of PhD Students 463

Bukvisova Zuzana, Wannous Kinan, Toman Petr, Orsagova Jaroslava, Alves Pedro Henrique de Paula, Topolanek David,
The Influence of Ripple Control Signal on Protection Relay Operation **464**

Janíček František, Poničan Ján, Sadloň Matej
Modified Discounted Payback Period of Various Small-Scale Photovoltaic Installations **468**

Urbanský Jakub, Špes Michal, Beňa Ľubomír, Pál Daniel
Determination of Overhead Power Lines Ampacity Based on CIGRE 207 Brochure **474**

Motyka David, Kajanova Martina, Bracinik Peter, Tomasov Marian, Roch Marek
Analysis of Batteries Impact on a Distribution Network with Distributed Generation **479**

Novak Marek, Belamy Pavol, Siranec Marek, Roch Marek
Analysis of a Power Looses and Local Factors affecting the Performance of the Photovoltaic Power Plant **483**

Bakhteev Kamil, Fedotov Alexander, Chernova Nataliya, Misbakhov Rinat
Methodological Approaches to the Choice of Energy Storage and Optimization of Their Parameters to Improve the Electric Power Quality in Various Types of Electric Power Systems **488**

Belamy Pavol, Novak Marek, Siranec Marek, Roch Marek
The Impact of the Retrofitted Lighting System to the Electrical Power Grid **494**

Hrbek Jan, Ebel Wladimir
Mathematical Model to Design Experiment Arrangement of Induction Furnace with Cold Crucible **499**

Siranec Marek, Hoger Marek, Otcenasova Alena
Evaluation of Power Line Parameter Estimation Accuracy Based on Data from Airborne LiDAR **504**

Pál Daniel, Urbanský Jakub, Beňa Ľubomír, Špes Michal
The Impact of the Placement of a Renewable Energy Resources on Power Losses in the Electricity Distribution System ... **510**

Margitová Anastázia, Kanálik Martin, Kolcun Michal
Verification of Synchronous Generator Time Constants Given by Manufacturers Using the Short-Circuit Current Calculation **515**

Sefl Ondrej, Prochazka Radek
XLPE Ageing Under Non-Standard Voltage Stresses **521**

Ivančák Michal, Kurimský Juraj
The Impact of the Smart Grid on the Distribution **525**

Knenicky Martin, Prochazka Radek
High-Frequency Resonance Overvoltages in Distribution Network with Large Photovoltaic Power Plant **529**

Hörömpöli Balázs, Rácz Ervin
Investigation of State-of-the Art Solar Cell Using Colour Filters **535**

Oliinyk Maksym, Džmura Jaroslav
Power Loss in Smart Grids **541**

Gorjani Ojan M., Vanus Jan, Kubicek Jan, Bilik Petr
Prediction of CO2 Concentration in Smart Homes Using IBM SPSS Software Tool **546**

Baleja Richard, Dolejší Ondřej, Běčák Petr, Sokanský Karel, Novák Tomáš
The Visibility of Pedestrians in the Traffic Area, Using Different Light Source Technologies of Car Headlights **551**

Blazek Vojtech, Fulnecek Jan, Petruzela Michal, Prokop Lukas, Misak Stanislav, Seidl David
Smart Crash Barrier Based on Off-Grid Technology **557**

Vycital Vaclav, Ptacek Michal, Topolanek David, Toman Petr
Optimisation of Earthing System Design through Touch Voltage Minimisation Using Ansys Maxwell **562**

Janik Daniel, Batora Branislav, Toman Petr
Analysis of Daily Load Diagrams for Active and Reactive Energy **567**

Blazek Vojtech, Vysocky Jan, Petruzela Michal, Misak Stanislav, Prokop Lukas
An Estimation of Appliance Impact on Frequency and Power Factor Parameters in Off-Grid Environment **571**

Vysocky Jan, Blazek Vojtech, Petruzela Michal, Misak Stanislav, Prokop Lukas
Review of Radial Distribution Network Reconfiguration Algorithms Based on Mathematical Optimization Theory **577**

Kohan Vladimír, Čonka Zsolt
The Influence of FACTS Devices on Power System **582**

Index of Authors

A

Ackva Ansgar *400*
Aleksandrov Nikolay M. *183, 280*
Altuhov Igor *363*
Alves Pedro Henrique de Paula *464*
Arndt Bernhard *400*
Artyukhov Ivan *133*

B

Bachratá Barbora *310*
Bakhteev Kamil *488*
Balakh Eduard *258*
Baleja Richard *551*
Batora Branislav *567*
Beňa Ľubomír *474, 510*
Belamy Pavol *483, 494*
Beláň Anton *383*
Belik Milan *116*
Bemš Július *62*
Beneš Martin *406*
Beranovský Jiří *434*
Bernát Milan *242*
Berzina Kristina *147, 264, 458*
Bilik Peter *444*
Bilik Petr *92, 97, 102, 546*
Běčák Petr *551*
Blazek Vojtech *557, 571, 577*
Bělík Milan *139*
Bojda Peter *67*
Bracinik Peter *479*
Bukvisova Zuzana *464*
Bulatov Yuri *38*
Buryanina Nadezhda *269*

C

Český Lukáš *315*
Chernova Nataliya *488*
Chernykh Ilya V. *43*
Chesalkin Artem *448*
Chladová Miloslava *373*
Chukreyev Iurii *329, 333, 338*
Chukreyev Mikhail *329, 333, 338*
Chusovitin Pavel *258*
Cialla Björn *310*
Cimbala Roman *247*
Čonka Zsolt *357, 582*

D

Danylchenko Dmytro *229*
Dekhtiar Sergei *258*
Dolejší Ondřej *551*
Dolník Bystrík *215*
Dvořáček Lukáš *429*
Dvorsky Emil *368*
Džmura Jaroslav *193, 197, 242, 541*

E

Ebel Wladimir *499*

F

Fedor Pavol *352*
Fedotov Alexander *284, 488*
Filasová Anna *377*
Frolov Vladimir *143*
Fulneček Ján *440, 557*

G

Gamboš Peter *383*
Gawlak Anna *159, 163*
Gbelec Ján *67*
Gdovin Dominik *75*
Georgiev Georgi *53, 58*
Gerasimov Dmitriy *363*
Glazunova Anna *411*
Gorec Ilia *233*
Gorjani Ojan M. *444, 546*
Grackova Larisa *284*
Grega Michal *347, 383*
Grishkevich Andrey *289*
Gubin Pavel Y. *122*
Gusev Sergey A. *122, 147*
Gyurov Valentin *168, 172*

H

Hejtmankova Pavla *368*
Hájnik Samuel *215*
Hocko Pavol *357*
Hoger Marek *504*
Holá Michaela *453*
Horák Martin *107*
Hormádka Aleš *32*
Hrabčák Ondrej *247*
Hrbek Jan *499*
Hörömpöli Balázs *535*
Hulec Martin *139*

I

Ilenin Stanislav *71*
Ivančák Michal *189, 525*
Ivanov Dmitrij *143*

J

Janíček František *315, 415, 468*
Janik Daniel *567*
Jedinák Martin *383*
Jiřičková Jana *32*
Jirinec Jakub *421, 425*
Jirinec Stanislav *425*

K

Kajanova Martina *479*
Kanálik Martin *87, 515*
Kasembe Andrew *342*
Kaspirek Martin *111*
Khrennikov Alexander Yu. *183, 280*
Kim Dmitriy *53, 58*
Kinsht Nikolay *178*
Kireev Pavel *297, 302, 306*
Kirov Rumen *168*
Knenicky Martin *529*
Knápek Jaroslav *62, 434*
Kohan Vladimír *582*
Kolcun Michal *515*
Kolcunova Iraida *211, 251*
Kolosok Irina *48*
Kolychev Alexander. V. *221*
Komarnicki Przemyslaw *363*
Korkina Elena *48*
Korolyuk Yuriy *269*
Koryakina Maya *269*
Koval Ludvik *97, 444*
Kováč Zoltán *415*
Krejci Petr *92*
Kret Miroslav *383*
Krištof Vladimír *293, 373*
Králik Tamáš *406*
Krokavec Dušan *377*
Kryukov Andrey *38*
Kubicek Jan *546*
Kurbatsky Victor *389*
Kurimský Juraj *207, 251, 525*

L

Lachmanová Michaela *394*
Lesnykh Elena *269*
Lomane Tatjana *264*
Lombardi Pio *363*
Lubkov Alexander N. *221*

M

Mahnitko Anatolijs *48, 122, 128, 147*
Malinina Tatiana *325*
Malinowski Andrzej *163*
Margitová Anastázia *87, 515*
Martinek Radek *444*
Martínek Zbyněk *32*
Medveď Dušan *75, 81*
Menaem Abdel Amir *128*
Minarik Daniel *448*
Misak Stanislav *440, 557, 571, 577*
Misbakhov Rinat *488*
Molčan Patrik *215*
Moldrik Petr *448*
Morozenko Nikita A. *43*
Motyka David *479*
Máslo Karel *26, 293, 342, 373*

Mukhlynin Nikita D. *320*
Mužík Václav *225, 274*

N
Nagay Ivan *297, 302, 306*
Nagay Vladimir *297, 302, 306*
Nagay Vladimir Jr. *302, 306*
Negnevitsky Michael *16, 53, 58, 389*
Noháč Karel *274*
Noháčová Lucie *139, 274*
Novak Marek *483, 494*
Novikova Olga *325*
Novák Tomáš *551*

O
Oboskalov Vladislav *122, 128, 147*
Oeser David *400*
Olexa Richard *352*
Oliinyk Maksym *189, 541*
Olkhovskiy Mikhail *139*
Orsagova Jaroslava *464*
Otcenasova Alena *504*

P
Papkov Boriss *147*
Pavlík Marek *189, 211*
Pavlov Ľuboš *67*
Pazderin Andrey A. *43*
Pazderin Andrey V. *43, 320*
Perdukova Daniela *352*
Petráš Jaroslav *193, 197*
Petrunko Natalia *178*
Petruzela Michal *557, 571, 577*
Pál Daniel *474, 510*
Poničan Ján *415, 468*
Prieložný Stanislav *383*
Prochazka Radek *202, 521, 529*
Proikov Mladen *168*
Prokop Lukas *440, 557, 571, 577*
Ptacek Michal *562*

R
Rakova Lenka *368*
Rácz Ervin *535*
Roch Marek *479, 483, 494*
Rot David *421, 425*
Rubcovs Sergejs *264*
Rudolf Ladislav *242*

S
Sadloň Matej *415, 468*
Santarius Pavel *102*
Sarry Sergey *297, 302, 306*
Schnierer Tomáš *347*
Sedláček Jan *225*
Sefl Ondrej *202, 521*
Seidl David *557*
Senkyr Vojtech *421*
Shamarova Natalia *38, 269, 363*
Shevchenko Sergey *229*
Shushpanov Ilia *363*
Sikora Tadeusz *440*
Simeonova Neli *168, 172*
Simeonov Stanislav *172*
Šimon Pavel *453*
Siranec Marek *483, 494, 504*
Sirový Martin *32*
Sivaev Alexander *143*
Skudřík Filip *315*
Skurčák Ľuboš *67*
Sokanský Karel *551*
Sotirov Sotir *168, 172*
Sowinski Janusz *153*
Špes Michal *474, 510*
Stepanov Sergey *133*
Suslov Konstantin *38, 269, 363*

T
Tadjibaev Alexey *233*
Tashchilin Valeriy *258*
Tavlincev Alexander *147*
Tesarova Miloslava *111*
Tikhonov Alexander *48*
Titkov Vasiliy V. *221*
Toman Petr *464, 562, 567*
Tomasov Marian *479*
Tomin Nikita V. *389*
Topolanek David *464, 562*
Trotsenko Igor *297, 306*
Tukeev Pavel D. *221*

U
Urbanský Jakub *87, 474, 510*

V
Vašíček Jiří *62, 406*
Vagapov Georgii *284*
Valiev Rustam T. *122, 128, 258*
Vantuch Tomáš *440*
Vanus Jan *444, 546*
Varfolomejeva Renata *122, 128*
Varga Ladislav *71*
Vecka Jiří *406*
Velicka Jan *92, 97, 102*
Velicka Richard *92, 97, 102, 444*
Veremiichuk Yurii *458*
Vostracký Zdeněk *225*
Vávrová Kamila *62*
Vycital Vaclav *562*
Vysocky Jan *571, 577*
Vysogorets Svetlana *233, 238*

W
Wannous Kinan *464*

Z
Zak Frantisek *139*
Zamulko Anatolii *458*
Zbojovský Ján *189, 207, 211, 251*
Zemtsov Artem *133*
Zicmane Inga *147, 264, 458*
Ziegler Andreas *400*
Zuščak Jozef *315*
Žvanda Peter *81*

Invited Lectures

High Renewable Energy Penetration and Power System Security: New Challenges and Opportunities

Michael Negnevitsky
School of Engineering
University of Tasmania
Hobart, Australia
michael.negnevitsky@utas.edu.au

Abstract — **The word "security" in the context of a power system implies its security against a complete collapse, or a blackout. Secure operation involves practices aimed to keep the system operating normally when contingencies occur. An increasing penetration of intermittent renewable energy generation introduces additional uncertainties in power systems. However, the impact of variable generation on the system security is often exaggerated. On average, no significant mitigation measures are required until the wind and solar penetration reaches 20 per cent. The main challenge facing a power system with high penetration of renewables is the displacement of conventional synchronous generation by non-synchronous generation. Kinetic energy stored in the rotating masses of synchronous generators provides the system rotational inertia. Wind power generators are mostly doubly-fed induction or full-converter machines. Because these machines are either partially or completely decoupled from the grid by electronic converters, they do not provide inertia to the system. This reduces the total system inertia, and as a result, the system becomes more vulnerable to contingencies. Traditionally security assessment is performed based on deterministic criteria. The *N*-1 security criterion requires a power system to withstand an outage of any single system component without violating any system operating limits. This is based on the worst-case scenario criterion and provides a simple rule in the system design and operation. It has satisfied the needs of the power industry for decades. However, the deterministic approach to security is not adequate in modern power systems with market driven dispatch and high penetration of renewable energy and distributed generation. In this paper, security is defined as the risk in the system's ability to withstand random contingencies without interruption to customer service. The higher the risk the lower the security, and vice-versa. System operational risk is defined as the sum of products of the probabilities of random contingencies that may occur in a particular system state and the expected cost of load interruptions caused by these contingencies. In calculating the operational risk, we consider not just the likelihood of contingencies, but also uncertainties in load variability and renewable energy generation. In risk-based security assessment, we generate contingencies at random, based on their probabilities. Then, we assess the consequences of these contingencies to determine whether loads are disconnected following voltage violations, overloads and significant imbalance between load and generation.**

Keywords: power system, security, condition, high renewable energy penetration.

I. Introduction

The word "security" in the context of a power system implies its security against a complete collapse, or a blackout. Because blackouts cause massive economic losses and disruptions in communication and transport, heating and cooling, water supply, emergency services and financial trading, secure operation of power systems is essential. Secure operation involves practices aimed to keep the system operating normally when contingencies occur.

We can now define a secure power system in terms of operating states. A system is classified as secure if it operates in the normal state and will remain in this state following any single likely contingency. However, if a system operates in the normal state but will not remain in this state following any single likely contingency, the system is said to be in the alert state and classified as insecure. Power system security is its ability to withstand sudden disturbances or unanticipated loss of system components.

Power systems are designed and operated to withstand contingencies selected on the basis of their probabilities. In practice, these contingencies are usually defined as the loss of any single major component in a power system. For example, if the largest generator of the system is suddenly taken out of service, the remaining generators must make up the deficit to avoid load shedding. Similarly, if the most heavily loaded transmission line of the system is switched off, the remaining lines must share the increased loading without exceeding their constraints. We cannot stop contingencies from happening, and we cannot predict when they will occur. But we can model potential contingencies and analyse their consequences. Because the specific times of failures are unpredictable, the system is operated at all times in such a way that it will not be left in a dangerous condition when any *credible* contingency occurs.

Credible contingencies are most probable contingencies. For example, we know from experience that the probability of failure of a single component of the system is much higher than simultaneous failures of multiple components. Thus, a failure of a single generator, transformer or transmission line would be considered more credible than a simultaneous failure of three 500 kV lines serving the same substation. We can also estimate the consequences of each contingency (*what if it really happened?*) through simulations and case studies.

Some less probable contingencies may lead to unacceptable consequences, so we must plan and operate the system so as to minimise the effect of these contingencies when they do happen. For example, in the case of failure of three 500 kV serving the same substation, we might consider as acceptable automatic or manual load shedding in order to preserve the system integrity. Clearly, a relatively short customer outage is an acceptable consequence if we avoid losing the system integrity or damaging power system equipment, which might require weeks or even months to repair.

The list of credible contingencies varies from control area to control area and depends on operational practices in a particular power system. A common practice is to define a *credible contingency* as the loss of any single component in a power system either spontaneously or preceded by a single-, double-, or three-phase fault [1]. This practice is referred to as the *N-1 criterion* because it examines the consequences of the loss of any single major component of the *N*-component power grid. In some cases, the list of credible contingencies may include more complex situations, and even a sequence of events leading to cascading outages. Consideration is also given to so-called "critical" contingencies. For example, the failure of a 230/138 kV transformer may become critical for the system integrity when some transmission lines are out of service due to maintenance. Load shedding and controlled islanding may be used to cope with such critical contingencies and prevent widespread blackouts.

We must plan and operate power systems so that even the worst credible contingency will not result in an unacceptable performance of the system. We must also ensure that after the first credible contingency occurs, the system operator will be able to adjust the system and prepare it for the next credible contingency.

Power systems cannot be operated securely without real-time system monitoring, contingency analysis, and corrective action analysis. These are major security functions carried out in network operations and control centres. System monitoring provides power system operators with on-line information about the system conditions, contingency analysis detects potential network problems, and corrective action analysis identifies actions to prevent network problems before they actually arise.

II. Contingency Analysis

There are two important aspects in contingency analysis: static and dynamic. Static contingency analysis involves steady-state analysis of post-disturbance conditions. This analysis must ensure that after contingency occurs, the system will settle to a normal state and all system components will operate within their constraints. However, we also need to ensure that the system survives the transition to this state, and therefore, we need to examine the dynamic performance of the system. This includes studies of rotor angle stability, voltage stability and frequency stability. We will discuss power system stability later in the book. In this chapter, however, we will concentrate on static contingency analysis.

Let us consider a simple example. Suppose, System A and System B are interconnected via two tie lines with the maximum capacity of 200 MW each. The total generation capacity of System A is 2500 MW with 2000 MW installed on hydro power plants and 500 MW on wind farms. System A has significant excess capacity and is exporting power to System B. System B is generating power on a coal-fired power plant with the total installed capacity of 500 MW (five generators, 100 MW each). Let us also assume that both systems operate on the market through a centrally-coordinated economic dispatch that pools generators from the systems and delivers required quantities of electricity from the pool to the wholesale consumers. The objective of economic dispatch is to meet the system load at the lowest possible cost, subject to transmission and operational constraints. Let us now assume that the load of System B is 600 MW. Because cost of electricity production in System A is lower than in System B, economic dispatch maximises power transfer via the tie lines. Figure 1 (a) shows power flows via the tie lines and generation in System B under economic dispatch. Is this state secure?

A simple "what if" test reveals that it is not. If one of the two running generators of System B suddenly fails, the total transfer via the tie lines will become 500 MW, as shown in Figure 1 (b). Both lines are now overloaded and will be taken out of service by protection. Imbalance between load and generation in now isolated System B will lead to load shedding and probably blackout. Thus, to make this state secure, we need to reduce the power transfer from System A to System B. This means that we need to run three generators on the coal-fired power plant, as shown in Figure 1 (c). However, is the system really secure? If one of the tie lines is disconnected due to forced outage, the other line will take on an additional 150 MW, as shown in Figure 1 (d), and become heavily overloaded. The line will be switched off by protection. As in the previous case, System B will be isolated with a significant imbalance between load and generation. Therefore, to ensure a secure operation of System B, we actually need to run four generators on the coal-fired power plant.

To ensure the required level of security, a power system must have sufficient reserves. These reserves are needed to balance short-term variations in supply and demand, mitigate the effects of load forecasting errors, handle peak demand, and manage fluctuations in renewable energy generation. But most of all, reserves are needed to withstand contingencies.

Power system equipment is subject to random failures, and thus additional generation capacity is needed so that it can be called upon when a large generator or a heavily loaded transmission line is suddenly taken out of service. This additional capacity is referred to as *operating reserve*.

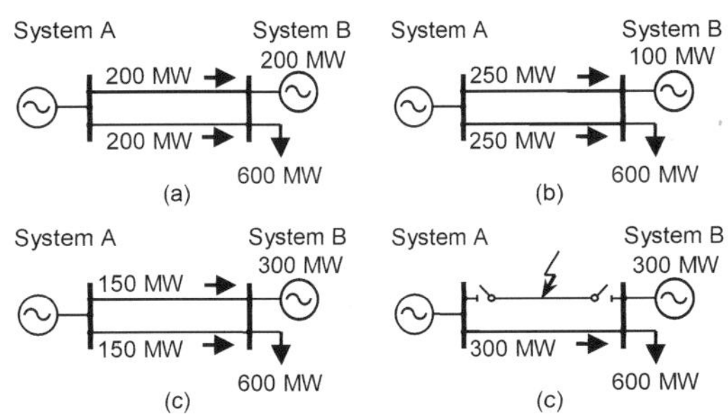

Figure 1. Contingency analysis for System B.

The North American Electric Reliability Corporation (NERC) defines operating reserves as [2]: *"That capability above firm system demand required to provide for regulation, load forecasting error, equipment forced and scheduled outages and local area protection. It consists of spinning and non-spinning reserve".*

Figure 2 shows a typical system response to a large loss of generation. The frequency disturbance event was recorded in the Western Interconnection, North America [3]. At the beginning of this event the frequency was at about 59.98 Hz (point A). At the very instant of the generation loss, synchronous generators that are still connected to the grid release kinetic energy stored in their rotating masses (turbines, shafts and rotors), and thereby slow down the frequency decline. This type of response is called *inertial* response. Large fast rotating generators have larger inertial response than smaller or slowly rotating machines. The system inertia, or the cumulative inertial response of all rotating machines, determines the initial slope of the frequency decline. In our example, it took just 5 s for the frequency to drop to 59.85 Hz (point B).

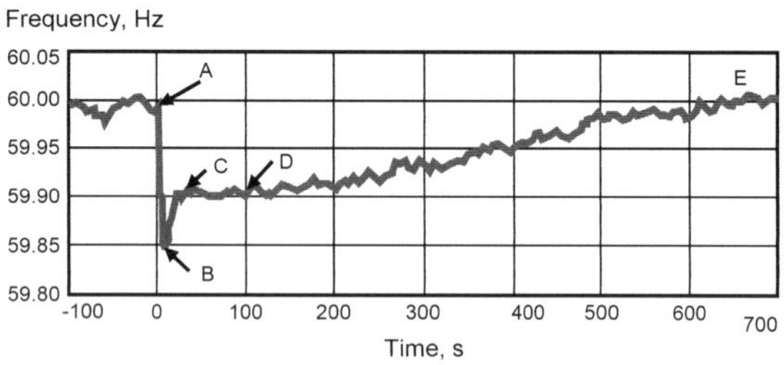

Figure 1. Contingency event and a typical system response.

Following a disturbance, turbine governors detect the frequency decrease and additional power is provided to balance the load. This type of response is called *primary frequency* response. It is used to restore power balance, and thus stabilise the frequency. The primary frequency response is usually delivered within 30 s. However, even when power balance is restored, the frequency can still be off the nominal value. In our case, the primary response was delivered in 15 s, and the frequency stabilised at about 59.90 Hz (point C). The frequency stayed at this level for about 80 s (point D) until the *secondary frequency response* took over and brought the frequency back to its nominal value (point E). The secondary frequency response is delivered by AGC. Generally, the AGC action can take anywhere from seconds to minutes. In our case, it took about 10 min for the frequency to recover fully.

Maintaining sufficient reserves of active power is necessary for preventing under-frequency load shedding and frequency collapse, but it does not guarantee voltage stability. To prevent voltage violations and voltage collapse, we also need to have sufficient reserves of reactive power. And this task is becoming increasingly challenging. In fact, almost all recent blackouts have been caused by problems associated with voltage instability [4].

Maintaining voltage stability has always been a challenging problem in operating large interconnected power systems. In the past, voltage problems were mostly related to power transfers over long distances from remote generation sites to load centres. These problems were adequately addressed by specific control and protection schemes. However, more recent voltage problems are different. They are caused by operating transmission systems closer to their limits than ever before, the changing nature of loads, and an increasing penetration of intermittent renewable energy sources (wind and solar). Voltage problems are now emerging over much wider areas of power systems.

A typical voltage collapse *scenario may* begin with a gradual voltage drop in one or several areas due to, for example, high air conditioning load on a hot summer day. This pushes transformer taps towards their maximum. Generators reach their reactive power limits. Then a generator or a transmission line may trip due to a fault. Since reactive power cannot be transmitted over long distances due to high losses, the sudden tripping of the transmission line results in the increased need for the local reactive power (MVAr) support. If this support is not available, the voltage drops rapidly and may cause collapse of the entire system.

III. Impact of Renewable Energy Generation on Power System Security

An increasing penetration of intermittent renewable energy generation introduces additional uncertainties in power systems. However, the impact of variable generation on the system security is often exaggerated. A common misconception is that significant additional reserves must be kept to be available when the output from wind farms is drastically reduced. In fact, however, a modest amount of variable generation requires a relatively small amount of additional reserves. For example, in the Ireland and Northern Ireland power system, the system stability degrades only when non-synchronous variable generation exceeds 50 per cent of demand [5]. On average, no significant mitigation measures are required until the wind and solar penetration reaches 20 per cent.

Variable energy generation causes changes in the power available from one hour to the next. The maximum expected hourly rate of change provides an indication of the amount of reserves required to cover generation shortfalls. Wind speed variations within the range of 15-25 m/s do not cause any changes in power generation because in this range, wind turbines operate at full output. On the other hand, wind variations within the range of 4-15 m/s can result in a substantial change of the turbine output. This problem can usually be mitigated by geographical dispersion of wind farms – the decrease of wind power generation at one side is often compensated by the increase at another. For example, in Germany, a single wind farm can frequently exhibit hour-to-hour power swings of up to 60 per cent of the installed capacity while aggregated power swings of wind farms over entire Germany do not exceed 20 per cent [6]. This figure provides the basis for estimating the ramping capability requirement for conventional generation.

The main challenge facing a power system with high penetration of renewables is the displacement of conventional synchronous generation by non-synchronous generation.

Kinetic energy stored in the rotating masses of synchronous generators provides the system rotational inertia. Wind power generators are mostly doubly-fed induction or full-converter machines. Because these machines are either partially or completely decoupled from the grid by electronic converters, they do not provide inertia to the system. This reduces the total system inertia, and as a result, the system becomes more vulnerable to contingencies – even contingencies that previously were considered "safe" can now lead to frequency violations and the system's instability.

During a frequency disturbance caused by the loss of a generator, the system frequency decreases and energy stored in rotating masses of all synchronous machines connected to the grid is released to the system. This assists in maintaining both the speed of remaining machines and the frequency of the system. The relationship between the stored kinetic energy and inertia is expressed as [7]:

$$E = \frac{1}{2} J \omega^2 \quad (1)$$

where E is the kinetic energy stored in the rotating masses, [MW·s]; J is the moment of inertia around the axis of rotation, [kg·m^2]; and ω is the angular velocity of the rotor [rad/s].

Based on the amount of energy stored in the rotating mass, we can determine how long the generator can supply its rated power solely from the stored kinetic energy:

$$H = \frac{E_0}{S} = \frac{J\omega_0^2}{2S} \quad (2)$$

where H is the inertia constant, [MW·s/MVA]; E_0 the kinetic energy measured at the rated angular velocity of the rotor ω_0, [MW·s]; and S is the rated apparent power of the generator, [MVA].

The inertia constant is expressed in seconds. It indicates the duration that the generator can supply its rated power to the system using only its kinetic energy. For example, the inertia constant of 6 s means that the generator can supply its rated power to the system for 6 s using only energy stored in its rotating masses. The value of H represents the combined inertia of the generator and turbine. For large machines, the inertia constant lies within the range of 2-10 s. Typical values for thermal and hydraulic machines are presented in Table 1 [7].

Table 1. Inertia constant for conventional machines.

Machine type	Inertia constant, MW·s per MVA
Thermal	
3600 r/min (2-pole)	2.5 – 6.0
1800 r/min (4 pole)	4.0 – 10.0
Hydraulic	2.0 – 4.0

In power systems integrating wind energy on a large scale, inertia becomes imperative in determining the rate of change of frequency after a contingency event. In other words, the system inertial response becomes a dominant factor in contingency analysis. Power systems with higher penetration of renewables will require more robust frequency control.

Until recently, it was generally assumed that the aggregated rotational inertia of a system was constant. However, this assumption may not be true any longer. For example, Figure 2 shows that the aggregated inertia of the German power system can fluctuate between its nominal value of 6 s when only conventional generators are dispatched and much lower values of 3–4 s when significant wind and PV generation is deployed [8]. The lowest level of rotational inertia was recorded during the Christmas period when demand was at its lowest, while wind power generation was unusually high. We should also note that results of this example are also valid for other power systems with high penetration of renewable energy generation.

In security analysis, we want to know the system frequency response to a contingency. We want to ensure that the system frequency will remain within acceptable limits. Therefore, we need to develop a model describing the frequency dynamics of a power system.

Let us assume that the system experiences the loss of a large generator. In milliseconds following the loss of the generator, the imbalance between load and generation is covered by kinetic energy stored in the rotating masses. In these first milliseconds, no power increase can be delivered by steam or hydro turbines due to slower reaction time of the valves or wicket gates that control the amount of steam or water entering the turbines. Thus, we can express this situation by

$$P_m - P_e = \frac{dE}{dt} \quad (3)$$

where P_m is the mechanical power supplied via the generator's shaft and P_e is the electrical power demand.

The total kinetic energy of the system varies as the square of frequency and we can, therefore, write

$$E = \left(\frac{f}{f_0}\right)^2 E_0 \quad (4)$$

where E_0 is the total kinetic energy measured at nominal frequency f_0, and f is the instantaneous frequency of the system.

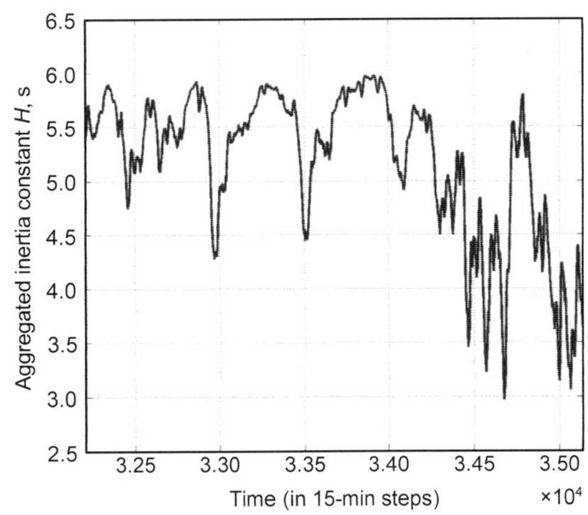

Figure 2. Aggregated rotational inertia in the German power system (December 2012). It is assumed that conventional generators provide inertia of 6 s and wind and PV generators do not contribute any inertia [8].

We can represent the instantaneous system frequency as

$$f = f_0 + \Delta f \quad (5)$$

and because Δf is small, we can rewrite *Eq.* (4) as follows:

$$E = \left(\frac{f_0 + \Delta f}{f_0}\right)^2 E_0 \approx \left(1 + 2\frac{\Delta f}{f_0}\right) E_0 \quad (6)$$

Substituting *Eq.* (6) into *Eq.* (3) results:

$$P_m - P_e = \frac{2E_0}{f_0}\frac{d\Delta f}{dt} \quad (7)$$

Now we can express frequency dynamics of the system as:

$$\frac{d\Delta f}{dt} = \frac{f_0}{2E_0}(P_m - P_e) = \frac{f_0}{2H}(P_m - P_e) \quad (8)$$

where H is the total system inertia, P_m is the total mechanical power, and P_e is the total electrical power demand.

The total inertia of the system is defined as:

$$H = \frac{\sum_{i=1}^{n} H_i S_i}{S_{base}} \quad (9)$$

The total mechanical power is defined as:

$$P_m = P_{m0} + \Delta P_m \quad (10)$$

where P_{m0} is the total mechanical power produced by the generators in steady state before the contingency occurred, and ΔP_m is the deviation from that value after the contingency.

The total electrical power demand is equal to the sum of the system loads and transmission losses, i.e.

$$P_e = P_{load} + P_{loss} \quad (11)$$

If we rewrite *Eq.* (11) in the same way as *Eq.* (10), we obtain

$$P_e = (P_{e0} + \Delta P_{load}) + (P_{loss0} + \Delta P_{loss}) \quad (12)$$

where P_{e0} and P_{loss0} are the total electrical power demand and the total transmission losses, respectively, before the contingency occurred, and ΔP_{load} and ΔP_{loss} are the changes in the system load and the system losses, respectively, after the contingency.

In s steady state, mechanical power is equal to electrical power demand. Thus,

$$P_{m0} = P_{e0} \quad (13)$$

Since any frequency change is caused by changes either in power demand or power generation, it is convenient to rewrite *Eq.* (8) in the following form:

$$\frac{d\Delta f}{dt} = \frac{f_0}{2H}(\Delta P_m - P_{load} - \Delta P_{loss}) \quad (14)$$

We can assume that transmission losses before and after the contingency are equal. Thus,

$$\Delta P_{loss} = 0 \quad (15)$$

and

$$\frac{d\Delta f}{dt} = \frac{f_0}{2H}(\Delta P_m - \Delta P_{load}) \quad (16)$$

Case Study 1

Let us consider a simple case study. Suppose, ten 100 MVA synchronous generators are supplying a total load of 800 MW. The inertia constant of each unit is 4.0 s on the 100 MVA base. Let us determine the rate of frequency change following a sudden loss of one of the generators assuming that there is no speed-governing control.

The rate of change of frequency (ROCOF) following a sudden loss of generation or load is directly proportional to the active power imbalance ΔP between the load and generation:

$$\text{ROCOF} = \frac{\Delta P \cdot f_0}{2H} \quad [\text{Hz/s}] \quad (17)$$

Note that ΔP must have the same MVA base as the total system inertia constant H.

The total system inertia following the loss of the 100 MVA unit, on the 100 MVA base, is

$$H = \frac{\sum_{i=1}^{9} H_i S_i}{S_{base}} = \frac{9 \times 4.0 \times 100}{100} = 36 \text{ s}$$

$$\Delta P = -\frac{100}{100} = 1 \text{ pu}$$

Thus,

$$\text{ROCOF} = -\frac{1 \times 60}{2 \times 36} = 0.83 \text{ Hz/s}$$

The value of ROCOF provides the initial slope of the frequency deviation vs. time. The greater the ROCOF, the lower the frequency is expected to dip following the loss of generation. The primary frequency response cannot stabilise the frequency instantaneously even when the system has sufficient rotating reserves. It usually takes a few seconds before turbine governors can adjust positions of the valves or gates and deliver the required additional power to the system. Therefore, we can assume that there is no governor action during the first 2 s after the contingency, and determine the frequency decline as

$$\Delta f = -0.83 \times 2 = -1.66 \text{ Hz}$$

The frequency dip increases as the system inertia decreases. Let us consider a case when four 100 MVA synchronous generators are displaced (or decommitted) by wind and solar PV generation. We can assume that the system still has sufficient rotating reserves when the contingency occurs. The total system inertia following the loss of the 100 MVA unit:

$$H = \frac{5 \times 4.0 \times 100}{100} = 20 \text{ s}$$

and

$$\text{ROCOF} = -\frac{1 \times 60}{2 \times 20} = -1.5 \text{ Hz/s}$$

Thus, in 2 s after the contingency, the system frequency would potentially reach 58.5 Hz, and under-frequency load shedding would be activated.

In security analysis, we are interested in the combined effect of synchronous machines on the system frequency dynamics in the case of contingency. If we assume the coherent response of all machines, we can represent them as a single equivalent generator driven by the combined mechanical outputs of the individual turbines. The speed of the equivalent generator represents the system frequency.

Case Study 2

Let us consider another example. Suppose, six 200 MVA and ten 100 MVA synchronous generators are supplying a total load of 2000 MW. The inertia constant of each 200 MVA unit is 5.0 s on the 200 MVA base, and the inertia constant of each 100 MVA unit is 4.0 s on the 100 MVA base. The governor speed regulation of each 200 MVA unit is 5%, and each 100 MVA unit is 4%. The speed governor time constant for the 200 MVA units is 0.2 s, and for the 100 MVA units is 0.1 s. The turbine time constant for the 200 MVA units is 0.8 s, and for the 100 MVA units is 0.6 s. The load damping constant is 1%. Let us etermine the frequency deviation following a sudden loss of one of the 200 MVA units. Examine the frequency dynamics of the system when five 100 MVA synchronous generators are displaced by non-synchronous renewable generation.

Let us assume S_{base} = 1000 MVA. Then, we can determine the total system inertia constant following the contingency:

$$H = \frac{5000 + 4000}{1000} = 9 \, s$$

The load damping of the system on the 1000 MVA base:

$$D = 1\% \times \frac{2000}{1000} = 2\%$$

The loss of generation can be represented as a step load increase. Thus,

$$\Delta P_{load} = \frac{200}{1000} = 0.2 \text{ pu}$$

Since there is a sudden reduction in generation, the system frequency decreases, and all generating units connected to the grid respond according to their droop characteristics.

The total spinning generation capacity of the five remaining 200 MVA units is 1000 MVA, or 1.0 pu. A regulation of 5% means that a 5% change in frequency causes 100% change in power generation. In p.u., we obtain:

$$\frac{1}{R_{200}} = \frac{1.0}{0.05} = 20$$

Similarly, for ten 100 MVA units, we have:

$$\frac{1}{R_{100}} = \frac{1.0}{0.04} = 25$$

Then, the total speed regulation:

$$\frac{1}{R} = \frac{1}{R_{200}} + \frac{1}{R_{100}} = 20 + 25 = 45$$

The equivalent governor time constant:

$$T_G = \frac{0.2 \times 1000 + 0.1 \times 1000}{1000 + 1000} = 0.15 \text{ s}$$

The equivalent turbine time constant:

Now we can build a model of the system frequency dynamics in MATLAB Simulink, as shown in Figure 3. Simulation results are shown in Figure 4 (Case 1).

The steady-state frequency deviation:

$$\Delta f_{\infty} = -\frac{\Delta P_{load}}{D + \frac{1}{R}} = -\frac{0.2}{2 + 45} \times 60 \text{ Hz} = 0.255 \text{ Hz}$$

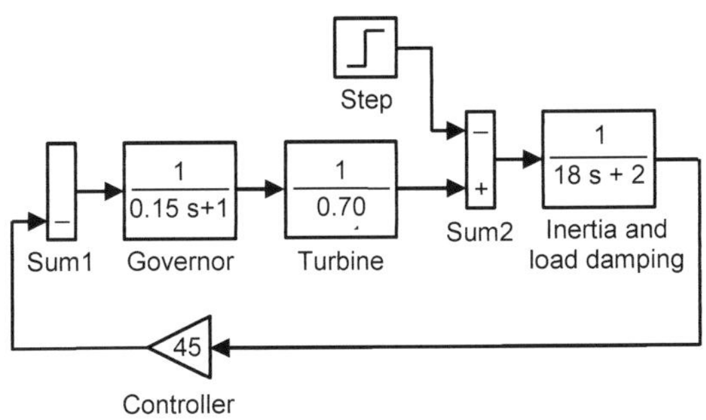

Figure 3. The MATLAB Simulink model of the system frequency dynamics.

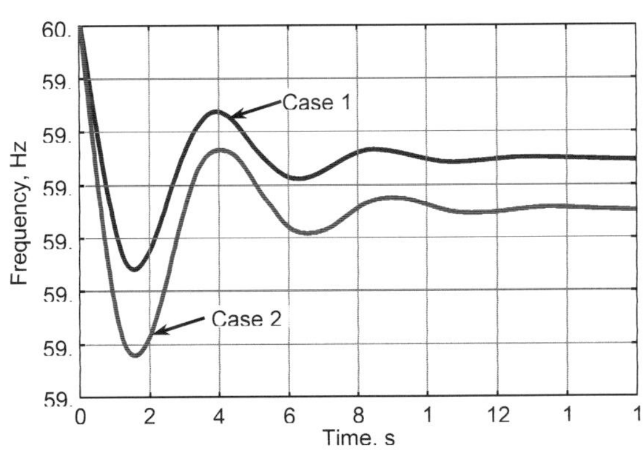

Figure 4. Frequency response of the power system.

Let us now consider the system frequency dynamics when five 100 MVA synchronous generators are displaced by non-synchronous power generators, assuming that they do not provide any inertia to the system.

The total inertia constant:

$$H = \frac{5000 + 2000}{1000} = 7 \text{ s}$$

The total speed regulation:

$$\frac{1}{R} = \frac{1}{R_{200}} + \frac{1}{R_{100}} = \frac{1.0}{0.05} + \frac{0.5}{0.04} = 20 + 12.5 = 32.5$$

The equivalent governor time constant:

$$T_G = \frac{0.2 \times 1000 + 0.1 \times 500}{1000 + 500} = 0.167 \text{ s}$$

The equivalent turbine time constant:

$$T_T = \frac{0.8 \times 1000 + 0.6 \times 500}{1000 + 500} = 0.733 \text{ s}$$

Now we can modify the system frequency dynamics model, as shown in Figure 5. Simulation results are shown in Figure 4 (Case 2). As we can see, frequency in the "lighter" case declines much faster, although in both cases, the system has sufficient rotating reserves to cover for the loss of the largest generator.

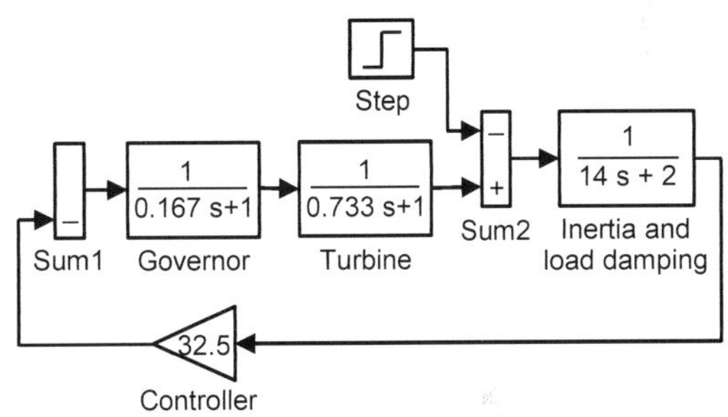

Figure 5. The MATLAB Simulink model for Example 9.3 when five 100 MVA synchronous generators are displaced by wind and solar PV generation.

The combined effect of the lower rotational inertia and reduced governor response results in deeper frequency excursions following contingencies. This increases the risk of under-frequency load shedding and cascading outages.

In general, faster frequency dynamics raises the question whether frequency control of a power system is fast enough for mitigating contingency events before a critical frequency drop occurs.

The frequency response of wind turbines is dependent on active controls. These controls are designed to provide a response, which is *functionally* similar to the frequency response of synchronous machines. For example, General Electric MW-class wind turbines are provided with WindINERTIA^{TM} control, which can deliver "inertial" frequency response [9]. For large under-frequency events, WindINERTIA^{TM} can increase the turbine output by up to 10 per cent of its rated power and supply this power to the grid for several seconds. The delivery of extra power from wind turbines reduces the rate of frequency decline, thus allowing time for governors of synchronous generators to take action.

Modern wind turbines are also capable of providing "governor" frequency response similar to the primary frequency response of synchronous machines. This is accomplished through blade pitch control. In order to provide "governor" response to low frequency events, wind turbines must operate below the level of maximum power output possible for a given wind condition. In other words, wind farms that contribute to frequency regulation in a power system must be curtailed. Curtailment, however, has obvious economic consequences in terms of lost energy production. Therefore, it is important that the level of curtailment is estimated and rules and commercial arrangements are in place.

Recent developments in wind turbine and battery storage technologies offer a *hybrid* solution when a battery is incorporated into the turbine electric and control system. This solution can potentially enable operators of large wind farms to provide a guaranteed amount of power output for up to an hour. In general, hybrid wind farms can provide frequency response to meet or even exceed the NERC requirements. Hybrid wind farms would be particularly valuable in the systems with very high penetration of renewable energy and limited or slow primary reserves.

IV. RISK-BASED SECURITY ASSESSMENT

Traditionally security assessment is performed based on deterministic criteria. The *N*-1 security criterion requires a power system to withstand an outage of any single system component without violating any system operating limits. This is based on the worst-case scenario criterion, and provides a simple rule in the system design and operation. It has satisfied the needs of the power industry for decades. However, the deterministic approach to security may not be adequate in modern power systems with market driven dispatch and high penetration of renewable energy and distributed generation [10]. Increasing installation of unobservable distributed and PV generation with unknown operating characteristics increases forecasting errors in both demand and generation. This, together with large-scale integration of wind power generation, increases system uncertainties and creates a number of technical challenges in the power system planning and operation. Thus, there is a need to investigate new security assessment methods, which can cope with the uncertain behaviour of wind power generation as well as the probabilistic nature of system conditions and events in the operation planning studies.

Deterministic security criteria define a set of "credible" contingencies that the system should be able to withstand. This set includes a loss of any component or a group of components that could be taken out of service as a result of a single fault. Losing two or more independent components simultaneously is considered "not credible" because the probability of two or more independent faults occurring at the same time is very low.

We should note, however, that the deterministic contingency analysis does take the probabilities of contingencies into account – the selection of "credible" contingencies implicitly implies that these contingencies are more likely to occur in reality. Indeed, the probability of a single independent contingency is much higher than the simultaneous outage of two or more components. However, probabilities of single contingencies are not equal. For example, a fault on a long power transmission line is much more likely to occur than a contingency of a large generator.

Security can be defined as the risk in the system's ability to withstand random contingencies without interruption to customer service [11]. The higher the risk the lower the security, and vice-versa. Although risk cannot be eliminated fully due to unexpected faults and probabilistic behaviour of a power system, it can be assessed and managed within an acceptable level in power system planning, design and operation [12].

Risk-based security analysis is concerned with voltage violations, overloads and frequency response adequacy. Frequency response adequacy is defined as the capability of frequency response resources to prevent frequency from dropping below a certain limit [13]. As we discussed earlier, in security analysis, it is necessary to assess the primary frequency response because of the risk of under-frequency load shedding, particularly in small systems with low inertia.

In risk-based security assessment, we do not use a predefined list of contingencies, but generate contingencies at random based on their probabilities. Then, we assess the consequences of these contingencies in order to determine whether loads are disconnected following voltage violations, overloads and significant imbalance between load and generation. This allows us to measure the impact of random contingencies in terms of loads not served. Using a model of the restoration process, we can then obtain the time needed to restore power, and thus, estimate the cost of the generated contingency. Finally, we repeat generating contingencies over and over, each time using a different set of random values. This process is called Monte Carlo simulation – a problem solving technique used to calculate the probability of certain outcomes by running multiple trials, called simulations, using random variables.

In risk-based security assessment, we need to evaluate the system's future operating conditions. In other words, we need to have an appropriate model of the future. Such a model is based on forecasts of the loading conditions and renewable energy generation at some future time period. Any forecast is uncertain, the actual loading conditions and renewable energy generation may, and most likely will, differ from the forecasted values.

To model these uncertainties, we need to create random operating conditions. First, we select the network topology, loading conditions and power generation based on the expected operating conditions. Then, we add a normally distributed random variable, α_{load}, with mean zero and pre-selected deviation δ_{load} to the load forecast. In power systems with high wind penetration, we also add a normally distributed random variable, α_{wind}, with mean zero and pre-selected deviation δ_{wind} to the wind generation forecast. Although, at a regulation time scale (minute-to-minute) the impact of wind power generation on the power system operational risk is relatively small, it becomes significant at the hourly time scale and beyond. Therefore, in power systems with high wind penetration, we have to consider uncertainties in demand as well as unexpected variations in power generation caused by the intermittency of wind.

Load at time t, P_{load}, is the sum of the load forecast, $P_{f\,load}$, and the forecasting error, $e_{f\,load}$

$$P_{load} = P_{f\,load} + e_{f\,load} = P_{f\,load} + \alpha_{load}(0, \delta_{load}) \quad (18)$$

Similarly, wind generation at time t, P_{wind}, is the sum of the wind generation forecast, $P_{f\,wind}$, and the forecasting error, $e_{f\,wind}$

$$P_{wind} = P_{f\,wind} + e_{f\,wind} = P_{f\,wind} + \alpha_{wind}(0, \delta_{wind}) \quad (19)$$

Normally distributed forecasting errors with mean zero reflect that smaller errors occur more frequently than large ones. Although, in practice, forecasting errors in wind power generation do not necessarily fit a normal distribution, this assumption is still valid due to a large number of geographically dispersed wind turbines.

In power system analysis, it is common to assume that the failure rate of a power system component is constant, and the probability of failure depends only on the length of time being considered. This assumption allows us to use an exponential distribution. Thus, the probability of failure of the system component i, $p_i(t)$, is given by:

$$p_i(t) = 1 - e^{-\lambda_i t} \quad (20)$$

where λ_i is the failure rate of the system component i. The term $\lambda_i t$ represents the probability of the component failing during the interval t.

Random outages are simulated assuming that the system is in its normal state at the beginning of the simulation period. If a random number corresponding to the component i is less than $p_i(t)$, the component i is considered to have failed, otherwise, it remains in service.

To compute the value of risk, we sum over all possible outcomes the product of the outcome probability and its severity [14]:

$$Risk = \sum_{n=1}^{N} \sum_{m=1}^{M} p(S_n) \cdot p(C_{nm}) \cdot Sev(C_{nm}) \quad (21)$$

where $Risk$ is the risk index, $p(S_n)$ is the probability of the pre-contingency operating state S_n, $p(C_{nm})$ is the probability of contingency C_m occurring in the state S_n, $Sev(C_{nm})$ is the severity of the contingency C_m in the state S_n, N is the total number of pre-contingency operation states, and M is the total number of contingencies considered in each state.

Severity is introduced to quantify post-contingency security violations in terms of economic consequences of a contingency. These consequences can be evaluated using a unit interruption cost multiplied by expected energy not supplied (EENS). In other words, we need to estimate the cost of a contingency.

However, it is difficult to quantify the interruption cost. We can, for example, estimate this cost based on utility damages such as revenue reduction and additional costs associated with emergencies (e.g., installation and running diesel generators in an outage situation), or we can assess customer damages due to interruption of power supply based on customer surveys. However, the customer damages do not reflect the utility damages and thus, cannot be used to balance the capital investment of a utility. On the other hand, customers eventually pay all the costs through rates and therefore, the customer interruption cost should be considered in system planning [12]. This is clearly a complicated issue; the unit interruption cost is country-, region-, and system specific. For example, the average unit interruption cost in Canada is between $4/kWh and $10/kWh, and in Australia it is $12.5/kWh.

To quantify consequences of the steady-state voltage violations and overloads, the post-disturbance system condition is checked via power flow solutions. If any equipment such as a transmission line or a transformer is overloaded and/or voltage at any bus is out of the allowable range, the load located at or near the violated element is disconnected in small blocks until these violations are cleared. In practice, in response to violations of the system operational limits, an operator can reschedule generation, change set-points of voltage regulators and transformer tap ratios, switch on or off capacitor banks, and transfer loads from one part of the system to another. In fact, emergency load shedding is used as the last resort when there is no other way to save the system. However, because a general indicator is needed to quantify the consequences of a system contingency, load shedding is commonly used in probabilistic benchmarking studies [10].

To quantify consequences of the frequency response inadequacy, we need to develop a mathematical model to represent frequency trajectories during frequency excursion events. Then, using the frequency trajectories, we can determine the consequences of the frequency event in terms of the load curtailment. By using the analytical method, the frequency response adequacy assessment can be run simultaneously with the steady-state voltage and overload evaluations [15].

For calculating the value of EENS, we need to multiply the total amount of interrupted load (in kW) by the duration of the load interruption (in hours). Thus, we need to develop a model of the load restoration process. This model is not required to be exhaustive, because we only need to capture the duration of interruption. This duration can be estimated with a simple rule-of-thumb, which suggests that the time needed for restoration is directly related to the number of disconnected buses and severity of the problem. Table 2 represent a heuristic model of the restoration process [16]. The model assumes that the restoration rate increases with time.

Table 2. Load restoration rates.

Restoration rate, MW per min	Time, min
10.0	0 – 30
33.3	30 – 60
66.6	60 – 90
83.3	> 90

The risk-based security assessment for power system operation planning consists of the following steps [15]:

Step 1:
Choose the simulation period (for example, one hour), and specify failure rates for the system components. Assign the total number of pre-contingency operating states N, and the total number of contingencies M to be considered in each state.

Step 2:
Initialise the base state by selecting the network topology, loading conditions and power generation based on expected system conditions in this state. In the base state, the loads are represented by their forecasts, all conventional generators produce power according to the specified dispatch and renewable energy generation is specified by the forecast.

Step 3:
Create a random operating condition by adding normally distributed random variables to the load forecast at each bus, and the wind forecast for each wind farm.

Step 4:
With associated failure rates, generate a random outage of a system component.

Step 5:
Check the system integrity and the load-generation balance. If the system is islanded or the load-generation balance is violated, calculate ROCOF and the steady-state frequency deviation. If the system frequency drops below the maximum under-frequency set point, shed 10 per cent of the system load, and re-check the load-generation balance. If the balance is not restored, shed an additional 10 per cent and a further 10 per

cent if frequency continues to decline and reaches the minimum under-frequency set point. The 30 per cent limit is commonly accepted as the maximum for under-frequency load shedding (UFLS). In the event when generation suddenly drops by more than 30 per cent, the system collapse cannot be prevented by UFLS.

Step 6:
Perform power-flow analysis to determine bus voltages and power flows in the system. Check for any violations of operating limits. If no limit is violated, go to Step 4 and repeat the process until the total number of contingencies, M, is considered. If any element, such as a transmission line or a transformer, is overloaded or voltage at any bus is out of the allowable range, shed the load located at or near the violated element. The load is shed in small blocks until all violations are cleared.

Step 7:
Calculate the cost of the contingency as the product of the unit interruption cost and EENS.

Step 8:
Calculate the value of risk associated with the contingency, go to Step 3 and repeat the process until the total number of pre-contingency operating states, N, is considered.

Step 9:
Calculate the risk index associated with the simulated period.

The probabilistic method represents an alternative to the conventional deterministic approach to power system security assessment within the context of operational planning and decision making. This method takes into account uncertainties associated with operating conditions and outages, and estimates potential effects of these outages. Given the high frequency of stressed conditions experienced by power systems today and increasing uncertainties in demand and generation, it is clear that the probabilistic risk-based security assessment will play a significant role in operating sustainable power systems.

V. Conclusions

The main challenge facing a power system with high penetration of renewables is the displacement of conventional synchronous generation by non-synchronous generation. Because renewables do not provide inertia to the system, the total system inertia reduces, and as a result, the system becomes more vulnerable to contingencies. Even contingencies that previously were considered "safe" can now lead to frequency violations and the system instability. In power systems integrating wind energy on a large scale, the system inertial response becomes a dominant factor in contingency analysis. The combined effect of the lower rotational inertia and reduced governor response results in deeper frequency excursions following contingencies. This increases the risk of under-frequency load shedding and cascading outages.

Recent developments in wind turbine and battery storage technologies offer a hybrid solution for mitigating the effect of faster frequency dynamics in power systems with high penetration of renewable energy. This solution can potentially enable operators of large wind farms to provide a guaranteed amount of power output for up to an hour.

Traditionally security assessment is performed based on deterministic criteria. The N-1 security criterion requires a power system to withstand an outage of any single system component without violating any system operating limits. This approach has satisfied the needs of the power industry for decades. However, the deterministic approach to security may not be adequate in modern power systems with market driven dispatch and high penetration of renewable energy and distributed generation.

In risk-based security assessment, we generate contingencies at random, based on their probabilities. Then, we assess the consequences of these contingencies in order to determine whether loads are disconnected following voltage violations, overloads and significant imbalance between load and generation. This allows us to measure the impact of random contingencies in terms of loads not served. Using a model of the restoration process, we can then obtain the time needed to restore power, and thus, estimate the cost of the generated contingency. Finally, we repeat generating contingencies over and over, each time using a different set of random values.

References

[1] P. Kundur, J. Paserba, V. Ajjarapu, G. Andersson, A. Bose, C. Canizares, N. Hatziargyriou, D. Hill, A. Stankovic, C. Taylor, T. Van Cutsem, V. Vittal, "Definition and Classification of Power System Stability", *IEEE Transactions on Power Systems*, Vol. 19, No. 2, May 2004, pp. 1387–1401.

[2] North American Electric Reliability Corporation, "Glossary of Terms Used in NERC Reliability Standards", October 1, 2014.

[3] E. Ela, M. Milligan, B. Kirby, "Operating Reserves and Variable Generation", *National Renewable Energy Laboratory (NREL), Technical Report NREL/TP-5500-51978*, August 2011.

[4] Report: ENTSO-E, UCTE Operational Handbook Policy 1, Load-Frequency Control and Performance, March 2009. [Online] Available: https://www.entsoe.eu/publications/system-operations-reports/operation-handbook

[5] A. Rogers, J. O'Sullovan, I. Dudurych, "Operational tools and policies for integrating large amounts of variable generation in Ireland and Northern Ireland", *Proceedings of the IEEE/PES General Meeting,* Detroit, MI, USA, 24-28 July, 2011.

[6] F. Van Hulle, "Large scale integration of wind energy in the European power supply: analysis, issues and recommendations", *Report by EWEA*, December 2005.

[7] P. Kundur, *Power system stability and control*. New York: McGraw-Hill, 1994.

[8] A. Ulbig, T. S. Borsche, G. Andersson, "Impact of Low Rotational Inertia on Power System Stability and Operation", *IFAC World Congress 2014*, Capetown, South Africa, pp. 7290-7297.

[9] N. W. Miller, M. Shao, S. Venkataraman, "Impact of Frequency Responsive Wind Plant Controls on Grid Performance", *Cigre Session 2012*, Paper C4-113, Paris, France, 26-31 August 2012.

[10] D. S. Kirschen, D. Jayaweera, "Comparisons of risk-based and deterministic security assessments," *IET Generation, Transmission and Distribution*, vol. 1, no. 4, pp. 527-533, July 2007.

[11] CIGRE Working Group C4.601, "Review of on-line dynamic security assessment tools and techniques," 2007.

[12] Li, W., *Risk assessment of power systems: Models, methods, and applications*, New York: Wiley, 2014.

[13] Electric Power Research Institute (EPRI), "Frequency response adequacy and assessment: Global industry practices and potential impact of changing generation mix," 2012.

[14] **J.** McCalley, S. Asgarpoor, L. Bertling. et al., "Probabilistic security assessment for power system operations", IEEE PES Task Force on Probability Aspects of Reliability Criteria, In *Proc. IEEE PES General Meeting, Denver, CO, USA, June 2004*, pp. 212-220.

[15] M. Negnevitsky, D.H. Nguyen, M. *Piekutowski,* "Risk Assessment for Power System Operation Planning with High Wind Power Penetration", *IEEE Transactions on Power Systems*, 30 (3), May 2015, pp. 1359 – 1368.

[16] D. S. Kirschen, K. R. W. Bell, D. P. Nedic, D. Jayaweera, R. N. Allan, "Computing the value of security," *IEE Proceedings Generation Transmission and Distribution*, 150 (6), pp. 673-678, November 2003.

Dynamic Models in the Context of the EU Commission Regulations

Karel Máslo
Transmission System Analysis Department.
ČEPS, a.s.
Prague, Czech Republic
maslo@ceps.cz

Abstract — **This paper presents dynamic models which are suitable to simulate the compliance of power generating modules with the network code on requirements for grid connection of generators requirements, namely the fault-ride-through capability and power system stabilizer function for the synchronous generating modules and the capabilities of providing synthetic inertia and fast fault current activation for asynchronous generating modules.**

Keywords: Dynamic models, Transmission System Operator (TSO), compliance tests and simulations

I. Introduction

The previous paper [1] was focused on dynamic models suitable for the frequency stability assessment. This paper deals with dynamic models more comprehensively in relation to European Union regulations.

According to the guideline on electricity transmission system operation (so called SOGL [2]) Article 41 paragraph 4a) each TSO shall exchange with the other TSOs of the same synchronous area or of its relevant part data concerning power generating modules (which are so called significant grid users), that are relating to, but not limited to voltage and speed controller models and prime movers models and excitation system models suitable for large disturbances. For TSOs to be able to fulfill these obligations, another EU regulation (so called network code RfG [3]) defines the requirements for power-generating facility owners dealing dynamic models.

The network code RfG defines in Article 15 (**General requirements for type C power-generating modules**) paragraph 6c) requirement for the power-generating facility owner to provide simulation models which properly reflect the behavior of the power-generating module in both steady- state and dynamic simulations (50 Hz component) or in electromagnetic transient simulations. These are sub-models of:

1. alternator and prime mover,
2. speed and power control,
3. voltage control, including, if applicable, power system stabilizer ('PSS')
4. function and excitation control system,
5. power-generating module protection models (as agreed between the relevant system operator and the power- generating facility owner) and
6. converter models for power park modules.

The power-generating facility owner shall ensure that the models provided have been verified against the results of compliance tests and shall notify the results of the verification to the relevant system operator or relevant TSO. These requirements are valid for the type D power generating modules.

The power-generating facility owner shall ensure that each power-generating module complies with the requirements applicable under the the RfG network code. This shall be demonstrated by test and / or simulation calculations in dependency on the power generation module type (A, B, C or D and synchronous or asynchronous).

The network code RfG specifies the following functions and capabilities, which must be proven by tests or simulations:

1. LFSM-O and LFSM-U responses
2. FSM response
3. Reactive power capability
4. Frequency restoration control
5. Black start capability
6. Quick re-synchronization capability
7. **The fault-ride-through capability**
8. Island operation capability
9. **Power System Stabilizer function**
10. Reactive power control
11. **Capability of providing synthetic inertia**
12. **Fast fault current activation**
13. Power oscillations damping control.

For example, items 7 - 9 for synchronous modules will be demonstrated by simulations only, while items 1 – 3 have to be proven by both tests and simulations and items 4 – 6 by test only. The items 11 -13 for asynchronous generating modules will be demonstrated by simulations only, while the item 10 by tests only.

The paper presents dynamic models which are able to simulate the compliance of power generating modules with the network code RfG requirements, namely the items 7 and 9 for the synchronous generating modules and 11 and 12 for asynchronous generating modules.

The network simulator MODES was used as a tool to present examples of simulation results.

The remainder of the paper is organised as follows: Section II describes simulations of the fault-ride-through capability and power system stabilizer function for the synchronous generating modules. Section III presents the capabilities of providing synthetic inertia and fast fault current activation for asynchronous generating. Section IV summarizes the paper.

II. Synchronous Generating Modules

A. *Fault-ride-through capability*

With regard to the fault-ride-through of type B synchronous power-generating modules, this capability shall be demonstrated by simulation according to the RfG (Article 51 - Compliance simulations for type B synchronous power-generating modules). These simulations are also mandatory for types C and D according to the Article 52 and 53.

The RfG defines in Article 14 a shape of voltage-against-time-profile at the connection point, which describes the conditions in which the power-generating module is capable of staying connected to the network and continuing to operate stably after the power system has been disturbed by secured faults on the transmission system. Figure 1 shows examples of such profiles for different power-generating modules types:

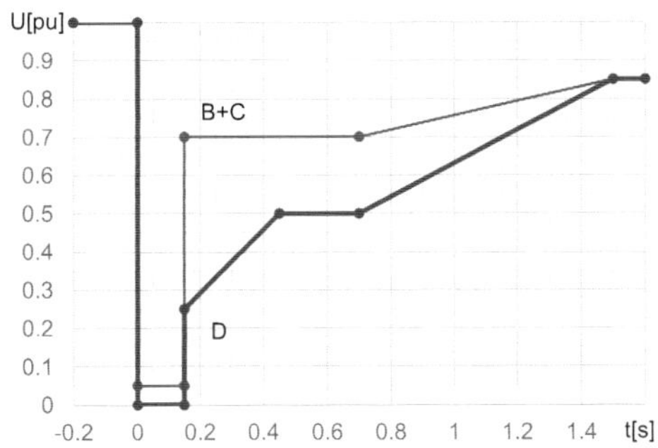

Figure 1. Voltage-against-time-profile for fault-ride-through

If the voltage is above this profile during the fault, the power-generating module must remain connected to the network.

We will examine this capability on a simple network model for modules equipped with three different excitation systems. The network model is based on the basic test grid model that was developed by a working group ENTSO-E [5]. This model was slightly modified by replacing the grid side switch by a transmission line and it is depicted in the following figure:

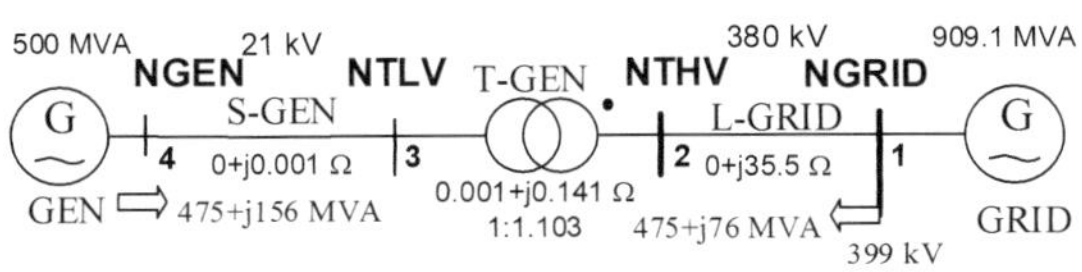

Figure 2. Scheme of the network model for fault-ride-through testing

The generator GEN has parameters described in the report [5]. The original excitation model SEXS was replaced by more complex model depicted in the following figure:

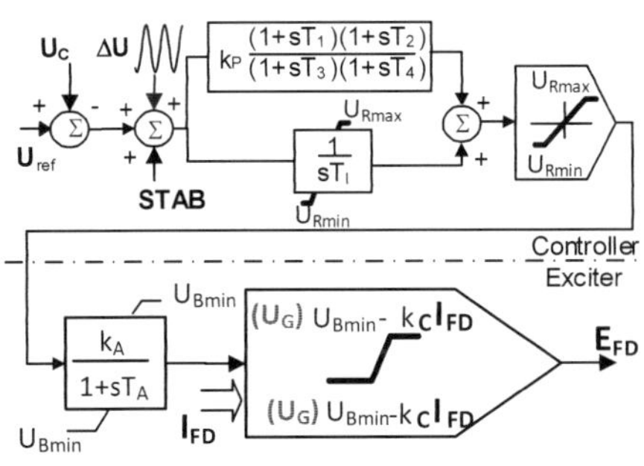

Figure 3. Block scheme of the excitation model

This scheme enables to simulate controlled rectifier exciters supplied from auxiliary alternator (e.g. IEC model ExcAC4A or ExcAC6A from [4]) or potential source controlled rectifier exciters supplied from the generator terminal or auxiliary bus (e.g. IEC model ExcST1A or ExcST7B from [4]). For the first group of exciters (so called independent excitation) the output E_{FD} is not limited by the terminal voltage U_G, while for the second group is (upper and lower limit is multiplied by U_G).

Three types of excitation are selected for the simulation:

1. Type1: independent excitation with PI controller,
2. Type2: potential source excitation with P controller and
3. Type3: potential source excitation with PI controller.

All types of excitation are equipped by a power system stabiliser (PSS) models with scheme corresponding to a dual-input stabilizer (e.g. IEC model PssIEEE2B), which uses combination of the power P_G and the speed deviation s_G to derive the stabilizing signal STAB:

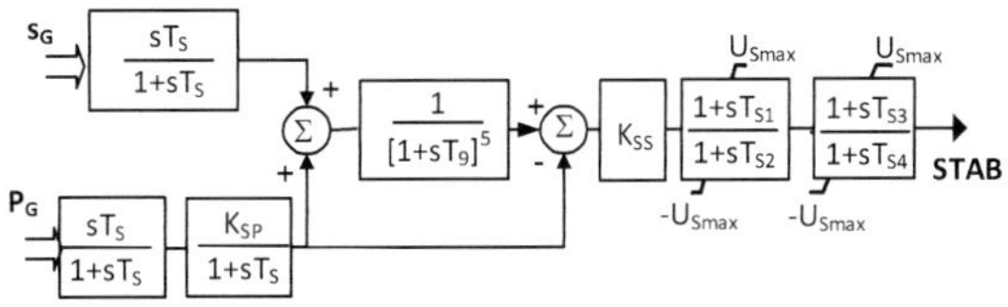

Figure 4. Block scheme of the dual-input stabilizer model

Parameters of these models are in the following tables:

Table 1. Parameters of regulators

	K_P	$T_1 = T_2 = T_3 = T_4$	T_I	U_{Rmin}	U_{Rmax}
Type1	8	0	0.33	-0.825	1
Type2	210	0	0	-99	99
Type3	40	0	0.075	-99	99

Table 2. Parameters of exciters

	k_A	T_A	K_C	U_{Bmin}	U_{Bmax}	Model
Type1	8	0	0	-6.5	8	AC_4
Type2	1	0	0	-6	6.63	ST_1
Type3	1	0	0	-4.5	5	ST_1

Table 3. Parameters of PSS

	T_S	K_{SP}	K_{SS}	T_{S1}	T_{S2}	T_{S3}	T_{S4}	T_9	U_{Smax}
Type1	5	0.4	13	0.15	0.0125	0.125	0.01	0.1	0.1
Type2	2	0.167	11.56	0.27	0.09	0.27	0.09	0.1	0.1
Type3	2	0.2	12	0.12	0.02	0.3	0.02	0.1	0.1

Now that we have a dynamic model defined, we can proceed to a simulation calculation of dynamic behavior during a three phase short circuit in the connection node NGRID with a duration 150 ms. Figure 5 shows the simulation results:

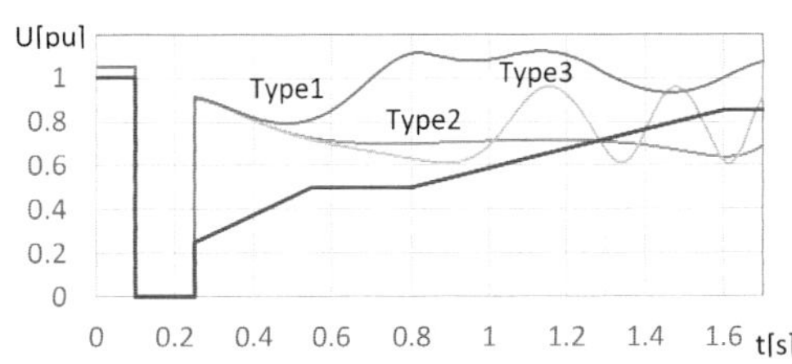

Figure 5. Voltage waveforms during fault and after its clearing

While the generator with the independent excitation (Type1) was stable the other two generators had lost stability, but they all meet the RfG requirements. In cases of Type2 and Type3 the voltage dropped below the profile and the generators can be disconnected from the network.

B. Power System Stabilizer (PSS) function

With regard to the power oscillations damping control simulation according to the RfG Article 53 (Compliance simulations for type D synchronous power-generating modules) paragraph 2a) it shall be demonstrated that PSS is capable of damping active power oscillations.

This capability can be easily demonstrated by simulation calculations: once with the stabilizer switched off and the second switched on. Figure 6 shows the active power waveforms with PSS switched off (the parameter $U_{Smax} = 0$ in the model), all types of excitation are unstable (without any disturbance).

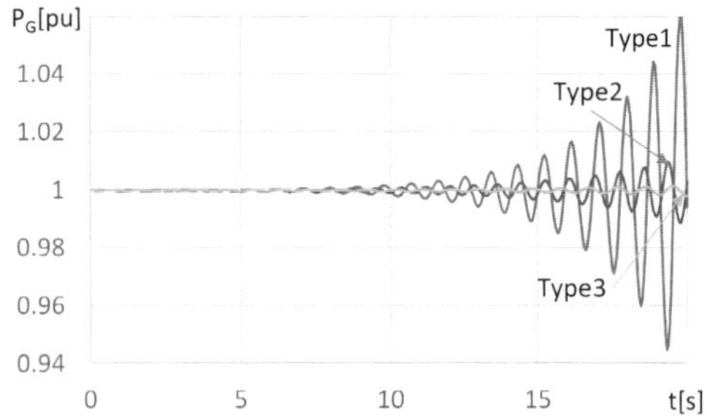

Figure 6. Active power waveforms for exciters without PSS

Since the waveforms are stable for the stabilizer switched on for all excitation types, damping capabilities are demonstrated.

If we decreased the reactance of the line L-GRID (in the scheme on Figure 2) to X=1 Ω) the waveforms stayed stable for the stabilizer switched off for the Type3. Then it is possible to demonstrate damping capabilities in another way. The excitation model (depicted in Figure 3) enables to add a harmonic signal to the summation point of the controller and emulate some network oscillations. The following figure shows responses on such forced signal ΔU with the amplitude 0.01 per unit and the frequency of 0.3 Hz (it corresponds to the typical interarea oscillation in the continental Europe synchronous area).

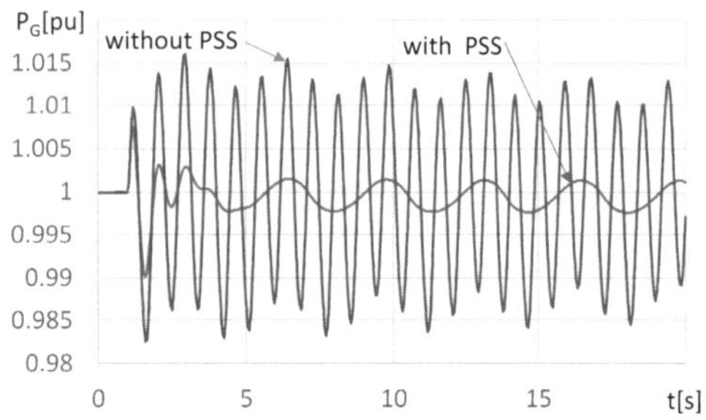

Figure 7. Active power waveforms for exciters without PSS

The signal connection at t=1 second will cause undamped oscillations with a frequency of 1.16 Hz and an amplitude of 0.013 pu in the case of the switched-off PSS. The oscillations were dampened to an amplitude of 0.0015 pu and the frequency of 0.3 Hz with PSS.

Frequency-response characteristics can be obtained in this way by using additional signals in the range of 0.1-2 Hz. The frequency-response characteristics (also called Bode diagrams) are relations between steady state sinusoidal inputs and the resultant steady state sinusoidal outputs (e.g. terminal voltage U_G and active power P_G) - see [6]. According to the methodology (published in [7]), the amplitude characteristic with the PSS switched on should always be under the characteristic without PSS so that the PSS does not worsen the oscillations damping.

III. ASYNCHRONOUS GENERATING MODULES

A. Capability of providing synthetic inertia

With regard to the synthetic inertia (the relevant TSO shall have the right to require it in the Article 21.2a), this capability shall be demonstrated by simulation according to the RfG (Article 55 - Compliance simulations for type C power park modules). These simulations are also mandatory for the type D according to the Article 56. Of course used models must contain circuits for the synthetic inertia emulation. Such model for wind farms is described in this chapter.

We consider the wind farms first. A model of wind turbine with full power converter for wind farms simulation was published and described in [8]. This model allows to simulate RfG requirements such as frequency sensitive mode, limited frequency sensitive mode for under-frequency/over-frequency and of course the synthetic inertia. We use single node models from the paper [8] to demonstrate the synthetic inertia capability. This single model with a large share of wind farms is depicted in the following figure:

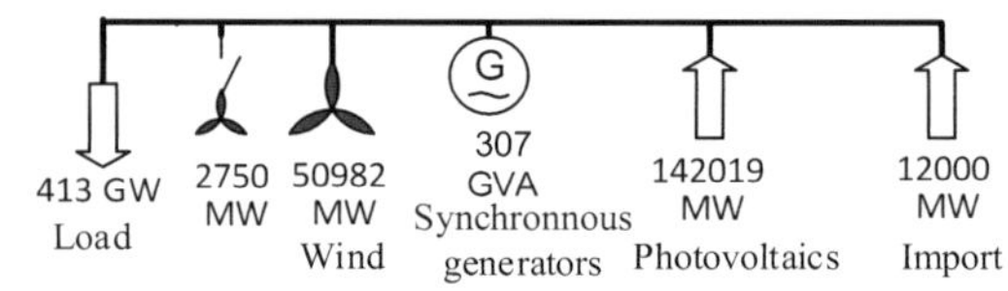

Figure 8. Scheme of single node model for synthetic inertia testing

The model corresponds to the 100% renewable energy sources scenario for summer low load period. This scenario was created in the framework of e-Highway 2050 project [9].

The following figure shows the frequency deviation after the wind farm outage with generation 2750 MW:

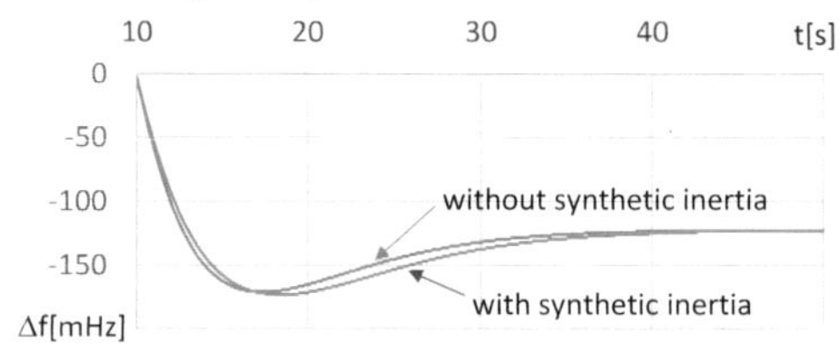

Figure 9. Frequency deviation for generation outage

In case that half the wind turbines is provided with a synthetic inertia, the rate of change of frequency is smaller compared to the case without the synthetic inertia. But the next frequency deviation course is worse because, after the turbine speed drops, its efficiency decreases and the power delivered is then smaller.

For a load outage, the situation is better as shown in the following figure for the loss of load 2250 MW:

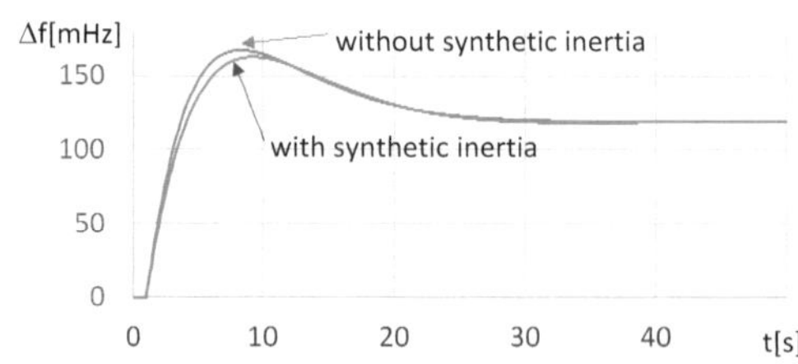

Figure 10. Frequency deviation for load outage

Excess energy accumulates in rotating masses of the turbine, which increase in speed. This will cause a decrease in efficiency and thus in turbine performance.

B. *Fast fault current activation*

RfG defines 'fast fault current' as a current injected by a power park module during and after a voltage deviation caused by an electrical fault with the aim of identifying a fault by network protection systems at the initial stage of the fault, supporting system voltage retention at a later stage of the fault and system voltage restoration after fault clearance.

With regard to the fast fault current injection (the relevant TSO shall have the right to require it in the Article 20.2b for type B power park modules) this capability shall be demonstrated by simulation according to the RfG (Article 54 - Compliance simulations for type B power park modules). These simulations are also mandatory for the types C and D according to the Articles 55 and 56.

Of course used models must contain circuits for the synthetic inertia emulation. Such model for photovoltaics (PV) plant is depicted in the following figure:

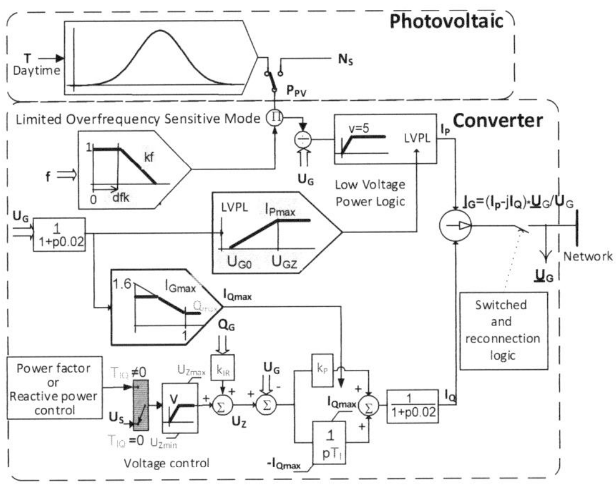

Figure 11. Block scheme of the photovoltaic plant model

The model consists of two parts. The first emulates the PV panel power P_{PV}. This power can have the constant value N_S for a short term dynamic (e.g. for transient stability calculations) or it can change in dependency on the day time for a long term dynamic (e.g. for frequency stability calculations – see [10] for more information).

The second part is a converter model that delivers the power P_{PV} to an AC network. This model is based on GE Energy models published in [11]. Two parts of this model contributes to the fault current injection during and after a voltage drop caused by an electrical fault (usually a short circuit). The Low Voltage Power Logic (LVPL) reduce the active power immediately after fault and enables to increase reactive current (it is supposed that PV plant is connected to a high voltage network so the reactive current is needed to restore the voltage and the delivery of reactive power has priority). At the same time a voltage dependent limit I_{Qmax} increases linearly as voltage drops. LVPL, voltage dependent limiter and voltage control parameters are in Table 4.

Table 4. Parameters of PV converter model

I_{Pmax}	I_{Gmax}	U_{G0}	U_{GZ}	Q_{max}	T_{IQ}	U_{Zmin}	U_{Zmax}	k_P	T_I	T_I
1.0	1.1	0.4	0.9	0.4	0	0.8	1.2	2	0.04	0.04

The test system from the benchmark simulations in [11] was used for the fast fault-current demonstration. The test system represents an aggregate model of PV plant modeled as one photovoltaic power module PVP connected through substation transformer TRAN to the 34.5 kV point of connection POI_34. One line diagram of the test system is shown on Figure 12.

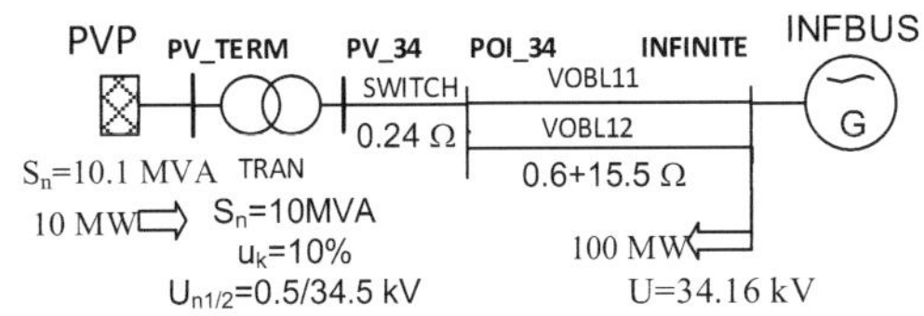

Figure 12. Scheme of the network model for fast fault-current testing

The same grid disturbance was simulated as in the report [11]: a three-phase short circuit on one of the 34.5 kV lines near the point of connection POI_34. The fault was cleared in 150 ms by the line disconnection.

The following figures shows the simulation results:

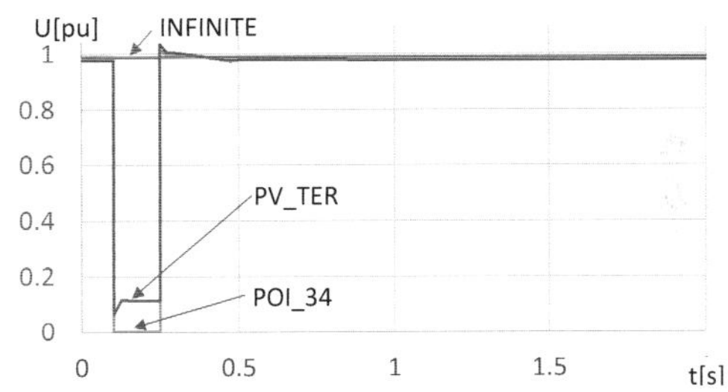

Figure 13. Voltage waveforms

The voltage in INFINITY node was constant – independent on the short circuit. The voltage in POI_34 dropped to zero due to the three-phase short bold circuit. The terminal voltage PV_TER decreased to 0.1 pu during the short circuit due to a fast fault current delivery (see the value I_G in Figure 14). Then the terminal voltage restored to the pre-fault value very fast due to the voltage control.

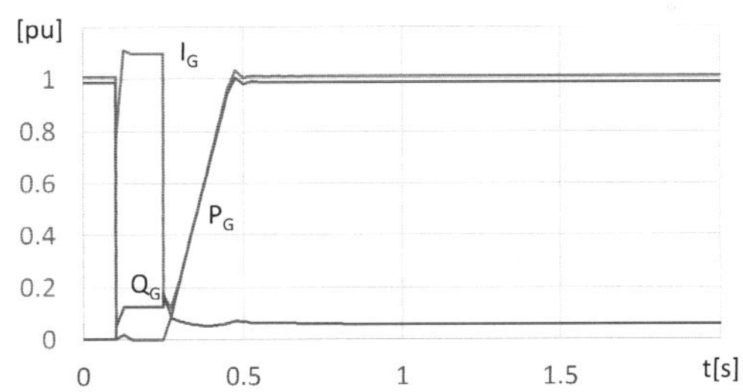

Figure 14. Active and reactive power and current waveforms

The effect of the low voltage power logic reduced the active power P_G to zero during the short circuit. The active power was restored in 250 ms after the short circuit was cleared.

The post-fault reactive power Q_G is slightly higher than the initial value, because of PV plant compensated for the greater voltage drop in the node POI_34 after one parallel line switching off.

IV. Conlusions

European Commission regulations (the guideline on electricity transmission system operation and the network code on requirements for grid connection of generators) impose an obligation for TSOs to exchange dynamic models but also to requirement for the power-generating facility owner to provide (on the request of the relevant system operator) the simulation models which properly reflect the behavior of the power-generating module in dynamic simulations.

This implies the importance of understanding these dynamic models and their use for simulation calculations. Especially if the network code allows or even requires dynamic simulations to confirm compliance with its requirements. Knowledge of dynamic models is therefore important not only for TSOs, but also for equipment owners, possibly third parties, which will provide dynamic models and appropriate simulation calculations.

This paper presents dynamic models which are suitable to simulate the compliance of power generating modules with the network code RfG requirements, namely the fault-ride-through capability and power system stabilizer function for the synchronous generating modules and the capabilities of providing synthetic inertia and fast fault current activation for asynchronous generating modules.

The paper shows what components the dynamic model must contain in order to be able to simulate the analysed capabilities or functionalities, eventually the influence of individual model parameters.

Results of the simulation calculations on simple test systems are also presented, including all necessary input data, so they can be easily reproduced in other computational tools than the used network simulator MODES.

References

[1] K. Máslo, "Frequency stability in the context of the European network codes," in Proc. 9th International Scientific Symposium on Electric Power Engineering, Sept. 2017, pp. 436-442

[2] *Commission regulation (EU) 2017/1485 of 2 August 2017 establishing a guideline on electricity transmission system operation*

[3] *Commission regulation (EU) 2016/631 of 12 April 2016 establishing a network code on requirements for grid connection of generators*

[4] *IEC 61970-457 Common Information Model (CIM) for Dynamics Profile*

[5] ENTSO-E SG SPD Report, "Documentation on controller tests in test grid configurations," 2013, *[On line] Available: https://www.entsoe.eu/fileadmin/user_upload/_library/publications/entsoe/RG_SOC_CE/131127_Controller_Test_Report.pdf*

[6] IEEE Std. 421.2-1990, IEEE Guide for Identification, Testing, and Evaluation of the Dynamic Performance of Excitation Control Systems

[7] K. Máslo, K. Witner, A. Kasembe, M. Škach, Z. Brettschneider, "Systémové stabilizátory – požadavky a jejich prověření," in Proc. 4th International Scientific Symposium Elektroenergetika, Sept. 2007

[8] K. Máslo, A. Kasembe, S. Moroni, R. Pestana, "Frequency stability modelling of the future continental Europe power system", in Proc. 52nd International Universities' Power Engineering Conference (UPEC 2017)

[9] e-Highway 2050 Deliverable 2.1, "Data Sets of Scenarios developed for 2050", [Online]. Available: http://www.e-highway2050.eu/fileadmin/documents/Results/D2_1_Data_sets_of_scenarios_for_2050_20072015.pdf

[10] K. Máslo, A. Kasembe, "Frequency Stability of Continetal Europe Synchronnous Area", in Proc. 10th International Scientific Symposium on Electric Power Engineering, Sept. 2019

[11] K. Clark, N.V. Miller, R. Walling, "Modeling of GE Solar Photovoltaic Plants for Grid Studies studies," GE International Inc., Schenectady, NY, Tech. Rep, v1.1, Apr. 2011

A: Generation of Electricity, Transmission, Distribution and Consumption of Electricity

Operation Assessment of Cogeneration Power Plant by Thermodynamic Model

Aleš Hromádka , Jana Jiřičková
Department of Electrical Power Engineering and Ecology and Regional Innovation Centre for Electrical Engineering
University of West Bohemia
Pilsen, Czech Republic
aleshrom@rice.zcu.cz, jjiricko@kee.zcu.cz

Zbyněk Martínek
Department of Electrical Power Engineering and Ecology
University of West Bohemia
Pilsen, Czech Republic
martinek@kee.zcu.cz

Martin Sirový
Regional Innovation Centre for Electrical Engineering
University of West Bohemia
Pilsen, Czech Republic
sirovy@rice.zcu.cz

Abstract — **This paper is focused on modeling of existing cogeneration power plant with backpressure turbine in Pilsen. The first motivation of the modeling is better understanding of all cycle features and its limitations. The second motivation is further investigation of some innovative implementations into the cycle. The main aim of the paper is to model the thermodynamic cycle with the turbine. The model has been created in Dymola software. Some innovative implementations for improvement of the cycle features are proposed in conclusion.**

Keywords: cogeneration power plant; backpressure turbine; thermodynamic model; Dymola software; Clara+ library

I. Introduction

Currently many cogeneration power plants (CPPs) with backpressure turbines (BPTs) are still in operation. These CPPs have been built in last five decades. The BPTs have had several advantages for cogeneration operation such as: [1]

The BPT generally reaches higher efficiency in cogeneration operation, i.e. up to 90%. The efficiency in cogeneration operation represents the sum of electrical and thermal efficiency. In other words, BPTs has a lower loss of thermal energy from the primary fuel than classical thermal power plants.

The BPTs have no problem with erosion of rotor blades due to the wetness at the output of the last low-pressure (LP) turbine stage. The problem has been often occurred during expansion in the last LP turbine stage. However, the BPT has had generally higher thermodynamic parameters at the output of the last LP turbine stage. It has meant that steam at the output of the turbine is not wet and does not cause the erosion.

The BTP does not need a condenser and a cooling tower. The higher thermodynamic parameters at the output of the BPT are used for a basic heating for heat demand. The transfer of heating energy from wet steam to hot water in heat pipeline enables condensation of full steam volume. Conventional condensing turbine (CT) has a condenser and cooling towers instead of a basic heater (BH).

On the other hand, this BPT has much more disadvantages than advantages for current cogeneration operation. These disadvantages are following:

The most significant disadvantage of the BPT is its strong dependence on heat demand. If the situation of a lack of heat demand occurs, the total efficiency will be very significantly dropped. This problem can be explained by the following fact. The water in the BPT is not able to purely condensate without additional components. That is reason why the operation of the BPTs must be quite often limited especially in summer season because of a lack of heat demand.

Moreover, currently there is a trend of decreasing energy consumption of all buildings. It means, that more customers have to be attached to the heating system to ensure crucial heat demand for condensing the complete steam volume at the output of the BPT. [3]

The heat production trend of the BPT can be described by the three areas in Figure 1. The first area can be covered only by the BPT. The second one must be covered by some other heat sources, which can cooperate with the BPT simultaneously. The BPT cannot cover the third one. The third area represents mostly summer season and some other period without heat demand.

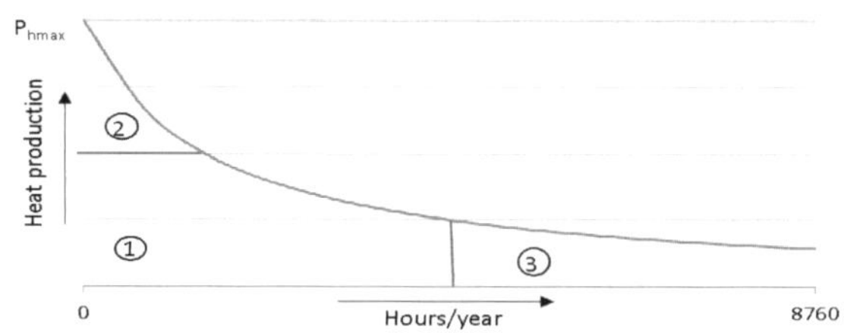

Figure 1. Coverage of the heat production by the BPT [1]

The second very significant disadvantage of the BPTs is generally lower electrical efficiency than CTs. It is tightly connected with the higher thermodynamic parameters at the output of the BPTs.

The BPTs need a certain heat demand for optimal operation. Heat energy, which is used for heat demand coverage of, ensures condensation of the complete steam volume in a primary circuit.

However, some BPTs offer an option of condensation by dry mechanical condenser (DMC). The DMC is a device, which enables fully or partly pure condensing operation at the BPT. The device can be described as a system of many electrical ventilators, which emit the heat energy from a primary circuit to the atmosphere.

However, the pure condensing operation at the BPT has two big disadvantages. The one of the disadvantages is big losses of thermal energy. It is connected with the drop of the total efficiency down to 20%. The second disadvantage is increasing own consumption of the CPP, because the system of ventilators consumes additional electricity. This way seems to be less effective and less convenient for the CCP economy.

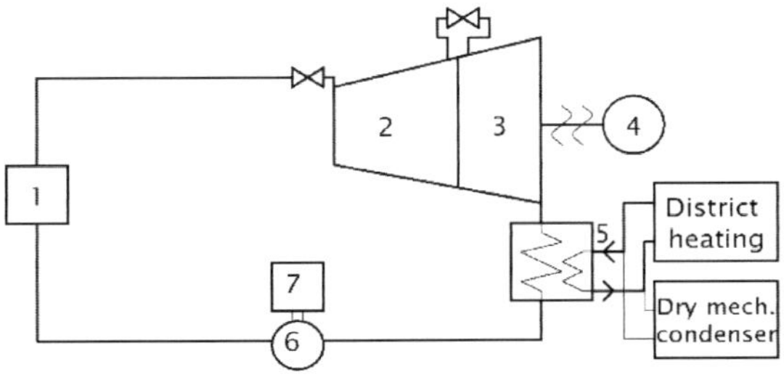

Figure 2. Principle of BPT with condensation by the DMC [1], [2]

The last disadvantage is the less flexible operation than CTs. This disadvantage can be demonstrated in Figure 3. and Figure 4. There are shown options of heat and electric production for both turbines.

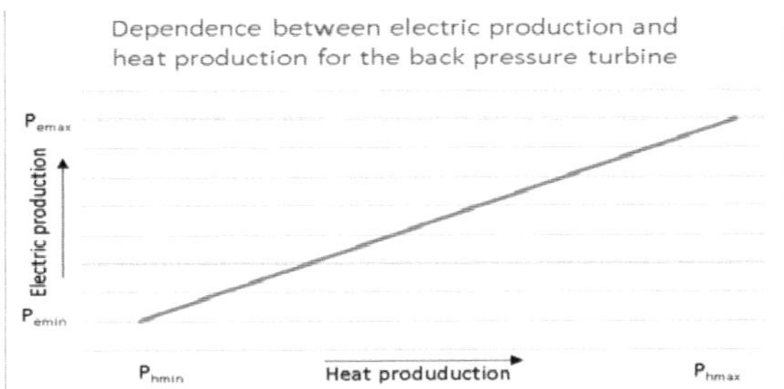

Figure 3. Dependence of electric and heat production of a BPT [1]

There is no regulation space for the BPT, which is shown in Figure 3. The dependence is strict without any options to production changes.

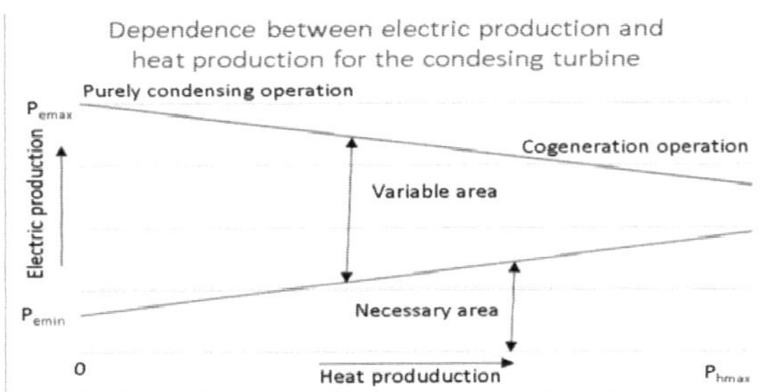

Figure 4. Dependence of electric and heat production of the CT [1], [5], [9]

The CT has completely different production trend. There is significant regulation area in Figure 4. The correct setup of the production can be chosen quite easily according to heat demand. Of course, the total efficiency is not optimal in all working points. However, the CPP with the CT can provide the flexibility, which is connected with better production incomes.

Therefore, there are contracts for primary, secondary and minute backup, which offer participation on the daily electricity market. The participation of the daily electricity market guarantees often better prices of electric energy for their producers. It means that electricity producers want to be as flexible as possible in the production. If important regulation requirements are met, the power plant obtains the supporting services contracts. These efforts are generally connected with better economy of the power plants.

The efforts are quite similar in CPPs. Therefore, the BPTs have become unsuitable, because of its limited regulation. That is reason why this paper is focused on BPTs. The BPTs are still used in the current CPPs. Our first aim is to assess its operability. The second aim is to improve its constraints for better sustainability of the BPT.

Therefore, the thermodynamic model of the BPT in Pilsen. has been created. The biggest BPT, which is currently in operation in Czech Republic, has been installed in Pilsen CPP. The nominal electric power of the BPT is 70 MW. [1], [2] The model of the BPT shows all thermodynamic parameters at variable operation. That was the first step for further investigation how to improve the current BPT and how to minimize its constraints.

II. Identification of Thermodynamic Parameters and Description of Particular Parts in the BPT

It is necessary to identify nominal thermodynamic parameters of the BPT for correct modeling of the cycle.

A. Description of the System of the Boilers

Firstly, it is necessary to describe the investigated heat source. Three boilers has been modeled as the one heat source. All of the three boilers has had the same heating power, which has been 128MWt. Two of the boilers have been identical supercritical solid fuel fired boilers. The last one has been fluidized bed combustion boiler. All boilers have worked as the heat source for the two turbines and reduction station for technological steam production. The main steam collector (MSC) has been fed by the boilers. Therefore, the boilers has been defined by thermodynamic parameters of the MSC.

Table 1. Nominal parameters of the MSC

Nominal parameters of the main steam collector		
Thermodynamic parameter	***Value***	***Unit***
Pressure of the MSC	133.5	bar
Temperature of the MSC	535	°C
Enthalpy of the MSC	3426.3	kJ/kg
Mass flow of the MSC	97.22222	kg/s

B. Desription of the BPT

The most important part of the model is already mentioned the BPT. The BPT, which has been divided into HP and LP turbine stages (TS), has been fed from the MSC. The HP TS has two non-regulated outputs for regenerative heating. One regulated output has been placed between HP TS and LP TS. The output has enabled obtaining technological steam for additional heating purposes or commercial purposes. The LP TS has had two non-regulated outputs for regenerative heating as well. The BH for district heating has been placed at the end of the LP TS of the BPT. The BH has offered suitable thermodynamic parameters for district heating.

Table 2. Nominal thermodynamic parameters of the modeled BPT

Nominal parameters of the backpressure turbine		
Thermodynamic parameter	***Value***	***Unit***
Pressure in the first partial arc	36.7	bar
Temperature in the first partial arc	361.5	°C
Enthalpy in the first partial arc	3130.5	kJ/kg
Mass flow in the first partial arc	6.47222	kg/s
Pressure in ther second partial arc	20.5	bar
Temperature in the second partial arc	291.6	°C
Enthalpy in the second partial arc	3003.8	kJ/kg
Mass flow in the second partial arc	5.44444	kg/s
Pressure in the end of HP TS	13	bar
Temperature in the end of HP TS	249.6	°C
Enthalpy in the end of HP TS	2930.5	kJ/kg
Mass flow in the third partial arc	11.25	kg/s
Mass flow to technology steam	9.66389	kg/s
Mass flow in the input of LP TS	73.54167	kg/s
Pressure in the fourth partial arc	7.8	bar
Temperature in the fourth partial arc	196.8	°C
Enthalpy in the fourth partial arc	2832.2	kJ/kg
Mass flow in the fourth partial arc	5.06944	kg/s
Pressure in the fifth partial arc	3.4	bar
Temperature in the fifth partial arc	137.9	°C
Enthalpy in the fifth partial arc	2690.9	kJ/kg
Mass flow in the fifth partial arc	3.75	kg/s
Pressure at the end of LP TS	1.2	bar
Temperature at the end of LP TS	104.8	°C
Enthalpy at the end of LP TS	2559.9	kJ/kg
Mass flow at the end of LP TS	64.75556	kg/s

C. *Description of the BH*

The next component in the cycle has been the BH for heat transfer to district heating. There have been primary and secondary side. The primary side has been used for the condensation of the complete steam volume at the end of the BPT. Additional water has been added into the BH primary side, because a part of feed water has been used for technological purposes. It has meant that the additional water represents the cycle loss of the feed water and the complete steam volume for technological purposes. The secondary side has been dedicated for basic heating for the district heating.

Table 3. Nominal thermodynamic parameters of the BH

Nominal parameters of the basic heater		
Thermodynamic parameter	***Value***	***Unit***
Pressure at the end of LP TS	1.2	bar
Temperature at the end of LP TS	104.8	°C
Enthalpy at the end of LP TS	2559.9	kJ/kg
Mass flow at the end of LP TS	64.75556	kg/s
Additional feed water	10.14167	kg/s
Pressure at the end of LP TS	3	bar
Temperature at the end of LP TS	76.4	°C
Enthalpy at the end of LP TS	320	kJ/kg
Mass flow at the end of LP TS	1777.778	kg/s

D. *Description of the LP Regenerative Heaters*

LP regenerative heaters (LPH) have been fed by two non-regulated arc from the LP TS. The system of the LP regenerative heating is that the steam from the fourth arc has been led to the LPH1. The steam has been condensed there into hot water and after that has been led to the LPH2. The hot water from the LPH1 have been mixed with steam from the fifth arc at the input to the LPH2. The complete volume of the wet steam has been condensed in the LPH2. After that, the water has been pumped into the feed water. This operation has increased the total cycle efficiency.

Table 4. Nominal parameters of the LPHs

Nominal parameters of the LP heaters		
Thermodynamic parameter	***Value***	***Unit***
Pressure in the LPH1	7.8	bar
Temperature in the LPH1	196.8	°C
Enphalpy in the LPH1	2832.2	kJ/kg
Mass flow in the LPH1	5.06944	kg/s
Pressure in the LPH2	3.4	bar
Temperature in the LPH2	137.9	°C
Enphalpy in the LPH2	2690.9	kJ/kg
Mass flow in the LPH2	3.75	kg/s

E. *Description of the Feed Water Tank*

The feed water tank (FWT) has worked more or less as a heater as well. However, there has not been any separated part only one space with the same pressure. In the other words, the pressure balance has been created in the FWT by the filling level regulation of the feed water. All the outputs except outputs from the LP TS have ended in the FWT. It has meant that the FWT has been fed by the water from the LP regenerative heating, the third output steam and condensed water from the HP regenerative heating. The output water from the FWT has been led to the main feed pump.

Table 5. Nominal thermodynamic parameters of the FWT

Nominal parameters of the feed water tank		
Thermodynamic parameter	***Value***	***Unit***
Pressure in the FWT	8	bar
Temperature of the FWT	160.3	°C
Input enthalpy of the feed water	677.1	kJ/kg
Input mass flow of the feed water	83.72222	kg/s
Temperature in the third output	240.73	°C

Nominal parameters of the feed water tank		
Thermodynamic parameter	***Value***	***Unit***
Enthalpy in the third output	2930.5	kJ/kg
Mass flow in the third output	1.58611	kg/s
Pressure in the HPH output	19.5	bar
Temperature in the HPH output	211.1	°C
Enthalpy in the HPH output	902.7	kJ/kg
Mass flow in the HPH output	11.91667	kg/s

F. Description of the System of Feed Pumps

Three feed pumps have worked in the cycle. The first one has been placed between the BH and LP regenerative heaters. The condenser feed pump (CFP) has increased the pressure for achieving balance in the FWT.

The second one has been placed at the output of the LPH. The LP feed pump (LPFP) has enabled that the condensed water from LPH has been fed to the feed water.

The last one has been placed at the FWT output. The main feed pump (MFP) has had to increase the pressure up to 168 bars. It has had to be reached, because that has corresponded with nominal pressure of the boilers.

Table 6. Nominal parameters of the feed pumps

Nominal parameters of the feed pumps		
Thermodynamic parameter	***Value***	***Unit***
Input pressure of the CFP	1.2	bar
Output pressure of the CFP	8	bar
Nominal mass flow of the CFP	74.897	kg/s
Input pressure of the LPFP	3.23	bar
Output pressure of the LPFP	8	bar
Nominal mass flow of the LPFP	8.8194	kg/s
Input pressure of the MFP	8	bar
Output pressure of the MFP	168	bar
Nominal mass flow of the MFP	97.222	kg/s

G. Identification Parameters and Description of the HP Regenerative Heaters

HP regenerative heaters (HPH) have been fed by two non-regulated arc from the HP TS. The system of the HPHs have been that the steam from the first arc has been led to the HPH1. The steam has been condensed there into hot water and has been led to the HPH2. The hot water from the HPH1 have been mixed with steam from the second arc at the input of the HPH2. The complete volume of the wet steam has been condensed in the HPH2. After that, the condensed hot water has been fed to the FWT. This operation has increased the cycle efficiency.

Table 7. Nominal parameters of the HPHs

Nominal parameters of the HP heaters		
Thermodynamic parameter	***Value***	***Unit***
Pressure in the HPH1	36.7	bar
Temperature in the HPH1	361.5	°C
Enphalpy in the HPH1	3130.5	kJ/kg
Mass flow in the HPH1	6.4722	kg/s
Pressure in the HPH2	20.5	Bar
Temperature in the HPH2	291.6	°C
Enphalpy in the HPH2	3003.8	kJ/kg
Mass flow in the HPH2	5.4444	kg/s

III. Thermodynamic Model of the BPT

The thermodynamic model of the entire cycle is described in that chapter. The entire model has been created in Dymola Software by its thermodynamic library Clara+, which has been changed for needs of CPP modeling. Dymola software offers really good environment for modeling of dynamic systems, which can be described by ordinary differential equations.

A. Explanation of Particular Models

Firstly, it is necessary to describe the model of the heat source. These has been used the changed model of the boiler from the ClaRa+ library. However, three boilers, which have been already installed in the CPP, are considered as one big heat source. That is the reason why just one model of the boiler has been used for our modeling.

The changed model of boiler is based on lumped balance equations for mass and energy and on transfer functions and characteristic maps for heat release and pressure losses. The aim of this model is to model the boiler's main dynamics without allowing deep insight into the process. The model contains the evaporation and superheat. This simplification does not allow detailed investigations of the control of steam temperature and feed water mass flow. There is applying an enthalpy correction.

The further modeled part is a turbine. The model is based on Stodola's cone law (law of the ellipse) with an optional efficiency degradation using the trend of consumption characteristics. Main equation of the Stodola's turbine model is following: [4], [7], [8]

$$\frac{m\sqrt{T_{01}}}{p_{01}} = k\sqrt{\left[1-\left(\frac{p_{02}}{p_{01}}\right)^2\right]}\ [-] \quad (1)$$

Where m represents mass flow rate, T_{01} represents input temperature, p_{02}/p_{01} represents a ratio between input and output pressure and k represents a typical constant for Stodola's equation, which is given by a type of turbine.

The next modeled part is a heater. The model of a heater is created by two heat exchangers with a wall between each other.

All the models of heat exchangers have been modeled as static heat exchangers using logarithmic mean temperature difference (LMTD). Main equation for the model of the static heat exchanger is following: [8]

$$Q = k_c * A_{heat} * \frac{dT_2 - dT_1}{\ln|dT_2| - \ln|dT_1|}\ [MWt] \quad (2)$$

Where Q represents transferred heat power, k_c represents heat transfer coefficient, A_{heat} represents transfer area, dT_1, dT_2 represent temperature differences on primary and secondary side. [8]

The next modeled part is a feed pump. The feed pump model is based on a volume flow rate, which is dependent on drive power and pressure difference only. The feed pump model is determined for VLE mixtures, which are defined in the library ClaRa+.

The last modeled part of the model is the feed water tank. This model is based on the mixed volume approach and level-dependent phase separation.

B. *Introduction of the Model of the BPT*

The model of the cycle has been created in Dymola by the model library ClaRa+. This model has been able to simulate the operation at various boundary conditions. The complex view at its operation of the model has been obtained. The parametrization of the model has been done at the nominal operation. The trend at partial operation has been used for all turbine stages of the BPT.

Many measurement points in the entire cycle model are shown in Figure 6. All these points have been correctly set according to data from nominal operation of the CPP.

The partial operation trend of the BPT has been set for each of the turbine stages according to the efficiency of the turbine stage at the nominal operation. The trend of overpressure turbine is shown in Figure 5. The trend has described the change of the thermodynamic efficiency of the overpressure turbine at various operation percentages. The estimation of the partial operation has been created for all the turbine stages by the trend. [9]

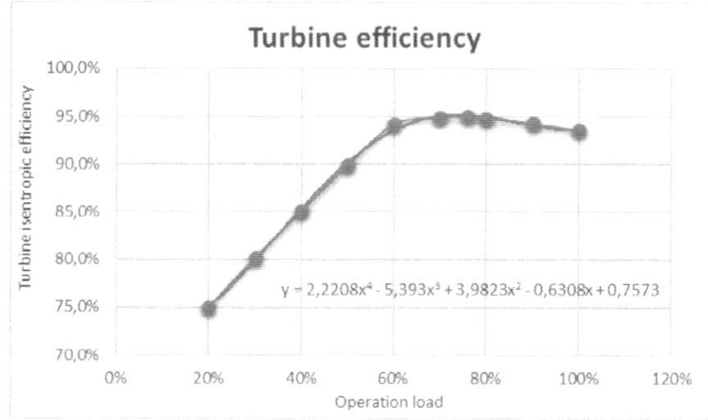

Figure 5. Trend for overpressure turbine at partial operation [9]

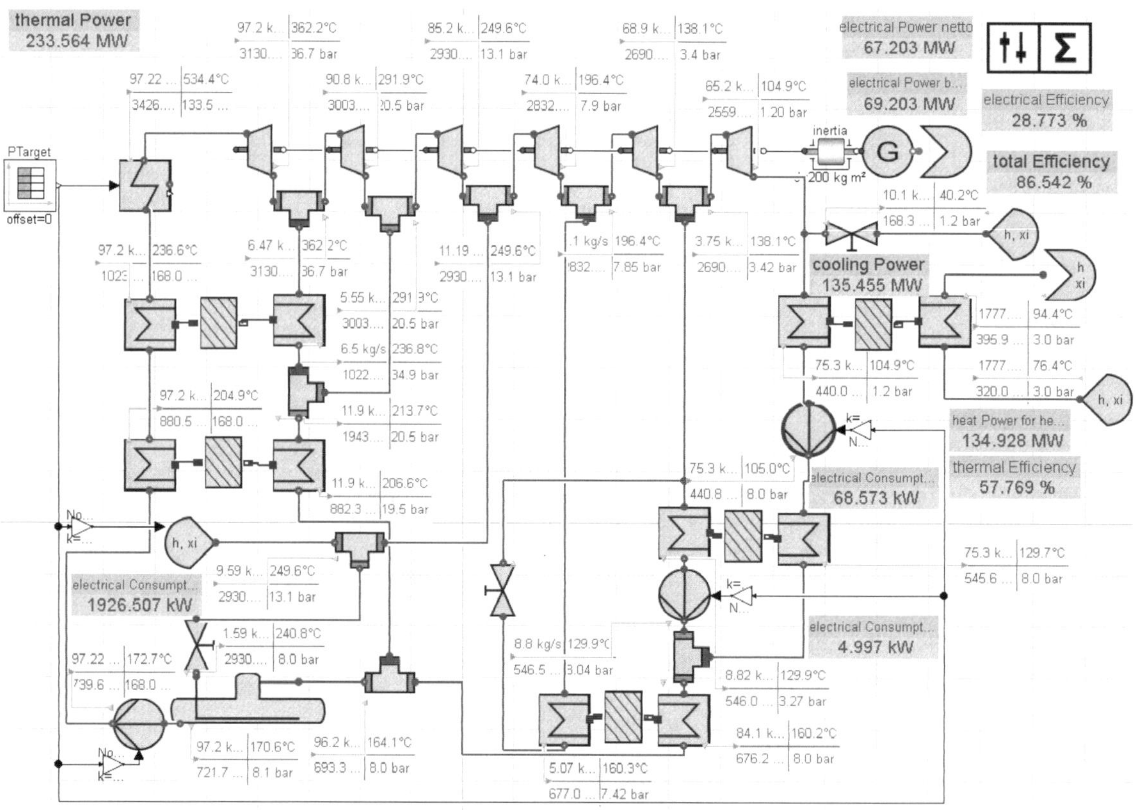

Figure 6. Thermodynamic model of the entire cycle at the nominal operation

IV. RESULTS OF THE MODEL

The results of the model have verified by the data from the nominal operation of the cycle.

Table 8. Verification of the modeled values

Verification of the modeled values		
Thermodynamic parameter	***Modeled***	***Real***
Electric power [MWe]	69.203	70
Electric efficiency [%]	28.773	29.114
Thermal efficiency [%]	57.769	57.659

According to achieved results, it is obvious that the model has worked quite well at the nominal operation. The accuracy of the entire model has been a little bit lower at partial operation, but it has still been sufficient for the aim of the paper.

The production trend, which is shown in Figure 7. , is the further result of the assessment.

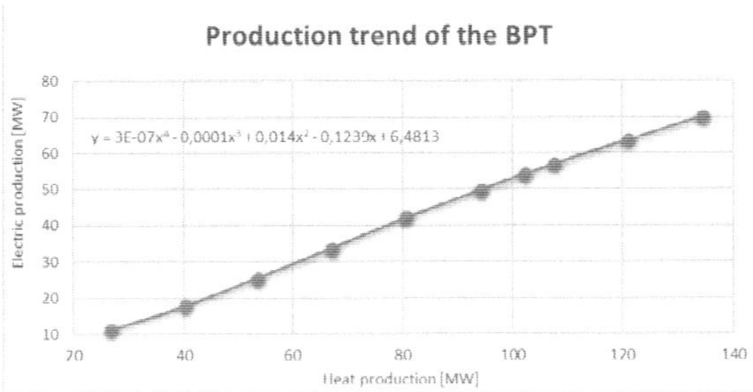

Figure 7. Production trend between electric and heat production

V. DISCUSSION

The entire created model will be used for further modeling of some innovative parts into the cycle. The model can be interconnected with the real time measuring system. It would mean that the model could be used for an optimal setup estimation of the boundary conditions at various operation. [10]

One of the possible way, how to improve current cycle could be adding a thermal energy storage inside to the district heating. The implementation could cover dynamic changes in the district heating. The innovation could improve the cycle flexibility.

Another utilization of a thermal energy storage could be in the MSC. Most of boilers have a problem with fuels consistency. Therefore, the thermodynamic parameters in the MSC could be different from operation requirements. In case of adding of a thermal energy storage, that problem could be probably fixed.

The last way how to improve the cycle could be utilization of an electrical energy storage for the primary regulation coverage. This innovation could increase the flexibility and decrease the loss, which is connected with dynamic changes of the cycle.

VI. CONCLUSION

This paper has concerned the thermodynamic model of the BPT. Firstly, it has mentioned the certain motivation of the paper. Next part has been focused on the description of all parts of the cycle and identification of the nominal parameters. Next part has included some details of the model. The last parts of the paper have informed about the assessment of the model accuracy and some suggestions of adding some innovative components into the cycle, which could be modeled by the model in Future.

Nowadays the BPTs is worse options than the CTs for utilization in CPPs. However, there are several options how to improve cycles with the BPTs. The worst BPTs features seem to be its impossibility of pure condensing operation and its bad flexibility. Further investigation must be focused on these features and how they could be improved.

ACKNOWLEDGMENT

This paper is supported by the student research project SGS-2018-023, and research project EPSILON TH02020798.

REFERENCES

[1] A. Hromádka, “Kombinovaná Výroba Tepla a Elektřiny v Plzeňské Teplárenské A.s.” DSpace at University of West Bohemia: NO TITLE, UNION Agency, 25 June 2013, dspace5.zcu.cz/handle/11025/7252.

[2] A. Hromádka, “Analýza a Optimalizace Procesního Řízení.” DSpace at University of West Bohemia: NO TITLE, UNION Agency, 26 May 2015, dspace5.zcu.cz/handle/11025/18815.

[3] P. Belany, et al. “Decreasing the Power Consumption of Buildings with Utilization of Lighting Control.” 2018 Elektro, 2018, doi:10.1109/elektro.2018.8398303.

[4] A. Hromádka, J. Jiřičková, Z. Martínek, M. Sirový, “Comprehensive mathematical model of the turbomachinery in the czech nuclear power plant Temelin : verification of the Thermal power library model for the Dymola software by real measurement data from the power plant,” In Proceedings of the 9th International Scientific Symposium on Electrical Power Engineering (ELEKTROENERGETIKA 2017). Košice: Technical University of Košice, 2017. s. 545-550. ISBN: 978-80-553-3195-9

[5] J. Zuwała, “Life Cycle Approach for Energy and Environmental Analysis of Biomass and Coal Co-Firing in CHP Plant with Backpressure Turbine,” Journal of Cleaner Production 35, 2012, s. 164–175.

[6] R. Dipippo, “Binary Cycle Power Plants.” Geothermal Power Plants, 2012, pp. 151–182

[7] S. I. Dixon, C.A. Hall. “Introduction.” Fluid Mechanics and Thermodynamics of Turbomachinery, 2014, pp. 1–37., doi:10.1016/b978-0-12-415954-9.00001-2.

[8] D. H. Cooke, “On Prediction of Off-Design Multistage Turbine Pressures by Stodola’s Ellipse.” Journal of Engineering for Gas Turbines and Power, vol. 107, no. 3, 1985, p. 596., doi:10.1115/1.3239778.

[9] T. G. Erhart, et al. “Influence of Condenser Conditions on Organic Rankine Cycle Load Characteristics.” Journal of Engineering for Gas Turbines and Power, vol. 135, no. 4, 2013, p. 042301., doi:10.1115/1.4023113.

[10] M. Novak, et al. “Off-Grid Model of Photovoltaic System with Real Time Monitoring Function.” 2018 IEEE International Energy Conference (ENERGYCON), 2018, doi:10.1109/energycon.2018.8398804.

Ensuring Postemergency Modes Stability in Power Supply Systems Equipped with Distributed Generation Plants

Yuri Bulatov
Bratsk State University
Country Bratsk, Russia
bulatovyura@yandex.ru

Andrey Kryukov
Irkutsk State Transport University
Irkutsk National Research Technical University
Irkutsk, Russia
and_kryukov@mail.ru

Konstantin Suslov
Natalia Shamarova
Irkutsk National Research Technical University
Irkutsk, Russia
souslov@istu.edu

Abstract — **Study results are presented aimed at emergency control methods development in power supply systems equipped with distributed generation plants. Control actions ensuring mode introduction into the stability region, were formed by changing the vector of controlled parameters along the preset path. Quality of the dynamic processes when implementing the control actions was ensured by harmonized setting of automatic voltage regulators (AVR) and automatic speed governors (ASG) of synchronous generators. It is shown that the efficient introduction of postemergency mode into the stability region can be implemented based on ratios obtained while using singular value decomposition of Jacobi matrix for steady-state mode equations. The application of the fuzzy technologies for AVR and ASG tuning allows to generate a qualitative dynamic transition for generators load shedding in the postemergency mode.**

Keywords: power supply systems, distributed generation plants, emergency control

I. Introduction

Introduction of smart grids implies a broader use of distributed generation (DG) plants connected to power consumers [1–7] power supply systems (PSS). The use of DG technologies in electrical power systems (EPS) requires development of new control algorithms in normal, emergency and postemergency modes [8–12].

One of the main issues resolved by emergency control (EC) systems is ensuring static dead-beat stability of postemergency modes (PEM) of EPS; in this case, control actions are generated for PEM introduction into the permissible region (stability region) along a certain path in the $D\mathbf{Y}$ space of the controlled parameters $\mathbf{Y}$ [11, 12]. As a rule, the path $D\mathbf{Y}$ is assumed to be linear, and can be determined proceeding from the following assumptions: to be set based on the preliminary calculations; to be determined based on the shortest distance to limiting hypersurface [11] or by minimal damage associated with cutting power sources and consumers [9–11].

Use of DG plants makes the task of modes introduction into the stability region up-to-date for distribution networks and power supply systems. This task acquires a special importance in PSS equipped with DG plants, implemented on the basis of non-traditional renewable energy sources. Such plants, as mini hydropower plants (HPP) and offshore windmill farms can be in a remote location from the consumption centers, which leads increasing the probability of going beyond the static dead-beat stability regions boundaries.

II. Formulation of the Problem

The EC task is illustrated in Fig. 1, where the cross section of the stability region is shown with the coordinate plane of the active generating powers P_i, P_j is shown; here and further in the text it is assumed that the stability and transmitted power limits coincide [12]. Curve 1 corresponds to the boundary of the stability region for the complete network diagram, curve 2 corresponds to a similar boundary when one of the main transmission lines is disconnected, and curve 3 corresponds to condition $\Im$ = const, where $\Im$ – the required value of the stability margin of the PEM.

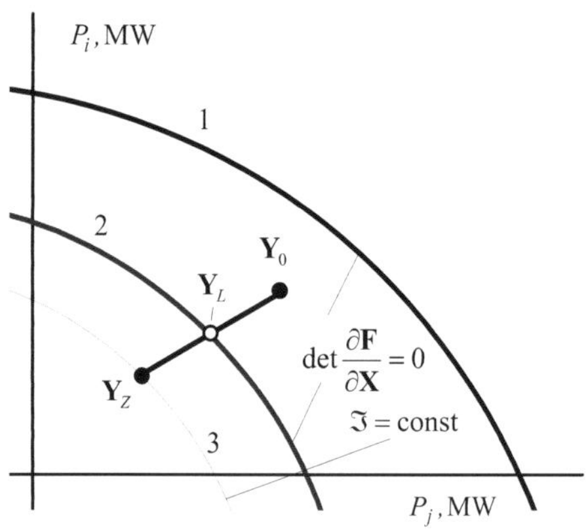

Figure 1. To the task of PEM introduction into the stability region: $\frac{\partial \mathbf{F}}{\partial \mathbf{X}}$ – matrix of Jacobi for steady-state equations (SSE); $\Im$ – stability reserve

The task of EC application is to reach the point $\mathbf{Y}_z$:

$$\mathbf{Y}_z = \mathbf{Y}_0 + D\mathbf{Y} = \mathbf{Y}_0 + t\Delta\mathbf{Y},$$

where $\Delta\mathbf{Y}$ – the direction of regime change in space $\mathbf{Y}$; t – a scalar parameter that determines powers change values in the relevant direction.

Generators power change in the direction chosen $\Delta\mathbf{Y}$ should be effected for an acceptable quality of the dynamic processes, which can be achieved through the use of automatic voltage regulators (AVR) and automatic speed governors (ASG) of DG plants synchronous generators. The relatively small power of DG plants and the low value of rotors inertia constant of their generators require taking into account the mutual influence of AVR and ASG when tuning them. It should also be noted, that optimal control requires AVR and ASG settings correction when significant changes are introduced in the operation modes of both DG and power supply systems (PSS). These requirements can be met by using the intelligent control algorithms [13–21].

Below are the results of studies aimed at development of methods of the mode introduction into stability region meeting the required stability margin and ensuring a quality transition when load shedding generators.

III. Method for Postemergency Mode Introduction into the Stability Region

Static stability margin is one of the factors determining admissibility of power system operating mode. Ensuring the margin is necessary to preserve EPS stability at irregular $D\mathbf{Y}$ fluctuations of $\mathbf{Y}$ parameters lying within hyperellipsoid boundaries [11]:

$$d\mathbf{Y}^{\mathrm{T}}\mathbf{S}^{-1}d\mathbf{Y} - C^2 = 0,$$

where $d\mathbf{Y} = \mathbf{Y} - M\mathbf{Y}$; M – the sign of mathematical expectation; $\mathbf{S} = M\left[(\mathbf{Y} - M\mathbf{Y})(\mathbf{Y} - M\mathbf{Y})^{\mathrm{T}}\right]$ – covariance matrix.

Therefore only those EPS operating modes are of practical importance which are at a sufficient distance from the stability region boarder, satisfying the condition: $\det\frac{\partial\mathbf{F}}{\partial\mathbf{X}} = 0$.

One of the possible admissibility criteria of the current power system mode characterizing its proximity to the boarder of static dead-beat stability, is a minimal singular number $\sigma_{\min}$ of Jacobi matrix for steady-state equations (SSE) (fig. 2).

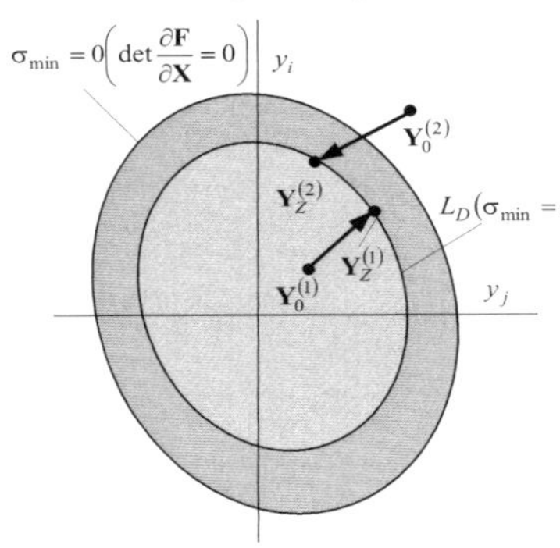

Figure 2. Determining permissible mode in the intended direction of weighting

The value is $\sigma_{\min}$ determined by the following equations:

$$\left(\frac{\partial\mathbf{F}}{\partial\mathbf{X}}\right)^{\mathrm{T}}\mathbf{K} = \mathbf{L};\ \left(\frac{\partial\mathbf{F}}{\partial\mathbf{X}}\right)\mathbf{L} = \sigma_{\min}\mathbf{K}$$

or

$$\left(\frac{\partial\mathbf{F}}{\partial\mathbf{X}}\right)^{\mathrm{T}}\left(\frac{\partial\mathbf{F}}{\partial\mathbf{X}}\right)\mathbf{L} = \sigma_{\min}^2\mathbf{L};\ \left(\frac{\partial\mathbf{F}}{\partial\mathbf{X}}\right)\left(\frac{\partial\mathbf{F}}{\partial\mathbf{X}}\right)^{\mathrm{T}}\mathbf{K} = \sigma_{\min}^2\mathbf{K}.$$

Hence, $\sigma_{\min}$ is a square root from the minimal own value of $\lambda_{\min}$ matrices $\left(\frac{\partial\mathbf{F}}{\partial\mathbf{X}}\right)^{\mathrm{T}}\left(\frac{\partial\mathbf{F}}{\partial\mathbf{X}}\right)$ or $\left(\frac{\partial\mathbf{F}}{\partial\mathbf{X}}\right)\left(\frac{\partial\mathbf{F}}{\partial\mathbf{X}}\right)^{\mathrm{T}}$.

To avoid algorithmic difficulties associated with differentiating of matrices product $\left(\frac{\partial\mathbf{F}}{\partial\mathbf{X}}\right)^{\mathrm{T}}\left(\frac{\partial\mathbf{F}}{\partial\mathbf{X}}\right)$ or $\left(\frac{\partial\mathbf{F}}{\partial\mathbf{X}}\right)\left(\frac{\partial\mathbf{F}}{\partial\mathbf{X}}\right)^{\mathrm{T}}$, one can use the following equations [11]:

$$\begin{aligned}
&\mathbf{F}(\mathbf{X},\mathbf{Y}_0 + T\Delta\mathbf{Y}) = \mathbf{0};\\
&\mathbf{V1}(\mathbf{X},\mathbf{K},\mathbf{L}) = \left(\frac{\partial\mathbf{F}}{\partial\mathbf{X}}\right)^{\mathrm{T}}\mathbf{K} - \sigma_{\min}\mathbf{L} = 0;\\
&\mathbf{V2}(\mathbf{X},\mathbf{K},\mathbf{L}) = \left(\frac{\partial\mathbf{F}}{\partial\mathbf{X}}\right)\mathbf{L} - \sigma_{\min}\mathbf{K} = 0;\\
&U(\mathbf{K}) = \mathbf{K}^{\mathrm{T}}\mathbf{K} - 1 = 0.
\end{aligned}\qquad (1)$$

Solving equations (1) using the Newton's method is related to the root-finding of the following linear equations system:

$$\begin{bmatrix} \frac{\partial\mathbf{F}}{\partial\mathbf{X}} & \mathbf{0} & \mathbf{0} & \frac{\partial\mathbf{F}}{\partial T}\\ \frac{\partial\mathbf{V1}}{\partial\mathbf{X}} & \frac{\partial\mathbf{V1}}{\partial\mathbf{K}} & \frac{\partial\mathbf{V1}}{\partial\mathbf{L}} & \mathbf{0}\\ \frac{\partial\mathbf{V2}_2}{\partial\mathbf{X}} & \frac{\partial\mathbf{V2}}{\partial\mathbf{K}} & \frac{\partial\mathbf{V2}}{\partial\mathbf{L}} & \mathbf{0}\\ \mathbf{0} & \frac{\partial U}{\partial\mathbf{K}} & \mathbf{0} & 0 \end{bmatrix}\begin{bmatrix}\Delta\mathbf{X}\\ \Delta\mathbf{K}\\ \Delta\mathbf{L}\\ \Delta T\end{bmatrix} = -\begin{bmatrix}\mathbf{F}\\ \mathbf{V1}\\ \mathbf{V2}\\ U\end{bmatrix}$$

or

$$\begin{bmatrix} \frac{\partial\mathbf{F}}{\partial\mathbf{X}} & \mathbf{0} & \mathbf{0} & \frac{\partial\mathbf{F}}{\partial T}\\ \frac{\partial\mathbf{V1}}{\partial\mathbf{X}} & \left(\frac{\partial\mathbf{F}}{\partial\mathbf{X}}\right)^{T} & -\sigma_{min}\mathbf{E} & \mathbf{0}\\ \frac{\partial\mathbf{V2}}{\partial\mathbf{X}} & -\sigma_{min}\mathbf{E} & \frac{\partial\mathbf{F}}{\partial\mathbf{X}} & \mathbf{0}\\ \mathbf{0} & 2\mathbf{K}^{T} & \mathbf{0} & 0 \end{bmatrix}\begin{bmatrix}\Delta\mathbf{X}\\ \Delta\mathbf{K}\\ \Delta\mathbf{L}\\ \Delta T\end{bmatrix} = -\begin{bmatrix}\mathbf{F}\\ \mathbf{V1}\\ \mathbf{V2}\\ U\end{bmatrix}$$

where $\frac{\partial\mathbf{V1}}{\partial\mathbf{X}} = \sum_{i=1}^{n}k_i\mathbf{\Gamma}_i$;

$$\frac{\partial\mathbf{V2}}{\partial\mathbf{X}} = \frac{\partial}{\partial\mathbf{X}}\left[\left(\frac{\partial\mathbf{F}}{\partial\mathbf{X}}\right)^{\mathrm{T}}\mathbf{L}\right] = \begin{bmatrix}\mathbf{\Gamma}_{11}\mathbf{L} & \mathbf{\Gamma}_{12}\mathbf{L} & \ldots & \mathbf{\Gamma}_{1n}\mathbf{L}\\ \mathbf{\Gamma}_{21}\mathbf{L} & \mathbf{\Gamma}_{22}\mathbf{L} & \ldots & \mathbf{\Gamma}_{2n}\mathbf{L}\\ \ldots & \ldots & \ldots & \ldots\\ \mathbf{\Gamma}_{n1}\mathbf{L} & \mathbf{\Gamma}_{n2}\mathbf{L} & \ldots & \mathbf{\Gamma}_{nn}\mathbf{L}\end{bmatrix},$$

where $\mathbf{\Gamma}_{ij}$ – *j-th* matrix line $\mathbf{\Gamma}_i$.

IV. The Modeling Results

Modeling was carried out with respect to PSS diagram shown in fig. 3. In the PSS under study, two mini-HPP with a capacity of 24 MW each, operate at an industrial point load connected at nodes 1 and 2. The enterprises operate in one shift and in the hours of the evening maximum of the EPS, a power output of 15 MW from each generator to the receiving system (node 3) is carried out. The network is composed of symmetrical electrical pathways. Power outage of lines 1–3 was considered as an emergency mode due short circuit and relay protection operation.

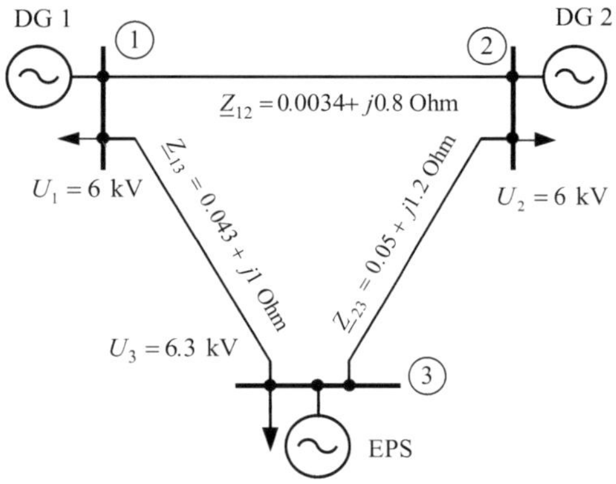

Figure 3. Diagram of the network under study

Results of postemergency modes (PEM) introduction into the stability region by the shortest path calculated on the basis of equations (1) are illustrated in fig. 4. As the initial loading mode of the DG plants generators a point with coordinates is considered $\mathbf{Y}_0 = [20 \quad 20]^T$; introduction into boundary of the stability region using the set path, is represented by a point $\mathbf{Y}_{Z2} = [15.8 \quad 13.6]^T$ (fig. 4).

A qualitative dynamic transition of PEM into the stability region can be ensured by harmonized setting of AVR and ASG of DG synchronous generators [18, 19]. Dynamic processes modeling was carried out in MATLAB system in the diagram under study (fig. 3) with provision for AVR and ASG, whose description is given in works [18, 20]. Using the method of concordant setting, regulators parameters were identified for three generators operating modes (modes of minimal, middle and maximal generator loads). Based on obtained AVR and ASG settings, base of the controllers fuzzy control system was compiled in the provided operating modes.

When DG plants operate, the system loses stability, in case one of the couplings is disconnected. To ensure PEM stability, DG plants generators should be load-shedded. In this case, the regulators harmonized tuning and functioning of the control fuzzy system used to change DG plants generators settings in different operation modes allows to significantly enhance the transient processes quality parameters.

The best effect in terms of improving the dynamic transition is observed for generators voltage, and parameters of a generator having immediate communication with the line disconnected. The corresponding oscillograms of DG plants voltage, rotor speed, and power in case of short circuit and shut-down of relay protection lines 1-3 are shown in fig. 5.

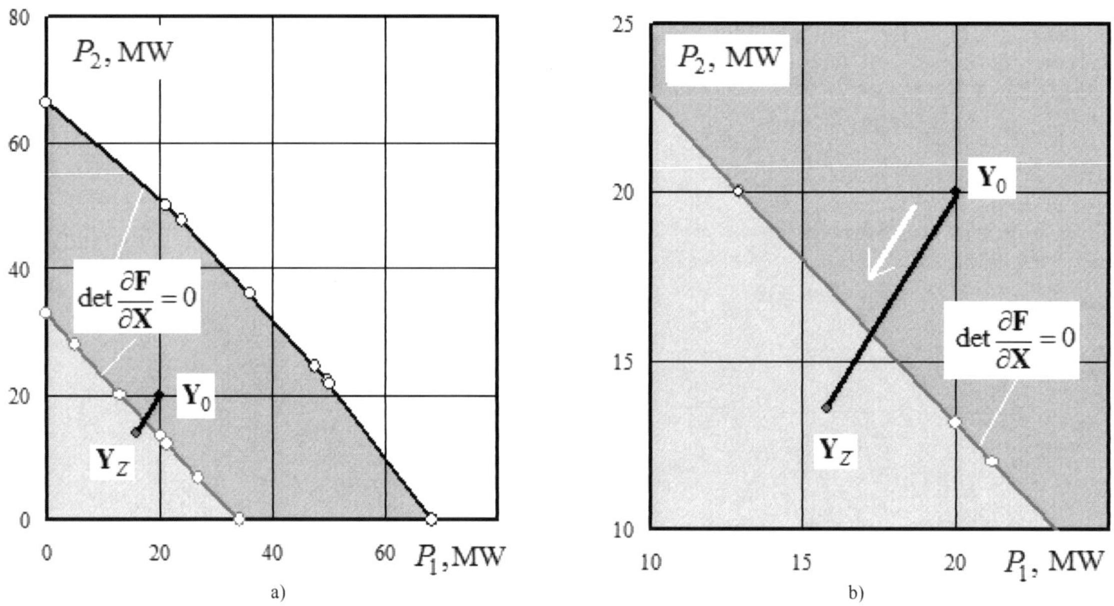

Figure 4. Mode introduction in the stability region using limit loads equations $\sigma_{\min} = 8$

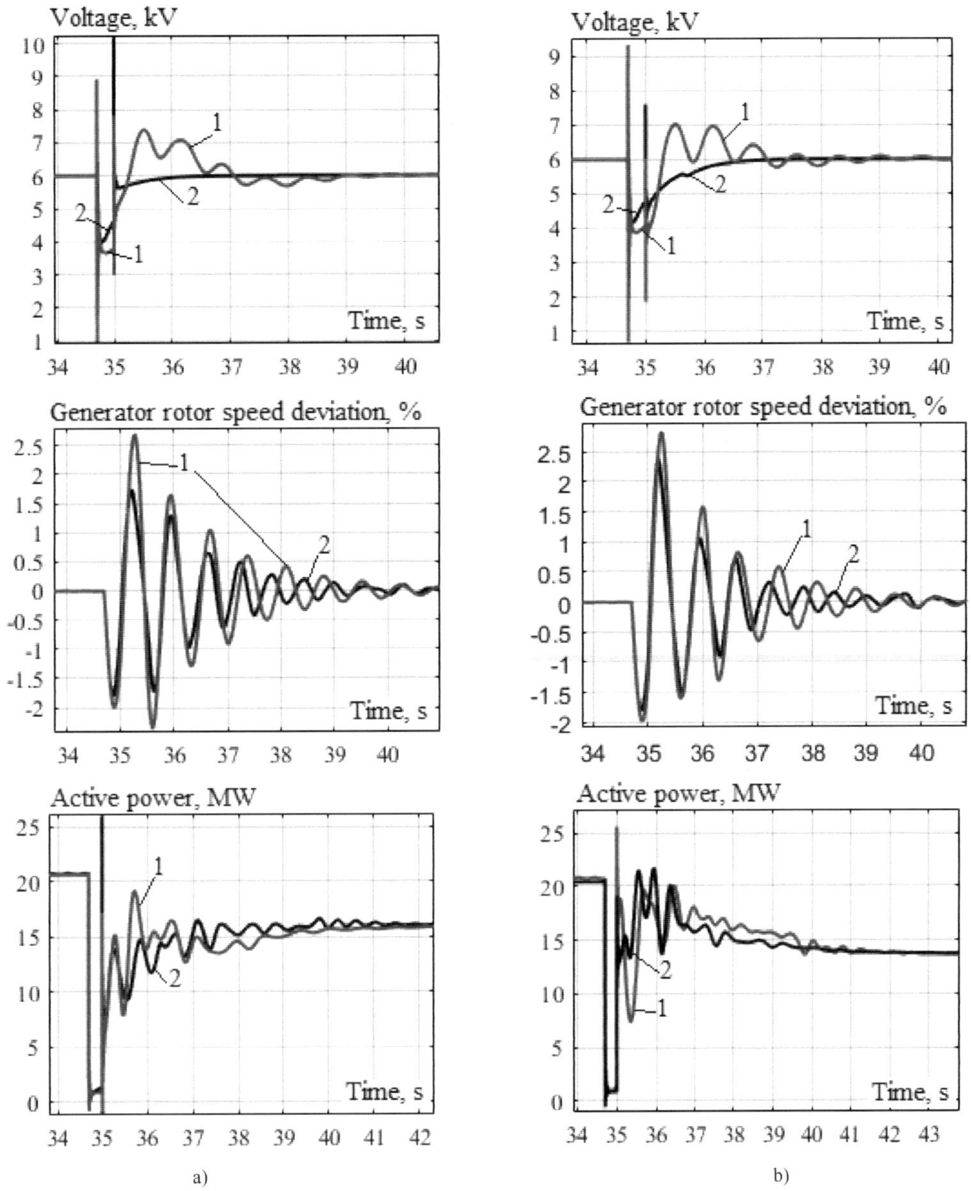

Figure 5. Oscillograms for voltage, rotor speed and power of DG plants in case of short circuit and line 1-3 disconnection: 1 – without changing the tuning factors of AVR and ASG; 2 – fuzzy control of AVR and ASG settings.

V. Conclusions

The computer modeling results allow us to make the following conclusions:

1. The efficient introduction of postemergency mode to the permissible region can be implemented based on ratios using singular values of Jacobi matrix for steady-state mode equations.

2. Generator regulators harmonized setting allows formation of quality dynamic transition while load shedding DG plants in the postemergency mode.

3. Application of fuzzy algorithms to control tuning of synchronous generators controllers allows to obtain additional effect with respect to enhancing quality of voltage, frequency and power transient processes while load shedding DG plants generators in postemergency mode. The best effect in terms of improving the dynamic transition is observed for generators voltage, and parameters of a generator having immediate communication with the line disconnected.

References

[1] T. Ackermann, G. Anderson, L. Söder, "Distributed generation: a definition", Electric Power Systems Research, No.57, 2001, pp. 195-204.

[2] J. Wang, A.Q. Huang, W. Sung, Y. Liu, B.J. Baliga, "Smart Grid Technologies", IEEE Industrial Electronics Magazine, vol. 3. No 2, pp. 16-23, 2009.

[3] N. I. Voropai, Z.A. Stychinsky, Renewable energy sources: theoretical foundations, technologies, technical characteristics, economics, Magdeburg: Otto-von-Guericke-Universität, 2010. 223 p.

[4] F. N. Mohsen, M.S. Amin, H. Hashim, "Application of smart power grid in developing countries", IEEE 7th International Power Engineering and Optimization Conference (PEOCO), 2013.

[5] O. Ellabban, H. Abu-Rub, F. Blaabjerg, "Renewable energy resources: Current status, future prospects and their enabling technology", Renewable and Sustainable Energy Reviews, vol. 39, 2014, pp. 748-764.

[6] B.M. Buchholz, Z.A. Styczynski, "Smart Grids – Fundamentals and Technologies in Electricity Networks", Berlin Heidelberg: Springer-Verlag, 2014, 396 p.

[7] M. S. Mahmoud, M. Fouad AL-Sunni, "Control and Optimization of Distributed Generation Systems," Cham: Springer International Publishing: Imprint: Springer, 2015. 578 p.

[8] N. I. Voropai, V.G. Kurbatsky, N.V. Tomin [and others], Complex of intellectual means for prevention of major accidents in power systems, Novosibirsk: Science, 2016, 332 p.

[9] I. Z. Gluskin, B.I. Iofyev, Emergency automation in power systems, Znak, Moskow, Vol.1, 2009, 568 p.

[10] I. Z. Gluskin, B.I. Iofyev, Emergency automation in power systems, Znak, Moskow, T.2, 2011, 528 p.

[11] A.V. Kryukov, Limiting regimes of electric power systems, Irkutsk, 2012, 236 p.

[12] Yu. N. Bulatov, A.V. Kryukov, "Emergency Control of Distributed Generation Plants", Electricity, No. 2. 2019. pp. 18-25. DOI: http://dx.doi.org/10.24160/0013-5380-2019-2-18-25

[13] A. N. Belyaev, S.V. Smolovik, "Designing adaptive automatic excitation controllers using neuro-fuzzy modeling", Electricity, No. 3, 2002, pp. 2-9.

[14] N. I. Voropai, P.V. Etingov, "Two-Stage Adaptive Fuzzy PSS Application to Power Systems", Proc. of International Conference on Electrical Engineering ICEE'2001, July 22-26, Xi'an, China, vol.1, 2001, pp. 314-318.

[15] K. Suslov, N. Solonina, V. Stepanov, "A principle of power quality control in the intelligent distribution networks", International symposium on smart electric distribution systems and technologies EDST 2015 Proceedings, 2015, pp. 260–264.

[16] Y. Chen, Y. Ma, W. Yun, "Application of Improved Genetic Algorithm in PID Controller Parameters Optimization", Telkomnika, Vol. 11, No. 3, 2013, pp. 1524-1530.

[17] A.Y. Jaen-Cuellar, R. de J. Romero-Troncoso, L. Morales-Velazquez, R.A. Osornio-Rios, "PID-Controller Tuning Optimization with Genetic Algorithms in Servo Systems", International Journal of Advanced Robotic Systems, Vol. 10, 2013, pp. 324-337.

[18] Yu. N. Bulatov, A.V. Kryukov, "Optimization of automatic regulator settings of the distributed generation plants on the basis of genetic algorithm", 2nd International Conference on Industrial Engineering, Applications and Manufacturing (ICIEAM). IEEE Conference Publications, 2016. pp. 1-6. DOI: 10.1109/ICIEAM.2016.7911456

[19] A.V. Kryukov, S.K. Kargapol'cev, Yu.N. Bulatov, O.N. Skrypnik, B.F. Kuznetsov, "Intelligent control of the regulators adjustment of the distributed generation installation", Far East Journal of Electronics and Communications, Vol. 17, No. 5, 2017, pp. 1127-1140.

[20] Yu. N. Bulatov, A.V. Kryukov, "Neuro Fuzzy Control system for Distributed Generation Plants", Proceedings of the Vth International workshop "Critical infrastructures: Contingency management, Intelligent, Agent-based, Cloud computing and Cyber security" (IWCI 2018), Atlantis Press, Advances in Intelligent Systems Research, Vol. 158, 2018, pp. 13-19. DOI: https://doi.org/10.2991/iwci-18.2018.3

[21] R. Wang, P. Wang, G. Xiao, "Intelligent Microgrid Management and EV Control Under Uncertainties in Smart Grid," Springer, 2018, 218 p.

Electric Losses Flow Distribution Method for Power Systems (utilities)

Andrey A. Pazderin
Ural Federal University
Yekaterinburg, Russia
pazderin-aa@ural.fsk-ees.ru

Andrey V. Pazderin
Ural Federal University
Yekaterinburg, Russia
a.v.pazderin@urfu.ru

Nikita A. Morozenko
Ural Federal University
Yekaterinburg, Russia
ingenirium@mail.ru

Ilya V. Chernykh
Ural Federal University
Yekaterinburg, Russia
i.v.chernykh@urfu.ru

Abstract — **The article considers the method of distribution of losses between nodes of electricity supply from the grid on the basis of the distribution of electricity flows. The approach is based on the balance equations for the nodes and branches of the grid scheme, which are written for electricity flows and loss flows. The method allows to obtain a loss distribution similar to the proportional sharing method without calculating the targeting coefficients. On the basis of distributed losses, correction factors are introduced for transmission tariffs that encourage consumers to further losses reduce.**

Keywords: energy flow model, energy transport costs, element costs, cost flows

I. Introduction

A review of the literature showed a great variety in approaches of countries to the formation of tariffs for the transmission of electric energy (TT) based on various technical indicators. The most common features are associated with the differentiation of TT by voltage levels, with separate payments for energy and power, with payment for EE losses [1-3]. Additional paid indicators in different countries can be reactive energy (power), geographical distance of consumers, time of day or season of the year, the form of the load pattern, indicators of reliability and quality of power supply. Reported EE losses are the main mode indicator, which utilities can influence on and reduce the cost of EE transfer. To reduce energy efficiency losses, it is necessary to create financial incentives for consumers within the framework of the tariff model. In this article, this is done by introducing correction factors to TT, taking into account the consumer's alignment of its load schedule (reduction in form factor), reduction in reactive energy consumption (reduction in power coefficient).

Many researches are devoted to the problem of distribution of transmission cost and losses allocation in the grid scheme [4-9]. According to [4] embedded cost transmission pricing methods can be divided into roll-in methods, like postage stamp and contract path methods [5]. In Russia the postage stamp method is used. It is very simple, but not sensitive to distance and direction. Grid-based Methods, like MW-Mile and MVA-Km methods [5-7] take into account distance when paying for EE. More complex and time consuming flow-based methods use the principle of tracing for the calculated grid mode [8, 9]. The main principle used in power flow tracing is proportional sharing [9]. The basis of the approach linking technical and economic parameters is the energy-cost flow model [10, 11]. The approach allows to represent the process of transferring EE in the form of two interrelated transport-balance tasks on the graph of the electric grid: the distribution of energy flow distribution and cost flow distribution (ECD). The basis of the technological subsystem is the energy-flow model (EFM), which allows to calculate the distribution of flows and losses of energy efficiency for each element of the grid scheme, based on EE measurements [12-14]. The economic subsystem of the model allows to calculate the distribution of cost flows in the same grid diagram and determine the full and specific cost of EE transmission to any point of the grid using the energy-flow mode [10, 11].

In accordance with the ECD model [10], the elemental cost of buses and branches is the cost of maintenance and the cost of loss of energy efficiency. The elemental cost of energy loss is determined by multiplying the elemental energy loss by the tariff used to compensate the losses. Elemental costs, including the cost of losses, are transmitted to EE supply nodes or consumption nodes based on the power distribution mode. Thus, within the framework of the ECD model, it is possible to solve the task of allocating responsibility for EE losses between consumption nodes based on the calculation of the loss distribution flow.

II. Energy-Losses Flow Distribution Model

Within the ECD, it is possible to distribute various types of electric power (power) losses between the supply nodes and everything is determined only by how the values of the elemental losses are formed. Within the framework of the developed ECD model, the elemental values are formed on the basis of the technical loss of energy efficiency of each element of the grid scheme, since technical losses of energy efficiency are known and are calculated values. Thus, if the elemental costs are formed by the only technical losses of each element, then technical losses will be distributed between the nodes of leave. If elemental costs are generated only on the basis of load losses, then pure load losses will be distributed. If elementary values are generated on the basis of shunt (static) losses, then shunt losses will be allocated.

The model can be used for a more reasonable definition of economic incentives in the form of increasing or decreasing coefficients to the tariffs for the transfer of energy efficiency. With its help, it is possible to more accurately distribute the losses between the load nodes and estimate the cost of these losses. Knowing the nodal losses, it is possible to determine the correction factors for the tariffs for the transmission of EE for a specific consumer for equalizing their power consumption schedule and reducing the consumption of reactive energy/power.

III. Energy Flow Model

The distribution of EE fluxes is tied to the time intervals at which electric meters are collected from EE counters. To obtain an algebraic system of equations of the ECD model, the equations describing the technological process of EE transfer are presented first [12, 13]. The basis of the mathematical model of EFM form the equations of EE balances, which can be written for all N buses and all M branches of the electric grid. The process equations in this section will be written only with respect to active EE fluxes. Such an assumption in the framework of the task is fully justified, since, in accordance with the current regulatory documents, when paying for services for the transfer of EE in Russia, only active power of EE is taken into account.

The basis of the system of equations for the problem of energy distribution is formed by the equations of EE balances for all buses and all branches of the computational model [12, 13]. Each equation of the bus energy balance determines that the sum of energy flows W_{ij} over all adjacent bus i-branches from the set ω_i is zero.

$$\sum_{j\in\omega_i} W_{ij} + W_i + \Delta W_i = 0, \qquad i = 1, 2...N \tag{1}$$

The second group of equations of the ER problem relates to the EE balances in the branches of the scheme. For each of the M branches, the sum of EE fluxes at the beginning of the W_{ij} branch, at the end of the W_{ji} branch and of the technical losses of the EE ΔW_{ij} of this branch is zero.

$$W_{ij} + W_{ji} + \Delta W_{ij} = 0, \qquad ij = 1, 2...M \tag{2}$$

The systems of equations (1) and (2) are written for a single computational time interval t, during which circuit-mode changes can occur and the directions of power flows on grid sections can be reversed. In the balance equations of EE for nodes and branches there are technical losses of EE ΔW_i and ΔW_{ij}, and the system of equations (1) - (2) must be supplemented with formulas for calculating technical losses of EE, which are described in [14]. At the same time, it is necessary to form a calculated equivalent circuit of the electrical grid so that it correctly takes into account the loss of EE in the longitudinal and transverse elements of the equivalent circuit. Within the framework of the ECD model under consideration, the process of distribution of loss flows should be carried out on the same electrical grid scheme for which the power distribution problem is being solved.

The balance model of loss distribution EE assumes that the distribution of losses between buses and branches of the scheme is known. As in the technological model, the elements of the scheme are the buses associated with electrical stations/substations, and the branches associated with power lines/transformers.

Further, the elemental losses of buses ΔW_i and branches ΔW_{ij} are transmitted (transferred) to subsequent elements of the grid diagram in accordance with the paths of flow of energy efficiency starting from the buses of energy supply to the grid and ending with the buses of energy supply from the grid in a cascade, absorbing all elemental costs of subsequent elements of the scheme. At the same time, on each element of the circuit new losses are formed, called "stream losses" of buses ∇W_i and branches ∇W_{ij}.

The flow distribution of losses in the electrical grid diagram can be described by a system of balance equations. Each equation of nodal balance determines that the sum of all losses flowing in and out of a bus is equal to zero, that is, the bus loss ∇W_i and all stream losses ∇W_{ij} along the adjacent branches of the set ω_i.

$$\sum_{j\in\omega_i} \nabla W_{ij} + \nabla W_i + \Delta W_i = 0, \qquad i = 1, 2...N \tag{3}$$

In addition, in the equation of nodal balance there is a value of nodal element losses ΔW_i of bus i. Similarly, we can write the balance equations for the loss flows for all M branches of the circuit. The sum of the flow losses at the beginning of the branch ΔW_{ij} and at the end of the branch ∇W_{ji} is equal to the elemental losses of this branch ΔW_{ij}. For all M branches of the equation of value balance can be represented as follows:

$$\nabla W_{ij} + \nabla W_{ji} - \Delta W_{ij} = 0, \qquad ij = 1, 2...M \tag{4}$$

At the same time, the flow losses at the end of the branch are more than at the beginning due to adding elemental losses of the branch, and the EE flux, on the contrary, is less at the end than at the beginning due to the presence of EE losses. As the elemental losses of all branches are positive, a minus sign is placed in front of the balance equations for the branches. By this, the flow distribution of losses differs from the flow distribution of EE, since in each branch the flow of EE is greater at the beginning than at the end, and the loss flow is vice versa, is less at the beginning than at the end.

Comparison of balance equations for EE flows (1) and (2) with balance equations for loss flows (3) and (4) shows their structural similarity. For both processes, the conditions of nodal and linear balances are fulfilled, that is, the sum of energy flows flowing into a bus and a branch is equal to the sum of flowing flows.

For the creation of a system of equations for the flow distribution of losses, it is necessary to establish a relationship between EE flows and loss flows. This relationship is determined by the fact that the amount of flow losses flowing into a bus completely translates into stream losses of adjacent branches in proportion to the resulting EE flows based on proportional sharing [9]. The flux losses ∇W_{ij} at any branch of the EE tempering (outflow) adjacent to node i, is determined by the equation:

$$\nabla W_{ij}^{-} = -(W_{ij}^{-} / \sum_{j\in\beta_i} W_{ij}^{-})(\sum_{j\in\alpha_i} \nabla W_{ij}^{+} + \Delta W_{ij}^{+}),\ ij = 1, 2...M \tag{5}$$

The sets β_i and α_i define the list of branches of the leave "-" and the reception "+" of EE for bus i. The distribution of outgoing flow losses from a bus in proportion to the outflows from the bus of the EE provides the receipt of value flows in the scheme similar to the address approach [8, 9]. However, there is no need to calculate the matrix of targeting coefficients.

The following part of the paper has an example of the distribution of technical losses of energy efficiency between load buses on the basis of the balance model on the previously considered fragment of the grid of technical losses in fig. 1a a fragment of the electrical grid considered in [11] is presented, with the calculated flows of EE and the losses of EE obtained on the power distribution stage applied to the grid diagram. All calculated flows and EE losses are balanced, that is, the EE balance equations (1) and (2) are performed for them, taking into account the signs. The distribution of technical losses between the load buses is carried out on the basis of the power distribution mode presented in Fig. 1a.

Fig. 1b shows the results of the distribution of losses between the buses of consumption (supply) EE from the grid based on the described balance model. The figured (red) arrows indicate the technical loss of energy efficiency for all buses and all branches of the scheme.

For EE flows (Fig. 1a) and loss flows (Fig. 1b), the following rule is accepted: flows flowing into the bus are positive, and flows flowing from the bus are negative. This rule allows not to set positive directions for the branches when forming balance equations for EE flows and loss flows.

Blue arrows indicate flow losses. The final values of EE losses in the load buses 4-7 of the scheme total technical losses are distributed, that is, the losses of EE that are associated with each specific consumer.

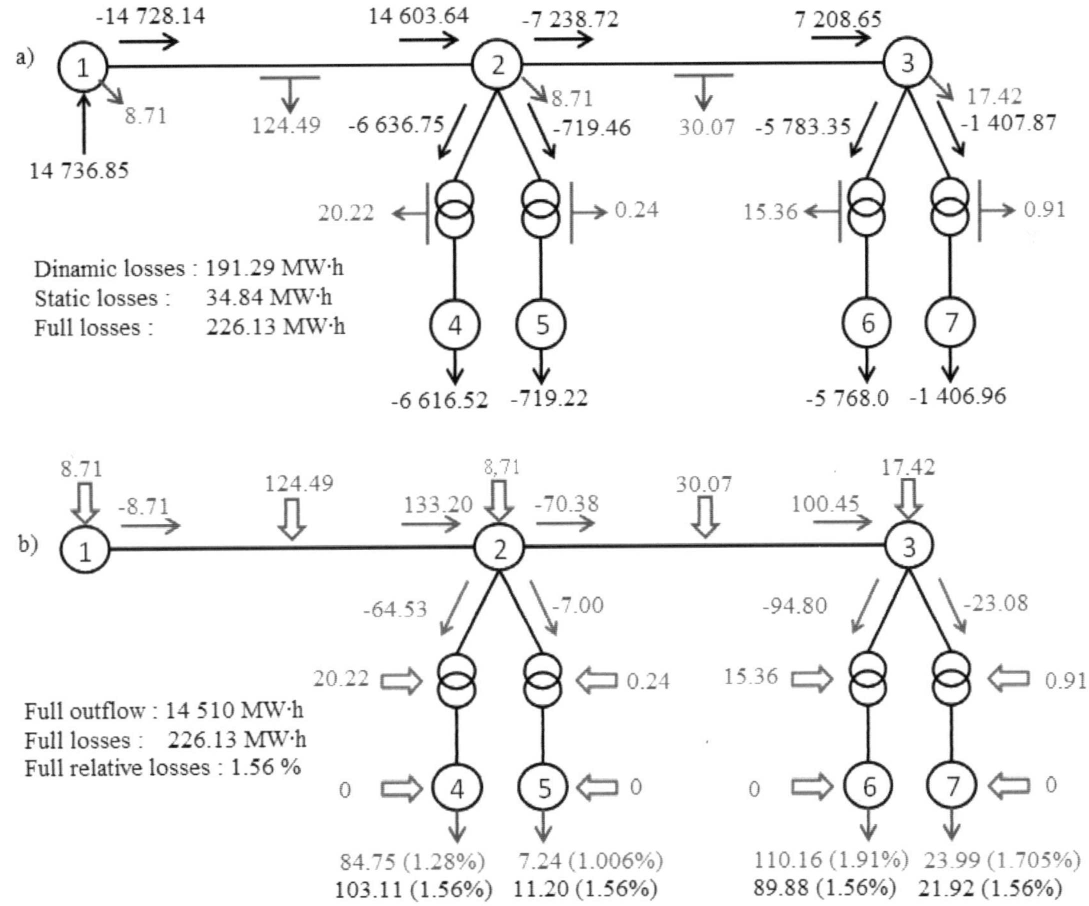

Figure 1. Distribution in the electrical grid scheme: estimated flow and losses of EE (a) and technical losses of buses and branches (b).

To EE losses distributed between the outgoing buses, it is advisable to put in line the term "outgoing losses" ΔW_{Pi}^{out}. In Fig. 1b temporal losses are located directly under the load buses (upper values). The sum of all estimated vacation losses is equal to the total technical loss of 226.13 MW·h.

Fig. 1b also shows the values of buses losses $\Delta W_{Pi}^{\%}$, which were obtained by distributing the total losses in proportion to the load of each node based on the average relative loss of energy efficiency, which is 226.13/14510=1.56 %. Such a distribution of losses in Russia is used in practice to pay for EE losses. It does not take into account the remoteness of the consumer and the mode of operation of the grid. The averaged values of the losses distributed between the nodes are marked black, and they are located below the values of the outgoing losses ΔW_{Pi}^{out}. It is sure, that the sum of all $\Delta W_{Pi}^{\%}$ is also equal to 226.13 MW·h.

It should be noted that in nodes 5 and 6, the estimated release losses of ΔW_{Pi}^{est} are less than the losses of $\Delta W_{Pi}^{\%}$ obtained on the basis of the proportional method, and in nodes of 7 and 8, on the contrary, more. The latter can be explained by the fact that the ECD model makes it possible to more equitably distribute the EE losses between the load buses, taking into account the mode of operation of the grid, taking into account the distance of the load buses from the power centers and the paths of the EE flows that these losses create.

Correction coefficients for the transfer tariff should encourage consumers to reduce grid losses. The approach is based on the proposition that, due to a change in its electricity consumption, there is a decrease in EE losses in the utilities grid and additional income generation. A certain share of income η remains at the disposal of utilities, and the remaining share of 1-η is spent on reducing the transmission tariff for this consumer.

Reducing the cost of load losses $\Delta C_{\Delta W}$ by reducing the load curve shape factor from k_{f1} to k_{f2} can be represented as:

$$\Delta C_{\Delta W} = (k_{f1}^2 - k_{f2}^2)\cdot\frac{W_P^2 + W_Q^2}{tU_{av}^2} RT_{\Delta W} \quad (6)$$

where $T_{\Delta W}$ is the tariff for loss compensation, R is the active resistance of the grid section, t is the time interval, U_{av} is the average voltage value.
For the introduction of the mechanism of allowances and discounts for the form of the schedule, it is necessary to establish the normative value of the form factor $k_{f\ norm}$, which corresponds to the correction coefficient to the tariff equal to one. If the consumer form factor k_{fi} exceeds k_{fnorm}, then such consumer makes an increased contribution to the loss and in relation to it a addition to TT should be introduced. In the reverse situation, the consumer helps to reduce losses and can count on a discount to TT. Taking into account the estimated vacation losses ΔW_{Pi}^{es}, the surcharge/discount to TT ΔT_i^{form} for the form factor can be determined by the expression.

$$\Delta T_i^{form} = \eta\alpha(k_{fnorm}^2 - k_{fi}^2)T_{\Delta W}\Delta W_{P_i}^{out} \quad (7)$$

The coefficient α determines the share of load losses in the composition of the estimated EE release losses. The correction factor to TT for the coefficient of the form of the load graph based on the estimated release loss of the consumer unit can be represented as:

$$K_i^{form} = 1 - \eta\alpha(k_{fnorm}^2 - k_{fi}^2)\cdot\frac{T_{\Delta W}\Delta W_{P_i}^{out}}{T_{form}W_{P_i}} \quad (8)$$

Special coefficient for the consumption of reactive energy. By analogy with (8), an expression is obtained for a correction coefficient that takes into account the deviation of the actual $tg\varphi_{fact}$ from the normal $tg\varphi_{norm}$.

$$K_i^{tg\varphi} = 1 - \eta\alpha(tg^2\varphi_{norm} - tg^2\varphi_{fact(i)})\cdot\frac{T_{\Delta W}\Delta W_{P_i}^{out}}{T_{form}W_{P_i}} \quad (9)$$

IV. Conclusion

The proposed balance model allows to distribute electricity losses in the grid diagram basing on the analysis of the energy-flow model. The distribution of losses in the grid is proportional to the energy flow. Based on the loss distribution model, methodologies have been developed for introducing correction factors to electricity transmission tariffs, which should encourage consumers to reduce EE losses in the grid. The benefits of reducing losses should be shared between consumers and grid organizations and contribute to further tariff reductions.

Acknowledgment

The reported study was funded by the Ministry of Science and Higher Education of the Russian Federation under Federal Targeted Programme according to the agreement № 075-02-2018-1219 (internal number 14.578.21.0226, project identifier: RFMEFI57817X0226).

References

[1] R. Green," Electricity transmission pricing: an international comparison," R. Green // Utilities Policy. – 1997. – Vol. 6. – №.3. – pp. 177-184.

[2] ENTSO-E Overview of Transmission Tariffs in Europe: Synthesis 2016, official site of Evropean community of electrical operators/ www.entsoe.eu.

[3] «Study on tariff design for distribution systems. Final Report» 20150313 Tariff report fina_revREF-E.PDF This is a document from the site ec.europa.eu

[4] B. Khan, G. Agnihotri, "A Comprehensive Review of Embedded Transmission Pricing Methods Based on Power Flow Tracing Techniques"/ Baseem Khan, Ganga Agnihotri // Chinese Journal of Engineering Volume 2013, Article ID 501587

[5] B. Khan, G. Agnihotri, "An approach for transmission usage &loss allocation by graph theory", WSEAS Transaction on Power Systems, vol. 9, pp. 44-53, 2014.

[6] Lo K.L Positive and negative aspects of MW-mile method for costing transmission transaction / K.L. Lo, M.Y. Hassan // 37th Int. Universities Power Engineering Conf. (UPEC). September 2002, – vol.1. – pp. 358-362.

[7] D. Avinash, "MW-Mile method considering the cost of loss allocation for transmission pricing," Conference on Power, Control, Communication and Computational Technologies for Sustainable Growth. PCCCTSG 2015. 7503892. – pp. 128-131.

[8] J. Bialek, "Tracing the flow of electricity," IEE Proceedings—Generation, Transmission and Distribution, vol. 143, no. 4, pp. 313–320, 1996.

[9] J. Bialek and P. A. Kattuman, "Proportional sharing assumption in tracing methodology," IEE Proceedings—Generation, Transmission and Distribution, vol. 151, no. 4, pp. 526–532, 2004.

[10] A. A. Pazderin, "Technical and Economic Model of Energy Transmission and Distribution Based on the Smart Metering Technologies, The 26th IEEE International Symposium on Industrial Electronics. 18-21 June 2017 Edinburgh, Scotland, UK. № 8001241. pp. 163-168.

[11] A. A. Pazderin, "Energy-cost flows model of electric energy distribution at an electric grid," The 11th IEEE International Conference on Compatibility, Power Electronics and Power Engineering. CPE-POWERENG 2017. Cadiz, Spain, 4-6 April 2017. № 7915188. pp. 308-312.

[12] A.V. Pazderin, E.A. Plesniaev, E.V. Mashalov An energy flow calculation method based on energy meters data // 3rd International Conference on Deregulation and Restructuring and Power Technologies, DRPT 2008; Nanjing; China; 6-9 April 2008. № 4523522. pp. 834-839.

[13] E. A. Plesniaev and A. V. Pazderin, "Analysis of the Computation Techniques for Energy Flow Problem Solving," *EUROCON 2005 - The International Conference on "Computer as a Tool"*, Belgrade, 2005, pp. 1469-1472.

[14] A. V. Pazderin V. O. Samoilenko, "Localization of non-technical energy losses based on the energy flow problem solution." Proceedings of the 6th IASTED Asian Conference on Power and Energy Systems, AsiaPES 20132013, Pages 100-103C. Gu, C. Yuan, F. Li and Y. Song, "Risk Management in Use-of-System Tariffs for Grid Users", IEEE Transactions on Power Systems, vol. 28, no 4, pp. 4683-4691, 2013.

Use of PMU Data for HVDC Link Modeling in the Power System State Estimation

Irina Kolosok, Elena Korkina
Melentiev Energy Systems Institute
Siberian Branch of Russian Academy of Science
Irkutsk, Russia
kolosok@isem.irk.ru

Alexander Tikhonov
Joint Stock Company, Electronetwork company
Irkutsk, Russia
tikhonovaleks@mail.ru

Anatolijs Mahnitko
Riga Technical University
RTU
Riga, Latvia
anatolijs.mahnitko@eef.rtu.lv

Abstract — **Updating of electric networks for raising or limiting the voltage level, for increasing the transmission capacity and for some other purposes is accomplished by installing FACTS (Flexible Alternative Current Transmission System) devices. Installation of FACTS devices in the energy company's networks enhances the modes of their operation. Modern methods of FACTS modeling in the task of electric power system (EPS) state estimation vividly prove that.**

Development of Wide Area Measurement Systems (WAMS) evidences the transition of EPS to a Smart Grid. Phasor measurement units (PMU) are used in different problems of on-line control primarily for monitoring the EPS modes, which allows more accurate modeling of processes occurring in the energy systems when solving the basic control problems.

The paper presents a model of High Voltage Direct Current (HVDC) device on voltage converters that was developed in Melentiev Energy Systems Institute (further MESI SB RAS). The model was included into an equivalent circuit for solving the state estimation problem. To ensure the HVDC model observability the substation with HVDC is proposed to be equipped with PMUs. Integration of PMUs ensures correctness of HVDC link operation monitoring.

Keywords— FACTS, HVDC link, PMU, state estimation

Introduction

Creation of a Smart Grid in Russia necessitates large-scale introduction of FACTS technology. Updating the existing electric networks for enhancing the voltage control, increasing the transmission capacity of lines and for some other purposes is accomplished by installing FACTS devices in the EPS networks.

HVDC is an inseparable part of modern EPS and of technologies for EPS control. HVDC links are used for harmonizing the operation of AC and DC electric networks, for harmonizing the operation of networks with different current frequencies, and for increasing the transmission capacity of the network elements with 'weak' links [1].

In present-day electric power systems, renewable energy sources are expected to appear, consequently, the volumes and modes of transmitted electricity will begin to change. Therefore, it is necessary to improve control of power systems while maintaining the reliability and security of the network. New technologies FACTS and HVDC are designed to provide flexible and safe operation of power systems.

HVDC link based on voltage converters and modern power electronic devices (Gate Turn-Off Thyristors or Integrated Gate Commutated Thyristors (GTO or IGCT) or Insulated Gate Bipolar Transistors (IGBT)) are currently more and more widely used.

Such HVDC link have a wide range of mode parameters control, ensure control both under the active and reactive power, and belong to FACTS devices. HVDC device in Russia is installed at 220kV substation in Mogocha for interconnecting the power systems of Siberia and the Far East.

In the practice of dispatching control two problems are used for generating the model of current mode: the steady-state calculations and EPS state estimation (SE) problem using the SCADA and WAMS data [2]. Issues of HVDC link modeling for calculating the steady-state conditions have been widely discussed in Russian and foreign scientific publications [3-7]. Studies on integrating the models of FACTS devices into the SE problems were started later, but in the latest years such studies have been actively carried on both abroad [8,9] and in Russia [10, 11]. Enhancement of electric network operating modes after integrating FACTS devices is demonstrated in [12] on the example of their modeling in the EPS state estimation problem.

Transition to Smart Grid without WAMS integrating is not possible [2]. WAMS record changes in the state parameters with high precision. Use of WAMS for monitoring different processes in EPS gives true representation of the EPS behavior and allows detailed modeling of some or other processes when solving the basic state control problems.

The advent of advanced monitoring device, phasor measurement unit, as a next-generation telemetry technology opens up great opportunities for

- control of power grid operation;
- monitoring of transient processes;
- PMU integration into emergency control facilities;
- as well as for optimal and reliable operation of second-generation FACTS devices (HVDC, static synchronous compensator (STATCOM), unified power flow controller (UPFC) and others), made on the basis of voltage converters (VC), to generate control actions for these devices.

Development of WAMS in Russia that started in 2005 is now accelerating: in 2005 only two foreign phasor measurement units (PMUs) were installed at energy facilities, whereas in 2018 their number was as high as 715 (mainly of Russian manufacture) [2]. Starting from 2020, PMUs are expected to be mandatorily installed at newly commissioned energy facilities. HVDC is a new substation to be engineered, constructed and equipped with different devices, PMU devices being among them.

This paper presents a developed in MESI SB RAS model of HVDC link that was included into the SE problem [11]. To ensure the HVDC link model observability and control the substation with HVDC link is proposed to be equipped with PMUs. It is shown that decisions on monitoring the HVDC mode can be much more grounded if WAMS data are available.

II. HVDC Link Modeling in the EPS SE problem

A. EPS state estimation problem

EPS state estimation assumes computation of state parameters based on data of measurements $\bar{y}$ [2].

SE algorithms are traditionally used with telemetry data, i.e., data coming from SCADA system. They are magnitudes of nodal voltages U_i, nodal injections of active P_i and reactive Q_i power, load flows in the lines and transformer links P_{ij}, Q_{ij}, and sometimes they are currents in nodes I_i and lines I_{ij}. A SE measurement vector has the form:

$$\bar{y} = (P_i, Q_i, P_{ij}, Q_{ij}, U_i, I_i, I_{ij}) \cdot \quad (1)$$

WAMS development gave way to measurements of new generation that include use of PMU for SE problem. They are magnitudes U_i and angles δ_i of nodal voltages, and currents $\dot{I}_{ij}$ in the lines that are incidental to the node where PMU is installed. A vector of measurements has the form:

$$\bar{y} = (\dot{U}_i, \dot{I}_i, \dot{I}_{ij}), \quad (2)$$

where $\dot{U}_i = U_{ai} + jU_{ri}$; $\dot{I}_i = I_{ai} + jI_{ri}$; $\dot{I}_{ij} = I_{aij} + jI_{rij}$ are vector measurements of nodal voltages, nodal currents, and currents in the lines.

A method of Test Equations for SE has been developed in ISEM SB RAS [2]. Test Equations (TE) are steady-state equations that include only measured variables of the state y :

$$w_k(y) = 0 \quad (3)$$

With the use of control equations the SE problem consists in minimization of the criterion:

$$J(y) = (\bar{y} - \hat{y})^T R_y^{-1} (\bar{y} - \hat{y}) \quad (4)$$

i.e., in calculating the estimates of measured variables $\widehat{y}$ under constraints in the form of a system of TEs (3). In formula (4) R_y is a covariance matrix of measurement errors. Estimates of the state vector $\hat{x}$ that includes magnitudes and angles of nodal voltages are calculated at the second step using the estimates of basic measurements that are further used for estimation of non-measured variables.

B. HVDC link modeling in the EPS state estimation problem

HVDC link is a substation designed for converting AC into DC and for subsequent conversion of DC into AC of initial or desired frequency. Both current and voltage converters can be used in HVDC link as converters [4, 13]. In this paper consideration is given to HVDC model on voltage converters (VC).

As it has been mentioned above, HVDC link using GTO, IGCT or IGBT structures have a wide range of control over the state parameters and in addition to the function of active power control both converters can be operated as STATCOM devices. Equations of steady-state HVDC link conditions are similar to equations describing the behavior of STATCOM devices [7,14].

Fig. 1,a shows a line with calculated parameters of steady-state conditions, HVDC link being included. Active power flows from node *j* to node *i*.

HVDC link can be represented by two voltage sources [4]. Each voltage source is connected to the AC system via a transformer. HVDC link is modeled at a substation of node *j* with two branches connected to it that simulate a transformer and a rectifier, respectively. For connecting a converter, k node is added to the circuit with two branches connected to it that simulate a transformer and an converter, respectively. Fig. 1,b shows an equivalent circuit after HVDC link connection.

There are high requirements for the quality of information received from the PMU, since it can be distorted for many reasons (cyber attacks, malfunctions of the device itself, etc.). The effects of solar radiation and magnetic storms on the GPS system, causing PMU timing problems, should also be taken into account. Therefore, it is imperative to execute the state estimation (SE) of the PMU data, because the estimated information will be used to generate control actions for the emergency control systems and commands for control of the FACTS settings, which will make it possible to control power flows and to localize the development of various transients and voltage asymmetries.

Solution of the SE problem is reduced to calculation of state vector estimates, i.e., phasors of nodal voltages that are further used for calculation of load flows by lines and power injections in the nodes. Equations for load flow calculation by lines (Fig. 1) are similar to equations for calculating the flows of a rectifier and converter at steady-state conditions calculation.

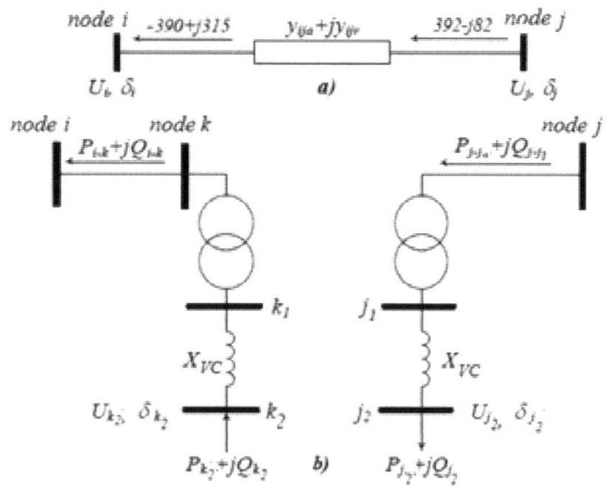

Figure. 1. A circuit of HVDC link connection for EPS SE

a) a line with design parameters; b) an equivalent circuit

When modeling some FACTS devices the additional equations occur that interrelate state parameters of those devices. For HVDC link an additional equation is introduced that describes the process of active power transfer via HVDC, i.e., equality of active power P_{j_2}, consumed by a rectifier and P_{k_2} that id output by a converter:

$$P_{j_2} + P_{k_2} = 0 \quad (5)$$

A controlled automatic DC link is a complex non-linear element that can impact the convergence of an iteration process when computing the steady-state conditions and at SE of EPS. This fact imposes strict requirements to adequacy of mathematical models of HVDC and accuracy of algorithms for modes calculation.

III. Results of Calculation

A. State Estimation with the HVDC model

A 19-node fragment of an actual network of 500kV power grid of Siberia was used as a test network for validation of the developed model [15]. HVDC link was simulated on the inter-system substation K. (node 12), Fig. 2. For the purpose of simulation node 22 was added to the circuit of the substation K. For modeling the step-down 630 MVA transformers 500/10.5kV, nodes 112, 122 and branches 12-112, 22-122, were added to the circuit; for simulating the voltage converters, nodes 1112, 1122 and branches 112-1112, 122-1122 were added.

The network integrated blocks composed of eight (8) voltage converters. Full rated capacity of each voltage converter was 56 MVA at 11kV voltage. Rated active power of each voltage converter made 50 MW, rated reactive power was 25 MVAr. Total rated capacity of a block composed of eight voltage converters (VC) was 448 MVA. Block VC-2 was working in the rectifier mode, and Block VC-1 in the converter mode. Depending on the direction of active power flow via Line 11-12 they may function either as a converter or as a rectifier. When HVDC is connected to the circuit used in the SE problem the circuit is divided into two parts along the branch denoting a DC line 1112-1122.

When a HVDC model is created the undesirable problems of lack of measurements appeared (see Table 1). The redundancy of measurements in HVDC link controllers can be effectively increased due to PMU installation in that very area. Thanks that fact, PMU-data, namely, voltage phasors and current phasors or even power flows based on PMU, are added to convinient set of measurements. Such a way the problem of critical measurements can be solved too. In practice the PMU installs exactly on voltage converters, thus the nodes 1112 and 1122 were chosen for this aim. Additionally, two PMU were installed at nodes 12 and 22.

Calculations are carried on for two subsystems simultaneously, a basic node with a zero voltage angle being assigned to each part of the circuit. Node 12 becomes the basic node in the subsystem separated from the main circuit. A node that is not shown on the network fragment (Fig. 2) is assigned to be the basic node in the main part of the circuit. Calculations for subsystems are given in Table 2.

As it has been shown, HVDC link is composed of voltage converters, i.e., a *rectifier* that converts AC energy from a generating system into DC energy, and a *converter* that receives energy from a rectifier, converts it into AC energy and forwards it to a receiving system. A *rectifier* here is VC2, whose voltage is higher than that in the grid:

$$U_{1112} > U_{112} \text{ (11.4kV > 11.0kV)} \quad (6)$$

A *converter* here is VC1, whose voltage is lower than that in the grid:

$$U_{1122} < U_{122} \text{ (9.3kV < 10.0kV).} \quad (7)$$

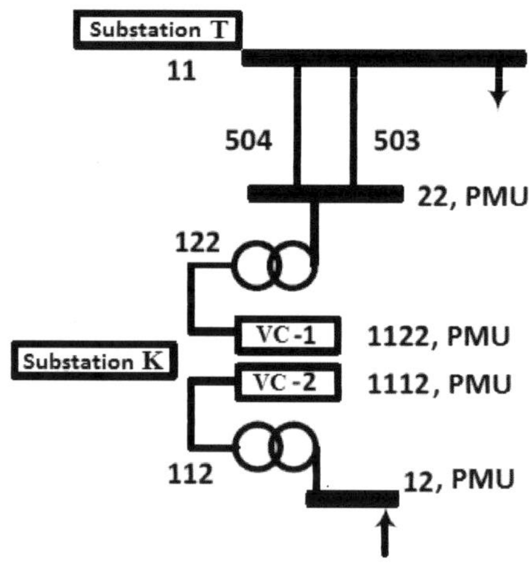

Figure. 2. A fragment of an actual power system.

VC2 outputs active power from the network and receives reactive power:

$$P_{1112} = -391 kW; \quad Q_{1112} = 166 MVAr$$

VC1 receives active power from the network and outputs reactive power:

$$P_{1122} = 391 kW; \quad Q_{1122} = -190 MVAr$$

Table 1 shows that condition (5) at active power flow via HVDC link is valid:

$$P_{1112} + P_{1122} = 0. \qquad (8)$$

B. Installation of PMUs over HVDC link

The problem of PMU installation is well known to researchers and engineers dealing with PMUs in power engineering [16-20 et al.]. All the PMUs are synchronized in time by signal receivers of global positioning system (GPS) to ensure adequate operation. Each option of PMU installation pursues its own purposes:

- on the most significant facilities of EPS [5],
- in the boundary nodes of adjacent EPS at SE problem decomposition [17],
- in the nodes enhancing the properties of SE problem (raising the redundancy of measurements, elimination of critical measurements and critical groups [18]),
- at linear SE in the areas fully observable by PMU [19, 20], etc.

The problem of PMU installation on the substations with FACTS devices is discussed in a number of papers [20-24 et al.].

This paper offers to install PMU on HVDC link primarily for obtaining accurate measurements of voltage magnitudes and angles in the nodes included into the HVDC link circuit. Inclusion of such measurement into the state vector of SE problems gives more accurate results of calculations. For validating the proposed approach, consideration is given to PMU installed in the reference node of circuit 12 and in the additional node 22 'cut in' line 12-11 for HVDC link simulation. This installation produces accurate PMU measurements of voltage magnitudes and angles.

Table 1.State Estimation With HVDC Link

	Node No.	U_{meas}	U_{SE}	$\delta°_{meas}$	$\delta°_{SE}$	P_{meas}	P_{SE}	Q_{meas}	Q_{SE}
Substation T	11	516	516	-	8.7	-386	-386.7	-86	-93.9
Substation K	12	512	512	-	0	-	391.5	-	-7.6
1-st Subsystem, VC2	112	-	10.8	-	-5.4	0	0	0	0
	1112	-	11	-	-10.8	-391	-391	85	85
2-nd Subsystem, VC1	22	512	511.8	-	11.8	0	0	0	0
	122	-	10	-	17.7	0	0	0	0
	1122	-	9.3	-	25	391	391	-	-187.6

Table 2.State Estimation With HVDC Link And PMU

	Node No.	U_{meas}	U_{SE}	$\delta°_{meas}$	$\delta°_{SE}$	P_{meas}	P_{SE}	Q_{meas}	Q_{SE}
Substation T	11	516	516	-	8.8	-386	-386.7	-86	-87.4
Substation K	**12, PMU**	**511.9**	**511.9**	**1.8**	**1.8**	**391**	**391.52**	**-85**	**-84.94**
1-st Subsystem, VC2	112	-	11.0	-	-3.6	0	0	0	0
	1112, PMU	**11.4**	**11.4**	**-9.0**	**-9.0**	**-391**	**-391**	**166**	**166**
2-nd Subsystem, VC1	**22, PMU**	**511.9**	**511.9**	**11.8**	**11.8**	**0**	**0**	**0**	**0**
	122	-	10.0	-	17.8	0	0	0	0
	1122, PMU	**9.3**	**9.3**	**25**	**25.0**	**391**	**391**	**-188**	**-188**

Table 2 presents PMU measurements - magnitude and angle of nodal voltages. The obtained voltage estimates equal to measurements of voltages in the node with PMU. Thanks to more precise voltages the resulting values of reactive power have been made more accurate. Conditions of HVDC link cut-in(6)-(8) have been met.

Along with improvement of the SE results, installation of PMU on HVDC allows monitoring of voltages on VC1 and VC2 and checking the coherence of their operation for correct HVDC operation.

IV. Conclusions

1. The paper proves the applicability of including the HVDC link models into the equivalent circuit when solving the SE problems.
2. Developed are the modified algorithms for including the HVDC link models into the SE problem.
3. For state estimation of the power system being synchronized it is necessary to install phasor measurement units on voltage converters (rectifier/inverter) at HVDC link, which allows controlling the currents through the HVDC link as well as localizing different emergency situations at early stages.
4. Calculations performed for a fragment of 500kV network demonstrated rapid response and reliable convergence of developed algorithms, which allows their use for on-line estimation.
5. PMU installation in HVDC link circuit ensures monitoring of HVDC operation and timely warning of operating

personnel of the station and dispatchers of the Grid about the frequency variation.

ACKNOWLEDGMENT

All The research was carried out under State Assignment, Project III.17.4.2 (reg. № AAAA-A17-117030310438-1) of the Fundamental Research of Siberian Branch of the Russian Academy of Sciences.

REFERENCES

[1] HVDC. URL: http://www.ntc-power.ru/innovative_projects/insert_dc/ (date of addressing: 20.12.2018).

[2] A. Gamm, A. Glazunova, Yu. Grishin, I. Kolosok, and E. Korkina, "Development of algorithms for state estimation of electric power system," in Electricity, 2009, No. 6, pp. 41-49.

[3] V. Kochkin, "New technologies of raising the transmission capacity of power lines. Controllable power transfer," in Power Engineering News, № 4 (46), 2007, pp. 2-6.

[4] N. Voropai, and A.Osak, "Development of equipment and systems for large energy systems control," in Scientific Report, code 2008-0-2.7-31-01-007. Irkutsk, 2009, 248 p.

[5] P. Erokhin, V. Neuimin, A. Aleksandrov, and D. Maksimenko, "Modeling the FACTS devices in the problems of power systems mode computation and optimization," in News of Research Center of Unified Grid, № 66(1), 2012, pp.22–28.

[6] N. Hingorani, and L. Gyugyi, "Understanding FACTS concepts and technology of flexible AC transmission systems," in IEEE Press, 2000, 452 p.

[7] E. Acha, C.R. Fuerte-Esquivel, H. Ambriz-Perez, and C. Angeles-Camacho, FACTS. Modelling and Simulation in Power Networks. England: JohnWilley&Sons, Ltd, 2004.

[8] T. Okon, and K. Wilkosz, "WLS state estimation in polar and rectangular coordinate systems for power system with UPFC: significance of types of measurements," "PowerTech'2011", Trondheim, Norway, 19-23 June, 2011.

[9] E. Zamora-Cárdenas, B. Alcaide-Moreno, and C. Fuerte-Esquivel, "State estimation of power systems containing FACTS controllers," in Electric Power Systems Research, #81, 2011, pp. 995-1002.

[10] I. Kolosok, and A. Tikhonov, "Algorithms for modeling the static thyristor equalizers at state estimation of power systems," in Industrial power engineering, №10, 2015, pp. 30-35.

[11] A. Tikhonov, "Modeling the FACTS devices at state estimation of modern power systems", Author's abstract of doctor's thesis. Irkutsk, 2017, 25 p.

[12] I. Kolosok, A. Tikhonov, and E. Korkina, "Study of the efficiency of the state estimation algorithms when connecting the FACTS devices to the substitution circuit," in Collected Works "Methodological issues of studying the reliability of large energy systems. Problems of energy systems reliability" Syktyvkar: Publishing House "Komi Republic Printshop", Issue 67, 2016, pp. 206-214.

[13] E. Ametistov, Foundations of modern power engineering. Modern power engineering. Moscow Publishing House of MEI, vol. 2, 2010, 632 p.

[14] I. Kolosok, and A. Tikhonov, "Modeling the FACTS devices (STC and STATCOM) when solving the EPS state estimation problems," Collected Works Methodological issues of studying the reliability of large energy systems. Problems of energy systems reliability", Minsk: Issue 66, 2015, pp. 176-182.

[15] I. Kolosok, A. Tikhonov, E. Korkina, A. Mahnitko, and A.Gavrilovs, "Modeling of FACTS devices for smart grid state estimation", "RTUCON-2018", Riga, Latvia, doi:10.1109/rtucon.2018.8659809.

[16] A. Zhukov, "Experience of using the technology of synchronized vector measurements (WAMS) for monitoring and control in power engineering," Report at NTS SO EPS Meeting. (October 10, 2018).

[17] I. Kolosok, E. Korkina,and A. Paltsev, "Bad Data Detection at Decomposition of State Estimation Problem," "PowerTech'2011", Trondheim, Norway, 2011, doi:10.1109/ptc.2011.6019459.

[18] A. Gamm, I. Kolosok, A. Glazunova, and E. Korkina, "PMU Placement Criteria for EPS State Estimation," in DRPT'08, Nanjing, China,2008, pp.645-649, doi:10.1109/drpt.2008.4523485.

[19] I. Kolosok, E. Korkina, and E. Buchinsky, "The Test Equation Method for Linear State Estimation Based on PMU Data," PSCC-2014, Poland, Wroclaw, 2014, doi:10.1109/pscc.2014.7038473.

[20] A. Phadke, and J. Thorp, "Synchronized Phasor Measurements and Their Applications", New York: Springer, 2008, 247 p.

[21] M. Dalali, and H. K. Kargar, "PMU placement for HVDC connected network considering united and islanded conditions," PSPC'2015, Tehran, Iran, pp.19-24.

[22] G. Yang, M. Gordon, A.H. Nielsen, and J. Østergaard, "PMU applications - from situation awareness to blackout prevention," 2009.

[23] B. Singh, N.K. Sharma, A.N. Tiwari, K.S. Verma, and S.N. Singh, "Applications of phasor measurement units (PMUs) in electric power system networks incorporated with FACTS controllers," International Journal of Engineering, Science and Technology, vol.3, #3, 2011, pp. 64-82.

[24] E.A. Zamora-Cárdenas, C.R. Fuerte-Esquivel, A. Pizano-Martínez, H.J. Estrada-García, "Hybrid state estimator considering SCADA and synchronized phasor measurements in VSC-HVDC transmission links", Electric Power Systems Research, #133, 2016, pp. 42–50.

Probabilistic Approximation of Voltage Values for Power Distribution Network with Large Number of Nodes and Solar Generation

Georgi Georgiev
Plethora Ltd, arna, Bulgaria
Email: georgidg@gmail.com

Dmitriy Kim
Narxoz University,
Almaty, Kazakhstan
E-mail: dmitriy.kim@narxoz.kz

Michael Negnevitsky
School of Engineering and ICT
University of Tasmania
Hobart, Australia
michael.negnevitsky@utas.edu.au

Abstract — **We study the problem of state estimating in the networks with distributed generation. Traditional state estimation methods provide the expected value of the estimated parameters. We consider the parameters of the system as random variables and study probability that voltage magnitudes belong to admissible set of values. We show that for large number of nodes we can apply central limit theorem and approximate the sought probability.**

Keywords: state estimation of energy system, distribution generation; distribution networks; voltage constraints; central limit theorem; probabilistic approach

I. Introduction

Due to the development of distributed generation, such as solar, the previously unidirectional power flows in distributed power system can become bidirectional [1]. This can lead to significant voltage and security concerns [2-5] and the distribution network operator needs information about the state of the network, which is provided by the state estimation (SE) process that derives the network state from the available set of measurements [6]. It is important to note that in distribution networks the number of available real-time measurements is usually very limited [7]. Therefore, pseudo-measurements with large margins of error are used as additional measurements. Traditionally state estimation methods aim to estimate the exact value of the estimated parameters [8]. This means that the estimated network state will contain a large amount of uncertainty [9-10].

In this work we suggest another approach to the problem of SE in distribution networks. We consider measurements as random variables with some probabilistic distribution. We use the fact that the principal concern of the distribution network operator is to keep the state of the network within its constraints. It means that it is not necessary for the network operator to know the exact values of the estimated parameters, but rather if these parameters remain within their respective constraints. According to this logic instead of finding of exact values of parameters we investigate the probability that the sought parameters belong to admissible intervals. Moreover, if we can find or assess such probability the next step might be to find such values of some input parameters that maximize probability.

Let us formulate our approach in more detail. The relationship between a vector of measurements Y and a vector of unknown parameters of the system X can be written as follows:

$$Y = h(X) + \varepsilon, \qquad (1)$$

where $h(\cdot)$ – vector of functions that relates the state vector and ε – errors of measurements. If we consider ε as the vector of random variables we can treat $Y - \varepsilon$ also as random vector. Therefore we can rewrite equation (1) as follows:

$$Y = h(X)$$

where Y is a vector of random variables with known probabilistic distribution. In this case the solution X is also random vector. Instead of exact estimation X we study the probability that values of X belong to some constraints:

$$\boldsymbol{P}(\,C_{min} < X < C_{max}\,).$$

If we know this probability we can analyze it, for example to study conditions which maximize it, study maximal values, etc.

In this work we consider distribution network with large number of nodes and random currents which depend on renewable generation and load of customer. We show how application of probabilistic methods such as central limit theorem can get approximation for the sought probability under general restrictions on the probabilistic distribution of currents. We investigate example of distribution network with linear topology. The achieved results allow us to understand better the network behavior and its properties. Moreover, we get exact probabilities where currents have Gaussian probabilistic distribution. These results might be compared with general case to "touch" the accuracy of approximation.

II. Problem Formulation

We consider distribution network with $m + 1$ nodes and their voltages

$$U_0, U_1, U_2, \dots, U_m.$$

Put that initial node is balancing, i.e. U_0 is known constant. Define the vector of voltages for each node except balancing

$$\boldsymbol{U} = (U_1, U_2, \dots, U_m)^T,$$

the vector of currents

$$\boldsymbol{I} = (I_1, I_2, \dots, I_m)^T$$

and active resistances of the network

$$R_1, R_2, \dots, R_m.$$

We suppose that we can neglect reactive parameters of the network. The relationship between the vector of voltages and the vector of currents might be expressed as

$$\boldsymbol{U} = \boldsymbol{V} + \boldsymbol{Z} \times \boldsymbol{I},$$

where $\boldsymbol{Z}$ – the matrix of nodal resistances and $\boldsymbol{V}$ – vector of known values.

We suppose that values of $\boldsymbol{I}$ could be expressed as follows:

$$I_i = -d_i \times SI + J_i, \quad i = 1,2, \dots, m,$$

where SI – the current from solar panels to network, d_i – coefficient, which characterize technical details of renewables (e.g. solar panels) in node i, J_i – the current from the node to the customer.

We suppose that SI and J_i are independent random variables with known probabilistic distribution, J_i are independent between each other, parameters d_i are known. Then the problem might be formulated in such way: to assess probability that voltages belong to admissible set of values, for instances from 0.9 to 1.1 of nominal voltage U_{nom},

$$\boldsymbol{P}\left(0.9\, U_{nom} < \min_{1\le i\le m} U_i, \max_{1\le i\le m} U_i < 1.1\, U_{nom}\right).$$

III. Currents with normal distribution

At the beginning suppose that SI and J_i have normal (or Gaussian) distribution, i.e. they might be written as

$$SI = e + \sigma_0 \times \xi_0,$$

$$J_i = a_i + \sigma_i \times \xi_i, \quad i = 1,2, \dots, m,$$

where $e, \sigma_0, a_i, \sigma_i$, $i = 1,2, \dots, m$, are known and ξ_0, ξ_i, $i = 1,2, \dots, m$, are independent and have standard normal distributions.

Then we can write

$$I_i = a_i - d_i e - d_i \sigma_0 \xi_0 + \sigma_i \xi_i, \quad i = 1,2, \dots, m,$$

in matrix form

$$\boldsymbol{I} = \boldsymbol{Ad} + \boldsymbol{W} \times \Xi,$$

where

$$\boldsymbol{Ad} = \begin{pmatrix} a_1 - d_1 e \\ a_2 - d_2 e \\ \dots \\ a_m - d_m e \end{pmatrix}, \quad \Xi = \begin{pmatrix} \xi_0 \\ \xi_1 \\ \dots \\ \xi_m \end{pmatrix},$$

$$\boldsymbol{W} = \begin{pmatrix} -d_1\sigma_0 & \sigma_1 & 0 & \dots & 0 \\ -d_2\sigma_0 & 0 & \sigma_2 & \dots & 0 \\ & & \dots & & \\ -d_m\sigma_0 & 0 & 0 & \dots & \sigma_m \end{pmatrix}.$$

In this case

$$\boldsymbol{U} = \boldsymbol{\Gamma} + \boldsymbol{A} \times \Xi,$$

where

$$\boldsymbol{\Gamma} = \boldsymbol{V} + \boldsymbol{Z} \times \boldsymbol{Ad}, \qquad \boldsymbol{A} = \boldsymbol{Z} \times \boldsymbol{W}.$$

Then the vector $\boldsymbol{U}$ has multivariate normal distribution with mean vector $\boldsymbol{\Gamma}$ and covariance matrix

$$Cov_m = \boldsymbol{A} \times \boldsymbol{A}^T.$$

Example 1. *Let us consider linear network with $m = 3$ nodes from Fig.1*

The node 0 is balancing and its voltage U_0 is known. Then

$$\begin{pmatrix} U_1 \\ U_2 \\ U_3 \end{pmatrix} = \begin{pmatrix} U_0 \\ U_0 \\ U_0 \end{pmatrix} +$$

$$\begin{pmatrix} -R_1 & -R_1 & -R_1 \\ -R_1 & -(R_1 + R_2) & -(R_1 + R_2) \\ -R_1 & -(R_1 + R_2) & -(R_1 + R_2 + R_3) \end{pmatrix} \times \begin{pmatrix} I_1 \\ I_2 \\ I_3 \end{pmatrix},$$

$$\boldsymbol{\Gamma} = \begin{pmatrix} U_0 \\ U_0 \\ U_0 \end{pmatrix} + \begin{pmatrix} -R_1 & -R_1 & -R_1 \\ -R_1 & -(R_1 + R_2) & -(R_1 + R_2) \\ -R_1 & -(R_1 + R_2) & -(R_1 + R_2 + R_3) \end{pmatrix} \times$$

$$\begin{pmatrix} a_1 - d_1 e \\ a_1 - d_1 e \\ a_1 - d_1 e \end{pmatrix},$$

$$\boldsymbol{A} = \begin{pmatrix} -R_1 & -R_1 & -R_1 \\ -R_1 & -(R_1 + R_2) & -(R_1 + R_2) \\ -R_1 & -(R_1 + R_2) & -(R_1 + R_2 + R_3) \end{pmatrix} \times$$

$$\begin{pmatrix} -d_1\sigma_0 & \sigma_1 & 0 & 0 \\ -d_2\sigma_0 & 0 & \sigma_2 & 0 \\ -d_3\sigma_0 & 0 & 0 & \sigma_3 \end{pmatrix}.$$

Let

$U_0 = 200\ V$, $U_{nom} = 222\ V$,
$a_1 = a_2 = a_3 = 10\ A$,
$e = 5\ A$,
$\sigma_0 = 2$,
$\sigma_1 = \sigma_2 = \sigma_3 = 3$,
$d_1 = d_2 = d_3 = 0.9$,
$R_1 = R_2 = R_3 = 0.01\ Om$.

Then

$$\boldsymbol{P}\left(0.9\, U_{nom} < \min_{1\le i\le 3} U_i, \max_{1\le i\le 3} U_i < 1.1\, U_{nom}\right) = 0.2012.$$

If we put $U_0 \geq 201\ V$, then

$$\boldsymbol{P}\left(0.9\, U_{nom} < \min_{1\le i\le 3} U_i, \max_{1\le i\le 3} U_i < 1.1\, U_{nom}\right) = 1.$$

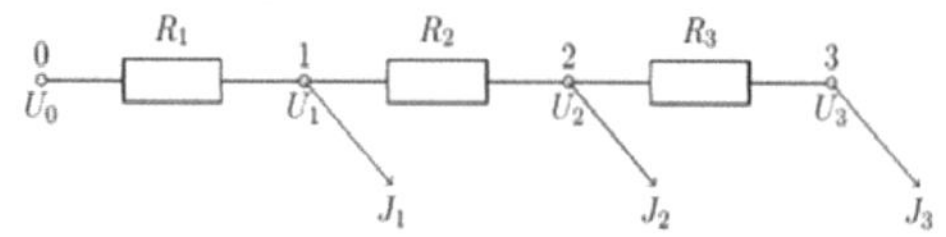

Figure 1. The 3 nodes line.

Example 2. *Let us consider linear network with* $m = 100$ *nodes from Fig.2*

Let

$$U_0 = U_{nom} = 230\ V,$$
$$a_1 = a_2 = a_3 = 5\ A,$$
$$e = 4/0.9\ A,$$
$$\sigma_0 = 0.4,$$
$$\sigma_1 = \sigma_2 = \sigma_3 = 1,$$
$$d_1 = d_2 = d_3 = 0.9,$$
$$R_1 = R_2 = R_3 = 0.01\ Om.$$

Then

$$\boldsymbol{P}\left(0.9\ U_{nom} < \min_{1 \le i \le m} U_i,\ \max_{1 \le i \le m} U_i < 1.1 U_{nom}\right) = 0.0746,$$

$$\boldsymbol{P}(0.9\ U_{nom} < U_1 < 1.1\ U_{nom}) = 1,$$

$$\boldsymbol{P}(0.9\ U_{nom} < U_m < 1.1\ U_{nom}) = 0.0747.$$

Remark that if

$$0.9\ U_{nom} = 207 < U_0 < 253 = 1.1\ U_{nom}$$

The maximal probability will be achieved with $U_0 = 253\ V$

$$\boldsymbol{P}\left(0.9\ U_{nom} < \min_{1 \le i \le m} U_i,\ \max_{1 \le i \le m} U_i < 1.1 U_{nom}\right) = 0.4018.$$

If we suppose that only first 10 nodes $i = 1,2,\dots,10$ *or last 10 nodes* $i = 91,92,\dots,100$ *in line have renewable generation then*

$$\boldsymbol{P}\left(0.9\ U_{nom} < \min_{1 \le i \le m} U_i,\ \max_{1 \le i \le m} U_i < 1.1 U_{nom}\right) = 0,$$

$$\boldsymbol{P}(0.9\ U_{nom} < U_1 < 1.1\ U_{nom}) = 1,$$

$$\boldsymbol{P}(0.9\ U_{nom} < U_m < 1.1\ U_{nom}) = 0.$$

If we suppose that last 80 nodes $i = 21,22,\dots,100$ *in line have solar panels then*

$$\boldsymbol{P}\left(0.9\ U_{nom} < \min_{1 \le i \le m} U_i,\ \max_{1 \le i \le m} U_i < 1.1 U_{nom}\right) = 0.0251,$$

$$\boldsymbol{P}(0.9\ U_{nom} < U_1 < 1.1\ U_{nom}) = 1,$$

$$\boldsymbol{P}(0.9\ U_{nom} < U_m < 1.1\ U_{nom}) = 0.253.$$

If $U_0 = 253\ V$ *for the case where only last 80 nodes have renewable generation*

$$\boldsymbol{P}\left(0.9\ U_{nom} < \min_{1 \le i \le m} U_i,\ \max_{1 \le i \le m} U_i < 1.1 U_{nom}\right) = 0.2405.$$

IV. Network with Large Number of Nodes

The proposition about normal distribution of currents is rather restricted and does not reflect the reality. Consider distribution network with large number m of nodes with

$$SI = e + \sigma_0 \times \xi_0,$$

$$J_i = a_i + \sigma_i \times \xi_i,\ \ i = 1,2,\dots,m,$$

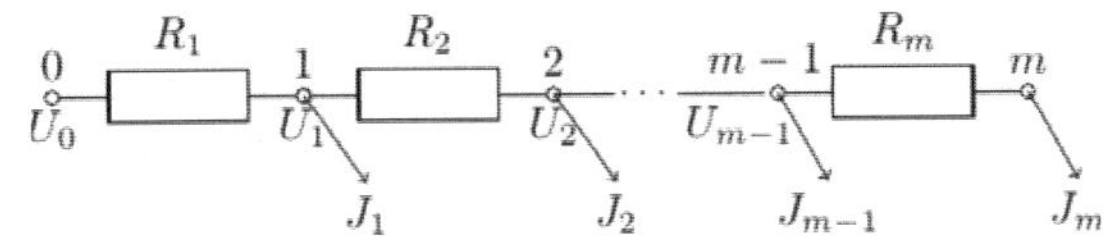

Figure 2. The $m = 100$ nodes line.

where $e, \sigma_0, a_i, \sigma_i,\ \ i = 1,2,\dots,m$ are known and $\xi_0,\ \xi_i,\ i = 1,2,\dots,m$ are independent random variables with $\boldsymbol{E}\xi_j = 0$, $\boldsymbol{Var}\xi_j = 1$, $j = 0,1,2,\dots,m$. We suppose that for every $j = 0,1,2,\dots m$

$$\boldsymbol{E}|\xi_j|^3 < \infty.$$

As in previous paragraph we can write that

$$\boldsymbol{U} = \boldsymbol{\Gamma} + \boldsymbol{A} \times \boldsymbol{\Xi}.$$

Due to known inequality Frechet

$$\boldsymbol{P}\left(0.9\ U_{nom} < \min_{1 \le i \le m} U_i,\ \max_{1 \le i \le m} U_i < 1.1\ U_{nom}\right) \le$$

$$\min_{1 \le i \le m} \boldsymbol{P}(0.9\ U_{nom} < U_i < 1.1\ U_{nom}).$$

If we put that a_{ij}, $i = 1,2,\dots,m$, $j = 0,1,2,\dots,m$, are elements of matrix $\boldsymbol{A}$ and γ_i are elements of vector $\boldsymbol{\Gamma}$ then

$$\boldsymbol{P}(0.9\ U_{nom} < U_i < 1.1\ U_{nom}) =$$

$$\boldsymbol{P}(0.9\ U_{nom} - \gamma_i < S_i(m) < 1.1\ U_{nom} - \gamma_i),$$

where

$$S_i(m) = a_{i0}\xi_0 + a_{i1}\xi_1 + \dots + a_{im}\xi_m.$$

It means that the problem reduces to the investigation distribution of sum of independent random variables. This theory is well-known. We can use either bounds or approximations for studying distribution of $S_i(m)$. In the next paragraph we will show how to use central limit theorem for analysis of linear topology of network.

A. Linear network with large number of nodes

Consider linear network from fig. 2 with large number m of nodes. For brevity we put

$$R_i = R,\ d_i = d,\ \sigma_i = \sigma,\ \ i = 1,2,\dots,m.$$

In such case for $i = 1,2,\dots,m$

$$S_i(m) = R\left(\left(mi - \frac{i(i-1)}{2}\right) d\sigma_0\xi_0 - \sigma\left(\sum_{k=1}^{i-1} k\xi_k + j\sum_{k=j}^{m} \xi_k\right)\right).$$

If we consider only first and last node then

$$S_1(m) = R\left(md\sigma_0\xi_0 - \sigma\sum_{k=1}^{m} \xi_k\right),$$

$$S_m(m) = R\left(\frac{m(m+1)}{2} d\sigma_0\xi_0 - \sigma\sum_{k=1}^{m} k\xi_k\right).$$

We can see that if every node has solar panels the dispersion of current from renewables (e.g. solar radiation) in md^2 times greater the dispersion of customer currents. It means that if for some constant $c > 0$

$$d \sim \frac{c}{\sqrt{m}},\ m \to \infty,$$

then the dispersion of current from renewable generation is proportional to the dispersion of customer currents. If $d \approx 1$, e.g. every node has solar panels producing current which is proportional to the customer currents, then for every $i = 1,2,\dots,m$

$$\boldsymbol{P}(0.9\ U_{nom} < U_i < 1.1\ U_{nom}) \approx$$

$$\boldsymbol{P}\left(0.9U_{nom} - \gamma_i < dR\sigma_0\left(mi - \frac{i(i-1)}{2}\right)\xi_0 < 1.1U_{nom} - \gamma_i\right).$$

Example 3. *Under the conditions of Example 2 with* $m = 100$ *if* ξ_0 *has uniform distribution on the interval* $[-\sqrt{3},\sqrt{3}]$, *then*

$$\boldsymbol{P}(0.9\ U_{nom} < U_m < 1.1\ U_{nom}) \approx 0.0633.$$

Let us study the case where only first $c\sqrt{m}$ nodes of the network have solar panels. If we consider only first and last node then

$$S_1(m) = R\left(c\sqrt{m}d\sigma_0\xi_0 - \sigma\sum_{k=1}^{m}\xi_k\right),$$

$$S_m(m) = R\left(c\frac{m}{2}d\sigma_0\xi_0 - \sigma\sum_{k=1}^{m} k\xi_k\right).$$

It is obvious that according to central limit theorem (see [11]) the distribution of

$$\frac{S_1(m)}{\sqrt{m}}$$

converges to normal distribution with zero mean and $R^2(c^2d^2{\sigma_0}^2 + \sigma^2)$ dispersion as $m \to \infty$. It means that under large m

$$\boldsymbol{P}(0.9\ U_{nom} < U_1 < 1.1\ U_{nom}) =$$

$$\boldsymbol{P}\left(\frac{0.9\ U_{nom} - \gamma_1}{\sqrt{m}R\sqrt{c^2d^2{\sigma_0}^2 + \sigma^2}} < \frac{S_1(m)}{\sqrt{m}R\sqrt{c^2d^2{\sigma_0}^2 + \sigma^2}} < \frac{1.1\ U_{nom} - \gamma_1}{\sqrt{m}R\sqrt{c^2d^2{\sigma_0}^2 + \sigma^2}}\right) \approx$$

$$\boldsymbol{\Phi}_{0,1}\left(\frac{1.1\ U_{nom} - \gamma_1}{\sqrt{m}R\sqrt{c^2d^2{\sigma_0}^2 + \sigma^2}}\right) - \boldsymbol{\Phi}_{0,1}\left(\frac{0.9\ U_{nom} - \gamma_1}{\sqrt{m}R\sqrt{c^2d^2{\sigma_0}^2 + \sigma^2}}\right).$$

Consider the last node. It is easy to see that

$$Var\left(\frac{S_m(m)}{m^{\frac{3}{2}}}\right) \to \frac{1}{3}R^2\sigma^2,\ m \to \infty.$$

The Lyapunov condition (see [11]) for the central limit theorem

$$\frac{1}{m^{\frac{9}{2}}}\sum_{k=1}^{m} k^3\ E{\xi_k}^3 \to 0,\ m \to \infty,$$

is satisfied and

$$\frac{S_m(m)}{m^{\frac{3}{2}}}$$

converges to normal distribution with zero mean and dispersion $\frac{1}{3}R^2\sigma^2$ as $m \to \infty$. Remark that limit distribution for $\frac{S_m(m)}{m^{\frac{3}{2}}}$ does not depend on ${\sigma_0}^2$. Therefore, for large m we can use approximation

$$\boldsymbol{P}(0.9\ U_{nom} < U_m < 1.1\ U_{nom}) =$$

$$\boldsymbol{P}\left(\frac{0.9\ U_{nom} - \gamma_m}{m^{\frac{3}{2}}R\sigma\frac{1}{\sqrt{3}}} < \frac{S_m(m)}{m^{\frac{3}{2}}R\sigma\frac{1}{\sqrt{3}}} < \frac{1.1\ U_{nom} - \gamma_m}{m^{\frac{3}{2}}R\sigma\frac{1}{\sqrt{3}}}\right) \approx$$

$$\boldsymbol{\Phi}_{0,1}\left(\frac{1.1\ U_{nom} - \gamma_m}{m^{\frac{3}{2}}R\sigma\frac{1}{\sqrt{3}}}\right) - \boldsymbol{\Phi}_{0,1}\left(\frac{0.9\ U_{nom} - \gamma_m}{m^{\frac{3}{2}}R\sigma\frac{1}{\sqrt{3}}}\right).$$

Example 4. *Under the conditions of Example 2 with* $m = 100$ *we consider the linear network in the case where only 10 first nodes have solar panels. Then*

$$\boldsymbol{P}(0.9\ U_{nom} < U_1 < 1.1\ U_{nom}) \approx 1,$$

$$\boldsymbol{P}(0.9\ U_{nom} < U_m < 1.1\ U_{nom}) \approx 0.$$

Consider the case where only last $c\sqrt{m}$ nodes of the network has solar panels. If we consider only first and last node then

$$S_1(m) = R\left(c\sqrt{m}d\sigma_0\xi_0 - \sigma\sum_{k=1}^{m}\xi_k\right),$$

$$S_m(m) = R\left(c\frac{m\sqrt{m}}{2}d\sigma_0\xi_0 - \sigma\sum_{k=1}^{m} k\xi_k\right).$$

Using the Lyapunov conditions one can prove as before that

$$\frac{S_m(m)}{m^{\frac{3}{2}}},\ m \to \infty,$$

converges to normal distribution with zero mean and dispersion

$$R^2\left(\frac{1}{4}c^2d^2{\sigma_0}^2 + \sigma^2\frac{m(m+1)(2m+1)}{6m^3}\right) \to$$

$$R^2\left(\frac{1}{4}c^2d^2{\sigma_0}^2 + \frac{1}{3}\sigma^2\right) = DM,\ m \to \infty.$$

In this case for large m we can use approximation

$$\boldsymbol{P}(0.9\ U_{nom} < U_m < 1.1\ U_{nom}) =$$

$$\boldsymbol{P}\left(\frac{0.9\ U_{nom} - \gamma_m}{R\ m^{\frac{3}{2}}DM^{\frac{1}{2}}} < \frac{S_m(m)}{R\ m^{\frac{3}{2}}DM^{\frac{1}{2}}} < \frac{1.1\ U_{nom} - \gamma_m}{R\ m^{\frac{3}{2}}DM^{\frac{1}{2}}}\right) \approx$$

$$\boldsymbol{\Phi}_{0,1}\left(\frac{1.1\ U_{nom} - \gamma_m}{R\ m^{\frac{3}{2}}DM^{\frac{1}{2}}}\right) - \boldsymbol{\Phi}_{0,1}\left(\frac{0.9\ U_{nom} - \gamma_m}{R\ m^{\frac{3}{2}}DM^{\frac{1}{2}}}\right).$$

Example 5. *Under the conditions of Example 2 with* $m = 100$ *if the linear network. has only 10 last nodes with solar panels. Then for* $c = 1$ *:*

$$P(0.9\, U_{nom} < U_1 < 1.1\, U_{nom}) \approx 1,$$

$$P(0.9\, U_{nom} < U_m < 1.1\, U_{nom}) \approx 0.$$

If we suppose that the network has only 80 last nodes with solar panels. Then for $c = 8$

$$P(0.9\, U_{nom} < U_1 < 1.1\, U_{nom}) \approx 1,$$

$$P(0.9\, U_{nom} < U_m < 1.1\, U_{nom}) \approx 0.0344.$$

If we put $U_0 = 253 = 1.1\, U_{nom}$, *then*

$$P(0.9\, U_{nom} < U_m < 1.1\, U_{nom}) \approx 0.2552.$$

V. Conclusion

The work considers the problem of analyzing the parameters of a distribution network with renewable energy sources in its structure, which allow households to not only reduce consumption from the network, but also generate surplus capacity to the network.

Instead of searching for estimates of unknown parameters, we studied the probability that the parameters satisfy constraints.

If we assume that the consumption and generation currents in the distribution network will be independent and normally distributed random variables, then the required probabilities can be obtained with little effort.

Without the assumption of normal distribution of currents we have to consider a network with a large number of nodes and apply known probabilistic methods to study and approximate the probability that the sought parameters are within acceptable limits. In particular, the central limit theorem for the linear network topology with a large number of nodes allows the following conclusions about node voltages:

- If each node with renewables generates current from node to network which is approximately as much as each node consumes, then to study the distribution of voltages we can neglect the uncertainty of consuming currents. If each node generates renewable current from node to network which is approximately proportional to $\frac{c}{\sqrt{m}}$, c > 0, of the current consumption, then the contribution of renewable's uncertainty to voltage distribution is approximately proportional to uncertainties of other consumption currents.

- If there are only m^{α} nodes in the network with generation of renewable energy sources, then for $\alpha < 1/3$ the influence of renewable energy sources can be neglected. If $1/3 < \alpha < 2/3$ then the "contribution"' of renewable energy sources to the uncertainty of voltage values is proportional to the "contribution" of uncertainty of current of consumption. If $\alpha > 2/3$ then the "contribution" of renewable current's uncertainty determines the distribution of voltage values.

References

[1] M. Eisenreich, G. Balzerl, J. Backes, B. Maurer, "Integration of PV systems into low voltage networks using standard load profiles", International Conference on Renewable Energies and Power Quality (ICREPQ'10), Granada (Spain), 23th to 25th March, 2010, pp. 972-976.

[2] J. Paoli, "Optimising low voltage transformer tap settings in distribution systems", Theses for obtaining bachelor degree, UTAS, Australia, pp. 1-84, 2018.

[3] A. Priye and K. Komla, "Voltage Rise Issue with High Penetration of Grid Connected PV", vol. 47, no. 3, pp. 4959-4966, IFAC, 2014.

[4] C. L. Masters, "Voltage rise: the big issue when connecting embedded generation to long 11 kV overhead lines," Power Eng. J., vol. 16, no. 1, pp. 5–12, 2002.

[5] M. J. E. Alam, K. M. Muttaqi, and D. Sutanto, "A comprehensive assessment tool for solar PV impacts on low voltage three phase distribution networks," 2012 2nd Int. Conf. Dev. Renew. Energy Technol., pp. 1–5, 2012.

[6] A. Abur and A. G. Expósito, "Power System State Estimation Theory and Implementation", New York: Marcel Dekker, Inc., 2004.

[7] D. G. Colome and C. E. D. L. Vega, "Voltage quality control at distribution systems applying a fuzzy state estimator", PES T&D LA, pp. 1-7, 2008.

[8] L. Mihet-Popa, X. Han, H. Bindner, J. Pihl-Andersen, J. Mehmedalic, "Grid Modeling, Analysis and Simulation of Different Scenarios for a Smart Low-Voltage Distribution Grid". Conference: ISGT 2013, Copenhagen, October 2013, 5p.

[9] B. Brinkmann, M. Negnevitsky, T. Yee, and T. Nguyen, "An Observability Index for Distribution Networks using Information Entropy," in Australasian Universities Power Engineering Conf. (AUPEC), Wollongong, NSW, 2015, pp. 1-6.

[10] B. Brinkmann, M. Negnevitsky, "A probabilistic approach to observability of distribution networks", IEEE Transactions on Power Systems, vol. 32, no. 2, pp. 1169-1178, 2017.

[11] I.I.Gihman, A.V. Skorohod, "Introduction to the Theory of Random Processes," W.B. Saunders Co., Philadelphia, 1969.

Probabilistic Approach to State Estimation in Electricity Networks with Distributed Generation

Georgi Georgiev
Plethora Ltd, Varna, Bulgaria
Email: georgidg@gmail.com

Dmitriy Kim
Narxoz University,
Almaty, Kazakhstan
E-mail: Dmitriy.kim@narxoz.kz

Michael Negnevitsky
School of Engineering and ICT
University of Tasmania
Hobart, Australia
michael.negnevitsky@utas.edu.au

Abstract — **We study the behavior of low voltage networks parameters during period of time. It is supposed that some parameters of the networks are random processes and the problem is to investigate the probability if the voltages satisfy some constraints during period of time. We use known probabilistic results and reduce this problem to the solution of partial differential equation. The numerical example is given at the end of the work.**

Keywords: distribution generation; distribution networks; voltage constraints; state estimation; probabilistic approach; partial differential equations.

I. Introduction

Due to the development of distributed generation the previously unidirectional power flows in distributed power system can become bidirectional [1]. This can lead to the problem of holding voltages in admissible set of values [2-4]. The distribution network operator needs information about the state of the network, which is provided by the state estimation (SE) process that derives the network state from the available set of measurements [5-6]. It is important to note that in distribution networks the number of available real-time measurements is usually very limited [7]. Therefore, pseudo-measurements with large margins of error are used as additional measurements. Traditionally state estimation methods aim to estimate the exact value of the estimated parameters at some moment of time [8].

In this work we suggest another approach to the problem of SE in distribution networks. First of all we consider the behavior of networks parameters during period of time. Secondly, we suppose that parameters of the network are random processes. Thirdly, instead of finding of exact values of parameters in time we investigate the probability that the sought parameters belong to admissible intervals during of time. We use the fact that the principal concern of the distribution network operator is to keep the state of the network within its constraints. It means that it is not necessary for the network operator to know the exact values of the estimated parameters, but rather if these parameters remain within their respective constraints. If we can find or assess such probability the next step might be to find such values of some input parameters that maximize this probability.

According to classical SE problem formulation the relationship between a vector of measurements Y and a vector of unknown parameters of the system X can be written as follows:

$$Y = h(X) + \varepsilon, \tag{1}$$

where $h(X)$ – vector of functions that relates the state vector and ε – errors of measurements. If we consider ε as the vector of random variables we can treat $Y - \varepsilon$ also as random vector. Therefore, we can rewrite equation (1) as follows:

$$Y = h(X),$$

where Y is a vector of random variables with known probabilistic distribution. If we plan to study the behavior of the network during period of time $t \in [0, T]$ we consider equation

$$Y(t) = h(X(t)),$$

where $Y(t)$ is random process. In this case the solution $X(t)$ is also random process. Instead of exact estimation X we study the probability that values of $X(t)$, $t \in [0, T]$, belong to some constraints:

$$\boldsymbol{P}\left(C_{min} < \inf_{t\in[0,T]} \mathrm{X(t)} \leq \sup_{t\in[0,T]} \mathrm{X(t)} < C_{max} \right).$$

If we know this probability we can analyze it, for example to study conditions which maximize it, to study maximal values, etc.

In this work we consider distribution network with random currents which depend on renewable generation and load of customers during period of time $t \in [0, T]$. We suppose that power consumption by households and renewable generation are random processes in time and consider the problem of probabilistic estimation that voltage values satisfy constraints during this period of time. We put that known random processes are Ito diffusion and we show how to transfer the sought probability to solution of partial differential equation. Thus, we will demonstrate the connection between SE problem in time and the well-developed theory of partial differential equations. At the end of the work we will give a simple example and find its solution numerically.

II. Problem Formulation and Main Result

We consider distribution network with $m+1$ nodes and their voltages

$$U_0(t), U_1(t), U_2(t), \dots, U_m(t),\ t \in [0,T].$$

Put that initial node is balancing, i.e. $U_0(t) = U_0$ is known constant. Define the vector of voltages for each node except balancing

$$\boldsymbol{U(t)} = (U_1(t), U_2(t), \dots, U_m(t))^T,$$

the vector of currents

$$\boldsymbol{I} = (I_1(t), I_2(t), \dots, I_m(t))^T$$

and active resistances of the network

$$R_1, R_2, \dots, R_m.$$

We suppose that we can neglect reactive parameters of the network. The relationship between the vector of voltages and the vector of currents might be expressed as

$$\boldsymbol{U}(t) = \boldsymbol{V} + \boldsymbol{Z} \times \boldsymbol{I}(t),\ t \in [0,T],$$

where $\boldsymbol{Z}$ – a matrix of nodal resistances and $\boldsymbol{V}$ – vector of known values.

We are interested in studying

$$\boldsymbol{P}(0.9\ \boldsymbol{U}_{nom} - \boldsymbol{V} < \boldsymbol{Z} \times \boldsymbol{I}(t) < 1.1\ \boldsymbol{U}_{nom} - \boldsymbol{V}) \qquad (2)$$

For every $t \in [0,T]$, where $\boldsymbol{U}_{nom} = (U_{nom}, U_{nom}, \dots, U_{nom})^T$.

We suppose that values of $\boldsymbol{I}$ could be expressed as follows:

$$I_i(t) = -d_i \times SI(t) + J_i(t),\ \ i = 1,2,\dots,m,$$

where $SI(t)$ – the current from renewables to network, d_i – coefficient, which characterize technical details of renewables in the node i, $J_i(t)$ – the current from the node i to customer, $t \in [0,T]$.

The problem of finding probability (2) is very difficult under general restrictions, therefore we will use the assumption that the currents are diffusion processes.

We put that $SI(t)$ and $J_i(t)$, $i = 1,2,\dots,m$, are Ito diffusion [6], which satisfy stochastic differential equation of the form:

$$d\,SI(t) = e(t)dt + r(t)dW_0(t),$$

$$d\,J_i(t) = a_i(t)dt + s_i(t)dW_i(t),\ \ i = 1,2,\dots,m,$$

where $W_j(t)$, $j = 0,1,2,\dots,m$ are independent Wiener processes. Let (t), $a_i(t)$, $r(t)$, $s_i(t)$ be continuous functions for $t \in [0,T]$.

Then $\boldsymbol{I}(t)$ is a vector with Ito diffusion:

$$d\boldsymbol{I}(t) = \boldsymbol{Ad}(t)dt + \boldsymbol{S}(t) \times d\boldsymbol{W}(t),$$

where

$$\boldsymbol{Ad}(t) = \begin{pmatrix} a_1(t) - d_1e(t) \\ a_2(t) - d_2e(t) \\ \dots \\ a_m(t) - d_me(t) \end{pmatrix},\ d\boldsymbol{W}(t) = \begin{pmatrix} dW_0(t) \\ dW_1(t) \\ \dots \\ dW_m(t) \end{pmatrix},$$

$$\boldsymbol{S}(t) = \begin{pmatrix} -d_1r(t) & s_1(t) & 0 & \dots & 0 \\ -d_2r(t) & 0 & s_2(t) & \dots & 0 \\ & & \dots & & \\ -d_mr(t) & 0 & 0 & \dots & s_m(t) \end{pmatrix}.$$

In our conditions, we can formulate the problem in the following form: let $\boldsymbol{\xi}(t) = (\xi_1(t), \xi_2(t), \dots, \xi_m(t))^T$ be m-dimensional diffusion process as a solution to a stochastic differential equation

$$d\boldsymbol{\xi}(t) = \boldsymbol{b}(t)dt + \boldsymbol{A}(t)d\boldsymbol{W}(t), \qquad (3)$$

$$\boldsymbol{\xi}(0) = \boldsymbol{Z} \times \boldsymbol{I}(0),$$

$$\boldsymbol{b}(t) = \boldsymbol{Z} \times \boldsymbol{Ad}(t),$$

$$\boldsymbol{A}(t) = \boldsymbol{Z} \times \boldsymbol{S}(t).$$

Remark that for this equation there exists unique solution, which is continuous diffusion process. Let v_i, $i = 1,2,\dots,m$, be the elements of vector $\boldsymbol{V}$. Consider m-dimensional domain

$$\boldsymbol{D_m} = (0.9U_{nom} - v_1, 1.1U_{nom} - v_1)x \times (0.9U_{nom} - v_2, 1.1U_{nom} - v_2) \times \dots \times (0.9U_{nom} - v_m, 1.1U_{nom} - v_m) \subseteq \mathbb{R}^m.$$

Denote by τ the first exit time of $\boldsymbol{\xi}(t)$ from rectangle $\boldsymbol{D_m}$:

$$\tau = \inf\{t > 0: \boldsymbol{\xi}(t) \notin \boldsymbol{D_m}\}.$$

We can rewrite the probability from (2) as

$$\boldsymbol{P}(\tau > T).$$

As a result, we can reformulate the problem statement in the following way: let $\boldsymbol{\xi}(t)$ be m-dimensional diffusion process (3) and τ is the first exit time of $\boldsymbol{\xi}(t)$ from rectangle $\boldsymbol{D_m}$. Our aim is to study

$$\boldsymbol{P}(\tau > T).$$

It is known that there is a connection between the distribution of functionals of the trajectories of the diffusion process and partial differential equations. In particular, in our case, the problem can be reduced to the well-known heat conduction equation (see [9]):

$$\frac{\partial}{\partial t}u(t,\boldsymbol{x}) + L\,u(t,\boldsymbol{x}) = 0,\ \ t \in [0,T], \qquad (4)$$

with boundary conditions for $i = 1,2,\dots,m$

$$u(t, x_1, x_2, \dots x_{i-1}, 0.9U_{nom} - v_i, x_{i+1}, \dots, x_m) = 0, \qquad (5)$$

$$u(t, x_1, x_2, \dots x_{i-1}, 1.1U_{nom} - v_i, x_{i+1}, \dots, x_m) = 0, \qquad (6)$$

$$u(t,\boldsymbol{x}) = 1, \qquad (7)$$

where $\boldsymbol{x} = (x_1, x_2, \dots, x_m)^T$, L – differential operator

$$L = \sum_{i=1}^{m} b_i(t)\frac{\partial}{\partial x_1} + \frac{1}{2}\sum_{i=1}^{m}\sum_{k=1}^{m} \sigma_{ik}(t)\frac{\partial^2}{\partial x_i \partial x_k},$$

$\sigma_{ik}(t)$ are elements of the matrix

$$\boldsymbol{\Sigma}(t) = \boldsymbol{A}(t) \times \boldsymbol{A}(t)^T.$$

It is known from the theory of partial differential equations that there exists a unique twice continuously differentiable function $u(t, \boldsymbol{x})$ satisfying the conditions (4-7). If we find such function $u(t, \boldsymbol{x})$ then

$$u(0, \boldsymbol{\xi}(0)) = \boldsymbol{P}(\tau > T).$$

In the next paragraph we find $u(t, \boldsymbol{x})$ numerically based on well-developed numerical methods for the heat equation.

III. Example

Let us consider linear network with $m = 2$ nodes from Fig.1. We put

$$U_{nom} = 220\ V,$$
$$R_1 = R_2 = R = 0.1\ Om,$$
$$T = 24,$$
$$d_1 = d_2 = d = 0.9.$$

The node 0 is balancing and its voltage U_0 is known. Then

$$\begin{pmatrix} U_1(t) \\ U_2(t) \end{pmatrix} = \begin{pmatrix} U_0 \\ U_0 \end{pmatrix} + \begin{pmatrix} -R & -R \\ -R & -2R \end{pmatrix} \times \begin{pmatrix} I_1(t) \\ I_2(t) \end{pmatrix}.$$

Let $\boldsymbol{E} J_1(t) = \boldsymbol{E} J_2(t)$. Graphs of the function $\boldsymbol{E} J_1(t)$ and $\boldsymbol{E}\, SI(t)$ in amps are shown in Fig. 2 and Fig. 3. Graphs of the function $e(t)$and $a(t) = a_1(t) = a_2(t)$ are shown in Fig. 4 and Fig. 5.

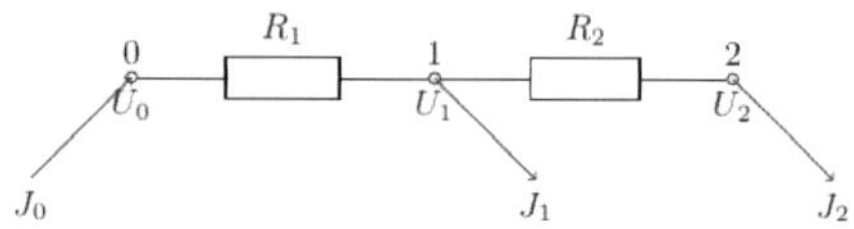

Figure 1. The two nodes line.

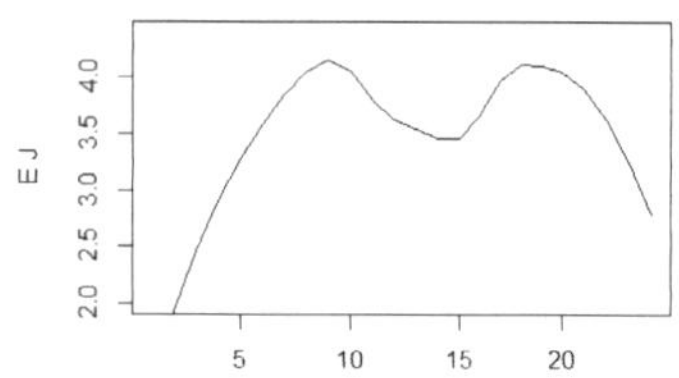

Figure 2. Function $\boldsymbol{E} J_1(t)$.

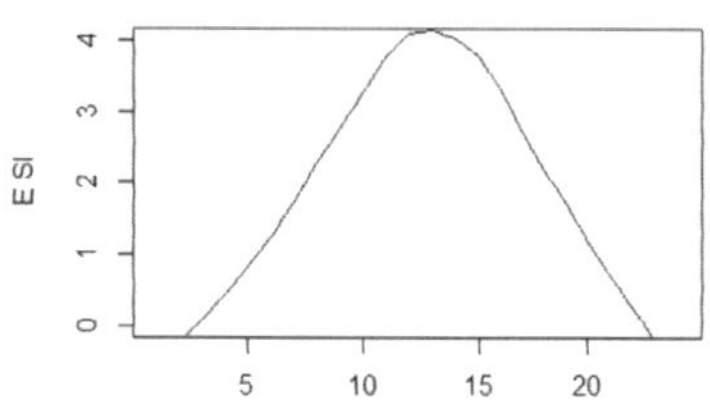

Figure 3. Function $\boldsymbol{E}\, SI(t)$.

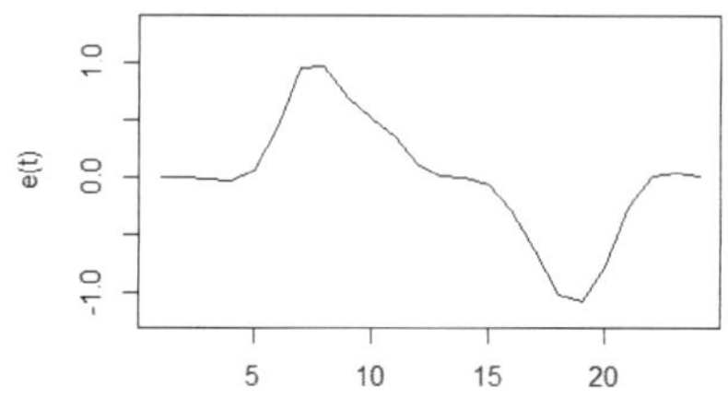

Figure 4. Function $e(t)$.

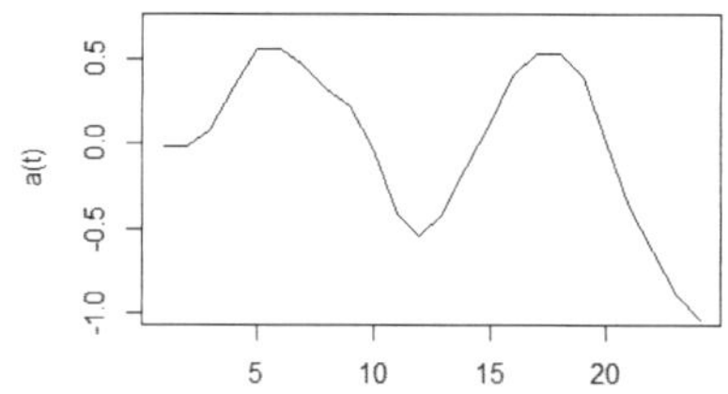

Figure 5. Function $a(t)$.

We put that

$$r(t) = 0.8 \times E J\_1(t), \quad s_i(t) = s(t) = 0.6 \times |a(t)|, i = 1{,}2.$$

It is easy to see that

$$J_1(0) = 2\text{ A}, \qquad J_2(0) = 2\text{ A},$$

$$d\boldsymbol{I}(t) = \boldsymbol{Ad}(t)dt + \boldsymbol{S}(t) \times d\boldsymbol{W}(t),$$

where

$$\boldsymbol{Ad}(t) = \begin{pmatrix} a(t) - d\, e(t) \\ a(t) - d\, e(t) \end{pmatrix}, \ d\boldsymbol{W}(t) = \begin{pmatrix} dW_0(t) \\ dW_1(t) \\ dW_2(t) \end{pmatrix},$$

$$\boldsymbol{S}(t) = \begin{pmatrix} -d\, r(t) & s(t) & 0 \\ -d\, r(t) & 0 & s(t) \end{pmatrix}.$$

Consider Ito diffusion $\boldsymbol{\xi}(t) = (\xi_1(t), \xi_2(t))^T$,

$$\xi_1(0) = -4R, \xi_2(0) = -6R,$$

$$d\xi_1(t) = -2R\big(a(t) - d\, e(t)\big)dt - d\, r(t)dW_0(t) + s(t)dW_1(t) + s(t)dW_2(t),$$

$$d\xi_1(t) = -3R\big(a(t) - d\, e(t)\big)dt - d\, r(t)dW_0(t) + s(t)dW_1(t) + s(t)dW_2(t).$$

Define

$$\boldsymbol{D_2} = \{(x_1, x_2) \colon 0.9U_{nom} - U_0 \le x_1, x_2 \le 1.1U_{nom} - U_0\},$$

$$\sigma_{11}(t) = 4\, R^2 d^2 r^2(t) + 2\, R^2 s^2(t),$$
$$\sigma_{22}(t) = 9\, R^2 d^2 r^2(t) + 5\, R^2 s^2(t),$$
$$\sigma_{12}(t) = 6\, R^2 d^2 r^2(t) + 3\, R^2 s^2(t),$$

$$\tau = \inf\{t > 0 \colon \boldsymbol{\xi}(t) \notin \boldsymbol{D_2}\}.$$

To find $\boldsymbol{P}(\tau > T)$ we consider partial differential equation:

$$\frac{\partial}{\partial t} u(t, x_1, x_2) -$$

$$R\left(a(t) - d\,e(t)\right)\left(2\frac{\partial}{\partial x_1}u(t,x_1,x_2) + 3\frac{\partial}{\partial x_2}u(t,x_1,x_2)\right) + \frac{\sigma_{11}(t)}{2}\frac{\partial^2}{\partial {x_1}^2}u(t,x_1,x_2) + \frac{\sigma_{22}(t)}{2}\frac{\partial^2}{\partial {x_2}^2}u(t,x_1,x_2) + \sigma_{12}(t)\frac{\partial^2}{\partial x_1 \partial x_2}u(t,x_1,x_2) = 0$$

with boundary conditions:

$$u(t, 0.9U_{nom} - U_0, x_2) = 0,$$
$$u(t, 1.1U_{nom} - U_0, x_2) = 0,$$
$$u(t, x_1, 0.9U_{nom} - U_0) = 0,$$
$$u(t, x_1, 1.1U_{nom} - U_0) = 0,$$
$$u(t, x_1, x_2) = 1.$$

We have reduced the problem of finding probability that voltages belong to admissible intervals to the well-known problem of heat propagation in a rectangular rod. We can apply well-developed numerical methods to obtain the numerical solution.

We can consider grid

$${x_1}^k = 0.9U_{nom} - U_0 + (k-1)h, \quad k = 1,2,\ldots,N+1,$$
$${x_1}^{N+1} = 1.1U_{nom} - U_0,$$
$${x_2}^k = 0.9U_{nom} - U_0 + (k-1)h, \quad k = 1,2,\ldots,N+1,$$
$${x_1}^{N+1} = 1.1U_{nom} - U_0,$$
$$t_j = (j-1)\tau, j = 1,2,\ldots,M+1, t_{M+1} = T,$$

and apply finite difference approximations for

$$\frac{\partial}{\partial t}u(t,x_1,x_2),$$
$$\frac{\partial}{\partial x_i}u(t,x_1,x_2), i = 1,2,$$
$$\frac{\partial^2}{\partial x_i \partial x_j}u(t,x_1,x_2), i = 1,2, j = 1,2,$$

with accuracy $O(\tau + h^2)$ to find approximation for $u(t,x_1,x_2)$. The 3d graph of probability $\boldsymbol{P}(\tau > T)$ for every initial value $\boldsymbol{\xi}(0) = (\xi_1(0), \xi_2(0))^T$ is shown in the Fig. 6.

IV. Conclusion

The work considers the problem of analyzing the parameters of a distribution network with renewable energy sources in its structure, which allow households to not only reduce consumption from the network, but also generate surplus capacity to the network.

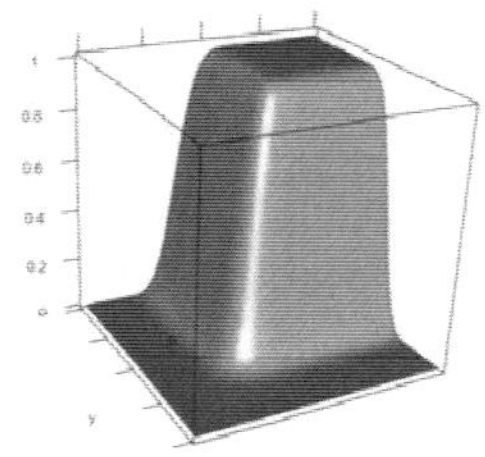

Figure 6. Function $\boldsymbol{P}(\tau > T)$.

We studied the behavior of network parameters not at a fixed point in time, but for a period of time, for example, a day. Instead of searching for estimates for unknown parameters, we considered the probability that the parameters would belong to compliance range ranges during some period of time.

If we assume that the consumption and generation of renewable energy currents can be described in time as diffusion processes, then the problem of estimating network parameters and the study of probabilistic distribution of many voltages functionals might be reduced to the problem of well-studied theory of partial differential equations.

In particular, the search for the probability that the network voltage values do not go beyond the permissible limits during the time interval is reduced to solving the heat conduction equation with certain boundary conditions. It means that the voltage dynamics of the network is similar to the propagation of heat in a rectangular rod. As an example, we used known numerical methods for the heat equation, and thus we found a numerical solution for the unknown probability.

References

[1] M. Eisenreich, G. Balzerl, J. Backes, B. Maurer, "Integration of PV systems into low voltage networks using standard load profiles", International Conference on Renewable Energies and Power Quality (ICREPQ'10), Granada (Spain), 23th to 25th March, 2010, pp. 972-976.

[2] A. Priye and K. Komla, "Voltage Rise Issue with High Penetration of Grid Connected PV", vol. 47, no. 3, pp. 4959-4966, IFAC, 2014.

[3] C. L. Masters, "Voltage rise: the big issue when connecting embedded generation to long 11 kV overhead lines", Power Eng. J., vol. 16, no. 1, pp. 5–12, 2002.

[4] M. J. E. Alam, K. M. Muttaqi, and D. Sutanto, "A comprehensive assessment tool for solar PV impacts on low voltage three phase distribution networks", 2012 2nd Int. Conf. Dev. Renew. Energy Technol., pp. 1–5, 2012.

[5] A. Abur and A. G. Expósito, "Power System State Estimation Theory and Implementation", New York: Marcel Dekker Inc., 2004.

[6] Joshua Paoli, "Optimising low voltage transformer tap settings in distribution systems", Theses for obtaining bachelor degree, UTAS, Australia, pp. 1-84, 2018.

[7] D. G. Colome and C. E. D. L. Vega, "Voltage quality control at distribution systems applying a fuzzy state estimator," PES T&D LA, pp. 1-7, 2008.

[8] L. Mihet-Popa, X. Han, H. Bindner, J. Pihl-Andersen, J. Mehmedalic. „Grid Modeling, Analysis and Simulation of Different Scenarios for a Smart Low-Voltage Distribution Grid". Conference: ISGT 2013, Copenhagen, October 2013, 5p.

[9] I.I.Gihman and A.V. Skorohod, "Introduction to the Theorv of Random Processes," W.B. Saunders Co., Philadelphia, 1969.

Effectiveness of Biogas Power Plant Transformation into Biomethane Production

Jaroslav Knápek, Jiří Vašíček, Július Bemš
Dept. of Economics, Management and Humanities
CTU in Prague, Fac. of Electrical Engineering
Prague, Technická 2, the Czech Republic
knapek@fel.cvut.cz, vasicek@fel.cvutt.cz,
bemsjuli@fel.cvut.cz

Kamila Vávrová
Department of Phytoenergy and Biodiversity
VÚKOZ, Publ. Res. Inst.
Květnové nám. 391, Průhonice, the Czech Republic
vavrova@vukoz.cz

Abstract — **The efficiency of biomass utilization in a biogas plant, where there is no possibility of using the heat generated by the cogeneration unit, is relatively low and most of the biomass energy content is lost unproductively. One of the ways to increase the efficiency of biomass utilization is to supplement the biogas plant with a technology for upgrading biogas to the quality of natural gas (biomethane production). This almost double the efficiency of biomass use. The paper presents a methodology for evaluating the transformation efficiency of a biogas plant to a biomethane plant in the half of the expected life of a biogas plant. The methodology for calculating the minimum price for biomethane to have an economic incentive for the operator to switch to biomethane production is based on the equality of IRR values for the continuation of the biogas station and for the transformation into biomethane production. For the typical technical and economic parameters of both technologies and for the conditions of support for electricity production in biogas stations, the minimum price of biomethane is based on approx. CZK 21.3 / m³.**

Keywords: biogas stations; biomenthane; economic effectiveness, RES support

I. Introduction

The total electricity production from renewable energy sources (RES) in the Czech Republic currently (2017) reaches 9.62 TWh, which is about 13.03% of the total domestic gross electricity consumption. The largest share of electricity production from RES was contributed by biogas plants – BPS (BPS using biomass from agriculture contributed by 93%) - 2.64 TWh (27.4% of RES), biomass burning - 2.2 TWh (22.9% of RES) and PV power plants - 2.2 TWh (22.9% of RES). The boom of BPS's development took place in 2008-2013, when biogas production increased from 0.091 TWh (2008) to 2.083 TWh (2013) [1]. Then there was a sharp braking, respectively stagnating further development of BPS. This was mainly due to a change in the support system from 2013 onwards, the complete stop of operating support for biogas resources. In 2017, total installed capacity in biogas stations was 323 MW, and electricity production (excluding sludge and landfill gas) was 2.46 TWh. Share of power generation based on RES reached 13.03% on gross power consumption – see Fig. 1.

The rapid growth of electricity generation from RES between 2006 and 2012 led to a sharp increase in the operating support (in the form of Feed-in Tariffs and Feed-in Premiums), which reached a total of CZK 43.2 billion in 2017 (plus app. CZK 2 billion for support of cogeneration and secondary sources).

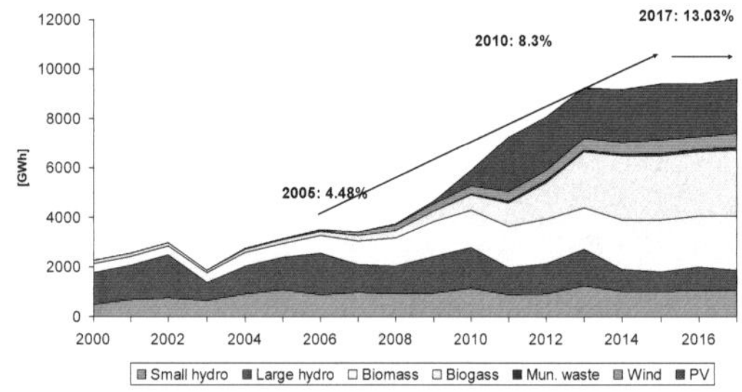

Figure 1. Development of RES power generation in the Czech Republic and shares of RES electricity on gross domestic power consumption [1], [2]

The largest share of the support for the production of electricity from renewable energy sources has a photovoltaic power plant - CZK 27 billion (62.5%), the second biggest part of the support was used by all forms of biogas (biogas stations using planted and residual biomass from agriculture, plants using landfill gas, sewage gas and gas from closed coal mines – CZK 8.2 billion (18.9%) od which biogas stations were CZK 7.4 billion (17.1%) – see Fig. 2. The average amount of operating support per kWh was in the year 2017 CZK 12.5/kWh for PV and CZK 3.3/kWh for biogas stations [3]. EU targets for RES development for 2030 foresee a further rapid development of RES for electricity generation and for cold and heat production, while their further development is expected in the transport sector.

A key aspect of further development of the use of RES is their energy and economic efficiency. Under the conditions of the Czech Republic, the use of biomass as well as biogas will play a key role in RES development in the next decade. At present, about 350-400 thousands ha of agriculture land is used for energy purposes, further developments in biomass and biogas will put pressure on increasing the efficiency of agricultural land utilization. Biogas plants using intentionally grown biomass on agriculture land only for electricity generation are characterized by the low efficiency of using raw biomass (and of land too). This can be documented on the example of the efficiency of utilization of agricultural land for the production of electricity in a biogas plant and production of biomethane. In the typical production conditions of the CR for the cultivation of maize (after counting all losses in the fuel cycle from raw biomass production to the production of biogas or biomethane) it is possible to estimate the energy supply from 1 ha of land in the amount of 40 GJ (electricity) and 75 GJ (biomethane) [4].

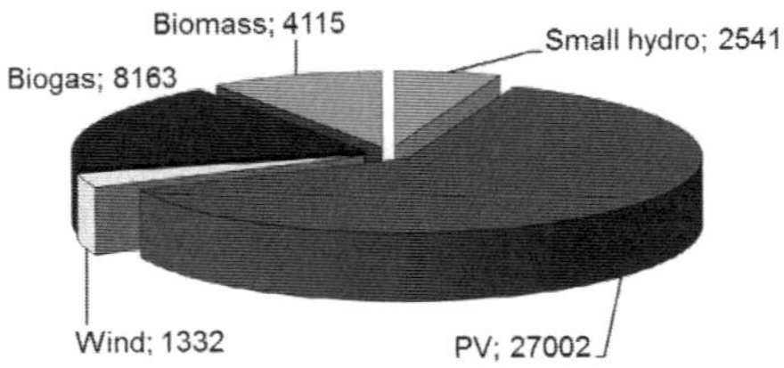

Figure 2. Structure of RES support for power generation in Czech Republic, year 2017, millions CZK [3]

The reason for almost half the energy efficiency of biogas plants compared to biomethane plants is mainly that there is usually no rational use of the heat generated by biogas combustion in a cogeneration unit. Part of the generated heat is used for heating the fermenter and for the heating of buildings by the BPS facilities, however, there is usually no possibility to supply heat for heating of residential or industrial buildings. In the Czech Republic, 10-30% of heat in BPSs is usually used, the rest is waste heat which is discharged to the environment.

The energy and economic efficiency aspect is further strengthened by the projected massive RES growth to the year 2030. National Energy Climate Plan (NECP) for the Czech Republic assumes rapid development of biomass utilization esp. for decentralized space heating (substitution of domestic brown coal with heat pumps, natural gas and biomass pellets and briquettes) and for transportation (biomethane). Especially in the transportation sector, a massive increase in the consumption of biofuels is expected - in the case of 1st generation biofuels it is about an increase of 2PJ, in the case of the 2nd generation biofuels is an increase of approximately 18 PJ. Biomethane will play a significant role here, according to the NECP, almost all expected biomethane production is expected to be directed to the transport sector. On the contrary, it is not expected, especially due to low energy efficiency, the further development of biogas plants producing only electricity.

II. METHODOLOGY

In the future, the development of biomethane production and the reduction of BPSs, which are typically between 6 and 10 years of operational lifetime, are expected in the Czech Republic.

The topic currently under discussion is the effectiveness of their potential upgrade to biomethane production. An upgrade to biomethane would allow both a reduction in high operating support paid to BPSs, while increasing the efficiency of the use of raw biomass. A key aspect for investing in biomethane transformation is the incentive to set the amount of support for biomethane production so that the overall economic effect for the investor outweighs the effect of continuing the operation of the biogas plant by the end of the support period (20 years).

A. *Existing support of BPSs*

Support for BPSs (pursuant to Act 180/2005 Coll.), but also for other types of RES-based power plants, was based on the principle of guaranteed return on the total invested capital for the plants put into operation between 2006 and 2012. The methodology of the so-called minimum price of the production unit (electricity) was used. This methodology was based on the simulation of cash flows associated with the construction and operation of a renewable electricity generation plant. The basic principles of this methodology were [5]:

- Purchase price guarantee for 20 years period (for small hydro power plants for 30 years).
- Differentiation of feed-in tariffs of electricity by type of RES, in case of BPSs also by type of material input - basically two basic types are distinguished here - BPS type AF1 processing predominantly cultivated biomass and BPS type AF2 processing mainly waste.
- The manufacturer could choose the form of support, either in the form of feed-in tariffs (FIT) or green bonus (FIP), which was a surcharge to the electricity market price. In the case of BPSs, almost all support in 2017 was provided in the form of green bonuses.
- Purchase prices of electricity for individual supported categories were determined using technical-economic data of so-called reference plants - the parameters of these production plants were derived from typical technology prices in the given period and from typical expected operational characteristics including the size of electricity generation.
- No other forms of support were foreseen when modeling cash flows of individual reference plants (eg investment subsidies from EU operational programs).
- The discount value was set at 7%, then from 2009 to 6.3%, the discount had the meaning of the so-called weighted capital price (WACC) - the method of financing was not addressed.

The FITs were based on the condition that the net present value (NPV) of the reference project is zero. In keeping with the expected technical and economic parameters expected by the reference project, the investor made a return on the invested capital in the amount of discount used to calculate the NPV. The guaranteed feed-in tariffs (FIT) then corresponded to the so-called minimum price of the production unit determined by the formula (1):

$$\sum_{t=0}^{T} CF_t . (1+r)^{-t} = \sum_{t=0}^{T} (c_{min,t} . Q_t - E_t) . (1+r)^{-t} = 0 \quad (1)$$

Where

T ... guaranteed period of support [years]

CF_t ... Net cash flow in year t [CZK]

r ... nominal discount rate [-]

t ... year of operation, year 0 means year of investment

$c_{min,t}$ ... minimum price in year t [CZK/kWh]

Q_t ... power delivered to the grid [kWh]

E_t ... expenditures in year t, including taxes [CZK]

The calculation of the minimum price was made at current prices, ie with the assumption of price development of individual operating expenses, the long-term average inflation was assumed to be 2.5%, the purchase price escalation was 2%. The reference models were updated over a two-year cycle, allowing both technology and price developments to be reflected.

The system of RES support under Act 180/2005 Coll. enabled the relatively rapid development of RES-based electricity generation and meeting the RES target of 8% of gross domestic electricity consumption by 2010. In 2009-2010, several factors combined (rapid decline in PV component prices, limited Energy Regulatory Office ability to respond flexibly to this by reducing the purchase prices of electricity from PV without legislative changes, the political crisis characterized by the fall of the government in March 2009, the impossibility to issue new elections earlier than June 2010, etc.), which led to the so-called photovoltaic boom and rapid growth in installed capacity, especially in large PV plants up to a total installed capacity of more than 2000 MW. As a result of the rapid increase in RES support - see Fig. 2, legislative changes were adopted leading to a drastic decrease in support for the new RES power plants (Act No. 165/2012 Coll.). Sources put into operation since 2013, have guaranteed only a 15-year payback period. This led to the stop of further development of RES use for electricity production, as shown in Fig. 1. However, for the sources put into operation by the end of 2012, the original support rules and the amount of support remained.

B. Methodology for biomethane economic analysis

Maintaining the original amount and duration of support for BPSs put into operation by the end of 2012 is a key factor in considering the possibilities of transforming current BPSs into biomethane plants. The methodology is based on the basic assumption that investors' decisions on the possible reconstruction of an existing BPSs to biomethane plants are based solely on the aspect of economic advantage and are not enforced by other external (legislative or regulatory factors). Thus, the price of the produced biomethane, which will provide the investor with at least the same economic effect as compared to the option of continuing the operation of the BPS with guaranteed purchase prices for electricity is searched. The biomethane price determined in this way can be considered the lower price limit from the investor's point of view. In practice, an investor would require a higher price to compensate for the higher risk of starting a new project compared to continuing an existing project - otherwise it would not have the economic incentive to make the transition from BPS to biomethane production.

The decision of the operator on a possible transformation of an existing BPS into a biomethane production plant can be documented on a model case, where at the end of the 10th year of operation the BPS is considering whether to supplement BPS with the technology of biomethane production or whether to carry out only a renewal investment for the continuation of BPS operation for another 10 year operation. Roughly halfway through the lifetime, a major refurbishment investment in a significant part of the technology (fermenter, pumps, etc.) is envisaged, in the reference models for the calculation of feed-in tariffs, this renewal investment of about 30% of BPS's initial total investment cost was considered. The investor is considering two options:

- Continue to operate BPS for the next 10 years and use the guaranteed feed-in price.
- Invest in biogas purification technology for natural gas quality and implement other induced investments (eg ensuring fermenter heating, covering waste heat from cogeneration units).

It is a decision-making role realized at a given moment (after 10 years of BPS operation). All economic effects that occurred before this point are irrelevant in terms of the logic of the task (investor's decision). Thus, from the first 10 years of operation, only depreciation from the original investment that exceeds 10 years of depreciation period is relevant to the task being solved. These depreciations relate to equipment that continues to operate in both variants (typically a construction part, etc.), are the same in both variants and are therefore not taken into account in the calculations.

The net present value criterion - NPV (see above) is used as the basic criterion for assessing the economic efficiency of projects and for investors' decision-making. However, the application of the NPV criterion is not suitable for the solved role of BPS transformation to biomethane production. The reason for this is that the investor decides to implement the transformation on the basis of expected future economic effects. In practice, the investor will therefore decide about whether to realize only a renewing investment in the old part of the BPS, or whether to invest a significantly higher amount in the transformation of BPS into BMS. This compares differently sized investments with the same estimated lifetime (10 years).

The NPV net present value criterion is the absolute value and expresses the effect of the assessed investment over the period of the assessment in CZK. If NPV is positive, the yield is greater than the specified discount and vice versa. The NPV criterion is thus suitable for expressing the effect of the investment being evaluated against the discount, ie the return on alternative investments, without evaluating these alternatives in a particular form.

For this task (finding the feed-in price, or the amount of biomethane support, which will be sufficiently motivating) it is advisable to start from the IRR criterion (2).

$$\sum_{t=1}^{T_{zb}} CF_t.(1+IRR)^{-t} = 0 \qquad (2)$$

Where:

T_{zb} ... remaining period of guaranteed support [years]

IRR ... internal rate of return [-]

Determination of the lower limit of the guaranteed purchase price of biomethane is then determined analogously by (1) so as to achieve the same IRR in both the continuation of BPS for the next 10 years and in the case of BPS transformation to the biomethane plant.

The IRR criterion is relative because the return is based on the initial investment being evaluated. Therefore, it is suitable for comparing the economic efficiency of differently big investments. This is the case of the evaluated options, where one of them assumes the continuation of the operation of the existing biogas plant, the second option assumes a major reconstruction of the biomethane production. In order to keep operators' motivation at least the same, both variants must have the IRR value the same. The use of the NPV criterion would not be appropriate since, with the same IRR, a larger investment would yield an absolutely higher return compared to alternative investments whose return is at the discounted rate (IRR).

C. Reference models

To demonstrate the use of the above methodology, a BPS reference model and a biomethane reference model were developed. For both reference models, the general assumptions used to calculate the feed-in tariffs for electricity between 2006 and 2012 have been used.

- 6.3% discount (in terms of WACC).
- 2.5% operational cost escalation.
- Zero salvage value of the device after the end of the service period (ie after the end of the guaranteed 20-year period).

All the technical and economic data for BPS and biomethane stations reflect the typical data used within the Czech RES support scheme [5]. Data on biomethane production and material inputs were collected with help of Czech Biomass Association – (CZ Biom).

For the BPS reference model, BPS with an installed capacity of 550 kWe was used (which corresponds to the most common type of BPS put into operation between 2006 and 2012). The assumed typical values of BPS investment and operating costs realized in 2006-2012, ie:

- Investment costs in the amount of 110 thousands CZK/kWe
- Mid-life renewal investment to ensure further operation of BPS as the 30% of the original investment value for BPS.
- The annual utilization of the installed capacity for the supply of electricity to the grid in the amount of 7800 hours, this corresponds to the annual electricity supply to the network of 4290 MWh/y, the total gross utilization of the installed capacity is 8500 hours.
- The proportion of heat utilization (excluding own technological consumption) of about 10% (ie a total of 1,200 GJ, which is about 2.5% of total BPS sales). We assume that usable heat displaces the heat generated from the natural gas, assuming heat payments of 300 CZK/GJ.
- Operating and maintenance expenses, operating costs and other expenses are expected to be 5.5% of the initial investment cost.

  The feedstock is expected to have an average composition per operated BPS - ie 40% corn silage, 20% grass silage, 30% manure of farm animals, 10% other (various types of residual and waste biomass). The price is approximately 710 CZK/t of combined input substrate. The total annual cost of the input substrate is about CZK 10.4 million (2018 price level, data obtained from the CZ Biom Association survey).
- The biogas yield of the defined substrate is approx. 150 m^3/h and biogas consumption of 255 m^3/h is assumed for full BPS power.

The biomethane reference model is based on the BPS reference model, but cogeneration is no longer considered (all biogas is converted to biomethane, which is fed to the natural gas system). The use of membrane technology to upgrade biogas to biomethane is expected. Since all biomethane is supplied to the gas network, there is a need for technological heat demand for plant operation (eg, fermenter heating). At the same time, sales of heat are no longer needed. The following technical and economic data are foreseen:

- The biomethane biogas technology has a lifespan of 10 years.
- The investment in the technology for biomethane production is about CZK 34 million, plus the investment in the renewal of the original BPS - this investment will be lower than in the case of BPS, this investment is expected to be about 20% of the original investment costs of BPS.
- The yearly operation at full capacity is the same as in the case of BPS, ie 8500 hours, which is then methane content in biogas of 52% corresponds to annual biomethane production of about 1100 thousands m^3.
- There is a change in electricity consumption (the cogeneration unit's own consumption is not needed; on the contrary, it is necessary to add the purchase of electricity to the compressor operation), which is due to the absence of cogeneration unit to be purchased from external suppliers, electricity consumption (without compressors) is estimated at approx. the original gross production of BPS, the average purchase price is about 3 CZK/kWh. The cost of purchasing electricity for the operation of compressors is estimated at about CZK 1.3 million.
- There is also a change in operating costs - there is no need for maintenance and repair of the cogeneration unit and, on the contrary, the operating costs of technology for biomethane production are increasing. It is expected that the operating expenses of the remaining part of the original BPS will decrease to about 2.5% (from the original total investment costs of the BPS).

III. RESULTS AND DISCUSSION

For the reference model of the continuation of BPS in operation, the IRR value is about 8.1%. It is not possible to directly compare this value with the return on the invested capital of 6.3% for which the guaranteed purchase price of electricity was calculated. The value of 6.3% relates to the entire guaranteed period of 20 years (and in relation to the total amount of the investment), while the value of 8.1% expresses only the efficiency of the renewal investment (which is significantly smaller than the original investment) while maintaining the original guaranteed feed-in price.

In order to achieve the same IRR value of 8.1% in the variant of BPS transformation to the biomethane production plant it is necessary to reach the purchase price of biomethane about 21.3 CZK/m^3.

The effectiveness of BPS reconstruction on biomethane production is fundamentally influenced by the following reference model items in order of severity:

- Investments for the necessary reconstruction including reconstruction of the existing BPS.
- Cost of raw materials.
- An increase in input prices, where there may be a shortage of some input raw materials that have alternative uses in other technologies.
- Annual growth in feed-in tariffs for biomethane.

Significant is the amount of investment costs for biomethane production - if the cost of biomethane production decreased by 10 million CZK, the minimum price of biomethane falls to about 19.8 CZK/m^3.

Significant influence on the minimum price of biomethane is also due to growth in feed-in tariffs for electricity and biomethane. In the case of feed-in tariffs for electricity from BPS, there has been no (inflation) increase in the last decade (as opposed to feed-in tariffs for electricity from small hydroelectric power plants, photovoltaic power plants and wind farms). Therefore, the calculations were carried out with a zero value of the redemption price of the electricity purchase price from BPS. Of course, the question is whether one can expect this in the future (ERO has an obligation to index the FIT by 2% only for power plants without a fuel component according to the rules used to set up feed-in tariffs). If the purchase price of electricity from BPS would increase in the future, it would have a significant impact on increasing the minimum price of biogas.

Also interesting is the impact of changes in input substrate prices. E.g. when its price drops by 20%, the efficiency of BPS's own operation is significantly increased. This then requires, in order to achieve the same IRR value, an increase in the minimum price of biomethane to a value of approximately CZK 24.50/m^3.

In general, all changes that increase the efficiency of existing BPS lead to an increase in the minimum price of biomethane. The reduction of the minimum price of biomethane can thus only be achieved by items affecting only the biomethane production option - in particular, the amount of investment costs.

If we were to consider the implementation of a new biomethane plant, it would be necessary to build a complete new biogas plant (without a cogeneration unit). Specific costs can then be estimated at approx. 84,000 CZK/kWe (BPS without cogeneration unit) and the cost of biomethane production technology (with a lifetime of 10 years). The minimum price of biomethane would then be (under otherwise identical conditions) higher by about 2 CZK/m^3. On the other hand, if the biogas upgrade technology is expected to last 20 years (and a renewal investment of about 30% - similar to the entire BPS) would be assumed - then the minimum price of biomethane would be around 21.3 CZK/m^3. A similar impact (reduction of the minimum price by 2 CZK/m^3) is expected to reduce the price of the input substrate by 20%.

IV. CONCLUSIONS

The development of biomethane production at the expense of biogas is an interesting opportunity to increase the efficiency of biomass utilization and thus the efficiency of agricultural land use. The development of biomethane production is one of the anticipated tools after fulfilling the objectives of using renewable energy sources in transport.

One of the options under consideration for the development of biomethane production is the transformation of the current biogas plant in about half of its useful life (or 20-year support period) to biomethane plants. This transformation is connected not only with investment in own technology for upgrading biogas to the quality of natural gas, but also with other induced costs associated with eg the need to provide heat for biogas production technology, purchase of electricity to cover own consumption, etc. The price of biomethane produced in this way is relatively high - for comparison, the price of natural gas as a commodity was around 4 CZK/m^3 in November 2018 and 2.2 CZK/m^3 in April 2019 [6]. At the same time, the final prices of natural gas for households of 6.5 CZK/m^3 (without a fixed payment of approximately 100-150 CZK/month) can also be stated for comparison. However, these prices include not only their own commodity but also the gas transmission costs and the gas trader margin.

The minimum price of biomethane is relatively high, however, if we convert the minimum price of biomethane to 1 kWh of calorific value (when the calorific value is about 9 kWh/m^3), then the minimum price of biomethane is about 2.3-2.5 CZK/kWh.

We can then compare the minimum price of biomethane with other alternatives to achieve the desired share of RES in transport and final energy consumption.

ACKNOWLEDGMENT

This article was performed in the frame of project no. TACR TK01010017, which was supported by the TAČR (Technology Agency of the Czech Republic).

REFERENCES

[1] MPO 2018: Renewable Energy Sources in 2017. https://www.mpo.cz/assets/cz/energetika/statistika/obnovitelne-zdroje-energie/2018/12/Obnovitelne-zdroje-energie-v-roce-2017-new.pdf (accessed 10 March 2019)

[2] ERU 2018: Yearly Report on the Operation of the Czech Electricity Grid for 2017. http://www.eru.cz/documents/10540/4580209/Yearly_report_electricity_2017.pdf/a4042a98-8750-4524-bb3c-f172ecba2f3b (accessed 10 March 2019)

[3] OTE 2018: Statistics on Supported Sources in the Czech Republic.https://www.ote-cr.cz/cs/statistika/statistika-poze/poskytnuta-podpora-2013-2017 (accessed 10 March 2019)

[4] J. Knápek, T. Králík, M. Valentová, T. Vořišek. "Effectiveness of biomass for energy purposes: a fuel cycle approach". Wiley Interdisciplinary Reviews: Energy and Environment. 2015, 4(6), 575-586. ISSN 2041-8396.

[5] Vasicek et al. Study on guaranteed purchase price and green bonuses for the electricity from RES power plant. ELEKTRA ČVUT FEL, study for the Energy Regulatory Office, 2012 (in Czech)

[6] Development of natural gas prices. https://www.kurzy.cz/komodity/zemni-plyn-graf-vyvoje-ceny/ (accessed 10 April 2019)

Evaluation of Unbalance in the Transmission System of the Slovak Republic

Ľuboš Pavlov, Ľuboš Skurčák, Ján Gbelec, Peter Bojda
VUJE a.s.
Trnava, Slovakia
lubos.pavlov@vuje.sk, lubos.skurcak@vuje.sk

Abstract — **Since 2015, the transmission system operator of the Slovak Republic (SEPS, a. s.) Has been operating 18 phasor measuring units (PMUs) in the most important PS SR nodes in continuous mode. The measurements are extensive databases containing three-phase values of voltage and current phasors, active and reactive component values of transmitted power and network frequency. The measured data are stored in the database for individual years. The database of measured data for 2017 was used to evaluate the development of the size of its own unbalance on selected power lines. The presented article presents the results of this evaluation.**

Keywords: Power system; unbalance; Phasor; Phasor Measurement Unit

I. Introduction

In transmission systems, there are operational states that generate an unforeseen emergency state, but the operational safety is maintained (Alert State), as well as states where operational safety is not met due to exceeding the limits of some operating parameter [10]. Many states are caused by the interconnection of transmission systems, the overall increase of complex electrical systems, the production of renewable energy in a pan-European context. The role of the transmission system operator is to guarantee the safe, reliable, quality and economically efficient operation of the transmission system and its sustainable development [1]. Compliance with a given task is only possible if the transmission system operator is informed of the current operating status. From this point of view, the installation of phasor measurement units (PMUs) at selected transmission system nodes is a powerful tool with wide application possibilities. The installation of several metering units in the transmission system is a system called "Wide Area Monitoring System" (WAMS). WAMS networks are a system of time-synchronized measurements of phase and current phase values with corresponding phasors, which can be used as a means of the solution not only the immediate reliability but also the problems of TS management as well as long-term system development planning [9].

II. Application of the PMUs in the Transmission System of the Slovak Republic and Their use

The gradual installation of PMU equipment in the Slovak transmission system has been in progress since 2011. In 2011, four PMU units were installed on the national lines V426, V406, V407 and V409 as part of the first pilot project. In 2014, another pilot project was launched to gradually implement the WAMS system in the transmission system. Within this project, 10 new PMU devices were installed in the point of connection of the line at the electrical substation of all cross-border lines of the TS SR. In 2015, the last 8 PMU devices were installed, equipped with important national lines V427, V492, V407 and V439. Phasor measuring units were installed at both ends of these lines. PMU measure phasors of phase voltages $\bar{U}1$, $\bar{U}2$, $\bar{U}3$, phasors of currents $\bar{I}1$, $\bar{I}2$, $\bar{I}3$, values of three-phase active (P) and reactive power (Q), frequency value (f) and its time change (df / dt). The sampling frequency of the measurement is 12 kHz with a total vector error (TVE) of <0,5%. The measured data is stored in a phasor data concentrator (PDC) every 20 ms [2].

At present time, the transmission system operator uses the WAMS (Figure 1) system in real operation and is looking for further application possibilities or use of the measured data provided by the given system [2].

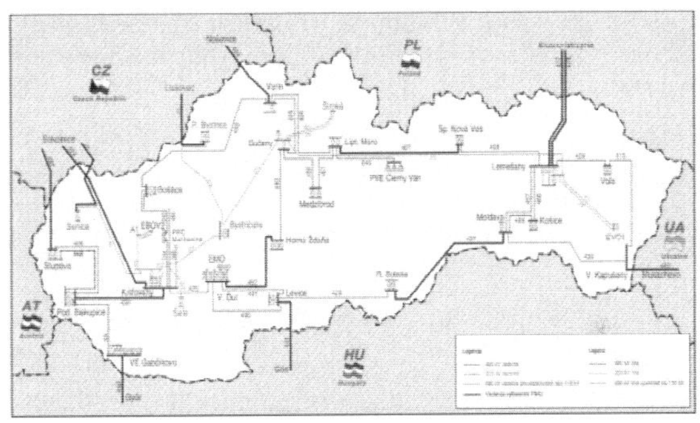

Figure 1. Diagram of TS SR with maked lines included in WAMS

In real operation, the WAMS system is used by dispatchers of the Slovak Electricity Power Dispatching Center and by specialized units of the TS SR operator, in particular: [2]
- measuring the phase angle,
- analysis of oscillations in the transmission system,
- management of an island, where appropriate, to stabilize the transmission system,
- Ex-post event analysis in the transmission system,
- monitoring of the interconnected transmission system.

III. The Risk of Unbalance in the Transmission System

Voltage unbalance is one of the quality parameters of electricity. It arises predominantly as a consequence of the unsymmetrical sampling, possibly due to the use of asymmetric elements. Typical asymmetric elements of the transmission system are non-transposed external high voltage and extra high voltage lines. The cause of unbalance is the longitudinal impedance inhomogenity caused by: change of mast configuration, phase sequence on mast brackets, change of type of phase conductors and ground wires in sections of line, the presence of runs with other lines, etc. The size of the unbalance of the outer conduit can also be influenced by the operating modes, which may be non-symmetrical.

Other elements of the transmission system, such as transformers and compensating chokes, cannot be omitted. These also belong to the group of unbalanced elements by their structural arrangement.

In the case of transformers, the size of the unbalance depends on the design of the magnetic circuit of their core (nuclear or shell construction), the design of the 3-phase primary and secondary windings, the clock angle and the corresponding internal current connections of the primary and secondary windings [3, 4].

Compensation chokes are designed as air coils, which can be connected separately for each phase as separate units arranged in a row (geometrically unsymmetrical arrangement) or in a triangle (geometrically symmetrical arrangement). The existence of choke imbalance is caused by a change in the size of the electromagnetic field, which is generated by the passage of the electric current through the end coils against the coil located in the middle of the three separate coils of the compensation chokes. The same applies if the coils are installed in a common metal container. If the connection of the three coils of the compensating choke in the common metal vessel were arranged in a row, higher unbalance values than in the case of the arrangement of the coils in the triangle can be expected. This is due to the higher magnetic coupling between the coils.

The calculation of unbalance can be done in several ways. Most accurately, however, it is possible to quantify the unbalance taking into account the difference in phasor values (amplitude, angle). In this case, the unbalance is determined from the ratio of the zero and negative component of voltage. It is based on the theory of unbalanced system distribution to symmetrical components: consecutive, reverse and non-rotating.

$$\text{Voltage unbalance} = \frac{\text{zero voltage component}}{\text{negative voltage component}} * 100\% \qquad (1)$$

The consecutive and backward voltage components can be obtained by distributing unbalanced voltages $UL1$, $UL2$, $UL3$ to 2 symmetric components U1 and U2. These symmetrical components are defined as follows [5]:

$$U_1 = \frac{\dot{U}_{L1} + \dot{a} * \dot{U}_{L2} + (\dot{a})^2 * \dot{U}_{L3}}{3} \qquad (2)$$

$$U_2 = \frac{\dot{U}_{L1} + (\dot{a})^2 * \dot{U}_{L2} + \dot{a} * \dot{U}_{L3}}{3} \qquad (3)$$

while $\dot{a} = 1\angle 120^\circ$, $\dot{a}^2 = 1\angle 240^\circ$.

The technical conditions of access and connection, the rules of operation of the transmission system, document C (referred to below as Technical Conditions) defines asymmetry as a state of a multi-phase network in which the effective values of phase voltages or phase angle differences between successive phases are not the same. Under normal operating conditions, 95% of the ten-minute mean effective values of the feed-back component must be within the range of 0 to 2% of the consecutive component of the primary harmonic supply voltage within one week [6].

IV. Unbalance Size in the Transmission System of the Slovak Republic

Two types of archived data from measurements of synchronous phasors from the WAMS system recorded in the period from 1st January 2017 to 3rd January 2018 were provided by the transmission system operator of the Slovak Republic. The first type is data with so-called. a short-term archive stored for 93 days with 50 Hz write frequency. The second type is data with so-called. a long-term archive stored for 365 days with a 10 Hz write frequency. Individual measured data are stored as csv. files.

A short-term archive is made up of files stored automatically in a cycle every 32 hours. Individual files are records of measured data in the number of 5864219 lines written in the time interval of 0, 02 s (50 Hz). The size of one saved file is 1, 5 GB. In the 2017 assessment, a total of 1260 files were included with a total data volume of more than 1, 8 TB.

A long-term archive is made up of files stored automatically in a cycle every 163 hours. Individual files are records of measured data in the number of 5881795 lines written in time interval 0,1 s (10 Hz). The size of one saved file is 1,0 GB. A total of 1,100 files were included in the 2017 evaluation. The total volume of data subject to assessment exceeded 2,9 TB.

Processing such an amount of data is time-consuming and requires appropriate hardware and software. Most attention was paid to choosing the appropriate software tool. Commonly used software tools have limited processing capabilities for large databases. These are mainly caused by the number of stored records in one file and the ability to load them. An important factor is the data format used, which in case of incompatibility needs to be transformed into a suitable form. Similarly, high demands are placed on the hardware used, especially with regard to the time it takes to perform the necessary analyses and to achieve the desired results.

The main criteria for choosing the right software tool are:

- Data retrieval options for large databases for post-processing,
- the ability to develop custom algorithms for data processing,
- possibility to define own calculation functions using bulk data processing,
- the possibility of graphical processing of achieved results.

In view of the above criteria, a programming environment specializing in scientific and numerical computations, modeling, algorithm designs, computer simulations, data analysis and presentation was used to process the unbalance results. An algorithm for editing the read data was created in the software. Subsequently, an algorithm for computation of consecutive and backward components for each record of measured phasors U and I was created, which were subsequently used for calculation of unbalance on selected national lines V427, V492, V407 and V439.

Table 1 shows the basic characteristics of selected electrical wiring from the point of view of the overall length of the line, the types of masts, conductors and ground wires used. From the power lines analyzed, three lines are realized as simple (V427, V439, V407) and one line (V492) is implemented as a single line and as a single line in the other.

In the case of this line, it can be expected that the resulting size of unbalance will be influenced by parallel electric lines.

Table 1. Basic characteristics of selected power lines

Line number	Length (km)	Electrical Towers type	Conductors	Earth Conductors
V427	93,705	Portal Macka Trojdriek Ypsilon	350 AlFe 6, 450 AlFe 6	70 Fe, AlFe 180/59
V439	45,767	Portal Macka Trojdriek	350 AlFe 6, 450 AlFe 6	70 Fe, 185 AlFe 3, Alcatel 187/75, 147/59
V492	62,903	Macka Donau Trojdriek	AlFe 450/52, 450 AlFe 6, 434-AL1/56- ST1A	185 AlFe 3, AlFe 180/59, 70 Fe
V407	85,785	Macka Portal, Ypsilon Kotveny V Trojdriek	350 AlFe 4	70 Fe, 185 AlFe 3

The results of the unbalance calculations for each selected line were then statistically processed. For better interpretation of the results, frequency functions have been developed indicating the relative abundance of the recorded values of calculated unbalance. Figures 2 to 5 show the resulting frequency functions of the individual power lines analyzed. All unbalance values calculated from the available database for 2017 ranged from 0,16 to 1,68%.

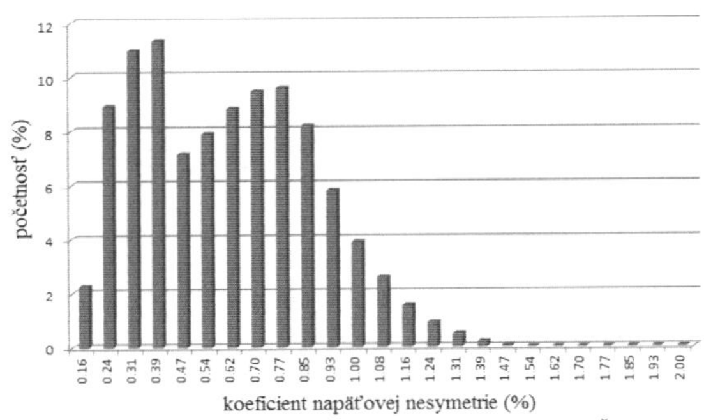

Figure 2. The unbalance size of the V492 line (ESt Veľký Ďúr - ESt Horná Ždáňa)

The unbalance size of the V492 line ranged from 0,16 to 1,54%. The average value for the monitored period was 0,61%.

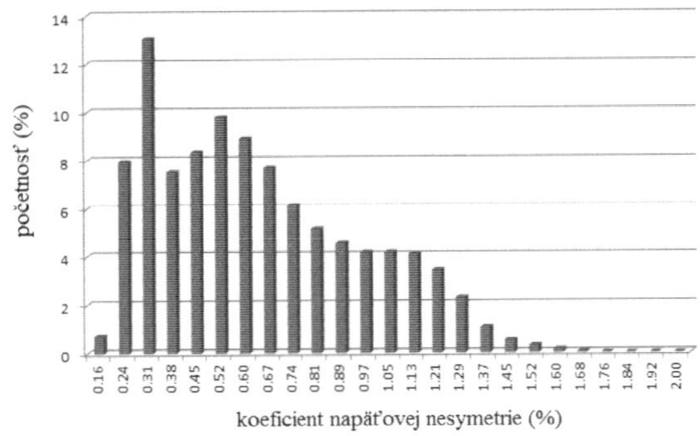

Figure 3. The unbalance size of the V439 line (ESt Križovany - ESt Podunajské Biskupice)

The size of the unbalance on the V439 line ranged from 0,16 to 1,68%. The average value for the monitored period was 0,62%.

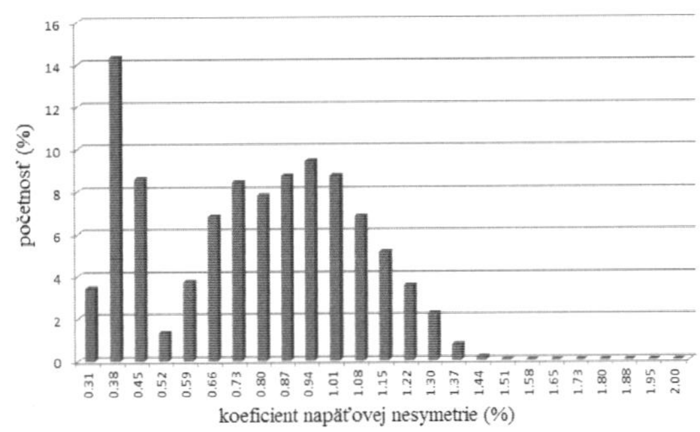

Figure 4. The unbalance size of the V427 line (ESt Rimavská Sobota - ESt Moldava nad Bodvou)

The size of the unbalance on the V427 line ranged from 0,31 to 1,51%. The average value for the monitored period was 0,84%.

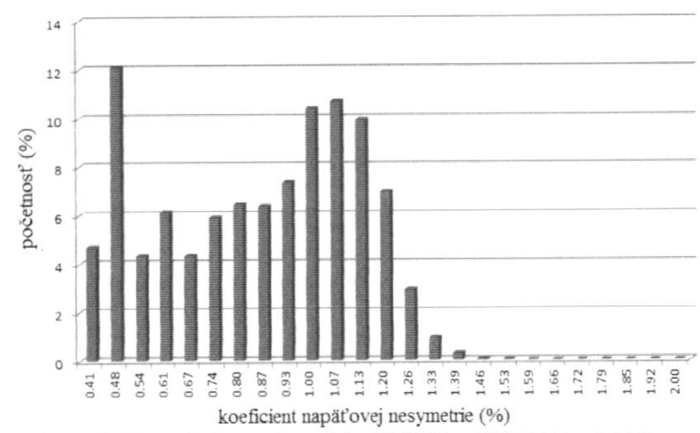

Figure 5. The unbalance size of the V407 line (ESt Liptovská Mara - ESt Spišská Nová Ves)

The size of the unbalance on the V407 line ranged from 0,41 to 1,53%. The average value for the monitored period was 0,84%.

The size of unbalance is influenced by many factors such as soil resistivity, terrain profile, geometric conductor layout and height, type and design, weather conditions etc. Despite the diversity of conditions that, in terms of theoretical assumptions, affect the magnitude of the inherent imbalance of the power lines, the lines under consideration are very similar in terms of both maximum and average unbalance levels. Based on the presented graphs, it can be stated that the proportional representation of the individual levels is different on the evaluated power lines. Due to the complexity of the whole issue, it is currently not possible to define precisely the weight of individual factors. This will be the subject of further investigation.

V. Conclusion

The results of synchronized system measurements generally have broad application possibilities. Data recorded with a high sampling rate is useful for analyzing transient events that occur in the system only at short time intervals, as well as for other specialized system operation analyzes. Some of the on-line data usage options are listed below:

- **assessing the steady running of the network** (traffic analysis and evaluation),
- **monitoring of network stability** (determination of amplitude and frequency of oscillations, monitoring of dynamic properties of the system allows sending of early warning against non-standard states, thus avoiding system oscillations),
- **determination of line parameters** (active resistance, reactance, susceptance, lead), prediction and fault localization (using neural networks),
- **identification of island operation** (management and connection of an existing island to the transmission system).

In the case of so-called. post-mortem (off-line) analyzes using a large database of stored data obtained by PMU can be implemented for example: - reliability assessment of operation - analysis of fault conditions, - quantification of unbalance. The presented paper was focused on demonstration of possible use of records from measurements of synchronous phasors for the purpose of assessing the unbalance of PS SR management, which can be implemented in future for all lines monitored by the WAMS system.

Based on the results presented in IV. th chapter it can be stated that the values of the calculated magnitude of the voltage unbalance comply with the requirements given in the technical conditions of access and connection, the rules of operation of the transmission system. The size of the asymmetry was assessed in a significantly shorter time interval (0,02 s and 0,1 s, respectively). In accordance with the technical conditions for assessing the unbalance of the supply voltage, a measuring interval of 10 min is sufficient. Although the quantification of the unbalance size was performed with a denser calculation step, 100% of the calculated values of the voltage unbalance factor (ρ_u) were in range from 0% to 2% of the Technical Conditions range. More than 88% of the voltage unbalance factor is less than 1%. Less than 0,7% of all values were calculated above 1,3% unbalance. In general, the obtained results indicate that transmission lines considered as the most unsymmetrical element in the Slovak transmission system (especially when they use Portal and Ypsilon masts) do not cause an increase in the voltage unbalance factor above 2%. **Possible exceeding of the limit value of the unbalance of the supply voltage in TS SR is therefore with a high probability caused mainly by the unsymmetrical load.** However, each case will need to be considered separately on the basis of a detailed analysis.

Acknowledgment

This work was supported by the Slovak Research and Development Agency under contract No. APVV-15-0464.

Special thanks to the Slovak transmission system operator - Slovenská elektrizačná prenosová sústava, a. s. for providing data and professional assistance.

References

[1] https://www.sepsas.sk/ZakladInfo.asp?kod=77

[2] M. Grega, S. Prieložný, M. Jedinák, „Implementation of Wams into transmission System of Slovakia", 12th International Scientific Conference, Control of Power Systems, May 31 - 2 June 2016, Slovakia.

[3] V. Mentlík, P. Prost, R. Polanský, J. Michalík, M. Brandt, " Instruments for On-line Monitoring of Transformers"

[4] Z. Paukner "On-Line Monitoring transformátoru, zkušenosti, modely"

[5] P. Szathmáry, M. Kanálik, J. Rusnák, M. Hvizdoš, "Nepriaznivé vplyvy nesymetrie napätia na elektrické zariadenia a možnosti ich eliminácie", AT&P journal, vol. 2, 2010, 51-53.

[6] https://www.sepsas.sk/Dokumenty/TechnickePodmienky/2019/02/01/TP-Dok.C_2018.pdf

[7] IEEE Standard for Synchrophasors for Power Systems. IEEE Std C37.118™-2005. USA.

[8] M. Ivanič, „Aspekty nesymetrie parametrov a veličín v elektrizačnej sústave", Dizertačná práca - STU FEI, 30.6.2017, Bratislava

[9] V. Krištof, S. Kušnír, M. Kolcún, „Základy technológie WAMS", Electric power engineering 2011

[10] Continental Europe Operation Handbook ENTSO-E; P5 – Policy 5: Emergency Operations; V 3.1;

Measurement the Ground Resistance of Masts of Overhead Power Lines in HV and EHV

Stanislav Ilenin
Department of Electric Power Engineering
Faculty of Electrical Engineering and Informatics
Technical University of Košice
Košice
stanislav.ilenin@tuke.sk

Ladislav Varga
Department of Electric Power Engineering
Faculty of Electrical Engineering and Informatics
Technical University of Košice
Košice
ladislav.varga@tuke.sk

Abstract — **The article deals with the measurement of resistance of grounding of masts of external power lines in HV and EHV. Conventional methods use measurement methods with disconnecting earthing ropes. There is also a measurement method without disconnecting the earthing ropes, which is difficult to use in practice. The article describes two new V-A methods that can be easier to use.**

Keywords: overhead power lines; resistance; grounding of mast; V-A methods

I. Introduction

The size of the grounding resistance of the metal poles is important both for operational reasons and for the safety of persons, respectively. animals moving in their vicinity.

STN EN 50341-1 specifies that the earthing resistance of individual poles with disconnected earthing cables should not exceed the following values under favorable soil conditions [1].

For lines with rated voltages of 220 kV and 400 kV and for lines with nominal voltages of 110 kV in areas with lower lightning strikes (less than 3 lightning / year.km^2):

- 15 Ω at the support points in the route
- 10 Ω at support points within 800 m of the power station

For lines with a rated voltage of 110 kV in areas with a higher lightning strike density (greater than 3 lightnings / year.km^2) and for lines with higher operational reliability requirements:

- 10 Ω at the support points in the route
- 7 Ω at support points within 800 m of the power station.

Higher grounding resistance values may only be allowed under adverse soil conditions. The highest grounding resistance values of the support points must then not be higher than the values given in Tab. 1

If, for cables with earthing cable, the individual support point does not have a ground resistance value according to Tab. 1 and cannot be achieved in a conventional manner, the earth resistance may be up to three consecutive support points of up to 50 Ω, provided that several adjacent support points have a ground resistance according to this table.

Thus, the grounding resistance of the metal masts must be permanently low and regularly measured. If the line does not have a grounding cable, the grounding resistance is relatively simple, because the grounding of a single pole is a simple and not a large earthing switch. To measure the overall earth resistance, it is advantageous to use conventional earth resistance measuring devices.

Table 1. The highest earthing resistance values [STN EN 50341-1]

Soil resistivity [Ω.m]		Earthing resistance [Ω]
up	to	
500	1000	20
1000	2000	30

It is no longer easy to measure the grounding resistances of MV, HV and MV grounding poles. The outer conductor with the earthing cable is, in terms of measuring the earthing resistances, an extensive set of earthing resistances of the individual poles connected in parallel by the earthing cable.

II. Tagg Earth Ground Resistance Method

Mr. Tagg provides a method for determining the grounding resistance of individual masts connected by a ground wire from the measured values of three pole mast resistances [2].

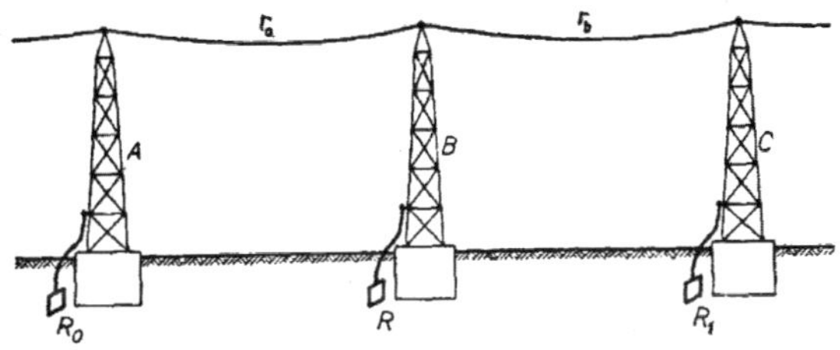

Figure 1. Earthing resistance of steel poles with earthing wire

The grounding rope is grounded on each mast. R_0, R, R_1 are the actual grounding resistances of the *A*, *B*, *C* masts.

Using the measurement circuit arrangement as shown in Figure, the apparent resistances of individual poles with the ground wire attached are measured. The measured values will differ very little, since the same set of ground resistors is always measured and the result of the measurement is more or less influenced only by the specific resistance of the soil in the vicinity of the measured pole.

If

X is the mast grounding resistance measured,

Y is the grounding resistance of the pole *B*,

Z is the ground resistance of the *C* pole.

The actual grounding ground resistance will then be

$$R = \frac{Q \cdot (M + r_a)}{M + r_a - Q} \quad (1)$$

Where

$$Q = \frac{r_b \cdot \left(2Y - r_b + \sqrt{r_b^2 + 4YZ}\right)}{2 \cdot (r_b - Y + Z)} \quad (2)$$

And

$$M = \frac{r_a \cdot \left(2X - r_a + \sqrt{r_b^2 + 4XY}\right)}{2 \cdot (r_a - X + Z)} \quad (3)$$

In this evaluation method, it is necessary to ensure a high accuracy of measurement and reading of the results, due to the little differing measured values. As seen in equations 1 and 2, the measured values in the fraction denominator are subtracted. Any small measurement errors reading results, can significantly affect the calculated value.

III. VOLT-AMPERE METHOD A

Substantially more accurate results, as reported by Mr. Tagg, are provided by the volt-ampere method of measuring the grounding resistances of masts in the arrangement of Fig. 2.

Measuring principle

It is not necessary to disconnect the earthing cable when measuring. Its disadvantage is that the measured overhead line must be turned off during the whole measurement period [3].

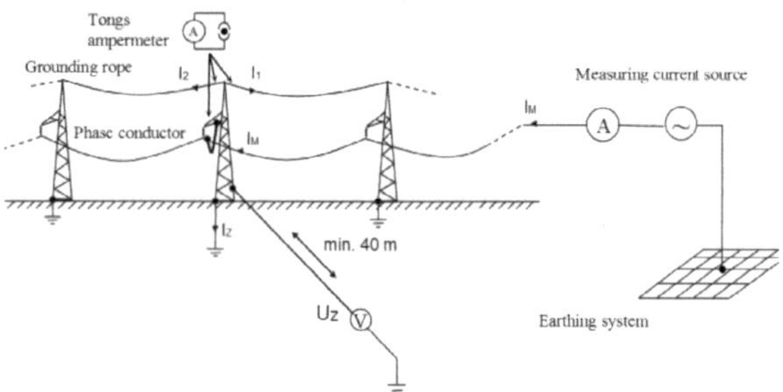

Figure 2. Volt-amper method of measurement

The source of the measuring current is connected via an ammeter at the station where the air conduction is measured, so that one pole is connected to the grounding system of the station and the other pole is connected to one phase conductor of the measured line. The measurement simulates a single-phase short circuit on each measured pole so that the powered phase conductor and the pole structure are connected via a conductor (shorting conductor wire). A part of the measuring current flows through the measured ground, ground, and ground station of the electrical station back to the source. This part of the current creates a voltage drop at the pole ground, which can be measured by a voltmeter connected between the structure of the measured pole and the auxiliary voltage electrode PE. The distance of the voltage electrode must be sufficient to capture almost all of the earthing voltage and depends on the grounding shape below the measured mast (typically 40 m). The grounding resistance can then be calculated by the equation [4,6].

$$R_Z = \frac{U_Z}{I_Z} \quad (\Omega) \quad (4)$$

While

U_Z grounding voltage,

I_Z current flowing through the measured ground,

In practice, it is quite difficult to directly measure a part of the current flowing through the measured ground. Therefore, to determine it, it is necessary to measure a part of the current that flows through the grounding cable after separation. Then, the current flowing through the pole ground can be determined as

$$I_Z = I_M - I_1 - I_2 \quad (5)$$

Where

I_M measuring current flowing through the short-circuiting set,

I_1 current flowing through the ground wire (according to wiring diagram),

I_2 current flowing through the ground wire (according to wiring diagram)

It is very advantageous to use a measuring current source with a frequency different from the 50 Hz industrial frequency. A voltmeter with high internal resistance is then required to measure the voltage. The advantage of using a different frequency is clearly reflected in the measurement of grounding resistances of poles of parallel air lines. Even with relatively short runs, several amperes flow in the earth line off the current line [5].

IV. VOLT-AMPERE METHOD B

The second method of measurement by a volt-ampere method, in which it is not necessary to disconnect the grounding cable and in addition to switch off the power line and can be measured during operation is shown below.

Principle of technical solution

Fig. 3 shows a circuit diagram for measuring the grounding resistance of power lines without disconnecting the ground wire.

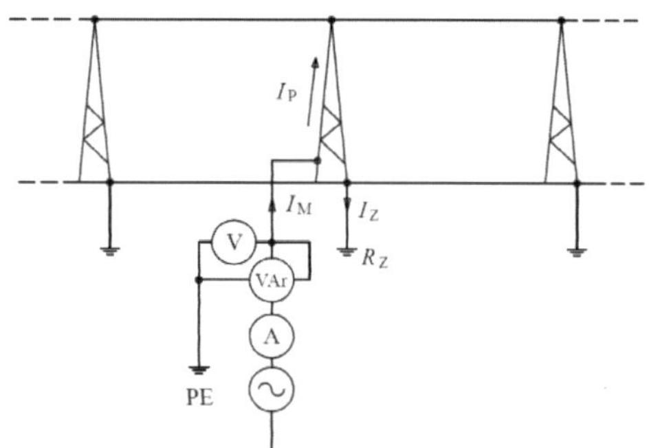

Figure 3. Wiring diagram of the measuring circuit

The measuring circuit is powered by a variable current measuring current source. The power supply is connected to the CE current electrode with one pole, the second pole through the measuring devices to the conductive structure of the measured mast. The ampermeter measures the total current of the measuring circuit I_M, which is the sum of the current flowing into the ground ground of the mast into the ground I_Z and the current flowing into the ground rope and adjacent grounding of the masts and the grounding of the electrical stations at both ends of the grounding cable I_P.

Voltmeter measures mast voltage U_M against reference electrode PE (remote ground). The VAr - meter is connected to the measuring circuit to measure the reactive power (inductive) flowing into the measuring circuit.

The same rules apply to the placement of the measuring electrodes CE and PE as in the case of classical earth resistance measurements by the VA method.

The replacement circuit of the measuring circuit is shown in Fig. 4

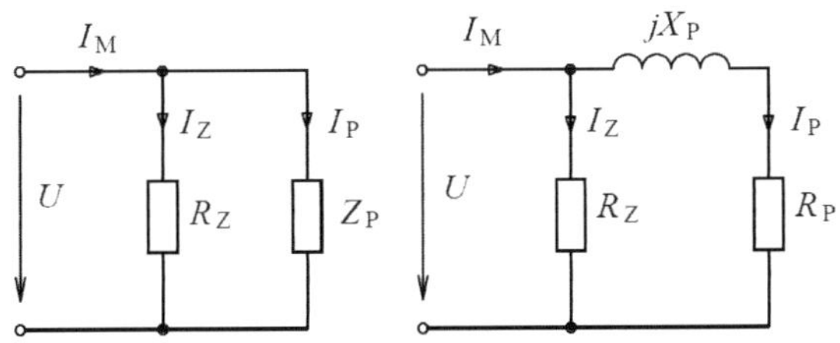

Figure 4. Replacement circuit of the measuring circuit

Where

R_Z the grounding resistance of the measured mast,

Z_P the resulting ground cable impedance and adjacent grounding resistances of the masts and electrical substations,

The impedance Z_P is a complex quantity and has a value

$$Z_P = R_P + j \cdot X_P \tag{6}$$

When powering the measurement circuit from a variable frequency source, the ohm resistors R_z and R_p, which will always have the same value, and the reactivity X_P will vary depending on the source frequency according to the relationship

$$X_P = \omega \cdot L_P = 2\pi f L_P \tag{7}$$

The following applies to the total current of the measuring circuit:

$$I_M = I_Z + I_P \tag{8}$$

Where

I_Z is the current flowing to the measured pole ground,

I_P is the current flowing through the earthing switches to ground the adjacent masts and electrical stations,

Next

$$I_M = \frac{U_M}{R_Z} + \frac{U_M}{(R_P + j\omega L_P)} \tag{9}$$

Where

U_M is the mast voltage to the PE reference electrode,

This relationship at a changing frequency is an equation of a circle in a complex plane (circle diagram, Fig. 5).

The end point of the total measuring point I_M describes the half circle in the complex plane. The center of the circle S lies on the real axis at the distance X with the radius of the circle r (at $f = 0$).

While

$$X = \frac{U_M}{R_Z} + \frac{U_M}{2R_P} \tag{10}$$

And

$$r = \frac{U_M}{2R_P} \tag{11}$$

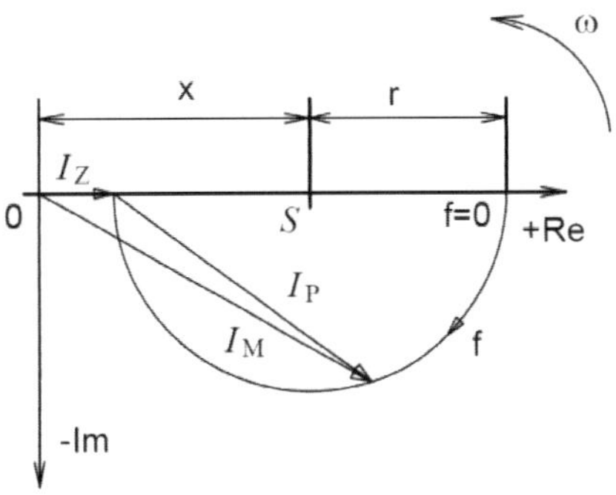

Figure 5. Circle diagram

To determine the current I_Z that flows into the ground of the measured pole, the maximum value of the reactive component is the I_M. That is, the maximum reactive power of the measuring circuit Q_{MAX}.

At this point in the circle diagram, the equality of the real and imaginary components of the impedance is Z_P, that is, $R_P = X_P$ (Fig. 6)

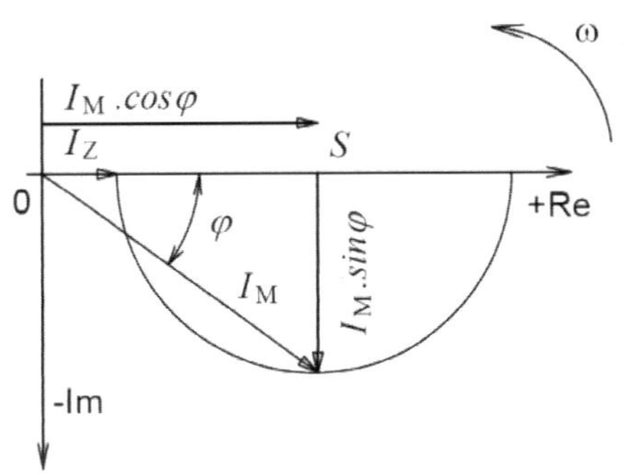

Figure 6. Circular diagram for equality of real and imaginary components

Valid formula: $Q_{MAX} = U_M . I_M . sin\varphi$ (measured by VAr-meter)

From this relationship (known values of Q_{MAX} U_M and I_M), sin and cos are determined

Thus:

$$X = I_M . cos\varphi , \tag{12}$$

resp.

$$r = I_M . sin\varphi \tag{13}$$

Then the current flowing to the measured ground of the mast:

$$I_Z = X - r \tag{14}$$

Then

$$I_Z = I_M \left(cos\varphi - sin\varphi\right) \tag{15}$$

Mast Ground Resistance:

$$R_Z = \frac{U_M}{I_Z} = \frac{U_M}{I_M \cdot (cos\varphi - sin\varphi)} \tag{16}$$

V. Conclusion

The advantage of this measurement is the usability of the method in operation using conventional measuring devices without major financial demands, such as e.g. using special measuring devices.

Acknowledgment

This work was supported by the Ministry of Education, Science, Research and Sport of the Slovak Republic and the Slovak Academy of Science under No. VEGA 1/0372/18 „Výskum pôsobenia SMART sietí na distribučnú sústavu“

Authors thank Ing. Zoltán Hogya for his expert advice and material remarks when drafting the article.

References

[1] STN EN 50341-1: 2013 (33 3300) Vonkajšie elektrické vedenia so striedavým napätím nad 45 kV. Časť 1: Všeobecné požiadavky. Spoločné špecifikácie.

[2] J. Osolsobě, M. Zapletal: Zemnění a bezpečnosť. Praha: Nakladatelství ČSAV, 1964

[3] L. Varga, D. Hlubeň: Meracie metódy v elektroenergetike. Vydavateľstvo PRO, Košice 2010, ISBN:978-80-89057-26-9.

[4] D. Medveď, M. Kolcun, Z. Conka et al.: Simulation of Faults in the Power System and Their Elimination in the EMTP-ATP, PROCEEDINGS OF THE 7TH INTERNATIONAL SCIENTIFIC SYMPOSIUM ON ELECTRICAL POWER ENGINEERING (ELEKTROENERGETIKA 2013) Pages: 484-487 Published: 2013

[5] Z. Conka, M. Kolcun, G. Morva: Utilizing of Phase Shift Transformer for increasing of Total Transfer Capacity, ACTA POLYTECHNICA HUNGARICA, Vol. 13, Issue:5 Pages: 27:37, DOI: 10.12700/APH.13.5.2016.5.2, 2016, ISSN: 1785-8860 .

[6] M. Kolcun, Z. Conka, et. Al, "Improvement of Transmission Capacity by FACTS devices in Central East Europe power system" IFAC PAPERSONLINE, Volume: 49 Issue: 27 Pages: 376-381 DOI: 10.1016/j.ifacol.2016.10.756

Solution of Electromagnetic Field Distribution Around the 22 kV Poles Embedded with Bird-barrier

Dušan Medveď, Dominik Gdovin
Department of Electrical Power Engineering
Mäsiarska 74
041 20 Košice, Slovak Republic
Dusan.Medved@tuke.sk , Dominik.Gdovin@student.tuke.sk

Abstract — **This paper deals with the solution of electromagnetic field distribution in the vicinity of 22 kV poles equipped with the bird barriers. However, in some cases, these barriers serve as birds' habitats. Bird barriers are mostly electrically and magnetically non-conductive. However, due to the weather conditions, the material properties of these barriers are changed. For the purpose of this paper, appropriate anti-bird barriers have been selected that have been placed in a high-polluted area. Due to pollution, electric and magnetic properties on the surface had changed. Since pollution was an electrically conductive material, it also affected the distribution of electrical and magnetic fields. Therefore, this paper presents the results of simulations of the distribution of electric and magnetic fields around 22 kV poles fitted with bird barriers, considering a high degree of pollution of the avi-barrier.**

Keywords: electric pole; bird barrier; 22 kV power line; electromagnetic field; electric intensity; magnetic induction

I. Introduction

This article analyzes and describes the electromagnetic field distribution around console of the 22 kV pole. The main initiative for the solution of the subject matter was the fact that the placement of bird barriers in areas with a high degree of contamination and pollution can affect the distribution of the electric and magnetic fields around the pole. A dirty barrier is essentially a high-ohmic load, and for a given high-voltage power line it is like a short-circuited turn. Pollution affects the change in the distribution of the electromagnetic field and it is therefore necessary to consider this change, which can affect eg. higher power losses, higher corona rates at the edges or edges of the soiled barrier, higher safety requirements for live work on these poles and the like.

For this reason, this article describes a computer simulation in the ANSYS Workbench environment of the electromagnetic field distribution around the 22 kV pole console. The simulation results for different barrier types were evaluated and compared.

II. Analysis of the Distribution of the Electromagnetic Field Around the HV Poles, Before and After Equipping the Bird-Barrier

A. *Analysis of magnetic field distribution*

The 22 kV pole, together with its base and conductor fasteners, eventually grounding cables, is the element of the support points that form the external power lines. Their shape, height, construction and material depend mainly on the voltage level and the number of systems. An iron-concrete pole of 22 kV power line with horizontal conductor arrangement was considered in this paper (Fig. 1).

Figure 1. View of considered (real) 22 kV pole

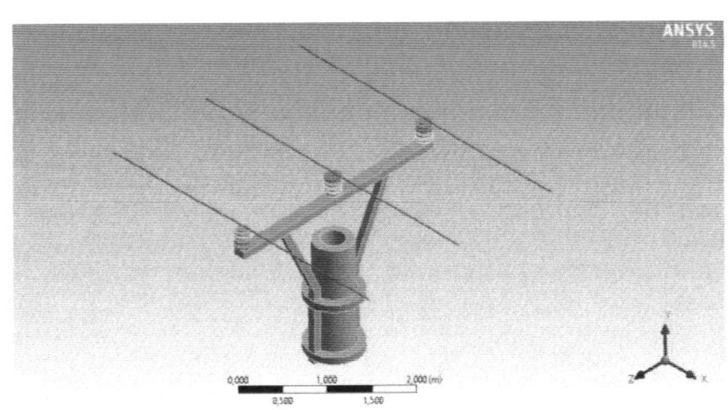

Figure 2. View of a modeled 22 kV pole

The following material constants listed in Table I [20–24] were used to simulate the magnetic field distribution. Pollution represents continuous solid contamination of 1 mm thickness and with electric and magnetic properties mentioned in Table I.

TABLE I. MATERIAL CONSTANTS FOR MAGETIC FIELD

Material	Relative magnetic permeability [–]	Electric resistivity [Ω/m]
Concrete	1,0013	400
Copper	0,99999	$1{,}75.10^{-8}$
Plastic (PVC)	1	1.10^{14}
Porcelain	1	1.10^{13}
Air	1,0000003	1.10^{11}
Iron	50	$1{,}3.10^{-7}$
Pollution	1,0006	34,84

Additional data were needed to be set up in the ANSYS Workbench, like current load of the conductors. Since the most unfavorable state was sought in this analysis, a full load of phase conductors, ie 300 A, was considered.

The simulations were performed on a server where the ANSYS software license was installed (12-core processor, 96 GB RAM). The calculation was quite fast (meshing about 4 hours (mesh size 1 cm), calculation of magnetic induction distribution, or electrical intensity: about 3 hours).

Pole without bird barriers

In order to compare the magnetic induction values around the contaminated bird barrier, it was necessary to first calculate the magnetic field distribution around the 22 kV pole console without using an avi-barrier (Fig. 3).

Figure 3. Magnetic field distribution around the 22 kV pole

The highest values of magnetic induction ***B*** are located in the center of the console because it is magnetized not only from the middle phase L_2, but also from the remaining (outer) phases L_1 and L_3 (Fig. 3). It can also be observed that, without the presence of bird barrier, and with a symmetrical load, the magnetic field is not deformed. The biggest influence on the magnetic field distribution is the metal pole console, which has relatively high relative permeability.

Pole with ridge bird barriers

Although there are no new ridge barriers at the moment, there are still plenty of poles where these ridge barriers are being installed. For this reason, these types of barriers were also considered in the simulations. The shape and fitting of the ridge barriers can be seen in Fig. 4.

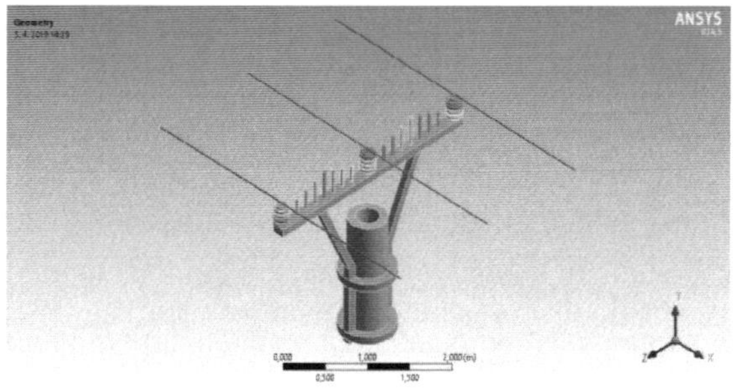

Figure 4. View of the 22 kV pole fitted with the ridge barriers

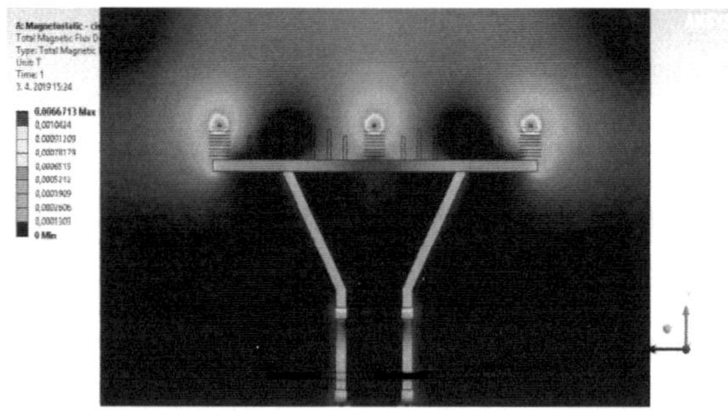

Figure 5. Magnetic field distribution around the 22 kV pole fitted with ridge barriers

The presence of ridge barriers does not have a significant effect on the magnetic field distribution (Fig. 5). This is due to the fact that these ridge barriers are made of a plastic material that has a relative magnetic permeability close to 1. For this reason, these ridges are very little magnetized. Furthermore, it is seen that the highest values of magnetic induction ***B*** are found on the carrier console, due to the fact that its relative permeability is high compared to other materials. It is also seen that the higest values of magnetic induction are located in the center of the console where the adjacent phases L_1 and L_3 are affected.

Since the barrier is relatively difficult to contaminate on its surface (from the top of the barrier, at the tips), the resulting magnetic field distribution, including contamination, was almost identical to the no-pollution variant.

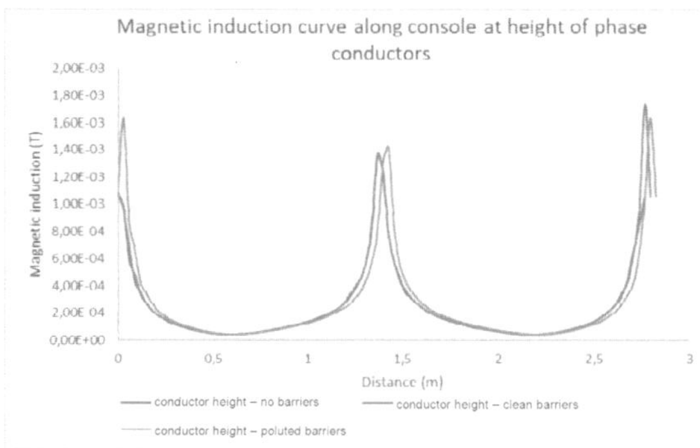

Figure 6. Comparison of the pole magnetic field and the magnetic field of the pole with ridge barriers at conductor height

Pole with bracket barrier against birds

The currently most widely used type of bird barrier used on 22 kV pole consoles is so-called bracket barrier. Its advantage, compared to other types of barriers, is that it allows the birds to short-time to sit on barrier, while electrically isolating the conductive parts from the possible contact of the bird with the conductor.

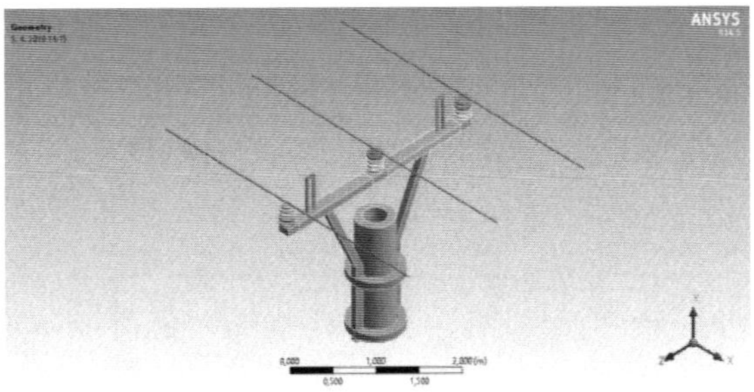

Figure 7. View of the 22 kV pole fitted with a bracket barrier

Figure 8. Magnetic field distribution around a 22 kV pole fitted with a bracket barrier

The largest values of magnetic induction appeared on the support metal console (Fig. 8). Thanks to the middle phase (L_2), a magnetic induction of 0.27 mT is induced on the surface of the bracket barrier. Based on the results of the simulation, it can be said that the presence of the bracket barrier itself does not affect the shape of the magnetic field, mainly because the bracket barrier is made of plastic material.

Figure 9. Magnetic field distribution around a 22 kV pole fitted with a bracket barrier and with pollution

The pollution caused an increase in the maximum magnetic induction value to 6.2 mT, because the electrical conductive contamination represented a quasi short-circuit turn (Fig. 9). As a result, magnetic induction values have increased in other locations around the console. This fact is due to the bracket barrier is fixed directly to the support console and also its top surface, which can become dirty, is quite large, therefore its influence on the magnetic field distribution is also greater than in other types of bird barriers.

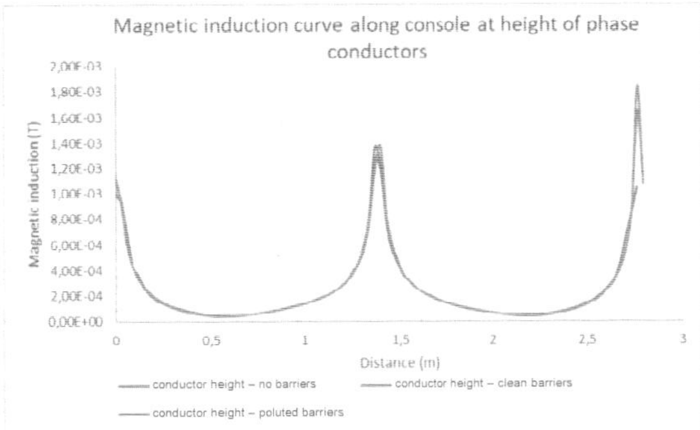

Figure 10. Comparison of magnetic field of pole without avi-barrier and fitted with bracket barrier at height of phase conductors

Pole with „H"-shape barrier against birds

Another protection against birds is the 4TECH „H" barrier. The ZSE distribution company mainly uses this bird barrier, but it can also serve as an alternative to, for example, a bracket barrer or ridge barrier (Fig. 11).

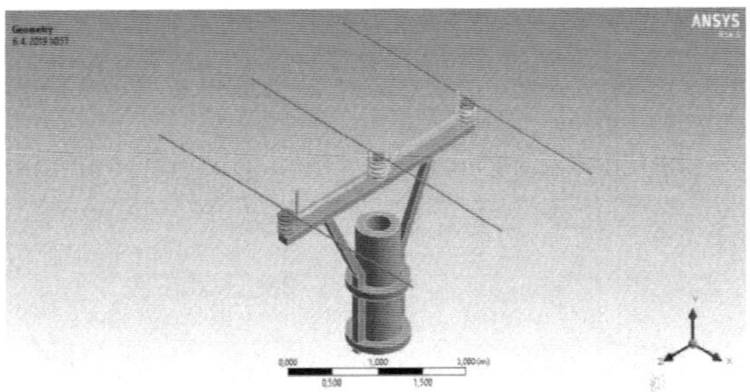

Figure 11. View of 22 kV pole fitted with avi-barrier of „H"-shape

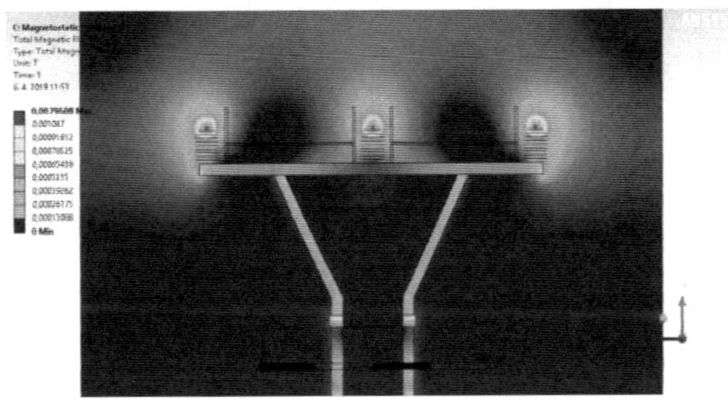

Figure 12. Magnetic field distribution around the 22 kV pole fitted with avi-barrier of „H"-shape and with pollution

From Fig. 12 it can be seen that if this barrier is contaminated, the maximum value of the magnetic induction rises to a value of nearly 8 mT, since the electrical conductive contamination represents a quasi short-circuit turn. Therefore, there is an increase in magnetic induction values. The shape of the magnetic field has not been greatly influenced by the fact that these barriers are made of plastic material and because the relative permeability of the pollution is in the range of paramagnetic substances.

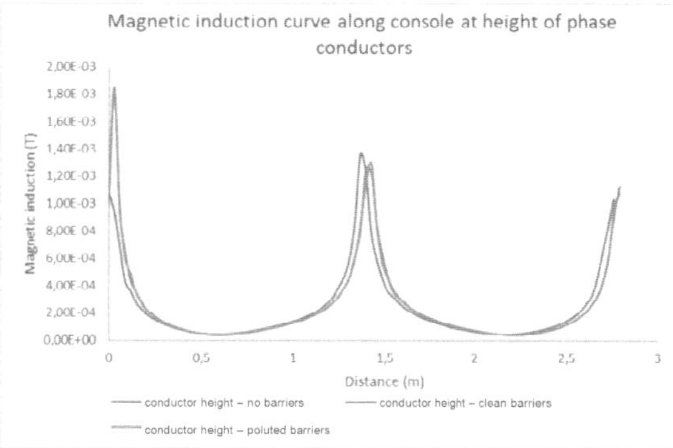

Figure 13. Comparison of magnetic field of pole without avi-barrier and fitted with „H"-shape barrier at height of phase conductors

B. *Analysis of the electric field distribution*

As mentioned above, the distribution of the electric field was solved separately in the ANSYS Workbench, not as a coupled field problem of electric and magnetic fields.

The material properties (electrical resistivity) of the respective pole components were selected according to Table I. Other input data in the simulations were values of the load. Since limit states were addressed for the purpose of this paper, a 100% phase conductor load (300 A) and a voltage of 23.6 kV were selected for the simulation.

The presentation of the results of the electric field will be in this paper in the form of graphs that were read on the selected curves (Fig. 14).

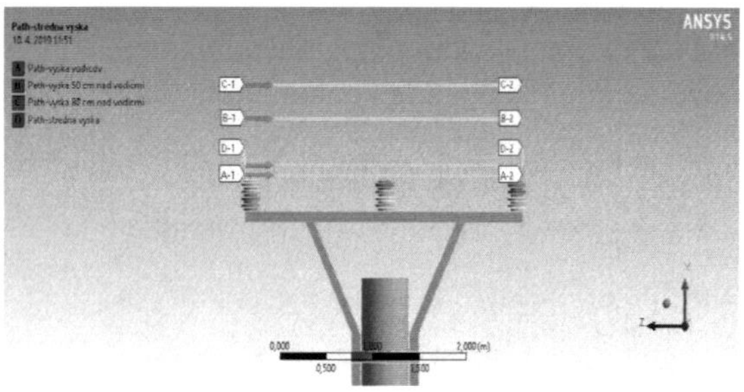

Figure 14. Curves along which the electric intensity was measured

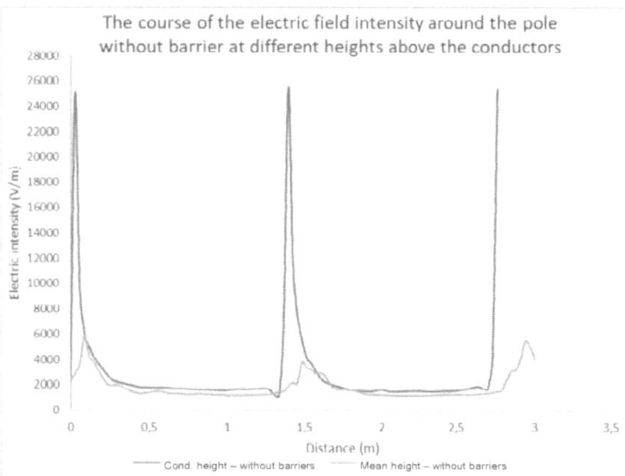

Figure 15. The course of the electric field intensity around the pole without barrier at different heights above the conductors

The graph in Fig. 15 shows that the highest values of the electric field intensity are at the hight of conductors (note: not inside the conductors, where $\boldsymbol{E} = 0$), where in the places where the phase conductors are located it reaches about 25 000 V/m. Furthermore, it can be seen that higher values are also in the middle. This mean height is needed to make further comparisons between individual bird barriers and to compare them at the same height. It can also be seen that the highest values occur around the phase conductors, with a value of about 6000 V/m. This rapid decrease in values was because the electric field intensity values are decreasing indirectly with the square of the distance from the source. It should be noted that the graph did not indicate values from inside the conductor and from near surroundings of the conductor (where the path curve passed through the conductors) because the values were oscilated, respectively, in the order of $1 \cdot 10^5$ V/m, what would lead to distortion of the graph and its visual opacity.

The following graphs show the results of electrical intensity courses around the selected types of avi-barriers.

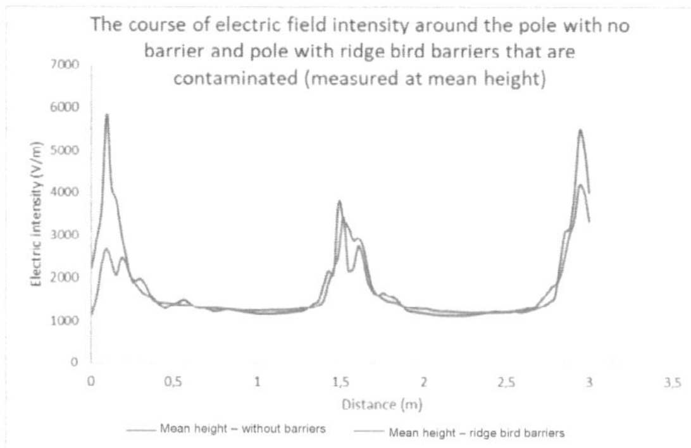

Figure 16. Comparison of the electric field intensity around the pole with no barrier and pole with ridge bird barriers that are contaminated

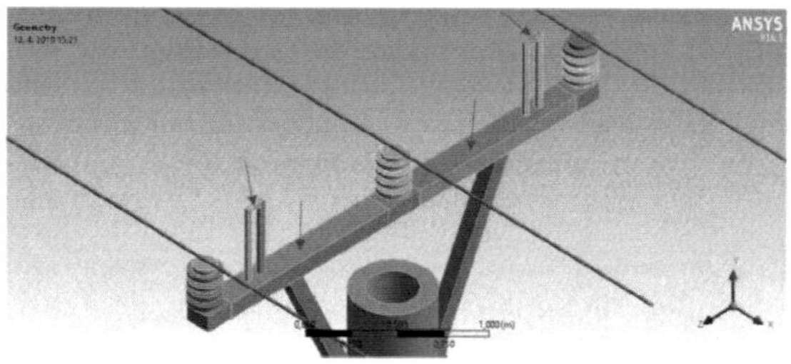

Figure 17. Detailed view of the sites where pollution was applied

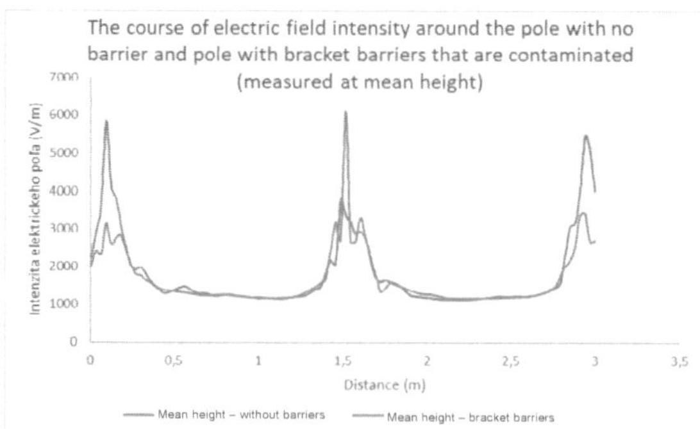

Figure 18. Comparison of the electric field intensity around the pole with no barrier and pole with bracket barriers that are contaminated

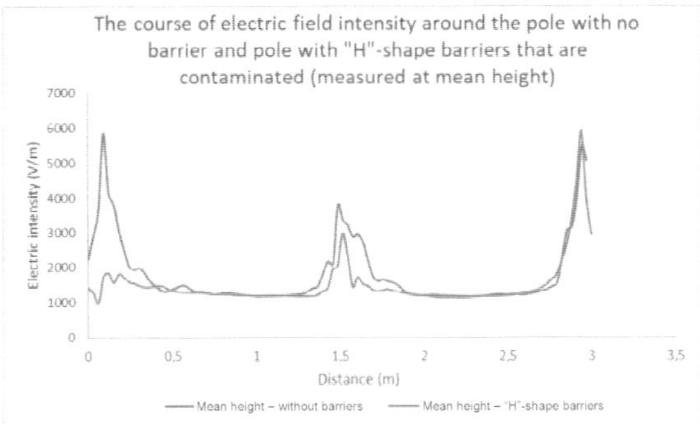

Figure 19. Comparison of the electric field intensity around the pole with no barrier and pole with „H"-shape barriers that are contaminated

From graphical dependence in Fig. 19 it can be seen that when the bird barrier is contaminated, the values of the electric field intensity at the conductor height decrease. As far as the electric field intensity in the middle height is concerned, the trend continues from the variant without pollution. The values of the electric field intensity at this height are comparable to those of a pole without barriers or even decrease below these values in some places. Exception is the location around the L_3 phase conductor. This trend is because the "path" curve along which the electric field intensity values are measured is high above the bird-placing where the bird barrier is polluted.

III. Conclusion

The aim of this paper was to analyze and describe the distribution of the electromagnetic field in the case of the application of a bird-barrier to the construction of 22 kV pole console. To compare the values, the EMF characteristic values were selected, namely the magnetic induction and the electrical intensity that are given in this paper.

It should be noted that, according to the simulated values, if the plastic material is used for barriers, their influence on the magnetic field is minimal. As far as the impact on the electric field is concerned, the electric field intensity values have increased mainly around the phase conductors. Therefore, it is also necessary to consider whether these elevated values can adversely affect birds in contact, and whether high electric field intensity can result in any injury or injury to birds and whether the operator cannot be compromised in the event of maintenance. Therefore, distribution system operators should address this issue, develop similar but more complex analyzes, complemented by real measurements and innovate bird-barriers so that their impact, whether on birds, distribution of electromagnetic fields or operators, is minimized.

Acknowledgement

This work was supported by the Ministry of Education, Science, Research and Sport of the SR and the Slovak Academy of Sciences under the contract No. VEGA 1/0372/18 and VEGA 1/0435/19.

References

[1] L. Deutschová, M. Galis, *„Vtáky a elektrina"*. [online]. Available at: < https://www.lifeenergia.sk/index.php/projekt/hlavne-informacie/item/406-vtaky-a-elektrina >

[2] Úrad priemyselného vlastníctva SR, *„Ochranný kryt elektrických rozvodných vedení"*, 2008. [elektronický dokument]. Available at: < http://skpatents.com/7-u5054-ochranny-kryt-elektrickych-rozvodnych-vedeni.html >

[3] D. Mayer, *„Aplikovaný elektromagnetizmus"*, Nakladatelství Kopp, České Budejovice 2012, 540 p. ISBN 978-80-7232-436-1.

[4] D. Mayer, J. Polák, *„Metody řešení elektrických a magnetických polí"*, Praha 1983. 450 p.

[5] D. Rot, J. Jiřinec, J. Kožený, A. Podhrázký, J. Hájek, S. Jiřinec, *„Induction system for hardening of small parts"*. In Proceedings of the 2018 19th International Scientific Conference on Electric Power Engineering (EPE). Piscataway: IEEE, 2018, p. 277-282. ISBN: 978-1-5386-4612-0, ISSN: 2376-5623.

[6] R. Cimbala, L. Kruželák, S. Bucko, J. Kurimský, M. Kosterec, *"Influence of Electromagnetic Interference on Time-Domain Spectroscopy of Magnetic Nanofluids"*, In: EPE 2016. Danvers: IEEE, 2016, p. 279 - 282. ISBN 978-1-5090-0907-7.

[7] J. Zbojovský, P. Liptai, M. Moravec, *"Modelling and calculating the shielding effectiveness of building materials"*, In: Technical sciences and technologies. Vol. 6, no. 4 (2016), p. 205-210. ISSN 2411-5363.

[8] M. Kanálik, M. Pavlík, M. Kolcun, *"The impact of multi-system overhead lines operation with different voltage levels to voltage unbalance"*, In: Elektroenergetika 2015. Košice: TU, 2015, p. 73-76. ISBN 978-80-553-2187-5.

[9] M. Kanálik, M. Kolcun, *"Computation of harmonic flows in three-phase systems"*. In: Acta Electrotechnica et Informatica. Vol. 7, No. 3 (2007), p. 41-45. ISSN 1335-8243.

[10] J. Zbojovský, M. Pavlík, Z. Čonka, L. Kruželák, M. Kosterec, *"Influence of shielding paint on the combination of building materials for evaulation of shielding effectiveness"*, In: EPE 2018. Brno: Un.ofTechn., p. 324-327. ISBN 978-1-5386-4611-3.

[11] J. Zbojovský, et al., *"Model of anechoic chamber for evaluating the shielding effectiveness of electromagnetic field"*, Acta Techn. Corviniensis. Vol. 10, no. 3, pp. 159-161, ISSN 2067-3809.

[12] M. Pavlík, et. al., *"The mapping of electromagnetic fields in the environment"*, In: Acta Technica Corviniensis: Bulletin of Engineering. Vol. 10, no.2 (2017), p. 107-110. ISSN 2067-3809.

[13] M. Pavlík, et. al., *"The impact of electromagnetic radiation on the degradation of magnetic ferrofluids"*, In: Arch. of El. Eng. Vol. 66, no. 2 (2017), p. 361-369. ISSN 1427-4221.

[14] V. Královcová, Z. Martínek, *"Evaluation of photovoltaic power plan"*, Proceedings of the 13th International Scientific Conference EPE 2012, Electric Power Engineering 2012, 23.-25.5.2012 hotel SANTON - Brno, Czech Republic, University of Technology Brno. IEEE, ISBN 978-80-214-4514-7.

[15] J. Kyncl, M. Novotny, *"Education of Digital and Analog Circuits supported by computer algebra system"*, (ISCAS) 2011 IEEE International Symposium on Circ. and Syst., p. 341 - 344, DOI: 10.1109/ISCAS.2011.5937572, ISBN: 978-1-4244-9473-6.

[16] M. Tesařová, R. Vykuka, *"Impact of Distributed Generation on Power Flows Along Parallelly Operated MV Feeders"*, Proceedings - 2018 IEEE International Conference on Environment and Electrical Engineering and 2018 IEEE Industrial and Commercial Power Systems Europe, EEEIC/I and CPS Europe 2018, DOI: 10.1109/EEEIC.2018.8493640.

[17] K. Nohac, M. Tesarova, L. Nohacova, J. Veleba, V. Majer, *"Utilization of Events Measured by WAMS-BIOZE-Detector for System Voltage Stability Evaluation"*, IFAC-PapersOnLine, Volume 49, Issue 27, 2016, Pages 364-369, DOI: 10.1016/j.ifacol.2016.10.747.

[18] Ž. Eleschová, A. Beláň, B. Cintula, B. Bendík, *"Smart grids analysis - View of the transmission systems voltage stability"*, In EPE 2018. Brno: University of Technology, 2018, p. 37-42. ISBN 978-1-5386-4612-0.

[19] D. Kaprál, P. Braciník, M. Roch, M. Höger, *"Optimization of distribution network operation based on data from smart metering systems"*, Electrical Engineering, Vol. 99, Issue 4, Springer, New York, USA, 2017, December, pp: 1417-1428, ISSN 0948-7921.

[20] Kaye&Laby, *"Magnetic properties of materials"*, 2017. [online]. Available at: < http://www.kayelaby.npl.co.uk/general_physics/2_6/2_6_6.html >.

[21] NGU, "*Magnetic Susceptibility Measuremets on* Concrete", 1998. [online]. Available at: < https://www.ngu.no/upload/Publikasjoner/Rapporter/1998/98_122.pdf >.

[22] Fyzikální cabinet, "*Měrný odpor a teplotní součinitel elektrického odporu*", 1988. [online]. Available at: < http://kabinet.fyzika.net/studium/tabulky/merny-odpor.php >.

[23] TechmaniaScienceCenter, "*Odpor vodiče*", [online]. Available at: < https://edu.techmania.cz/cs/encyklopedie/fyzika/elektricky-proud/odpor-vodice >.

[24] ResearchGate, "*Electrical resistivity and the dielectric constant of materials*", [online]. Available at: < https://www.researchgate.net/figure/Electrical-resistivity-and-the-dielectric-constant-of-materials-related-to-iron-rust-at_tbl1_276039770 >.

..

Electromagnetic Field Distribution Modeling and Measuring of the 110 kV Substation

Dušan Medveď, Peter Žvanda
Department of Electrical Power Engineering
Mäsiarska 74
041 20 Košice, Slovak Republic
Dusan.Medved@tuke.sk , Peter.Zvanda@student.tuke.sk

Abstract — **This article deals with the solution of electromagnetic field distribution inside an electric 110 kV substation. The selected variables, which define the distribution of the electric and magnetic fields, in the respective place, were electric intensity *E* and magnetic induction *B*. Measurements were performed at three selected heights of 0.3 m; 1 m and 1.8 m above the ground, consisting of 168 measuring points along the selected part of the power-line. Subsequently, simulation was performed in ANSYS, where the distribution of electric and magnetic fields was solved. The measurement and simulation results were evaluated and compared with the action values set out in Decree 534/2007 Coll. (for residents) and 209/2016 Coll. (for employees).**

Keywords: electric substation; electromagnetic field; electric intensity; magnetic induction

I. Introduction

This article analyzes and describes the distribution of the electromagnetic field (EMF) in a high voltage substation. The reason for addressing this issue was the fact that residents are increasingly appealing to a healthy environment and thus to avoiding harmful chemical or electromagnetic sources that cannot be registered by human sensors, but have an impact on the proper functioning of the organism itself. In Slovakia, the Decree of the Ministry of Health of the Slovak Republic no. 534/2007 Coll. (for residents) and government regulation 209/2016 Coll. (for employees), set exposure limit values and exposure action values for electromagnetic fields. These two documents are currently decisive for assessing whether a given source of electromagnetic radiation does not exceed the limit values, but since each living organism is different (in a professional language: it has different physical material properties), the real impact of an electromagnetic field on a living organism may be different.

For this reason, the present study and this paper were developed, which describes the measurement and computer simulation of electromagnetic field distribution in the 110 kV substation. The results of measurements and simulations were subsequently compared with the action values set out in Decree 534/2007 Coll.

II. Measurement of Electromagnetic Field Distribution Around the Very High Voltage Substation

A. *Description of electrical substation*

The measurement was carried out in a very high voltage substation in eastern Slovakia of a voltage level of 110 kV. The selected part of the electric array on which the measurement was made is shown in the following figure (Fig. 1). The measurement of the electromagnetic field was carried out separately for the electrical component and for the magnetic component. The selected array of the electric substation, where the measurement was carried out, contained 3-phase power-lines, consoles, supports, voltage instrument transformers, grounding, hence common equipment of electric array.

Figure 1. Illustration of the selected part of the power substation where the measurement was made

B. *List of measuring instruments*

The FW Bell model 8030 Magnetometer Gaussmeter Teslameter was used to measure the electromagnetic field and its ***magnetic component*** along with an external 3-axis probe. Using this instrument, it was possible to measure the magnetic component of EMF in three axes (x, y, z). In addition to these three axes, the instrument measures the resulting vector value, i.e., the vector sum of the three axes. This makes it possible to measure the resulting value of the magnetic induction $\boldsymbol{B}$ that affects humans. In addition to the instrument, the 3-axis probe Z0A83-3208-10-T was used for the measurement. This probe, along with Gauss/Teslameter, allowed the magnetic induction to be measured from 0 T to 1 T. The measuring of EMF lasted for about 3 hours during constant weather, so thermal instability was not considered. Since it was necessary to move the probe in 3 axes, a fixed fastening structure of non-ferromagnetic material was provided. Accuracy of this device was 0,5 %.

The Gigahertz Solutions ME 3840B was used to measure the electromagnetic field and its ***electrical component***. Its frequency range is from 5 Hz to 100 kHz. The measuring range is limited to 0÷2000 V/m. This equipment is used to measure electromagnetic fields around the overhead power lines. This instrument contains an internal measuring probe (for both electric intensity and magnetic induction, but in this case it was used only for measuring the electric intensity). Accuracy of this device was 2 %.

C. Electromagnetic field measurement

The electric substation consisted of a number of arrays, but one 110 kV array was chosen, where the load of this array was doubled compared to normal load, for measurement purposes. The marking of the measurement points is shown on the plan view of the substation array (Fig. 4).

The geometrical dimensions of the selected array were: length (x axis) 15 400 mm and width (y axis) 9 600 mm. An x-axis measurement step 2.2 m and y-axis 1.6 m were selected from the determined array dimensions. With this array size and the selected measurement steps, 56 measurement points were created. At each point, another three measurements were made at a height of 0.3 m; 1 m; 1.8 m. Together, 168 values of electric intensity $\boldsymbol{E}$ and 168 values of magnetic induction $\boldsymbol{B}$ were measured.

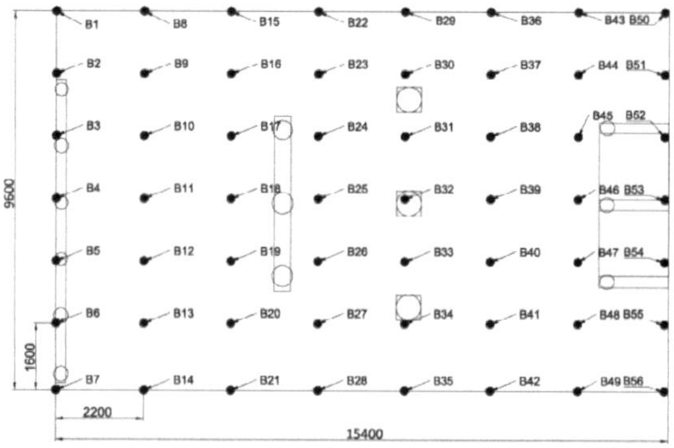

Figure 2. Plan view of the selected array in the electrical substation, indicating the measurement points

During measurement of $\boldsymbol{E}$ and $\boldsymbol{B}$, the ambient temperature was 19 °C and the humidity was 93 %. This humidity is relatively high and affected the measured results. The measurement of the electric field intensity was carried out in parallel with the measurement of the magnetic induction from 9:30 until 12:00. At the time of measurement, all three phases were loaded almost symmetrically. Mean effective current value in phases L_1, L_2, L_3 was $I_{L1} = 231.37$ A, $I_{L2} = 236.02$ A, $I_{L3} = 235.12$ A. The mean voltage at time of measurement was $U = 117{,}69$ kV. Later, the current and voltage values were used in the ANSYS simulation environment.

D. Measurement of magnetic induction $\boldsymbol{B}$

Since number of measurement points is 168, only some selected outputs will be given in this paper in order to visualize and then assess the distribution of magnetic induction in selected locations of the array.

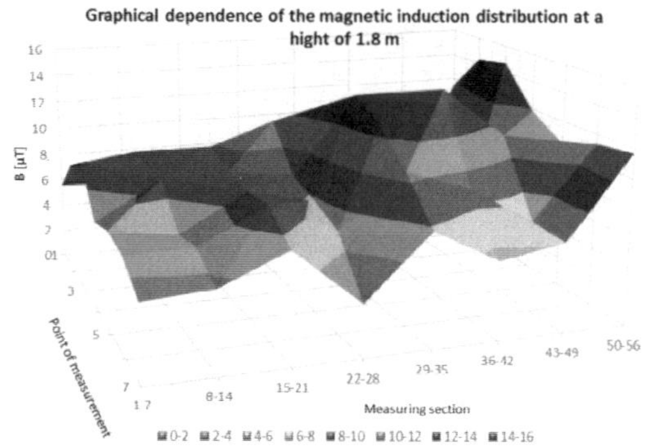

Figure 3. Graphical dependence of the magnetic induction distribution at height of 1.8 m from the ground

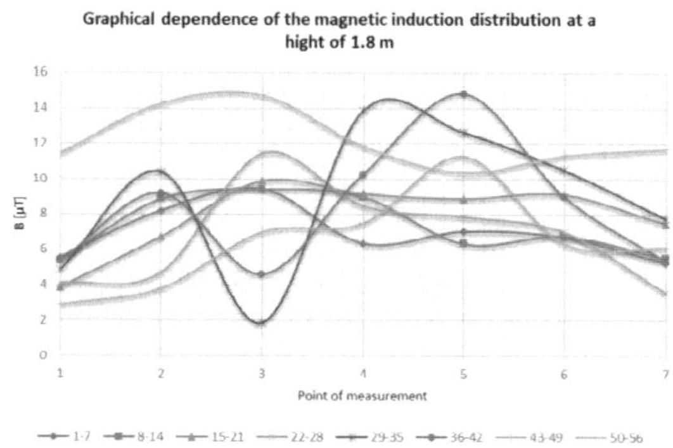

Figure 4. Graphical dependence of magnetic induction distribution in individual sections at a height of 1.8 m

A measurement at height of 1.8 m was made to detect the effect of magnetic induction on a human head. From the graphical dependence shown in Fig. 5, it is possible to see the large oscillations in all sections of the measurement. The largest oscillation was recorded in the measurement section 29-35, where the maximum value was at the fourth measurement point and was 13.89 μT (point B32, Fig. 4). Another oscillation was recorded in the measurement section 36-42 where the maximum value was at the fifth point of the measurement and was 14.81 μT (point B40). The overall increase in these values was in the section where the height of the cable was close above the ground due to the power switch that was located vertically (see Fig. 1). Therefore, when entering the power switch, the conductor's height above the ground was greater than that of the terminal where the height above the ground was less about 1.2 m. Therefore, the magnetic induction values also increased at all three stages. These values are comparatively as high as those measured in the last measurement section 50-56. In the measurement section 50-56, the magnetic induction along the whole section is the same as in the case of the magnetic induction measurement of 0.3 m and 1 m. The change occurs at the last point where the value of the magnetic induction no longer tends to rise, as it was at the previous heights, but stabilizes. The course of the magnetic induction curves in the individual sections can be observed in the Fig. 6. Each curve shows one section (cp. Fig. 4).

E. Measurement of electrical intensity $\boldsymbol{E}$

Similar to the measurement of magnetic induction, as well as the measurement of electrical intensity $\boldsymbol{E}$, it was performed at 168 points. For the purpose of this paper, however, only essential results will be listed, suitable for visual assessment of the electric intensity distribution in selected area locations.

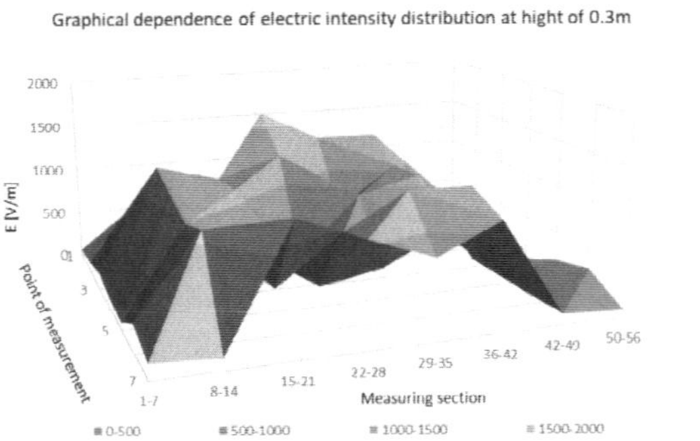

Figure 5. Graphical dependence of the electric intensity distribution at height of 0.3 m from the ground

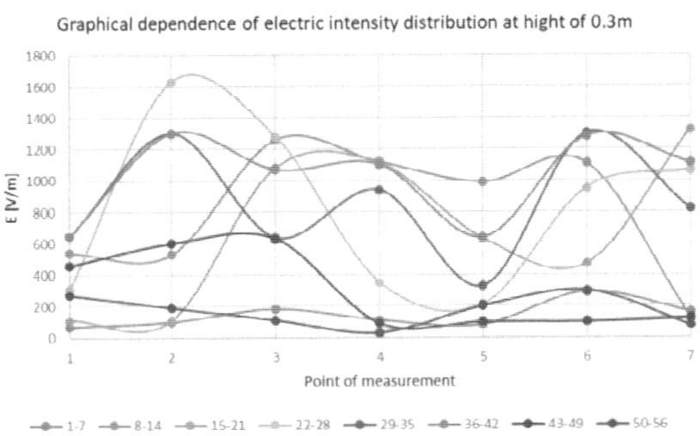

Figure 6. Graphical dependence of electric intensity distribution in individual sections at a height of 0.3 m

The graphical dependence in Fig. 8 shows the large fluctuations in all measuring sections. The smallest measured values of the electric field intensity were recorded in the first measurement section 1-7 and in the last measurement section 50-56. This is because in these sections, at the beginning and end of the array, is no large concentration of cables (phase conductors) of different constructions that would be the source of an electric field. The greatest oscillation was recorded in the measurement section 22-28, where the curve in this measurement section reached the greatest difference between the maximum and minimum values of the electric field intensity. The maximum value was at the second measurement point and was 1,630 V/m (point B23). This value is also the largest measured value at height of 0.3 m above the ground. In the other sections of the array, from the measurement section 8-14 to the measurement section 43-49, the measured values ranged from 400 V/m to 1300 V/m. In these sections, the instrumentation of the electric array is located. There are various consoles and high concentration of cables (phase conductors). The total electric intensity distribution throughout the array in this power substation is shown in Fig. 7.

Note: since the measuring instrument used did not allow the measurement of electrical intensity above 2000 V/m, the measurement at 1 m and 1.8 m was difficult and exceeded the stated measuring range of the instrument. Therefore, these values were verified and compared by simulation in ANSYS.

III. Modeling of the Electromagnetic Field Distribution in ANSYS Environment

A. *Design of geometry and meshing*

The simulation tool ANSYS Workbench was used to simulate the distribution of the electromagnetic field in which a 3D geometric model of the selected part of the array of 110 kV electric substation was created. In the first step, a detailed geometric model was created to provide more accurate EMF results. Due to the available computer technology (12-core processor, 96 GB RAM), where the ANSYS license is installed, it was found that it was not possible to obtain a meshed model after 2 weeks of operation. Therefore there was applied a gradual simplification of the geometric model, which was already applicable with given computer technology (for illustration: meshing of the model: 9 days (21 276 196 elements, mesh size 10 cm), magnetic induction calculation: 7 days, electric intensity calculation: 7 days).

B. *Analysis of magnetic field distribution*

For the functionality of the whole model and its calculation it was necessary to define the material properties of the individual elements. All material properties are listed in Table I. This table shows the resistivity values needed to solve the electromagnetic field distribution (also for electric and magnetic fields). It also contains the relative permeability required for the magnetic field calculation. The resistivity value for electric and magnetic fields calculation was identical.

TABLE I. Electromagnetic Properties of Materials

Material	Resistivity [Ω·m]	Relative permeability
Air	$2 \cdot 10^{14}$	1.00000037
Aluminum	$8.21 \cdot 10^{-5}$	1.000022
Concrete	$1 \cdot 10^{8}$	2
Porcelain	$1 \cdot 10^{12}$	1
Iron	$1.7 \cdot 10^{-7}$	100

The positioning of the probes in the model was the same (cf. Fig. 4) as in the case of real measurements, for the purpose of comparing the measured values with the simulated (at heights of 0.3 m, 1 m and 1.8 m above the ground).

Figure 7. Magnetic field intensity ***H*** distribution in the measurement array (front view)

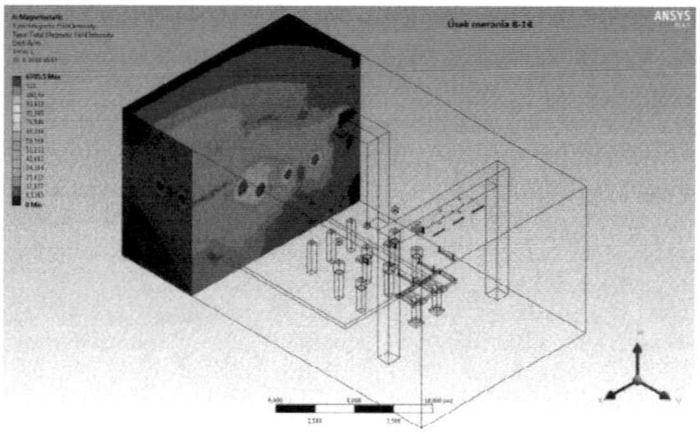

Figure 8. Visualisation of magnetic field intensity $\boldsymbol{H}$ distribution in selected measurement section 8-14

Similarly, it was possible to obtain a magnetic induction distribution $\boldsymbol{B}$ in the individual field locations (Fig. 11).

The magnetic induction distribution in the simulated array can be seen in Fig. 12 and Fig. 13. The magnitude of the magnetic induction distribution at height of 1.8 m is comparable to that of 1 m above the ground. As at a height of 1 m above the ground, as well as at this height of 1.8 m above the ground, the highest values of magnetic induction were recorded in the last measurement section 50-56. Values in this section were in the range of 40÷55 µT. In particular, the increase is seen at the beginning of the section at the point of measurement 2 and at the end of the section at the point of measurement 6. At these points there is a steel structure which serves to lead the conductors (and also power) out of the array. Therefore, in these locations, increased values arise due to the magnetic induction radiating from the phase conductors, (see Fig. 11). In the remaining sections, the curves of magnetic induction distribution are nearly the same. Values are in the range of 15÷45 µT.

It should be noted that the magnetic induction values at height of 0.3 m above the ground were considerably increased due to the grounding effect, which consisted of steel and concrete. The material properties of this grounding and geometric simplification (considered full ground, not latticed, and geometrically closer to the consoles for model simplifying and minimizing) had great impact to the simulation results.

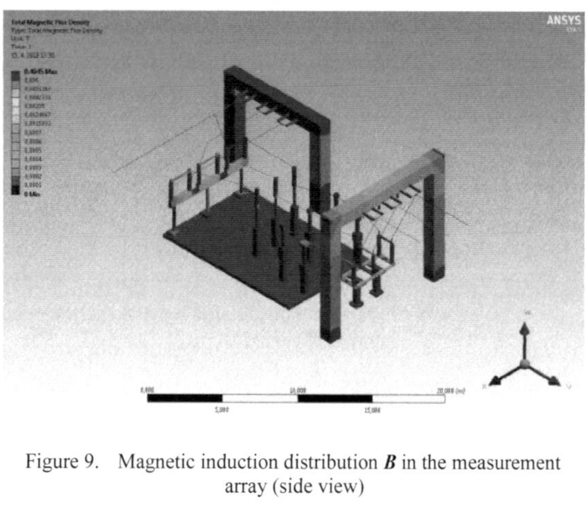

Figure 9. Magnetic induction distribution $\boldsymbol{B}$ in the measurement array (side view)

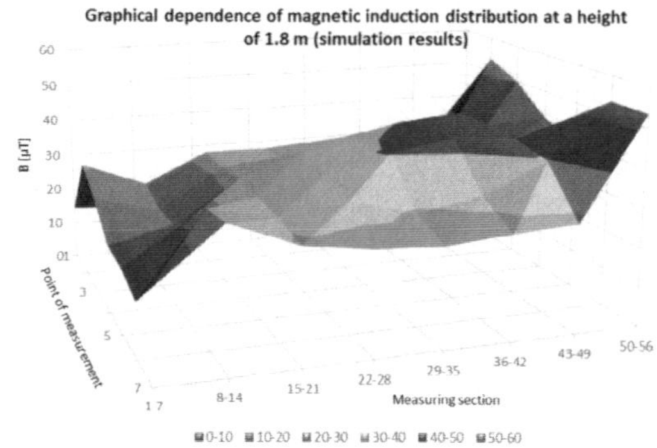

Figure 10. Graphical dependence of the magnetic induction distribution at height of 1.8 m from the ground

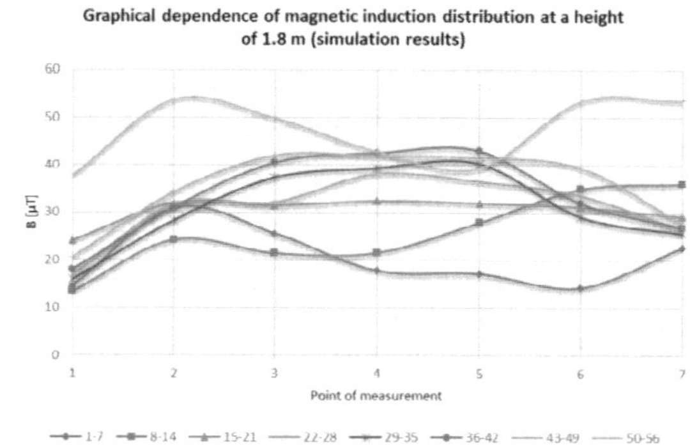

Figure 11. Graphical dependence of magnetic induction distribution in individual sections at a height of 1.8 m from the ground

C. *Analysis of the electric field distribution*

For the purpose of simulation of the electric intensity distribution $\boldsymbol{E}$, the same mesh model was used as in the case of magnetic field. The properties that influence the electric field distribution are electric resistivity and relative permittivity.

The placement of the probes in the model was the same as in the case of real measurements (Fig. 4), for the purpose of comparing the measured values with the simulated (at height of 0.3 m, 1 m and 1.8 m). The results of the distribution of electrical intensity $\boldsymbol{E}$ at the selected locations and heights can be seen in the following graphs.

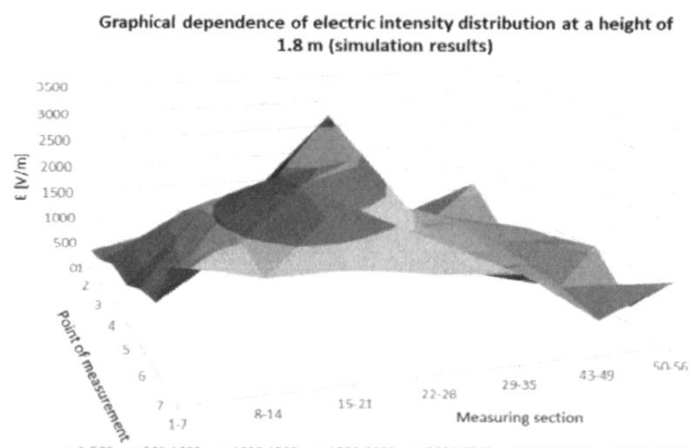

Figure 12. Graphical dependence of the electric intensity distribution at height of 1.8 m from the ground

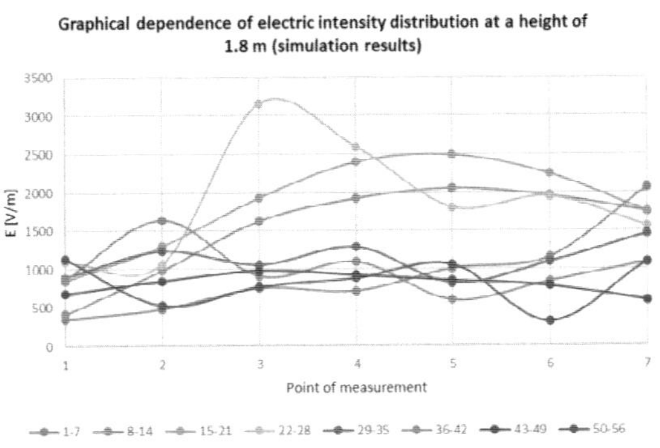

Figure 13. Graphical dependence of electric intensity distribution in individual sections at a height of 1.8 m above the ground

The graphical dependencies of electric intensity at height of 1.8 m above ground are in Fig. 14 and Fig. 15. It can be seen that in the most part of the array the electric intensity ***E*** is from 313.71 V/m to 1 282.5 V/m. However, there are also increased values in this height. Specifically, in the measurement range of 22-28, the value rapidly increases at the measurement point 3 to 3 152.4 V/m. Such an increase was due to a change in phase conductor height. The conductor (phase line) passes through the power switch, which is stored vertically. Therefore, the outgoing conductor from the power switch is positioned approximately 1.2 m lower than the input conductor. Therefore, such a decreasing of conductor height and increasing of measurement height caused an increase in the electric field intensity value. In the other two measurement sections, 8-14 and 15-21, a similar behavior of electric field intensity distribution is seen. These increases are caused by crossing conductors from the supplying power line and conductors in the array. Therefore, there are two sources that are the source of the electric field and therefore these values are also higher than the remaining values.

IV. Comparison of Measurement and Simulation Results

Here it is worth mentioning that both the measurement results and the simulation results may have been influenced by several factors and thus have some accuracy. For example, the measurement probe of the instrument may be affected by the measurement, whether by EMF sources themselves or by the climatic conditions in which it was measured. Similarly, in simulations, results could be influenced, for example, by simplifying geometry, material constants of elements of electric substation, and so on. It follows from the above that the presented results have a certain accuracy (respectively measurement/simulation error) and serve only for illustrative comparison of individual characteristic values of the electromagnetic field.

A. Comparison of results for magnetic field

TABLE II. Comparison of Maximum Magnetic Induction Value by Measurement/Simulation With National Legislation

	Maximum measured / simulated value	Action value for employees (209/2016 Coll.)	Action value for residents (534/2007 Coll.)
B (μT) (0,3 m above ground)	12.8 / 78.156	1 000	100
B (μT) (1 m above ground)	13 / 59.154	1 000	100
B (μT) (1,8 m above ground)	14.81 / 53.792	1 000	100

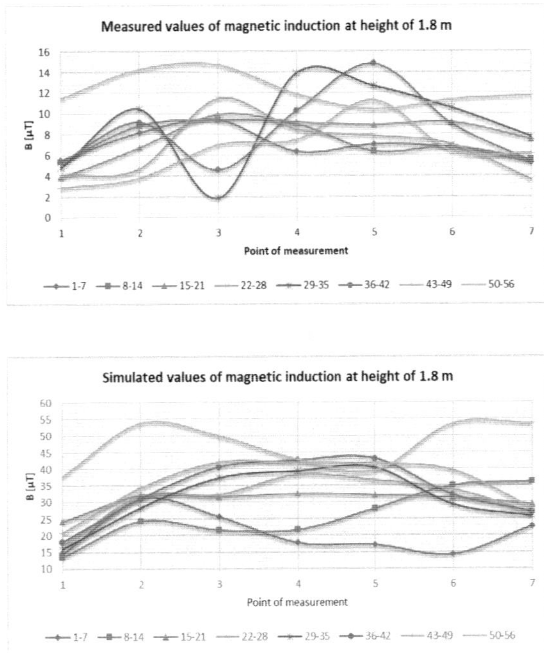

Figure 14. Measured and simulated values of magnetic induction at a height of 1.8 m above the ground

B. Comparison of results for electric field

TABLE III. Comparison of Maximum Electric Intensity Value Obtained by Measurement/Simulation with National Legislation

	Maximum measured / simulated value	Action value for employees (209/2016 Coll.)	Action value for residents (534/2007 Coll.)
E (V/m) (0,3 m above ground)	1630 / 1 802.7	10 000	5 000
E (V/m) (1 m above ground)	NA / 2 214.2	10 000	5 000
E (V/m) (1,8 m above ground)	NA / 3 152.4	10 000	5 000

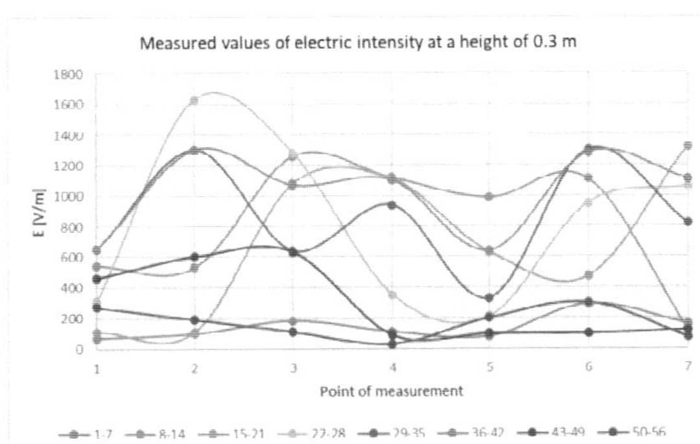

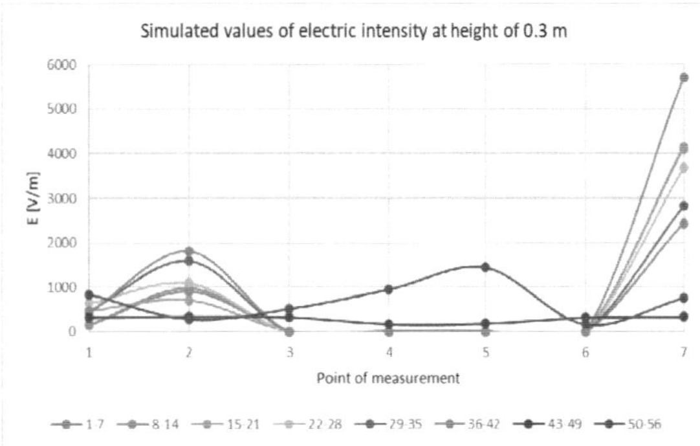

Figure 15. Values of electric intensity E at a height of 0.3 m above the ground obtained by measurement and simulation

V. Conclusion

The aim of this paper was to analyze and describe the distribution of the electromagnetic field inside the very high voltage power substation. The results of measurement and computer simulation of electromagnetic field distribution in 110 kV electrical substation are described in this paper. The results of measurements and simulations were compared with the action values set out in Decree of the Ministry of Health of the Slovak Republic no. 534/2007 Coll. (for residents) and government regulation 209/2016 Coll. (lower electric field intensity action values (RMS) were considered and compared in this case. For employees.), that set exposure limit values and exposure action values for electromagnetic fields. Other standards that are relevant for dealing with electromagnetic field are STN EN 50499; STN EN 50647; STN EN 62110.

When performing the measurement at the electrical substation, it was measured at three heights, in (0.3; 1; 1.8) m above the ground. The 168 measuring points were created along the whole array at the power substation. At a height of 0.3 m above the ground, the maximum electric field intensity of 1 630 V/m was measured. At subsequent heights, the measurement was limited due to an improperly selected measuring instrument that had a maximum measuring range of up to 2 000 V/m. The maximum measured magnetic induction value was 12.8 μT at a height of 0.3 m above the ground, 13 μT for a height of 1 m and 14.81 μT at a height of 1.8 m. In each measured height, neither the electric field intensity nor the magnetic induction it was not exceeded the action value in comparison with the action values found in Decree no. 534/2007 Coll. (for residents) and in regulation no. 209/2016 Coll. (for employees).

The simulation was realized in ANSYS environment. For the simulation, the measurement points were equally spaced as was the case with physical measurements. As one can see from the obtained results and the necessity of simplification of geometry and processing/calculation time, this software is not very useful for such a difficult geometry and there are necessary to accept lower accuracy for a long time calculation time. For this reason, we suggest to use different software for calculation of EMF in such a large object. The maximum simulated value of the magnetic induction was 78.156 μT at a height of 0.3 m above the ground, 59.154 μT at a height of 1 m and 53.792 μT at a height of 1.8 m above the ground. From this finding, it can be said that the action value of magnetic induction was not exceeded for both employees and ordinary citizens. For the electric field, the electric field intensity of 0.3 m above the ground was 1 802.7 V/m, for the height of 1 m it was 2 214.2 V/m and for the height of 1.8 m it was 3 152.4 V/m. Even the electric field has not exceeded the action values for employees and for ordinary people according to national legislation. It can be said that electric field modeling and its measured maximum values coincide with theoretical knowledge.

Acknowledgement

This work was supported by the Ministry of Education, Science, Research and Sport of the SR and the Slovak Academy of Sciences under the contract No. VEGA 1/0372/18 and VEGA 1/0435/19.

References

[1] D. Mayer, J. Polák, "*Metody řešení elektrických a magnetických polí*". SNTL/ALFA, Praha 1983.

[2] D. Rot, J. Jiřinec, J. Kožený, A. Podhrázký, J. Hájek, S. Jiřinec, „*Induction system for hardening of small parts*". In Proceedings of the 2018 19th International Scientific Conference on Electric Power Engineering (EPE). Piscataway: IEEE, 2018., p. 277-282. ISBN: 978-1-5386-4612-0, ISSN: 2376-5623.

[3] R. Cimbala, L. Kruželák, S. Bucko, J. Kurimský, M. Kosterec, *"Influence of Electromagnetic Interference on Time-Domain Spectroscopy of Magnetic Nanofluids"*, In: EPE 2016. Danvers: IEEE, 2016, p. 279 - 282. ISBN 978-1-5090-0907-7.

[4] M. Pavlík, *"Prediction of electricity price using RSI mechanism"*, In: Scientific Letters of Academic Society of M. Baludansky, Vol. 4, No. 6A (2016), p. 82-84. ISSN 1338-9432.

[5] J. Zbojovský, P. Liptai, M. Moravec, *"Modelling and calculating the shielding effectiveness of building materials"*, In: Technical sciences and technologies. Vol. 6, no. 4 (2016), p. 205-210. ISSN 2411-5363.

[6] M. Kanálik, M. Pavlík, M. Kolcun, *"The impact of multi-system overhead lines operation with different voltage levels to voltage unbalance"*, In: Elektroenergetika 2015. Košice: TU, 2015, p. 73-76. ISBN 978-80-553-2187-5.

[7] M. Kanálik, M. Kolcun, "*Computation of harmonic flows in three-phase systems*". In: Acta Electrotechnica et Informatica. Vol. 7, No. 3 (2007), p. 41-45. ISSN 1335-8243.

[8] J. Zbojovský, M. Pavlík, Z. Čonka, L. Kruželák, M. Kosterec, "*Influence of shielding paint on the combination of building materials for evaulation of shielding effectiveness*", In: EPE 2018. Brno: Un.ofTechn., p. 324-327. ISBN 978-1-5386-4611-3.

[9] J. Zbojovský, et al., "*Model of anechoic chamber for evaluating the shielding effectiveness of electromagnetic field*", Acta Techn. Corviniensis. Vol. 10, no. 3, pp. 159-161, ISSN 2067-3809.

[10] M. Tesařová, R. Vykuka, "*Impact of Distributed Generation on Power Flows Along Parallelly Operated MV Feeders*", Proceedings - 2018 IEEE International Conference on Environment and Electrical Engineering and 2018 IEEE Industrial and Commercial Power Systems Europe, EEEIC/I and CPS Europe 2018, DOI: 10.1109/EEEIC.2018.8493640.

[11] K. Nohac, M. Tesarova, L. Nohacova, J. Veleba, V. Majer, "*Utilization of Events Measured by WAMS-BIOZE-Detector for System Voltage Stability Evaluation*", IFAC-PapersOnLine, Volume 49, Issue 27, 2016, Pages 364-369, DOI: 10.1016/j.ifacol.2016.10.747.

[12] D. Kaprál, P. Braciník, M. Roch, M. Höger, "*Optimization of distribution network operation based on data from smart metering systems*", Electrical Engineering, Vol. 99, Issue 4, Springer, New York, USA, 2017, December, pp: 1417-1428, ISSN 0948-7921.

[13] Ž. Eleschová, A. Beláň, B. Cintula, B. Bendík, "*Smart grids analysis - View of the transmission systems voltage stability*", In EPE 2018. Brno: University of Technology, 2018, p. 37-42. ISBN 978-1-5386-4612-0.

Calculation of Power Transformer Losses due to Harmonic Current Flow

Martin Kanálik, Anastázia Margitová, Jakub Urbanský
Department of Electrical Power Engineering
Technical University of Košice
Košice, Slovakia
martin.kanalik@tuke.sk, anastazia.margitova@tuke.sk, jakub.urbansky@tuke.sk

Abstract — **Nowadays, when the power quality term is more frequently used, new questions comes about the solutions of both technical and theoretical problems associated with this phenomenon. One of these questions is the approach of power losses determination in power transformers due to harmonic currents flowing through their windings. This article describes the method of calculating the active power losses due to harmonic current flows through transformers based on IEEE standard C57.110-2018 and analyzes the effect of harmonic current flow on the power loss of a real transformer with different size of load.**

Keywords: power quality; transformer power losses; harmonic current flow

I. Introduction

IEEE standard C57.12.90 and IEEE standard C57.12.91 categorize transformer losses as no-load losses (excitation losses), load losses (short-circuit losses), and total losses (the sum of no-load losses and load losses) [1]:

$$P_{TL} = P_{NL} + P_{LL}\,, \tag{1}$$

where

P_{TL} are total losses in watts,

P_{NL} are no-load losses in watts,

P_{LL} are load losses in watts.

Load losses are subdivided into RI^2 losses and stray losses. Stray losses are determined by subtracting the RI^2 losses (calculated from the measured resistance) from the measured load losses (impedance losses). Stray losses can be defined as the losses due to stray electromagnetic flux in the windings, core, core clamps, magnetic shields, enclosure or tank walls, and so on. Thus, the stray losses are subdivided into winding stray losses and stray losses in components other than the windings. The winding stray losses include winding conductor strand eddy-current losses and losses due to circulating currents between strands or parallel winding circuits. All of these losses may be considered to constitute winding eddy-current losses. The total load losses can then be stated as follows in (1) [1]:

$$P_{LL} = P + P_{EC} + P_{OSL}\,, \tag{2}$$

where

P are RI^2 losses in watts,

P_{EC} are winding eddy-current losses in watts,

P_{OSL} are other stray losses in watts.

If the RMS (root mean square) value of the load current is increased due to harmonic components, the RI^2 losses will be increased accordingly. Winding eddy-current losses in the frequency range under consideration (power frequency and associated harmonics) tend to be proportional to the square of the load current and approximately proportional to the square of frequency (see [2] to [5]). It is this characteristic that can cause excessive winding losses and hence abnormal winding temperature rise and hottest spot temperatures in transformers supplying non-sinusoidal load currents. It is recognized that other stray losses in the core, clamps, and structural parts will also increase at a rate proportional to the square of the load current. However, these losses will not increase at a rate proportional to the square of the frequency, as in the winding eddy losses. Studies by manufacturers and other researchers have shown that the eddy-current losses in bus bars, connections, and structural parts increase by a harmonic exponent factor of 0.8 or less. Therefore, as a conservative estimate, an exponent of 0.8 is used throughout this article. The effects of these losses also vary depending on the type of transformer [1].

Several authors deal with the issue of harmonic impact on power transformers in their publications. In [6] to [8], authors compare simulation and real measurements on transformers, using older version of IEEE standard C57.110 from 1998. They consider the influence of temperature on operation and life of transformers. In [6], [7] and [8] are used various approaches to calculate transformer losses, these approaches differ from the calculation procedure chosen in this article. Interesting results of transformer oil properties which can be used for future development of the standard are presented in [9] to [12].

II. Transformer Losses Calculation

To calculate given transformer losses (P, P_{EC}, P_{OSL}) with considering the influence of the load and harmonics, it is necessary to determine the rated losses (P_r, P_{ECr}, P_{OSLr}) without considering harmonics (only 1st harmonic), and know the following transformer parameters and other data:

- number of phases,
- type (dry-type or liquid-filled),
- rated power,
- rated voltages,
- short-circuit losses,
- no-load losses,
- winding connection,
- DC resistances of high-voltage (HV) and low-voltage (LV) winding,
- transformer load,
- frequency spectrum and RMS values of individual harmonic currents flowing through the transformer relative to the fundamental frequency current.

Since in the case of no-load losses their value is independent of load and harmonics, it can be considered as constant from this point of view.

For the RI^2 losses under rated frequency and rated load conditions P_r in watts following applies [1]:

$$P_r = K\left(R_1 I_{1r}^2 + R_2 I_{2r}^2\right), \tag{3}$$

where

I_{1r} is the high-voltage RMS fundamental line current under rated frequency and rated load conditions in amperes,

I_{2r} is the low-voltage RMS fundamental line current under rated frequency and rated load conditions in amperes,

R_1 is the DC resistance measured between two HV terminals in ohms,

R_2 is the DC resistance measured between two LV terminals in ohms,

K is the constant dependent on the number of phases: 1.0 for single-phase transformers and 1.5 for three-phase transformers.

Resistances R_1 and R_2 may be calculated as follows [1]:

- delta winding: R_1 or $R_2 = 2/9$ of three-phase DC resistance (resistance is the sum of the three phases in series),
- wye winding: R_1 or $R_2 = 2/3$ of three-phase DC resistance.

As mentioned above, a portion of the stray losses is taken to be eddy-current losses. For dry-type transformers, the winding eddy-current losses under rated conditions P_{ECr} in watts is assumed [1]:

$$P_{ECr} = 0.35 P_{TSLr}, \tag{4}$$

for liquid-filled transformers, the winding eddy-current losses P_{ECr} under rated conditions in watts is assumed

$$P_{ECr} = 0.50 P_{TSLr}. \tag{5}$$

P_{TSLr} are the total stray losses under rated conditions in watts [1]:

$$P_{TSLr} = \Delta P_k - P_r, \tag{6}$$

where ΔP_k are short-circuit losses (load losses under rated conditions) in watts.

The rated other stray losses P_{OSLr} can then be determined from the equation [1]:

$$P_{OSLr} = P_{TSLr} - P_{ECr}. \tag{7}$$

For any nonsinusoidal load currents, the equation for the RMS current in per-unit form is:

$$I = \sqrt{\sum_{h=1}^{h=h_{max}} I_h^2}, \tag{8}$$

where

h is the harmonic order,

h_{max} is the highest significant harmonic number,

I_h is the RMS current at harmonic h in with respect to the RMS fundamental frequency current in per units.

For the RI^2 losses formed by any load and harmonics in watts following applies:

$$P = P_r K_{load}^2 I^2 = P_r K_{load}^2 F_{HL1}, \tag{9}$$

where

K_{load} is the load coefficient with respect to the rated load value,

F_{HL1} is the harmonic loss factor for RI^2 losses.

For the winding eddy-current losses formed by any load and harmonics in watts following applies [1]:

$$P_{EC} = P_{ECr} K_{load}^2 \frac{\sum_{h=1}^{h=h_{max}} \left(I_h^2 h^2\right)}{\sum_{h=1}^{h=h_{max}} I_h^2} = P_{ECr} K_{load}^2 F_{HL2}, \tag{10}$$

where F_{HL2} is the harmonic loss factor for winding eddy currents.

For other stray losses formed by any load and harmonics in watts following applies [1]:

$$P_{OSL} = P_{OSLr} K_{load}^2 \frac{\sum_{h=1}^{h=h_{max}} \left(I_h^2 h^{0.8}\right)}{\sum_{h=1}^{h=h_{max}} I_h^2} = P_{OSLr} K_{load}^2 F_{HL3}, \tag{11}$$

where F_{HL3} is the harmonic loss factor for other stray losses.

III. Real Transformer Losses at Different Loads

In this chapter, the results of the power transformer losses calculation considering different non-linear transformer loads are presented. The basic transformer parameters are shown in Table 1. The calculation is based on real measured harmonic currents at transformer LV terminals. Considered (measured) harmonic currents related to the 1st harmonic current are shown in Table 2.

Table 1. Rated parameters of analyzed power transformer

Quantity	Value
Number of phases	3
Type of transformer	Dry
Rated power (MVA)	1.25
Primary rated voltage (kV)	22
Secondary rated voltage (kV)	0.42
Short-circuit voltage (%)	5.87
No-load current (%)	0.256
Short-circuit losses (W)	**9 954**
No-load losses (W)	**1 716**
Primary winding connection	Delta
Secondary winding connection	Wye
DC resistance between two HV terminals (Ω)	2.352
DC resistance between two LV terminals (Ω)	0.000719

Table 2. Measured values of harmonic currents

Harmonic order	Magnitude (%)	Harmonic order	Magnitude (%)
1	100	26	0.11
2	3.47	27	0.21
3	20.32	28	0.09
4	3.22	29	0.50
5	9.45	30	0.07
6	2.21	31	0.41
7	8.12	32	0.08
8	1.15	33	0.08
9	6.19	34	0.06
10	0.77	35	0.24
11	4.55	36	0.04
12	0.90	37	0.24
13	3.84	38	0.05
14	0.75	39	0.05
15	3.30	40	0.04
16	0.83	41	0.15
17	4.86	42	0.04
18	0.56	43	0.12
19	2.47	44	0.04
20	0.49	45	0.04
21	0.78	46	0.04
22	0.27	47	0.08
23	2.17	48	0.04
24	0.14	49	0.07
25	0.88	50	0.03

There are calculated rated losses considering only 1st harmonic (according to the equations (3), (4) and (7)) and with considering all harmonics (according to Table 2) shown in Table 3. The sum of the rated RI^2, eddy-current and other stray losses is equal to the short-circuit losses (9 954 W). The no-load losses are constant (1 716 W) and they are not dependent on load and harmonics. Total rated losses with considering harmonics according to Table 2 are 1.35 times larger than losses with considering only 1st harmonic.

Table 3. Rated losses of analyzed power transformer

Losses (W)	First harmonic	All harmonics	Harmonic loss factor
RI^2 losses	6 980.8	7 484.4	1.0241
Eddy-current losses	1 040.6	4 261.8	4.0955
Other stray losses	1 932.6	2 287.4	1.1836
Load losses	9 954	14 033.6	
No-load losses	1 716	1 716	
Total losses	11 670	15 749.6	

Fig. 1 and Fig. 2 show the comparison of active power losses with and without considering harmonics depending on size of transformer load. The 100 % load represents the rated conditions (rated losses of analyzed transformer are listed in Table 3). As the transformer load increased, load losses (RI^2 , eddy-current and other stray losses) increased, and the load losses was zero at zero load. No-load losses increased the total loss by a constant value 1 716 W (Fig. 2). The influence of harmonics was most evident in the case of eddy-current losses, which are also characterized by the highest harmonic loss factor (Table 2). The difference between total losses with and without harmonics consideration is increasing with the load increasing (Fig. 2), in the case of rated conditions it was approximately 4 kW.

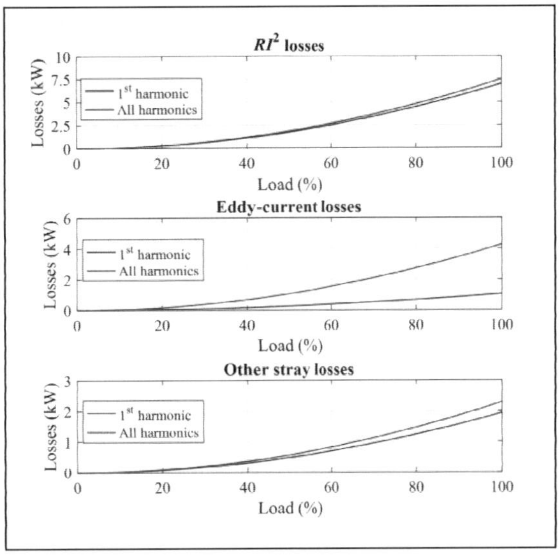

Figure 1. RI^2, eddy-current and total stray losses with and without considering harmonics depending on transformer load

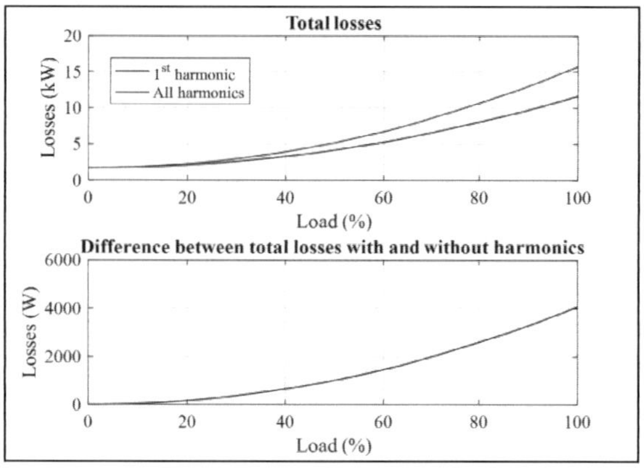

Figure 2. Total losses with and without considering harmonics depending on transformer load

Fig. 3 and Fig. 4 show the comparison of active losses with and without considering the real measured harmonics (harmonic currents) of analyzed transformer for one week. The transformer load ranged from approximately 7 to 22 % of the rated load. At each time the transformer was loaded differently, so there were also various losses at each time. The highest loads, and hence losses, were achieved on weekdays (in the middle of the day). For the weekend, the losses dropped significantly compared to the first five days of the week.

In the case of RI^2 and other stray losses, the difference with and without consideration harmonic current is minimal. Again, the highest difference was achieved with eddy-current losses (maximum losses of 51.70 W considering the 1st harmonic only, maximum losses of 97.46 W considering all harmonics, Fig. 3).

Since the transformer was low loaded, a large part of the total losses were no-load losses (1 716 W, Fig. 4). The maximum total losses value considering only the first harmonic was 2.21 kW and, in the case of considering all harmonics 2.27 kW. Both peaks were reached on Tuesday at around 1 pm. In the case of the difference between total losses with and without harmonics, the maximum difference was only 75.76 W achieved on Tuesday around 16 pm.

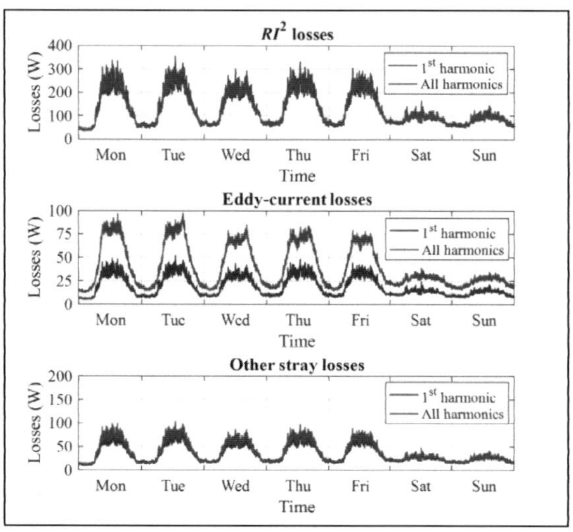

Figure 3. RI^2, eddy-current and total stray losses with and without considering real measured harmonics for one week

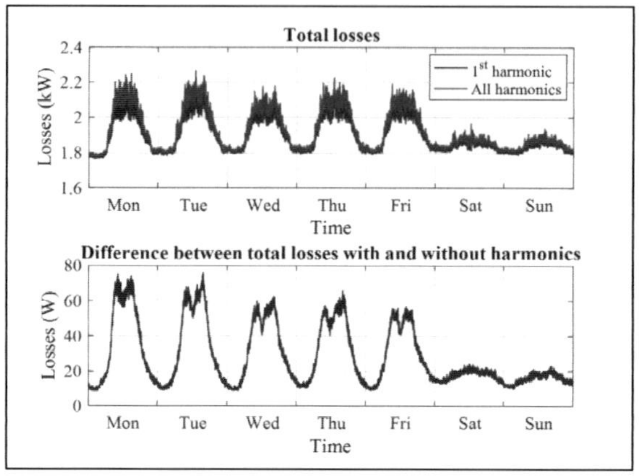

Figure 4. Total losses with and without considering real measured harmonics for one week

IV. Conclusion

The study of harmonic impact on power transformers is important because power transformers are among the most expensive equipment and the most critical plant equipment in electric utilities especially in transmission and distribution network. Harmonics cause additional losses in transformers. In the case of transformer losses calculation, due to neglect of harmonic current effects, it is possible, for example, an incorrect design or overloading of the transformer.

This article deals with the calculation of real power transformer active losses. The paper examines the effect of load and harmonic currents that can actually occur in the analyzed transformer. The calculation is based on the latest version of the IEEE standard C57.110-2018, which shows some differences compared to earlier versions of this standard.

The active losses of the analyzed transformer were calculated from the real measured currents (with considering only the 1st harmonic and also considering harmonics up to 50th order). The maximum difference between total losses (with and without harmonics consideration) was only 75.76 W, but the transformer was loaded to 22 % of the rated load at that time. Loss analysis at rated conditions of the analyzed distribution transformer (100 % transformer load) leaded the difference between total losses with and without harmonics consideration to about 4 kW, as the harmonic effect is proportional to the square of the load with respect to the rated load value.

Acknowledgment

This work was supported by the Ministry of Education, Science, Research and Sport of the Slovak Republic and the Slovak Academy of Sciences under the contract No. VEGA 1/0372/18.

References

[1] IEEE Std C57.110-2018, "IEEE Recommended Practice for Establishing Liquid Immersed and Dry-Type Power and Distribution Transformer Capability when Supplying Nonsinusoidal Load Currents".

[2] M. T. Bishop, C. Gilker, "Harmonic Caused Transformer Heating Evaluated by a Portable PC-Controlled Meter," 37th Annual Rural Electric Power Conference, 1993.

[3] L. F. Blume, et al., Transformer Engineering, 2nd ed. New York: Wiley, 1951, pp. 56-65.

[4] S. Crepaz, "Eddy current losses in rectifier transformers," IEEE Transactions on Power Apparatus and Systems, vol. PAS-89, no. 7, pp. 1651-1656.

[5] H. B. Dwight, „Electrical Coils and Conductors," New York: The McGraw-Hill Companies, 1945, Ch. 3.

[6] D. M. Said, K. M. Nor, "Effects of Harmonics on Distribution Transformers," Australasian Universities Power Engineering Conference (AUPEC'08), vol. P-107, 2008.

[7] J. Singh, S. Singh, A. Singh, "Effect of Harmonics on Distribution Transformer Losses and Capacity," International Journal of Engineering Technology Science and Research (IJETSR), vol. 4, 2017, pp. 48-55.

[8] S. B. Sadati, H. Yousefi, B. Darvishi, A. Tahani, "Comparison of Distribution Transformer Losses and Capacity under Linear and Harmonic Loads," 2nd IEEE International Conference on Power and Energy (PECon 08), December 1-3, 2008, Johor Baharu, Malaysia.

[9] J. Zbojovský, L. Kruželák, M. Pavlík, "Impedance Spectroscopy of Liquid Insulating Materials," Proceedings IEEE International Conference AND Workshop in Óbuda on Electrical and Power Engineering, New York (USA): Institute of Electrical and Electronics Engineers, 2018, pp. 249-253.

[10] S. Bucko, M. Krchňák, R. Cimbala, L. Kruželák, J. Zbojovský, "Abrasive Properties of Transformer Oil-Based Magnetic Nanofluid," EPE 2018, Brno, University of Technology, 2018 pp. 320-323.

[11] D. Medveď, L. Mišenčík, "Design of Dry Transformer Cooling," Elektroenergetika, vol. 6, 2013, pp. 28-31.

[12] D. Medveď, M. Beluščák, "Influence of Oil Transformer Cooling on Its Lifetime" Elektroenergetika, vol. 6, 2013, pp. 25-27.

Behavior of Electricity Meters and Quality Analyzers Using Equipment to Optimize Subscriptions in Small Photovoltaic Power Plants

Richard Velicka
Department of cybernetics and biomedical engineering
VSB – Technical university of Ostrava
Ostrava, Czech Republic
richard.velicka@vsb.cz

Petr Bilik
Department of cybernetics and biomedical engineering
VSB – Technical university of Ostrava
Ostrava, Czech Republic
petr.bilik@vsb.cz

Petr Krejci
Department of Power Engineering
VSB – Technical University of Ostrava
Ostrava, Czech Republic
petr.krejci@vsb.cz

Jan Velicka
Department of cybernetics and biomedical engineering
VSB – Technical university of Ostrava
Ostrava, Czech Republic
jan.velicka@vsb.cz

Abstract — **There are currently many users of photovoltaic power plants that have devices to optimize their consumption or supply of electricity. These devices use the aggregation interval lengths to measure the electricity meter to redistribute the energy flow so that the balance is zero at this time. The electricity meters are unable to record these fast changes in the direction of the power flow, but the modified quality analyzers do. Therefore, tests were performed to simulate the different algorithms used by these optimization devices. In the laboratory, the behavior of electricity meters and modified quality analyzers in the measurement of the test progress was tested. The result is a comparison of the calculated values of the energy consumed and delivered, over time, for all devices tested.**

Keywords: Power Quality, Wattrouter, Electricity meter, Quality analyzer, Non-harmonic signal, Photovoltaic power plant.

I. Introduction

With the rapid increase in small photovoltaic power stations in family houses, problems are starting to appear with regard to connection to the energy network. Most users of these power stations are connected in a manner that does not transfer excess electrical energy back into the grid. Many, however, are trying to use the properties of the energy meter to achieve a resulting balance of zero. From the point of view of measured consumed and supplied energy, this evaluated state is only truly zero if the direction of energy is manipulated within the space of a single second. This interval is equal to the aggregate interval of the energy meter. The problem, therefore, isn't in the faulty billing of the consumed energy but in the supplied energy in terms of excess energy, which is inadmissible. Users of these devices, who regulate the flow of energy, make use of the energy grid for short-term accumulation of produced energy. In doing so, however, they are overloading the distribution power lines and in extreme cases can even lower the quality of electricity (PQ) for other consumers. In order to measure this behavior in the network in places where photovoltaic sources are connected, a device is required that can measure energy much more quickly than a standard energy meter. In our research, we compared the properties of two special, quality analyzers modified for these measurements to conduct tests in a laboratory environment. The results show how energy meters behaved during the regulated mode of energy consumption and supply described above [1], [2].

II. Materials and Methods

A. The tested devices

Two specially modified, quality analyzers underwent testing. One of the analyzers was custom designed in our department of Cybernetics and Biomedical Engineering. It was built on the hardware platform analyzer DEWE-561 by the company DEWETRON, the main difference being the software, which was our own implementation. The second tested analyzer was designed by an external company that also modified the software to measure energy in smaller intervals. For the comparison of results, we used three four-quadrant electricity meters, which are standardly applied where photovoltaic power stations are installed for measuring energy consumption and supply. Two of them were the same type and one was different. The two same types were selected to rule out production faults during testing. The third type eliminated the properties of the second manufacturer [3].

B. Tailored PQ analyzer from VSB – TU Ostrava

A PC-based PQ analyzer was used for the measurement of all the phenomena described in this paper (Figure 1.). The authors used hardware of commercial PQ analyzer DEWE – 561 from DEWETRON, which contains the following main blocks: a PC running a Windows operating system, electronic circuits for voltage and current inputs signal conditioning and the National Instruments data acquisition board. DEWE - 561 has direct voltage inputs and Chauvin-Arnoux MN39 current clamps (two ranges 20A/200 A) were used as a current sensor [4].

The standard off-the-shelf PQ analyzer functionality is defined in CSN IEC 61000-4-30. Special analyzer firmware has been developed to detect the potential cause of a high flicker value due to very fast power changes; LabVIEW graphical programming language using the Electrical Power Toolkit library was used as a development environment to make an instrument prototype. In addition to the 10 min and 200 ms aggregation intervals, the firmware developed can record active power for an aggregation interval of 10 ms (synchronized half a period of 50 Hz fundamental frequency) and an instantaneous flicker for an aggregation interval of 20 ms [2], [3], [5].

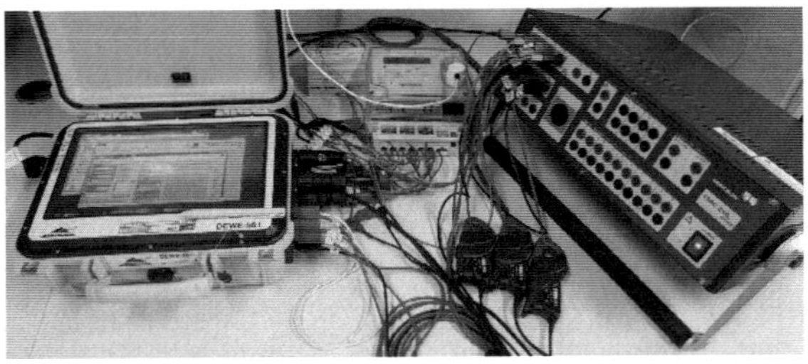

Figure 1. PQ analyzer DEWE - 561 and calibrator OMICRON CMC 256+

The flickermeter provides two quantities: each 10 minutes, short time flicker P_{st}, each 2 hours, long time flicker P_{lt}.

Firmware measures developed and records following quantities:

• Voltage, current, active and reactive power with a 10 ms aggregation interval (each half-period of signals with a fundamental frequency of 50 Hz).

• flicker P_{st} (10 min) and P_{lt} (2 hours).

The functionality of the firmware and the precision of the entire PQ analyzer were checked by a source OMICRON CMC 256+ calibrator, see Figure 1..

C. Voltage and current testing

It is not easy to design the course of currents for laboratory conditions so that the properties correspond to applied regulators. Voltage is continuously sinusoidal, but the current has to change depending on the type of regulator. Sometimes regulators change the direction of the current for 3, 4 or 5 periods. There are also regulators that can detect a change in the amplitude of the current within one period or oscillation so that the resulting energy in a single second is null. The type of regulator to use under certain conditions and how often regulation is active is unknown. The course must therefore be designed as generally as possible in order to define both the behavior of different types of regulator and include a certain level of randomness. The testing course was designed from three different intervals with differing current courses, there are shown in Figures 2. – 4.. This testing course was periodically and simultaneously generated with a sinusoidal voltage waveform for a period of approximately 80 minutes. The values of consumed and supplied electrical energy per hour were then taken from the main segment of voltage and current courses. Any errors arising because of inconsistencies in time were minimal in terms of the number of generated segments [6], [7], [8].

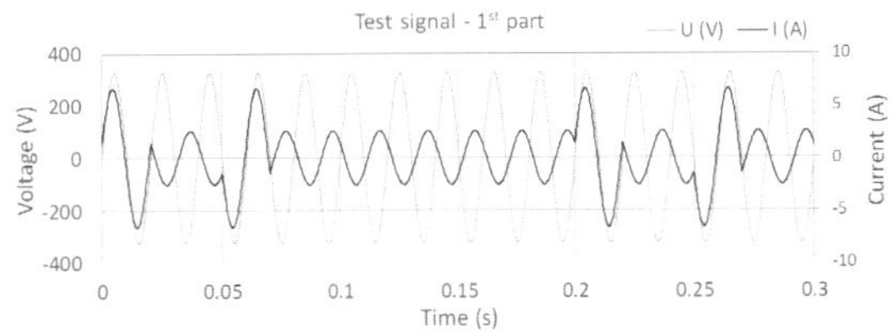

Figure 2. Example of the first part of the test signal

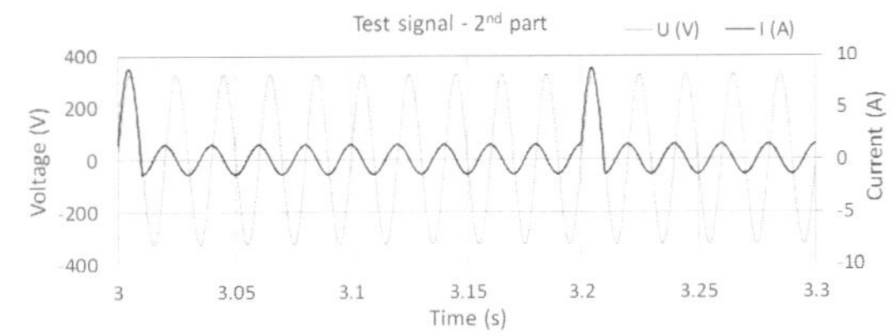

Figure 3. Example of the second part of the test signal

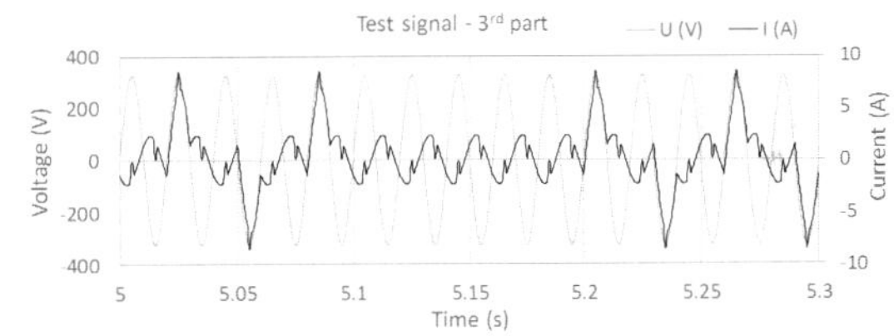

Figure 4. Example of the third part of the test signal

Generating such a long record with a so many repetitions brought a certain level of risk that the source would cease working as the courses were generated. Measurements were repeated often, and supply was not generated for the entire expected time. Mostly though, it only ceased just before the end, which did not influence the results. It was necessary to evaluate the accumulated energy supplied and consumed over one hour. To simplify evaluation, the signal was designed so that the value of both the positive and negative energy was the same. To create the waveform, our colleagues from Faculty of Electrical Engineering and Communication (FEEC), Brno University of Technology created a program that calculated the courses and then generated a file with data that could be used to program the OMICRON CMC 256+ power supply. As a result of the length of the main course, only segments are shown in Figure 2. – 4..

D. Measuring setup

All testing equipment was connected in parallel to the source of the signal for all measurements. The OMICRON 256+ programmable source was used as voltage and current supply. It can generate voltage and current courses independently. Measurements were made in single-phase, although the supply permits up to three-phase outputs. Single-phase generated signals are easier, and problems with synchronising differing phase currents on individual phases are eliminated. Real connections are also frequently single-phase regulators. If a device can measure the supplied and consumed energy correctly in single-phase, then it can do the same in three-phase. The courses were prepared in MS Excel, which also generated a text file to assign the courses to the OMICRON CMC 256+ power supply. In the software module, Advanced TransPlay was applied to the OMICRON CMC 256+, power supply, which allowed it to generate its own voltage and current courses according to time signatures and values. With respect to the power supply's internal storage, individual signals were generated with a sampling frequency of 5 kHz. In other words, one second of a waveform recorded 5000 samples of time, voltage and current. If the length of the main signal is seven seconds, then 35,000 rows of values are used, the values of time, voltage and current always being recorded. These 35,000 rows then periodically repeat according to the set amount so that generation lasts a minimum of one hour. The OMICRON CMC 256+ power supply's advantage is its capability of supplying a so-called "phantom power", which means that the generated voltage is completely independent of the generated current and vice versa. Large amounts of power can therefore be generated without them even existing in reality, but they are real to the measuring equipment. All the tested equipment was connected to the source in parallel to the voltage, and the current was shared between all devices. The analyzers were input with current from their own current probes. The electricity meters shared current input from a current transformer. These methods of connecting current did not influence recording (Figure 5.).

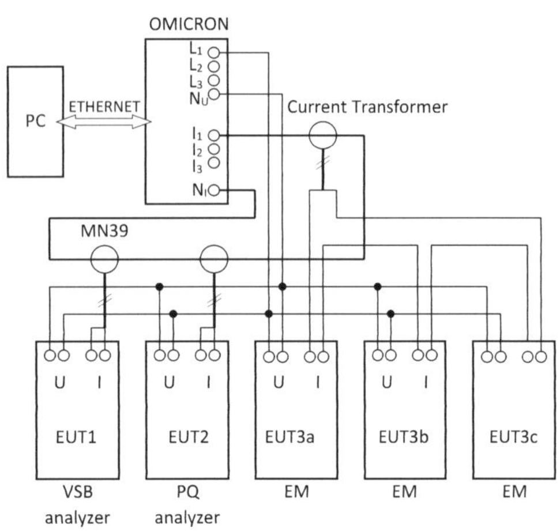

Figure 5. Principle circuit diagram for tests

III. Results

To compare the results of the measured active and reactive energy, their theoretical values for one hour were calculated. As mentioned above, the courses were designed so that the values would be very similar. The analytically calculated energy of the testing waveform was as follows:

+A = 1142,379 Ws (positive active energy)
-A = 1151,458 Ws (negative active energy)

The values were over seven seconds. If the values of energy are recalculated to one second or one hour, then we generate courses with the energy:

+A = 163,197 Wh (one-hour positive active energy)
-A = -164,494 Wh (one-hour negative active energy)

All the other recorded values from the testing equipment relate to the values above.

The VSB analyzer (VSB – Technical University of Ostrava) recorded energy in two speeds. The quicker record was every 200 ms, the slower every 15 minutes (this interval also corresponded to the intervals between recordings in the electricity meters). The difference is that the 15-minute value was calculated here from the recorded half-period values. The parameters of the analyzer were described above. The following Table 1. shows the recorded values from the VSB analyzer [2], [3], [9].

Table 1. Values of EUT1 – VSB analyzer

VSB Analyzer - EUT1	Interval **200 ms**		Interval **15 min**	
One-hour energy	+A (Wh)	-A (Wh)	+A (Wh)	-A (Wh)
Measured values #1	165.504	166.338	165.504	166.338
Analytic values	163.197	164.494	163.197	164.494
Relative error	1.41%	1.12%	1.41%	1.12%
VSB Analyzer - EUT1	Interval **200 ms**		Interval **15 min**	
One-hour energy	+A (Wh)	-A (Wh)	+A (Wh)	-A (Wh)
Measured values #2	165.477	167.253	166.555	166.112
Analytic values	163.197	164.494	163.197	164.494
Relative error	1.40%	1.68%	2.06%	0.98%
VSB Analyzer - EUT1	Interval **200 ms**		Interval **15 min**	
One-hour energy	+A (Wh)	-A (Wh)	+A (Wh)	-A (Wh)
Measured values #3	165.516	167,369	166.584	166.137
Analytic values	163.197	164.494	163.197	164.494
Relative error	1.42%	1.75%	2.08%	1.00%
VSB Analyzer - EUT1	Interval **200 ms**		Interval **15 min**	
One-hour energy	+A (Wh)	-A (Wh)	+A (Wh)	-A (Wh)
Measured values #4	165.373	167.479	165.975	166.456
Analytic values	163.197	164.494	163.197	164.494
Relative error	1.33%	1.81%	1.70%	1.19%
VSB Analyzer - EUT1	Interval **200 ms**		Interval **15 min**	
One-hour energy	+A (Wh)	-A (Wh)	+A (Wh)	-A (Wh)
Average Meas. values	165.467	167.023	166.155	166.261
Analytic values	**163.197**	**164.494**	**163.197**	**164.494**
Average relative error	**1.39%**	**1.54%**	**1.81%**	**1.07%**

The second instrument was also a modified but commercial analyzer. The measured values are shown in Table 2..

Table 2. Values of EUT2 – PQ analyzer

PQ Analyzer - EUT2	Interval **20 ms**	
One-hour energy	+A (Wh)	-A (Wh)
Measured values #1	166.686	169.647
Analytic values	163.197	164.494
Relative error	2.14%	3.13%
PQ Analyzer - EUT2	Interval **20 ms**	
One-hour energy	+A (Wh)	-A (Wh)
Measured values #2	166,768	169,657
Analytic values	163,197	164,494
Relative error	2,19%	3,14%
PQ Analyzer - EUT2	Interval **20 ms**	
One-hour energy	+A (Wh)	-A (Wh)
Measured values #3	166.861	169.756
Analytic values	163.197	164.494
Relative error	2.25%	3.20%
PQ Analyzer - EUT2	Interval **20 ms**	
One-hour energy	+A (Wh)	-A (Wh)
Measured values #4	166.201	169.013
Analytic values	163.197	164.494
Relative error	1.84%	2.75%
PQ Analyzer - EUT2	Interval **20 ms**	
One-hour energy	+A (Wh)	-A (Wh)
Average Meas. values	166.583	169.472
Analytic values	**163.197**	**164.494**
Average relative error	**2.07%**	**3.03%**

The EUT2 analyzer shows that the relative error in all the measurements was a little higher than in EUT1. However, it stored 10× more values and recorded almost identical values in each measurement. The scattering of relative errors in individual tests was minimal. Only the 20 ms records shown were evaluated. Table 3. shows the values obtained from the three electricity meters in only three tests. Here, only the 15-minute interval records are shown.

Table 3. Values of EUT3 – Electricity meters

EUT3	**EUT3a – 15 min**		**EUT3b – 15 min**		**EUT3c – 15 min**	
One-hour energy	+A (Wh)	-A (Wh)	+A (Wh)	-A (Wh)	+A (Wh)	-A (Wh)
Measured values #1	9.79	18.34	13.09	19.43	13.14	11.88
Analytic values	163.20	164.49	163.20	164.49	163.20	164.49
Relative error	-94.00 %	-88.85 %	-91.98 %	-88.19 %	-91.95 %	-92.78 %
EUT3	**EUT3a – 15 min**		**EUT3b – 15 min**		**EUT3c – 15 min**	
One-hour energy	+A (Wh)	-A (Wh)	+A (Wh)	-A (Wh)	+A (Wh)	-A (Wh)
Measured values #2	6.58	18.03	14.43	20.76	14.56	13.34
Analytic values	163.20	164.49	163.20	164.49	163.20	164.49
Relative error	-95.97 %	-89.04 %	-91.16 %	-87.38 %	-91.08 %	-91.89 %
EUT3	**EUT3a – 15 min**		**EUT3b – 15 min**		**EUT3c – 15 min**	
One-hour energy	+A (Wh)	-A (Wh)	+A (Wh)	-A (Wh)	+A (Wh)	-A (Wh)
Measured values #3	3.66	16.51	14.95	20.76	12.60	11.36
Analytic values	163.20	164.49	163.20	164.49	163.20	164.49
Relative error	-97.76 %	-89.96 %	-90.84 %	-87.38 %	-92.28 %	-93.09 %
EUT3	**EUT3a – 15 min**		**EUT3b – 15 min**		**EUT3c – 15 min**	
One-hour energy	+A (Wh)	-A (Wh)	+A (Wh)	-A (Wh)	+A (Wh)	-A (Wh)
Avg. Meas. values	6.68	17.63	14.15	20.32	13.43	12.19
Analytic values	163.20	164.49	163.20	164.49	163.20	164.49
Avg. relative error	**-95.91 %**	**-89.29 %**	**-91.33 %**	**-87.65 %**	**-91.77 %**	**-92.59 %**

The recorded values show that the electricity meters make such enormous mistakes in recording values that they cannot even be considered measuring devices. The average relative mistake in measurement of the test courses was around 90%, which is an unacceptable level for measuring [10].

IV. Conclusions

The results of the measurements show that we have the technical means for measuring even small changes in the change of energy flow. The tested analyzers were suitable for measuring with sufficient accuracy [7]. By contrast, the standard electricity meters installed as the billing equipment in subscribing areas are entirely unsatisfactory if we consider quick regulation of change in energy flow for the purpose of optimizing consumption. If the device fulfils the condition of evaluating energy in a one half-period, then the results are sufficient. To do this though, standard electricity meters need greater processing power. Table 4. shows a comparison of all the tested devices.

The values in Table 4. show that the best results were achieved by the EUT1 device. The results from EUT2 were only slightly worse, and the worst were the EUT3 electricity meters. A medium relative error was observed in all electricity meters (EUT3a, EUT3b and EUT3c). The relative errors also demonstrate that the analyzers measured slightly higher values, while the electricity meters measured noticeably lower values, shown by the symbols on values in Table 4.. Times shown in Table 4. show the storage speed, or the intervals between individually stored values. The EUT1 and EUT2 devices recorded every 10 ms but only stored the middle value in the stated intervals. The electricity meters had a standard aggregate interval of one second and stored the middle value in 15-minute intervals. In such a long aggregate interval, rapid changes cannot be observed [11], [12].

Table 4. Average relative error – final table

Average relative error	EUT1 - VSB		EUT2 - PQ		EUT3 - EM	
Time / Energy	+A (kWh)	-A (kWh)	+A (kWh)	-A (kWh)	+A (kWh)	-A (kWh)
20 ms	-	-	2,07%	3,03%	-	-
200 ms	1,39%	1,54%	-	-	-	-
15 min	1,81%	1,07%	-	-	- 93,00 %	- 89,84 %

If electricity meters are to react to the processes in the distribution network, their entire concept must be changed. Much more powerful measuring devices similar to analyzers need to be used, though of course with limited functions aimed solely at quicker evaluation of measured energy. Even this would bring an enormous increase in their price, however. Their development would also need to be considered. Perhaps these kinds of precautions cannot be avoided in the future and manufacturers will have to adapt, although there may be another way.

If we keep the contemporary concept of electricity meters, the distribution network will have to adapt to the excess flow of energy from the user to the distributor's side, that is, to the excess flow of energy in very small intervals. Thus far, the second way seems more conceivable.

ACKNOWLEDGMENT

This work was supported by the Student Grant System of VSB – Technical University of Ostrava [grant number SP 2019/118].

REFERENCES

[1] R. Bodnar, A. Otcenasova, M. Regula, M. Repak. "Methodology for determination of the number of equipment malfunctions due to voltage quality." In: *2016 IEEE 16th International Conference on Environment and Electrical Engineering (EEEIC)* [online]. IEEE, 2016, 2016, pp. 1-6 [cit. 2019-04-14]. DOI: 10.1109/EEEIC.2016.7555685. ISBN: 978-1-5090-2320-2. Available at: http://ieeexplore.ieee.org/document/7555685/, [2016].

[2] M. Regula, A. Otcenasova, M. Roch, R. Bodnar, M. Repak. "Software for power quality monitoring in model smart grid with using LabView." In: *2016 ELEKTRO* [online]. IEEE, 2016, 2016, pp. 355-358 [cit. 2019-04-14]. DOI: 10.1109/ELEKTRO.2016.7512096. ISBN 978-1-4673-8698-2. Available at: http://ieeexplore.ieee.org/document/7512096/.

[3] *IEC 61000-4-15 ed. 2. Electromagnetic compatibility (EMC):* Part 4-15: Testing and measurement techniques – Flickermeter – Functional design specifications. Geneva : International Electrotechnical Commission, 2010. 83 p. ISBN 978-2-88912-076-5.

[4] M. Repak, A. Otcenasova, M. Regula, A. Boli, P. Belany. "Design of power quality analyzer." In: *2018 ELEKTRO* [online]. IEEE, 2018, 2018, pp. 1-6, [cit. 2019-04-14] DOI: 10.1109/ELEKTRO.2018.8398305. ISBN:978-1-5386-4759-2 Available at: https://ieeexplore.ieee.org/document/8398305/, [2018].

[5] T. Spielböck, A. Belan. "Design of a one-sided, impedance-based transmission line fault locator using line topology and source impedances.", *Electric Power Systems Research.*, 2018, vol. 161, p. 123-138. ISSN 0378-7796.

[6] K. Gorecki, M. Szmajda. "The power quality in low-power solar off-grid system.", Proceedings of: *16th International Conference on Harmonics and Quality of Power, ICHQP*, 2014, 244-248, ISBN: 978-146736487-4.

[7] D. Chathurangi, U. Jayatunga, M. Rathnayake, A. Wickramasinghe, A. Agalgaonkar, S. Perera. "Potential power quality impacts on LV distribution networks with high penetration levels of solar PV", Proceedings of: *18th International Conference on Harmonics and Quality of Power, ICHQP*, 2018, Ljubljana, Slovenia, ISBN: 978-153860517-2.

[8] IEC/TR 60725:2012 Consideration of reference impedances and public supply network impedances for use in determining the distrurbance characteristics of electrical equipment having a rated current ≤75A per phase, IEC, ISBN 978-2-83220-161-9.

[9] J. Drapela, R. Langella, J. Slezingr, A. Testa. "A Tunable Flickermeter to Account for Different Lamp Technologies." *IEEE Transactions on Power Delivery* [online]. 2017, 32(2), 872-880 [cit. 2019-04-14]. DOI: 10.1109/TPWRD.2016.2537699.,ISSN:0885-8977. Available at: http://ieeexplore.ieee.org/document/7423823/, [2016].

[10] J. Drapela, R. Langella, J. Slezingr, A. Testa. "Generalized lamp model for light flicker studies." *Electric Power Systems Research* [online]. 2018, 154, pp. 413-422, [cit. 2019-04-14]. DOI: 10.1016/j.epsr.2017.09.007.,ISSN:03787796. Available at: https://linkinghub.elsevier.com/retrieve/pii/S0378779617303747 .

[11] J. Novotny, J. Drapela, D. Topolanek. "Frequency response of revenue meters in measured active energy". In: *2016 17th International Conference on Harmonics and Quality of Power (ICHQP)* [online]. IEEE, 2016, 2016, pp. 524-529 [cit. 2019-04-14]. DOI: 10.1109/ICHQP.2016.7783309., ISBN:978-1-5090-3792-6. Available at: http://ieeexplore.ieee.org/document/7783309/, [2016].

[12] J. Moravek, J. Drapela, V. Wasserbauer, P. Mastny. "Power quality issues related to power flow control in systems with renewable energy micro sources." In: *2016 17th International Scientific Conference on Electric Power Engineering (EPE)* [online]. IEEE, 2016, 2016, pp. 1-6, [cit. 2019-04-14], DOI: 10.1109/EPE.2016.7521784. ISBN: 978-1-5090-0908-4. Available at: http://ieeexplore.ieee.org/document/7521784/, [2016].

Testing Behavior of Electricity Meters when Measuring Non-Standard Electricity Consumption and Deliveries

Richard Velicka
Department of Cybernetics and Biomedical Engineering
VSB – Technical University of Ostrava
Ostrava, Czech Republic
richard.velicka@vsb.cz

Petr Bilik
Department of Cybernetics and Biomedical Engineering
VSB – Technical University of Ostrava
Ostrava, Czech Republic
petr.bilik@vsb.cz

Ludvik Koval
Department of Cybernetics and Biomedical Engineering
VSB – Technical University of Ostrava
Ostrava, Czech Republic
ludvik.koval@vsb.cz

Jan Velicka
Department of Cybernetics and Biomedical Engineering
VSB – Technical University of Ostrava
Ostrava, Czech Republic
jan.velicka@vsb.cz

Abstract — **To test the correct function of meters, standard tests are determined according to valid legislation. However, there are conditions where the meter may not be maintained correctly, even if it is legally correct. Such situations occur, for example, in a one-phase is consumption (import) and delivery (export) in other phases or in fast changes between power consumption and supply. These are both static and dynamic states. A set of different tests have been proposed for these atypical states. The results were compared with the calculated values. Different manufacturers have been tested and connected for both direct and indirect measurements.**

Keywords: Electricity meter, Electricity meter tests, OBIS codes, Active power, Reactive power, Consumption, Deliveries, Imported energy, Exported energy.

I. Introduction

As devices designed to measure electrical energy consumption or supply, electricity meters are standardly checked in accordance with the methodology stated in the legislation. Testing is also conducted in accordance with this legislation. If we disregard single-phase electricity meters, we have a number of tests for three-phase electricity meters. These tests, however, specify consumed or supplied electricity in one direction and in static mode. In real use though, conditions may occur in which the system simultaneously supplies and consumes. For example, in a single phase, electricity is supplied while the other two consume electricity. The question is how the measuring system reacts to this condition. Similar conditions can occur even with phase shifts of current in all phases. This shift can be inductive or capacitive. Finally, an interesting state for measuring electrical energy is during relatively quick changes in current direction, whether in all phases or only some. The distribution system can react in many ways at the location where the electricity meter takes its measurements. In these unusual situations, their behavior must be tested. In cooperation with an external partner, this study involved these kinds of tests. The aim was to compare the behavior of a several types or several numbers of the same type of electricity meter under the non-standard conditions mentioned above. In order to conduct these measurements, testing and measurement of individual variables were specified by OBIS codes [1], [2], [3].

II. Materials and Methods

A. Testing methods

We designed twelve tests divided into four groups, in which two groups always consisted of static tests, the other two groups of dynamic tests. The structure of the tests was as follows:

1. Static tests:
   - a – Power consumption in all phases, test reference (**1a**)
   - b – Supply in all phases, test reference **1b**
   - c – Combined consumption and supply:
     - Two phase consumption, single phase supply (**1c1**)
     - Single phase consumption, two phase supply (**1c2**)
   - d – Interphase: consumption between phases L1 a L3 (**1d**)
2. Dynamic tests – alternating consumption and supply in one-minute intervals, in single or all phases:
   - a – Alternating in phase L1, phase L2 and L3 constant consumption, test reference (2a)
   - b – Alternating in all phases, test reference (**2b**)
   - c – Alternating in all phases simultaneously instead of one-minute intervals: consumption of three periods / supply of two periods, test reference (**2b**)
3. Static tests with phase shift:
   - a – Consumption in all phases with phase shift +20°, test reference (3a)
   - b – Consumption in all phases with phase shift -20°, test reference (**3b**)
4. Dynamic tests with phase shift – alternating consumption and supply in one-minute interval. Simultaneously in all phases:
   - a – Consumption in all phases with phase shift +20°, test reference (**4a**)
   - b – Consumption in all phases with phase shift -20°, test reference (**4b**)

B. Selection of observed variables – OBIS codes, registries

The values of energy are stored inside the electricity meter in registries that are pre-determined by OBIS codes (Figure 1.). The variables selected for evaluation were the following registries:

- Registers for active energy - Export:
  - 1.8.0 Positive active energy (A+) total 3~
  - 21.8.0 Positive active energy (A+) L1
  - 41.8.0 Positive active energy (A+) L2
  - 61.8.0 Positive active energy (A+) L3
- Registers for active energy - import:
  - 2.8.0 Negative active energy (A-) total 3~
  - 22.8.0 Negative active energy (A-) L1
  - 42.8.0 Negative active energy (A-) L2
  - 62.8.0 Negative active energy (A-) L3
- Registers for total reactive energy:
  - 5.8.0 Imported ind. reactive energy (I. Quad.) total 3~
  - 6.8.0 Imported cap. reactive energy (II. Quad.) total 3~
  - 7.8.0 Exported ind. reactive energy (III. Quad.) total 3~
  - 8.8.0 Exported cap. reactive energy (IV. Quad.) total 3~

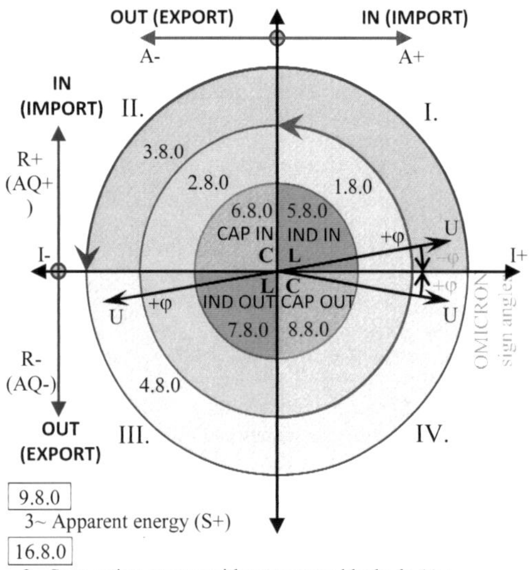

Figure 1. Energy registration diagram, OBIS codes

The diagram in Figure 1. shows the overall recorded energy in terms of current and voltage in the same manner as electricity meters in their registries. Because of the different labelling of angle symbols on electricity meters and in the OMICRON CMC 256+ power supply, this had to be taken into account in the tests. The conventional angle symbols of the power supply are also shown in the diagram [4], [5], [6], [7].

C. Measurement of the selected registries

The measurement tests were divided into three groups according to the parameter settings as follows: electricity meters for direct measurement, indirect measurement for a voltage of 230 V and indirect measurement for 57.7 V. The groups were designated with the letters A, B, and C and are referred to in the tables in this manner. Measurements were conducted by supplying the electricity meters with defined values according to the relevant tests described above. An Omicron CMC 256+ calibrator was used as a power supply to individually generate voltage and current in all three phases in the required values. The tests were programmed on a PC, which was connected to the OMICRON CMC 256+ power supply via Ethernet. Before commencing the tests, the condition of the registries was checked. After a specified time spent measuring, the values of the registries were checked again. Readings were conducted on an optical sensor and service software, which stored the values in text form in a folder. Each test was run for a minimum of four hours and each group separately. Forty-eight successful four-hour recordings were made. More were conducted, but some were repeated because the power supply had failed, or they had taken much longer as a result of other technical conditions. The block diagrams of connections are shown in the following figures. For direct measurements, current was increased from 5A to 20A using laboratory transformers (Figure 2.). For indirect measurements, current was measured directly at the power supply output (Figure 3.) [7], [8].

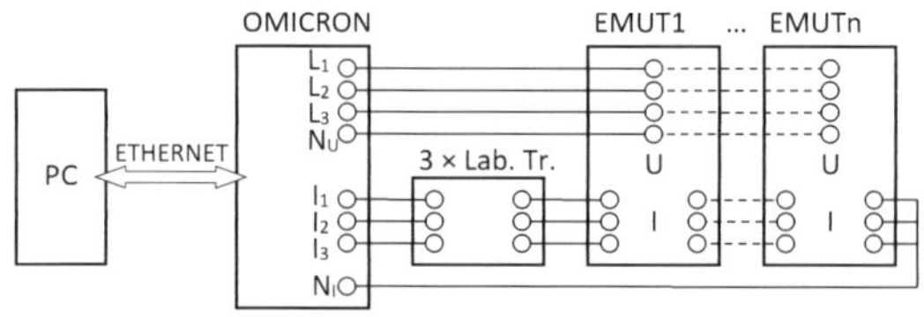

Figure 2. EMUTs direct measurement schematic diagram

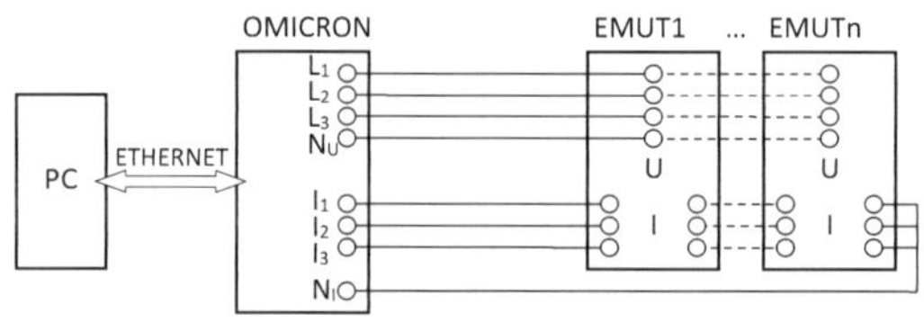

Figure 3. EMUTs non-direct measurement schematic diagram

The connections shown are principal connections. For some electricity meters, connections were selected with regard to the impossibility of separating the voltage component from the current component [9], [10].

D. Referential energy values

For each of the three groups of devices (A, B, and C) and the corresponding twelve tests, referential values were calculated for the registries. Deviations to these values were then evaluated in the form of absolute and relative errors. The aim was not to specifically identify all the errors in the measurement chain, but to compare the properties of the electricity meters. The following results should therefore be viewed from a relative point of view. The following equations define the calculation of referential values of power for the tests. Because of the parameter settings in the electricity meters, the constants were modified to produce resulting values in kWh or kvarh.

Tests: 1a, 1b, 1c1, 1c2, 1d, 3a, 3b

$$A+ = U_p \cdot (I+) \cdot \cos(\varphi)/1000 \cdot t \text{ (kWh)} \quad (1)$$

$$A- = U_p \cdot (I-) \cdot \cos(\varphi)/1000 \cdot t \text{ (kWh)} \quad (2)$$

Tests: 1a, 1b, 1c1, 1c2, 1d, 2a, 2b, 2c, 3a, 3b, 4a, 4b

$$\text{total } A+ = (\text{L1 } A+) + (\text{L2 } A+) + (\text{L3 } A+) \text{ (kWh)} \quad (3)$$

$$\text{total } A- = (\text{L1 } A-) + (\text{L2 } A-) + (\text{L3 } A-) \text{ (kWh)} \quad (4)$$

Tests: 1d

$$\text{L1 } A+ = \frac{1}{2}\cdot U_l \cdot (I+) \cdot \cos(\varphi)/1000 \cdot t \text{ (kWh)} \quad (5)$$

$$\text{L2 } A+ = \frac{1}{2}\cdot U_l \cdot (I+) \cdot \cos(\varphi)/1000 \cdot t \text{ (kWh)} \quad (6)$$

Tests: 2a, 2b, 4a, 4b

$$A+ = \frac{1}{2}\cdot [U_p \cdot (I+) \cdot \cos(\varphi)/1000 \cdot t] \text{ (kWh)} \quad (7)$$

$$A- = \frac{1}{2}\cdot [U_p \cdot (I-) \cdot \cos(\varphi)/1000 \cdot t] \text{ (kWh)} \quad (8)$$

Test: 2c

$$A+ = \frac{3}{5}\cdot [U_p \cdot (I+) \cdot \cos(\varphi)/1000 \cdot t] \text{ (kWh)} \quad (9)$$

$$A- = \frac{2}{5}\cdot [U_p \cdot (I-) \cdot \cos(\varphi)/1000 \cdot t] \text{ (kWh)} \quad (10)$$

Tests: 3a, 3b, 4a, 4b

$$Q+ = 3 \cdot U_p \cdot (I+) \cdot \sin(\varphi)/1000 \cdot t \text{ (kvarh)} \quad (11)$$

$$Q- = 3 \cdot U_p \cdot (I-) \cdot \sin(\varphi)/1000 \cdot t \text{ (kvarh)} \quad (12)$$

The positive variables in the equations were given a "+" sign, and similarly negative variables a "-" sign. Phase voltage was given the symbol U_p, and phase-to-phase voltage (line voltage) U_l.

The referential results of the tests of the electricity meters in an Aron connection (group C) were calculated by our software in order to calculate the energy for static power direction. The other two tests, (1c3) and (1c4), also added here, enabled improved control of the reactive power registries.

Table 1. Reference values of group A

konst. k = 1						OBIS register											
Group	Reference values - inputs					1.8.0	21.8.0	41.8.0	61.8.0	2.8.0	22.8.0	42.8.0	62.8.0	5.8.0	6.8.0	7.8.0	8.8.0
A	k × t	Up	Uc	I	cos φ	total A+	L1 A+	L2 A+	L3 A+	total A-	L1 A-	L2 A-	L3 A-	IN IND	IN CAP	OUT IND	OUT CAP
TEST	(h)	(V)	(V)	(A)	(-)	(kWh)	(kWh)	(kWh)	(kWh)	(kWh)	(kWh)	(kWh)	(kWh)	(kvarh)	(kvarh)	(kvarh)	(kvarh)
A 1a	4	230	-	20	1.00	55.20	18.40	18.40	18.40	0.00	0.00	0.00	0.00	0.00	0.00	0.00	0.00
A 1b	4	230	-	20	1.00	0.00	0.00	0.00	0.00	55.20	18.40	18.40	18.40	0.00	0.00	0.00	0.00
A 1c1	4	230	-	20	1.00	36.80	18.40	18.40	0.00	18.40	0.00	0.00	18.40	0.00	0.00	0.00	0.00
A 1c2	4	230	-	20	1.00	18.40	18.40	0.00	0.00	36.80	0.00	18.40	18.40	0.00	0.00	0.00	0.00
A 1d	4	-	400	20	1.00	32.00	16.00	0.00	16.00	0.00	0.00	0.00	0.00	0.00	0.00	0.00	0.00
A 2a	4	230	-	20	1.00	46.00	9.20	18.40	18.40	9.20	9.20	0.00	0.00	0.00	0.00	0.00	0.00
A 2b	4	230	-	20	1.00	27.60	9.20	9.20	9.20	27.60	9.20	9.20	9.20	0.00	0.00	0.00	0.00
A 2c	4	230	-	20	1.00	33.12	11.04	11.04	11.04	22.08	7.36	7.36	7.36	0.00	0.00	0.00	0.00
A 3a	4	230	-	20	0.94	51.87	17.29	17.29	17.29	0.00	0.00	0.00	0.00	0.00	0.00	0.00	18.88
A 3b	4	230	-	20	0.94	51.87	17.29	17.29	17.29	0.00	0.00	0.00	0.00	18.88	0.00	0.00	0.00
A 4a	4	230	-	20	0.94	25.94	8.65	8.65	8.65	25.94	8.65	8.65	8.65	0.00	9.44	0.00	9.44
A 4b	4	230	-	20	0.94	25.94	8.65	8.65	8.65	25.94	8.65	8.65	8.65	9.44	0.00	9.44	0.00

The values shown in Table 1. show the results of tests of electricity meters for direct measurement. Although the electricity meters used a larger current range, 20 A was selected for a greater accuracy during measurement. Higher currents showed a greater current distortion because the current transformers were overloaded.

Table 2. Reference values of group B

konst. k = 1						OBIS register											
Group	Reference values - inputs					1.8.0	21.8.0	41.8.0	61.8.0	2.8.0	22.8.0	42.8.0	62.8.0	5.8.0	6.8.0	7.8.0	8.8.0
B	k × t	Up	Uc	I	cos φ	total A+	L1 A+	L2 A+	L3 A+	total A-	L1 A-	L2 A-	L3 A-	IN IND	IN CAP	OUT IND	OUT CAP
TEST	(h)	(V)	(V)	(A)	(-)	(kWh)	(kWh)	(kWh)	(kWh)	(kWh)	(kWh)	(kWh)	(kWh)	(kvarh)	(kvarh)	(kvarh)	(kvarh)
B 1a	4	230	-	5	1.00	13.80	4.60	4.60	4.60	0.00	0.00	0.00	0.00	0.00	0.00	0.00	0.00
B 1b	4	230	-	5	1.00	0.00	0.00	0.00	0.00	13.80	4.60	4.60	4.60	0.00	0.00	0.00	0.00
B 1c1	4	230	-	5	1.00	9.20	4.60	4.60	0.00	4.60	0.00	0.00	4.60	0.00	0.00	0.00	0.00
B 1c2	4	230	-	5	1.00	4.60	4.60	0.00	0.00	9.20	0.00	4.60	4.60	0.00	0.00	0.00	0.00
B 1d	4	-	400	5	1.00	8.00	4.00	0.00	4.00	0.00	0.00	0.00	0.00	0.00	0.00	0.00	0.00
B 2a	4	230	-	5	1.00	11.50	2.30	4.60	4.60	2.30	2.30	0.00	0.00	0.00	0.00	0.00	0.00
B 2b	4	230	-	5	1.00	6.90	2.30	2.30	2.30	6.90	2.30	2.30	2.30	0.00	0.00	0.00	0.00
B 2c	4	230	-	5	1.00	8.28	2.76	2.76	2.76	5.52	1.84	1.84	1.84	0.00	0.00	0.00	0.00
B 3a	4	230	-	5	0.94	12.97	4.32	4.32	4.32	0.00	0.00	0.00	0.00	0.00	0.00	0.00	4.72
B 3b	4	230	-	5	0.94	12.97	4.32	4.32	4.32	0.00	0.00	0.00	0.00	4.72	0.00	0.00	0.00
B 4a	4	230	-	5	0.94	6.48	2.16	2.16	2.16	6.48	2.16	2.16	2.16	0.00	2.36	0.00	2.36
B 4b	4	230	-	5	0.94	6.48	2.16	2.16	2.16	6.48	2.16	2.16	2.16	2.36	0.00	2.36	0.00

The values for group B shown in Table 2. correspond to the generated values directly from the OMICRON CMC 256+ power supply. Phase voltage and currents were generated independently of each other. Their mutual relationship is given by the software-controlled power supply. A large amount of power could be generated in this manner without exposing it to a real load [3], [7].

Table 3. Reference values of group C

konst. k = 20						OBIS register											
Group	Reference values - inputs					1.8.0	21.8.0	41.8.0	61.8.0	2.8.0	22.8.0	42.8.0	62.8.0	5.8.0	6.8.0	7.8.0	8.8.0
C	k × t	Up	Uc	I	cos φ	total A+	L1 A+	L2 A+	L3 A+	total A-	L1 A-	L2 A-	L3 A-	IN IND	IN CAP	OUT IND	OUT CAP
TEST	(h)	(V)	(V)	(A)	(-)	(kWh)	(kWh)	(kWh)	(kWh)	(kWh)	(kWh)	(kWh)	(kWh)	(kvarh)	(kvarh)	(kvarh)	(kvarh)
C 1a	80	57.7	-	5	1.00	69.28	23.09	23.09	23.09	0.00	0.00	0.00	0.00	0.00	0.00	0.00	0.00
C 1b	80	57.7	-	5	1.00	0.00	0.00	0.00	0.00	69.28	23.09	23.09	23.09	0.00	0.00	0.00	0.00
C 1c1	80	57.7	5A,-,2.5A		5.8.0	51.96	23.09	17.32	11.55	0.00	0.00	0.00	0.00	10.00	0.00	0.00	0.00
C 1c2	80	57.7	-2.5A,-,5A		8.8.0	28.84	0.00	5.74	23.09	11.55	11.55	0.00	0.00	0.00	0.00	0.00	30.00
C 1c3	80	57.7	2.5A,-,-5A		6.8.0	11.55	11.55	0.00	0.00	28.84	0.00	5.74	23.09	0.00	30.00	0.00	0.00
C 1c4	80	57.7	-5A,-,2.5A		7.8.0	11.55	0.00	0.00	11.55	28.84	23.09	5.74	0.00	0.00	0.00	30.00	0.00
C 1d	80	-	100	5	1.00	40.00	20.00	0.00	20.00	0.00	0.00	0.00	0.00	0.00	0.00	0.00	0.00
C 2a	80	57.7	-	5	1.00	46.19	11.55	11.55	23.09	11.55	11.55	0.00	0.00	20.00	0.00	0.00	0.00
C 2b	80	57.7	-	5	1.00	34.64	11.55	11.55	11.55	34.64	11.55	11.55	11.55	0.00	0.00	0.00	0.00
C 2c	80	57.7	-	5	1.00	41.57	13.86	13.86	13.86	27.71	9.24	9.24	9.24	0.00	0.00	0.00	0.00
C 3a	80	57.7	-	5	0.94	65.10	21.70	21.70	21.70	0.00	0.00	0.00	0.00	0.00	0.00	0.00	23.70
C 3b	80	57.7	-	5	0.94	65.10	21.70	21.70	21.70	0.00	0.00	0.00	0.00	23.70	0.00	0.00	0.00
C 4a	80	57.7	-	5	0.94	32.55	10.85	10.85	10.85	32.55	10.85	10.85	10.85	0.00	11.85	0.00	11.85
C 4b	80	57.7	-	5	0.94	32.55	10.85	10.85	10.85	32.55	10.85	10.85	10.85	11.85	0.00	11.85	0.00

In group C (Table 3.), we can observe that the generated phase voltages corresponded to the indirect measurements, as with current. For tests in an Aron connection, the currents were adjusted to half in some phases. (i.e. to 2.5 A). Test (1c) added other tests with a combination of currents. The behavior of electricity meters for measuring reactive energy in all summary registries in all quadrants could therefore be observed.

III. Results

The results of the analysis are given in several tables comparing the recorded referential values. Nine electricity meters underwent testing and were separated according to their parameters into groups. The results were sorted into group tables according to the electricity meter group, OBIS codes and the respective test. It is not possible to provide all the results, and only samples of individual evaluations are given. Table 4. shows an evaluation of errors according to group and respective test. The tables of results are numerous, only one with the results of tests (4a) and (4b) in group B was selected as a representative example.

Table 4. Example for results by group

TEST	Group		OBIS registrers											
4a, 4b	B		1.8.0	21.8.0	41.8.0	61.8.0	2.8.0	22.8.0	42.8.0	62.8.0	5.8.0	6.8.0	7.8.0	8.8.0
			total A+	L1 A+	L2 A+	L3 A+	total A-	L1 A-	L2 A-	L3 A-	IN IND	IN CAP	OUT IND	OUT CAP
			(kWh)	(kWh)	(kWh)	(kWh)	(kWh)	(kWh)	(kWh)	(kWh)	(kvarh)	(kvarh)	(kvarh)	(kvarh)
B 4a	REFERENCE VALUES		6.48	2.16	2.16	2.16	6.48	2.16	2.16	2.16	0.00	2.36	0.00	2.36
	EMUT 3	Increment in registry	6.43	2.14	2.14	2.14	6.43	2.14	2.14	2.14	0.00	0.00	0.00	2.33
		Absolute error	-0.06	-0.02	-0.02	-0.02	-0.06	-0.02	-0.02	-0.02	0.00	-2.36	0.00	-0.03
		Relative error (%)	**-0.88**	**-0.89**	**-0.85**	**-0.85**	**-0.88**	**-0.94**	**-0.89**	**-0.89**	-	**-100**	-	**-1.31**
	EMUT 7	Increment in registry	6.46	2.15	2.15	2.15	6.46	2.15	2.15	2.15	0.00	2.36	0.00	2.35
		Absolute error	-0.03	-0.01	-0.01	-0.01	-0.02	-0.01	-0.01	-0.01	0.00	0.00	0.00	0.00
		Relative error (%)	**-0.39**	**-0.39**	**-0.41**	**-0.36**	**-0.37**	**-0.38**	**-0.39**	**-0.34**	-	**-0.19**	-	**-0.21**
B 4b	REFERENCE VALUES		6.48	2.16	2.16	2.16	6.48	2.16	2.16	2.16	2.36	0.00	2.36	0.00
	EMUT 3	Increment in registry	6.42	2.14	2.14	2.14	6.42	2.14	2.14	2.14	2.35	0.00	2.35	0.00
		Absolute error	-0.06	-0.02	-0.02	-0.02	-0.06	-0.02	-0.02	-0.02	-0.01	0.00	-0.01	0.00
		Relative error (%)	**-0.97**	**-0.99**	**-0.99**	**-0.94**	**-0.97**	**-0.99**	**-0.99**	**-0.99**	**-0.63**	-	**-0.63**	-
	EMUT 7	Increment in registry	6.46	2.15	2.15	2.15	6.46	2.15	2.15	2.15	2.35	0.00	2.35	0.00
		Absolute error	-0.03	-0.01	-0.01	-0.01	-0.03	-0.01	-0.01	-0.01	-0.01	0.00	-0.01	0.00
		Relative error (%)	**-0.41**	**-0.42**	**-0.43**	**-0.39**	**-0.39**	**-0.39**	**-0.41**	**-0.37**	**-0.57**	-	**-0.56**	-

Table 5. shows the relative errors as an example in evaluating tests (4b) on all tested electricity meters. The dark grey areas represent the registries, which the electricity meters did not measure. These results were organized into twelve tables (14 tables if tests (1c3) and (1c4) in an Aron connection are included). The resultant values were organized into tables of values from the registries and processed using conditional formatting. This enabled the values that exceeded limits to be highlighted. In the presented Table 5., the limit was set at ±10 %. Only one value deviated from this.

Table 5. Example for results by test

TEST	OBIS registers											
4b	1.8.0	21.8.0	41.8.0	61.8.0	2.8.0	22.8.0	42.8.0	62.8.0	5.8.0	6.8.0	7.8.0	8.8.0
Group A,B,C	total A+	L1 A+	L2 A+	L3 A+	total A-	L1 A-	L2 A-	L3 A-	IN IND	IN CAP	OUT IND	OUT CAP
EMUT	(%)	(%)	(%)	(%)	(%)	(%)	(%)	(%)	(%)	(%)	(%)	(%)
1	-3.61	4.10	4.10	4.10	0.25	4.10	4.10	-7.46	-	-	-	-
2	-0.08	-0.07	-0.12	-0.06	-0.05	-0.03	-0.11	-0.03	-	-	-	-
3	-0.97	-0.99	-0.99	-0.94	-0.97	-0.99	-0.99	-0.99	-0.63	-	-0.63	-
4	-1.70	1.38	1.38	1.38	-1.70	1.38	1.38	1.38	-7.16	-	-7.16	-
5	0.25	-7.46	4.10	-7.46	0.25	4.10	-7.46	4.10	-4.66	-	5.93	-
6	0.25	-	-	-	-	-	-	-	-	-	-	-
7	-0.41	-0.42	-0.43	-0.39	-0.39	-0.39	-0.41	-0.37	-0.57	-	-0.56	-
8	-1.16	-	-	-	-1.11	-	-	-	-1.27	-	-1.25	-
9	**99.5**	-0.37	-0.07	-0.33	-0.28	-0.45	-0.13	-0.28	-	-	-	-

Table 6. shows an evaluation of results concerning the OBIS registries. Because twelve of these registries were measured, there are twelve corresponding tables in total. The cells in each table indicating that the electricity meter did not react or did not support the given registry are highlighted. Evaluation was done automatically according to the pre-determined format. The values that did not correspond to any criteria are highlighted. The values suggest that the electricity meters struggled the most with test (2c), which quickly changed current direction in the given range of periods. The limit was set at ±5 %.

Table 6. Example for results by OBIS register

OBIS	TEST													
61.8.0	1a	1b	1c1	1c2	1c3	1c4	1d	2a	2b	2c	3a	3b	4a	4b
L3 A+ (%)	Y/N	Y/N	Y/N	Y/N	Y/N	Y/N	Y/N	Y/N	Y/N	Y/N	Y/N	Y/N	Y/N	Y/N
EMUT	(%)	(%)	(%)	(%)	(%)	(%)	(%)	(%)	(%)	(%)	(%)	(%)	(%)	(%)
1	3.26	-	-	-	-	-	0.00	3.26	-2.17	**-67.79**	-1.68	-1.68	**-7.46**	4.10
2	0.02	-	-	-	-	-	0.49	0.03	0.11	**-50.14**	0.05	0.00	-0.05	-0.06
3	-0.07	-	-	-	-	-	-0.08	**-100.0**	-1.00	**-66.68**	-0.01	-0.11	-0.85	-0.94
4	-0.41	-	**-13.40**	-4.74	-	3.92	0.00	-0.41	3.92	**-99.96**	1.38	1.38	1.38	1.38
5	-2.17	-	-	-	-	-	0.00	-2.17	-2.17	**-67.79**	-1.68	4.10	4.10	**-7.46**
6	-	-	-	-	-	-	-	-	-	-	-	-	-	-
7	0.01	-	-	-	-	-	0.00	**-100.0**	-0.40	**66.70**	-0.01	0.02	-0.36	-0.39
8	-	-	-	-	-	-	-	-	-	-	-	-	-	-
9	-0.27	-	-	-	-	-	0.31	-0.20	0.37	**7.84**	-0.22	-0.25	-0.04	-0.33

IV. Conclusions

The final result of all tests was the table of successful tested electricity meters. Each electricity meter underwent all the tests, even though it was already known that some did not support some of the registries. These results were not included in the overall evaluation. If a method of evaluation is designed with regard to the capabilities of the electricity meters, the success rate of the electricity meters must be included with regard to their capabilities, relative error limits and referential values. A Table 7. was created for the 5% relative error that shows the relative success of the tested electricity meter with regard to the number of performed tests. At the end, the demonstrated average success rate per number of performed tests was shown. The capabilities of the electricity meters under non-standard conditions, which, however, occur typically in the network, are somehow quantified in this manner. It can be concluded that the electricity meters were not designed for quick changes in energy flow. These conditions though occur more and more frequently in order to best balance consumption and supply at the given location of consumption. The electricity meters are built for the aggregate one-second interval, which might seem quite a long time today for measurements to be correct, especially when regulators handle changes in the current flow at the level of a few periods and the readings on electricity meters cannot be relied on as correct. Some of the tested electricity meters were not successful in dealing with different directions of energy flow in individual phases. Again, this is a consequence of the electricity meter's construction, which does not consider these conditions at the time of their design. Nevertheless, it is good to know that some of the electricity meters were capable of correctly registering these states [3], [11].

Table 7. Measurement success of EMUTs at tests

TESTS	OBIS registry												MEAS. Success Total
1a - 4b	1.8.0	21.8.0	41.8.0	61.8.0	2.8.0	22.8.0	42.8.0	62.8.0	5.8.0	6.8.0	7.8.0	8.8.0	
REG. Success	total A+	L1 A+	L2 A+	L3 A+	total A-	L1 A-	L2 A-	L3 A-	IN IND	IN CAP	OUT IND	OUT CAP	
EMUT	(%)	(%)	(%)	(%)	(%)	(%)	(%)	(%)	(%)	(%)	(%)	(%)	(%)
1	90.9	90.9	77.8	77.8	87.5	50.0	66.7	57.1					**75**
2	90.9	81.8	88.9	88.9	87.5	83.3	83.3	85.7					**86**
3	81.8	90.9	77.8	77.8	75.0	83.3	83.3	85.7	50	0	100	100	**75**
4	92.3	81.8	80.0	83.3	88.9	75.0	85.7	83.3	100	50	50	100	**81**
5	72.7	90.9	88.9	77.8	87.5	83.3	66.7	85.7	100	0	0	50	**67**
6	63.6												**64**
7	81.8	90.9	77.8	77.8	62.5	83.3	83.3	85.7	100	100	100	100	**87**
8	61.5				44.4				100	100	100	100	**84**
9	45.5	90.9	88.9	87.5	87.5	83.3	83.3	85.7					**82**

An evaluation of the overall results presented in Table 7. show that none of the electricity meters were 100% successful in measuring in all tests. The results of the dynamic tests were important factors, especially the fast test (2c), in which none of the meters succeeded. The meters that achieved a success rate of more than 80% can be considered very good. However, this might not be sufficient in the future and a re-evaluation of the electronics design in electricity meters may become necessary in order to adapt them to measurements in an energy network that is also continually developing.

Confidentiality of undisclosed facts does not permit the specific devices or their parameter settings to be shown.

Acknowledgment

This work was supported by the Student Grant System of VSB – Technical University of Ostrava [grant number SP 2019/118].

References

[1] R. Bodnar, A. Otcenasova, M. Regula, M. Repak. "Methodology for determination of the number of equipment malfunctions due to voltage quality." In: *2016 IEEE 16th International Conference on Environment and Electrical Engineering (EEEIC)* [online]. IEEE, 2016, 2016, pp. 1-6 [cit. 2019-04-14]. DOI: 10.1109/EEEIC.2016.7555685. ISBN: 978-1-5090-2320-2. Available at: http://ieeexplore.ieee.org/document/7555685/.

[2] M. Regula, A. Otcenasova, M. Roch, R. Bodnar, M. Repak. "Software for power quality monitoring in model smart grid with using LabView." In: *2016 ELEKTRO* [online]. IEEE, 2016, 2016, pp. 355-358 [cit. 2019-04-14]. DOI: 10.1109/ELEKTRO.2016.7512096. ISBN 978-1-4673-8698-2. Available at: http://ieeexplore.ieee.org/document/7512096/

[3] *EN 50160 Ed.3 Voltage characteristics of electricity supplied by public distribution systems.* Brussels: Eurepean Committee for Electrotechnical Standardization, 2010. 20 p..

[4] M. Konicek, A. Belan, B. Cintula, D. Viglas. "Reactive power regulation in photovoltaic power plants from the view of voltage in the distribution network." (Conference Paper), *Proceedings of the 8th International Scientific Symposium on Electrical Power Engineering, ELEKTROENERGETIKA 2015*, Stara Lesna; Slovakia; 16th September 2015 through 18th September 2015, pp. 472-475; ISBN: 978-80-553-2187-5.

[5] *IEC 61000-4-15 ed2.0. Electromagnetic compatibility (EMC)*: Part 4-15: Testing and measurement techniques – Flickermeter – Functional design specifications. Geneva: International Electrotechnical Commission, 2010. 83 p. ISBN 978-2-88912-076-5.

[6] T. Spielböck, A. Belan. "Design of a one-sided, impedance-based transmission line fault locator using line topology and source

impedances.", *Electric Power Systems Research.*, 2018, vol. 161, p. 123-138. ISSN 0378-7796.

[7] M. Regula, A. Otcenasova, M. Roch. "SCADA system with power quality monitoring in Smart Grid model.", *Proceedings of IEEE 16th International Conference on Environment and Electrical Engineering (EEEIC)*, 2016, Florence, Italy, ISBN: 978-1-5090-2320-2.

[8] M. Repak, A. Otcenasova, M. Regula, A. Boli, P. Belany. "Design of power quality analyzer." In: *2018 ELEKTRO* [online]. IEEE, 2018, 2018, pp. 1-6 [cit. 2019-04-14]. DOI: 10.1109/ELEKTRO.2018.8398305. ISBN 978-1-5386-4759-2. Available at: https://ieeexplore.ieee.org/document/8398305/ .

[9] V. Kus, T. Josefova. "Education strategy regarding the electromagnetic compatibility at low-frequency.", *Revue Roumaine des Sciences Techniques Serie Electrotechnique et Energetique*, vol. 62, pp. 48-52, 2016, ISSN: 00354066.

[10] V. Kus, T. Josefova. "Study of the Input Current Hharmonic Distortion of Voltage-Source Active Rectifiers." *Revue Roumaine des Sciences Techniques Serie Electrotechnique et Energetique*, Vol. 62, Issue: 2, pp. 185- 191, Published: 4/2017.

[11] IEC 61000-4-30 ed.2 Electromagnetic compatibility (EMC): Part 4-30: Testing and measurement techniques – Power quality measurement methods. Geneva: International Electrotechnical Commission, 2008, 134 p., ISBN: 2-8318-1002-0.

Verification of Electrical Properties of Photovoltaic Panels

Richard Velicka
Department of cybernetics and biomedical engineering
VSB – Technical university Ostrava
Ostrava, Czech Republic
richard.velicka@vsb.cz

Pavel Santarius
Department of cybernetics and biomedical engineering
VSB – Technical university Ostrava
Ostrava, Czech Republic
pavel.santarius@vsb.cz

Jan Velicka
Department of cybernetics and biomedical engineering
VSB – Technical university Ostrava
Ostrava, Czech Republic
jan.velicka@vsb.cz

Petr Bilik
Department of cybernetics and biomedical engineering
VSB – Technical university Ostrava
Ostrava, Czech Republic
petr.bilik@vsb.cz

Abstract — **Photovoltaic panels available on the market are produced in most cases in China. Their documentation lists panel parameters such as maximum power and maximum current. We decided to verify these parameters. Verification equipment has been built up to enable these parameters to be verified with respect to the actual panel illumination intensity. Different ways of loading panels, such as dependence on constant power output, constant supply current or constant load resistance, have been measured. The results were compared with the manufacturer's data.**

Keywords:* *Solar module, Photovoltaic power plant, Electronic load, Solar module testing, Solar cells, P-V curves, I-V curves

I. Introduction

Currently, most photovoltaic panels imported into the European Union (and therefore also the Czech Republic) come from China. These panels generally lack the documentation that specifies the power information, voltage limits and current requirements of the panels or information about the various electrical values. Usually no information is provided about the light intensity of the panel. In some cases, detailed information can be found on the manufacturer's website. We decided to verify the manufacturer's stated parameters for the photovoltaic panels in terms of solar radiation intensity. In order to check this information, electrical parameters and the power per unit of area surface of the photovoltaic panel were measured. The results obtained were compared with the values given by the panel's manufacturer. Specific performance values were compared at different levels of power load on the panel in terms of light intensity, i.e. the power supplied from the sun. A method using electronic load was used to facilitate accurate measurement of the operating points. Because electronic loading permitted multiple modes of use during performance-testing, these modes were also then compared. The resultant values were then compared with the parameters obtained from the manufacturer.

II. Materials and Methods

A. The tested photovoltaic panel

A polycrystalline photovoltaic panel was used for testing. The photovoltaic panel, or solar panel, comprised 72 solar cells. The manufacturer supplies these solar panels with a power range of 330 W up to 335 W, graded in 5 W intervals. Table 1 lists the parameters of the selected 330 W solar panel.

Table 1. The solar module details

Electrical Data		
Peak Output Tolerance	0 – +5	
Peak Power – Pmax	**330 W**	**1000 W/m²**, 25°C Cell
Maximum Power Voltage – Vmp	37.8 V	
Maximum Power Current – Imp	8.73 A	
Open Circuit Voltage – Voc	46.1 V	
Short Circuit Current – Isc	9.24 A	
Module Efficiency	17 %	
Peak Power – Pmax	**239 W**	**800 W/m²**, 20°C Cell
Maximum Power Voltage – Vmp	34.6 V	
Maximum Power Current – Imp	6.91 A	
Open Circuit Voltage – Voc	41.3 V	
Short Circuit Current – Isc	7.49 A	
Mechanical Data		
Solar Cells	Polycrystalline 156.75 x 156,75 mm, 6"	
Cell Orientation	72 cells (6 x 12)	
Module Dimens.	1960 x 992 x 40 mm	
Weight	22.5 kg	
Frame	Silver Anodized Aluminium Alloy	
Junction Box	IP67 or IP 68 rated	
Connector	MC4	
Temperature and Maximum Ratings		
NOCT	44°C (±2°C) – Nominal Oper. Cell Temp.	
OT	-40 – +85°C – Operational Temperature	
Max Fuse	15A	

B. Testing the photovoltaic panel (Solar module)

A connection supplying an electronic load was the most suitable method for measuring the various characteristics of the solar module. Electronic load enables voltage, current resistance and power to be measured. Some of the values can be set according to the load's mode and others will adapt to this established mode. The modes used in this test were constant resistance (CR), constant current (CC), constant power (CP) and constant voltage (CV). In order to best measure the parameters given by the manufacturer, it was also important to measure external variables. These were measured using a special module with sensors for an ENA-520 analyzer, which measures light intensity with a pyranometer and ambient temperature and relative humidity. This module provides information about these variables, including the solar panel's current and voltage and the current on the voltage after the changer on the output side of the photovoltaic power station. Only one input measuring single-phase current and voltage was used on the solar panel to provide readings.

The values were displayed on a measuring module with a LEM sensor. The electrical parameters measured with the analyzer were single-phase only for testing purposes and compared with the values measured directly on the electronic load (type: Aim TTi LD400P, Programmable DC Electronic Load). Figure 1. shows the measuring diagram of the experiment. It is important to note that measurement should be done with continuous and a sufficient amount of sunlight. It is best to measure on a clear sunny day around noon, when intensity is the highest. We were able conduct measurements in the right conditions, even in terms of intensity, which was 800 – 1000 W/m^2. Most measurements were conducted under an intensity of around 950 W/m^2. Measuring intensity, however, depended on correct placement of the pyranometer in the sun, and likewise the photovoltaic panel had to be tilted to face the sun as accurately as possible. Otherwise, a noticeable decrease of power from the panel could have occurred [1],[2].

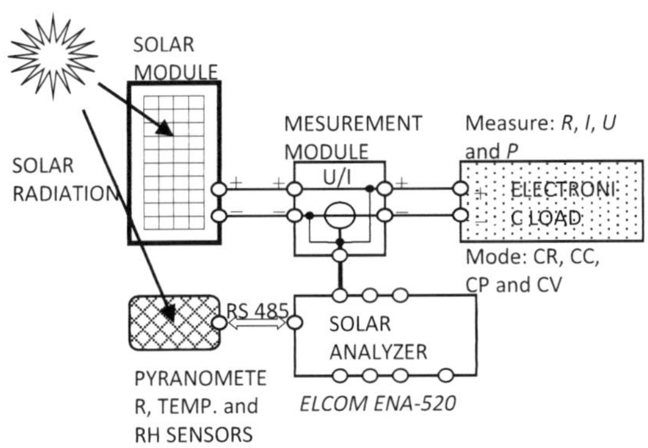

Figure 1. The solar module measurement

C. *Electronic load mode – constant resistance (CR)*

In constant resistance mode, the amount of the resistance load can be set from the maximum value (Hi) to a value of 2 Ω. If the resistance is set too high, it is idle, and the load does not draw any current. As resistance decreases, current and the supplied power rises. However, the voltage in the module also decreases and the relationship, therefore, is not linear. In this mode, the full power potential of the panel cannot usually be reached. Table 2. and the graphs in Figures 2., 3. and 4. show the measured relationships in CR mode.

Table 2. Data for CR mode

R	*U*(*R*)	*I*(*R*)	*P*(*R*)	*E*
Ohms	Volts	Amps	Watts	W/m^2
2.00	32.90	7.38	240.00	876.00
2.00	36.84	7.34	255.00	962.00
3.00	37.51	5.77	218.40	973.00
3.00	37.40	5.84	220.00	964.00
4.00	38.82	4.68	181.40	960.00
5.00	39.94	3.96	156.40	963.00
6.00	39.55	3.88	153.50	962.00
10.00	41.39	2.14	88.00	962.00
100.00	43.65	0.24	10.20	974.00

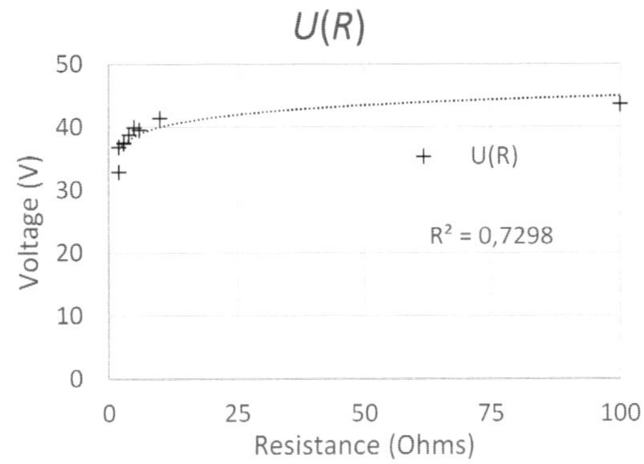

Figure 2. Voltage – Resistance characteristic

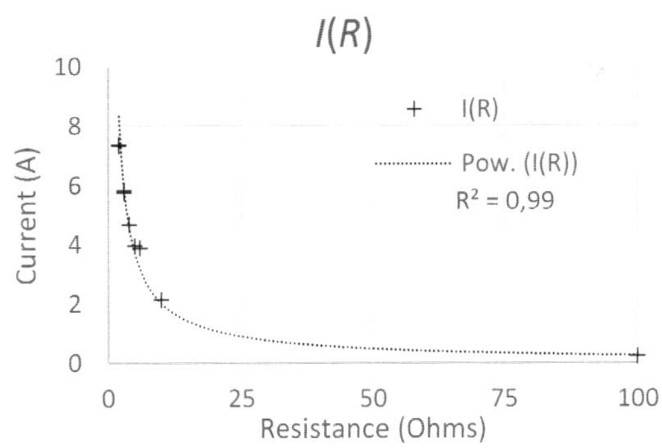

Figure 3. Current – Resistance characteristic

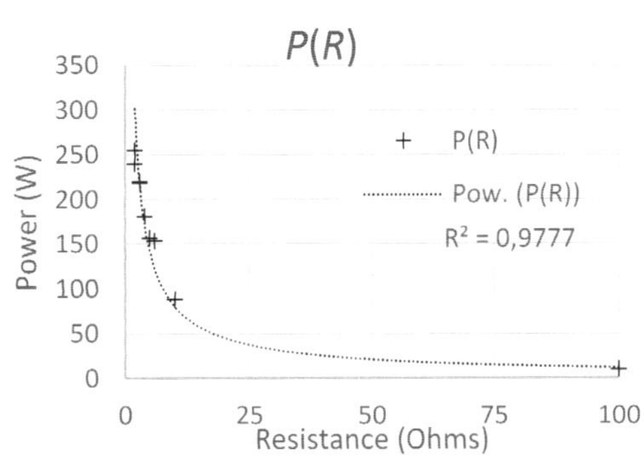

Figure 4. Power – Resistance characteristic

D. *Electronic load mode – constant current (CC)*

In constant current mode, measurements are very stable and the maximum power value for the given immediate light intensity can be reached. However, if a higher value of current is set than the panel is capable of supplying at the given moment, the electronic load is automatically disconnected. The following measurements are always conducted in steady increments of current. The parameters can be measured equally across the whole load current range. Table 3. and the graphs in Figures 5., 6. and 7. show the plotted measured values in CC mode.

Table 3. Data for CC mode

I	*R(I)*	*U(I)*	*P(I)*	*E*
Amps	Ohms	Volts	Watts	W/m²
1.00	42.2	42.21	42.1	956.00
2.00	20.6	41.21	82.4	956.00
3.00	13.40	40.27	130.70	957.00
4.00	9.80	39.19	156.70	953.00
5.00	7.61	38.03	190.00	952.00
6.00	6.11	36.63	219.70	947.00
6.50	5.52	35.57	233.00	948.00
7.00	4.98	34.86	244.00	943.00
7.50	4.50	33.65	252.70	946.00
8.00	3.98	31.82	254.50	946.00

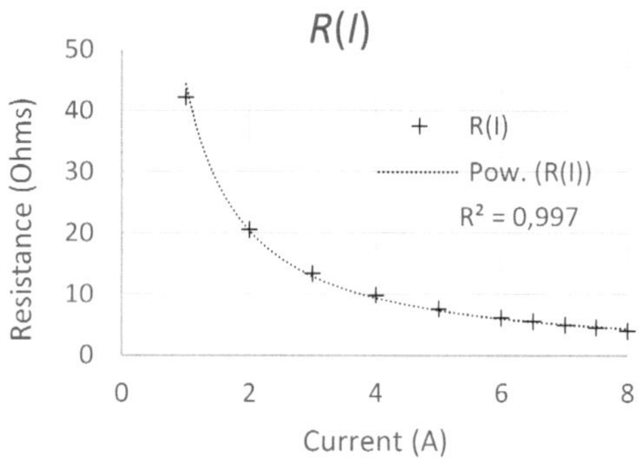

Figure 5. Resistance – Current characteristic

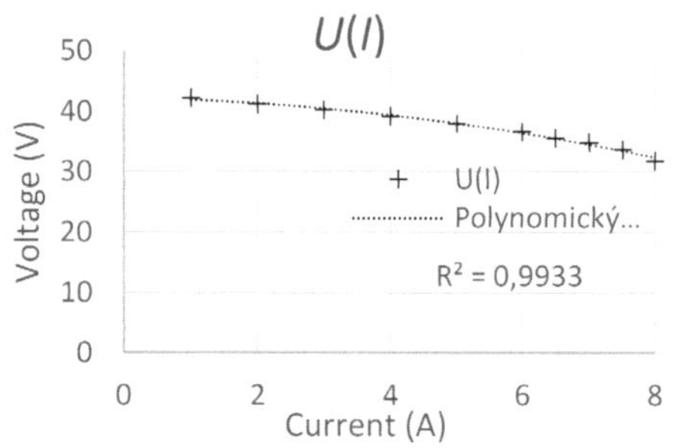

Figure 6. Voltage – Current characteristic

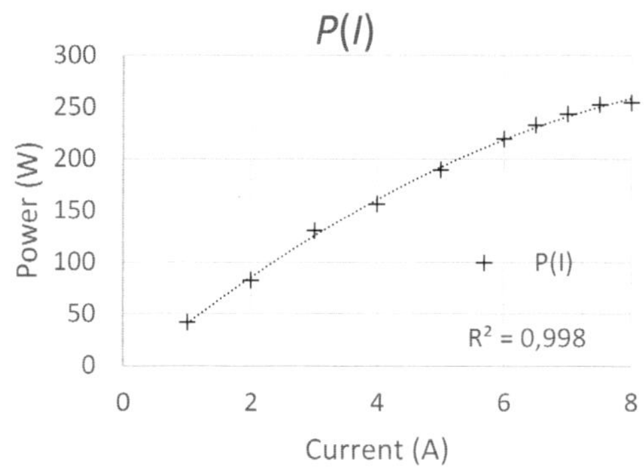

Figure 7. Power – Current characteristic

E. *Electronic load mode – constant power (CP)*

The constant power mode enables power to be set incrementally up to the maximum possible value with regard to light intensity. However, after it achieves this value, the load disconnects. It is difficult to determine the right maximum power value, especially as light intensity on the panel continuously changes [3], [4]. The measured relationships do not evenly correspond to the measured values. This results from the effort to make measurements near peak power. Table 4. and the graphs in Figures 8., 9. and 10. show the measured relationships in CP mode.

Table 4. Data for CP mode

P	*R(P)*	*U(P)*	*I(P)*	*E*
Watts	Ohms	Volts	Amps	W/m²
100.00	16.40	40.53	2.47	940.00
150.00	10.20	39.05	3.84	942.00
200.00	6.96	37.27	5.37	943.00
220.00	5.98	36.25	6.07	945.00
240.00	4.97	34.55	6.95	934.00
241.00	4.90	34.37	7.01	931.00
245.00	4.63	33.76	7.25	937.00
246.00	4.54	33.53	7.33	924.00
247.00	4.45	33.12	7.44	920.00
248.00	4.45	33.17	7.47	921.00
249.00	4.37	32.90	7.56	923.00
250.00	4.55	33.73	7.41	948.00

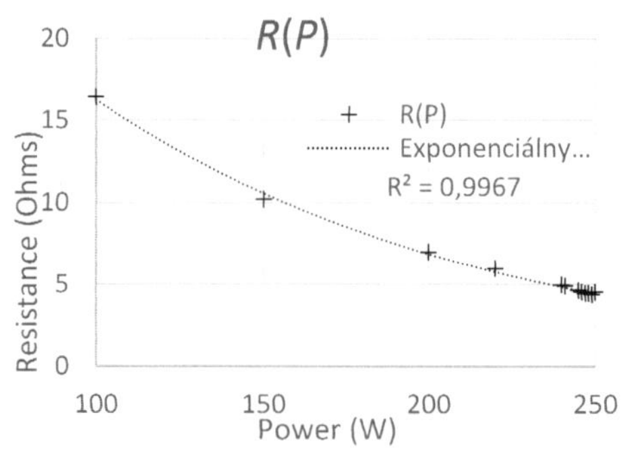

Figure 8. Resistance – Power characteristic

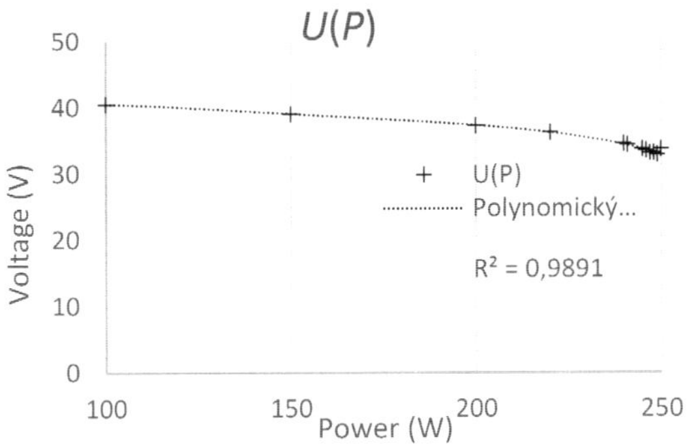

Figure 9. Voltage – Power characteristic

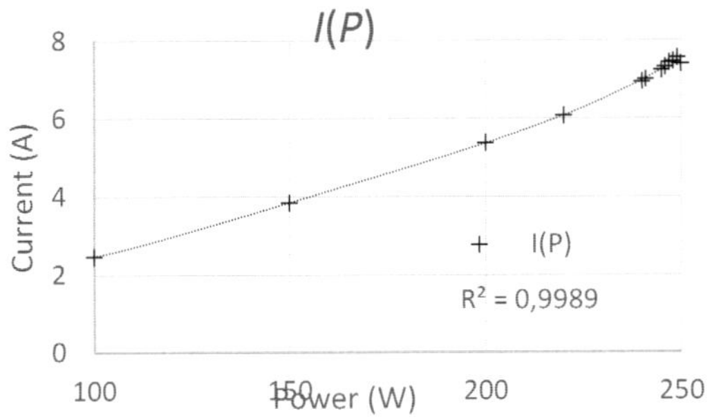

Figure 10. Current – Power characteristic

F. Electronic load mode – constant voltage (CV)

In constant voltage mode, all the characteristics of the solar panel as provided by the manufacturer can be consistently measured. The measurements though, are very sensitive to changes in light intensity. No conditions arise that disconnect the load because of insufficient power from the panel, but the data are generally distorted. A comparison of the power and current graphs with the graphs from the manufacturer reveal interesting information. In this mode, a load can be exerted even with low resistance and acquire the maximum value of current. This value is during an almost null voltage state, however, which also means no supplied power. Table 5. and the graphs in Figures 11., 12. and 13. show the measured relationships in CV mode.

Table 5. Data for CP mode

U	*R(U)*	*I(U)*	*P(U)*	*E*
Volts	Ohms	Amps	Watts	W/m²
0.10	0.02	8.25	1.30	913.00
5.00	0.63	8.18	42.30	902.00
10.00	1.26	8.21	85.00	908.00
20.00	2.54	8.17	169.00	911.00
25.00	3.22	8.05	210.00	899.00
30.00	3.95	7.85	244.50	902.00
35.00	5.95	6.10	220.80	903.00
40.00	33.20	1.26	52.00	897.00
42.69	-	0.00	0.00	-

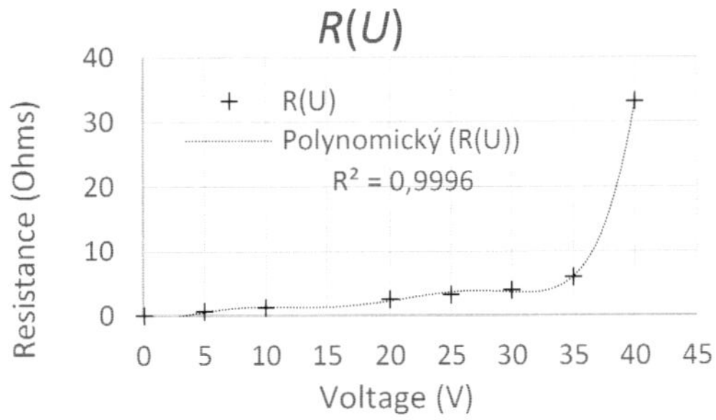

Figure 11. Resistance – Voltage characteristic

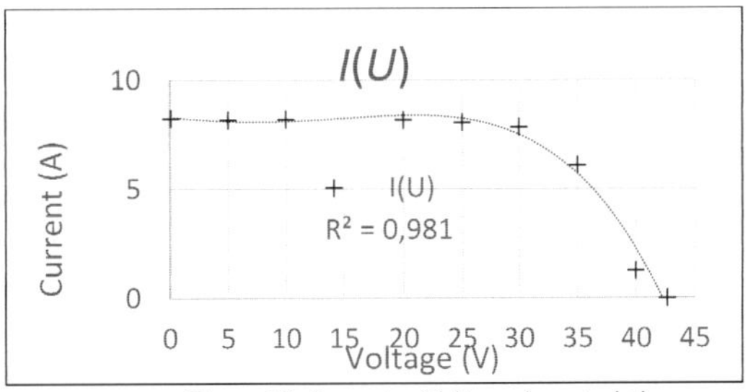

Figure 12. Current – Voltage characteristic

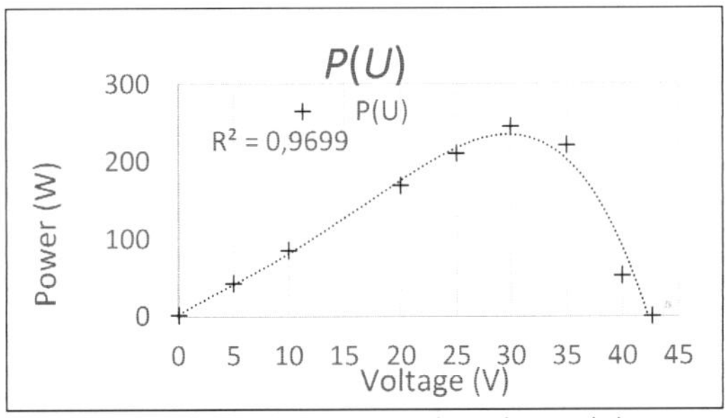

Figure 13. Power – Voltage characteristic

III. Results

If we compare the measured parameters with the parameters given by the manufacturer, we can infer that their shapes are very similar [5], [6]. The measured values are somewhat lower than the manufacturer claims, however. Figures 14. and 15. chart the voltage and power as stated by the manufacturer. Figures 16. and 17. depict the same parameters measured in our test.

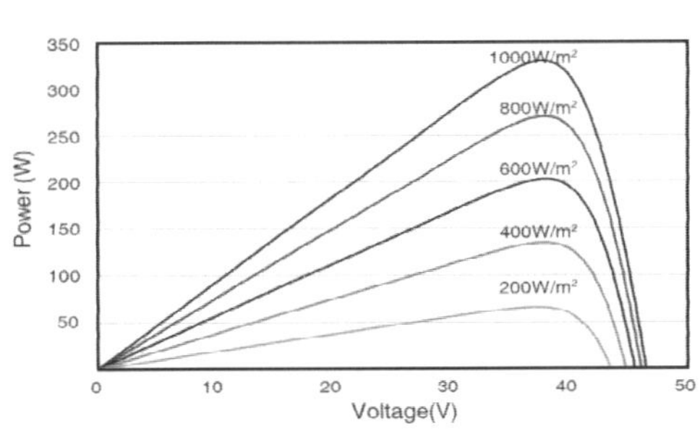

Figure 14. Producer P-V curves

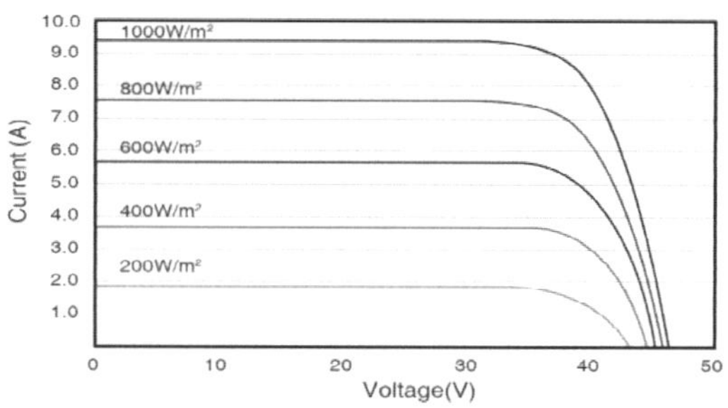

Figure 15. Producer C-V curves

The following characteristics were measured during a high light intensity of approximately 950 W/m². According to the manufacturer, a current of approximately 8.6 A and power of more than 300 W should be attainable in the panel. This could not be achieved, however. According to our measurements, a current of approximately 8.2 A and power of 250 W was reached, as shown in the graph in Figures 16. and 17. [4].

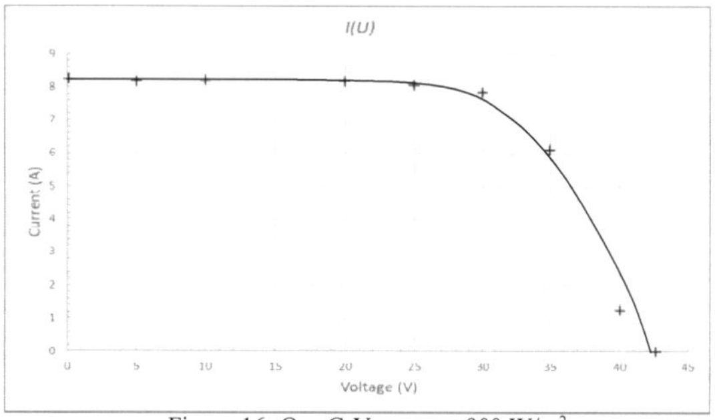

Figure 16. Our C-V curve at 900 W/m²

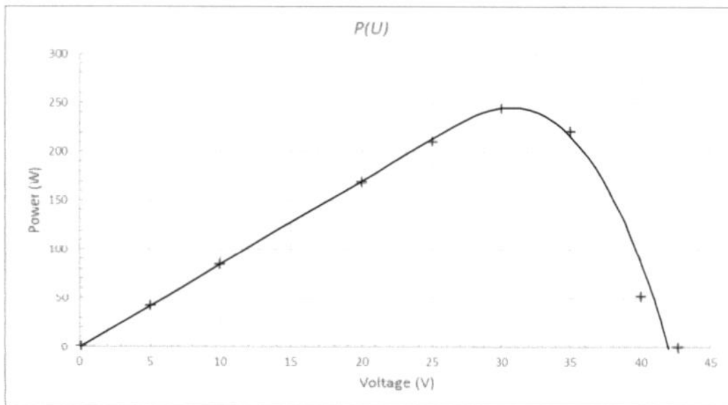

Figure 17. Our P-V curve at 900 W/m²

IV. Conclusions

Electronic load is a suitable method for measuring the characteristics of solar panels. The correct mode with regard to the measured parameter should be selected for conducting measurements. The results can then show several characteristics that tell us interesting information about the panel's properties. Finally, it is also interesting to compare the measured values of parameters with the parameters supplied by the manufacturer. The test was not conceived as a comparative measurement test in terms of accuracy, but to discover the respective relationships between individual variables and the possibility of using an electronic load to test the properties of solar panels [5]. Some properties and results from the solar panel confirmed our assumptions, while some were surprising. Testing is difficult though, because the intensity of solar radiation cannot be influenced and must be constant for the duration of measurements in order to be accurate [3], [6].

Different results may be due mainly to other measurement conditions, but also to the instruments used and the real panel. The measurement was into a real outdoor environment, where the intensity of the radiation changes significantly. Also, the panel direction directly to the sun has a great effect. Laboratory conditions are not the same as in real environment, results must be different too.

Acknowledgment

This work was supported by the Student Grant System of VSB – Technical University of Ostrava [grant number SP 2019/118].

References

[1] M. Repak, A. Otcenasova, M. Regula, R. Bodnar (2016). "Design of power converter for model of photovoltaic power plant." In: *2016 ELEKTRO* [online]. IEEE, 2016, 2016, s. 359-363 [cit. 2019-04-14]. DOI: 10.1109/ELEKTRO.2016.7512097. ISBN 978-1-4673-8698-2. Available at: http://ieeexplore.ieee.org/document/7512097/ [2016].

[2] P. Janiga, M. Liska, A. Belan, V. Volcko, M. Ivanic. "Power quality measurement of small solar off-grid system.",(Article), Department of Electrical Power Engineering, Slovak University of Slovakia, ISSN: 17905079.

[3] J. Novotny, J. Drapela, D. Topolanek. "Frequency response of revenue meters in measured active energy." In: *2016 17th International Conference on Harmonics and Quality of Power (ICHQP)* [online]. IEEE, 2016, p. 524-529 [cit. 2019-04-14]. DOI: 10.1109/ICHQP.2016.7783309. ISBN 978-1-5090-3792-6. Available at: http://ieeexplore.ieee.org/document/7783309/ [2016].

[4] J. Moravek, J. Drapela, V. Wasserbauer, P. Mastny. "Power quality issues related to power flow control in systems with renewable energy micro sources." In: *2016 17th International Scientific Conference on Electric Power Engineering (EPE)* [online]. IEEE, 2016, 2016, p. 1-6, [cit.2019-04-14]. DOI:10.1109/EPE.2016.7521784. ISBN:978-1-5090-0908-4. Available at: http://ieeexplore.ieee.org/document/7521784/ [2016].

[5] EN 50160 Ed.3 Voltage characteristics of electricity supplied by public distribution systems. Brussels: European Committee for Electrotechnical Standardization, 2010. 20 p..

[6] IEC 61000-4-30 Ed.2.0 Electromagnetic compatibility (EMC). Part 4-30: Testing and measurement techniques – Power quality measurement methods. Geneva : International Electrotechnical Commission, 2008, 134 p., ISBN: 2-8318-1002-0.

Prediction of Household Consumption Based on the Model

Martin Horák
Dep. Economic, Management and Humanities,
Faculty of Electrical Engineering
Czech Technical University
Prague
horakm11@fel.cvut.cz

Abstract — **In the electric power grid, an influence of intermittent sources starts not to be negligible. Increasing the percentage of production from renewable energy sources may increase demands for power regulation in the network. However, the power regulation is now mostly ensured through the control of fossil power plants. Exiting from the use of fossil fuels will further reduce the possibility of regulation. One of the solutions can be regulation on the customer side where they can actively manage their consumption. This solution requires the development of infrastructures such as smart meter and communication systems, which have been, however, uneconomic so far.**

Until the cost of technology is reduced, the way to develop a decentralized energy grid is by using the predictions of user consumption. Such estimate must be much more accurate than the currently available load profile of distribution tariffs. The present research, therefore, evaluates the changes in diagrams of households` electricity consumption based on usage of the different household electric appliances, prediction of lifestyle patterns, improving thermal and the technical properties of the buildings leading to a model to estimate the future consumption profile of households. All research is carried out on data from the Czech Republic.

Keywords—Energy load profile; Smart city; Zero-energy-building; Decentralization

I. Introduction

With the use of new types or more efficient appliances and the installation of decentralized power sources, e.g., PV and (micro)cogeneration, the households are no longer just consumers and become so-called "prosumers" (producer and consumers). They produce electricity on their own or sell the surpluses return to the energy grid [1] In the electric power grid, an influence of intermittent sources starts not to be negligible. To achieving the RES EU target of 2030, the RES share for electrical generation will increase from this year's 13% to 18 % in 2030 and further increase according to vision EU 2050 [2] [3] [4]

However, electricity as a commodity differs from others in the difficulty of storing a large amount of energy for a long time. For short-term balance in the grid, the active management of a power plant system has to be used. At the time, the power inflowing to the grid from power plants must always correspond to the amount of energy outflowed, consumed, in the grid. If there is an imbalance in the network, the frequency in the grid is changing, later the overload and blackout may occur.

The original consumption pattern that is used for current electric profile prediction follows only the use of household's amenities like a water heater and central electrical boiler. This model has been so far sufficient for group hour load profile but not for single household prediction.

The consumption may vary due to minor influences like the behavior of householders or so-called rebound effect. This effect describes the fact that low consumption of new appliances can lead to more often use, and thus the total household consumption can increase. However, two more challenges in future load profile changes are; growing interest in electromobility that can cause the short-term high consumption electricity demand from the grid that distribution network may not be able to satisfy, and the increasing number of PV panels whose an intermittent production cause abnormality in consumer load profile.

The solution to this problem may be implementing battery repositories (or other technologies) so that the maximum energy produced in small decentralized units (PVs) is consumed at the area of production and do not increase the distribution grid load. Thanks to cost-effective technologies for storing energy (i.e., battery storage, domestic water heating, storage tanks for heating), deviations could be eliminated at the level of buildings, groups or small district. To properly design these local repositories, it is necessary to create a sufficiently accurate consumption pattern for individual buildings or units.

This project tries to evaluate the changes in diagrams of households' electricity consumption due to using another household electric appliances, changing lifestyle. The results of the analysis will fit into the draft of the updated load profile.

II. Electricity Tariffs

All participants electricity trading are responsible for consuming all the energy they have purchased. If the trader does not take the purchased energy or vice versa, the trader is penalized for deviation. For small consumers would not be technically advantageous to buy energy yourself. For this reason, traders provide purchase for end users. The traders associate end consumers into groups with similar grid requirements. These groups are described by STP (Standard load profile) diagram issued by the market operator (OTE a.s. for the Czech Republic). The model is generally divided into two parts; for household and industry. The research deals with only the household sector; the description of the industry part will be omitted. The household sector contains a total of 10 distribution tariff groups in four TDD load profiles. [1] (a brief description in Table 1).

Table 1 Distribution tariffs

STP	Distribution tariffs	Appliances	Num. cheaper hours
SINGLE-RATES OF ELECTRICITY TARIFF			
4	D01d	XXX	0
4	D02d	XXX	0
TWO-RATES ELECTRICITY DISTRIBUTION TARIFF			
5	D25d	boiler (lower)	8
5	D26d	boiler (higher)	8
5	D27d	electro-car	Min. 8
6	D35d	Storage heaters	16
7	D45d	Electric heaters	20
7	D56d	Heat pump	22
7	D57d	Heat pump and electric heater	20
4	D61d	XXX	Weekend

Estimation of the shape of the loading profile is calculated from the values of the previous years. These data are obtained from AMR or smart meters. After gathering, data are statistically processed and recalculated by the temperature profile of last year. The goal of this process is to create a standardized diagram for each TDD type. By using the weather correlation coefficients forecast, the further load profile is estimated. Each TDD is calculated for the whole distribution network of the Czech Republic, except the TDD5, whose data are strongly dependent on the district. Therefore, consumption patterns include seven national TDDs and eight district TDD5 diagrams.

III. Statistical Survey

To estimation future changes in consumption, the current state of household equipment and consumer behavior has to be analyzed. For this purpose, the latest statistical survey ENERGO 2015 [5] gathered between 7 July 2015 and 15 January 2016. The sample contains statistical data about 20,000 households, i.e., 0.5% of all permanently inhabited dwellings. The survey is divided into several modules mapping the energy demands of the population.

- **Housing module:** Energy performance of the selected apartment/house, residential area, number of floors, year of construction, a way of house insulation, and other additional questions.
- **Fuel module:** For heating, water heating, cooking, lighting, cooling
- **Lighting module**: The structure of the light sources.
- **Appliance module**: Only appliances with the most significant share of the total household electricity consumption were included (refrigerators, freezers, automatic washing machines, clothes dryers, dishwashers, televisions, computers, and air conditioning). Their energy class, age, and frequency of use.
- **Transport module:** Use of cars and their characteristic parameters

At the end of the questionnaire, participants were asked about the total household income for further analysis.

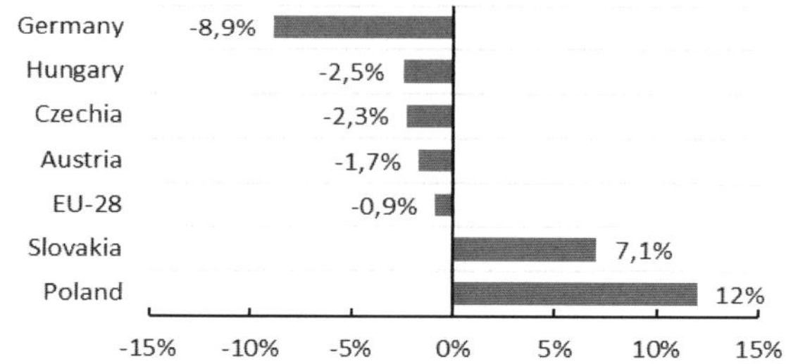

Figure 1 Changes in households electricity consumption between the years 2005-2015 [6]

To estimate further changes in household amenities the values from the ENERGO 2015 are compared with the Odyssee-Mure statistic data for neighboring countries. During the ten years from 2005 to 2015, the consumption of electricity by households fell in the whole EU-28 by 0.9 % in the Czech Republic by 2.3%; the seventh largest decline in the EU-28 (Figure 1). [6]

A significant influence on electricity consumption has household electrical appliances growth. In households, there is a growing number of new types of appliances with specific consumption, such as entertainment technology, dryers, smart home appliances.

From figure 2 showing the coverage of household appliances in the Czech Republic can be assumed that the coverage will approach the level of EU and Austria and Germany. Moreover, the natural replacement of the appliances will increase the average cost-effectiveness of the main appliances. A lifetime of appliances depends on many factors, but according to ENERGO 2015 statistic data from the Czech Republic in Table 3 and the proportion of appliance age in Table 2 indicates that approx. 50% of refrigerators, washing machines, and dishwashers older than six years will be exchanged for Class A+, or better devices and consumption can decrease by about 30%. [5] [7]

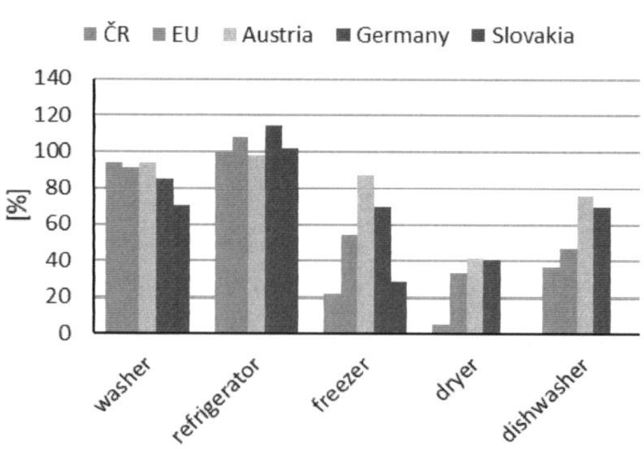

Figure 2 Home appliances [6]

Table 1. A share of energy classes in household appliances owned by households in the year 2015 [5]

Energy class (%)	A+++	A++	A+	A	B-D
refrigerator combined	8,9	20	25,6	33,2	12,3
refrigerator	6,3	13,7	19,6	32,9	27,5
freezer	5,3	12,6	19,3	33,8	29
dryer	23,4	32,3	23,8	14,8	5,7
dishwasher	11,9	26	27,4	27,3	5,4
washer	8,7	22,2	26,6	30,8	11,7

Table 2 Average age of the appliances in households in the year 2015 [5]

Age (%)	2 yrs	2 -5 yrs	6-10 yrs	more
refrigerator combined	6,3	36,3	37,1	18,3
refrigerator	4,9	28,5	36,9	29,7
freezer	2,8	25,6	34,2	34,4
dryer	25,2	55,7	16,5	2,6
dishwasher	10,5	51,3	31,1	7,1
washer	9	40,5	36,5	14

IV. Smart Metering Data Analysis

The smart meters data has been used for consumption estimation — the data contained ten annual load profile from households within five different distribution tariffs. Data from the smart meter contained consumption value for every 15 minutes and a timestamp for each entry. All the data comes from the region of Central Bohemia, namely the distribution power grid under the management of the company PREdistribuce, a.s.

This analysis aims to determine the predictability of consumption based on real household data. The data are processed regarding the assumption that the activities of the inhabitants related to the annual cycle and the natural weekly period.

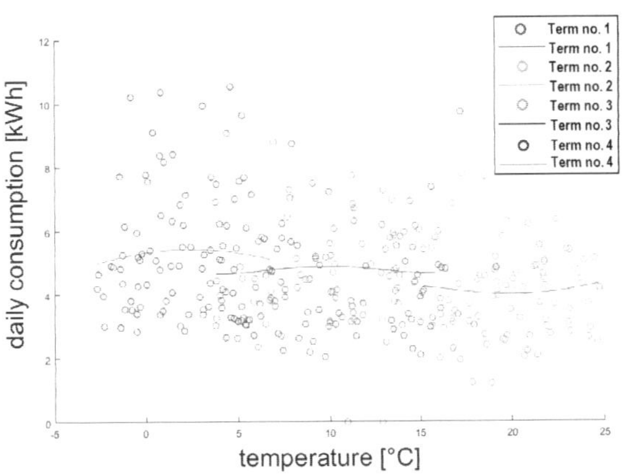

Figure 3 Temperature correlation

The annual consumption period is mostly associated with weather, ambient temperature. As an example, consumption during the summer months can be higher as usual by thanks to working air conditioning. On the contrary, in winter the consumption might be higher due to higher lighting requirements. For the annual period reason, the data has been correlated with ambient temperature in the year as shown in Figure 3.

The weekly period is mainly influenced by the householder's daily routine. Therefore, the analysis focuses on the similarity of the daily consumption diagrams during the same weekday. Since it cannot be eliminated the sessional dependency of consumption, the calculation was always created within a shorter period than a year, usually one month. Figure 4 shows the mean consumption and standard variation for March Monday of 2017.

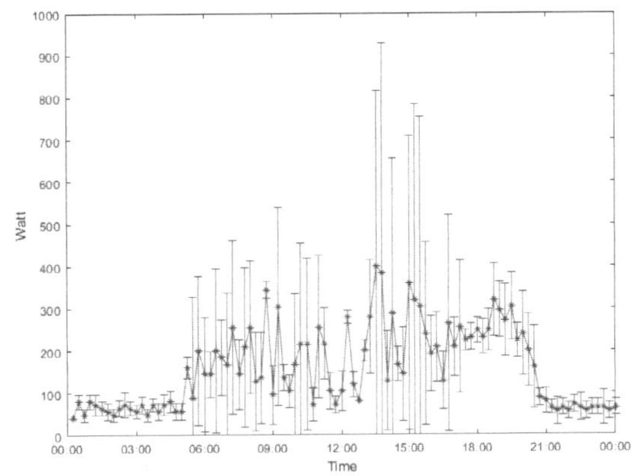

Figure 4 Five-week Monday consumption

V. Discussion

The research uses both statistical data (ENERGO 2015), and real data from already installed smart meters. The survey shows that nearly half of household appliances have an energy class worse than the minimum value that can be sold in the EU after 2013. Assuming the average replacement of appliances is about 7 - 10 years, it can be expected that approximately half of the major appliance will be exchanged for class A+ or better devices and consumption from these appliances decrease approximately about 20%. Also, due to using of new facilities in households, we can expect significant changes in consumption pattern after 2020.

The pilot analysis of the real-time data from households shows the correlation among consumption, ambient temperature, and periodic living pattern. Analyzed data, for instance, showed that a family with the parent on maternity leave consumes about 70% of daily energy during the lunchtime unlike households with residents with a full-time job whose consumption is moved to afternoon or evening hour. Still, the identification of individual smaller appliances contribution in total consumption profile remains a challenge for further research.

With the incoming entry ban for diesel vehicles into European city centers, partial replacement by electric cars can be expected. Assuming an average annual mileage of Prague driver is about 8 thousand kilometers [8], and average consumption of an electric car is 15 kWh / 100 km [9], average household consumption can grow by a non-negligible 2.7 MWh per year. This amount can be a part of household consumption in case of household in a house. However, as it is not possible to charge the car in flats, in those cases the direct consumption would not be affected.

VI. Conclusion:

Developments in the consumption of electrical appliances cause changes in the total consumption of households. Together with the effort to reduce energy losses in flats and houses, household dependence on electricity supplies is growing. Moreover, with the EU regulation on the minimum standard for newly built houses from 2020, greater deployment of heat pumps, recuperation unit, and smart heating system deployed can be expected. When comparing household amenities in neighboring countries, we can also expect the use of dryers or dishwashers that are not so common in the Czech Republic.

Acknowledgments:

Thanks to prof. Ing. Jaroslav Knápek, CSc. for help and suggestions for this work. This work was supported by the Grant Agency of the Czech Technical University in Prague, grant No. SGS18/136/OHK5/2T/13.

References:

[1] "Egú Brno," 20 09 2018. [Online]. Available: https://www.egubrno.cz/wpcontent/uploads/2018/10/TZ_Energetika_2018.pdf. [Accessed 23 03 2019].

[2] E. COMMISSION, "Policy 2030," [Online]. Available: http://data.consilium.europa.eu/doc/document/ST-169-2014-INIT/cs/pdf.. [Accessed 13 3 2018].

[3] E. COMMISSION, "Roadmap2050," 2011. [Online]. Available: http://eur-lex.europa.eu/legal-content/EN/TXT/PDF//?uri=CELEX:52011DC0885&from=EN.. [Accessed 13 3 2018].

[4] M. o. o. a. trade, "Státní energetická koncepce," [Online]. Available: https://www.mpo.cz/dokument158059.html. [Accessed 23-Mar-2019].

[5] Č. s. ú. |. ČSÚ, "Energo 2015," 12 2014. [Online]. Available: https://www.czso.cz/csu/czso/energo-2015.. [Accessed 13 3 2018].

[6] Odyssee-mure, "Energy Efficiency Trends and Policies in Buildings," 2015. [Online]. Available: http://www.odyssee-mure.eu/publications/br/energy-efficiency-in-buildings.html. [Accessed 13 03 2019].

[7] S. A. E. B. Anette Michel, "Energy efficiency of White Goods in Europe:monitoring the market with sales data – Final report. ADEME," 2016.

[8] Odyssee-mure, "distance travelled by car," 2015. [Online]. Available: http://www.odyssee-mure.eu/publications/efficiency-by-sector/transport/distance-travelled-by-car.html. [Accessed 2019].

[9] U. D. o. E. -. E. E. a. R. Energy, "fueleconomy.gov," [Online]. Available: https://www.fueleconomy.gov/feg/PowerSearch.do?action=noform&path=1&year1=1984&year2=2019&vtype=Electric. [Accessed 20 03 2019].

Evaluation of Short Interruptions of Supply Voltage in Distribution System

Miloslava Tesarova
Department of Electrical Power Engineering and Environmental Engineering
University of West Bohemia
Plzen, Czech Republic
tesarova@kee.zcu.cz

Martin Kaspirek
Management of grid
E.ON Distribution
Ceske Budejovice, Czech Republic
martin.kaspirek@eon.cz

Abstract — **Continuity of supply can be regulated by a national regulation authority to maintain or improve its level. Short interruptions are monitored and regulated much less frequently than long interruptions. They can occur as single event or in sequences cased by a single power system event or by independent events. Frequency of their occurrence is mainly characterized by the indicator MAIFI. Method used for counting interruption sequences greatly affects the values of the indicator. In regulatory framework, it is important to define continuity of supply indicators and to establish procedures for counting of interruptions. The presented work is initial step in preparation of methodology for MAIFI calculation in the Czech Republic. In the paper, long-term monitoring of events in selected MV/LV substations is evaluated. Number and percentage of short and long interruptions during monitoring period is shown. Values of MAIFI indicator are compared with values of modified indicator that besides short interruptions includes also severe voltage dips. The effect of various aggregation methods and time-aggregation intervals for counting of short interruptions is illustrated on the case of a MV/LV substation with extreme occurrence of short interruptions.**

Keywords: distribution system; power quality; continuity of supply; voltage quality; interruption; aggregation

I. Introduction

Continuity of supply refers to the availability of electricity to network users. It concerns interruptions in electricity supply and focuses on the events during which the voltage at the supply terminals of a network user drops to zero or nearly zero (according to EN 50160) [1]. The European technical standard EN 50160 defines interruptions that last more than 3 minutes as *long interruptions*, and others as *short interruptions*. Contrary to CEER/WGQES (Council of European Energy Regulators / Working Group on Quality of Electricity Supply), the UNIPEDE considers short interruptions to be voltage quality problem, although they may produce equipment damage like long interruptions. Long interruptions are monitored in all European countries involved in the 6th CEER benchmarking report on the quality of electricity, but less than half of the countries collect separate data on short interruptions [1].

Current and future continuity levels result from investment decisions of network operators. To maintain or improve of continuity levels, the continuity of supply can be regulated by a national regulation authority. The regulator publishes continuity data to facilitate benchmarking the performance of the networks operators and implements reward or penalty schemes or incentives to optimize the continuity of supply. Regulation measures can be implemented at system level (overall standard) and single-user level (guaranteed standards). Overall standards relate to the overall provision of supply and services measured and regulated at the system level, and frequently include quality incentives. Guaranteed standards of performance relate to individual service delivery and carry a penalty payment. Regulation of the continuity of supply on the system level is applied in more than half of the countries involved in the 6th CEER benchmarking [1]. Nearly two thirds of countries offer individual compensation to network users when standards focused on duration of long interruption are not met. Although short interruptions may have a strong economic impact on industrial customers, fewer European countries have involved short interruptions into regulation [1].

In the Czech Republic, continuity of supply is regulated both at system level and at single-user level. Guaranteed standard for individual customers relates to individual duration of very long unplanned interruptions. Since 2013, continuity of supply has been also regulated at system level. Continuity regulation for distribution system operators (DSOs) is based on SAIDI and SAIFI indicators. The mechanism of incentive-based regulation is described in [1]. At present, the word "continuity of supply" is very frequently mentioned by Czech DSOs and requirement on improving of continuity levels has been implemented into grid development [2], [3]. In the last year, the discussion to introduce the regulation of short interruptions has started in the Czech Republic. But no recommendations or rules are introduced for monitoring and evaluation of short interruptions in the Distribution Code.

In regulatory framework, it is important to define continuity of supply indicators and to establish their limits and procedures for data collection and counting of interruptions. In the paper, long-term monitoring of events in selected MV/LV substations is evaluated. Indicator MAIFI is compared with modified indicator that involved short interruptions and severe voltage dips. Various aggregation methods and time-aggregation intervals are applied for counting of short interruptions recorded in a LV substation with extreme occurrence of interruptions. The number of short interruptions, counted by various methods, differs in wide range.

II. Counting of Short Interruptions

A. Continuity of supply indicators

Continuity of supply is mainly characterized by the number and duration of interruptions. Commonly used indicators are SAIDI (System Average Interruption Duration Index) and

SAIFI (System Average Interruption Frequency Index), usually weighted by the number of network users. Other used indicators, mostly used for transmission networks, are unavailability (interrupted minutes) and energy not supplied due to interruptions (ENS) per year. Comparison of continuity levels in various countries is difficult because different approaches to calculating these parameters exist, e.g. weighting to number of network users or to their size, counting exceptional events. Short interruptions are mainly characterized by the indicator MAIFI (Momentary Average Interruption Frequency Index). Some countries differentiate between short (T≤3 min) and transient (T≤1 s) interruptions. According to EN 50 160, an interruption occurs when the voltage drops below 5% of the reference voltage (nominal voltage) on all the phases.

B. *Indicator MAIFI*

MAIFI is the average number of momentary interruptions that a customer would experience during a reporting period (year) and is expressed as:

$$MAIFI = \frac{\sum NM_i}{N_T} \tag{1}$$

where ΣNM_i is the total number of momentary interruptions to customers during reporting period, and N_T is the total number of customers served.

The short interruptions results from transient short-circuit faults (e.g. caused by lightning strokes or vegetation contacting), automatic reclosing operation or switching. They can occur as single event or in sequences cased by a single power system event or by independent events. For example, multiple interruptions due to automatic reclosing may be counted as one interruption or as multiple interruptions. Method used for counting of interruption sequences should respect their real impact on customers. Time aggregation counts all interruptions during a defined time interval as a single interruption. The time aggregation can be based on the time elapsed since the end/start of the previous interruption and range between a few cycles and a few tens of minutes. The aggregation rules can also reflect the fact that a long interruption can be followed or preceded by a single short interruption or a sequence of short interruptions. If the calculation of MAIFI is based on counting of momentary interruption events (not interruptions), the indicator is being called $MAIFI_E$.

$MAIFI_E$ is used in practice in most countries [4]. In France, the aggregation interval is 60 minutes for a short interruption after a long interruption, and 2 minutes for a short interruption before a long interruption and for a short interruption after another short interruption. In Italy, 60 minutes aggregation (either after long or after short) is possible only if same interruption cause and same origin voltage level. In Hungary, a short interruption after a transient interruption is counted as 1 if the automatic reclosing action was successful. Further, a short interruption before a long interruption is counted as 1 short interruption, whereas short interruptions after long or short interruptions are not counted. The method used for counting interruption sequences greatly affects the values of the indicator [5].

C. *Interruption monitoring in the Czech Republic*

In the Czech Republic, voltage quality has been continuously monitored on the secondary side of all EHV/110 kV and 110 kV/MV transformers since 2006 and 2010 respectively. Few MV/LV substations, as well as important customers are also continuously monitored. The interruptions are divided according to their duration on long, short and transient ones (Table 1). The system and as well as voltage level indicators SAIDI and SAIFI are calculated for long interruptions [5]. The system indicators SAIDI and SAIFI are regulated. Short interruptions are only monitored.

Table 1. Definition of long, short and transient interruptions in the Czech Republic

Transient interruptions	*Short interruptions*	*Long interruptions*
10 ms < T ≤ 1 s	1 s < T ≤ 3 min	T > 3 min

III. Case study

Occurrence of interruptions is evaluated for selected MV/LV substations where events are continuously monitored from 2006. The number and percentage of short and long interruptions is evaluated and continuity indicators as calculated. To show their effect, various aggregation methods and time-aggregation intervals are applied for counting of short interruptions recorded in a LV substation with extreme occurrence of interruptions.

A. *Input data*

Interruption data have been collected from six MV/LV distribution substations (DTSs) for monitoring period 2006-2018, a total of 78 monitor-years. Monitors are placed on secondary side of the MV/LV transformers. Types of MV feeders and supplied LV network (overhead lines, cables, mixed) are different for particular DTSs (Table 2). Data contains both planned and unplanned interruptions. Both types of interruptions are included in the analysis, although the calculation of continuity indicators, used for regulatory purposes in the Czech Republic, includes only interruptions, which are within the system operator's control (planned interruptions and unplanned failure-related interruptions caused by failures originating in the system operator's system or in the operation thereof under usual weather conditions).

Table 2. Characteristics of monitored distribution substations

DTS	*MV network*	*LV network*	*Transformer output*	*Number of customers*
DTS 1	mixed	mixed	250kVA	123
DTS 2	mixed	mixed	400kVA	210
DTS 3	OHL	mixed	400 kVA	118
DTS 4	OHL	OHL	160kVA	65
DTS 5	cables	mixed	630 kVA	353
DTS 6	cables	cables	400 kVA	382

B. *Occurrence of short and long interrtuptions*

Number and percentage of long and short interruptions was evaluated. Results are summarized in Table 3 and Fig. 1. Almost 70% of recorded interruptions lasted less than 3 minutes. Transient interruptions occurred rarely. Significantly higher number of interruptions was recorded in DTSs that are supplied from a MV network with overhead lines (DTS 3 and DTS 4). Higher percentage of short interruptions was also recorded in DTS 3 and DTS 4. Most of recorded short interruptions were probably caused by automatic reclosing. Because automatic reclosing is not used in LV networks, the short interruptions were originated mainly from MV or HV networks. On the contrary, percentage of long interruptions was higher in DTSs supplied from MV cable networks (DTS 5 and DTS 6). Connection between temporary or permanent faults and number of short or long interruptions is evident.

Table 3. Occurrence of interruptions in the LV networks (2006-2018)

Number of interruptions			
10 ms < T ≤ 1 s	***1 s < T ≤ 3 min***	***T > 3 min***	***total***
4	433	195	632
Percentage occurrence of interruptions			
0.6%	68.5%	30.9%	100%

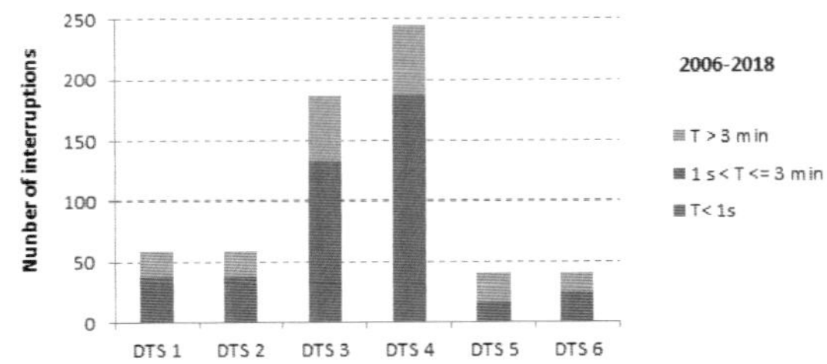

Figure 1. Total number and percentage of long and short interruptions recorded in DTSs during monitoring period 2006-2018

Although percentage of long and short interruptions recorded varies from year to year (Fig. 2), it is evident that higher number of long interruptions is expected in cable networks.

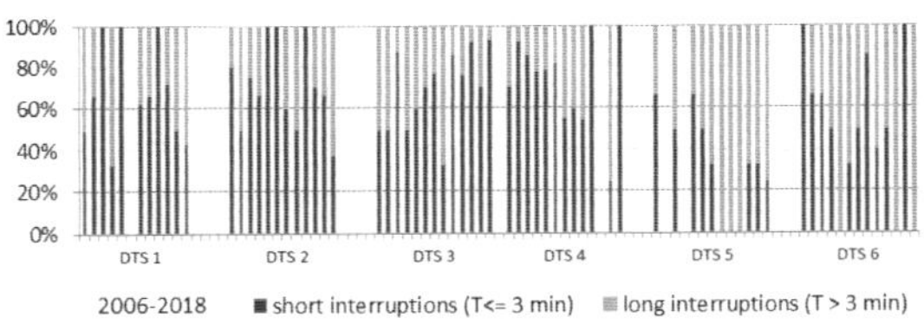

Figure 2. Percentage of long and short interruptions recorded in DTSs in individual years

C. *Indicators SAIFI and MAIFI*

SAIFI and MAIFI indicators of individuals DTSs were evaluated for each year. Because the indicators are not weighted by total number of customers served, but only customers served by given DTS, SAIFI and MAIFI are equal to a number of long and short interruptions respectively, recorded in given DTS.

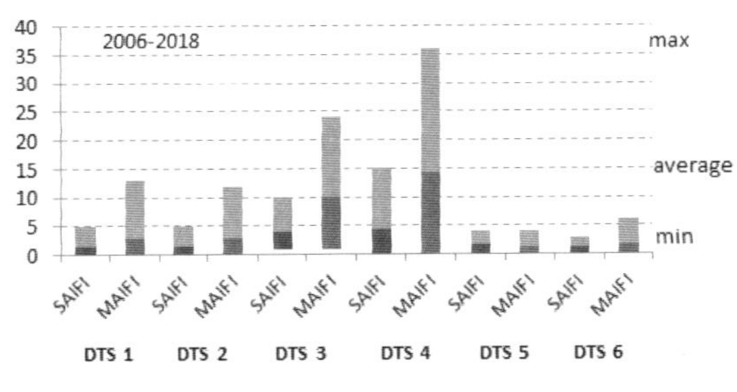

Figure 3. Range of SAIFI and MAFI calculated for individual DTSs and years during monitoring period 2006-2018

Fig. 3 shows the range of indicator's values calculated for individual years. Average value of indicators during monitoring period 2006-2018 is added in the graph. Lighter part of the columns corresponds with indicator's values above the average value and darker part below the average value. Significantly wider range of the above-average values than range of the under-average values indicates extreme occurrence of interruptions in given year or years. For example, 13 short interruptions were recorded in DTS 1 in 2015, although number of interruption recorded in other years does not exceed 4 events. In case of DTSs supplied from cable MV networks (DTS 5 and DTS 6), value of MAIFI is comparable to SAIFI because of similar occurrence of long and short interruptions (see Fig. 1).

Annual voltage level indicators SAIFI and MAIFI were also calculated for the group of DTSs according to (1). Total number of interrupted customers during a year $\sum NM_i$ is expressed as:

$$\Sigma NM_i = \Sigma ID_i \cdot N_i \quad (2)$$

where ID_i is the number of interruptions recorded in the DTS i during reporting period, and N_i is the number of customers served by the DTS i. SAIFI and MAIFI calculated for individuals years are shown in Fig. 4. The curves of 2-year moving average of indicators are added. In the Czech Republic, 2-year moving average of indicators, i.e. average value of actual performance indicators in the past 2 years, is compared with the required value which was set by the Regulator. In the current regulatory period, the required value of system indicator $SAIFI_Q$ is 1.547 interruptions per year and per a customer for the EON company. SAIFI for the selected group of 6 DTSs is higher than the required value of the system indicator $SAIFI_Q$ which is "whole-system" index, relating to the system operator's entire distribution system without differentiating between voltage levels and counting only events being within the system operator's control.

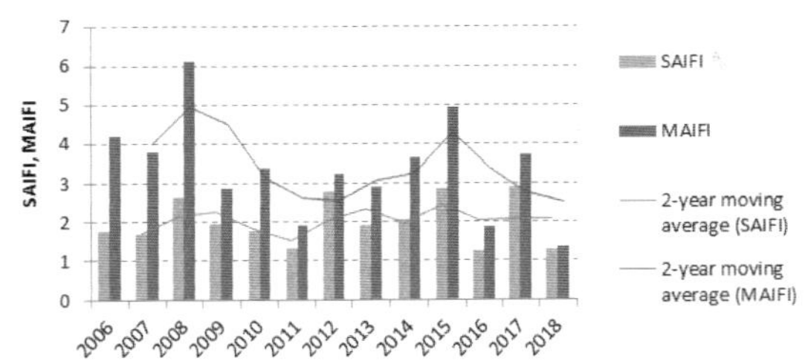

Figure 4. Voltage level indicators SAIFI and MAFI calculated for individual years during monitoring period 2006-2018

D. *Indicator MAIFI-dip*

Not only interruptions, but also voltage dips may produce equipment damage. The reference curves to delimit severe dips are often based on Class 2 and Class 3 of equipment dip immunity [6]. Because no rules are introduced for evaluation of voltage events in our country, we introduce a new indicator *MAIFI-dip* that involves short interruptions and voltage dips below a reference curve. The reference curve is derived from the upper reference curve used for systems above 45 kV for regulation purposes in Sweden [4]. Area below this reference curve is called Area C. In the analysis, the Area C is extended by dips lasting more than 1 minute (cells C6-D6-X6 and C7-D7-X7 in magnitude-duration table in the Czech Distribution Code, appendix no. 3), see Table 4.

Table 4. Reference curve and area used for counting of severe dips

U (%)	Duration of dips t (s)				
	0.01≤t<1	1≤t<5	5≤t<60	60≤t<180	180<t
80≤U<90	A1-A3	A4	A5	A6	A7
70≤U<80	B1-B3	B4	B5	B6	B7
40≤U<70	C1-C3	C4	C5	C6	C7
5≤U<40	D1-D3	D4	D5	D6	D7
0<U<5	X1-X3	X4	X5	X6	X7

Percentage of interruptions and severe voltage dips is shown in Fig. 5. Besides dips with duration up to 3 minutes, the graph shows also severe dips lasting more than 3 minutes. These dips are not reported in magnitude-duration table in the Czech Distribution Code, but they have been recorded. Their occurrence in some DTSs is substantial, e.g. DTS 4.

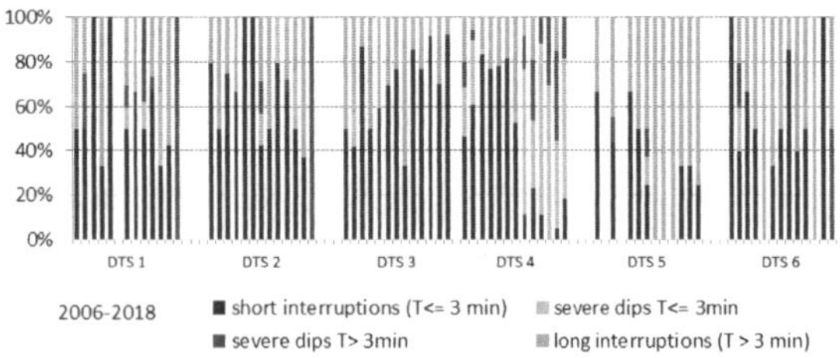

Figure 5. Percentage of long, interruptions, short interruptions and severe dips recorded in DTSs in individual years

Excluding DTS 4, only a few severe dips have been recorded in DTSs during all monitoring period. One third of annual event records in DTS 4 contains more than 13 severe dips, which is the same number as severe dips recorded during all monitoring period in other DTSs. The substation DTS 4 has a significant impact on values of *MAIFI-dip*, see Fig. 6. Fig. 6 compares values of MAIFI indicators. Indicator *MAIFI-dip* includes short interruptions and severe voltage dips with duration T≤3 min. Indicator *MAIFI-dip T>3 min* includes short interruptions and severe voltage dips with unlimited duration. Magnitude and duration of polyphase voltage dips are based on polyphase aggregation rule.

Figure 6. Voltage level indicators MAIFI and MAIFI-dip calculated for individual years during monitoring period 2006-2018

The substation with extreme occurrence of voltage events may have a significant impact on values of MAIFI and modified MAIFI indicators.

E. Aggregation rules for counting of short interruptions

Records of events show that short interruptions often occur in sequences or their occurrence is associated with long interruptions. Because the case of events can be the same, aggregation rules are applied to respect their real impact on customers. A sequence of events within a defined time are counted as a single event. Furthermore, various ways how to count of short interruptions associated with long interruptions are used in European countries, see the chapter II-B. Method used for counting of interruption sequences greatly affects the results. The effect of various aggregation methods and time-aggregation intervals is illustrated on the case of the substation DTS 4 where extreme occurrence of short interruptions and voltage dips has been recorded (years 2006, 2007 and 2009 are evaluated).

Methods used for counting sequences of short interruptions:

1. Method 1 - the aggregation interval is 3 minutes either after long or after short interruption (Fig. 7a)
2. Method 2 - like the method 1, but the aggregation interval is 60 minutes
3. Method 3 - the aggregation interval is 3 minutes before and 60 minutes after a long interruption (Fig. 7b)
4. Method 4 - like the method 1, but T1= 60 minutes and T2= 60 minutes
5. Method 5 - like the method 3, furthermore the aggregation interval is 3 minutes after a short interruption
6. Method 6 - like the method 4, furthermore the aggregation interval is 3 minutes after a short interruption.

Aggregation interval 3 minutes corresponds to automatic reclosing delay and interval 60 minutes to recovery time of production processes.

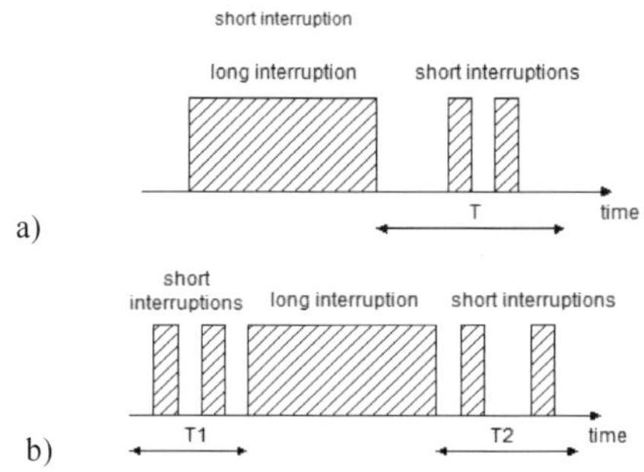

Figure 7. Aggregation methods for counting of short interruptions

The aggregation methods were applied on events recorded on 21/06/2007 (Table 5) to demonstrate how they affect counting of short interruptions.

Table 5. Events recorded in the DTS 4 on 21/06/2007

Start	End	Duration	L1 (%)	L2 (%)	L3 (%)
15:41:19,590	15:41:20,690	1,10 sec	44,9	44,9	42,8
15:43:04,000	15:43:06,000	2,00 sec	0	0	0
15:43:19,000	15:43:19,100	0,10 sec	77,5	37	
15:46:28,000	15:46:30,000	2,00 sec	0	0	0
15:47:09,000	15:47:11,000	2,00 sec	0	0	0
15:49:48,610	15:49:48,700	0,09 sec		78,3	
15:51:03,000	15:51:05,000	2,00 sec	0	0	0
15:51:44,000	15:51:46,000	2,00 sec	0	0	0
15:51:56,810	15:54:59,000	3 min 2,19 sec	0	0	0
15:59:14,000	15:59:16,000	2,00 sec	0	0	0
15:59:53,000	15:59:55,000	2,00 sec	0	0	0
16:00:50,000	16:00:52,000	2,00 sec	0	0	0
16:01:07,980	16:01:10,000	2,02 sec	0	0	0
16:01:45,000	16:01:47,000	2,00 sec	0	0	0
16:05:22,000	16:05:24,000	2,00 sec	0	0	0
16:17:03,790	16:17:03,890	0,10 sec	49,6		80,3
16:29:10,080	16:29:10,310	0,23 sec	58,4	59	
16:35:07,440	16:35:07,670	0,23 sec		56,3	61,3
16:37:23,040	16:37:23,140	0,10 sec	60		85
16:37:42,000	16:42:55,000	5 min 13,00 sec	0	0	0

Table 6. Results of aggregation methods used on record in Table 5

Number of short interruptions					
M1	*M2*	*M3*	*M4*	*M5*	*M6*
5	1	3	0	2	0

Results are summarized in Table 6. The one-hour record contains 2 long interruptions, 11 short interruptions and 7 dips. According to the method used the number of short interruptions counted varies from 0 to 5. The aggregation methods were applied on events recorded of three selected years with extreme number of recorded interruptions (Fig. 8). Number of short interruptions recorded in given year was: 36, 36 and 24 respectively. The lowest number of short interruptions was counted by using the method 2 that does not count all short interruptions occurring in the interval 60 minutes either after long or after short interruption are not counted. The method 3 and 4 have limited the number of counted short interruptions for minimum. The reason is that they limit number of short interruptions associated only with a long interruption. Although aggregation interval associated with a long interruptions differs in the methods 5 a 6, their results are very similar. Occurrence of long interruptions is less frequent than of sequences of short interruptions that are aggregated within 3 minutes after the first short interruption.

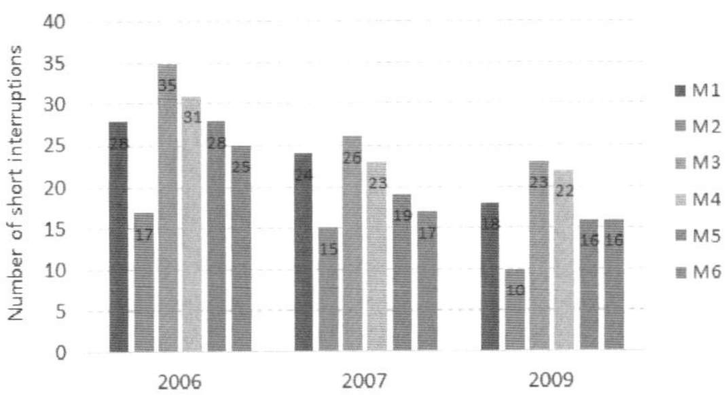

Figure 8. Comparison of various aggregation methods for counting short interruptions

In event records you can find also sequences of voltage dips followed by an interruption or surrounding a swell. The authors dealt with the counting of voltage dips associated with other events in [7]. Question how to count sequences of different events remains unanswered.

IV. CONCLUSIONS

The presented work is initial step in preparation of methodology for reporting and regulation of short interruptions in the Czech Republic. The national regulation authority, Energy Regulatory Office, has planned to introduce reporting of short interruptions in the near future. Before that it is important to define indicators and to establish their limits and procedures for data collection and counting of interruptions, especially when continuity of supply is regulated. But no rules have been introduced for monitoring and evaluation of short interruptions in the Distribution Code.

Methods for counting of short interruptions are presented in the paper and they results are illustrated on the case of a MV/LV substation with extreme occurrence of short interruptions. Proposed methods differ in aggregation method and time-aggregation interval used. Method used for counting of short interruptions greatly affects the number of reported short interruptions. For regulatory purposes, it should respect real impact of interruption sequences on customers.

Because not only interruptions but also voltage dips can cause equipment damage or production outage, the modified indicator MAIFI, so called MAIFI-dip, was introduced. The indicator MAIFI-dip involves short interruptions as well as voltage dips below a reference curve that define severity of voltage dips.

Despite the proposed procedures the discussion on the counting of short interruptions remains open.

ACKNOWLEDGMENT

This work was supported by the project of the University of West Bohemia [SGS-2018-023].

REFERENCES

[1] Council of European Energy Regulators (CEER) - 6th CEER Benchmarking Report on the Quality of Electricity and Gas Supply, 2016.

[2] D. Kouba, D. Mezera, M. Kopt, F. Brož, "Smart improvement of distribution grid reliability," Proceedings of the 23rd International Conference on Electricity Distribution (CIRED 2015), Lyon, France, paper 1240.

[3] D. Kouba, L. Kolář, J. Čeleda, J. Jurík, "MV grids development and automation," Proceedings of the 24th International Conference on Electricity Distribution (CIRED 2017), Glasgow, UK, paper 0750.

[4] Council of European Energy Regulators (CEER) - 5th CEER Benchmarking Report on the Quality of Electricity Supply, 2011.

[5] K. Procházka, F. Brož, J Šefránek, M. Konc, M. Kašpírek, J. Hradecký, "Evaluation of long and short interruptions indices of power supply in the Czech Republic," Proceedings of the 24th International Conference on Electricity Distribution (CIRED 2017), Glasgow, UK, paper 0336.

[6] CIGRE TB 412: Voltage Dip Immunity of Equipment and Installations, CIGRE/CIRED/UIE working group JWG C4.110, April 2010.

[7] M. Tesařová, M. Kašpírek, "Voltage dip assessment in context of voltage quality regulation," Proceedings of the 25th International Conference on Electricity Distribution (CIRED 2019), Madrid, Spain, paper 542, in press.

Emergency Island Grids with Small Hydro Power Stations

Milan Belik
Department of power engineering and ecology
University of West Bohemia
Pilsen, Czech Republic
belik4@kee.zcu.cz

Abstract— **This article focuses on alternative islands powered from small hydro power stations as an emergency power source for specific communities. Grids under consideration should backup regions of small communities. These areas are usually connected to main grid, but are provisionally disconnected from the grid due to some unexpected situations in the main grid. Mentioned small local grids in island mode are not capable to cover entire energy demand in the community, but it has probably the capacity to power main crucial facilities in the community (hospitals, public buildings, traffic lights, street lighting, water stations, broadcasting systems, etc.). Specific sequential coupling to or shut-off from the local grid must be defined. The main task is to discuss and define features of the local island, operating conditions, proper local power sources and particular source cooperation in dependency on the local power capacity, energy demand and duration of main grid outage. Although the complete research deals with particular types of small or distributed power sources, this article focuses mainly on hydro power plants from the category small hydro and thus utilization of adequate hydroengineering structures.**

Keywords: small island grids, temporary power grid outage, power demand and consumption, distributed power sources, small hydro power plants.

I. Introduction

If we focus on emergency situations in power distribution grids, we can simply identify some unexpected situations such as overloads or malfunctions that can disintegrate the main grid and split it into several local islands.

This research deals with local grids covering territory of a small community usually connected to main grid, but temporarily disconnected from the grid due to some technical malfunctions in the distribution grid. Studied island operation should not cover complete energy demand in considered community, but it should be capable to power main crucial public services in the community such as public buildings, traffic lights, street lighting, water stations, broadcasting systems and hospitals. Eventual sequential connection to or shut-off from the island grid must be defined in dependency on the local power sources capacity and duration of the island operation [1].

The blackout situation is more plausible than the society is ready to expect. Fig. 1 shows number of DTS belonging to CEZ Distribuce without power and delay of the blackout as the result of the largest storm winds in the country - Kyril 2007 and Emma 2008 [2].

The paper was supported by SGS-2018-023 and TK01030082.

Other interesting example is from recent past. Heavy snows and storm winds at the beginning of 2019 evoked 20 simultaneous malfunctions on high voltage power lines and plenty of failures on low voltage power lines. More then 30 000 consumers were out of the power throughout the entire area of Czech republic [2].

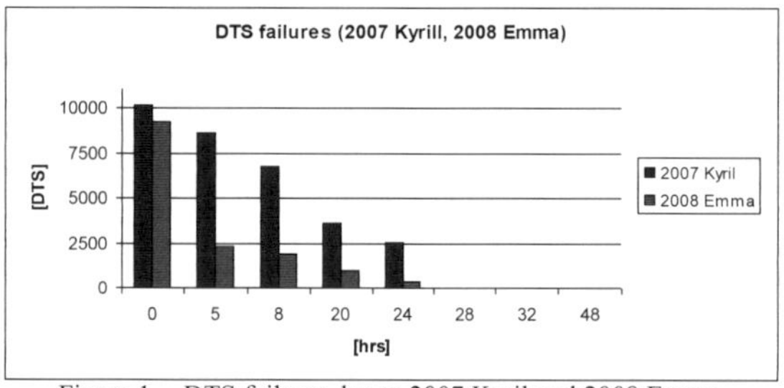

Figure 1. DTS failures due to 2007 Kyril and 2008 Emma.

Traffic chaos, public disturbances, eventually violence and plunder can occur in extreme situations. Some studies anticipate that 160 hours blackout could completely paralyse the society. After 5 days citizens will probably avoid to go out after sunset, because the crime will increase after the dusk. Outage aftermaths could be significantly reduced, if at least some crucial public services would be in operation (traffic lights, water stations, wireless broadcasting, etc.) [1].

The aim is to concretize and discuss features of the local island, conditions of emergency operation, definition of proper power sources and their eventual cooperation.

Although the complete research speculates with all kinds of small local or distributed power sources from conventional plants to territorial renewable energy sources, this article focuses mainly on hydro power plants from the category small hydro.

The study evaluates possibilities of small hydro power plants operating as main power source for the small local grid in island mode. In contrast of conventional energy sources, scanned small hydro power stations have almost zero accumulation capability and thus no regulation abilities. Possible cooperation of 2 or more hydro generators for grid regulation is also discussed. Main features of basic hydraulic engines (water wheels and water turbines) are elaborated with the view of island mode operations [1].

Practical aspects are demonstrated on real generation facilities and communities. The article elaborates situation on the rivers mostly in central and western Bohemia (Berounka, Javornice, Sazava, Uhlava, Uslava, etc.). Existing power stations (e.g. 27 facilities on Berounka river) are coupled with virtual islands covering territories of particular communities

(Beroun, Cernosice, Revnice, Chrast, Plzen, etc.). The problem is discussed from geographical, electrical, technical, ecological and urban point of view. Abandoned facilities capable for revitalization and recommissioning are also analyzed in relevant situations, because this is the only reasonable economical variant for any further development. Emergency operation, grid stability, power generation and consumption of each particular island are explained and feasible upgrade possibilities are illustrated. Cooperation between diverse power sources in island mode is deeply itemized in case study of Plzen island.

Particular variants of the island power supply and grid topology features could be compared. For example, total power installed inside the territory has enough capacity to supply all consumers, but if conventional sources (Pilsen heat plant or Pilsen power station) fail, the island has no abilities for black start, because even if particular small hydro power plants can supply enough energy, they are not able to keep the grid in operation. Features of this emergency island depending only on hydro plants and impact on power supply for Faculty hospital are also summarized [3].

II. Small Hydro Power Stations

Small hydro power station is defined as a power station up to 10 MW. Approximately 30 000 small hydro power plants were in operation during thirtieth of 20th century in the Czechoslovakia territory [4].

Majority of them was decommissioned and abandoned during central electrification process in post war era. Further revitalization of these sources started after the velvet revolution 1989. At present time Czech power grid utilizes except 9 large hydro plants (total installed power 753 MW) also 1614 small hydro power stations with total installed power 348 MW.

Some suitable surviving structures can be reconstructed and recommissioned. Fig. 2 demonstrates conditions of Hubalov plant on Jizera river after privatization in 1992. Together with other local power source could be interesting components of evaluated islands [8].

Figure 2. Conditions of Hubalov plant before reconstruction.

Installed power or theoretical power in the location, flow duration curve in the location and size of the island define conditions for particular plant usage in the project. The purpose allows a sequence of single facilities on the stream to be considered as one complex source.

Rigorous coordination of the group components could allo specific regulation of the island although small hydro power stations have often no reservoir and are operated as nonpondage run of the river plants with no regulation abilities. Regulation of the turbine flow is usually the only control, but the water bypassing turbine can not be accumulated for future utilization. By contrast conventional and accumulation power plants have some reservoir and can be utilized either for power regulation or for stream regulation. But these plants do not belong to the small hydro category and can not be used for purposes of this project. Analogous pumped storage hydro power plants serve as the accumulator and do not directly produce electricity, so that also can not be used in studied islands [8].

Stream energy potential (1) is defined as the work necessary for transportation of a liquid between standings A and B and is equal to the potential energy difference in A (W_A) and B (W_B) depending on the vertical interval (head) h and the mass m or volume V of the liquid. Constant g represents the gravitation field and ρ is the liguid density.

$$W = W_A - W_B = m \cdot g \cdot h = V \cdot \rho \cdot g \cdot h \quad \text{[J]} \qquad (1)$$

The work during the period τ defines the theoretical power of the stream (2). Average period volume depends on the flow Q and is calculated for average flow $Q_{50\%}$ and 95% from maximum flow $Q_{95\%}$.

$$P = \frac{E}{\tau} = \frac{V \cdot \rho \cdot g \cdot h}{\tau} = \frac{V}{\tau} \cdot \rho \cdot g \cdot h = Q \cdot \rho \cdot g \cdot h \quad \text{[W]} \qquad (2)$$

Only communities in locations with sufficient energy potential can utilize purposeful island. Fig. 3 presents sample flow duration curve of Jizera river in the location of reconstructed hydro plant Hubalov.

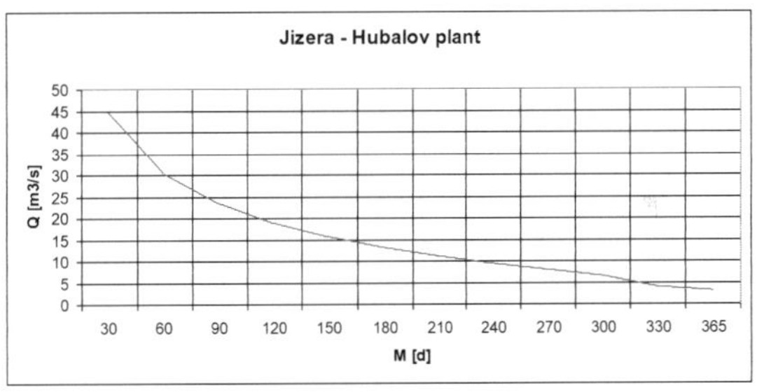

Figure 3. Sample flow duration curve of Jizera river.

The flow duration curve is mathematically a cumulative frequency curve presenting the time percentage with the flow reaching or exceeding given amount during given period. The flow duration curve is only the integral representation of the flow and does not show the chronological sequence of immediate flow.

In fact, it is not allowed to utilize the entire flow. Hydrological and ecological aspects prescribe sanitation flow to bypass the power plant all the time. Sanitation flow is usually defined as Q_{330}, so the utilizable flow (3) is the maximum flow to power the turbine.

$$Q_{util} = Q - Q_{330} \quad [\text{m}^3/\text{s}] \qquad (3)$$

Technical parameters Qmin and Qmax further affect the flow utilizable in the turbine and thus produced power. Data in further calculations come from water level gages operated by CHMI in certain points. Flows in remaining locations are interpolated.

Subsequent question is the selection of proper hydraulic engine and its stability. Mechanical behavior and machinery vibrations spectrum can be simulated from (4). All machines in small hydro plants are used in turbine mode, so the vibration spectrum d_j depends on D_0 (background value of wavelet coefficient), v_j, p_j, q_j (wavelet coefficient of the jth frequency band at particular time caused by the hydropower unit's own mechanical faults), P (hydrogenerator active power) and upper and lower level H1, H2 [5, 7].

$$\left|d_j\right| = D_0 + d_{j-1} + v_j P + \frac{1}{p_j + q_j (H_1 - H_2)^2} \quad \text{[m]} \quad (4)$$

Water turbines are usually high speed reaction machines utilizing two sets of blades. Complicated design makes them very sensitive on water level alternations and mechanical impurities in the water. Efficiency can be higher than 85 %. Many turbine designs have existed during the history, but nowadays only 3 or 4 basic types are being utilized in small hydro power stations [6].

Francis turbine is the oldest reaction radial-axial turbine. Utilizable head varies from meters to hundreds meters and it can be operated in reverse mode as a pump, what makes it the most universal engine.

Kaplan reaction axial turbine is the youngest. Adjustable distributor and runner blades provide wide regulation capability and high efficiency in variable flow rate, but complicated mechanical design increases purchase costs. Small hydro plants often use cheaper modification with fixed runner blades (propeller turbine).

Pelton turbine is the most utilized tangential impulse turbine. High pressure water is distributed through jets onto dish shaped runner blades. The need of high pressure water (high head) limits usage in small hydro plants.

Banki turbine is an impulse turbine with double discharge through the runner and is being successfully used on smaller flows.

Several techniques allow the calculation of the proper turbine type such as the specific power revolutions (5). Real revolutions n, efficiency η, flow Q and the specific energy E (derived from the head h) are the input parameters.

$$n_s = 1200 \cdot n \cdot \frac{\sqrt{Q}}{\sqrt[4]{E^3}} \cdot \sqrt{\eta} \quad [-] \quad (5)$$

The turbine must be verified against cavitation. Very complex and unpleasant mechanical process can irretrievably damage partial components of the turbine. To avoid the cavitaion specific pressure energy must be higher than specific energy of saturised water steam. Maximum sucking height (6) can be approximately calculated from the air pressure p_a, utilized head H and Thom cavitaion factor σ_T.

$$H_S = \frac{p_a}{\rho \cdot g} - \sigma_T \cdot H \quad \text{[m]} \quad (6)$$

The segment of small hydro plants utilizes not only conventional water turbines but also often profits from obsolete water wheels or rather exotic hydraulic piston engines and hydraulic spirals or propellers. Fig. 4 demonstrates usage of non conventional Archimedes spirals in Kacerov hydro plant on river Berounka.

These rather primitive impulse engines provide interesting features for small hydro plants. Above all they can utilize very low head (only few centimeters or less), that turbines are not capable. Next, robust low speed construction with simple runner is neither sensitive for mechanical impurities (sticks, leaves, ice drift, etc.) nor for oscillating water head during floods droughts Decreased water flow during the dry season is partially compensated with increased head while the head loose during floods is vice versa compensated through the higher flow. Although the efficiency is rather low, it can also reach 85 % [6, 8].

Figure 4. Archimedes spirals in Kacerov plant (Berounka river).

Small hydro plant basic features can be calculated from (1) and (2). Table 1 shows sample calculations of theoretical power from actual flow and head.

TABLE I. THEORETICAL POWER [KW]

	Q [m³/s]					
H[m]	1	2	3	4	5	7
1	9,8	19,6	29,4	39,2	49,1	68,7
2	19,6	39,2	58,9	78,5	98,1	137,3
3	29,4	58,9	88,3	117,7	147,2	206,0
5	49,1	98,1	147,2	196,2	245,3	343,4
10	98,1	196,2	294,3	392,4	490,5	686,7
	Q [m³/s]					
H[m]	10	20	30	50	70	100
1	98	196	294	491	687	981
2	196	392	589	981	1373	1962
3	294	589	883	1472	2060	2943
5	491	981	1472	2453	3434	4905
10	981	1962	2943	4905	6867	9810

Conventional weirs utilize head between 2 and 5 m. Table 1 shows that minimal flow must not be lower than 1 m^3/s. Because small streams usually do not reach this value, we must focus on rivers such as Berounka, Sazava, or Jizera.

III. Definitions Of The Local Island

This research focuses on midsize communities. Problems in small urban areas (< 1000 citizens) during emergency situations are relatively not so serious and these habitations have usually only basic public services so they do not meet criteria of this project. Likewise metropolitan areas such as Prague, Brno and Ostrava are excluded because they have own complex emergency strategies. Hospitals are also not included, because they have own backup generators (e.g. Faculty Hospital in Pilsen has 2x 1 MW and 3x 0,8 MW diesel generators) [4].

Island mode means, that some particular object, territory or part of the grid operates independently. Transmission grid disintegration defines the conditions. The island can consist from entire Czech republic or from one or more regions or urban areas. It must respect urban structure, but also geography, grid topology and other technical aspects [5].

Local power island has some specific charactestics:

1. continuous power supply,
2. unstable key quantities (U, f,...),
3. variable grounding conditions,
4. stability of local public services,
5. economical benefits.

Stability and quality of U and f is defined in standards, such as protection settings. Undervoltage and overvoltage protections are to be set to 0,7 and 1,1 Un with speed 0,3 s while frequency protections are usually set to 49 and 51 Hz with reaction time 0,1 s.

Ideal situation of local island can be demonstrated on the Pilsen area. Power consumption consists from industry (24 %), households (33 %), public services (41 %), transportation (10 %) and agriculture (10 %). Power generation is comprised of 2 conventional plants (Pilsen heat plant – 90 MW and Pilsen power station – 111,6 MW), solar power plant (11,2 MW), waste incineration plant Chotikov (10,5 MW), small cogeneration units (3,6 MW) and 9 small hydro power plants (731 kW). Fig. 5 demonstrates sample of historical dependencies between power generation and consumption in this territory. Several situations have proved, that Pilsen can operate in an island mode without any limitations, because the power production in the island is equal or higher than the consumption [4].

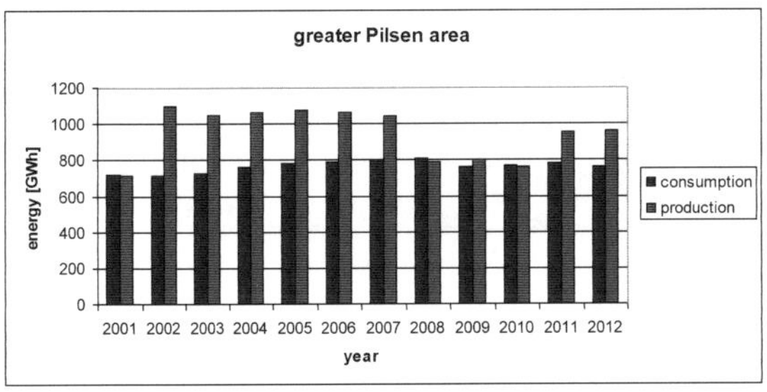

Figure 5. Power production and consumption in greater Pilsen area.

Balanced power production and consumption is the critical condition for the island stability. It also defines features of small hydro plants that can be utilized for that purposes. Pilsen is again ideal example. Although the island has enough capacity to power all consumers in optimal situation, if Pilsen heat plant or Pilsen power station fails, fails also the island. Both sources are not capable to black start and even if particular hydro plants can supply enough energy for that, they are not configured to keep the grid in operation. In fact, they have capacity only to power local emergency island.

IV. Small Hydro Power Stations For Local Islands

Two different situations can be identified. Several existing power structures could be utilized to power the emergency island or some abandoned facility can be reconstructed and recommissioned to supply the island.

First situation can be demonstrated on Berounka river. It is 139 km long, originates in Pilsen at elevation 300,5 m.a.s.l. and flows into Vltava river near Prague at elevation 188,2 m.a.s.l. Average flow varies between 23 m^3/s in Pilsen and 36 m^3/s in the confluence with Vltava. Theoretical power potential P_{teor} = 27 541,63 MW can be calculated according to (1) and (2). Nowadays, there are 23 operating hydro plants with installed power 5, 83 MW utilizing total head 32,11 m.

Respecting geographical situation, grid topology, technical aspects and urbanization around Berounka river, we can find 6 or 8 possible islands using 11 hydro stations. Table 2. summarizes basic parameters of these islands.

TABLE II. Islands On Berounka River

island	citizens	units	head [m]	Q_{aver} [m3/s]	P_{inst} [kW]
Cernosice	7 000	2	3,5	35,6	340
Dobrichovice	3 600	1	1,85	35,6	84
Revnice	3 700	1	0,77	35,6	190
Beroun	19 000	2	3,51	35,6	945
Sykorice	1 100	1	1,3	32,8	15
Roztoky	1 200	1	1,49	31	50
Chrast	1 900	1	3	23,95	138
Pilsen	171 000	2	3,07	23,95	1170

First island on Berounka river (Cernosice) covers area 9 km^2 of setlements Cernosice, Vraz and Mokropsy with total population 7 000. Fig. 6 shows flow chart and power generation for Cernosice island. 2 particular hydro plants (Cernosice and Mokropsy) are connected to 22 kV grid and would operate in parallel. If Cernosice plant (250 kW) fails, Mokropsy plant (90 kW) can not cover entire demand. Local conditions allow power upgrade up to 262 kW in Mokropsy.

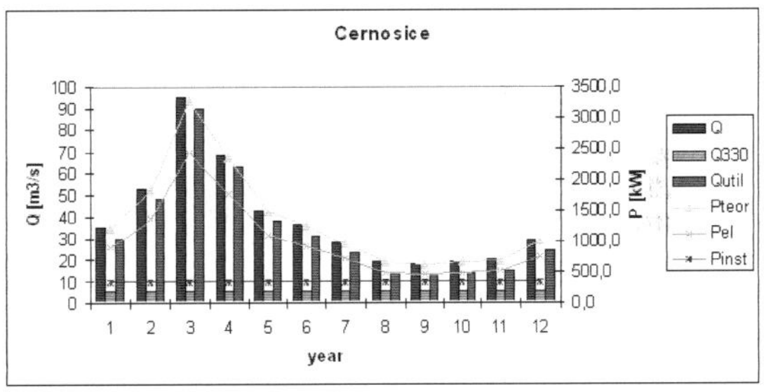

Figure 6. Cernosice island.

Last island on Berounka (Pilsen) covers area 138 km^2 of greater city Pilsen with population 171 000. Fig. 7 shows flow chart and power generation for this island from river Berounka. 2 individual power stations (Bukovec 540 kW and Papirna 630 kW) are connected to 22 kV grid and operate in parallel mode.

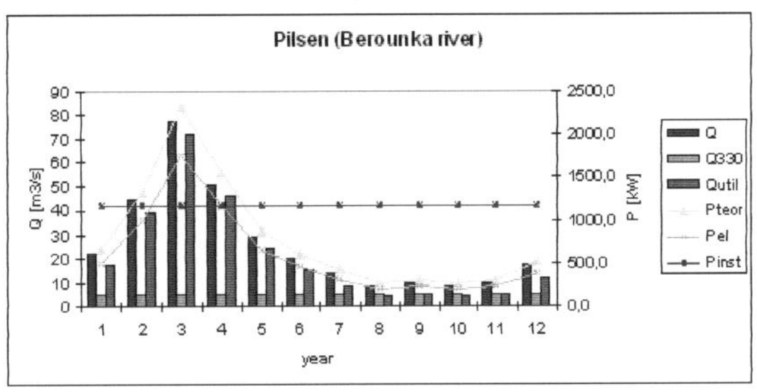

Figure 7. Pilsen island.

Second situation can be demonstrated on reconstruction of Hubalov Plant on Jizera river. It is 164,6 km long, originates at elevation 980,2 m.a.s.l. under Smrk mountain and empties into Labe river near Lazne Tousen at elevation 169,3 m.a.s.l. Average flow varies between 5 m^3/s in Jablonec (km 131) and 24 m^3/s in the confluence with Labe. Theoretical power potential is P_{teor} = 15 911,82 MW. Power plant Hubalov is an excellent example of reconstructed facility to power an island.

Fig. 8 demonstrates power production in Hubalov before reconstruction. Drops in particular year are results from unexpected repairs, maintenance and floods.

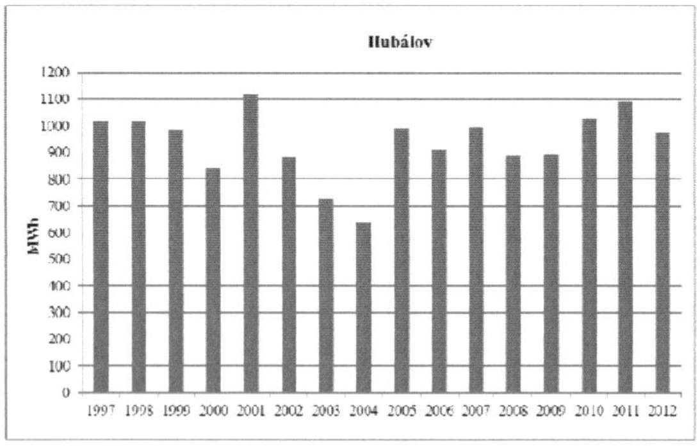

Figure 8. Hubalov plant production before reconstruction.

RETScreen software was used to simulate optimal state of Hubalov plant. Results are displayed on Fig. 9. Flow duration curve is based on data from CHMI.

Simulation and recommended reconstruction is based on 2 turbines (112 kW and 82 kW) allowing wide regulation range and utilization of variable flow between 2,7 m^3/s and 14,0 m^3/s.

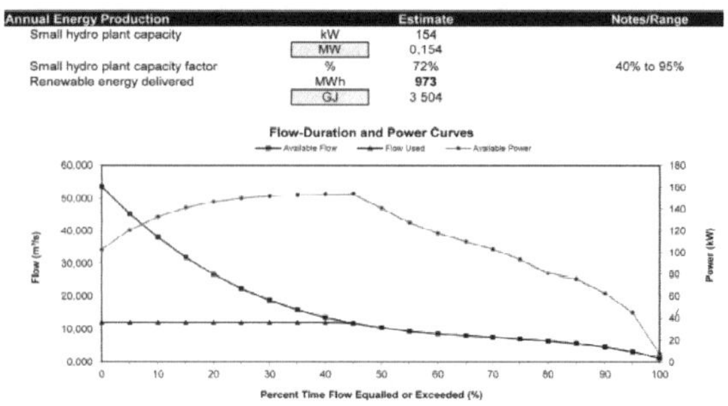

Figure 9. Simulation of Hubalov plant in RETScreen.

Subsidiary problem of the plant was oscillating flow as the result of insufficient machinery of trash racks. The bar was often clogged as evident from Fig. 10.

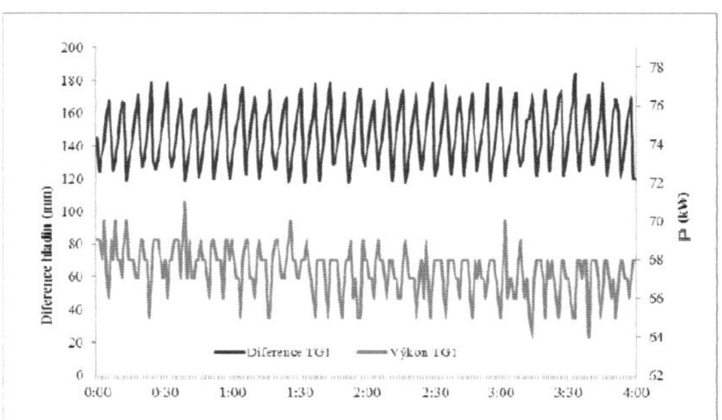

Figure 10. Oscilating flow after trash rack in Hubalov plant.

Fluctuating power generation can cause also problems with power quality in the island. The power factor should stay between 0,95 and 0,98 inductive. Although mentioned flow fluctuations do not increase requirements on fast compensation, the compensation should be adjustable. The voltage lift as the result of power source connection or reconnection should not exceed 2 % in HV grids and 3 % in LV grids. The condition is assessed according to rate of short circuit powers k (7).

$$k = \frac{S_{kV}}{S_{A\max}} \quad [-] \qquad (7)$$

S_{kv} indicates grid short circuit power while S_{Amax} stands for the sum of short circuit power for connected or reconnected sources. Voltage fluctuation in the junction point generate no false influence on grid when the voltage change does not exceed the limits and the period is not shorter than 1,5 min.

If the fluctuation or switching frequency is low, the voltage changes can be higher. The change Δu_{max} can be estimated according to grid short circuit power S_{kV} and the source short circuit power S_n (8).

$$\Delta u_{\max} = k_{iMAX} \frac{S_n}{S_{kV}} \quad [-] \qquad (8)$$

Island Loukovec covers territory 11 km2 of communities Loukovec, Hubalov, Loukov and Koryta with population 1 600. Fig. 11 shows flow chart and power generation for this island. 1 reconstructed hydro plant 194 kW connected to 22 kV grid can cover the emergency demand.

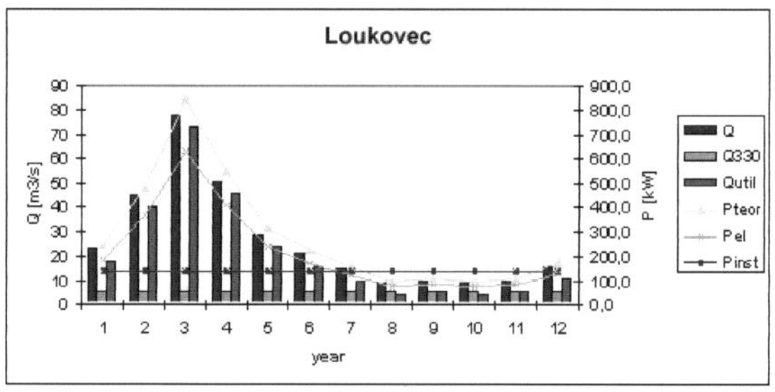

Figure 11. Loukovec island.

V. Results And Conclusions

Presented calculations and simulations proved that small hydro power plants are capable to power emergency island in specific communities. Even if the theoretical stream potential is high enough, urban dispositions, grid topology, technical aspects and geographical situation in the region can disappoint these plans.

Usage of existing facilities is demonstrated on the river Berounka with 6 – 8 emergency islands using 11 plants with 2 932 kW from total 22 plants with 5 097 kW installed along entire river length, while theoretical hydro potential is 10 998 kW.

Usage of reconstructed, revitalized or recommissioned facilities is presented on Hubalov plant on Jizera river powering island on territory of 4 communities with 1 600 residents.

References

[1] M. Balaban, J. Ludvik, and L. Stejskal, "Strategické trendy globálního vývoje," Roman Karpas RK, Prague, 2012.

[2] https://www.cezdistribuce.cz/cs/pro-media/tiskove-zpravy/302.html, 2019.

[3] L. Bicek, "Nouzove zdroje v nemocnici, bakalarska prace," ZCU, Pilsen, 2012.

[4] V. Muzik, " Connection of electric power sources in city of Pilsen – island performance – Master thesis," ZCU, Pilsen, 2014.

[5] V. Kukharchuk, S. Katsyv, V. Hraniak, "Vibro-forecasting of fault development in hydropower units," Bulletin of the Karaganda University, Karaganda, 2018.

[6] O. A. Kovalchuk, V. V. Kulyk, O. B. Burykin," Optimisation of joining circuits of extended power sources to local electric systems ," Technical Electrodynamics, 2012, ISSN 1607-7970.

[7] V. V. Kukharchuk, V. V. Bogachuk, V. F. Hraniak, "Method of magneto-elastic control of mechanic rigidity in assemblies of hydropower units," Proc. SPIE, Optical Fibers and Their Applications, 2017.

[8] M. Kubin, "Rozvoj energetiky v Ceskoslovensku," CEZ, Prague, 1989.

Adequacy Analysis in Bulk Power Systems Based on the Analytical-Probabilistic Method

Vladislav P. Oboskalov
Department of APS Ural Federal University
SEC "RSLSM" of Ural Branch of RAS
Ekaterinburg, Russia
vpo1704@mail.ru

Anatolijs Mahnitko, Renata Varfolomejeva
Electrical power systems dept.
Riga Technical University
Riga, Latvia

Pavel Y. Gubin, Sergey A. Gusev, Rustam T. Valiev
Department of automated power systems
Ural Federal University
Ekaterinburg, Russia
p-tul@yandex.ru, s.a.gusev@urfu.ru,
r.t.valiev@urfu.ru

Abstract — **This paper presents an application of the probability-based analytical procedure in the framework of the adequacy assessment problem for bulk power systems. The proposed procedure includes two stages. At the first stage, generation capacity allocates between load buses in proportion with their power. Secondly, real power flows are distributed proportionally to the voltage angle difference, which helps to account for power losses in the power system adequacy indices assessment. The results show that power losses significantly affect the resultant adequacy indices of a bulk power system, therefore power losses analysis should be considered as an essential part of the adequacy assessment process. The proposed mathematical models and algorithms have been tested with a sample power system. For comparison purposes, the Monte Carlo simulation technique have been used as a benchmark. The comparison of the case study results has shown a high computational efficiency and an acceptable accuracy of the proposed probabilistic-analytical method (PAM).**

Keywords: Power systems, adequacy indices, power imbalance, Monte Carlo simulation.

I. Introduction

The Monte Carlo simulation (MCS) technique is the most widely used method for calculating adequacy indices of bulk power systems in practice [1 – 5]. The technique allows to get relatively accurate results with account for a lot of practical constraints, but its computational efforts are very high usually.

One of the most important aspects which have to be considered in the process of adequacy indices calculating is a load shedding strategy. Mathematically, the dispatch strategy is often formulated as a large-scale Mixed-Integer Linear-Programming problem. In case of the MCS, the dispatch strategy for each random state is considered in a deterministic way, and linear [6] and non-linear [7, 8] procedures can be used to find least-cost solution in which no operational constraints (e.g. ramping, minimum up/down time, transfer capacity limits, etc.) are violated.

The liberalization of power and energy markets along with increased focus on renewable energy sources which is motivated by both ecological and economical incentives, enlarges the number of parameters in a power system modelling process [9-11]. As it is mentioned in the [12], each step towards increasing the level of data details significantly increases the complexity of the mathematical problem to be solved and thereby the development of new fast-calculating approaches for generation adequacy indices (GAI) problem becomes urgent.

Currently, the complexity of a market simulation process requires to take certain assumptions. One of the key assumptions in the adequacy assessment is that an electrical power system is considered as a number of concentrated power systems (i.e. power systems with unlimited transfer capacities) which are connected in a power pool. This allows to avoid usage of mathematical methods for solving system of non-linear power flow equations and finally to employ only linear programming and flow-oriented models (Ford-Fulkerson algorithm etc.).

Mean values of power losses in power systems and load shedding mean values are of the same order of magnitude. The excluding of power losses could potentially result in a significant GAI error [5, 7 - 8]. It is well known that power losses depend on the square of the tie line current which, in turn, defines in accordance with the Ohm's and the Kirchhoff's laws. Considering power losses in adequacy assessment requires additional constraints in the power system model to be included.

The additional constraints consider distribution of active and reactive power flows. In accordance with the fast-decoupled load flow model, per-unit reactive flows are in proportion to a difference in per-unit voltages at EPS buses and per-unit real power flows are in proportion to the voltage angular difference across EPS buses. It results in necessity of power system steady-state parameters analysis which should be formulated as a probabilistic problem (load and generation are considered as stochastic parameters) due to characteristic features of adequacy calculations. This problem is also known as the probabilistic optimal power flow problem [13-17]. This paper presents the combined probabilistic analytical method for calculating adequacy indices with respect to EPS power losses and dependency of active power flows upon voltage vector angles at transmission ends.

II. The Mathematical Formulation of the Problem

The main aim of the paper is to calculate generation adequacy indices taking into account power losses in a power system. The load and the available capacity of generators in the power system are considered as random values with predetermined mean values and variances which are obtained from statistical data. The distribution of per-unit real power flows is in proportion to the voltage angular difference across the electrical network buses and is considered as an additional constraint. The main assumptions of the problem can be formulated as follows:

- Reactive power balances and stochastic nature of tie-lines state are not considered.
- Load and available capacity of generators are described by the normal probabilistic distribution.
- Voltage magnitudes are constant and equal to a rated power system voltage

A. A power flows distribution

The strategy of proportional load shedding distribution can be expressed as the equality of power surplus indices in the nodes (1):

$$\bar{G}+\Delta\bar{G}=\mu\bar{L}, \tag{1}$$

where: $\bar{L}$ is the vector of nodal active loads, $\bar{G}$ is the vector of available active capacity of generators, $\Delta\bar{G}$ is the vector of nodal active power injections from power system, and μ is generation power surplus index, which can be calculated as shown in (2).

$$\mu=G_\Sigma/L_\Sigma=\left(\sum G_i\right)/\left(\sum L_i\right) \tag{2}$$

$\{G_i, L_i, i=1,\dots n\}$ – components of vectors $\bar{G}$, $\bar{L}$

According to (1) vector of nodal active power injections $\Delta\bar{G}$ can be written as:

$$\Delta\bar{G}=\mu\bar{L}-\bar{G}. \tag{3}$$

If all electrical parameters are expressed in per-units, the nodal voltage equations can be presented as [17, 18]:

$$B\bar{\delta}=\Delta\bar{G}=\mu\bar{L}-\bar{G}, \tag{4}$$

where: B – the imaginary component of nodal conductance matrix; $\bar{\delta}$ – the vector of voltage angles.

In accordance with (4), voltage angles can be calculated as follows:

$$\bar{\delta}=B^{-1}\left(\mu\bar{L}-\bar{G}\right) \tag{5}$$

The power flows vector, that corresponds both to the load shedding strategy and the additional constraint can be calculated by equation (6):

$$\bar{X}=D_{B_{br}}\hat{M}^T\bar{\delta}=A\left(\mu\bar{L}-\bar{G}\right), \tag{6}$$

where:

$\dim(M)=n\times k; \dim(A)=k\times(n-1), D_{B_{br}}, \dim\left(D_{B_{br}}\right)=k\times k$ – branch reactive conductivity, diagonal $\left(\dim(B_{br})=k\times 1\right)$; $\hat{M}^T$ – the matrix of incidence without one row, n- the number of nodes, k– the number of branches

B. Power flows mean values

The mean value of the random values product equal to the product of their mean values. This mathematical feature allows to write the next equation:

$$\bar{m}_X=-A\left[m_{G_\Sigma}E\left(\frac{\bar{L}}{L_\Sigma}\right)-\bar{m}_G\right], \tag{7}$$

where: $m_{L_\Sigma}=\sum_{i=1}^{n}m_{L_i}$, $m_{G_\Sigma}=\sum_{i=1}^{n}m_{G_i}$.

Numerator and denominator in the ratio $E\left(\bar{L}/L_\Sigma\right)$ are dependent values. Division of two dependend stochastic parameters forms a new probabilistic relationship and requires an additional probabilistic analysis. For this purpose, let's consider the equation:

$$\bar{\psi}\left(\bar{L}\right)=\frac{\bar{L}}{L_\Sigma};\quad\left\{\psi_i=\left(\frac{L_i}{L_\Sigma}\right),\quad i=1,\dots,n\right\}.$$

The quadratic approximation of the function $\psi_i\left(L_i,L_\Sigma\right)$:

$$\psi_i=\psi_i\left(\bar{m}_L\right)+\sum_{j=1}^{n}\frac{\partial\psi_i\left(\bar{m}_L\right)}{\partial L_j}\tilde{L}_j=+$$
$$+\frac{1}{2}\sum_{j=1}^{n}\sum_{k=1}^{n}\frac{\partial^2\psi_i\left(\bar{m}_L\right)}{\partial L_j\partial L_k}\tilde{L}_j\tilde{L}_k,\quad i=1,\dots,n \tag{8}$$

where: $\tilde{L}=\bar{L}-\bar{m}_L$ is the centered random variable.

After taking derivations and collecting terms, (8) can be written as:

$$\frac{1}{2}\sum_{j=1}^{n}\sum_{s=1}^{n}\frac{\partial^2\psi_i(\bar{m}_L)}{\partial L_j\partial L_s}\tilde{L}_j\tilde{L}_s=-\frac{1}{m_{L_\Sigma}^2}\tilde{L}_i\tilde{L}_\Sigma+\frac{m_{L_i}}{m_{L_\Sigma}^3}\tilde{L}_\Sigma\tilde{L}_\Sigma$$

Mean value of this function:

$$E(\psi_i)=\frac{m_{L_i}}{m_{L_\Sigma}}+\frac{m_{L_i}}{m_{L_\Sigma}^3}D_{L_\Sigma}-\frac{1}{m_{L_\Sigma}^2}\operatorname{cov}\left(L_iL_\Sigma\right).$$

At the same time the correlation moment of the loads L_i, L_Σ is:

$$\operatorname{cov}\left(L_i,L_\Sigma\right)=\operatorname{cov}\left(L_i,\sum_{j=1}^{n}L_j\right)=D_{L_i}$$

where: D_{Li} - load variance in the node i.

According to the equations presented above, mathematical expectation of (8) can be formulated and thus, the power flows mean value is obtained as follows:

$$E(\bar{\psi})=\frac{1}{m_{L_\Sigma}}\left(1+\frac{D_{L_\Sigma}}{m_{L_\Sigma}^2}\right)\bar{m}_L-\frac{1}{m_{L_\Sigma}^2}\bar{D}_L.$$

Taking into account the equations presented above, power flows' mean values can be represented as:

$$\bar{m}_X = -A\frac{m_{G_\Sigma}}{m_{L_\Sigma}}\left[\left(1+\frac{\sigma^2_{L_\Sigma}}{m^2_{L_\Sigma}}\right)\bar{m}_L - \frac{1}{m_{L_\Sigma}}\bar{D}_L\right] - A\bar{m}_G.$$

C. The power flows correlatin matrix

The strategy of proportional load shedding distribution results in a covariance between tie line power flows. In order to find the power flows correlation matrix, the nodal power injections function has to be considered:

$$\Delta\bar{G} = \bar{\varphi}(\bar{L},\bar{G}) = \frac{G_\Sigma}{L_\Sigma}\bar{L} - \bar{G}$$

In accordance with (6) the power flows correlation matrix can be defined as $K_X = AK_\varphi A^T$. The matrix K_φ can be calculated using the load and generation correlation matrices and linear approximation of the function $\bar{\varphi}(\bar{L},\bar{G})$ shown in (9):

$$\varphi(\bar{L},\bar{G}) = \varphi(m_L, m_G) + (J_L \quad J_G)\begin{pmatrix}\bar{L}\\ \bar{G}\end{pmatrix}. \qquad (9)$$

where: J_L, J_G are Jacobian matrices obtained for the case of $\bar{L} = \bar{m}_L$, $\bar{G} = \bar{m}_L$:

$$J_L = \left[\frac{\partial\varphi_i}{\partial L_j}\right] = m_\mu R; \quad J_G = \left[\frac{\partial\varphi_i}{\partial G_j}\right] = -R; \qquad (10)$$

$$\frac{\partial\varphi_i}{\partial L_i} = \mu_m\left(1 - \frac{L_i}{L_\Sigma}\right); \quad \frac{\partial\varphi_i}{\partial L_j} = -\mu_m\frac{L_i}{L_\Sigma}, \; j \neq i;$$

$$\frac{\partial\varphi_i}{\partial G_i} = \frac{L_i}{L_\Sigma} - 1; \quad \frac{\partial\varphi_i}{\partial G_j} = \frac{L_i}{L_\Sigma}, \; j \neq i;$$

$$\mu_m = m_{G_\Sigma} / m_{L_\Sigma}$$

$$R = \left[R_{ii} = 1 - \frac{m_{L_i}}{m_{L_\Sigma}}; \quad R_{ij} = -\frac{m_{L_i}}{m_{L_\Sigma}}, j \neq i\right]$$

or in the matrix form:

$$R = I - \frac{1}{m_{L_\Sigma}} diag(m_L)[1]_{n,n};$$

where: I – unity matrix; $diag(m_L)$ – diagonal matrix of load mean values; $[1]_{n,n}$– n-degree quadratic matrix consisting of identities.

Nodal power injections correlation matrix could be calculated as follows:

$$K_\varphi = (J_L \quad J_G)\begin{pmatrix}K_L & 0\\ 0 & K_G\end{pmatrix}(J_L \quad J_G)^T = J_L K_L J_L^T + J_G K_G J_G^T.$$

where: K_{LL}, K_{GG} –diagonal covariance matrices for load and generation respectively.

The matrix K_φ can be calculated by the following equation:

$$K_\varphi = (\mu_m R \quad -R)\begin{pmatrix}K_L & 0\\ 0 & K_G\end{pmatrix}(\mu_m R \quad -R)^T = RK_N R^T$$

Power imbalances correlation matrix can be represented by:

$$K_N = \mu_m^2 K_L + K_G;$$

The equation (6) allows to obtain the power flows correlation matrix:

$$K_X = AK_\varphi A^T = (AR)K_N(AR)^T.$$

D. Power losses

One of the possible solutions for the problem of accounting power losses in the process of power systems adequacy assessment is the using of power losses indices [1 – 8]. For extra high voltage networks ($U \geq 500$ kV) power losses is approximately k_π=0,5-1% of whole power delivery to a power system [19]. The sum of all positive nodal power injections (3) can be considered as the overall power delivery to the power system. Overall power losses can be calculated as follows:

$$\pi_\Sigma = k_\pi \sum_{i=1}^{n}(\Delta G_i > 0)\cdot$$

In case of the normal probability distribution function of nodal power injections, it is possible to calculate mean values $E(\pi_\Sigma)$ and variance $D(\pi_\Sigma)$ of overall power losses by means of mean value $M(a,b,m,D)$ and variance $D(a,b,m,D)$ of reduced distribution function [20]:

$$E(\pi_\Sigma) = k_\pi \sum_{i=1}^{n} E(\Delta G_i > 0) = k_\pi \sum_{i=1}^{n} M(0,\infty,m_{\Delta G_i},D_{\Delta G_i});$$

$$D(\pi_\Sigma) = k_\pi^2 \sum_{i=1}^{n} D(\Delta G_i > 0) = k_\pi^2 \sum_{i=1}^{n} D(0,\infty,m_{\Delta G_i},D_{\Delta G_i}),$$

where:

$$M(a,b,m,D) = aF(a) + b[1-F(b)] + m[F(b)-F(a)] + D[f(a)-f(b)]; \qquad (11)$$

$$D(a,b,m,D) = a^2F(a) + b^2[1-F(b)] + m^2[F(b)-F(a)] + 2mD[f(a)-f(b)] + D[(m+a)f(a)-(m+b)f(b)] - m_p^2; \qquad (12)$$

where: $F(X)$ and $f(x)$ – cumulative distribution function and probability density distribution function respectively.

In practice, real power losses could vary considerably from the power loss indices and to increase the accuracy of the power losses accounting it is necessary to consider them another way. Active power losses could be expressed through active and reactive power flows:

$$\pi_\Sigma = \sum_{j=1}^{k} R_j \frac{P^2_{j(s)} + Q^2_{j(s)}}{U_s^2} \qquad (13)$$

where: s – number of an incidental node of tie line j.

In the adequacy assessment problem, there is no calculation of reactive power flows distribution due to considerable uncertainty of power system parameters. The load structure of concentrated power systems differs slightly from each other, consequently, power factors $\{tg\phi_i, i=1,\ldots,n\}$ for each system could be set equal. Under such an assumption the distribution of reactive power flows is in proportion to the active power flows distribution:

$$\pi_\Sigma = (1 + tg^2\varphi)\sum_{j=1}^{k} R_j x_j^2 \qquad (14)$$

where: x_j – active power flow through the tie-line j.

Power line losses are proportional to square of a power line flows that are defined as follows:

$$x_j^2 = \left(\tilde{x}_j + m_{x_j}\right)^2 = \tilde{x}_j^2 + 2\tilde{x}_j m_{x_j} + m_{x_j}^2 \quad (15)$$

Taking into account (15), equation (13) can be written as:

$$\pi_\Sigma = \left(1 + tg^2\varphi\right)\left(\pi\left(m_{\tilde{X}}\right) + 2\sum_{j=1}^{k} R_j m_{x_j} \tilde{x}_j + \sum_{j=1}^{k} R_j \tilde{x}_j^2\right).$$

The vector of power line losses could be obtained by the next equation:

$$\bar{\pi} = \left(1 + tg^2\varphi\right)\left(\bar{\pi}\left(m_{\tilde{X}}\right) + 2\left[R_i m_{X_i}\right]\tilde{X} + \left[R_i\right] diag(\tilde{X}\tilde{X}^T)\right) \quad (16)$$

where: $[R_i]$ – diagonal matrix of elements $\overline{R_i}$,$diag(A)$ – the vector of diagonal elements of matrix A.

Mean value of the lines power losses:

$$m_{\bar{\pi}} = \left(1 + tg^2\varphi\right)\left(\bar{\pi}\left(m_{\tilde{X}}\right) + \left[R_i\right] diag(K)\right)$$

where: $diag(K)$ – vector of diagonal elements of the power flows correlation moment matrix.

The matrix of correlation moments of overall power losses can be defined by linear component of (16):

$$\text{cov}(\bar{\pi}) = \left(1 + tg^2\varphi\right)^2 \left[R_i m_{X_i}\right] K_X \left[R_i m_{X_i}\right]^T$$

Mean value of overall power losses:

$$E(\pi) = \left(1 + tg^2\varphi\right)\left(\pi\left(m_{\tilde{X}}\right) + \sum_{j=1}^{k} R_j \sigma_j^2\right);$$

Quadratic representation of (14) can be used for obtaining variance of overall power losses:

$$\pi_\Sigma = \left(1 + tg^2\varphi\right)\bar{X}^T \left[R_i\right]\bar{X}$$

In assuming of the normal probability distribution function of power flows, mean value of total power losses can be calculated as follows [21]:

$$\text{var}\left(\pi_\Sigma\right) = \left(1 + tg^2\varphi\right)^2 \left[2\text{tr}\left(\left(R_D K_X\right)^2\right) + 4\bar{m}_X^T R_D K_X R_D \bar{m}_X\right]$$

where tr(.) –the trace of matrix.

Obtained mean values and variances of overall power losses can be used to correct the equation (2) in the following way:

$$\mu = G_\Sigma / \left(L_\Sigma + \pi_\Sigma\right) = \left(\sum G_i\right) / \left(\pi_\Sigma + \sum L_i\right)$$

The overall power losses can be distributed across the system by attributing it to each load node as an additional load in accordance with the strategy of proportional load shedding.

In general, the correction of power losses is an iterative process, however test results have shown that single iteration is sufficient for practical calculations.

E. *Probabilistic characteristics of the power shortage of a bulk power system*

The surplus index defined by (2) can be used for calculating probability, mean value and variance of power shortage in a bulk power system and for calculating the same parameters for each concentrated power system in case of using proportional power imbalance distribution strategy.

The fact that:

$$\frac{\partial \mu}{\partial L_i} = -\frac{G_\Sigma}{L_\Sigma^2}; \quad \frac{\partial \mu}{\partial G_i} = \frac{1}{L_\Sigma}; \quad \frac{\partial^2 \mu}{\partial L_i^2} = \frac{2G_\Sigma}{L_\Sigma^3};$$

$$\frac{\partial^2 \mu}{\partial L_i \partial G_j} = \frac{\partial^2 \mu}{\partial G_i \partial L_j} = -\frac{1}{L_\Sigma^2}; \quad \frac{\partial^2 \mu}{\partial G_i^2} = 0,$$

and quadratic approximation of (2) allow to represent the surplus index in the following way:

$$\mu(G_\Sigma, L_\Sigma) = \frac{m_{G_\Sigma}}{m_{L_\Sigma}} - \frac{m_{G_\Sigma}}{m_{L_\Sigma}^2}\tilde{L}_\Sigma + \frac{1}{m_{L_\Sigma}}\tilde{G}_\Sigma - \\ -\frac{1}{m_{L_\Sigma}^2}\sum_{i=1}^{n}\sum_{j=1}^{n}\tilde{L}_i\tilde{G}_i + \frac{m_{G_\Sigma}}{m_{L_\Sigma}^3}\sum_{i=1}^{n}\tilde{L}_i^2. \quad (17)$$

The Mean value of the power surplus index can be written as follows (G_Σ, L_Σ – independent random values):

$$m_\mu = \frac{m_{G_\Sigma}}{m_{L_\Sigma}}\left(1 + \frac{\sigma_{L_\Sigma}^2}{m_{L_\Sigma}^2}\right);$$

By using the linear component of (17) the variance of the power surplus index can be represented as:

$$\sigma_\mu^2 = \frac{1}{m_{L_\Sigma}^2}\left(\frac{m_{G_\Sigma}^2}{m_{L_\Sigma}^2}\sigma_{L_\Sigma}^2 + \sigma_{G_\Sigma}^2\right).$$

Probabilistic characteristics of the power shortage in a bulk power system can be calculated based on the mean value and the variance of the power surplus index. In particular, the probability of power shortage can be calculated as:

$$p(\mu < 1) = F_\mu(1, m_\mu, \sigma_\mu)$$

The mean value and the variance of the power shortage of a bulk power system can be obtained by using (11) and (12):

$$E(def) = m_{L_\Sigma}\left(1 - M\left(0;1;m_\mu;D_\mu\right)\right); \\ D(def) = m_{L_\Sigma}^2 D\left(0;1;m_\mu;D_\mu\right), \quad (18)$$

It should be noted that the equations (18) are deduced in the assumption of the normal probability distribution function of the surplus index. The ratio of two normally distributed random variables is asymmetrically distributed due to the asymmetrical probability distribution of the hyperbolic function $1/x$ as shown in Fig. 1. The using of the gamma distribution provides more accurate results in comparison to the results obtained by using (18) but an error is relatively small.

The relationship between the power shortage and the power surplus index allows using power imbalance $N_\Sigma = L_\Sigma - G_\Sigma$ in the mathematical equations presented above:

$$E(def) = M\left(0, \infty, m_N, D_N\right);$$

$$D(def) = D(0, \infty, m_N, D_N).$$

When power shortage is positive, it can be corrected only by load shedding although the whole power balance includes a generation component also. Numerical integration methods and equations of reduced distributions provide more accurate calculation results when the load variance is used:

$$E(def) = M(0, \infty, m_N, D_L);$$

$$D(def) = D(0, \infty, m_N, D_L).$$

III. Test calculations

The test scheme that have been used for calculations is shown in Fig. 2. The data for buses and tie lines are presented in Table 1 and Table 2 respectively.

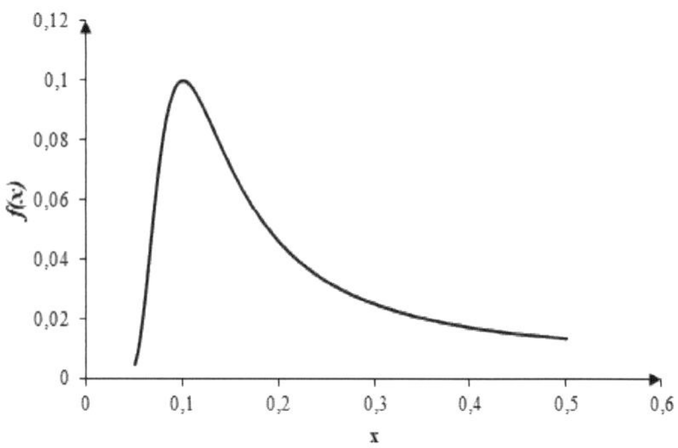

Figure 1. Probability density z=1/x for mx=10,σx=4

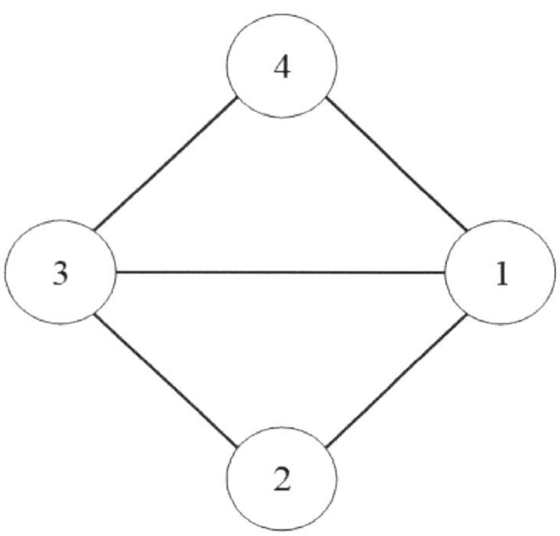

Figure 2. Test system

The Monte-Carlo simulation technique has been used as a benchmark method (the number of samples - 100000). The table 3 represents the power flows mean values (MP_{ij}), the power losses mean values ($M\pi_{ij}$) and the probabilities of the power flows exceeding the tie-lines capacities. The results obtained by using the combined probabilistic analytical method (PAM) in comparison to the results obtained by Monte-Carlo simulation technique is accurate enough, but the computation burden of the first method is far less. (In case of using the MATLAB (2017a) software package, Intel core i7 CPU, 0,08 sec. for combined method vs. 11,8 sec. for the Monte-Carlo simulation technique).

Table 1. load and generation Parameters of bulk power system

Node	m_L, MW	σ_L, MW	m_G, MW	σ_G, MW
1	1000	100	1500	225
2	2000	200	4000	600
3	500	50	1200	180
4	5000	500	4000	600
Bulk Pow. System	8500	550	10700	896

Table 2. Parameters of the power system tie-lines

№	Fr	To	R, Ohm	X, Ohm	B, Mho	Transfer Capability, MW
1	1	2	10	100	0.10	500
2	1	3	5	50	0.02	500
3	1	4	5	50	0.02	500
4	2	3	3	33	0.03	1500
5	3	4	1	10	0.10	2000

Table 3. Power flows test results

	Monte-Carlo Method			Prob. Analytical Method		
№	MP_{ij}, MW	$M\pi_{ij}$, MW	Pr over	MP_{ij}, MW	$M\pi_{ij}$, MW	Pr over
1	-329	4.31	0.112	-327	4.30	0.107
2	101	0.20	0.000	101	0.20	0.000
3	465	4.30	0.409	463	4.30	0.401
4	1151	15.90	0.198	1145	15.70	0.193
5	1816	13.20	0.340	1813	13.10	0.334
	Σ	37.90		Σ	37.60	
	RMS	22.20		RMS	22.00	

The table 4 shows the probabilistic characteristics of the power shortage for test scheme in case of considering it as a bulk power system. Three main cases have been considered in test calculations:

Case 1: without considering reactive power flows and power losses;

Case 2: $tg\varphi = 0$ – without considering reactive power flows but considering power losses;

Case 3: $tg\varphi = 1$ – considering reactive power flows which correspond to $tg\varphi$=1 and power losses. It can be seen that power losses ($m\pi$, $\sigma\pi$) strongly affect resulting probability $\mathcal{P}(def)$ of system power shortage and loss of energy mean value M_{DP}. It should be noted that it is not sufficient in terms of power losses calculations to consider only active component of a power flow through the tie-lines (second case $tg\varphi = 0$) and reactive power flows have not less effect on total power losses over bulk power system (the third case $tg\varphi \neq 0$).

Table 4. Indices Results of bulk power system

Loss status	Method	P(def)	M_{DP}, MW	m_μ	σ_μ	m_μ, MW	σ_μ, MW
Without losses	MC	0.0179	6.93	1.264	0.135	0.0	0.0
	CPA	0.0182	6.95	1.264	0.133	0.0	0.0

Loss status	Method	P(def)	M_{DP}, MW	m_μ	σ_μ	m_μ, MW	σ_μ, MW
tg =0	MC	0.0196	7.54	1.255	0.133	42.2	23.1
	CPA	0.0203	7.83	1.257	0.132	41.8	22.9
tg [illegible]	MC-100000	0.0217	8.40	1.251	0.132	84.2	46.4
	MC-1000	0.019	7.29	1.260	0.130	81.9	45.7

IV. Conclusions

A combined analytical probabilistic method is proposed in this paper for calculating bulk power system adequacy indices based on power flows distribution strategy. The strategy depends on active power flows distribution among tie lines in proportion to angle difference between their end points and reactive power flows proportionally to nodal voltage absolute values considering linear dependencies of angles and values of nodal powers. It provides an opportunity to define probabilistic characteristics of transfer angles and power flows through tie lines and so, determine the power losses in power systems and risk of power flows exceeding tie-lines capacity constraints.

The results have shown that power losses considerably affect adequacy indices and supported a high efficiency of the proposed method for calculating bulk power system adequacy indices.

References

[1] Yu. N. Rudenko and M.B. Cheltsov," Reliability and reservation in power systems," Novosibirsk: Science, 1974.

[2] R. Billinton and L. Wenyan, "Reliability assessment of electric power systems using Monte Carlo methods," 1st ed. New York, Springer, 1994. 361 c.

[3] Yu.N. Rudenko, "Reliability of power systems and their G," Reliability of power systems, Vol. 4, Under the general ed.. T.2., Moscow: Energoatomizdat, 2000, p. 568.

[4] Nu. Y. Chukreev and M. Yu. Chukreev," Models for assessing adequacy indices in managing the develop-ment of electric power systems," Syktyvkar, Komi Scientific Center, Ural Branch of the Russian Academy of Sciences, 2014. p. 207.

[5] G.F. Kovalev and L. M. Lebedeva, "Reliability of electric power systems. Novosibirsk, Siberian Publishing Company "Science" of the Academic Publishing House, 2015.

[6] Chukreev Yu. Ya. Models for ensuring the reliability of electric power systems. Syktyvkar: Komi Scientific Center, Ural Branch of the Russian Academy of Sciences, 1995.

[7] V. I. Zorkaltsev, G.F. Kovalev, L. M. Lebedeva and S. M. Perzhabinsky, "Minimizing the power shortage in an EPS in the light of power losses in power transmission lines," Electricity, 2010, No. 9, pp. 56-60.

[8] V. I. Zorkaltsev and S. M. Perzhabinsky, "Models for assessing the power shortage of the electric power system," Siberian Journal of Industrial Mathematics 2012, Volume XV, No.1 (49)

[9] Mid-term adequacy forecast. – ENTSO-E, 2016, p.144.

[10] CIGRE Technical brochure on review of the current status of tools and techniques for risk-based and probabilistic planning in power systems. Working group 601 of study committee C4. International conference on large high voltage electric systems. March 2010.

[11] N. Hary, "The electricity generation adequacy problem: Assessing dynamic effects of capacity remuneration mechanism," Energy Policy, 2016, issue 91, pp.113-127, ISSN 0301-4215.

[12] V. Z. Manusov, " Probabilistic problems in electric power industry," Novosibirsk, NETI, 1981., p.119.

[13] B. Borkouska," Probabilitic load flow," IEEE Paper No, T73 485-0 presented at the summer Meeting, Vancouver. February 1973.

[14] R.N Allan, A-M. Late da Silva and R.C. Burchett," Evaluations methods and accuracy in probabilistic load flow solutions," IEEE Transactions on Power Apparatus and Systems. PAS-100(5):2539-2546, May 1981.

[15] A.M. Leite da Siva, V.L. Arienti and R.N Allan, "Probabilistic load flow considering dependence between input nodal powers," IEEE Transactions on Power Apparatus and Systems, PAS-10s(6): 2539- 2546, June 1981.

[16] M.Th. Schilling, A.M. Leite da Silva, R. Billinton and M.A. El-Kafy," Bibliography on power system probabilistic analysis (1962-1988)," IEEE Transactions on Power Systems, 5(1):1-11, February 1990.

[17] B. I. Ayuyev, V. V. Davydov, P. M. Erokhin and V. G. Neuimin," Computational models of flow distribution in electrical networks, Moscow: Flint: Science, 2008. 256 p.

[18] B. M. Weedy and B. J. Cory, "Electric Power Systems, John Wiley & Sons Ltd. 1998., p.545.

[19] Handbook for the design of electrical networks / ed. D.L. Faibisovich, Moscow: ENAS, 2012. 376 p.

[20] V. P. Oboskalov, " Reliability of ensuring the power balance of electric power systems," Ekaterinburg: USTU. 2002, p.217.

[21] Seber. Linear regression analysis, World. 1980, p.456.

Mathematical Methods for Probabilistic Estimation of Power Shortage in Concentrated Electric Power Systems

Vladislav Oboskalov[1,2], Amir Abdel Menaem[1]and Rustam Valiev[1]
[1] Ural Federal University,
Yekaterinburg, Russia
[2] SEC of Ural Branch of RAS,
Yekaterinburg, Russia
v.p.oboskalov@urfu.ru

Anatolijs Mahnitko, Renata Varfolomejeva
Institute of Power Engineering,
Riga Technical University
Riga, Latvia
mahno@eef.rtu.lv

Abstract — **As an integral part of assessing the reliability of electric power systems, the calculation of probabilistic indices for power deficit between load and generation are needed considering the stochastic nature of generation and load. Therefore, the mathematical basis and estimation procedure for different analytical and approximate probabilistic techniques such as convolution method, equivalent method, cumulants combined with Gram-Charlier expansion and cumulant combined with Von Mises functions required to define the probability distribution of generation system is introduced in this paper. Also presented are numerical results for the expected value, standard deviation, and probability of the power shortage, which are calculated under a condition of system load loss and compared with the numerical method, i.e., Monte Carlo simulation, in two generation cases (identical – diverse generation unit sizes). Since the convolution of the probability series of generating power method provides the most accurate solution, it is taken as a benchmark method. The results show that with a large number of heterogeneous generator groups for practical application, the von Mises method is recommended as a more accurate method on average and equivalent method as a fairly accurate, but mathematically more simple.**

Keywords: power system adequacy; power shortage; probability distribution function; Gram-Charlier expansion; von Mises method

Abbreviations and Acronyms

LOLP: Loss of Load Probability;
LOLE: Loss of Load Expectation;
SAIFI: System Average Interruption Frequency Indices;
SAIDI: System Average Interruption Duration Indices;
CAIFI: Customer Average Interruption Frequency Index;
CAIDI: Customer Average Interruption Duration Index;
PDF: Probability Distribution Function;
PMF: Probability Mass Function;
CDF: Cumulative Distribution Function.

I. Introduction

The adequacy assessment problem for bulk power system includes the following steps: power system equipment status modeling, forming the system of constraints and assumptions, allocation of load between the parallel generators, the load shedding distribution and finally calculating adequacy indices of the concentrated power systems [1-3]. This work compares the efficiency and accuracy of the mathematical methods for calculating the resultant adequacy indices for concentrated power systems.

The probability, the mean value and variances of power shortage are considered as the main adequacy indices. The other indices such as LOLP, LOLE, SAIFI, SAIDI, CAIFI, and CAIDI etc. can be calculated through the main ones [**Chyba! Nenašiel sa žiaden zdroj odkazov.**]. In practice, combined probabilistic analytical or statistical methods are used for calculating the indices listed above. The type of a PDF of load, generation and of power shortage in the system plays a crucial role in the process of calculating adequacy indices. The parameters of a power system are represented by different PDFs (normal, gamma, binominal etc.). The normal PDF can be used for the load representation, while the binominal PDF is accurate enough for the generation, and low probability cases (e.g. generator failure) are represented by the Poisson distribution. A large number of generators of different types connected in parallel requires normal or gamma distribution. The convolution method can be used for getting the equivalent model that describes many generators of different types [5-6]. Probabilistic analysis of a system that includes several parameters described by different PDFs can also be made by the convolution method [7]. The applicability of using one or another PDF can be tested by the Monte-Carlo simulation technique. The accuracy and efficiency of the combined probabilistic analytical methods for calculating adequacy indices of concentrated power systems are analyzed in the article.

II. Groups of the Same-Type Generators

The group k of n generators of the same type, with the rated power of a generator – P_{bl} and probability of no failure operation – p, can be defined by the binominal PDF: $G \sim B(n,p)$ that is presented as the PMF $\{p_k, G_k\}$where G_k is the combined rated power. The normal PDF $f_L(x)$ and the CDF $F_L(x)$ are sufficiently accurate for load description. The power shortage $def = L - G$ is non-zero in case of $L > G$, and it is zero for $L \leq G$. Generation is a discrete stochastic variable, and load is a continuous stochastic variable. In such a representation, the probabilistic characteristics of power shortage can be calculated via the following equations [7]:

$$Pr_{def} = \sum_{k=0}^{n_G} p_k(1 - F_L(G_k))$$

$$m_{def} = \sum_{k=0}^{n_G} p_k \int_{G_k}^{\infty}(x - G_k)dF_L(x)$$

$$m_{2,def} = \sum_{k=0}^{n_G} p_k \int_{G_k}^{\infty} (x - G_k)^2 dF_L(x)$$

$$\sigma_{def}^2 = m_{2,def} - m_{def}^2$$

where: $m_{1,def}$ – the first moment of power shortage (mean) and σ_{def}^2 is the power shortage variance.

For $L \sim N(m_L, \sigma_L)$ the right parts of the equations can be written in the form below:

$$m_{def} = \sum_{k=0}^{n_G} p_k[(m_L - G_k)(1 - F_L(G_k)) + \sigma_L^2 f_L(G_k)]$$

$$m_{2,def} = \sum_{k=0}^{n_G} p_k\{[(m_L - G_k)^2 + \sigma_L^2][1 - F_L(G_k)] + \sigma_L^2(m_L - G_k)f_L(G_k)\}.$$

The equations presented above allow calculating the power shortage probabilistic characteristics for a relatively small generators group in a mathematically simple way.

III. GROUPS OF THE DIFFERENT-TYPE GENERATORS

In practice, the generators of different types are used in power systems $\{G_i \sim B(n_i, p_i), i = 1, \dots, K\}$. One of the most widely used methods, for representing the PDF of the generation with a complex structure, is the PMF convolution method [1-3, 8-12 etc.]. The method gives advantages over other methods. First of all, the table form of initial data is convenient for representing nonlinear and discrete PDF in software. Secondly, the PMF can be used for a nonstandard PDF description. Thirdly, the PDF of the random variables sum with different PDF set by PMF can be calculated in a mathematically simple way.

Mathematically, the convolution of the generators PMF is the product of probabilities related to the possible operational conditions of the generators. In particular, for the convolution of a PMF of two different generators groups given as $\{G_{A,i}, p_{A,i}, i = 0, \dots, n_A\}$, $\{G_{B,j}, p_{B,j}, j = 0, \dots, n_B\}$, The two matrices $\{G_{\Sigma,ij} = G_{A,i} + G_{B,j},\ p_{\Sigma,ij} = p_{A,i}p_{B,i},\}$, which represent the PMF of equivalent generator unit, are to be formed. The matrices are algorithmically transformed into two vectors which represent the equivalent PMF and can be used in the next convolution procedures if required. Other forms of convolution procedures do exist and they are not in contradiction with the ones presented above [2,9-10].

The representation of the multiplication of a PMF in matrix form can be useful in case of using software products such as Matlab, MS Excel etc. In case of representing the probabilities and rated power of the generators operational conditions in the vector form, the resultant matrices can be expressed as follows: $p_\Sigma = \bar{p}_A\, \bar{p}_B^T$, $G_\Sigma = \bar{G}_A \bar{e}_B^T + \bar{e}_A \bar{G}_B^T$, where $\bar{e}_A, \bar{e}_B$ the unit vectors with the dimensions corresponding to the vectors $\bar{G}_A, \bar{G}_B$. Test calculations have shown that the convolution method provides accurate results. The dimension of equal PMFs expands exponentially with the number of generators groups, and the computational burden becomes significant.

IV. USING THE NORMAL DISTRIBUTION FOR GENERATION REPRESENTATION

According to the de Moivre–Laplace theorem, the normal distribution may be used as an approximation of the binomial distribution if $n \to \infty$. It can be assumed that, when the number of generators in the systems amounts to hundreds, the unite «load-generation» variable can be described by the normal distribution. Let's consider the power imbalance variable $H = L - G$. The power shortage is nonzero when $H > 0$. The mean value and the variance of the power imbalance are:

$$m_H = m_L - m_G$$
$$\sigma_H^2 = \sigma_L^2 - \sigma_G^2.$$

The probability that the power shortage will be more than zero can be calculated through PDF or CDF:

$$Pr_{def} = \int_0^{\infty} f_H(x)dx = 1 - F_H(0)\,.$$

The mean value of H is defined by the reduced distribution [8]:

$$m_{def} = \int_0^{\infty} x dF^p{}_H(x)dx \qquad (1)$$

There is removable discontinuity in point $x = 0$ of function (1) $\Delta F_H^p(0) = F_H(0)$. The discontinuity is not considered in the process of calculating of the first and the second moments since $\text{def}(x = 0) = 0$. In assumption that $H \sim N(m_H, \sigma_H)$ and according to [8]:

$$m_{def} = \int_0^{\infty} x f_H(x)dx = m_H(1 - F_H(0) + \sigma_H^2 f_H(0)$$
$$\sigma_{def}^2 = (m_H^2 + \sigma_H^2)[1 - F_H(0)] + \sigma^2 m_H f_H(0) - m_{def}^2.$$

Test results have shown that the error brought by the assumption can be significant if a rated power of generators of the highest capacity is comparable to a load of a power system.

V. GRAM-CHARLIER EXPANSION APPROACH

The using of Gram-Charlier expansions [13,14] reduces the error caused by the assumption of the power imbalance normal distribution. In addition, the Gram-Charlier expansion allows to partially consider the difference of generators types a power system. The CDF and the PDF of the normalized power imbalance $\tilde{H} = (H - m_H)/\sigma_H$ can be calculated as follows:

$$F_{\tilde{H}}(x) = \sum_{i=0}^{r} \frac{c_i}{i!} \Phi^{(i)}(x) \qquad (2)$$

$$f_{\tilde{H}}(x) = \sum_{i=0}^{r} \frac{c_i}{i!} \varphi^{(i)}(x) \qquad (3)$$

where: $\Phi\,(x)$ and $\varphi\,(x)$ are the CDF and PDF of the reference distribution (e.g. normal), $\Phi^{(i)}(x)$, $\varphi^{(i)}(x)$ – i-th derivative of the reference distribution; r – accepted order of the Gram-Charlier expansions. The Gram-Charlier expansions indices:

$$c_0 = 1 \quad c_1 = c_2 = 0 \quad c_3 = -\frac{\mu_3}{\sigma^3} = -\frac{k_3}{\sigma^3}$$
$$c_4 = \frac{\mu_4}{\sigma^4} - 3 = \frac{k_4}{\sigma^4} \quad c_5 = -\frac{\mu_5}{\sigma^5} + 10\frac{\mu_3}{\sigma^3} = -\frac{k_5}{\sigma^5} \qquad (1)$$
$$c_6 = \frac{\mu_6}{\sigma^6} - \frac{15\mu_4}{\sigma^4} + 30 = \frac{k_6 + 10k_3^2}{\sigma^6}$$

where: μ_j, k_j the central moment and the semi invariant of order j for variable H respectively. The division of k_j by σ^j is due to H normalization. Taking into account that $\Phi^{(1)}(x) = \varphi(x)$, (2) can be rewritten as:

$$F_{\tilde{H}}(x) = \Phi(x) + \sum_{i=3}^{r} \frac{c_i}{i!} \varphi^{(i-1)}(x).$$

The first six derivatives of the normal CDF are:

$$\varphi(x) = \frac{1}{\sqrt{2\pi}} e^{-\frac{x^2}{2}}$$
$$\varphi^{(1)}(x) = -x\varphi(x)$$
$$\varphi^{(2)}(x) = (x^2 - 1)\varphi(x)$$
$$\varphi^{(3)}(x) = (3x - x^3)\varphi(x)$$
$$\varphi^{(4)}(x) = (x^4 - 6x^2 + 3)\varphi(x)$$
$$\varphi^{(5)}(x) = -(x^5 - 10x^3 + 15x)\varphi(x)$$
$$\varphi^{(6)}(x) = (x^6 - 15x^4 + 45x^2 - 15)\varphi(x).$$

The correction of the normal PDF by the Gram-Charlier expansion considerably reduces the error in comparison to the original case, but still the error is significant in absolute terms. Another disadvantage of the Gram-Charlier expansion is the necessity of calculating of semi-invariants of order higher than 4 and it can be difficult for the majority of the distribution functions.

VI. von Mises method

The normal distribution using for the group of generators of different types does not take into account the individual influence of the generators that can be considered as «impulses». As proposed by Richard von Mises [16], the equivalent PDF of the generators group can be considered as the weighted sum of variables represented by the normal distribution:

$$f_{rez}(x) = \sum p_k(x_k) f(x - x_k)$$

where: p_k is the impulse x_k weight. The method allows to find the set of the equivalent impulses $\{x_k, p_k\}$ if there exist n+1 moments of the random variable. The determinants of the matrices, which are made of moments to be determined for obtaining the equivalent impulses are in the following form:

$$D_0 = |m_0|; D_1 = \begin{vmatrix} m_0 & m_1 \\ m_1 & m_2 \end{vmatrix}; \ldots D_r = \begin{vmatrix} m_0 & m_1 & \ldots & m_r \\ m_1 & m_2 & \ldots & m_{r+1} \\ \ldots & \ldots & \ldots & \ldots \\ m_r & m_{r+1} & \ldots & m_{2r} \end{vmatrix}.$$

The maximum order of the positive determinant defines the maximum number of impulses. In practice, the number of used impulses is not more than 5. But the test calculations have shown that the error is significant if the number of used impulses is less than 7. Solving the linear system with the structure of the matrix of coefficients similar to the D_r, allows finding the $\{c_i\}$ indices of the $\sum c_i x^i = 0$ equation. The roots of the equation determine the coordinates $\{x_i\}$of the equivalent impulses. The c_i corresponds to the equation:

$$\begin{pmatrix} m_0 & m_1 & \ldots & m_r \\ m_1 & m_2 & \ldots & m_{r+1} \\ \ldots & \ldots & \ldots & \ldots \\ m_r & m_{r+1} & \ldots & m_{2r} \end{pmatrix} \begin{pmatrix} c_0 \\ c_1 \\ \ldots \\ c_r \end{pmatrix} = \begin{pmatrix} -m_{r+1} \\ -m_{r+2} \\ \ldots \\ -m_{2r+1} \end{pmatrix}.$$

The probabilities of the equivalent impulses can be calculated through the next equation:

$$\begin{pmatrix} 1 & 1 & \ldots & 1 \\ x_1 & x_2 & \ldots & x_r \\ \ldots & \ldots & \ldots & \ldots \\ x_1^r & x_2^r & \ldots & x_r^r \end{pmatrix} \begin{pmatrix} p_1 \\ p_2 \\ \ldots \\ p_r \end{pmatrix} = \begin{pmatrix} 1 \\ m_1 \\ \ldots \\ m_r \end{pmatrix}.$$

The high order moments which are to be calculated restrict the von Mises's method range of applications. As a rule, the sum of several probabilistic variables which are described by different PDFs with the moments or semi-invariants that can be calculated, form the required probabilistic variable. The high order moments of the required variable can be calculated by means of the semi-invariants of the elements that form the variable.

The binominal distribution is usually used for describing generation in a power system. The structure of equations used for calculating the high order moments of the binominal PDF is complex. The problem of calculating high order moments of variables described by PDFs with a complex structure can be potentially alleviated by considering each generator separately. Assuming that there are only two possible operational conditions of generator (on and off), it is possible to describe the available capacity of a generator X_i by the Bernoulli distribution. The k-th moment $m_{k,i}$ of a variable X_i can be calculated as:

$$m_{k,i} = p_i P_{bl,i}^k$$

where: p_i *is* the generator availability factor, $P_{bl,i}^k$ – available capacity related to p_i. The semi-invariant $k_{r,i}$ of the r-th order, $r > 1$ of the variable X_i are defined by the next equation [15]:

$$k_{r,i} = m_{r,i} - \sum_{j=1}^{r-1} C_{r-1}^j m_{j,i} k_{r-j,i}.$$

The equal generators influence $X_\Sigma = \sum X_i$ characterized by semi-invariants:

$$k_{r,\Sigma} = \sum k_{r,i}.$$

The moments of the variable X_Σ can be calculated based on the equations presented above [15]:

$$m_{r,\Sigma} = k_{r,\Sigma} + \sum_{j=1}^{r-1} C_{r-1}^j m_{j,\Sigma} k_{r-j,\Sigma}.$$

VII. Equivalization Procedure for The Different-Type Generator Groups

The convolution of generation and load PMFs reduces the generation discrete nature influence. The convolution allows to represent a group of generators of the different types by an equivalent group of generators [6]. The available capacity, the cutoff power mean values and the variances related to initial and equal group of generators have to be equal. The number, the probability of failure and the rated power $\{n_e, q_e, P_e\}$ of the equal generators group are considered as initial data. Binominal PDF is considered as the most accurate one for the representation of the equivalent generators group cutoff power. Poisson's, gamma, normal and exponential distributions can also be considered as the possible ones for the cutoff power representation.

A. The binominal distribution

It is assumed that failures of real and equivalent generators are random events and can be described by binominal PDF. According to the equality assumptions mentioned above, the relationships between the installed capacity and the cutoff power probabilistic characteristics of initial and the groups of generators can be described by the next equations:

$$P_{inst} = \sum n_i P_{g,i} = n_e P_{g,e} \quad (5)$$

$$M_{cutoff} = \sum n_i q_i P_{g,i} = n_e q_e P_{g,e} \quad (6)$$

$$D_{cutoff} = \sum n_i q_i p_i P_{g,i}^2 = n_e q_e p_e P_{g,e}^2 \quad (7)$$

where: $p_e = 1 - q_e$ is the availability factor of an equivalent group generator. The probability of failure for an equivalent group generator can be obtained based on (5) and (6):

$$q_e = \frac{M_{cutoff}}{P_{inst}} = \frac{\sum n_i q_i P_{g,i}}{n_e P_{g,e}} = \frac{\sum n_i q_i P_{g,i}}{\sum n_i P_{g,i}}. \quad (8)$$

The rated power of an equivalent group generator can be calculated based on the equations (6) and (7):

$$\frac{D_{cutoff}}{M_{cutoff}} = \frac{n_e q_e p_e P_{g,e}^2}{n_e q_e P_{g,e}} = p_e P_{g,e}$$

and therefore:

$$P_{g,e} = \frac{D_{cutoff}}{p_e M_{cutoff}} = \frac{\sum n_i q_i p_i P_{g,i}^2}{p_e \sum n_i q_i P_{g,i}}. \quad (9)$$

The number of generators in the equivalent group are defined by (5):

$$n_e = \frac{\sum n_i P_{g,i}}{P_{g,e}}. \quad (10)$$

The number of generators calculated using (5) can be a fractional number and in this case, it has to be round up $n_{ee} = [n_e]$. The rated power of a generator in the equivalent group also has to be corrected $P_{g,ee} = P_{inst}/n_{ee}$. The probability of failure for the corrected number of generators is equal to the probability calculated using (8). Rounding up of the number of generators results in the cutoff power variance error. The correction of the variance can be made according to the next equation:

$$D_{cutoff}^{new} = D_{cutoff}^{old} \frac{n_{ee} q_e p_e P_{g,ee}^2}{n_e q_e p_e P_{g,e}^2} = D_{cutoff}^{old} \frac{P_{g,ee}}{P_{g,e}} = D_{cutoff}^{old} \left(1 + \frac{\Delta P_{g,e}}{P_{g,e}}\right)$$

where: $\Delta P_{g,e} = P_{g,ee} - P_{g,e}$ .The ratio error of the variance is defined by the equation:

$$\varepsilon D_{cutoff}^{new} = \frac{\varepsilon D_{cutoff}^{new}}{\varepsilon D_{cutoff}^{old}} - 1 = \frac{\Delta P_{g,e}}{P_{g,e}}.$$

Considering that:

$$P_{g,ee} = \frac{P_{inst}}{n_{ee}} \approx \frac{P_{inst}}{n_e} - \frac{P_{inst}}{n_e^2}(n_{ee} - n_e) = P_{g,e}\left\{1 - \frac{(n_{ee} - [n_e])}{n_e}\right\}$$

and due to the fact, that $|n_{ee} - [n_e]| \leq 0.5$, the ratio error of the variance can be calculated as:

$$\varepsilon D_{cutoff}^{new} \leq \frac{0.5}{n_e}.$$

In practice, the number of different types of generators is quite large, and the procedure of obtaining an equivalent group of generators of the same type results in variance error that is about 2-3%. The error is small enough for practical purposes, but in calculations it affects both the mean value and the probability of power shortage. For reducing the error mentioned above, it is proposed to consider the generation as two subsystems: equal generators group $\{n_{ee}, q_{ee}, P_{ee}\}$ and additional deterministic load $L_{add} = -(P_{inst} - n_{ee}P_e)(1 - q_e)$. The additional deterministic load is summarized with the initial power system load. Test calculations have shown that such an approach considerably reduces the error.

VIII. Test Calculations

For comparison purposes, the concentrated power system that includes three groups of generators of the different types and the load described by the normal distribution have been considered. The initial data for the generators groups is the number of generators, the rated power of a generator in the group, the failure probability of a generator: {5; 200; 0,08}; {4; 100; 0,05}; {6; 50; 0,05}]. The mean value and variance of the load are $m_L = 1000\ MW$ and $\sigma_L = 100\ MW$ respectively. Test calculations results are presented in the Tables 1 and 2. The Table 2 considers the particular case of the single generators group {8; 200; 0,08} .The convolution method is set as a benchmark method. Table 1 and 2 show that the increase in the number of samples in the Monte Carlo simulation results in the significant error decrease. The computation burden of the Monte-Carlo technique is huge, and for comparison purposes the number of samples should not be less than 10^6 for obtaining accurate results. Using the normal PDF for generation has led to significant errors {-89,9%, -94%, -81,6%}. The accuracy can be considerably improved by using the Gram-Charlier expansions for the binominal PDF correction {11,6%, -5,6%, -17,6%} and (0) {23,3%, 2,6%, -14,2%} (0). The binominal PDF is used as a benchmark (Since using the Poisson PDF results in significant error).

The von Mises method and the method of obtaining the equivalent group of generators are the most accurate ones among the considered. The von Mises method results {0,1%, -0,3%, 0%} (Table 1) and {0%, 0%, 0%} (Table 2) related to the case of 11 impulses. The reducing of the number of calculated impulses to 5 increases the error up to 2-3%. The errors of the same order {-0,3%, 0,9%, 1,4%} (Table 1) and {0%, 0%, 0%} (Table 2) correspond to the method of obtaining the equivalent group of generators. The computation time of the method is considerably less in comparison to the Monte Carlo technique. For the particular case presented in the Table 2 the results of the methods fully coincide with the convolution method (0% error). Therefore, the method of obtaining the equivalent group of generators is recommended for practical purposes since mathematically less complicated and preferable for calculations.

IX. Conclusions

- The separate considering of the load and generation parts (with the convolution method in use) is the most accurate way for calculating probabilistic characteristics of the capacity shortage.
- The Monte Carlo simulation technique requires a relatively huge number of samples (at least 10^6) for obtaining accurate results.

Table 1. The Probability and Mean Values of Power Shortage in Concentrated Power System for Different Generators Groups

Method	Probability	Probability error	Mean value	Mean value error	Standard deviation	Standard deviation error
Convolution	0.002	0.0%	0.143	0.0%	4.388	0.0%
Monte Carlo 10^6	0.002	-1.5%	0.140	-2.1%	4.315	-1.7%
Monte Carlo 10^4	0.0032	60.5%	0.226	58.6%	5.702	30.0%
Normal PDF	0.0002	-89.9%	0.008	-94.2%	0.806	-81.6%
Binominal PDF+ Gram-Charlier expansion	0.0022	11.6%	0.135	-5.6%	3.616	-17.6%
von Mises	0.0020	0.1%	0.142	-0.3%	4.388	0.0%
Equivalent	0.0020	-0.3%	0.144	0.8%	4.449	1.4%

Table 2. The Probability and Mean Values of Power Shortage in Concentrated Power System for Single Generator Group

Method	Probability	Probability error	Mean value	Mean value error	Standard deviation	Standard deviation error
Convolution	0.0043	0.0%	0.338	0.0%	7.102	0.0%
Monte Carlo 10^6	0.0041	-3.6%	0.329	-2.6%	7.053	-0.7%
Monte Carlo 10^4	0.0034	-20.0%	0.236	-30.2%	5.244	-26.2%
Normal PDF	0.0005	-87.6%	0.022	-93.4%	1.327	-81.3%
Binominal PDF+ Gram-Charlier expansion	0.0052	23.3%	0.346	2.6%	6.097	-14.2%
von Mises	0.0043	0.0%	0.338	0.0%	7.102	0.0%
Equivalent	0.0043	0.0%	0.338	0.0%	7.102	0.0%

- Using the normal PDF for generation representation is not sufficiently accurate and cannot be recommended for practical calculations.
- The correction of the power imbalance PDF by the Gram Charlier expansions reduces the error associated with representing generation by the normal PDF.
- Both the von Mises method as the more accurate one and the method of obtaining the equivalent group of generators as sufficiently accurate one and less complicated mathematically can be recommended for cases with a large number of generator units of different types

References

[1] Y. N. Rudenko and M. B. Cheltsov, Reliability and reservation in power systems. Novosibirsk: Science, 1974.

[2] R.Billinton and R. Allan, Reliability Evaluation of Power Systems. New York : Plenum Press, 2006.

[3] R. Billinton and R. Allan, Reliability Assessment of Large Power Systems. Boston-MA: Kluwer, 1988.

[4] B. Bagen, J. Moura and K. Jefferies, Probabilistic reliability assessment of North American Electric Power Systems, 2014 International Conference on Probabilistic Methods Applied to Power Systems (PMAPS), Durham, 2014, pp. 1-6.

[5] J. Endreny, Reliability modeling in electric power systems. New York: Wiley, 1979, pp. 338-450.

[6] O. Terno and M. Valdma, Optimization of operating reserves in power systems, Oil Shale, Vol. 22, pp. 153–160, 2005.

[7] A. Leon-Garcia, Probability, statistics, and random processes for electrical engineering. Third edition, 2017.

[8] C. DeSieno and L. Stine, A Probability Method for Determining the Reliability of Electric Power Systems, *IEEE Transactions on Reliability*, Vol. R-14, Issue: 1, pp. 30-35, 2009.

[9] V.P. Oboskalov, Structural reliability of electric power systems: tutorial, Publisher: Ural Federal University, 2012.

[10] L. Krumm, U. KurreL, and O. Terno, Development of theory and methods of complex optimisation of control and operational reliability of interconnected power systems , Oil Shale, Vol. 24, pp. 377–388, 2007.

[11] O. Dzobo, C.T. Gaunt and R. Herman, Investigating the use of probability distribution functions in reliability-worth analysis of electric power systems, *International Journal of Electrical Power & Energy Systems,* Volume 37, Issue 1, pp. 110-116, May 2012.

[12] A. I. Khalyasmaa, Y. V. Sandakov and O. M. Kotov, "Special models in the structural reliability calculations of electric grids," 2017 11th IEEE International Conference on Compatibility, Power Electronics and Power Engineering (CPE-POWERENG), Cadiz, 2017, pp. 88-93.

[13] M.Kendall, Kendall's Advance Theory Statistic. New York: Oxford Univ, 1987.

[14] H.Cramer, Numerical Methods of Statistics. Princeton: N.Y. Princeton Univ. Press, 1946.

[15] W.D.Tian, D. Sutanto, Y.B.Lee and H.R.Quthred, Cumulant based probabilistic power systems simulation using Laguerre polynomials, *IEEE Trans. Energy conversion*, vol. no 4, pp 567-574, Dec. 1989.

[16] Von Mises, Mathematical Theory of Probability and Statistics. N.Y. Academic, 1964.

Transient Processes in the Microgrid Based on a Group of Autonomous Operating Synchronous Generators

Ivan Artyukhov, Sergey Stepanov
Power Supply and Electrical Technology Department
Yuri Gagarin State Technical University of Saratov
Saratov, Russia
ivart54@mail.ru

Artem Zemtsov
Power Supply for Industrial Enterprises Department
Samara State Technical University, Syzran Branch
Syzran, Russia
artex283@mail.ru

Abstract — **The current level of power electronics development allows creating of microgrids based on synchronous generators, working at different speeds of the shaft rotation. The energy flows' control in such networks and the specified parameters provision of the generated electricity is implemented via using conversion devices. The paper presents the results of the study on transient processes in a microgrid, in which the energy flows are combined on a DC bus. Modeling microgrid can be performed using MATLAB+Simulink software. We established that to ensure the system stability, it is necessary to carry out an operative correction of the automatic regulator parameters.**

Keywords: microgrid; synchronous generator; power conversion; transient processes; simulation modeling

I. Introduction

For the modern stage of development of the electric power engineering, an intensive process of creating and developing microgrids is characteristic. This term, the authorship of which belongs to Professor Robert H. Lasseter, designates an integrated low-power energy system with distributed generators and energy consumers [1]. Such energy system, as a rule, is characteristic for rather small areas. Currently, about 90% of existing electric microgrids occupy an area of up to 1 km^2, and the total power of energy generators in a single microgrid does not exceed 1 MW.

The feasibility of using microgrids is determined by the fact that they have a number of advantages compared with the classical methods of generating, transmitting and distributing energy. In a microgrid, generated energy is mainly used by local consumers, which ensures a reduction in losses associated with transmission and distribution of energy via electrical grids. Reliability of electricity supply within a microgrid is much easier to ensure than in larger energy systems. Energy consumers can participate in the process of power balancing by regulating their loads, generating, accumulating and delivering electricity to a microgrid [2].

There are various options of microgrids. They can operate autonomously or be connected to main utility grid. Combining individual local grids into a microgrid can be conducted on both AC and DC buses. At the same time, DC voltage microgrids have several advantages [3]. At present, the prospects for creating microgrids are considered mainly in connection with the possibility of integrating renewable energy sources, such as solar and wind energy [4, 5]. However, microgrids can also be built on the basis of conventional energy sources, such as diesel generators, gas turbine generators, etc.

The most frequently used sources of energy are synchronous electric machines, which operate at a fixed shaft rotation speed to ensure the stability of the output voltage frequency. However, the operation of an electric machine in this mode is not always possible, e.g., when the source of mechanical energy is represented by a wind turbine. In other cases, stabilizing the rotational shaft speed in a drive motor reduces an energy efficiency of the generation system. In particular, when using a diesel engine as a prime engine, actual fuel consumption when operating the system for a variable load is significantly higher than its declared value.

With development of power electronics, new opportunities emerged for improving generation of electricity. We could use electric machines with variable rotation speed, both synchronous and asynchronous. It is worth noting that rotational speed of an electric machine shaft can be set by force, in accordance with specified settings of the fuel supply regulator in a diesel generator plant, or determined by external operating conditions of the power generation system, e.g., when regulating the performance of gas pumping units via changing shaft speeds.

Conversion devices provide required parameters of the generated electricity (its effective voltage value, frequency and a curve shape) in accordance with the requirements of contemporary standards for the quality of electrical energy in conditions of changing the shaft speed of an electric machine and the load [6].

II. Structural Diagram and Features of Operating the Microgrid

Fig. 1 depicts a microgrid circuit formed by combining N autonomously functioning modules across a DC bus [7]. An energy source in each module is a synchronous generator with an engine. The generator shaft rotates with a speed n_k. Voltage and frequency have values of U_k and f_k, respectively.

The AC-DC and DC-DC converters along with a smoothing filter LD,C convert AC electrical energy at the output of synchronous generator into DC electrical energy with required parameters. The modules include electrical consumers of direct and alternating current. To obtain AC energy, DC-AC converters are used. They are equipped with filters to provide the required voltage curve shape. If there is an excess of power in the microgrid, it can be transferred to external consumers or to an energy storage system.

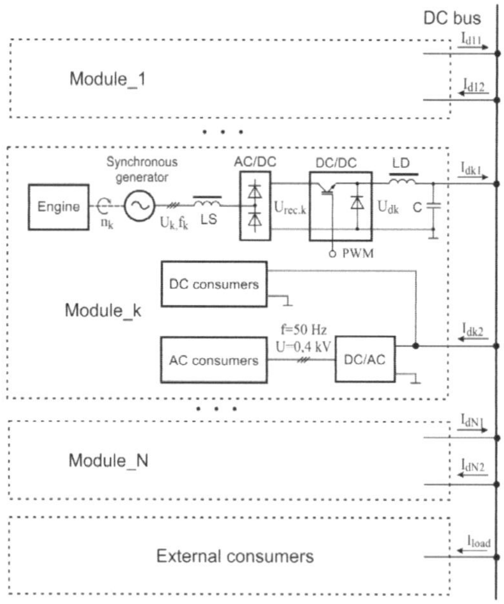

Figure 1. Microgrid structural diagram.

To study the processes in the microgrid under consideration, an equivalent circuit was used. It is depicted in Fig. 2 on the assumption of infinitely small resistance of the conductors forming the DC bus. The energy source of each module is presented in the form of an equivalent two-pole device with electromotive force (EMF) E_{dk} and internal resistance R_{dk}. The diode simulates the unidirectional nature of the current I_{dk1} . DC load of the module is represented by the equivalent resistance R_{mk} , through which the current I_{dk2} flows. External load of a microgrid is modeled by resistance R_{load}.

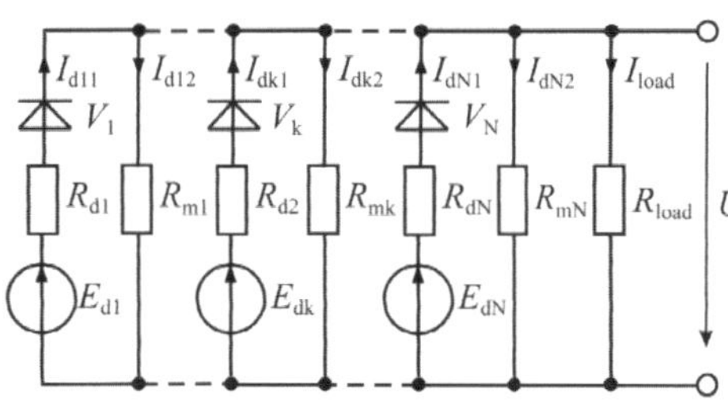

Figure 2. Equivalent circuit of a microgrid.

Based on the equivalent circuit, the formula for the voltage on the DC bus can be written as follows:

$$U_d = \frac{\sum_{k=1}^{N} F_k E_{dk} R_{dk}^{-1}}{\sum_{k=1}^{N} \left(R_{dk}^{-1} + R_{mk}^{-1} + R_{load}^{-1} \right)}, \tag{1}$$

where F_k designate logical functions reflecting the possibility of the flow of currents I_{dk1} just in one direction.

Equivalent EMFs of energy sources are output voltages of DC-DC converters in off-load mode. The values of these voltages depend on the rectifiers' output voltages and the control signals for DC-DC converters

$$E_{dk} = D(\mu_k) U_{rec.k} , \tag{2}$$

where $D(\mu_k)$ is the control characteristic of the k-th module of DC-DC converter. The type of control characteristic is determined by specific implementation of the converter.

For a three-phase bridge rectifier, the following expression is used to determine an average value of an output voltage:

$$U_{rec.k} = \frac{3\sqrt{6}}{\pi} \cdot E_{0k} - \frac{3 \cdot \omega L_{0k} \cdot I_{dk}}{\pi} - 2 \cdot R_{0k} \cdot I_{dk} , \tag{3}$$

where E_{0k} is EMF of the k-th generator; it depends on the main magnetic flux Φ_k in the air gap and the rotor speed n_k; I_{dk} is an average value of the rectifier output current; ω is a cyclic frequency; L_k , R_k are inductance and resistance circuit at the rectifier input, respectively.

EMF value is computed according to the formula:

$$E_{0k} = C_e \Phi_k n_k , \tag{4}$$

where C_e is a constructive coefficient of electric machine.

If a synchronous generator has an excitation winding, then by changing its current, it is possible to adjust the magnitude of a magnetic flux. In generators with permanent magnets, this is not possible.

An output current of the k-th module and the current consumed by electrical equipment of this module are calculated according to the formulas:

$$I_{dk1} = F_k (E_{dk} - U_d) / R_{dk} ; \tag{5}$$

$$I_{dk2} = U_d / R_{dk} . \tag{6}$$

External load current of a microgrid can be computed from the expression:

$$I_{load} = U_d / R_{load} . \tag{7}$$

The expressions (1) - (7) imply that, for a given set of electrical equipment characteristics, the voltage value on DC bus and the distribution of currents between the modules depend on the speeds at which the rotors of synchronous generators rotate and on the parameters of the loads. Excitation currents and control signals of DC converters provide control actions for synchronous generators with electromagnetic excitation.

Thus, by changing excitation currents of synchronous generators and control signals of DC converters, it is possible to ensure specified distribution of energy flows between the microgrid modules.

Microgrid modules can give power to external consumers:

$$P_{load} = \sum_{k=1}^{N} P_{dk} = \sum_{k=1}^{n} U_d \cdot (I_{dk1} - I_{dk2}) , \tag{8}$$

where P_{dk} is a power delivered by the k-th generator. It is necessary to point out that P_{dk} does not exceed generator's nominal power.

III. MICROGRID SIMULATION MODEL

The microgrid (Fig. 1) includes electric machines, semiconductor converters and a large number of inductive and capacitive elements. Studying the processes in such electrical complex is a daunting task. The mathematical model describing such intricate object is a system of nonlinear high-order differential equations.

Changing operation mode of synchronous generators, load parameters and microgrid structure leads to appearance of the transient processes. The study of these processes can be conducted solely using modern software products, such as MATLAB+Simulinlk.

When simulating a microgrid, it was assumed that changing the load on the module would not change the rotation speed of the synchronous generator shaft. This assumption is true when the drive power is many times greater than the generator power. Such situation occurs, e.g., in the power supply system of a gas pumping unit, in which a 250 kW synchronous generator is driven by a 10 MW turbine.

A diagram of a simulation model for studying transient processes in a microgrid with two modules is shown in Fig. 3. The power section of the microgrid is represented by Subsystem blocks (Fig. 4). Each of those models a synchronous generator with AC-DC and DC-DC converters in the form of a two-terminal circuit with EMF E_{dk} and internal resistance R_{dk} .

The validity of this approach was justified in [8, 9], among other publications. In addition, it was assumed that the DC-DC converter is represented by an aperiodic link of the first order under an assumption that the power switch of the converter is controlled with a sufficiently high frequency [10].

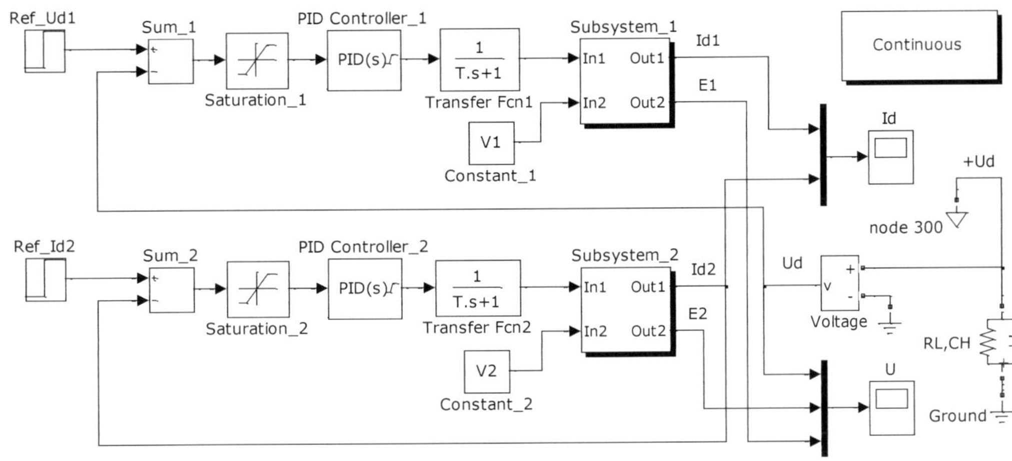

Figure 3. Simulink model for two modules.

The basis of the Subsystem is a Controlled Voltage Source, connected in series with a RS resistor and a Diode, which ensures the unidirectional nature of the module output current. Resistor RS simulates an internal resistance of the energy source.

The voltage value of the Controlled Voltage Source is set by a signal formed by multiplying signals at the information inputs of the Subsystem.

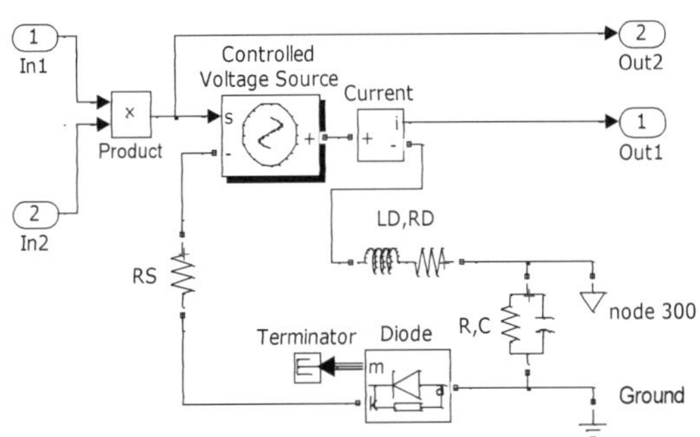

Figure 4. Subsystem diagram.

The control signal from the regulator (PID Controller block) is sent to the input 1 via the Transfer Fcn block, while the constant V1 numerically equal to the output voltage of AC-DC converter in off-load mode is sent to the input 2. As a result, at the output of the Controlled Voltage Source, voltage is generated at the values corresponding to EMF of the two-terminal circuit.

Subsystem has also information outputs with the signals indicating the values of EMF and current.

The first module operates in the voltage stabilization mode on DC bus. To ensure this, the signal from the voltage meter (Voltage block) is fed to the inverse input of the summing element Sum_1, to the direct input of which a reference signal from Ref_UD1 block is fed. The difference of signals is processed by the regulator (PID Controller_1) until the voltage on DC bus reaches specified value.

Sometime after switching on first module, the second one is supposed to enter the stabilization mode of the output current. For this purpose, a feedback loop is formed, which transfers information about the output current of module 2 to the inverse input of the comparison element Sum_2. The direct input of the element Sum_2 receives the signal of the current setting device (Ref_ID2 block).

The difference of signals is processed by the regulator (PID Controller_2), which changes the voltage at the output of the Subsystem_2 block. The maximum voltage at the output of this block is set by the constant V2.

IV. Microgrid Simulation Results

Fig. 5 presents the results of a microgrid simulation with the following parameters: rectifier output voltage – 500 V; internal resistance of sources – 0.1 Ω; equivalent load resistance – 3 Ω; total capacitance of capacitors – 6600 μF; inductance of smoothing chokes – 1 mH.

Time constants of the Transfer Fcn blocks were 0.1 s. The controller settings are shown in Table 1.

Table 1. Controller settings

Controller type	Regulator coefficients		
	KP	*KI*	*KD*
PID Controller_1	0.002	0.014	0
PID Controller_2	0	0.05	0

The assumed sequence of operations is as follows. First, the module 1 initiates and goes into the voltage stabilization mode on DC bus at 300 V. Then, after two seconds, the module 2 starts up and goes into the current stabilization mode at 40 A.

From our graphs, it is clear that the first stage of the transient process is developing successfully. The coefficients of the module 1 regulator are chosen so that the voltage on DC bus reaches a predetermined level without overshoot.

However, after the command to start module 2 arrives, and after EMF of this module reaches the level of EMF of the module 1, self-oscillations take place, accompanied by alternating switching the modules on and off.

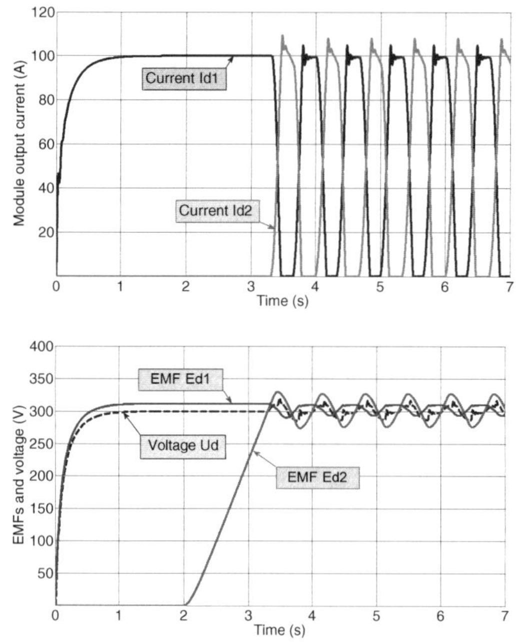

Figure 5. Results of transient processes simulation in a microgrid with two modules.

To eliminate this phenomenon, we proposed to reduce the transfer coefficient of the second regulator in those modes, when the difference of EMF modules becomes less than the specified value. Diagram of a microgrid simulation model with a transfer coefficient correction in a current stabilization circuit is shown in Fig. 6.

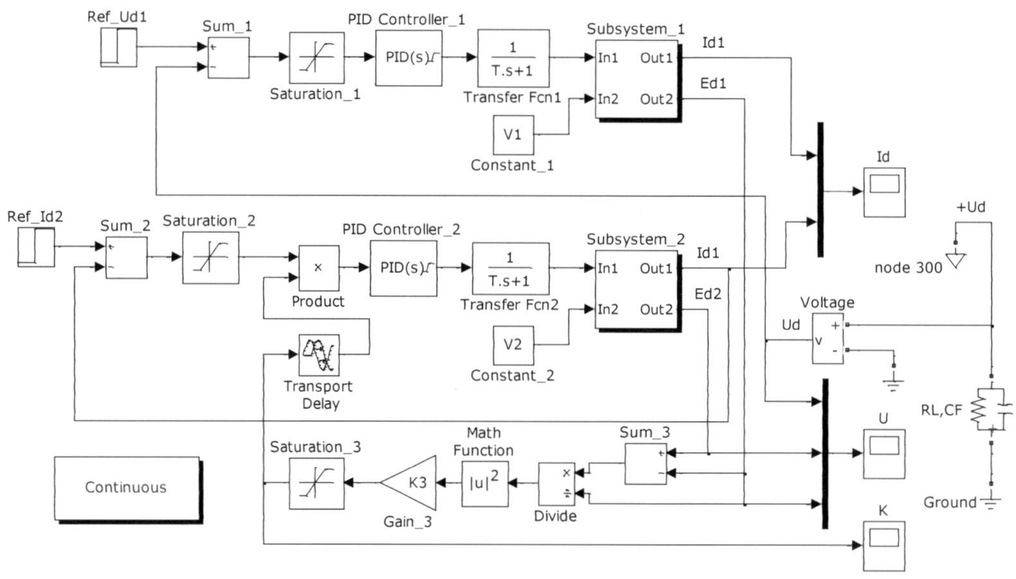

Figure 6. Simulink model with a transfer coefficient correction.

The Summator_3, Divide, Math Fuction, and Gain_3 blocks are introduced into the model diagram. They implement the following function:

$$A = K_3\left(\left(E_{d1} - E_{d2}\right) / E_{d1}\right)^2, \tag{9}$$

where K_3 is coefficient, which is set in the Gain_3 block.

The parameter A controlling a transfer coefficient in the current stabilization circuit, is limited by the range A_{min}...1 via the Saturation_3 block.

Simulation results of the transients in a microgrid with correction of a transmission coefficient in the current stabilization circuit are presented in Fig. 7.

Parameter A_{min}=0.02; coefficient K_3=20. With such value of K3, the parameter A remains equal to 1 until the modulus of the relative difference of EMFs is less than 0.22.

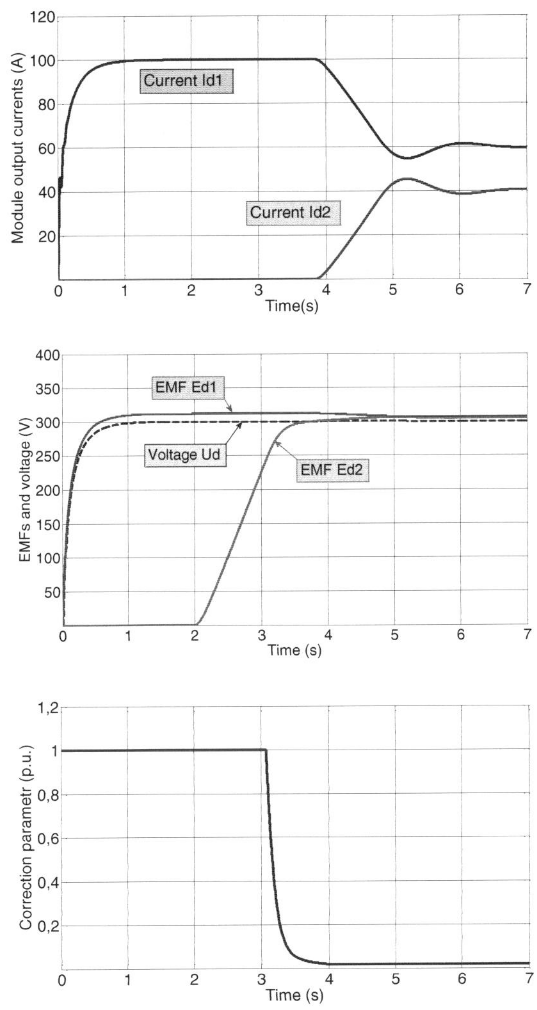

Figure 7. Results of transient processes simulation in the system with transfer coefficient correction.

From the graphs in Fig. 7 it can be seen that after three seconds of the transient process, the rate of change in EMF E_{d2} declines by decreasing the parameter A value, which leads to the stability of the system. Hence, predetermined redistribution of the modules' loads occurs: the current of the module 1 decreases to 60 A, while the current of the module 2 increases to 40 A.

As a result of the simulation, it was established that the parameter A_{min} strongly influenced the character of transient processes. This was confirmed by the graphs presented in Fig. 8. They were obtained for different values of the parameter A_{min} in modeling the scenario, which was a development of the previous sequence of operations.

After switching on the module 1 in the voltage stabilization mode at 300 V and module 2 in the current stabilization mode at 40 A, a command was given to increase the current of the second module up to 60 A. At parameter A_{min} = 0.02, a predetermined redistribution of currents occurred with a slight overshoot. At parameter A_{min} = 0.06, an oscillatory process was observed. Value A_{min} = 0.12 resulted in self-oscillations.

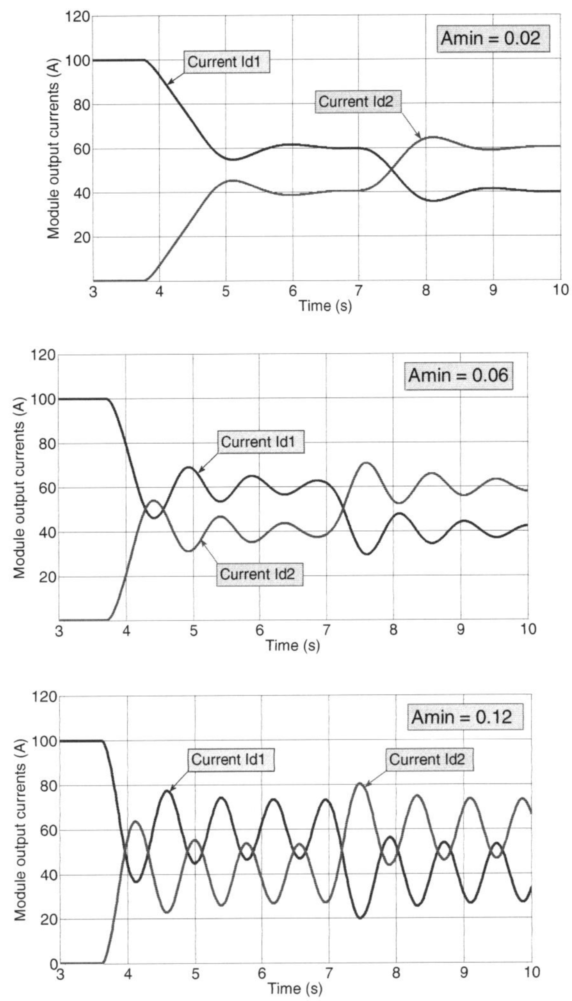

Figure 8. Transient processes for different parameter values of A_{min}.

Conclusion

Currently, there is an intensive process of creating and developing microgrids, which has a number of advantages in comparison with the classical methods of generating, transmitting and distributing energy. At the same time, the modern level of developing power electronics allows creating microgrids based on generators, the drive shafts of which rotate at different speeds.

A microgrid based on a group of autonomously operating synchronous generators is characterized by a complex dynamics. When changing the operating mode of the generators, the parameters of the load and the structure of the microgrid, transient processes arise, for the study of which it is necessary to use modern software products.

The study of dynamic modes in a microgrid of two modules with autonomous operating synchronous generators, conducted using the MATLAB + Simulinlk software, showed the possibility of self-oscillations, which are accompanied by alternately switching the modules on and off.

It was proposed and confirmed by numerical experiments on a microgrid simulation model that, in order to ensure the stable operation of the system, it is necessary to carry out operational correction of the parameters of the automatic controller in the module, which operates in the current stabilization mode.

Acknowledgment

This research was financially supported by the Ministry of Science and Higher Education of the Russian Federation (project number 8.7075.2017/8.9).

References

[1] R.H. Lasseter, "MicroGrids." *Proceedings IEEE Power Engineering Society Winter Meeting*," 2002, pp. 305-308.

[2] T. Dragicevic, X. Lu, J.C.Vasquez, and J.M. Guerrero, "DC Microgrids–Part II: A review of power architectures, applications, and standardization issue," *IEEE Transactions on Power Electronics*, 2016, vol. 31, № 5, pp. 3528-3549.

[3] Y. Ito, Z. Yang, and H. Akagi, "DC microgrid based distribution power generation system." *Proceedings of Power Electron. and Motion Control Conference*, 2004, vol. 3, pp. 1740–1745.

[4] J. Dudiak and M. Kolcun, "Integration of renewable energy sources to the power system," *Proceedings of 14th International Conference on Environment and Electrical Engineering*, 2014, pp. 148-151, doi: 10.1109/EEEIC.2014. 6835854.

[5] M. Vojtek, M. Kolcun, M. Mikita and M. Špes, "Integration of photovoltaic and wind power plants in low voltage networks using energy storages," *Proceedings of the 9th International Scientific Symposium on Electrical Power Engineering*, ELEKTROENERGETIKA, 2017, pp. 602-607.

[6] I.I. Artyukhov, D.A. Bochkarev and E.T. Erbaev, "Voltage stabilization on the power supply system based on the synchronous generator with variable rotor speed," *Proceedings of International Conference on Actual Problems of Electron Devices Engineering*, 2014, doi: 10.1109/APEDE.2014.6958206.

[7] I.I. Artyukhov, S.F. Stepanov, D.A. Bochkarev, G.N. Tulepova and A.I. Zemtsov, "Microgrid based on a group of autonomous operating synchronous generators," *Urban construction and architecture*, 2017, vol.7(4), pp. 127-131, doi: 10.17673/Vestnik.2017.04.22.

[8] J. Jatskevich, S.D. Pekarek and A. Davoudi, "Parametric average-value model of synchronous machine-rectifier systems," *IEEE Transactions on Energy Conversion*, 2006, vol. 21, issue 1, pp. 9-18, doi: 10.1109/TEC.2005.847974.

[9] F.M. Gonzalez-Longatt, P. Wall and V. Terzija, "A simplified model for dynamic behavior of permanent magnet synchronous generator for direct drive wind turbines," *Proceedings IEEE Trondheim PowerTech*, 2011, doi: 10.1109/PTC.2011.6019425.

[10] E.P. Popov. *The Dynamics of Automatic Control Systems*. Pergamon, 1961.

The influence of load on the Transformers on the size of their own losses

Lucie Nohacova[1)], Frantisek Zak[2)] ,Martin Hulec[3)], Milan Bělík[4)], Mikhail Olkhovskiy[5)]

[1, 3, 4, 5)] Department of Electrical Engineering and Environmental Engineering, University of West Bohemia, Plzen

[2)] Ceske Budejovice

Czech Republic

nohacova@kee.zcu.cz, ee.zak@seznam.cz, hulecm@students.zcu.cz, belik4@kee.zcu.cz, mikhail@students.zcu.cz

***Abstract* — The topic of the article is new insights into the load of transformers depending on their own losses. Transformers are among the most common, but important elements of the transmission and distribution networks. This is a reliable device with a guaranteed lifetime at the level of 25 to 30 years. New materials and new technologies for the production of transformers are leading to significantly reduce their own losses. Loss of transformers can be further influenced by both the change in their load and the flow of reactive power over transformers. Managing the overflow of reactive power through the transformers can then also influence the voltage conditions in the network.**

Keywords: transformers own losses; electrical network operation

I. Introduction

The concept of the contemporary alternating-current electrical networks was put through by N. Tesla (1856-1943) to triumph over that of T.A. Edison (1847 – 1931) in this respect. In 1896, Westinghouse Electric Corporation commissioned a big power plant at Niagara Falls to generate alternating electric current as proposed by N. Tesla. The fundamentals of this concept are electrical energy transmission at higher voltage levels, whereby reducing transmission losses and making it possible to transmit electrical energy over great distances with the voltage level being reduced again in the place of consumption. In this philosophy, a significant role is played by transformers. Transformers are considered to be the crucial elements within the alternating-current networks. As in all other fields, transformers also experience a constant development. Owing to the introduction of new technologies and materials, transformer losses are reduced, as is the consumption of the transformer production materials.[1],[3],[4]

Transformers are classified among the stationary electric machines consisting of coils and magnetic circuits. The material of the conductors used to produce the coils remained unchanged over the years. Copper and Aluminum are used for their good electrical conductivity. The coil design itself has changed both due to the use of new insulation materials and due to changes and new variants in the coil winding technology. Using computers in designing coil structures and in modeling the magnetic flux generated by coils facilitates making the designs more precise and effective. Conductors of different cross-sections as well as metal foils are used for windings. The coil design influences especially the short-circuit transformer losses. Short-circuit losses are affected by the current load of the transformer, irrespective of whether under an active or reactive power (see Fig. 2) [2], [5]

Figure 1. 220 kV/ 22 kV, 25 MVA Transformer

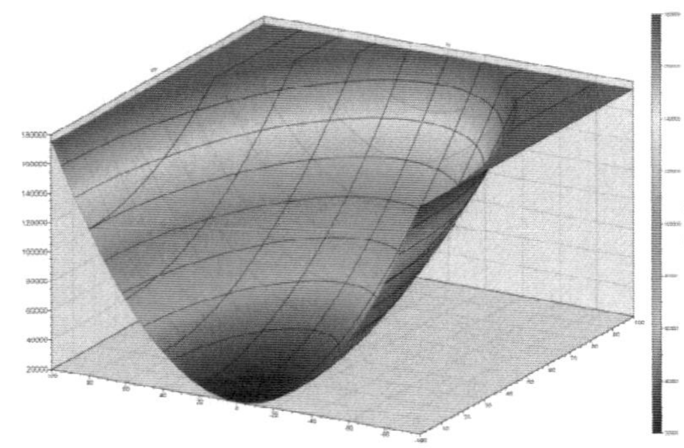

Figure 2. Illustration of 110/22 kV transformer losses (Sn 40 MVA, uk = 11.2%, Pk = 155 kW, Po = 20 kW) when loaded by both active and reactive power of inductive and capacitive character

Scale on the axes in the figure: for transformer losses from 20 kW to 180 kW.

II. Transformer Losses

Another important structural element of the transformer is its magnetic circuit, where emphasis is put on eliminating losses in the magnetic circuit caused by eddy currents and magnetic circuit re-magnetizing. Losses in the magnetic circuit are dependent on its saturation. Modern magnetic materials facilitate higher saturation even with lower losses. This makes it possible to reduce no-load transformer losses and also the weight of the whole transformer. The no-load losses are independent on transformer loading. They may, however, be affected by the network's higher harmonics or by stray direct currents from the electric traction, for instance. [7]

Thus, the new transformer designs make it possible to reduce not only the material demands but also the losses, both the no-load and the short-circuit ones. For example, a 110 kV/22 kV 25 MVA transformer produced approximately 50 years ago showed no-load losses of 55 kW and short-circuit losses of 200 kW. At present, such losses are achieved in transformers of outputs around 63 MVA.

The transformer design changes owing to the development of new materials and technologies. One has, however, also change one's approach to transformers in designing new networks and when replacing existing obsolete transformers.[6]

In general, transformers belong to electrical network elements providing high efficiency. In spite of this, transformer losses should be monitored. In our country, the highest emphasis was put on transformer losses at the time of the construction boom of photovoltaic power plants (PVPP). The subsidized price of electrical energy supplied from PVPP was a multiple of the usual price of electrical energy. For this reason, also, energy losses in delivery to the transmission point had to be assessed in these price proportions. In designing, demands were then put for not only greater conductor cross-sections but also for low loss transformers. The higher investment costs of PVPP construction usually returned within one to three years of PVPP commissioning.

With the low prices of electrical energy, however, the increased investments into economies do not pay much. Replacement of older transformers with higher losses by low loss ones usually brings a long period of return on the investments spent. In some load levels, the investment payback period can be as long as the transformer's service life. A different situation, however, occurs where existing obsolete transformers are to be replaced. Here, the transformer output can be optimized to contribute to reducing the total network losses. To design a suitable transformer output, one has to know its parameters, price, and load. To determine transformer's rated power supplied to a specific network, the expected load is just of crucial importance. Where an existing transformer is to be replaced, determining transformer load usually is no problem. The expected development in transformer power overflow should always be considered. [8]

III. Transformer Load

In newly constructed networks, determining the load poses a greater problem, and it usually depends on the designer's experience of how well he is able to estimate power overflow through the transformer. As a rule, losses in the current path are dependent on the square of the current flow. Therefore, to determine transformer losses, one has to allow for the arithmetic quadratic mean value of transformer's power overflow (Fig. 3). As a standard, quarter-hour values of transformer load obtained from electricity meters are used for the purpose. The arithmetic quadratic mean of load can be calculated using the formula below:

$$\bar{S} = \sqrt{\frac{\sum_{i=1}^{n} S_i^2}{n}} \qquad (1)$$

Where:

$\bar{S}$ is the arithmetic quadratic mean of load

S_i is the load in the particular time interval

n is the number of time intervals

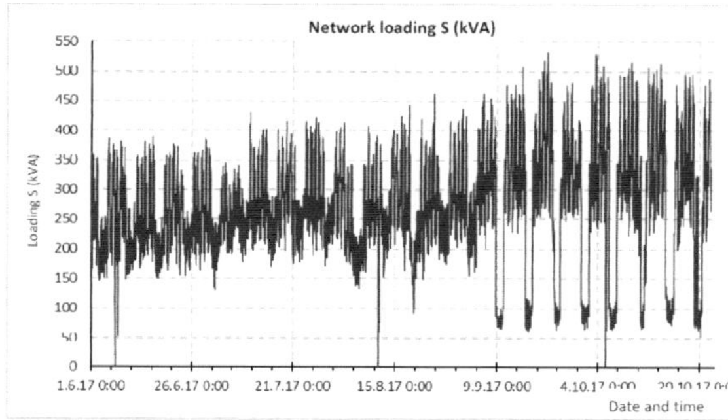

Figure 3. Chronological record of ¼-hour transformer load

The transformer shows the lowest efficiency at a zero load, when the losses reach 100 %. In the absolute value, however, the transformer losses are least here and they correspond to the no-load losses. When at the nominal load, the transformer losses comprise the no-load and the full short-circuit ones. The efficiency of advanced transformers at nominal loads usually ranges around 99 %. The optimum transformer load can be determined with the knowledge of transformer losses (Fig. 4):

$$S\,(\%) = 100 * \sqrt{\frac{P_0}{P_k}} \qquad (2)$$

Where:

S (%) is the optimum percentage of the load of a transformer of a given output

P_0 are transformer's no-load losses

P_K are transformer's short-circuit losses

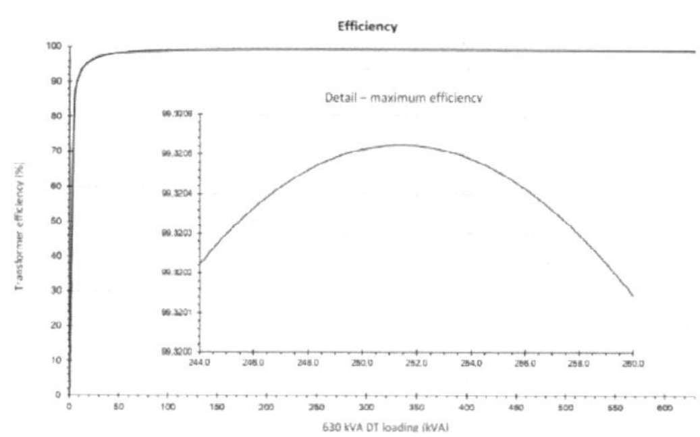

Figure 4. 630 kVA distribution transformer efficiency curve with respect to load

Transformer's optimum output can be determined by comparing the arithmetic quadratic mean load of the transformer with the value of its optimum load. The optimum load values of old transformers ranged from 60 % to 70 % of their nominal output. Transformers with magnetic circuits produced from oriented transformer sheets used to have the optimum of their loading at about 50 % of their rated outputs. The present-day transformers show an optimum load at approximately 30 % of their rated outputs. This is due to the decrease in transformer losses, especially in no-load losses. The decrease in no-load losses is caused by using new magnetic materials and new technologies of sheet processing and magnetic circuit setup. In distribution transformers, for instance, the no-load losses take up approximately 7 % to 10 % of the short-circuit ones. Short-circuit losses are influenced especially by the cross-section and length of the winding conductors. Reducing the coil dimensions affects the length of the coil conductors. A shorter conductor, even if keeping the same cross-section, causes a decrease in resistance and, thereby, also in losses. As the short-circuit (winding) losses are decreased to a lesser extent in comparison with the no-load ones, the value of optimum transformer loading decreases as well.

As a result of the decrease in the value of transformer optimum load, the currently designed transformer outputs for the given load are usually higher than they used to be in the older transformers. And despite higher transformer outputs are chosen now, they provide an absolute decrease in losses, with a higher output reserve for transformer loading being achieved.

Losses in transformation depend especially on the magnitude of the current load of the transformer. Where a certain area can be power-supplied using more transformers, by dividing consumption among them one can also influence the magnitude of losses in transformation (Fig. 5). The optimum value making the division of consumption among transformers economical is influenced by the parameters of the transformers, with the no-load losses again playing an important role here. In new transformers, for instance, consumption division between two identical distribution transformers pays back economically with the total load exceeding 42 % of the rated output of the single transformer.

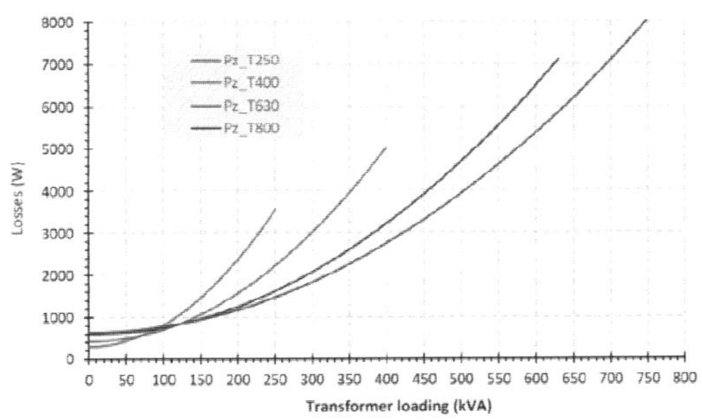

Figure 5. Losses of distribution transformers of various outputs

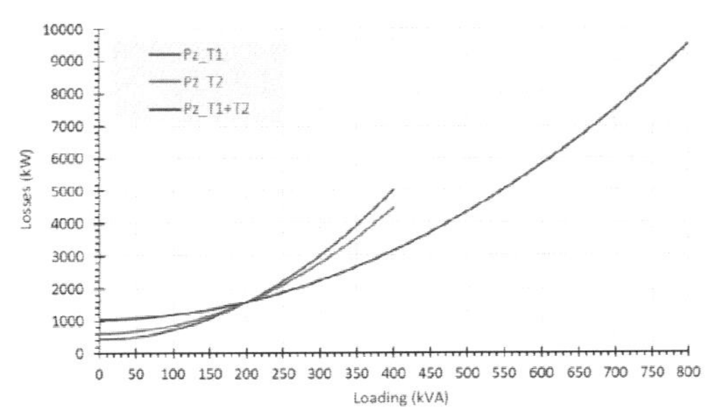

Figure 6. Consumption division between two different transformers

(The operation of T1 alone is only optimal up to the load of approx. 200 kVA, above that, an optimum is achieved by splitting the load between the two of them).

Transformer losses are quadratic-dependent on transformer loading, which means that one can influence the total losses by controlling loading. As a rule, the options to control the active power overflow are limited considerably. Where higher loading occurs, one can choose to increase the transformer output or to distribute the loading among a number of transformers. However, such measures need not be beneficial economically, and they may even cause technical problems. Reducing transformer's current loading can frequently be achieved by eliminating the reactive power overflow. At the time of high loading, transformers usually are loaded by both active and reactive power of an inductive character. At present, however, one can often find transformers loaded by a reactive power of a capacitive character, which is most frequently caused by the charging capacity currents of the extensive cable networks occurring mainly at the time of lower loading. At the same time, with the voltage level of the cable lines increasing, the charging power increases considerably. In the 22 kV networks, as a rule, the charging power of the cable line ranges from 40 kVAr/km to 60 kVAr/km, and in the 35 kV cable line it reaches ca. 100 kVAr/km (Fig. 7). The 110 kV cable networks provide charging powers exceeding 500 kVAr/km up to 1.5 MVAr/km. The charging capacity of the line is influenced by the cable type in use and its cross-section. The cabling projects in the urban and suburban areas, however, contribute significantly to the increase in capacitive reactive power overflow, which influences losses especially at the time of lower network loading. Another aspect increasing the reactive power overflow is the change in the character of appliances. Many appliances are fitted with power electronics which, at the time of low loading, contributes to capacitive reactive power overflow. The present-day light sources (Fig. 8) and also idle inverters, for instance, are the considerable sources of capacitive reactive power. All these aspects very frequently cause the capacitive character of the reactive power overflow through transformers.

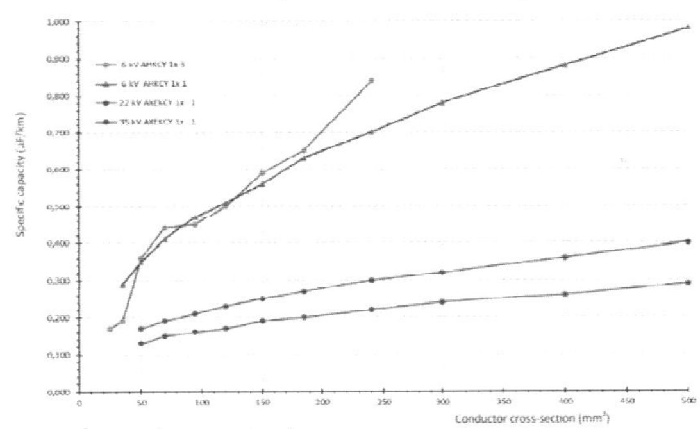

Figure 7. Specific capacity of high-voltage cables – illustrative graphs

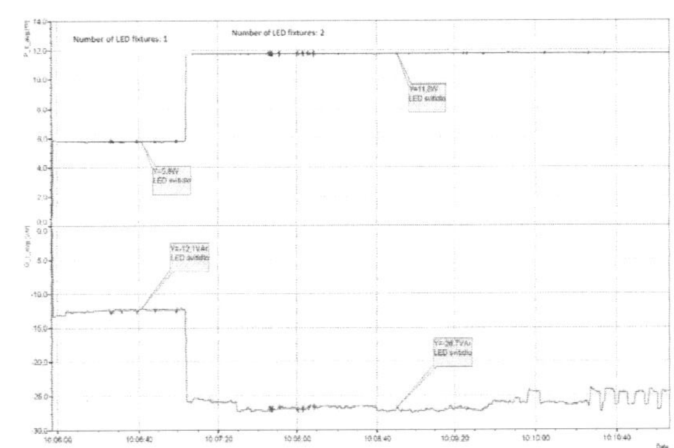

Figure 8. Active and reactive power consumption by LED lighting fixtures

By eliminating the reactive power overflow through the transformer one can influence the transformer losses and, at the same time, contribute to voltage stabilization in the network. To eliminate the charging power overflow in extensive cable networks, one can use both decompensate reactors and the VSS system (Fig. 9). The latter is a system which enables eliminating the capacitive charging current, especially in large networks, the value of the ground capacitive current in case of a ground fault, as well as the phase voltage unbalance in the network.

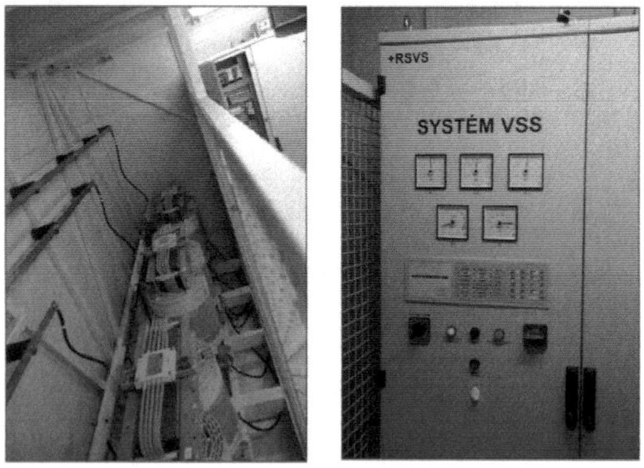

Figure 9. VSS system used to eliminate capacitive charging currents, ground capacitive currents and phase voltage unbalance

IV. Conclusion

The focus of the transformer manufacturers on transformer loss reduction is motivated by the effort to achieve advantages in the market. Also, the investors and transformer operators make calculations to evaluate the benefits of increasing their investments in purchasing transformers featuring lower losses. The price of electrical energy and, in the case of the electrical network operators, also the allowable losses in electrical energy distribution play the important roles here. The concern of losses is likely to grow in importance with the growth of the electrical energy prices. The advantage of transformers is, in addition of low losses, their high service reliability. At present, efforts are focused on replacing the conventional distribution transformers with state-of-the-art switched sources. Such sources can provide characteristics never achievable by the common distribution transformers due to their physical substance. The advancement of the electronic substitution for transformers is so far prevented by the high price, the amount of losses and lower service reliability. In spite of this, one can expect that power electronics will succeed also in this respect as it already has in the power supply of small appliances.

Today, such power electronics can be found in distribution transformers where it is used as a device to deal with certain network problems. However, it is possible that the issue of the concept of the alternating-current electrical networks will be reconsidered again in the future, not excluding return to the direct-current distribution of electrical energy. Using the direct-current network would mean the application of semi-conductor power elements. Such power electronics would ensure voltage level change instead of the standard transformers used in the alternating-current network. The advantage of electrical energy direct-current transmission lies in the elimination of reactive power overflow and in the consequent reduction of losses associated with electrical energy transmission.

It is possible to approach the problem of reducing losses in the transmission of electrical energy in different ways. This always requires certain measures to be reflected in the investment costs. Investing in savings usually has a long payback period. It is therefore good to assess the individual elements of the electricity system, determine their loss rate and, if these elements also show other benefits, they can be well used in the transmission and distribution of electricity.

This article has been written based on own knowledge and longtime experience acquired by the processing of measurements in the distribution network.

Acknowledgment

This paper was supported by the student research project SGS-2018-023 (WBU in Plzen, Czech Republic).

References

[1] P. Hocko, M. Novak and M. Kolcun, "Vplyv obnoviteľných zdrojov na dlhodobú dynamiku v ES SR," Proceedings of 13th International Scientific Conference on Electric Power Engineering (EPE 2012). Brno, Czech Republic, May 2012, pp. 687-691, ISBN 978-80-214-4514-7.

[2] K. Nohac and L. Nohacova, "Detection of nonstandard states in networks with renewable energy sources," Proceedings of 7th International Scientific Symposium Elektroenergetika 2013, September 2013, Stara Lesna, Slovakia, pp. 188-191.

[3] F. Zak, „Measuring of the capacitive earth-fault current in a 6 kV network using the indirect method," research Report, May 2013.

[4] L. Nohacova, F. Zak and J. Mertlova, "Elimination of Impact of Cross Parameters Unbalance," Proceedings of 14th International Scientific Conference on Electric Power Engineering (EPE 2013). Ostrava, Czech Republic, May 2013, pp. 479-483.

[5] D. Kouba and L. Nohacova, "Two Simultaneous Faults in Middle Voltage Distribution Network," Proceedings of 22th International Expert Meeting Power Engineering. Maribor 2013, Slovenia, May 2013. pp. 1-8.

[6] F. Zak, "Reducing energy intensity in the field of energy transfer (transmission, distribution) of energy", research Report, December 2018.

[7] A. Otcenasova, A. Bolf, J. Altus and M. Reguľa, "The Influence of Power Quality Indices on Active Power Losses in a Local Distribution Grid," Energies 2019, Volume 12, 1389, SWITZERLAND, p. 32, ISSN 1996-1073.

[8] Z. Martinek, J. Holy and T. Klor, "Reliabity of the electrical power system," Proceedings of 14th International Scientific Conference on Electric Power Engineering (EPE 2013). Ostrava, Czech Republic, May 2013, pp. 479-483, DOI: 10.1109/EPE.2014.6839427

Simulation of Arc Extinction in Discharge Chambers of Lightning Protection Equipment

Vladimir Frolov, Dmitrij Ivanov
Dep. of Electrical Power Engineering and Equipment
Peter the Great Saint-Petersburg Polytechnic University
Saint-Petersburg, Russia
eie@spbstu.ru

Alexander Sivaev
Streamer Electric AG
Saint-Petersburg, Russia
office@streamer-electric.com

Abstract — **This article describes theoretical investigations of new generation of surge arresters designed for protection of critical power objects. The results of mathematical modeling, namely, the arc temperature distributions are given. The influence of the chamber design on the arc extinction process is shown.**

Keywords:* *lightning protection, arrester, arc extinction

I. Introduction

A protection of critical energy facilities from direct lightning strike requires a development of special lightning protection devices.

Currently, a new promising way to protect overhead power lines from lightning overvoltage is the use of multi-chamber arresters with rod electrodes of steel or copper placed in silicone rubber [1-3].

The multi-chamber arrester consists of a large number of series-connected chambers (Fig. 1, 2) in which an arc discharge occurs during breakdown.

Such a discharge is accompanied by erosion of the electrodes material and ablation of the discharge chamber material. In this case, an increased pressure occurs in the chamber, leading to formation of the plasma jet from the discharge chamber and the extinction of the electric arc.

The purpose of the paper was to carry out theoretical investigation allowing to reasonably choose the design of the discharge chamber of a multi-chamber arrester.

Figure 1. Multi-chamber arrester

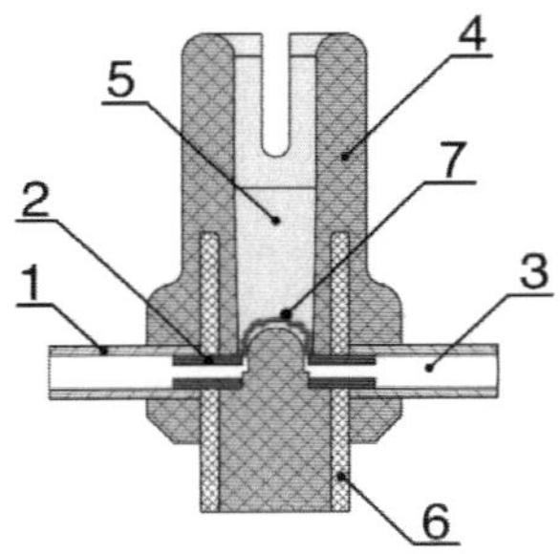

Figure 2. The design of a single discharge chamber of a multi-chamber discharger

Numbers in Fig. 2 denote: 1 – outer tube, 2 – inner tube, 3 – cavities, 4 – silicone rubber, 5 – discharge gap, 6 – fiberglass tube, 7 – arc.

II. Mathematical model

A three-dimensional non-stationary mathematical model has been developed for modeling the processes in the discharge chamber of the multi-chamber arrester. The model includes the following equations [4-9]:

- the energy equation:

$$\frac{\partial}{\partial t}(\rho h) + \nabla \cdot (\rho \vec{v} h) = \sigma E^2 - u_{rad} + \nabla \cdot \left(\frac{\lambda}{c_p} \nabla h\right); \quad (1)$$

- the motion equation:

$$\frac{\partial}{\partial t}(\rho \vec{v}) + \nabla \cdot (\rho \vec{v} \vec{v}) = -\nabla p + \left[\vec{J} \times \vec{B}\right] + \nabla \cdot (\mu \nabla \vec{v}); \quad (2)$$

- the continuity equation:

$$\frac{\partial}{\partial t}(\rho) + \nabla \cdot (\rho \vec{v}) = 0; \quad (3)$$

- the electromagnetic equation on the basis of Maxwell's system:

$$\begin{cases} \nabla \times \vec{H} = \vec{J} \\ \nabla \times \vec{E} = -\frac{\partial \vec{B}}{\partial t} \\ \nabla \cdot \vec{D} = q \\ \nabla \cdot \vec{B} = 0 \end{cases}. \quad (4)$$

Here: ρ is the density, h is the enthalpy, υ is the velocity, σ is the electrical conductivity, E is the electric field strength, u_{rad} is the specific radiation power losses, λ is the thermal conductivity, c_p is the specific heat, p is the pressure, J is the current density, B is the magnetic induction, μ is the viscosity, H is the magnetic field intensity, D is the electric displacement vector, q is the specific volume electric charge.

The turbulence of plasma flow was taken into account by the shear stress transport (SST) model which is a combination of k-ε and k-ω turbulence models: k-ε model are used for calculation in the free flow zone while k-ω model are used in the zones near walls.

The use of polygonal mesh accelerates the convergence of the solution in comparison with the conventional triagonal mesh.

An electrical potential difference is defined on electrodes surfaces as boundary condition for the electromagnetic problem so that a discharge current through the chamber corresponds to the experimentally obtained dependence at each moment of time (see Fig. 3) [5].

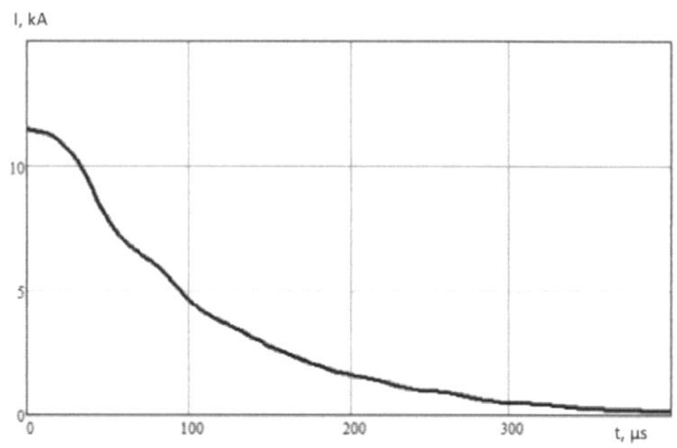

Figure 3. Discharge current dependence on time

The properties of gas (plasma) was calculated based on erosion of the electrodes and ablation of material of walls of the discharge chamber.

A method of calculation of the composition, thermodynamic and transport properties of plasma is described in sufficient detail in the references [10–14].

Silicone rubber has the following chemical formula:

$$(C_2H_6SiO)_n$$

Using this information and considering possible cases of the electrode material (Cu, W, Fe) the 85 components were taken into calculation.

The equilibrium constants for the equations of the mass action law were taken from [15, 16], and the partition functions of atoms and ions were calculated based on NIST data [17]. Enthalpies of individual components were taken from [15, 16] to calculate the thermodynamic properties.

To calculate the transport properties, it is necessary to find data on collision cross sections for each pair of species. The book [18] was used as the main source for specifying collisions between two neutral species and between neutral and charged components. If the required cross section was not found in the book [18] then we acted as follows.

When considering collisions between two neutral components it was assumed that the cross section can be described by the Lennard-Jones interaction potential [19], the main interaction parameters were taken from [20, 21] using rules given in [21].

When considering collisions between neutral and charged components it was assumed that the cross section can be described by the polarization potential, the parameters of which were taken from [20, 21].

The collision of two charged components was described by a screened Coulomb potential [22].

A series of calculations were carried out for a pressure range of 0.3–1.0 *atm* and a temperature range of 1000–30 000 *K* with the following proportions of basic chemical elements included into the system:

Si:O:C:H=1:1:2:6;   Fe:O=1:10.

These ratios were determined using a condition that the walls of the discharge chamber are made of silicone rubber and the electrodes are made of steel.

The ratio Fe:O=1:10 was made on the basis of experimental studies of the plasma composition at a discharge in the chamber of the multi-chamber arrester.

The obtained dependences of plasma properties on the temperature were used in the further mathematical model.

Two cases of the discharge chamber design shown in Figs. 4, 5 were calculated. The main difference between cases is the design of the electrodes. A design with tube electrodes forming cavities connected to the main volume of the discharge chamber is shown in Fig. 4. A design with rod electrodes (without cavities) is shown in Fig. 5.

Figure 4. Design of the discharge chamber with tube electrodes

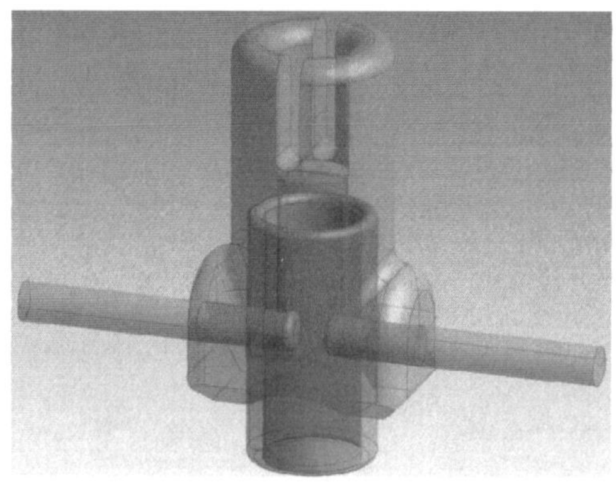

Figure 5. Design of the discharge chamber with rod electrodes

III. Results of Calculation

Results of simulation i.e. distributions of plasma temperature at different points of time are shown in Figs. 6, 7 for two cases of the discharge chamber design.

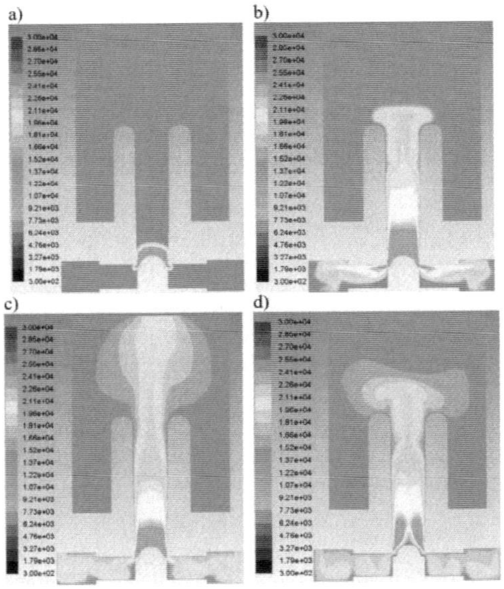

Figure 6. Distributions of temperature (T, K) in the discharge chamber with tube electrodes at different points of time: a – 2 μs, b – 30 μs, c – 140 μs, d – 210 μs

The obtained results show that:

1) In the case of a discharge chamber with tube electrodes the plasma jet reaches the exit of the chamber gap at time point of 10 μs approximately, and the chamber outlet – at time point of 30 μs;

2) In the case of a chamber with rod electrodes the plasma jet reaches the exit of the chamber gap at time point of 5 μs approximately, and the chamber outlet – at time point of 15 μs. It means that motion of plasma to the outlet is approximately twice as high as in the case with tube electrodes;

3) The maximum temperature at the outlet of the discharge chamber is 15-20 thousand K, in the first 100-120 μs the plasma jet increases in size, then its size decreases;

4) The high-pressure zone reaches the outlet of the discharge chamber earlier than the high-temperature zone. It means that a plasma heated to a high temperature moves towards the exit and "pushes" cold gas before itself;

5) The presence of a longitudinal cut in the cylindrical part of the discharge chamber leads to the ejection of a part of plasma in this direction and to a decrease in the velocity and temperature of the plasma jet;

6) Analysis of the simulation results shows that the presence of cavities joined to the main volume of the discharge chamber by small openings leads to the fact that at the initial stage of discharge, when pressure inside the discharge chamber exceeds pressure inside the cavities, plasma begins to spread into these cavities. The pressure inside the cavities is increased as long as it begins to exceed the pressure in the discharge chamber, which is lowered due to propagation of the discharge towards the outlet of the discharge chamber. It occurs in about 140-150 μs after the start of discharge. Thereafter, the gas accumulated in the cavities with lower temperature than the discharge core, begins to move in the opposite direction, i.e. from cavities to the discharge chamber. After 160 μs it leads to cooling of the central part of the discharge, which undoubtedly contributes to the extinction of the electric arc.

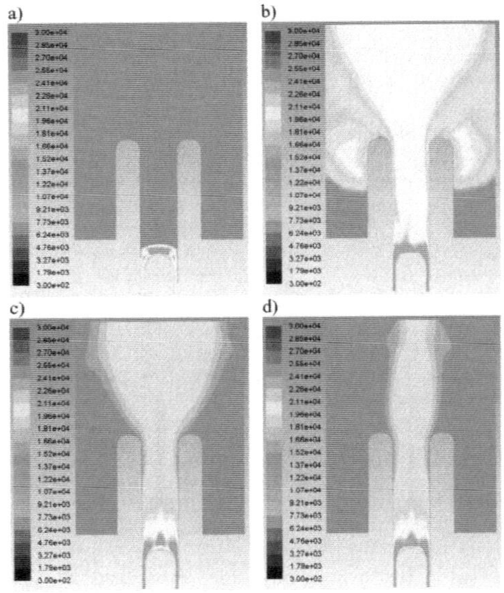

Figure 7. Distributions of temperature (T, K) in the discharge chamber with rod electrodes at different points of time: a – 2 μs, b – 30 μs, c – 140 μs, d – 210 μs

IV. Conclusion

The developed mathematical model allows to adequately simulate the processes occurring in the discharge chamber of the multi-chamber arrester and in the surrounding space.

Further studies will be devoted to taking into account in the model the following effects: deformation of the chamber due to high pressure, transport of plasma radiation, deionization processes near the zero of the pulse current.

References

[1] G. V. Podporkin, "Development of Long Flashover and Multi-Chamber Arresters and Insulator-Arresters for Lightning Protection of Overhead Distribution and Transmission Lines", Plasma Physics and Technology, vol. 2, no. 3, pp. 241-250, 2015.

[2] R. Kozakov, A. Khakpour, S. Gorchakov, D. Uhrlandt, D. Ivanov, I. Murashov, G. Podporkin, V.Frolov, "Investigation of a Multi-Chamber System for Lightning Protection at Overhead Power Lines", Plasma Physics and Technology, vol. 2, no. 2, pp. 150-154, 2015.

[3] V.Ye. Pilshchikov V.Ya. Frolov R.A. Belsky, "Investigation of the electrical strength and electrical conductivity of plasma near the chamber outlet during the pulse current through the discharge chamber", Proceedings of V Russian Conference on Lightning Protection", pp. 243-246, 2016 (in Russian).

[4] V.Ya. Frolov, D.V. Ivanov, Yu.V. Murashov, A.D. Sivaev, "Calculation of the composition of plasma of an arc pulsed discharge in a multichamber arrester", Technical Physics Letters, vol. 41, no. 4, pp. 310–313, 2015.

[5] V.Ya. Frolov, D.V. Ivanov, Y.V. Murashov, A.N. Chusov, "Numerical simulation of processes in discharge chamber of multi-chamber arrester for lightning protection", Proceedings of V Russian Conference on Lightning Protection", pp. 334-337, 2016 (in Russian).

[6] A.N. Chusov, G.V. Podporkin, M.E. Pinchuk, D.V. Ivanov, Y.V. Murashov, V.Ya. Frolov, "Modeling of discharge in multi-chamber systems", Proceedings of V Russian Conference on Lightning Protection", pp. 351-357, 2016 (in Russian).

[7] V. Frolov, D. Ivanov, I. Murashov, A. Sivaev, "Mathematical simulation of processes in discharge chamber of multi-chamber system for lightning protection at overhead power lines", Proc. 2016 IEEE North West Russia section young researchers in electrical and electronic engineering conference, ELCONRUSNW 2016, pp. 562-565, 2016.

[8] V.Y. Frolov, D.V. Ivanov, R.A. Belsky, "Iincreasing of operation security and of breaking capacity of surge arresters", Proc. 2017 IEEE Russia Section Young Researchers in Electrical and Electronic Engineering Conf., ElConRus 2017, pp. 1520-1523, 2017.

[9] V.Y. Frolov, D.V. Ivanov, G.V. Podporkin, A.D. Sivaev, "Development of mathematical model of processes in multi-chamber arrester for identification of criteria of arc extinction", 2017 Int. Symp. Lightning Protection, XIV SIPDA 2017, pp. 240-243, 2017.

[10] M.I. Boulos, P. Fauchais, E. Pfender, Thermal Plasmas: Fundamentals and Applications, Vol. 1. Boston, MA: Springer US:,Imprint: Springer, 1994.

[11] M. Capitelli, G. Colonna, A. D'Angola, Fundamental Aspects of Plasma Chemical Physics: Thermodynamics, Springer Verlag, 2011.

[12] Thermodynamic and Thermophysical Properties of Combustion Products, Ed. by V. P. Glushko, Vol. 1, Moscow: AN SSSR-VINITI, 1971 (in Russian).

[13] A.L. Suris, Thermodynamics of High-Temperature Processes, Moscow:Metallurgiya, 1985 (in Russian).

[14] V.Ya. Frolov, D.V. Ivanov, "Calculation of a plasma composition and its thermophysical properties in cases of maintaining or quenching of electric arcs", J. Phys.: Conf. Ser., vol. 1058(1), p. 012040, 2018.

[15] NIST Atomic Spectra Database: http://physics.nist.gov/PhysRefData/ASD/index.html

[16] Thermodynamic Properties of Individual Substances: Reference Book, Ed. by V. P. Glushko et al., 3rd ed. Nauka, Moscow, 1978–1982, Vols. 1–4 (in Russian).

[17] M.W. Chase, Jr., Editor. NIST-JANAF Thermochemical Tables, J. Phys. Chem. Ref. Data. Monograph No. 9 (Part I and II), New York: American Chemical Society and the American Institute of Physics for the National Institute of Standards and Technology, 1998.

[18] M. Capitelli, D. Bruno, A. Laricchiuta, Fundamental Aspects of Plasma Chemical Physics: Transport, Springer Verlag, 2013.

[19] J.O. Hirschfelder, C.F. Curtiss, R.B. Bird, Molecular theory of gases and Liquids, J. Wiley, New York, 1964.

[20] Y. Cressault, R. Hannachi, Ph. Teulet, A. Gleizes, J.-P. Gonnet, J.-Y. Battandier, "Influence of metallic vapours on theproperties of air thermal plasmas", Plasma Sources Sci. Technol., vol. 17, p. 035016, 2008.

[21] P. Andre, J. Aubreton, S. Clain, M. Dudeck, E. Duffour, M.F. Elchinger, B. Izrar, D. Rochette, R. Touzani, D. Vacher, "Transport coefficients in thermal plasma. Applications to Mars and Titan atmospheres", Eur. Phys. J. D, vol. 57, pp. 227-234, 2010.

[22] E.A. Mason, R.J. Munn, F.J. Smith, Phys. Fluids, "Transport Coefficients of Ionized Gases", vol. 10(8), pp. 1827-1832, 1967.

Analysis of the Quality Problem of Electric Power and the Management of Reliability of Power Supply

Boriss Papkov
Nizhny Novgorod State Engineering and Economic Institute,
Knyaginino, Russia boris.papkov@gmail.com

Anatolijs Mahnitko, Inga Zicmane, Kristina Berzina
Faculty of Power and Electrical Engineering
Riga Technical University Riga, Latvia
mahno@eef.rtu.lv, Inga.Zicmane@rtu.lv,
Kristina.Berzina@rtu.lv

Vladislav Oboskalov, Sergey Gusev
Department of Automated Power Systems
Ural Federal University
Ekaterinburg, Russia
s.a.gusev@urfu.ru, s.a.gusev@urfu.ru

Alexander Tavlincev
Department of Automated Power Systems
Ural Federal University
Ekaterinburg, Russia
s.a.gusev@urfu.ru

Abstract — **The paper discusses the main indicators of the quality of electricity (frequency and voltage), determining the balances of active and reactive power in the EPS. The analysis of the dependence of the quality of electricity on changes in the generated and consumed power has been done. The deterioration of technical and economic indicators of EPS and PSS when there is a deviation of electricity from its nominal values is observed.**

The task of ensuring the level of consumer reliability is proposed to be solved in several stages, the implementation of which is possible only under complex organizational and economic development. At the same time, the reliability of power supply to consumers should be considered as a commodity that should be paid at the market price and as an obligation of the EPS to its customers.

Keywords — power supply system, power quality, power balance, reliability

I. Introduction

Modern development of electric power systems (EPS) and power supply systems (PSS) of consumers, aimed at the creation of intelligent electrical networks, imposes increased requirements to ensure reliable power supply to consumers of electricity with standardized quality, quantity and continuity [1-2]. The fulfilment of these requirements leads to an increase in the operating efficiency of electricity consumers and EPS. The solution to the problem of power quality should be based on a feasibility comparison of the effect of measures to improve the quality of electricity and the inevitable additional costs. According to the LEK standard, the frequency of the alternating current f in the electrical network and the voltage U at its nodes are the most important indicators of the quality of electricity.

The technical and economic indicator for assessing the quality of electricity is the damage (technological and electromagnetic) of the EPS and consumers from poor-quality electricity.

There are a large number of papers dealing with various aspects of power quality and reliable power supply to consumers [2-10]. Most of them consider the issues of assessing both regime and structural [2-9, 18] reliability of electrical networks. The problem of power quality and the provision of reliable power supply to consumers is of particular importance when creating intelligent energy, operating on the principles of the Smart Grid, involving the participation of the end user of electricity in the system of planning, functioning and control of EPS [11-20]. The latter is explained by the fact that the intelligent electrical network is represented as the integration of existing technologies of generation, transmission and distribution of electricity with modern information and communication computer technologies associated with external information networks. The emergence in these networks of many additional devices for control, registration, management, etc., various kinds of data only complicate the problem under consideration. This is explained both by the fact of the increase in the number of these devices in the EPS and PSS, as well as by the increased requirements of these devices to the quality of electricity due to its sensitivity, for example, to the frequency of alternating current.

It should be noted that there still remains the problem of formalizing the analysis of the reliability of power supply to consumers for solving the system reliability problems. Currently, there are three main approaches to formalizing the reliability of PSS consumers: a) analysis of structural schemes: c) analysis of functional integrity schemes or using a failure tree; c) statistical modelling [5, 7, 18].

The article provides an approach to solving the problem of ensuring the reliability of power supply to consumers, based on the principle of mutual benefit. The latter implies that EPS and consumers should receive an economically reasonable monetary reimbursement of their costs to achieve the required level of reliability of power supply. The practical implementation of this model of power supply reliability management is relevant in the modern conditions of a liberalized power industry, the introduction of intelligent electrical networks and distributed energy.

II. Power Balance and Power Quality

At each moment of the steady-state regime in the EPS, balances of active and reactive powers should be observed, represented respectively by the equations

$$\sum P_G = \sum P_L + \sum \Delta P_S + \sum \Delta P_{O.N}, \quad (1)$$

$$\Delta Q_G = \sum Q_L + \sum \Delta Q_S + \sum \Delta Q_{ON} \pm \sum Q_{CD} - \sum Q_{PL}, \quad (2)$$

where $\sum P_G, \sum Q_G$ - total active and reactive powers of the generating sources; $\sum P_L, \sum \Delta Q_L$ - total active and reactive power of loads; $\sum \Delta P_S, \sum \Delta Q_S$ - total power losses in the elements of power supply systems and EPS; $\sum \Delta P_{ON}, \sum \Delta Q_{ON}$ - total power consumption for own needs of power plants; $\sum \Delta Q_{CD}$ - the total power of the compensating devices (the sign "+" corresponds to the devices consuming reactive power, the sign "-" to the generating ones; $\sum Q_{PL}$ - total (charge) reactive power generated by overhead power transmission lines.

Power balances in EPS exist in steady-state regimes, characterized by certain values of the basic parameters of the quality of electricity - the frequency of alternating current f and voltage at the nodes of the network U, including the consumer's buses. In case of power imbalances in the EPS, disturbances of the steady-state regime occur, and transient processes begin, which end in new steady-state regimes with new values of f_1 and U_1. The frequency value at any time of the normal mode is the same in all nodes of the electrical network and is a single indicator of the quality of electricity. At the same time, the voltage levels at different points in the network can vary greatly. In this case, in some nodes the voltage levels may meet, while in others they may not meet the requirements of LEK.

Capacities consumed from the network by electrical receivers, as a rule, depend on the frequency of the network and the voltages on the buses of the electric receivers. Groups of power receivers, receiving power from the EPS, form load nodes. These dependences (under different complex composition of power receivers in the load nodes) are called static load characteristics. A general view of such characteristics for a complex load node is shown on Fig.1.

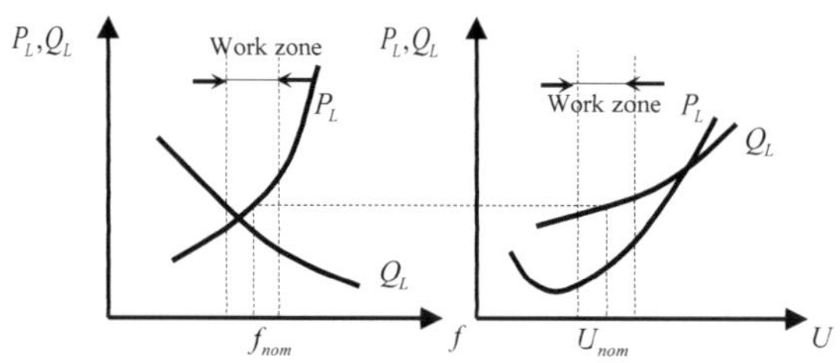

Fig. 1. Static Load Characteristics: a - frequency; b - voltage

The characteristic of the active power P_L in frequency is a steep curve with a positive slope. Accordingly, as the frequency f increases in the EPS, the active power consumption of the load increases and vice versa - the active power of the load decreases as the frequency f decreases. Thus, the partial derivative $\frac{\partial P_L}{\partial f}$ of active power in frequency (the slope ration $P_L(f)$) is positive and has a rather large value compared to the derivative of the active power in voltage $\frac{\partial P_L}{\partial U}$, which is also positive, but close to zero. Reactive load power decreases with increasing of frequency $\left(\frac{\partial Q_L}{\partial f} < 0\right)$ and increases with increasing in voltage on the load buses $\left(\frac{\partial Q_L}{\partial U} > 0\right)$.

It is known that to determine the relationship between the parameters of the quality of electricity (frequency and voltage) and power balances can be most clearly done using the example of the EPS equivalent to one generator ($\sum P_G + j\sum Q_G$) and one load ($\sum P_L + j\sum Q_L$), Fig. 2.

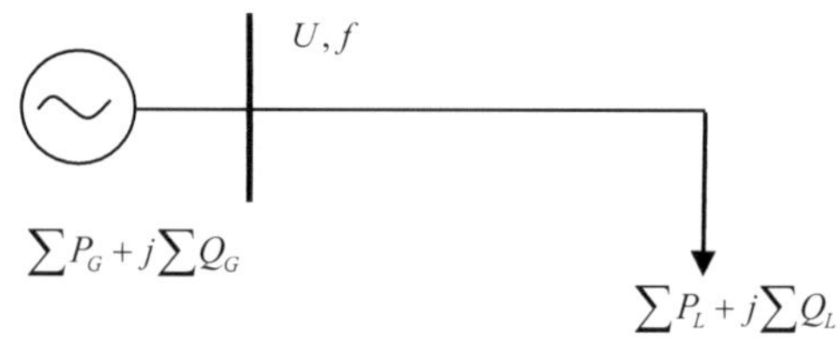

Fig. 2. Schematic Equivalent Power Energy System

Imbalance in such an EPS, and, consequently, a change in the parameters of power quality occurs under the change in load $\sum P_L, \sum \Delta Q_L$ and the change in generation $\sum P_G, \sum \Delta Q_G$. To assess the effect of the power balance on the change in frequency f and voltage U on consumer buses, you can use the linear part of the expansion of functions in the Taylor series (1) and (2) near the point of initial state (f_0, U_0) of power quality parameters.

If power balance in such an equivalent EPS is violated due to a change in the output of the generated active power $\sum P_G$ by the value ΔP_G and reactive power $\sum Q_G$ by the value ΔQ_G, the frequency will change by the value Δf and on the consumer's buses by the value ΔU.

As a result, it is possible to determine linear dependencies between changes in generating capacities $\sum P_G$ and ΔQ_G, and changes in the parameters of power Δf and ΔU quality and under constant load composition in the form of a system of linear equations:

$$\left.\begin{aligned} \Delta P_G &= \frac{\partial \sum P_L}{\partial f}\Delta f + \frac{\partial \sum P_L}{\partial U}\Delta U, \\ \Delta Q_G &= \frac{\partial \sum Q_L}{\partial f}\Delta f + \frac{\partial \sum Q_L}{\partial U}\Delta U. \end{aligned}\right\} \quad (3)$$

Based on these dependencies, unknown frequency Δf and voltage ΔU deviations can be determined by expressions:

$$\left.\begin{aligned} \Delta f &= \frac{\Delta P_G \dfrac{\partial \sum P_L}{\partial U} - \Delta Q_G \dfrac{\partial \sum P_L}{\partial U}}{D}, \\ \Delta U &= \frac{\Delta Q_G \dfrac{\partial \sum P_L}{\partial f} - \Delta P_G \dfrac{\partial \sum Q_L}{\partial f}}{D}, \end{aligned}\right\} \quad (4)$$

where D is the numerical value of the system identifier (3).

Analysis of relations (4) allows you performing an analysis of the quality of electricity with known parameters of the initial state of the EPS and changes in generated power. From these relationships, for example, it follows that under a constant generation of reactive power $(\Delta Q_G = 0)$, the frequency in the EPS increases, if the generation of active power increases $(\Delta P_G > 0)$ and vice versa, it is decreased, if the release of generating active power $(\Delta P_G < 0)$ occurs. With the constancy of the generation of active power $(\Delta P_G = 0)$, the following considerations are true.

Considering that the steepness of the characteristic $\frac{\partial \sum P_L}{\partial U}$ is much less than the steepness $\frac{\partial \sum P_L}{\partial f}$, a change in the mode of reactive power affects mainly the change in voltage levels in the EPS and has little effect on the change in frequency. At the same time, the reduction ΔQ_G leads to a decrease in voltage on the buses of the generator and in the system as a whole, but the increase ΔQ_G leads to the increase in voltage.

Also, it follows from (4) that an increase ΔP_L leads to a decrease in frequency, but an increase ΔQ_L, mainly, decreases voltage.

The requirements for the parameters of power quality determine the need for measures to regulate the balance of power. Since the frequency is a system-wide parameter and in the steady-state regime at any point of the EPS has the same value, it makes no difference at which point of the EPS to perform frequency control, the level of which depends on the active power balance throughout the system.

The balance of reactive power in the system determines the average level of voltages in the EPS. However, the voltage level in the EES as a whole does not guarantee the required voltage levels in all nodes of the electrical network. Consequently, the balance of reactive power should be checked not only for the system as a whole, but also for its individual areas. In order to maintain the balance of reactive power, it is necessary, besides centralized voltage regulation, to use the local one.

III. IMPACT OF FREQUENCY DEVIATIONS ON THE WORK OF EPS AND POWER SUPPLY SYSTEMS

The frequency deviation from the nominal one negatively affects both the work of the EPS and PSS.

For example, blades of steam turbines are designed for shock-free steam at the nominal speed of turbines. If the frequency deviates, the steam enters the turbine with a blow, which can lead to a beating and even destruction of the turbine blades. Reducing the frequency greatly affects the operation of the feed pumps of boilers of power plants, which have a steep dependence of performance on frequency. In turn, a reduction in the feed water supply to the boilers leads to a decrease in steam generation, which as a result reduces the generation of active power by the generators and a further decrease in the frequency.

With a decrease in the frequency of consumers in the PSS, there is a deterioration in the operation of electric drives and a decrease in production.

Maintaining the frequency at the level regulated by the LEK standard is carried out by powerful EPS generators by changing the output of active power to the electrical network.

IV. IMPACT OF VOLTAGE ON THE WORK OF ELECTRIC RECEIVERS AND EPS

The deviation of the voltage from the optimum one leads to a change in the technical and economic indicators of electricity receivers. When the voltage changes, the indicators of the electrical network itself also change, mainly due to changes in power and energy losses. Thus, voltage deviations in individual points of the network affect the entire PSS of the industrial enterprise. It should be noted that now the most common receivers of electricity in the industry are asynchronous motors used to drive a wide variety of mechanisms. When the voltage of the network changes compared to the nominal one, the active power on the shaft of the asynchronous motor remains almost constant, but the losses of active power in it change, which can cause over-consumption of electricity.

Reducing the voltage in the electrical network increases the power loss in the EPS, which, as a result, increases the fuel over-burn to compensate for these losses. In addition, the performance of the engines of own needs falls, which can lead to a decrease in electricity generation.

As the voltage increases, the rate of aging of the insulation of electrical equipment increases, which may be the cause of emergency situations. When the voltage rises, losses in the steel of power transformers increase, which leads to an increase in the temperature of the steel. Increasing the voltage increases the generation of reactive power by overhead transmission lines and batteries of static condensers, which can lead to large voltage problems in the network and the danger of insulation breakdown of electrical equipment.

V. VOLTAGE REGULATION

To limit the deviations of voltages from their nominal values, a set of well-known technical measures should be taken on the buses of electrical consumers. The implementation of these measures allows adjusting the voltage within the limits required by LEK standard. With centralized regulation in the supply node, the tolerable limits are generally maintained for the group of consumers in the nearest area at the same time. Local regulation involves maintaining the required voltage levels directly on the consumer's buses. Voltage regulation can be carried out:

1. Generators at power plants; transformers with voltage control devices under load (on-load changing) and without load (off-circuit tap-changer); booster transformers and linear regulators; compensating devices that produce (condensers batteries, synchronous compensators in under-excited mode) and consume (reactors, synchronous compensators in under-excited mode) reactive power.

As an example of voltage regulation on consumer buses, we present the problem of choosing the power of a compensating device necessary to ensure acceptable levels to the consumer (Fig. 3). The consumer load power is set to maximum. To simplify the calculations, the power losses will not be taken into account.

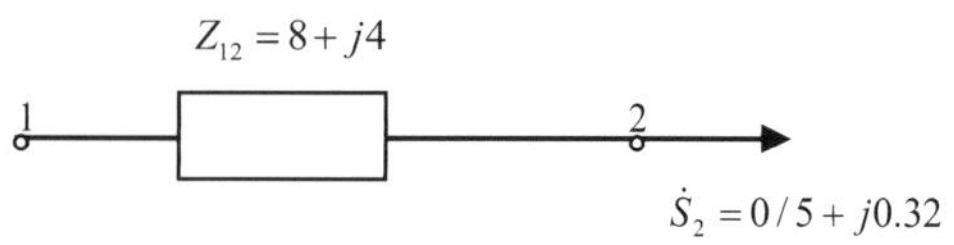

Fig. 3. Network Circuit

To choose the power of the compensating device, must first you make sure that it is necessary. To do it, you need to determine the voltage on the consumer's buses and compare it with the voltage levels permissible by the standard, in a network of 6 kV in normal modes correspond to the range from 6.3 to 5.7 kV ($\pm 5\%$ from U_{nom}).

Given the no-loss condition, the power flow is determined at the beginning of the network section:

$$S_{12} = S_2 = 0.5 + j0.32MV \cdot A.$$

The voltage drop in the 6 kV network is mainly due to the longitudinal component of the voltage drop, then:

$$U_2 = U_1 - \Delta U_{12} = U_1 - \frac{P_{12}R_{12} + Q_{12}X_{12}}{U_1} = 6 - \frac{0.5 \cdot 8 + 0.32 \cdot 4}{6} = 5.12kV.$$

Therefore, the voltage at node 2 does not meet the condition of the quality of electricity at the consumer; thus, it is necessary to install a compensating device that generates reactive power to increase the voltage at node 2. Such a device may be a battery of static condensers. Reactive power generated by a battery of static condensers is determined by the expression

$$Q_{CD} = \frac{\Delta U \cdot U_1}{X},$$

where ΔU - the difference between the rated voltage at node 2 and the minimum allowable one. We have

$$\Delta U = U_2 - U_{\min\ ac} = 5.12 - 5.7 = -0.58kV;$$

allowing you to determine the power of the required compensating device

$$Q_{CD} = \frac{-0.58 \cdot 6}{4} = -0.87MVar.$$

Here, the minus sign means that to ensure the quality of electricity regulated by LEK at node 2, the power of the compensating device must be generated. Given that the scale of the nominal capacity of the compensating devices is discrete, you should choose the nearest large standard power.

2. By consumer loads reducing the consumption of active and, especially, reactive power while reducing the voltage on their buses (regulating the effect of load).

3. By the change the configuration of the electrical network.

VI. RELIABILITY

Ensuring the reliability of EPS and PSS for consumers is now becoming one of the most topical problems of energy development. One of the most obvious ways to solve this problem is to back up or duplicate the power supply of a large number of consumers. This way is associated with significant material costs, which must be properly justified. Assessing the damage caused to consumers by interruption of power supply, damages associated with emergency repairs, as well as the cost on improving reliability, one can raise the question of the optimal level of reliability of electric power equipment, facilities and systems.

For the PSS consumers, as well as for the EPS ones, it is important to ensure the sustainability of operation under various (internal and external) disturbances, since they can contribute to the cascade development of the accident. In this connection, the question of survivability arises, i.e. preservation of the efficiency of production after the manifestation of disturbances leading to negative consequences.

The possible consequences of unreliability become so significant that it requires constant improvement of methods for modelling the development, design, construction, installation, operation and diagnostics of PSS and EPS, allowing for a fuller consideration of reliability and the most economical use of funds allocated for its provision.

It should be noted that the reliability of decisions made to ensure the reliability of systems largely depends on the completeness, accuracy and timeliness of the information taken as the basis. It is never absolutely accurate, even for the reason that for making future-oriented decisions, it is possible to use only a limited amount of historical data and take into account only the present state of the controlled object. Even under the complete absence of retrospective information about the conditions of operation of highly reliable structurally complex systems in extreme situations, the value of comparative calculations is obvious, since they make it possible to obtain data for further interaction with power supply organizations [5, 6, 18, 21-25].

Obviously, with the further development of market relations, intelligent networks and distributed generation, the solution of problems associated with the assessment and minimization of consumer damage becomes even more relevant.

The qualitative composition and magnitude of losses for consumers can be completely different: from zero to unacceptably large quantities, from material losses to social ones. The difficulty of obtaining sufficient and reliable information about the possible losses of consumers, as well as the lack of universal methods for calculating it, does not allow to adequately solving actual problems of determining the real level of reliability of electricity supply to consumers. The issues of ensuring a universal information and methodological basis for registering the information about consumer losses in any practical and theoretical problems of energy have been studied and are being studied by many researchers. The analysis of these researches leads to the conclusion about the prospects of the developments outlined in [6] allowing to solve the problem of ensuring the reliability of power supply. To solve it, principles for the conclusion of contractual relations between the subjects of the electric power industry for the implementation of an alternative system of relations between the subjects (on an economic basis) have been proposed. The alternative system of relations between the subjects differs from the existing ones by universality, simplification of implementation in the conditions of the market electric power industry. Its implementation will allow the subjects of the electric power industry, who are interested in managing the reliability of electricity supply to mutually beneficially and targetedly provide the necessary level of reliability with costs close to the objectively reasonable minimum.

When analysing the reliability of power supply and power consumption management, various approaches are used. The three most common approaches to formalize the reliability of PSS consumers are the most widespread: a) analysis of structural schemes; b) analysis of functional integrity schemes or using fault trees; c) statistical modelling. All these approaches are quite versatile and interchangeable; they are widely used in practice, but the problem of formalizing the analysis of the reliability of power supply to consumers remains.

VII. THE ENSURING THE BASIC LEVEL OF RELIABILITY OF ELECTRICAL SUPPLY OF CONSUMERS

The problem of organizing the control of the level of reliability of power supply arises mainly for economic reasons. In particular, the cost for maintaining or improving reliability (strengthening and upgrading the electrical network, building and commissioning new capacities) is covered by the power company, and damages related to failures of the EPS elements mainly occur to the end user as a result of power supply disruption. At the same time, the consumer and the power company are rigidly connected by the electrical network, which at the current stage of development of the EPS does not allow the change of the service provider. Damage from a power failure is inversely related to the cost of reliability and is largely determined by them.

At present, the relationship between the consumer and the power company is such that the power company has no motivation to increase the cost on reliability, but the consumer can hardly influence them.

In this regard, the task of ensuring the level of reliability can be represented in the form of several subtasks:

1) determining the level of reliability that must be provided to the consumer;

2) the definition of a group of measures due to which the customer's required level of reliability will be ensured;

3) determining the source of funding for a group of measures to ensure reliability;

4) determining the rules for the distribution of responsibility for power supply disruptions due to non-compliance with the contractual terms of reliability.

Today there is an acute shortage of solutions to the first, third and fourth tasks. At the same time, there are many proposals for solving the second task. The latter circumstance creates the illusion of ease of solving the problem of reliability. It is believed that it is sufficient to implement several technical measures and the required level of reliability will be obtained. However, any measures cannot be implemented until clear solutions to the three other tasks are identified.

This imbalance may explain the long existence of the problem of reliability and its aggravation in the conditions of the market electric power industry. A well-developed and well-known mathematical apparatus of the theory of reliability is aimed at the technical support of the structural and elemental reliability of EPS and PSS [8, 16]. For this reason, most of the proposed solutions are aimed at ensuring the existing and (or) contractual level of reliability in a purely technical sense, without proper comprehensive organizational and economic development. These methods of reliability control are located between two diametric approaches to its assessment [4]. On the one hand, reliability is considered as a commodity that must be paid at a free market price, and on the other, as an obligation of energy companies to their customers. With all the practical opposite of these approaches, only their simultaneous and balanced implementation will solve the problem. Energy companies should receive fair reimbursement of the cost on maintaining reliability, but the consumers should get reliable electricity supply at a reasonable price. Violation of this balance leads to a decrease in reliability either due to the irresponsibility of monopolized energy companies (the current situation), or due to their bankruptcy. In the event of a decrease in reliability from the "ordered" (basic) level due to the fault of the power company, the consumer must receive financial compensation.

VIII. CONCLUSION

The quality of electricity determines the economic efficiency of consumers and EPS in general. Deviations of the most important indicators of the quality of electricity, namely, the frequency of alternating current and voltage at the nodes of the EPS from their nominal values adversely affect the operation of EPS and PSS.

Existing requirements for the parameters of power quality necessitate the implementation of measures for their regulation. Taken into account the relationship of power quality with the reliability of power supply to consumers, it is advisable to have a consistent understanding of the relationship and distinction between these indicators. It is quite obvious that with low reliability of electricity supply, it's impossible in general to provide consumers with electricity of high quality.

Therefore, to ensure the required level of PSS reliability, an appropriate problem solving strategy must be implemented. The essence of this strategy should be to ensure that energy companies can receive a fair reimbursement of their cost for maintaining reliability, but the consumers - reliable power supply at a reasonable price. The implementation of this strategy is of particular relevance in the creation of intelligent energy, operating on the principles of the Smart Grid, providing for the participation of the end user of electricity in the system of planning, functioning and control of EPS.

REFERENCES

[1] Likumi.lv-Republic of Latvia legislation, https://likumi.lv/doc.php?id=260769, Likumi.lv-Latvijas Republikas tiesību akti, https://likumi.lv/doc.php?id=260769

[2] "The concept of reliability in the power industry", Moscow, LLC ID "Energy", 2013, 304 p. "Концепция обеспечения надежности в электроэнергетике", Москва, ООО ИД " Энергия ", 2013, 304 с.

[3] N.I. Voropai, P. V. Etingov, A.S. Udalov, A. Germont and R. Sharkawi, "Coordinated emergency control of load and FACTS devices," Electricity, 2005, No 10, p. 25. Н.И. Воропай, П.В. Этингов, А.С. Удалов, А. Жермон, Р. Шеркави, "Координированное противоаврийное управление нагрузкой и устройствами FACTS", Электричество, 2005, № 10, с. 25.

[4] V. Nepomnyaschiy, J. Gerhards, A. Mahnitko and T. Lomane, "Reliability of Latvian power systems 330 kV substations," Latvian . Journal of. Physics and Technical Sciences, 2014, vol.51(3), pp. 15 – 23, ISSN 0868-8257.

[5] B. Papkov, "Formalization of damage assessment from power supply interruptions of structurally complex technical systems," Methodical issues of the study of the reliability of large energy systems. Issue 56. Problems of reliability of reformed energy systems and methods for their solution, Irkutsk: ISEM SB RAS, 2006. p. 259 - 263. Б.В. Папков, „Формализация оценки

ущерба от нарушений электроснабжения структурно-сложных технических систем", сб. Методические вопросы исследования надежности больших систем энергетики. Вып. 56. Задачи надежности реформируемых систем энергетики и методы их решения, Иркутск: ИСЭМ СО РАН, 2006. с. 259 - 263.

[6] M.V. Sharygin, "Assessment of consequences of power outage," Methods and models, Nizhny Novgorod, 2014, 202 p. М.В. Шарыгин, "Оценка последствий отключения потребителей электроэнергии", Методы и модели, Н. Новгород, 2014, 202 с.

[7] B. Papkov, J. Gerhards and A. Mahnitko,"Systems Problems of Power Suply Reliability Analysis Formalisation," Proceedings. 2015 IEEE 5 th International Conference on Power Engineering, Energy and Electrical Drives (POWERENG), May 11-13, 2015, Riga, Latvia.

[8] G.K. Moscov,"Reliability of adaptive systems," Moscow, Soviet Radio, 1973, p. 104. .Москатов Г.К. "Надёжность адаптивных систем", М, Сов. Радио, 1973, с 104.

[9] IEEE Standards 1366-2003 IEEE Guide for Electric Power Distribution Reliability Indices // 2004, 44 c.

[10] A.E. Chernov, A.V and Akimov, "Quality and reliability of electrotechnical complexes of autonomous objects," News of MSTU "MAMI", No 1 (13), 2012, pp 105-112. А.Е., Чернов, А.В. Акимов, "Качество и надежность электротехнических комплексов автономных объектов", Известия МГТУ «МАМИ» № 1(13), 2012, с. 105-112.

[11] Amin S. M., Wollenberg B. F., "Toward a Smart Grid", IEEE P&E Magazine, September/October, 2005.

[12] European SmartGrids Technology Platform, "Vision and Strategy for Europe's Electricity Networks of the Future", Luxembourg: Office for Official Publications of the European Communities, 2006, p.44. – URL: http://www.youblisher.com/p/166842-SmartGrids-European-Technology-Platform-European-Commission/

[13] Liserre M., Sauter T. and Hung J.Y., "Future energy systems: Integrating renewable energy sources into the smart power grid through industrial electronics," IEEE Industrial Electronics Magazine: – March, 2010. – Vol. 4, Is. 1. – p. 18–37. – URL: http://dx.doi.org/10.1109/MIE.2010.935861

[14] Gellings C. W., "The Smart Grid. Enabling Energy Efficiency and Demand Response," CRC Press, 2010.

[15] Analysis of Distributed Generation Systems, Smart Grid Technologies and Future Motivators Influencing Change in the Electricity Sector / School of Engineering and Science, 2011. URL: http://file.scirp.org/pdf/SGRE20110300001_32428639.pdf.

[16] A.F. Dyakov, V.A. Stennikov, S.M. Senderov and others, "Reliability of energy systems: Problems, models and methods for solving them," rep. ed. N.I. Whisper, Novosibirsk: Science, 2014., 284 p. А.Ф. Дьяков, В.А. Стенников, С.М. Сендеров и др., "Надёжность систем энергетики: Проблемы, модели и методы их решения", отв. ред. Н.И. Воропай, Новосибирск: Наука, 2014., 284 с.

[17] Oboskalov,V., Gerhards, J. and Mahnitko, A. Structural Reliability of Electrical Power Systems, Riga: RTU Press, 2015. -202 p.

[18] B. V. Papkov and V. L. Osokin, "Power consumption management in the conditions of the market and intellectualization of power supply systems," Knyaginino, NGIEI, 2016, 218 p. Б.В. Папков, В.Л. Осокин, "Управление электропотреблением в условиях рынка и интеллектуализации систем электроснабжения", Княгинино, НГИЭУ, 2016, 218 с.

[19] 16. IEC TR 63097: 2017 Smart grid standardization roadmap. Edition 1.0. 2017.

[20] A. V. Ivanov, Yu. N. Kucherov, V. M. Samkov and D. A. Corlew, "Development of standardization of intelligent power supply systems of the future," Energy of a single network, No 3 (38) June-July 2018, pp 70-79. А.В. Иванов, Ю.Н. Кучеров, В.М. Самков и Д.А. Корев, "Развитие стандартизации интеллектуальных систем электроснабжения будущего", Энергия единой сети, № 3 (38) июнь-июль 2018, с. 70-79.

[21] A.N. Borisov, A.V. Alekseev, G.V. Merkurjeva and others, "Processing of fuzzy information in decision-making systems," Radio and Communications, 1989, p. 304. А.Н. Борисов, А.В. Алексеев, Г.В. Меркурьева и др., "Обработка нечёткой информации в системах принятия решений", Радио и связь, 1989, с. 304.

[22] V. Oboskalov, S. Timashev, A. Bushinskaya and S.Gusev, S, "Stochastic network reduction technique for calculations of electrical power system structural reliability," Proceedings of the 2nd International Symposium on Stochastic Models in Reliability Engineering, Life Science, and Operations Management, SMRLO 2016.

[23] V. Oboskalov, R. Valiev, S. Gusev, A. Mahnitko and I. Moshkin, "An analytical approach for calculations of power system generation adequacy indices," Proceedings of the 58th Annual International Scientific Confererence on Power and Electrical Engineering of Riga Technical University, RTUCON 2017.

[24] A.S. Tavlintsev, A.A. Suvorov, S.A. Gusev, E.D. Staymova, I. Zicmane and K. Berzina,"Search for the Single-type Load Schedules of the Power Facitily," Proc. of the 2018 IEEE 59th International Scientific Conference on Power and Electrical Engineering of Riga Technical University (RTUCON), November 12-14, 2018, Riga, Latvia. 5 pp.

[25] A. Pazderin and A. Tavlintsev,"Static load characteristic of complex load node with allowance for automation," (2015) MATEC Web of Conferences, 37, DOI: 10.1051/matecconf/20153701040

Modeling of Electricity Demand in the Region under Uncertainty

Janusz Sowinski
Faculty of Electrical Engineering
Czestochowa University of Technology
Czestochowa, Poland
jansow@el.pcz.czest.pl

***Abstract*—The changes on the electricity market affecting consumers increase uncertainty concerning demand for electricity and force the development of new methods of the analysis and the forecast of electricity demand in the regions understood in the paper as the province. A method for forecasting electricity demand in the region based on the idea of an end-use model is presented. Such models use regional electricity consumption rates. The data published by GUS (Statistics Poland) form the Local Data Bank. The data are grouped according to the division of the country into units and according to the category, such as electricity consumption in the sectors of economy, ie. the industrial sector, power sector, transport, households, agriculture and other consumption. Using the above data, demographic data and regional economic data, electricity consumption rates are created. The time series of electricity consumption rates are treated as dynamic processes with random perturbations and can be modeled by means of stochastic equations of Ito type. The model deploys stochastic differential equations (SDE) for simulating time-variation of electricity consumption rates by means of the Euler method. On the basis of available statistical data, the results of a forecast of the electricity demand in the region in a medium-term horizon are presented.**

Keywords: electricity; regional demand; uncertanty;

I. Introduction

Contemporary energy management is facing a period of rapid technological changes in the power industry [1]. Changes in technologies of obtaining and processing energy [1] are related to the development of distributed generation, mainly using renewable energy sources. The above changes force distribution system operators to plan the development of electricity networks [2], guaranteeing adequate reliability of electricity supply [3]. Structural changes on the electricity market and technological progress are also observed on the consumer side. The above phenomena increase the uncertainty of the future electricity demand, even in the short time horizon. It is necessary to develop new forecasting techniques for electricity demand under conditions of uncertainty.

II. Statistical Data of Electricity Consumption in the Regions

A. Energy Management in the Regions

Analyzing the energy consumption in the region, it is impossible to omit changes in the regional electricity market. In this context, prosumers' behavior and their impact on the electricity demand are considerable important. The development of energy storage technologies is an important element supporting the development of unstable electricity generation technologies. The amendment to the Act on Renewable Energy Sources (June 22, 2016) introduced a system of discounts in prosumer settlements, but the real innovation was the introduction of energy clusters. The agreement within the energy cluster assumes the generation and balancing of the demand, distribution or trade in energy from RES or other sources in the distribution network with rated voltage below 110 kV, geographically limited, as the cluster's area of activity can not exceed one district (poviat) or 5 communes. Implementation of the energy policy of the cluster, but also of the commune, requires analysis of the demand for energy and electric power. It is necessary to build appropriate regional forecasting models.

In addition, the Energy Law imposes on the municipalities the obligation to develop assumptions for the plan of supplying heat, electricity and gas fuels for the commune area for at least 15 years and its update at least every 3 years. The development of assumptions for the plan should include variant forecasts, among others electricity based on the municipal development program. The method for forecasting electricity demand proposed in the article will easily be adopted to fulfill the above obligation as well. The market character of the regional energy management causes that the functions of energy balance models have to change. Drastic modifications to the prognostic methods are required (e.g. [4-5]), because current trends break down. The models require taking into account the uncertainty and various forms of risk.

The GUS (Central Statistical Office – Statistics Poland) statistics on regional energy data [6] cover electricity consumption broken down into basic sectors: industry, power sector, transport, agriculture, households and other consumption as well as basic energy intensity factors, e.g. total consumption per capita. Analysis of coefficients is necessary to monitor energy management and to manage energy in a way that takes into account sustainable development. The analysis of energy intensity coefficients with different aggregation is indispensable for predicting the demand for energy with end-use models [4, 5, 7- 11].

Time series of energy intensity coefficients are dynamic processes with random disturbances that can be modeled using stochastic differential equations [12]. The article presents an analysis for regions, understood as provinces (voivodeships). The essence of the end-use model is reduced to the analysis of the energy intensity and the reference variable. The end-use model presented in the article has been extended to the sectoral analysis of the regional economy, understood both as a voivodeship, as well as a poviat, a commune or a cluster.

B. Input Data of the Models

The Central Statistical Office (GUS) publishes data forming the Local Data Bank [6]. Data are grouped according to the arrangement of territorial units and relate to a number of areas, including material and fuel-energy market. Data include installed and available capacity in power plants, electricity production by sources and energy consumption by economic sectors.

In Fig.1. the number of inhabitants in particular regions is given. Only small increases in the number of inhabitants in individual regions can be observed. Higher growth is the domain of the Mazowieckie Voivodeship at the expense of population migration, mainly from the Silesia Voivodeship, in which the number of inhabitants dropped.

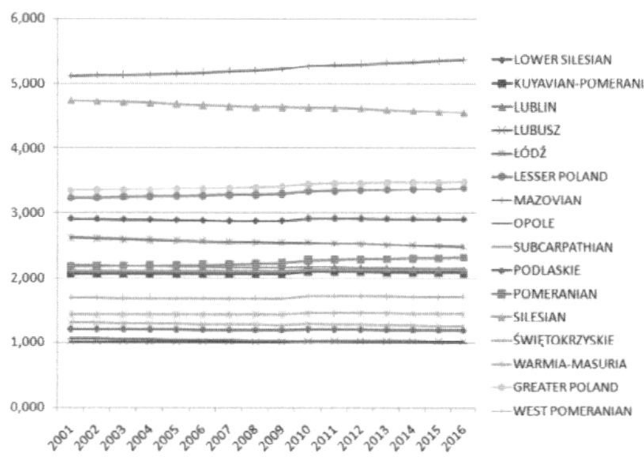

Figure 1. Population in Polish regions in millions of inhabitants [6]

The electricity consumption of economic sectors per capita for territorial units was analyzed. The trends of the sectors' electricity consumption coefficients per capita for 2001-2016 are shown in Figs. 2÷7.

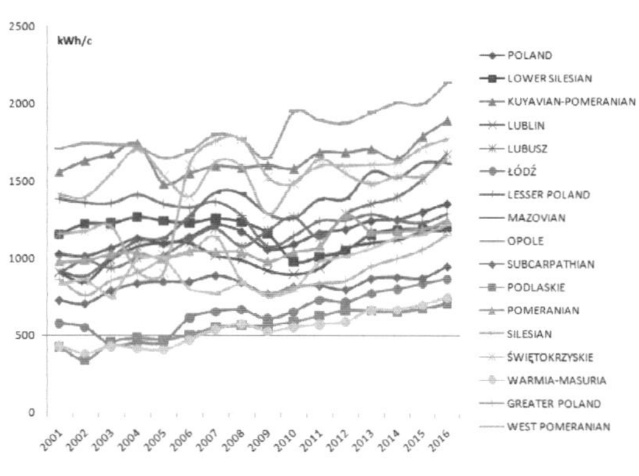

Figure 2. Electricity consumption coefficients per capita in industrial sector [6]

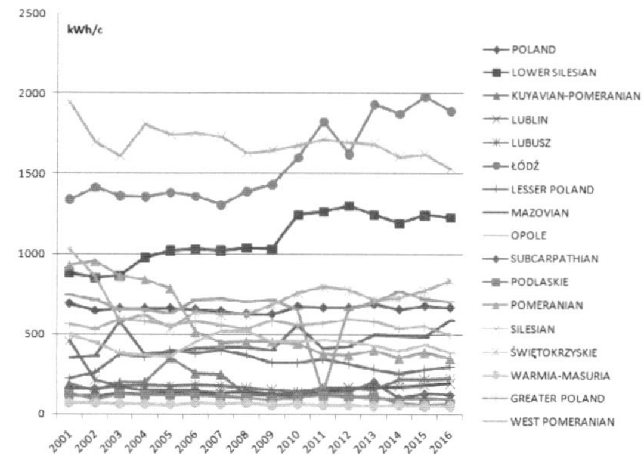

Figure 3. Electricity consumption coefficients per capita in power sector [6]

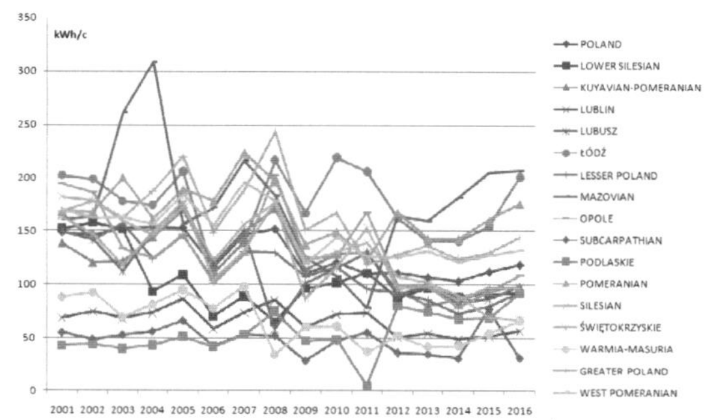

Figure 4. Electricity consumption coefficients per capita in transport [6]

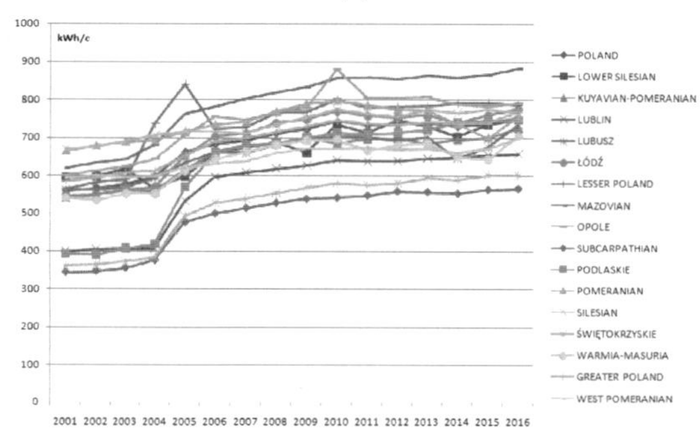

Figure 5. Electricity consumption coefficients per capita in households [6]

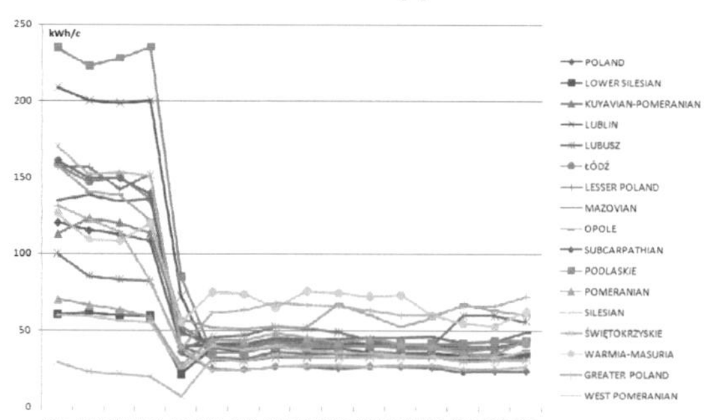

Figure 6. Electricity consumption coefficients per capita in agriculture[6]

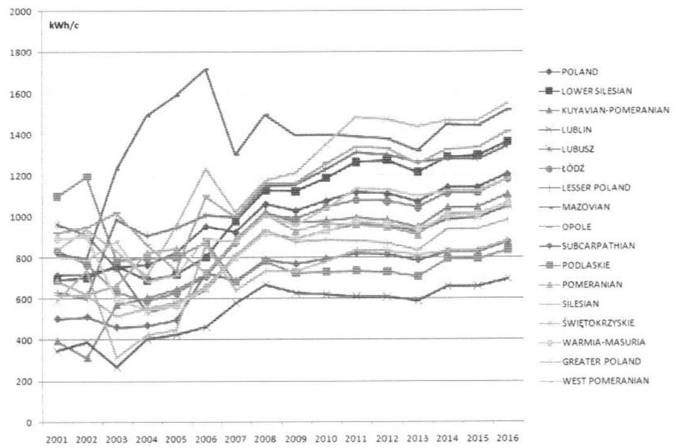

Figure 7. Electricity consumption coefficients per capita in other consumption [6]

The coefficients of total electricity consumption per capita both for Poland and for each region, broken down by sectors, show clear upward trends in recent years. The large distribution of their values is the result of differences in the economic structure. A division into agricultural voivodeships and industrial provinces is revealed.

III. End-use Model of Electricity Demand

The end-use model allows analyzing of energy demand at the end-user level. This concept can be assigned to the consumer at a different hierarchical level. In typical end-use models of the electricity demand forecast, the end user of electricity is a consumer, which may be a physical or legal person, having electricity receivers. Such a definition allows to understand the consumer both as a person living in a single- or multi-family building, as well as an industrial or a service facility consumer. The consumption of electricity per customer is analyzed. Depending on the detail of the disaggregation of the energy economy, the end user in the analysis of energy intensity may be the economic sector, branch of industry, or specific technology, but also, for example, the household. Such design of the model enables relatively easy modification and expansion of the model, and at the same time provides the possibility of introducing some assumptions about the structure and changes of energy consumption of the end user. The end-use model depends mainly on the research capabilities limited by the acquisition of detailed statistics. If the research is limited to the basic statistics available for the regions, understood as voivodeships, then we are able to determine annual coefficients of electricity consumption per capita:

$$e_{ij}(t) = \frac{E_{ij}(t)}{P_i(t)}, \qquad (1)$$

where: $E_{ij}(t)$ - total electricity consumption in the i-th voivodeship, the j-th sector of the economy in the t-th year, $P_i(t)$ - population in the i-th voivodeship in the t-th year, $e_{ij}(t)$ - coefficient of electricity consumption per capita in the i-th voivodeship, the j-th sector of the economy in the t-th year.

The essence of the end-use model is based on the assumption that the energy demand is proportional to the energy intensity and the value of the input variable, which in this case is the population. The model of forecasting energy demand in regions requires the construction of an input scenario, which could be a demographic forecast. This scenario may be the result of external models or the result of heuristic analysis. On the other hand, the key task of model construction is to forecast the coefficients of electricity consumption per capita in the medium-term horizon.

Modeling of dynamic processes, in which there are random disturbances, using stochastic differential equations SDE (the Ito equations) [13] gives good results both in economics and in technology. The specificity of modeling of energy processes allows the use of stochastic differential equations for their forecast.

The use of the Euler method [12] to solve stochastic differential equations simulates the behavior of energy intensity coefficients in the future. The stochastic process of many time-continuous variables, in this case the coefficients of electricity consumption for regions, can be described by stochastic differential equations with the general form:

$$de_{ij}(t) = F(t, e_{ij}(t))dt + G(t, e_{ij}(t))dW_t, \qquad (2)$$

where: $e_{ij}(t)$ - state variable (coefficient of electricity consumption in the i-th region and the j-th sector of the economy), W - variable of Wiener's process (Brownian motion), F - trend-determining function, G - diffusion function.

Based on the general form (2), it is possible to define special variations of the model. The simplest model is a linear trend model with a random component of Brownian motion (BM - Brownian Motion) containing two components: determined and random:

$$de_{ij}(t) = A_{ij}(t)dt + V_{ij}(t)dW(t). \qquad (3)$$

Another form of this model is a model of relative changes of the state variable with a random component described by Brownian motion (GBM - Geometric Brownian Motion):

$$de_{ij}(t) = B_{ij}(t)e_{ij}(t)dt + V_{ij}(t)e_{ij}(t)dW(t). \qquad (4)$$

The simulation of the behavior of energy intensity coefficients, taken as state variables, relies on the numerical solution of stochastic differential equations. For this purpose, Euler's numerical method of solving differential equations was used, based on the geometric interpretation of the differential equation.

Modeling electricity consumption coefficients per capita in the region [12, 13] with stochastic differential equations, determination of solutions using the Euler method and analysis were done by programming own scripts using the Matlab tools (software by MathWorks) and Microsoft Excel spreadsheets.

The analysis uses the GBMC model, which is one of the SDE models, taking into consideration correlations between electricity consumption coefficients in Wiener's processes. The annual relative increments of the coefficients of electricity consumption per capita in the region for the economy sector is the state variable in the model:

$$de_{ij}(t) = \mu_{ij}\, e_{ij}(t)dt + \sigma_{ij}\, e_{ij}(t)dW(t). \qquad (5)$$

where: μ - mean value of variable, σ - standard deviation of variable.

In the presented model we have a determinate component describing the trend of the variable, and a random component in which the measure of uncertainty is the standard deviation of the variable. The energy consumption per capita is identified with the state variable. The uncertainty is understood here in economic interpretation (in mathematical terms it is a risk), because the probability distribution of the state variable is known.

In the GBMC model (5), in which the state variable is an annual relative change of the coefficient of electricity consumption per capita in the region, it is necessary to estimate two parameters: mean value and standard deviation. The value of the standard deviation σ is the risk measure in the model.

IV. Simulations of Electricity Demand in the Regions

For further analysis, the results of simulation of electricity demand per capita in the regions and for the specific sector made by the GBMC model were selected as the most adequate to the economic development. Figures 8÷13 present the simulation results.

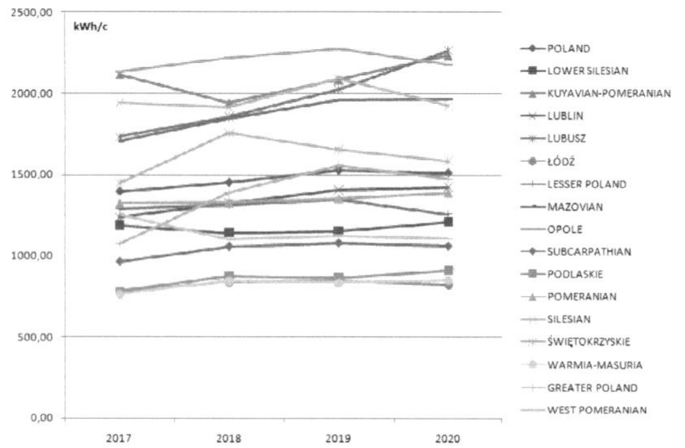

Figure 8. Simulation of electricity demand coefficients per capita in industrial sector in 2017÷2020

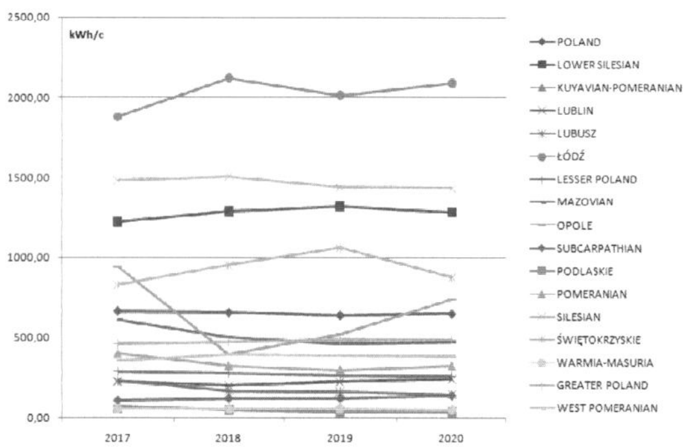

Figure 9. Simulation of electricity demand coefficients per capita in power sector in 2017÷2020

Simulations of the total electricity demand in the regions based on electricity demand coefficients require also a scenario of future population projection. The regional demographic forecast of the Central Statistical Office was used (Table 1).

An exemplary simulation of electricity demand by 2020 for Poland is presented in Table 2, while results for regions in Fig.14. In Table 2, the first row presents the results being a forecast determined from the simulation of electricity demand factors in the country and the national demographic forecast. In the second row of Table 2, a forecast calculated as the sum of regional forecasts is provided. The differences between the two forecasts are relatively small.

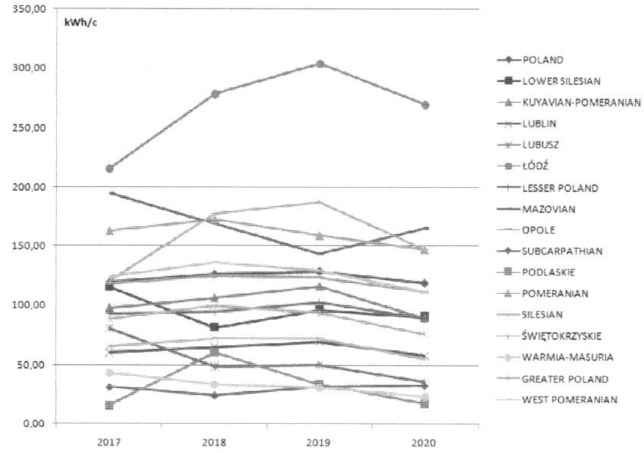

Figure 10. Simulation of electricity demand coefficients per capita in transport in 2017÷2020

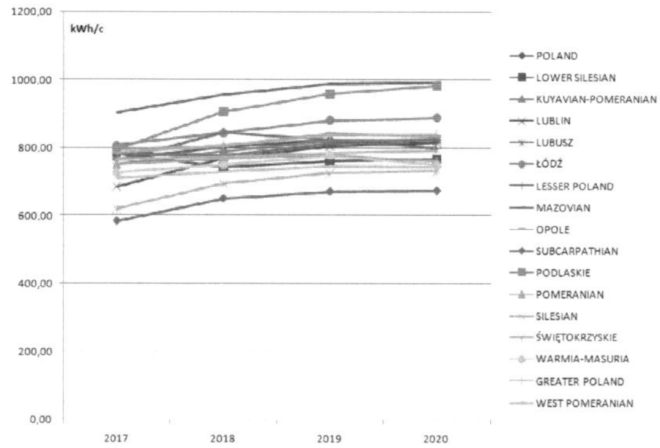

Figure 11. Simulation of electricity demand coefficients per capita in households in 2017÷2020

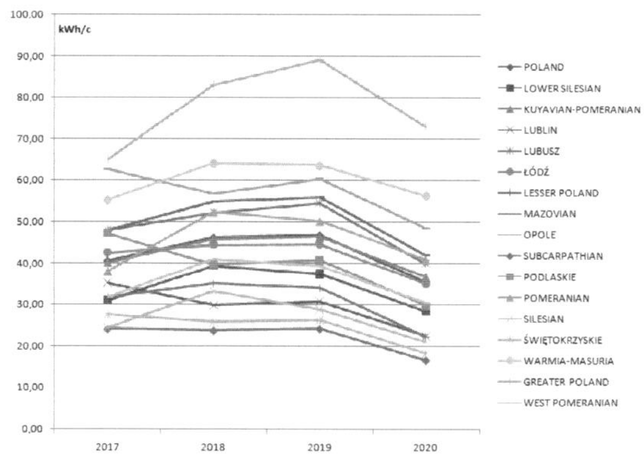

Figure 12. Simulation of electricity demand coefficients per capita in agriculture in 2017÷2020

When performing simulations using the GBMC model, the standard deviation values were assumed at the level of estimation from the course of historical relative changes in energy consumption coefficients, thus assuming a similar level of risk as in the past. The adoption of a higher value of the standard deviation rate results in a higher level of risk. The discussion of risk quantification was presented in [13]. The results of CGBM forecasts can be considered as a scenario of trend continuation.

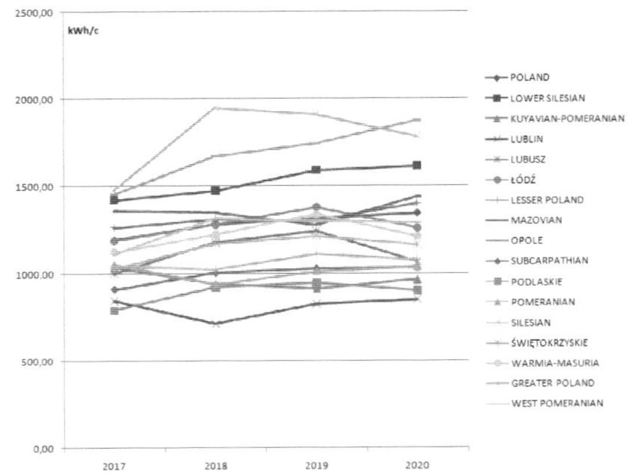

Figure 13. Simulation of electricity demand coefficients per capita in other consumption in 2017÷2020

Table 1. Demographic forecast for Poland and regions in 2017-2020

Region	Demographic forecast in millions inhabitants			
	2017	*2018*	*2019*	*2020*
POLAND	38,315	38,260	38,201	38,138
LOWER SILESIAN	2,887	2,880	2,873	2,867
KUYAVIAN-POMERANIAN	2,078	2,074	2,070	2,065
LUBLIN	2,123	2,114	2,105	2,096
LUBUSZ	1,016	1,014	1,012	1,010
ŁÓDŹ	2,468	2,457	2,446	2,434
LESSER POLAND	3,383	3,388	3,392	3,396
MAZOVIAN	5,361	5,371	5,380	5,388
OPOLE	0,984	0,978	0,972	0,967
SUBCARPATHIAN	2,123	2,120	2,118	2,115
PODLASKIE	1,180	1,176	1,172	1,168
POMERANIAN	2,313	2,317	2,321	2,324
SILESIAN	4,534	4,516	4,497	4,478
ŚWIĘTOKRZYSKIE	1,244	1,238	1,232	1,226
WARMIA-MASURIA	1,433	1,429	1,425	1,421
GREATER POLAND	3,483	3,486	3,488	3,490
WEST POMERANIAN	1,705	1,701	1,697	1,692

Source: [6]

Table 2. Forecast of electricity demand for Poland (GBMC model) in 2017-2020

Years	-	*2017*	*2018*	*2019*	*2020*
Electricity demand	TWh	160.2	167.0	171.3	171.0
Electricity demand as the sum of regional forecasts	TWh	159.5	166.4	170.3	168.9

The results of the electricity demand forecast in 2020 (Table 2) were compared with the government forecast equal to 169.3 TWh [14]. The difference between the electricity demand forecast and the government forecast is less than 1 %.

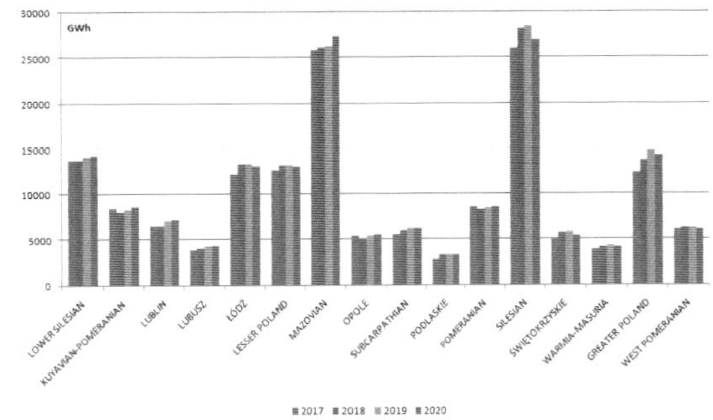

Figure 14. Forecast of regional electricity demand (in GWh) in 2017÷2020

V. Conclusions

The values of electricity consumption coefficients determine the demand for electricity in regions and in the country, because the demographic forecast is characterized by low volatility.

The performed simulations of electricity consumption coefficients in the horizon up to 2020 depend on the adopted model. The results of each of the analyzed models can be treated as a scenario of the energy demand in the future. The presented simulations are only an example of the SDE models results. The models determine the energy demand using the electricity demand factors and the end-use methodology. The presented tools support the process of analyzing the country's energy balance and enable taking into account expert assessments by applying demographic scenarios and verifying model coefficients.

As this study indicates, the Euler method is adequate for carrying out simulations of stochastic variables described by differential equations, taking into account the uncertainty of processes occurring in power industry, resulting from the impact of the external environment on the models.

Acknowledgment

The project financed under the program of the Minister of Science and Higher Education under the name "Regional Initiative of Excellence" in the years 2019 - 2022 project number 020/RID/2018/19, the amount of financing 12,000,000 PLN.

References

[1] I. Pavlova-Marciniak, "Rozwój energetyki jądrowej w Polsce – nowe wyzwania," *Przegląd Elektrotechniczny*, vol. 3, 2009, pp. 216-219, ISSN 0033-2097 .

[2] M. Kornatka and A.Gawlak, "Comparative analysis of operating conditions in Polish medium-voltage and 110 kV networks," *Proceedings of The 8th International Scientific Symposium ELEKTROENERGETIKA* 2015, Stara Lesna, Slovak Republic, 2015, ISBN 978-80-553-2187-5.

[3] M. Kornatka, "The weighted kernel density estimation methods for analysing reliability of electricity supply," *in: 17th International Scientific Conference on Electric Power Engineering (EPE)*, Prague, 2016, ISBN 978 - 1 - 5090 - 0907 - 7.

[4] G.R. Newsham and C.L. Donnelly, "A model of residential energy end-use in Canada: Using conditional demand analysis to suggest policy options for community energy planners," *Energy Policy*, vol. 59, 2013, pp. 133-142, ISSN: 0301-4215.

[5] L.G. Swan and I.V. Ugursal, "Modeling of end-use energy consumption in the residential sector: A review of modeling techniques," *Renewable and Sustainable Energy Reviews*, vol. 13 (8), 2009, pp. 1819-1835, ISSN: 1364-0321.

[6] www.stat.gov.pl (16.09.2018), GUS, 2018.

[7] B. Chateau and B. Lapillonne, "Term energy demand forecasting, a new approach," *Energy Policy,* vol. 6 (2), 1979, pp. 140-157 ISSN: 0301-4215.

[8] Commission of European Communities, The MIDAS Energy Model, Bruxelles, EC, 1985.

[9] International Atomic Energy Agency (IAEA), Model for Analysis of the Energy Demand (MAED), Vienna, IAEA, 1986.

[10] J. Sowiński, ""End-use"-model prognozy zapotrzebowania bezpośredniego na energię," In: Dobrzańska I. (red), *Prognozowanie w elektroenergetyce. Zagadnienia wybrane*, WPCz, 2002.

[11] J. Sowiński, "Model typu „end-use" prognozy zapotrzebowania na energię pierwotną i finalną w warunkach ryzyka," *Rynek Energii*, Vol. 112, No 3, 2014, pp. 39-43, ISSN 1425-5960.

[12] B.K. Øksendal, Stochastic Differential Equations: An Introduction with Applications, Berlin Springer, 2003.

[13] J. Sowiński, Inwestowanie w źródła wytwarzania energii elektrycznej w warunkach rynkowych, Wydawnictwa Politechniki Częstochowskiej, Częstochowa, 2008.

[14] Ministry of Economy, Polityka energetyczna Polski do 2030 roku (Energy Policy of Poland until 2030), Warszawa, 2009

Cost-efficiency of Constructing Medium Voltage Cable Lines

Anna Gawlak
Institute of Electrical Power Engineering
Technical University of Częstochowa
Częstochowa, Poland
gawlak@el.pcz.czest.pl

Abstract — **The paper analyses cost-efficiency of replacing MV overhead lines by cable lines. The analysis is carried out by means of the IRR and NPV methods. The impact resulting from the difference in investment and operating costs was taken into account. Also the influence of line damage intensity and the cost of energy not delivered to the recipients.**

Keywords: medium-voltage network; cost-efficiency; cable lines

I. Introduction

The power network has to meet requirements concerning the quality of energy, its reliability of supply to consumers as well as safety for maintenance crew and users. It should also be adaptable to serving a growing number of consumers [1,2].

The condition of the distribution network is of vital importance for the economy and for ensuring the safety of power supply. The development of the power network infrastructure has to be strictly correlated with economic development, which causes increase in demand for power, with investments in distributed power generation, especially wind farms, with Poland's priorities as well as EU strategies [3,4,5]. Fig. 1 presents dependence of the average duration of unplanned power interruptions on the percentage of cable lines in selected European countries [6].

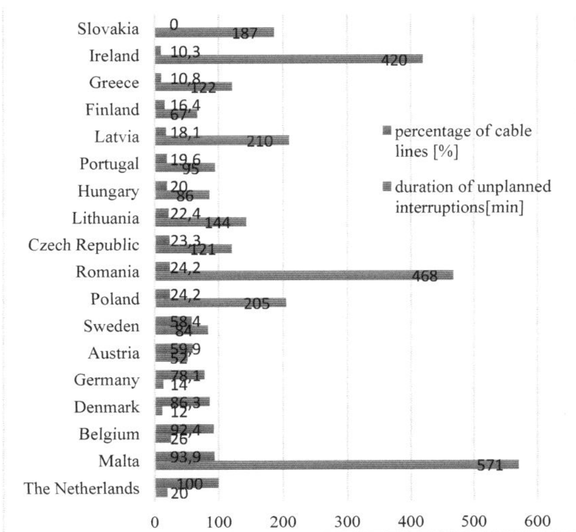

Figure 1. Average duration of unplanned power interruptions vs. the percentage of cable lines in selected European countries.

As the graph indicates, there exists a correlation between the duration of unplanned interruptions and the share of cable lines. Fig.2 presents the length of MV cable lines and their percentage in the total MV lines for the five greatest distribution companies in Poland, and Table 1 presents the age structure of MV lines.

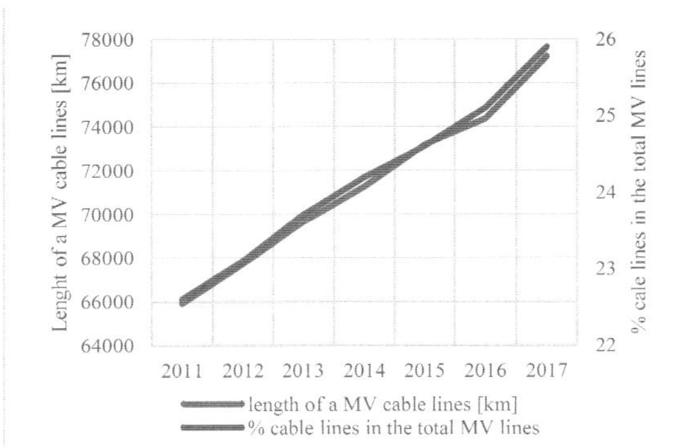

Figure 2. The length of MV cable lines and their percentage in the total MV lines for the five greatest distribution companies in Poland.

The length of MV cable lines and their share in MV lines increases with every year. At the end of 2017 the share of cable lines in total MV lines was 25.9 %, with increase by 0.8 % with respect to the preceding year, which was the highest year-to-year increase noted. Before that the increase was at the level of 0.5 % year-to-year. It appears that reaching the average European percentage of cable line share in MV networks, which exceeds the Polish average by 20 % will take between a decade and two decades. The average in the ten countries with the highest share was 75% in 2017.

Table 1. Age structure of MV lines [%]

MV lines	older than 40 years	25-40 years	10-25 year	below10 years
Overhead	37	39	17	7
Cable	16	24	28	31

The age structure presented in Table 1 indicates the directions of network development adopted in distribution companies. The small percentage of new overhead lines (7%) and the relatively high percentage of cable lines (31%) means that cable lines are preferred, despite being more costly to construct. The decisive factor is higher reliability and less complicated preparations of the investment project with respect to legal requirements.

Overhead lines are directly or indirectly exposed to atmospheric conditions, which causes numerous breakdowns, possibly long-lasting and extending over large areas. This results in power interruptions affecting many customers. The cost of interruptions for power distribution companies is also very high, mainly due to significant cost of repairs, which often have to be carried out in challenging conditions [7,8,9].

Analysing many disasters, it was concluded that their causes could be largely attributed to unreliability of overhead lines. This is another argument that the share of cable lines should be increased, the operation of which is largely not affected by extreme weather conditions. Since the reliability indices are mostly affected by the performance of MV networks, distribution companies undertake actions accordingly. Apart from being resistant to adverse weather condition, cable lines have other advantages: they do not generate noise or harmful electromagnetic field (or to a significantly lesser extent) and they do not have negative impact on the landscape.

II. COST-EFFICIENCY ANALYSIS OF CONSTRUCTING MV CABLE LINES

A cost-efficiency analysis of constructing MV cable lines was performed on the basis of data for the year 2017 obtained from 8 distribution companies (OSD). Table 2 presents percentage of lines of a given cross-section, percentage of cable lines and an average cross-section of MV lines.

Table 2. Percentage of MV lines of a given cross-section

Cross-section [mm2]	**≤ 35**	**50**	**70**	**95**	**120**	**% of cable lines**	**Average cross-section**
OSD1	50.2	16.2	33.4	0.1	0.2	24.6	49.30
OSD2	62.9	8.0	2.4	0.5	0.2	15.9	46.67
OSD3	73,9	11.2	14.2	0.4	0.4	19.5	42.20
OSD4	62.5	12.7	24.6	0.0	0.2	15.4	45.71
OSD5	60.3	7.5	31.5	0.1	0.5	19.3	47.67
OSD6	50.8	21.2	27.0	0.0	1.1	10.4	48.55
OSD7	54.0	16.9	28.1	0.7	0.4	14.6	48.11
OSD8	53.9	15.8	28.3	0.4	1.5	61.5	48.82

As can be seen in Table 2. over 50% of MV lines (50.18 for OSD1 and 73.9% for OSD3) has a cross-section of 35 mm^2 or smaller. The average cross-section of MV lines in the area under scrutiny is 43.89 mm^2. with the lowest being attested in OSD3 (42.20 mm^2) and the highest in OSD1 (49.30 mm^2). Cable lines constitute from 15.39% in OSD4 to 24.55% in OSD1. Figs 3 and 4 show the rates of utilizing the capacity of MV cable and overhead lines. respectively.

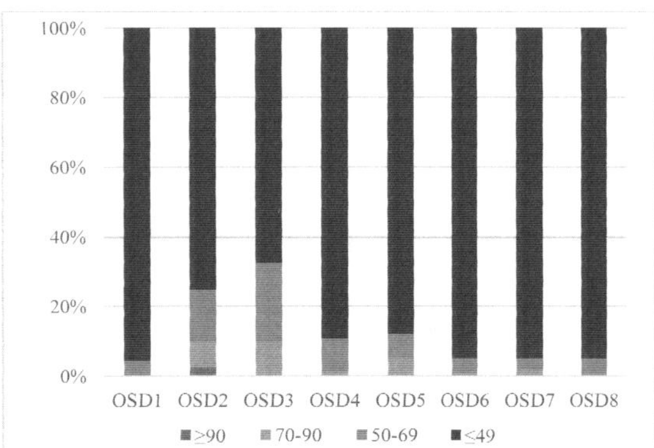

Figure 3. Rates of utilizing the capacity of MV cable lines.

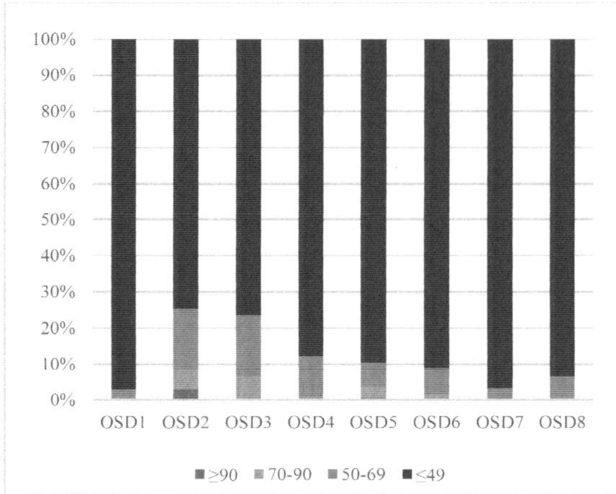

Figure 4. Rates of utilizing the capacity of MV overhead lines.

In over 88% of overhead lines and in 87% of cable lines the rates of utilizing their capacity are below 49%. Only in 0.31% of cable lines and in 0.64% of overhead lines the rates are higher than 90%. This indicates that the cross-sections are appropriate and could be increased only in a few cases.

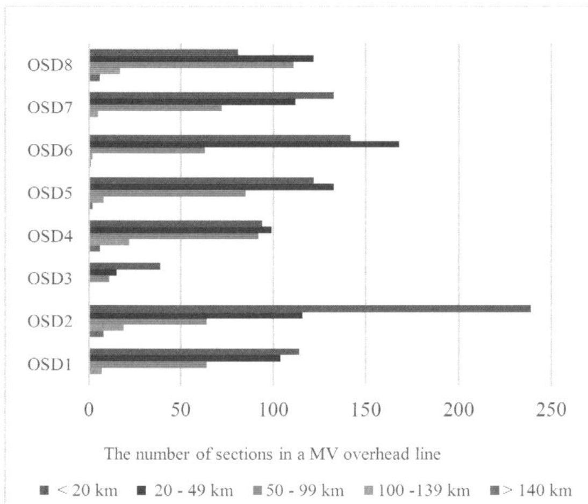

Figure 5. The number of sections in a MV overhead line.

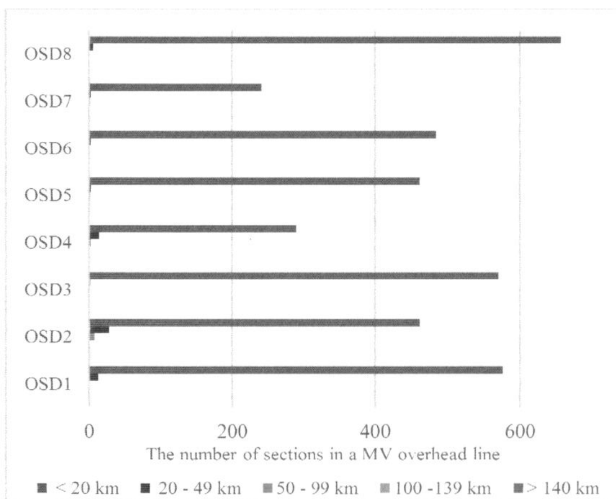

Figure 6. The number of sections in a MV cable line.

As Figs 5 and 6 show. the length of over 97% MV cable line sections is below 20 km and there are no cable lines with sections exceeding 140 km. Contrary to cable lines. overhead lines have only 38.59% of sections below 20 km. and 4.12% of sections are over 100 km long.

The economic criterion is often considered to be the most important when decisions concerning investment projects are taken [3,4,5]. A unitary cost of constructing a cable line depends on a number of factors. but in any case it is significantly higher than the corresponding cost of overhead lines. With faultless operation of cable lines. their exploitation cost is lower than that of overhead lines (it is not necessary to cut trees and branches. there are no repairs due to extreme weather conditions). but if faults occur in cable lines. repairs are more costly that those of overhead lines.

III. METHODS OF COST-EFFECTIVENESS ANALYSIS

The cost-effectiveness analysis of constructing cable lines will be carried out by means of the most frequently used valuation methods. i.e. NPV and IRR [10,11].

The net present value (NPV) is method of assessing economic effectiveness of an investment project as well as a rate obtained on the basis of this method. It is a dynamic method based on the analysis of discounted cash flows when the discount rate is known. It can be defined as the difference between the discounted cash flows and the invested cash and is described by the formula:

$$NPV = \sum_{t=1}^{n} \frac{CF_t}{(1+r)^t} - I_o,$$

where:

NPV- Net Present Value,

CF_t - net cash low at time t,

r - discount rate,

I_o - invested cash,

n- consecutive periods (e.g. years) of exploitation of the investment project.

The value of NPV indicates the profitability of investment for a company from the perspective of a present evaluation. The investment is profitable if NPV>0. A positive value of NPV means that the return on the investment is greater than the threshold rate corresponding to the discount rate assumed as profitable. Each investment with a positive value of NPV should be implemented because it is likely to generate profit for the company. Investments with a negative value of NPV should not be considered for implementation.

Internal Rate of Return (IRR) represents a real return on an investment project. It can be defined as a discount rate for which the Net Present Value (NPV) is equal to zero (NPV=0). i.e. the present value of outflowing cash is equal to inflowing cash. Based on IRR. an investment project is profitable. if IRR is higher than a threshold rate assumed by the investor as the minimal profitability rate.

To assess the profitability of replacing overhead lines by cable lines. the following assumptions were made:

- the cost of building an overhead line is 184,000 PLN,
- average cost of repairs in overhead lines – 2,542 PLN in cable lines-3,452 PLN,
- exploitation costs of overhead lines from 2 to 4.5% of the value of the investment project,
- the cost of the investment project is equal to the difference between the cost of constructing the cable line and the overhead line,
- the profit is equal to the difference between the exploitation costs and repair costs of the overhead lines and cable lines,
- reliability indices are those from 2017 r.

Fig. 5. presents NPV calculated on the assumption that the cost of constructing a cable line exceeds the cost of constructing an overhead line by 10 - 50%. and the exploitation cost of an overhead line exceeds that of a cable line by 50%.

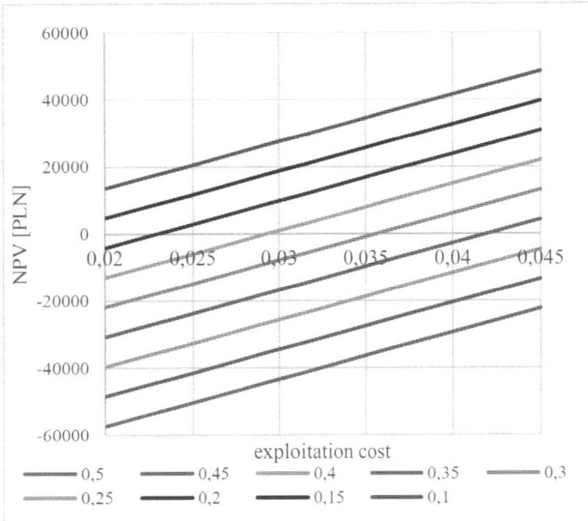

Figure 7. NPV vs. exploitation cost.

The analysis indicates that the investment in cable lines is profitable if the cost of their construction is at most 30% higher than the construction cost of overhead lines. with the difference in exploitation costs exceeding 3.5%. If the construction cost of cable line does not exceed that of building an overhead line by more than 20%. then the investment in cable lines is profitable regardless of the exploitation cost. Cable lines are more reliable and the average interruption time is shorter.

In order to assess the impact of exploitation costs on the profitability of the investment. a comparative analysis for cable and overhead lines was performed on the assumption that the construction cost of a cable line exceeds the construction cost of an overhead line by 30% and the total exploitation cost is equal to 3% of the investment project. The results are presented in Table 3.

Table 3. Impact of the difference in exploitation costs on NPV and IRR

	The cost of ownership ratio: the cable to the overhead line						
	0.8	***0.7***	***0.6***	***0.5***	***0.4***	***0.3***	***0.2***
NPV	17223	8835	447	-7941	-16330	-24718	-33106
IRR	6.75%	5.41%	3.98%	2.43%	0.72%	-1.24%	-3.63%

A distribution company will achieve profit if the exploitation cost of a cable line is lower by at least 40% that the exploitation cost of an overhead line.

Incidence of faults in overhead and cable lines as well as average repair time are random. To asses the influence of the incidence of faults on the profitability of constructing a cable line. an analysis was carried out for the period of two years. on the assumption that the construction cost of a cable line is higher by 30% that that of an overhead line. The results are shown in Table 4.

Table 3. Influence of fault incidence on NPV and IRR

Exploitation depreciation	2017		2016	
	NPV [PLN]	*IRR [%]*	*NPV [PLN]*	*IRR [%]*
0.02	-21921.7	-0.55%	-25045.3	-1.32%
0.025	-14931.6	1.02%	-18055.2	0.34%
0.03	-7941.48	2.43%	-11065.1	1.82%
0.035	-951.372	3.73%	-4074.98	3.16%
0.04	6038.733	4.94%	2915.129	4.41%
0.045	13028.84	6.09%	9905.234	5.58%

Despite the fact that the incidence of faults was relatively low in 2016. the difference between overhead lines and cable lines still indicated advantage of building cable lines in 2017 by about 0.5%.

The amount of energy not supplied to customers due to interruption could also potentially affect the profitability of the investment project. The results of an analysis taking into account this cost are shown in Table 5.

Table 5. Influence of the cost of energy which was not supplied due to interruptions

Exploitation depreciation	Without the cost of energy which was not supplied		With the cost of energy which was not supplied	
	NPV [PLN]	*IRR [%]*	*NPV [PLN]*	*IRR [%]*
0.02	-21921.70	-0.55%	-20158.96	-0.13%
0.025	-14931.60	1.02%	-13168.86	1.39%
0.03	-7941.48	2.43%	-6178.75	2.77%
0.035	-951.37	3.73%	811.35	4.04%
0.04	6038.73	4.94%	7801.46	5.24%
0.045	13028.84	6.09%	14791.56	6.37%

The cost of energy not supplied due to interruptions does not affect the investment in a significant way.

IV. Concluding remarks

The cost-efficiency of replacing overhead lines by cable lines depends on a number of factors. Among these factors the cost of building cable lines is of vital importance. As was demonstrated. with this cost exceeding the cost of constructing overhead lines by more than 20%. the construction of cable lines is economically unjustified. The difference in exploitation costs is also important. If such costs are at least 40% lower in cable lines than in overhead lines. the investment turns out to be profitable. The repair cost of a cable line is higher by about 74% than the repair cost of an overhead line. but this difference is compensated by a lower incidence of faults in cable lines. The cost of energy not supplied due to interruptions is practically irrelevant from the viewpoint of the profitability of an investment.

Acknowledgment

The project financed under the program of the Minister of Science and Higher Education under the name ["Regional Initiative of Excellence" in the years 2019 - 2022 project number 020/RID/2018/19], the amount of financing 12,000000 PLN.

References

[1] A. Gawlak, "The Influence of Investment on Reducing Energy Losses in Distribution Networks," Proc. IEEE International Scientific Conference Electric Power Engineering (EPE 2015), IEEE Press, May 2015, pp. 315-319.

[2] M. Kolcun, M. Kornatka, A. Gawlak and Z. Čonka, "Benchmarking the reliability of medium-voltage lines," Journal of Electrical Engineering vol. 68 (3), 2017, pp. 212-215, DOI: https://doi.org/10.1515/jee-2017-0031.

[3] A. Gawlak, "Technological aspects of electrical energy distribution," Proc. IEEE International Scientific Conference Electric Power Engineering (EPE 2014), IEEE Press, May 2014, pp. 45-48, DOI: 10.1109/EPE.2014.6839420.

[4] M. Goňo, M. Kyncl, R. Goňo and I. Kłosok-Bazan, "Experience with the production of electricity from biogas at sewage treatment plant in the Czech Republic," Przeglad Elektrotechniczny, Volume 89, Issue 11, 2013, pp. 12-15.

[5] A. Gawlak, "Noninvestment Forms of Reducing Energy Losses in Distribution Networks," Proc. IEEE International Scientific Symposium on Electrical Power Engineering (EPE2015), IEEE Press, May 2015, pp. 61-64.

[6] www.ptpiree.pl/examples/raport_2018

[7] A. Gawlak, „Podział środków inwestycyjnych na rozwój sieci rozdzielczych przy zastosowaniu metody taksonomicznej," Przegląd Elektrotechniczny, R.85 nr 3, 2009, pp. 157-160.

[8] M. Kornatka and A. Gawlak, "Comparative Analysis of Operating Conditions in Polish Medium-voltage and 110 kV Networks," Proc. IEEE International Scientific Symposium on Electrical Power Engineering (EPE2015), IEEE Press, May 2015, pp. 57-60.

[9] A. Gawlak, "Analysis of technical losses in the low and medium voltage power network," Proc. IEEE International Scientific Conference on Electrical Power Enginering (EPE2010), IEEE Press, May 2010, pp. 119-123.

[10] J. Sowiński, "Comparison of RAINS and Fisher's models for calculating sulphur deposition in Poland," Atmospheric Environment Vol.29, No.22, 1995, pp. 3385-3389.

[11] J. Paska., Ekonomika w elekktroenergetyce, Oficyna Wydawnicza Politechniki Warszawskiej, 2007.

Renewable Sources of Energy and Hybrid Electrical Wiring in Individual Houses

Possibilities and Restrictions

Anna Gawlak
Institute of Electrical Power Engineering
Technical University of Częstochowa
Częstochowa, Poland
gawlak@el.pcz.czest.pl

Andrzej Malinowski
A.Malinowski Projekt
Częstochowa, Poland
a.malinowski.projekt@gmail.com

1

Abstract — **In the article are presented basic issues related to the build, in individual houses, installations producing the electrical energy from sources of renewable energy with hybrid electrical installations.**

Keywords: ***the energy independence; the energy self-sufficiency; renewable sources of energy.***

I. Introduction

For purposes of the article is the approximation and the description of basic issues related to the energy production of electrical from renewable sources of energy and a build of electrical hybrid installations in individual houses with manners of the utilization of obtained at command the energy. The article deals with objects of the one-family building in the aspect of the possible obtainment, already currently, and most interesting of owners of houses, the state of the energy self-sufficiency, marking the lack of charges for the energy received from the network. Contained issues can concern the same small service [1,2] works. Built or existing individual house, will be touched inevitably, with the matter related to growing costs of his maintenance, and this will concern, first of all, heating costs, preparations of the useful hot water and the electrical energy on vital purposes, at the exclusion of the use from the coal. Soon will appear, on the mass-scale, the new and very essential kind of the load of electrical domestic installations. This will be the necessary electrical energy on the download of electrical [3] cars. This fact substantially will come in the entire energetics, from the level of final consumers, across networks, till for power stations. If after the ending era of the affluence of oil our civilization is supposed to exist and to function, though on level 50% all currently possessed possibilities within the range mobilities, are the professional energetics and the electricity produced by themselves users, constitute a very serious and real part for solutions which will permit the fulfilment of our expectations. The script of events, in the global economy which will follow violently, or calmly, related to steering itself on the electrical energy, after the era of oil, already in the present time is a great challenge and a crop of issues to urgent solution, both in the sphere of the professional energetics as and in the project of general and everywhere, where this only will be possible, the own electrical energy production. The article shows the fragment of results from accomplished calculations and analyses within the range possibilities and conditionings technical, in planned yield of the energy from domestic installations of renewable sources of energy and reveals possibilities and restrictions resulting from their use. Issues for the professional energetics, resulting from planned incorporation to the network, on a large scale, new objects with hybrid installations, are for distributional networks a problematic set of tasks to solution in projects preparatory, investment and technical [4,5,6]. These matters will be an object of the separate article.

II. Individual house near at hand as the receiver of the electrical energy

An object of calculations and analyses is the model of the house, in the individual building, for 4 resident persons, about the surface approx. 137m2, to be built in the geographical latitude from approx. 450 to 650 in which the necessary electric power for purposes useful, heating in the system of the thermal pump, and on the download of electrical cars from the usual nest will amount about 17.2 kWh, and prognosed annual consumption of energy of electrical will amount at least 18.5 MWh.

III. What independence and the energy self-sufficiency is possible and how to programme, systemically, building of hybrid installations?

The answer to this question is following: becoming independent from the association to the network of the professional energetics, conditioned the average individual house, is excluded. However is possible, already currently, the obtainment of the self-sufficiency in the balance of the collection and the production and the return of the energy to the network of the professional energetics. This kind of the self-sufficiency will completely satisfy owners of houses, as far as they will be in a position to overcome the investment economic barrier and will build proper installations. They will not have, in that case, any costs for used electrical energy, and at arising of appropriate building conditions and the good configuration of the installation of renewable sources of energy will receive the return for the sold energy to the network. It is necessary to have the consciousness that this will be also dependent from State-settlements legally - administrative in relations: distributors of the energy - domestic producers of the energy [7,8,9]. The incorporation of the installation of renewable sources of energy to the network of the professional energetics state the large problem, because hybrid installations mostly give back the made electrical energy to the network when it is least necessary, and they consume energy in time, when this is least favourable for the network. This is particularly severe for the installation with the most of photovoltaic installations and this happens during the large insolation. The large wind power plants create transmission serious problems in networks, in time unforeseeable. The great problem came to light to solution for the network of the professional energetics [10,11]. Simultaneously arose the main chance for designers and producers of domestic hybrid installations which are installations at the suitable construction and the character, will be able to greatly to reduce the collection of the energy from the network in the top of the demand and to reduce the collection of the energy at all [12,13]. Results of calculations and analyses show as maybe to function this in the reality and which is great and positive meaning. The question appears: what are other options for presented examples and solutions?

IV. Energy Balance

The energy self-sufficiency will follow:

$$E_z - E_w + E_s = 0$$

Where: E_z – the energy consumed, E_w – the own energy produced in renewable sources of energy, E_s – the energy received from the network of the professional energetics.
In case of when:

- $E_z - E_w + E_s < 0$ it is the positive balance to house owner and goes the transmission of the energy to the network and it sales.
- $E_z - E_w + E_s > 0$ it is the negativ balance to house owner and follows the purchase of the energy from the network.

Table 1 shows energy data of analysed technical of the house for the case (2).

TABLE I. Technical characteristics of the house (2)

Useful surface	137 m^2
Quantity of inhabitants	4
Heating and the hot water	thermal pump
Electrical preparation of meals and other	Yes
Load of 2 electrcial cars [total]	18000 km
Own photovoltaic power plant	5 kWp
Own wind power plant	5 kW
Own store of the electrical energy	25 kWh
Energy ordered in the year	18.57 MWh
Electric power orderer in the object	17.8 kW
Balance of the collection on the energy production	+425 kWh

Table 2 shows results for the case (2), the calculations within the range the energy: ordered, produced and received from the network and the balance these of three sizes.

TABLE II. Demand, the own production and the receiving of the energy from the network (2)

The term	I	II	III	IV	V	VI	Total
Requirement [kWh]	2315	2015	1735	1405	995	955	
Own production [kWh]	1500	1550	1650	1725	1550	1625	
Receiving from the net [kWh]	815	465	85	-320	-555	-670	
The term	**VII**	**VIII**	**IX**	**X**	**XI**	**XII**	
Requirement [kWh]	945	900	1245	1735	2015	2315	18575
Own production [kWh]	1625	1550	1525	1650	1550	1500	19000
Receiving from the net [kWh	-680	-650	-280	85	465	815	-425
Final balance result to owner [kWh]							+425

Fig.1 shows, for the case (2), diagrams of the value of the energy: ordered, produced and received from the network.

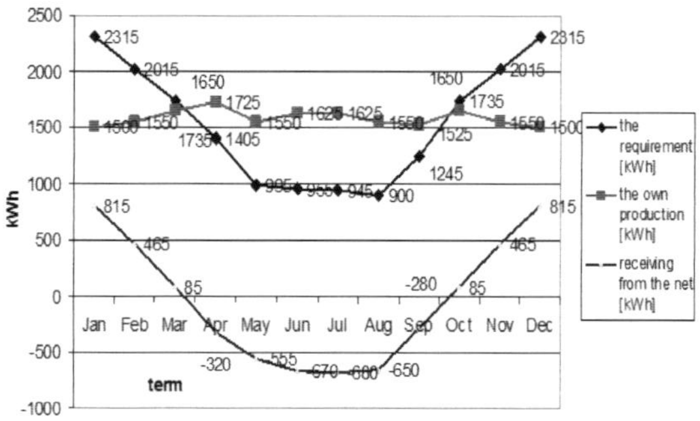

Figure 1. Energy ordered, produced and received from the network. Own elaboration.

Fig. 2 shows, for the case (2), the diagram of the hybrid installation at individual home.

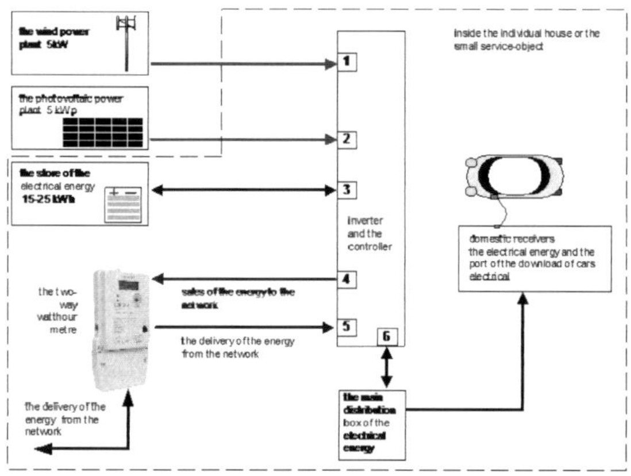

Figure 2. Drawing of the hybrid, house electric wiring.Own elaboratin.

From performed calculations and shown diagram results that within a period of from January to March and from October to December will have to be received, in short supply, the electrical energy from the network. In the residual term will follow the feed of the installation of the object with the own made energy and prosument sales of the energy to the network. The huge part performs the store of the energy which causes that the diagram of the load of the network from such object will be very gentle and stable. This is the irrefutable advantage, to be seen from the point of view of the network to which the object, or a great deal such objects will be connected [14,15,16,17,18]. Obvious is the fact that will happen situations of the collection of the energy from the network within a period of from April to September in unforeseen moments, caused with weather conditions, with failures of the network or other. On such critical situations must be the prepared distribution network, what will be an object of the separate article.

V. Possibilities

Shown above results of calculations and graphical records for the chosen model and the kind of installation are the case (2), where follows the positive energy balance and the owner of the object, at the enough substantial consumption of energy on the heating and vital purposes for 4 persons, together with loading of electrical cars on running over total approx. 18

thousand km, will pay nothing for the energy from the network and will receive still the return payment for the sold of energy. This is the most desirable situation for every one-family home-owner, or the service-object. On the obtainment of the so positive effect are necessary: the own photovoltaic power plant about the power approx. 5 kWp, the own wind power plant about the power 5 kW and the own store of the energy with the energy approx. 15-25kWh. Every from these elements is necessary. For the work of such arrangement is the required modern controller whose algorithm of the work also will take into account the download of the store of the energy and her rendition, to the installation, at assorted programming temporary, proper to applied network tariffs of the distributor of the energy from the network.

VI. Restrictions

Presented and performed calculations for the model (2) are not fitted for automatic copying in other installation configurations with other parameters of devices of renewable sources of energy. One ought each time to analyse it. It is performed also calculations and analyses for the model (3) where will be the negative balance of the energy in the object, to the disadvantage of the home-owner. In the table 3 and on fig. 3. are showed results of calculations for the case (3). This is the case with parameters of the model (2) in which one did not install the wind power plant. This creates the completely different situation which shows how great and positive meaning have wind power plants in the electrical energy production in domestic hybrid installations. Table 3 shows energy data of analysed technical of the house for the case (3).

Table 1. Technical characteristics of the house

Useful surface	137 m^2
Quantity of inhabitants	4
Heating and the hot water	thermal pump
Electrical preparation of meals and other	Yes
Load of 2 electrcial cars [total]	18000 km
Own photovoltaic power plant	5 kWp
Own wind power plant	no
Own store of the electrical energy	13.2 kWh
Energy ordered in the year	18.57 MWh
Electric power orderer in the object	17.8 kW
Balance of the collection on the energy production	-14.5 kWh

Fig.3 shows, for the case (3), diagrams of the value of the energy: ordered, produced and received from the network.

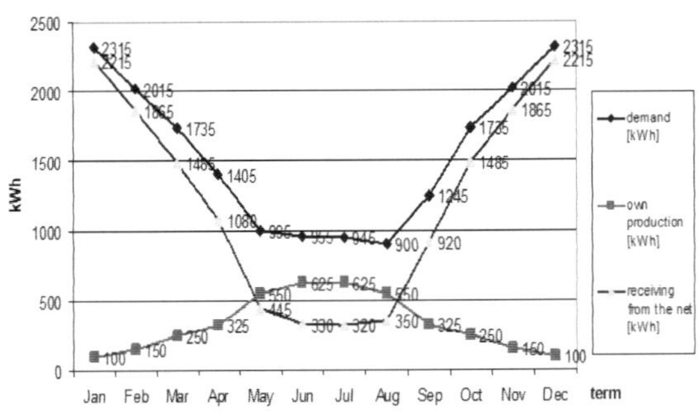

Figure 3. Demand, the own production and the receiving of the energy from the network. Own elaboration.

VII. Results of analyses of other configurations of the installation

The case in shown previous configurations into which owners should have invested enough large funds, what will not be too general. In this chapter, for comparison and realizings in the scale of the issue, are presented results of calculations and analyses for other, but a very probable configuration requiring with a lot of smaller financial expense and giving the effect of the partial energy self-sufficiency. The table No. 4 shows energy data of analysed house for the case of the build of the installation containing the photovoltaic power plant, without the wind power plant, but with the store of the energy 13.2 kWh with the option no-load electrical cars. This is the example which one of these odd-come-shorties will begin to be commonly used, and in the nonce differentiates him from installed photovoltaic power plants this, that is in him the store of the energy whose application will spread because of the considerable saving on charges for the electrical energy from the network.

Table 2. Technical characteristics of the house

Useful surface	137 m^2
Quantity of inhabitants	4
Heating and the hot water	gaseous
Electrical preparation of meals and other	Yes
Load of 2 electrcial cars [total]	No
Own photovoltaic power plant	5 kWp
Own wind power plant	No
Own store of the electrical energy	13.2 kWh
Energy ordered in the year	4 MWh
Electric power orderer in the object	12.8 kW
Balance of the collection on the energy production	0 kWh

Fig.4 shows, for the case of other configuration, diagrams of the value of the energy: ordered, produced and received from the network.

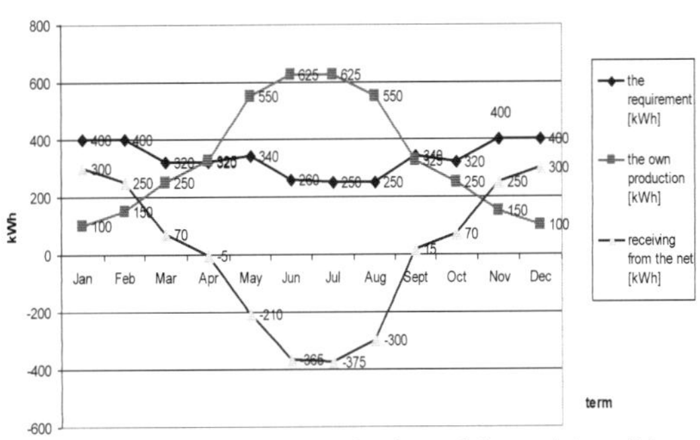

Figure 4. Demand, the own production and the receiving of the energy from the network. Own elaboration.

VIII. Stores of the energy

The storage and the way of the use of the produced own electrical energy at home, are the same important as it production. In this chapter we show in an evident way the storage of the energy it influences, in the fundamental way, on the economics of the work of the entire hybrid installation. An inseparable element of the hybrid installation is the proper controller of the flow of the energy which will be programmed „brain" of the energy management in our house. To controllers of the flow of the energy is dedicated the next chapter. The table V shows the example of calculations of operating costs for the hybrid installation in the month October, in the average tariff G13 of the Polish operator, for the object with the daily power requirement in the height 13,2 kWh and with the own daily energy production in alt. 10,4 kWh and with the store of the energy about the capacity 13,2 kWh.

Table 3. The calculation of costs for the energy in Poland (The standard of prices 12/2018) in the installation with the store of the energy

Hours in 24 h/October	7-13	13-16	16-21	21-7	Total per/day	Payment euro/day
Requirement [kWh]	1.5	2.5	6.0	3.2	13.2	
Own production [kWh]	4.4	6.0			10.4	
Posible sales energy to the net [kWh]	-2.9	-3.5			-6.4	-1.116
Low tariff energy from the net incl. load magazine [kWh]				13.2	13.2	1.381
Hight tariff energy from the net [kWh]					0.0	0.0
Energy from the magazine to home installation [kWh]	1.5	2.5	6.0	3.2	13.2	
Balance of the payment from the owner to the net						0.265

The results from the table are showing that at the use of the store of the energy, in this specific installation, in this specific month, will not be required consuming energy from the distributional network during the duration of hours of high charges, and this will be necessary receiving only during the duration of hours of low charges. We assumed, for simplicity's sake that the made energy, from e.g. the photovoltaic bodybuilding, will be in full directed for prosument sale in alt. daily 6,4 kWh (by work of the programmed controller of the flow of the energy - the higher rate for prosument sales than for the collection from the distributional network, at forced limits and ponds of the tariff G13), and the necessary energy to the work of the installation in hours of high charges will be received to the store of the energy in hours of the download at low price charges (21-7h) and will be given back to the installation in the top of the demand and high charges (the work of the controller of the flow of the energy). Here does not matter whether the store of the energy will be a stationary device, or this will be e.g. the electrical car with the technology of the rendition of the energy to the network in already functioning system V2G. Other extremely essential argument is, that this programmed work, so constructed installation, will cause the lack of the collection of the energy from the distributional network, in peak hours, what at the more and more greater quantity of the installation will have a very serious influence on her stability and the certainty of the action.

The table VI shows the example of calculations of costs for the same hybrid installation in October, in the tariff of the Polish operator G13, for the object with the daily power requirement in the height 13,2 kWh and with the own energy production in alt. 10,4 kWh but without installed store of the energy.

Table 4. Calculation of costs for the energy in Poland (The standard of prices 12/2018) in the installation without the store of the energy

Hours in 24 h/October	7-13	13-16	16-21	21-7	Total per/day	Payment euro/day
Requirement [kWh]	1.5	2.5	6.0	3.2	13.2	
Own production [kWh]	4.4	6.0			10.4	
Posible sales energy to the net [kWh]	-2.9	-3.5			-6.4	-1.116
Low tariff energy from the net incl. load magazine [kWh]		2.5		3.2	5.7	0.597
Hight tariff energy from the net [kWh]	1.5		6.0		7.5	1.221
Balance of the payment from the owner to the net						0.265

The table shows how great economic meaning has an application of stores of the energy. The difference in borne daily costs for the electrical energy from the distributional network, for both shown cases, to the disadvantage of the installation without the store of the energy, carries out till above 2,6 times. All calculations for other configurations of the installation and other seasons will be different and a task of the planner and the designers will be the preparation of calculations and forecasts and optimal solutions. For the character of the installation will have also the influence of State-provisions in the matter of rules of sales of the prosument energy.

IX. Controller of the flow of the energy

The controller of the flow of the energy in the electrical, domestic hybrid installation will be a crucial element and the technical properties they will depend and give possibilities of the installation and the steward of the exploitation. In the nonce there are no standardized rules of the construction of controllers and they are in the state to lead selective tasks provided by their producers. Controllers of the flow of the energy should have, at given processing powers DC/AC, possibilities of the incorporation of several renewable sources of energy, and in this of photovoltaic bodybuilding, wind bodybuilding and other and stores of the energy. The quantity of annexation ports should be standardized and foreseen for several series of types of sent powers. Annexation points should have possibilities of programming of the work in reference to requirements of the installation and the opening time/of the lock of ports depending on tariffs and charges for the electrical energy received/given back to, or from distributional networks. The driver of the controller should have the interface to programmed setting of the algorithm of the work of the entire installation. Together with controllers should become compulsorily delivered devices of the bypass to the dissociation of domestic installations from remaining elements of the hybrid installation for cases of any failures, threatening the fire-safety, the electric shock with the electric current from the network and from atmospheric discharges. In primitive hybrid installations, in case of the failure of some devices, the lack of systems the bypass deprives their users, for many hours, the access to the energy from the network, to the time of the arrival of service agents, nevertheless that the installation is joined to the object with the efficient mains supply. One actually designed and built switch - the bypass - for the group of switches (relays) can cut off defective devices from the domestic installation and that can provide the normal and safe work of the network energy in the object.

X. Conclusions

The superior purpose should be the promotion of the own electrical energy production from renewable sources of energy and the build of electrical hybrid installations, in individual houses, for the obtainment of presented, in the article, the energy self-sufficiency, everywhere where this only will be possible.

- Every quantity of the made energy from renewable sources of energy, even when one will not reach described energy self-sufficiency, because an obstacle will be economic matters, or architectural, for e.g. lack of the possibility of the build of wind power plantss, is priceless for the saving of our environment, through the reduction of the emission CO_2.
- Not picayune is the way of the use of the made energy. In this can help prepared informational programmes and, particularly, heating technologies being able to accept the made energy from renewable sources of energy in the intelligent way, unpermissive

on any losses and complications of installation systems.

The large part in this intention will perform intelligent and unified controllers of the flow of the energy being able in the optimum way and safe „to lead" the energy management of our houses. One ought urgently to prepare for investors and designers multifaceted guidelines concerning of principles of conduct at the designing in individual houses and small service-works, hybrid electric wirings and renewable sources of energy and related to they heating systems and feeds of other energy-consuming domestic receivers [18,19].

ACKNOWLEDGMENT

The project financed under the program of the Minister of Science and Higher Education under the name ["Regional Initiative of Excellence" in the years 2019 - 2022 project number 020/RID/2018/19], the amount of financing 12,000,000 PLN.

REFERENCES

[1] A. Malinowski and J. Szkutnik, "Hybrid low voltage networks as the element of the lifting of the efficiency and the independence in the distribution of electricity," Przegląd Elektrotechniczny, R.95 Nr 4/2019.

[2] Creating on energy efficiency mortgage for Europe : Towards a new market standard, World Green Buiding Council, 9th Annual World Green Buiding Week, 24-30 September 2018.

[3] P. Birkner, "Electrical Grids of Tomorow, Seminar on Present Challenges of European Power Engineering," Częstochowa University of Technology, May 2010.

[4] K.Billewicz, "Smart Grids Inteligentne sieci elektroenergetyczne," cz.I, cz.II, IMD Anna Korba 2015.

[5] Dyrektywa 2006/32/WE Parlamentu Europejskiego i Rady z dnia 5 kwietnia 2006 r. w sprawie efektywności końcowego wykorzystania energii i usług energetycznych oraz uchylająca dyrektywę Rady 93/76/EWG(EC Directive of the European Parliament and of the Council of 5 April 2006 on Energy end-use efficiency and Energy services, 2006/32/EC Official Journal L 114 0f 27/04/2006 (ESD).

[6] A. Gawlak, "The Influence of Investment on Reducing Energy Losses in Distribution Networks," Proc. IEEE International Scientific Conference on Electric Power Engineering (EPE2015), IEEE Press, May 2015, pp. 315-319.

[7] Green Building Standard – Nowy standard budownictwa, Wytyczne Green Building Standard, Ogólnopolskie Stowarzyszenia Wspierania Budownictwa Zrównoważonego Warszawa 2017.

[8] Doing more with less, Green paper on energy efficiency, Office for Official Publication of the European Communities, Luxembourg 2005.

[9] A. Gawlak, "Technological aspects of electrical energy distribution," Proc. International Scientific Conferenc Electric Power Engineering (EPE2014), IEEE Press, May 2014, pp.45-48.

[10] A. Gawlak, "Noninvestment Forms of Reducing Energy Losses in Distribution Networks," Proc. IEEE International Scientific Symposium on Electrical Power Engineering (EPE2015), IEEE Press, May 2015, pp. 61-64.

[11] A. Gawlak, "Analysis of technical losses in the low and medium voltage power network," Proc. IEEE International Scientific Conference on Electrical Power Enginering (EPE2010), IEEE Press, May 2010, pp. 119-123.

[12] A. Malinowski and J. Szkutnik, "Studium wykonalności Ekologiczne, oszczędne osiedla mieszkaniowe. Wybrane zagadnienia z zakresu programowania inwestycji w zakresie energetycznym", Częstochowa, luty 2019 r.

[13] BPTS 5 Hybrid The intelligent storage solution Robert Bosch.

[14] Smart Future, Folder Ml System, www.mlsystem.pl.

[15] M. Kornatka and A. Gawlak, "Comparative Analysis of Operating Conditions in Polish Medium-voltage and 110 kV Networks," Proc. International Scientific Symposium on Electrical Power Engineering (EPE2015), IEEE Press, May 2015, pp. 57-60.

[16] B. Soroczyński and J. Szkutnik, "Wpływ magazynów energii elektrycznej na pracę sektora energetycznego," Rynek Energii nr.2 kwiecień 2016 r.

[17] The battery futur is modular, Solarwatt, www.solarwatt.com.

[18] APASMART; Nazca, Smart City, Smart Building, https://apagroup.pl.

[19] World green building trends, Smark market report, Dodge Data & Analitycs, November 2018. .

Study on Interconnections Between Electromagnetic Compatibility and Techno-Economic Characteristics of Industrial Enterprises from Chemical and Mining Industry in Republic of Bulgaria

Valentin Gyurov
Department of Electric Power Supply and Electrical Equipment
Technical University of Varna
Varna, Republic of Bulgaria
valentin.giurov@tu-varna.bg

Mladen Proikov
Department of Electronic, Electrical Engineering and Machine Knowledge
Bourgas, Republic of Bulgaria
Bourgas University "Asen Zlatarov"
m_proykov@digicom.bg

Neli Simeonova
Department of Electronic, Electrical Engineering and Machine Knowledge
Bourgas, Republic of Bulgaria
Bourgas University "Asen Zlatarov"
neli_simeonova@btu.bg

Sotir Sotirov
Department of Electronic, Electrical Engineering and Machine Knowledge
Bourgas, Republic of Bulgaria
Bourgas University "Asen Zlatarov"
ssotirov@btu.bg

Rumen Kirov
Department of Electric Power Supply and Electrical Equipment
Technical University of Varna
Varna, Republic of Bulgaria
valentin.giurov@tu-varna.bg

Abstract - **The report is based on multi-year research in industrial enterprises of the chemical and mining industry in Republic of Bulgaria and presents the definition of the relationship between techno-economic performance, power quality and electromagnetic compatibility (EMC). The influence of the voltage deviation on the specific energy consumption at different levels of loads of the objects is shown. The resulting quantitative ratios give opportunity for establishment of the relationship between power quality, EMC and energy efficiency (EE). The study formulates recommended values of different indicators and characteristics for increasing of EMC in the particular branch of industry.**

Keywords: Reactive power compensation; electromagnetic compatability; power and energy losses; electric power supply systems

I. Introduction

The reduced electricity consumption in industrial enterprises affects a number of technical and economic indicators of electric power supply systems (EPSS) and has a negative impact on EE and EMC. The power distribution transformers (PDT) in reduced load regime significantly alter their electrical parameters such as reactive and active losses, voltage deviation, cosφ, etc., which affects their EMCs with the LV electrical equipment and consumers. The slowdown in industrial growth in the Republic of Bulgaria over the last 30 years has led to a significant drop in electricity consumption in all branches of the economy. The biggest decrease in electricity consumption was observed in machine building and the electrical industry - over 28% [1]. This reduction in transport is about 20%. For large structurally significant enterprises the decrease is in the range (11÷12)%, for the light industry is about 25% and for the small and medium enterprises in the sphere of services and trade is 9%. The chemical and mining industry is characterized by wide dynamic and profound changes in the structure and specific characteristics of the industrial enterprises. For a large part of the enterprises in the sector there is a reduction of the electricity consumption and a deterioration of the energy efficiency due to technological reorganizations and changes in the subject activity.

II. Experimental Study

The surveys were carried out for a period of 25 years in 19 power plants with a total installed capacity of the branch $S_{BRANCH} = 665$[MVA], annual average $\cos\varphi_{AVG} = 0{,}843$, annual average utilization $T_{AVG.YEAR} = 8205$hours, annual average active load $P_{AVG.YEAR} = 5732$[MW] and annual electric energy consumption $W_{AVG.YEAR} = 4703.103$ [MWh/year]. The studies are based on a theory of Design of Experiment (DOE) and represents passive multifactor experiment, consist of two complex energy parameters (target functions) - specific energy consumption per production unit and balance equation of power.

1. Specific energy consumption per production unit

The specific energy consumption per unit of production is determined on the basis of information from the "Production and energy department" of the surveyed companies, which includes the quantity of product M and the electricity consumption E [kWh] [2].

$$\Delta E = \frac{E}{M} \qquad \Delta E[\%] = \frac{\Delta E(U_{MAX})}{\Delta E(U)} \cdot 100 \qquad (1)$$

where: $\Delta E(U_{MAX})$ and ΔE specific energy consumption for U_{MAX} - maximum recorded voltage and for U - voltage values under the experiment conditions. With the use of expressions (1), mean values for ΔE and ΔE [%] are determined on the basis of data obtained from measurements for each object with a duration of 41 to 55 working days for the summer and winter season. The research was carried out with a variety of measuring equipment from various manufacturers - ABB Power Plus, Iskra PNA760, Fluke 190-504/S; UMG 96S; KEW6310; Siemens SCADA, Schneider Electric SCADA and others. Voltage regulation is performed with steps voltage switches of the distribution transformers and tap changing transformers. The measured data is grouped into two voltage ranges - lower - $U=(0.9\div1)U_n$ and higher - $U=(1\div1.1)U_n$, which correspond to position 1 and 3 of the steps switches. The load varies into three levels - low (b1), medium (b2) and high (b3), and reactive power modes are two - without reactive power compensation(RPC) (Mode A) and with RPC (Mode B). The total number of measurements for this branch of industry is $n = 430.10^3$. The results for the reduction of the specific electricity consumption ΔE for mode A and mode B (ΔE_A and ΔE_B) in the branch "Chemical and mining industry" (CMI) are presented in Table 1. The values of the dispersion of voltage deviation δU - $\sigma(U)_A$, $\sigma(U)_B$ and the average $\cos\varphi$ for the two modes - $\cos\varphi_A$ and $\cos\varphi_B$ modes are also shown. For all measurements, mathematical statistics were applied, including calculations for the stationarity and ergodicity of the probability process for all individual samples.

Table 1. Reduction of specific energy consumption ΔE[%]

	Position of Step-Switch 1 U=(208÷219)V		
	$\beta_1=(0.16\div0.38)$	$\beta_2=(0.39\div0.57)$	$\beta_3=(0.58\div0.8)$
ΔE_A	-1.8	-0.3	0.8
ΔE_B	-1.3	0.2	1.7
$\cos\varphi_A$	0.79	0.81	0.83
$\cos\varphi_B$	0.9	0.9	0.91
$\sigma(U)_A$	3.5	3.9	3.7
$\sigma(U)_B$	3.8	3.5	3.5
	Position of Step-Switch 3 U=(228÷239)V		
	$\beta_1=(0.13\div0.36)$	$\beta_2=(0.37\div0.59)$	$\beta_3=(0.6\div0.82)$
ΔE_A	-0.2	1.8	4.2
ΔE_B	0.7	2.7	5.2
$\cos\varphi_A$	0.77	0.79	0.83
$\cos\varphi_B$	0.9	0.91	0.9
$\sigma(U)_A$	3.15	2.83	2.9
$\sigma(U)_B$	2.87	2.98	3.82

Graphical interpretation of the reduction of the specific energy consumption for the sector is shown in Figure 1, as a result of which the following conclusions can be drawn:

- If the voltage drops is in the range of $\delta U = (0.95\div1)U_n$, (position 1 of the steps switch), then negligible positive and more significant negative changes of ΔE are observed. For low loads level $\beta1 = (0.16 \div 0.38)$, ΔE is changed in the range $\Delta E = (-1.3 \div -1.8)\%$. At levels $\Delta U = (1\div1.05)U_n$ there is an increase in equipment productivity and ΔE decreases within the range of $\Delta E = (1.8\div5.2)\%$. This is due to the presence of users specific to this industry, such as powerful electro-systems, thermal installations, powerful specialized drives with synchronous and asynchronous motors, etc. which operate inefficiently at $U < U_n$. Therefore, for the CMI sector at lower loads, it is more appropriate to work with higher voltages to improve EMC and increase EE.

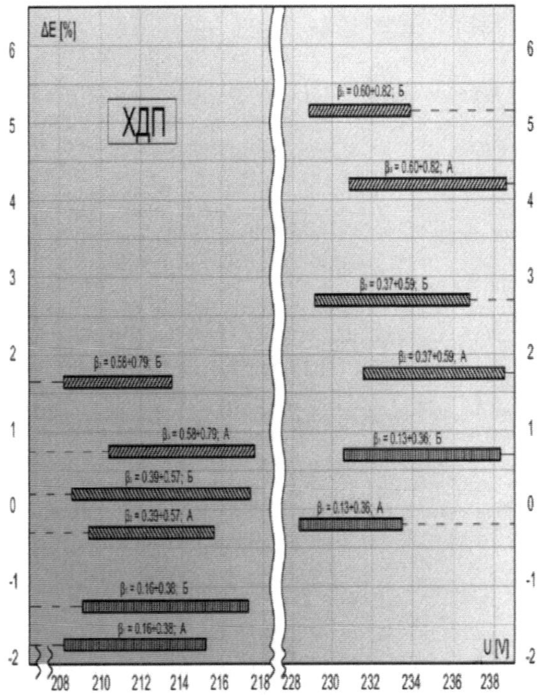

Figure 1. Reduction of specific energy consuption ΔE[%] for branch Chemical and Mining Industry

- The studies also show that in position 1 of the steps switch of distribution transformers, RPC contributes to increasing of the ΔE, i.e., at lower voltages RPC is less efficient. This finding is not confirmed in the rest branches of the industry.
- Analysis of the research shows that there is a strong relationship between EE, estimated by the specific EE consumption and the voltage deviation δU, as a key indicator of power quality. This correlation determines the level of EMC of the electrical equipment and the EPSS in the industry. In order to achieve a higher level of EMC during the operation of the electrical equipment, the technical event is recommended to regulate the supply voltage in the workshop stations without load by using the voltage in the range $\delta U = (1\div1.05)U_n$ corresponding to the position 3 of the steps switch or to regulate the voltage under load in the same range with the use of tap-changing transformer in the 110/20, 10, 6 kV power substations.

2. Balance Equation of Power

Compensation of reactive loads at different voltage levels in EPSS is accompanied by an increase in the voltage at the measurement point, resulting in a positive regulating effect on the active and reactive load. At the same time, there is a negative regulating effect on the losses of active power. Therefore, there are two opposing trends on the power balance δP. In order to reduce the electrical consumption, it is necessary to regulate the voltage. The optimization of such a mode of operation can be realized by simultaneously with RPC and voltage regulation [3]. The mathematical model for δP* is represented by Formula 2 [3,4].

$$\delta P^* = \frac{\delta P}{P_n} \cdot 100[\%] = f(\beta, \delta U^*, \cos\varphi, H_K) \tag{2}$$

where: δU * - relative variation of voltage due to RPC and its regulation; H_K = f(p,q) - complex indicator determined by the simultaneous influence of characteristic coefficients for active (p) and reactive (q) load.

The main criterion for economical operation of the EPSS of each industrial enterprise is the fulfillment of the condition δP * <0 and this can be achieved by appropriately combining the factors in the target function (2). This can be accomplished by applying a complex multifactor approach using the Design of Experiment theory (DOE) [3.5], based on the possibility for simultaneous modification of various significant SF factors and their impact on output parameters or target function. In this case, the output parameter is δP*, and the SFs are four, justified by the expressions (2) - β, δU*, cosφ and H_K. The DOE was carried out with the use of a second-order plan Bm (m = 4) and SFs were defined on three levels. The mathematical model and corresponding graphical characteristics are represented with MathCAD software in Figure 2.

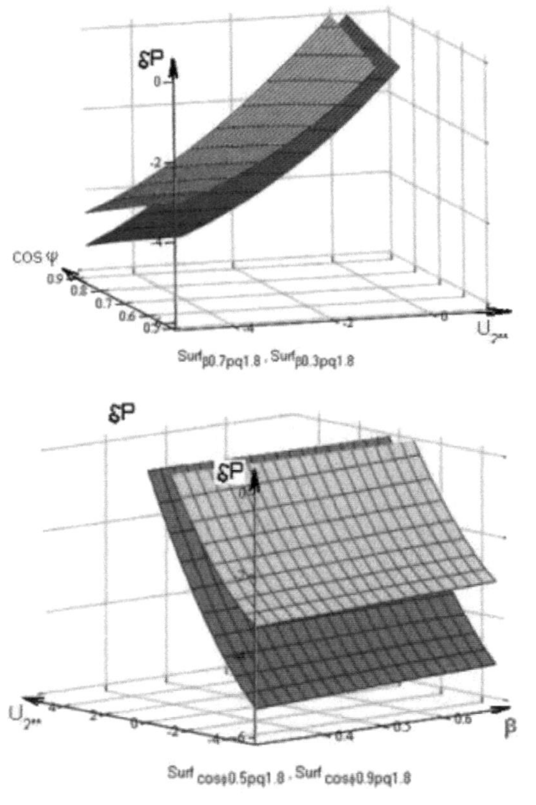

Figure 2. Segment δP*<0 for branch Chemical and Mining Industry

$$\begin{aligned} \delta P^* &= 3.282 + 0.675 \cdot \beta + 2.136 \cdot \delta U^* - 3 \cdot \cos\varphi + \\ &+ 0.047 \cdot H_K - 0.009 \cdot \beta \cdot H_K - 0.001 \cdot \delta U^* \cdot H_K + \\ &+ 0.0037 \cdot H_K \cdot \cos\varphi - 0.0048 \cdot \beta^2 + 0.197 \cdot (\delta U^*)^2 - \\ &- 0.0697 \cdot \cos^2\varphi + 0.058 \cdot H^2 \end{aligned} \tag{3}$$

The Table 2 shows the optimal SFs ranges by the dP*<0 criterion for the branch CMI .

Table 2. Optimal areas of SF by criteria δP*<0 for branch Chemical and Mining Industry

β=0.3	*β=0.5*	*β=0.7*
cosφ≥0.7	cosφ≥0.8	cosφ≥0.9
0.98≤δU*≤1.03	0.98≤δU*≤1.04	0.98≤δU*≤1.05
H_K≤ 1.2	H_K≤ 1.2	H_K≤ 1.8

On the basis of these and other results, the following conclusions can be drawn:

- The fulfillment of the criterion δP*<0 occurs when cosφ increases and this can also start from quite low cosφ = 0.7.
- This low threshold for optimal cosφ values is due to the effect of mutual compensation and the positive influence of other SFs each one.
- For the CMI, higher EE and improved EMC is obtained at δU* values in the range δU* = (1÷1.05)U_n both in low load mode and medium and high load, which is characteristic of other industries from the industry. This is due to the influence of specific consumers in this industrial branch, including powerful chemical and thermal installations.
- Significant fluctuations of the voltage δV = (1.2÷ 5.4) % at the frequency of fluctuations (30÷80) fluctuations per hour were registered in the sector. The coefficients of voltage unbalance ε_U and current unbalance ε_I are in the range of ε_U = (1.4÷3.8)% and ε_I= (4.7÷17.9)%. The coefficients of zero sequence unbalance of voltage α_U and current α_I are respectively within the limits α_U = (1.35÷3.12)% and α_I = (3.25÷ 11.12)%. The generalized three-phase load-weighted indicators of voltage and current TUW_U and TUW_I , which varies within the range TUW_U = (1.08 ÷ 2.93)% and TUW_I = (3.18÷14.51)%, are also defined. The measured average values for THD_U and THD_I are within the range THD_U = (2÷9)% and THD_I = (6 ÷ 35)%, with a high probability of occurrence of current and voltage resonance. The ineffective RPC leads to an increase of this probability, and at the same time leads to a shift of the resonant frequencies from higher to lower. Harmonic distortions are evaluated by the three-phase load-weighted indicators $THDW_U$ and $THDW_I$, which varies within the range $THDW_U$ = (1.63÷ 7.12)% and $THDW_I$ = (4.2÷23.7)%. For some industrial sites in the sector, the intensities of the electric field (E) and the magnetic field (H) were measured and they varied in the range E = (3.4 ÷ 23.7) [kV/m] and H = (50 ÷ 2000)[A/m]. The high values are recorded in electrolyze systems with overlap of constant and variable field. For switching mode of individual 20 kV circuit breakers, E = (0.8÷3.5) [kV/m] and H = (13÷19) [A/m] were measured. The over voltages occurred with a small duration of the impulse front and a high frequency of fast-pacing fluctuations.

The multi-annual studies and their results presented above provide the opportunity to formulate the optimal indicators, characteristics and guidelines for their achievement which are presented in a synthesized form in Table 3.

Table 3. Recommended levels of parameters for achievement of EMC for Branch Chemical and Mining Industry

β	$\cos$	$\delta U^*[\%]$
≥ 0.45	≥ 0.9	0.98÷1.05
H_K	$\delta V[\%]$	$\varepsilon_U[\%]$
≤ 1.8	1÷2	0.5÷2
$\varepsilon_I[\%]$	$TUW_U[\%]$	$TUW_I[\%]$
3÷10	0.4÷1.6	2÷7
$\alpha_U[\%]$	$\alpha_I[\%]$	$THD_U[\%]$
0.5÷1.5	2÷5	2÷6
$THD_I[\%]$	$THDW_I[\%]$	$E[kV/m]$
4÷16	2.5÷12	5÷10
$H[A/m]$	$\delta P^*[\%]$	$\Delta E[\%]$
10÷400	3÷8	2÷8

The reasons for the deterioration of EMC are the operating modes of specific thermal and chemical installations electrolyze systems, and powerful electrical motors.

These results are intended for designers, researchers and operators of electric power supply systems and electrical equipment in order to achieve clear, complete and accurate information on the status of processes and phenomena in electric power supply systems in this branch of industry.

III. Conclusion

For the CMI industry, appropriate voltage variation ranges were established within the range $\delta U = (1 \div 1.05)U_n$ for which it is possible to achieve high EE and EMC. It has been shown that in this range of variation of δU, RPC is more effective. Modeling of energy processes under reduced load was made using the defined complex multifactor approach, based on the theory of planning the experiment (DOE). According to the criterion $\delta P^* < 0$, the mathematical model for output parameter $\delta P^* = f(\beta, \delta U^*, \cos\varphi, H_K)$ is defined, and plan B_4 is applied. The results are presented in graphical form on the basis of which optimal areas for SF and output parameters are identified. Research provides opportunity to formulate recommended values and guidelines for achieving EMS in the industry by defining specific values for different indicators and characteristics.

Acknowledgment

This work was supported by European Regional Development Fund through the Operational Programme "Science and Education for Smart Growth" under contract UNITe № BG05M2OP001-1.001-004 (2018-2023)

References

[1] A. Geotgiev, "Statistical data for electricity consumption in the country", Journal of Power Engineering, Bulgaria, 1989-2018.

[2] R. Kirov, I. Iliev, N. Naydenov, "Study on power efficiency by sectors for the management of the voltage mode of power supply systems", Proc. Symp. Electrical Power Engineering , 2014.

[3] R. Vasilev, "Exploring the possibilities to improve the efficiency of power supply systems of industrial sites in low-load mode", PhD Thesys, Technical University of Varna, 2003.

[4] V. Orlov, "Reduction of energy consumption with reactive power compensation of industrial networks", Journal of Energetics, Vol. 4, Russia, 1989.

[5] A. Mitkov, S. Kardashevski, "Statistical methods in agricultural technologies", Sofia, 1977.

Study on Electromagnetic Compatibility of Powerful Industrial Loads in Different Working Regimes

Valentin Gyurov
Department of Electric Power Supply and Electrical Equipment
Technical University of Varna
Varna, Republic of Bulgaria
valentin.giurov@tu-varna.bg

Stanislav Simenonov
Department of Electronic, Electrical Engineering and Machine Knowledge
Bourgas, Republic of Bulgaria
Bourgas University "Asen Zlatarov"
stanislav_simeonov@btu.bg

Neli Simeonova
Department of Electronic, Electrical Engineering and Machine Knowledge
Bourgas, Republic of Bulgaria
Bourgas University "Asen Zlatarov"
neli_simeonova@btu.bg

Sotir Sotirov
Department of Electronic, Electrical Engineering and Machine Knowledge
Bourgas, Republic of Bulgaria
Bourgas University "Asen Zlatarov"
ssotirov@btu.bg

Abstract — **The study analyzes basic characteristics related to the electromagnetic compatibility (EMC) of some powerful industrial objects under different operating modes. The level, nature and impact of EMC from the load and processes related to the quality of electrical energy (EE) and the reactive power compensation (RPC) in the electric power supply systems (EPSS) of the objects under consideration have been established. The results demonstrate the effect of RPC and confirm its suitability in low-load mode as well. The conditions and prerequisites for deterioration of the EMC of the investigated sites have been studied.**

Keywords: Reactive power compensation; electromagnetic compatability; power and energy losses; electric power supply systems

I. Introduction

The EMC is explored as a complex concept related to the identification of the sources of disturbances, the technical and methodological means for their investigation, the identification of their negative influence and the possibilities for their minimization [1,2]. For the electric power supply system, electromagnetic compatibility represents the ability of individual users to operate normally in an electromagnetic environment and do not create unacceptable disturbances, affecting the performance of other consumers. Electromagnetic processes in the EPSS can create negative consequences in two directions - on the one hand, to degrade the properties and characteristics of the transmission grids, thereby reducing their throughput and, on the other, deteriorating the parameters of EE, respectively its quality. For control and limit of these negative phenomena a number of standards have been created. The main ones are the IEC 61000-X, IEEE 519 series of standards. The IEC standards are more important because they place limits not only on the harmonic composition of the voltages but also on the harmonics of the current. The IEC and CENELEC regulate through standards the limitation of odd harmonics of currents from 3rd to 39th. Similar requirements are set by the US (ANSI), Russia (GOST), China (CNS) and Japan (JIS) standards, but they are not mandatory. Different approaches are used in the development of national standards for EMC [2]. For example, in the New Zealand standard, regulatory requirements for the quality of electrical energy are regulated without taking into account the specificities of the sources of disturbances, while UK and Australian standards takes into account the negative influence on the quality of electrical energy caused by the connection of powerful users. The system's short-circuit power is taken into account by some standards, with higher requirements of EMC for high-power consumers. The IEEE 519 standard regulates the permissible higher harmonics (VH) of the current to be 50% higher at startup modes, and p/6 higher for rectifiers (where p is the number of rectifier pulses. The South African Standard (ESLOM) examines three types of networks depending on their sensitivity to electromagnetic disturbances. At Japanese Standard, the permissible levels of harmonic current are given in % of the consumer power [mA/kW]. The Swedish National Standard SS4211811 regulates $THD_U \leq 6$ [%] for all modes of operation. There are possibilities to determine the energy impact of the quality of electrical energy by using aggregated three-phase load-related indicators. The harmonic distortions and asymmetry are quantified by *THDW* (Total harmonic distortion weighted) and *TUW* (Total unbalance weighted) indicators. The studies shows [1,2,4,6] that the most correct approach is to determine and limit the impact of each harmonic of voltage and current. Evaluation of EMC only through the THD_U and THD_I does not allow the probability of resonance phenomena of voltage and current to be predicted. In addition, a more adequate approach is recommended for the global ESM assessment by using the *THDW* and *TUW* indicators.

II. Experimental Study

The analysis of normative documents shows the existence of direct relation between EMC and power quality. In addition, in a number of studies [3,5,7] it has been shown that reactive load compensation (CRT) has not only an impact on the EPSS energy parameters, but also has a strong impact on power quality and EMC, and the level of this influence depends heavily by the electrical equipment consumption modes and load coefficient. In order to establish the interdependence of these factors, a study was carried out in three powerful industrial sites in the Republic of Bulgaria - "Alcomet" AD, Shumen town, "Solvay Sodi", Devnya town, "Varna Shipyard", Varna town.

1. Research on power quality and EMC in EPSS of "Alcomet"AD, Shumen town

The main activity of the company is the processing of aluminum and its alloys. The main electrical equipments are rolling mills, equipped with winding and unwinding mechanisms driven by DC motors. The winding and unwinding machines are types 8206, 500 kW, 440 V, 1225 A and MAF 7304, 400 kW, 440 V, 984 A respectively. The electric motors are produced by the French company "Control Techniques" and they are mechanically connected with shaft to the main gearbox. The excitation of DC motors is accomplished by a controlled rectifier consisting of 4 thyristors, and the control pulses are generated by operational amplifiers. The DC motors are controlled by Mentol II DC Drivers 25A, 1850A drivers, and the power supply unit is a 6-pulse controlled rectifier with pulse generator control. In addition to the rolling mills, other powerful electrical equipment in the company are furnaces, straightening machines, separators, frequency controlled fans, etc. The energy management of the company is realized with a microprocessor system Yanitza Electronics – Germany. The research was conducted for 60 days in winter and summer season at two control points of winding and unwind machines with the use of UMG 96S universal meter device, equipped with RS232/485 communication, Profibus DPVC. The measured real-time parameters are currents (I), voltages (U), power, cosφ, frequency, harmonic components, etc. at sampling time of 1 min. The passive experiment enables data from the studies to be grouped for two levels of cosφ (respectively in the absence and presence of RPC) and two levels of load β, which are presented in Table. 1.

Table 1. Range of grouping of data for β and cosφ

Control points	β	cosφ
winding machine	β_1=0.31÷0.63	$cos\varphi_1$=0.31÷0.63
	β_2=0.63÷0.92	$cos\varphi_2$=0.89÷0.93
unwinding machine	β_1=0.27÷0.59	$cos\varphi_1$=0.31÷0.63
	β_2=0.59÷0.88	$cos\varphi_2$=0.31÷0.63

The volume of data generated is 3.10^4, while the surveyed sample is about 20% of all data. The results of the studies and their processing make it possible to draw the following conclusions:

- The voltage deviation (1) is in the range of δU = (-4÷8) % and the unbalance coefficient (2) is in the range of ε_U = (0.31÷0.51) % for both objects and they are within the values allowed by the EN50160 standard. For the coefficient of current asymmetry (3), high values in the range ε_I = (1.39÷13.18) % were recorded. In Table 2 the correlation dependencies R(β,ε_U) and R(β,ε_I) are presented (4). For both objects, they have negative values, therefore the variance of the values is inversely proportional. Therefore, at lower load the current and voltage asymmetry are more pronounced, and EMC is worsened by the influence of these indicators. The generalized three-phase weighted indicators for the voltage and current unbalance TUW_U and TUW_I, which are changed respectively in TUW_U = (0.35÷ 0.47) % and TUW_I = (1.54÷8.27) %, are defined. These ranges of variations shows a lower influence of the unsymmetrical regimes on the EMC, i.e. the impact of the load on these indicators is taken into account and such an assessment is more appropriate.

Table 2. Corelation coefficients R(β, ε_U) and R(β, ε_I)

	R(β, ε_U)	R(β, ε_I)
winding machine	-0.47	-0.62
unwinding machine	-0.52	-0.75

- For both machines it was found that $THD_U \leq$ 8%, i.e. meet the requirements of EN50160 while THD_I significantly exceeds the IEC 61000-3-4 value $THD_I \leq$ 25%, with higher values being recorded at higher loads and low *cosφ* values. Consequently, it is confirmed that RPC is an effective measure to achieve good EMC at low loads β.
- The generalized three-phase weighted indicators $THDW_U$ and $THDW_I$ have values in the range $THDW_U$ = (2.78÷4.31) % and $THDW_I$ = (28.6÷41.7) %, finding that good EMC at low loads can be achieved by implementing an effective RPC .
- The voltage harmonics are smaller than the EN50160 requirements, confirming the tendency for lower U_v values to be observed at lower loads and higher cosφ. In accordance with IEC 61000-3-4, the I_v values are: $I_3 \leq$ 21.6 %, $I_5 \leq$ 20.7 %, $I_7 \leq$7.2 %, $I_{11} \leq$ 3.1 % and $I_{13} \leq$ 2 %. The measured values for both aggregates significantly exceed the regulated of the standard. This indicates that the EMC is worse when the two units work together. The probable reasons for that are the inadequate characteristics and parameters of the drivers of DC motors and power rectifiers.

$$\delta U = \frac{U - U_n}{U_n} \cdot 100 \tag{1}$$

where: U – line voltage [V]; U_n – nominal voltage [V].

$$\varepsilon_U = \frac{U_2}{U_1} \cdot 100 \tag{2}$$

where: U_2 – sequence 2 in Fortesque transform [V]; U_1 – sequence 1 in Fortesque transform [V].

$$\varepsilon_I = \frac{I_2}{I_1} \cdot 100 \tag{3}$$

where: I_2 – sequence 2 in Fortesque transform [A]; I_1 – sequence 1 in Fortesque transform [A].

$$R(\beta,\varepsilon_I) = \frac{n \cdot \sum_{i=1}^{n}(\beta_i . \varepsilon_{Ii}) - \sum_{i=1}^{n}\beta_i . \sum_{i=1}^{n}\varepsilon_{Ii}}{\sqrt{\left[n . \sum_{i=1}^{n}(\beta_i)^2 - \left(\sum_{i=1}^{n}\beta_i\right)^2\right] \cdot \left[n . \sum_{i=1}^{n}(\varepsilon_{Ii})^2 - \left(\sum_{i=1}^{n}\varepsilon_{Ii}\right)^2\right]}} \tag{4}$$

2. Research on power quality and EMC in EPSS of "Solvay Sodi", Devnya town

The company produces light and heavy calcined soda and soda bicarbonate. The EPSS consists of 1 main substations 21/6.3 kV, 1 main distribution substation, 25 transformer substations with power transformers (1 MVA – 55 pcs, 1.6 MVA – 8 pcs, 2 MVA – 5 pcs) and operate in continuous mode. The EPSS include SCADA POWER CC Siemens dispatcher system for control and management of various power processes and regimes. The main users of the company are non-linear loads in the range of 65 ÷ 87 % of the total installed load and includes electric drives with frequency control, rectifiers, pulse power supplies, gas discharge lamps with electronic drivers, arc welding devices, computer systems,

The experimental research was carried out on a medium voltage side 6.3 kV with the use of the information on the energy processes from the SCADA I/O module modules type 6MD 6130 and on the low voltage side 0.4 kV with the use of Kyriotsu KEW6310 mobile analyzer. With the help of these two metering systems, five substations (SS06-1, SS07-1, SS07-2, SS14-2 and SS31-1) were analyzed by determining the currents and voltages, all types of power, current harmonics and voltage and other magnitudes required to evaluate EMC. The RPC is realized with capacitor banks on the LV side of the transformer substations. The required compensation power Q_c[kVA] is determined (5).

$$Q_c = \frac{k_T \cdot U_{0.4} \cdot U_{6.3}}{X_{PT}} + P \cdot \left(\frac{R_{PT}}{X_{PT}} + tg\varphi \right) \quad (5)$$

where: K_T – transformer ratio; $U_{0.4}$ and $U_{6.3}$ – measured voltages on LV and MV side; P [kW] – active power; R_{PT} and X_{PT} – active and inductive resistance [ohm].

Table 3. Results for 5 transformer substations

Parameter	Substation									
	06-1		07-1		07-2		14-2		31-1	
	with RPC	without RPC	with RPC	without RPC	with RPC	without RPC	with RPC	without RPC	with RPC	without RPC
S_n [kVA]	1000	1000	1000	1000	1000	1000	1000	1000	1600	1600
β	0.26	0.94	0.23	0.69	0.27	0.81	0.19	0.77	0.21	0.78
$cos\varphi$	0.87	0.89	0.74	0.91	0.68	0.9	0.81	0.92	0.73	0.91
Q_c [kVAr]	-	210	-	400	-	530	-	600	-	650
f_r [Hz]	567.5	358	738.4	307.5	667.7	248.7	869	234.6	544.6	216
ν	11	7	15	6	13	5	17	5	11	4
$I_{nw}(\nu)$ [A]	26.4	102.6	29.4	65	31.5	179	42.6	138.6	46.8	152.8
$I_{ns}(\nu)$ [A]	35.4	163.8	39.8	73.2	42.8	196	76.3	143.9	63.1	186.4
$U_{nw}(\nu)$ [A]	81.8	138	57.2	82.7	63.1	156	135.6	125.6	118.7	122.7
$U_{ns}(\nu)$ [A]	106.8	220.3	77.4	93.13	85.7	170.8	242.8	130.4	160	149.3

For the five substation, the required Q_c, β, $cos\varphi$, the resonance frequency f_r and the corresponding harmonics to the non-linear load currents and voltages of the respective resonant frequency with and without RPC for working and start-up regime were calculated ($I_{nw}(\nu)$, $I_{ns}(\nu)$, $U_{nw}(\nu)$, $U_{ns}(\nu)$. The results are presented in Table 3.

The following conclusions can be drawn from the results:

- Due to the fact that the non-linear load can be capacitive and the remaining load is inductive, there is a possibility of a current resonance occurring in the range 11 to 17 harmonics;
- The probability of resonance with presence of RPC shifts to lower frequencies (4-7 harmonic), and can be observed with even harmonics (4 and 6).
- The most significant overload of harmonic current is registered for SS07-2. In presence of CRT the current of the 5th harmonic increases from 31.5 to 179A, i.e. about 6 times, which is not proportional to the change of β = 0.27 to β = 0.81, i.e. about 3 times.

3. Research on power quality and EMC in EPSS of "Varna shipyard", Varna town

The main workshop of the company is a "Hulls manufacturing", where the main consumers are welding units, rectifiers, motor generators, welding systems, flow welding lines, etc. with a total installed power of about 9200 kVA. The electric power supply of the workshop consist of 9 substations with total installed power S_n=14310 [kVA]. The remaining load of the company includes mainly electrical machines with a installed power of about 6350 kW (rolling machines, single-column, two-column and horizontal presses, guillotines, production lines, etc.). Research shows that the quality of the electrical energy in the ESS of the company is considerably degraded [5]. Significant deviations from the normative requirements for unbalance and non-linear currents and voltages are observed. This is caused by some specific technical solutions, such an existence of power transformers with Y-Yo connection circuits, which allows the passage of harmonics to the MV side. The voltage fluctuation δV varies widely in the range of δV = (13.5÷19.3)%. The power factor PF is very low (PF = 0.34÷0.67) and for its improvement and for minimizing the δV was built a system for serial RPC with total power 1380 kVA connected on LV to a rectifying group with 6 pcs. welding machines. Table 4 presents the results for some indicators without and with serial RPC. It can be seen that with the connection of serial RPC, the PF increases significantly, the voltage oscillation δV decreases abruptly (about 3 times), the voltage unbalance ε_U reduce to $\varepsilon_U \leq 2\%$, i.e. the requirements of standard EN50160 are met. As a consequence of the reduction of currents, the power losses ΔP are reduced from 18.4% to 7.9%.

The study includes measurement of the voltage and current harmonics U_v and I_v at different loads β over 18 working days for a production line with 18 pcs. welding transformers with a total power of 1080kVA and 20 pcs. rectifiers with total power 1600kVA. The results are presented in Table 5.

The analysis of the results of the table shows that there is a tendency with the increase of the load of the power units, the higher harmonics U_v and I_v to decrease their values. This fact is in contradiction to results from other research made for powerful industrial sites in the chemical and metallurgical industries, where EMC deterioration has been reported with an increase in load [6,7]. Furthermore, even harmonics are observed in the current and voltage spectrum, which is due to the asymmetry of the thyristor control impulses.

Table 4. Parameters for group 3A with S=1380[kVA] with and without RPC

Parameters without RPC			
K_P	δV	ε_U	ΔP[%]
0.43	15.98	4.27	18.4
Parameters with RPC			
0.958	4.82	1.84	7.9

III. Conclusion

In the surveyed industrial enterprises with the increment of load, there are prerequisites for worsening of the EMC. The implementation of efficient RPC, both parallel and serial, helps to improve the harmonic composition of voltage and current, and the use of special RPC schemes can improve non-symmetrical modes of operation. In low-load modes, the probability of current resonance arises in the range of higher frequencies, and efficient RPC helps displacement of resonance phenomena in the lower frequency range. It has been shown that RPC is an effective technical solution for improving electrical energy efficiency and EMC not only at medium and high but also at low loads. In real studies, the efficiency of the serial CRT is confirmed. The improvement of the EMC being expressed in a sharp decrease of the voltage fluctuation, substantial decrease of the asymmetry and significant increase of the power factor, respectively reduction of the losses of active power.

Acknowledgment

This work was supported by European Regional Development Fund through the Operational Programme "Science and Education for Smart Growth" under contract UNITe № BG05M2OP001-1.001-004 (2018-2023)

Table 5. Results for voltage harmonics U_v and curret harmonics I_v

ν	2	3	4	5	6	7	8	9	10	11	12	13	14	15
β=0.2÷0.45														
U_v	4.82	4.32	2.83	2.38	0.93	1.68	0.66	0.87	0.51	0.61	0.58	0.65	0.45	0.32
I_v	7.59	14.58	5.98	8.46	3.11	3.02	1.71	1.93	0.94	0.87	0.89	0.87	0.61	0.64
β=0.45÷0.7														
U_v	4.57	3.85	2.14	2.11	0.86	1.44	0.54	0.78	0.44	0.54	0.55	0.58	0.44	0.31
I_v	7.19	13.28	4.25	9.52	2.83	2.73	1.49	1.58	0.91	0.89	0.86	0.83	0.6	0.55
β=0.7÷0.8														
U_v	4.31	3.27	1.62	2.07	0.75	1.32	0.39	0.56	0.38	0.43	0.45	0.46	0.42	0.29
I_v	6.58	12.54	3.96	8.32	2.68	2.38	1.52	1.31	0.85	0.82	0.77	0.79	0.6	0.51

References

[1] M. Proykov, G. Georgiev, R. Panchev, V. Gyurov, "Impact of electrical switching processes and resonance phenomena on electromagnetic compatibility and reliability of electric power supply", Proc. Symp. Unitech 2018, pp.48-52.

[2] I. Iliev, R. Panchev, M. Proykov, R. Kirov, "Concerning the extensions of the definition area of the electric power category "Electromagnetic Compatibility" and its relation to the reliability of the electric power supply", Proc. Symp. Unitech 2018, pp.42-47.

[3] M. Proikov, G. Georgiev, H. Panchev, G. Ivanova, V. Gyurov, "Study of parameters and characteristics in the explotation of the elecabels systems of Elkabel Inc., Bourgas", Journal of Mechanics, Transport, Communications, vol. 16, Issue 3/2, pp. 68-74, Bulgaria, 2018, Paper №1725.

[4] R. Vasilev, R. Kirov, V. Gyurov, M. Proikov, H. Panchev, "About the formulation of approaches and concepts for the evalution of electromegnetic compatibility in electricity supply systems", Journal of Mechanics, Transport, Communications, vol. 16, Issue 3/2, pp. 98-104, Bulgaria, 2018, Paper №1729.

[5] I. Illiev, M. Proykov, R. Kirov, "Improvement of electromagnetic compatibility in power supply systems with serial compensation of reactive loads", Proc. Symp. Energy Forum 2017, Bulgaria, pp.156-163.

[6] I. Illiev, M. Proykov, R. Kirov, "Investigation and analysis of electromagnetic compatibility in the operation of rolling mills" Proc. Symp. Energy Forum 2017, Bulgaria, pp. 392-402.

[7] R. Kirov, V. Gyurov, V. Chikov, N. Makedonski, "Study on electromagnetic compatability in reactive power compensation systems", Proc. Symp. Energy Forum 2017, Bulgaria, pp. 341-347.

B: High Voltage Technique and Diagnostics in Power Engineering

Multiple Partial Discharges and Set of Their Sources

Nikolay Kinsht
Tech. Diagnostics Lab.
IACP FEBRAS
Vladivostok, Russia
kin@dvo.ru

Natalia Petrunko
Tech. Diagnostics Lab.
IACP FEBRAS
Vladivostok, Russia
pnn@dvo.ru

Abstract — **The subject of the discussion is the problem of multiple partial discharges (PD) in high voltage equipment insulation. It is declared PD set has a double character. First, PD multiplicity can be connected with the few defects in the HV insulation active space. At the other hand, it connects with a set of PD occurs in the solitary insulation defect as a PD source. The International standard IEC 60270:2000 and other national standards declare methods of measurement of partial discharges and ignore the problem about the multiple PD sources as potential defects. From point of view of the "apparent charge" conception, it's impossible to learn this problem. The problem of generating the PD sequence by any PD source is considered in the context of learning the integral pattern of the PD. The question about the number of PD sources which initiate this PD integral pattern is investigated. It is shown, the way of the investigation of the set of the PD pulse instants can clarify the problem. The problem of distinguishing of the set of defects in insulation on the basis of the set of generated PD is formulated and solved. An illustrative example is shown.**

Keywords: partial discharge, defect, PD source, PD set, PD integral pattern

I. Preface

Partial discharges (PD)can be considered from several points of view: they have been studied for a long time as an electro physical phenomenon, as a technical phenomenon, and as a property of potential defects in the insulation of the high-voltage equipment (HVE). The state and quality of HVE insulation is associated with the presence, intensity and distribution of partial electrical discharges. There occur as during normal operation, and when degradation of the insulation and other structural elements.

It is fundamental clarity in terms of electro physics theory, but there are still many unclear places and contradictions in their study. The question arises as to what set of basic PD properties is necessary (and sufficient, if strictly speaking) to assess their danger from the point of view of prevention of an accident. For a long time the main criterion by evaluating PD is used "apparent charge". This criterion still prevails in the practice of assessing the performance of the HVE insulation. Let us turn to the international standard IEC 60270:2000 [1] and the corresponding National standards (e.g. [2, 3]). They declare the general principles of partial discharge measurements in high-voltage tests and "apparent charge" is considered as the main normative parameter.

The apparent charge model creates the illusion of technical simplicity and correctness, based on an equivalent circuit with a pair (or three) capacitors. It actually relies on the assumption of the homogeneity of the electric field, in which there is a small inhomogeneity (bubble). Naturally, the location i.e. the coordinates of this inhomogeneity located in a homogeneous electric field does not affect the external manifestations of the PD arising in it.

We can say that diagnostic methods based on the "apparent charge" in simple cases meet the needs of practice. However, experts note that the interpretation of the results of the PD observation is still not sufficiently clear. The problem of the PD multiplicity becomes be increasingly the focus of attention and its solution perspectives can be associated with high-precision high-frequency measurements [4-16]. These problems can not be solved you within the framework of the paradigm of apparent charge.

II. PD Set and PD Sources Ratio

Thus, the purpose of the PD study is diagnosis or to assess the technical condition (health index) of the equipment on-line and on-site. In other words, the task is to search for PD sources which are potential defects and to assess their danger from the point of view of further operation of the equipment. Especially note that the expression "PD source" (or its equivalent) does not occur in the text of IEC. There are expressions: "source of stress" "sources of error", "sources of interference", but not "PD source". The concept of "PD" and "defect" is in some sense synonymous when it comes to a single PD [1,2]. When (potential) defects are not single, their locations are different and these concepts must be clearly distinguished.

The image of the PD integral pattern (IP) represents usually by a three-dimensional function of the set of PD pulses in the coordinates 'time'-'apparent charge'-'pulse repetition rate' or two-dimensional color image.

Different interpretations of this picture are used to obtain some meaningful assumptions about the PD sources or quantitative assessments of the hazards of these phenomena. First, let us the main criterion of danger is the maximum impulse among all PD. Here it is implicitly assumed that there are many PD sources, but they are all insignificant in comparison with the "greatest". Second, if the whole set of apparent charges is integrated, it is implicitly assumed that all of PD are generated by a single defect. And what version actually takes place?

The problem can be summarized in the following couple of questions. How many PD pulses are there? And how many PD sources this integral pattern reflects? For example, let suppose series of (approximately) identical PD is revealed during the operating voltage cycle. The question arises whether all of these PD correspond to the set of small inclusions scattered in the insulation volume or they occur in a single developed local inclusion. In both cases, the apparent charges may be equivalent, but if all PD occur in a single defect, the development of discharge processes is much more dangerous and the probability of equipment failure is much higher. The solution of such simple problem is the key to the problem of analyzing the complex PD integral pattern.

However, the question of the number of real defects in the insulation and the danger of each of them is still not reflected in the mentioned Standards.

III. Parameters of the Certain PD Series

The transition from the classical model with a single apparent charge to the PD set requires a complex formalization of the process. Let us consider supervision of PD relaxation process in a single defect for one circle (Fig.1).

We introduce into consideration the voltage at the defect $v(t)$, PD current I PD(t), and some parameters. There are forced component of the voltage $v_f(t)=V_m\sin\omega t$, the PD ignition voltage V_b, the PD extinction voltage V_e, and the initial condition of the process $V0$. We consider the PD process to be periodic [17, 18]. PD current is a series of exponential pulses uniquely associated with jumps of the relaxation voltage $v(t)$. The amplitude of the pulse "...it is impossible to measure directly" [1], p.3.1.1). But, there is a real technical possibility to accurately measure the instant set $T=(t_1, t_2,...)$ of PD pulses occurrence, tying them to the phase of the operating voltage. The totality of these instants carries the most reliable information about the whole process. The analysis shows a unique relationship (accurate to a constant factor) between the set of "energy" parameters (V_m, V_b, V_e) with the initial condition V_0, and the instant set T of the PD series (Fig.1). On the other hand, the energy parameters depend on the set of the defect electro physical parameters. There are the size of the inclusion and its shape, geometric coordinates, the structure of the operation field at the neighborhoods given point in the high-voltage device active space, the dielectric properties of the main insulation and inclusion, the polarization parameters of the insulation, and the properties of the PD volt-ampere characteristic.

An analysis of the considered PD model in a separate defect allows us to draw the following conclusions. The triple of values of the energy parameters (V_m, V_b, V_e) with the initial condition V_0 can be considered as generalized coordinates of the separate defect, individualize the defect and can serve as a basis for discrimination of the PD sources.

Without loss of generality, taking as the basic value of the ignition voltage V_b, it is possible to go to the triple independent normalized energy parameters

$$\left(V_m^*, V_e^*, V_0^*\right) = \left(V_m, V_e, V_0\right)/V_b.$$

(For brevity, then the upper asterisk index will be omitted).

It is easy to evaluate number of the PD in the series for the first half-cycle of the operating voltage:

$$n_{0.5} \approx (V_m - V_b)/(V_m - V_e). \tag{1}$$

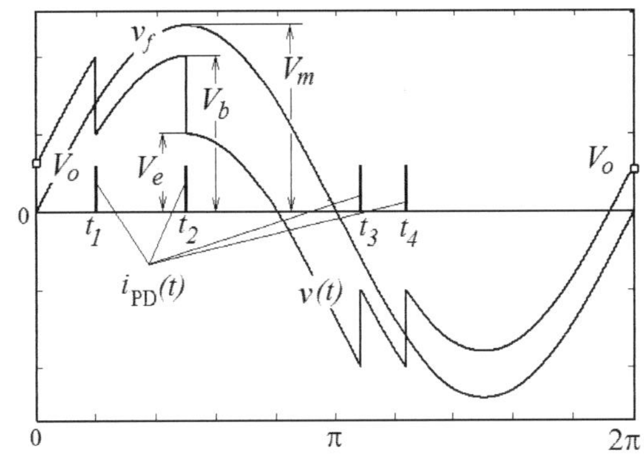

Figure 1. The designation of the parameters of the PD relaxation process

Let illustrate the effect of energy parameters variation on the appearance of the PD series by conducting mathematical modelling for specific numerical parameters. The simulation results are summarized on the Fig. 2. The simulation was performed on a cycle of the operating voltage with a discreteness of 10^4 points per cycle. For the parameter values (according to Fig.1)

$$(V_m, V_e, V_0) = (1.2, 0.5, 0.3),$$

length of series is 4 PD; this option is taken as the baseline. Let increase V_m of forced voltage to 1.5:

$$(V_m, V_e, V_0) = (1.5, 0.5, 0.8),$$

and length of series increases here to 6. (We don't pay attention to the initial condition yet).Increasing the extinction voltage to $V_m = 0.8$:

$$(V_m, V_e, V_0) = (1.2, 0.8, 0.3)$$

also causes to an increase length of series to 6. Let us turn to the initial condition influence of the process. Let change the initial condition compared to the baseline option:

$$(V_m, V_e, V_0) = (1.2, 0.5, 0.5).$$

It causes to a nonlinear shift of the entire series of PD in time. Here we can't be made yet additional definite conclusions about influence of the initial condition to PD formation. It value is random, but it can not be neglected. Finally, if at the same time to increase the amplitude of the forced voltage and the extinction voltage it causes to a multiplicative increase the PD series length:

$$(V_m, V_e, V_0) = (1.5, 0.8, 0.7).$$

Let us estimate maximal length of the PD series. If assume that the maximum electrical strength in the oil equal to E_o=60 kV/cm, the breakdown electrical strength in the gas bubble equal to E_b=30 kV/cm, the maximum forced voltage in the gas turns out in gas bubble V_m>2V_b,and in accordance with [17] the ratio between the PD extinction voltage and the PD ignition is in the range of $V_e/V_b \approx$ (0.1...0.9).Thus, in accordance with (1), the length of the PD series for the half cycle of the operating voltage approximately is $n_{0.5} \approx 10$.

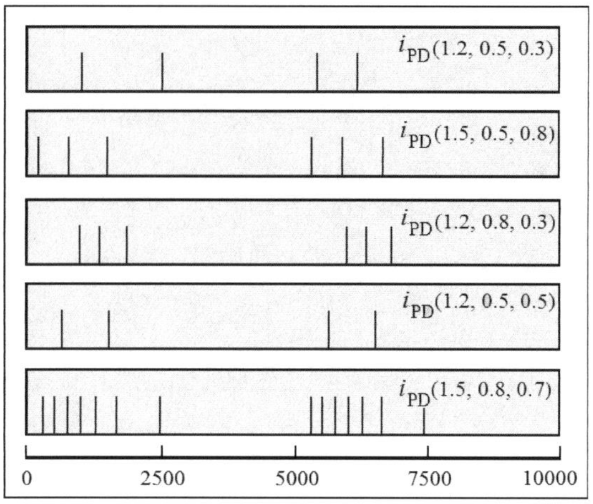

Figure 2. Simulation of PD series i_{PD} (V_m, V_e, V_0) for different values of energy parameters

IV. PD Integral Pattern as a Join of the Certain Series

The set of observed PD is caused both by the multiplicity of defects and by the possibility of multiple repetition of pulses in the form of a series of PD in each of them.

The general view of the set (integral pattern) of the registered PD appears as the sum of the PD from the whole set of sources as the integral pattern IP:

$$i_{\mathrm{IP}} = \sum_{k=1}^{N} h_k i_k, \ \left(j = 1, ..., n_k\right) \tag{2}$$

where the coefficients h_k characterize the attenuation of the signals from the pulses of the PD on the path of propagation of a defect monitoring system.

Each single j-th PD in the k-series of inclusion occurs at the time t_{kj} and creates an exponential current pulse with amplitude I_k, continuing for a time t_{PD}. PD current of the k-th defect is the sum of a series of n_k pulses:

$$i_k = \sum_{j=1}^{n_k} i_{kj}(t) = |I_k| \sum_{j=1}^{n_k} \exp\left(\left(t_{kj} - t\right)/\tau\right)\left[\mathbf{1}\left(t - t_k\right) - \mathbf{1}\left(t - t_{kj} - t_b\right)\right] \tag{3}$$

where are:

n_k– the number of pulses in the k-series;

τ– the discharge time constant;

$\mathbf{1}(t)$ is a function of a single jump;

the distribution of the instants of individual pulses of the PD is a unique series of pulses

$$\mathbf{T}_k = (t_{k1}, t_{k2}, ...), \ k = (1,..., N). \tag{4}$$

As shown above, the defect is uniquely individualized when it is associated with a set of parameters (V_{mk}, V_{ek}, V_{0k}). All of the k-th defects make a contribution to integral PD pattern; it can be illustrated in the form shown on the Fig. 3.

When interpreting the PD ID, the inverse problem arises: to distinguish the correspondence of a subset of the PD (of the full set of recorded PD) to individual sources.

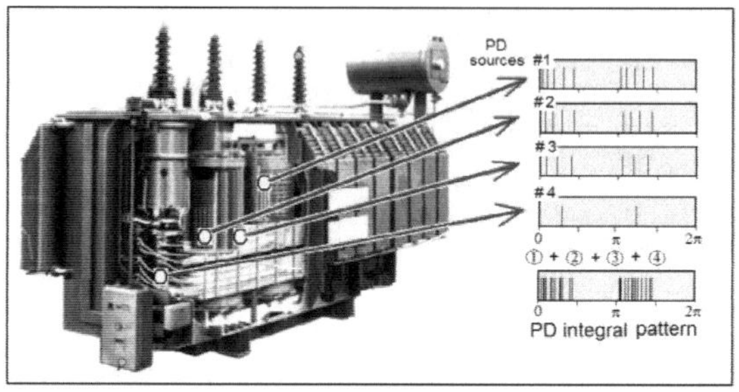

Figure 3.Formation of the PD IP

So, for the j-th defect, the triple of the parameters

$$\mathbf{V}_j = (V_{jm}, V_{je}, V_{j0})$$

uniquely determines the series of PD instants as a $\mathbf{T}_j = (t_{1j}, t_{2j}, ...)$. If the set $\mathbf{T}_j$ is caused by $\mathbf{V}_j$, let be named set $\mathbf{T}_j$ as a Right set (or Right instant series).

And if the set $\mathbf{T}_j$ is caused by $\mathbf{V}_j$, $\mathbf{T}_j$ and $\mathbf{V}_j$ are corresponded one-to-one each other, it necessary to pay attention, not every random set of time points $\mathbf{T}_x = (t_{1x}, t_{2x}, ...)$ can be a Right instant series. It is easy to prove this fact. For example, let's like Fig. 2 operation period contains 10^4 points, and observed PD sequence contains of 3 instants at the first quarter of period:

$$\mathbf{T}_x = (1250, 1251, 1252).$$

It's enough to examine PD series $\mathbf{T}_x$ and to make sure to impossibility to create the function $v(t)$ like the Fig. 1.

V. Possibility of the PD Integral Pattern Decomposition

As it shown above the IP is the set of N_{IP} PD current pulses initiated from N sources,

$$N_{IP} = \sum_{k=1}^{N} n_k$$

and the problem is to decompose the set of N_{IP} instants to N fragments $\mathbf{T}_k = (t_{k1}, t_{k2}, ...)$ associated with separate defects. So, we ask ourselves: how many sources of PD reflect this integral picture? Answering this question, we (in any case-theoretically) can as a solution to the problem arrange a full search of points (with suitable for accuracy discreteness) in the three-dimensional space of relative power parameters (V_m, V_e, V_0) and get a discrete set of all Right sequences $\mathbf{T}_s$.

Formally to solve this problem it's simple. It is need to enumerate all of the N_{IP} PD pulses subsets and to examine every of them for "Right or Not-Right". All of Right subsets are present independent PD source. But this way is non-constructive, since a number of all subsets of the set of N_{IP} elements is $N_{IP}!$ (N_{IP} factorial), it is excessively large.

The description of the PD process in analytical form and the formalized algorithm are given in [19]. We do not stop at solving the whole problem and let us use simple option. We pay attention here, the simplest solution is to select pulses by amplitude. However, since the attenuation coefficients of h_k are unknown, then (especially in the presence of noise) the amplitudes of the PD current pulses from individual sources are weakly informative. More hard is the analysis of the IP by pulses with approximately equal amplitudes.

The simplest approach is to prepare a database of PD series (Right sets of time points). Let's see what the practical possibilities for such a search are. Let the interval of values of each parameter (V_m, V_e, V_0) be represented by 100 points, so there are 10^6 points in this three-dimensional space.

It's easy to compile a database of a million records, and it allows solving the problem of correspondence or discrepancy of an arbitrary set of moments of time of the PD to some potential defect.

VI. Computing Example

We illustrate the analysis of the integral picture of the PD obtained as a result of computer simulation of the process. There was generated the set of the 28 PD, which are caused from several defects. Results of the mathematical modeling are based on PD quasideterministic model [17]. The accuracy of the calculations along the time axis was 10^4 points for the period of the operating voltage; below we will operate with numbers of instants as a time points. For simplicity when modeling it is assumed that the value of the relative extinction voltage is given (V_e=0.75).

The PD integral pattern as an instant sequence has an appearance of the time points of the PD pulses number list:

IP={462, 629, 769, 834, 979, 1112, 1253, 1297, 1397, 1535, 1846, 2019, 2058, 2500, 5403, 5629, 5705, 5912, 5979, 6039, 6098, 6350, 6397, 6439, 6660, 7039, 7058, 7074}.

At the Fig.4 is shown the set of pulses and $v_{op}(t)$.Let us assume that a priori the ratio of ignition and discharge extinction voltages is set (or experimentally obtained):

$$V_e/V_b = 0.75.$$

For the decomposition this example series we create an auxiliary data base for V_m=1.1, 1.2, 1.3, 1.4, 1.5;V_b=1,V_e=0.75, V_0= 0.1, 0.2, …, 0.9. (The fragment of the database presented as the Table 1). Any PD sequence is declared as a Right sequence if it corresponds to any string in the data base.

All of the PD strings one by one are inspected to presence all of the string elements in the IP set. First of suitable string corresponds to pair of defect parameters (V_{1m},V_{10}) = (1.1, 0.2). It is Right sequence:

T$_1$= (1297, 2019, 6098, 6660).

And the search process was repeated. (For shortness, only strings which are neighbourhood with the Right sequences we show in the Table 1).So, we have more 3 Right sequences:

T$_2$= (834, 1253, 1846, 5912, 6350, 7039)
T$_3$= (629, 979, 1397, 2058, 5629, 5979, 6397, 7058)
T$_4$=(462,769, 1112, 1535, 2500, 5403, 5705, 6039, 6439 7074);

(V_{2m}, V_{20})=(1.2 0.4), (V_{3m}, V_{30})=(1.3 0.5), (V_{4m}, V_{40})=(1.4 0.6).

As a result of PD decomposition we divide all of 28 PD to 4 series which associate with different defects. The result of decomposition at the first quarter of period is shown in the Fig. 5. The each of defects is characterized by individual pair parameters (V_m, V_0).Note that the values of the initial conditions V_{10},...,V_{40} are random parameters and they are important to the solution of the problem and parameter individualization. So, 4 sources of PD are revealed, and they are located in the different places of the active device volume with different electric field intensity.

Summing up the currents of the PD pulses of individual sources, and guided by a quadratic assessment, it is possible to assess the dangers of the identified PD sources as:

$$n_1^2:n_2^2:n_3^2:n_4^2=4^2: 6^2: 8^2: 10^2 = 16: 36: 64: 100.$$

It is obvious that the 4-th defect has the greatest danger.

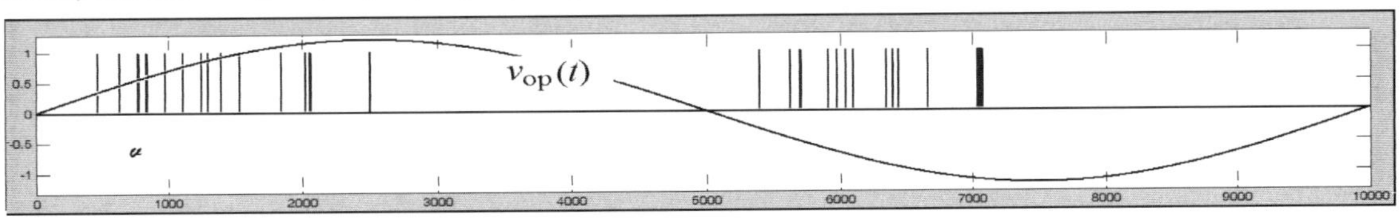

Figure 4. Example of illustration PD set on the operating voltage cycle

Table 1. Fragment of the table of Right PD sequences

V_m	V_0	Right PD sequences (marked strings correspond to example answer)									
1.1	0.1	1526	6406	7500							
1.1	0.2	1297	2019	6098	6660						
1.1	0.3	1098	1660	6297	7019						
		…									
1.2	0.3	992	1454	2500	5758	6162	6696				
1.2	0.4	834	1253	1846	5912	6350	7039				
1.2	0.5	685	1075	1568	6075	6568					
		…									
1.3	0.4	764	1135	1606	5834	6218	6728				
1.3	0.5	629	979	1397	2058	5629	5979	6397	7058		
1.3	0.6	498	834	1218	1728	5764	6135	6606			
		…									
1.4	0.5	582	900	1267	1757	5582	5900	6267	6757		
1.4	0.6	462	769	1112	1535	2500	5403	5705	6039	6439	7074
1.4	0.7	344	643	969	1350	1895	5521	5834	6188	6639	

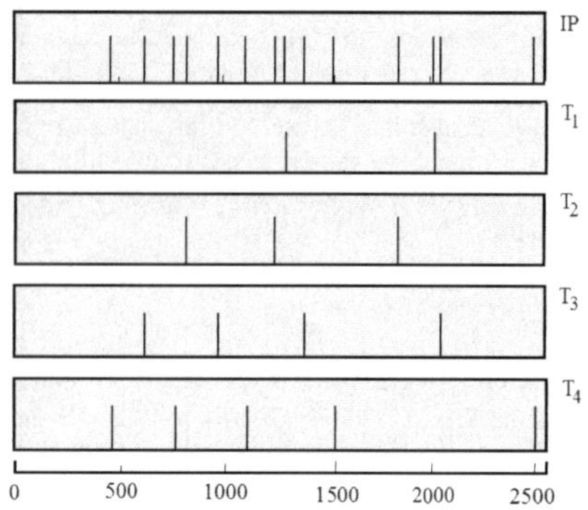

Figure 5.The result of IP defragmentation at the first quarter of operation voltage period

VII. CONCLUSION

The article shows the methodological principles of the analysis of the integral picture of the PD, when there is reason to believe that the set of registered PD is generated by a variety of sources – potential defects. It is postulated, the application of the idea of apparent charge to the case of multiple charges is unproductive. It is introduced the idea of Right PD sequences which belongs to individual defect as a PD source. An identification of these sources makes it possible to assess with greater certainty the real danger of the many registered PDs reflected in their integral picture of the PD.

ACKNOWLEDGMENT

This work was partly supported by the Russian Foundation of Basic Research [grant number 17-08-00693].

REFERENCES

[1] International IEC Standard 60270 – 2000.

[2] IEEE-Std.-C57.104-2008-IEEE-Guide-for-the-Interpretation-of-Gases-Generated-in-Oil-Immersed-Transformers.

[3] IEEE Std. 400.3-2006 IEEE Guide for Partial Discharge Testing of Shielded Power Cable Systems in a Field Environment

[4] S. D. Mitchell, Siegel, M., M. Beltle, and S. Tenbohlen, "Discrimination of Partial Discharge Sources in the UHF Domain", IEEE Trans. on Dielectrics and Electrical Insulation, vol. 23,no. 2, pp.1068-1075, April 2016, doi: 10.1109/TDEI.2015.005015

[5] I. Kolcunova, M. Sipos, "Measurement of Partial Discharge Picture in disc-shaped voids", Proc. of the 8th International Scientific Symposium on Electrical Power Engineering (ELEKTROENERGETIKA 2015), Stará Lesná, Slovakia, 2015,pp. 322-325.

[6] N. H. Nik Ali, P. Rapisarda, P. L. Lewin, "Separation of Multiple Partial Discharge Sources Within a High Voltage Transformer Winding using Time Frequency Sparsity Roughness Mapping", Proc. of the 2016 Int. Conf. on Condition Monitoring and Diagnosis (CMD2016), Xi'an, China, 2016,pp.225-228.

[7] N. H. NikAli, W. Goldsmith, J. A. Hunter, and P. L. Lewin, "Comparison of Clustering Techniques of Multiple Transformer Windings", Proc. of the IEEE 11th Int. Conf. the Properties Appl. Dielectr. Mater., 2015, pp.256–259.

[8] S. Grzybowski, "Optimal features selected by NSGAII for partial discharge pulses separation based on time-frequency representation and matrix decomposition", IEEE Trans. on Dielectrics and Electrical Insulation, vol. 20, no. 3, pp. 825–838, Jun. 2013.

[9] N. H. Nik Ali, J. A. Hunter, P. Rapisarda, and P. L. Lewin, "Identification of Multiple Partial Discharge Sources in High Voltage Transformer Windings", Proc. of the IEEE Conf. Electr. Insul. Dielectr. Phenom., 2014, pp. 188–191.

[10] J.C. Chan, H. Ma, T.K. Saha, and C. Ekanayake, "Self-adaptive partial discharge signal de-noising based on ensemble empirical mode decomposition and automatic morphological thresholding", IEEE Trans. on Dielectrics and Electrical Insulation, vol. 21, no. 1, pp. 294–303,Feb. 2014.

[11] N. H. Nik Ali, M. S. AbdRahman, J. A. Hunter, P. Rapisarda, and P. L. Lewin, "Wavelet and Mathematical Morphology as the de-noising methods for PD analysis of high voltage transformer windings", Proc. of the IEEE Electr. Insul. Conf., 2015, pp. 214–217.

[12] J. C. Chan, H. Ma, and T. K. Saha, "Time-frequency sparsity map on automatic partial discharge sources separation for power transformer condition assessment", IEEE Trans. on Dielectrics and Electrical Insulation, vol. 22, no. 4, pp. 2271–2283, 2015.

[13] Xi Li, Xiaohua Wang, Mingzhe Rong, Dingli Xie, Yuwei Fu, Qingqing Gao, Nan Yin, "Relationship Between Time-frequency Representation of PD-induced UHF Signal and PD Current Pulse", Proc. of the 2016 Int. Conf. on Condition Monitoring and Diagnosis (CMD2016), Xi'an, China, 2016, pp.229-232.

[14] C. Suwanasri, T. Sangpakdeejit, N. Vipulum, P. Fuangpian, S. Ruankon and T. Suwanasri, "Investigation on Partial Discharge Inception Voltage and Discharge Pattern of Simulated Defect Cable System", Proc. of the 2016 Int. Conf. on Condition Monitoring and Diagnosis (CMD2016), Xi'an, China, 2016, pp.233-236.

[15] WenboFan, Shilei Guan, Jinwei Fu, Lisheng Li, Jianxiu Li,Wei Wang, Xiaoyu Yan,"Comparison Study of Partial Discharge Detection Methods for Switchgears", Proc.of the 2016 Int. Conf. on Condition Monitoring and Diagnosis (CMD2016), Xi'an, China, 2016, pp. 318-322.

[16] G. Li, M. Rong, X. Wang, X. Li, and Y. Li, "Partial dischargepatterns recognition with deep convolutional neural networks," Proc. of the 2016 Int. Conf. on ConditionMonitoring and Diagnosis(CMD2016), Xi'an, China, 2016, pp. 323-326.

[17] N. V. Kinsht and N. N. Petrunko, "Some possibility of constructing of the PD processes quasi-deterministic model," Proc. of the 2014 Int. Conf. on Condition Monitoring and Diagnosis(CMD2014), Jeju, Korea, 2014, pp. 371-374.

[18] N. V. Kinsht, and N. N. Petrunko, "Regularities of the formation an integrated partial discharges picture as a set of quasideterministic processes Proc. of the 8th International Scientific Symposium on Electrical Power Engineering (ELEKTROENERGETIKA 2015), Stará Lesná, Slovakia, 2015, pp. 315-318.

[19] N. V. Kinsht and N. N. Petrunko, Method of control of technical condition of high-voltage equipment elements, RU 2604578, 2016

Diagnostics of High Voltage Electrical Equipment in Power Engineering

Alexander Yu. Khrennikov, PhD of El. Eng.
Moscow Energy Institute (Scientific Research University)
Moscow, Russia
ak2390@inbox.ru

Nikolay M. Aleksandrov,
SPE "Dynamics"
Cheboksary, Russia
nickdynamics@gmail.com

Abstract - **paper presents an application's experience of LVI and SFRA testing to check the condition of transformer windings, infra-red control (IR-control) results of electrical equipment, inspections of the SF6 insulation current transformers by thermal-vision control, monitoring of the electric activity of the partial discharges (PD) into the insulation, the gas pressure in the housing, data of electrical tests, the results of analysis of the quality of SF6 by different methods. IR-control was used for detection of overheating of disconnector contacts, bushing condition of power transformers, condition of circuit breakers, arresters and limiters of overvoltages, coupling transformers, measuring current transformers, measuring voltage transformers and other. Methods and equipment for testing XLPE cables by voltage on the ULF (0.1-0.01 Hz.) are also considered in the article.**

Keywords: IR-control; low voltage impulse method; sweep frequency response analysis; short-circuit inductive reactance measurement; SF6 insulation; quality of SF6; Partial discharges (PD); digital registrar-analyzer; XLPE cables; ULF for testing.

I. INTRODUCTION

A large part of the electrical equipment of electrical stations, substations and power transmission and distribution systems has developed a service resource, but continues to operate, because its replacement requires more funding. Therefore, every year the increasing cost to undertake comprehensive surveys and diagnostics. Assessment of the actual status of power electric equipment based on diagnostic measurements is very difficult and pressing challenge.

Infra-red (thermal) control is one of the main directions of development of the system of technical diagnostics, which provides the ability to control the thermal state of electrical equipment and substation's buildings without taking them out of work, identify defects at an early stage of development, reducing the cost of technical surveys by predicting the timing and extent of the repairs [1, 2, 3].

Electrical equipment, in particular power transformers, is one of the basic parts in the circuitry of power transmission and delivery. Therefore the interest to perfection of electrical equipments' fault diagnostic methods is being increased. The repairs of power transformers and other electrical equipment are carried on, using diagnostic measurement results [1, 2, 3, 5-7].

LVI-testing and short-circuit inductive reactance measurements are sensitive for detecting such transformers winding faults as buckling, axial shift and other. The 70 units of 25-240 MVA 110-500 kV power transformers were checked by low voltage impulse (LVI) method. A few power transformers were detected with winding deformations after short-circuit with aperiodical short-circuit current.

Also this paper presents inspection of insulation current transformers by infra-red (thermal-vision) control, measuring the activity of the partial discharges (PD) into the insulation of transformers, data of electrical tests, the unreliable fastening of protective hoods on the diaphragm devices, the defects of installation.

II. INFRA-RED AND ULTRAVIOLET CONTROL RESULTS.

Examples of efficient application of infra-red control for electrical equipment faults diagnostic are in the fig. 1.

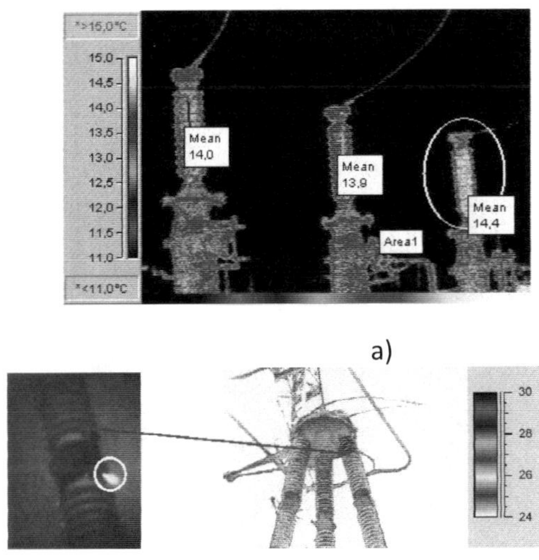

Figure 1. An examples of efficient application of infra-red control for electrical equipment faults diagnostic : a) - decrease of insulation resistance of 110 kV limiter of overvoltage to 300 MΩ, moistening and the entry of moisture inside limiter of overvoltage, overheating (ΔT=0,5°C); b) - ultraviolet control, glow of surface PD in the cracks of stand-off insulators (defective insulator in the 500 kV disconnector with the worsened electrical properties of porcelain).

III. LVI-TESTING OF TRANSFORMER'S WINDING AND FRA (SFRA) METHOD.

The LVI method is very sensitive to small local changes of winding geometry: turn-to-turn and coil-to-coil capacitances, mutual inductances between transformer windings. The LVI oscillograms, which contains basic resonance frequencies of transformer winding, are a "fingerprint" or condition state of transformer. Generally, windings of large power transformers have three basic resonance frequecies. Frequency Response Analysis (FRA) showed presence of 110 kHz, 320 kHz and 550 kHz frequencies for 250 MVA /220 kV transformer. An amplitude of these resonance frequencies changed 1,3-2 times after detection of radial buckling in LV winding at short-circuit tests (fig.2) [5-7].

Figure 2. Typical example of deformation due to radial buckling in the C phase LV internal winding of 250 MVA /220 kV transformer (Xk= +3,1%).

Inductive reactance deviation was Xk= +3,1% in this case.

Radial buckling in MV 220 kV winding of 167 MVA/500 kV/220 kV autotransformer after three short-circuits in service is in the fig.3.

Figure 3. An example of radial buckling in MV 220 kV winding of 167 MVA/500 kV/220 kV autotransformer after three short-circuits at "Bugulma-500 kV" substation in service.

Diagnostics experience was showed that the transformer plant documentation should be include: normograms of LVI-testing, data bases of partial discharge technique, normograms of infra-red diagnostics (for first models) during heat testing, data bases of winding pressure by vibration measurements of transformer. Short-circuit reactance measurements (Zk) and LVI-testing should be carry out together. The necessity of LVI-normograms of all new manufactured transformer with capacity over 2,5 MVA, which are produced at the transformer manufacturers, is came. It's necessity for data bases of the mechanical winding conditions for the future LVI-testing at the energy system after probable short-circuit.

The analysis of mechanical state of power transformer's windings is based on comparison of normogramms with defectogramms. For easier interpretation of results oscillogramms are usually transformed from time scale to frequency scale with the use of Fourier transformation. As a result it becomes possible to provide further analysis to define the actual internal defect of transformer. The drawback of such approach is a low sensitivity in the area of high frequencies. So the Sweep Frequency Response Analysis method was presented for mechanical condition diagnosis of power transformers winding. It is based on measuring of attenuation of a signal on different frequencies. At each step of frequency the signal's value stays constant ant it allows to get the same sensitivity for the whole frequency range.

Recent researches proved that changes in different parts of frequency range can indicate the fault in some transformer's part. Table 1 indicates frequency ranges to interpret the condition from the SFRA results.

Table 1. Frequency ranges used in SFRA results interpretation

Frequency Ranges	**Elements**
Below to 20 kHz	transformer core and magnetic circuits
20 kHz to 400 kHz	radial relative geometrical movements between winding
Above 400 kHz	deformations of each single winding

The special device RETOM-SFRA was developed for frequency response measurements. It has generator with 2.88 Vp-p, 10 Hz - 2.5 MHz frequency range, 2 measurement inputs. The software for this device can simultaneously display 3 responses and calculate cross-correlation factor (CCF) (1) for all 3 responses in 3 different frequency ranges.

$$CCF = \frac{\sum_{i=1}^{n}(X_i - \bar{X})\cdot(Y_i - \bar{Y})}{\sqrt{\sum_{i=1}^{n}(X_i - \bar{X})^2}\cdot\sqrt{\sum_{i=1}^{n}(Y_i - \bar{Y})^2}} \quad (1)$$

Recommended boundary values for frequency response's CCF are given in table 2.

Table 2. CCF

	CCF
Good match	0.95-1
Close match	0.9-0.94
Poor match	<0.89
No match	≤ 0

Figure 4 presents phase-to-phase frequency responses for 10 MVA 35 kV power transformer. The results of CCF calculation indicate radial relative geometrical movements between winding especially in phase B.

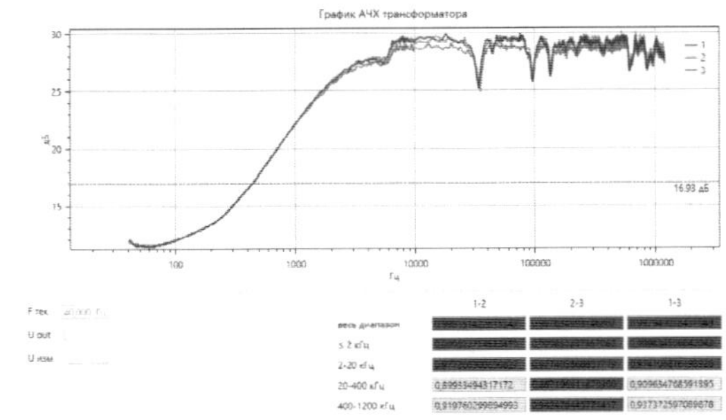

Figure 4. Frequency responses for 10 MVA 35 kV power transformer

IV. DIAGNOSTICS METHODS OF INSTRUMENT TRANSFORMERS WITH SF6 INSULATION

In fig. 5 are given the basic methods of complex diagnostics of instrument transformers with SF6 insulation [3]. These are thermal-vision inspection, the measurement of the acoustic activity of the partial discharges (PD) into the insulation and the analysis of the quality of SF6 by different methods [3, 7].

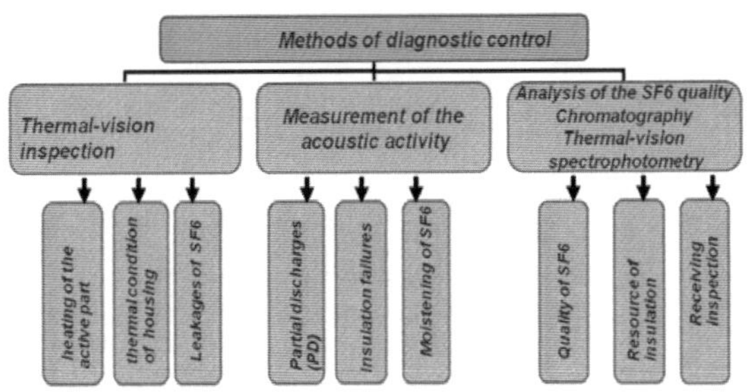

Figure 5. Basic methods of complex diagnostics of instrument transformers with SF_6 insulation.

Thermal-vision inspection is very effective for predictive diagnostic for SF6 insulation current and voltage transformers, and it is achieved in accordance with national standard RD 153-34.0-20.363-99 "The basic condition of the procedure of infrared diagnostics of electrical equipment and transmission airlines" (Fig. 6).

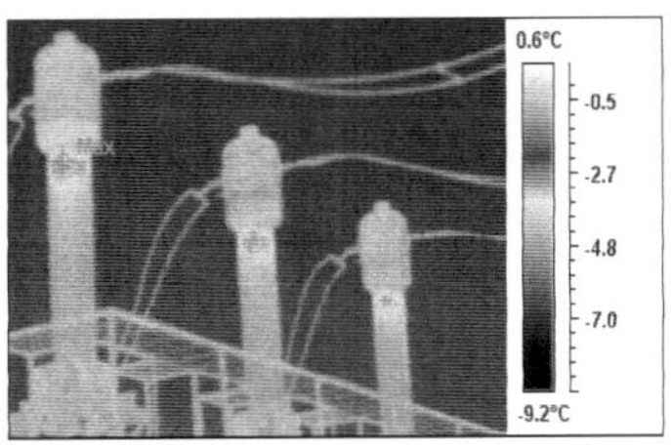

Figure 6. Heat-gram of SF6 insulation current transformer (T=2,1°C).

An electrical activity of the partial discharges (PD) into the insulation of 80 MVA/110 kV power transformer are in the fig. 7. Color shows intensity PD per second (from the dark-blue to the red).

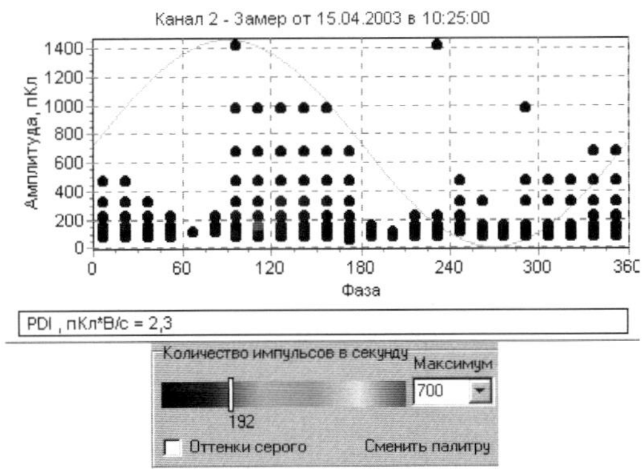

Figure 7. An electrical activity of the partial discharges (PD) into the insulation of 80 MVA/110 kV power transformer.

Normative documentation during PD measurements are National standard GOST-20074-83 "Method of measuring the PD characteristics and IEC-60270 "The measurements of PD characteristics".

Compositional analysis of SF_6 is carried out by spectrophotometry by portable dissolved multi-gas analyser TRANSPORT X.

Analysis of a transformer oil sample for dissolved gases by a laboratory is an established technique recognized as the most important test for monitoring power transformers.

Normative documents and national standard is RD-16.066-05 "Sulfur hexafluoride electrotechnical equipment technical requirements for the production" and IEC 60480-2004.

V. Partial Discharges (PD) for the Insulation Monitoring. Breakdown of the Turn-to-turn Insulation of the Power Transformers as a Result of the Action of Partial discharges (PD)

The internal short-circuits of transformer windings can be caused by the breakdown of the turn-to-turn insulation as a result of the destruction of insulation under the action of operational factors and by the action of the partial discharges in the place of future breakdown, as initiator of which can serve commutation, thunderstorm and other increased actions on the transformer insulation. The tracks of partial discharges during the insulating construction are represented in Fig. 8.

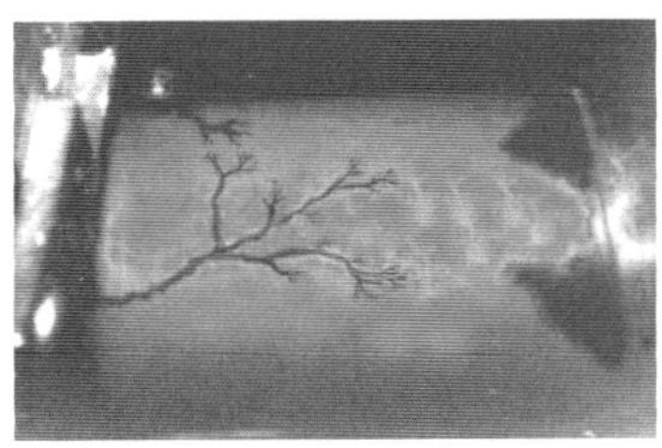

Fig. 8. The tracks of partial discharges during the insulating construction.

Thus, for instance, on the substation 500 kV of Far-Eastern region was switching off by the action of gas protection of the 167 MVA/500/220 kV autotransformer. It was established with the chromatographic analysis by dissolved gas analysis (DGA) of transformer oil that the concentration of the dissolved gases exceeds boundary values several times. The damage of the 4th turns of regulating winding was revealed with the internal inspection of autotransformer windings.

All results of the traditional preventive insulation routine tests of autotransformer according to National testing Standard, the preceded moment of damages, gave conclusion about satisfactory condition state of insulation.

The specialists of the Novosibirsk specialized testing laboratory carried out the unplanned measurements of PD intensity on the autotransformer. Although the PD measurements on the autotransformer were carried out not fully, only on the 220 kV winding side as a result of turning off of sensors along the side 500 kV, that it did not make it possible to accurately localize defect, also in this case were discovered discharge phenomena, were given recommendations about conducting of revision of tap-changer. Revision was not carried out.

The results of the measurements of PD intensity on the autotransformer before the damage are given below [by 1-4].

VI. Procedure of PD Measurements and Measuring Means.

The determination of PD intensity was produced with the aid of the electrical method of PD registration. As the measuring means was used digital oscillograph TDS-2204 "Tektronix", and also the eight-channel digital registrar-analyzer of the partial discharges «R2000» of production "Vibro-center", Perm' city. The diagram of PD registration in each phase AT is given to fig. 9.

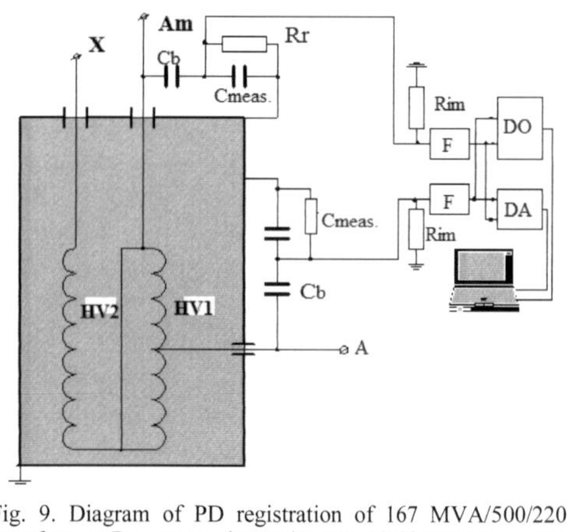

Fig. 9. Diagram of PD registration of 167 MVA/500/220 kV autotransformer: Rr – measuring resistance, 50 Ohm; Rim – matched impedance, 50 Ohm; F – the filter of high frequency; Cb – the capacity of 500 and 220 kV bushings; Cmeas. – the capacity of the measuring facing of 500 and 220 kV bushings; DO – digital oscillograph; DA – digital analyzer.

It was used as the sensors the stationary devices of connection of the type UP-6. It was impossible to use 500 kV sensors, since continued the process of the agreement of the joint operation of the device of the connection UP-6 with KIV-500, and 500 kV sensors were switching off.

The procedure of PD intensity measurement with the aid of the digital oscillograph (DO) consisted of the following: the measured signals from the sensors of different phases were investigated. It was produced the filtration of signals by series connection into the chain of high-pass filters in this case, the agreement of the wave drag of cable with the load, changes in polarity and level of the starting of development, scales along the axes of oscillograms and etc.

PD selection was produced on the specific criteria, which it is discussed below. With the use of a digital analyzer was produced the more thorough selection of PD pulses from the noise signals. Recorded in the digital signal aspect underwent the subsequent the processing by special software for determining the PD fundamental characteristics.

VII. Results of Partial Discharge Measurements.

The signals from the 220 kV sensors during several periods 50 Hz are represented in fig. 10.

Figure 3 depicts characteristic PD oscillograms, obtained from the measurement sensors of the 220 kV bushings of all autotransformer phases. PD distinguishing features are the steep front of signals in this case, the absence "of the prepulse" of another polarity, which characteristic for the electrostatic influence of charge of streamer corona, and, furthermore, the symmetry of PD signals on both half-waves of 50 Hz voltage, which indicates the appearance of discharge processes in the region of uniform field.

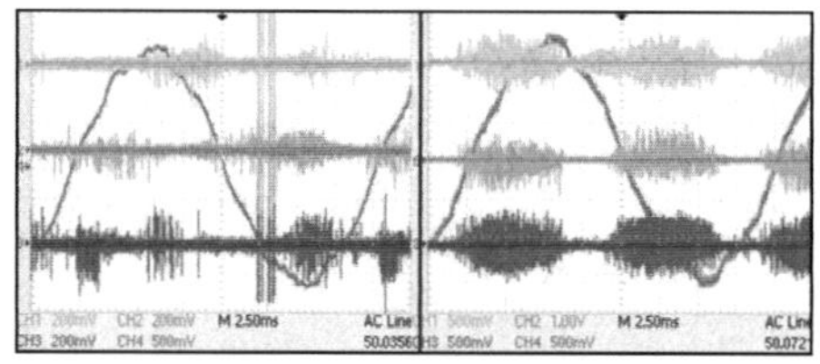

Fig. 10. The common form of the distribution of PD signals on the autotransformer phases (side 220 kV) during several periods 50 Hz: 1 – signal from measurement bushing of 220 kV "A" phase; 2 – signal from measurement bushing of 220 kV "B" phase; 3 – signal from measurement bushing of 220 kV "C" phase; 4 - 50 Hz signal from measurement bushing of 220 kV "A" phase.

As can be seen from obtained PD oscillograms, in the transformer insulation of phase "B" is registered the partial discharges, which correspond to the level of amplitude in 10 nC. The form of amplitude-phase diagrams (AFD) makes it possible to assume by possible PD center in tap-changer of phase "B" [by 4-6].

According to PD Standards the maximum permissible apparent charge of partial discharges with the manufacturer tests of transformers with 150-500 kV voltage is defined as 0,3 by nC with the greatest operating voltage. For the transformers 110 kV and above for the operating conditions and aged equipment maximum magnitudes of the PD charge it is possible to accept value 3-5 nC (GOST-20074-83 "Method of measuring the PD characteristics and IEC-60270 "The measurements of PD characteristics").

Therefore as a whole according to the results of inspection it was possible to estimate the state of main autotransformer insulation and bushing insulation as satisfactory. Were at the same time discovered discharge phenomena in the tap-changer device of phase "B". Therefore it was recommended to estimate the state of oil in tap-changer and, during exceeding of Standards to gas- and moisture content, the increased tangent (tan δ) for 700 and 900 temperatures, high acid number and to conduct revision of tap-changer into the spring-summer season with the replacement of transformer oil.

As a result the internal inspection of autotransformer windings after emergency switching off was revealed the damage and short-circuit of the forth turns of regulating winding (fig.11).

Fig. 11. Internal short-circuit of 167 MVA/500/220 kV autotransformer regulating windings.

As in the report for the sums of the commission work the suppositional reason for the damage of 167 MVA/500/220 kV autotransformer was reflected the entry of moisture into the internal volume during the replacement of the bushing. This version became determining and it was proposed by the representative of transformer manufacturer [3-7].

Normative documentation during PD measurements are national standard GOST-20074-83 "Method of measuring the PD characteristics and IEC-60270 "The measurements of PD characteristics" [by 1-2, 5-7].

VIII. Advantages of Cables with XLPE Insulation

Cable construction with XLPE insulation is in the fig. 8. In recent years, cables with XLPE insulation have become widespread on the Russian market. The accompanying documentation provided by the manufacturers lists the advantages of cables with XLPE insulation before cables with impregnated paper insulation (BIP):

• higher reliability in operation; • Lower costs for the reconstruction and maintenance of cable lines;

• low dielectric losses (dielectric loss factor 0.001 instead of 0.008); • high resistance to damage;

• Greater throughput due to the increase in the permissible heating temperature of the veins: long (90C instead of 70C), with overload (130C instead of 90C);

• higher thermal stability current in case of short circuit (250C instead of 200C); • Low permissible temperature for laying without preheating (-20C instead of 0C);

• low moisture absorption; • Lighter weight, diameter and bending radius, which facilitates laying on difficult routes;

• possibility of laying on routes with an unlimited difference in levels; • more ecological installation and operation (absence of lead, oil, bitumen) ; • Large construction length.

Cable construction with XLPE insulation is in the fig. 12, where is:

1. Round multiwire sealed conductor core material - Aluminum, Copper;
2. Screen for the core of extruded semiconducting cross-linked polyethylene;
3. XLPE insulation (Pv);
4. Screen for insulation from extruded semiconducting cross-linked polyethylene;
5. Separating layer;
6. Screen made of copper wires fastened with copper tape of different sections;
7. Separating layer of two ribbons of crepe paper or rubberized fabric or polymer tape;
8. Separating layer;
9. Sheath: polyethylene (P), polyethylene, increased thickness (Pu), plastic (B); plasticity of reduced fire hazard.

Fig. 12. Cable construction with XLPE insulation.

Disadvantages:

• In the case of an electrical breakdown of a solid dielectric, a XLPE cable will not be able to restore its electrical strength, and any single-phase ground fault (SPGF) will lead to a stable emergency mode.

• Cable BIP in 60-70% (SPGF) cases is self-destructing and the operating personnel do not record these emergency modes.

IX. Methods and Equipment for Testing XLPE Cables by Voltage on the ULF (0.1-0.01 Hz.)

Underground cables have to be tested before on-site commissioning or recommissioning. Although cables are tested by the manufacturer before delivery, they maybe damaged while being laid. As well joints or terminations may be defective.

Therefore laying construction companies and network operators are obligated to carry out standard tests. In the BIP insulated cables have been tested with DC voltage with the leakage current being an indication of the circuits quality.

Modern cables with EPR/XLPE insulation require a sinusoidal voltage and should be tested in accordance with the most recent national or international specifications. Such specifications define the voltage level and time for which the user performs the test. If at the end of the test time the circuit has withstood the applied voltage, then it gives the user a very high degree of confidence that when he switches it in it will operate satisfactorily.

Such a test is limited in that it gives no information about the quality of the circuit. The test result is a simple "pass/fail" information. Some test equipment can be used on both traditional and modern type cables.

X. Diagnostics of Cables with XLPE Insulation.

Cable diagnostics provide the user with information about quality and condition of the circuit. This goes far beyond the "pass/fail" information from cable testing. The diagnostic information that is available with some equipment gives "present and future analysis" of the insulation status of cable systems (fig. 13).

The target is to perform maintenance measures only if necessary and not to plan purely based on operating time or error probability of a cable circuit. Based on VLF test voltage, the dissipation factor measurement tan δ and partial discharge diagnostics give the necessary information without destructive testing, which can be put into a database and provide trend analysis. Diagnostics technology can be used to test all cable types like plastic cables or paper-mass cables [7].

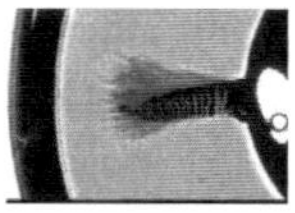

Fig. 13. Water triing of the type " fan".

XI. Conclusion

Infra-red control was effective for detection electrical equipment's faults, defects and weaknesses: overheating of disconnector contacts, bushing condition of power transformers, condition of circuit breakers, arresters and limiters of overvoltages, coupling transformers, measuring current transformers, measuring voltage transformers, control of connecting heads soldering quality of turbo-generator stator windings.

The low voltage impulse testing is a very sensitive and reliable method of deformation's detections of transformer windings. The LVI oscillograms are a "fingerprint" of transformer.

This winding "fingerprint" are defined by major resonance frequencies (a winding spectra). The 250 MVA 220 kV winding transformer's FRA spectrum was contained a 110 kHz, 320 kHz and 550 kHz frequencies, which are changed 1.3-2 times after the mechanical radial winding deformations. The FRA spectrum of 80 MVA 110 kV transformer changed after short-circuit. The original 300 kHz, 500 kHz, 700 kHz resonance frequencies disappeared and a new 400 kHz, 800 kHz resonance frequencies.

The results of the carried out complex inspection showed that the basic reasons for the failure of instrument transformers they are the low quality of SF6 (sulfur hexafluoride), design deficiencies in the transformers, the absence of absorbers for adsorbing of moisture and decomposition products of SF6, the application of not corrosion-resistant materials for manufacturing the transformers and the low quality of service maintenance [3-7]. The quality of SF6 must correspond to the norms, indicated in IEC 60480-2004.

The internal short-circuits of transformer windings can be caused by the breakdown of the turn-to-turn insulation as a result of the destruction of insulation under the action of operational factors and by the action of the partial discharges in the place of future breakdown, as initiator of which can serve commutation, thunderstorm and other increased actions on the transformer insulation.

The determination of PD intensity was produced with the aid of the electrical method of PD registration. As the measuring means was used digital oscillograph, and also digital registrar-analyzer of the partial discharges.

As a result the internal inspection of autotransformer windings after emergency switching off was revealed the damage and short-circuit of the forth turns of regulating winding.

The suppositional reason for the damage of 167 MVA/500/220 kV autotransformer was reflected the entry of moisture into the internal volume during the replacement of the bushing.

References

[1] IEC-60270 "The measurements of PD characteristics".

[2] A.Yu Khrennikov, R.V. Mazhurin, P.S. Radin, "Infra-red and ultraviolet control, LVI-testing, partial discharges and another diagnostic methods for detection of electrical equipment's faults, defects." Journal of Multidisciplinary Engineering Science and Technology (JMEST) (ISSN: 3159-0040), Volume1-Issue 4 Nov. – 2014. http://www.jmest.org/vol-1-issue-4-november-2014/

[3] A.Yu. Khrennikov, R.V. Mazhurin, P.S. Radin, "Diagnostic Methods for Detection of Electrical Equipment's Faults and Defects Including with Sulfur Hexafluoride Gas Insulation." Energy and Power Engineering Science, EPES 2015, 2(1): pp. 13-19. http://manuscript.sciknow.org/uploads/epes/pub/epes_14139061671.pdf

[4] A.Yu. Khrennikov, "Smart Grid technologies for Detecting Electrical Equipment Faults, Defects and Weaknesses. Workshop on Mathematical Modelling of Wave Phenomena with applications in the power industry", Linnaeus University, Växjö 23-24 April 2013. http://lnu.se/mmwp

[5] A.Yu. Khrennikov, T. V. Ryabin, N.M. Aleksandrov, "Diagnostic methods for detection of electrical equipment's faults, defects." International Conference on Condition Monitoring, Diagnosis and Maintenance CMDM 2017. September 25th -27th 2017, Bucharest, Romania.

[6] A.Yu. Khrennikov, N.M. Aleksandrov, I.N. Zhelbakov, T.A. Khrennikova, "Smart Grid for increasing the reliability of electric power supply of green cities by detecting electrical equipment faults and defects", 2017 2nd International Conference on the Applications of Information Technology in Developing Renewable Energy Processes & Systems (IT-DREPS). Petra University. Amman, Jordan. December 6th & 7th 2017.

[7] A.Yu. Khrennikov, "Diagnostics of Electrical Equipment Faults and Power Overhead Transmission Line Condition by Monitoring Systems (Smart Grid)", Short-Circuit Testing Of Power Transformers, Nova publishers, New York, 2016, USA, p. 165 https://www.novapublishers.com/catalog/product_info.php?products_id=56561

Compare of Shielding Effectiveness for Building Materials with Different Air Gap

Marek Pavlík, Ján Zbojovský, Maksym Oliinyk, Michal Ivančák
Department of Electric Power Engineering
Technical University of Košice
Košice, Slovak Republic
marek.pavlik@tuke.sk, jan.zbojovsky@tuke.sk, maksym.oliinyk@tuke.sk, michal.ivancak@tuke.sk

Abstract — **During the 1990s, most electromagnetic field research focused on extremely low frequency exposures stemming from conventional power sources, such as power lines, electrical substations, or home appliances. Now, in the age of mobile telephones, wireless routers, and portable GPS devices (all known sources of EMF radiation), concerns regarding a possible connection between EMFs and adverse health effects still persist. Quality shielding materials are determined by three coefficients, shielding effectiveness SE, absorption coefficient A and a reflection coefficient R. The shielding effectiveness SE is closely related with coefficient R, and A. This paper deals with compare the shielding effectiveness of two bricks with different air gaps. The air gaps constitute 60% of Sample 1 and the air gaps constitute 80% of Sample 2. Measurements were performed in the frequency range from 1GHz to 9 GHz. Many devices that people use in dailylife and are exposed to are working in this frequency range. Measurements were performed according to the IEEE Standard, Method for Measuring the Effectiveness of Electromagnetic Shielding Enclosures. Measurements were performed in the non-reflection chamber in order to avoid the exterior influence. Measurements point to the shielding effect of building material (brick). The article also compares the shielding effectiveness of electromagnetic field for different air gaps.**

Keywords: shielding effectiveness, electromagnetic radiation, electromagnetic field

I. Introduction

After longer thinking about the development of electromagnetic equipment, it is possible to conclude that the amount of electromagnetic field resources is significant at present. Mobile phones, wireless internet is just a fraction of what affects the human body. In doing so, it is necessary to realize that it is almost impossible to avoid this influence. This is mainly because there is almost no place where one can hide in normal life so that the effect of the electromagnetic field would be zero. It is also important to realize that the electromagnetic field of higher frequencies penetrates materials better. It is possible to prevent electromagnetic field penetration into a human being by shading, but on the other hand, it is necessary to realize whether a person really wants this. On the one hand, there is convenience in the form of wireless internet on laptops, tablets or making calls from virtually any location, and on the other hand there are potential concerns about the influence of the electromagnetic field [1] [2] [3].

The effects of the electromagnetic field can be divided into thermal and non-thermal. The magnitude of the thermal effects of an electromagnetic field depends on the frequency, and with increasing frequency the range of thermal effects of the electromagnetic field increases. The essence is the fact that electromagnetic radiation is converted into heat upon absorption into tissues. The total electromagnetic radiation absorbed depends on the amount of water content in the tissue, the size of the exposed surface exposed to radiation, and the field strength. Because the body cannot absorb the heat it receives, it overheats. The organs most susceptible to temperature increase are the lens of the eye, the brain, and the testes. Non-thermal effects, in contrast to thermal effects, are related to the long-term exposure of the human body to electromagnetic radiation.[4] The non-thermal effects of the electromagnetic field are related to the electrical properties of biological systems, which can be manifested after prolonged exposure by body weakness, exhaustion, increased fatigue, memory impairment, sleep disorders and headaches [2][4].

Already in the 1970s, the term electrosmog was introduced. The electrosmog has been explained as "electricity pollution" and it includes every electromagnetic wave that is created by man and is not a natural source. Unlike conventional smog, electrosmog cannot be seen with the naked eye, and so the fears of it are smaller for the population. Thus, even an elevated level of electrosmog will not be noticed by an ordinary person. On the contrary, the increased level of electrosmog causes the following problems: chronic headache, sleep disturbances, heart rhythm disorders, chronic fatigue and exhaustion, internal restlessness, hyperactivity in children, tinnitus, nerve and soft tissue pain, concentration and memory disorders, depressions, blood pressure fluctuations, susceptibility to infections, decreased immunity, cancer diseases such as leukemia and brain tumors [1] [5] [6].

Various studies, measurements and researches warn about these effects. Recent researches published in study [5] have partly confirmed that diseases such as fatigue, irritability, loss of appetite, sleep disturbances, depressions and lack of concentration were mainly related to the exposure of the electromagnetic field of GSM mobile phones. However, on the other hand, the authors also admitted that these problems did not need to be caused only by this radiation. In conclusion, Indian research draws attention to the effects of electromagnetic fields from mobile towers and recommends that these towers be located outside densely populated areas. It also describes that in the case of already installed mobile towers, it would be appropriate to at least increase the height of these towers, or change the direction of the antennas [2][5][7] [8][9].

II. Protection Against Electromagnetic Field Effects

Due to the previously mentioned thermal and non-thermal effects of the electromagnetic field, attention should be paid to protection from the effects of the electromagnetic field. The penetration of the electromagnetic field is closely related to the physical properties of the material through which the electromagnetic field passes. By using suitable material (shielding) it is possible to completely prevent the penetration of the electromagnetic field. However, electromagnetic field penetration cannot be completely prevented at present. Conversely, using wireless technologies (mobile phones and Wi-Fi devices) requires that the transmitted and received signals be as stable as possible. Finding a compromise between protection against electromagnetic fields and the convenience of using wireless technologies is oftentimes difficult [8][10][11].

It has been almost thirty years since the Maes Baubiology institute has begun to structure various aspects of the Building Biology methods. Based on testing experience, a standard has been developed in collaboration with scientists, experts, doctors and colleagues. Building Biology guidelines for relaxation and sleeping areas were soon published. Today, standards are used to guide professional and independent home testing worldwide, including Europe, the US, Canada, Australia, and New Zealand. Building Biology consultants, associations, institutes, laboratories and test equipment manufacturers leave their recommendations in them. Affected people, medical facilities, doctors, clinical ecologists, consumer associations, and civic organizations find in them the proper direction. Building Biology measurements are carried out on site, e.g. in bedrooms, living areas or in the workplace. Appropriate remediation recommendations are issued for any increased values [10][12][13][14].

III. Shielding Effectiveness

Most people spend (as opposed to previous generations) up to 90% of their time in rooms, often in dark and poorly ventilated buildings. The group of these people is growing, leading to the emergence of so-called Indoor Generation. Also, the number of electromagnetic field sources is increasing, both inside and outside. It is therefore important to address the issue of shielding the electromagnetic field of buildings in which we work and live. This post deals with shielding electromagnetic field bricks, which are commonly used as building material for residential buildings. Bricks contain air gaps for filling and strengthening. Different brick thicknesses have different air gaps. When electromagnetic radiation strikes the interface between two environments (e.g., air and brick), the incident beam is split into a reflected beam and transmitted beam. By the law of reflection, the reflection angle α 'of the reflected beam is equal to the angle of incidence α of the incident beam. The direction of the beam transmitted is not the same as the direction of the incident beam. This phenomenon is called refraction. The angle of refraction of the transmitted beam β depends on the physical properties of both environments on their interface. Reflecting the waves from the interface of two environments changes their direction without changing their speed. In refraction wave changes its direction when moving from one environment to another, accompanied by a change in speed. The transmission of the electromagnetic wave (and also the absorption) depends on its wavelength. The substance may transmit a wave of a certain wavelength while a wave with other wavelength may be absorbed. Radiation transfer occurs when the material has no absorption properties. The diffraction phenomenon occurs with all types of waves (and thus also with electromagnetic). Diffraction allows the bending of electromagnetic waves, their passage through small openings, and movement around small particles of matter. The smaller the hole or the particle size, the greater the bending of the light. Diffraction occurs when the propagating wave hits an obstacle (a hole or a particle) but is most noticeable in the case of waves with wavelength whose dimensions are of the same order of magnitude as obstacle dimensions [3][5][15][16][17][18].

Shielding efficiency is an indicator of shielding quality. In general, the shielding efficiency *SE* is calculated as the sum of the absorption coefficient *A* and the reflection coefficient *R* [2][4][6][7] [17].

$$SE = A + R \qquad (1)$$

$$SE = 8{,}69\frac{t}{\sqrt{\frac{2}{\omega\mu\sigma}}} + 20\log\left(\frac{1}{4}\sqrt{\frac{\sigma}{\omega\mu_r\varepsilon_0}}\right)$$

where t is material thickness, σ is electrical conductivity of shielding material, σ_R is the relative electrical conductivity, μ_R is the relative permeability of shielding material, μ is the magnetic permeability of shielding material, f is frequency, δ is depth of penetration, μ is permeability which included permeability of the shielding material, ε_0 is electric permittivity of vacuum. From equation (1) we can see, the relative permeability and relative conductivity affect shielding effectiveness SE [7][10][11][12].

Reflection depends on the characteristics of the environment in which the non-ionizing radiation impacts. One of these is the smoothness of the surface of the material compared to the wavelength of the incident radiation. Rough surface of the material affects the relative direction and phase coherence of the reflected wave. This characteristic determines the amount of radiation that is reflected back into the first environment and also determines the preservation of the accuracy of the information carried by the reflected wave. In the absence of reflection, the reflective environment does not retain the transmitted image information and scatters the radiation in all directions. When the radiation passes through the environment, it loses its coherence through scattering. A reflective environment that is unable to create an image is considered a diffuse environment [10][11][17].

Absorption has both a particle and a wave nature. In the interaction of electromagnetic radiation with a substance, this radiation can be absorbed and its energy passed to the substance. The energy of the electromagnetic wave in the environment is partly used to generate electron oscillations. Free electrons, even in a very thin layer of material, absorb all the light energy, so metals are opaque to light. In dielectric substances, light absorption is always small. Absorption of electromagnetic waves (light) in a substance, i.e. intensity varies according to Lambert's law of absorption [7][10].

IV. Experiment

Previous experiments have pointed to the shading ability of bricks. This experiment was focused on the effect of the air gap on the shielding efficiency value. For the purpose of this experiment, bricks of the same thickness were selected, but the air gap dimensions were different. The experiment was aimed at a frequency range of 1 GHz to 9 GHz with a 0.1 GHz step.

The measuring workplace consists of the Agilent N5181A pulse generator, which supplies the transmitting antenna RF spin DRH18-E. There is a measured shield between the antennas. The output from the receiver antenna of the R&S HF907 leads to the Agilent N9038A MXE EMI spectrum analyzer. The measuring workplace block diagram is shown in Fig.1. You can see the measured samples (thickness 30 cm) in the Fig.2. Two bricks with different air gap were measured.

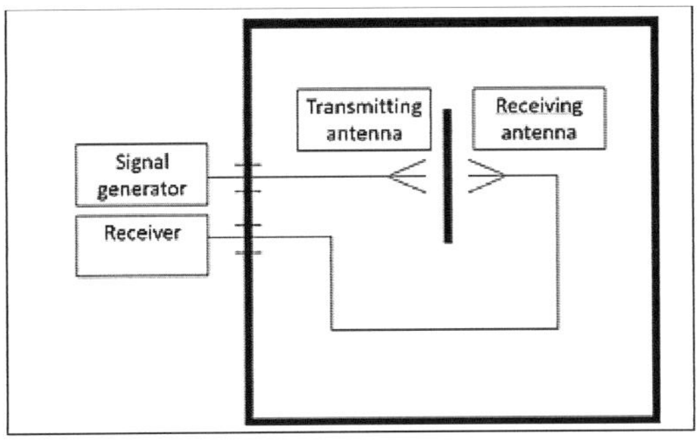

Figure 1. Measuring workplace for shielding effectiveness experiment

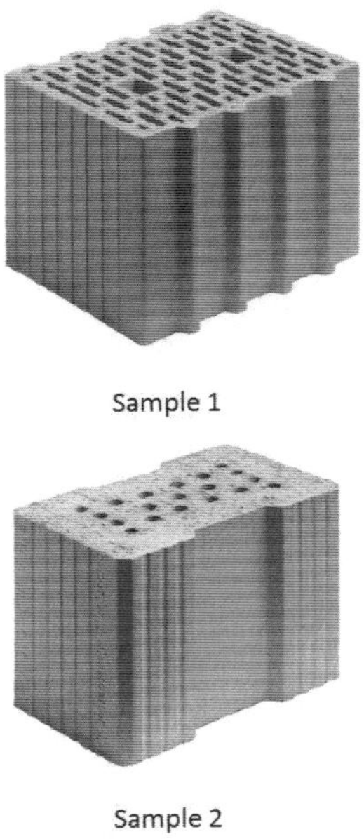

Figure 2. Measured samples – brick with different air gap

Before the actual measurement the measurement site was calibrated. Calibration consists in adjusting the position of the transmitting and receiving antenna to the position that the receiving antenna captures the electromagnetic field transmitted by the transmitting antenna. It needs to be realized that the receiving antenna will capture the electromagnetic field even if the receiving and transmitting antennas are not oriented exactly opposite each other. However, if it were not so, the receiving antenna may not capture the emitted electromagnetic field in the main lobe. This could directly affect the shielding efficiency where reflections from the material are already occurring and the incident electromagnetic field may be at a different angle. The calibration consisted of several steps.

First, it is necessary to determine the transmitted signal and to do this it is necessary to calculate the signal that is displayed on the spectrum analyzer. Calibration is performed by moving the receiving antenna in the vertical and horizontal directions - Fig. 20. When moving in the horizontal and vertical directions it is necessary to monitor the value of the electromagnetic field on the spectral analyzer.

Calibration of the measurement site was based on a radio-communication equation that is defined as follows:

$$P_P = P_V - L_0 + G_V + G_P \tag{2}$$

where P_P is received power, P_V is power transmitted, L_0 is free space loss, G_V is the gain of the transmitted antenna, and G_P is the receiving antenna gain.

The experiment was repeated 10x and the resulting shielding efficiency was calculated as the average of the individual frequencies. The shielding effectiveness was calculated by this formula: [8][10][12]

$$SE = P_{1\log} - P_{2\log} \text{ [dB]} \tag{3}$$

V. Results

Fig. 2 shows the shielding efficiency relationship for brick material with different air gap dimensions. The results show that bricks with the same thickness but different air gap dimensions have different shielding efficiency values. The experiment shows that if the air gap size is more, the shielding efficiency is fewer. For example, for a frequency of 3 GHz: shielding efficiency for air gap of Sample 1 is SE = 5.8 dB and shielding efficiency for air gap of Sample 2 is SE = 7.8 dB. Shielding efficiency consists of absorption and reflection of the electromagnetic field. However, this experiment does not confirm that the absorption/reflection ratio of the electromagnetic field must also change. The shielding efficiency value was changed, the ratio of R and A could remain constant. This claim could confirm or disprove further measurement of the absorption or reflection of the electromagnetic field. As the absorption inside the brick is complicated, measuring the reflection of the electromagnetic field would be sufficient. If the electromagnetic field reflection measurement was accurate, the absorption could be calculated as the difference of shielding efficiency and electromagnetic field reflection.

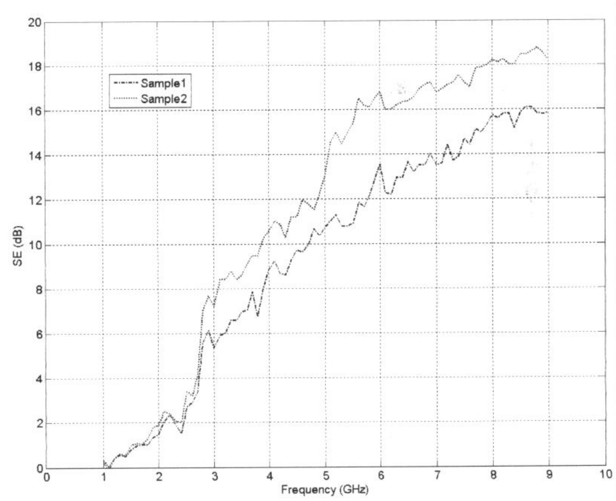

Figure 3. Shielding effectiveness for brick with different air gap

VI. Conclusion

The proposed contribution presents the results of research into the shielding of a construction object towards the electromagnetic field. Measured object was two brick wall with different air gap. The wall thickness was the same - 30cm.

The experiment pointed to the shading ability of the brick wall. The experiment shows that if the air gap size is more, the shielding efficiency is fewer. The Sample 1 (more air gap) has lower shielding efficiency as Sample 2 (fewer air gap).

The shielding effectiveness consists of absorption coefficient and the reflection coefficient. However, this experiment does not confirm that the absorption/reflection ratio of the electromagnetic field must also change. The shielding efficiency value was changed, the ratio of *R* and *A* could remain constant. This claim could confirm or disprove further measurement of the absorption or reflection of the electromagnetic field.

Acknowledgment

This work was supported by the Scientific Grant Agency of the Ministry of Education of Slovak Republic and the Slovak Academy of Sciences by the projects VEGA No. 1/0435/19.

This work was supported by the Cultural and Educational Grant Agency of the Ministry of Education, Science, Research and Sport of the Slovak Republic (KEGA) under the project No. 008TUKE-4/2019.

This work was supported by project: Influence of material on electromagnetic field penetration, FEI-2018-54.

References

[1] K. Kuznetsov, A. Zakirova and U. Averyanov, "Specific energy of 50 Hz electromagnetic field," 2017 International Conference on Industrial Engineering, Applications and Manufacturing (ICIEAM), St. Petersburg, 2017, pp. 1-4., doi: 10.1109/ICIEAM.2017.8076222.

[2] E. Yavolovskaya, G. Gabriadze, G. Chiqovani and R. Jobava, "Human exposure analysis for several different human voxel models," 2017 XXIInd International Seminar/Workshop on Direct and Inverse Problems of Electromagnetic and Acoustic Wave Theory (DIPED), Dnipro, 2017, pp. 183-186. doi: 10.1109/DIPED.2017.8100594.

[3] D. Medveď, "Investigation of influence of the additional electromagnetic shielding on domestic appliances," Elektroenergetika 2017, Košice: TU, 2017, pp. 196-201. - ISBN 978-80-553-3195-9.

[4] Z. Anane and F. Aouabed, "Electromagnetic fields analysis on OTL with presence of human body model using comsolmultiphysics," 2017 5th International Conference on Electrical Engineering - Boumerdes (ICEE-B), Boumerdes, 2017, pp. 1-6. doi: 10.1109/ICEE-B.2017.8191985

[5] J. Zbojovský, A. Mészáros, P. Kurimský, "Modelling the high frequency electromagnetic field propagation through the polystyrene," Elektroenergetika 2015, Košice, 2015 pp. 556-559., ISBN 978-80-553-2187-5.

[6] J. Besnoff, M. Chabalko, D. Ricketts, "A Frequency Selective Zero-permeability Metamaterial Shield for Reduction of Near-field Electromagnetic Energy," Antennas and Wireless Propagation Letters, IEEE, 2015, Volume: PP, Issue: 99, pp.1-4, ISSN : 1536-1225.

[7] D. Bambynek, A. Jakubas and P. Jablonski, "Examination of the possibilities to shield the electromagnetic field by selected polymer composites," Przegląd Elektrotechniczny, Vol. 2017, No. 01, pp. 121-124, ISSN 0033-2097.

[8] IEEE 299-2006 Standard: Method for Measuring the Effectiveness of Electromagnetic Shielding Enclosures, EMC Society, New York 2006, p.39.

[9] B. Dolnik, M. Rajnak, R. Cimbala, I. Kolcunova, J. Kurimsky, J. Balogh, J. Dzmura, J. Petras, P. Kopcansky, M. Timko, J. Briancin, M. Fabian, "The Response of a Magnetic Fluid to Radio Frequency Electromagnetic Field", ACTA PHYSICA POLONICA A, Vol. 131, n. 4, 2017, pp. 946-948. ISSN 0587-4246.

[10] C. Hao and L. Denghua, "Simplified Model of Shielding Effectiveness of a Cavity with Apertures on Different Sides," IEEE Transactions on Electromagnetic Compatibility, Vol.56, No.2, 2014, s.615-621, ISSN: 0018-9375.

[11] Y. Guang, "Analysis of Electromagnetic Field on the Permeability Test of Concrete," Power and Energy Engineering Conference (APPEEC), Asia-Pacific, Shanghai, 27-29 march, 2012, pp. 1-2, ISSN:2157-4839.

[12] P. Liptai, M. Moravec and M. Badida, "Analysis and comparison of electromagnetic fields radiated by the selected equipment," In: SGEM 2013 : 13th International Multidisciplinary Scientific Geoconference : Ecology, Economics, Education and Legislation : conference proceedings : Volume 1 : 16-22 June, 2013, Albena, Bulgaria. - Albena : STEF92 Technology Ltd, 2013 P. 17-20. - ISBN 978-619-7105-04-9.

[13] M. Pavlík, I. Kolcunová, B. Dolník, J. Kurimský, A. Mészáros, D. Medveď, M. Kolcun and J. Zbojovský, "Measuring of Dependence of Shielding Effectiveness of Wet Materials on The Frequency of Electromagnetic Field in the High Frequency Range," Acta Electrotechnica et Informatica. Vol. 13, No. 3 (2013), s. 12-16. - ISSN 1335-8243.

[14] S. Tumański, "Modern magnetic field sensors – a review," Przegląd Elektrotechniczny, Vol. 2013, No. 10, pp. 1-12, ISSN 0033-2097.

[15] M. Bereš and D. Kováč, "Digital pulse generator for multiphase bost converter,"Acta Electrotechnica et Informatica. Vol. 14, No. 4 (2014), s. 46-51. - ISSN 1335-8243.

[16] D. Kováč, D., et al., "An Automated Measuring Laboratory (VMLab) in Education," International Journal Of Engineering Education 32.5 (2016): 2250-2259.

[17] M. Češkovič, S. Bucko, M. MIĽO and J. Labun, "Antennas for UAV application," Aeronautika 17. - Lublin : University College of Enterprise and Administration, 2017 P. 230-237. - ISBN 978-83-60617-49-6.

[18] Z. Čonka, M. Kolcun and G. Morva, "Utilizing of Phase Shift Transformer for increasing of Total Transfer Capacity," ACTA POLYTECHNICA HUNGARICA, Volume: 13, Issue: 5, Pp. 27-37, 2016, ISSN:1785-8860, DOI: 10.12700/APH.13.5.2016.5.2.

Pollution Ash Powder Electric Parameter Measurement

Jaroslav Petráš
Department of Electric Power Engineering
TU of Košice, FEI
Košice, Slovakia
jaroslav.petras@tuke.sk

Jaroslav Džmura
Department of Electric Power Engineering
TU of Košice, FEI
Košice, Slovakia
jaroslav.dzmura@tuke.sk

Abstract — **The measurement of the electrical properties of ash from thermal power stations is described in this paper. The experiments were focused on the measurement of the relative permittivity, conductivity measurement and measurement of ash charging. These parameters should be part of input for the design of a suitable type of separator.**

Keywords: dust electrical properties; relative permittivity measurement; conductivity measurement; charging measurement

I. Introduction

The electrical separation of macroscopic sized an-organic particles is important both from technological and environmental points of view. It is one of the most important stages in the thermal processing of coal for electric power purposes. The effectiveness of separation depends of course on electrical parameters of the ash. The electrical parameters also influence the resistivity of pollution deposit on electric insulator surface. The in depth analysis of these electric parameters should help to design more efficient separator and it should also help us in the process of pollution sediment resistivity monitoring to prevent device damages caused by partial discharges on electric insulators [1], [2].

Electrical properties of such ash are influenced also by other physical parameters of the environment where the ash occurs. They are especially the environment temperature, humidity, the specific weight of the particles, size of particles and the chemical composition of the ash. Therefore, basic dielectric measurements have to be made such as the measurement of relative permittivity, the specific electric resistivity and the chargeability of macroscopic sized particles [3], [4].

II. Relative Permitivity Measurement

Basically, the relative permittivity measurement is a heterogeneous system measurement which consists at least of two components: solid state part (the ash) and gas (air). The gas fills the space between particles and it is also absorbed by a thin layer on the particle surface [5]. The solid part has bigger volume and weight. The parameters of the layer are determined by the kind of additional material and gas, by granule-metric and morphological composition of additional material and by outer conditions of layer sedimentation [6], [7], [11].

During in-depth analysis we can assume that the element with h thickness and dS surface can be defined as C_C capacity and R_C resistivity. The capacity of air layer is C_V, the border capacity is C_P and the transition resistivity is R_P. After simplification we get a circuit substitute of condenser and resistor connected in parallel which is precise enough from macroscopic point of view:

$$R = \rho_V \cdot \frac{h}{S}, \; C = \frac{\varepsilon_0 \cdot \varepsilon_r \cdot S}{h} \quad (1)$$

Where ρ_V – the specific resistance of the layer,
ε_r – the relative permittivity of the particle.

The ash layer for measurement purposes can be modelled by resistor-capacity model as seen on Fig. 1.

The relative permittivity of the mixed dielectric material can be calculated under condition that we assume that the particles have a sphere shape and that they are placed in a simplified way [8].

Most often Lorentz – Lorentz equation is applied:

$$\frac{\varepsilon_r - 1}{\varepsilon_r + 2} = \frac{\varepsilon_{r1} - 1}{\varepsilon_{r1} + 2} \cdot V_1 + \frac{\varepsilon_{r2} - 1}{\varepsilon_{r2} + 2} \cdot V_2 \quad (2)$$

Where ε_r – relative permittivity of the ash,
ε_{r1}, ε_{r2} – relative permittivity of the components,
V_1, V_2 – volume parts of the components.

For the measurement of permittivity of ash particles with high specific resistance a high precision high voltage Schering bridge was used in our experiments with the tested voltage with frequency 50 Hz and regulated voltage in the range 0 – 2000 V. During thermal dependency measurement, we used the specific

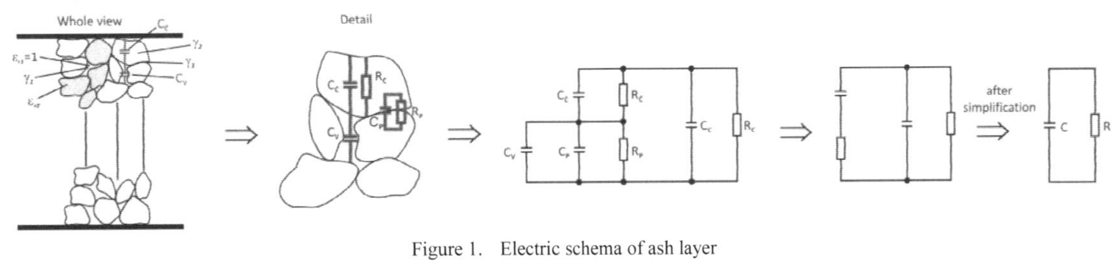

Figure 1. Electric schema of ash layer

measurement capacitor TETTEX with internal heating of tested sample.

Generally used 3-electrode system has also a protector electrode, which eliminates the influence of surface leak current and determines the electric field on the periphery of measuring electrode.

The measuring capacitor TETTEX 2904 used in our experiments is dedicated for permittivity measurements of solid dielectrics. As it is not suitable for powder consistency material measurement as it is, we had to modify it slightly. We modified the lower part of the electrode system in order to make a base for measured ash powder. An intermediate ring was used to keep the insulation parameters (Fig. 2). The lower part of the electrode was made of high glazed material to keep constant thickness of the measured ash layer after it is pressed by protector electrode. Thus, the measurements were reproducible. The measurement capacitor was dedicated for thermal dependency measurements. The heating elements were supplied from the power supply network and the temperature was regulated by thermal regulator TETTEX with ± 0.5°C precision.

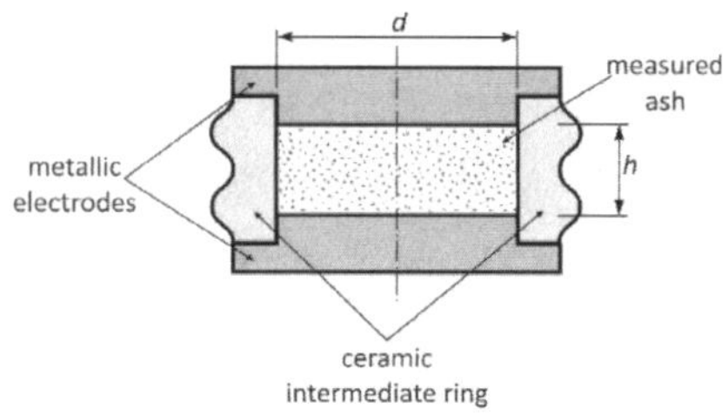

Figure 2. Electrode system for ash electric parameter measurement

Special attention is dedicated to the fact that the measured ash could be wet which interferes with specific conductivity measurement and especially with relative permittivity measurement which is very sensitive to the humidity of the measured sample. The value of relative permittivity can be calculated form C – measured capacity and capacitor capacity value ratio for the same geometrical size and air dielectrics. When the measured ash contains too much humidity at lower temperatures before ash drying the relative permittivity has very high value. Therefore, the ash powder must be dried thoroughly before measurement in order to eliminate hysteresis of the measurement process. Thanks to the drying and the stabilization of the sample parameters, the measurements are reproducible. The tempering during one measurement lasts 2 hours for one capacity value. Then the evaluation is made according to [2], [12]:

$$C = \frac{C_3}{C_4} \cdot C_N \quad (3)$$

For our TETTEX measuring bridge the capacity values C_3 and C_N are constant and the bridge is balanced by C_4 capacity change. Then the measured capacity can be calculated from the equation:

$$C_3 \cdot C_N = 100 \rightarrow C = \frac{100}{C_4} \quad (4)$$

In addition, the relative permittivity value is determined form

$$\varepsilon_r = \frac{C \cdot h \cdot 4 \cdot \pi \cdot 0.8988}{S} = 0.5647 \cdot C \cdot h \quad (5)$$

Where C – capacity in pF,
h – thickness of the measured sample in cm,
S – surface of measurement electrode in cm^2.

We used up-down type measurements i.e. in order to eliminate the sample humidity the temperature was changed from basic temperature until 150°C and after that the temperature decrease was checked by a regulator. The overall process of measurement lasted 24 to 36 hours for one sample. The voltage connected to the sample was in the range 100 – 1000 V with 100 V step and at each step 10 values of relative permittivity was saved. Same process of measurement was repeated for 18 test samples. An example of relative permittivity of the ash powder dependency on temperature is depicted on Fig. 3.

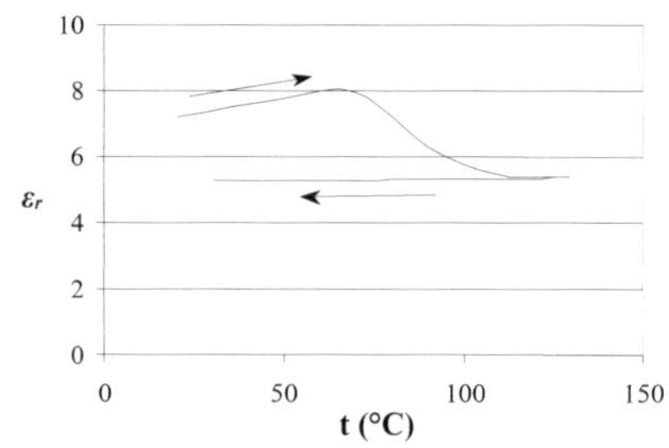

Figure 3. Relative permitivity dependency on temperature

The high peak of the relative permittivity value at the beginning temperature is caused by polarized water molecules, which were absorbed by the powder (water molecule size is $2.7 \cdot 10^{-10}$m). The powder drying lowers the relative permittivity value.

We verified our experiments by ash powder samples with high specific resistance in the range $10^7 \div 10^9\,\Omega$m. The ash samples were taken directly from high voltage separator input.

Some limitations for this method were samples of ash with high conductivity (low specific electric resistivity), in this case the method was not applicable. Another limitation was the higher frequency of tested voltage.

III. Specific Resistivity Measurement

Another important parameter for separation and insulation resistance measurement is the specific resistance of the ash powder, often specified as specific conductivity [1].

The specific conductivity is defined by:

$$\gamma_P = \frac{i}{E} \quad (6)$$

The overall value of conductivity depends not only on volume and surface resistance but also on contact resistance of two or more particles in touch. This is also a function of granularity and the shape of particles. To make experiments reproducible following parameters must remain constant [3]:

- powder particle density,
- environment relative humidity,
- temperature,
- electric field intensity.

To measure inner and surface resistance we had to measure the current flowing through sample and the current flowing on the surface of the sample separately. This is achieved by electrode configuration. To simplify the calculations, the used geometry should be as simple as possible.

We found optimal electrode system configuration – circle shaped electrodes in plan parallel configuration for ash with low specific resistance. In this case the specific resistance can be calculated from:

$$\rho = R \cdot \frac{S}{h} \tag{7}$$

Where R – resistance measured by the device,
h – thickness of the ash layer,
S – intersection of opposite electrodes.

Experiments were made at 30-130°C temperature range with 20°C steps at low voltages in order to prevent the creation of conductive channels which would deteriorate the measured results when the voltage reaches a critical value. The temperature was stabilized by thermal regulator during each measurement of open hysteresis loop (Fig. 4).

We have seen that all measured samples had low ρ values, which of course is a disadvantage for separator process. After temperature stabilization we have made at least 5 measurements and the arithmetical average was recorded. The Fig. 4 shows the $\rho = f(T)$ dependency. Measurements were very time consuming.

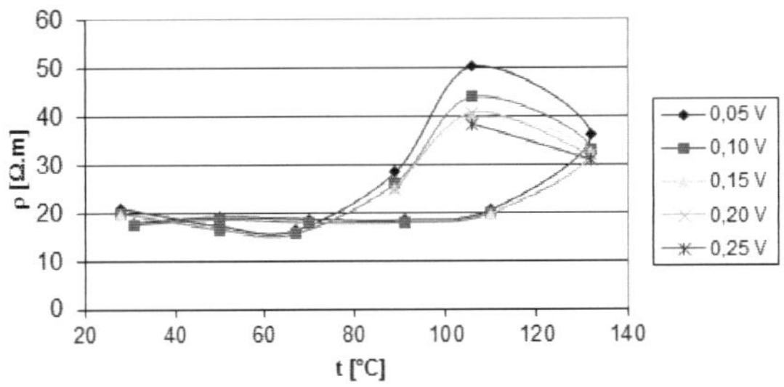

Figure 4. Dependency of specific resistance on temperature

IV. Chargeability Measurement

As seen above, the value of specific resistance ρ and the value of resistance R for actual ash layer are determined also by its chargeability [4]. If the specific resistance is low the leak of charge from macroscopically large particle is very quick. This fact can be expressed by an exponential dependence graph with negative exponent [9], [10]. Therefore, a new method of chargeability was introduced.

Electrically charged ash falls onto a tank made from conductive material though a narrow hole. This tank is placed in another shielding container which creates a capacitor object together with inner tank. The electrically charged particles invoke charge due to influence and this can be measured by a device depicted on Fig. 5.

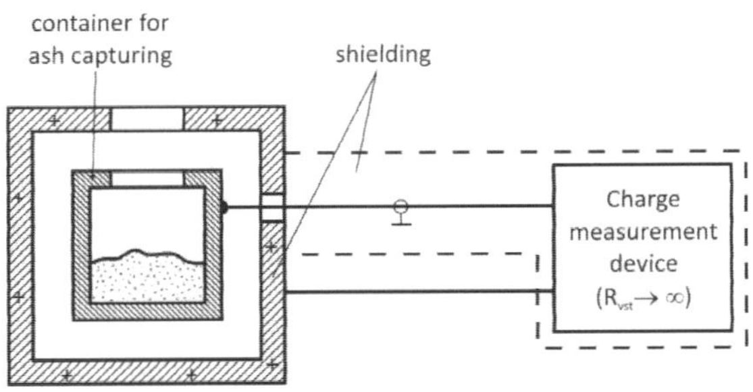

Figure 5. Device for ash chargeability measurement

With known capacity of the device it is possible to determine the charge by [5] if we know the voltage against the earth:

$$Q = C \cdot U \tag{8}$$

Where Q – the amount of charge in the ash powder,
C – overall capacity (container capacity + conductor capacity + additional capacity),
U – voltage drop on measuring capacitor.

We used a measuring capacitor consisting of two coaxial valves. The covers of the inner valve are freely removable. On the upper cover, there is a hole in the middle for exhausting air. On the lower cover there is a pin in the middle. The outer valve is fastened to a wooden tray and it is insulated by Teflon material. The inner valve is centered by the above-mentioned pin in the Teflon tray ensuring thus perfect centering.

During experiment the powder particles of dried ash were poured into the measuring capacitor on a metallic transport lead in order to avoid electro-static charge creation. The time of pouring was constantly 5 s.

The polarity of charge was determined by experiment on the touching surfaces. The materials were categorized by the value and the polarity of the charge.

V. Conclusion

The dielectric parameter analysis of ash powder helped us to decide which design and construction type is most suitable for high voltage separator. Furthermore, the experiments helped us to develop methods for parameter measurement during insulator pollution analysis. We examined materials with complex dielectric parameters with combination of organic and an-organic elements.

During separator construction we have to take into account these parameters as examined materials had complex polarization and conductivity properties as these materials have limit values of some parameters, which would classify them as semi-conductors.

Acknowledgment

The authors would like to acknowledge Scientific Grant Agency of the Ministry of Slovak Republic and Slovak Academy of Science for funding experimental works in the frame of VEGA No. 2/0141/16, VEGA No. 1/0340/18, VEGA No. 1/0435/19.

This work was supported by the Slovak Research and Development Agency under the contract No. APVV-15-0438 and No. APVV-17-0372.

References

[1] K. Marton. "Príspevok k štúdiu mechanizmu nanášania práškov". *Statická elektřina 78*, DT České Budejovice 1978, pp. 27–38.

[2] A. D. Moore. *Electrostatics and its Applications*. John Wiley and Sons, New York 1973.

[3] D. A. Lloyd. *Electrostatic precipitator handbook*. J W Arrowsmith Ltd., Bristol 1988, ISBN 0-85274-492-7.

[4] D. M. Taylor and P. E. Secker. *Industrial electrostatics*. Research Studies Press ltd., Somerset 1994, ISBN 0-86380-158-7.

[5] K. R. Parker. *Applied Electrostatic Precipitation*. Blackie Academic & Professional, London 1997, ISBN 0-7514-0266-4.

[6] J. Zbojovský, M. Pavlík. "The propagation of electromagnetic fields through the various materials". *Scientific Letters of Academic Society of Michal Baludansky*, Vol. 4, No. 6A, 2016, pp. 107-109, ISSN 1338-9432.

[7] J. C. Ge, S. K. Yoon, N. J. Choi. "Application of Fly Ash as an adsorbent for removal of air and water pollutants," *Applied Sciences*, Vol. 8, No. 7, art. no. 1116, ISSN: 20763417.

[8] M. Pavlík, I. Kolcunová, B. Dolník, J. Kurimský, A. Mészáros, D. Medveď, M. Kolcun, J. Zbojovský. "Measuring of Dependence of Shielding Effectiveness of Wet Materials on The Frequency of Electromagnetic Field in the High Frequency Range". *Acta Electrotechnica et Informatica*. Vol. 13, No. 3, 2013, pp. 12-16, ISSN 1335-8243.

[9] Z. Čonka, M. Kolcun and G. Morva. "Utilizing of Phase Shift Transformer for increasing of Total Transfer Capacity". *ACTA POLYTECHNICA HUNGARICA*, Vol. 13, Iss. 5, pp 27-37, 2016, ISSN:1785-8860.

[10] D. Medveď, M. Kolcun and Z. Čonka. "Simulation of Faults in the Power System and Their Elimination in the EMTP-ATP". *Proceedings of the 7th international scientific symposium on electrical power engineering (Elektroenergetika 2013)*, pp. 484-487, 2013.

[11] I. Kolcunová, M. Pavlík, Ľ. Beňa, Z. Čonka, S. Ilenin, M. Kanálik, M. Kolcun, A. Mészáros, D. Medveď and J. Zbojovský. "Influence of electromagnetic shield on the high frequency electromagnetic field penetration through the building material". *Acta Physica Polonica A*. Vol. 131, No. 4, pp. 1135-1137, ISSN 0587-4246.

[12] M. Kosterec, J. Kurimský, M. Foľta, S. Bucko, Z. Conka, "Investigation of effects of non-ionizing electromagnetic fields interacting with biological systems," Proceedings of the 8th international scientific symposium on electrical power engineering, P: 564-567, Sep. 16-18, 2015, Stará lesná, Slovakia ISBN 978-80-553-2187-5

New Methods of De-Noising Partial Discharge Signal by Morphological Filters

Jaroslav Petráš, Jaroslav Džmura
Department of Electric Power Engineering
Technical University of Košice, FEI
Košice, Slovakia

jaroslav.petras@tuke.sk, jaroslav.dzmura@tuke.sk

Abstract— **We study the partial discharge signal de-noising by morphological filters which are used mainly in signal processing. These filters have the advantage of not corrupting original useful signal during filtering process. However, these filters efficiently remove noise pulses from signal corrupted by noise and acquired during partial discharge measurement on electric power devices. We analyze the efficiency of morphological filtering for different noise type occurring in environment where partial discharge activity is monitored.**

Keywords*— *partial discharge, acoustic emission, morphological filter

I. Introduction

Electric energy supply reliability has nowadays a high importance and when considering electric energy supply continuity and quality it is one of the observed parameters [6, 7]. The operation of electric power devices without dropouts is the main aspect influencing electric energy supply reliability. Thus continual generation and distribution of electric power energy (by devices such as generators, power transformers device transformers) is ensured. An insulation system in such device in an important aspect and it considerably determinates the lifetime of electric power device. The methods for status determination of insulation system are very important using expert systems. Improving such monitoring methods is an important issue for electric power device manufacturers as well as for their operators, because of lower maintenance costs and costs caused by servicing.

A non-destructive method of insulation system diagnostics is the partial discharge measurement method and it is widely used.

Partial discharge activity in electric power device uses monitoring by direct galvanic, indirect acoustic method and there are other methods.

II. Noise and Its Filtering Methods

Measured signal measured during partial discharge activity measurement is obtained by direct galvanic method or indirect acoustic method in most cases. In addition, in real operation conditions there are many interfering noise signals. Depending on method used for partial discharge activity measurement, these noise signals have different character. For acoustic method, the noise is most disturbing when it has the same spectrum as the useful signal originating from partial discharge emission. In the case of direct galvanic method, the electrical noise generated by environment is superimposed to useful partial discharge electric signal [1, 5, 8].

In both cases, it is often very hard or even impossible to eliminate this interference. One of the solutions can be the usage of narrow band sensors, which deteriorates also the shape and characteristics of useful signal.

Usually, electric signal from useful source is filtered or shielded [2, 3, 4]. The possibilities of filtering are numerous. The selection of the filter has to be done carefully and it depends on effective signal characteristics and on the characteristics of noise signal. Band filters is a common choice. These filters can be built as analog filters (discrete components) or they can have digital character, i.e. they are applied after sampling and quantifying the measured signal.

One of the methods under research in the field of partial discharge measurement is the usage of Wavelet transform as a filtering tool. The Wavelet transform is used together with thresholding. The disadvantage of these filters (Wavelet transform, band filters) is the fact that filtering deteriorates also the shape of useful signal disabling thus the extraction of partial discharge type markers from measured signal. E.g., the distortion of leading edge of partial discharge pulse is a typical case, which is caused by higher frequency removing from useful signal, called as smoothing effect. The determination of pulse delay comparing to referential time is ruled out in this case. However, this is a very important parameter for acoustic measurement method when estimating a partial discharge source position in electric power device construction.

Morphological filters do not have these disadvantages and these filters do not change the shape of original useful signal. At the same time, they efficiently remove noise signals with pulse characteristics. The only condition is that noise pulses have shorter duration than useful signal. The higher frequencies in useful signal spectrum are left untouched.

III. Noise Sources Interferring with Partial Disacharge Signal

During partial discharge activity measurement by direct galvanic method in real operation conditions the useful signal can usually be deteriorated by pulse noise signals, which can in addition change their characteristics during measurement.

Most common noise sources interfering with electric signals of partial discharge activity during measurement are depicted on Fig. 1.

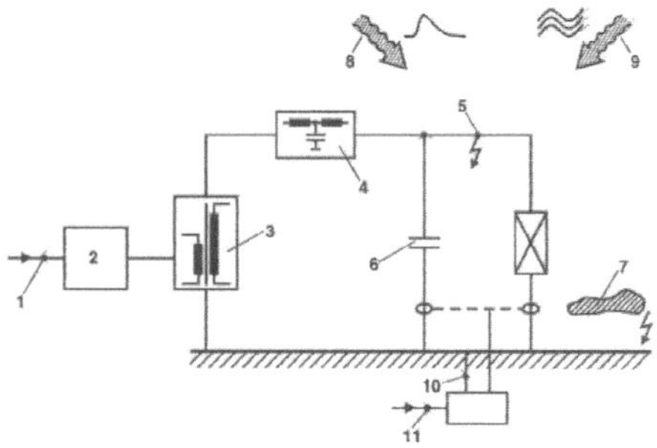

Fig. 1. Most common noise sources

1- Measurement source circuit generating testing voltage
2- Regulation unit
3- High voltage source
4- Filter in high voltage part of the circuit
5- Wiring of the high voltage circuit
6- Coupling capacitor
7- Near conducting objects without fixed potential
8- Pulse noise
9- Electro-magnetic interference from radio frequency transmitters
10- Noise currents in earthing system
11- Measurement device supply circuits

Noise signals can be divided according to their characteristics into:

- Pulse noise, in most cases caused by random pulse signals with very wide spectrum. These signals have very similar shape as the useful signal of partial discharge emission.
- Harmonic noise originating from power regulators, thyristor regulators etc.

Noise signals can be divided also according to the source type:

- External noise sources caused processes that are independent from testing voltage in measurement circuit. These noise sources can be detected by different methods and they are identified by characteristic pulse shape. Typical noise sources of this category can be e.g. telecommunication transmitters, rotary machines, switching devices etc.

- Internal noise sources, which are dependent on testing voltage in measurement circuit. Typically, with higher testing voltage internal noise increases also. Main sources of internal noise can be e.g. partial discharges on physical parts of measurement circuit, un-sufficiently earthed parts of the circuit or near object without fixed potential.

Using high voltage filters internal noise can be filtered out. In case of corona partial discharges sharp peaks and edges should be removed from measurement circuit and its environment.

IV. White Noise

A random noise process with equal power in the whole spectrum is called white noise. This type of noise can be achieved only theoretically as indefinitely high power is needed to generate such noise. Practically in some cases, we can consider noise with limited flat spectrum as white noise.

V. Pulse Noise Type

Noise with the pulse shape waveform consists of one or more short randomly generated pulses. Typically, they have different amplitude, duration and other signal characteristics depending on noise source. In addition, their shape is changed by transmission channel through which pulses travel from source to the sensor.

VI. Dot Noise

Dot noise can be seen e.g. in systems using radar wave technology, in image processing systems etc. It is called also „salt and pepper" noise due to white or black dots in image. The term image stands for real two-dimensional image or a one-dimensional signal considered as one-dimensional image with superimposed pulse noise creating thus dots in original partial discharge waveform. Conventional filters, as stated above with all of disadvantages can do removing such noise. Morphological filtering can use a much more progressive method.

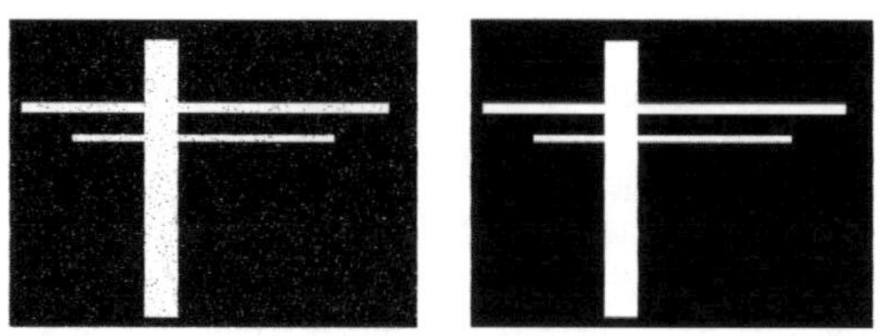

Fig. 2 Image with dot noise (left) and morphologically filtered image (right)

VII. Morphological Filtering

Mathematical morphology exploits the theory of sets, integral geometry and topology. Morphological operators used for filtering an n-dimensional image are local non-linear operators using masks or so called structuring elements.

By a set of dots, images of any dimension can be modelled. Euclidean space E2 and a system of its subsets is a natural definition domain for planar shape description. For binary image, the dot set consists of a set of integer number pairs Z2 and in the case of greyscale image these become triplets Z3.

E.g.

$$X=\{(0,2),(1,1),(1,2),(1,3),(2,0),(2,1),(3,1),(4,1)\}$$

Is a set of whole number integers describing black and white image on fig. 3.

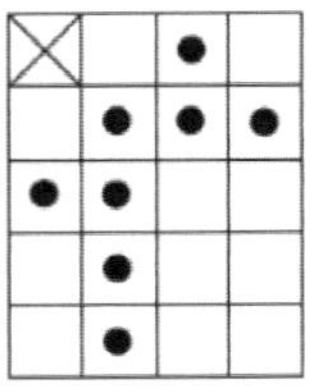

Fig. 3 Black and white image

A morphological transform of dot set A can be defined as an interaction with other smaller set of dots B, which is called a structuring element. This set is used to go through the A set and locally transforms the A set according to predefined rules. For n-dimensional A set a n-dimensional B set is needed.

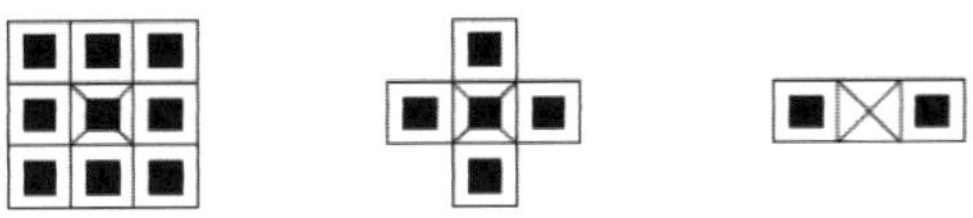

Fig. 4 Examples of planar two-dimensional structuring elements

According to rules used during filtering process by a structuring element, we can differentiate between morphological filters:

Dilation searches for maximum value of dots in X set within the area of the structuring element B. This maximum value is then assigned to the referential coordinates of X set. The dilation is denoted as:

$$X \oplus B$$

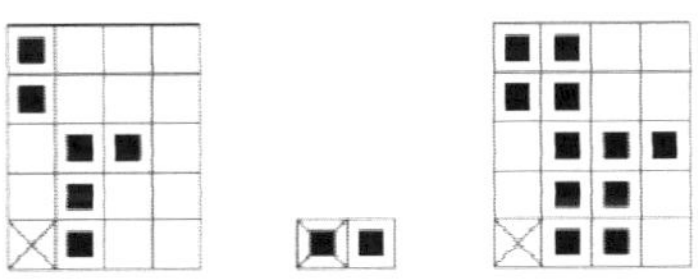

Fig. 5 An example of dilation filter, A set (left), structuring element (middle), filtered A set (right)

Dilation can be used to fill small gaps in original signal and it serves as a basic operation for more complex morphological operators.

Fig. 6 An example of dilation operator by 3x3 structuring element

Erosion is an operator dual to the dilation, enlarging small gaps in the original signal.

Fig. 7 An example of erosion operator by 3x3 structuring element

Dilation and erosion are inverse operations and by combining them in determined order we get the operators of opening and closing, which result in simplified but much less sharp signal - image.

The definition of opening morphological operator:

$$X \circ B = (X \oplus B) \oplus B$$

Fig. 8 An example of opening operator by 3x3 structuring element

The definition of closing operator:

$$X \circ B = (X \oplus B) - B$$

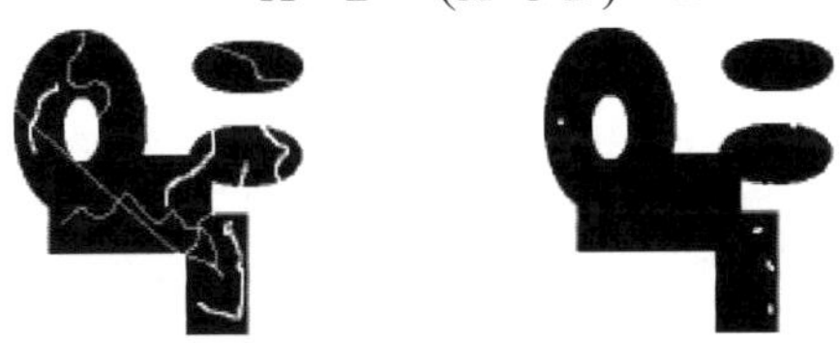

Fig. 9 An example of closing operator by 3x3 structuring element

These operators still have the disadvantage of corrupting the useful original signal, though removing efficiently short pulse noise signals.

Therefore, more complex morphological filters such as opening-closing, closing-opening, opening-closing by reconstruction and closing-opening by reconstruction are used for noise filtering which in addition reconstruct the original shape of useful signal.

The pulse noise signal with duration shorter than the structuring element is filtered out from the useful signal of partial discharge emission.

VIII. APPLICATION OF MORPHOLOGICAL FILTERS ON PARTIAL DISCHARGE SIGNALS

During our experiments, we have acquired a large set of partial discharge activity signal. We have made both experiments by direct galvanic method and indirect acoustic method. For both cases, signals were recorded and archived for later analysis. As they were influenced by different types of noise signals, we tried to apply morphological filtering to reduce noise interference.

Below waveforms of recorded partial discharge activity signals can be seen in time domain. We applied morphological filtering implemented in Scilab environment.

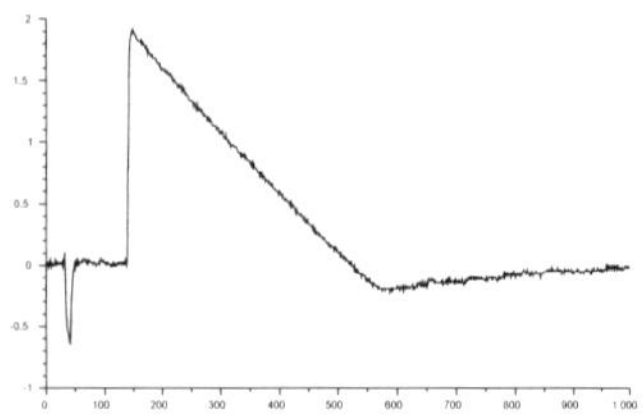

Fig. 10 Partial discharge pulse waveform with superimposed pulse noise

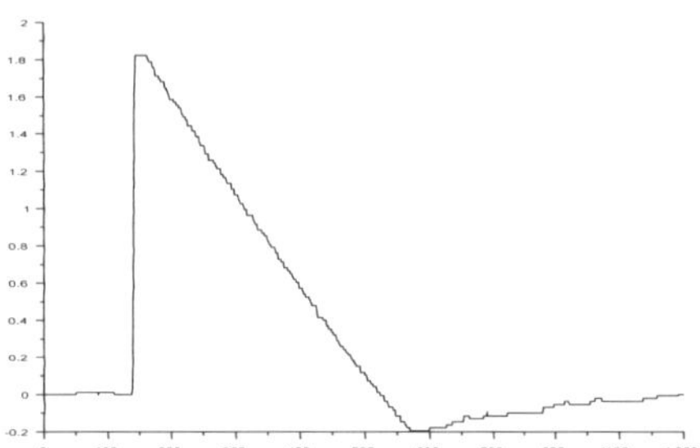

Fig. 11 Partial discharge pulse waveform with superimposed pulse noise after morphological filtering with structuring element of size 5

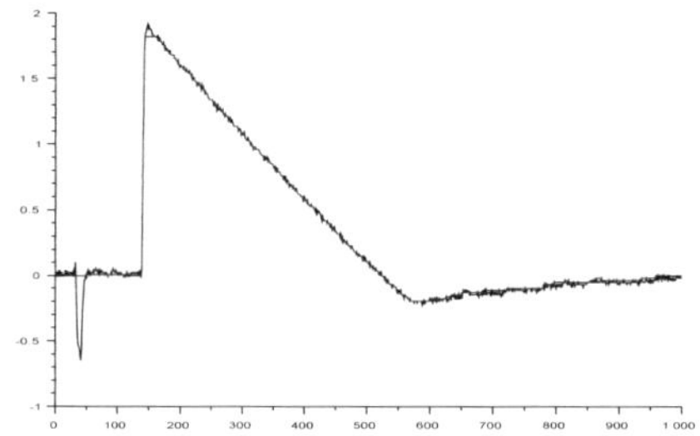

Fig. 12 Partial discharge pulse waveform with superimposed pulse noise and filtered signal after morphological filtering with structuring element of size 5

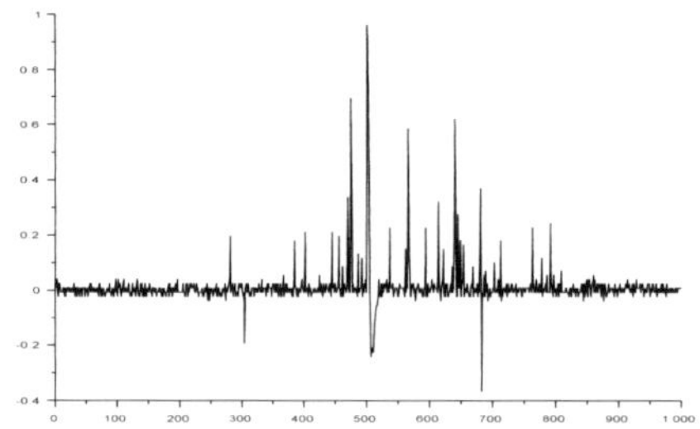

Fig. 13 Partial discharge pulse waveform with superimposed pulse noise – partial discharge in stator coil

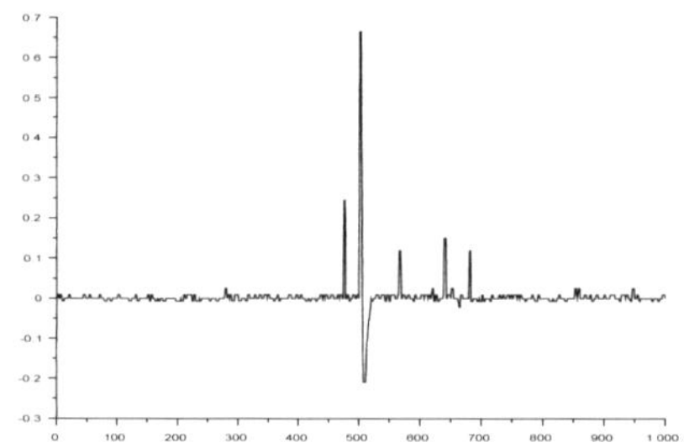

Fig. 14 Partial discharge pulse waveform with superimposed pulse noise – partial discharge in stator coil, after morphological filtering – structuring element with size of 5 samples

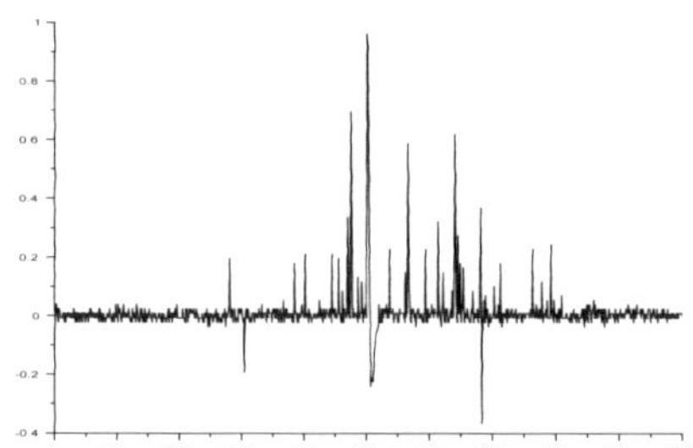

Fig. 15 Partial discharge pulse waveform with superimposed pulse noise – partial discharge in stator coil and the signal after morphological filtering – structuring element with size of 5 samples

IX. Conclusion

As demonstrated on figures above, morphological filtering removes efficiently pulse shape noise peaks from partial discharge activity signals, no matter if they are measured by direct galvanic or indirect acoustic method. At the same time these filters leave the original useful signal untouched (they reconstruct it). By setting appropriate size of structuring element we can adjust the width of noise pulses to remove from useful signal.

Acknowledgment

The authors would like to acknowledge Scientific Grant Agency of the Ministry of Slovak Republic and Slovak Academy of Science for funding experimental works in the frame of VEGA No. 2/0141/16, VEGA No. 1/0340/18, VEGA No. 1/0435/19.

This work was supported by the Slovak Research and Development Agency under the contract No. APVV-15-0438 and No. APVV-17-0372.

References

[1] J. Zbojovský, M. Pavlík. "Calculating the shielding effectiveness of materials based on the simulation of electromagnetic field". Elektroenergetika 2017. Košice TU, 2017 pp. 369-372. ISBN 978-80-553-3195-9

[2] J. Zbojovský, A. Mészaros, P. Kurimský. "Modelling the high frequency electromagnetic field propagation through the polystyrene". Elektroenergetika 2015. Košice: TU, 2015 pp. 556-559. ISBN 978-80-553-2187-5

[3] J. Zbojovský, M. Pavlík, Zs. Čonka, L. Kruželák, M. Kosterec. "Influence of shielding paint on the combination of building materials for evaulation of shielding effectiveness" EPE 2018. Brno: University of Technology, 2018 pp. 324-327. ISBN 978-1-5386-4611-3

[4] M. Pavlík, I. Kolcunová, B. Dolník, J. Kurimský, A. Mészáros, D. Medveď, M. Kolcun, J. Zbojovský. "Measuring of Dependence of Shielding Effectiveness of Wet Materials on The Frequency of Electromagnetic Field in the High Frequency Range". Acta Electrotechnica et Informatica. Vol. 13, No. 3, 2013, pp. 12-16, ISSN 1335-8243.

[5] I. Kolcunova, M. Pavlik, L. Bena, Z. Conka, et al., Influence of electromagnetic shield on the high frequency electromagnetic field penetration through the building material, Acta Physica Polonica A. Vol. 131, No. 4, pp. 1135-1137, ISSN 0587-4246

[6] M. Kosterec, J. Kurimský, M. Foľta, S. Bucko, Z. Conka, "Investigation of effects of non-ionizing electromagnetic fields interacting with biological systems," Proceedings of the 8th international scientific symposium on electrical power engineering, P: 564-567, Sep. 16-18, 2015, Stará lesná, Slovakia ISBN 978-80-553-2187-5

[7] Z. Čonka, M. Kolcun, G. Morva. "Utilizing of Phase Shift Transformer for increasing of Total Transfer Capacity". ACTA POLYTECHNICA HUNGARICA, Vol. 13, Iss. 5, pp 27-37, 2016, ISSN:1785-8860.

[8] D. Medveď, M. Kolcun, Z. Čonka. "Simulation of Faults in the Power System and Their Elimination in the EMTP-ATP". Proceedings of the 7th international scientific symposium on electrical power engineering (Elektroenergetika 2013), pp. 484-487, 2013.

Simulation Model of Partial Discharges in Internal Void

Radek Prochazka
Department of Electrical Power Engineering
Faculty of Electrical Engineering, CTU in Prague
Prague, Czech Republic
xprocha3@fel.cvut.cz

Ondrej Sefl
Department of Electrical Power Engineering
Faculty of Electrical Engineering, CTU in Prague
Prague, Czech Republic
seflondr@fel.cvut.cz

Abstract — **This paper deals with partial discharges inside the internal voids of solid insulating materials. First, the physical background of partial discharges is introduced. The streamer inception, charge deployment, and charge distribution on the void surface are described. Conclusions of physical description are then used for the construction of a mathematical model which can respect dynamic behavior of applied voltage. The simulation model is based on the finite element method calculations with variable parameters that respect the current condition of the insulation system. The output of the simulation model is in the form of phase-resolved partial discharge diagrams. Model parameters were determined from experimental results measured on an artificial void in epoxy resin sample. The simulated results show good agreement with the results gathered from the experimental measurements.**

Keywords: partial discharges, insulation systems, numerical models

I. Introduction

A partial discharge measurement is a standard tool for insulation state assessment in high voltage systems. A partial discharge (PD) is usually defined as localized electric discharge in gaseous voids which exist within a solid insulating material. The presence of PDs causes acceleration of insulation degradation, reduction of insulation quality and finally shortening its lifetime. Hence, PD phenomena have become one of the leading scientific and research activities in the area of high voltage insulating materials and their diagnostics. The simulation models can provide a better understanding of all effects related to PDs. Since PD is a very complex phenomenon characterized by stochastic and non-stationary behavior, including memory effect, i.e., when the space charge left from previous discharge event influences PD inception conditions of next PD, searching for a similar generalized model is challenging work. An attempt to summarize all these aspects and proposals of numerical models has been made in fundamental works [1], [2]. Simulation models assume two main conditions which have to be fulfilled for a PD to occur in an internal void. First, the electric field in the void must reach the PD inception field and second, a free electron must be available to start the ionization process. Then the streamer structure is developed. The time needed for discharge development is in the range of nanoseconds. The electric field inside the void consists of the local enhancement of the external field by the defect and electric field related to the space charge on the void's surface which has been left from previous discharges. The local field enhancement can be expressed by potential difference over void using dimensionless factor determined as the average of enhanced field distribution in the void. Usually, it is assumed, that the PD in a void is characterized by streamer discharge type, which can be mathematically described by continuity equations including diffusion [3], [4]. For the cases when the phase and amplitude of a discharge are the only model outputs, the detailed description of streamer propagation is not needed [5]. Moreover, the time of the streamer discharge development within the void is in the range of nanoseconds, which is too short in comparison with the applied voltage to be taken into account.

The proposed model is a simplified dynamic model of PD in the internal void, which allows simulating phase resolved partial discharge (PRPD) patterns. The mathematical model was simplified to decrease the calculation time. Thus, the model does not consider conductivity and temperature dependencies. Model parameters could be determined from an optimization process based on experimental results from the measurement on insulation sample with an artificial void.

II. Physical Background

As was already mentioned, two main conditions must be fulfilled to start the PD process inside a void: sufficient electric field and an initial free electron. The description of electric field distribution inside the void can be determined from the following fundamental equations:

$$\nabla \cdot \overline{E} = \frac{\rho}{\varepsilon}, \tag{1}$$

$$\nabla \cdot \overline{j} + \frac{\partial \rho}{\partial t} = 0, \tag{2}$$

where E is the electric field intensity, ρ is the free charge density, j is the free current density and ε is the permittivity of the material. The equations (1) and (2) can be combined and written as:

$$\nabla \cdot (-\sigma \nabla V) - \nabla \cdot \varepsilon \nabla \frac{\partial V}{\partial t} = 0, \tag{3}$$

where σ is the electrical conductivity, and V is the electric potential. This equation can be solved by the finite element method, if the geometry, material constants, and initial and boundary conditions are known.

The electric field inside of void has two components. Except for the contribution of the background field, which is caused by the void, there is also the contribution of the electric field associated with local space charge distribution from previous discharge activity. The value of local charge directly depends on the electric field which caused the partial discharge. The deployed charge is decreased with the time constant which depends on the void diameter and surface conductivity. This effect is usually called "memory effect" and has a significant influence on the time delay of PDs.

The streamer initiation criterion is given by a minimal stability charge and for spherical voids can be written in the form [6]:

$$E_i = \left(\frac{E}{p}\right)_C p\left[1 + \frac{B}{\sqrt{2pd}}\right] \quad (4)$$

where E_i is the inception electric field strength, B is the characteristic constant for gas, p is the pressure, d is the void diameter, and $(E/p)_C = 25.2\ \mathrm{Vm^{-1}\ Pa^{-1}}$ is the critical ratio between the electric field required to sustain the discharge and pressure p.

The generation of first electrons that potentially start the avalanche ionization process in the void is from two sources, volume and surface processes [1]. The volume generation is caused by a photoionization process started by high-energy photons or by electron detachment from negative ions. For both cases, the electron rate is a function which mainly depends on the kind of gas, electric field, gas volume, and the probability that a single electron can develop the avalanche process. The photoionization process is more important for internal voids, which are enclosed by insulating material. However, the contribution of volume generation is sufficiently small compared to the surface generation and usually can be neglected. Electron production from an internal void surface comes from detrapping of electrons from insulation material, electron release by ion impact or photoemission. It was experimentally proven that surface electron emission can be approximately described by Richardson-Schottky scaling [1]. It is assumed, that the number of electrons, which can be detrapped from the surface, is proportional to the PD charge q:

$$N_{d0} = k\frac{q}{e} \quad (5)$$

where k is the proportional constant, and e is the elemental charge. The charges decay with a time constant τ_d as they are moving into deeper traps. The number of detrappable electrons in time t after the previous PD can be expressed as:

$$N_d = N_{d0}e^{\frac{-t}{\tau_d}}. \quad (6)$$

The surface emission increases with the applied electric field. For the PD occurrence likelihood determination, some simplifications introduced in [5] will be further used. If it is assumed that the electron surface emission rate is exponentially dependent on the electric field, thus:

$$N_e = N_d e^{\left|\frac{E_v}{E_i}\right|}. \quad (7)$$

where E_v is the electric field on void surface. The likelihood of PD occurrence is dependent on the value of N_e and the time interval after the previous PD.

III. Experimental Measurement

Fig. 1 shows the electrode arrangement with the test sample which represents spherical void inside the insulating material. The test sample was prepared from epoxy resin material with artificially created spherical void placed in the middle of the sample. The spherical void is prepared by injection of air bubble during the hardening process. The following sanding and polishing processes adjust the sample thickness and position of the air bubble. The final sample thickness is 3.5 mm and the void diameter 1.4 mm. The whole electrode system with the test sample is immersed in natural insulating oil (Envirotemp™ FR3™ natural ester dielectric fluid) to avoid surface discharges.

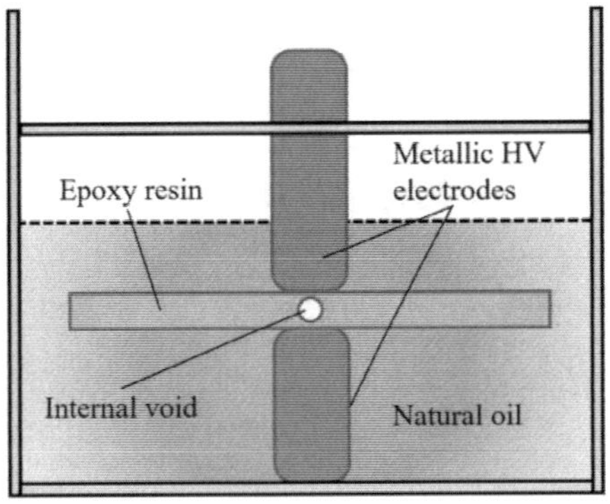

Figure 1. Electrode and test sample arrangement

The high voltage circuit for PD measurement is shown in Fig. 2. The high voltage can be controlled by the supply transformer up to the voltage level 100 kV. The PD signal and voltage signal are transferred to the PD detector via optical cables and evaluated in a personal computer. OMICRON MPD600 system was used for PD data gathering. The measurement circuit must be first calibrated. After that, the background noise is measured and potential sources of disturbing PD discharges must be eliminated. Subsequently, the applied voltage is being continuously increased until the inception voltage is reached. The test sample is then stressed for 30 minutes to ensure stable conditions. If the apparent charge and the number of PDs are almost constant in time, the PRPD diagrams are recorded.

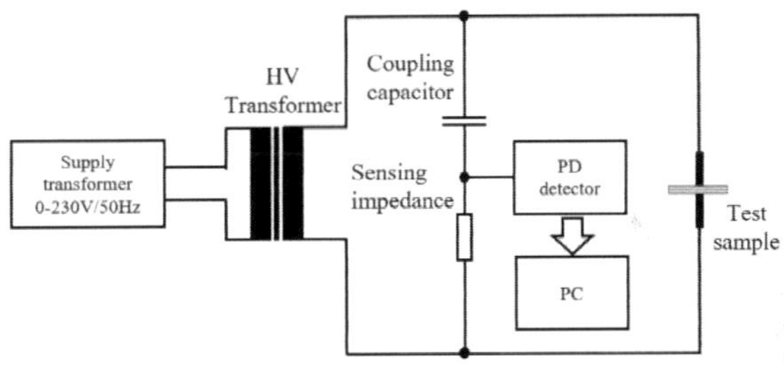

Figure 2. Schematic diagram of test high voltage circuit for PD measurement

IV. Numerical Model of PDs in the Internal Void

A finite element method (FEM) numerical model of PD activity inside a void was created in Wolfram Mathematica software. The model was designed to work with any void shape or voltage waveform, and therefore could be employed for simulation of PD activity inside the sample from the experimental part and compared with the measured data. The modus operandi of this model is shown as a flowchart in Fig. 3; the individual processes are explained in detail below.

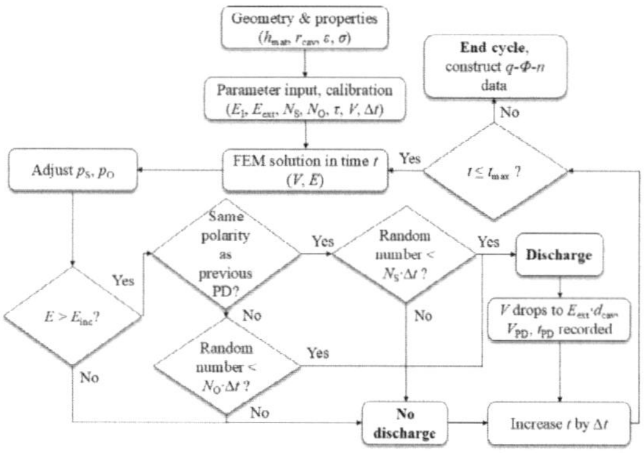

Figure 3. Schematic diagram of test high voltage circuit for PD measurement

A. Geometry & Properties

Model geometry corresponded to the dimensions of the sample (sample diameter, thickness, void diameter). As the sample was practically axisymmetric, a significant simplification of the model could have been made to increase its calculation speed. In this manner, the standard 3D configuration was turned into a 2D axisymmetric one, assuming that the discharge activity could be modeled as an axisymmetric problem with less than a marginal error. The relative permittivity and conductivity of both the void and bulk material were estimated for the employed sample material and can be seen in Table I. These properties remained constant throughout the simulation, as opposed to those in other models, such as in [5].

TABLE I. MODEL PARAMETERS

Parameter	Value
V	$20\ 000 \cdot \sin(2\pi \cdot 50 \cdot t)$ V
h_{mat}	3.5 mm
r_{cav}	0.7 mm
d_{mat}	5.0 mm
$\varepsilon_{r,cav}$	1.0
$\varepsilon_{r,bulk}$	4.4
σ_{cav}	$1 \cdot 10^{-15}$ S/m
σ_{bulk}	$1 \cdot 10^{-12}$ S/m
E_{inc}	3.012 kV/mm
E_{ext}	0.842 kV/mm
τ_d	2 ms
N_O	10
N_S	3 000
q-n MSE	1 126
ϕ-n MSE	1 209
Total n	1 528 (1 578)

B. Model Parameters, Calibration

Each of the model parameters listed in the flowchart determined the shape of a resulting phase resolved partial discharge (PRPD) pattern. The inception electric field strength was unequivocally given by Equation (4) as 3.01 kV/mm, while the extinction field strength had to be determined from the voltage amplitude and the ratio between the maximum and minimum magnitude of discharges of experimental results. Free electron availability probability was divided into two terms: N_S for the same and N_O for the opposite polarity as/to the previous discharge. These parameters changed throughout the calculation in agreement with the theoretical assumptions listed above. Initial values of both probabilities were estimated from the resulting patterns by finding the minimal error between calculated and measured q-n and φ-n distributions. The employed model utilized a time constant that represented the momentary build-up and subsequent decay of availability of free electrons after a discharge occurred. In agreement with the theory, this time constant is denoted as τd, and its value was chosen as $\tau_d = 2$ ms [5]. Voltage V and its waveform were similar to the experimental ones - a 50 Hz sinusoidal waveform with an amplitude of 20 kV. Regarding the size of the time step, a total number of 3 000 steps per one period was deemed sufficient to provide solid accuracy while maintaining reasonable calculation speed. Taking into account the frequency of the employed voltage, 3 000 steps per cycle yielded a time step size of about 6.67 μs.

C. FEM Model Solution & Main Cycle

The finite element method was used for the simulation of the electric field inside the sample. The electric field distribution was determined from the solution of Equation (3). Since partial discharges were regarded as sharp drops of voltage in the void, the model parameters had to be changed during the calculation. In agreement with the theoretical assumptions, several discrete variables were introduced, such as the time of previous PD occurrence, its magnitude, or space charge electric field inside the void. A partial discharge could only occur when the void internal electric field was equal or larger than the inception electric field strength and when there was a free electron available. The former could be easily determined from the field model, while the latter required additional calculation using free electron availability probabilities. At the start of each step, both probabilities were recalculated from the new values of electric field strength and time since the previous PD. Next, a random number was generated and compared with the respective probability depending on the polarity of the previous discharge. If the random number was lower than the probability multiplied by the time step size and if the internal electric field was stronger than the inception field, a PD occurred. As the duration of a partial discharge is much shorter than the employed time step, the electric field was reduced instantly to the value of extinction electric field strength and the difference was recorded and later recalculated as discharge magnitude. Afterward, time t was increased by Δt, and the calculation was repeated until the maximum time was reached.

The electric field strength difference was also used as an auxiliary parameter in the calculation of the next time step, in both previous discharge inception field strength and adjustment if space charge on the void surface. Each PD altered the space charge by a value equal to its magnitude. The initial value of space charge was considered to be zero, and the magnitude of each PD was either added to or subtracted from it, depending on the discharge polarity.

D. Pattern Construction

After the main cycle had concluded, PRPD patterns were constructed from the solved variables. Phase value φ was gathered simply by using modulo on times of PD occurrences. Next, discharge magnitude q was determined from the difference between the void internal electric field before and after a PD and calibrating it to the real measured values. Finally, the number of PDs for a given phase and discharge magnitude values *n* was calculated by sorting individual PDs into a 500 by 400 φ-q table. Such classification corresponded to the one used by the measuring device used in the experimental part. The number of PDs inside each box of the grid represented the searched variable *n*.

V. MODEL RESULTS AND EXPERIMENTAL VERIFICATION

Using the parameters defined in the previous chapter, both initial electron availability probabilities were estimated as N_O = 10 and N_S = 3 000 by finding the minimum of mean squared error between the calculated and measured results for both q-n and φ-n data. Table I shows the model parameters used for the best estimation of availability probabilities along with the mean squared errors (MSE) between the model and experiment.

A PRPD pattern of the calculated model is shown in Fig. 4. The same type of pattern constructed from experimental results can be seen in Fig. 5. For comparison, both q-n and φ-n data plots from model and experimental results can be seen in Fig. 6 and Fig. 7, respectively.

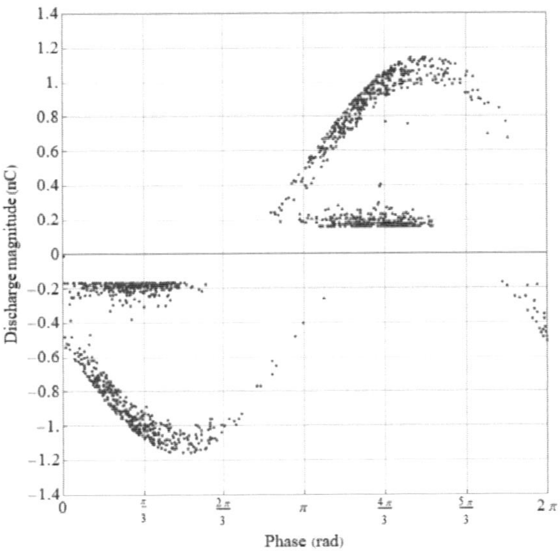

Figure 4. PRPD pattern of numerical model results

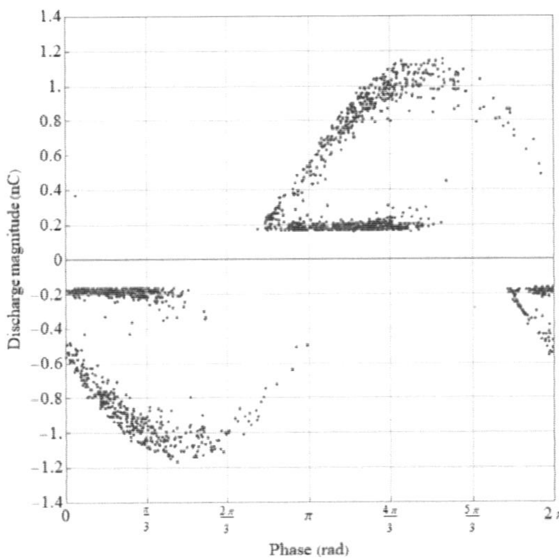

Figure 5. PRPD pattern of experimental results

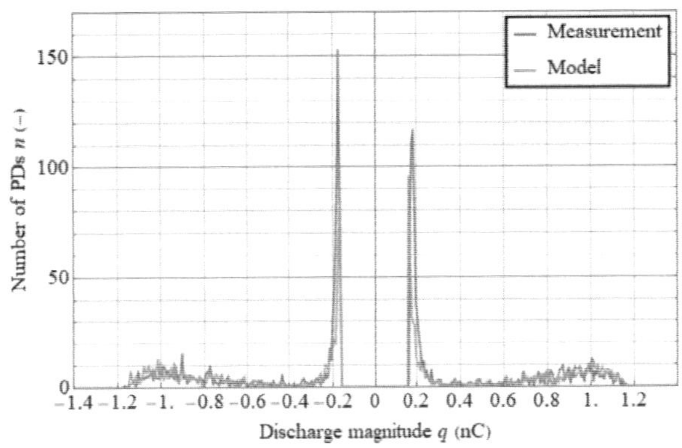

Figure 6. Comparison of q-n data between the model and experiment

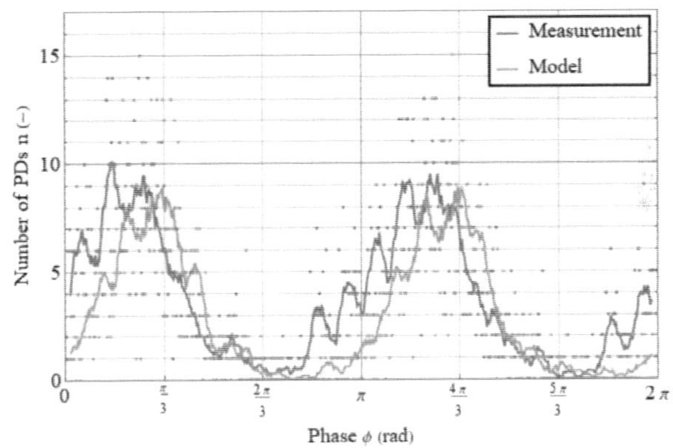

Figure 7. Comparison of φ-n data between the model and experiment. Moving averages with a period of 10 samples are shown as solid lines

VI. CONCLUSION

A numerical dynamic model for PD activity inside an insulating material void has been constructed. The model is The generated PRPD pattern resembles the pattern gathered from PD measurement on the sample well enough. From the comparison of q-n data, it is evident that the PD activity at lower discharge magnitudes is more intensive in case of the experimental measurement. A small phase shift is present between both trends in φ-n data plots. Nevertheless, the proposed model is based on the physical processes that occur inside and in close vicinity of the void when a sufficient voltage is applied. Several simplifications were made to enhance the calculation speed of the model. A sample with a single artificial void was prepared from epoxy resin. The sample's dimensions were recorded and PRPD patterns were measured at 20 kV. These patterns were compared to those generated by the numerical model. Model input parameters were chosen to represent the experimental sample as well as possible. Some parameters had to be calibrated from the measured patterns, such as the extinction electric field strength and discharge magnitudes. numerical model seems viable for use in the simulation of PD activity inside the given cavity, and therefore applicable for various shapes of voids, materials and voltage waveforms due to its generic nature.

ACKNOWLEDGMENT

This work was supported by the Grant Agency of the Czech Technical University in Prague, grant No. SGS18/133/OHK3/2T/13.

REFERENCES

[1] L. Niemeyer, "A generalized approach to partial discharge modeling," in IEEE Transactions on Dielectrics and Electrical Insulation, vol. 2, no. 4, pp. 510–528, Aug. 1995.

[2] R. J. Van Brunt, "Physics and chemistry of partial discharge and corona. Recent advances and future challenges," in IEEE Transactions on Dielectrics and Electrical Insulation, vol. 1, no. 5, pp. 761–784, Oct. 1994.

[3] I. Gallimberti, "A computer mode for streamer propagation," in Journal of Physics D: Applied Physics, vol. 5, no. 12, pp. 2179–2189, Dec. 1972

[4] R. Morrow, "Theory of negative corona in oxygen," in Americal Physical Society, vol. 32, no. 3, pp. 1799–1809, Sep. 1985

[5] H. Illias, G. Chen and P. L. Lewin, "Partial discharge behavior within a spherical cavity in a solid dielectric material as a function of frequency and amplitude of the applied voltage," in IEEE Transactions on Dielectrics and Electrical Insulation, vol. 18, no. 2, pp. 432–443, April 2011.

[6] G. C. Crichton, P. W. Karlsson and A. Pedersen, "Partial discharges in ellipsoidal and spheroidal voids," in IEEE Transactions on Electrical Insulation, vol. 24, no. 2, pp. 335–342, April 1989.

Towards to Thermal Stability of Equivalent Series Resistance and Capacity of Flexible Sensor

Ján Zbojovský, Juraj Kurimský
Department of Electric Power Engineering
Technical University of Košice, FEI
Košice, Slovakia
jan.zbojovsky@tuke.sk, juraj.kurimsky@tuke.sk

Abstract—**The paper presents the thermal stability of flexible (microstrip) sensor. A brief characteristic of flexible sensor is presented and the equivalent series resistance and capacity was measured. The temperature was changed in the range from -10 to 90 °C with step 5 °C. The frequency during measurement start from 100 kHz to 1 MHz.**

Keywords: flexible sensor, thermal stability, frequency, resistance

I. Introduction

Regularly changing, spaced side by side electrodes are a frequently used electrode arrangement. They are used in the field of non-destructive testing (NDT), acoustics, biotechnology, electrochemistry, telecommunications, etc. Although the same physical principle is used, different names have been used over time for this structure, such as: nested or interdigital, periodic, comb, or microstrip. When designing their specific application, it is important to solve geometry, microstrip manufacturing process, material selection, analytical or numerical modeling, system integration and processing, addressing and data analysis. [1][2]

Common properties

The microstrip sensor can be easily placed in the measured sample and penetrated by electrical, magnetic or acoustic fields from one side of the sample. The other side may be open, allowing the absorption of gas, moisture, or chemicals that can change the electrical properties of the sample being measured. A sensitively sensitive layer of chemical or biological nature located above the electrodes may interact with the gaseous or liquid medium. Thus, it is allowed the monitoring concentration of chemicals in materials such as air, transformer oil or even the human body.[3][4]

The level of the measured signal in the microstrip sensor can be controlled by changing the size of the sensor area, adjusting the number of electrodes and the distance between them. The signal-to-noise level and minimum area of the microstrip sensor is designed on the requirements by the specific application. Microstrip sensor is also used in frequency dielectrometry. This technology uses pairs of electrodes in a ridge arrangement, with the spatial periodicity of arrangement is variable.

The penetration depth of the quasi-static electric field lines into the material is independent of the frequency and directly proportional to the spatial wavelength. This is defined as the distance between the center axes of adjacent electrodes of the same polarity.

The dielectric properties may be related to many other physical properties, such as porosity, density, structural integrity, but also the chemical composition of the test materials. The microstrip sensor can be placed on the surface of the insulator to measure surface conductivity. The working temperature on the surface of the insulators ranges from cca -20 ° C to 90 ° C. Due to the temperature range it is necessary to observe whether the temperature affects the sensor properties. [5][6]

II. Properties of Sensor

The electrodes are placed on a flexible PEN (Polyethylene naphthalate) PEN film can be mechanically stressed up to 160 ° C, electrically up to 180 ° C. The electrode material is silver Ag. Resistance of the Ag paste is generally less than 100 mΩ /sq. The thickness of the substrate is 0.050 mm, glass transition temperature of 155 ° C (class F use possible), dielectric strength up to 300 kV / mm (25 μm).

An example of an arrangement of mutually overlapping the active parts of the electrodes are shown in Figure 1. Geometric characteristic consists of the number of electrodes, the length of the electrode *L*, width *W* and inter-electrode gap *G*. In English literature, these types of arrangements are called "inter-digited."capacitors ". [7] They are widely used as chemical and biological sensors where capacitance or impedance is changing as a result of interaction between analytical and sensor layer [3]. At present, this arrangement is made in different scales, normally they also use micro-devices known as MEMS (Micro-Electro-Mechanical Systems). [8]

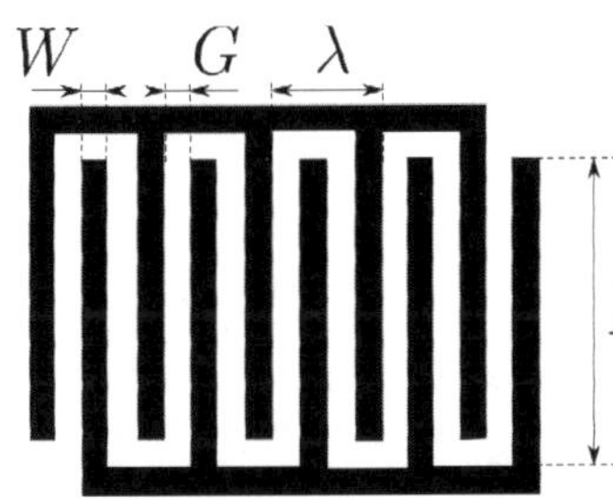

Figure 1. Example the errangement of electrodes

The sensor's geometric layout is as follows:

Width, W= 0.5 mm

Gap, G= 0.5mm

Length L= 49 mm

Lambda λ = 2 mm

III. On Resistance and Capacity

Surface resistivity refers to the resistance that occurs at the leakage current along the surface of the insulating material. It can also be defined as an electrical resistance that exists between two parallel electrodes that contact the surface. The electrodes must be separated by a distance equal to the length of the contacts of the electrodes. The four ends of the electrodes form a square, so the resistance is expressed in ohms and sometimes in ohms per square meter. Surface resistivity is calculated as the ratio of the potential gradient with units V/m and current per unit length of electrodes, A/m. This is summarized as the ratio of the DC voltage drop per unit length of the electrode to the current per unit, width. The surface resistance determines the guidance of any given material. The higher the surface resistivity, the lower the conductivity of the material because it will have a low leakage current [8]. The surface resistance, R_S (1), is defined as the ratio of the direct voltage U to the current I_s passing between the two electrodes with the specified configuration that are in contact with the same side of the test material (Fig. 2) [9].

$$R_s = \frac{U}{I_S} \tag{1}$$

The surface resistivity ρ_s, on the other hand, is determined by the ratio of the DC voltage U falling to the ratio unit of the length L and to the surface current I_s to the ratio unit of the width D.

$$\rho_s = \frac{UD}{I_S L} \tag{2}$$

Surface resistivity is a property of a material. Theoretically, it should remain constant regardless of the method and arrangement of the electrodes used to measure surface resistivity. The result of measurement of surface resistance depends on material and geometry electrodes used in measurement [9].

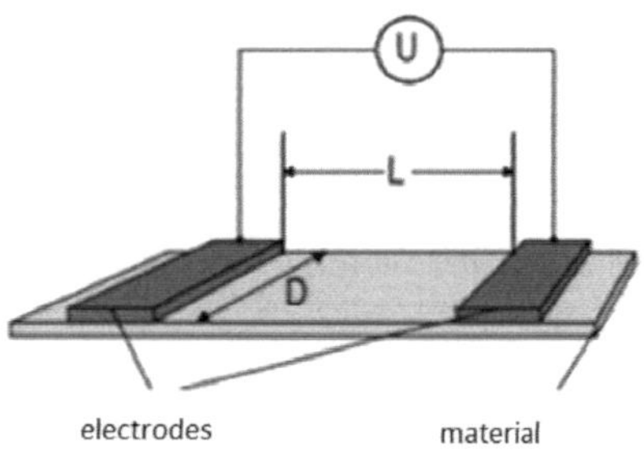

Figure 2. The basic configuration of electrodes for the measuring of the surface resistivity (source [9])

A. Surface resistivity

Surface resistivity refers to currents flowing in the surface layer of the material. It is determined indirectly by measuring the surface resistance, taking into account the effective length of the measuring electrode and the gap width between the electrodes. This measurement is particularly difficult because of the surrounding environment influence the surface conductivity. This fact can be seen on materials with very large resistivity [10] [11] [12].

B. Volume resistivity

The volume resistivity refers to the current flowing through the insulating material under the influence of a constant electric field. It is always determined indirectly, by measuring the volume resistance, taking into account the effective surface electrode and the thickness of sample. Measurements of this type are very simple, especially with regard to very small currents. Therefore, when determining the internal material resistance, a large number of samples should be examined and then an average value calculated, especially if the material is inhomogeneous [11].

It is recommended to measure the internal resistivity using three electrodes: the measuring and protection electrodes on one side and the voltage electrode on the other side. The same number of electrodes is also recommended in case of surface resistivity, but here the electrode arrangement is different. The measuring and voltage electrodes are located on one side of the dielectric, the protective electrode on its opposite side [13].

C. On determining the capacity

To determine/calculate the capacity of sensor, a replacement scheme is provided that contains two types of capacities: C_I is the half capacity value of the internal electrode to grounded electrode, C_E is the capacity of the outer electrode of the adjacent grounded electrode, Fig.3

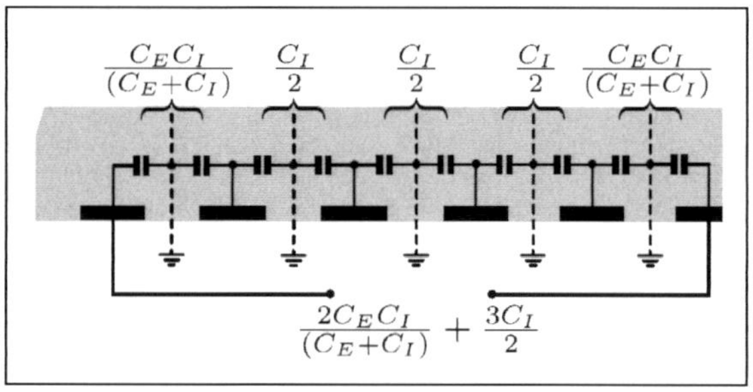

Figure 3. Equivalent cirquit for the calculation of the capacity

The resulting capacity of the microstrip sensor is:

$$C = (N-3)\frac{C_I}{2} + 2\frac{C_I C_E}{C_I + C_E} \tag{3}$$

Where N>3 is the number of electrodes.

For the application of a sensor for external contamination, it is useful to have two substances applied with different permittivity. The capacities C_E and C_I for multiple dielectric layers which has different thicknesses and permittivity are calculated by the so-called the partial capacitance method, Fig. 4,. ε_1 ε_2 are permittivity of the layers and their thickness are h_1 and h_2. The resulting sensor capacity is then:

$$C_u = C_{h=\infty} + (\varepsilon_1 - 1)C_{h=h_1} + (\varepsilon_2 - \varepsilon_1)C_{h=h_2} \tag{4}$$

C_h is the geometric capacity of one layer, which is dependent on its thickness and on the concrete geometry of electrode.

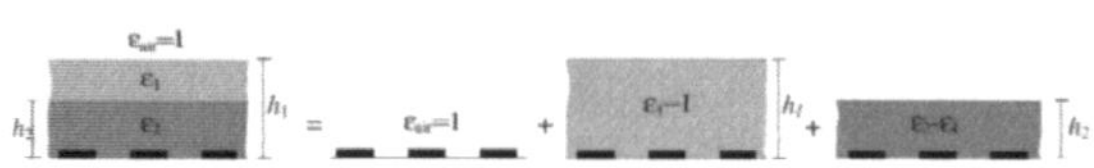

Figure 4. Calculating the capacity of sensor for two dielectric layer

The surface capacity of sensor depends on the properties of the surrounding substances and thus also the substances present in the fallout pollution that covers its active area.

IV. Method of Measurement

The equivalent series resistance and capacity of the capacitor model where measured. On the following picture is shown whole sensor.

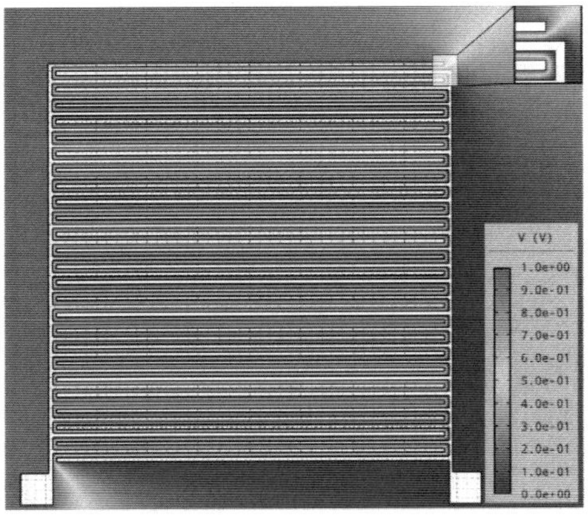

Figure 5. Construction of sensor

Behavior of the capacitor can be modeled according to a serial equivalent model, Fig. 6

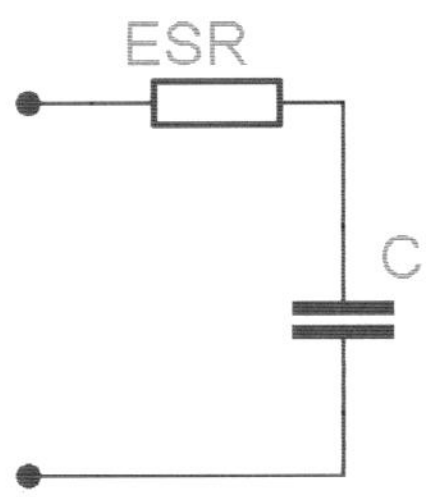

Figure 6. Serial equivalent model

From an equivalent circuit model, the relationship between loss factor and ESR applies:

$$\tan \delta = ESR \cdot \omega \cdot C \qquad (5)$$

Measurement parameters:

- Start Frequency: 100 kHz
- Stop Frequency: 1 MHz
- Sweep Mode: linear
- Number of Points: 51
- Source level: 13dBm
- Bandwidth: 10 Hz

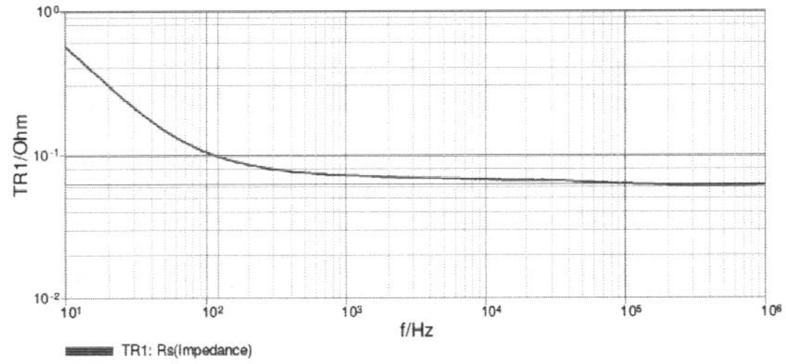

Figure 7. Graphical dependence of impedance from frequency

V. Results

On the following figures are described the results from measurement. At first the changes of capacity in frequency spectrum where measured. Temperature was changed from -10 to 90 °C with step 5°C. It can be seen that the capacity oscillated with changing the temperature in range aproximately 0,18 in comparison with measurement at 0 °C. The capacity was increasing from the frequency about $1{,}0*10^6$Hz.

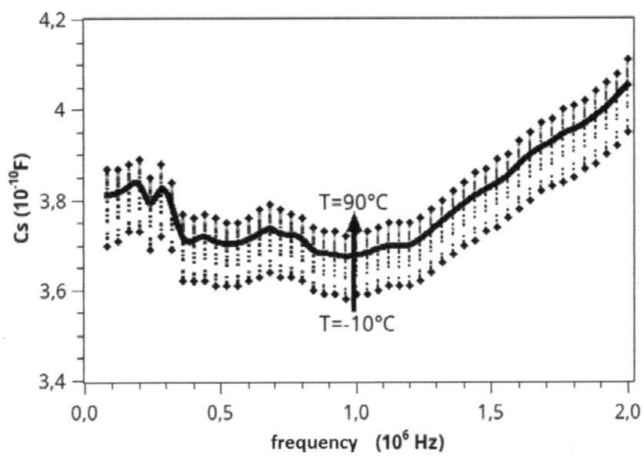

Figure 8. Changes of the capacity in frequency spectrum

Next experiment was focused on the measurement of ESR in frequency spectrum. The temperature was also was changed from -10 to 90 °C with step 5°C. The resistance in dependence from the frequency was decreasing. It can be seen that the resistance oscillated with changing the temperature in range aproximately 27 Ohm from the lower frequencies and gradualy was stabilized.

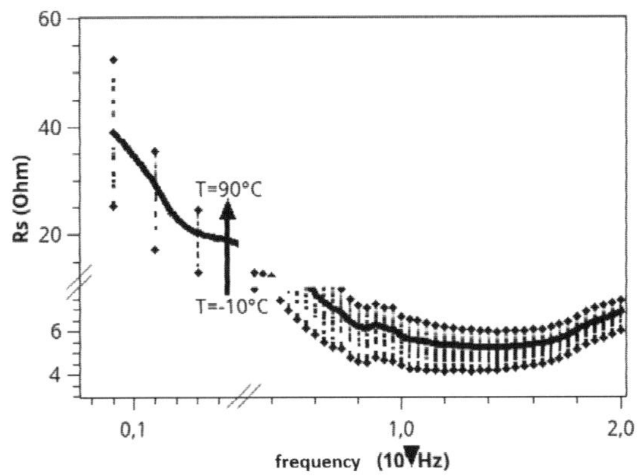

Figure 9. Changes of the ESR in frequency spectrum

VI. Conclusion

The goal of this paper was to analyze the thermal stability and the capacity of flexible sensor. It was measured the changes of capacity and the resistance in dependence from frequency and temperature. The output is graphical dependencies based on which the sensor properties can be determined.

Acknowledgment

This work was supported by the Slovak Research and Development Agency under the contract No. APVV-15-0438.

References

[1] J. Kim, M. Chaudhury, and M. Owen, "Hydrophobicity loss and recovery of silicone HV insulation", IEEE Trans. Dielectr.Electr. Insul., Vol. 6, pp. 695–702, 1999.

[2] William A. Maryniak, Toshio Uehara, Maciej A. Noras, :Surface Resistivity and Surface Resistance Measurements Using a Concentric Ring Probe:, Technique Trek Number 1005.

[3] A. Pietrikova, "Potentiality of LTCC for sensor applications," in 24th International Spring Seminar on Electronics Technology. Concurrent Engineering in Electronic Packaging. ISSE 2001. Conference Proceedings (Cat. No.01EX492), Calimanesti-Caciulata, Romania, 2001, pp. 112–116. DOI: 10.1109/ISSE.2001.931025

[4] González, G., Kolosovas-Machuca, E.S., López-Luna, E., Hernández-Arriaga, H., González, F.J., „Design and Fabrication of Interdigital Nanocapacitors Coated with HfO2", Sensors 15, 1998–2005, 2015, doi:10.3390/s150101998.

[5] A. Pietrikova and M. Kravcik, "Investigation of rheology behavior of solder paste," in 2012 35th International Spring Seminar on Electronics Technology, 2012, pp. 138–143. DOI: 10.1109/ISSE.2012.6273124

[6] Abdul Rahman, M.S., Mukhopadhyay, S.C., Yu, P.-L., "Novel Planar Interdigital Sensors", in: Novel Sensors for Food Inspection: Modelling, Fabrication and Experimentation. Springer International Publishing, Cham, pp. 11–35, 2014.

[7] M. Pavlík, I. Kolcunová, B. Dolník, J. Kurimský, A. Mészáros, D. Medveď, M. Kolcun and J. Zbojovský, "Measuring of Dependence of Shielding Effectiveness of Wet Materials on The Frequency of Electromagnetic Field in the High Frequency Range," Acta Electrotechnica et Informatica. Vol. 13, No. 3 (2013), s. 12-16. - ISSN 1335-8243.

[8] Solymar, L., Walsh, D., " Electrical properties of materials", Oxford university press, Issue 8, 2010, ISBN 978-0-19-956592-4

[9] M. L. Clingerman, J. A. King K. H. Schulz J. D. Meyers, "Evaluation of Electrical Conductivity Models for Conductive Polymer Composites", Journal of Applied Polymer Science, Vol. 83, 1341–1356 (2002), © 2002 John Wiley & Sons

[10] J. Džmura, J. Petráš, J. Balogh and M. Bernát, "Modeling and Computer Simulation of Electrical Separation," 17th International Scientific Conference on Electric Power Engineering (EPE), MAY 16-18, 2016, Prague, CZECH REPUBLIC, p. 537-542, ISBN:978-1-5090-0908-4.

[11] D. Medved, M. Kolcun, Z. Conka, „Simulation of Faults in the Power System and Their Elimination in the EMTP-ATP" In: Proceedings of the 7th international scientific symposium on electrical power engineering (Elektroenergetika 2013) pp. 484-487, ISBN 978-80-553-1441-9, 2013.

[12] M. Špes, Ľ. Beňa, M. Kosterec, L. Kruželák, Z. Müller, H. Wachta, M. Márton, "Impact of Environmental Conditions on the Capacity of Power Lines", Acta Electrotechnica et Informatica. 17 (2017) 3–7. DOI:10.15546/aeei-2017-0018.

[13] A. V. Mamishev, K. Sundara-Rajan, Fumin Yang, Yanqing Du, and M. Zahn, "Interdigital sensors and transducers," Proceedings of the IEEE, vol. 92, no. 5, pp. 808–845, May 2004. DOI: 10.1109/JPROC.2004.826603.

Physical Characteristics Influence of Shielding Coating on the Electromagnetic Field Penetration

Ján Zbojovský, Marek Pavlík, Iraida Kolcunová
Department of Electric Power Engineering
Technical University of Košice, FEI
Košice, Slovakia
jan.zbojovsky@tuke.sk, marek.pavlik@tuke.sk, iraida.kolcunova@tuke.sk

Abstract —**The paper deals with the shielding effectiveness evaluation of experimentally created coatings. Many studies have shown potential biological and thermal effects of GSM electromagnetic fields, therefore people are concerned about their health. Electromagnetic wave does not penetrate the wall of the building as a whole. A small part of the wave is reflected and a small part is absorbed by the building material.**

Keywords: shielding coating, electromagnetic field, shielding effectiveness, frequency

I. Introduction

In recent years, there has been rapid growth in the use of technologies in which the electromagnetic radiation of different frequencies is produced. The sources of this radiation include 50 Hz electrical lines, local power lines, as well as cables located in homes, offices and many other buildings. In addition, all machines and equipment connected to the network create an electric field. Similarly, a rapid increase began in technologies using radio frequency. This includes radios, TV sets, telecommunication lines, satellite connections, mobile phones and wireless internet. [1] Electromagnetic fields emitting from electronic devices can be harmful for organism or devices. [2][3]The only known natural source of electromagnetic radiation is the negligibly weak cosmic rays from space, but several radiation sources exist in radio frequencies due to atmospheric phenomena such as the Northern Lights and Storms. By the 1950s, the frequency of electrosmogs remained outside the boundaries of microwaves, but in the 1960s television stations began broadcasting at these frequencies. Portable phones were added in the 1980s and WiFi came in the 1990s. Both use microwave frequencies considerably. Electromagnetic fields are often used in medicine, such as magnetic resonance imaging (MRI), which gives a very good and clear picture, making it much easier to diagnose a patient. [4][5]

In addition to the positive results generated by the electromagnetic field, they are also negative. Therefore, the impact of the electromagnetic field has become the subject of a great deal of research, much of which has been focused on the long-term effects of the electromagnetic field. It is much more difficult to determine because human testing is not possible, and so it is not possible to speak accurately about the diseases caused primarily by this field. However, the electromagnetic field may have a negative impact on the individual, depending on his / her state of health and other conditions. The discovery, that an electromagnetic field could have a direct negative impact on the emergence of some types of diseases could negatively affect the current lifestyle dependent on electricity.

II. Evaluating the Shielding Effectiveness

Shielding is the ability of a material to prevent electromagnetic field penetration. Shielding is used in various kinds of devices from different areas of everyday life, for example in mobile phones and stations, Wi-Fi devices and the Internet, through various medical devices, communication networks and so on. For electromagnetic fields of lower frequencies, the electrical and magnetic components of the electromagnetic field are considered separately [6 [7] This term is related to electromagnetic compatibility, which can be defined as the ability of a device to coexist in one electromagnetic environment. To determine the quality of shielding, the Shielding effectiveness (SE) indicator, defined by the relationship [8], is used.

$$SE = A + R \tag{1}$$

where A represents the absorption of the electromagnetic field and R represents the reflection of the electromagnetic field. For the absorption coefficient>

$$A = 8{,}69 \cdot \frac{t}{\delta} \tag{2}$$

where μ represents the permeability of the enviroment in which the permeability of the shielding material is also included and σ represents the specific conductivity of the shielding material

For the reflection coefficient [9] [10]:

$$R = 20 \cdot \log\left(\frac{1}{4} \cdot \sqrt{\frac{\sigma}{\omega \cdot \mu_r \cdot \varepsilon_0}}\right) \tag{3}$$

If the transmitted signal values are set in logarithmic units, the shielding effectiveness can be determined by the following relationships [11 [12]:

$$SE = |E_1| - |E_2| \tag{4}$$
$$SE = |H_1| - |H_2| \tag{5}$$
$$SE = |V_1| - |V_2| \tag{6}$$
$$SE = P_1 - P_2 \tag{7}$$

where E_1 represents the intensity of the electric field impinging on the barrier, E_2 is the intensity of the electric field at a certain point of the shielded space, H_1 represents the intensity of the magnetic field impinging on the barrier, H_2 is the intensity of the magnetic field at a certain point of the shielded space, V_1 is the wave voltage of the electromagnetic field incident on the barrier V_2 is the wave voltage of the electromagnetic field at a certain point in the shielded space, P_1 is the power of the electromagnetic field incident on the barrier, P_2 is the power of the electromagnetic field at a certain point in the shielded space.[11][12]

III. Measurement Setup

Measurement of the shielding effectiveness of a building element during the penetration of electromagnetic waves took place in an anechoic chamber. The measurement workplace consisted of the Agilent N5181A signal generator, which was powered by the RF antenna DRH18-E. Between the antennas there was placed a building element with a shielding coating. The output from the receiver antenna R&S HF 907 was connected to the Agilent N9038A MXE EMI spectrum analyzer. A block diagram for measuring the shielding effectiveness of an electromagnetic field is shown in Fig. 1 and the workplace on the Fig.2.

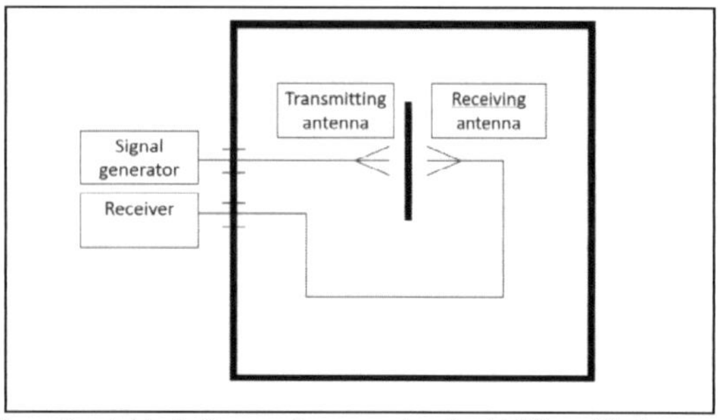

Figure 1. A block diagram for measuring the shielding effectiveness

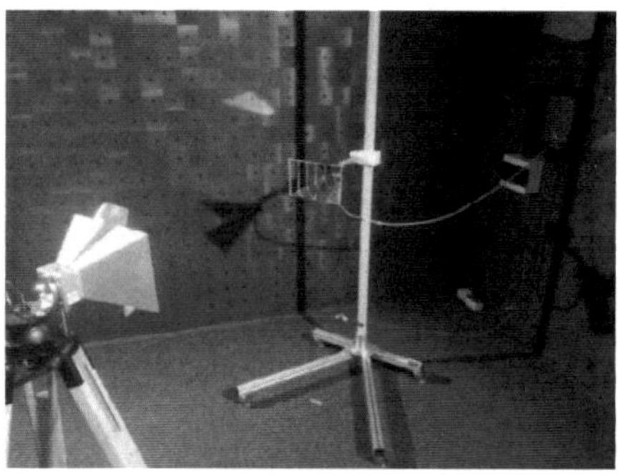

Figure 2. Measurement workplace

The whole measurement can be divided into two parts. The first part was to measure the attenuation of a building element without a shielding coating. The second part was to measure the attenuation of the building element on which the shielding coating was located.

An OSB board was chosen to measure the shielding effectiveness of the electromagnetic field as a building block. Subsequently, the following shielding coatings were applied to this building element:

- YShield HSF 74
- THOMSIT R762
- experimental coating with ferrite powder 15g

YShield HSF 74 is a professional shielding paint containing a preservative-free silicate paint. Capable of shielding electrical and magnetic field components. The manufacturer gives a shielding effectiveness of 39 dB using a single coating, in the case of two coatings the manufacturer gives 46 dB. THOMSIT R762 is a solvent-free conductive surface coating used to provide electrical conductivity before laying antistatic floor coverings. It is also necessary to realize that uncertainty of measurement occurs during this measurement. Calculation was done based on the document of Slovak national accreditation service. The overall uncertainty of measurement equals 0.35 dB.

IV. Results

At first was measured wooden material OSB 1000x1000x4mm without shielding coating. On the next table are shown the results for shielding effectiveness. Shielding effectiveness measurements were performed in the 0.9 GHz-6 GHz frequency range with 0.1 GHz step.

Table 1. Shielding effectiveness for the OSB wooden board.

Frequency f (MHz)	The value of received signal power (dBm)	Average value of the received signal after shielding (dBm)	Shielding effectiveness SE (dB)
900	-13,2117	-16,7361	0,1667
1100	-13,2117	-14,2856	-0,0467
1800	-15,8398	-17,4296	0,1167
1900	-16,0334	-16,5362	0,3867
2100	-17,8676	-17,5662	0,2733
2400	-17,3782	-17,7289	0,1667
2600	-20,0195	-18,8472	1,0067
4000	-24,6216	-24,3191	2,2433
5000	-29,7374	-30,5096	2,8400
6000	-31,1677	-31,7274	2,3000

OSB has reached numerically higher values of shielding effectiveness against electromagnetic radiation at certain frequencies, but has a significantly greater number of frequencies at which the effectiveness is close to zero and in some cases equal to zero. Therefore, it can be concluded that the OSB shielding effectiveness is largely dependent on the used frequency. Graphical dependence the shielding effectiveness from frequency is shown on the following Figure.

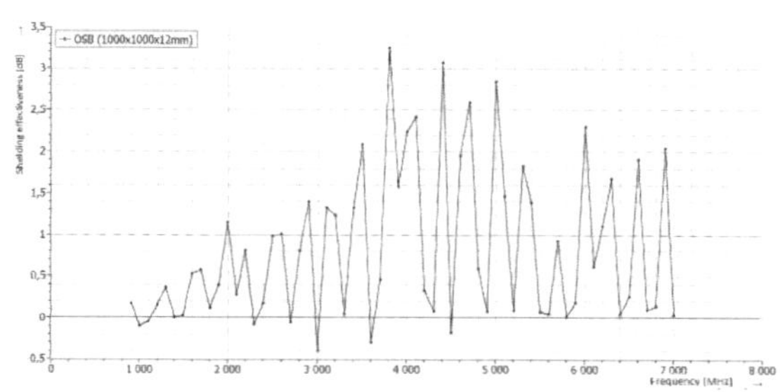

Figure 3. Graphical dependence of shielding effectiveness from frequency, OSB wooden material

In order to compare the shielding effectiveness of different types of coatings, the shielding effectiveness of experimental coating with YSHIELD is compared. For the purpose of the experiment, HSF 74 was selected as described above. This coating is intended mainly for indoor use and its primary use is primarily in the area of higher frequencies of the EM field or electric field and lower frequencies, where the magnetic component of low-frequency EM fields is not able to shield. The coating is based on the use of carbon and therefore does not contain any metallic particles, therefore the coating has a long life and is resistant to oxidation. Next experiment represents the OSB material on which was applied HSF74 coating.

Table 2. Shielding effectiveness for the OSB board with applied HSF74

Frequency f (MHz)	The value of received signal power (dBm)	Average value of the received signal after shielding (dBm)	Shielding effectiveness SE (dB)
900	-15,4293	-19,7773	4,3479
1100	-13,3421	-16,5706	3,2286
1800	-14,8187	-23,3061	8,4873
1900	-14,2043	-20,8792	6,6749
2100	-16,4593	-22,5355	6,0763
2400	-18,0802	-23,0460	4,9658
2600	-18,6003	-24,8674	6,2671
4000	-25,6524	-35,3226	9,6702
5000	-29,0993	-38,9244	9,8250
6000	-31,0988	-37,6276	6,5287

For the purpose of the experiment, the thinnest possible thickness of the shielding coating was applied evenly over the entire surface of the material, thus explaining the reduced shielding effectiveness proposed by the manufacturer in the documentation. It is possible to observe a gradual increase in shielding effectiveness with increasing frequency, but as in previous measurements with a certain oscillating character.

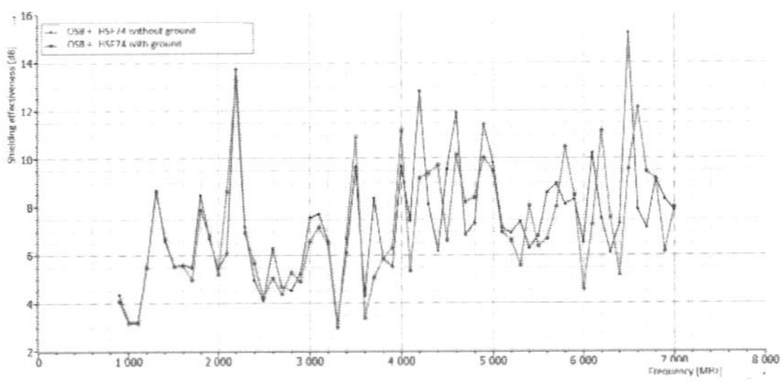

Figure 4. Graphical dependence of shielding effectiveness from frequency, OSB + HSF74 shielding coating

On the next table are shown the results from the application of the second layer on OSB material.

Table 3. Shielding effectiveness for the OSB board with 2x HSF 74 coating.

Frequency f (MHz)	The value of received signal power (dBm)	Average value of the received signal after shielding (dBm)	Shielding effectiveness SE (dB)
900	-15,1790	-24,0317	8,8527
1100	-12,7438	-37,6979	24,9541
1800	-16,4138	-31,6381	15,2243
1900	-14,0199	-30,0789	16,0590
2100	-17,0208	-36,5099	19,4891
2400	-16,4416	-42,3682	25,9266
2600	-18,1922	-42,1360	23,9438
4000	-26,1127	-39,6655	13,5528
5000	-29,2103	-44,7922	15,5819
6000	-29,9617	-49,1808	19,2191

In view of the values reported by the manufacturer of HSF74, where the shielding effectiveness value is estimated to be 45dB, which is the 99.997% shielding value quoted, the measurement was carried out to verify that the professional paint shielding capability met the present values in the attached documentation. According to the results achieved, it is possible to see a significant improvement in the shielding effectiveness of the SE compared to using only one HSF 74 coating layer. The highest effectiveness value in the measured frequency range (0.9 - 6 GHz) was achieved at the frequency of 4.2 GHz, 31.912 dB.

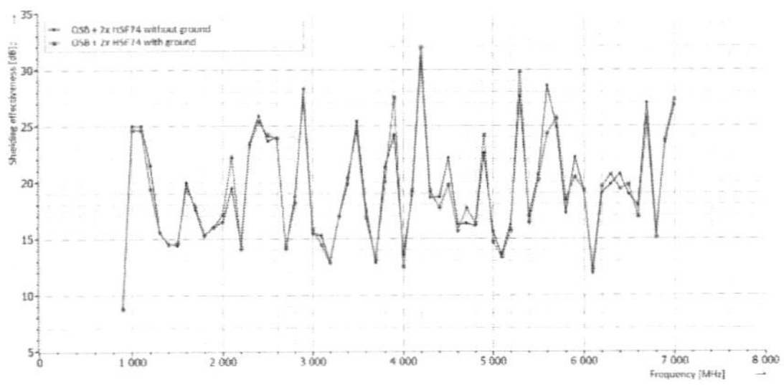

Figure 5. Graphical dependence of shielding effectiveness from frequency, OSB +2x HSF74 shielding coating

Next experiment was focused on the conductive coating R 762. The conductive coating R 762 serves primarily to achieve increased electrical conductivity before laying conductive floor coverings. In the case of performing the experiment, it is used as an electrically conductive medium to serve the purpose of combining with additional admixture. The conductive coating consists of a polyacrylate-based synthetic resin dispersion with conductive properties, and the leakage resistance value is less than 3.10^{5} Ohms. For the purpose of determining the possible ability to shield the electromagnetic field, the measurement of the board was made primarily using solely R 762 pure coating without any additional admixture. The results are included in Table 4

Table 3. Shielding effectiveness for R762 applied on the OSB board..

Frequency f (MHz)	The value of received signal power (dBm)	Average value of the received signal after shielding (dBm)	Shielding effectiveness SE (dB)
900	-17,0270	-17,0289	0,0019
1100	-14,2353	-14,2390	0,0037
1800	-14,8839	-15,3124	0,4284
1900	-16,9076	-16,8348	-0,0728
2100	-16,1405	-17,5332	1,3927
2400	-17,5624	-18,0452	0,4828
2600	-18,0780	-18,5199	0,4419
4000	-24,4158	-25,5878	1,1720
5000	-27,9717	-27,9588	-0,0129
6000	-30,2348	-31,2786	1,0439

From the results obtained using a clean conductive coating, it is possible to observe a minimum to almost zero shielding capability at certain frequencies, regardless of whether the object was grounded or not when measured, so a clean coating of R 762 does not have a separate use in the electromagnetic field shielding and is necessary additional coating enhancement with additives. The measured values of effectiveness will be used as reference values for the next experiment continuation, where the next paint will be prepared by mixing of R762 with ferrite powder.

The use of ferrite is wide-spread in electrical engineering because it is a kind of ferromagnetic material that has the necessary physical properties as low eddy current losses and can be used, for example, as a magnet. However, in the shielding area, it is often used to prevent the propagation of radio frequency interference in power cables in the form of a ferrite toroid or cylinder. In view of this, ferrite can also be tested for the ability to shield electromagnetic field propagation in combination with a conductive coating of R 762. For the purpose of the experiment, several pieces of ferrite cores were pulverized and 15g of ferrite powder was mixed into a conductive coating and applied on the OSB board, and the results are follows:

Table 4. Shielding effectiveness for R762 and 15 g of ferrite powder

Frequency f (MHz)	The value of received signal power (dBm)	Average value of the received signal after shielding (dBm)	Shielding effectiveness SE (dB)
900	-15,2780	-17,8455	2,5674
1100	-13,3252	-16,3048	2,9796
1800	-15,2281	-17,1739	1,9458
1900	-14,9824	-16,0603	1,0780
2100	-16,7895	-18,9338	2,1444
2400	-18,5915	-19,8059	1,2143
2600	-17,4376	-19,6363	2,1987
4000	-25,9014	-27,6718	1,7704
5000	-29,9736	-32,0279	2,0542
6000	-32,9333	-35,5503	2,6170

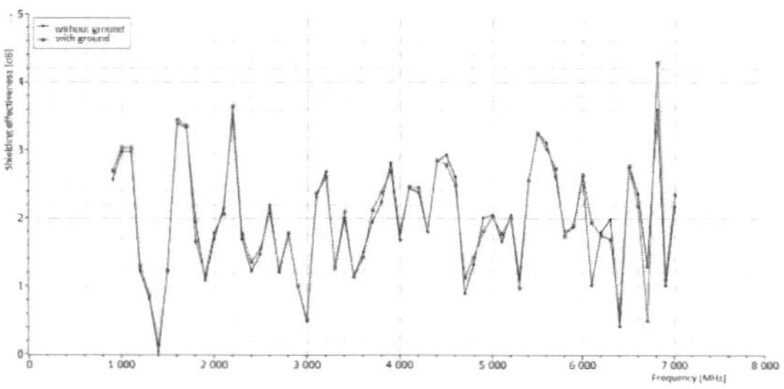

Figure 6. Graphical dependence of shielding effectiveness from frequency, R762 + 15 grams of ferrite powder

V. Conclusion

In this paper was done the experimental evaluating of shielding effectiveness of some materials. At first was analyzed the shielding coating HSF 74. The best values of the shielding effectiveness achieved by the application on the second layer and it was similar as described in documentation from the manufacturer.

I the experiment was as the first measured R762 conductive paint. The measured values of effectiveness were used as reference values for the next experiment continuation, where the next paint was prepared by mixing of R762 with 15 grams of ferrite powder. The ferrite powder achieved some shielding properties. The highest values of the shielding effectiveness were measured in the frequency range up to 3 GHz and the area above 6 GHz. The potential for better utilization of the ferrite powder is to improve the crushing of the ferrite cores to a finest powder and to use a different method for combining materials such as the conductive coating used, where conventional mixing is not an appropriate choice.

Next experiments will be focused on the improving of the mixing such as constant mixing with the conductive coating or using ultrasound or magnetic mixing, and the increasing of the amount of powder in the conductive coating.

Acknowledgment

This work was supported by the Scientific Grant Agency of the Ministry of Education of Slovak Republic and the Slovak Academy of Sciences by the projects VEGA No. 1/0435/19.

This work was supported by the Cultural and Educational Grant Agency of the Ministry of Education, Science, Research and Sport of the Slovak Republic (KEGA) under the project No. 008TUKE-4/2019.

This work was supported by project: Influence of material on electromagnetic field penetration, FEI-2018-54.

References

[1] Health protection agency: Static magnetic fields: Report of the independent advisory group on Non-ionising radiation. Centre for radiation, Chemichal and Enviromental Hazards, Oxfordshire pp 143. ISBN 978-0-85951-616-7, 2017

[2] M. Pavlík, I. Kolcunová, B. Dolník, J. Kurimský, A. Mészáros, D. Medveď, M. Kolcun and J. Zbojovský, "Measuring of Dependence of Shielding Effectiveness of Wet Materials on The Frequency of Electromagnetic Field in the High Frequency Range," Acta Electrotechnica et Informatica. Vol. 13, No. 3 (2013), s. 12-16. - ISSN 1335-8243.

[3] S. Bucko, M. Češkovič, J. Labun, P. Kurdel, "Mobile phone effect on quality of results of the sensitive measurements" In: MOSATT 2017. - Košice:TU, LF, pp. 22-25. - ISBN 978-80-553-2864, 2017

[4] IEEE 299-2006 Standard, Method for Measuring the Effectiveness of Electromagnetic Shielding Enclosures, EMC Society, New York 2006, p.39.

[5] D. Medveď and O. Hirka O, "Investigation of Electromagnetic Fields in Residential Areas," In: Acta Electrotechnica et Informatica. Vol. 17, No. 3 (2017), pp. 48-52. - ISSN 1335-8243.

[6] C. Erbas, S. Kent, "Shielding effectiveness of a rectangular cavity with aperture between 1-3 GHz ", IEEE International Symposium on Electromagnetic Compatibility, , Istanbul, pp.343 - 346 Vol.1, ISBN: 0-7803-7779-6, 2013.

[7] J. Labun, S. Fábry, M. Češkovič, P. Kurdel, "Mechanical demodulation of aircraft antenna signal", Transportation Research Procedia Volume 28, pp 149-155, ISSN: 2352-1465 , 2017.

[8] J. Džmura, J.Petráš, J.Balogh, M. Bernát, „Modeling and Computer Simulation of Electrical Separation" 17th International Scientific Conference on Electric Power Engineering (EPE), Prague, CZECH REPUBLIC, p. 537-542, ISBN:978-1-5090-0908-4, 2016

[9] D. Medved, M. Kolcun, Z. Conka, „Simulation of Faults in the Power System and Their Elimination in the EMTP-ATP" In: Proceedings of the 7th international scientific symposium on electrical power engineering (Elektroenergetika 2013) pp. 484-487, ISBN 978-80-553-1441-9, 2013

[10] IEEE Guide--Adoption of IEC/TR 61000-3-7:2008, „Electromagnetic compatibility (EMC)--Limits--Assessment of emission limits for the connection of fluctuating installations to MV, HV and EHV power systems" IEEE Std 1453.1-2012 (Adoption of IEC/TR 61000-3-7:2008) , pp 78, E-ISBN : 978-0-7381-7285-9, 2012.

[11] B. Dolník, "Electromagnetic compatibility. (Elektromagnetická kompatibilita)," TU of Košice, monography. ISBN 978-80-8086-221-3, 2013

[12] M. Špes et al., "Impact of enviromental condition on the capacity of power lines," In: Acta Electrotechnica et Informatica. Vol. 17, No. 3 (2017), pp. 3-7. - ISSN 1335-824

The Changes of Electrical Properties of ZnO-based Varistors Initiated with Current Pulses

Bystrík Dolník, Patrik Molčan, Samuel Hájnik
Department of Electrical Power Engineering
Technical University of Košice, FEE&I
Košice, Slovakia
bystrik.dolnik@tuke.sk

Abstract — **Electronics used in various industries and household appliances is threatened by the overvoltage. The protection of electrical equipment against the effects of the overvoltage is provided by the surge protection devices. The ZnO-based varistor represents an important element used in surge protection devices. In this paper, we focus on the degradation effects of current pulses to ZnO varistors. Standard current pulses 8/20 shape were applied to varistor samples of commercial production. The electro-physical properties of ZnO varistors were monitored by measuring the current-voltage characteristic and the dielectric parameters by the spectroscopy in the frequency range from 20 Hz to 2 MHz. It refers to the temperature dependence of the ZnO varistor's current-voltage characteristic in the leakage current region and to the reduction of the coefficient of non-linearity of the varistor at elevated temperature.**

Keywords: varistor; ZnO; ceramics; current pulse; non-linearity coefficient; degradation

I. Introduction (Heading 1)

Today, extreme electronics production continues to be used in various industries as well as in households. At the same time, there are threats that adversely affect electronics. Overvoltage is one of the most serious threats to electronic devices. Overvoltage is the voltage that exceeds the maximum permissible voltage value in the electrical circuit. It can also be said that it is extremely high voltage, lasting in the order of microseconds to milliseconds. Overvoltage may have a natural origin, such as atmospheric overvoltage (lightning strike and induced overvoltage) and electrostatic discharge (ESD), or are generated artificially, i.e. by human intervention or technical equipment, generally called switching overvoltage [1-7].

Overvoltage effects occur or immediately (destructive effects on electrical devices) or gradually, with accelerated degradation of electrical equipment [8-10]. To protect electrical equipment and electronic equipment against overvoltage, surge protection devices (SPD) are used also referred to as surge arresters. The purpose of the SPD is to limit the extreme effects of overvoltage to the isolation of electrical equipment, thereby avoiding power outage caused by electrical equipment failures, while avoiding economic damage. There are different types of SPDs, which are constantly upgraded [10].

The basic types of SPDs used in low voltage network include the protective spark gap, gas discharge tube (GDT), and varistor [11]. Protective spark gaps provide a protective temporary potential equalization at the moment of overvoltage, mostly bonding to the ground. The gas discharge tube is actually a spark gap housed in a cylindrical shape made housing mostly of glass or ceramic. The housing interior of the GDT is filled with an inert gas at low pressure. Metal oxide varistors are voltage-dependent resistors, the resistance of which decreases substantially with increasing voltage and vice versa, with the voltage decreasing the varistor goes into a non-conductive state and the resistance increases. In operation varistors are stressed by surges, resulting in the deterioration of their electro-physical properties. The changes in the electro-physical properties of ZnO based varistors can be monitored under laboratory conditions by injection of current pulses mimicking atmospheric overvoltage.

II. Zinc Oxice Based Varistors

Zinc oxide varistors are produced by sintering zinc oxide ZnO at high temperatures (typically from 1100 °C to 1400 °C) with a precisely defined temperature course. The metal oxide ceramic has a strongly non-linear volt-ampere characteristic which, in addition to the sintering process, is dependent on the additives added in the manufacturing process. The chemical composition consists of 95 % to 98 % zinc oxide mixture and the rest consists of additives such as Bi_2O_3, TiO_2, CoO, MnO, and Sb_2O_3 and others [12]. Thanks to these additives, the nonlinear response and the stability of the varistors are further improved. In addition to the aforementioned ingredients, various studies and testing of the effects of other additives are carried out to improve the properties of varistors [13-17].

A. The microstructure of ZnO varistor

In the production of a varistor by sintering zinc oxide powder with other metal oxide additives, a polycrystalline ceramic is formed, which, like the added ingredients, has a great influence on the electro-physical properties of the varistor.

Zinc oxide grains, see Fig 1, are highly conductive, while a non-conductive layer that is formed by other oxides is highly non-conductive. In the areas of contact of zinc oxide grains, microvaristors are formed during sintering. The electro-physical properties of microvaristors are comparable to Zener diodes. The breakdown voltage at the ZnO grain boundaries is approximately 3 V to 3.5 V [18].

Typically, ZnO grains are 5 µm to 20 µm in size, with grain sizes depending on the composition of the material, temperature and sintering time [19]. Varistors with different voltage gradients must be produced to accommodate different levels of protection. The voltage gradient represents the basic indicator in the whole volume of the varistor. When the voltage gradient value increases, the effect of the avalanche effect that causes the breakdown increases. At the ZnO grain boundaries, the microstructure is complicated and divided into three

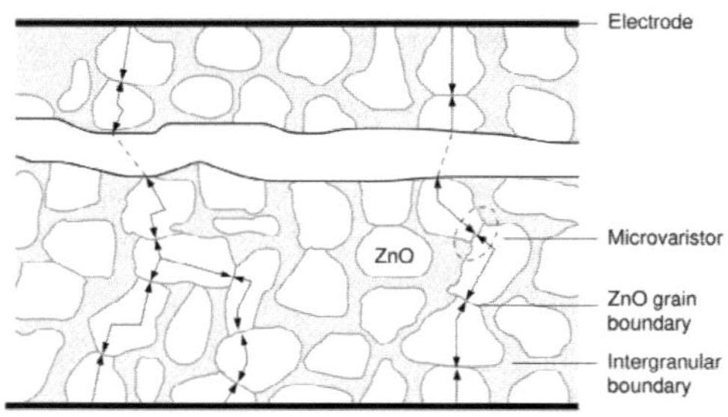

Figure 1. The microstructure of ZnO based varistor.

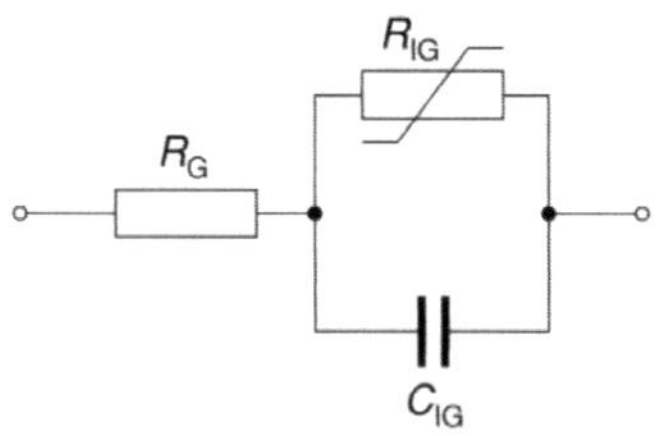

Figure 2. The simple equivalent circuit model of MOV.

groups. The layer between ZnO grains formed by Bi_2O_3 at grain boundaries marked as type I has a thickness of 0.1 μm to 1 μm. For the type of layer referred to as II, the layer between ZnO grains becomes thinner because of the space between the ZnO grains. The layer III type is characterized by the content of elements such as Bi, Co and by an excessive amount of ions at the grain boundaries of several nanometers [12].

A simple equivalent electrical circuit for describing ZnO material consists of a $R_{IG}C_{IG}$ parallel electrical circuit in series with R_G resistor and inductance [20]. R_{IG} is defined as a highly non-linear, voltage-dependent resistor. Similar to the non-linear behavior of ZnO layers at grain boundaries, we have the capacity, C_{IG}. R_G is the electrical resistance of conductive ZnO grains. The capacitance of ZnO grains is very small and can be in most cases neglected. The serial inductance represents the inductance of the wire considered only for steep responses. The resulting equivalent circuit model of MOV is shown in Fig. 2.

B. *Electrical properties of ZnO varistor*

The electrical properties of the ZnO based varistor (MOV) are characterized by the voltage and current (VI) characteristic. For different types of varistors, the VI characteristic is different. The VI characteristic varies not only from the varistor species but also from the percentage of additives added to the sintered powder. The VI characteristic of the MOV is almost symmetrical, and the working area can be described by the equation

$$I = C\,V^{\alpha}\,, \tag{1}$$

where I is the current flowing through the varistor, V is the voltage applied to the varistor, C is the constant dependent on the varistor type, and α is the non-linearity coefficient of the MOV.

The non-linearity coefficient α for MOV varies from 25 to 60, some MOV also have a higher value of the non-linearity coefficient [16]. SiC based varistors have a non-linearity coefficient in the range of 3 to 7. The non-linearity coefficient is determined either from the VI characteristic or by the equation

$$\alpha = \frac{\log\left(\frac{I_2}{I_1}\right)}{\log\left(\frac{V_2}{V_1}\right)}\,, \tag{2}$$

where V_1 and V_2 ($V_1 < V_2$) are the voltages at the currents of I_1 and I_2 respectively. The relationship in (2) can be written as

$$\alpha = \frac{\log I_2 - \log I_1}{\log V_2 - \log V_1}\,. \tag{3}$$

In the case of specified points on the VI characteristic, for example $I_1 = 0.1$ mA and $I_2 = 1$ mA, the non-linearity coefficient is calculated as

$$\alpha = \frac{1}{\log V_2 - \log V_1}\,. \tag{4}$$

The VI characteristic of the ZnO varistor is divided into three regions. The first region is called the leakage region. In this area, varistor acts as a linear resistor and has a high resistance value, of the order of 10^9 Ω. The second region is the work region defined by (1). In this region, the varistor resistance decreases, varistor acts as a variable resistor. The last region is a region of extreme currents, characterized by a very low resistance value, in the order of ohms. In this area, the overvoltage is limited to the desired protection level.

It is recommended to continuously monitor the varistor state by measuring the selected point on the VI characteristic. Practically, it is sufficient to check more defined points on the VI characteristic. The most frequently used control point on the VI characteristic is the so-called milliampere point. This point represents the DC voltage that causes a 1 mA current flow through the varistor.

C. *The aging of ZnO varistor*

Aging, respectively varistor degradation is the process by which MOV impairs its electrical and physical properties. The degradation of electrical properties is caused mainly by the deterioration of the microstructure at the ZnO varistor grain boundaries. The most appropriate way to observe degradation is to continuously monitor changes in VI characteristics.

The processes that cause MOV degradation can be divided into three groups. The first group represents the degradation of the MOV by the operating voltage. The second group includes high-amplitude current pulses due to overvoltage that adversely affects the VI characteristic of the MOV. The deflection of the VI characteristic is mainly manifested in the leakage current region. The greatest change in VI characteristic can be observed around the operating voltage. The degradation of the microstructure of MOV caused by chemical reactions belongs to the third group, which differs significantly from the degradation caused by current pulses. The degradation of the MOV microstructure caused by chemical reactions can be caused, e.g. partial discharges, resulting in an increase in leakage current.

A large number of experiments have been performed to better understand the aging processes of MOV. In the experiments, the methods of accelerated aging by the electric field, the elevated temperature, and injection of current pulses are generally used. The combined stresses are often used, e.g. the DC at elevated temperature [12], [21], [22].

III. Experiment

In the experiment, the temperature dependence of MOV of commercial production in the first and second region of the VI characteristic have been investigated, and also the impact of current pulses on the degradation of MOV non-linearity coefficient.

A. Temperature dependence measurements

The VI characteristic of the ZnO varistor was measured with KEITHLEY 2410 source meter in the temperature range from room temperature 23 °C to 100 °C. After measuring at room temperature, successive measurements were made from 23 °C to 100 °C with a 10 °C step. Similarly, the dielectric parameters of the MOV were measured using the Agilent E4980A Precision LCR meter in the frequency range from 20 Hz to 2 MHz.

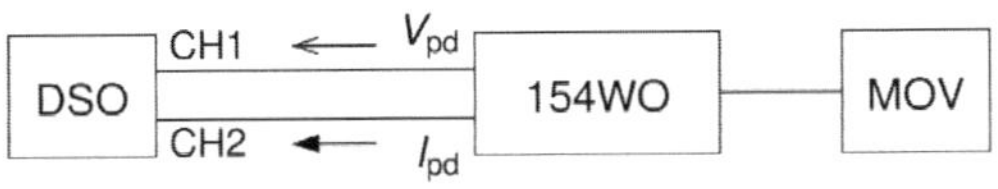

Figure 3. The block giagram of the measuring circuit.

The measured sample was placed in an electric furnace, with the VI characteristic being started after 10 minutes from the time the oven temperature had stabilized. During the VI characteristic measurement, the temperature in the electric furnace varied with a maximum deviation of ±0.3 °C from the desired temperature. The temperature in the electric furnace was measured with a K-type thermocouple with a device with a temperature resolution of ±0.1 °C and a declared accuracy of ±(0.5 % + 0.5 °C).

The non-linearity coefficient was calculated according to (4) for each temperature of the varistor measured. The current values of $I_1 = 0.1$ μA and $I_2 = 0.1$ mA, were chosen as the current reference values

B. Surge current injection

The degradation of the MOV with current pulses was performed using HAKEL current pulse generator 154WO. A normalized 8/20 current pulses were injected into the MOV sample. The time course of voltage on the MOV and the corresponding current flowing through the MOV were measured using a two channel digital storage oscilloscope (DSO) GDS806S. The measured data in the DSO memory was stored in a text file using the script written in the Python programming language. In Fig 3 the block diagram of the measuring circuit is shown.

In Fig 3, the V_{pd} and the I_{pd} are the acronyms for varistor voltage and the current passing through the varistor measured with voltage divider and current shunt, respectively. The energy absorbed by the test sample was calculated numerically using a trapezoidal method. The numerical calculations of the energy absorbed by the MOV were programmed in GNU Octave version 4.0.0.

C. Test procedure

The test sample of the MOV was measured at the start of the experiment to obtain initial data on the non-degraded varistor. The VI characteristic and also the dielectric parameters on the sample at elevated temperatures were measured.

After the initial measurement, the test sample was degraded by 8/20 current pulses to accelerate the aging of the measured sample. After the injection of current pulses into the sample, the VI characteristic and dielectric parameters of the sample were measured again. The test procedure was then repeated until the test sample was completely destroyed.

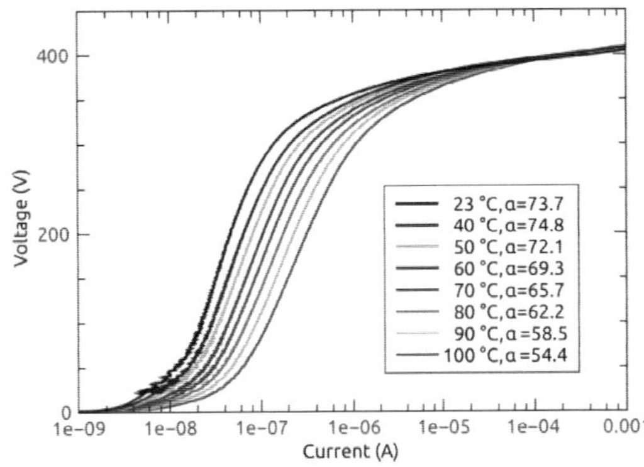

Figure 4. The tempereture dependence of the VI characteristics of the MOV sample tested.

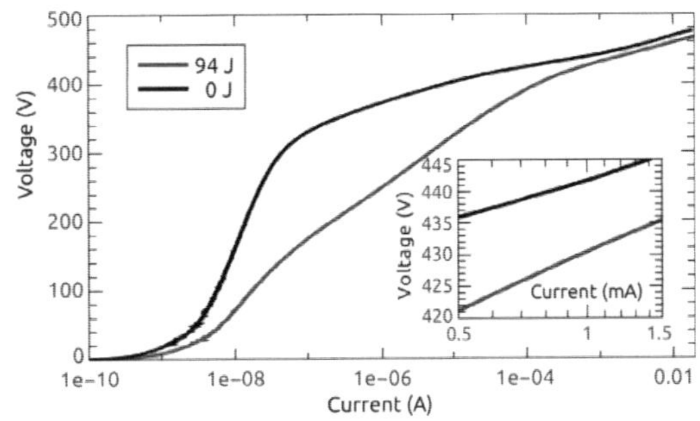

Figure 5. The VI characteristics of the MOV sample tested.

D. Evaluation

As mentioned above, we measure at the beginning of the experiment the initial state. From the measured data, the VI characteristics, the non-linearity coefficient and the complex permittivity were calculated. Subsequently, the changes in VI characteristics, the temperature dependence of 1 mA point on VI characteristic, the changes in non-linearity coefficient, relative permittivity and dielectric loss factor were compared. Changes in electrical parameters are interpreted using graphical dependencies and tables.

IV. RESULTS AND DISCUSSION

In Fig 4. the temperature dependence of the VI characteristics of the MOV sample tested in the temperature range from 23 °C to 100 °C is depicted. From Fig. 4, the temperature dependence in the leakage region is evident. With increasing temperature, the leakage current increases at a constant test voltage. This increase in leakage current has a significant impact on the degradation of the VI characteristic of the MOV and thus on its service life.

The changes in VI characteristics of the MOV sample at the initial state and after degradation with current pulses with 94 J energy absorbed are shown in Fig. 5. As can be seen from the Fig. 5, significant changes in VI characteristics occurred already after absorbing relatively small energy. Figure 5 also shows a detail of the change in the 1 mA point on the VI characteristic of the MOV sample. The voltage value at 1 mA point reduces from 441.4 V at the beginning of the experiment to 430.3 V after absorbing 94 J energy. The change in the 1 mA point on the VI characteristic is due to the degradation of the individual microvaristor at the ZnO grain boundary [23].

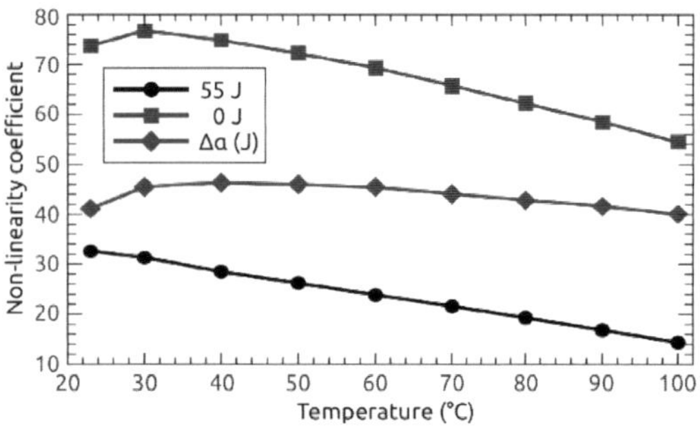

Figure 6. The temperature dependence of the non-linearity coefficient of the MOV sample tested.

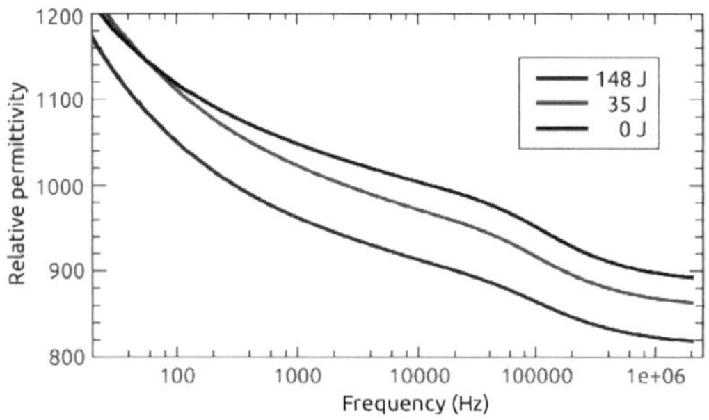

Figure 7. The change of the relative permittivity in frequency domain.

Figure 6 shows the temperature dependence of the varistor non-linearity coefficient. The parameter is the energy absorbed by the MOV sample. It is clear from the calculated values of the non-linearity coefficient that, with increasing sample temperature, the non-linearity coefficient of the MOV decreases. By linear approximation, the decrease in the MOV non-linearity coefficient in the temperature range from 23 °C to 100 °C can be expressed by a linear function

$$a(t) = -0.323t + 87.77\,, \tag{5}$$

where $\alpha(t)$ is a temperature-dependent MOV non-linearity coefficient and t is the MOV sample temperature. The coefficient of determination is $R^2 = 0.991$. As can be seen from graphical dependence shown in Fig. 6, after absorbing the energy of 55 J, the MOV non-linearity coefficient was significantly reduced. The temperature dependence of the non-linearity coefficient over the whole measured temperature range was also confirmed. Also in this case the decrease in the non-linearity coefficient of the MOV with increasing temperature can be expressed by the linear equation

$$a(t) = -0.237t + 38.13\,, \tag{6}$$

where $\alpha(t)$ and t have the same meaning as in the previous equation 5. In this case, the coefficient of determination is $R^2 = 0.999$. In Fig. 6, the dashed line represents the difference between the MOV non-linearity coefficient in the initial state and after the 55 J energy absorption

$$\Delta a(t) = a_{0J}(t) - a_{55J}(t)\,, \tag{7}$$

where $\alpha_{0J}(t)$ is the temperature-dependent MOV non-linearity coefficient in the initial state and $\alpha_{55J}(t)$ is the corresponding temperature-dependent MOV non-linearity coefficient after the 55 J.

It can be seen from Fig. 6 that the MOV non-linearity coefficient difference $\Delta a(t)$ is almost constant. The calculated average value of the MOV non-linearity coefficient difference is 43.7, and the corresponding calculated standard deviation is 2.3. In other words, after absorbing the energy of 55 J, the varistor non-linearity coefficient decreased by an average of 43.7 in the measured temperature range from 23 °C to 100 °C.

Table 1 shows the average leakage currents of the MOV tested at selected stages of the experiment, the multiples of the leakage current (LC) to the initial state, and the percentage change in LC to the initial state. The initial LC was determined to be $0.8V_{1\,mA}$, where the $V_{1\,mA}$ is defined as the voltage corresponding to 1 mA direct current passing through the MOV sample.

In Table 1, the LC multiples are calculated as the ratio of the LC of the degraded MOV to the LC of the MOV in the initial state. The percentage change in the LC of the MOV is calculated as an increment of the LC to the initial state.

Table 1. The change in leakage current

State	Leakage current (μA)			
	Mean	*STD**	*Multiple*	*Change (%)*
Initial	0.314	0.185	–	–
Degraded (94 J)	2.810	0.196	8.960	759.9
Degraded (113 J)	3.581	0.306	11.416	1041.6
Degraded (438 J)	9.198	1.050	29.326	2832.6

*Standard deviation

Based on these results, it can be concluded that the injected current pulses significantly reduce the MOV non-linearity coefficient even at low energy absorbed values, in the order of tens of Joules. Moreover, current pulse degradation results in a significant increase in the leakage current, resulting in an increase in the Joule losses of the MOV and an increase in the MOV temperature.

The changes in relative permittivity of the MOV sample in the frequency domain at the initial state and after degradation with current pulses with 35 J and 148 J energy absorbed are shown in Fig. 7. As can be seen, the relative permittivity of the MOV decreases with increasing frequency, both for the initial state of the MOV and for the MOV degraded by the current pulses.

At frequencies greater than 100 Hz, the decrease of the relative permittivity is not too sharp compared with the lower frequency region. For test frequencies greater than 100 Hz, the MOV's relative permittivity is shifted to lower values with increasing absorbed energy. However, for test frequencies less than 100 Hz, the relative permittivity of the MOV was shifted towards lower values after energy absorption of 148 J.

According to the equivalent circuit model depicted in Fig. 2, one can argue that the shift of the relative permittivity of the MOV after energy absorption to lower values is due to the reduction of the space charge at the ZnO grain boundaries [23] and changes in the depletion region, since the relative permittivity of the ZnO grain itself is about 8.5 [24-26].

In Fig. 8, the change of the dielectric loss factor of the MOV sample in the frequency domain is presented. As can be seen from the dielectric loss factor versus frequency dependence in Fig. 8, the dielectric loss factor decreases with the increasing

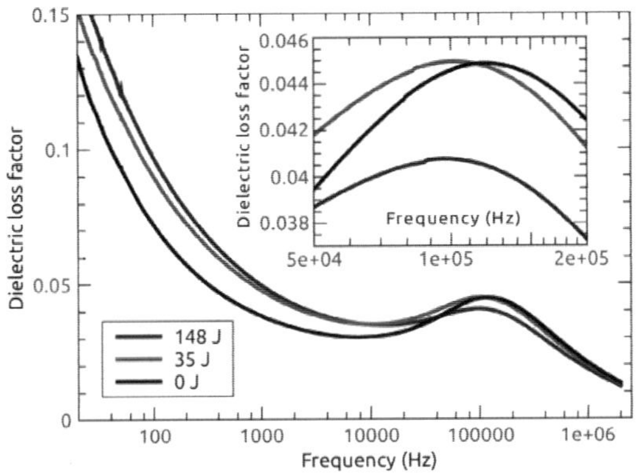

Figure 8. The change of the dielectric loss factor in frequency domain.

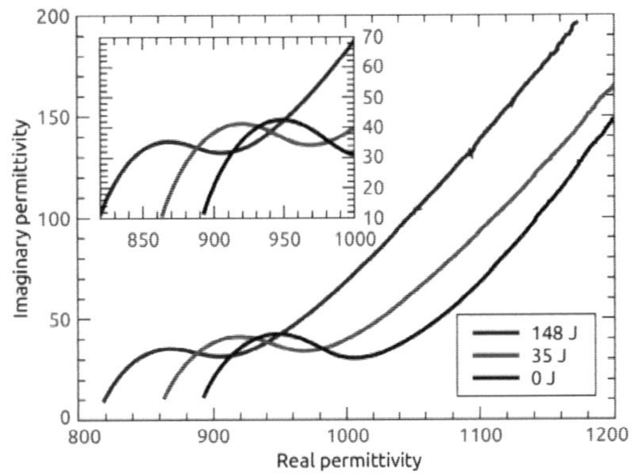

Figure 9. The Cole-Cole plot of the MOV sample tested.

frequency, with a relaxation peaks tgδ_m (where the dielectric loss factor reach its maximum) around 100 kHz.

In a more detailed view of Fig. 8, a change in the amplitude of the relaxation peak Δtgδ_m and also the relaxation peak shift to lower frequencies with increasing energy absorbed by the MOV is evident. Table 2 shows the relaxation peak amplitudes as well as the corresponding frequency f_r and a change in the frequency of the relaxation peak Δf_r for the individual varistor degradation steps. The shift of the relaxation peak to lower frequencies (a negative value of Δf_r) is due to the degradation of the depletion layer at the ZnO grain boundaries [23] caused by injection of current pulses into the MOV sample.

Table 2. The change in the relaxation peak of the MOV sample

State	tgδ_m	Δtgδ_m	f_r	Δf_r
	(-)	(%)	(Hz)	(%)
Initial	0.04487	–	119130	–
Degraded (35 J)	0.04495	0.192	100240	–15.86
Degraded (148 J)	0.04076	–9.156	95726	–19.65

Finally, Fig. 9 shows a Cole-Cole plot of the MOV sample gradually degraded by current pulses. As can be seen in Fig. 9, the Cole-Cole plot of the ZnO varistor contains depressed semicircles in the high frequency range, with the center of the semicircle shifted below the axis of relative permittivity. The shift of the semicircle below the relative permittivity axis demonstrates that the dielectric relaxation nature of the porous ZnO ceramic is not an ideal mono-dispersive relaxation, but poly-dispersive relaxation [23]. From the Cole-Cole plot in Fig. 9, the shift of the center of the semicircle toward the higher frequencies with increasing absorbed energy of the MOV sample is evident. The Maxwell--Wagner relaxation is a dominant factor in the dielectric properties of ZnO ceramics at low frequencies [23]. The corresponding Cole--Cole plot is investigated by the following relationship

$$\varepsilon^* = \varepsilon' - \mathrm{j}\varepsilon'' = \varepsilon_\infty + \frac{\varepsilon_s - \varepsilon_\infty}{1 + (\mathrm{j}\omega\tau)^{1-\alpha}}, \quad (8)$$

where α is the distribution of relaxation time (depression parameter), and the value of $\alpha\pi/2$ denotes the angle between the horizontal axis and the line defined by semicircle center and high frequency intercept.

In Table 3, the depression parameter α at selected stages of the experiment are listed. The relaxation time values are in the range between 0.36 and 0.45 corresponding to previous published results for the other samples [23]. Figure 10 shows the Cole-Cole plot of the MOV sample tested with semicircles according to Eq. 8.

Table 3. The change in the relaxation time of the MOV sample

State	depression parameter α
Initial	0.36
Degraded (35 J)	0.37
Degraded (148 J)	0.45

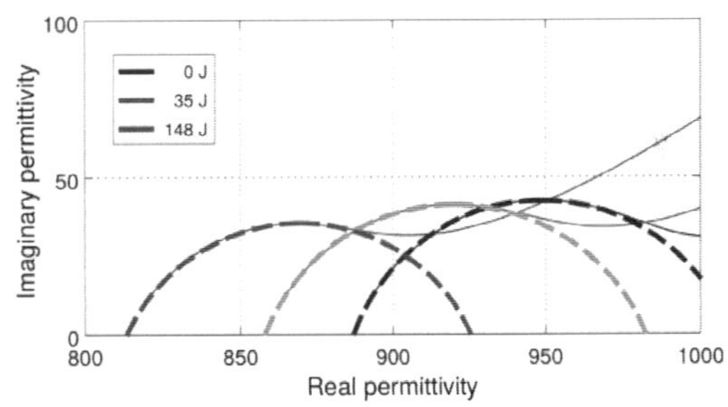

Figure 10. The detailed Cole-Cole plot of the MOV sample tested with circles according to (8).

V. Conclusion

The temperature dependence of the VI characteristic of the ZnO varistor of commercial production in the leakage region and the temperature dependence of the non-linearity coefficient of ZnO varistor is pointed out. It is confirmed, that with the increasing temperature of the varistor the leakage current increases at a constant test voltage, and it is also pointed out, that with the increasing temperature of the ZnO varistor the non-linearity coefficient of the varistor decreases linearly in the temperature range from 23 °C to 100 °C. For this temperature range, a linear approximation equation of the temperature dependence of the non-linearity coefficient was calculated. We assume that at higher sample temperatures, the non-linearity coefficient is even more reduced.

Dielectric spectroscopy and the description of VI characteristics have shown that injection of current pulses significantly degrades ZnO varistor even at relatively low values of absorbed energy. The observed changes were confirmed by increasing the leakage current, reducing the non-linearity coefficient and relative permittivity, and increasing the dielectric losses in the low frequency range. The non-Debay conducting process in ZnO varistor sample tested was confirmed.

Acknowledgment

The author would like to thank to the Scientific Grant Agency of the Ministry of Education of the Slovak Republic and the Slovak Academy of Sciences under the project VEGA No. 1/0340/18. This work was supported by th Cultural and Educational Grant Agency of the Ministry of Education, Science, Research and Sport of the Slovak Republic (KEGA) under the project No. 008TUKE-4/2019.

References

Authors should gave DOI numbers, where applicable [8]. Use strictly unique Reference style list for all items.

[1] A. I. Nousdilis, G. C. Christoforidis, and G. K. Papagiannis, "Active power management in low voltage networks with high photovoltaics penetration based on prosumers' self-consumption," *Applied Energy*, vol. 229, pp. 614–624, 2018.

[2] M. M. Haque and P. Wolfs, "A review of high pv penetrations in LV distribution networks: Present status, impacts and mitigation measures," *Renewable and Sustainable Energy Reviews*, pp. 1195–1208, 2016, doi:10.1016/j.rser.2016.04.025.

[3] L. M. Anderson and K. B. Bowes, "The effects of power-line disturbances on consumer electronic equipment," *IEEE Transactions on Power Delivery*, vol. 5, no. 2, pp. 1062–1065, 1990, doi:10.1109/61.53122.

[4] Z. Čonka, M. Kolcun, and G. Morva, "Utilizing of phase shift transformer for increasing of total transfer capacity," *ACTA POLYTECHNICA HUNGARICA*, vol. 13, no. 5, pp. 27–37, 2016.

[5] M. Kolcun, Z. Čonka, L. Beňa, M. Kanálik, and D. Medveď, "Improvement of transmission capacity by facts devices in central east europe power system," *IFAC-PapersOnLine*, vol. 49, no. 27, pp. 376–381, 2016.

[6] V. Volokhin, I. Diahovchenko, V. Kurochkina, and M. Kanálik, "The influence of nonsinusoidal supply voltage on the amount of power consumption and electricity meter readings," *Energetika*, vol. 63, no. 1, pp. 1–7, 2017.

[7] M. Kanálik, J. Buša, and J. Kyncl, "Analysis of voltage transients caused by single-line to ground fault in medium voltage distribution networks," in *ELEKTROENERGETIKA* 2015: proceedings of the 8th International Scientific Symposium on Electrical Power Engineering, 2015, pp. 380–383.

[8] H. Wang, C. Zhuang, R. Zeng, S. Xie, and J. He, "Transient voltage measurements for overhead transmission lines and substations by metal-free and contactless integrated electro-optic field sensors," *IEEE Transactions on Industrial Electronics*, vol. 66, no. 1, pp. 571–579, 2019, doi:10.1109/TIE.2018.2826455.

[9] C. Bodeker and N. Kaminski, "Implementation and investigation of the dynamic active clamping for silicon carbide mosfets," in *2016 18th European Conference on Power Electronics and Applications (EPE'16 ECCE Europe)*, 2016, pp. 1–7.

[10] S. B. Smith and R. B. Standler, "The effects of surges on electronic appliances," *IEEE Transactions on Power Delivery*, vol. 7, no. 3, pp. 1275–1282, Juli 1992, doi:10.1109/61.141841.

[11] N. Kularatna, A. S. Ross, J. Fernando, and S. James, Eds., *Design of Transient Protection Systems*. Elsevier, 2019, ch. Components Used in Surge Protection Circuits, pp. 29–42.

[12] C. Cheng, J. Hu, and J. He, "Characterization of dielectric behavior in ZnO electroceramic: Superior grain boundary, inferior grain boundary and grain," *Materials Letters*, pp. 240–242, 2014, doi:10.1016/j.matlet.2014.06.100.

[13] C.-W. Nahm, "Aging characteristics of ZnO–$PrO_{1.83}$-based semiconducting varistors for surge protection reliability," *Microelectronics Reliability*, vol. 54, no. 11, pp. 2417–2422, 2014, doi:10.1016/j.microel.2014.05.008.

[14] B. M. Abu-Zied, Y. A. Mohamed, and A. M. Asiri, "Fabrication, characterization, and electrical conductivity properties of Pr_6O_{11} nanoparticles," *Journal of Rare Earths*, vol. 31, no. 7, pp. 701–708, 2013, doi:10.1016/S1002-0721(12)60345-7.

[15] C.-W. Nahm, "Nonlinear behavior of Tb_4O_7-modified $ZnOPr_6O_{11}$-based ceramics with high breakdown field," *Journal of Rare Earth*, vol. 31, no. 3, p. 276, 2013.

[16] P. Meng, S. Lyu, J. Hu, and J. He, "Tailoring low leakage current and high nonlinear coefficient of a Y-doped ZnO varistor by indium doping," *Materials Letters*, vol. 188, pp. 77–79, 2017.

[17] C.-W. Nahm, "Highly stable nonlinear behavior against impulse current of ZnO-Pr_6O_{11}-CoO-Cr_2O_3-Y_2O_3 varistors," *Materials Letters*, vol. 64, no. 23, pp. 2631–2634, 2010.

[18] M. A. Alim, S. Li, F. Liu, and P. Cheng, "Electrical barriers in the ZnO varistor grain boundaries," *Physica Status Solidi (A) Applications and Materials Science*, vol. 203, no. 2, pp. 410–427, 2006, doi:10.1002/pssa.200521084.

[19] M. Matsuoka, "Nonohmic properties of zinc oxide ceramics," *Japanese Journal of Applied Physics*, vol. 10, no. 6, pp. 736–746, 1971.

[20] M. A. Alim, A. Batra, M. Aggarwal, and J. R. Currie, "Immittance response of the SnO_2–Bi_2O_3 based thick-films," *Physica B: Condensed Matter*, vol. 406, no. 8, pp. 1445–1452, 2011, doi:10.1016/j.physb.2011.01.046.

[21] C. Nahm, "Microstructure, electrical and dielectric properties, and aging behavior of ZPCCA varistor ceramics with Er_2O_3 doping," *Journal of Rare Earths*, vol. 33, no. 4, pp. 403–410, 2015, doi:10.1016/S1002-0721(14)60433-6.

[22] P. Meng, J. Hu, and J. He, "Low-residual-voltage ZnO varistor ceramics improved by multiple doping with gallium and indium," *Materials Letters*, vol. 195, pp. 209–212, 2017.

[23] X. Li, L. Xu, L. Liu, Y. Wang, X. Cao, Y. Huang, C. Meng, and Z. Wang, "High pressure treated ZnO ceramics towards giant dielectric constants," *Journal of Materials Chemistry A*, vol. 2, no. 39, pp. 16 740–16 745, 2014, doi:10.1039/c4ta03434a.

[24] T. Chen, Q. Fu, D. Zhou, Y. Hu, Z. Zheng, C. Gao, and Y. Liu, "Effects of bn on phase compositions and electrical properties of ZnO–Bi_2O_3 varistors prepared by the reduction–reoxidation method," *Journal of the European Ceramic Society*, vol. 36, no. 3, pp. 609–614, 2016, doi:10.1016/j.jeurceramsoc.2015.09.035.

[25] B. Efafi, M. S. Ghamsari, M. Aberoumand, M. M. Ara, and H. H. Rad, "Highly concentrated zno sol with ultra-strong green emission," *Materials Letters*, vol. 111, pp. 78–80, 2013.

[26] S. Nezhadesm-Kohardafchahi, S. Farjami-Shayesteh, Y. Badali, S. Altindal, M. Jamshidi-Ghozlu, and Y. Azizian-Kalandaragh, "Formation of ZnO nanopowders by the simple ultrasoundassisted method: Exploring the dielectric and electric properties of the Au/(ZnO-PVA)/n-Si structure," *Materials Science in Semiconductor Processing*, vol. 86, pp. 173–180, 2018, doi:10.1016/j.mssp.2018.06.03.

Several Approaches to Development of Dynamic Predictive Model Thermal Modes of Underground Power Transmission Cable Lines

Alexander V. Kolychev, Vasiliy V.Titkov, Pavel D.Tukeev
Energetics and transport Institute
Peter the Great Polytechnic University
Saint-Petersburg, Russia
alkoly4ev@yandex.ru, titkovprof@yandex.ru, paultuk@yandex.ru

Alexander N. Lubkov
High voltage testing lab
Power transmission by DC Institute
Saint-Petersburg, Russia
lubkoff@pisem.net

Abstract — **In this study the ways of creating dynamic digital twins of underground power transmission cable lines, able to formulate its mathematical model parameters based on a cable shell temperature monitoring data, has been considered. Possibility of Green's function and standard differential equation systems application for creating compact digital model along with the problems and mathematical procedures of dynamical parameter adjustment, have been discussed.**

Keywords: power cable transmission line; heating, fiber thermal sensor; thermal mode prediction

I. Introduction

The development of measuring instrument, power equipment diagnostics in operating mode and digital technologies of data processing set a number of tasks in the field of operating and control of power engineering objects technical conditions. At the same time computer technology development now allows to create computer models, also known as digital twins, allowing to analyze and control parameters and condition of object in real-time.

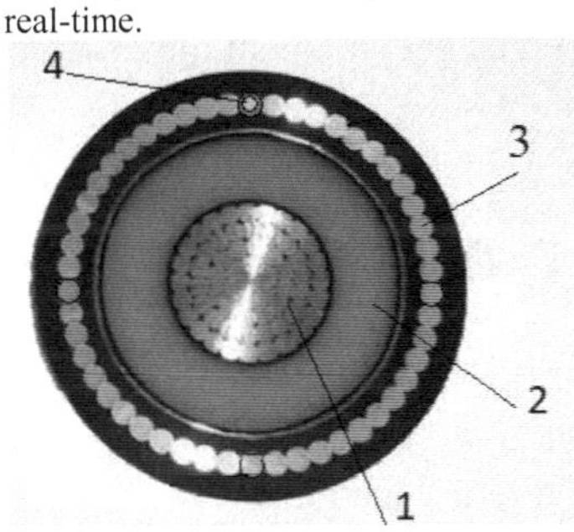

Figure .1. XPLE-cable cross-section: 1 – core, 2 –main insulation, 3-metallic shield, 4 – fiber-optic in steel tube

In this study the dynamic models of underground three-phase cable line heating mode, allowing to perform the prognosis of its heating mode given planned capacity graph for some period of operation, have been considered. The distinction of underground cable heating mode is almost complete absence of stationary temperature field. It happens due to considerable, up to tens of hours, time constants of cable line temperature transients. Given daily and seasonal capacity adjustments stationary mode is almost unreachable. Estimations of cable line load capacity, especially 35 kV or higher, regulatory documents recommendations are based of stationary field calculations [1]. In addition, when cable line capacity grows temporary for a limited time stationary mode hasn't been reached and actual cable temperature proves lower than stationary estimation. That's why exploiter of cable line have the ability to temporally overload his cable line. In order to make such a decision it seems viable to apply a model, operatively predicting cable temperature for some period of time (one day, for example). The second field of thermal mode prediction models is predicting this modes when amount and composition of power consumers may vary. For example, thermal mode of cable lines, operating in metropolises, tends to change due to surrounding infrastructure changes (building new roads, heating mains etc), and with this thermal condition of cable operation change too. It is advisable to manage such changes using temperature monitoring of cable line. In this case prediction model is an effective mean of planning and control of technical condition of power cable transmission lines.

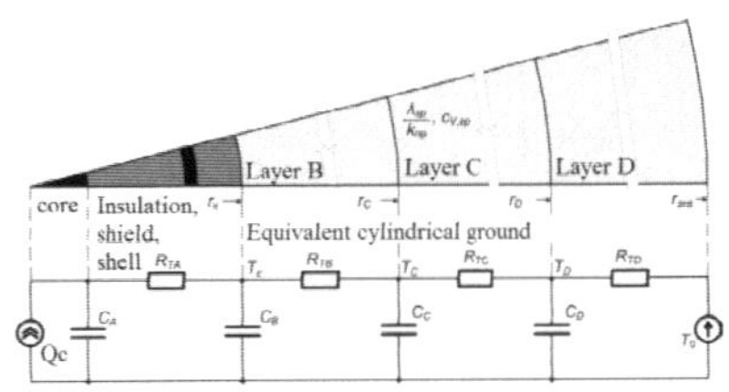

Fig .2. Equivalent network of axysymmetrical thermal model of cable

Modeling of cable line thermal mode is now performed using finite-element analysis [2]. However, finite-element models are usually time- and computation power-consuming and are normally used only for design. Taking into account aforementioned operation condition changes the application of finite-element analysis, requiring exact thermo physical parameters of surrounding ground, cable specifications, cable shield connection method etc, becomes difficult. Thereby it is advisable to develop the simplified models of thermal modes, using temperature and electrical parameters monitoring. The elements of this approach are used en-masse by fiber-optic sensors in cable shields (fig. 1).

II. The Simplfied Models of Cable Line

A. *Electro thermal analogy*

The electro thermal analogy method had a wide application in thermal mode analysis in a days before finite-element analysis had become widespread. For instance, the Van Wormer method is well-known in cable systems calculation [3,4]. An idea of electrothermal analogy is based on parallel between Bio-Fourier thermal flow and Ohm law differential form equations along with equations for solid body internal energy and electrical capacity charge. Based on aforementioned analogies real body can be split in finite volumes, each of them will be characterized by specific heat value and thermal resistance, wherein the connections of these volumes are set by equivalent electrical circuit. For instance, in [3] the electrical equivalent circuit of cable thermal mode is shown on Fig. 2.

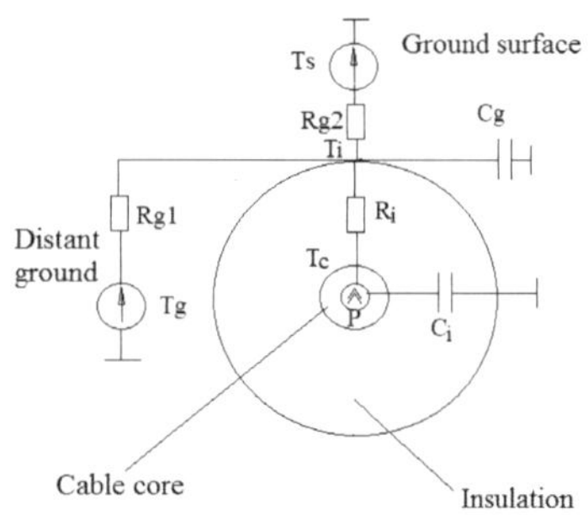

Figure. 3. Equivalent circuit for heat transfer between cable and ground

The ground in this model is represented by three axysymmetrical layers concentrated around cable. "The current source" Q_c – the heating power of cable core. Circuit parameters – thermal resistances (R_T) and specific heat values (C_V), defined by Van Wormer method [4]. Skipping the details of equivalent electrical circuit parameters calculations, note that transient in it is described by system of ordinary differential equations for cable and layers temperatures. Since the output is temperature of outer cable shell only, in order to dynamically adjust such a model not only R_T and C_V, but initial values of layer temperatures are required. Thus, the number of model parameters will significantly increase when being deeply detailed, as in [5].

Based on aforementioned it is advisable to assume the simplest equivalent circuit when using electrothermal analogy method. Let's build a circuit, taking into account heating in cable core, heat transfer from core surface to shell surface, heat transfer from shell surface to ground surface with T_s temperature and to distant ground with T_g temperature (fig. 3).

Writing the Kirchhoff equation for Fig. 3, we get

$$C_i \frac{dT_c}{dt} + \frac{T_c - T_i}{R_i} = P$$

$$C_g \frac{dT_i}{dt} + \frac{T_i - T_s}{R_{g2}} + \frac{T_i - T_g}{R_{g1}} = \frac{T_c - T_i}{R_i} \qquad (1)$$

where T_c and T_i – temperatures of core and insulation respectively, P – linear Joule heating power. If monitor with relatively big time steps of 30-60 minutes, the system (1) can be simplified even further, because in this case

thermal transient in cable can be neglected, wherein the math model will consist of one equation only

$$C_g \frac{dT_i}{dt} + \frac{T_i - T_s}{R_{g2}} + \frac{T_i - T_g}{R_{g1}} = P(t) \qquad (2)$$

where using, for example, daily or even larger capacity graph P(t) and cable temperature T(t), is possible to choose parameters of latter equation. The equation (2) can be written down in more simple form

$$\frac{dT_i}{dt} + \frac{T_i}{\tau} = \alpha P(t) + \beta \qquad (3)$$

Selection of equation parameters α, β, τ is performed by the minimizing relative standard deviation of observed and calculated by (3) dependence of temperature from time

$$F(\alpha, \beta, \tau) = \frac{\int_0^{t_1} [\tilde{T}(t) - T_{obs}(t)]^2 dt}{\int_0^{t_1} T_{obs}^2(t) dt} \rightarrow min \qquad (4)$$

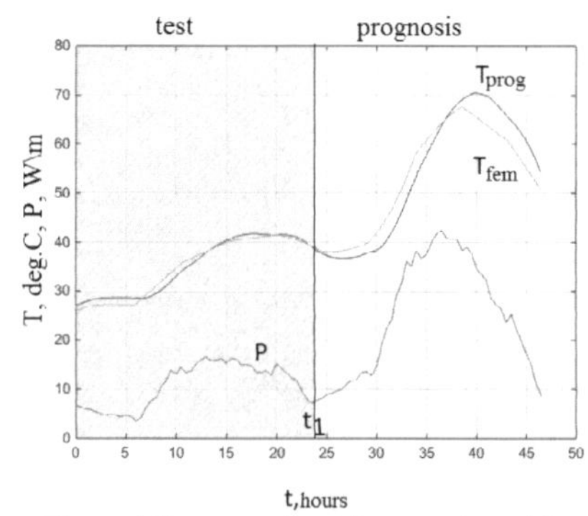

Figure. 4.The comparison of prognosis model temperature (T_{prog}) and finite-element method (T_{fem}) under real capacity graph P

The minimizing of (4) functional is performed by numerical methods. After defining of α, β, τ parameters the reasonably accurate approximation of dependence of temperature from time can be calculated using (3) with parameters defined on prediction period. An example of test time period analysis and prognosis, made for real 6 kV cable line capacity graph, is shown on Fig. 4. Note that despite reasonable accuracy and high speed of algorithm performance this methods shows a drawbacks too, which are related mainly to initial values selection to obtain parameters for (3) – initial values α_0, β_0, τ_0, which couldn't be automated due to big polymodality of functional (4) relief. Thus, the problem of initial approximation selection while using ODE technique should have been solved by manual combination sorting.

B. Green function method

The template is used to format your paper and style the text. All margins, column widths, line spaces, and text fonts are prescribed; please do not alter them. You may note peculiarities. For example, the head margin in this template measures proportionately more than is customary. This measurement and others are deliberate, using specifications that anticipate your paper as one part of the entire proceedings, and not as an independent document. Please do not revise any of the current designations.

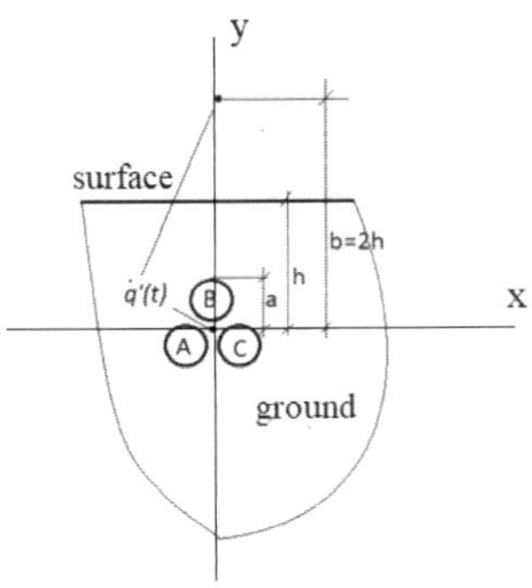

Figure 5. Calculation field of cable line thermal field model

As a second approach to problem of cable line thermal mode prediction a Green function for non-stationary heat transfer equation will be used [6]. In case of 2D thermal field, typical for underground power cable lines, the non-stationary thermal field of linear source q' in infinite space with uniform thermo physical parameters distribution – thermal conductance coefficient (λ), density (ρ) and specific heat (C) is defined by a formula

$$T(x,y,t) = T_0 + \frac{1}{\rho c}\int_0^t \dot{q}'(\tau)\left(\frac{1}{\sqrt{4\pi D(t-\tau)}}\right)^2 exp\left[-\frac{x^2+y^2}{4D(t-\tau)}\right]d\tau$$

where D= λ/(ρc)

For taking into account the ground surface, the latter counts as thermal-insulated in first approximation, assuming that the main mechanism of cable cooling is heat transfer to deep ground (Fig. 5.). In that case superposition of main source, located at depth of h, and source, mirrored from ground surface, allows to fulfill that boundary conditions. The expression for thermal mode in coordinate system with base point in cable cross-section centre point takes the form

$$T(x,y,t) = \frac{1}{\rho c}\int_0^t \dot{q}'(\tau)\left(\frac{1}{\sqrt{4\pi D(t-\tau)}}\right)^2 exp\left[-\frac{x^2+y^2}{4D(t-\tau)}\right]d\tau + \frac{1}{\rho c}\int_0^t \dot{q}'(\tau)\left(\frac{1}{\sqrt{4\pi D(t-\tau)}}\right)^2 exp\left[-\frac{x^2+(y-2h)^2}{4D(t-\tau)}\right]d\tau + T_0$$

In order to estimate shell B surface temperature (Fig. 6), using this formula, let's set an observation point *a* to a cable surface, assuming $x=x_a=0$, $y=y_a$. The equivalent linear heat source will be placed in cable line center of symmetry – in base point *x,y* (Fig. 6), where in

$$\dot{q}'(t) = R'(I_A^2 + I_B^2 + I_C^2)$$

where I_A, I_B, I_C – phase currents, R' – equivalent active phase resistance per length.

$$T(0,y_a,t) = \frac{1}{\rho c}\int_0^t \dot{q}'(\tau)\left(\frac{1}{\sqrt{4\pi D(t-\tau)}}\right)^2 exp\left[-\frac{{y_a}^2}{4D(t-\tau)}\right]d\tau + \frac{1}{\rho c}\int_0^t \dot{q}'(\tau)\left(\frac{1}{\sqrt{4\pi D(t-\tau)}}\right)^2 exp\left[-\frac{(y_a-2h)^2}{4D(t-\tau)}\right]d\tau + T_0 \quad (5)$$

Changing *h* in that expression for another parameter *h'* will result in appearance of heat flow on ground surface directed from ground when *h'*>2*h* and to ground when *h'*<2*h*. Adjusting the parameter *h'* value allows to effectively taking into account heat transfer between ground and the environment. Thus, the model tuning is performed by minimizing the object function

$$F(\rho c,\lambda,h,T_0) = \frac{\int_0^{t_1}[\tilde{T}(t)-T_{obs}(t)]^2 dt}{\int_0^{t_1} T_{изм}^2(t)dt} \to min \quad (6)$$

The minimizing of (6) functional is performed by numerical methods. After defining of $\rho c, \lambda, h, T_0$ parameters the reasonably accurate approximation of dependence of temperature from time can be calculated using (5) with parameters defined on prediction period. An example of test time period analysis and prognosis, made for real 6 kV cable line capacity graph, is shown on Fig. 6.

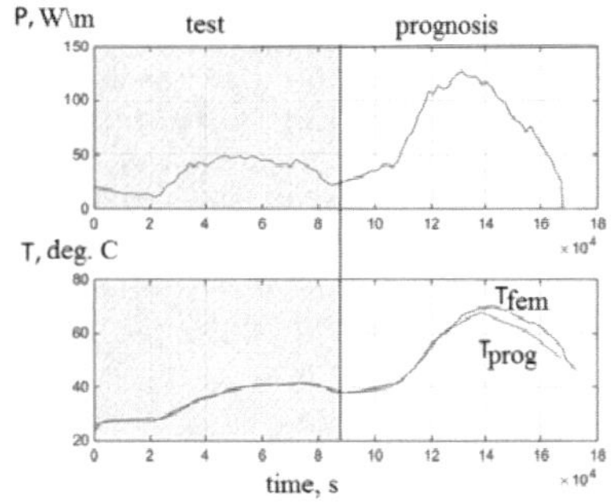

Fig. 6. The result of 6 kV cable line thermal mode prediction for a case of triangle phases composition under real capacity graph

The advantages of Green function method are:

- easy and physically reasonable selection of initial approximation point in Green function parameters space.
- high-quality approximation to prediction model in case of "high-contrast" load graphs.

The disadvantage is relatively long duration of algorithm operation time caused by the specifics of numerical calculation of convolution integrals

C. Conclusions

In order to solve cable line mode prediction problem two algorithms, using estimated capacity on prediction time period graph, have been developed. Both algorithms have been tested on examples, calculated using detailed finite-element analysis model. The algorithms proposed can be realized while using common math computation computer systems

References

[1] IEC Standard 60287–1–1. Electric Cables – Calculation of the Current Rating. Part 1: Current Rating Equations (100% Load Factor) and Calculation of Losses – Section 1: General. – 2006.

[2] S. Dubitsky, G. Greshnyakov, N. Korovkin, "Comparison of Finite Element Analysis to IEC 60287 for Predicting Underground Cable Ampacity", 2016 IEEE International Energy Conference (ENERGYCON) 4-8 April 2016, Leuven, Belgium, Conf. Proc., pp. 1411-1416 DOI: 10.1109/ENERGYCON.2016.7514107

[3] R.J. Millar, "Real-Time Transient Temperature Computation of Power Cables Including Moisture Migration Modelling", 15th Power Systems / Computation Conference. – Liege. – Belgium. – 2005. Vol. 1. – Session 39. – Paper 4. – pp. 1388–1395, ISBN: 9781617388569

[4] F.H. Van Wormer, "An Improved Approximate Technique for Calculating Cable Temperature Transients" ,AIEE Transactions. – 1955. – Vol. 74. – Part 3. – pp. 277–281, DOI: 10.1109/AIEEPAS.1955.4499079

[5] R. Zaytsev, "E. Development of Calculation Methodology of Thermal Equivalent Circuit's Parameters for Predicting Temperature of High Voltage Cable Lines 110 – 500 kV", Applied Mechanics and Materials. – Switzerland: Trans Tech Publications, 2015. – Vol. 792. – pp. 272–279, DOI:10.4028/www.scientific.net/AMM.792.272

[6] Y. A.N.Tikhonov, A.A.Samarskii, "Equations of mathematical physics", Pergamon Press Ltd., New York, 1963, 754 pp

Coupled Tests Including Parameters TF and SLF of High Voltage SF6 Circuit Breakers

Zdenek Vostracky, Jan Sedlacek, Vaclav Muzik
Faculty of Electrical Engineering
University of West Bohemia
Plzen, Czech Republic
zdenekv@kee.zcu.cz, sedlacek@ntc.zcu.cz, vmuzik@kee.zcu.cz

Abstract — **Standard short circuit tests of HV SF6 circuit breakers need high power sources and special synthetic facilities. Tests consist of many standard disciplines. Each of short fault disciplines requires high parameters from high power sources. They also require exact measurement and accurately control consequence of switching high current and high voltage circuit of synthetic method. Developing tests are therefore very expensive. The aim of this article is to introduce new point of view on testing by coupled terminal fault and short line fault while reducing high cost of developing tests.**

Keywordss: HV circuit breaker; interaction interval; dielectric interval; reliability; coupled tests

I. Introduction

This contribution deals with an analysis of interruption process especially from the point of view of time characteristics parameters. This attempt enables classification of interrupting process and understanding individual phenomena and influence of parameters not only in circuit breaker but also of the testing method. Testing synthetics methods are very complicated and capability of circuit breakers is extremely sensitive on even small change of parameters in the surrounding of current zero. Analyses of interrupting processes had been done by dividing on the partial time interval. This also responds to basic theories of Cassie and Mayr. This background enables to easily understand interrupting process. It is also possible to assess reliability from the point of view limiting characteristics.

II. Interrupting Process

It is possible to divide the interrupting procedure into three main time intervals according to characteristics of physical phenomena (Fig. 1) [1]. All intervals consequently create initial conditions for the next one. High current interval influences long life of quenching chambers, especially contacts. Interruption capability depends very much on parameters in surroundings current zero which is represented in the greater scale on Fig. 2 [6]. Great attention is therefore payed on interaction interval and dielectric interval. Scientific work is concentrated especially on theoretical models derived from models Cassie of high current interval and Mayr of interaction interval. There exit further theories so called black-box models combine both above mentioned basic theories. Input parameters for theoretical models are obtained from very difficult and expensive tests and measurement by short fault experiments. It has been topic for CIGRE Study Committee 13.

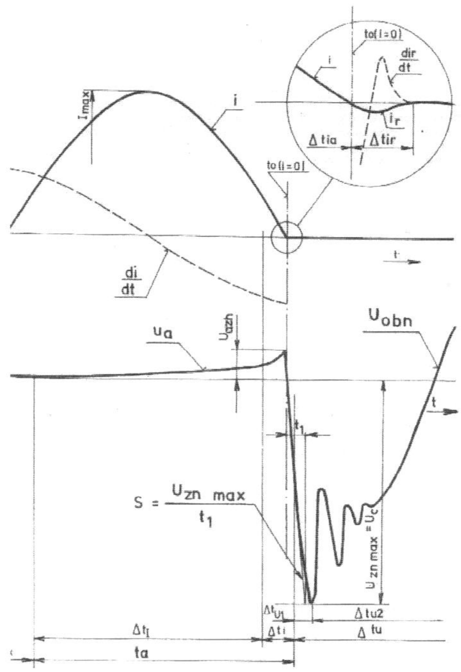

Figure 1. Three basic intervals in an interruption process: ts - high current interval, t_i - interaction interval, t_u - dielectric interval [1].

A. Interaction Interval

The interaction interval is short period of time vicinity of current passing zero. Here exists power interaction between arc and circuit. Failure in this interval is called thermal failure. The most important parameters are time delay t_d, current derivation di/dt and voltage derivation du/dt and time delay t_d.

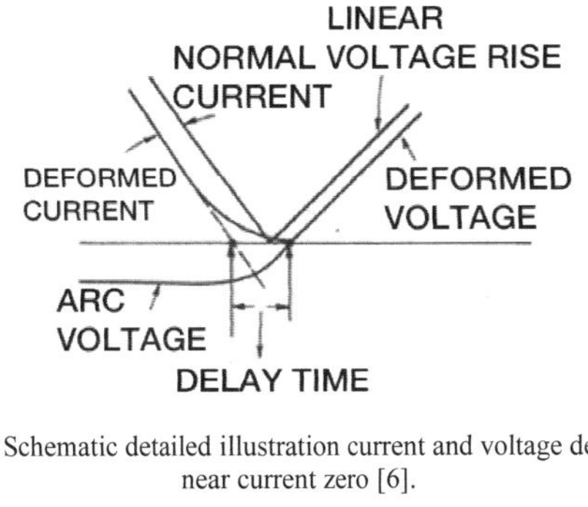

Figure 2. Schematic detailed illustration current and voltage deformation near current zero [6].

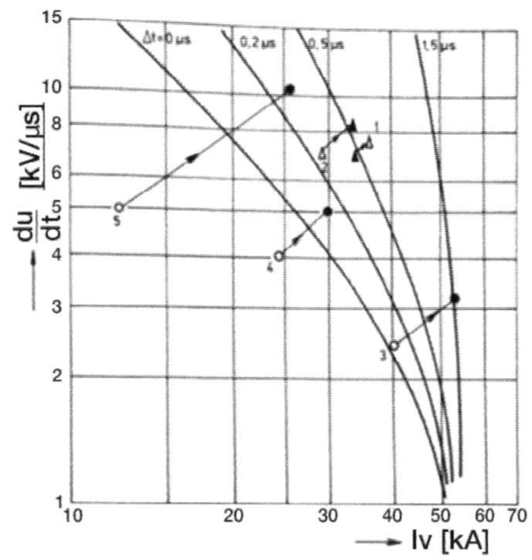

Figure 3. Experimental results of du/dt on interrupting current for parameters of time delay td [4].

In the short time in the surrounding current passing zero are decisive parameters, derivation current and derivation voltage (rate of rise of recovery voltage - RRRV) that is described theoretically by differential equations based on Cassie or Mayr equations. These theoretical conclusions are also confirmed by experimental results in Fig. 4.

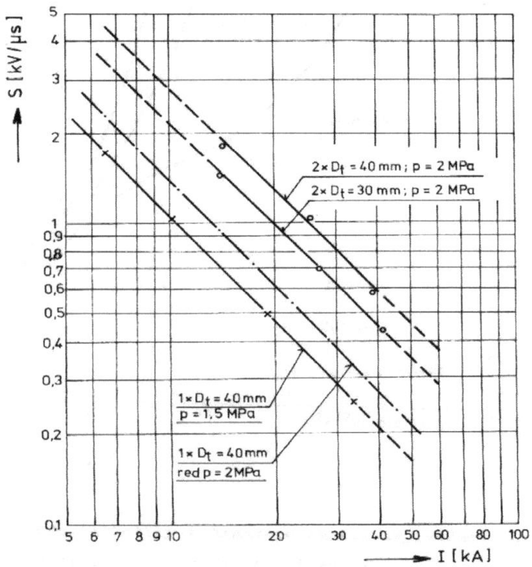

Figure 4. RRRV vs. interrupting current: different pressure and design parameters [1].

These experimental results correspond to basic theoretical equations Dependence du/dt on di/dt is derived as an example in [1], [7] and many others as

$$\frac{du}{dt} = S = \frac{\frac{di}{dt}}{\left|\frac{di}{dt}\right|} \cdot konst \cdot \left|\frac{di}{dt}\right|^{-m} \tag{1}$$

Dependence of these three parameters could be expressed by [4] and others.

$$S_{\lim} = S_0 \cdot e^{\varepsilon \cdot t_d} \cdot I_B^{-C_1 \cdot e^{v \cdot I_B}} \tag{2}$$

Because of t_d is more or less given by standards we can express limiting recovery-voltage for terminal fault from the curve for short line fault

$$S_{\lim} = S_1 \cdot \left(\frac{I_B}{I_1}\right)^{-m} \tag{3}$$

where S_1, I_1 are experimentally assigned. Argument "m" lays on the extension from 2 up 3. It is chosen experimentally for the given quenching chamber and depends on the design of the breaker.

$$S_{\lim} = S_1 \cdot e^{\varepsilon(t_{dT} - t_{dL})} \cdot \left(\frac{I_B}{I_1}\right)^{-m} \tag{4}$$

where t_{dT} is time delay for terminal fault and t_{dL} is the time delay for short line fault an also

$$S_{\lim T} = k \cdot S_{\lim L} = e^{\varepsilon(t_{dT} - t_{dL})} \cdot S_{\lim L} \tag{5}$$

Constant "k" can be derived from

$$k_{Rt} = \frac{S_{\lim}}{S_N} \cong \frac{const_{\lim} \cdot I_B^{-m}}{0{,}2 \cdot I_B} = const \cdot I_B^{-(m+1)} \tag{6}$$

$$const_{\lim} = S_0 \cdot e^{\varepsilon \cdot t_d} \tag{7}$$

It is obvious that below reaches high steepness of reliability. Theoretical relationship and practical experience gives value $0{,}9 I_{BN}$ as the most severe discipline (I_{BN} is nominal value of short current). Therefore was enabled the cancel standard test SLF 0,6 I_{BN}. There are now only two disciplines SLF 0,9 I_{BN} and 0,75 I_{BN}. If we compare terminal fault test (TF), we can see that due to smaller du/dt and also due to longer t_d we need not be afraid that after SLF the test circuit breaker could failure below 20 kA.

This is valid up to high current without blocking gas flow through nozzle that is not thermal blocked. When a nozzle is blocked even after current passed zero, ionized gas remains in the contact space. Consequence is then a comparatively sharp bend down of this curve on lower voltage.

B. *Dielectric Iinterval*

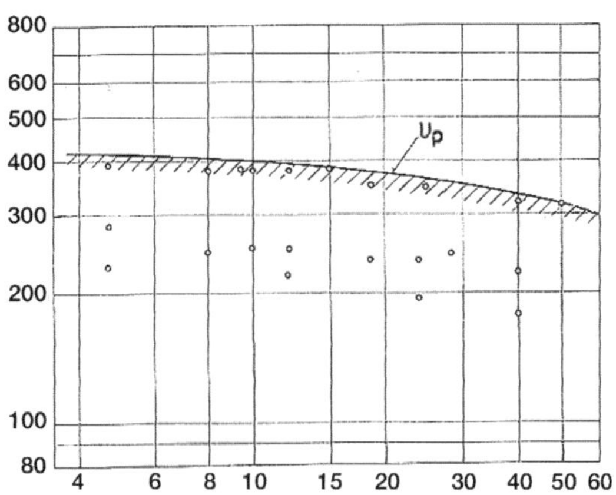

Figure 5. Breakdown voltage on interrupting current with RRRV and 8 kV/microsecond [8].

The peak of recovery-voltage U_C in the dielectric interval does not change too much. Dependence breaking voltage on interrupting current I_B could be expressed by equation

$$U_{BD} = konst \cdot I_B^{-\delta} \tag{8}$$

Where δ is approximately in the interval between 0,1 and 0,15 and it is possible to derive equation

$$U_{BD} = U_1 \cdot \left(\frac{I_B}{I_1}\right)^{-\delta} \tag{9}$$

III. Breaking Capability vs Standards

Critical parameters for the range of 123 kV and 420 kV are obvious from the characteristics of breaking capability of CB and standards Fig. 6 and Fig. 7

If we divide limit parameters to standardized parameters, we obtain so called coefficient of safety limit on each individual current. From the Fig.8 we can see the critical disciplines are for interaction interval SLF 40 kA and for 420 kV it is and dielectric interval TF 0,3 or special discipline fault by transformer 10 kA. Quenching of SLF is not so important because circuit breaker 420 kV has two quenching chambers and recovery voltage and RRRV is divided by "2" but has on the contrary voltage on one chamber "210kV ".

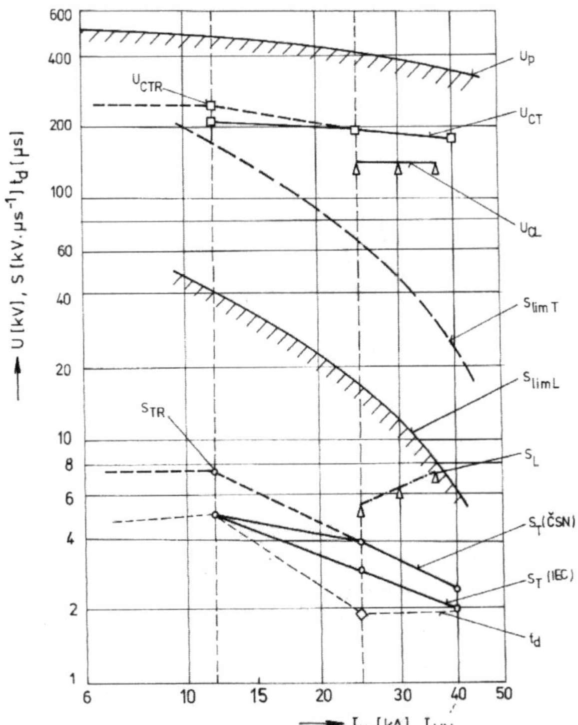

Figure 6. Characteristics of interrupting capability and standardized parameters of circuit breaker U_N=123 kV, I_{BN}= 40 kA. – where: U_{BD}-breakdown voltage, Slim- RRRV, U_c-maximum value of recovery-voltage td-time delay, T-terminal fault, L-short line fault, TR-fault near transformer, time delay t_d=1,5 µs)

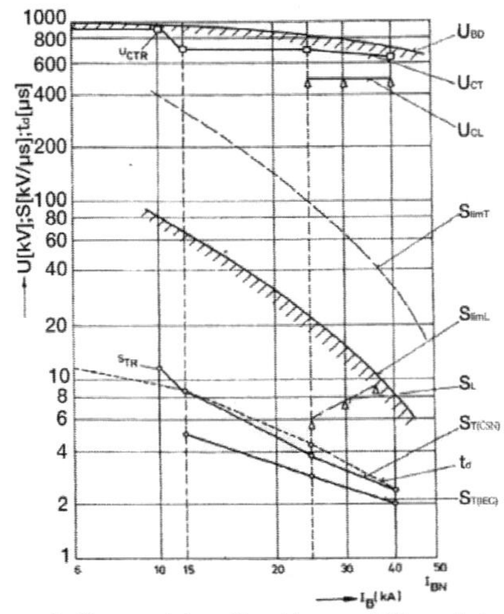

Figure 7 .Characteristics of breaking capability and standardised parameters for circuit breaker U_N=420 kV, I_{BN}= 40 kA. Legends see Fig. 6.

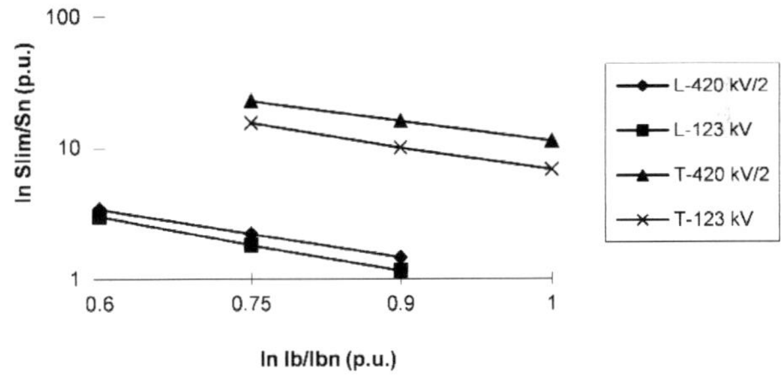

Figure 8. Rate of limiting of recovery voltage in the logarithmic scales: T-terminal fault, .L – short line faults

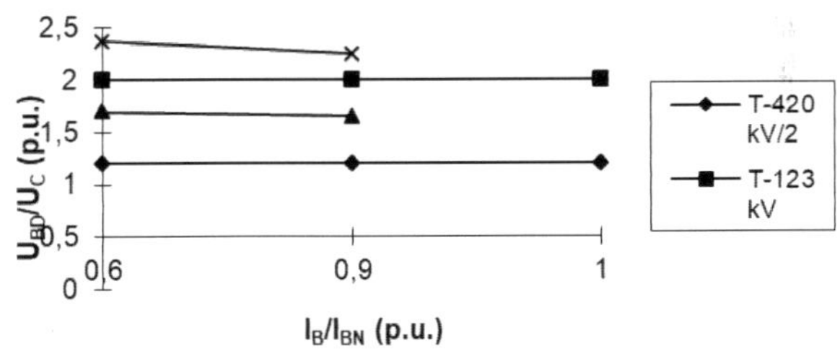

Figure 9. Rate of break down voltage to the recovery voltage in the logarithmic scales for breakers U_N=123 kV, I_{BN}= 40 kA and, U_N=420 kV, I_{BN}=40 kA with two breaking units:

IV. Coupled Developing Tests - Example

The previous analysis shows the possibility by developing period to couple some disciplines of interrupting capabilities and to get valuable results. There are two basic points of view – interaction (thermal) and dielectric interval. It is possible to couple some disciplines that will cover by an envelope standard stresses. An envelope is used for spectrum of recovery voltage in IEC. Example is introduced below for circuit breakers 123 kV (witlh one interruption units) and 420 kV (with two interruption units – suppose distribution 50%on one unit). Discussions of the spectrum interrupting disciplines are graphically documented in Fig.6 and Fig.7.

First we will consider the heavy current interval. It creates suitable condition for next intervals – interactive – in the period before current zero by thermal activity increase gas SF6 pressure, in compression space and close proximity current zero the temperature of the flowing gas decreases less than 2000 K. That procedure can be evaluated by the oscillograms Fig. 1 and Fig. 2 on which is typical increasing arc voltage before current zero. If current starts or continues to flow after, e.g. three arc time constants are 20 microsecond or more, occurs a top on the arc voltage and then the failure is dielectric nature.

Second we will consider the interaction interval. It is obvious that from the point of view of TF (terminal fault) is rated short circuit current 40kA and relevant RRRV (rated rise of recovery voltage) RRRV is 2 kV/microsecond and for lower current 60% is 3 kV/microsecond, for relevant current 30% is 5 kV/microsecond.

Interrupting capability is tested for SLF (short line fault). The current difference of rated TF and SLF is only TF 40kA- SLF 36kA= 4kA (10 %). That small difference would not play decisive role from the point of view of heavy current interval. This is possible to assume and to derive from the equation (2),(4) and Fig. 6 and Fig.7. It is obvious that circuit breaker 123 kV with one interrupting unit has crucial stress at 7.2 kV/microsecond. Circuit breaker 420 kV is stressed on half level 3.6 kV/microsecond because of two interrupting units and it is tested by circuit breakers 123 kA and covers also RRRV of the spectrum of TF for both rated 123 kV and 420 kV.

Third we will consider the dielectric interval – voltage withstand – during an interrupting process. It deals in the time of maximum recovery voltage. Voltage withstand in dielectric interval is possible to test with related voltages for a 420 kV/2 interrupting units. This tested voltage is higher than for circuit breaker 123 kV and covers it.

Classification of type failure is not often clear. There are the influences of all intervals. Within approximately 3 arc time constant, that is 15 microseconds, is not clearly to say which type is the cause of failure.

In developing period is possible to couple some disciplines to create envelops covering the most important points of interrupting capability even in one coupled discipline. In this example are derived disciplines for the circuit breakers of 123 kV and 420 kV and interrupting current 40 kA. Decisive role plays from the point of view of interaction the interrupting interval4 90% SLF of 40 kA=36 kA and 7.2 kV/microsecond. Dielectric interval is than tested on level of the voltage 420kV/2 e.g. on the voltage level 210 kV. It is possible to test these parameters by one tested synthetic circuit.

Total price of short circuit breaking current test consist of more than 9 disciplines and including auxiliary breakers is about 700 000 EURO. It is possible to expect by coupled developing testing save app. 15 % costs.

It emerges especially for circuit breakers 420 kV with two interrupting units and also for self-blast circuit breakers. There is no great difference between U_{BD} for I_{BN} and 0,9 I_{BN} and their ratio is approximately in the range from 1,05 to 1,1. We can state that if we will test for 0,9 I_{BN} with higher U_{CL}=1,1 U_C (I_{BN}) a circuit breaker will pass successfully also the test TF with I_{BN} and U_C.

We can compare breaker 420 kV with two chambers with breaker 123 kV with one chamber is for circuit breaker 123 kV critical short line fault and for circuit breaker terminal faults.

V. Conclusion

The testing is technically very difficult and expensive. This article introduces an idea of how to combine testing parameters, terminal fault and short line fault in one coupled discipline to effectively reduce test cost. This attempt is possible also to use for the circuit breakers of the same design but different voltage. Spectrum of recovery voltage for different waveforms of recovery voltage is by IEC defined as an envelope. This attempt could be basically used also for the testing of other parameters but with theoretical and experimental background.

Switching has statistical distribution and statistics is by decision procedure needed to be taken account. The frequency of occurrence of short circuits is from the word wide regarding on systems and for 300 kV and above are introduce as: faults per station and year 0,21, and 1,92 faults per 100 km o fine per year. This has an influence on clarifying coupled tests and also reliability and supporting idea on coupled testing [9]. The decision has to be made also from the point of view of frequency of switching off and the parameters with respect to reserve on limiting capability. There is a difference of reliability where voltage is near voltage withstood by nearly every switching on the contrary switching off SLF 90 % which is near to the boundary limit but occurs very rarely. The illustration of reserves for individual disciplines is in the Fig. 8 and Fig. 9.

Coupled (combined) tests enable to verify behaviour of circuit breakers in the interaction interval as well as in the dielectric interval. It gives opportunity to simplify very expensive testing of circuit breakers. This tendency of cost effectiveness evidences also confirms the step CIGRE and IEC in simplifying testing by omitting one discipline short line fault (SLF) for SLF60% I_{bn} and keeps only procedure disciplines SLF 90% and 75%.

Acknowledgment

This paper was supported by the student research project SGS-2018-023.

References

[1] Z. Vostracký, *Charakteristiky vypínací schopnosti vypínačů a jejich užití praxi.* Doktorská disertační práce. ČVUT Praha, 1986, pp. 150 str. (*Charakteristics of interrupting capability of HV circuit breakers and their practical use.* Doktoral Thesis, Czech Technical University in Prague, 1986, pp. 150.

[2] J. Sedláček, Z. Vostracký, H. Schramm, H. Knobloch and C. Wiesinger, "Optimization of h.v. self-blast interrupter by gas flow and electric field computations". IEEE Transactions on Power Delivery, 2003, Vol.. 18, No. 4, s. 1228-1235. DOI: 10.1109/TPWRD.2003.817536.

[3] K. Bárta and Z. Vostracký. *Spínací přístroje velmi vysokého napětí*, (*HV Circuit Breakers Switching apparatus*), Prague SNTL 1983.

[4] M. Hrabovský, V. Mastný and Z. Vostracký, "Use of dynamic arc model for extrapolation of short-circuit test results and determination of performance characteristics of circuit breakers". CIGRE Session, Paris 1984, Report 13-01.

[5] Nakanishi Kunto (editor), *Switching Phenomen in High-Voltage Circuit Breakers*. Marcel Dekker, Inc.New York, ISBN 0-8247-8443-6, 1991.

[6] G. Frind and J. A. Rich, "Recovery Speed of Axial Flow Gas Blast Interrupter: Dependence on Pressure and dI/dt for Air anf SF6," IEEE Transactions on Power Apparatus and Systems, vol. PAS-93, no. 5, pp. 1675–1684, Sep. 1974. DOI: 10.1109/TPAS.1974.293900

[7] W. Hermann and K. Ragaller, "Theoretical description of the current interruption in hv gas blast breakers," IEEE Transactions on Power Apparatus and Systems, vol. 96, no. 5, pp. 1546–1555, Sep. 1977. DOI: 10.1109/T-PAS.1977.32483

[8] J. Kopainski, "Influence of the Arc on Breakdown Phenomena in Circuit Breakers". In: Ragaller K. (eds) Current Interruption in High-Voltage Networks. Earlier Brown Boveri Symposia. Springer, Boston, MA, pp. 329-354. DOI: 10.1007/978-1-4757-1685-6_12

[9] M. Erche, "Switching Surges", In: Ragaller K. (eds) Surges in h.v. networks. Proc of Brown Boveri Symposium, 1979. pp. 63-95

Method of Determination of the Current-Voltage Characteristics of Surge Arresters in the Area of Leakage Current

Sergey Shevchenko
dept. «Electrical Power Transmission»
National Technical University
"Kharkiv Polytechnic Institute"
Kharkiv, Ukraine
syurik42@gmail.com

Dmytro Danylchenko
dept. «Electrical Power Transmission»
National Technical University
"Kharkiv Polytechnic Institute"
Kharkiv, Ukraine
danylchenko.e@khpi.edu.ua

Abstract—**The article is intended to substantiate the possibility of determining the parameters of the voltage-ampere characteristics of overvoltage limiters of non-linear ones under the influence of the maximum permissible voltage of the network by statistical analysis of experimental studies, for the use of the obtained models in the choice of apparatus in the presence of higher harmonic voltage components in the network at the design stage.**

Keywords: surge arresters, mathematical models, volt-ampere characteristic, leakage current.

I. Introduction

Non-linear surge arrester (SA) are in normal operation under the action of the operating voltage of the electrical network. In this mode, the body of the SA flows a current, the value of which is determined by the parameters of the apparatus. In the presence of any type of surge current in the network through the SA increases due to the nonlinearity of its resistance [1]. The main property of an SA is its ability to dissipate overvoltage energy, which essentially depends on the initial conditions in which the SA operates. One of the parameters that affects the ability of an SA to efficiently dissipate the energy of an overvoltage is the temperature of its components [2]. With an increase in the temperature of the varistors from which the SA is composed, its ability to maintain a thermal balance, which can lead to heat damage, varies greatly. Increasing the temperature of the varistors may be caused by distortions in the voltage in the electrical network if there are higher voltage harmonics in it. For the estimation of energy of losses of active power a method of its determination was developed [3].

In order to determine the losses of active power in the SA, at low quality of electric energy, it is necessary to have information about its volt-ampere characteristic (VAC) under the influence of the maximum allowable voltage of the network. Unfortunately, such information is not provided by manufacturers in directories. Therefore, we made an attempt to summarize the experimental results of the VAC at the operation of the SA in the zone of leakage.

Typically, studies and calculations of energy absorbing the SA are related to the varistor's VAC workspace and are described in detail by many writers [4-14]. This is due to the main function of the SA - to protect the equipment of electrical networks from over-voltage. However, the analysis of the work of an array at the impact of the maximum allowable voltage of the network researchers pay very little attention and publications on this topic are almost very rare. At the same time, such analysis is necessary to determine the ability of an SA to remain under the influence of the largest operating voltage of the electrical network without disturbing the thermal balance. This question is especially relevant for networks with low quality electrical energy. This is due to the fact that in the presence of higher voltage harmonics in the network, the loss of active energy in the SA can become significant and exceed the allowable values

Previous studies of the approximation examples of VACs of SAs have shown that various segments of VAC have different approximation functions. All attempts to approximate VACs with one mathematical expression did not receive positive results. The main attention was paid to the working segment of VAC, which characterizes the ability of the SA to perform the main function of work - overvoltage protection. The study of the work of an array at the influence of the maximum permissible voltage of the network, in which they work most of the time devoted enough attention. The determination of the VAC of the SA under the influence of the maximum allowable voltage of the network is an important task for ensuring the normal operation of the SA during the entire operation time.

II. Definition of VAC in the Leakage Current Zones

Typically, VAC SA is defined by the expression (1) or (2):

$$U = A \cdot I^{\alpha} \quad (1)$$

where α is the coefficient of nonlinearity of the material (the value of $\beta = 1\text{-}\alpha$ is called the ventilation coefficient);

A is a constant, which depends on the material and size of the varistor sample;

I - current, *A*;

U - voltage, *V*.

$$I = K \cdot U^{\lambda} \quad (2)$$

However, attempts to use these expressions, when exposed to the maximum permissible voltage of the network, give very large errors in relation to the experimentally obtained. This state of affairs makes it very difficult to analyze the work of SA under the influence of the maximum permissible voltage of the network and makes it practically impossible to assess the SA's ability to operate without accidents during operation under the influence of higher voltage harmonics.

A typical VAC, which is usually referred to by manufacturers in the directories for the SA has the form shown in Fig. 1 [1].

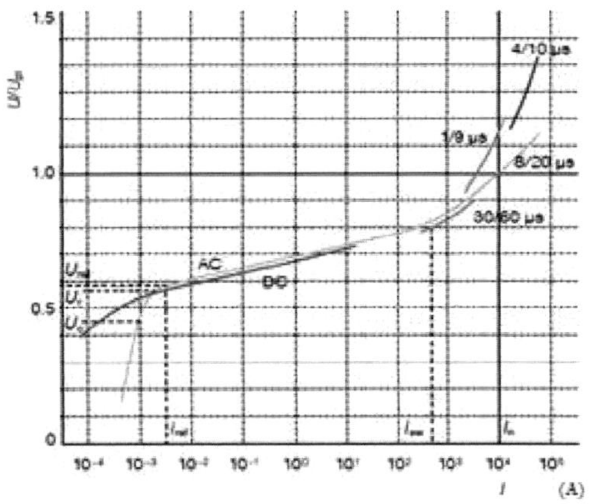

Figure 1 - Typical VAC SA

As can be seen from the figure of the VAC, the SA has three distinct sections. For the analysis of the operation of an overhead arc, when exposed to the maximum permissible voltage of the network, it is necessary to consider the first two of them. On the first segment of the SA, leakage currents are due to the properties of the ceramics from which the varistors are made. On the second section through the SA, conduction currents flow and it becomes a conductor. Each of these segments VAC of SA has its approximating function, which differs substantially from the approximating function of the adjacent segment of the VAC of the SA. This circumstance makes it necessary to use different coefficients in terms (1) and (2) for each segment. However, from the experimental data presented in Fig. 1 [1] we can conclude that on each of the segments of the VAC, the SA can be displayed using straight lines. Indeed, manufacturers of SA have a coefficient α of 1 in the zone of leakage currents VAC. If we substitute this value in expressions (1) and (2), we obtain an expression corresponding to the curve of the first order - a straight line beginning at the origin. Nonlinearities are expressed more significantly in the second segment of the VAC (VAC working area), however, with a certain error, one can assume the approximation of this segment of VAC with the help of a linear function. The search for such functions will be carried out by means of a dispersion analysis of the experimentally obtained VAC of SA and varistors in the zone of leakage currents.

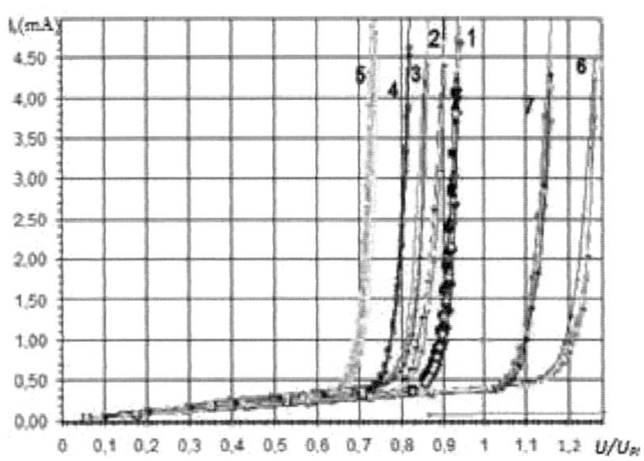

Figure 2 - Experimental VAC varistors and varistor pairs of different manufacturers:

1 - a pair of varistors producing ABB 6 *kV*; 2 - a pair of varistors of 5 *kV* of ABB production, aimed at meeting each other; 3 - 6 *kV* varistors of ABB production; 4 - a pair of varistors 5 *kV* of ABB production; 5 - *5 kV* varistors manufactured by ABB; 6 - a pair of varistors of 6 *kV* of ABB production, aimed at meeting each other; 7 - pair of varistors 3 and 5 *kV* produced by EPCOS [1]

The varistors for which the VACs are obtained, under the influence of the maximum permissible voltage of the network, have different nominal, residual and qualification voltages and permissible active power, which are determined by the manufacturer and are of a random nature. So the difference between varistors of one type is due to the physical properties of oxide-zinc ceramics, from which they are made and technology of production. In order to study the statistical dependences of the obtained results, according to the well-known method [15], we performed a dispersion analysis of the experimental data obtained. The complete dispersion expansion was performed using the KOVAN program in Borland Delphi 7.0. The calculation results are shown in Table 1.

Table 1 Results of covariance analysis of experimental VAC varistors

Statistical criterion	The obtained value of the statistical criterion	The critical value of the statistical criterion
Fo	316.38194	3.8415
Fwg	5.040007	3.8415
Fg	0.321196	1.3826
Fw	0.113681	1.3788
Fα	0.450113	1.3788
Fso	0.271062	1.485

In Table 1 shows the results of calculating the Fisher criteria F_0 (total angular regression coefficient), F_α (total), F_w (among the angular coefficients in groups inside), F_{s0} (remainder), as well as the F_{wg} (the angular coefficient of the group average in comparison with the average weighted coefficient) and F_g (relative to the linear regression of group mean) criteria, which can be used to estimate the uniformity of the scattering of additive deviations, with respect to the mean value of the free linear regression term by expression [15]:

$$M\left[\alpha_t\right] = \overline{Y} - B_0 \cdot \overline{T} \qquad (3)$$

To assess the dynamics of changes in the values of the decomposition criteria at all stages of the preliminary processing of data, the decomposition of the total sum of squares S deviations of the values of standardized measurement results, was carried out for the received data array. Completed decomposition of the total sum of squares S deviations of the values of standardized measurement results allows to estimate additive and multiplicative changes in the processes of parametric shift of mathematical expectations of the values of VAC currents. To do this, consider two criteria F_α and F_w. The F_α criterion reflects the random changes of the free member a_i. regression model, and F_w - random variations of the angular coefficient of the b_i model.

Comparing the results of calculating the values of the criteria F_w from Table. 1 with critical points F - distribution [15] taking into account the number of degrees of freedom, it is easy to conclude that for the analyzed data array, the hypothesis H_0 (1) is not rejected. This indicates the absence of a multiplicative shift, which may be due to different physical properties of the varistor structure, which, incidentally, is confirmed by a small excess of the values of the criteria F_{wg}, the values of the critical points of the F criterion. Such an excess testifies to the practical equality of the angular coefficients of linear regression, that is, the hypothesis $H_0^{(3)}$– is not rejected.

By comparing the values of the F criteria from Table 1 with critical points F - distribution, taking into account the number of degrees of freedom, hypothesis $H_0^{(4)}$ was also confirmed. From Table 1 it is evident that the criteria F_w and F_α inequality $F_w < F_\alpha$, indicating the presence of small systematic differences in the characteristics of the varistors studied.

The hypothesis that the group averages lie on the straight line $H_0^{(2)}$ has been confirmed. In addition, small values of the F_g criterion suggest a practical lack of influence of the unevenness of the distribution of the initial values of VAC currents, which indicates the variability of varistor characteristics varies depending on their manufacturers and nominal voltages.

The performed analysis of the values of the criterion $F_{\Sigma 0}$ did not reveal any significant effect of the total changes in the parameters a_i and b_i models for the investigated data array. As for the systematic changes in the mathematical expectation of indicators, the value of the criterion F_0 is significantly greater than the critical value obtained from the F-distribution, taking into account the number of degrees of freedom. This indicates a significant change in the mean values of the investigated VACs, that is, the presence of non-linearity of VAC in the zone of leakage currents.

Divide the obtained VAC varistors into sections where they can be represented by straight lines. On the basis of the well-known method [16], an analysis was made for deviation from the linearity for each of the varistor groups. Groups were determined based on the data shown in Fig. 2. The first group combines all investigated varistors on a segment of VAC to a voltage equal to 0.65 of the highest operating voltage of the network, all the following groups are determined by the type of the investigated varistor. The results obtained during the analysis for deviations from the linearity are given in Table 2. The notation is taken in the Table 2: M1 - a common array of all kinds of varistors to a voltage of 0.65 from the maximum permissible; M2 - 5 *kV* varistors manufactured by ABB; M3 - 5 *kV* varistor pairs of ABB production; M4 - 6 *kV* varistors manufactured by ABB; M5 - vaporizers of production of ABB 6 *kV*; M6 - a pair of varistors 6 *kV* of ABB production, aimed at meeting each other; M7 - pair of varistors 3 and 5 *kV* produced by EPCOS.

The obtained results showed that the hypothesis of the linearity of the curve corresponds to reality for all without variance of VAC segments of varistors of different groups, which makes it possible to conclude about its linearity on the considered segments. Such a result, the analysis for deviation from linearity, confirmed the possibility of representing VAC varistors with two straight lines. The first of these straight lines reflects the segment of leakage currents, and the second working area of the VAC varistors. The approximation of the VAC varistors, such lines, makes it possible to calculate the operating modes of the SA, when exposed to them, the voltage is close to the highest operating voltage, which allows to significantly improve the quality of their choice at the design stage.

To obtain mathematical models of VAC segment varistors, a regression analysis of the experimental data was performed. On the basis of the results of regression analysis, mathematical models of SA on the VAC segments for all groups of varistors studied were obtained. To determine the transition point from one segment of the model to the second we use that they have a single point of intersection. Having determined the coordinates of this point, we obtain the condition for the transition from one segment of the VAC to the second. We write the system of equations for the specified condition:

$$\begin{cases} I = A_0 + B_0 \cdot U \\ I = A_i + B_i \cdot U \end{cases} \quad (4)$$

Having solved this system relative to U and I, we obtain the coordinates of the point of transition from one VAC to another:

$$U_n = \frac{A_i - A_0}{B_i - B_0} \quad (5)$$

$$I_n = A_0 + B_0 \times \frac{A_i - A_0}{B_i - B_0} \quad (6)$$

The obtained expressions for the coordinates of the transition point allow us to obtain the mathematical model of the VAC of the SA in the zone of leakage currents:

Table 2 - Results of analysis of deviation from linearity of VAC varistors

Data	***k***	***N***	**Additive dispersion expansion**				**Number of degrees of freedom**		***F* - statistics**		**Decision**
			Q_{total}	$Q_{linear\ regression}$	Q_{discon}	Q_{remain}	v_{discon}	v_{remain}	$F_{calculated}$	$F_{critical}$	
M-1	3	227	2.4425	1.9041	0.2037	0.3346	1	224	136.39	1018.3	H_0 not rejected
M-2	5	50	84.0081	78.9789	2.3746	2.6543	3	45	13.419	14.025	H_0 not rejected
M-3	4	40	48.5804	44.9636	0.7568	2.8599	2	36	4.7633	39.469	H_0 not rejected
M-4	4	17	22.0648	18.3518	0.4648	3.2481	2	13	0.9302	39.4203	H_0 not rejected
M-5	4	19	14.2857	10.7371	0.2362	3.3123	2	15	0.5349	39.43	H_0 not rejected
M-6	6	26	23.1969	22.0358	0.5721	0.5891	4	20	4.8559	8.5599	H_0 not rejected
M-7	5	23	25.6715	23.6713	0.3329	1.6673	3	18	1.1983	14.201	H_0 not rejected

$$\begin{cases} U \le \dfrac{A_i - A_0}{B_i - B_0} \to I = A_0 + B_0 \cdot U \\ U_i \ge \dfrac{A_i - A_0}{B_i - B_0} \to I = A_i + B_i \cdot U \end{cases} \quad (7)$$

where i - number of varistors group.

The obtained expressions of mathematical models of varistors give somewhat overestimated values of currents, however, the maximum error is not more than 10%, which is sufficient for the engineering calculations of the operating modes of the SA in the zone corresponding to the largest operating voltage of the network.

The values of the coefficients A_0, B_0, A_i, and V_i for each of the varistor groups are given in Table 3. Designations adopted in Table 3: MM1 - 5 *kV* varistors manufactured by ABB; MM2 - 5 *kV* varistor pair of ABB production; MM3- 6 *kV* varistors manufactured by ABB; MM4 - 5 *kV* varistors of ABB production, aimed at meeting each other; MM5 - Variors of production of AVK 6 *kV*; MM6 - a pair of varistors of 6 *kV* of ABB production, aimed at meeting each other; MM7 - pair of varistors 3 and 5 *kV* produced by EPCOS.

Table 3 - Coefficients of approximating lines for varistor groups.

No. groups	Coefficients of approximating straight lines			
	A_0	B_0	A_i	B_i
MM-1	- 0.00139	0.54196	- 52.5938	62.8399
MM-2	- 0.00139	0.54196	- 71.6242	92.5351
MM-3	- 0.00139	0.54196	- 73.1479	90.2913
MM-4	- 0.00139	0.54196	- 77.3415	110.1748
MM-5	- 0.00139	0.54196	- 81.4229	114.4777
MM-6	- 0.00139	0.54196	- 90.4633	101.5741
MM-7	- 0.00139	0.54196	- 116.1008	128.3104

Mathematical models obtained as a result of the dispersion analysis of experimental data can be used as the basis for the method of determining the VAC of the SA in the zone of leakage currents. The essence of the method is the separation of the VAC of the SA into segments and their replacement on these segments by curves of the first order (straight lines). Such a method of representing the VAC of the SA will simplify the definition of their characteristics when choosing at the design stage.

Conclusions

1. The performed covariance analysis revealed the presence of non-linearity of VACs under the influence of the maximum allowable voltage of the network.

2. Mathematical models obtained as a result of the dispersion analysis of experimental data can be used as the basis for the method of determining the VAC of the SA under the influence of the maximum permissible voltage of the network. The essence of the method is to divide the VAC of the SA into segments and replace them with these segments with curves of the first order (straight lines).

3. The method of representing the VAC of the SA will simplify the definition of their characteristics when selecting their parameters at the design stage.

References

[1] S. Yu. Shevchenko, P.V. Petrov, G.M. Katrenko, O. M. Pasko, BF Yermolenko, O.M. Bunches"Limiters of overvoltages nonlinear: application, installation and selection Bunches edited by Shevchenko S.Yu. The monograph Kharkiv "Fort" 2015 - 286 pp.

[2] S. Shevchenko, S. Khlomko, O. Berchuk, "Influence of energy characteristics of surge arresters on their selection" ,The East European Journal of Advanced Technologies. - Kharkiv: UkrDUZT, PE "Technological Center". - 2017; No. 4/8 (88): P. 48-55.

[3] S.Yu Shevchenko, "Peculiarities of thermal modes of overvoltage limiter of nonlinear in conditions of low quality of electric energy The East European Journal of Advanced Technologies." - Kharkiv: UkrDUZT, PP "Technological Center" - 2015; No. 4/8 (76): P. 11-16.

[4] S.S.D., Dmitriy, "Experimental studies of current voltage characteristics of the arrester leakage currents in the area" International Young Scientists Forum on Applied Physics (YSF). - Dnipropetrovsk: Oles Honchar Dnipro National University. – 2015. – 1-4 pp. DOI: 10.1109/YSF.2015.7333227

[5] A. de Conti, F. H. Silveira, S. Visacro, "On the role of transformer grounding and low-voltage surge arresters on protecting loads in complex distribution networks" Electric Power Systems Research, vol 113, august 2014, pp. 204-212 DOI:10.1016/j.epsr.2014.02.035

[6] A. Christodoulou, G. Perantzakis, G. E. Spanakis, P. Karamelas, "Evaluation of the lightning performance of transmission lines protected by metal oxide surge arresters using artificial intelligence techniques",Energy Systems 2012; 3 (4) pp. 379-399.

[7] Overvoltage limiters are nonlinear voltage 6-35 kV. Instruction on the selection and application in switchgear installations: СОУ-Н МЕВ 40.1-00100227-67: 2012 .- Officer. kind. - K.: GRIFRE: Ministry of Fuel and Energy of Ukraine, 2012, 50 p. - (Regulatory document of the Ministry of Fuel and Energy of Ukraine. Instruction.).

[8] ABB MWK Surge Arrester Datasheet, Medium Voltage Products & Systems Catalog 2014

[9] K. Aodsup, T. Kulworawanichpong, "Numerical Modeling and Very Fast Transient Simulation of MOV Surge Arresters", IEEE 2012 IEEE PES Asia-Pacific Power and Energy Engineering Conference (APPEEC) - Shanghai, China (2012.03 .27-2012.03.29)] 2012, - 1-4. DOI: 10.1109/APPEEC.2012.6307335

[10] H. Sugimoto, K. Shimasaki, H. Kado,"Distribution Surge Arrester Due to Winter Lightning and Measurement of Energy Absorption Capability of Arresters", IEEJ Transactions on Power and Energy, - 2012; - 132 (6), - P. 554-559.

[11] H. Daman-Khorshid, F. Ajri, R. Shariatinasab, "Probabilistic evaluation of the failure risk of transmission line surge arresters caused by lightning flash", IET Generation Transmission & Distribution, - 2014, - 8 (2), - P.193- 202

[12] K. Aodsup, T. Kulworawanichpong, "Numerical Modeling and Very-Fast Transient Simulation of MOV Surge Arresters", Power and Energy Engineering Conference (APPEEC), 2012 Asia-Pacific 1-4 (2012). DOI: 10.1109/APPEEC.2012.6307335

[13] G. D. Peppus, I. A. Naxakis, Ch. T. Vits, E. C. Pyrgioti," - Surge arresters for fast transients" [IEEE 2012 International Conference on Lightning Protection (ICLP) - Vienna, Austria (2012.09.22.09.09)] 2012 International Conference is Lightning Protection (ICLP) -1-6 (2012).

[14] K. Demyanenko,"Stability of highly nonlinear zinc oxide arrestors under the prolonged action of a commerial frequency potential" Sov. Eltectr Eng. (USA) Vol. 55, No 9, p.46-51, 2013

[15] AM. Andronov, E.A.Kopitov, L.Ya.Gricllaz," Probability Theory and Mathematical Statistics", A Textbook for High Schools. St. Petersburg: Peter, 2004.

New express-method for liquid dielectric quality valuation

Svetlana Vysogorets, Alexey Tadjibaev
Department of Electrical equipment Diagnostics
FSEE "PEIPK"
Saint Petersburg, Russia
s-151075@yandex.ru

Ilia Gorec
School of Humanities and Social Sciences
NRU "HSE SPb"
Saint Petersburg, Russia
gorecil@ya.ru

Abstract — **Presented the results of researches, aimed to develop methods for express-analysis of transformer oil quality indicator "Acid Value". Developed an algorithm of preparation for and conducting measurement. Proposed implementation of mixed indicator. Outlined the valuation criteria for results of "AV" semiquantitative measurement via express-method. Defined the criteria for RGB measured values distribution by AV groups.**

Keywords: valuation, transformer oil, quality indicators, method

I. Introduction

In modern conditions, the issue of increasing operational efficiency is acute for Russian energy companies. Maintenance and repair (hereinafter referred to as MRO) are traditionally a "black hole" for the budget, while the current level of wear and tear of equipment of the Russian Federation electric grid organizations exceeds 60%.

In the conditions of a market economy, to ensure competitiveness, the tactics of minimum expenditures came. The increase in the efficiency of energy production is inextricably linked with the improvement of the quality of repairs. The search for innovative technologies and techniques in the performance of repair work is oriented both to improving the quality of work (correspondingly, increasing the in-between overhauls period) and to reducing costs.

The aging equipment park requires medium and capital repairs, which are followed by regenerative work to restore the exploitational and resource characteristics of the liquid dielectric. The physico-chemical analysis of the oil is an extremely reliable tool that allows to influence the efficiency of these works: to control the technological cycle, to determine the point of the procedures completion. At the same time, carrying out control chemical measurements involves the need to transport samples to chemical laboratories with considerable time losses. The search for a minimum set of control parameters and the development of methods for their rapid analysis will improve the quality and efficiency of repair work in oil-filled equipment..

II. Determination of oil quality key parameter to value liquid dielectric resource

All the processes taking place in the insulation of oil-filled equipment affect, above all, the impregnating mineral oil and spread to solid cellulose materials (paper, cardboard). At the present time, there have been developed theoretical provisions on the fundamental aging processes of individual insulation components - oil and paper [1].

Damage to insulation is the result of complex physico-chemical processes that occur under the influence of electric and magnetic fields, gas and hydrodynamic processes, temperature, as well as other factors - humidification, pollution, etc. [2]

In [3], there is revealed a considerable influence of the oxidation processes occurring in the transformer oil and the aging products formed in it on the state of solid insulation, affecting, eventually, its electrical insulation characteristics and durability. Timely and sufficient removal of aging products from the oil and subsequent restoration of its resource allows maintaining the characteristics of both isolation of power transformers and liquid dielectric stable and reliable.

The carried scientific experiment [3] statistically proved a significant relationship between the oil antioxidant stability (oil resource) and a number of quality indicators which are indicative also for aging products, among which are the indicators "Acid value" (further AV), "Water-soluble acids and alkalis" (further WAA), "Dielectric loss angle tangent" (hereinafter Tangent). By methods of correlation-regression analysis, a significant direct relationship between AV and antioxidant stability (hereinafter Stability) was established [3]. Along with this, the main Stability grouping factor of transformers was found to be AV [3].

Thus, AV is a key indicator of quality, allowing with a high degree of reliability to indirectly assess the change/deterioration of the oil resource. Accordingly, in case of monitoring the trend of the AV during regeneration work it is permissible to conduct a test valuation of the performed regeneration quality with the subsequent determination of the works completion point.

A significant part of the oil resource regeneration work is carried out at the place of oil-filled equipment installation, accordingly, there is a need to develop an express-method of operational control of the regeneration works quality. Thus, the appearance of the express-methods for estimating AV in the arsenal of exploitational divisions will eventually result in:

- reducing the loss of time for the delivery of intermediate samples to the laboratory, and, accordingly, will affect the reduction in the duration of direct capital repairs;
- reducing the cost of capital repairs: reducing the idle time of crews before obtaining the results of control measurements, increasing the efficiency of regeneration machines by sorbent timely replacement based on the results of control measurements, reducing costs by eliminating the excessive carrying out of expensive laboratory measurements;

- improvement of the quality of directly regenerative work due to the provision of operational control without loss of organizational procedures, and accordingly to increase the period in-between overhauls of transformers, increase the reliability and durability of their isolation.

III. Development of a laboratory experiment

One of the recognized modern methods of measuring transformer oil AV is the method [4] providing for volume titration (titrimetric method) using an "Pigment Blue 61" indicator. In this case, unlike the more common method [5] of determination of AV using the indicator of nitrazine yellow, the method [4] does not contain boiling procedures. Accordingly, the method [4] is defined the optimal as the basis for the development of the express-analysis of the oil quality index AV.

The developed technique is an empirical (semi-quantitative) method for assessing the compliance of the AV of the analyzed mineral transformer oil with a given initial value. To assess the results, a categorical assessment is used: "less than the set value" and / or "more than the set value". A list and composition of necessary solutions was determined, the interaction of which with the defined volume of the transformer oil sample, according to the adopted rules, by the colour shifts the range of AV values characteristic for a given liquid dielectric is determined.

All experimental measurements are carried out on the basis of the chemical laboratory of PJSC "IDGC of the North-West" "Komienergo".

During the trial experimentations, it was found that for a number of oils there is no more or less bright colour transition, which makes it difficult to determine the end point of titration/the category of evaluation.

Justification of the choice of indicators [6]:

1. "Pigment Blue 61" has a pH range of the colour transition (in aqueous solutions): from pH 9.4 (blue) to pH 14.0 (red):

2. Curcumin has the first range of the pH colour transition (in aqueous solutions): from pH 7.4 (yellow) to pH 9.2 (brown-red); The second range of the pH colour transition (in aqueous solutions): from pH 10.2 (brown-red) to pH 11.8 (orange-yellow):

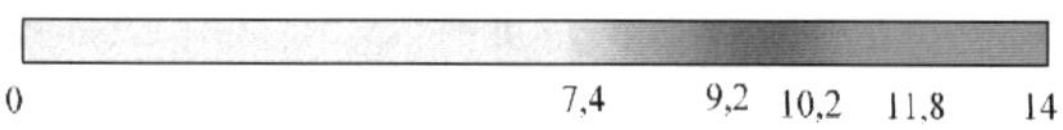

When using the above-mentioned pair of indicators at the same time, it is expected that the yellow colour of the curcumin indicator in an acidic region will not affect the initial colour of the analyzed mixture, given the original yellow colour of the oil. At the same time, in combination with the blue colour of the indicator of "Pigment Blue 61" in a mixture with oil, a strengthening of green shades is expected. When approaching the equivalence point, curcumin begins to give the red-brown colour of the mixture before the "Pigment Blue 61", and then strengthen the red colour it.

Thus, the use of an additional indicator made it possible to make the green tint of the colour more pronounced and hence the (more clearly observed) disappearance of the green tint is considered an achievement of the end point of titration.

IV. Research results

The key objectives of the study are:

- Development of the methodology for express analysis of transformer oil AV: determination of conditions and algorithm for conducting measurements, a list of necessary reagents, the order of evaluation of measurement results;
- search for the dependence of the change in the colour component composition of the reaction mixture according to the RGB model from the measured value of the AV.

In the course of experimental studies, 54 samples of transformer oils with different AV were measured. The results of the measurements are grouped into four ranges of analyzed values of AV (less than 0.02 mgKOH/g; 0.02-0.03 mgKOH/g; 0.03-0.05 mgKOH/g; 0.05-0.10 mgKOH/g). For the analysis of one working sample of transformer oil, a list of necessary solutions, an algorithm for preparing and conducting an express-analysis of the AV quality index were developed. Based on the results of the test measurements, it is envisaged to form a conclusion about the compliance of the AV to one of the set values.

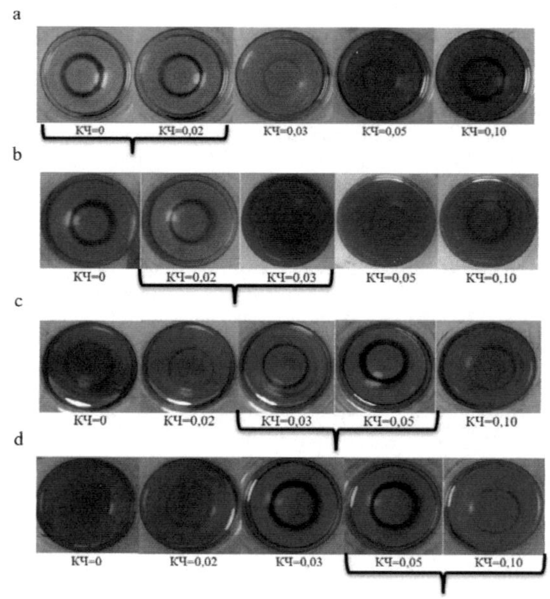

Figure 1. Photos of solution colour variation in test-tubes at chemical reaction:
a) Visible shift of coloration: definite range of AV value – no more than 0,02 mgKOH/g (measured value - 0,0003 mgKOH/g);
b) Visible shift of coloration: definite range of AV value – no more than 0,03 mgKOH/g (measured value - 0,022 mgKOH/g);
c) Visible shift of coloration: definite range of AV value – no more than 0,05 mgKOH/g (measured value - 0,048 mgKOH/g);
d) Visible shift of coloration: definite range of AV value – no more than 0,10 mgKOH/g (measured value - 0,056 mgKOH/g). (figure caption)

In the developed method of rapid analysis of the AV, the following interpretation of the results of the test analysis is provided without taking into account the numerical representation / change in the component composition of the colour of the solution during a chemical reaction:

- if the colour of the solution in the test tube with an aliquot of the sample retains the main colour / shade (red with tints) and significantly differs from the colour / shade of the test solution (green with tints), this means that the substances causing the oil AV are not sufficient to neutralize all the amount of KOH contained in Solution 2. Accordingly, it can be concluded that the AV of the oil sample does not exceed the preset AV in Solution 2;

- if the colour of the solution in the test tube with an aliquot of the sample changed the main colour / shade became green and / or close to the colour / shade of the test tube with the reference solution, this means that the substances causing the oil AV in the mixture are contained in excess, all KOH is neutralized. Accordingly, it can be concluded that the AV of the oil sample exceeds the preset AV value in Solution 2.

The evaluation of the reliability of the first developed AV express-analysis method was carried out by comparing the ratio of the obtained by the newly developed method analysis results to the results of the oil samples under study, the measurement of the AV in which was carried out using a control certified method [4]. Based on the results of the comparative analysis of the new method with the control one, it was established that in most of the analyzed samples, an admissible visualization for determination of the titration end point is found (see Fig.1)

After carrying out in the laboratory conditions of oil regeneration with the subsequent control measurement of the AV by the express-method, the dynamics of the colour change is also noticeable.

V. Statistical Analysis of Empirical Results

To eliminate the influence of the human factor in determining the end point of titration / the decision point, quantitative criteria are searched for.

A hypothesis has been put forward on the existence of a dependence of the change in the solution component colour composition the during chemical reactions on the quantitative content of acids in the analyzed oil sample (values of the AV).

The analysis of the colour component constitution of the solution in tubes is carried out according to the RGB model using the «Цветоанализатор ColorAnalyzer» software: the data is read in three channels - R (red), G (green) and B (blue), followed by its fixation. During the research, the results of measurements of each stage of titration were recorded:

- 1 stage of titration - an aliquot of the oil sample without adding an additional volume of KOH (reference solution);

- 2 stage of titration - an aliquot of the oil sample with the additional calculated volume of KOH in the amount necessary to completely neutralize the substances causing the AV less than 0.02 mgKOH/g;

- 3 stage of titration - an aliquot of the sample of oil with the additional calculated volume of KOH in the amount necessary to completely neutralize the substances causing the AV equal to 0,02-0,03 mgKOH/g;

- 4 stage of titration - an aliquot of the oil sample with the additional calculated volume of KOH in the amount necessary to completely neutralize the substances causing the AV equal to 0,03-0,05 mgKOH/g;

- 5 stage of titration - an aliquot of the sample of the oil with the additional calculated volume of KOH in the amount necessary to completely neutralize the substances causing the AV to be 0.05-0.10 mgKOH/g.

The dependence of the AV and RG-values for each stage of titration (1-5) should be statistically analysed, where RG is the corresponding value of the ratio of the measured green (G) to red (R) solution colour for each of the five stages of titration: RG1, RG2, RG3, RG4, RG5. Also, the dependence of the AV and B-values for each titration step (1-5) is statistically analysed, where B is the corresponding value of the blue colour of the solution in each of the five stages of titration: B1, B2, B3, B4, B5.

In the course of the a priori analysis, a generalization of the initial data was carried out: variational series were constructed for each of the traits studied. In the first series of analysis, the factor is the AV, the response - RG1, RG2, RG3, RG4, RG5, respectively, in the second series of analysis, the factor is the AV, and the response is B1, B2, B3, B4, B5, respectively.

To determine the presence of a linear relationship between the AV factor and the RG and B responses for each stage of titration, as well as an assessment of its tightness and statistical significance, data were checked on the basis of the Pearson[7] correlation criterion (see Table 1).

Table 1.

№ of titration stage	**Pearson coefficient (r) for every stage of titration, for RG response**	**Pearson coefficient (r) for every stage of titration, for B response**
1	0,14330881	-0,17888
2	0,11850754	-0,10838
3	0,25350886	-0,10486
4	0,42049732	-0,29418
5	0,56998923	-0,44897

Thus, 10 correlation coefficients were obtained, on the basis of which it was established:

- a statistically significant relationship between the AV factor and the RG response on the fourth and fifth titrations: the value of the coefficient for RG4 $r = 0.4$, the tightness / correlation strength is moderate; the value of the coefficient for RG5 $r = 0.6$, the tightness / the correlation strength is noticeable;

- inverse statistically significant relationship between the AV factor and the B response on the fifth titration: the value of the coefficient for B5 $r = -0.4$, the tightness / correlation strength is moderate.

Statistically significant dependencies are analyzed using scatter plots in the PowerBI utility (see Fig. 2).

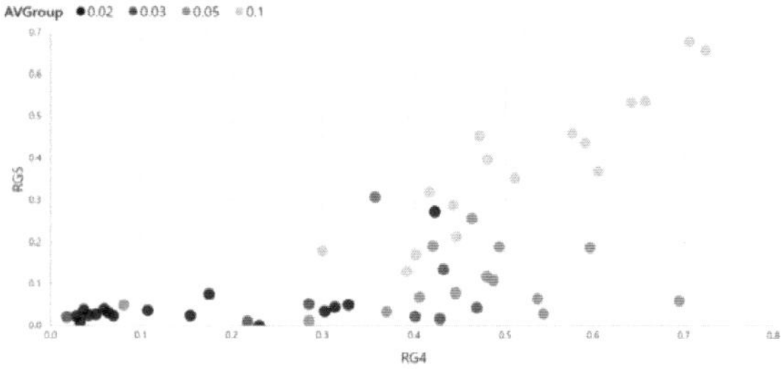

Figure 2. Scatter plot of RG4 factor dispersion on RG5 response by AV Group:

- ● AV value for each oil sample in group no more than 0,02 mgKOH/g;
- ● AV value for each oil sample in group 0,02-0,03 mgKOH/g;
- ● AV value for each oil sample in group 0,03-0,05 mgKOH/g;
- ● AV value for each oil sample in group 0,05-0,1 mgKOH/g.

When considering the scatter plot (Fig. 2), a stratification into clusters according to the AV group was found. The formed clusters allow us to distinguish the boundaries of the RG4 and RG5 indices for each AV group, the measurement of which in the course of the express-analysis will allow to attribute the analyzed sample to a certain AV group. At the same time, it is difficult to outline the cluster boundaries for the AV group of 0.02-0.03 mgKOH/g. Thus, the indicated AV group of 0.02-0.03 mgKOH/g is scattered between the AV groups less than 0.02 mgKOH/g and 0.03-0.05 mgKOH/g. Accordingly, it can be concluded that the use of parametres RG4 and RG5 is not enough. To eliminate the above drawback, the use of the B indicator (Blue) to evaluate the obtained measurements was proposed.

The introduction of an additional indicator B5 (indicator B for stage 5 of titration) allows to improve the accuracy of determining the quantitative criterion of the measured RGB indicators ratio to a certain group of AV. Thus, it has been established that different groups of AV have different B5 indices, which make it possible to separate the AV group samples 0.02-0.03 mgKOH/g from other groups (see Fig. 3).

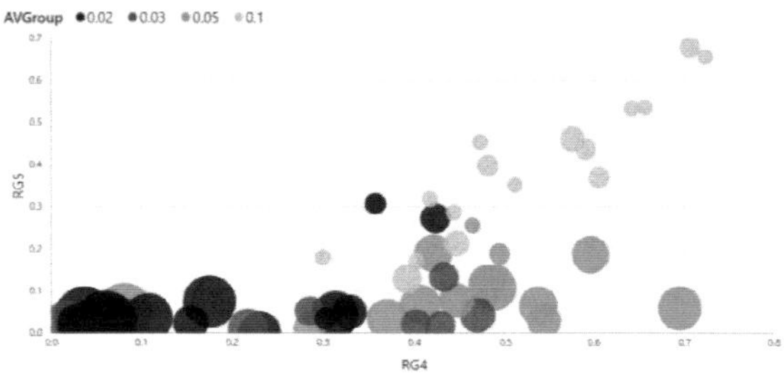

Figure 3. Scatter plot of RG4 factor dispersion on RG5 response by AV Group, sized by B5

- ● AV value for each oil sample in group no more than 0,02 mgKOH/g;
- ● AV value for each oil sample in group 0,02-0,03 mgKOH/g;
- ● AV value for each oil sample in group 0,03-0,05 mgKOH/g;
- ● AV value for each oil sample in group 0,05-0,1 mgKOH/g.

On the basis of the conducted analysis, the criteria for distributing the measured RGB values by the AV groups are defined (Table 2).

Table 2.

AV Group	Indicator	Value range
Less than 0,02	RG4	0.00-0.35
	RG5	0.00-0.1
	B5	5-11
0,02-0,03 mgKOH/g	RG4	0.00-0.5
	RG5	0.00-0.15
	B5	3-5
0,03-0,05 mgKOH/g	RG4	0.35-0.7
	RG5	0.00-0.2
	B5	5-7
0,05-0,1 mgKOH/g	RG4	0.35-0.75
	RG5	0.1-1
	B5	1-3

Based on the results of statistical analysis, one can state the presence of a statistically significant dependence of the change in the solution component colour composition (in the RGB system) in chemical reactions on the quantitative content of all free acids in the sample of the liquid dielectric (AV values). The criteria for the distribution of measured RGB values by the AV groups are determined.

VI. Conclusions

1. In works [3] it is justified that the key parameter providing a reliable operational control of the process of oil regeneration (resource restoration) is the AV.
2. Substantiated the necessity of the development of express analysis of the AV oil quality parameter as a step allowing to increase the efficiency of capital repairs of power transformers.
3. The method of AV express analysis is developed, the algorithm of preparation and carrying out of measurements along with criteria for visual valuation of the AV semi-quantitative express-measurement results are presented. A mixed indicator is proposed.
4. Based on the results of the statistical analysis, the hypothesis of the dependence of the change in the solution component colour composition (RGB) in chemical reactions on the quantitative content of all free acids in the sample of the liquid dielectric (values of the AV) is proved. The criteria for the distribution of measured RGB values by the AV groups are determined.

References

[1] G.G. Lysakovskii, "Research of transformer oil aging in paper-oil insulation in close to exploitational circumstances," in Liquid fuel and oil in energetics: treatise collection. - M.: Energoizdat, 1982

[2] G.Ya. Shneider, Electic insulation of high-voltage - M.: Znak, 2009, 160 p.

[3] S.P.Vysogorets, Development of the new methods and quality valuation algorithm of 35-110kV power transformer exploitational oil, Dissertation for the degree of Candidate of Engineering Sciences SPb., 2012, 260 p.

[4] MKHA 05-09 Measurement methods determination of transformer oil acid number by titrimetric method, SPb.: Scientific and Production Association «Electrum» 2009 (2014 edition), 15 p.

[5] GOST 5985-79 (ST SEV 3963-83) Petroleum products. Method of acidity and acid number determination, M.: Izdatelstvo Standartov, 1994, 9 p.

[6] P.P. Korostelev, Reagents for technical analysis: B.R., M.: Metallugria, 1988, 384 p.

[7] K. Pearson, "Notes on regression and inheritance in the case of two parents" - Proceedings of the Roy. Soc. London, vol. 58, pp. 240–242, June 1895.

Development of a Predictive Model for Estimating the Liquid Dielectric Resource

Svetlana Vysogorets
Department of Electrical equipment Diagnostics
FSEE "PEIPK"
Saint Petersburg, Russia
s-151075@yandex.ru

Abstract — **Proposed a measurement of the "antioxidation stability" quality parameter for the valuation of exploitational transformer oil resource. Defined the rules for the classification of transformers that allow to predict the aging of the oil based on the measured regulated quality parameters.**

Keywords: valuation, transformer oil, quality indicators, prognosis

I. Introduction

In a rapidly aging electrical equipment park, the most important is the choice of effective approaches to the organization of maintenance and repairs (hereinafter referred to as MRO) that can reduce exploitational costs. To solve this problem, it is necessary to develop and implement methods of technical diagnostics, which allow the organization of MRO to take the path of preventive measures that affect the reduction of emergency blackouts and reduce exploitational costs.

II. Influence of a Liquid Dielectric Quality on Reliability and Durability of Power Transformers

During exploitation, the oil in the transformer changes its physico-chemical and exploitational properties, which affects its quality parameters shifts, while the oxidation of oil is a key process that determines the change in its properties. A significant deterioration in the quality of exploitational transformer oils leads to the need for complex, expensive repairs [1].

Residual resource of solid insulation is closely related to the residual resource of transformer oil, since the products of oil "aging" (oxidation) heavily influence the "aging" of solid insulation [2, 3]. Timely removal of aging products from exploitational oil is a preventive measure that allows to ensure the required level of reliability of oil-filled transformers and affect their durability.

Effective monitoring of the transformers condition based on the results of a physico-chemical analysis of a liquid dielectric makes it possible to determine the time when the above procedures will be necessary and effective at minimum exploitational costs.

The complex chemical composition of transformer oils explains the ambiguity of their "behavior" in exploitation, which ultimately causes difficulties in predicting the early stages of oil aging due to the diversity of the formed oxidation products. The diversity of the chemical composition of different groups of transformer oils indicates the need to find methods for assessing the quality of exploitational oils, taking into account their individual characteristics and working conditions.

III. Method for Estimating the Resource of Exploitational Oils of Power Transformers

An indicator of oil quality, determining its resistance to aging factors, is "antioxidant stability". Stability against oxidation of oils is determined by the following quality parameters: the acid value of the oxidized oil (hereinafter referred to as "*AVOO*") and the amount of sediment formed after oxidation (hereinafter "*Sediment*"). The following values are taken as threshold values, measured during determining the antioxidant stability [4]: "Acid number of oxidized oil", mgKOH / g - no more than 0.1; "Content of sediment formed after oxidation", % of mass - no more than 0.01.

Timely implementation of measures to maintain the stability of the insulation by continuously removing aging products from the exploitational oil and timely restoring its resource is a preventive measure to ensure the required level of reliability of oil-filled transformers. Effective chemical monitoring of the state of power transformers is a "tool" that allows to determine the point in time when the implementation of the above measures will be necessary and effective with minimal operating costs.

"Residual resource of transformer oil" is the total exploitational time of transformer oil from the moment of control of its technical condition to the transition of oil quality to the extreme state. "The extreme oil state" is a condition of the oil in which its further exploitation is unacceptable or impractical, or the restoration of its exploitational state is impossible or impractical. Thus, the quality management of transformer oil is probably the only element through which it is possible to control the reliability (quality) of a power transformer. [5].

The method of exploitational oil resource valuation is justly attributed to the "antioxidation stability" parameter measurement, standardized by GOST 981-75 "Petroleum Oils. The method for estimating the antioxidation stability"[6]. By measuring the antioxidant stability of the exploitational transformer oil in conditions of accelerated oxidation of the oil sample in the laboratory, it is estimated how much the condition of the oil under examination, at a given time, is resistant to the influence of key factors that determine the course of oxidation processes, which take place in real circumstances. Accordingly, the worse the antioxidant stability indicators, the greater the initial resource proportion oil has lost [1].

The method of measuring oil antioxidant stability is rather complicated and laborious process, therefore its use is rational in the presence of oil aging signs - exactly, when the deterioration of the oil quality is registered according to a number of controlled parameters by the results of the physico-chemical analysis.

A number of measured oil quality parameters are indicative to oxidation products and allow, with a certain confidence probability, to determine the stage of oxidation process development in oil. On the basis of the theoretical studies, it has been established that the following quality parameters are indicative for the oil aging processes: acid value (further as *AV*), water-soluble acids and alkalis (hereinafter referred to as *WAA*), tangent of the dielectric loss angle of oil (hereinafter *Tgδm*), content of antioxidant additive agidol-1 (*"Additive"*). The theoretical possibility of the relationship between the change in the above-mentioned oil quality indicators and the degree of expenditure of its resource is found [5]. The carried out experimental studies confirmed the theoretical possibility of changing the values of the above-mentioned indicators of the operational transformer oil quality from the degree of its aging - antioxidant stability [1].

IV. Search for a way to forecast a residual resource of transformer exploitational oil on the basis of the measured regulated oil quality parameters

Correlation of the quality parameters and the degree of oil aging was studied to construct a model that predicts a decrease in the oil resource in the early stages, and, accordingly, timely determine the optimal time for the necessary exploitational procedures.

During statistical data processing, the degree of oil aging was represented by categorical variables: satisfactory and unsatisfactory stability. In case the antioxidant stability is not reduced (satisfactory), the stability class is set to 0, if the antioxidant stability is reduced (unsatisfactory), the stability class is set to 1.

Statistical modeling of the relationship between oil quality parameters and the degree of its aging was carried out using a nonparametric method - classification trees. The purpose of constructing a classification tree is to construct the decisive rules that allow you to classify objects. Any decision tree is essentially a tree-like graph consisting of nodes connected with each other by edges. Each level in the tree can be considered as one of the solutions; the decision-making node provides a condition check, and each edge indicates one of the possible options. At mathematical modeling of dependence of oil quality parameters from a degree of its aging, algorithm of a classification tree CART is used. CART (Classification And Regression Trees) builds a binary tree (with a separation on only two nodes at each branching step) based on a complete enumeration of tree variants with one-dimensional branching.

Figure 1 presents the rules for classification into groups by Stability based on the values of various factors. The zero node of the tree in Figure 1 shows that out of 52 transformers (100%), 16 transformers (30.8%) actually belong to the Stability class = 0 and 36 transformers (69.2%) actually belong to the Stability class = 1. Graphically, the proportion of transformers of each class is shown by a bar chart. Node 2 shows that Stability = 1 unambiguously includes all transformers whose AV is > 0,0195 mgKOH/g. If, on the other hand, the value of AV does not exceed 0,0195 mgKOH/g, then the next grouping attribute is the indicator "oil dielectric loss tangent" (Tgδm): if this indicator does not exceed 0.78%, then with probability 0.846 (84.6%) transformer can be classified as Stability = 0.

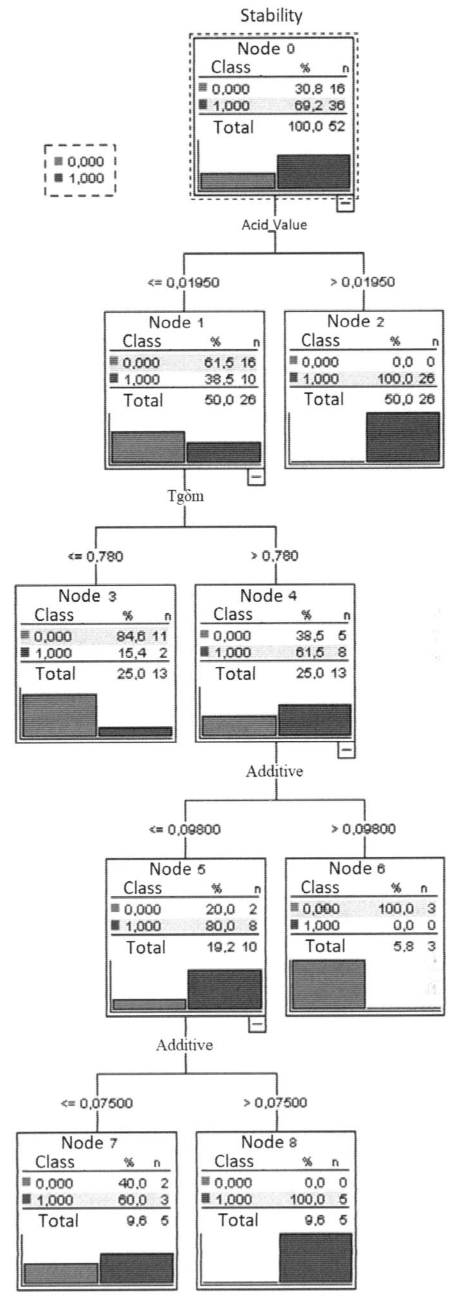

Figure 1. The classification tree of transformers into groups according to Stability classes based on three factors: 0.000 - Stability satisfactory = 0; 1,000 - Unsatisfactory stability = 1.

If Tgδм > 0,78%, then the following grouping attribute is the indicator "content of antioxidant additive" (Additive): if the value of this indicator is more than 0,098%, then the transformer belongs to the Stability class = 0 with probability one (100%). If the Additive is > 0.075% of weight, but does not exceed 0.098%, then the transformer is classified as Stability = 1 with a probability of one (100%). If the Additive does not exceed 0.075% of weight, then with a probability of 0.6 (60%) the transformer can be assigned to the class Stability = 1 and with a probability of 0.4 (40%) the transformer can be assigned to the class Stability = 0. It should be noted that the verification of branching for the factors Tgδm and Additive is difficult due to the insufficient number of transformers in the subgroups.

The classification according to three criteria gives 92.3% of the correct classifications, while the proportion of properly classified objects in the class Stability = 1 is 94.4%.

The risk (the average proportion of incorrectly classified objects) during cross-validation of the classification is 32.7%, which exceeds 30%, which indicates that the quality of the classification is insufficient. The reason for this is the insufficient number of objects in the subgroups based on the factors TgδM and Additive.

Since the verification of branching by Tgδm and Additive factors is difficult due to the insufficient number of transformers in the subgroups, it is advisable to shorten the tree - to classify only by one factor - AV (the first level of this tree).

Classification by one factor by the method of trees is shown in Fig.2. The zero node of the tree in Fig. 2 shows that of all 52 transformers (100%) 16 transformers (30.8%) actually belong to the class Stability = 0 and 36 transformers (69.2%) actually belong to the class Stability = 1. Graphically, the specific weight of the transformers of each class is shown by a bar chart. Node 2 shows that the class of Stability = 1 definitely refers to all transformers for which the value of AV > 0.0195 mgKOH/g.

If the value of AV does not exceed 0,0195 mgKOH/g, then with a probability of 0.62 (61.5%) transformers can be classified as Stability = 0, and with a probability of 0.38 (38.5%) transformers can be classified to class Stability = 1

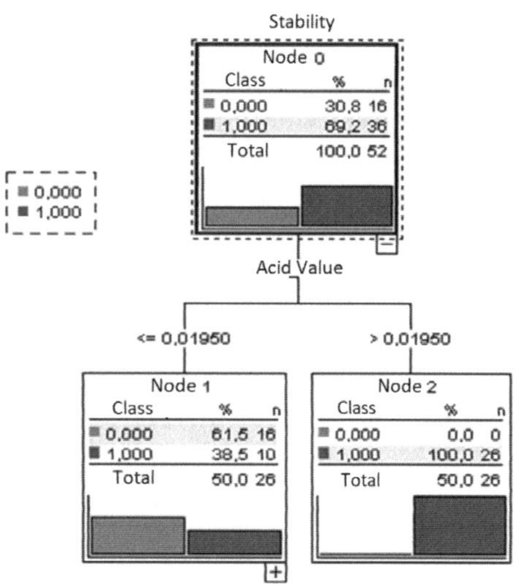

Figure 2. The classification tree of transformers into groups according to Stability classes based on factor AV: 0.000 - Stability satisfactory = 0; 1,000 - Unsatisfactory stability = 1.

Classification of transformers into groups according to the classes of Stability only on the basis of the value of factor AV gives 80.8% of correctly classified objects.

Thus, the classification tree method makes it possible to classify transformers into groups according to the Stability class based on the values of the AV factor. The threshold (grouping) AV is 0.02 mgKOH/g.

Additionally, the transformers were classified into groups according to the Stability class by discriminant analysis. The best way of classification based on discriminant analysis is determined by the method of enumeration of various variants. Simulation was carried out with the exception of the WAA factor, which is a multicollinear AV. The coefficients obtained from the results of discriminant analysis allow us to obtain two discriminant functions (DF), which determine to which group in the Stability class should the transformer be included.

Stability equals 0:

$$DF = -11{,}537 + 52{,}876 * AV - 0{,}221 * Tg\delta m + 113{,}540 * Additive + 0{,}377 * Age \quad (1)$$

Stability equals 1:

$$DF = -11{,}942 + 130{,}586 * AV - 0{,}007 * Tg\delta m + 95{,}230 * Additive + 0{,}364 * Age \quad (2)$$

Thus, discriminant analysis allowed to correctly classify 84.6% of transformers into Stability classes. The most important factor for classification into groups according to the Stability class is AV, then Tgδm and Additive. Age is not a grouping variable, but using it increases the proportion of properly classified objects.

Transformers were also classified into groups according to the Stability class based on logistic regression. The probability of attributing a transformer to a class Stability equal to 1 is determined by the equation:

$$P = \frac{1}{1+e^{(2{,}350-193{,}608*AV)}} \quad (3)$$

where P is the separating value (probability of attribution). If the separating value of probability P > 0.3, then the transformer will be assigned to the Stability class equal to 1. With an increase in the AV of the transformer, transformer's chance to be assigned to the class Stability 1 increases.

Thus, 78.8% of transformers were classified using the logistic regression method. The probability of assigning a transformer to Stability class equal to 1 depends only on the AV.

Comparison of the results of the transformers classification into groups according to the Stability classes by various non-parametric methods showed that in any method, the main grouping factor is AV. When using any classification method, there is a proportion of incorrectly classified transformers in the Stability class. The average proportion of correct classifications is higher when applying discriminant analysis, however, the proportion of correct classifications in the Stability group is 1 higher in the case of applying logistic regression, while method of classification trees has more understandable decisive classification rule.

The analysis confirmed the presence of a correlation between the indicators of the quality of the oil (AV, WAA, Tgδm, Additive) and the degree of its aging, assessed by measuring the stability against oxidation.

Respectively, based on the results of the study, there firstly established the rules for classifying transformers into groups of antioxidant stability, which allow predicting the oil aging according to the measured quality parameters. The rules are based on the developed single-factor mathematical model with a key parameter of the oil resource probabilistic estimation - the value of the AV. The threshold value of the single-factor mathematical model of the AV value is 0.02 mgKOH/g. On the basis of the obtained model, the "Algorithm for selecting a system of measures aimed at maintaining a satisfactory quality of exploitational oils" was developed for the first time [5]

V. CONCLUSIONS

1. The influence of a liquid dielectric quality on the reliability and durability of power transformers is shown.
2. The measurement of the "antioxidation stability" quality parameter for the estimation of transformer exploitational oil resource is proposed.
3. The rules for the classification of transformers have been defined that allow to predict the oil aging according to the measured regulated quality parameters.

REFERENCES

[1] S.P. Vysogorets, "Forecasting the residual life of power transformers oil on the basis of analysis of the the liquid dielectric quality," in Scientific and Technical Journal «Safety and Reliability of Power Industry №1(32)2016, Moscow, 2016, pp. 50-54.

[2] V.V. Sokolov and V.A. Lukaschuk, Issues of valuation and guaranteeing reliability of power transformers, Ekaterinburg: Publishing House «Autograph», 2010, pp. 22-30.

[3] I.S. Aptov and M.V. Homyakov, Chemical materials in electrichold, M. : Energy, 1969, 280 p.

[4] GD 34.45-51.300-97. Volumes and standarts of electrical equipment testing, 6th ed. with changes and additional, M.: ENAS, 2007, 256 p.

[5] S.P.Vysogorets, "Development of the new methods and quality valuation algorithm of 35-110kV power transformer exploitational oil," Dissertation for the degree of Candidate of Engineering Sciences, SPb., 2012, 260 p.

[6] GOST 981-75 Petroleum Oils. The method for estimating the antioxidation stability, M.: Izdatelstvo standartov, 1992, 9p.

Computer-Aided Design in Technological Process of Applying Textile Fibres and Flocks Using the Force Effect of Electric Field

Jaroslav Džmura
Department of Electric Power Engineering
Technical University of Košice, Faculty of Electrical Engineering and Informatics
Košice, Slovakia
jaroslav.dzmura@tuke.sk

Milan Bernát
Department of Physics, Mathematics and Techniques
University of Prešov, Faculty of Humanities and Natural Sciences
Prešov, Slovakia
milan.bernat@unipo.sk

Ladislav Rudolf
Department of Technical and Vocational Education
University of Ostrava, Faculty of Education
line 3: City, Country
ladislav.rudolf@osu.cz

Abstract — **Flocking is a technological process of applying textile fibers and flocks using the force effect of electric field to transport charged fibres and flocks. The computer models presented can be applied by designers as well as engineers in terms of a better detailed PC supported design of electrostatic technologies that require the use of computer simulation, for instance to predict and forecast the critical operation states, as presented.**

Keywords: flocking; electrostatic; mathematical modelling; computer simulation

I. Introduction

Flocking is a technological process of applying textile fibers and flocks using the force effect of electric field to transport charged fibers and flocks. The technology of flock process can be basically divided as follows [1].

A flock process with corona charging (modification powered by direct and alternating voltage is distinguished according to discharge electrode charging). When charged with direct (hereinafter DC) voltage, we speak of a direct corona. The particles – textile fibers and flocks – are being charged when being brought into the space of homopolar charge of direct corona. The charged particles are then transported by the electric field together with gravity vertically into substrate following virtual lines of force. A layer of adhesive is applied to the substrate. This is the technological essence of flock process with charging inside the direct corona (Fig. 1). At modification powered by alternating (hereinafter AC) voltage, the discharged electrode is adjusted so that pointed insulating termination develops at its tip with generation of homopolar space charges. Those can be used at macroscopic particle charging for their transport to the opposite electrode (Fig. 2) [2].

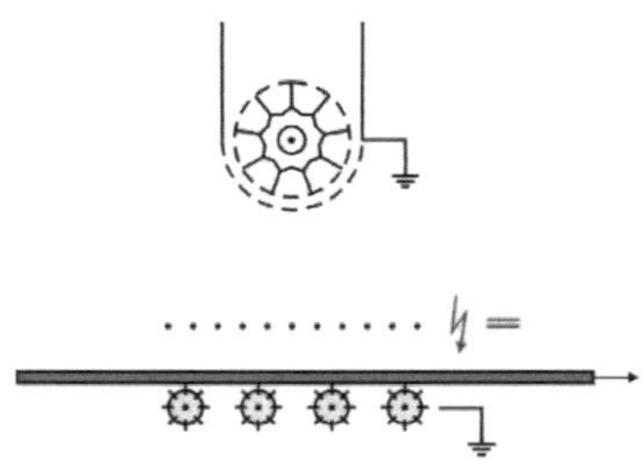

Figure 1. The principle of flocking with DC charging system of forces

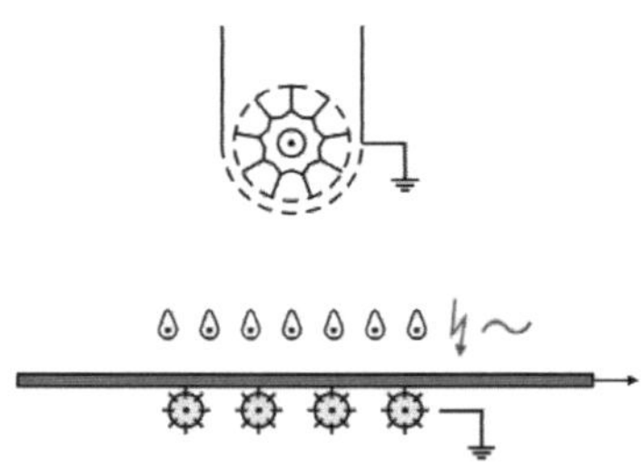

Figure 2. The principle of flocking with AC charging system of forces

A flock process with direct charging. Direct charging (more often called contact charging in the flock technology) is implemented by means of immediate contact with a HV electrode. The lower part of the technological system (narrowed opening – covering) is also a homopolar HV electrode.

Electrostatic technologies utilize the interaction of macroscopic particles and electric field, particles of various shapes and sizes. The electric fields are used to manipulate particles in variable conditions [3]. In some cases, the laws of rotational motion of macroscopic particle, directing the particle at the electric field lines of forces, must be also taken into account. An example of such technology would be a flock process with textile fibers are applied on the substrate coated with a layer of adhesive (our reason due to the need for acceptance of the rotational motion is that the final orientation of fibers fixates in the layer of adhesive).

II. Synthesis of Mathematic and Computer Model of Rotational Motion of Macroscopic Cylinder-Shaped Particle

A. Physical Model Description Layout

Let us examine the rotational motion of the cylinder-shaped particle with diameter D and the length l ($l \gg D$) placed into a homogeneous electric field produced by direct voltage according to Fig. 3. Let's assume that the particle – fiber – undergoes an almost instant dielectrical polarization, therefore, electric dipole results comprise $+q$ and $-q$ identical polarizing charges spaced apart by the l length of fiber.

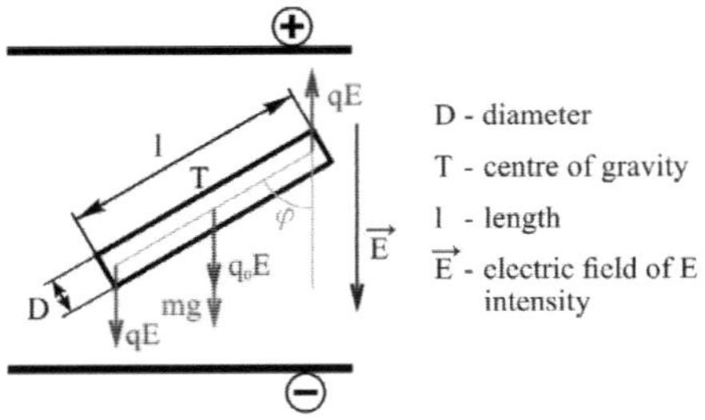

Figure 3. Effects of selecting different switching under dynamic condition

Besides that, let us suppose that the fiber contain q_0 charge (gained e.g. at the contact with the upper electrode or in the space charge of corona discharge) placed at T centre of fibre in simplified case.

B. Mathematic model synthesis

The q_0 charge causes that the advance motion of fibre towards the lower electrode accelerates. At first, let's assume that there is no air resistance at the rotation of the particle with the moment of inertia I.

Next, consider only the rotational motion of the particle and derive its kinetic equation. To solve the problem, we will use the methods of "generalized" analytical mechanics.

The potential energy of electric dipole with p moment in electric field of E intensity is

$$W_p = -(\vec{p} \cdot \vec{E}) = -p \cdot E \cdot \cos\varphi = -q \cdot l \cdot E \cdot \cos\varphi \qquad (1)$$

where φ is the angle between the direction of $\vec{E}$ and $\vec{p}$ vectors.

The moment of this couple acts on the dipole [41] is

$$\vec{M} = \vec{p} \times \vec{E} \qquad (2)$$

The kinetic energy for a body going through rotational motion is

$$T = \frac{1}{2} \cdot I \cdot \dot{\varphi}^2 \qquad (3)$$

Then the Lagrange system function [1] for our system would be in this form

$$L = T - U = T - W_p = \frac{1}{2} \cdot I \cdot \dot{\varphi}^2 + q \cdot l \cdot E \cdot \cos\varphi \qquad (4)$$

Since the kinetic system under consideration is a conservative system, the Lagrange kinetic equation (of the second order) adopts the form below

$$\frac{d}{dt}\left(\frac{\partial L}{\partial \dot{\varphi}}\right) - \frac{\partial L}{\partial \varphi} = 0 \qquad (5)$$

If we substitute into the equation (5) for the particular expressions according to (4)

$$\frac{\partial L}{\partial \varphi} = -p \cdot E \cdot \sin\varphi \quad \frac{\partial L}{\partial \dot{\varphi}} \cdot I \cdot \dot{\varphi} \quad \frac{d}{dt}\left(\frac{\partial L}{\partial \dot{\varphi}}\right) = I \cdot \varphi'' \qquad (6)$$

we will get

$$I \cdot \frac{d^2\varphi}{dt^2} + q \cdot E \cdot l \cdot \sin\varphi = 0 \qquad (7)$$

Next, to simplify our reasoning as well as mathematical operations, let us assume that orientation only takes place in the region of small φ, thus it is possible to $\sin(\varphi) \approx \varphi$

$$I \cdot \frac{d^2\varphi}{dt^2} + q \cdot E \cdot l \cdot \varphi = 0 \qquad (8)$$

It is a homogenous linear differential equation of the second order with constant coefficients. The solution of the differential equation (8) for Cauchy's initial conditions $t = 0$; $\varphi(0) = \varphi_0$; $\varphi'(0) = 0$ is the equation form (the detailed solution is in [1]).

$$\varphi = \varphi_{max} \cdot \cos(\omega \cdot t + \varphi_0) \qquad (9)$$

As far as the theory of oscillation and undulation is concerned, this equation form is characterized by a harmonic oscillating motion (the motion of a physical pendulum in a particular instance).

Since technological conditions present a real environment; that is an environment with aerodynamic drag against the motion of a particle, it is necessary to examine the possibility for a mathematical description of course of fiber orientation within a homogeneous field under these conditions. It is an example of non-conservative kinetic system, the since forces of aerodynamic drag, those are classed as are dissipative forces, exist there.

Unlike in case described under item (8), the kinetic equation (10) will furthermore include the M_S term – a moment of air resistance force preventing rotation of the particle, thus

$$\vec{M}_{rot} - \vec{M}_S = I \cdot \frac{d^2\vec{\varphi}}{dt^2} \qquad (10)$$

The dependency of M_S on textile fibre particle has been derived in [1] provided course of action in the Stockes area (subcritical), i.e. concerning small numbers $Re \ll 1$. Suppose that M_S quantity, that is the moment of resistance force of environment preventing rotational motion, is proportional to the first derivation of the angular velocity of φ' particle. The derivation of M_S is implemented in detail in [5], whereas the result is presented here:

$$M_S = W_S \cdot \frac{d\varphi}{dt} \qquad (11)$$

where W_S is the coefficient including dependence on air viscosity, fiber shape and parameters. The values of W_S obtained through experiment are shown in [5].

$$I\cdot\frac{d^2\varphi}{dt^2} = -q\cdot E\cdot l\cdot sin\,\varphi - W_S\cdot\frac{d\varphi}{dt} \qquad (12)$$

$$\frac{d^2\varphi}{dt^2}+2\cdot\delta\cdot\frac{d\varphi}{dt}+\omega_0^2\cdot\varphi=0 \qquad (13)$$

It is an ordinary differential equation of the second order with constant coefficients which, describes any damped oscillating motion, in physics (our instance of the physical system can be regarded as a special modification of a physical pendulum). The equation solving and labeling of its coefficients (2δ, ω_0) have been borrowed from [1].

When examining the rotational motion of the macroscopic cylinder-shaped particle placed into a homogeneous electric field produced by AC voltage (the flock process with corona charging in modification powered by the AC voltage – see the principal scheme in Fig. 2), we proceed from the physical situation presented in Fig. 3. subject to a modification with electrodes are powered by AC voltage. The equation for harmonically changing field intensity will be in the following form

$$\vec{E}(t)=E_m\cdot cos(\omega\cdot t+\phi_0) \qquad (14)$$

The derivation of the kinetic equation of particle proceeds from the basic equation for rotational motion already mentioned that is in this form

$$I\cdot\frac{d^2\vec{\varphi}}{dt^2}=\sum\vec{M}_{rot}-\vec{M}_S \qquad (15)$$

where I refers to the moment of particle inertia (it is represented by $I=ml^2/12$ for the cylinder-shaped particle in our case), M_{rot} is the sum of the moments of forces initiating the rotational motion, M_S is the moment of Stokes force causing resistance to the motion of particle.

The basic equation (8) takes the form

$$I\cdot\frac{d^2\varphi}{dt^2}=q\cdot E(t)\cdot l\cdot sin\,\varphi - W_S\cdot\frac{d\varphi}{dt} \qquad (16)$$

As we have mentioned, low values of φ can be substituted with $\sin(\varphi)\approx\varphi$; then we use the relation (16) and express q based on the relation (17)

$$q=\frac{\alpha\cdot E}{l} \qquad (17)$$

The above-mentioned modified equation (16) with the acceptance of (14) and (17) will gradually produce (18) and (19)

$$I\cdot\frac{d^2\varphi}{dt^2}=\alpha\cdot E(t)^2\cdot\varphi - W_S\cdot\frac{d\varphi}{dt} \qquad (18)$$

$$\frac{d^2\varphi(t)}{dt^2}+W_S\cdot\frac{d\varphi(t)}{I\cdot dt}+\frac{\alpha\cdot E_m^2}{I}\cdot sin^2(\omega\cdot t+\phi_0)\cdot\varphi(t)=0 \qquad (19)$$

It is an ordinary non-linear differential equation of the second-order. So far, mathematics does not know any analytical way to solve this; although the equation looks very complicated, numerical solution for Cauchy´s initial conditions $\varphi(0)=\varphi_0$; $d\varphi/dt=0$ is possible. The equation is a mathematical model of our examined physical system. Its solving by computer (using the Runge-Kutta method 3 in the Matlab environment) is presented in Fig. 4.- Fig. 7.

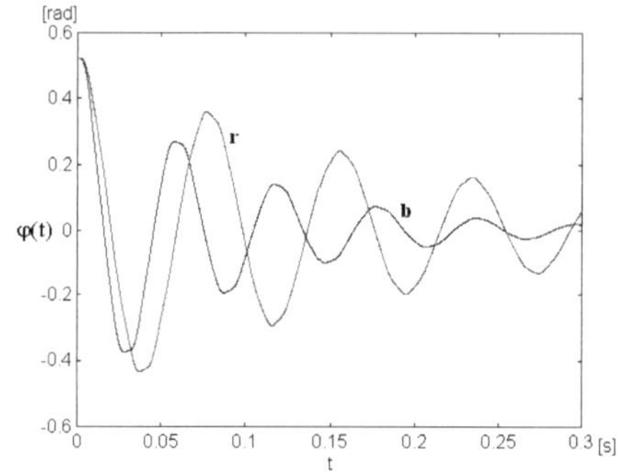

Figure 4. Graphical representation o $\varphi=f(t)$f dependence for particle 1 (b) and 2 (r)

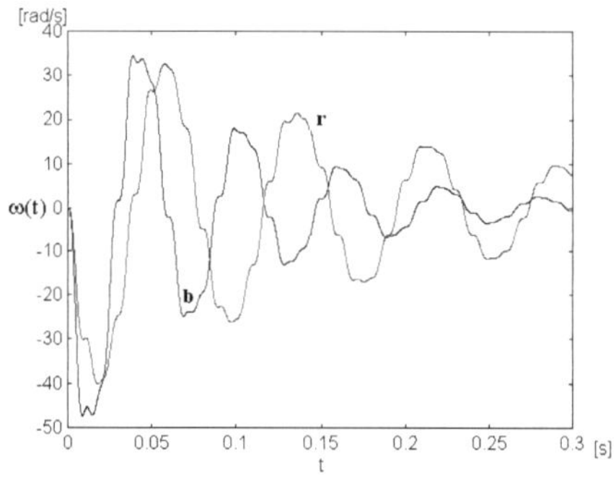

Figure 5. Graphical representation of $\varphi'=f(t)$ dependence for particle 1 (b) and 2 (r)

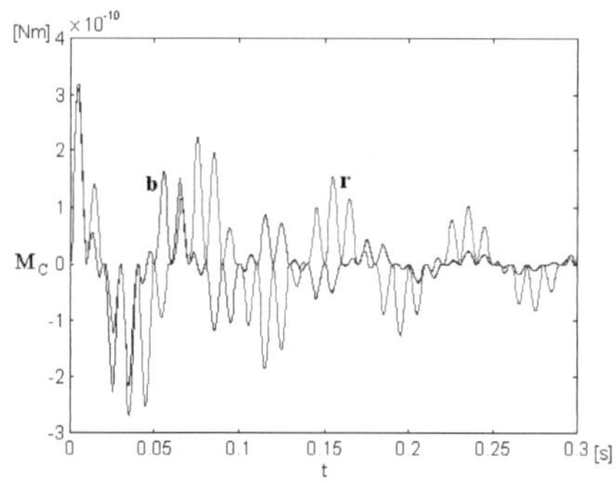

Figure 6. Graphical representation of $M_C=f(t)$ dependence for particle 1 (b) and 2 (r)

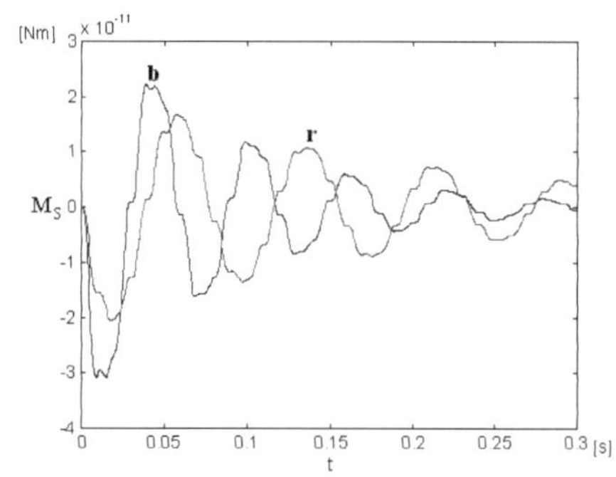

Figure 7. Graphical representation of $M_S=f(t)$ dependence for particle 1 (b) and 2 (r)

III. Computer simulation on a model, its results (analysis and discussion), their interpretation

The factor computer experiment (equation solving and construction of some dependences between state values) proceeded from the following system values-parameters:

Field: $E_m = 10^6$ V; $\omega = 314$ rad/s; $\phi_0 = 0$;

Particle 1: $\alpha = 0.6 \cdot 10^{-21}$; $m = 5.5 \cdot 10^{-8}$ kg; $l = 2.5 \cdot 10^{-3}$ m

Particle 2: $\alpha = 0.64 \cdot 10^{-21}$; $m = 6.6 \cdot 10^{-8}$ kg; $l = 3 \cdot 10^{-3}$ m

Environment: $W_{S1} = 0.295$ kg $\cdot$ m^2; $W_{S2} = 0.511$ kg $\cdot$ m^2;

these values are chosen according to [5].

The computer program we have implemented for solving the differential equation is a mathematical model of our physical system for implementation of the factor computer experiment (with the purpose of presenting the effect of some chosen quantities on the form of the presented $\varphi = f(t)$ dependence).

Factors influencing behavior of the system that we examine can be divided into the following categories defined below. The criterion for such division comprises the dependency on:

a – particle (α, m, l); the choice of values of these factors follows the rules described in [1];

b – electric field (E_m, ω, ϕ_0);

c – environment (W_S);

d – Cauchy (principle value) ($\varphi(0)$, $d\varphi(0)/dt = 0$).

Due to the time consumed, it is not practical to implement the full multifactor experiment because of a great number of factors within our system and it is not even necessary for the purpose of our system. Hence we have implemented the factor computer experiment only as a single-factor experiment and that involves the factors relevant to the initial conditions only.

Fig. 4 – Fig. 9 show the outcome of our computer simulation on the model (19); that is on a model representing the rotational motion of the macroscopic cylinder-shaped particle in a homogenous field produced by the AC voltage according to Fig. 3.

The single-factor experiment comprised the implementation technology of the said computer experiments and the graphic construction of the particular state quantities within state space (the construction of the particular operation kinematic-dynamic features of motion characterizing our physical system) was following the purpose of experimental efforts. It is also the presentation of the Matlab possibilities for simultaneous examination of kinematics and dynamics of motion. Besides the experiments already mentioned, we have employed the method of computer simulation to perform the examination (by means of an optimizing experiment); as well as the dependence between amplitude of rotational oscillating motion maximum value and voltage frequency on electrodes (having a mutual electric field in between). In our case, this is literally required for our methodological research, since, as we have already stated, the non-linear differential equation which cannot be solved analytically, describes the rotational motion of the particle within the AC electric field and thus it is not possible to determine dependence between the angular position and time using the analytical function for a particular frequency.

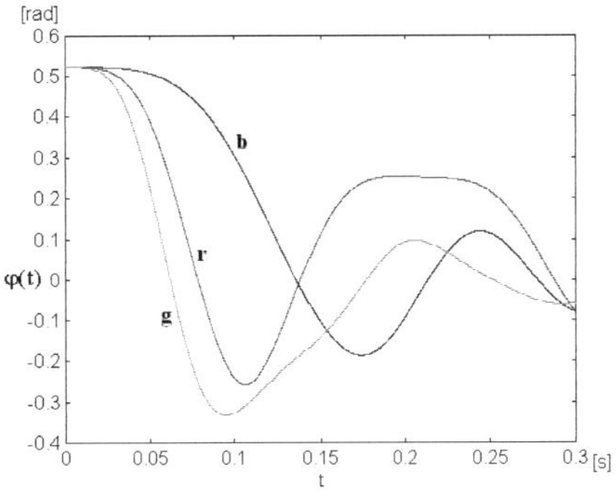

Figure 8. Graphic representation of $\varphi = f(t)$ dependence for f = 5 Hz (b), f = 15 Hz (r), f = 25 Hz (g)

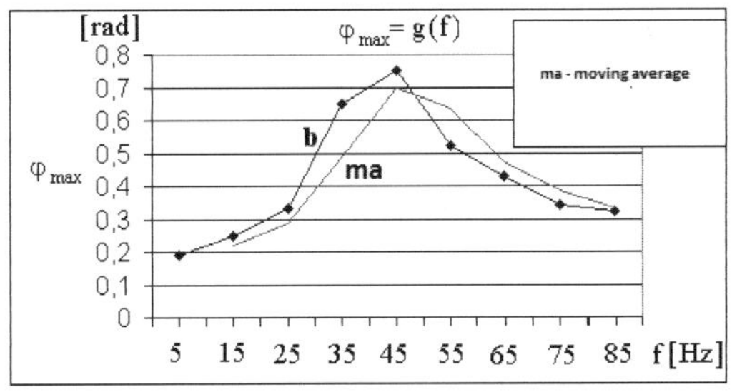

Figure 9. Graphical representation of $\varphi_{max} = g(f)$ dependence for particle 1 (b)

IV. Conclusion

The most important research contributions are:

- The improvement of the existing research methods in the technology of flocking (the emphasis has been so far put on constructional and operational aspects with empirical knowledge being the main and practically the sole source of information),
- the parallel solution of the aspects of the kinematics, the dynamics, and of the electric field of a moving macroscopic particles (in the real environment of electrostatic technologies).

Unlike the past, the proposed method offers a more complex solution to the issue. It solves the issue not only from the viewpoint of kinematics but also from the viewpoint of dynamics and of the theory of the electromagnetic field taking into account all the major factors affecting the ballistics of a particle. According to it, the then used empirical methods were capable of solving only the kinematics of a particle but not its dynamics, therefore the method proposed seems to be at the moment the only complex solution.

Flocking is a technological process of applying textile fibers and flocks using the force effect of electric field to transport charged fibers and flocks.

In conclusion, it is important to draw attention to the following: in the case of the existence of a great number of the particles (the more identical model to the real model) and their interactions (due to which they influence each other) it is recommended to use a statistical approach (statistic physics, Monte Carlo method) and stochastic computer modelling of the above-mentioned processes. The creation of the stochastic computer model is described in the dissertation thesis [1].

The computer models presented can be applied by designers as well as engineers in terms of a better detailed PC supported design of electrostatic technologies that require the use of computer simulation, for instance to predict and forecast the critical operation states, as presented [1].

ACKNOWLEDGMENT

This work was supported by the Slovak Research and Development Agency under the contract No. APVV-17-0372.

REFERENCES

[1] M. Bernát, Dynamics of space charges in highly non-homogheneous DC and AC fields. PhD. thesis, FEI TU Košice 2015.

[2] M. Bernát, Visualization of some electro-physical processes through computer for didactic purposes and its application in teaching electrotechnical subjects. PhD. thesis, PdF UKF Nitra 2005.

[3] A. Mikkelsen, J. Wojciechowski, M. Rajňák, J. Kurimský, K. Khobaib, A. Kertmen, Z. Rozynek, "Electric Field-Driven Assembly of Sulfonated Polystyrene Microspheres", Materials. 10 (2017) 329. doi:10.3390/ma10040329.

[4] M. Bernát, "Abut Applets Application in Natural Science in 1st Level of Basic School", in International Review on Computers and Software (IRECOS) Vol 4, No 5, 2009, pp. 621-626.

[5] W. Radowicki, B. Strelcov, Elektro-Aeromechanik textiler Faserstoffe. VEB Fachbuch Verlag Leipzig, 1975, pp.1-225.

Dielectric Properties of Segmented Sensor

Roman Cimbala
Department of Electrical Power Engineering,
Faculty of electrical Engineering and Informatics .
Technical University in Košice
Košice, Slovakia
roman.cimbala@tuke.sk

Ondrej Hrabčák
Department of Electrical Power Engineering,
Faculty of electrical Engineering and Informatics .
Technical University in Košice
Košice, Slovakia
ondrej.hrabcak@student.tuke.sk

Abstract — **This paper deals with the problem of monitoring the dielectric properties of a sensor for measuring the surface resistance of insulating materials. The influence of moisture on capacity and dielectric loss factor is analyzed. The paper describes an experiment for obtaining selected qualitative parameters describing the surface condition of the sensor insulation system. The dependencies for clean and dirty sensor surfaces are evaluated.**

Keywords: capacity; dielectric loss factor; outdoor insulator; industrial polution

I. Introduction

Electricity consumption is increasing worldwide. New and new power stations have been being built with ever-increasing performance. It is an extensive direction of development. However, there is a visible effort to use these devices to the maximum extent possible within their technical parameters. Thus, the issue of security and reliability is becoming one of the most important challenges facing current practice.

The most important elements of any electrical equipment include an insulation system. Its failure, resp. reaching the limit state affects the operation of a state-of-the-art facility, but has a secondary impact due to the technological interconnection of the machines and the operation of the entire production process.

Today's knowledge in the art does not allow unambiguously to determine the residual lifetime of the investigated insulation system, so manufacturers prescribe the replacement of the old insulation system after a defined period of operation (of course, if no failure occurs earlier). If a comprehensive insight into the aging of the insulation system and the resulting diagnostic method with high efficiency and accuracy in predicting the residual life of the insulation, there would be no need to replace the entire system after the manufacturer's prescribed time, but after a period that would be determined by objective isolation status. The financial costs that would be saved are huge.

II. Atmospheric Corrosion

Atmospheric corrosion is one of the main causes of electrical equipment faults in dirty outdoor environments. The basic condition for its formation is the existence of critical moisture, which causes the formation of a liquid layer of water on the surface of the insulating material. Critical relative humidity is mainly influenced by air pollution by gaseous and solid exhalates, the surface quality of the equipment, the temperature and airflow velocity in the surroundings and the likes.

The most significant air degradation component is SO2. It is water soluble and, to a greater or lesser extent, it also contains seemingly pure air. Other ingredients that are contained in industrially contaminated air are chlorine, hydrogen chloride, hydrogen sulphide, ammonia, and various chlorides. Compounds HNO3, HCL or Cl2 together with atmospheric moisture form strong electrolytes.

The very unfavorable effect on the state of the insulating system and in particular on its surface resistance has dust particles depositing on its surface. Solid particles not only cover the upper surfaces of the equipment, but also their entire surfaces. E.g. the upper layer of the insulator may be covered with a substantially thicker layer of adsorbed particles than the entire insulator. In dry conditions, this layer is poorly conductive, but its conductivity increases by sorption of water vapor and various gases from the atmosphere. The space charge can be created and it has influence on electric field intensity distribution [1]. This process is relatively fast and reversible depending on the temperature and partial pressure of water vapor or gas. However, the surface insulation resistance must not fall below a critical value that leads to a jump on the surface of the insulation system. Therefore, the periods with the occurrence of so-called "Dense" mists with a large liquid water content that occur in our climatic conditions in spring and autumn.

In the case of electric stations located near coal-fired power plants, fly ash from chimneys, mills and tailings is added to existing exhalations.

III. Humidity Influence on Surface Insulation Resistance Dielectric

The electrical conductivity on the dielectric surface is predominantly ionic. The surface resistance is mainly affected by a layer of contaminated water adsorbed on the surface of the insulator. The resistance varies little with the change in ambient temperature but varies significantly with the change in relative humidity. In the vicinity of the insulator, changes in surface temperature will cause changes in relative humidity and consequently in the thickness of the adsorbed aqueous film. This creates, depending on the relative humidity of the mono or polymolecular layer of water.

Specific surface conductivity increases with temperature [2]

$$\gamma_p = C_1.\exp\left(-\frac{a_m}{T}\right). \qquad (1)$$

- C_1 - conductive film constant on the surface of the insulator,

- a_m - a constant proportional to the activation energy of the film molecules on the surface
- γ_p - specific surface conductivity,
- T - absolute temperature.

In a small temperature range, equation (12.11) can be written in the form

$$\gamma = C_2.\exp(a_m.\upsilon). \quad (2)$$

As the temperature rises above 60 ° C, the increase in specific conductivity due to temperature is mitigated by the evaporation of the adsorbed water layer on the surface.

For the specific surface insulation resistance of an insulator with an imperfect dry surface [3],

$$\rho_p = f(p_{H_2O}, \upsilon_t). \quad (3)$$

- p_{H2O} - water vapor partial pressure
- υ_t - temperature in °C.

The aqueous film has a specific insulation resistance. The thickness of the water film on the surface of the insulator depends on the relative humidity of the ambient air and also on the surface properties. The surface resistance consists of the surface resistance of the insulating material itself and the resistance of the aqueous film, which are in parallel.

The specific surface insulation resistance is given

$$\rho_p = \frac{R'.b'}{l'}. \quad (4)$$

- R '- total surface insulation resistance,
- b '- length of electrodes
- l '- distance of electrodes.

The aqueous film has a specific insulation resistance

$$\rho = R''.b'.\frac{d'}{l'}. \quad (5)$$

- d '- water film thickness.

The sum of these two parallel resistances is the resulting surface resistance of the material

$$\rho' = \rho_p.d'. \quad (6)$$

Even traces of impurities that are difficult to detect analytically affect the surface conductivity. The sorption proceeds with increasing concentration of water vapor in the surface layers and is still slower with the decreasing pressure difference of the water vapor and the inner volume of the insulator. Gradually, the total insulation resistance of the insulator is determined by its internal resistance and less superficial.

The surface electrical conductivity is predominantly due to the presence of impurity ions in the water adsorbed on the surface of the insulator.

In the case of damp heat cycling, the surface resistance first decreases rapidly, after a period of time it increases considerably and then finally decreases. This is due to the influence of dissolvable water-soluble substances which are gradually washed away from the surface of the insulator during prolonged exposure to cyclic wet heat. The more polar the substance, the greater the amount of water it attaches to its surface.

IV. Influence of Moisture on Dielectric Dielectric Loss Factor

The dependence of the insulation resistance on the amount of water in the dielectric channel is defined by equation

$$R_c = \rho_{H_2O}.\frac{k_k^{\,2}.l^2}{V_k}. \quad (7)$$

- ρ_{H_2O} - water resistance,
- l - channel length,
- k_k - coefficient respecting that channel does not connect electrodes in the shortest way,
- V_k - volume of water in channels.

and the dielectric loss factor is

$$\tan\delta = \frac{1}{\omega.C.R_c}. \quad (8)$$

$$\tan\delta = \frac{V_k}{V}.\frac{1}{\omega.\varepsilon_0.\varepsilon_r.k^2.\rho_{H_2O}}. \quad (9)$$

By increasing the amount of water absorbed, the insulation resistance decreases, the factor to respect the shortening of the conductive paths between the electrodes decreases, the dielectric loss factor increases. If the humidity changes from 0 to 90%, the tan δ can vary from 10 to 300%.

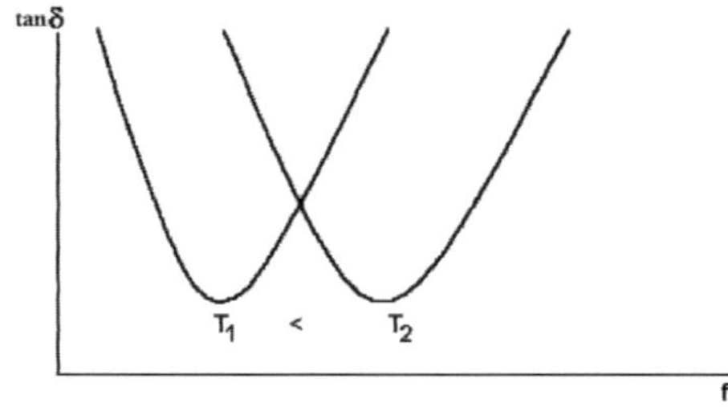

Figure 1. General Dependence of dielectric loss factor on frequency

In some wetted insulators, the dielectric losses are highly dependent on frequency and temperature, as shown in Fig. 1. Water has a considerable ionic conductivity and thus a large tan, which decreases with an initial frequency increase. When a certain frequency value is exceeded, the dielectric loss factor increases. This is due to the loss of energy due to the friction of the dipoles in oriented movements forced by the alternating electric field. The solid particles in electric field on the insulator surface assembly in the bulk of liquid droplets and have a significant influence on dielectric properties [4]. There is an influence of non-ionizing electric field on polarization phenomena observed via applied voltage in frequency dependency [5].

The large gradient of moisture concentration creates a dielectric with two different layers - considerably moistened and relatively dry according to Fig.2.

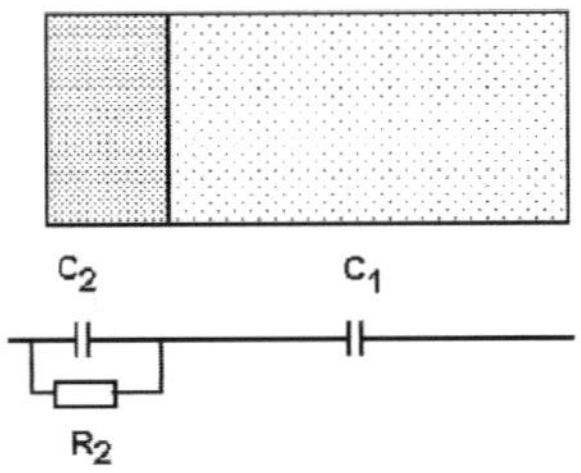

Figure 2. Substitute scheme of wetted dielectric [3]

Then it applies to tan δ

$$\tan\delta = \frac{\omega^2.C_1.C_2}{\frac{1}{G} + \omega^2.(C_1 + C_2).C_2} . \qquad (10)$$

V. Experiment Design

The task of the experimental part of the measurement was to find out the influence of pollution and relative humidity on the surface resistance of the sensory layer of industrial glass insulator. According to sensor design, it was necessary to measure it sensitivity to higher telecommunication frequencies, that can probably have influences of measured signal [6].

The design of the sensor topology is currently in the patent and therefore no further details on its shape and material composition will be given.

The sensory layer has been formed by a multi-segment comb electrode system whose application is currently in patent. Therefore, this paper does not give details of its material composition and topology.

Two sets of measurements were designed and performed to determine the effect of pollution and relative air humidity:

1. Measurement on a pure sensor layer insulator (hereinafter referred to as a clean insulator).
2. Measurement on an insulator with a contaminated sensor layer (hereinafter referred to as contaminated insulator). The relative humidity levels were determined to be 60% RH, 70% RH, 80% RH, and 90% RH. The insulator was polluted by its placement in the real environment on the open area Mäsiarska 74, Košice). Pollution exposure time was set to 2 months (17.7.2018 - 19.9.2018).

A drying oven was used for the measurement, which was adjusted so that the required air humidity can be adjusted and maintained automatically in its space.

Capacity and dissipation factor were measured with Agilent LCR meter E4980A. It provides impedance measurements in the 20 Hz - 2MHz frequency range.

The next devices were in the measurement set (Fig. 3):

- heating element,
- water container (in the bottom of oven),
- humidity probe for regulator,
- Moisture Meter Probe,
- antiparallel air circulation fans and cooling fan,
- multi-segment electrode system,
- extending the electrical output from the sensor surface.

Figure 3. Air oven with humidity regulator andoutdoor insulator

VI. Results and Discussion

With a clean insulator surface, a larger amount of charge is created due to the increasing thickness of the water film, which helps to charge the sensor surface. The higher content of water results in higher differences in capacity in low frequencies. In higher frequencies the value differences are decreasing. This phenomenon occurs at frequencies higher than 200 kHz.

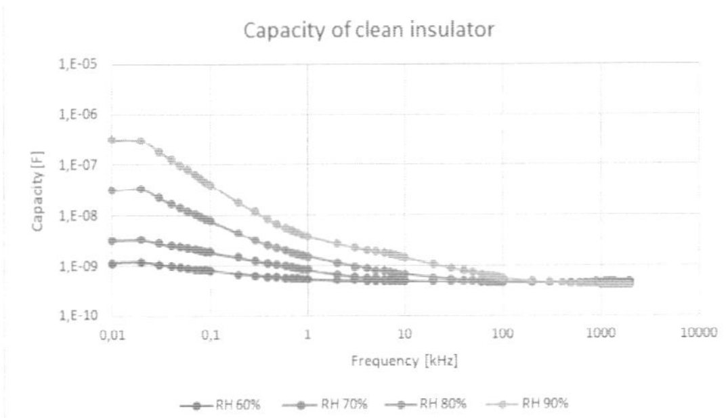

Figure 4. Capacity dependence on frequency of clean insulator

As the frequency increases, the possibility of charge movement increases, and thus the total capacity values are equalized at different humidity, as shown in Fig. 4

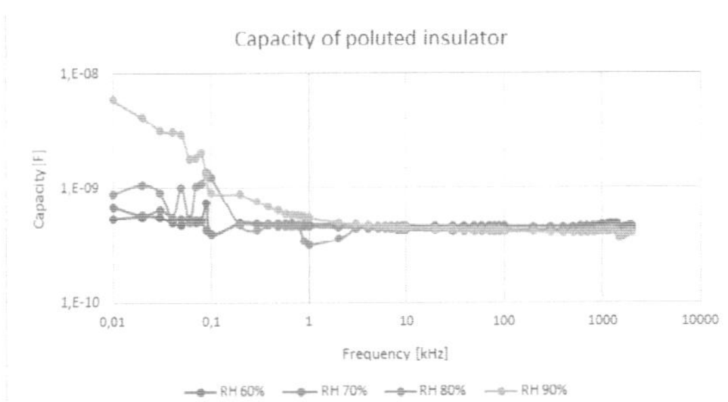

Figure 5. Capacity dependence on frequency of contaminated insulator

Due to the contamination of the insulator surface, there is a greater amount of charge that helps to charge the sensor surface and cause less capacity differences for different air humidity. As the frequency increases, there is a greater possibility of charge movement [7], so the total capacity is reduced, as shown in Fig.5. This phenomenon occurs at frequencies higher than 4 kHz.

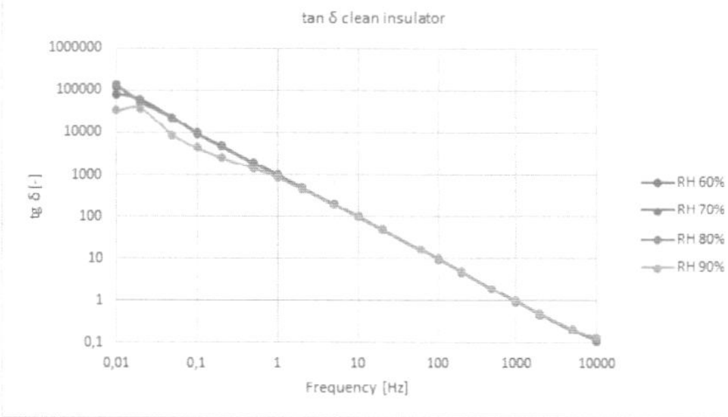

Figure 6. Dissipation factor dependence on frequency of clean insulator

The tan δ values for all relative humidity values are very similar on almost the whole range of the measured frequencies of 0.01 Hz - 10 000 Hz. Small differences tg δ can be observed at frequencies below 1 Hz at 90% relative humidity. At the remaining frequencies, this difference is observable to a lesser extent at frequencies below 0.05 Hz as it is shown in Fig. 6.

Measurements show that the resulting tan δ is not sensitive enough to determine the relative humidity value over the entire measured frequency range.

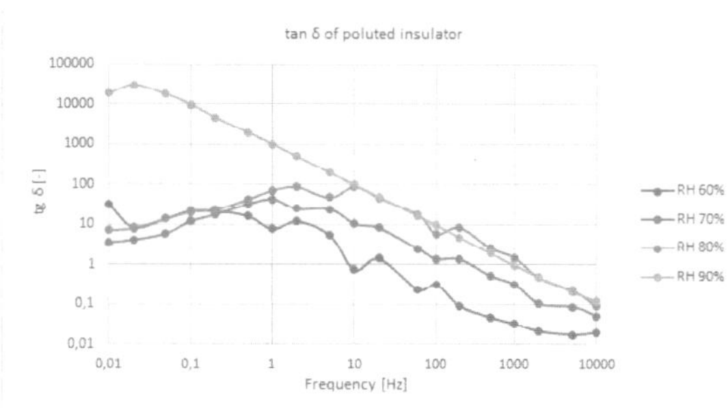

Figure 7. Dissipation factor dependence on frequency of contaminated insulator

Tan δ on a contaminated insulator is different for each relative humidity over the entire frequency range. The trend change to tan δ occurs at a frequency of 5 Hz, which is caused by an increase in the surface resistance value. According to theory of dielectrics, the complex permittivity can be calculated[8].For a relative humidity of 90%, a conductive layer formed by a combination of moisture and contamination is formed on the surface of the contaminated insulator, so that the trend to the melt occurs at 0.02 Hz as it is shown in Fig.7. It is possible, that the solver joints can have influence on dissipation factor behavior especially in higher temperature and humidity [9].

VII. CONCLUSION

Measurement of the influence of pollution and relative humidity on the surface resistance of the industrial glass insulator sensory layer was carried out as part of the research of the mentioned projects and its task was to determine the electrophysical properties of the surface conductivity sensor monitoring the contamination. The results of the measurements showed that the sensor is capable of capturing changes in surface conductivity when exposed to external influences over a period of 2 months. The changes are significant and the influence of relative humidity on their size can also be determined.

With increasing relative air humidity, with increasing frequency values, the capacity values for a clean and contaminated insulator begin to approximate each other. Measurements to date have shown that the determination of capacity patterns is complicated for this type of measurement and requires further research and fuzzy logics of artificial inteligence can be used [10].

For all relative humidity, the tan δ values for the clean and contaminated insulator begin to converge with increasing frequency. However, the approach speed with increasing relative humidity decreases.

ACKNOWLEDGMENT

Slovak Academy of Sciences and Ministry of Education, projects VEGA No. 1/0340/18 and 2/0141/16; Ministry of Education Agency for structural funds of EU projects No. 26220120055; Slovak Research and Development Agency project No. APVV-15-0438 and APVV-17-0372.

REFERENCES

[1] K. Marton, P. Kapáš, R. Cimbala and M. Kolcun, "The contribution of space charge in a high non-homogenous electric field to the creation of negative differential conductivity", Journal of Electrostatics. Vol. 71, no. 3 (2013), pp. 276-280

[2] B. Bartáková and J. Vařák, Elektrická zařízení v chemických a jiných agresívních prostředích, SNTL Praha, 1975

[3] M. Rychtera and B. Bartáková, Tropikalizace elektrických zařízení, NCSAV, Praha 1960

[4] A. Mikkelsen, J. Wojciechowski, M. Rajňák, J. Kurimský, K. Khobaib, A. Kertmen, Z. Rozynek, Electric Field-Driven Assembly of Sulfonated Polystyrene Microspheres, Materials. 10 (2017) 329.

[5] R. Cimbala, J. Kiraly, M. German-Sobek, S. Bucko, J. Kurimsky and J. Dzmura, "Polarization Phenomena in Magnetic Liquids", Chemické listy, Vol. 109, Issue: 2 pp. 117-124 Published: FEB 2015

[6] J. Zbojovský, M. Pavlík, D. Medveď, Z. Čonka, and L. Kruželák, "Shielding of the electromagnetic field in the range from 0.9 to 6 GHz by experimental shielding coatings", Proceedings IEEE International Conference ans Workshop in Óbuda on Electrical and Power Engineering. - New York (USA) pp. 273-276., ISBN 978-1-7281-1153-7

[7] I. Kolcunová, M. Pavlík and L. Lisoň, "Assessment of Long Thermal Ageing on the Oil-Paper Insulation", Advances in Electrical and Electronic Engineering. Vol. 14, no. 5 (2016), pp. 506-511, ISSN 1336-1376

[8] J. Kudelcik, S. Hardon and L. Varacka, "Measurement of Complex Permittivity of Oil-Based Ferrofluid in Magnetic Field, Acta Physica Polonica A, Volume: 131, Issue: 4, pp. 931-933

[9] A. Pietriková, J. Ďurišin, Ľ. Livovský and J. Urbančík, "Microstructure analysis and measurement of nonlinearity of vapour phase reflowed solder joints," 2008 31st International Spring Seminar on Electronics Technology, Budapest, 2008, pp. 363-366.

[10] D. Perduková and P. Fedor, "Design rules of a fuzzy controller using genetic algorithms", *Proc. of the 7th Int. Scientific Symposium on Electrical Power Engineering* Stará Lesná, 2013, pp. 513-516. ISBN 978-80-553-1441-9

Measurement of Corona Discharges Using UV Camera

Iraida Kolcunová , Ján Zbojovský, Juraj Kurimský
Department of Electric Power Engineering
Technical University of Košice, FEI
Košice, Slovakia
iraida.kolcunova@tuke.sk, jan.zbojovsky@tuke.sk, juraj.kurimsky@tuke.sk,

Abstract — **One of the non-destructive diagnostic methods of measuring partial discharge in high voltage equipment is measurement by using a UV camera. Camera DayCor Classic is the instrument for detection of the discharge activity in UV spectrum. With UV camera it is possible to observe this activity in daylight. This camera captures activity of the discharge in the UV spectrum from 240nm to 280nm. During of the discharge activity in the space between the electrodes occurs the different acts. One of them is negative differential conductivity. This article is focused to the research of this anomalous phenomenon using UV camera.**

Keywords: corona discharges, UV camera, DC positive voltage, AC voltage, negative differential conductivity.

I. Introduction

During the operation of power equipment, the presence of partial discharge activity leads to various disorders. Partial discharges have negative influence to the operation of power equipment and they cause lower reliability of supply electricity. The occurrence of partial discharge is inadmissible in the normal operation. Term of corona discharges is a designated for partial discharges in atmospheric air, which appear near the sharp electrodes. They are the first form of self-sustained discharges in nonuniform field gap [1]. The presence of corona discharges in high voltage test setup usually indicates an inadequate insulation and provides an early warning of possible insulation failure. The current pulses produced by streamer discharges also generate electromagnetic interferences in the hundreds megahertz range, affecting radio and TV reception and measuring instrumentation used in partial discharge detection [2, 3].

Corona discharge as a positive phenomenon is used in a wide variety of industrial applications such as electrostatic precipitation, electrically separating from each other two or more solid species in an air ambient, electrostatic application of coatings and so on [4-6].

When a free electron in gaseous dielectric acquires enough kinetic energy from the applied electric field, it makes a series of successful impact ionization colliding with neutral molecules. This process develops in a very systematic manner and leads to the formation of innumerable number of avalanches. In the case of extremely nonuniform field electrode configurations in gaseous dielectrics a local breakdown phenomenon known as corona discharge takes place. The complete breakdown in extremely nonuniform fields is always preceded with stable corona discharges [7-9].

Nowadays the research of partial discharges development in insulating system is very important. For detection the corona discharge serves UV camera that can work in daylight. These cameras observe the discharge activity in ultraviolet spectrum of electromagnetic radiation [10, 11].

The partial discharge activity causes the various phenomena in strongly non-homogeneous electric field. One of these phenomena is negative differential conductivity that is main subject of this article. This phenomenon is manifested like an area with negative differential resistance on current-voltage characteristic and it occurs at the discharge activity in strongly non-homogeneous electric field that is represented by system of electrode point - plate. Electrodes are connected to a source of DC voltage, where the point electrode has a positive polarity. The measurements of the current and the light emitted by the corona show that, as the voltage increases, the current first reaches a local maximum, then decreases to a local minimum after which is starts to steadily increase again (Fig.1). This current-voltage anomaly in the case of a positive polarity was first observed by prof. Karol Marton in 1980's [12]. Based on researches of professor Marton, this abnormal phenomenon is characterized by the following facts [12, 13]:

•The discharge process unfolds in the air at normal atmospheric conditions

•Inception voltage of corona discharges is a function of configuration of electrodes, distance between electrodes and physical properties of the electrode material

•Directive $dI/dU < 0$ depends of the distance between electrodes hence it depends on intensity of electric field in this configuration of electrodes

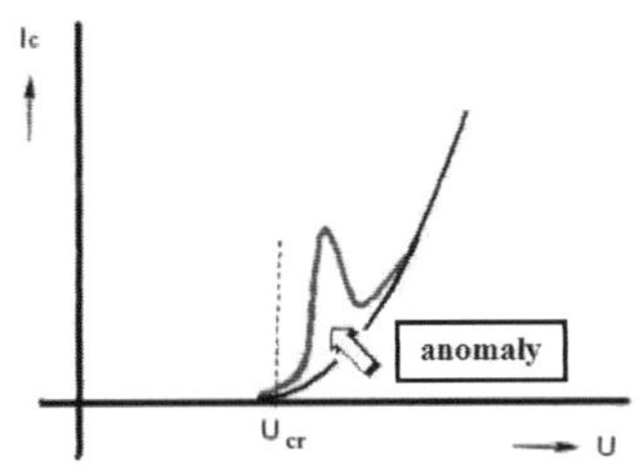

Figure 1. Negative differential conductivity [12]

Here are a several factors that have negative influence to this abnormality [11]:

•The UV radiation in place between electrodes

•The electrochemical activity of surface of negative electrode

•The influence of internal electric field

•The increasing of temperature of tip electrode toward environment.

Negative differential conductivity is characterized that from certain point the discharge activity is decreasing with increase of voltage (Fig.1). The abnormality is caused by the negative space charge (negative ions) that is amassed in certain distance from electrode. This space charge acts as a barrier. After overcoming this barrier the discharge activity is increasing with increase of voltage. [13, 14]

II. Measuring Set-up

Characteristics of corona discharge were measured using a point-plate electrode system. Distance between the high voltage electrode with corona and the earthed electrode was 4 cm. Copper electrodes with 2.5 mm radius were used as a point electrode with three different electrode endings (Fig.2):

-point tip electrode – with a 100 μm apex of the cone,
-hemispherical electrode – with radius 2.5 mm,
-cut off electrode.

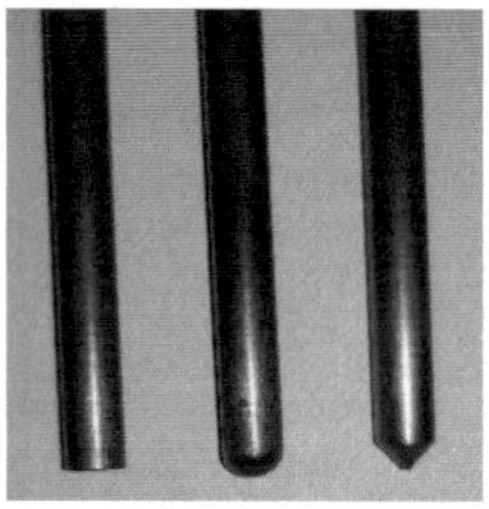

Figure 2. Measuring point electrodes

For discharge activity measurements a working setup was built as shown in Fig.3. DC high voltage was generated by BS222A source, where the voltage polarity positive or negative could be chosen. AC high voltage was generated by TuR – GPT 3/80 source. The applied voltage value on the point electrode could vary from 0 to 60 kV for DC and from 0 to 80 kV for AC.

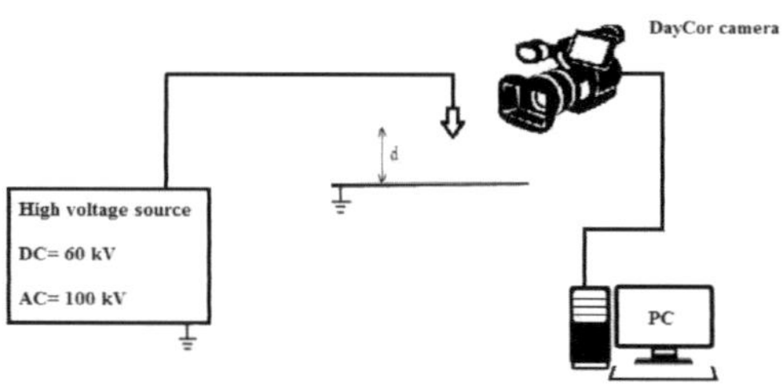

Figure 3. Set-up for optical measurement of corona discharges

The experiments were conducted using the following procedure: the voltage, with DC positive polarity or AC, was gradually increased on the point electrode until the first appearance of corona discharge. At this point, corona discharges was measured. Next, the voltage was increased in 1 kV steps up to breakdown. These procedure was repeated for various types of point electrodes. Optical measurement of corona activity near electrode was done by UV Daycor Classic made by OFIL Company. This camera detects UV radiation in 240 nm to 280 nm bandwidth. It is a bi-spectral device connected to a PC, which records images for various voltage levels.

III. Measuring Results

After finishing measurement qualitative pictures for different voltage level were recorded by using DayCor camera. Camera for discharge activity measurement was set to a UV mode and the camera gain stayed unchanged during measurement. An external card AVerTV M036 was used to connect the camera to a PC, i.e. converting an analog video signal from the camera to a digital form. To evaluate the intensity of a discharge activity a methodology was developed that evaluates the UV camera recordings. This method is based on ROI method used in a computer tomography to determine sizes of internal organs [14, 16]. So quantitative values of discharge activity were obtained.

Qualitative pictures of corona discharge for positive DC applied voltage are depicted in Fig. 4 near cut-off electrode, in Fig. 5 near hemispherical electrode and in Fig. 6 near point tip electrode.

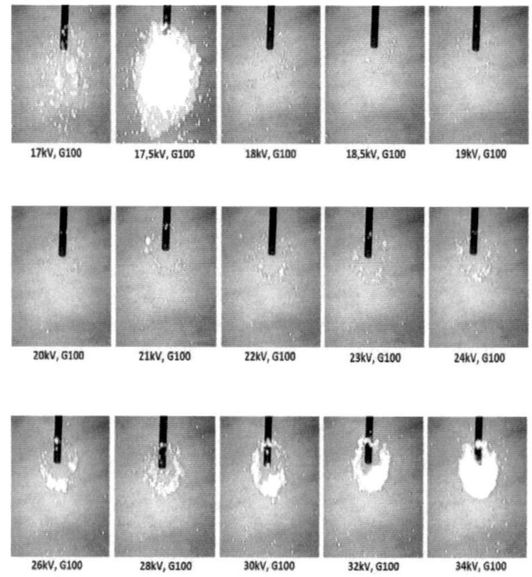

Figure 4. Corona discharges near cut-off electrode for positive DC

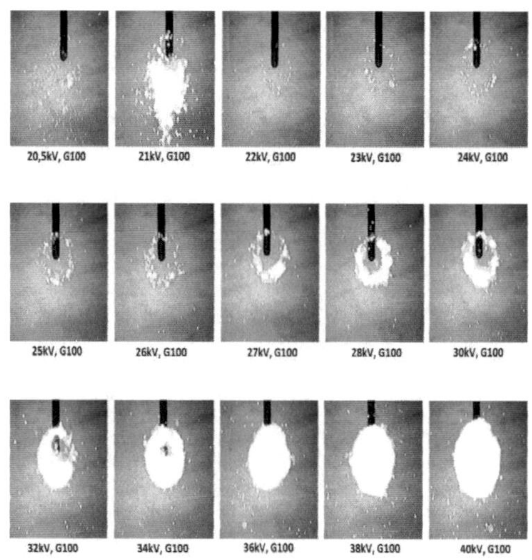

Figure 5. Corona discharges near hemispherical electrode for positive DC

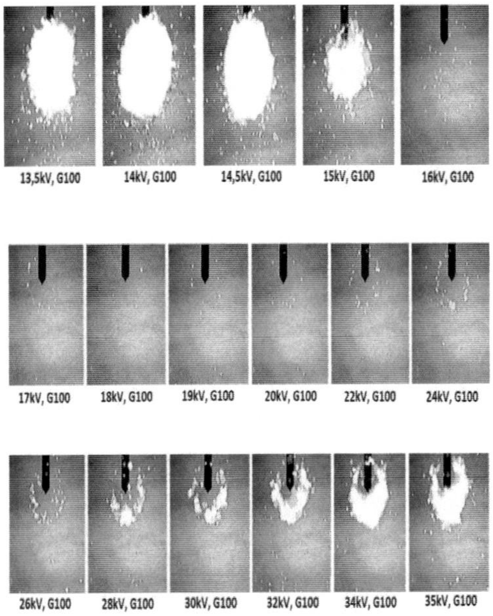

Figure 6. Corona discharges near point tip electrode for positive DC

Figure 4 shows a record of the discharge activity near cut-off copper electrode with positive DC voltage polarity. On this record, it can be seen that the discharge activity started at a critical voltage U_{cr} = 17kV. Subsequently, the UV activity of corona discharges increased and reached a maximum at 17.5kV. After that UV activity quickly dropped at 18.5kV. On this type of electrode it is possible to observe a sharp increase in discharge activity and a subsequent rapid decrease. This decrease in discharge activity is called negative differential conductivity and run on voltage extant 1 kV in this case. From a voltage of 18.5kV, the discharge activity fluently increases with increasing voltage up to breakdown at 38kV.

A record of the discharge activity on a hemispherical copper electrode is in Fig. 5. The record indicates that the discharge activity started at critical voltage 20.5kV. Then the discharge activity sharply raised until it reached a maximum at 21kV. From this voltage up to 23kV, the discharge activity decreased. It is again a phenomenon of the negative differential conductivity that had the duration 2kV. From a voltage of 23kV, the discharge activity gradually increases to the breakdown at 45kV.

Figure 6 shows a record of the discharge activity on a point tip copper electrode at a DC positive voltage. On this record it is possible to see that the discharge activity started at the critical voltage 13.5kV. Subsequently, the discharge activity increased until it reached a maximum of 14.5 kV. From this point on, the discharge activity fallen down to 18kV. Thus, negative differential conductivity takes 3.5kV for this type of electrode. From the voltage of 18kV, the discharge activity gradually increases up to 35kV, when breakdown occurred.

It is seen from the qualitative records that the critical voltage of the corona discharges and the breakdown voltage value are different for each type of electrodes. This is related to the geometry of the electrode arrangement and the intensity of the electric field around the tip of corona electrode. It should be noted, however, that negative differential conductivity has occurred in all cases. The voltage at which the discharge activity reaches its maximum and subsequently decreases is different for each type of electrodes and also the duration of the negative differential conductivity. It also relates to the geometry of the corona electrode.

Qualitative pictures of corona discharge for AC applied voltage are depicted in Fig. 7 near cut-off electrode, in Fig. 8 near hemispherical electrode and in Fig. 9 near point tip electrode.

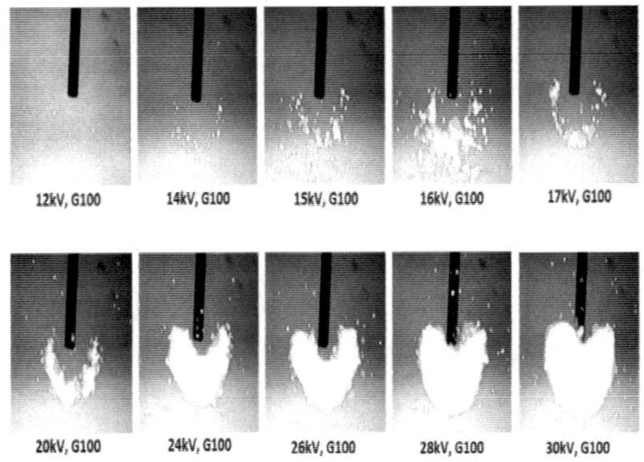

Figure 7. Corona discharges near cut-off electrode for AC

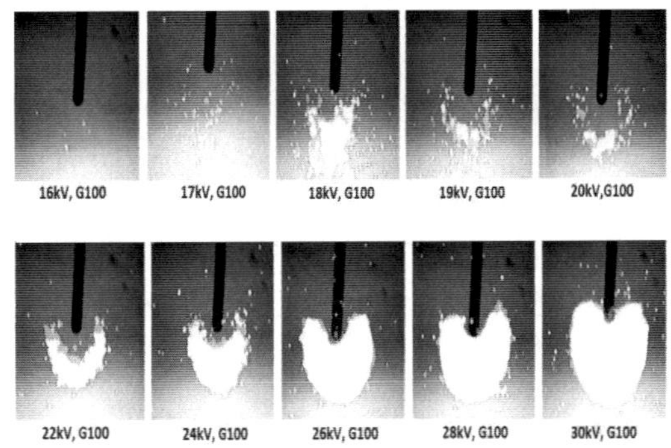

Figure 8. Corona discharges near hemispherical electrode for AC

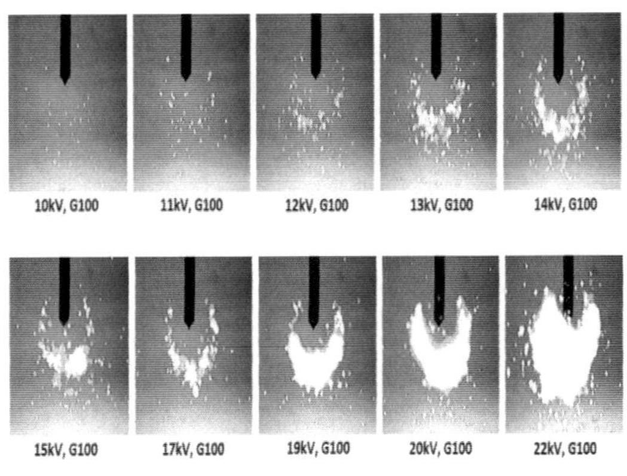

Figure 9. Corona discharges near point tip electrode for AC

In Fig. 7, it is visible that the discharge activity started on a cut-off electrode at the critical voltage 12kV. Subsequently, UV activity increased until it reached a local maximum at 16kV. From this point on, the discharge activity dropped to a local minimum at 17kV. At this type of electrode, the anomaly (negative differential conductivity) was observed in the voltage range 1kV. From a voltage of 17kV, the discharge activity gradually increased until breakdown voltage 38kV.

Figure 8 shows that the discharge activity started at the critical voltage 16kV for the copper hemispherical electrode and increased until it reached a local maximum at 18kV. From this point on, the discharging activity weakened until the voltage reached to 20kV. Thus, at this type of electrode, the anomaly was within voltage interval 2kV. From a voltage of 20kV, UV activity gradually increases up to breakdown at 42kV.

Figure 9 shows a recording of the discharge activity on the point tip electrode at the alternating voltage. The record shows that the discharge activity started at the critical voltage 10kV, then increased until it reached a local maximum at 13.5kV. From this point on, the discharge activity weakened to the voltage value of 17kV, so the anomalous phenomenon existed in the voltage range of 3.5kV. From the voltage value of 17kV, the discharge activity gradually increased until breakdown occurred at 22kV.

The qualitative records of the discharge activity indicate that the phenomenon of negative differential conductivity also exists at the applied AC voltage. The intensity of the UV radiation at the local maximum is less than was detected under the applied DC voltage of the positive polarity, but after exceeding the minimum, the intensity of the radiation increases faster under AC voltage.

Quantitative recording was done using UV mode on DayCor Camera and a program that evaluated the area of white dots on a black background [14, 15, 16]. The method is based on a gradual signal processing of UV signal from camera, where an image with a 24-bit colour depth is transformed to an image with a 1-bit colour depth. This process and evaluation was developed in Matlab. Such a recording is shown in Fig. 10 (for DC voltage) and Fig. 11 (for AC voltage).

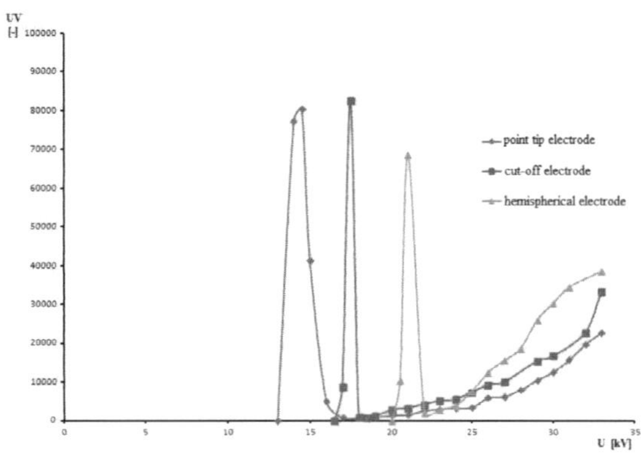

Figure 10. Graphical dependence of UV activity on the individual electrodes at DC positive voltage

The graphical dependencies indicate that the discharge activity first began on the point tip and at the latest on the hemispherical electrode. The negative differential conductivity at the both DC and AC were observed within the same voltage range, namely 1kV at the cut-off electrode, 2kV at the hemispherical electrode and 3.5KV at the point tip electrode. The difference between maximum and minimum UV activity at DC voltage was much greater than at the alternating voltage, thus the anomaly phenomenon is easier to observe at DC positive voltage than at AC.

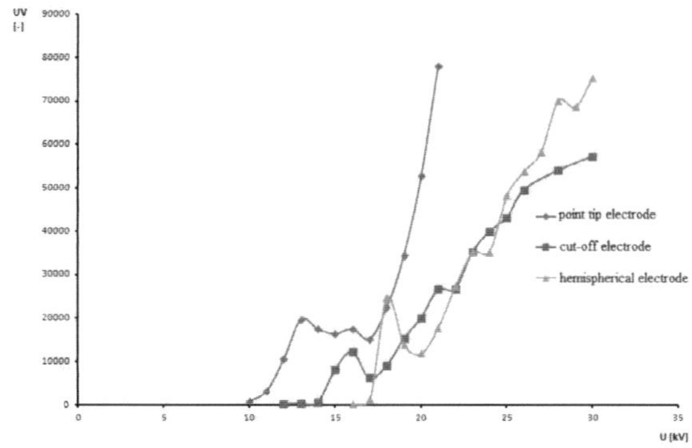

Figure 11. Graphical dependence of UV activity on the individual electrodes at AC voltage

IV. Conclusions

The aim of this paper was to qualitatively monitor the corona discharge activity and its anomaly phenomenon using a UV camera. Negative differential conductivity or anomaly phenomenon was both qualitatively and quantitatively recorded. Measuring results shown that the anomaly phenomenon occurs even at the alternating voltage, but in a smaller range of UV activity. The negative differential conductivity was the most evident in the point tip copper electrode, and therefore this type of electrode is ideally used to investigate this anomaly. The UV camera is a great detection device. Using this device it will be possible to obtain important information in the research of negative differential conductivity.

Acknowledgment

This work was supported by the Scientific Grant Agency of the Ministry of Slovak Republic and Slovak Academy of Science for funding of experimental works in the frame of VEGA-1/0340/18, VEGA-1/0435/19, VEGA-2/141/16. This work was supported by the Slovak Research and Development Agency under the contract No. APVV-15-0438, APVV-17-0372.

References

[1] F. Rizk and G. Trinh. High voltage engineering, CRC Press Taylor &Francis Group, 2014, ISBN 978-1-4665-1376-1.J. Clerk Maxwell, A Treatise on Electricity and Magnetism, 3rd ed., vol. 2. Oxford: Clarendon, 1892, pp.68–73.

[2] P. N. Mikropoulos and V. N. Zagkanas, "Threshold inception conditions for positive DC corona in the coaxial cylindrical electrode arrangement under variable atmospheric conditions", IEEE Transactions on Dielectrics and Electrical Insulation Vol. 22, No. 1; February 2015, DOI 10.1109/TDEI.2014.004641.

[3] D. Medveď and O. Hirka, "Investigation of electromagnetic fields in residential areas," Acta Electrotechnica et Informatica. Vol. 17, No. 3 (2017), pp. 48-52. - ISSN 1335-8243.

[4] R. Hackam and H. Akiyama, "Air pollution control by electrical discharges", IEEE Trans. Dielectr. Electr. Insul., Vol. 7, pp. 654-683, 2000.

[5] A. Pietrikova and M. Kravcik, „Investigation of rheology behaviour of solder paste", Proceedings of 35th International Spring Seminar on Electronics Technology - Power Electronics

(ISSE), Bad Aussee, Austria, 2012, pp.138-143. ISBN 978-1-4673-2239-3

[6] A. Fridman, A. Chikorov, and A. Gutsol, "Non-thermal atmospheric pressure discharges", J. Phys. D: Appl. Phys., Vol. 38, pp. R1-R24, 2005.

[7] R. Arora and W. Mosch. High voltage and electrical insulation engineering, IEEE Press Editorial Board, 2011, ISBN 978-0-470-94970-6.

[8] N. Vagapova, A. Fedotov, A. Sidorov and G. Vagapov, "Mechanisms of effective management in electric power companies", Polish Journal of Management Studies, 13 (2), 2016, pp. 183-191.

[9] J. Lii, „Breakdown in non-uniform fields". [online]. [citované 2012-12-28]. Dostupné na internete: http://www.cee.cqu.edu.cn/myweb/uplife/20110315173249371.pdf

[10] D. Perduková and P. Fedor, „ Design rules of a fuzzy controller using genetic algorithms",: Proc. of the 7th Int. Scientific Symposium on Electrical Power Engineering (EE), Stará Lesná, 2013, pp. 513-516, ISBN 978-80-553-1441-9.

[11] P. Linder, „Inspection for corona and arcing with the daycor camera", INMR, Roč. 2006, s.1.

[12] K. Marton. Štúdium unipolárneho korónového výboja a jeho využitie v praxi, Doktorská dizertačná práca, 1989(in Slovak).

[13] P. Kapás. Analýza priestorového náboja v slabých a silných elektrických poliach. Dizertačná práca. Košice, 1994 (in Slovak).

[14] I. Kolcunová, I., R. Cimbala, J. Petráš, J. Kurimský, B. Dolník, J. Džmura and J. Balogh, „Experimental observation of negative differential characteristic of corona discharge in ultraviolet spectrum", Journal of Electrostatics, Volume 88, 2017, Pages 139-147, ISSN 0304-3886.

[15] P. Duda. Meranie výbojovej činnosti vo vzduchu a po povrchu pevných dielektrík. Bakalárska práca, Košice, 2014 (in Slovak).

[16] M. Šipoš. Meranie výbojovej aktivity v nehomogénnych poliach. Bakalárska práca, Košice, 2013 (in Slovak).

C: Power Systems Relaying, Faults Analysis

Off-grid Distribution System Logical Protection for Resilience Enhancement

Pavel Chusovitin, Eduard Balakh, Sergei Dekhtiar, Rustam Valiev, Valeriy Tashchilin
Automated Electrical System department
Ural Federal University
Yekaterinburg, Russia
pvchus@gmail.com

Abstract — **The paper addresses resilience of real isolated power system. In the system, resilience may not be achieved due to induction motor slowing down during the fault. Whether transient instability will occur or not depends on fault clearance time. In the paper, fault location method is proposed based on logical signals exchange using GOOSE messages in accordance with IEC 61850. The method allows decreasing fault clearance time in the area of the network, which is critical from a resilience point of view. The algorithm is applied in physical intelligent electronic devices (IED) and tested with a model of the real system under consideration. Simulations and testing were carried out using the hardware-in-the-loop (HIL) approach with real-time dynamic simulations**

Keywords: isolated power system, resilience, transient stability, fault location, IEC 61850, hardware-in-the-loop, real-time dynamic simulation

I. Introduction

Isolated power systems play an essential role in supplying remote customers where there are no options for connection to the utility. The specific dynamic behavior of isolated systems depends upon energy source, which is used to produce electrical energy. Most of such systems can be characterized by small inertia [1] and limited active and reactive power reserves. As a result, the small isolated grid is subjected to losing stability in case of the various events.

While the isolated operation of smart grid may be an option for power system resilience improvement [2], the isolated system itself requires revision of protection systems [3] from a resilience point of view. Resilience is defined as the ability of the system to withstand significant and low-probability events without major disruption [4]. There are various approaches for resilience improvement of isolated power systems.

Minimizing of emergency event influence on electrical energy supply is a key element of resilience maintaining. For isolated power systems, one of the major events, which may lead to system shut down, is a fault. In [5], optimal switchgear allocation is considered in terms of resilience improvement in case of natural disasters. Mechanisms for fault location, isolation and supply restoration based on multiagent approach are considered in [6], [7]. Described fault location and isolation considers the post-disturbance operation and may not provide resilience due to transient issues. In isolated power systems, fault may lead to instability even with typical fault clearance time [8]. For distribution systems with power electronic-based generation/storage, transient stability improvement may be achieved with proper inverter control. In [9], storage control strategy for distribution system transient stability improvement is shown. In [10] application of renewable energy source inverter is proposed for isolated system support during transients. In the research, real system was considered which has issues with transient stability. During the fault, the system resilience may fail due to generators frequency increase [11] or induction motor slowing down. A similar problem is described in [12] regarding induction motor starting in an isolated power system. In the system under consideration, power electronic starting device is used, but during the fault, motors are slowed down and may not go back to nominal speed after fault clearance. The same issue is addressed in [8] by adjusting control strategies of different energy sources.

Another way of solving the problem is fast fault location. In the paper, the approach is proposed for logical fault location based on the application of IEC 61850. Introduction of IEC 61850 results in the development of new protection algorithms, which can improve transient stability of power systems. There have been developed a number of various approaches for IEC 61850 implementation especially in the field of protection device communication up to the new concept of virtual power system protection concept [13]. The research that considers the application of IEC 61850 for distance protection is presented in [14]. In [15] it is shown that distance relay selectivity can be improved using IEC 61850 GOOSE trip/block messages exchange between IEDs. Some accelerated protection schemes employing GOOSE messages exchange are presented in [16]. Various applications of GOOSE messages for distribution bus protection improvement are presented in [17, 18].

A proposed algorithm is based on the same GOOSE blocking signals exchange, but unlike abovementioned approaches, is not limited by the protection of one specific network element. It is extended to the coordinated protection of network fragment. A similar approach is applied in [19] for sensitivity improvement of phase directional protection and earth fault directional protection of distribution system. In [20] GOOSE coordination is used for overcurrent protection sensitivity improvement in presence of large amount of power electronic-based generation. Research [21] proposes centralized algorithm for the same aim.

The main feature of the presented research is that it addresses isolated system transient stability improvement. It is achieved by minimal information exchange, which improves the reliability of the protection system. The overcurrent protection acceleration using GOOSE messages exchange between IEDs is proposed for fast fault location. Proposed scheme allows decreasing fault clearance time thus increasing the resilience of the system under consideration.

II. The Isolated System Description

The presented research considers real isolated power system, which supplies mine. Total consumption of the system is about 6000 kW. For the research, snapshot of power system state with maximum load (6200 kW) was taken.

The system is supplied by seven diesel generators with a nominal power of 1591.6 kW each. Typically, six generators maintain power system normal conditions while one generator is in standby reserve. Such operation mode provides necessary reliability level for the isolated power system. Fig. 1 represents the system. Two 6 kV induction motors are connected through cables to 6 kV bus and thus raising high power quality and reliability requirements.

Dynamic characteristics of the diesel generator were simulated by a system of differential equations that presents characteristics of real engines acceptable for power system modeling [22].

Diesel Generator model comprises speed governor, actuator and engine. The inputs of the speed governor are speed deviation and load demand Load Reference obtained from secondary frequency control block. The output is a fuel demand signal that is obtained according to the first order transfer function. Actuator dynamics can be presented as first-order transfer function with proportional gain K2 and actuator time constant T2 [23]. The output of diesel engine model is connected to the input Pm of Three-Phase Synchronous Machine that simulates dynamic characteristics of a synchronous generator in accordance with Park-Gorev equations. In the research, model "ST1A" form [24] was used for modeling automatic voltage regulator.

III. Resilience Issue Description

In isolated power system problem may occur with induction motor start due to lack of reactive power support [12], [8]. In the system under consideration similar problem occurs, which leads to poor resilience. Fig. 2 shows the transient process (rotation speed of generator 1, the rotation speed of induction motors and the voltage at terminals of induction motor) after a three-phase fault at bus 9 (see Fig. 1). The motor is slowed down during the fault due to voltage decrease. After the fault is cleared, voltage is not recovered instantly since slowed motor consumes a large amount of reactive power. In the most cases, the motor is continuously speeded up, which provides less reactive power consumption, further voltage recovery and speed up. In case the fault is long enough, motor collapse occurs.

Fig. 2 shows the unstable process of that kind. The fault is cleared in time determined by overcurrent protection time delay, which, in turn, is determined by selectivity requirements. In this case, fault clearance time was 0.5 s. One can conclude by observing Fig. 2 that correct overcurrent protection settings do not provide motor stability after the fault. In Fig. 3, stable process is demonstrated for the same conditions, but fault clearance time is equal to the critical value of 0.46 s.

One can observe that critical fault clearance time allows stable induction motor speed up.

The phenomenon of motor collapse impacts adversely on the system resilience. Further research was intended on the localization of the problem.

Fig. 4 shows how critical fault clearance time is changed through the network. Induction motor dynamic instability was considered as a criterion of critical fault clearance time. It can be seen that critical fault clearance time increases with fault point moving away from motors.

When a fault occurs at low-voltage side of the transformers, it does not lead to a motor collapse at all. Large critical fault clearance time for the fault at motors terminals is explained by that faulted motor is switched off which ease reactive support requirements.

In the paper method is proposed to improve resilience by using IEC 61850 for communication between protection devices.

IV. Algorithm Description

A. Distribution network overcurrent protection

Application of overcurrent protection is worldwide spread way for distribution lines protection. Selectivity of this protection is achieved by different time delay settings for different devices. In the investigated system, protection with code ANSI 50/51 in accordance with [25] is applied. The coordination diagram of protection devices is presented in Fig 5-(a) for feeder 1-2-9-14 (see Fig. 1). Conventional time delay of such overcurrent protection is 0.25 s [26]. Due to selectivity requirements time difference between two consequent relays is also 0.25 s. Hence, fault clearance time for the head device is 0.75 s. Increased clearance time may lead to instability of the power system.

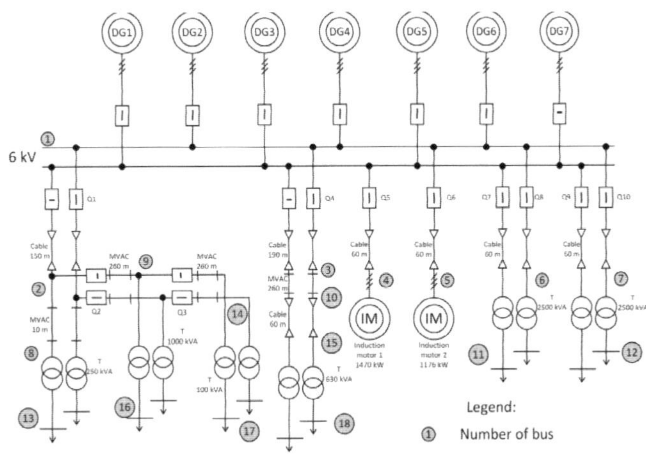

Figure 1. Scheme of the system under consideration

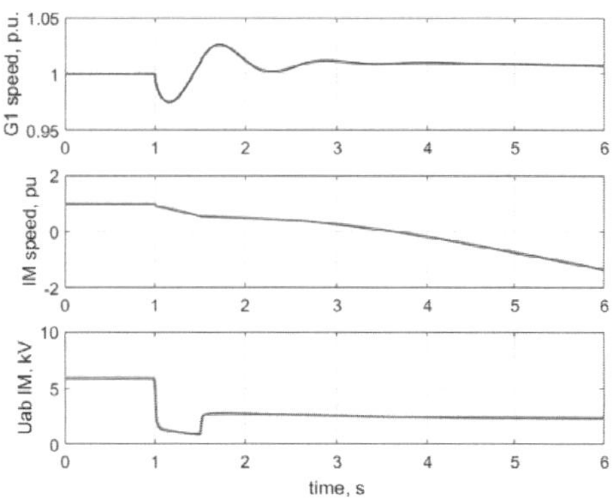

Figure 2. An unstable transient process in the system under the fault

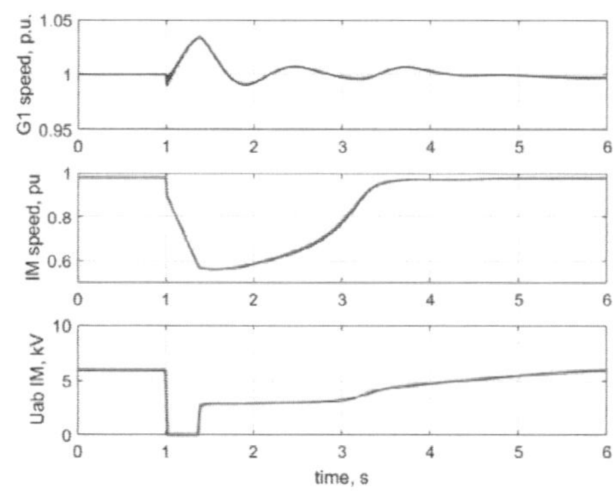

Figure 3. A stable transient process in the system under the fault

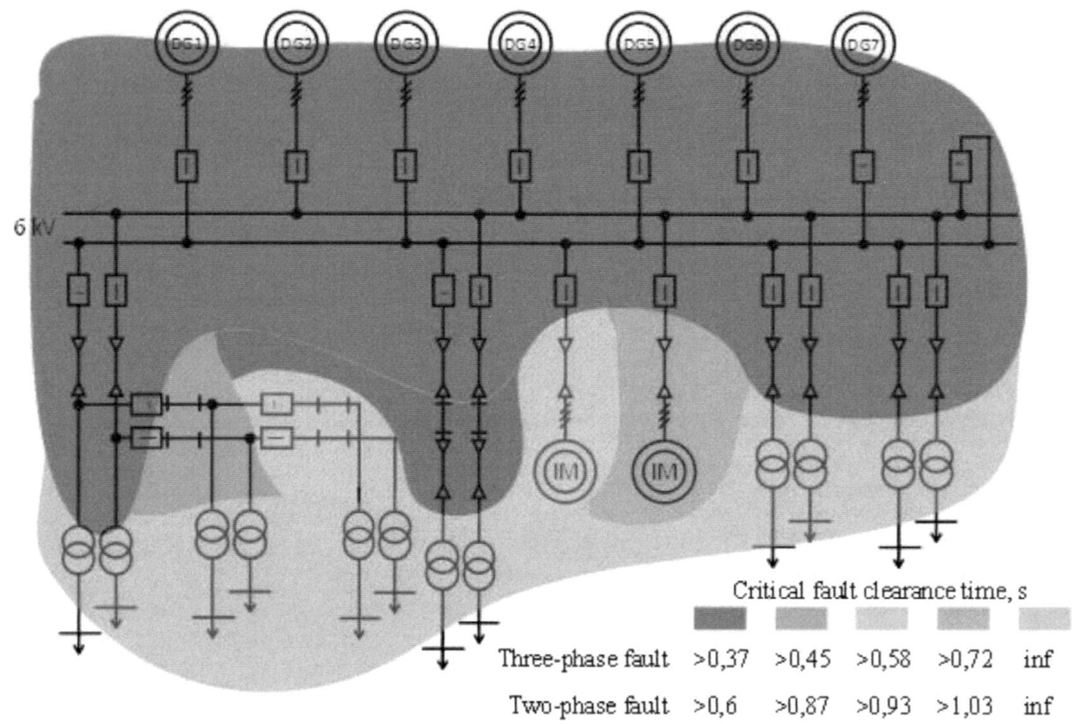

Figure 4. Fault clearance time for the system under consideration

(a)

(b)

Figure 5. Coordination of protection devices

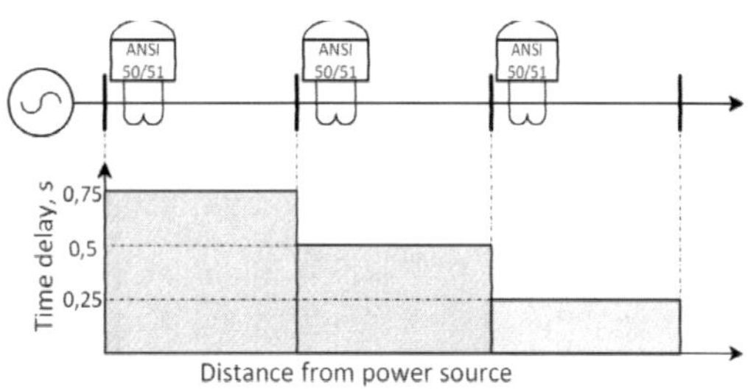

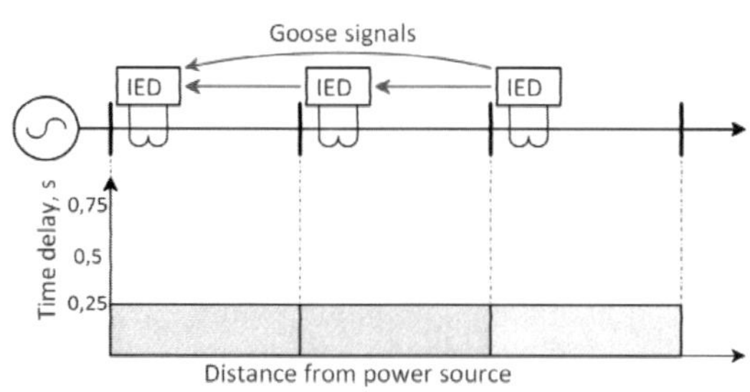

Application of IEC 61850 GOOSE communication between installed relays allows avoiding the necessity of adding time difference between various relays. In this case, relay coordination diagram becomes as it is shown by Fig 5 - (b). In this case, all consequent relays have the same time lag, which results in deceasing fault clearance time. As it was shown in the previous section, such decreasing of clearance time can be crucial in terms of keeping the stable operation of isolated power system.

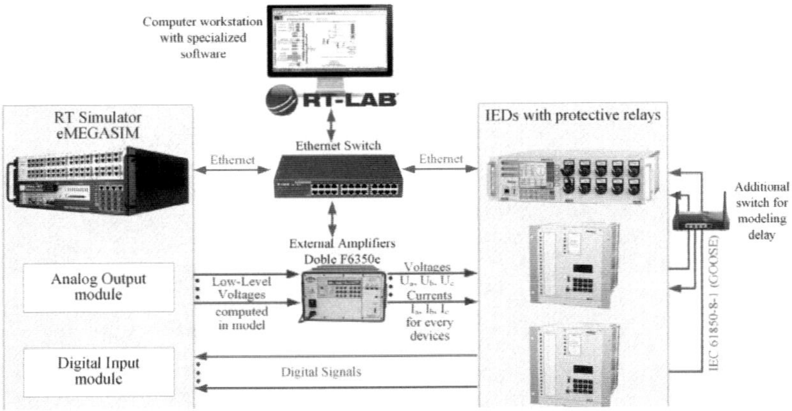

B. Proposed algorithm

The advent of digital protections and the development of standards such as IEC 61850 have led to the emergence of the logical type of selectivity (zone selectivity). The developed algorithm improves the efficiency of traditionally used overcurrent protection and uses the logical type of selectivity. The input algorithm parameters include line voltages, phase A and C currents, GOOSE messages from adjacent devices which allow or prevent the operation of a circuit breaker.

While working in a normal operation mode, measured voltages and currents are within predetermined limits and algorithm is off-duty. As soon as there is a fault in the power system, the limits for input parameters are violated and it is necessary to clear fault as fast as possible. For the sake of reliable operation of the logic, time delay for expecting GOOSE messages from the adjacent devices is applied. Signals from those devices allow or prevent sending the signal for circuit breaker operation. Delay time for GOOSE is much less, than that is used for conventional overcurrent protection and for that reason less time is necessary to clear the fault. Eventually, the basic principle of the algorithm is as follows:

- Each device send GOOSE signals to devices, which are closer to the source if it detected fault;
- This signal is used in those devices for forbidding circuit breaker operation.

It should be mentioned that, in order to clear a fault in case of communication channels damage that there is a signal in the algorithm, which is formed like in traditionally used overcurrent protection.

The algorithm described above was applied in intelligent electronic devices (IED) and was tested with a real-time dynamic simulation system.

V. Description of Simulation Platform

Fig. 6 shows hardware-in-the-loop real-time simulation platform, which was applied for the research. Computation system is presented by eMEGASIM simulator. The simulator allows developing power system models based on Matlab/Simulink software, which includes SimPowerSystems (SPS) library, and then to compute the dynamic behavior of the mathematical model in real-time. The system described in section 2 was modeled with eMEGASIM.

The output of the simulator was the set of voltages and currents as if it was measured by overcurrent protection devices connected to breakers Q1, Q2 and Q3 (see Fig. 1). Then these signals are amplified by DOBLE F6350e to the level required for IEDs installed in the real power system. IEDs used for the research were programmed to perform the proposed algorithm. Digital outputs of IEDs are connected to simulator's digital input, which, in turn, change circuit breaker status and trip protected line in the model. Additional communication switch is used to model communication network of the substation.

VI. Algorithm Testing Results

Fig. 7 demonstrates signals recorded by IED installed at Q1 during the two-phase fault at bus 2. Fig. 7 - (a) shows the operation of conventional overcurrent protection and Fig. 7 - (b) presents operation of the proposed algorithm. With the algorithm, the fault is cleared with minimal time allowed for overcurrent protection since no GOOSE signals were detected forbidding circuit breaker operation.

Fig. 9 demonstrates a comparison of power system response for the conventional and proposed approach. One can observe power system recovery after fault clearance with the proposed algorithm, which improves power system resilience.

Fig. 8 shows signals recorded by IED installed at Q1 and Q2 during three-phase fault at bus 9. Power system response for the proposed algorithm as well as conventional protection is presented in Fig. 10.

Although IED installed at Q1 detects a fault, it does not operate since forbidding GOOSE signal comes from IED installed at Q2. Similarly, with the previous experiment, the stable transient process is achieved by application of the proposed algorithm, which is demonstrated by Fig. 10.

VII. Conclusion

In the research, the resilience of the real isolated power system is considered. Resilience of the system may not be achieved after the fault due to induction motors slow down. Transient stability of the system was investigated and the area of network identified where fault may lead to transient instability. The algorithm was developed to avoid the problem using fast fault location. It uses GOOSE messages exchange based on IEC 61850 concept. The algorithm was applied in physical intelligent electronic devices (IED) and tested with HIL approach using real-time dynamic simulations.

Testing shown the reliability of the proposed approach. Algorithm successfully located all faults during the testing and cleared it with expected small time. The goal of improving the system resilience was achieved. It was proved the application of the algorithm allows eliminating transient instability events.

Figure 6. Coordination of protection device

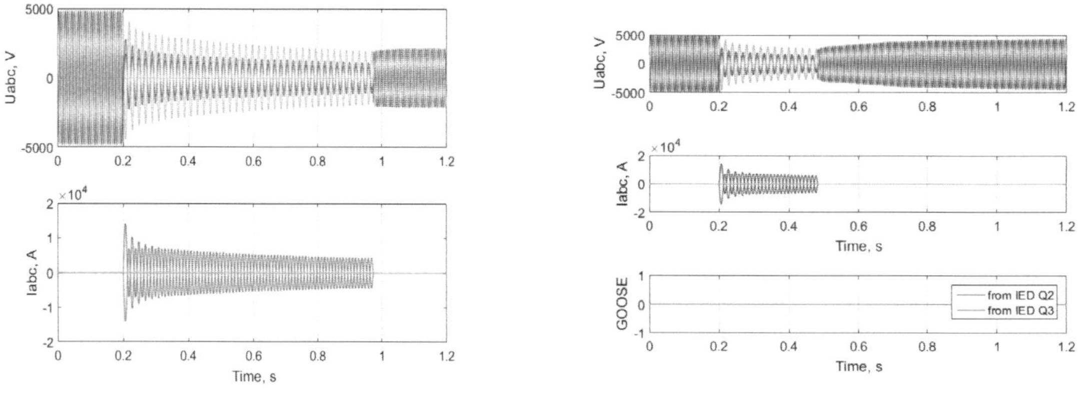

(a) (b)

Figure 7. Signals recorded by IED Q1 during the fault at bus 2

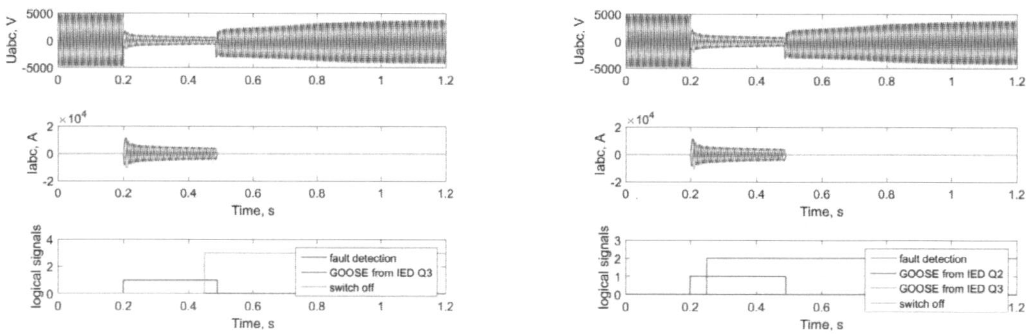

(a) (b)

Figure 8. Signals recorded by IED Q1 and Q2 during the fault at bus 9

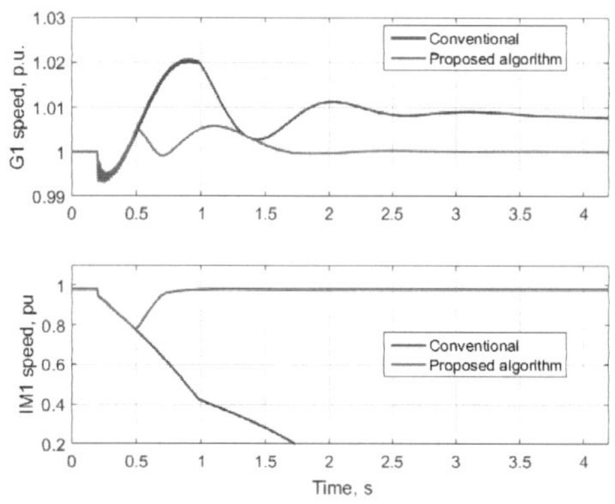

Figure 9. Power system response to two-phase fault at bus 2

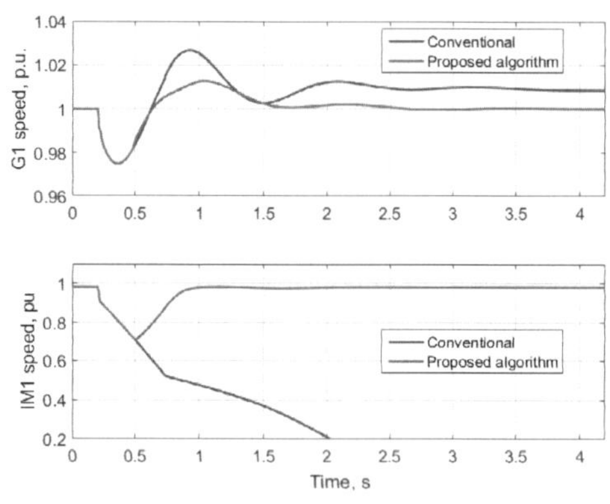

Figure 10. Power system response to three-phase fault at bus 9

ACKNOWLEDGMENT

The reported study was funded by the Ministry of Science and Higher Education of the Russian Federation under Federal Targeted Programme according to the agreement № 075-02-2018-1219 (internal number 14.578.21.0226, project identifier: RFMEFI57817X0226).

REFERENCES

[1] X. Tang, W. Deng, and Z. Qi, "Investigation of the dynamic stability of microgrid," IEEE Transactions on Power Systems, vol. 29, pp. 698-706, March 2014.

[2] C. Chen, J. Wang, F. Qiu, and D. Zhao, "Resilient distribution system by microgrids formation after natural disasters," IEEE Transactions on smart grid, vol. 7, no. 2, pp. 958-966, 2016.

[3] G. Jimenez-Estevez, A. Navarro-Espinosa, R. Palma-Behnke, L. Lanuzza, and N. Velzquez, "Achieving resilience at distribution level: Learning from isolated community microgrids," IEEE Power and Energy Magazine, vol. 15, pp. 64-73, May 2017.

[4] M. Simonov, "Dynamic partitioning of dc microgrid in resilient clusters using event-driven approach," IEEE Transactions on Smart Grid, vol. 5, pp. 2618-2625, Sept 2014.

[5] M. Zare-Bahramabadi, A. Abbaspour, M. Fotuhi-Firuzabad, and M. Moeini-Aghtaie, "Resilience-based framework for switch placement problem in power distribution systems," IET Generation, Transmission Distribution, vol. 12, no. 5, pp. 1223-1230, 2018.

[6] A. Zidan and E. F. El-Saadany, "A cooperative multiagent framework for self-healing mechanisms in distribution systems," IEEE transactions on smart grid, vol. 3, no. 3, pp. 1525-1539, 2012.

[7] M. J. Ghorbani, M. A. Choudhry, and A. Feliachi, "A multiagent design for power distribution systems automation," IEEE Transactions on Smart Grid, vol. 7, no. 1, pp. 329-339, 2016.

[8] A. H. K. Alaboudy, H. H. Zeineldin, and J. Kirtley, "Microgrid stability characterization subsequent to fault-triggered islanding incidents," IEEE Transactions on Power Delivery, vol. 27, no. 2, pp. 658-669, 2012.

[9] M. Ayar, S. Obuz, R. D. Trevizan, A. S. Bretas, and H. A. Latchman, "A distributed control approach for enhancing smart grid transient stability and resilience," IEEE Transactions on Smart Grid, vol. 8, no. 6, pp. 3035-3044, 2017.

[10] F. Katiraei, M. R. Iravani, and P. W. Lehn, "Micro-grid autonomous operation during and subsequent to islanding process," IEEE Transactions on power delivery, vol. 20, no. 1, pp. 248-257, 2005.

[11] D. Vangulick, D. Ernst, and T. V. Cutsem, "Resilience of the distribution system operator network near to 50.2 hz," CIRED - Open Access Proceedings Journal, vol. 2017, no. 1, pp. 2520-2524, 2017.

[12] M. Falahi, K. L. Butler-Purry, and M. Ehsani, "Induction motor starting in islanded microgrids," IEEE Transactions on Smart Grid, vol. 4, no. 3, pp. 1323-1331, 2013.

[13] R. Wojtowicz, R. Kowalik, and D. D. Rasolomampionona, "Next generation of power system protection automation - virtual power system protections," IEEE Transactions on Power Delivery, vol. PP, no. 99, pp. 1-1, 2017.

[14] K. Wannous and P. Toman, "Iec 61850 communication based distance protection," in Proceedings of the 2014 15th International Scientific Conference on Electric Power Engineering (EPE), pp. 107-112, May 2014.

[15] D. D. Giustina, A. Dede, G. Invernizzi, D. P. Valle, F. Franzoni, A. Pegoiani, and L. Cremaschini, "Smart grid automation based on iec 61850: An experimental characterization," IEEE Transactions on Instrumentation and Measurement, vol. 64, no. 8, pp. 2055-2063, 2015.

[16] A. Apostolov and B. Vandiver, "Iec 61850 goose applications to distribution protection schemes," in 2011 64th Annual Conference for Protective Relay Engineers, pp. 178-184, April 2011.

[17] A. Apostolov, "Impact of iec 61850 on the interoperability and reliability of protection schemes," in 2013 IEEE Power Energy Society General Meeting, pp. 1-5, July 2013.

[18] P. K. Naik, M. Bahadornejad, N. K. C. Nair, and V. Vyatkin, "Iec 61850 based smart distribution protection: Solutions for sympathetic tripping," in 2011 IEEE PES Innovative Smart Grid Technologies, pp. 1-7, Nov 2011.

[19] A. Silos, A. Senis, R. M. de Pozuelo, and A. Zaballos, "Using iec 61850 goose service for adaptive ansi 67/67n protection in ring main systems with distributed energy resources," Energies, vol. 10, no. 11, p. 1685, 2017.

[20] M. H. Cintuglu, T. Ma, and O. A. Mohammed, "Protection of autonomous microgrids using agent-based distributed communication," IEEE Transactions on Power Delivery, vol. 32, no. 1, pp. 351-360, 2017.

[21] M. Y. Shih, A. Conde, Z. Leonowicz, and L. Martirano, "An adaptive overcurrent coordination scheme to improve relay sensitivity and overcome drawbacks due to distributed generation in smart grids," IEEE Transactions on Industry Applications, vol. 53, no. 6, pp. 5217-5228, 2017.

[22] S. Roy, O. Malik, and G. Hope, "A k-step predictive scheme for speed control of diesel driven power plants," IEEE Transactions on industry applications, vol. 29, no. 2, pp. 389-396, 1993.

[23] R. Pena, R. Cardenas, J. Proboste, J. Clare, and G. Asher, "Wind-diesel generation using doubly fed induction machines," IEEE Transactions on Energy Conversion, vol. 23, no. 1, pp. 202-214, 2008.

[24] IEEE Std 421.5-2005, "IEEE Recommended Practice for Excitation System Models for Power System Stability Studies," IEEE Std 421.5-2005 (Revision of IEEE Std 421.5-1992), pp. 1-93, April 2006.

[25] IEEE Std C37.2-2008, "Electrical Power System Device Function Numbers, Acronyms, and Contact Designations," IEEE Std C37.2-2008 (Revision of IEEE Std C37.2-1996), pp. 1-48, October 2008.

[26] ABB, "Protection criteria for medium voltage networks," Technical Guide, pp. 1-50, 2016.

Estimation of Earth Fault Distance Protection Algorithms Effectiveness for High Voltage Cable Line

Tatjana Lomane
Faculty of Power and Electrical Engineering
Riga Technical University
Riga, Latvia
loman@eef.rtu.lv

Sergejs Rubcovs
Faculty of Power and Electrical Engineering
Riga Technical University
Riga, Latvia
Sergejs.rubcovs@ast.lv

Kristina Berziņa
Faculty of Power and Electrical Engineering
Riga Technical University
Riga, Latvia
k.berzina@inbox.lv

Inga Zicmane
Faculty of Power and Electrical Engineering
Riga Technical University
Riga, Latvia
Inga.Zicmane@rtu.lv

***Abstract* — In cable lines, there are problems of ensuring the effective operation of earth faults distance protection (DP) through the transient resistance due to the relatively small values of the specific parameters of the cable and the difference in the characteristics of the active and reactive resistances of the positive and zero sequences. The paper presents the efficiency estimations of distance measuring units (DMU) different algorithms operation in cables lines during the earth faults through transient resistance and analysis the impact of different zero sequence currents compensation coefficients to calculate the impedance to the fault location. To improve the efficiency of the DMU operation in the high-voltage cable line, it is proposed to synthesize an algorithm for measuring the line impedance to the fault location on the basis of a refined emergency mode model and use search optimization methods. The results of the analysis showed that the efficiency of such an algorithm is higher than the classical one and less depends on the value of the transient resistance in the fault location.**

Keywords: earth faults; distance measuring units algorithm; high-voltage cable; optimization methods; effectiveness

I. Introduction

Modern development of electric networks is characterized by the use of high-voltage cable inserts and the trend of construction of cable lines instead of overland lines [1-3]. This is a consequence of a more stringent environmental policy, the increase in the cost of land, increasing the density of built-up areas or the presence of water obstacles to the electricity transmission lines. In high-voltage networks, the most common protection of power lines from all types of damage is distance protection (DP), which is based on the use of the classical algorithm for determining the impedance to the fault location [1-12]. In cable lines, the operation of the earth fault DP is significantly influenced by the presence of transient resistance in the short circuit. The fact is that the AC cable inserts are very short. For example, in 330kV networks do not exceed 30 km. In this case, the specific parameters of the cable are very small and at close to the short-circuit resistance of the line to the fault location can be much less than the value of the transient resistance. In addition to cable networks to the problem of ensuring the efficient operation of earth fault distance protection caused by differences of characteristics of active and reactive resistance positive and zero sequence. This necessitates separate the calculating the compensation of voltage drop from zero sequence currents along the active and reactive resistances axes. The article considers the earth fault distance protection operation in the high-voltage cable line by three algorithms of distance measuring units. The influence of different methods of calculating the zero sequence currents compensation coefficients on the efficiency of distance measuring units (DMU) by different algorithms in cable lines during the earth faults through the transient resistance is analyzed. According to the results of statistical estimates, it is proposed to synthesize an algorithm for measuring the line impedance to the fault location on the basis of a refined emergency mode model and use search optimization methods. In cable networks, the efficiency of such an algorithm is higher than the classical one and the dependent on the magnitude of the transient resistance in the place of fault location is less.

II. Characteristics of Cable and Overhead Lines

The high-voltage cables specific technical parameters and features determines the DMU operation. Table 1 presents the main electrical characteristics of high-voltage cable and overhead lines.

Table 1. 330 kV overhead lines and cables positive and zero sequences parameters

330kV HVTL	X1,Om/km	X0/X1	R1,Om/km	R0/R1
Overground Line	0.3-0.42	2.5-3.5	0.02-0.1	3 - 5
Underground Cable	0.1-0.25	0.2-0.6	0.01-0.025	10-15

Cable lines are characterized by significantly lower values of active and reactive resistances and significantly higher values of capacitive conductivity in comparison with air. At the same time, the ratio of active and reactive parameters of the direct and zero sequence scheme is fundamentally different. Thus, in air lines the values of reactive and active resistances of the zero sequence are 2.5 - 4 times greater than the resistances of the direct sequence depending on the design of the wires, and the angles of the resistance vectors of the direct and zero sequences are very close. In cable lines, the reactive resistances of the zero sequence are 2-5 times less than the direct sequence, and the values of the active resistances of the zero sequence are 10-15 times more than the direct sequence. Therefore, the angles of resistance of the straight and zero sequences are significantly different, and the values of the resistance modules of the straight and zero sequences are close.

Figure1 shows the resistance vectors of the direct and zero sequences for overhead lines (Fig.1a) and cable (Fig.1b) lines used in the 330 kV network. Different characteristics of the resistance parameters of air and cable lines create problems in the coordination of distance protection in combined networks.

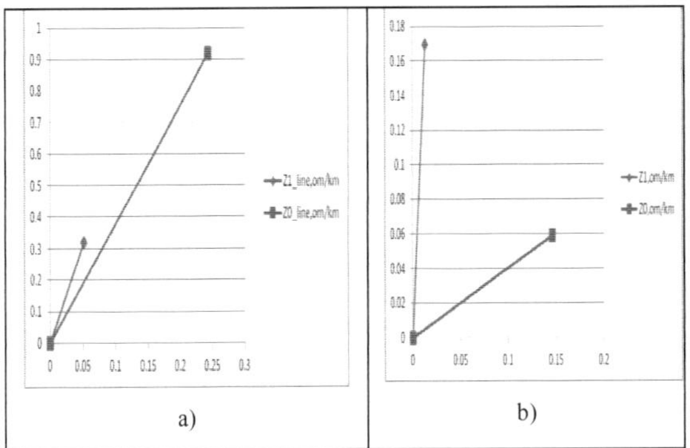

Figure1. Positive and zero sequences impedances for overhead – a) and cable – b) lines.

In contrast to the overhead lines in the cable lines, the ratio of the resistance of the direct and zero sequences for the active and reactive components are different and opposite. So the active component of the resistance of the zero sequence increases 5-10 times compared to the positive sequence, and the reactive component of the resistance of the zero sequence decreases 3-4 times compared to the positive sequence. If the angles of the resistance vectors of the positive and zero sequence of the overhead line are close in values, then for cable lines these vectors are in different quadrants, since the active and reactive components of the resistance of the zero sequence have different dependences on the components of the positive sequence.

III. The Earth Faults DMU Algorithms

Two algorithms of the DMU operation using different approaches to calculation of the compensation coefficient of zero sequence currents and one algorithm using the search method of optimization were considered.

A. *Algorithm 1*

The classical algorithm for calculating the resistance vector of a positive sequence from the place of distance protection installation to the fault location uses the expression [1-12]:

$$\dot{Z}1_m = \left[\frac{\dot{U}_{ph}}{\dot{I}_{ph} + \dot{K}_N \cdot \dot{I}_E} \right] , \qquad (1)$$

Where

$\dot{K}_N$ is the classical compensation coefficient of the voltage drop from the zero sequence currents in the line is calculated by the specific parameters of its line:

$$\dot{K}_N = \frac{\dot{Z}0_L - \dot{Z}1_L}{3 \cdot \dot{Z}1_L} \qquad (2)$$

B. *Algorithm 2*

Algorithm 2 uses the separate calculation of active and reactive positive sequence impedances between the distance protection installation place and earth fault point as follow [4]:

$$X1_m = \frac{Im\{\dot{U}_{ph} / \dot{I}_{ph}\}}{1 + K_x} , \quad R1_m = \frac{Re\{\dot{U}_{ph} / \dot{I}_{ph}\}}{1 + K_r} \qquad (3)$$

where

K_X , K_R are the compensation coefficients of the voltage drop from the zero sequence currents in the active and reactive line impedances, respectively:

$$K_X = \frac{X0 - X1}{3 \cdot X1} . \quad K_R = \frac{R0 - R1}{3 \cdot R1} \qquad (4)$$

C. *Algorithm 3*

The Algorithm 3 is based on the use of search optimization methods to improve the accuracy of the positive impedance vector calculation. The equation of the bus voltage balance at the place of protection installation is accepted as the objective function. The voltage balance is defined as the difference between the $\dot{U}_{A,m}$ measured voltage on the control bus A and the $\dot{U}_A$ residual voltage on the monitored bus calculated by the using the relay input currents and the specified coefficients. Then the vector of unbalanced voltages on the controlled buses can be written in the form [7]:

$$\dot{U}_{nb,A} = \dot{U}_{m,A} - \dot{U}_A \qquad (5)$$

where

$$\dot{U}_A = \dot{Z}1_{AF} * (\dot{I}_A + \dot{K}_N \cdot \dot{I}0_A) + \dot{I}_{EF} \cdot R_F \qquad (6)$$

$$\dot{K}_N = \frac{\dot{Z}0_L - \dot{Z}1_L}{3 \cdot \dot{Z}1_L} ;$$

$$\dot{I}_A = \dot{I}1_A + \dot{I}2_A + \dot{I}0_A ;$$

$$\dot{I}_{EF} = 3 \cdot \dot{I}0_F ;$$

$$\dot{I}0_F = \dot{K}_{cur} \cdot \dot{I}0_A ;$$

$$\dot{K}_{cur} = \frac{\dot{Z}0_B + (\dot{Z}0_L - \dot{Z}0_{AF})}{\dot{Z}0_B + \dot{Z}0_L + \dot{Z}0_A} , \qquad (7)$$

Where

$\dot{Z}1_L$, $\dot{Z}0_L$ are respectively positive and zero sequence impedance of the line;

$\dot{Z}1_A$, $\dot{Z}0_A$ are respectively the positive and zero sequence impedance of A system;

$\dot{Z}1_{AF}$ is the line positive sequence impedance to the fault location;

R_F is a fault resistance at the fault place; $\dot{I}1_A, \dot{I}2_A, \dot{I}0_A$ are respectively positive, negative and zero sequence currents;

$\dot{I}0_F$ is the zero sequence current in the fault location; $\dot{Z}0_B$ is the zero sequences impedance of B system.

The complex equation (5) is nonlinear with respect to the required variables and can be solved by various optimization methods. In the process of finding the minimum value of the objective function, the equation of the form is solved:

$$\dot{U}_{nb}(\overline{\Pi}) = 0 \qquad (8)$$

Where

$\overline{\Pi} = \{R_{AF}, X_{AF}, R_F\}$ is unknown parameters vector;

R_{AF} , X_{AF} are respectively active and reactive impedance of the line section from the protection installation place to the fault location.

In order to simplify the task, the numbers of unknown parameters have been reduced to a minimum if possible. Since the ratio of the specific parameters of the line active and reactive impedances is a constant value equal to the cotangent angle of the line. Then the active resistance can be expressed through the reactive resistance equation of the form:

$$R_{AF} = X_{AF} * ctg(\varphi_L) . \qquad (9)$$

Then the number of unknown parameters is reduced to two parameters and the unknown parameters vector is rewritten:

$$\overline{\Pi} = \{X_{AF}, R_F\} .$$

The complex vector of the voltage unbalance (8) can be divided into imaginary and real parts and it is obtained a system of real equations of the form:

$$\vec{U}_{nb}(\overline{\Pi}) = \left[U'_{nb}(\overline{\Pi}), U''_{nb}(\overline{\Pi})\right] = 0 \qquad (10)$$

One of the most rapidly convergent methods for solving the nonlinear equations system is Newton's method. In Newton's method, the initial approximation of the solution vector $\overline{\Pi}^0$ is chosen first by using the classical algorithm 1. Then the nonlinear function (5) is approximated by a linear function in the form:

$$\vec{U}_{nb}(\overline{\Pi}) = \vec{U}_{nb}(\overline{\Pi}^0) + \frac{\partial \vec{U}_{nb}(\overline{\Pi}^0)}{\partial \overline{\Pi}} * (\overline{\Pi} - \overline{\Pi}^0) , \qquad (11)$$

Where

$\dfrac{\partial U_{nb}(\overline{\Pi}^0)}{\partial \overline{\Pi}}$ is Jacobi matrix ;

$\overline{\Pi}^0$ is a vector of initial approximation.

$$\frac{\partial U_{nb}(\overline{\Pi})}{\partial \overline{\Pi}} = \begin{bmatrix} \dfrac{\partial U'_{nb}}{\partial R_F} & \dfrac{\partial U'_{nb}}{\partial X1_{AF}} \\ \dfrac{\partial U''_{nb}}{\partial R_F} & \dfrac{\partial U''_{nb}}{\partial X1_{AF}} \end{bmatrix} ,$$

At each step of the iterative process, the system (11) is solved with respect to the correction vector:

$$d\overline{\Pi}^k = (\overline{\Pi}^k - \overline{\Pi}^{k-1}) ,$$

which determines the transition to the value of the desired vector:

$$\overline{\Pi}^{k+1} = (\overline{\Pi}^k + d\overline{\Pi}^k) .$$

IV. Analysis DMU Efficiency

For the analysis of the different compensation coefficient influence on DMU efficiency the 330kV power transmission diagram on Fig. 2 was used.

Table 2 shows the values of different compensation coefficients calculated by equations (2) and (4) for overland line and cable. It is obvious, that the compensation coefficient vectors for the air and cable lines are in different quadrants of the complex plane.

Table 2. Compensation coefficients

330 kV	$\dot{K}_N$, (2)	K_R / K_X (4)
Land Line	$0.64e^{-j9^0}$	1.2 / 0.62
Cable	$0.34e^{-j125^0}$	3.4 / -0.22

Fig. 3 and 4 show estimations of the earth fault DMU operation in the cable line with a length of 30 km during earth faults along the line from the beginning of the line to the end of the first zone. Due to the small resistance values of cable lines, the length of DP first zone is 80% of the line length, which is 24 km. It is simulated earth faults with the variable fault resistance value in diapason equal to $R_F = 0 - 5\Omega$.

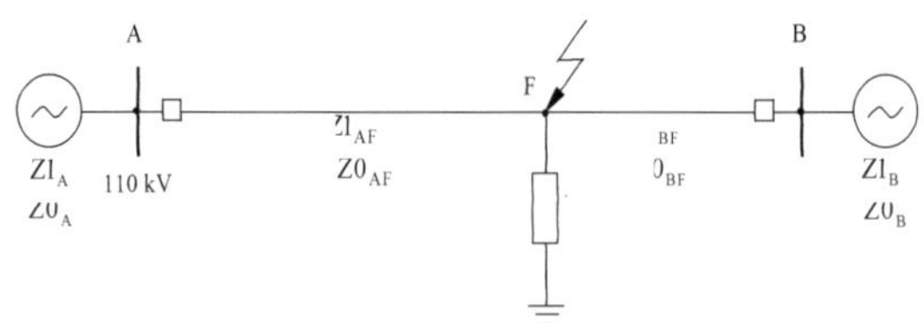

Figure 2. HVTL single diagram.

Fig. 3a and 3b show the complex impedance plane with the tripping characteristics of the DP first zone and estimations of DMU operation by the algorithm 1 and the algorithm 2 during earth fault along the line from the beginning to the end of the first zone in the absence and presence of fault resistance from 1 to 5 ohms.

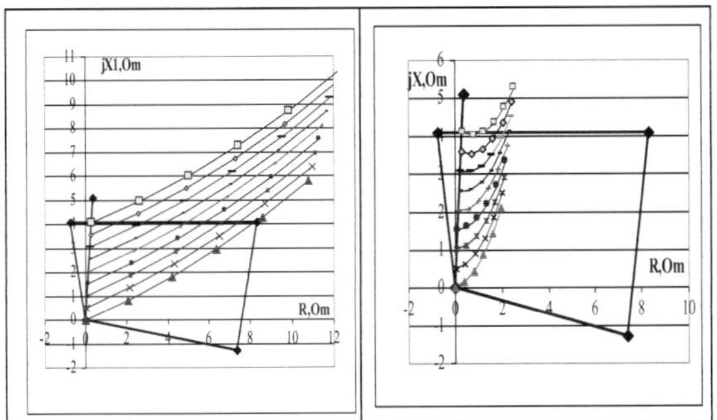

a) b)

Figure 3. Tripping characteristics of DP first zone and estimations of DMU operation by algorithm 1 - a) and algorithm 2 – b).

Both algorithms are synthesized on the base of a simplified model of the protected object for earth faults with taking into account the transient resistance at the fault location. So in the absence of the transient resistance there is practically no impedance calculation error. There are significant errors in the DMU operation by the algorithm 1 (Fig. 3a) in the cases of transient resistance presence. If its value exceeds the value of the cable active resistance to the fault location the DP may be out of operation.

Fig. 3b shows estimations of the DMU operation by the algorithm 2. It uses the separate compensation coefficients and the error of the active and reactive impedance to fault location calculation are reduced. In these cases the DP may be out of operation only at the earth faults through the big value of transient resistance at the end of first protection zone.

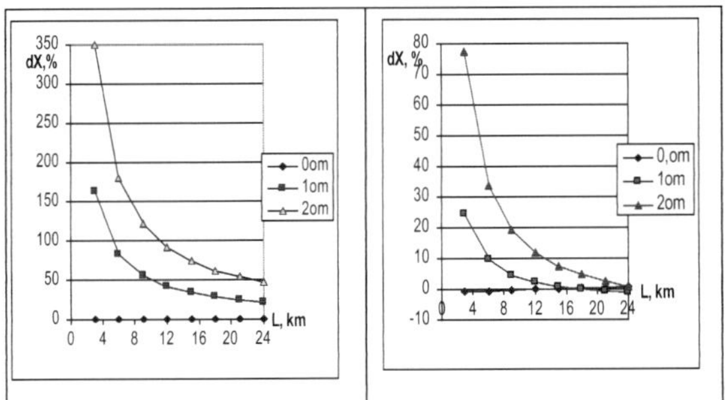

a) b)

Figure 4. The mathematical expectations of the impedance calculation error by Algorithm 1 – a) and Algorithm 2 – b).

Fig. 4a and 4b show the mathematical expectations of the calculation error of the impedance to fault location, expressed as a percentage, during earth faults along the line to the end of the first zone. In the presence of transient resistance the calculation error of the impedance to fault location for the classical algorithm 1 (Fig. 4a) is much greater than for the algorithm 2 (Fig. 4b).

In the considered cases, the errors of the algorithms 1 and 2 lead to an increase in the calculated parameters of the reactive and active resistances. This means a decrease in the sensitivity of protection and possible failure of the DP first zone.

Fig. 5 and 6 show estimations of the earth fault DMU operation by the algorithm 3. The application of optimization methods allows solving the nonlinear system of equations of the bus voltage balance at the place of protection installation. Algorithm (3) is based on Newton's method. The optimization objective function (7) contains a nonlinear voltage calculation function on the monitored buses (6), which takes into account the voltage drop on the transient resistance at the fault location.

The high voltage cable has small values of active and reactive impedance, so the presence of transient resistance in the fault location greatly affects on the impedance calculation. Fig. 5 shows the tripping characteristics plane and the values of the calculated impedance vectors, which in all cases coincide with the line impedance vector. The use of algorithm (3) reduces the probability of DP out of operation during earth faults through fault resistances and does not require the expansion of the tripping characteristic.

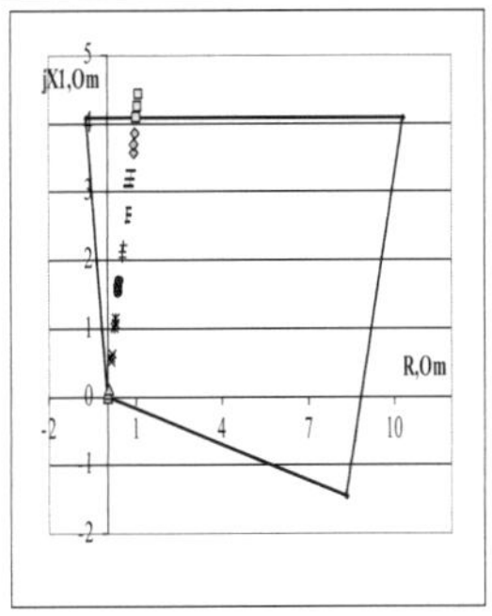

Figure 5. Tripping characteristics of DP first zone and estimations of DMU operation by algorithm 3

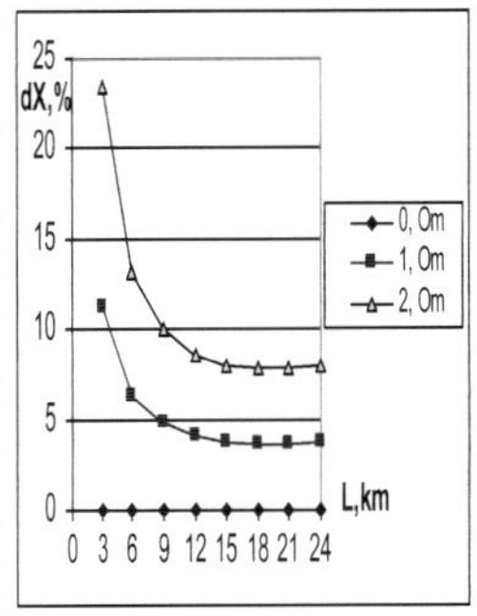

Figure 6. Mathematical expectations of impedance calculation error by algorithm 3

Fig. 6 shows the dependence of the calculation errors of the positive sequence impedance to the fault location from the distance to the fault location and the magnitude of the transient resistance.

Table 3 shows the algorithms errors of DMU operation by algorithms 1, 2 and 3 during the earth faults at the end of the first zone, since the most stringent requirements in terms of selectivity apply to the DMU operation during faults at the end of the protected zone.

Table 3. Algorithms Errors during the earth faults at the end of the first zone.

	Algorithm 1	Algorithm 2	Algorithm 3
Rf,Om	del X, %	del X, %	del X, %
0	0	0.7	0
1	21	1	3.7
2	47	2	8

VI CONCLUSION

The efficiency of DMU operation in the cable line by the algorithm 1 largely depends on the presence of transient resistance in the fault location, because it is based on a simplified model of the protected object without taking into account the transient resistance and with using a complex compensation coefficient, which is calculated by using the values of the positive and zero sequence line impedance vectors.

In the cable line, the efficiency of DMU operation by the algorithm 2 during the earth faults through the transient resistance is higher, because the algorithm 2 uses a separate compensation for the voltage drop from the zero sequence currents along the active and reactive resistances axes.

The efficiency of DMU operation by the algorithm 3 is higher than by the algorithm 1, since the objective function takes into account the voltage drop from the zero sequence currents on the transient resistance. In this case, the transient resistance becomes an optimization variable that is not related to the line active resistance. However, algorithm 3 uses a complex compensation factor, so its efficiency of DMU operation is lower than that of algorithm 2.

It is recommended to synthesize the algorithm based on the algorithm 1 and 3 in such a way that the calculated resistance value according to the algorithm 1 is used as the initial value in the optimization process of the algorithm 3. In this case, the tripping characteristic of DMU can use a narrow tripping characteristic along the active resistance axis. This is especially important in heavily loaded lines.

REFERENCES

[1] Application Guide on Protection of Complex Transmission Network Configurations, CIGRE, 1991.

[2] T. Lomane, S. Rubcovs, S. Kovalenko, K. Berziņa, "The Effectiveness Increasing of Earth Fault Distance Protection Operation for high-voltage cable line", 9th International Scientific Symposium on Electrical Power Engineering ELEKTROENERGETIKA 2017, EE2017, September 12-14, 2017, High Tatras - Stara Lesna, Slovak Republic.

[3] T. Lomane, S. Rubcovs. "Estimation of Earth Fault Distance Protection for High Voltage Mixed Network", Proceeding of the 7th International Scientific Symposium on Electrical Power Engineering ELEKTROENERGETIKA 2013, EE2013, September 16-18, 2013, High Tatras - Stara Lesna, Slovak Republic.

[4] G. Ziegler, "Numerical Distance Protection, Principles and Applications", SIEMENS, Germany, 2006

[5] V. Fabrikant, "Distance Protection", Moscow, Russia, 1978, p. 214.

[6] SIPROTEC 7A522 Distance Protection, Manual, SIEMENS AG, 1999.

[7] T. Loman, A.Voinovska, "New algorithm of distance relay", CP&HS''02, Zlin, Czech Republic, May 21-22, 2002, p.116-119.

[8] J. Izykowski, E. Rosolowski, M. Saha, "Adaptive Digital Distance Algorithm for Parallel Transmission Lines", 2003 IEEE Bologna Power Tech Conference, June 23-26, Bologna, Italy.

[9] E. Rosolowski, J. Izykowski, P. Pierz, M. M. Saha, P. Balcerek, M.Fulczyk, "Optimization of Distance Protection Algorithm for Series-Compensated Transmission Line", 2011 IEEE Trondheim PowerTech, 2011, pp. 1-7.

[10] T. Loman, V. Belugina, "The Distance Protection Effectiveness Increasing For The Tapped Lines", RTU, EN UN EL, vol. 20, part 4, Riga, Latvia, 2007, pp. 198-204.

[11] T. Loman, A. Sauhatas, A. Voinovska, "Statistical Analysis of Digital Distance Protection", paper BPT99-40-22 in Proceeding of IEEE PowerTech'99 Conf., Budapest, Hungary, 1999.

[12] REL 511, Line distance protection terminal, User's Guide, ABB Relays AB, ABB Network Control & Protection, Sweden, 1995.

Using a New Algorithm of Current Protection for Substations

Nadezhda Buryanina
Yuriy Korolyuk
Maya Koryakina
Chukotka branch of
North-Eastern Federal University
Anadyr, Russia
bns2005_56@mail.ru

Elena Lesnykh
Siberian Transport University
Novosibirsk, Russia
abbiel@mail.ru

Konstantin Suslov
Natalia Shamarova
Irkutsk National
Research Technical University
Irkutsk, Russia
dr.souslov@istu.edu

***Abstract*—Modern switches disable short circuits for two to three periods of frequency, so the requirements for algorithms implementing emergency control are very strict. Therefore, the development of algorithms capable allocate sinusoidal component during a half period and less relevant. The paper proposes equations for finding the current parametric values and their derivatives based on three samples. These equations reduce the effect of non-periodic components on how accurately their periodic counterparts are isolated from the current samples received by digital power-system protection devices. No calculation errors occur when parameters are sinusoidal. We have developed an algorithm to correct current data when the current transformers have their magnetic cores saturated.**

***Index Terms*—SmartGrid, digital power-system protection, current values of currents and the derivatives thereof, non-periodic current components, periodic-component calculation errors.**

I. Introduction

Computational devices used to control power facilities fall into two groups: those for slow (seconds-long) processes and those for fast (milliseconds-long) processes. The latter include power-system protection (PSP) and fault localization (FL), whereby faults include short circuits (SC) and open-phase faults. In case of FL, SC can be eliminated as a part of the automatic re-activation cycle; or the line can be switched off for a longer period until a repair team handles the fault. Given that SC have to be switched off within two or three frequency periods, PSP should operate within one or two periods. Thus, emergency control algorithms have to meet very stringent requirements. In such cases, we need algorithms that non-industrial-frequency currents and voltages can't interfere with. Modern digital PSP algorithms are triggered after analog or digital filtering is complete [1] – [3]. Filtering undergoes the transient-response stage, which is time-consuming. This is especially true when it comes to filtering such SC currents that contain both periodic and non-periodic components. Dev-proposed PSP provide for a protection delay for at least one period [2], [4]. The situation is exacerbated by the possible saturation of current-transformer magnetic cores. Saturation-induced errors and other errors can be reduced by means of least squares [5], [6] or by discrete Fourier transformation [7] – [11]. Cosine filters are often used in microprocessor protection [10] – [17]. Cosine filters are practically insensitive to non-periodic components. However, these methods take long to implement. This is why it is important to develop algorithms capable of isolating the sinusoidal components within half a period or less [18]. This is what our research is about.

II. Methods of Research

The transfer of relay protection to digital equipment has not yet changed the approach to the structure of relay protection. With the transfer of protection to digital equipment the protection of each object is still implemented in the same way with collecting data from their own measuring elements. Therefore, the fundamental shortcomings of protection, implemented with analog equipment, remained at the digital protection.

Let us consider the directional current protection of the power line. The first stage of protection in case of short circuit (SC) near the place of installation of protection may fail due to the low voltage value (in the limit to zero) used to determine the direction of power.

In order for the directional current protection to work correctly in case of a SC near its installation, it is proposed to replace the voltage removed from the protected object with the voltage of another object electrically connected with the protected one. Let us turn to the diagram of the part of the electrical network shown in Fig. 1. We can use the voltage U_2, U_3, and even U_4 instead of the voltage U_1. Similarly, with a SC on the side of a transformer winding with a voltage U_2, the voltage U_1 can be entered into the measuring element of directional protection of the line connecting the power system G_3 and the transformer T_1.

The above allows us to offer new directional protections as another basic protection for transformers and generators. In transformer protection, one should use the winding current of one voltage, and the voltage - windings of other voltages. Protection of a two-winding transformer will have two measuring units, protection of a three-winding – three. The voltage regulation under load and non-uniformity of current transformers are non-critical for directional protection of the transformer.

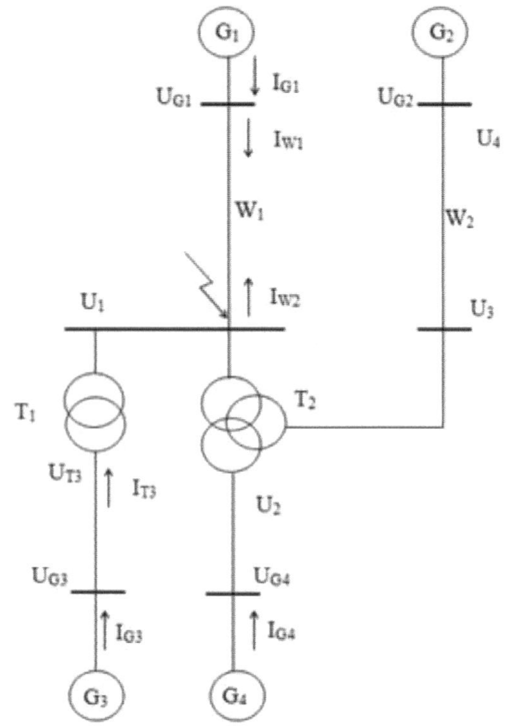

Figure 1. Power system diagram as an example of using adjacent elements mode parameters for relay protection of "their" element.

In order not to keep away from load currents, it is proposed to use voltages and currents of negative sequence. The protection will not respond to three-phase short circuits. But with the introduction of digital protection with a short circuit detection time of 1.5 - 2 ms. (milliseconds) every three-phase short circuit is first fixed as either single-phase or two-phase. Therefore, the beginning of a three-phase SCis defined by the protection as unbalanced. Therefore, the first stages of protection of main equipment should be performed as protection against asymmetrical SC.

Voltage for the directional protection of the generator can be used from the nearest equipment. For example, for directional protections of G_3 and G_4 generators, we can use voltage U_1 or U_2. The protection settings of the generator and transformer can be at the level of 10 - 12%.

Let us consider the part of the power system shown in Fig. 1. The power station with G_1 and G_2 generators is connected to 220 kV (U_1) buses through transformers T_1 and T_2. T_2 transformer has a winding of 35 kV, supplying buses of 35 kV, which are connected to the power station with G_2 generators through W_2 line. The 220 kV line connects the buses of 220 kV station to the G_1 generators. It is required to develop algorithms and select the main protections of generators, transformers, power lines and develop algorithms for determining the location of short circuits on power lines. All algorithms should be based on processing information about the currents and voltages of the fundamental frequency [19].

The currents of short circuits consist of periodic and aperiodic components. Let us consider how to eliminate aperiodic components. There are high-frequency components in the voltages during the first half period.

It is easy to eliminate high-frequency components during the first period of industrial frequency. A more serious problem is the elimination of aperiodic components from the currents.

Let us consider how to separate the periodic components of the industrial frequency from the currents for three sampling intervals.

Let us consider the method on the example of a SC current equal to

$$i(t) = \frac{E_{eq}}{\sqrt{R_{eq}^2 + X_{eq}^2}}(\sin(\omega t + \psi - \varphi) - \\ -\sin(\psi - \varphi) \cdot e^{\frac{-t}{T_N}}) \quad (1)$$

where $i(t)$ – instantaneous value of current, t – current time, E_{eq} – equivalent electromotive force, R_{eq} – equivalent active resistance, X_{eq} – equivalent inductance resistance, ω – the angular frequency, T_N – time constant equal to L/R, L and R – inductance and active resistance of the network relative to the location of the short circuit, ψ - branch EMF phase, φ – the initial phase of current.

It is proposed to calculate the current values of the currents by the equation:

$$i(\Delta t)_{cal} = \frac{2i(n \cdot \Delta t) - i((n-1) \cdot \Delta t) - i((n+1) \cdot \Delta t}{4sin^2\left(\omega \cdot \frac{\Delta t}{2}\right)} \quad (2)$$

Table 1. Current values of time currents, varying sinusoidally with an amplitude equal to one

The current values of the currents	
Calculated as sin(ωt)	*Calculated according to (2)*
0	0,19509
0,19509	0,38268
0,38268	0,55557
0,55557	0,70711
0,70711	0,83147
0,83147	0,92388

The data in the table confirm that the calculated current values of the currents correspond to the sinusoidal components of the currents, but are lagging behind them by the sampling interval. This allows us to calculate the aperiodic components of the currents as the difference between the measured value and that calculated one according to (2).

We use this property and determine the aperiodic component of the current, the time constant and, accordingly, the periodic component.

Let us calculate from (2) the actual values of the currents at times (Δt) and ($2\Delta t$) and subtract i(Δt) and i($3\Delta t$)from them respectively. Network constant will be defined as:

$$T_N = \frac{-\Delta t}{ln\left(\frac{i_{cal}(2\Delta t) - i(3\Delta t)}{i_{cal}(t) - i(2\Delta t)}\right)} \quad (3)$$

Angle φ determine through constant networking and angular frequency:

$$\varphi = arctan(T_N \cdot \omega) \quad (4)$$

The aperiodic component is calculated as:

$$i_A(t) = \left[E_{eq}/\sqrt{R_{eq}^2 + X_{eq}^2}\right][-\sin(\psi - \varphi)] \cdot e^{\frac{-t}{T_N}} \quad (5)$$

The sinusoidal component:

$$i_{sin}(t) = i(t) - i_A(t) \quad (6)$$

It is possible to replace the short transmission lines (up to 300-400 km) with the longitudinal active and reactive resistances in the calculation of short circuits. Regarding the SC point in the calculations, the line with two-way power can be divided into two independent with the source of EMF at this point, equal to the pre-emergency voltage. The time constants of lines diverging from the point of SC in different directions are determined by the ratio of active resistances to the inductances of these lines. SC current in each branch changes in time as

$$i(t) = \frac{U_{eq}}{\sqrt{R_{eq}^2 + X_{eq}^2}}(\sin(\omega t + \psi - \varphi) - \\ -\sin(\psi - \varphi) \cdot e^{\frac{-t}{T_N}}) \quad (7)$$

where U_{eq}, R_{eq}, X_{eq} – pre-accident stress in the place short circuits, active and reactive resistance EMF of a branch to the point of SC, ψ – the phase EMF of the branch, φ – the angle between the EMF of branches and current.

The sinusoidal component of the current is:

$$i_{sin}(t) = \frac{U_{eq}}{\sqrt{R_{eq}^2 + X_{eq}^2}}(\sin(\omega t + \psi - \varphi)) \quad (8)$$

The derivative of the current –

$$i'_{sin}(t) = \frac{U_{eq}}{\sqrt{R_{eq}^2 + X_{eq}^2}}(\cos(\omega t + \psi - \varphi)) \quad (9)$$

The amplitude and phase of the current are defined as:

$$I_B(t) = \sqrt{i_{sin}^2(t) + i'^2_{sin}(t)}, \\ \varphi_B(t) = arctg(i'_{sin}(t)/i'_{sin}(t)) \quad (10)$$

Active P and reactive Q power are defined as

$$P = (U \cdot I)/2 \cdot \cos(\varphi), Q = (U \cdot I)/2 \cdot \sin(\varphi) \quad (11)$$

where U – voltage, I – current.

or (the sinusoidal component of the voltage and its derivative are determined similarly to the definition of these parameters for currents).

$$S(t) = \frac{(u_{sin}(t) + u'_{sin}(t) \cdot j)}{2} \cdot \frac{(i_{sin}(t) - i'_{sin}(t) \cdot j)}{2} \quad (12)$$

Active power is equal to the real part of the complex S (t), reactive - to the imaginary part.

As an example, let us consider the selection of sinusoidal component of the SC current given by the equation

According to (2), (3), we find T = 0.01 s., and using (4) we get ϕ = 72,3° . According to (10) we have the amplitude value of the current I = 2 units. Fig. 2 shows dependences of the currents: 1 — at the input of the measuring element, 2 — the aperiodic component calculated according to (5), 3 — the periodic component obtained according to (6), 3 — the amplitude calculated by means of (10).

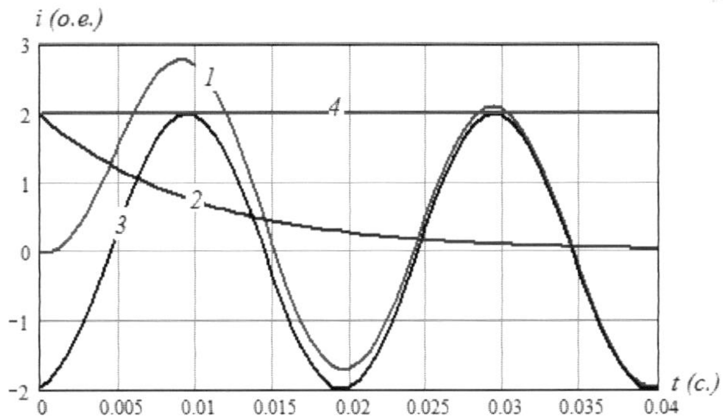

Figure 2. Currents at the input and output of the measuring element. (1 - at the input; 2 - aperiodic component at the output; 3 - periodic component at the output; 4 - amplitude value of the periodic component at the output).

In fig. 3. active and reactive powers are given according to (12) provided that the voltage is sinusoidal, and the active and reactive components of the currents are calculated according to (2).

The advantage of the four-sampling method is that only four values of instantaneous currents and voltages are recorded, separated by three sampling intervals. With a sampling interval of 0.625 ms. in less than two milliseconds, we obtain sinusoidal components of currents and voltages at any time, which frees digital protections from constantly calculating the necessary information in the accident process. Four samples are enough to implement any protections that respond to the periodic components of the mode parameters.

Active and reactive powers calculated: 5 and 7 - according to (12); 6 and 8 - the voltage is sinusoidal, and the active and reactive components of the currents are calculated according to (2).

The given power dependencies on the information processing method used show that errors can reach 100% or more over the time interval until half a period from the moment of occurrence of a short circuit, which can lead to false protection operation.

Given the fact that the four-sampling method allows us to calculate the mode parameters in "future" time, we can determine the instantaneous currents of the direct and inverse sequences at any time, and determine their complex values through the instantaneous values. The instantaneous values of the direct and inverse sequence currents can be calculated via phase currents without zero components as [17]:

$$i_1(t) = [i_A(\omega \cdot t) + +(i_B\left(\omega \cdot t + \frac{\pi}{2}\right) - i_C\left(\omega \cdot t + \frac{\pi}{2}\right))/\sqrt{3}]/2 \quad (13)$$

$$i_2(t) = [i_A(\omega \cdot t) - -(i_B\left(\omega \cdot t + \frac{\pi}{2}\right) - i_C\left(\omega \cdot t + \frac{\pi}{2}\right))/\sqrt{3}]/2 \quad (14)$$

We get the same results if we use the equations:

$$i_1(t) = \begin{matrix} ([i_A(\omega \cdot t) + i_B\left(\omega \cdot t + \frac{2\pi}{3}\right)]/3) + \\ + ([i_C\left(\omega \cdot t - \frac{2\pi}{3}\right)]/3) \end{matrix} \quad (15)$$

$$i_2(t) = ([i_A(\omega \cdot t) + i_C\left(\omega \cdot t + \frac{2\pi}{3}\right)]/3) + +([i_B(\omega \cdot t - \frac{2\pi}{3})]/3) \quad (16)$$

Only sinusoidal components of phase SC currents are used in equations (13), (14), (15), (16).

After calculating the mode parameters by the method of four samples, the angular frequency ω is present in all equations. Since the frequency in the network can vary, and the Δt interval is constant, the angular frequency needs to be periodically adjusted. For example, after 0.1 s.

By selecting the sinusoidal components of currents and voltages in the "end" elements of electrical networks (in the diagram of Fig. 1, the end elements are power systems, replaced by G_1, – G_4 generators), we can get the parameters of the equivalent circuit of these sequences during a short circuit. This will allow the automation to make decisions promptly at the pace of the process.

Equations (2) and (5-14) and others, respectively, can be used for the needs of automation, which turns on, as a rule, after the attenuation of aperiodic components. The operating time of the automation is from fractions of a second to seconds. Therefore, in equations (2) it is proposed to take the Δt interval equal to a quarter of the period of the industrial frequency, but at the same time use five samples [18].

The sampling interval of a quarter of the frequency period can be obtained by fixing the zero-voltage transitions on the active and inductive resistances of the RC – link, as shown in Fig. 4 [20].

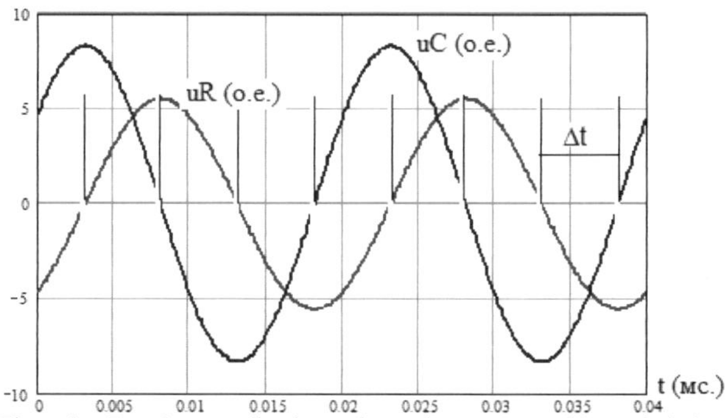

Figure 3. Ensuring sampling intervals equal to a quarter of the period of the network frequency.

III. Conclusions

The method that makes it possible to isolate the periodic components of short-circuit currents from currents equal to the sum of the periodic and aperiodic components has been developed.

The algorithm that implements the method uses a total of four samples, separated by three sampling intervals, which allows not to take into account the possibility of saturation of the magnetic circuits of current transformers, collecting information in areas of sufficiently accurate transformation.

The method allows to calculate currents of the direct, inverse sequences and any combinations thereof, resistance, power used in the relay protection and emergency control at any time.

Directional protection is proposed for generators and transformers. To exclude "dead" zones with close three-phase short circuits, it is proposed to use voltages of adjacent elements of the electrical network that are connected to protected objects.

Transferring the method to reading information for the operation of automation allows us to increase the sampling interval to a quarter of the period of industrial frequency.

ACKNOWLEDGMENT

The study was carried out with the financial support of the Russian Foundation for Basic Research and the Subject of the Russian Federation - the Republic of Sakha (Yakutia) № 18-48-140 010.

REFERENCES

[1] L.A. Trujillo Guajardo, "Prony filter vs conventional filters for distance protection relays: An evaluation." Electric Power Systems Research, vol. 137, pp. 163-174, Aug. 2016, ISSN: 0378-7796

[2] W. Rebizant, J. Szafran, and A. Wiszniewski, Digital Signal Processing in Power System Protection and Control. London: Springer, 2011, ISBN 978-0-85729-802-7

[3] J. A. de La O., "New family of digital filters for phasor computation [power system relay protection]." IEEE Transl. on Power Delivery, vol. 15, pp 86-91, 2000, ISSN: 1937-4208

[4] A. G. Phadke, and J. S. Thorp, Computer Relaying for Power Systems. 2 edition. Chichester: John Wiley & Sons, 2009.

[5] D. Barbosa, R. M. Monaro, D.V. Coury, and M. Oleskovicz, "Digital frequency relaying based on the modified least mean square method." International Journal of Electrical Power & Energy Systems, vol. 32, pp 236-242, March 2010.

[6] M. S. Sachdev, and M. A. Baribeau, "A New Algorithm for Digital Impedance Relays," IEEE Transactions on Power Apparatus and Systems, vol 6, pp. 2232-2240, 1979.

[7] S.R. Samantaray, and P.K. Dash. "Transmission line distance relaying using a variable window short-time Fourier transform," Electric Power Systems Research, vol. 78, Issue 4, pp. 595-604, April 2008.

[8] M. Monadi, M. A. Zamani, J. I. Candela, A. Luna, and P. Rodriguez, "Protection of AC and DC distribution systems Embedding distributed energy resources: A comparative review and analysis," Renewable and Sustainable Energy Reviews, vol. 51, pp. 1578-1593, November 2015.

[9] A. G. Phadke, and J. S., Thorp, Synchronized Phasor Measurements and their Applications. London: Springer Science & Business Media, 2008.

[10] B. Grcar, J. Ritonja, and B. Polajzer, "Estimation methods using dynamic phasors for numerical distance protection," in Proc. 2008 IEEE International Conference Industrial Technology, ICIT, 2008, pp. 1-6.

[11] K.V.Suslov, V.S.Stepanov, and N.N.Solonina, "Smart grid: Effect of high harmonics on electricity consumers in distribution networks," in Proc. 2013 IEEE International Symposium on Electromagnetic Compatibilit, 2013

[12] H. O. Pascual, and J. A. Rapallini, "Behaviour of Fourier, cosine and sine filtering algorithms for distance protection, under severe saturation of the current magnetic transformer," in Proc. IEEE Porto Power Tech, 2001.

[13] E. O. Schweitzer III, and D. Hou, "Filtering for Protective Relays," in Proc. IEEE Conference on Communications, Computers and Power in the Modern Environment, WESCA-NEX 93, pp. 15-23, 1993.

[14] M. B. Duong, Shi-Lin Chen, Keng-Yu Lien, Yung-Ruei Chang, Yih-Der Lee, and Jheng-Lun Jiang, "Investigation on transient behaviours of a uni-grounded low-voltage AC microgrid and evaluation on its available faultprotection methods: Review and proposals." Renewable and Sustainable Energy Reviews, vol. 75, pp. 1417-1452, August 2017.

[15] B. J. Brearley, and R. R.Prabu, "A review on issues and approaches for microgrid protection.". Renewable and Sustainable Energy Reviews, vol. 67, pp. 988-997, January 2017

[16] S.A. Hosseini, H.A.Abyaneh, S.H.H.Sadeghi, F.Razavi, and A.Nasiri, "An overview of microgrid protection methods and the factors involved." Renewable and Sustainable Energy Reviews, vol. 64, pp. 174-186, October 2016

[17] M. Singh, and P.Basak, "Fractionalization of microgrid protection system through detection of zero sequence component of fault current," in Proc. 7th India International Conference on Power Electronics (IICPE), 2016.

[18] E.V. Greshnov, and Y.F. Korolyuk, "On the choice of the sampling interval of analog parameters entered into the computer for programmable protections." Electronic modeling, vol. 5, pp. 71-74, 1982.

[19] N. Buryanina, Y. Korolyuk and E. Lesnyh, "Method for determining the instantaneous values of the parameters (currents and voltages) of the forward and reverse sequences," Russia Patent 2016137597, Sept. 20, 2016.

[20] N. Buryanina, J. Koroluk, M. Koryakina, E. Lesnykh, and K. Suslov, "Determination of the Place of Short-Circuit on Power Lines with a Simultaneous Rupture of a Phase," in Proc. 2018 International Multi-Conference on Industrial Engineering and Modern Technologies, FarEastCon, 2018.

Actual Possibilities of Open-Source Applications for Analysis of Power Systems

Karel Noháč, Lucie Noháčová, Václav Mužík
Department of electrical power and environmental engineering
University of West Bohemia
Pilsen, Czech Republic
nohac@kee.zcu.cz; nohacova@kee.zcu.cz; vmuzik@kee.zcu.cz

Abstract — **The paper recapitulates the current state and the assumed near future view of the possibility using open-source free software for complex simulation analysis of power systems. The declared general properties and capabilities of significant representative current tools for modeling complex power grid systems in both steady state and dynamic behavior analysis will be presented. Additionally the contribution of article authors for the development of conceptually differently targeted instruments will be briefly presented.**

Besides the public available information, especially authors' practical creative experiences in solving practical and research tasks were used for objective comparison of applications.

Keywords: power systems analysis; software application; open source

I. Introduction

Computational analyzes of energy systems are now an integral part of the routine work of not only a scientific specialist but also of many practical engineering activities. With modeling and simulation of electrical grids is dealing a relatively large number of tools in the commercial and also freely available categories. Some of the second one's are also called "open source", which means not only that the use of such applications is free of charge in non-profit and commercial sphere, but also that the user custom tools and existing support libraries are available for viewing and possible changes, including all source code. The concept of using free and open tools has wide support not only in terms of cost reductions but also with regard to the need to perform routine but above-standard analyzes by a wide range of technicians and the need to share and compare obtained results using maximum standardization of input and output data and component solution. We can observe this trend both at the level of transmission system tasks and local problems in distribution systems.

There are many reference lists of tools capable solving power systems electrical grid problems, for example, the "Power Systems Analysis Software" on the "Open Electrical" [1] or "Open Source Software (OSS) for Electricity Market Research, Teaching, and Training" [2].

In the context of solving practical application tasks, the authors performed many practical comparing of capabilities and performance of particular software in steady state and dynamic analysis categories. The essential part of the article is therefore an analysis of the experience gained and the evaluation of the qualities of selected tools.

The conclusion of the paper is a summary of the potential benefits of the wide application of modern free tools for simulation analyzes in electrical power engineering with regard to technical practice and scientific research.

II. Overview of open source tools

According to relative large number of existing open source software (OOS) application tools for power engineering authors divided them according to the platform used. The spectrum of possible candidates become limited, especially when not only the application itself, but also the platform and operation system on which the tool is capable to function should be at least completely free or better open source. As a result, an alternative option to use the LINUX-based operating system is also required.

Applications were created mostly according to their specific purpose. Although this article focuses only on tools analyzing electrical networks, there exists, of course, also many other software for particular problems. Regardless of the limited view, it is possible to distinguish between subcategories based on the tasks that the applications are capable able to solve:

- Power Flow (PF), also known as Load Flow (LF), is analysis of power system steady-state operating characteristics for a given load, generator power and voltage conditions. Once we have this information, we can calculate easily real and reactive power flow in all branches together with power losses.
- Continuation Power Flow (CPF) is extension of PF suited for voltage stability assessment. It is a cyclic calculation of PF and gradual overloading of the network according to a certain algorithm and strategy.
- Optimal Power Flow (OPF) is again based on PF. It calculates parametric optimization of a certain parameter, usually of an economic nature.
- Small Signal Stability Analysis (SSSA), also known as Static Stability (SS), is analysis which determines ability of the system stay under current conditions with constant system variables.
- Time Domain Simulation (TDS) explores course of all important system parameter in a specific time range. It is usually investigation of electromechanical processes as opposed to previous cases, where stabilized electromagnetic systems have been observed.

- Electro-Magnetic Transients analysis (EMT) is analyzing electromagnetic systems if time detail.
- Short Circuit Calculation (SCC) is investigation of short circuit current size at fault state.
- Graphic User Interface (GUI) is a form of user interface that allows users to interact with computing application through graphical icons and other visual indicators instead of text-based user interfaces or command-line interface.
- Computer Aided Design (CAD) is understood in context of applications for power engineering as a graphical input for entering electrical network topology and parameters of each components used.

Next, the basic characteristics of interesting representative applications will be described, according to their guest platforms:

A. *MATLAB, respectively GNU Octave platform*

Most of the contributions to the family of applications were built on the very popular MATLAB [3] platform, respectively on open source GNU Octave [4] platform. Although MATLAB is a purely commercial product, over time, its free open source alternative GNU Octave has become enough capable of running some of tools originally created for MATLAB on this independent open source platform. The best-known and most significant representatives are: MATPOWER [5], Power System Analysis Toolbox (PSAT) [6], MatDyn [7] and DiSC [8]. While MATPOWER is just a package of MATLAB-language M-files for solving steady-state simulation and optimization problems such as power flow (PF), continuation power flow (CPF) and extensible optimal power flow (OPF), the others are also capable of some additional functionalities. DiSC concerns on verifying voltage control approaches in European power distribution systems, MatDyn on dynamic analysis and PSAT on small signal stability analysis (SSSA), time domain simulations (TD), WAMS placement, contingency analysis and includes sophisticated special models of components.

Table 1. Features of MATLAB tools running on GNU Octave

Capability	Application			
	MATPOWER	*PSAT on GNU Octave*	*MatDyn*	*DiSC*
PF	♦	♦	♦	♦
CPF	♦	♦		
OPF	♦	♦		♦
SSSA		♦		
TDS		♦	♦	♦
EMT				
SSC				
GUI				
CAD				

B. *Python platform*

Python [9] is one of the most progressively evolving modern object programming languages. It is implemented on almost all operating systems and, thanks to its extremely sophisticated range of expansion libraries, it also creates a platform for technical and scientific computing and efficient application development. Therefore many new projects are moving from the previous platform to this one because of the advanced features of Python, of application type versatility and finally also of its performance. Thanks to the rapid development, the situation here is less clear than the previous platform. Many projects arise and disappear. However, the most representative examples are: PYPOWER [10], DOME [11], GridCal [12], PyPSA [13], Pylon, PandaPower, Minpower, Mosaik and PowerGAMA. While PYPOWER is simple port of MATPOWER to Python, Mosaic extends it for smart grid scenarios solution, GridCal adds many features (time series, blackout, three-phase short circuit calculation, numeric method options and optimization, beautiful GUI, data import and export in many formats). DOME is a direct follower of PSAT and contains improved all its computational capabilities, but without a GUI. Power GAMA is just a lightweight simulation tool of electric grid and market for high level analyses of renewable energy integration in large power systems. PyPSA is a very ambitious project which, apart from static power flow also includes linear optimal power flow, total energy system least-cost investment optimization and strong graphical output presentation features, unfortunately is still in low development phase.

Table 2. Features of tools using Python language and libraries

Capability	Application			
	PYPOWER	*DOME*	*GridCal*	*PyPSA*
PF	♦	♦	♦	♦
CPF	♦	♦		
OPF	♦	♦		♦
SSSA		♦		
TDS		♦	♦	♦
EMT				
SSC			♦	
GUI			♦	♦
CAD			♦	

C. *Modelica platform*

Modelica [14] is a non-proprietary, object-oriented, equation based programming and simulation language to conveniently model complex physical systems containing, e.g., mechanical, electrical, electronic, hydraulic, thermal, control, electric power or process-oriented subcomponents. Modelica introduces simulation environment and libraries with a large set of models from many providers. Some of these environments and libraries are open source and power system oriented with different approach to solution. Many of them are currently under development, the most popular of which are: ObjectStab [15], Spot [16], PowerSystems [17], iTesla Power System Library - iPSL [18] and Open-Instance Power System Library - OpenIPSL [19]. All these libraries are now natural part of standard installation package of mostly used Modelica implementation OpenModelica [20]. Focus of each library is different. The ObjectStab package is library for power systems voltage and transient stability simulations, voltages and currents are assumed to be sinusoidal and symmetrical. Spot library is intended to treat transient simulation and simulation of large systems in full generality within one common framework and to bring the power-flow approximation closer to the general case. PowerSystems library was created to model electrical power systems even more general at different levels of detail both in transient and steady-state mode and goes over level of Spot. The iPSL library contains a set of power system component models for phasor time domain simulations and use experiences gained by some proven commercial applications and PSAT mentioned earlier. This gives a high level of computing power, comfort and reliability to this library and its successor OpenIPSL.

As mentioned, the Modelica language platform is implemented in many products. The most important open source are OpenModelica, JModelica.org [21] and SimulationX [22]. With very few corrections all libraries of power systems are compatible to all these platforms, although they were originally mostly optimized just for OpenModelica or some commercial Modelica language platform. Each Modelica implementation has its own user interface. OpenModelica with GUI OMEdit and activated OpenIPSL library shows Fig. 1.

Table 3. Features of tools created for Modelica platform

Capability	Application			
	ObjectStab	*Spot*	*PowerSystems*	*iPSL/ OpenIPSL*
PF				
CPF				
OPF				
SSSA				
TDS	♦	♦	♦	♦
EMT	♦	♦	♦	
SSC	♦	♦	♦	♦
GUI	♦	♦	♦	♦
CAD	♦	♦	♦	♦

D. Other and special platforms

In the past also other platforms were used. For example the predecessor of iPSL – Pegase library was based on Scilab/Xcos. DynastPowerLibrary [23] on Dynast [24] etc. There is growing number of tools based on Java platform (DCOPFJ, InterPSS, RAMSES), on Java/Python cooperation (AMES) or are pure web-based (RPowerLABS). Many tools were written from scratch in non-high level modeling platform (GridPACK, OpenDSS, SmartGridToolbox) or are using combination of low level programming language and some of mentioned platforms (Dynawo [25]). Such products can be very sophisticated in terms of performance optimization, but they seem to be unpromising in the present-day requirements if we consider the ability to transparently exchange data, models and even processing methods across products and platforms much more important. Only combination of low and high level approach can be successful in the future, but such projects are still in very early stage of development.

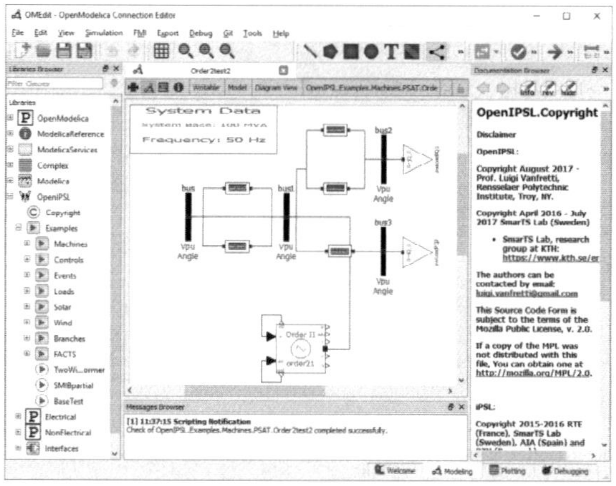

Figure 1. Example of OpenModelica's environment OMedit and using of OpenIPSL library

III. Contribution of Article Authors

Some contributions to the family of tools for analyzing power systems have been developed at the authors' department and outweighed by their participation or leadership. Although the academic and scientific application has always been a fundamental goal, focus and platforms of these applications are quite different:

- SimEPS - Steady-State Simulations of Electric Power Systems [26] is complete MATLAB application for power flow problems solution with many additions and improvements. It is mainly focused on load flow analysis, voltage stability analysis (with var limits) and shortest distance to voltage instability analysis of power systems with var limits and OLTC transformers. For this purpose, optimized and sparsely programmed iterative load flow methods and other codes were designed. It is not intended as open source, but computation core can be ported to GNU Octave and experiences gained by its development were important for next projects. GUI design of this application illustrates Fig. 2.

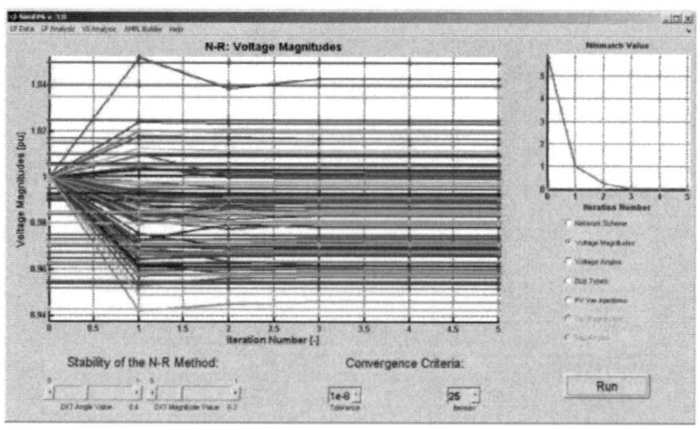

Figure 2. Example of SimEPS - Load flow analysis using the N-R method - voltage magnitudes of the IEEE 118-bus system

- MVA Schematics [27] is web based very fast power flow analyzing tool with comfortable GUI and Python computing core. Again it is not fully open source, but it documents a wide range of diverse approaches to the solution that the authors have explored. Example of user interface shows Fig. 3.

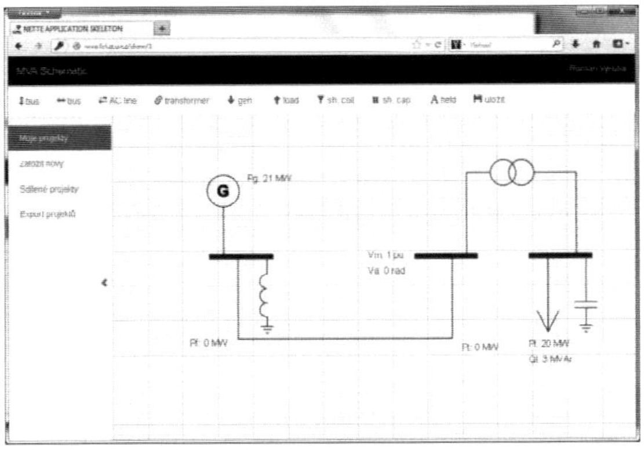

Figure 3. Example of MVA Schematics – Building of simple test grid with web user interface

- DPL - Dynast Power Library is collection of optimized power system components for detailed transient analysis created by using modelling platform DYNAST. Two years ago, DYNAST, a former fully commercial traditional modeling tool, has become completely free and open source. According to concept of DPL, compute with instantaneous values instead of phasor of efficient values, is DPL suitable especially for analyzing fast electromagnetic transients in limited electrical network topology. Example of DYNAST modeling environment and using of Dynast Power Library shows Fig 4.

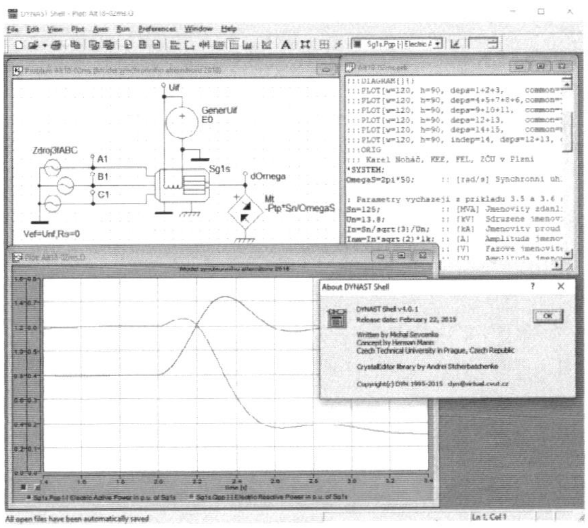

Figure 4. Example of DYNAST and Dynast Power Library

- PSAT-Octave-GUI is a new graphic environment and expansion of PSAT. Due to undeniable computational qualities, a wide range of possible solved tasks and a long-term proven reliability the PSAT tool has become author's point of interest. For full user friendly use in the open source mode, it was necessary to add a graphical interface to the existing computational core for GNU Octave system as well, since the original version has GUI only for MATLAB platform implemented. Developed application in Lazarus programming environment runs on both Windows and Linux operational systems. Overview of user interface and output graphs shows Fig 5.

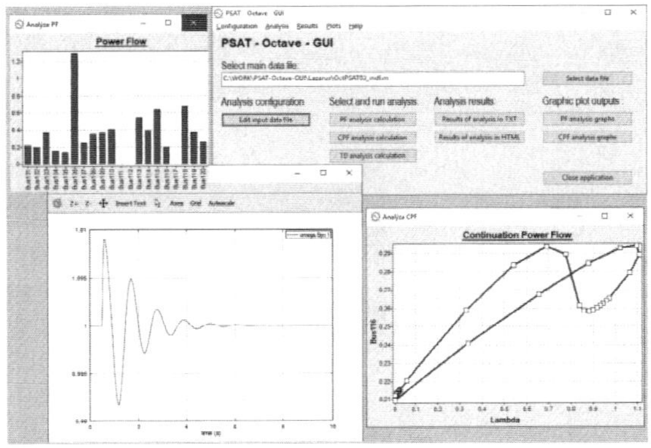

Figure 5. Example of PSAT-Octave-GUI – Overview of user interface and output graphs

Table 4. Features of tools with some contribution of authors

Capability	Application			
	SimEPS	*MVA Schematics*	*DPL*	*PSAT-Octave-GUI*
PF	♦	♦		♦
CPF	♦			♦
OPF				♦
SSSA				♦
TDS			♦	♦
EMT			♦	
SSC			♦	
GUI	♦	♦	♦	♦
CAD	♦	♦	♦	

IV. COMPARISON OF APPLICATIONS FEATURES

In the previous text, the applications were divided with regard to their specific purpose. It corresponds to the fact, that traditionally, power systems tools have always treat differently problems depending on the size. Whilst smaller systems were investigated using general transient methods, larger systems were restricted just to PF approximation. So different simulation tools were used in order their necessary different methods. The main difference was usually representation of harmonic parameters. For large system investigated in long time range just effective values with phase shifts (phasors) were used. On the contrary, the instantaneous parameter values were used for compact systems with the necessary detailed analysis in a short time range. So the first approach is suitable for analysis of type PF, CPF, OPF, SSSA and slow version TDS and SCC. Second approach is necessary suitable for analysis EMT and detailed TDS plus SCC. New applications and libraries promise common approach for both cases. This is especially offered by new packages created for Modelica platform. The legitimacy of such a solution in terms of reliability and performance was examined. For testing, the traditional problem stability of a simple alternator operating in a transmission network, which links the problem of PF and slow TDS with the ambition to use an EMT-capable tool, was chosen. Therefore, for the direct comparison, the very different tools PSAT, DPL and OpenIPSL were chosen. Detailed description of this SMIB (Single-Machine Infinite-Bus) reference problem is in [28]. Following pictures show different problem implementation in platform CAD:

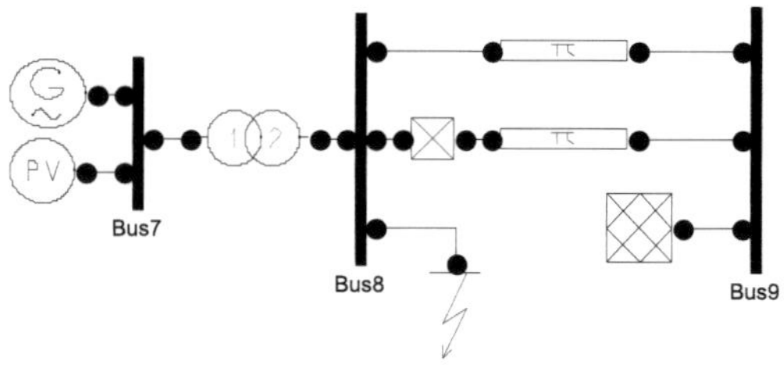

Figure 6. SMIB problem implemented in PSAT tool

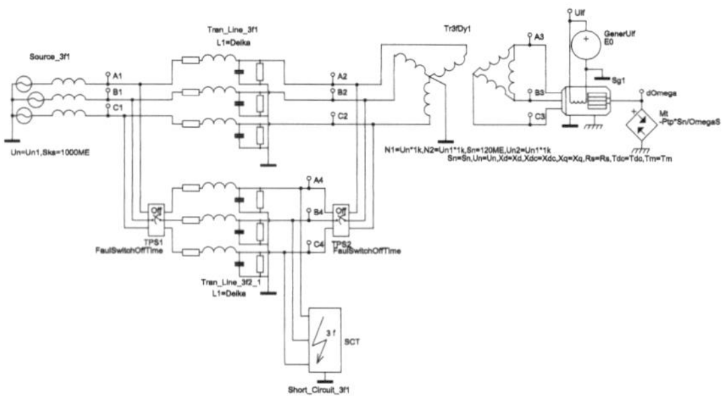

Figure 7. SMIB problem implemented in DPL library

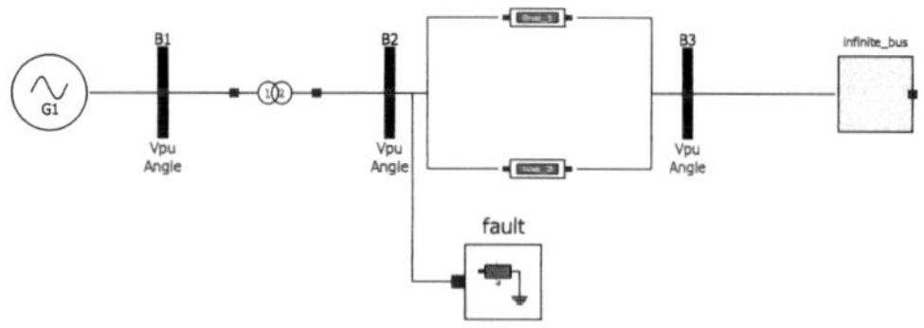

Figure 8. SMIB problem implemented in OpenIPSL librrary

Although the outputs of all tested applications have proven to be almost the same, there is a noticeable difference between phasor applications and applications using instantaneous parameters. Differences within the first category were quite negligible, on the other hand much more significant were differences versus the second category. The differences were most often due to other type of component models implementation, especially of alternators. All this clearly show following pictures:

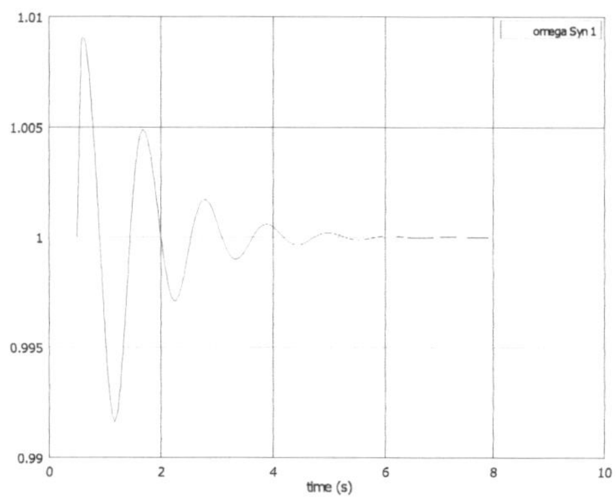

Figure 9. Angular velocity in p.u. of synchronous machine – output of PSAT and OpenIPSL

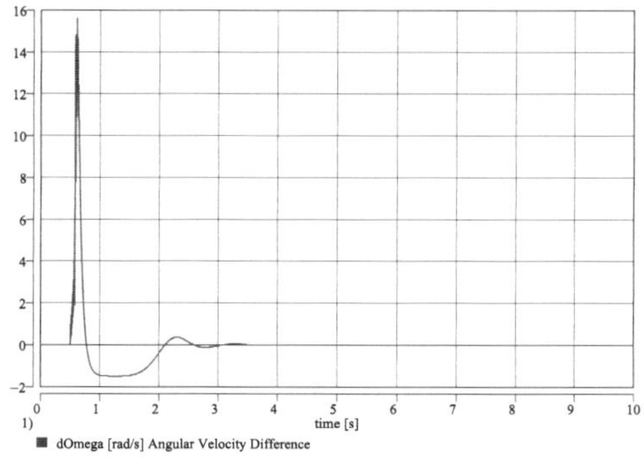

Figure 10. Angular velocity difference in rad/s of synchronous machine – output of DPL on DYNAST

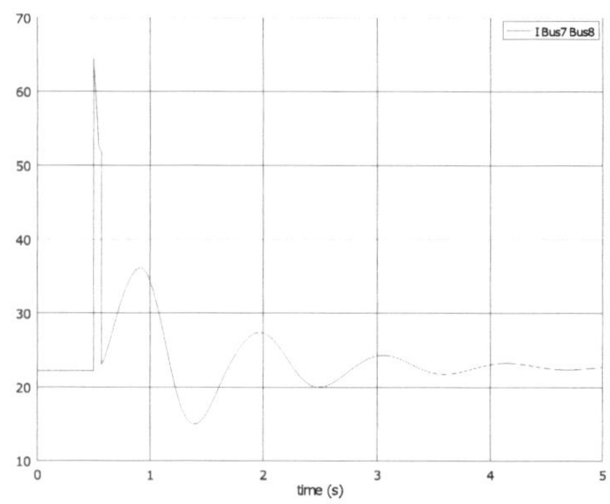

Figure 11. Current from Bus 7 to Bus 8 in A – output of PSAT and OpenIPSL

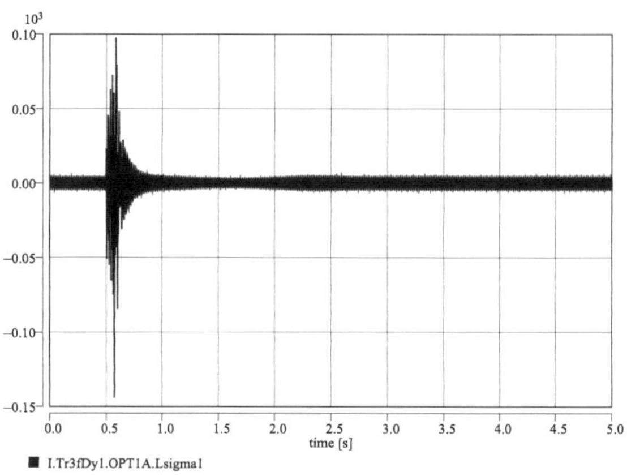

Figure 12. Current from Bus 7 to Bus 8 in A – output of of DPL on DYNAST

However, the reliability, accuracy and speed of the results obtained, user friendly interface and other traditional attributes are not the only features that are currently being investigated. Very important to support use of open source products are also other capabilities mainly related to collaboration between users working on different platforms and with different tools:

- Easy of standard data import describing power system.
- Model definition exchange according to standards Functional Mock-up Interface (FMI) and Functional Mock-up Unit (FMU) [29].

- Common modeling language and principle of modeling for all power system problems categories.
- Support of modern possibilities of computing power improvements such as cloud computing, multiprocessor and big data systems.

PSAT tool was clearly the most appropriate in point of view ability to retrieve data in different standard formats. On the other hand, libraries based on the Modelica platform excelled in unity of approach and openness for both creation of new models, their standardization for exchange and optimization for new computing technologies.

V. Conclusion

New progressive features of open source applications may be the basis for the gradual enforcement of these instruments in the future. Especially the user-friendliness can be crucial for the decision of instrument choice for using. The importance of model interoperability will increase strongly according to need for power engineering analyzes with cooperation of connected disciplines. Universal open source tools are still lagging behind in the performance assessment and accuracy versus specialized professional applications with long tradition and, so they still have long development process ahead. This applies especially in necessary stabilization of standards, as the rapid development suppresses model compatibility transparency and user orientation across versions is not easy.

Acknowledgment

This paper was funded and supported by the student research project SGS-2018-023.

References

[1] Open Electrical "Power Systems Analysis Software." Internet: wiki.openelectrical.org/index.php?title=Power_Systems_Analysis_Software, Jun. 15, 2018 [Apr. 10, 2019].

[2] "Open Source Software (OSS) for Electricity Market Research, Teaching, and Training." Internet: www2.econ.iastate.edu/tesfatsi/ElectricOSS.htm, Jan. 1, 2019 [Apr. 10, 2019].

[3] MathWorks "MATLAB." Internet: www.mathworks.com/products/matlab.html, [Apr. 10, 2019].

[4] GNU Octave, Gnu.org, 2019. Internet: www.gnu.org/software/octave, [Apr. 10, 2019].

[5] Ray D. Zimmerman, Carlos E. Murillo-Sánchez & others "MATPOWER.". Internet: www.pserc.cornell.edu/matpower. Feb 21, 2019 [Apr. 10, 2019].

[6] Power System Analysis Toolbox (PSAT), Faraday1.ucd.ie, 2019. Internet: faraday1.ucd.ie/psat.html, [Apr. 10, 2019].

[7] MatDyn, Esat.kuleuven.be, 2019. Internet: www.esat.kuleuven.be/electa/teaching/matdyn, [Apr. 10, 2019].

[8] DiSC - A Simulation Framework for Distribution System Voltage Control." Internet: kom.aau.dk/project/SmartGridControl/DiSC. [Apr. 10, 2019].

[9] "Python.". Internet: www.python.org, Apr 12, 2019 [Apr. 20, 2019].

[10] "PYPOWER.". Internet: github.com/rwl/PYPOWER. [Apr. 10, 2019].

[11] "DOME.". Internet: faraday1.ucd.ie/dome.html. Feb 21, 2019 [Apr. 10, 2019].

[12] "GridCal.". Internet: github.com/SanPen/GridCal. Mar 20, 2019 [Apr. 20, 2019].

[13] T. Brown, J. Hörsch and D. Schlachtberger. (2018) "PyPSA: Python for Power System Analysis", *Journal of Open Research Software*, [Online]. 6(1), Available: doi.org/10.5334/jors.188 [Apr. 20, 2019].

[14] "The Modelica Association.". Internet: www.modelica.org. [Apr. 20, 2019].

[15] "ObjectStab Library.". Internet: github.com/modelica-3rdparty/ObjectStab. Sep 23 2015 [Apr. 20, 2019].

[16] "Spot.". Internet: github.com/modelica-3rdparty/Spot. Dec 16 2015 [Apr. 20, 2019].

[17] "PowerSystems.". Internet: github.com/modelica/PowerSystems. Dec 16 2015 [Apr. 20, 2019].

[18] "iPSL: iTesla Power System Library.". Internet: github.com/itesla/ipsl. May 30 2017 [Apr. 20, 2019].

[19] "OpenIPSL: Open-Instance Power System Library.". Internet: github.com/OpenIPSL/OpenIPSL. Mar 14 2019 [Apr. 20, 2019].

[20] "OpenModelica.". Internet: openmodelica.org. Jan 31 2019 [Apr. 20, 2019].

[21] "JModelica.org.". Internet: jmodelica.org. [Apr. 20, 2019].

[22] "SimulationX.". Internet: www.simulationx.com. [Apr. 20, 2019].

[23] L. Nohacová and K. Nohac "New models for Power Engineering of simulation tool DYNAST." in *Proc. 13th International Scientific Conference – Electric Power Engineering 2012*, 2012, pp. 201-204.

[24] L. Nohacova and K. Nohac. "Dynast – Computer Program used in Power Engineering", *Applied Mechanics and Materials*, vols. 284-287, pp. 1035-1038, 2013.

[25] "Dynaωo - An hybrid C++/Modelica time-domain simulation tool for power systems.". Internet: github.com/dynawo/dynawo. Apr. 20, 2019 [Apr. 20, 2019].

[26] Nohac K., Sitar V. and J. Veleba. "Load flow analysis in alternative tools using equations, physical and block diagrams." in *Proc. 13th International Scientific Conference – Electric Power Engineering 2016*, 2016, pp. 11-16.

[27] L. Nohacova, M. Tesarova, R. Vykuka and K. Nohac. " Use of WAMS-BIOZE-Detector and MVA-Schematic applications in networks with connected RES", *Energetika*, vol. 65, pp. 627-630, 2015.

[28] P. Kundur. "Numerical integration methods," in *Power System Stability and Control*, N. J. Balu, M. G. Lauby Ed. New York: McGraw-Hill, 1994, pp.836-848.

[29] "What is FMI.". Internet: www.modelon.com/what-is-fmi, 2019 [Apr. 20, 2019].

Approaches and Experience in Testing of Digital Substation Primary Equipment and Relay Protection

Alexander Yu. Khrennikov
Moscow Energy Institute (Scientific Research University)
Moscow, Russia
ak2390@inbox.ru

Nikolay M. Aleksandrov
SPE "Dynamics"
Cheboksary, Russia
nickdynamics@gmail.com

Abstract — **It is necessary to take a new look on testing of the equipment due to active implementation of digital devices in power utilities, transmission and distribution stations and substations. This article describes new approaches for testing of primary equipment such as non-conventional CTs and PTs, CBs, as well as modern digital protection devices (IED), merging units and other equipment installed in the digital substations. Also the experiences and real cases of testing of IEC 61850 supported devices is presented, because nowadays there are a lot of new digital substations are being implemented in Russia.**

Keywords: testing, IEC 61850, IED, conventional CT, nonconventional CT,

I. Introduction

Currently, the electric power complex of the Russian Federation is experiencing a significant increase in the number of digital substations using the IEC 61850 standard. A few years ago these were either pilot projects or partial implementation of separate functions, now in many cases all levels of a digital substation are implemented (station level, bay level and process level) with station buss and process bus (fig 1).

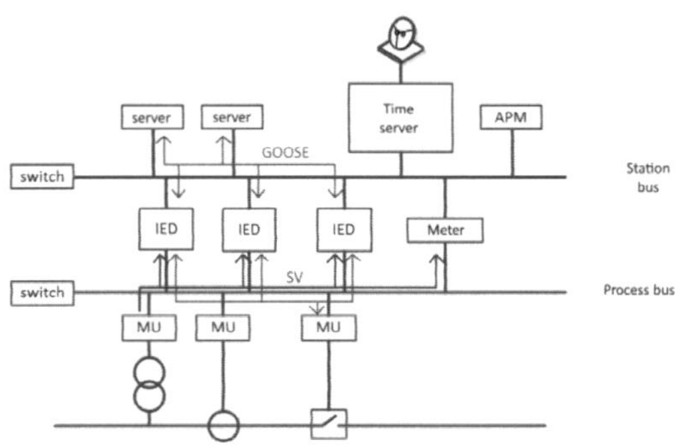

Figure 1. Digital substation

This article describes the approaches to testing equipment installed at the bay and station level.

II. IED Testing

In the first edition of the IEC 61850 standard there was no answer to the question of checking the IED, namely, to disconnect it from the information network or not? In this regard, different options were considered related to duplication of the IED, program simulation of its work, etc.[1] With the release of the second edition of the standard, a clear understanding emerged because the Behavior attribute appeared, which can have three meanings:

• "ON" is the normal operation mode of the device, it rejects all incoming data with the quality bit "q" with the value "test";

• "TEST" - test mode of the IED logical unit. The application is executed, only commands with the quality bit "q" with the value "test" are accepted, all data objects issued to the network will also have the "test" flag;

• "TEST / BLOCKED" - the test mode / blocking of the IED logic node. The application is executed, only commands with the quality bit "q" with the value "test" are accepted. The trusted node does not issue any commands, but can issue a notification about the issued command.

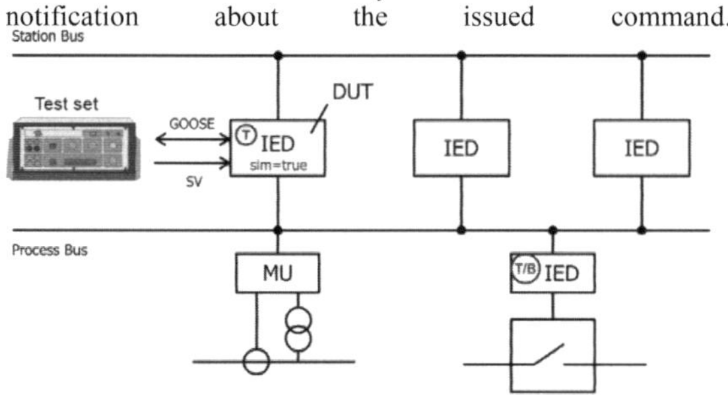

Figure 2. IED Testing

III. CT and VT Testing

It is possible to present a small classification of digital instrument transformers and give features for checking each of them.

A. Conventional Measuring Transformers with Remote merging unit (AMU)

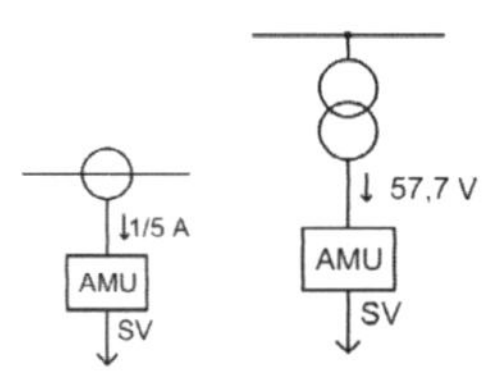

Figure 3. Conventional Measuring Transformers with Remote merging unit (AMU)

"Conventional" means electromagnetic transformers. Despite many years of experience in operating, manufacturing and designing electromagnetic current transformers, their main disadvantage is still the saturation of the magnetic circuit during short circuits with the presence of an aperiodic component. The use of magnetic cores with a gap can reduce the effect of saturation, but reduces the accuracy of both the level and the angle [2].

In this version, it is most convenient to divide the test object into two objects and check them separately: a measuring transformer and a converter. Then the verification schemes will look as follows (Figure 4).

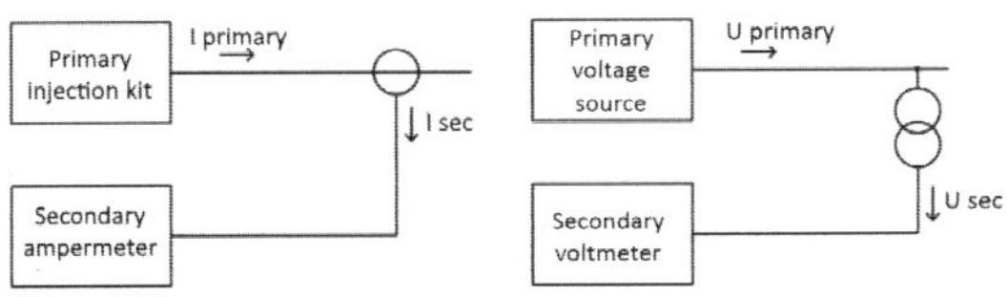

Figure 4. Verification of the transformation ratio and the angular error of traditional CT and VT

For electromagnetic current transformers, an important test is to check the properties of its magnetic core, namely, the saturation curve (Figure 5).

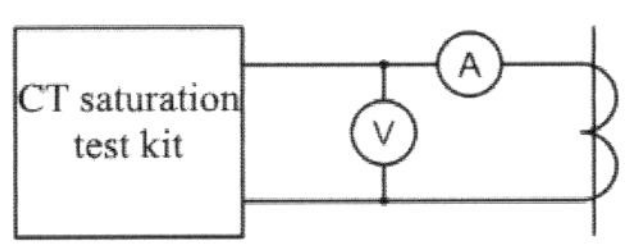

Figure 5. CT saturation curve testing

To verify a digital combining device, appropriate digital testing equipment is required that supports the IEC 61850 standard and its local profiles used in transducers. The test installation consists of a generator of secondary currents and voltages, as well as a digital stream recorder Sampled Values. The verification evaluates the accuracy of conversion of current and voltage values, conversion delay, data loss and distortion, as well as the dynamic characteristics of the converter. The user also needs to check the parameters of the generated Sampled Values flow (svID, AppID and MAC address).

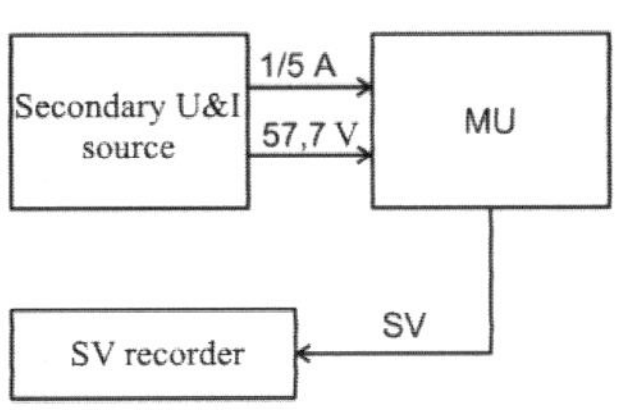

Figure 6. MU testing

In case the generator and the recorder are constructively two separate devices, or for some reason there is no internal synchronization of these devices, it is necessary to use external synchronization via PPS or PTP.

B. Conventional Instrument Transformers with Integrated AMU

In this type of the measuring CT it is no longer possible to divide the digital CT into two different test objects. Therefore, the composition of the test facility includes generators of primary currents and voltages, as well as a digital stream recorder Sampled Values.

This type of test is more preferable than the separation of digital transformers into components, since the possibility of erroneous connection after testing is excluded.

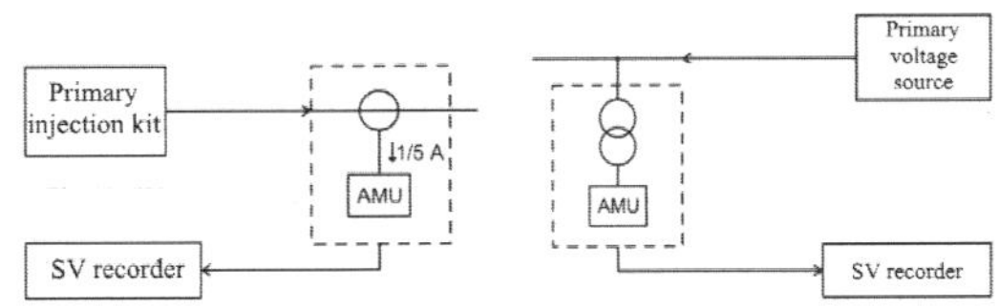

Figure 7. Testing CT and VT with built-in transducers

C. Digital optical CTs and VTs (OCT and OVT)

The use of optical instrument transformers is justified by the following advantages:

• lack of influence of EM effects, and therefore lack of saturation at KS;

• one-time conversion of information from analog to digital with the possibility of its unlimited replication;

• synchronization of all measurements of the protected object;

• other design benefits.

In fact, testing of OCT and OVT is no different from testing of traditional CTs with built-in digital converters. The advantages of OCT testing can also include the ability to use current generators with a lower output current during testing, but running several turns in the OCT window, as described in [2].

Due to the fact that almost all digital CTs and VTs have high accuracy (at the class level of 0.2 and above), a current generator capable of delivering current with the required accuracy (5 times better than the accuracy of the tested object, according to recommendations [3]) is selected quite challenging. In such cases, it is recommended to use reference CTs with the required declared accuracy.

IV. CB Testing

The digital part of the verification of switching primary equipment is limited by the possibility of sending a GOOSE message to it with a shutdown command and receiving a similar message with the "Disabled" status. For this test, it is sufficient to use a small digital device with a real-time clock to estimate the response time of the switch.

In case opening of the primary breaker is not desired, the breaker IED can be set to TEST-BLOCKED, so the trip GOOSE will be received and processed but outputs to the breaker will not operate

V. Communication Testing

Network testing is crucial on communication-based substations. That is why special tool was developed for this purposes. In the software, the corresponding functionality has been implemented for testing special algorithms of the relay protection IED, allowing to simulate the following effects [3]:

- loss of one or several network packets (samples) of digital data of SV streams in a fixed or random mode;

- offset of network packets with data of instantaneous values of one or several SV-streams relative to each other (due to delays introduced by the communication equipment of the local-computer network of the object);

- mixing (changing the correct order of samples) network packets within the SV-stream in a fixed or random mode.

In accordance with [1], for a relay protection device, the SV packet of a network of 80 samples / period, contains one sample of the instantaneous values of three phase currents and voltages (iA, iB, iC, uA, uB, uC), current and voltage zero sequences (neutral) 3i0, 3u0. The network packet of an SV stream of 256 samples / period intended for measurement tasks contains 8 samples of instantaneous values.

Evaluation of the behavior of the I & O RZA in the distortion modes of the SV-flows will allow to evaluate the work of special algorithms:

- approximation of lost data;

- buffering the values of incoming samples;

- blocking the relevant functions;

- fault signaling (unreliability of received data).

Next, we consider the features of testing with imitation of distortion modes with the RETOM-61850 testing system (Fig. 1). Distortion modes are set for any number of SV streams (out of 60 available) individually and do not depend on the settings of the remaining SV streams. All distortion modes can be combined with each other. The RETOM-61850 software and hardware complex has high-precision hardware synchronization using PTP and 1PPS protocols, which guarantees high accuracy of the characteristics of outgoing SV streams (interpacket intervals, time stamps, guaranteed distortion characteristics).

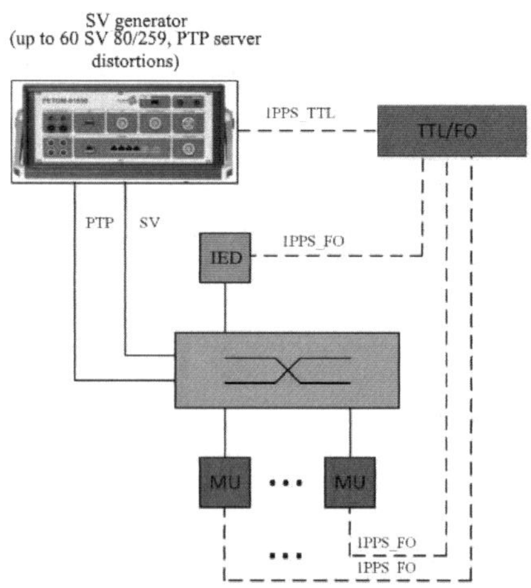

Fig. 8. Test scheme of relay protection IED

A. Packet Offset Mode

The "Offset Packet" distortion mode offsets all the SV stream data (the location of the network packet with the initial digital sample rate smpCnt = 0) relative to the 1PPS signal (the beginning of the second) by a positive or negative number of packets (samples).

The time of a given offset is:

$T_{CM} = T_{\Pi} \cdot N_{CM}$, where T_{Π} – time between packets (250 μs for an SV stream with a sampling rate of 80 samples / period and 625 μs for an SV stream with a sampling frequency of 256 samples / period and 8 instantaneous measurements in the packet); N_{CM} – number of network packets shifted (samples).

If the specified offset is positive, the delay is simulated, and if it is negative, the SV stream advance is simulated (Fig. 9).

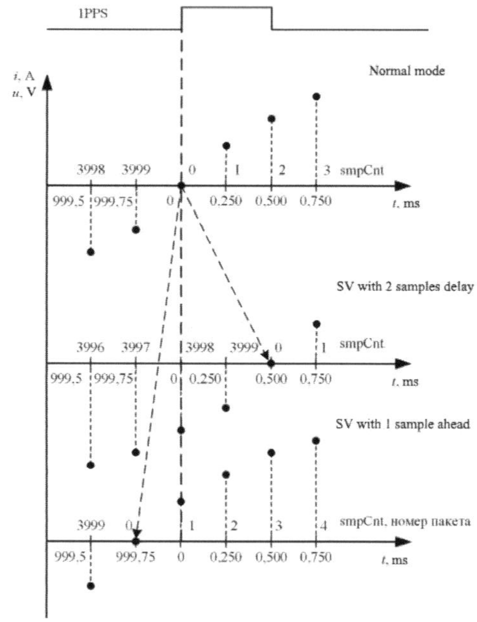

Fig. 9. An example of packet offset for an SV stream of 80 samples / period

The offset number of packets depends on the sampling rate and the number of instantaneous measurements in the SV stream. For an SV stream with a sampling frequency of 80 samples / period with one instant measurement in a packet, the offset range is from -3999 to 3999. For an SV stream with a sampling frequency of 256 samples / period and 8 instant measurements in a packet, the range from -1599 to 1599.

This mode of the test system RETOM-61850 allows you to test the performance of the sample alignment algorithm in the RZA device that receives SV streams from several signal sources when the transducers do not synchronize.

In the SV delay mode, an increase in the value of smpCnt can measure the buffer time of sample values, which compensates for uneven arrival of digital samples [4], which is quite relevant, since the buffer increases the response time of the measuring elements of the I / O RZA. The size of the compensation buffer in the I / O RZA is units of ms (2 ÷ 4), respectively, the accuracy of its measurement will be ± 250 μs or 6.25 ÷ 12.5%, which is sufficient to assess the performance of the buffering algorithm.

When simulating an SV flow advance (reducing the smpCnt value), in which network packets arrive "from the future", as a rule, the DUT algorithm should instantly block the function and signal the unreliability of the received data.

B. Package Order Mode

In the "Order of packets" distortion mode, the order of network packets inside a window of a specified duration changes from a given digital count and with a certain window repeat interval.

The size of the window mixing samples:

$T_{MIX} = T_{\Pi} \cdot N_{\Pi}$, where N_{Π} – specified size of the window in packages, not less than 2.

Only adjacent samples or random network packets can be mixed within a window. In fig. Figure 4 shows an example of randomly swapping samples starting from the 3rd packet from the 1PPS signal in a 6-packet mix window with a 3-packet repetition interval. The relay protection IED algorithm should rebuild the packets in the normal order.

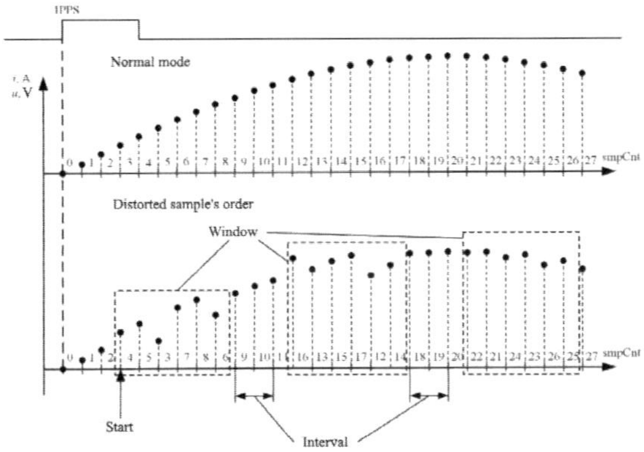

Fig. 10. An example of packet mixing for an SV stream of 80 samples / period

C. *Missing Packet Mode*

In the "Missing Packet" distortion mode, the absence of network packets in the SV stream due to interference or communication channel overload is simulated.

The disappearance of the samples is as follows. From a given digital sample, the size of the transmission window of the network packets (samples) and the window repeating interval are specified. In this case, the mode algorithm provides for the ability to skip all, or a random number of network packets inside the window. In fig. Figure 5 shows an example of randomly skipping samples starting from the 2nd packet from the 1PPS signal in a 3-packet mixing window with a repetition interval of 7 packets.

Packet skip window time:

$T_{\Pi P} = T_{\Pi} \cdot N_{\Pi P}$, where $N_{\Pi P}$ – The specified number of network packets to skip (samples).

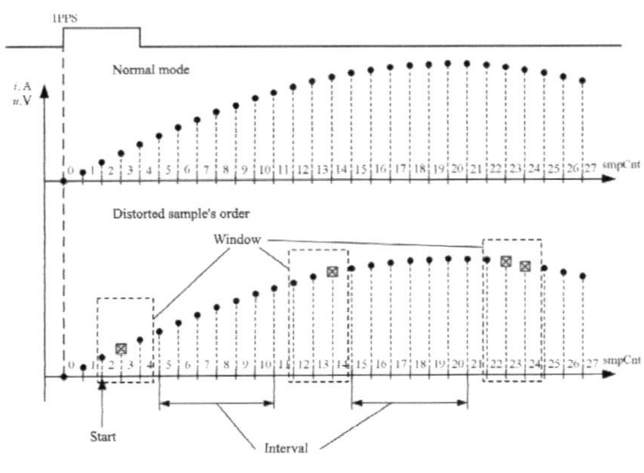

Fig. 11. An example of the mode of passing packets of the SV stream 80 samples / period

This mode allows you to test the functionality of the algorithms for recovering lost values with a single sample loss (approximation) and the corresponding blocking and signaling of invalid data in received SV streams when several samples are lost.

VI. Conclusion

With the widespread use of the IEC 61850 standard at energy facilities, personnel have to deal with the maintenance and testing of new, previously unused equipment. These devices include new unconventional instrument CTs and VTs, merging units, switches, network, optical lines. To test this equipment, it is necessary to develop new methods and means of testing, since traditional methods and devices do not allow testing of new digital devices with the support of IEC 61850.

The principles of testing also change when there is no need for physical isolation of the device under test, and only software tuning is sufficient for limiting its effect on other devices that are in the network.

The approaches and methods given in the article, which can be used when testing equipment of digital substations, will allow to evaluate the operation of equipment, adjust it if necessary and thereby improve the reliability of power supply

References

[1] A.N. Podshivalin, I.N. Nikolaev, N.M. Aleksandrov "Testing of a Digital Substation Based on IEC 61850", CIGRE B5 Colloquium, sub.№120, 2013

[2] A.Yu. Khrennikov, "Digital current transformers," Elteh.news, vol. 6(96), pp. 40–49, 2015.

[3] A.S.Shalimov, Testing of special algorithms for ied with iec 61850-9-2le process bus implementation.

[4] IEC 61850-9-2 LE (Lite Edition). Implementation Guideline for Digital Interface to Instrument Transformers using IEC 61850 9-2

[5] A.Yu. Khrennikov, T.V. Ryabin, N.M. Aleksandrov "Diagnostic methods for detection of electrical equipment's faults, defects." International Conference on Condition Monitoring, Diagnosis and Maintenance CMDM 2017 (4th edition) Bucharest, Romania

[6] A.Yu. Khrennikov, T.A. Khrennikova, I.N. Zhelbakov, N.M. Aleksandrov "Smart Grid for increasing the reliability of electric power supply of green cities by detecting electrical equipment faults and defects" 2nd International Conference on the Applications of Information Technology in Developing Renewable Energy Processes & Systems (IT-DREPS). Petra University. Amman, Jordan, 2017.

[7] A.Yu. Khrennikov, "Diagnostics of Electrical Equipment Faults and Power Overhead Transmission Line Condition by Monitoring Systems (Smart Grid): Short-Circuit Testing Of Power Transformers", Nova publishers, New York, 2016, USA, p. 174.

[8] A. Apostolov, B. Vandiver, "Functional testing of IEC 61850 based IEDs and systems" In: IEEE PES Power Systems Conference and Exposition, pp. 1683-1688, 2004.

[9] M. S. Thomas, D. P. Kothari, A. Prakash, "Design, Development, and Commissioning of a Substation Automation Laboratory to Enhance Learning" IEEE Transactions on Education. 2011, vol. 54, issue 2, pp. 286-293, vol. 54, issue 2, pp. 286-293.

Monitoring of Overhead Power Lines and Detection of Damages

Aleksander Fedotov
Professor of department
"Methods of Increase of Power Supply Reliability and Power Quality in Distributive Electrical Grids.
Kazan State Power Engineering University
Kazan, Russia
fed.ai@mail.ru

Georgii Vagapov
Researcher of department
"Methods of Increase of Power Supply Reliability and Power Quality in Distributive Electrical Grids.
Kazan State Power Engineering University
Kazan, Russia
vagapov@list.ru

Larisa Grackova
Institute of Physical Energetics
Laboratory of Energy System Analysis and Optimisation
Riga, Latvia
larisa.grackova@inbox.lv

Abstract—**At the present time, the requirements to reliability and continuity of power supply in electrical grids are continuously becoming stricter. The overhead power lines are the least reliable components of power supply systems. Early diagnosis and early elimination of damages on overhead power lines is one of the priority tasks for power supply company. This paper reviews theoretical and practical points of using different kind of characteristics during the damage for early diagnostic.**

Keywords: power system simulation; power system harmonics; transmission lines

I. Introduction

Electric power system development strategy of the Russia provide a reference point from which to view the progress achieved and the new targets to development the Unified Energy System safety and reliability [1]. Key targets that are set include:

- Increasing of electric grids transmission capacity.
- Reducing of restrictions on the flow of electricity in the transmission system.
- Progressively decreasing the greenhouse gas emissions produced by electricity generation.

The combination of natural factors, such as wind, snow, rain and ice, negatively affects the transmission capacity of overhead transmission lines with a voltage of 110 kV and above and, as a result, the quality supply of electricity to the end-user [2].

Before proceeding to the this study, the authors has studied modern practical, technical and theoretical studies about early detection and reduction of the negative impact of wind and ice-load on the wires of overhead power lines to ensure the reliability of the electrical network. It is important to indicate that all the scientific research has shown that electricity generation and distribution of electricity are of considerable theoretical and practical importance in all regions of the world. Particular attention has been developed to the theoretical methods used to predict and control a distribution network in adverse weather conditions (wind, snow, etc.).

The analysis has revealed there is wide range of models, algorithms and original studies already implemented in practice and in theory. They reflect different positions and approaches and suggest actions to ensure the reliability of the electrical network and national energy safety. However, these problems are remaining highly relevant.

Thus, the aim of the study is evaluate classical theoretical methods and propose a mathematical model for determining at the initial stage the parameters of an overhead power line with a voltage of 110 kV when exposed to wind and ice-load.

II. Theoretical Methods for Determining the Condision of the Wires the Influence of Wind and Ice Concentration

The classical theoretical methods used to determine the state of the wire in the span when exposed to wind and ice-load is based on parameters such as the sag of the wire and the mechanical stress in the wire. In order to avoid exceeding the permissible voltage, all these parameters are calculated using a maximum wire load and lowest ambient temperature [14].

At the same time, a modern monitoring system of overhead power line involves implementing more complex tasks. In general, the main objective of the state of the wire in the span under the influence of natural factors is determination of the parameters of an overhead power line at the initial stage. In order to fulfill the required reliability of power supply, the whole complex of parameters is necessary when planning organizational and technical solutions. Accordingly, during monitoring of the line, it is necessary to use highly sensitive sensors that monitor both the sag angle of the wire and the angle of its lateral deviation. Thus, the method for assessing the effect of ice deposits on the wire is based on measurements of geometrical parameters at a specific location using an installed sensor device on the wire, taking into account the elongation of the wire during operation [13, 19].

Figure 1 presents the general case of wire sag in a span: the sagging points of the wire are at different heights; the ordinate axis passes through the lower point of the wire sag [13, 19, 20 and 21].

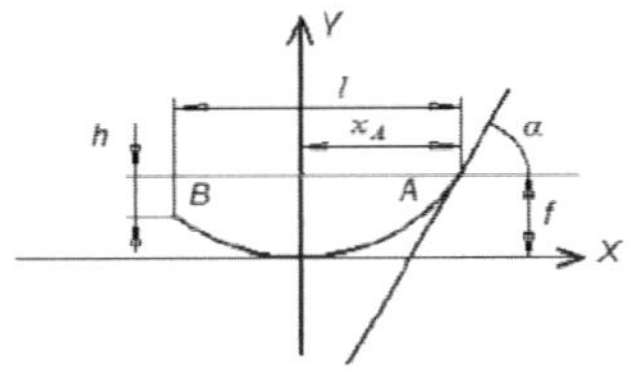

Figure 1. Wire sag in a span (general case)*.

*where: A and B – coordinates of the wire suspension points on the supports in the span; x_A – the distance between point A and the axis of ordinates; f – sag of wire; l – distance between pylons in the span; α – angle of inclination of the tangent at the point of suspension of the wire in the span; h – vertical distance of wire suspension points.

The conditions of wire in the span are based on equation (1).

$$y = a(ch\frac{x}{a} - 1), \tag{1}$$

where a – equation canonical parameter chain.

Equation (2) is determines the tangent of the angle of inclination of the tangent α at the point of suspension of the wire, indicating the dependence on the parameter a.

$$tg\,\alpha = \text{sh}\frac{x_A}{a}. \tag{2}$$

Experiments in practice have shown that in order to reduce the error and the definition of the sagging arm f of the 110 kV-overhead power line (OPL) wire, the angle α should be determined with an accuracy of hundredths of degrees with a span of OPL of more than one hundred meters using equation (3).

$$f = a\cdot\left(\text{ch}\frac{x_A}{a} - 1\right). \tag{3}$$

Also, Equation (1) was transformed as: $y = a(2sh^2\frac{x}{2a})$.

In Fig.1, the parameter h has the following dependence on the values of a and x_A :

$$h = a(ch\frac{x_A}{a} - ch\frac{l-x_A}{a}). \tag{4}$$

A system of equations (5) was obtained to determine a and x_A by combining equations (2) and (4): the angle α is determined according to the sensor device, x_A is determined from equation (5), h and l - according to the design data or the results of practical research OPL. A programming platform Matlab is used to solve the system of nonlinear equations (5).

$$\begin{cases} tg\,\alpha = \text{sh}\frac{x_A}{a} \\ h = a(ch\frac{x_A}{a} - ch\frac{l-x_A}{a})' \end{cases} \tag{5}$$

The results of equation (5) allow us to determine the length of the wire L in the span by equation (6) and the sag of the wire according to equation (3):

$$L = a\cdot\left(\text{sh}\frac{x_A}{a} + \text{sh}\frac{l-x_A}{a}\right), \tag{6}$$

The example of equation (6) in practice is presented below in the next section of this article.

Further, having determined the length of the wire in the span, a transition is made to the determination of the existing conditions of the OPL wire condition through their initial and final value. This transition from geometric to physical parameters is justified by the need to determine the degree of influence of external climatic parameters (wind, ice, etc.). This will allow determining the thickness of the ice on the wire with equation (7):

$$L_n = L_m\cdot\left[1 + \frac{1}{E}\cdot(\sigma_n - \sigma_m)\right]\cdot[1 + \alpha\cdot(t_n - t_m)], \tag{7}$$

where E – elastic modulus; σ_n – voltage in the wire at the point of wire suspension in the final state; σ_m – voltage in the wire at the point of wire suspension in the initial state; α – coefficient of temperature linear expansion; t_n – wire temperature in the final state; t_m – wire temperature in the initial state; L_m – the length of a wire in the initial state; L_n – the length of a wire in the final state.

A similar approach has already been presented in the studies of other authors, but the process of elongation of the wire during operation was not considered. [13, 19, 20, 21].

In equation (7), authors of this article, introduced a new multiplier $(1+\sigma_m{*}T)$, which takes into account the process of elongation of the wire during operation, and also takes into account the ground wire. Thus, the following equation (8) is:

$$L_n = L_m\cdot\left[1 + \frac{1}{E}\cdot(\sigma_n - \sigma_m)\right]\cdot[1 + \lambda\cdot(t_n - t_m)]\cdot[1 + \sigma_T\cdot T], \tag{8}$$

where T – the number of whole years of lightning protection cable in the span; λ – temperature coefficient of wire elongation; E – elastic modulus; Δl_T – length elongation of ground wire for the year. Δl_T is determined according to statistical data, or by calculation using specialized software.

Equation (7) is contains the parameters of large (wire length in the span) and small (changes in the length of the wire in the monitoring mode, within a few centimeters) orders. This leads to significant errors in determining the unknown components. To reduce the error, it is advisable to reduce the equations to a single order, which is usually performed by reducing equation (8) to the so-called classical wire state equation.

This article proposes this method of conversion, as the sensor devices allow you to calculate of a and x_A values with high accuracy. For this, equation (6) is converted as follows: the left-hand side is written as a Taylor series based on equation (5), while retaining only the first two terms of the expansion. As can see in equation (9):

$$L_{nx} \approx a_n\cdot\left[\frac{x}{a_n} + \frac{1}{3!}\cdot\left(\frac{x}{a_n}\right)^3\right] = x + \frac{x^3}{6\cdot(a_n)^2}, \tag{9}$$

In this case, the length of each line is calculated separately: for the first line $L_{n1} = x_{An} + \frac{x_{An}^3}{6\cdot(a_n)^2}$ and for the second line $L_{n2} = l - x_{An} + \frac{(l-x_{An})^3}{6\cdot(a_n)^2}$. Thus, the total line length is:

$$L_n = L_{n1} + L_{n2} = x_{An} + \frac{x_{An}^3}{6(a_n)^2} + l - x_{An} + \frac{(l-x_{An})^3}{6(a_n)^2} = l + \frac{x_{An}^3 + (l-x_{An})^3}{6(a_n)^2}$$

The right side of the equation is written similarly (8) and equation (10) is follower:

$$L_m = l + \frac{x_{Am}^3 + (l-x_{Am})^3}{6(a_m)^2} \tag{10}$$

By using equations (9) and (10) to equation (8), has been obtained an equation (11), while retaining members of the first order only.

$$l+\frac{x_{An}^3+(l-x_{An})^3}{6(a_n)^2}\approx L_m(1+\lambda\Delta t)(1+E^{-1}\Delta\sigma)(1+\sigma_T T)\approx \quad (11)$$

$$\approx l+\frac{x_{Am}^3+(l-x_{Am})^3}{6(a_m)^2}+L_m(\lambda\Delta t+\sigma_T T+E^{-1}\Delta\sigma)$$

As a result of shortening the span length l, equation (12) was obtained to first-order variables ordinary differential equation:

$$\Delta\sigma=E\left(\frac{x_{An}^3+(l-x_{An})^3}{6(a_n)^2 L_n}-\frac{x_{Am}^3+(l-x_{Am})^3}{6(a_m)^2 L_m}-\lambda\Delta t-\sigma_T T\right) \quad (12)$$

The Equation (12) is used to estimate the mechanical effect on the wire, but it does not allow us to make a conclusion about the thickness of the ice, since there are no mechanical loads on the wire.

This is the main difference between the classical equation of wire state [13, 14].

Thus, the next step, two different methods of ice-load are considered, but without exposed to wind.

The first method: in the district where ice is possible, measurements were made of the geometrical parameters of the wire, provided that there is no ice. The fact of absence is established by visual observation. Measurement dates: wire temperature + 20°C and fixed angle α_m. Calculations are performed according to equation (12), and arrays of reference values $[\Delta\sigma_e]$ are formed. In the monitor mode, it is necessary to control the mechanical stress in the wire, according to equation (12) and compared with the value of the array and selected according to the condition of the near temperature of the wire. The condition when the calculated $[\Delta\sigma_e]$value is higher than the reference value means the fact of ice on the wires.

The second method is based on the interaction of mechanical loads on the wire with its geometrical characteristics, namely, the sag of the wire, equation (13):

$$f=\frac{\gamma x_A^2}{2\sigma}, \quad (13)$$

where γ – specific mechanical load on the wire.

For the basic mode - m, by use formula (6):

$$\sigma_m=\frac{\gamma_m x_A^2}{2 f_m} \quad (14)$$

For the current mode n

$$\sigma_n=\frac{\gamma_n x_A^2}{2 f_n} \quad (15)$$

In equations (14) and (15), the f_m and f_n values are determined according to equation (3); a and x_A - according to equation (4). Further, $\Delta\sigma$ is calculated by the equation (13), and applied to equations (14) and (15). As a result, two equations (16) and (17) are obtained:

$$\sigma_n-\sigma_m=\Delta\sigma=\frac{\gamma_n x_A^2}{2 f_n}-\frac{\gamma_m x_A^2}{2 f_m} \quad (16)$$

$$\Delta\gamma=\gamma_n-\gamma_m=\gamma_m\left(\frac{f_n}{f_m}-1\right)+\frac{2\Delta\sigma f_n}{x_A^2} \quad (17)$$

After calculation, if $\Delta\gamma > 0$, it is means an additional mechanical load on the wire. Since the calculations were made without exposure to wind, this load means the presence of ice-load on the wire. The presence of wind does not on fundamentally change the proposed method. The effect of wind is taken into account when observing the angle of deflection of the wire from the vertical, Fig.2.

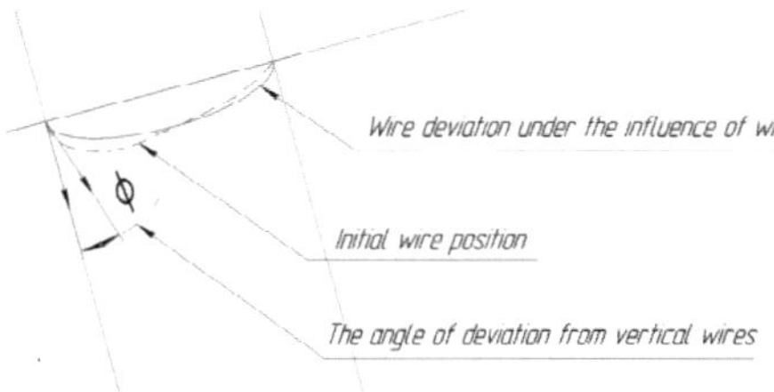

Figure 2. Visualization of wire deviation under the influence of wind.

In the case when there is an effect of ice-load and wind on the wire, the value of σ_n have been calculated according to equations (14) and (15):

$$\sigma_n=\frac{\gamma_n(1+tg\phi_\Pi)x_A^2}{2 f_n}$$

$$\Delta\gamma=\gamma_n-\gamma_m=\gamma_m\left(\frac{f_n}{f_m(1+tg\phi_\Pi)}-1\right)+\frac{2\Delta\sigma f_n}{x_A^2(1+tg\phi_\Pi)}$$

It should be noted that the f_m and f_n values are calculated by use equation (6), where in the presence of wind the angle α is different, it is not the one that was measured by the sensor. To make it more understandable, in Fig.3, can be seen that the wire is suspended on supports at points O-O' without a height difference, i.e. $h = 0$.

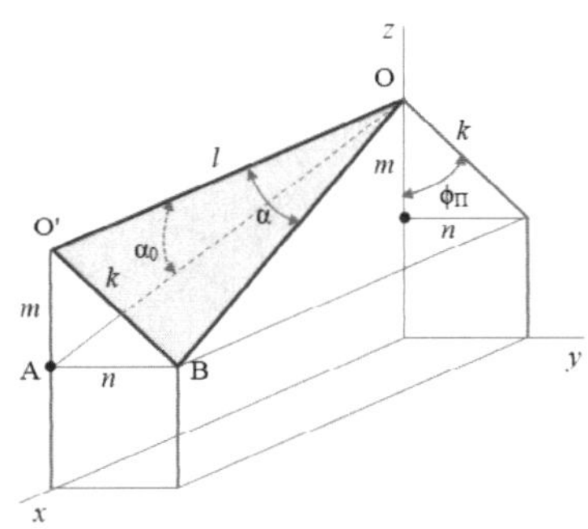

Figure 3. Geometric model of wire rotation.

The plane where the wire is located, when rotated by an angle ϕ_Π around the axis O-O' under the action of the wind, forms a right-angled triangle OO'B, where the OB line corresponds to the tangent to the sagging curve of the wire at the point O, which is the projection of the triangle OO'B on the xz plane. In accordance with Fig.3, an angle α could provide a suitable solution:

$$tg\alpha = \frac{k}{l};\ k = \frac{m}{\cos\phi_{\Pi}},\ \text{and}\ m = l * tg\alpha_0 .$$

As result:

$$tg\alpha = \frac{tg\alpha_0}{\cos\phi_{\Pi}} . \qquad (18)$$

If the drop in height of the wire suspension in the span, Fig.1, takes place, then the wire rotation under the action of the wind will vary with respect to the axis, which is also located in the xz plane. In this case, it will no longer be parallel to the x axis. By completing the necessary steps for Fig.3, the following equation was obtained:

$$\alpha = \alpha' + \varphi, \qquad (19)$$

where $tg\varphi = \frac{h}{l}$ and $tg\alpha' = tg\alpha_0 \sqrt{\frac{\cos^2\varphi}{\cos^2\phi_{\Pi}} + \sin^2\varphi}$.

Thus, from this data we can conclude that theoretical method is possible to organize energy saving melting of ice. Equation (17) allows the value to control. Thus, ice on the wires is not observed, $\sigma_m = \sigma_n$, when $\Delta\gamma = 0$.

III. Practical Implementation of Method

Currently, there is a wide selection of sensory equipment. As a practical application, primary tilt sensors were selected, which are described in detail in the article [18]. The considered theoretical method was applied in the monitoring system on the phase conductor of the existing 110 kV overhead transmission line. The values of the sag angle of the lightning protection cable from the sensor are determined every 20 minutes in normal mode. It is also possible to switch them to the interval of 10 seconds in the period of ice-load.

Analysis of the transverse angle of the wire also demonstrates the presence of a correlation between the temperature of the wire and the angle of the transverse wire rotation, Fig.4. At $\alpha = 0.05$ (confidence probability $p = 0.95$), the critical value t_{cr} is defined in Matlab, where $t_{cr} = t_{inv}$ (0.975, N-2); (0.975 = 1 - α/2).

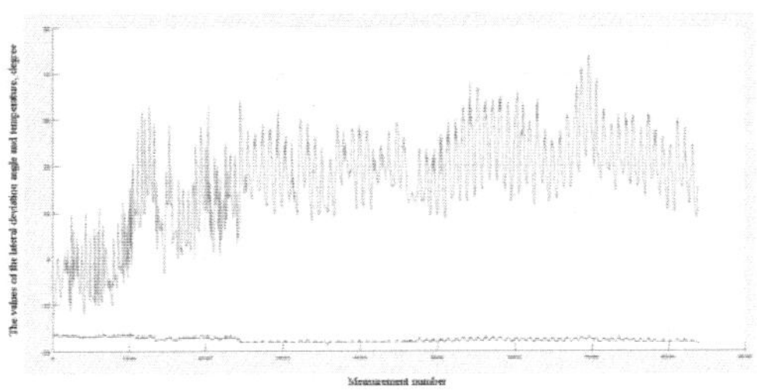

Figure 4. Comparison of changes in the lateral angle and temperature

The correlation coefficient is considered significant, therefore, there is a relationship between the data of the temperature and the values of the longitudinal change of the angle if $|\ t\ | > t_{cr}$. The theoretical method considered in practice has the following values: $t = 30.97$, $t_{cr} = 1.96$. The desired condition is performed with a large margin Fig.5 and Fig.6. A correlation between the temperature of the wire and the transverse angle of the sensor exists, since a change in temperature affects the change in the length of the wire and, therefore, a change in the tension of the wire is observed. Correlation analysis provides data indicating that the degree of temperature influence on the angle of transverse rotation is determined by the value $\Delta\beta / \Delta t = 0.016$ °/°C. Thus, $\Delta\beta$ is the change in the angle of rotation and Δt is the change in temperature.

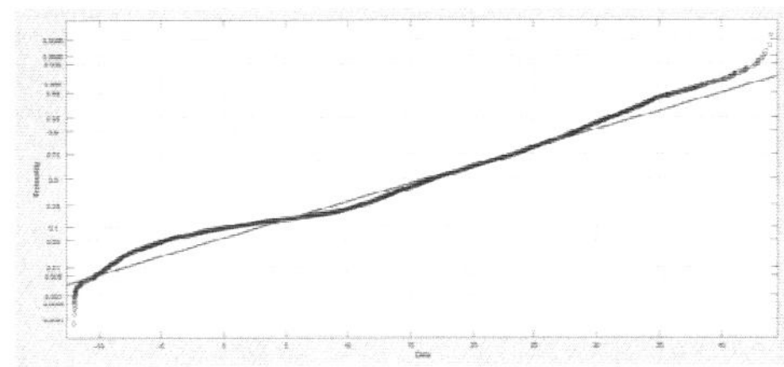

Figure 5. Probabilistic graph of temperature values distribution.

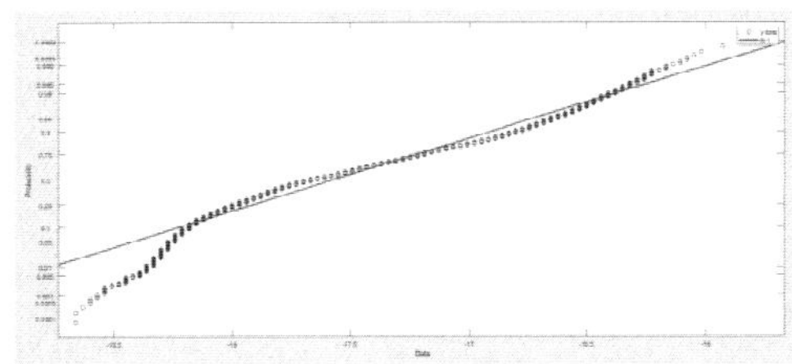

Figure 6. Probabilistic graph of the distribution of the transverse angle.

Changing the temperature of the wire leads to a change in the length of the wire in the span, which affects the sag angle of the wire. Thus, based on the proposed algorithm, it is possible to determine not only the boundary conditions of the wire, but also its current values.

IV. Conclusions

This research shows methodology for forecasting the ice on wire of overhead power lines. This methodology provides data on the value of the wire temperature, as well as the value of the longitudinal angle of the wire sag.

The obtained data allows the operating organization to implement the monitoring process and implement a set of emergency response measures at an early stage.

References

[1] Federal law dated March 26, 2003 N 35-FL " About power industry ". Web site: https://base.garant.ru/185656/

[2] Preparation and passing of autumn-winter period 2018-2019. Web site: https://minenergo.gov.ru/node/11896

[3] Regulation «Russian grid» «About complex uniform technical policy in the power grid». Web site: https://www.rosseti.ru/investment/science/tech/doc/tehpolitika.pdf

[4] Ministry of Energy of the Russian Federation. Министерство энергетики РФ. Web site: https://minenergo.gov.ru/node/7822

[5] E.I. Sacuk. Software and hardware for monitoring overhead power lines and power system control in extreme weather conditions. Doctoral Thesis. Novocerkassk, 2011, p. 295.

[6] I.I. Levchenko, A.S. Zasypkin, A.A. Alliluyev, E.I. Sacuk. Diagnostics, reconstruction and operation of overhead power lines in icy areas. MEI, 2007, p. 448.

[7] A.F. Dyakov. Prevention and elimination of ice accidents in electrical grids. Pyatigorsk: « Yuzhenergotekhnadzor », 2000, p. 284.

[8] L. Qimao, L. Peng, Z. Qing, R. Wenping, C. Min G. Shangfei, "Icing load prediction for overhead power lines based on SVM," Proceedings of 2011 International Conference on Modelling, Identification and Control. 2011. pp. 104-108. DOI: 10.1109/ICMIC.2011.5973684.

[9] P. Li, L. Qimao, C. Min, G. Shangfei, H. Haiyan, "Time series prediction for icing process of overhead power transmission line based on BP neural networks, " Proceedings of the 30th Chinese Control Conference. 2011. pp. 5315 – 5318.

[10] D.A. Yaroslavsky, M. F. Sadykov. Development of a device for monitoring and quantitative control of ice formation on overhead power lines. News of higher educational institutions. Energy problem. – 2017. – № 3-4. – p. 69-79.

[11] Minutes of the meeting of Volga Federal District to the passage of autumn-winter period 2018-2019 (№AH-431). Web site: https://minenergo.gov.ru/node/11896

[12] F. Reinhard: Freileitungen: Planung, Berechnung, Ausfiihrung; mit 86 Tabellen / R. Fischer; F. Kiessling.- 4. Aufl. - Berlin; Heidelberg; New York; London; Paris; Tokyo; Hong Kong; Barcelona; Budapest: Springer, 1993. ISBN-13: 978-3-642-97925-5. DOI: 10.1007/978-3-642-97924-8.

[13] V.I. Idelcik. Electrical systems and networks. – M.: Energoatomizdat, 1989. 592 p.

[14] A.A. Glazunov. Basics of the mechanical part of overhead power lines, Gosenergoizdat, 1956. 192 p.

[15] A.D. Boshnyakovich. Mechanical calculation of wires and strike protection wires of power lines. – L.: Gosenergoizdat, 1962. 254 p.

[16] K.P. Kryukov, B.P. Novgorod. Structures and mechanical calculation of power lines. –L.: Energy, Leningrad Branch, 1979. 312 p.

[17] D.R. Merkin. Introduction to the mechanics of a flexible thread. - M.: « The science ». – 240 p.

[18] A.Fedotov, S.Kurth, S.Voigt, G.Vagapov. A concept for an ice detection system on overhead power lines, theory and practical results // Proceedings of the 9th International Scientific Symposium on Electrical Power Engineering, ELEKTROENERGETIKA 2017 9. 2017. pp. 297-300.

[19] IEEE 738-2006. IEEE Standard for calculating the current-temperature of bare over-head conductors. https://www.ieee-pes.org/presentations/td2014/td2014p-000114.pdf

[20] V.G. Makarov, A.I. Fedotov, R.Sh. Basyrov, G.V. Vagapov. Simulation of an overhead power line in Matlab/Simulink. Bulletin of the University of Technology. 2017. V. 20. № 13. p. 93-96.

Usage of Empirical Statistical Distributuions in Simulation Models to Estimate Structural Reliability Indices of a Complex Systems

Andrey Grishkevich
Faculty of Mathematics, Physics & Technical Science
Kazimierz Wielki University
Bydgoszcz, Poland
andreygrishkevich@ukw.edu.pl; grishkev.gmai@gmail.com

***Abstract*—The method for generating pseudorandom numbers with a given empirical distribution law based on a numerical representation is presented. The usage of the numerical distribution law in structural reliability calculations for power supply systems is investigated. Recommendations on the numerical representation of the distribution functions are presented in order to ensure acceptable accuracy while determining the probability distributions of states of electric power systems using mathematical modeling methods.**

Keywords*: *Monte Carlo methods; empirical distribution function; probability distribution of states; reliability of electric power systems

I. Introduction

Calculation of the power supply system structural reliability [1] is based on the summation of failure states contribution (for highly reliable systems simultaneous failures of one, two and three elements are usually considered) of this system in the resulting reliability indicators of the system.

The usage of digital technologies has led to a wide distribution and availability of computing resources. Multicore technologies and integration of distributed computing resources into a system with parallel calculations has led to a dramatic increase in the performance of computing systems and cost reduction of individual mathematical operations. Simplest computing cluster creation does not require the purchase of additional equipment, and is performed fairly quickly and simply on the basis of the existing computer network. If the processing power of the configured cluster is not sufficient, it is possible to use a more powerful cluster or a supercomputer. Computational efficiently of calculation methods stops being crucial under such conditions. The simulation methods are gaining practical relevance [2-4]. These methods require a significant number of mathematical operations and have a number of advantages. In such conditions the simulation methods (estimating reliability indexes using a sufficiently large series of tests requiring a considerable number of mathematical operations) in practical applications are increasingly competing with the analytical methods.

The advantages of the simulation method include the possibility of numerical assignment of random variables distribution laws, an assignment of a rather complex logic of emergency and preventive repairs organization, and obtaining distributions of the states durations and their sets.

The numerical assignment of the distribution laws allows, in particular, the usage of empirical estimates of the electric power systems elements distribution laws - on the basis of statistical data on their functioning.

II. Simulation Model of the Electric Power System Functioning

Each element E of the power system (transformer, switch, ...) can be in one of four states [1]. Supposing that En is the state of normal operation of the element, Es is the state between the failure of the element and the completion of the operational switching, Er is the state of emergency repair of the element, Em is the state of preventive repair (deliberate shutdown) of the element; t_{Exy} is the random time of transition of the element E from the state Ex to the state Ey. Transitions between states are set

$$En \to Em \to En \to \dots,$$
$$En \to Es \to Er \to En \to \dots.$$

The state of the system Si is determined by the state of each element of the system. Only one element can be in the state of preventive maintenance (deliberate shutdown).

Simulation of the system operation period `T_Model` requires time T for calculations. The simulation result [3] is obtaining the probabilities of states

$$P_{Si} = \Sigma P_{Si,k},$$

where $P_{si,k}$ – is the probability that the random time of being in the state Si will lie in the interval $[t_k, t_{k+1})$, $k = 0, 1, 2, 3, \dots, M$-1.

III. Generation of a Random Variable with Given Numerical Distribution in the Simulation Program for Modeling Structural Reliability

The distribution law $F_\xi(t)=P(\xi<t)$ of a random variable ξ is numerically specified by the vector `CDFI` of dimension `N` in computer memory or in a file `CDFI.txt` on the hard disk. The specified vector contains the values of the inverse function of distribution law

`CDFI[i]=`F^{-1}`(i/N), i=0,1,2,...,N-1.`

The presented data format allows storing information about `N` points of the distribution law function

`(i/N, `F^{-1}`(i/N)), i=0,1,2,...,N-1,`

in one array (the values i/N are calculated arithmetically).

The exponential distribution with the parameter L is numerically set by the vector

```
#define N number
double EXPL[N];

for (int i=0; i<N; i++)
EXPL[i]=SV_Exp((double)i/(double)N, L);

double SV_Exp(double y, double L){
double x;

                    x=-log(1-y))/L; //              (1)

return x;}//.
```

Pseudorandom numbers are generated on the basis of random numbers library (defined in the header file `<random>` of C++11/14/17) [5]. The random number library provides a set of pseudo-random number generators (in particular, `mt19937_64 generator (number)` – 64-bit Mersenne's Twister) that generate integer sequences with uniform distribution (depending on the `number` parameter)

```
#include <random>
number=1;
std::mt19937_64 gen(number); //.
```

The library also includes the standard classes of random numbers distributions: the uniform distribution in the interval [0;1)

```
std::uniform_real_distribution
<double>urd(0,1);
```

and exponential distribution with the parameter L

```
std::exponential_distribution
<double>EXP(L); //.
```

The code for generating the pseudo-random value y of a random variable with a uniform distribution law in the interval [0;1) is represented by the following code in the C++ programming language using the `<random>` library

```
                    y=urd(gen); //                  (2)
```

and the exponentially distributed quantity by code

```
                    y=EXP(gen); //.                 (3)
```

In the inverse function method [6], the value y of a random variable with a uniform distribution law is converted to the value $t=F_{\xi}^{-1}(y)$ of a random variable ξ with the distribution law $F_{\xi}(t)=P(\xi<t)$.

In accordance, the exponential distribution (3) represents the procedure implemented by means of the `<random>` library of C++ 11/14/17

```
               y= SV_Exp(urd(gen), L)); //,         (4)
```

where analytical representation (1) of function F^{-1} is used.

The code in the C++ programming language using the distribution representation in the form of a `CDFI` array

```
          t=SV_Tab(urd(gen), CDFI); //        (5)

double SV_Tab(double y, double* ROZKLAD){
double t;

           int n1 = (int)(y *(double)N); //    (6)

int n2 = n1 + 1;
        if (n1>=N-1){

       t=-(log(1-y))*ROZKLAD[N-1]/log(N);//(7)

        return t;}

        else {
double x2, x1, y2, y1;
x1 = ROZKLAD [n1];
x2 = ROZKLAD [n2];
y1 = n1/(double)N;
y2 = n2/(double)N;

           t=(x2-x1)*(y-y1)/(y2-y1)+x1; //    (8)

return t;
}}
```

As the distribution function approaches 1, the numerical piecewise linear approximation F^{-1} leads to a significant inaccuracy increase. On the interval $[1-\alpha,1]$ for small α it is recommended to use the analytical expression, which provides $F^{-1}(y)\to+\infty$ with $y\to1-0$. As a consequence, a linear approximation is used on the interval $[y_i, y_{i+1}] = [i/N, (i+1)/N]$, $i = 0, 1, 2, \dots, N-2$, (8) (Fig.1), the exponential one – on the interval $[y_{N-1}, y_N] = [(N-1)/N, 1)$ (7) (Fig.2).

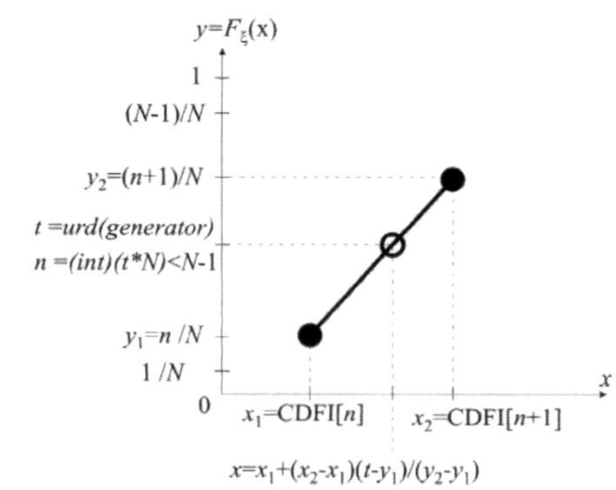

Figure 1. Calculating the value of the random variable ξ based on the value of the random variable y with equal distribution law in the interval [0; $(N-1)/N$].

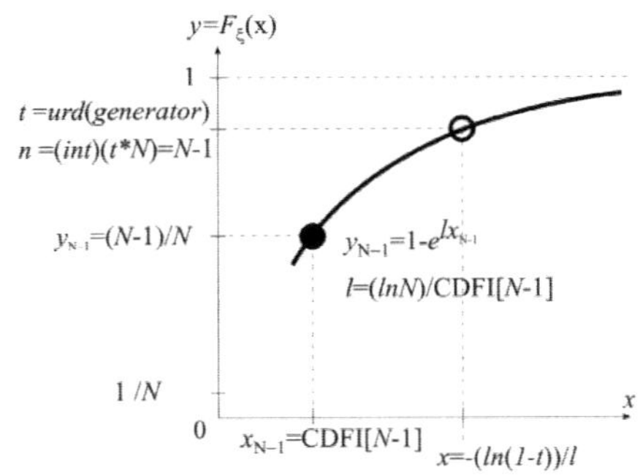

Figure 2. Calculating the value of the random variable ξ based on the value of the random variable y with equal distibution law in the interval $[(N\text{-}1)/N;\ 1)$.

The generation of next pseudo-random number is performed by the repeated execution of the corresponding procedure (2) – (5).

The random variable generation code is very efficient. The search for the interval $[y_n, y_{n+1}] = [n/N, (n+1)/N]$, such that $y \in [y_n, y_{n+1}]$, is performed in one operation (6).

IV. Example of Obtaining Numerical Representations for the Distribution Law

The distribution law $F_\xi(x)=P(t<x)$ of the random variable $\xi \in [0;+\infty)$ (failure function, unreliability function) is numerically given [7] in the form of number arrays x_i (array `CDFX[]` of dimension `NIn` = 7, column F of Fig.3) and $y_i = F(x_i)$ (array `CDFY[]` of dimension `NIn` = 7, column G of Fig.3)). The fragment of the EXCEL sheet and the corresponding graph are shown in Fig.3.

The values $x_i = F^{-1}(i/N)$, $N = 4$ (array `CDFI[]`, column A of Fig.3.) can be obtained on the basis of a value set $(x_i, F(x_i))$ (arrays `CDFX[]` and `CDFY[]`, columns F and G of Fig.3) by the linear approximation method. The values of column A are stored in the file `CDFI.txt` and are used in modeling.

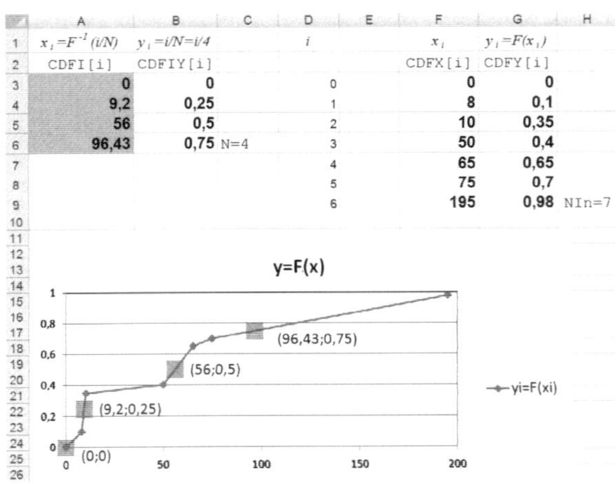

	A	B	C	D	E	F	G	H
1	$x_i=F^{-1}(i/N)$	$y_i=i/N=i/4$		i		x_i	$y_i=F(x_i)$	
2	CDFI[i]	CDFIY[i]				CDFX[i]	CDFY[i]	
3	0	0		0		0	0	
4	9,2	0,25		1		8	0,1	
5	56	0,5		2		10	0,35	
6	96,43	0,75	N=4	3		50	0,4	
7				4		65	0,65	
8				5		75	0,7	
9				6		195	0,98	NIn=7

Figure 3. Approximation of values $F^{-1}(i/N)$ of the random variable distribution law.

Fig.4 shows the numerically given function of the distribution law $F_\xi(x)$ (dashed line) and its approximation, which is used to obtain the inverse function when generating a random variable by the inverse function method.

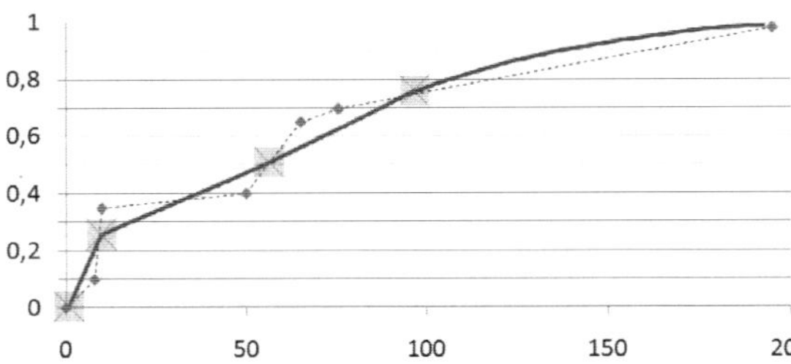

Figure 4. The approximation $F_\xi(x)$, which is used as an inverse function when generating a random variable.

Generation of `CDFI` array based on a set of values $(x_i, F(x_i))$ is implemented using the EXCEL macro of the MS OFFICE package. EXCEL visualization tools allow you to control the correctness of the numerical representation of the empirical distribution law.

V. Studying Numerical Assignment of the Distribution Law

In the study of numerical assignment of the distribution law, test numerical reliability indicators were used [2]. Analytical (3) and numerical (9)

```
t=SV_Tab(urd(gen), EXPL);//
```
(9)

was compared, when N = 100, 1000, 10000, the exponential distribution function set when modeling the states of one element.

The assumption of the exponentiality of the time distribution laws of transitions between the states of the system allowed us to obtain the exact values of the steady states probabilities from the Kolmogorov equations and compare them with the results of numerical simulation. The maximum relative error ε of determining the state probability and the time of computer simulation T are presented in Table 1.

Table 1. Comparison of the analytical (3) and table (9) assignments of the distribution function when modeling the states in the model of a single element

Options			Results	
Formula	*N*	`T_Model`	ε %	*T* [s]
(3)	-	10^9	0,04	431
(9)	100	10^9	0,03	356
(9)	1000	10^9	0,01	363
(9)	10000	10^9	0,10	377

With a permissible error fluctuation (relative error less than 1%), the tabular representation of the distribution function allowed us to slightly (1.1 times) speed up the simulation.

The probabilities $P_{Si,k}$ of the duration of staying (when choosing intervals $t_k = k$ [h], k = 0, 1, 2, 3, ...) in the state $Si = Ir$ for the analytical and tabular (N = 100) tasks of the exponential distribution are presented in Fig.5. Fig.6 shows the relative error of the states probability.

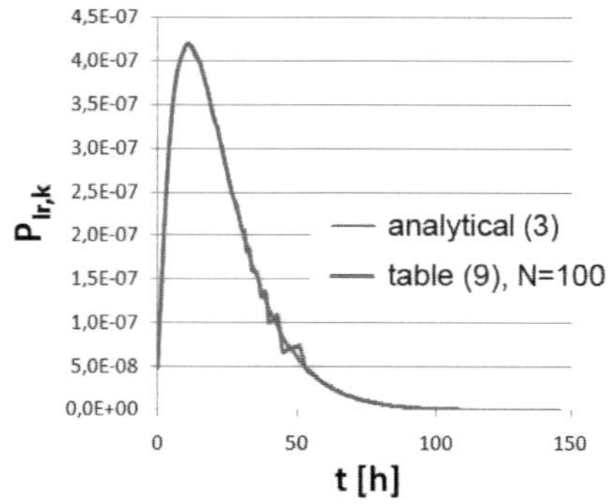

Figure 5. Probabilities $P_{Ir,k}$ of stay durations $[k, k+1)$ [h], $k = 0, 1, 2, 3, \ldots$ in the Ir state for analytical and tabular ($N = 100$) assignments of distribution functions.

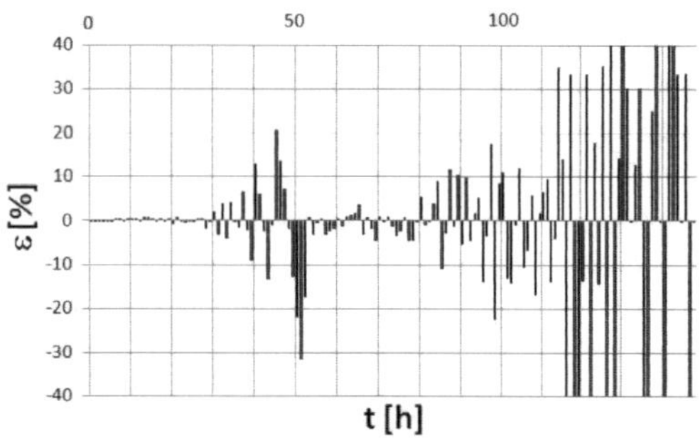

Figure 6. Relative error ε of the state Ir,k probability for tabular ($N = 100$) assignment of the distribution function.

The relative errors of the tabular ($N = 100$, 1000, 10000) assignment of the distribution function are presented in Fig.7.

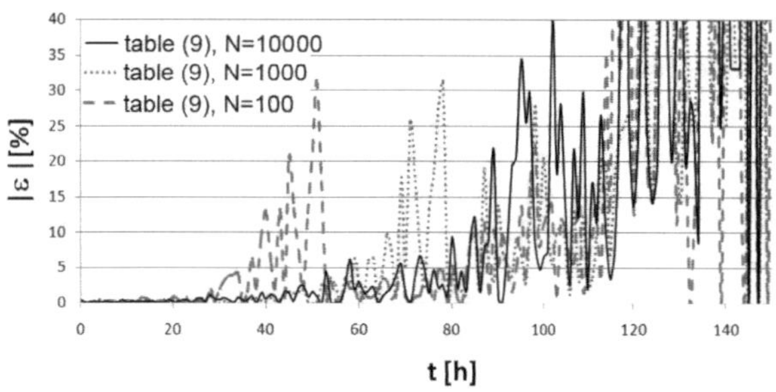

Figure 7. Relative error ε of state probability Ir,k for tabular ($N = 100$, 1000, 10000) assignment of distribution functions.

An increase in the accuracy of the distribution law representation (an increase in the dimensionality of the array N) allows to increase the time interval over which the analytical and numerical assignment of the distribution laws coincide with a relative error of 10%. Obviously, the piecewise linear approximation (8) of the distribution function is improved.

The increase in error for large time values is due to the small probability of such events. The hit rate for such states is small even with large modeling intervals. In accordance, the probabilities estimates based on a small number of events are imprecise and random.

VI. Conclusion

A method for obtaining the numerical representation for the empirical distribution function of the random transition time between the states of the electric power system is proposed for usage when calculating the reliability of power systems using mathematical modeling methods with the help of EXCEL MS Office macro.

The numerical representation of the distribution functions provides reasonable accuracy and time when determining the state probabilities with mathematical modeling methods for structural reliability calculation of power supply systems, which creates prerequisites for using statistical data to describe, in particular, emergency and preventive repairs of system elements.

The usage of 100–1000 (from 100 to 1000) values of the state transition time distribution function provides an acceptable accuracy in calculation of the states probability distribution with an interval of 1 hour with up to 50 hours duration, which is quite enough to study the long-term failures with very serious systemic consequences and for risk assessments for commercial customers with non-linear cost functions.

References

[1] A. Grishkevich, V.I. Hudym, A.M. Kruczinin, and A. Sawicki, Zagadnienia energetyczne wybranych współczesnych urządzeń i systemów elektrostalowniczych. Polska, Częstochowa: Wydawnictwo Politechniki Częstochowskiej, 2010.

[2] A.Grishkevich, "Simulation models using empirical statistical distributions to estimate structural reliability indices of electric power systems," Przegląd Elektrotechniczny (Electrical Review), vol. 93, pp. 103-106, nr. 12, 2017.

[3] A. Grishkevich, "Distributed simulation models to estimate indices of structural reliability of electric power systems," Proc. of the 9th International Scientific Symposium on Electrical Power Engineering LEKTROENERGETIKA 2017, 12.-14. 9. 2017, Stará Lesná, Slovak Republic, 2017, pp. 422-427.

[4] H. Ge, S. Asgarpoor, "Parallel Monte Carlo simulation for reliability and cost evaluation of equipment and systems," Electric Power Systems Research, vol. 81, pp. 347–356, 2011.

[5] Microsoft. Dokumentation: <random> (Internet: https://docs.microsoft.com/en-us/cpp/standard-library/random?view=vs-2019) [April 12, 2019].

[6] R. Saucier, "Computer Generation of Statistical Distributions," Army Research Lab Aberdeen Proving Ground MD., Storming Media, 2000.

[7] M. Kornatka, "Automatyzacja pracy sieci średniego napięcia a poziom ich niezawodności," Przegląd Elektrotechniczny (Electrical Review), vol. 90, pp. 109-112, nr. 8, 2014.

Simlified Criterium for Transient Stability Assessment

Karel Máslo
Transmission System Analysis Dpt.
ČEPS, a.s.
Prague, Czech Republic
maslo@ceps.cz

Vladimír Krištof
Gridman s.r.o.
Kosice, Slovakia
vladimir.kristof@gridman.sk

Abstract — **The paper examines the validity of a simple criterion for verifying the transient stability of synchronous generators connected to the network. The criterion is based on the contribution of the network to the short-circuit power and the sum of rated active powers of all generating units connected to a given node.**

Keywords: Rotor angle stability, Transient stability

I. Introduction

A power system dynamic stability assessment is (beside compliance with the criteria N-1 and the system adequacy) an important part of a secure power system operation. It is possible to examine various aspects of the dynamic stability, e.g. paper [1] deals with simplified method of the rotor angle (or transient) stability based on a Lyapunov function.

An even simpler method is proposed in paper [2]. The method is based on short circuit power S_k'' criterion. With consideration that short circuits in transmission network are cleared within time not exceeding 150 ms, the transient angle stability is maintained when the following condition is met:

$$S_k'' \geq 6 * P_n \qquad (1)$$

where P_n is the sum of rated active powers of all generating units connected to a given node of the transmission network and S_k'' is the contribution of the network to the short-circuit power calculated for a 3-phase short circuit at the given node (contribution of the generating units connected to the given node is not taken into account in the value of S_k'').

Short-circuit power is fictive value which determines „the hardness" of power system at given node. Level of short-circuit power is important value for e.g.:

- voltage sag propagation and analysis,
- analysis and calculation of power quality indicators (e.g. flicker, upper harmonics etc.)
- unwanted effects of specific loads (e.g. carbide furnace) or new connected sources to power system.

We show using of mentioned criterion based on S_k'' on a test case and compare it with a conventional method of the time domain simulation in this paper. The MODES network simulator was used as a computation tool.

The paper is structured as follows. A brief description of transient stability concept is listed in section II. The test cases are introduced in the section III. The results of time domain simulation are shown and compared with the criterion (1) in the section IV. Finally, the section V concludes the paper.

II. Transient stability concept

The simplest way to explain transient stability problem is on single machine-infinity bus model (Figure 1) with using equal area criterion.

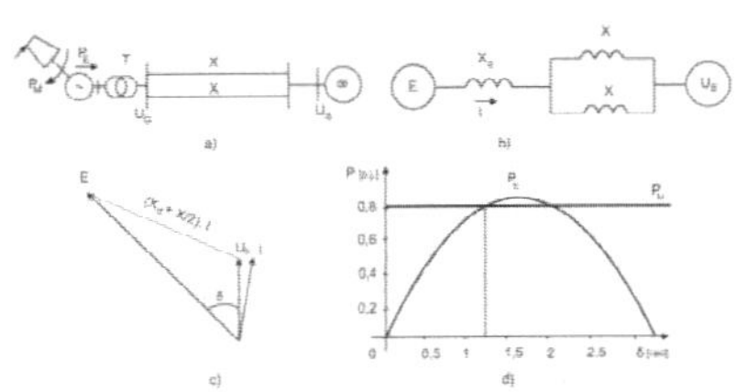

Figure 1. Scheme of single machine-infinity bus model

Single-line diagram a) shows the basic arrangement of the model. A synchronous generator driven by turbine supply electric power P_E through the double line with reactance X to so called infinity bus characterised by constant voltage U_S. The electric power P_E is equal to the mechanical turbine power P_M in the initial state (we neglect mechanical and electrical losses for simplicity). We can substitute synchronous generator by an electromotoric voltage E behind synchronous reactance X_d showed on scheme b) with corresponding phasor - diagram c). Generated active power transmitted through the line to the infinity bus is equal:

$$P_E = \frac{E * U}{X_\Sigma} \sin\delta \qquad (2)$$

where X_Σ is summary reactance.

Equation (2) is graphically showed in per unit (related to the nominal apparent power) on diagram d). Amplitude of the output characteristic is inversely proportional to the summary reactance.

Transient stability of power system is mostly threatened by short-circuits. Let´s assume that on one of parallel power lines will occur three-phase bolt short-circuit according to following Figure 2.

Transient process in this case consists from two phases:

- short-circuit occurrence (diagram "a"),
- switching off short-circuit and faulted power line (diagram "b").

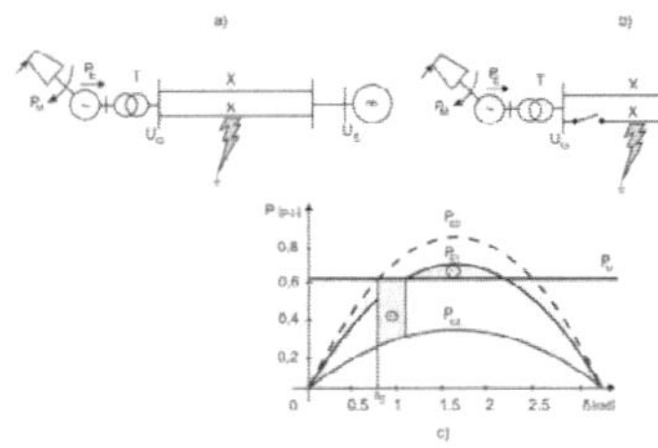

Figure 2. Transient stability concept (Single machine – infinity bus model)

Each phase has its own power-angle characteristic (see diagram "c"). During short-circuit power-angle characteristic decreases from P_{E0} to P_{EZ}. Rotor is accelerating and accumulating kinetic energy equal to yellow area marked with ⊕. After switching off short-circuit and faulted power line the power-angle characteristic will change to P_{E1}. Together with power-angle characteristic rotor angle is also changing – in first phase increase from δ_0 to δ_1. As long as the breaking area (marked by minus sign) is greater or equal to the accelerating area, the rotor decelerates (it moves back in respect of synchronously rotating reference frame) and it stabilises finally in a new equilibrium point δ_2. If the accelerating area is greater than the breaking area, the rotor is still accelerating, rotor angle δ is increasing and the machine losses synchronism and transient stability. Ensuring that accelerating area will be smaller than breaking is done mainly with short time of fault switching time. Modern digital protective relays and circuit breakers can switch off the fault under 100 ms.

III. Test cases

Two simple test systems were selected to verify the criterion (1). The first test system comes from the book [3] (example 13.1) and it corresponds to the to the simplest power system dynamic model – the single machine infinite bus.

The following figure shows the initial scheme of this system.

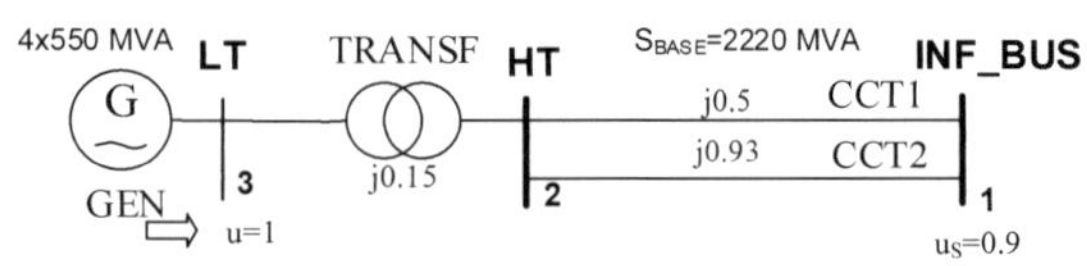

Figure 3. Single –line scheme of the first test system (according to [3])

This scheme in per units (related to $S_{BASE} = 2220$ MVA and $U_{BASE} = 24$ kV) has been converted to named values and infinite bus was modelled by a generator with very large power – see Figure 4.

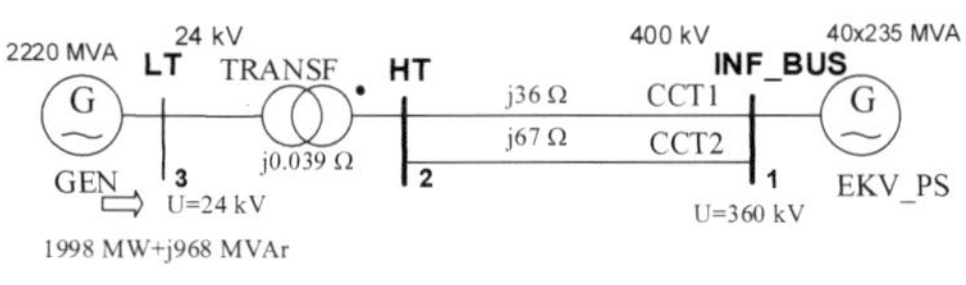

Figure 4. Adapted scheme for initial load flow calculation

The classical model (constant E' behind subtransient reactance x_d') was considered for the generator GEN with $x_d' = 0.3$ p.u. and with the inertia constant H = 2.915 MWs/MVA. This constant was somewhat smaller compared to the book [3] because of nominal frequency was considered 50 Hz instead of 60 Hz.

The second test system represents a high voltage system (380/110/30/10 kV) from the Technical report [4]. The test system was used in the paper [5], where are the details of dynamic models. Figure 7 shows the scheme of the modelled test case.

IV. Comparision of results

In first test system we will consider three phase ideal short-circuit on the line CCT2 near the node HT. Then the per unit short circuit power can be simply calculated for our case as follows:

$$S_k'' = \frac{u_S}{x_{CCT1}} = \frac{0.9}{0.5} = 1.8 \text{ p.u.} \qquad (3)$$

Since $6 * P_n$ is 5.4 p.u. it is obvious that the condition (1) is not met $(1.8 < 5.4)$. It means that the system was not stable according to criterion (1) for the clearing time 150 ms. We can check this by the calculation of the critical clearing time by conventional method (time domain simulation of the short circuit with different clearing times). The following figure shows results of the time domain simulation carried out by the MODES network simulator.

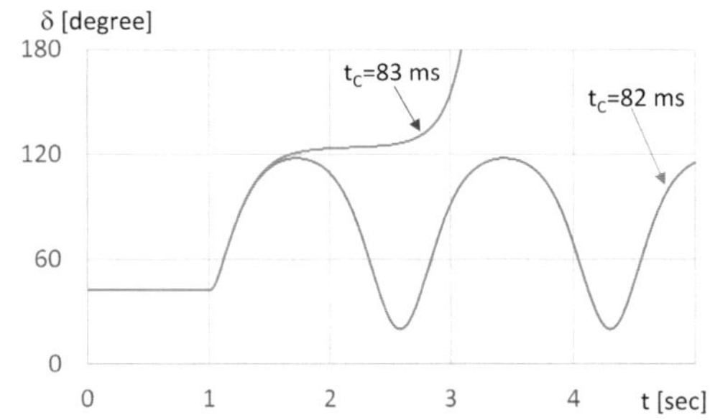

Figure 5. Waveforms of rotor angle for two clearing times

It is seen that for the fault clearing time $t_C = 83$ ms is the system already unstable. It means that the critical clearing time is $t_{CC} = 82$ ms, so the validity of the criterion (1) is confirmed.

The simplicity of the relationship (1) allows to determine the value of the line reactance x_{CCT} at which the system would remain stable for the $t_C = 150$ ms and the given short circuit.

$$x_{CCT} \cong u_S/6/P_n = 0.9/6/0.9/= 0.167 \text{ p.u.} \qquad (4)$$

The following figure shows results of the critical clearing time calculation.

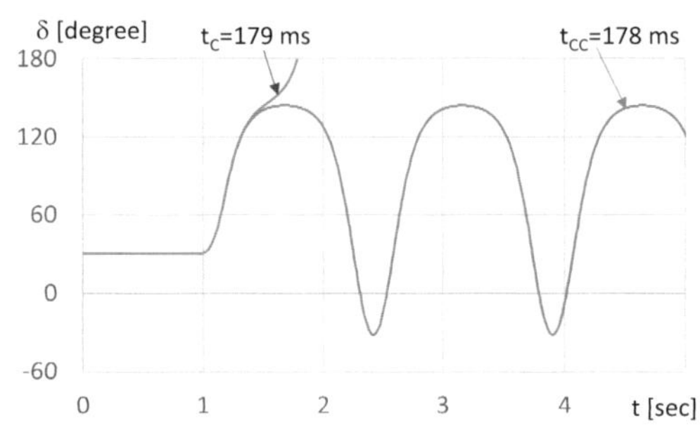

Figure 6. Waveforms of rotor angle for smaler line reactance

Since the critical clearing time is $t_{CC} = 178$ ms, which is more than the expected clearing time 150 ms, and the validity of the criterion is confirmed again.

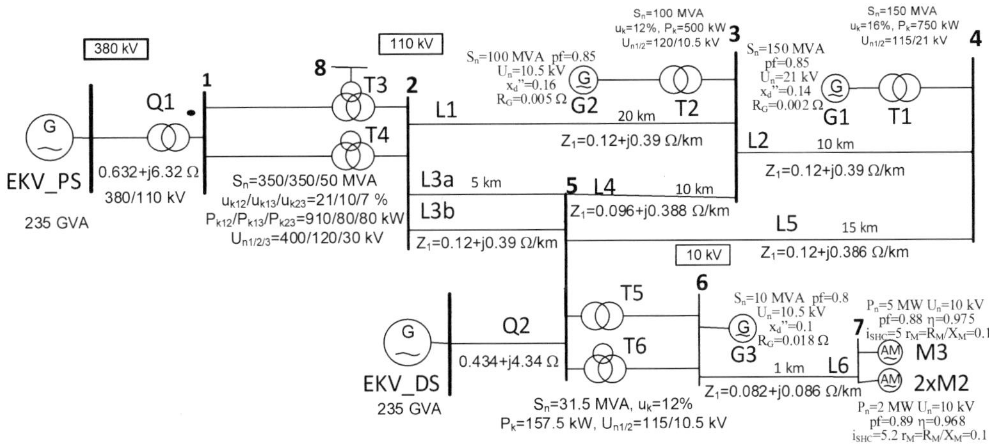

Figure 7. Single –line scheme of the second test system

Second test system is more complex. Again let´s assume that on power line L2 (near node 4) occurs three phase ideal short-circuit. Short-circuit contribution from the network could be calculated from the short circuit current of the line L5 during this short circuit (see Figure 8). Since this short-circuit power is $S_k'' = \sqrt{3} * 110\ \text{kV} * 7\ \text{kA} = 1333\ \text{MVA}$.

On the other hand six times rated active power of generator G1 connected to node 4 is $6 * P_n = 765$ MW. The equation of criterion $S_k'' \geq 6 * P_n \Rightarrow 1333\ \text{MVA} \geq 765\ \text{MW}$ denotes that system should remain stable when the fault will be cleared in 150 ms.

We check this claim again by calculation of critical clearing time on dynamic model created in MODES. The value of critical clearing time $t_{CC} = 274$ ms was estimated by series of calculations. The value of critical clearing time is much higher than clearing time defined by criterion (274 ms > 150 ms), so according to the criterion when the fault will be cleared in 150 ms, system remain stable.

Dynamic calculations confirmed correctness of proposed criterion. On Figure 9 and Figure 10 are shown waveforms of rotor angle of generator G1 for three cases:

- fault is cleared in 150 ms,
- fault is cleared in critical clearing time 274 ms,
- fault is cleared after critical clearing time 275 ms.

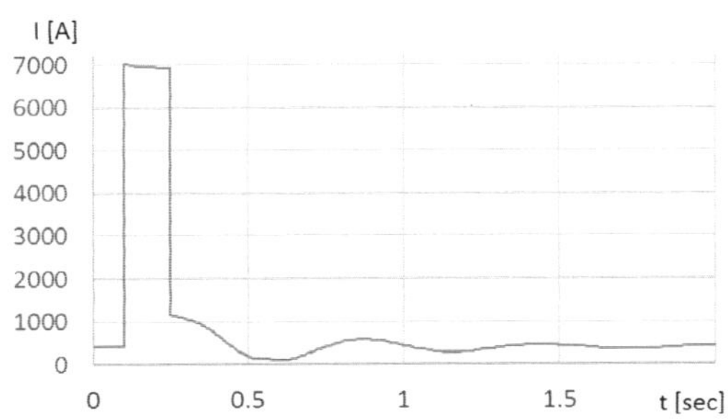

Figure 8. Waveforms of short-circuit current flowing through power line L5

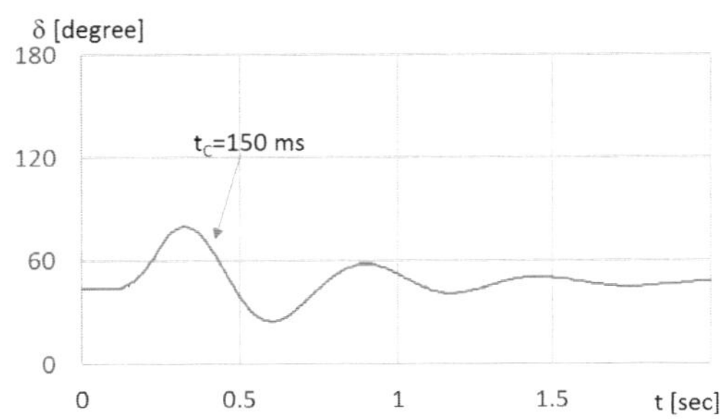

Figure 9. Waveform of rotor angle for clearing time 150 ms

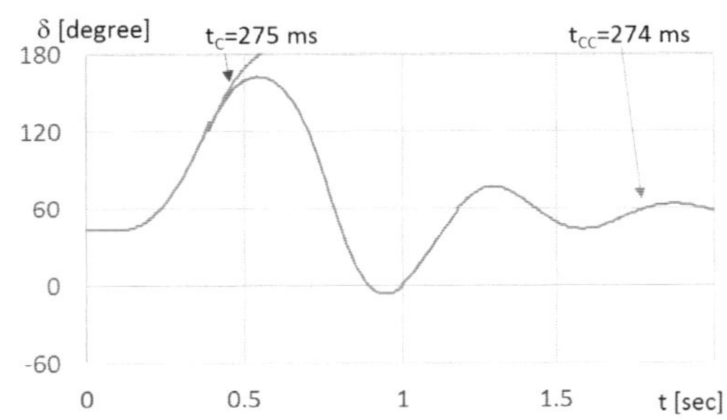

Figure 10. Waveforms of rotor angle for two clearing times

V. CONCLUSIONS

Dynamic stability calculations are fundamental standard of power system operational security. These calculations are performed in various stages:

- in phase of design of new source or future power system development,
- in preparation phase of power system operation (short-term operation, middle-term operation and so on),
- in real time,

- in fault analysis ex-post.

The simulation calculations confirm the validity of the transient stability criterion based on the network's short-circuit power $S_k^{''}$ and the rated active power P_n of the synchronous generator. The criterion is suitable for a cursory assessment of the transient stability in the stage of system development, when the details of system operation are not yet known, but these two basic values are already determined.

According to the European network code on requirements for grid connection of generators [6] relevant system operator should provide values of minimum short circuit power at each connection point. Depending on this value, it is possible to estimate the maximum size of connected units while ensuring the transient stability.

Analogical, it is possible this criterion adjusts for cursory assessment of steady-state stability or voltage stability.

References

[1] K. Máslo, P. Toman, B. Bátora, and J. Koudelka, "Methods for Dynamic Stability Assessment," Proc. 20th International scientific conference Electric Power Engineering (EPE 2019), May 2019

[2] J. Machowski, S. Robak, P. Kacejko, P. Miller and M. Wancerz, "Short-circuit power as important reliability factor for power system planning," Proc. 18th Power Systems Computation Conference, Aug. 2014

[3] P. Kundur, Power system stabilityand control, McGraw-Hill, 1994, pp. 843

[4] TR 60909-4, Technical report: Examples for the calculation of short-circuit currents

[5] K. Máslo, "Short circuit curret calculations," Proc. 13th International Scientific Conference Control of Power Systems (CPS 2018)

[6] *Commission regulation (EU) 2016/631 of 14 April 2016 establishing a network code on requirements for grid connection of generators*

Intellectual control system of switching devices of high-voltage substations

Sergey Sarry, Vladimir Nagay, Pavel Kireev, Ivan Nagay, Igor Trotsenko
Dept. Electric power stations and electric power systems
Platov South-Russian State Polytechnic University (NPI)
Novocherkassk, Russian Federation
sv@sarry.ru, nvi53@mail.ru, kireevps@yandex.ru, nagayiv@mail.ru, tiv110865@yandex.ru

Abstract—**The operational control systems (interlocks) of switching devices (disconnectors) are used to exclude the wrong actions of personnel during operational switching. A variant of designing an intellectual control system for switching devices using digital technologies is considered. A feature of the system is the use of inductive position sensors of switching devices. The use of specialized controllers and digital information exchange between them allows increasing the informational perfection of the system, reducing the total length of the cable network at the substation and reducing the costs of its operation. It is proposed to use standard communication protocols, including IEC 61850 to integrate the locking system into the substation automated process control system.**

Keywords: interlocks; switching device; disconnector; automated process control system.

I. Introduction

The modern concept in the field of automation of substations of electric power systems is focused on the creation of modern integrated automated process control systems based on intellectual electronic devices (IEDs) using microprocessors and a developed information and computing structure [1].

The development of modern information technologies and their application in the tasks of automation and management of electric power substations allows the transition to the creation of new type substations - digital substations [2] due to the possibilities of remote access to equipment and its integration into a single active-adaptive network [3].

The features of the power generation and substation control systems of the new generation are: the presence of embedded the IED in the primary equipment, the use of local computer networks for communications, a digital method of access to information, its transmission and processing, automation of operation and control processes of substations. New regulatory documents and standards are being developed on the designing of such modern types of power plants and substations based on microprocessor IEDs. Currently, the standard is the standard for network systems and communications of IEC 61850 substations [4], which regulates not only the transfer of information between individual devices, but also the description of schemesof substation, protection, automation and measurement, device configuration. The standard introduces the concept of an information model, which describes the connections to the primary equipment, information about the outputs, connections and feeders, logical blocks, through which digital data exchange between control and protection algorithms performed by different IEDs is implemented.

Information communications on the modern substations are performed by digital ways. This opens up the possibility of a quick direct exchange of information between devices, which ultimately makes it possible to reduce the number of copper cables and the number of devices, as well as their more compact placing [5]. The use of digital exchange significantly changes the approaches to the implementation of many schemes, in particular, operational interlocks schemes (OI), ensuring the safe operation of switchgears when switching devices (SD) are operating [6]. In the framework of modern substations, OI are implemented on the algorithmic level [7].

II. OI System Requirements

OI systems should prevent accidental actions of men on dutyduring operations with switching devices, reducing the likelihood of fault situations and increasing the safety and efficiency of power supply. The OI withelectromagnetic lock iswidespreadin high-voltage switchgear for the SD [8]. It should be noted that the classical electromagnetic OI of the SD, due to the presence of numerous connections within the bay of switching devices and between bays, due to the large number of contacts used does not allow for a sufficient level of operational reliability and is not recommended for use at new generation substations. When designing modern OI systems with digital control principles, the basic functioning algorithms remain unchanged and are supplemented with elements for solving interface problems.

Based on the analysis of existing technical solutions, it is possible to formulate technical requirements for the intellectual control system of the SD. The system should provide:

- getting information about the status of the SD drive and its position at the current time with the registration of the astronomical or relative time of occurrence of the operation;
- control of SDdrivers using remote control for motor drives or issuing control actions on electromagnetic locks for manual operation;
- ensuring the logic of the drive of the device and the logic of blocking of the drive control;
- reception and transmission of information by the digital communication channels;
- self-diagnosis of devices and the entire system as a whole;
- action logging;

- blocking of drive control in the event of a communication link break;
- monitoring of controlled equipment, including control of the number of commutation by drives;
- the possibility of including the system in an automated process control system (APCS) at the substation.

The transmission of digital information can be carried out both via optical and wired communication channels using standard protocols. To create a control system, it is necessary to use microprocessor controllers operating under the control of an intellectual control device.

The intellectual control device OI (AKB on schemes) controls the position of SD on the power object and, in accordance with the specified logic, transmits information to the controllers, which in turn apply the supply voltage to the locks of those disconnectors that are can be switched at the current time.

To make decisions in AKB, it is possible to use mathematical methods with information redundancy, which allows performing the control of the veracity of the initial information using a binary model of an object with a network structure. The undoubted advantage of the binary model is the representation of links of any complexity [9]. Structural and functional redundancy implies an increase in the cost of equipment and information transmission channels. At the same time, it is sufficient to implement temporary redundancy in the system, which involves the use of multiple repetitions of messages to controllers via communication channels. Logical methods of reliability control, which are semantic character, are based on the analysis of the technological scheme and the logic of the data being processed and are used in conjunction with digital codes for the detection and correction of information transmission errors.

The general principle of the execution of an OI is to block the closing of earthing switch that connected to equipment which is not separated by a visible gap, and actions with disconnectors, through which load current flows. A mathematical model of an intellectual OI system that is part of the digital substation structure [10] can be represented as a set of input values $D=\{d_1, d_2,...d_n\}$, depending on the state of sensors that control the state of switching devices and their grounding blades, and a set of output values $Y=\{y_1,y_2,...y_n\}$ that express overreach and block signals in discrete form to the block locks of electromagnetic keys.

It is possible to increase the reliability and veracity of results on the position of disconnectors and grounding blades by using contactless position sensors. In order to research the prospects for use in intellectual OI instead of unreliable block of contacts type KSA, the authors analyzed the characteristics and capabilities of contactless position sensors with different principle of operation: optical, induction and capacitive. Features of the propagation of infrared radiation of optical sensors limit their use in the open air due to the influence of dust, smoke, and precipitation. When frost appears on the surface of the optical sensor at sub-zero temperatures, its response is not guaranteed; therefore measures are needed to prevent the appearance of frost. Capacitive contactless sensors do not meet the response requirements when used on open switchgear. The most convenient for use in systems of OI induction position sensors. In this case, the plate associated with the shaft of the switching apparatus can be made of any metal. In this case, the distance to the sensor, at which it operates, can be chosen experimentally, taking into account the design of the disconnector drive and its grounding blades. The test results showed that in operation, the distance reduction of an object to an induction sensor, as a rule, is made not along the relative axis, but perpendicular to it. In this case, the region of operation within the guaranteed interval of operation depends on the removal of the object from the active surface.

When the software is implemented in the control device, the OI logic circuits can be replaced by logical expressions with bit variables reflecting the position of the substation's SD. For example, the diagram in Fig. 1 matches the expression (1).

$$YQS1 = \overline{Q1} \wedge \left(\left(QS1 \wedge \overline{QSG1} \wedge \overline{QS2} \right) \vee \left(\overline{Q2} \wedge QS3 \wedge \overline{QSG2} \right) \right) \quad (1)$$

where $YQS1$ — is a bit variable that allows/blocks switching of the disconnector (1 "allowed", 0 "blocked"), $Q1$-$Q2$, $QS1$-$QS3$, $QSG1$-$QSG2$).

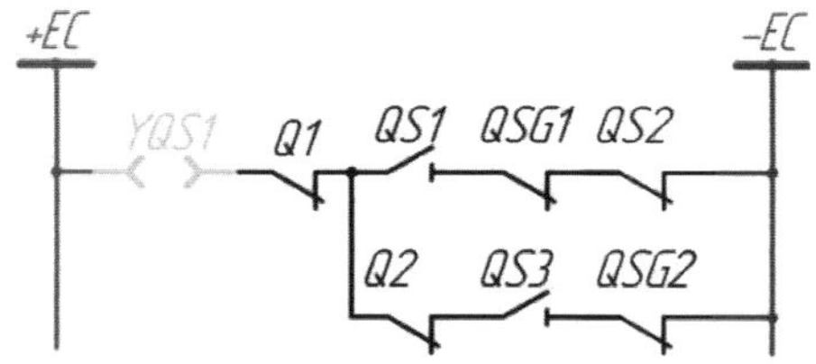

Figure 1. An example of an logicscheme of electromagnetic interlock

On the Fig.1 $YQS1$ - electromagnetic interlock of disconnector, $Q1$ $Q2$, $QS1$ $QS3$, $QSG1$ $QSG2$ - switches corresponding to the position of high-voltage switches, disconnectors and their grounding blades, respectively.

Theoretically, for the formation of bit variables in the random access memoryof the microprocessor of the OI control device, it is sufficient to install one "on" or "off" position sensor on each switching device (Fig. 2). Such solution has a minimal cost for the purchase and installation of sensors, but it has limited self-diagnostics capabilities, which leads to a high risk of the OI false operating as a result of sensor false or wire breakage.

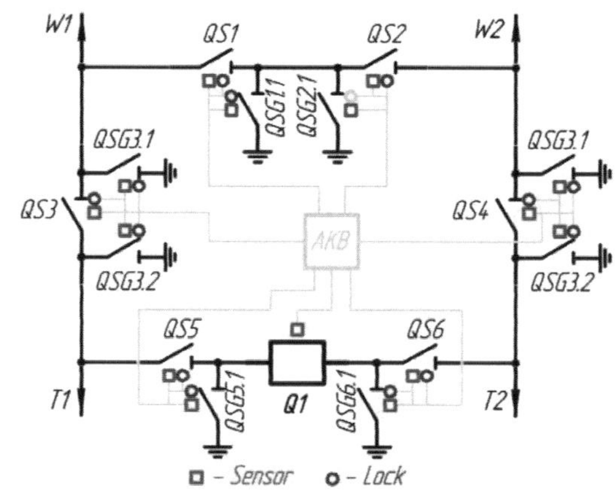

Figure 2. An example of the OI scheme with one position sensor and one lock for each switching device

Installing two position sensors (Fig. 3) on each switching device (one position sensor is "on" and one position is "off") allows us to separate the states of the switching device "closed", "opened", "intermediate position", and also some cases of sensor failure or cable connections.

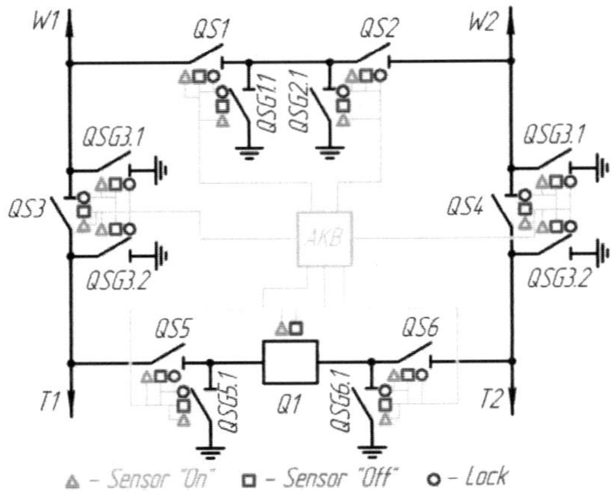

Figure 3. An example of the OI scheme with two position sensors and one lock on each switching device

The correspondence of the states of the sensors to the position of the switching device is given in Table 1. It should be noted that the state "intermediate position" may occur in cases of failure of the drive of the switching device or its sensors, as well as in the switching process of the switching device. In order to prevent blocking of the switching device's drive in the process of switching, it is necessary to introduce a time delay for fixing the "intermediate position" of the SD and system failure, exceeding the maximum switching time of the disconnector. We can also use additional features to differstate "switching" and state "failure".

Table 1. Correspondence of sensor states to the position of the switching apparatus

Sensor status "on"	*Sensor status "off"*	*Disconnectorstate*
non-operated	operated	"opened"
operated	non-operated	"closed"
non-operated	non-operated	" intermediate position"
operated	operated	"sensor failure"

Such sign can be:

1. The presence of current flowing through the electromagnetic lock of the disconnector with a manual drive. This current is present in the presence of a key in the lock, and, therefore, indicates a switching process.
2. The presence of voltage on one of the coils of the starters of the drive of the motor of the disconnector, which is present only when circuiting the complete chain of control of the specified starters since the moment of the start of switching to its completion.

It should be noted that due to the high switching speed of the switches and the absence of blocking of their operation, there is no need for an additional feature to recognize the switching mode of the switches.

III. Development and implementation experience

The authors have developed an intellectual control system for switching devices with the implementation of the operative blocking function for disconnectors BREM-01M with inductive position sensors [11]. The structural system for the substation with the main electrical circuit N-5 type (Russia) is shown in Fig. 3. The system includes (Fig. 4): a central control unit (CCU) that getting information from the state controllers of switching devices and bay controllers, as well as the formation of control actions for switching devices based on hard algorithms linked to the main electrical scheme of the substation; state controllers of switching devices (*A*), to which inductive sensors and electromagnetic interlocks are connected; bay controllers (*B*) for the interaction of controllers of switching devices, as well as monitoring the position of switches. The information communication channel between the system components is implemented using the RS-485 interface. To communicate with the automated process control system (APCS) of substations, another RS-485 port was added to the central control unit. The IEC 61850-8.2 protocols can be optionally implemented, however, in general, for simple substations this solution significantly increases the cost of the system. The increase in cost is due to the need to purchase additional communication adapters and IEC 61850 protocol drivers, the use of high-performance equipment certified for compliance with the standard and an increase in the complexity of project implementation due to the novelty of the technologies used.

With the help of the control system, the distributed [12] and centrally distributed [13] OI systems can be implemented.

The controllers of switching devices and bay controllers are installed on an outdoor switch-gear and near the high-voltage current-carrying parts. An experimental assessment of the electromagnetic environment at the equipment installation sites was performed using a PZ-60 electromagnetic field meter.

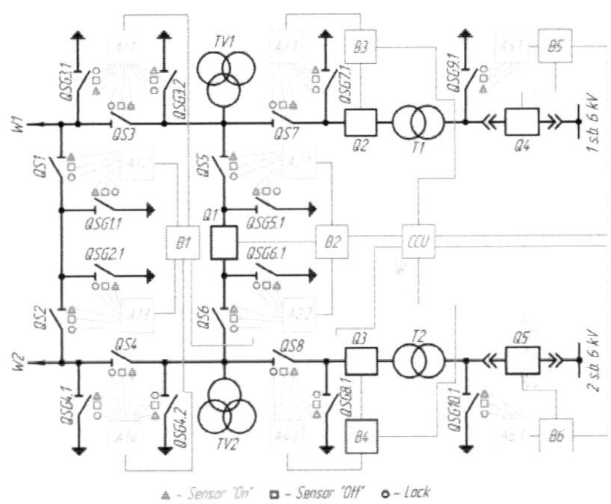

Figure 4. Structural scheme of BREM-01M

Measurements of electric and magnetic fields were carried out at three control points near each drive of the switching device: at the level of the drive at the intended location of the sensors; 0.5 meters higher and lower of it. The induction of the magnetic field did not exceed 3 μT, the magnetic field strength was 200 V/m. When designing an assessment of electromagnetic compatibility of controllers can be performed according to the method proposed in [14].

The connection between the controllers should be based on the deterministic principle: the master controls the operation of all slave devices in a predetermined order. In this case, all devices in the system must be registered, has their own self-control and reset function when the controller hangs when exposed by external electromagnetic interference when commutating disconnectors with a motor drive, followed by an arc discharge. The implementation of the system protocol should support the standard data transfer protocols currently adopted in the process control systems of substations, including those implementing digital substation protocols. For modern digital substations, the IEC61850 protocols should be supported [15], for OI reconstructions on the operating substations that do not have APCS systems, using MODBUS [16] with RS485 [17] or CAN [18] is sufficient.

The goals set in the work are implemented in software and hardware complexes of the intellectual control system of switching devices OI-type BREM-01M. Software algorithms for the operation of controllers implemented in C language AVR.

The advantages of the system include the possibility of supplementing the interlock logic with the help of the developed software "Diagnostic and Maintenance" during the reconstruction of the substation, with the input and output of equipment for repair. The software allows you to configure the equipment of the system, taking into account the primary scheme of the substation, to create the logic of operating and monitor the system in the construction and operation. Due to the possibility of distribution of functions among the devices of the system and free access to the information of all digital devices connected to the system's network, the efficiency of OI increases.

The software package for maintenance and diagnostics of the intellectual OI control system allows you to create configuration files in accordance with the main electrical scheme of the substation, perform a full diagnostics and check the operability of bay controllers and switching devices and display information about the position of switching devices, enabling signals on the block-locks and on an electromagnetic key inserted into the lock.

Works on the creation of an intellectual control system for switching devices of high-voltage substations wereimplemented in the framework of R & D with IDGC of the South, PJSC [19]. For the intellectual blocking system type BREM-01M, a reuse project has been prepared, recommended for universal use in Rossetti PJSC [20]. The developed systems were installed at 8 substations in branches of IDGC of the South.

IV. Conclusions

1. The use of induction position sensors and digital methods of information processing allows to increase the reliability of information about the position of the switching devices and, as a consequence, the reliability of the operational interlock system.

2. Increasing of the information base of the electromagnetic OI system by installing needy position sensors on each switching device and monitoring the state of controlled elements (electromagnetic locks and starter coils) allows identifying the state of a specific switching device with a high degree of veracity, which is of crucial importance of logic scheme functioning of OI and, consequently, the failure-free operation of the OI.

3. The intellectual control system OI can be a part of the equipment of a digital substation, as it has, besides the performance of its main functions, integration capabilities in an APCS, continuous monitoring, an advanced interface and advanced configuration functions.

4. Implementation open information transfer protocols in intellectual systems OI allows integrating them with terminals and controllers of various manufacturers.

The interaction of the operative interlocking system of the disconnectors BREM-01M with the APCS of the substation enables it to more effectively solve operational management tasks.

References

[1] The concept of the development of relay protection and automation of the power grid. Attachment No. 1 to the Protocol of the Board of JSC Rossetti dated 22.06.2-15 No. p.356, Relayshik, 2015. No3, pp 28-43.

[2] S.I. Chichev, V.F. Kalinin and E.I. Glinkin, "Methodology of designing a digital substation in the format of new technologies," Spectrum Publishing House, Moscow, 2014.

[3] V.V. Dorofeev and A.A. Makarov, "Active adaptive network - new quality of UES of Russia," vol. 4 (15), Energoexpert, 2009, pp. 29-34.

[4] A.O. Anoshin and A.V. Golovin, "IEC 61850 Standard. Document Structure," The news of electrical engineering,vol. 4(76), 2012.

[5] T.G. Gorelik and O.V. Kirienko, "Design issues for substation automation and control systems based on the IEC 61850 standard",International Scientific and Technical Conference "Modern directions of development of relay protection and automation of power systems," 4th International Scientific and Technical Conference "Actual Trends in Development of Power System Relay Protection and Automation",2013, Ekaterinburg.

[6] V.O. Tazin, A.A. Voloshin and B.K. Maksimov, "The use of multi-agent technologies for solving problems of adaptive automation of operative interlocking control of disconnectors and grounding blades at 110-750 kV substations," Vestnik MEI, vol. 6, 2015, pp. 48–55.

[7] The order of the organization of an operational interlocking on substations of new generation, the order №236p JSC FGC UES 05/05/2010, Publishing house of JSC FGC UES, 2010, p.10.

[8] A.A. Filatov, "Switching in electrical disconnect devices," Moscow, Energia, 1973, p.88.

[9] S. Russell and P. Norvig, "Artificial Intelligence: A Modern Approach," Williams, 2nd ed., 2007, p.1410.

[10] V.I. Dubrov, R.G. Oganyan, D.V. Shaykhutdinov, E.V. Kirievskiy, N.I. Gorbatenko and N.D. Narakidze, "Development of mathematical models of digital substations based on the analysis of the automated technological systems of electric power conversion," Modern high technologies, vol. 9, 2016, pp.36-40.

[11] S.V. Sarry, P.S. Kireev and A.V. Ukraintsev, "Intelligent interlocking of switching devices," 2018 International Multi-Conference on Industrial Engineering and Far Eastern Technologies, IEEE press, 2018, pp. 1-4. doi: 10.1109 / FarEastCon.2018.8602512.

[12] N.O. Kalinina, V.I. Nagay, S.V. Sarry, N.V. Goncharova and N.A. Nedosekov, "Distributed system of interlocking of switching devices of substations of electric networks," Russian Electromechanics, vol. 2, 2012 , pp.106 – 107, ISSN 0136-3360.

[13] N.O. Kalinin, V.I. Nagay and S.V. Sarry, "Centrally-distributed system of electromagnetic interlocking of switching devices," Russian Electromechanics, vol.1, 2013, pp.121-123, ISSN 0136-3360.

[14] E.D.Berezkin, V.I. Nagay, S.V Sarry, "Evaluation method of the spatial configuration of induction interference," 2016 2nd International Conference on Industrial Engineering, Applications and Manufacturing, ICIEAM 2016.

[15] A.A. Voloshin, Ya.L. Artsishevsky and A.V. Zhukov, "Principles of Creation of Intelligent Systems of Relay Protection and Automation", 5th International Scientific and Technical Conference "Actual Trends in Development of Power System Relay Protection and Automation," 2015, Electronic Resource. http://cigre.ru/activity/conference/relayprotect5/participants/materials/.

[16] Modbus messaging on TCP / IP implementation guide, V.1.0a, www.Modbus-IDA.org, June 4, 2004, 46 p.

[17] CAN open electronic data sheet specification for CANopen,CiA draft standard 306, v. 1.3, Jan. 01, 2005, p.30.

[18] TIA/EIA standard TIA/EIA-485-A, Electrical Characteristics of Generators and Receivers for Use in Balanced Digital Multipoint Systems, Telecommunications Industry Association, March 3, 1998.

[19] TIA/EIA standard TIA/EIA-485-A, Electrical Characteristics of Generators and Receivers for Use in Balanced Digital Multipoint Systems, Telecommunications Industry Association, March 3, 1998.

[20] Protocol of the meeting of PJSC "Rosseti" No.SM5-1000-3182, 27.07.2016.

Designing a backup system of relay protection for multi-end overhead lines

Ivan Nagay, Vladimir Nagay, Pavel Kireev, Vladimir Jr. Nagay, Sergey Sarry
Dept. Electric power stations and electric power systems
Platov South-Russian State Polytechnic University (NPI) – «RELDON»
Novocherkassk, Russian Federation
nagayiv@mail.ru, nvi53@mail.ru, kireevps@yandex.ru, wwn_1978@mail.ru, sv@sarry.ru

Abstract—**One of the numerous elements of modern distribution electric networks with a voltage of 6-110 kV in Russia are multi-ended overhead lines with branch substations with transformers rated at 2.5–25 MVA. Designing a backup protection system for these substations is difficult due to the comparability of short-circuit currents on the side of lower voltage with the current flowing between the supply substations in the normal regime. The authors proposes a backup system, the algorithms of which are based on multiparameter monitoring of signals from both the substation sides: same side and monitoring the parameters of the information signs of signals from other substations using device associated with signaling system for transmitting alarm signals and commands.**

Keywords: relay protection; short circuit; longitudinal-transverse asymmetry; recognition.

I. Introduction

Stable functioning of modern electric power systems is impossible without equipping them with relay protection and automation (RPA), providing shutdown of damaged elements of systems during short circuits (SC), open-phase regimes and unacceptable overloads. As a rule, relay protection (RP) is performed in the form of autonomous devices integrated into the protection system, although now with the development of microprocessor technology and the use of modern communication channels and communication protocols, for example, IEC 61850, the possibility of protecting several objects up to the entire substation by the one complex RPA. In this case, the relay protection system can be both centralized and distributed. However, the difference of the traditional approach is that the combination of individual protections into the system does not exclude their autonomous action. The principle of RP operation is chosen taking into account the exclusion of "dead" zones and the philosophy of designing relay protection involves multiple backup, which practically provides a practical guarantee for the elimination of a fault.

At the same time, backup or reserving (lat. Reservare - to save, save) is such protection that is provided for work instead of the main protection of the element in question in cases of failure or decommissioning (short-range backup) or instead of protection of adjacent elements in case of their failure or in case of failure of switching devices of the adjacent elements (remote backup). Backup is accomplished by introducing redundancy and / or using other principles for designing relay protection compared to protection with absolute selectivity to solve the problem of increasing the efficiency of relay protection.

Relay protection systems can be built on the basis of the following components: local backup (LB), breaker failure protection, remote backup (RB). Currently, in terms of backup protection of energy facilities, considerable international experience has been accumulated, many problems and problems has been solved. Improved backup protection contributed to the development of the theory of RP and adaptive algorithms of functioning, the active introduction of microprocessor technology, increased use and improvement of communication channels, manufacturer's competition.

Some challenges are the provision of due technical excellence backup protection with relative selectivity. The spread of new principles of designing electrical networks leads to the need to solve the problems of providing protective functions for networks with distributed generation, active adaptive networks and etc. However, some calls for backup protection are not caused by the current development and innovations of protection objects.

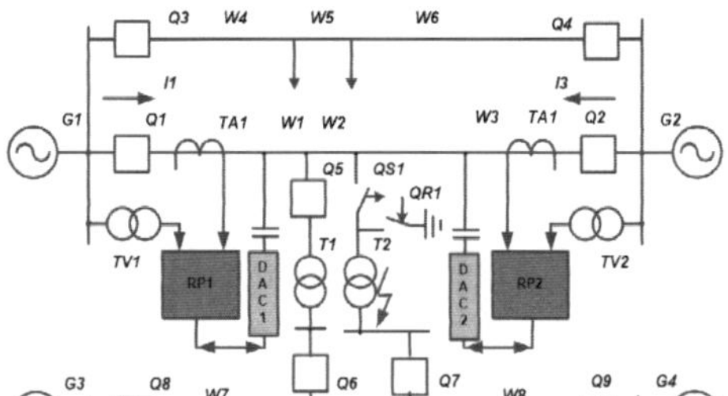

Figure 1. Scheme of the protected network

But they caused by existing technical solutions of the previous period of power industrydevelopment, in particular the form of substations with simplified schemes of primary and secondary connections. In conditions of limited investment in the power industry and the increasing maintenance time of the main equipment and RPA, the fault rate increase is possible, which requires correspondingly effective technical solutions to ensure backup of the protection and switching equipment of such substations. The problem of backup of relay protection is aggravated by electrical networks with transformers with a capacity of less than 10 MVA and on transit overhead lines.

This article addresses the issues of improving the efficiency of backup protection of branch and intermediate substations, which are one of the main power sources for consumers, which involves ensuring their high reliability and requires the use of a backup system with high technical excellence. Recognition of fault conditions for transformers *T1,T2* with a capacity of 2.5-25 MVA (Fig. 1) by remote backup protection (RB) *RP1,RP2* is difficult due to the comparability of short-circuit currents behind protected transformers with maximum load currents, magnetizing inrush (MI) when the transformer is turned on at idle, start-up and self-starting currents of high-power high-voltage electric motors (EM) [1-8], switching currents of capacitor banks (CB), the influence of transient resistances in the place of fault, low voltage levels of symmetrical components on buses in the place of installation of protection. Replacing of the fault throwing switches *QR1* and isolators *QS1* by the switches *Q5* (Fig. 1) on these substations does not lead to a fundamental solution to the problem of backup.

The sensitivity of the RB protection is determined by the maximum loads of the supply lines, the presence of the motor load on the low voltage side (LV) of the transformers, two- and multilateral power supply from both the highest voltage (HV) and the low (medium) voltage, which relevant in the implementation of distributed generation systems *G3, G4*. An illustration of the relevance of the problem of recognition of fault behind transformers *T1,T2* is the field of permissible and fault regimes of the considered substations connected to transit overhead lines *W1,W2,W3* (Fig. 2). At the same time, the intersection of the load and faultregimes in the case of a three-phase short circuit behind a 10.0 MVA transformer of the branch station both from the supplying substation (*Z1ld, Z1sc* areas) and from the receiving substation (*Z2ld, Z2sc* areas) was noted. In this case, the resistance of the transformer is taken as the base resistance *T2*. This makes it impossible to recognize the fault regime on the background of the permissible load regime.

II. Recognizable Regimes and Influencing Factors

As mentioned above, the recognized regimes of backup protection of transformers on the branch and intermediate substations are: maximum load regimes, start-ups (self-starts) of high-voltage electric motors, switching on the load in the auto reclosing regime, switching capacitor batteries, turning on transformers at idle, accompanied by MI, phase-to-phase short circuits (two-phase and three-phase), open-phase regimes (OPR), as well as a combination of the above regimes (Fig. 3).

In addition, it is necessary to take into account influencing factors, for example, the presence of arc resistance in the place of short circuit, errors of measuring current and voltage transformers. The transient resistance at the fault site leads to a decrease in the modulus and argument of the short-circuit current and the shift of the area of fault regimes towards the area of load regimes.

The presence at the substations of a powerful motor load leads not only to an increase in the modules and arguments of currents but to the turn of their vectors towards the area of fault regimes. So, the starting current module I_{*1} on the supplying substation in the basis of the current of the dead short circuit $I_{*T\,i}$ behind the protected transformer is equal to [1]:

$$I_{*1} = \sqrt{m^2 + 1 + 2m \cdot cos(\varphi_L - \varphi_T)} \qquad (1)$$

where $m = I_{S\,EM}/I_{T\,i}$, so

$$m = -k_U\, cos(\varphi_{S\,EM} - \varphi_T) + \sqrt{k_U^2\, cos^2(\varphi_{S\,EM} - \varphi_T) - (k_U^2 - 1)}$$

The arguments of the load currents I_L are in the range φ_L=-(10°÷30°), the arguments of the start-up currents $I_{S\,EM}$ and short-circuit behind transformers I_{Ti} respectively $\varphi_{S\,EM}$=-(50°÷70°) and φ_T=-85°, the voltage of the successful start (self-start) of the EM, $U_{EM} \geq k_U \cdot U_{nom}$, k_U=0.55÷0.7, U_{nom}- nominal voltage.

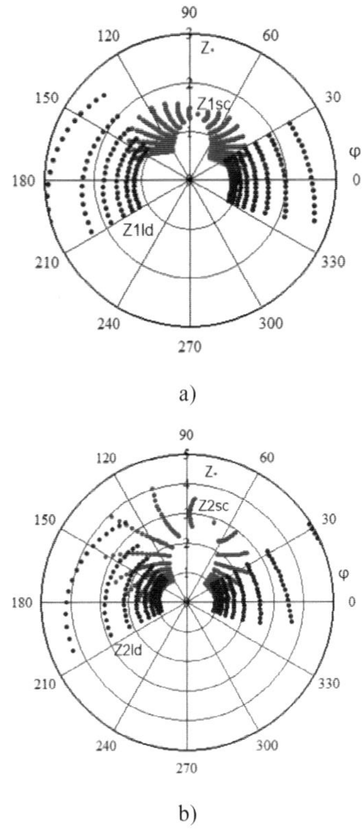

Figure 2. Areas of resistance of load (*Z1ld, Z2ld*) and fault (*Z1sc, Z2sc*) regimes of transit overhead line with branch stations for supply (a) and receiving (b) sides

The presence of transient resistance leads to the displacement of the area of fault regimes to the area of the starting regimes of EM and their possible intersection. The transient (arc) resistance at phase-to-phase short-circuit is determined by the parameters of the electric arc burning between the current-carrying parts.

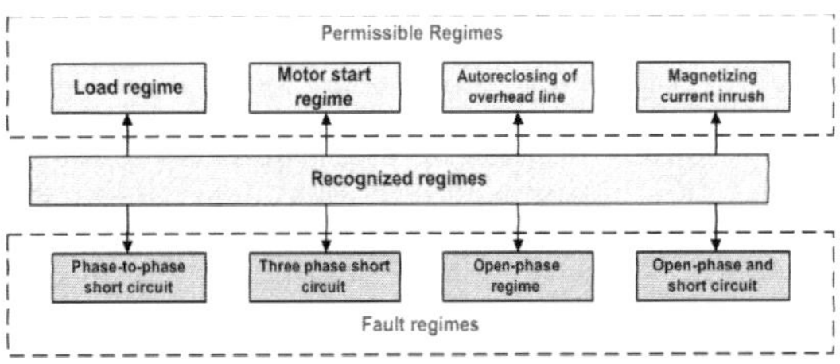

Figure 3. Permissible and fault regimes recognized by backup protection

As a rule, the voltage on it does not exceed U_a=(0.3-0.4)$\cdot U_{nom}$, and the resistance of a short-circuited circuit Z_{SC} with a transformer, taking into account the transient resistance of the electric arc [1], will be equal (2).

$$Z_{SC} = Z_T \left[\frac{j2A_a \sin\varphi_T + \cos\varphi_T (2A_a - 6U_{*a}^2) +}{+\sqrt{(6U_{*a}^2 \cdot \cos\varphi_T)^2 - 12A_a U_{*a}^2}} \right] \Big/ 2A_a \quad (2)$$

where Z_T, R_T, X_T – impedance, resistance and reactance of the transformer; $A_a=3U^2_{*a}-1$; $U_{*a}=U_a/U_{nom}$ – the voltage on the arc column referred to the nominal voltage.

In the networks under consideration, the regimes of longitudinal-transverse asymmetry (LTA) are also not excluded, due to the breaks of the phase wires of overhead lines and simultaneous faults in the network of higher or lower voltage. For example, a phase wire break with simultaneous short circuit from the side of the transformer of the branch substation is not accompanied by an increase in currents to levels comparable to short-circuit currents. At the same time, in most cases, relay protection of overhead lines with branching or intermediate substation does not have sufficient sensitivity to the considered regimes [1-8], despite the fact that this may lead not only to energy losses in distribution networks, but also cause damage to transformers due to for a long flow of currents exceeding their nominal currents by 2-4 times [1,4,6,8], electric motors, overvoltage in the ungrounded neutrals of power transformers, damage to measuring voltage transformers, possibly unnecessary action of RP on the adjacent parallel line. Phase break and even its short-circuit from the powered substation in which the fault throwing switch is installed may entail the impossibility of switching off the damaged substation from the supplying substation. This is due, on the one hand, as a rule, insufficient sensitivity of the RP on the supplying substation and a small change in the short-circuit current when the fault throwing switch is turned on, since in this case, its level changes by no more than 5-10% [1,7,8].

III. Ways to solve the problem

The analysis of fault and permissible regimes, conducted above, indicates the presence of a problem of their recognizability based on the use of traditional information signs (modules, arguments of currents and resistances, for example, Fig. 2). At the same time, the use of other informational signs, including fault components of currents, resistances (their increments) (Fig. 4) allows us to separate these areas of load (*Z1ld*, *Z2ld*) and fault ($\Delta Z1sc$, $\Delta Z2sc$) regimes. In this case, the area of fault regimes is narrowed and moves away from the area of permissible (normal) regimes. Based on this, we can conclude that the effective ways to increase the sensitivity of remote backup relay protection are the expansion of the information base (using the principle of multiparameter protection) and the implementation of its measuring units based on adaptation to the protected network regime and controlling the fault components of currents and resistances [1,2 ,7-9].

When implementing multiparameter protections (Fig. 5), it is supposed to expand the space of information signs (1-4) with the control of their parameters in interrelation with each other.In particular, this allows detecting developing faults (transition of two-phase short-circuit faults to three-phase (5)) and to provide correction of monitored signals (6) due to changes in load parameters both at the damaged substation and at adjacent substations, in particular with regard to static load characteristics.

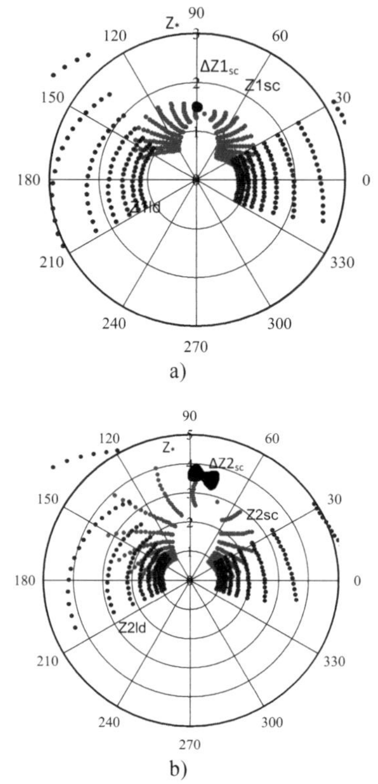

Figure 4. Areas of resistance load (*Z1ld*, *Z2ld*) and fault (*Z1sc*, *Z2sc*) regimes and their increments ($\Delta Z1sc$, $\Delta Z2sc$) transit overhead line with branch stations for supply (a) and receiving (b) of its sides

It is very important to monitor the state of switching devices and devices of adjacent relay protection devices at a protected substation (7), which allows for reducing the time of backup protection and, in some cases, increasing their sensitivity. The presence of information exchange between the substations of transit overhead lines with the help of alarm transmission devices and commands (DAC) (Fig. 1) also allows to achieve the above result [10]. In this case, there are two options for transmitting information about the direction of power on the sides of the protected line. In the first case, the transfer of information about the argument of a specially formed current on the sides of the substations, and in the second case, information on the state of the directional measuring units at the same substations. The use of these methods of information transfer is widely used in the use of differential-phase and high-frequency protections with blocking and permissive signals). However, for backup protections, these methods have so far practically not been used.

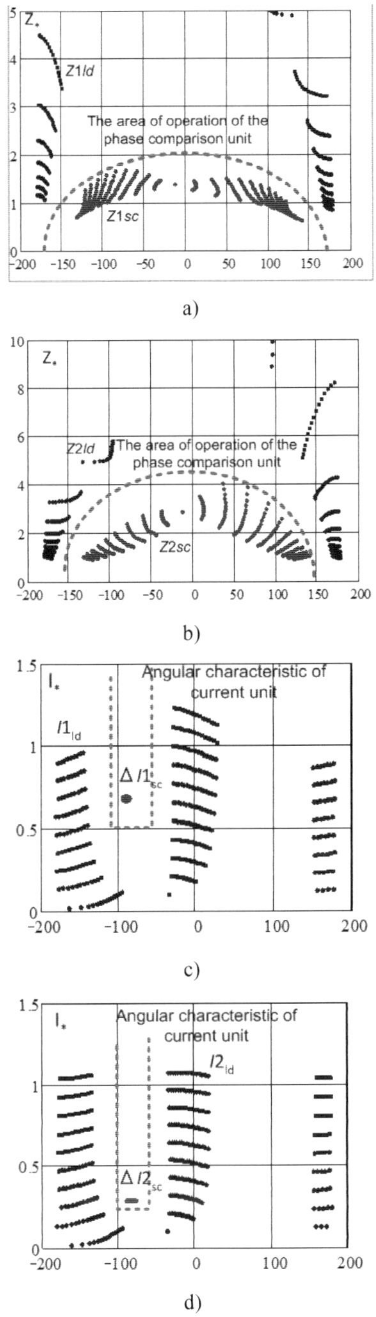

Figure 5. The area of operation of the phase comparison units depending on the resistance (a, b) and directional current units (c, d) of the transit overhead line with branching stations for the supply (a, c) and receiving (b, d) sides of it (load resistance areas ($Z1ld$, $Z2ld$) and increments ($\Delta Z1sc$, $\Delta Z2sc$) of fault regimes; areas of load current ($I1ld$, $I2ld$) and increments ($\Delta I1sc$, $\Delta I2sc$) of fault regimes)

The efficiency of using current argument control algorithms at the ends of the protected lines with such substations is illustrated in Fig.5. In this case, dotted lines indicate the areas of operation of the proposed protection, which clearly separate from each other the areas of fault and permissible regimes. According to the authors, the most preferred technical solution is the indirect principle of controlling the phase argument of currents at the ends of the protected line, i.e. use of protection with HF power line carrier blocking.

The considered approaches to designing a backup protection are implemented in a number of microprocessor-based protections developed by the authors of this work and have been tested in 110 kV electrical networks.

IV. Conclusions

1. To solve the problem of designing backup protection, ensuring recognition of electrically remote fault behind transformers of branching substations of transit overhead lines, the most effective are: multiparameter method, direct and indirect control of the phases of currents at the ends of the protected line, use of fault components of currents and resistances.
2. The response characteristics of the phase comparison and current directional unit should be based on the total or fault components of the resistances and currents.
3. To increase the efficiency of fault recognition behind transformers, increase sensitivity and reduce time exposures for backup relay protections of branch stations of transit overhead lines, it is proposed to use alarm transmission devices and commands

References

[1] V.I. Nagay, "Relay protection of the tapping substation in the electric networks," Moscow, Energoatomizdat, 2002, p.312.

[2] I.V. Nagay, "Forming response characteristics of backup overhead line protection with branches," News of Universities. Electromechanics,vol.2, pp. 56-61, 2011.

[3] I.F. Maruda, "Solving the problem of relay protection of dead-end 110-220 kV overhead lines with parallel induction with mutual induction," Energetik, vol.6, pp. 28-29, 2001.

[4] I.F.Maruda, "Relay protection of 110–220 kV lines during phase breaks," Power Technology and Engineering, vol.1, pp.40-42, 2002.

[5] I.F.Maruda, "Relay protection of walk-through substations without voltage transformers on the side of 110 kV",Powerl stations, vol.4, pp. 52-56, 1999.

[6] I.F.Maruda, "Relay protection of step-down transformers against short circuits on the overhead line during phase breaks,"Power Technology and Engineering, vol.2, pp.44-46, 2003.

[7] V.V. Nagay, "Analysis of the recognizability of asymmetrical short circuits behind the transformers of branch and feed-through substations," News of Uiversities. Technical science, Special edition, pp.46-49, 2003.

[8] I.V. Nagay, "Providing the functions of long-distance backup relay protection of transformers in conditions of longitudinal-transverse asymmetry," News of Universities. North Kavkaz region. Tech. science, vol.5, pp. 19-24, 2011.

[9] G.Kharun, B. Litash, "Innovative protection of long-range and near-redundant transformers," Power industry. Transfer and distribution, vol.3(29), pp.88-92, 2015.

[10] S.L. Kuzhekov, V.I. Nagay, A.A. Degtyarev, I.V. Nagay, D.B. Antonov, D.N. Kokulin, B.S. Litash, G.V. Kharun, "Phase comparision protection of overhead lines with the function of backup relay protection of power transformers and switches of substations connected to the line," Power industry. Transfer and distribution, vol.1(52), pp.108-115, 2019.

Recognition of longitudinal-transverse asymmetric faults in electrical networks

Vladimir Nagay
Dept. Electric power stations and electric power systems
Platov South-Russian State Polytechnic University (NPI)
Novocherkassk, Russian Federation
nvi53@mail.ru

Vladimir Jr. Nagay
Dept. Relay Protection and Automation
Branch of company "Energo-Yug" – "Yuzenergosetproekt"
wwn_1978@mail.ru

Ivan Nagay, Pavel Kireev, Sergey Sarry, Igor Trotsenko
Dept. Electric power stations and electric power systems
Platov South-Russian State Polytechnic University (NPI)
Novocherkassk, Russian Federation
nagayiv@mail.ru, kireevps@yandex.ru, sv@sarry.ru, tiv110865@yandex.ru

Abstract— **The article is devoted to improving the informational perfection of relay protection of electrical distribution networks (overhead lines, transformers). It shows the need to use special protection against longitudinal-transverse asymmetry (LTA) in networks with branch-line and intermediate substations, which will prevent damage of transformers and minimize energy losses. It is performed the estimation of recognizability of LTA regimes when relay protection uses modules and arguments of phase currents, currents of symmetrical components as information signs.**

Keywords: relay protection; short circuit; longitudinal-transverse asymmetry; recognition.

I. Introduction

Regimes of longitudinal-transverse asymmetry caused by the breaking of the phase wires of overhead lines (OHL) and simultaneous short circuit (SC) in the network with an effectively-grounded neutral in many cases are not accompanied by an increase in currents to levels comparable with short-circuit currents [1-4] in electrical distribution networks of 110-220 kV, which explains less attention to them.

At the same time, in most cases, relay protection (RP) of an overhead transmission line with branch substations do not have sufficient sensitivity to the considered regimes [1-3], despite the fact that it can lead not only to electric power losses in distribution electric networks, and cause damage of the power transformers due to the long flow of currents exceeding the nominal currents by 2-4 times [1-3], electric motors, overvoltage in the ungrounded neutrals of power transformers, the possible excessive effect of the RP of the adjacent parallel lines [2,3].

Phase wire rupture and even short circuit on the side of the powered substation, on which the switching device is installed (circuit-breaker, fault throwing switch with an automatic disconnector), may not cause disconnection of fault on the supply substation. It is due, as a rule, insufficient sensitivity of the relay protection on the supply substation and a small change in short-circuit current when the fault throwing switch of the damaged phase is turned on, because in this case, its value of current changes by no more than 5–10% [1,4,5].

Regimes of longitudinal-transverse asymmetry on the tapped sections of transit overhead lines are not calculated in this article, because in this case, the line will be disconnected on at least one side and the damaged line will be operating in radial scheme. In this case, the regime of operation of a parallel overhead line is of significant importance, which is especially evident when there is mutual induction between the chains of lines. Such regimes include: operation of one of the lines in the radial regime, disconnection and grounding of one of the lines on one or two sides.

It should be noted the effect of transient resistance in the place of the earth fault, when phase-to-phase short-circuit on the low voltage side of transformers of the branch substations, preceding the load regime, mutual induction between the circuits of the protected OHL.

Review of existing solutions. In spite of sufficiently in-depth researches of the regimes of longitudinal-transverse asymmetry [4, 5] in distribution electrical networks 6-110 kV, there is no special protection against these types of fault [1-3] and even the requirements for this protection. In networks with an effectively-grounded neutral, the main protection against these types of fault is current (directional) neutral current relay protection (DNCRP), the sensitivity of which is, as a rule, insufficient according to the conditions of technical implementation or the conditions of grading from LTA regimes on adjacent network elements. The overcurrent protection of the transformer measuring the difference of the phase currents, when one phase breaks and it closes with the other phase is insensitive. A number of technical solutions (relative measurement of zero and negative sequence currents, measuring of the difference of phase current modules, measuring of phase current increments and other methods) [1-3,6,7], allows in many cases to solve the problem of designing the protection against LTA, but according to the pointed earlier reasons are not widespread. The above arguments, as well as a number of researches on this problem [1-3] confirm the need to install a special RP in 110 kV networks with higher sensitivity by increasing its information perfection, which would ensure the selection of the type of fault and the formation of an appropriate operation on switching devices with different Time delays.

Determination of the area of existence of the regimes LTA. The considered approach in designing protection against LTA is shown on the example of a network with a radial overhead line (Fig. 1) supplying the branch substations, with parameters: Z_G=3.5+j30 Ohm - resistance of the $G1$ system; Z_{W1}=3.2+ j4.5 ohms, Z_{W2}=4.8 + j6.75 ohms - resistance of the first and second sections of the main overhead line; Z_{T1}=4+j225 ohms and Z_{T2}, taking the value from 46.6+j555 ohms to 4.38+j86.7- resistance of the transformers of branch stations $T1$, $T2$. The power of the transformer $T2$ varied from 2.5 to 16 MVA, and the load power of the transformers $T1$, $T2$ varied in the range (10 ÷ 100%) of their nominal power, the argument of the load current φ_{load}=-(20-40°), the resistance of the branches is an order of magnitude smaller than the resistances of the overhead lines of overhead lines, and even more transformers, and therefore in the calculations were not taken into account. The arc resistance at the site of fault varied in the range of 5-50 ohms.

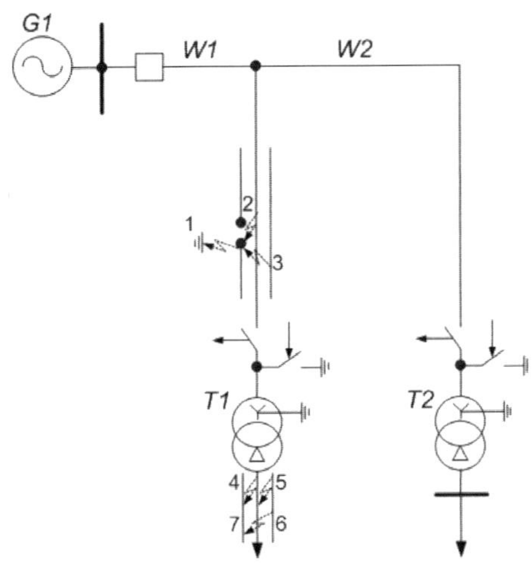

Figure 1. Radial OHL 110 kV with branches

The following LTA types are considered: a break of the phase wire and earth fault of phase A on the first overhead line $W1$ with a short-circuit on the side of the transformer $T1$ (1 - regime AY), a break of the wire of the phase A and short-circuit phases AB (2 - ABY), phases CA (3-CAY), a break wire of the phase A and short circuit on the low voltage side of the transformer $T1$ with the star-delta winding connection group of phases AB (4-ABΔ), BC (5-BCΔ), CA (6-CAΔ), ABC (7-ABCΔ). The first three types of complex faults are the development of one simple fault: 1 - falling of the broken wire to the ground, 2 and 3 - clashing (or breakdown of the air insulation) of the broken phase wire with another phase wire. The last four regimes can occur when the overhead line operates in an incomplete phase regime, as is practiced in some power systems, one of the poles of a switching device of the supplying, branching or intermediate substations, a phase wire, for example a stub line on an anchor tower, is breaked. Short circuits of phases B and C with simultaneous rupture of the wire of phase A, as well as phase-to-phase short-circuits are excluded from this list due to significant levels of currents and the sensitivity of the RP of the supply substation.

The asymmetric fault in the network causes the appearance of symmetrical components and a change in phase currents, their arguments, which is reflected in Fig. 2 (transformer $T2$ power is equal to transformer $T1$ power, neutral transformer $T2$ is not grounded). In this case, the nominal current of the transformer $T1\,I_{nom}$ is taken as the base current.

In this case, the three-phase short-circuit current behind this transformer is approximately 10 times higher than the nominal current (the impedance voltage of the transformer is equal U_{*k}=0.105). Phase currents in all cases exceed the nominal current of the protected transformer, which once again shows the inadmissibility of its long-term operation in LTA regime not only if there is fault on the low-voltage side, but also with LTA in the high-voltage network.

Evaluation of the selection regimes LTA. As starting units for the proposed protection, zero-sequence current units with an operating current value $I_{*set} \geq kI_{*sum}$ can be used. If its value $I_{*set} \leq (0.05 \div 0.1) I_{*sum}$, where I_{*sum} is the maximum load current of the supply line, then it is possible to use measuring units of current symmetrical component increment or phase current increment. For the designing of selective and highly sensitive protection, it is important to recognize LTA regimes with a selection of the type of fault. At the same time, it is necessary to form the set of informational signs [8-11], which allow one to obtain non-intersecting regions of the regimes considered above. If information signs of regimes are designated $I_{\varphi j}$ and j=1,2..,M, then each regime in M-dimensional space of signs can be represented as a vector $\underline{I}_{\varphi j}=(I_{\varphi 1}, I_{\varphi 2}, ..., I_{\varphi M})$, the coordinates of which characterize the properties of the protected object. To determine a measure of proximity or similarity between recognizing regimes in the M-dimensional space of signs, it is possible to introduce a Euclidean metric (1-4):

$$d^2(\omega_{pk}, \omega_{ql}) = \sum_{j=1}^{M} (I\varphi_{pk}^{j} - I\varphi_{ql}^{j})^2 \tag{1}$$

$$p,q=1,2\ldots,m \tag{2}$$

$$k=1,2\ldots, k_p \tag{3}$$

$$l=1,2\ldots, k_q \tag{4}$$

where $I\varphi_{pk}^{j}$ are the values of the j-th information sign of the k-th open-phase subregime of the p-th class of the longitudinal asymmetry, that is, the regime ω_{pk}; $I\varphi_{ql}^{j}$- the values of the j-th information sign of the l-th subregime with short-circuit and break wire of the q-th class of longitudinal-transverse asymmetry, i.e. the regime ω_{ql}. In this case, the required result can be obtained if the value of the measure of proximity between the regimes of a given class Ω_p is (5) and the measure of proximity between the regimes of the recognized classes Ω_p and Ω_q is (6), it means, that there is a "compression" of the regimes belonging to recognizing class and a "distancing" of objects belonging to other classes.

$$S(\Omega_p) = \sqrt{\frac{2}{k_p}\frac{1}{k_p - 1}\sum_{k=1}^{k_p}\sum_{l=1}^{k_p} d^2(\omega_{pk}, \omega_{pl})} \rightarrow \min \tag{5}$$

$$R(\Omega_p, \Omega_q) = \sqrt{\frac{1}{k_p k_q}\sum_{k=1}^{k_p}\sum_{l=1}^{k_q} d^2(\omega_{pk}, \omega_{pl})} \rightarrow \max \tag{6}$$

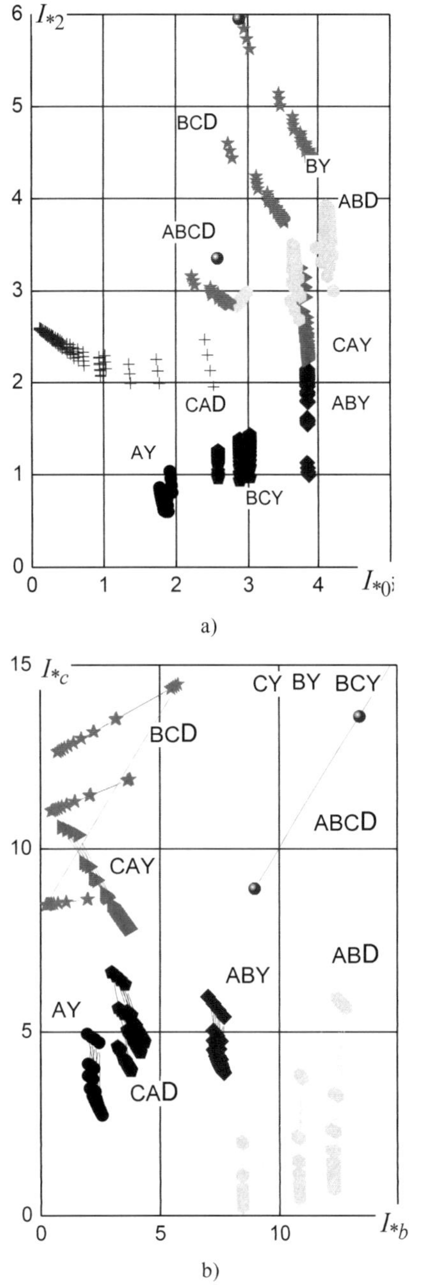

Figure 2. Portraits of longitudinal-transverse asymmetry regime in the "module-module" plane: (a) – I_2-I_1; (b) – I_c - I_b

As a criterion of the effectiveness of the designed RP, a functional depending on the functions $S(\Omega_p)$,$R(\Omega_p,\Omega_q)$ and a decision rule $L(\omega,\{\omega_g\})$, i.e. $J=F[S(\Omega_p),R(\Omega_p,\Omega_q),L(\omega,\{\omega_g\})]$. Then the measure of proximity between the recognizable regime ω and the class Ω_g with g=1,2...,m defined by its regimes $\{\omega_g\}$ (7).

$$L(\omega,\{\omega_g\}) = \sqrt{\frac{1}{k_g}\sum_{g=1}^{k_g} d^2(\omega,\omega_g)} \qquad (7)$$

$L(\omega,\{\omega_g\})$ determines the RMS distance between the regime ω and the class regime Ω_q. Improving the efficiency of the recognition system, in this case, the relay protection will be determined by equation the extremum of the decision rule $L(\omega,\{\omega_g\})$ and, as a result, of the functional J. The minimum RMS distance between the regime ω and the class regime Ω_q must be at least some threshold value determined by the capabilities of the recognition system, i.e. in this case, the measuring units of relay protection. The maximum unbalance values of currents and voltages of symmetrical components, maximum values of load currents and other can be selected as the threshold value.

As classes, you can take object regimes that require a special response of relay protection — a group of normal load regimes of an overhead line, a group of LTA regimes on an overhead line, a group of LTA regimes on an overhead line with simultaneous short circuit on the low-voltage side of the transformer and other regimes [12-18]. These classes are grouped according to the required response of the control system - here there is a relay protection: non-tripping or tripping with different time delays, signaling to service personnel and other. The same reaction is possible for different regimes, which are completely different by informational signs, so it is convenient to classify them as many subclasses of fault and normal regimes. For modern RP, it is also important to determine the type of fault, that makes it possible to more effectively analyze the response of the protection system to a fault. It means RP in this case is a chart recorder of faults.

To assess the recognizability of LTA regimes, we use the obtained results, shown in Fig.2. Zero and negative sequence currents and phase currents (in this case, currents I_{*b} and I_{*c}) can be used as the informational signs of the LTA. RMS distances between objects of different classes in the basis of the transformer nominal current for it are defined and its values are shown in Fig. 3. Only parts of the regimes are chosen for Fig. 3, due to the limited size of the article. The choice of a special phase in the LTA regime can be ensured by using the damaged phase selector, which measures phase currents and voltages, for example, if the phase A is interrupted, the current of this phase decreases and the voltage increases relative to other phases.

The smallest root-mean-square distance for currents I_{*2} and I_{*0} exists between regime with the breaks of the phase wire of the phase A and the interphase short circuit AB and BC on the low-voltage side of transformer and regime with the breaks of the phase wire of the phase A and interphase short circuit of phases AB and CA (shown on Fig.2 that areas of these regimes are getting closer to each other). It should be noted that these regimes belong to the same class and can be recognized based on the control of phase currents for which $L(\omega,\{\omega_g\})\geq 5I_{*nom}$. As a threshold value of the RMS distance for currents, we can take the value of the nominal current of the minimum power transformer. An additional informational sign can be the arguments of currents, which can also help the "distancing" of different regimes.

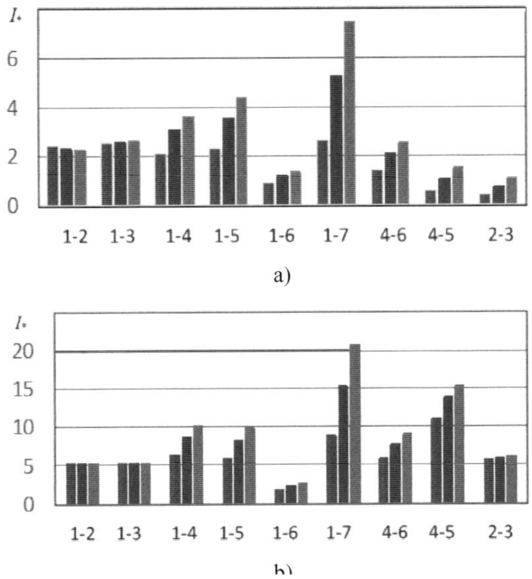

Figure 3. The values of the RMS distance between objects of different classes with longitudinal-transverse asymmetry in the network for symmetrical components (a) and for phase currents (b)

Monitoring the currents of symmetrical components provides for the selection of the type of fault and a possible section with a wire break, and monitoring the values and their changes over time (dynamics) of the phase currents and the current of a positive sequence on the parallel chain of OHL allows selective selection of the damaged circuit. This is true not only for the overhead line sections of the supply substation, but also for the line sections of the receiving substation. To increase the sensitivity of the proposed protection, their parameters should change depending on the previous load regime, especially since the almost linear nature of changes in fault regime currents depending on the load current of the previous regime simplifies the formation of response settings.

A similar approach is carried out in the remote backup protections developed at the Power Stations and Power Systems Department of the Platov South-Russian State Polytechnical University (NPI). At the same time, the areas of location of the listed regimes are formed depending on the previous load regime, and the parameters of the information signs are defined in the specified area.

II. Conclusions

1. It is necessary to develop special relay protection against these types of fault to exclude damage to electrical equipment and its non-selective disconnection in the regimes of longitudinal-transverse asymmetry.
2. Selection of regimes of longitudinal-transverse asymmetry on radial lines with branching substations based on monitoring of modules and arguments of phase currents and symmetrical component currents with the formation of areas of possible fault regimes taking into account the previous load regime allows us to perform adaptive relay protection, the response parameters of which (tripping currents and time delays) depends on the parameters of the current and previous regimes of line operation.
3. The overcurrent protection of the transformers on the branch and intermediate substations must be supplemented with zero-sequence protection or perform it in a three-phase version with the phase current measuring.

References

[1] V.I. Nagay, "Relay protection of the tapping substation in the electric networks," Moscow, Energoatomizdat, 2002, p.312.

[2] I.F.Maruda,"Relay protection of step-down transformers against short circuits on the overhead line during phase breaks," Power Technology and Engineering,vol.2, pp.44-46, 2003.

[3] I.F.Maruda,"Relay protection of 110–220 kV lines during phase breaks," Power Technology and Engineering, vol.1, pp.40-42, 2002.

[4] A.B. Chernin, "Electric values calculation and relay protection functioning in single-phase operation regimes in electric networks," Gosenergoizdat, Moscow, 1963.

[5] A.M. Averbuch,"Examples of single-phase operation regimes and short circuits calculation," Energy, Leningrad branch, Leningrad, 1979.

[6] V.I. Nagay, S.V. Sarry, V.V. Nagay and others, "Protection (signalling) against longitudinal-transverse asymmetry of electrical equipment 0.4-110 kV," pp. 158–159, April 2000 [Digests XIVth Annual Science-Technical Conf. Relay protection and automatic of the power systems 2000 Russia, 2000].

[7] V.I. Nagay,V.V. Nagay, S.V. Sarry and others, "Relay protection with longitudinal-transverse asymmetry regime recognition," pp. 48–50, May 2002 [Digests XVth Annual Science-Technical Conf. Relay protection and automatic of the power systems 2002 Russia, 2002].

[8] A.L.Gorelik and V.A. Skripkin, "Methods of recognition," Proc. allowance, Higher. HQ, Moscow, 1984.

[9] A.E. Lepsky and A.G. Bronevich, "Mathematical methods of image recognition," Lectures, Publishing house of the TTI SFU, Taganrog, 2009.

[10] A. Faure, "Perception et reconnaissance des formes," Transl. France, 1989, p. 272.

[11] J.T.Tou and R.C.Gonzales, "Pattern recognition principles," Moscow, Mir, 1978, p. 412

[12] D.S. Vasilyev, V.N.Kozlov and I.A. Rodionov, "Long-range backup protection of tapping substations in the open-phase mode," Relay protection and automation, vol. 3, pp. 62-64, 2011.

[13] I.V. Nagay, P.S. Kireev and G.N. Chmihalov, "Remote and local relay protection functions in the longitudinal-transverse asymmetry regimes," Russian Electromechanics, vol. 01, pp. 111-113, 2013.

[14] I.V.Nagay and V.I. Nagay, "Designing of multi-parameter backup protection of electric distribution grid 6-110 kV," Energetik, vol. 2, pp.18-21, 2013, in press.

[15] I.V. Nagai, "About perfection of protection from current networks open-phase modes," Global nuclear security, vol. 2. 2013, pp. 64-68.

[16] I.V.Nagay and P.S. Kireev, "Modeling of load regimes of branching substations," Russian Electromechanics, vol. 2, pp. 100-103, 2012.

[17] I.F. Maruda, "Relay protection of the overhead line 110-220 kV in the presence of the phase wire breaks," Power Stations, Moscow, vol.1, pp. 40-42, 2002.

[18] I.V.Nagay, V.I. Nagay, P.S. Kireev, S.V.Sarry and A.V. Ukraincev, "Recognition of Remote Short Circuits with the Transient Resistance of an Electric Arc," Proceedings of the 8th international scientific symposium on electrical power engineering Elektroenergetika 2015, Stara Lesna, SLOVAKIA, SEP 16-18, 2015, pp. 367-371, ISBN 978-80-553-2187-5.

System Testing of Protection Devices

Avoiding Faults and Increasing Testing Efficiency

Björn Cialla
Application Engineer
OMICRON Electronics Deutschland GmbH.
Erlangen, Germany
bjoern.cialla@omicronenergy.com

Barbora Bachratá
managing partner
AP Consulting Ltd.
Bratislava, Slovakia
bachrata@apcon.sk

Abstract - Testing protection devices is a vital procedure to make sure that the overall power system stays stable and therefore it ensures the security of energy supply. The methods of this important process always were derived from the capabilities of the protection devices. In the last couple of decades, the technology of protection devices changed drastically. Now digital processors and amplifiers with power electronics allow for computer operated testing of protection devices. The basic principal of testing has somehow stayed the same. We are still mostly testing thresholds and operating times and therefore test only if the parameters of the protection device have been set correctly. The question "Is this protection device suitable for the power system it should protect?" is typically not answered.

The system testing approach that is shown in this publication, does not focus on testing thresholds with artificial signals. Rather than that in consists of a simulation of the power system behavior under realistic fault conditions. This simulation is automatically send to a test set which generates these transient signals. The reaction of the protection device is then recorded.

The test system, described in this publication, overcomes these limitations by two main adaptations:

- **The base of the transient network simulation is an easy to use network calculation program, which is less complex than conventional network calculation tool.**
- **A closed loop simulation, which reacts to external stimuli while the simulation is running, is avoided. That reduces the necessary calculation power of the test system drastically. Therefore, the test system only needs to be able to replay transient signals.**

Furthermore, we found out that this solution can easily be used in lager systems like busbar differential protection with more than 10 feeders. It can operate multiple test sets at the same time. Therefore, the entire system could be tested in one test which drastically reduced the test time.

Seeing these benefits, we think that using this testing approach during conventional routine test will uncover mistakes and increase the safety of the protection system over all.

Keywords: protection testing; simulation of the power system; system testing approach; complex digital protection; safety;

I. History and Present of protection testing

A. The beginning of testing electromechanical relays

To understand todays methods of testing protection devices, it is necessary to know where their origin comes from. In this case it means testing of electromechanical protection relays. The early devices of this type were basically measuring currents or voltages. If the current or voltage exceeded a certain settable threshold, the device started a timer. After the timer had elapsed the trip contact was activated to operate the circuit breaker. Testing these relays meant to find out whether these pick-up thresholds and delay timers were set correctly. This meant to inject currents or voltages into these protection devices and to measure the delay time. The tests therefore consisted of two main parts:

- Pick-up test: A current or voltage was injected with a value well below the defined threshold. Then this value was gradually increased until the protection device operated. The operating value was then noted in the test report.
- Trip time test: A current or voltage well above the defined threshold – typically two times – was injected into the protection device as this should lead to a trip of the protection device. The time between the start of the current injection and the occurrence of the trip signal was then measured and noted to the test report.

As the protection devices had multiple measuring circuits (e.g. one for each phase) these tests had to be done for multiple fault loops. This made sure that all the phases were set correctly.

Furthermore, the tests had to be done exactly that way as the correct setting of threshold value and delay time are pivotal to the correct operation of the overall protection system. As the parts within an electromechanical protection relay are subjected to aging, their electrical parameters changed over time. This meant that the set thresholds and delay time are changing over the years. That meant that these protection relays must be tested on a periodical basis (typically every 1-4 years). This not only checks, if the defined values are still as expected by the overall protection system grading. It also gives the opportunity to readjust the parameters of the protection device and therefore ensures the overall functionality of the complete protection system. To ensure that, a competent tester, a variable transformer and a connected stop watch was enough.

B. Moving to digital relays

Today's digital processor protection relays behave very different from the electromechanical relays. Most notably they allow for much more complex protection functions due to advanced calculations. Also, their threshold values and timers do not change gradually over time. Their components, however, can fail. Which means if the CPU or the analog digital converter fails, the protection device will not be able to operate anymore. So, the goal of testing digital devices should not be the adjustment of threshold values anymore. Rather than that it should confirm the settings of more advanced protection functions as well as the proper operation of the components. Up to date test systems generate their signals with power electronics which are controlled by microprocessors. This helps handling the complexity of the new protection functions, but it does not necessarily influence the underlaying test philosophy.

In many utilities the basic principle of testing has not really adapted to the complexity of powerful digital protection devices. Most of the tests still confirm thresholds and trip times. With such tests the focus lies on confirming that the parameters are set according to the definition of the protection engineer. In the following this testing method shall be named parameter testing.

One example for an adaptation of older test routines to digital protection relays is the fault loop to be tested: For digital relays it is typically sufficient to test one phase-to-ground and one phase-to-phase loop. To be on the safe side a three-phase test is also recommended. This makes sure that differences in the ground and phase loop measurement are taken care of. In reality many digital relays are tested with all three phase-to-ground loops plus one phase-to-phase loop. This is mostly not critical to the quality of the protection test itself, but it is to some extend inefficient. Therefore, it showcases that rethinking the testing process might result in overall improvements of the protection test efficiency and quality.

II. MISOPERATIONS OF PROTECTION SYSTEMS

In order to define new ways of testing, the main issues of the protection systems need to be known. **Chyba! Nenašiel sa žiaden zdroj odkazov.**shows the main causes of a failing protection system protection in the USA. While there might be severe differences in the protection philosophy between America and Europe, it is safe to assume, that the fundamental mechanisms which cause the misoperations are similar. The following part will focus on the three most frequent causes:

A. Incorrect setting/logic/design errors

The misoperations in this category are often caused by the process of defining the parameter settings. The protection engineer who designs the protection system does this based on network calculations. The resulting parameters are then given to the engineer or technician who puts them into the protection relay. Then a protection test confirms, that the parameters in the protection relay are set as desired.

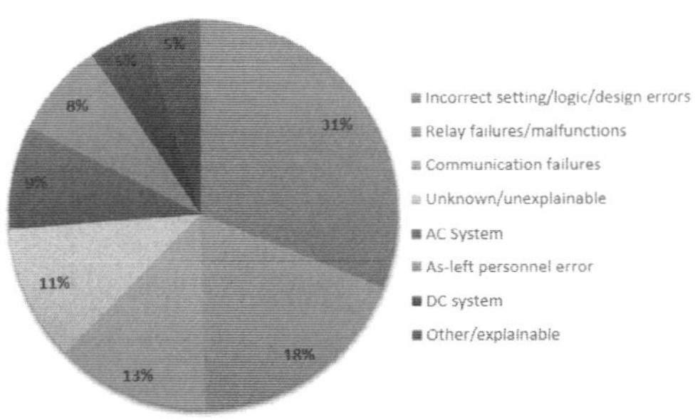

Figure 1. Causes for protection misoperations in the USA (2012-2014) [1]

This process, however, does not verify whether the calculated protection settings are feasible for the power system that should be protected. If a mistake happens during the design phase – for example when the calculation results are transferred to actual protection parameters – the parameter test will not detect it.

Furthermore, changes in the power system can make older parameter calculations irrelevant. So, if new overhead lines or generation plants are connected to the grid, the parameters in large parts of the overall protection system might need to change. If they are not adapted, the protection system could be inappropriate for the changed power system. This is also not detected by parameter tests.

The logic functions in modern protection devices can also become quite complex. Also, it is important to note that a logic of a protection system can include multiple protection devices. An automatic reclosing system for example can only work properly, if all terminals of an overhead line are switched of at the same time for a long enough duration. So even if the individual protection devices seem to be set up correctly, wrong timings could lead to cases, where the resulting dead time of the overhead line is not sufficient to clear the fault. Protection tests that focus on only one protection device will probably not detect these kinds of faults.

B. Relay failures/malfunctions

These misoperations are caused by defective components of the protection devices. A parameter test can and should find these defective devices. The standard intervals for protection tests, however, might not always be the best solution. For device types that show a tendency to fail it might make sense to reduce the interval of the protection tests.

C. Communication failures

This category reflects the fact that many modern protection systems (e.g. line differential protection, distance protection with teleprotection, decentralized busbar protection) rely on communication between separate protection devices to achieve a selective and fast disconnection of power system faults. Failures of the communication system generally lead to a limited emergency or backup operation of the protection system, thus increasing the operation time and/or reducing the likelihood of detecting the fault properly. A parameter test that focuses on the protection devices without including the communication system will not find these faults.

III. SYSTEM TESTING APPROACH

Rethinking the methodology of protection tests should focus on approaches to reduce the likelihood of the above mentioned misoperations. Therefore, the scope of a protection test should be widened. From focusing on singular parameters, elements and devices towards a more broader view of the protection system. It should also overlap with other steps in the process of designing a protection system (e.g. the calculation of protection parameters).

The system testing approach tries to achieve exactly this. It tries to verify whether a protection system fits to the power system it is meant to protect. That means that the ideal system test is a real fault in the power system [2]. Applying these real faults methodically to test the protection system is obviously not feasible. Rather than this, the behavior of the power system under nominal and fault conditions should be simulated in a sufficiently realistic manner. The results of such simulations can then be applied to the protection devices. To confirm correct operation, their behavior will subsequently be assessed.

Figure 2 shows a comparison of the test coverage between a system test and a parameter test.

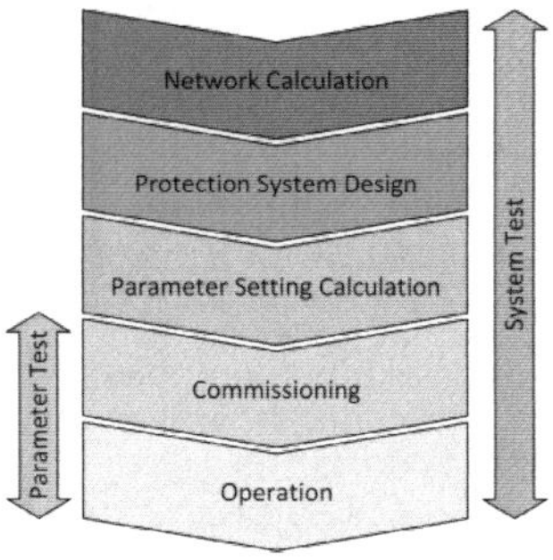

Figure 2. Test coverage of protection design process

Furthermore, the system testing approach does focus on a complete protection system which might involve multiple protection relays. Such a test should therefore be able to apply test quantities simultaneously on more than one location. That makes a proper time synchronization between the test locations necessary.

A. Testing with transient playback

The basic approach of system testing is not completely new. The most common way done is by using an existing network simulation tool which is able to generate transient traces – preferably in the COMTRADE format [3] . These traces can then be output to the relays by protection test sets. This process however has some major drawbacks:

- The simulation tool and the test tool are two independent pieces of software. Therefore, generating the transient signals and transferring them to the test set is manual work. With this, even handling a moderate number of different fault scenarios becomes cumbersome. Applying the efforts of distributed testing becomes even more complicated.
- The simulation tool often is quite expensive and needs in depth knowledge. If the right expert is not present during the test, it is almost impossible to adjust the test files to changes that might be necessary.
- The analysis of the results of such a test is to a certain degree manual work. That makes it susceptible for human mistakes. When it comes to distributed testing, the assessment is even more complicated. Here the results of multiple locations must be combined to one picture. Only after this is done an assessment is possible. Further than that, the whole test needs to be repeated, if the analysis shows, that the test was not sufficient to verify the behavior of the protection system.

All these points limit this approach to laboratory use. In the field it is typically too cumbersome to apply.

B. Testing with real time simulators

Another method for system testing that is already in use is a real time simulator. Here the simulation runs on a dedicated hardware which directly generates the test signals. This hardware also feeds back the reaction of the protection relay directly to the simulation. As these test systems have a lot of processing power, their simulation can immediately react on input signals and adapt the simulated network accordingly. E.g. if a trip signal occurs, the simulation will be able to react on this in time, open the circuit breaker and calculate the new resulting quantities properly. This makes it very convenient for an advanced analysis of the protection system behavior. But also, this approach has some drawbacks:

- The test hardware is very big, heavy and costly. Therefore, it is only used in laboratories.
- Distributed testing is usually not possible, but these systems can typically generate o lot of signals at the same time. That allows for testing larger and more complex protection systems.
- Setting up a simulation with a real time simulator can require a deep knowledge of the power system as well as of the operating software.

Real time simulation test systems are mostly used by relay manufacturers, big utilities and universities.

C. Testing with an integrated solution in the field

To make sure that a system test solution can be used in the field there are multiple requirements that need to be met:

- The test hardware needs to be portable.
- The simulation, test execution and analysis must be integrated in one software/solution.
- The simulation needs to be set up and adapted easily.
- Distributed testing must be set up easily
- Distributed testing must be controlled from one application
- The results of distributed tests must be shown and analyzed in one application

Due to the requirement of portability, the hardware typically will not be able to perform a real time simulation while executing the test.

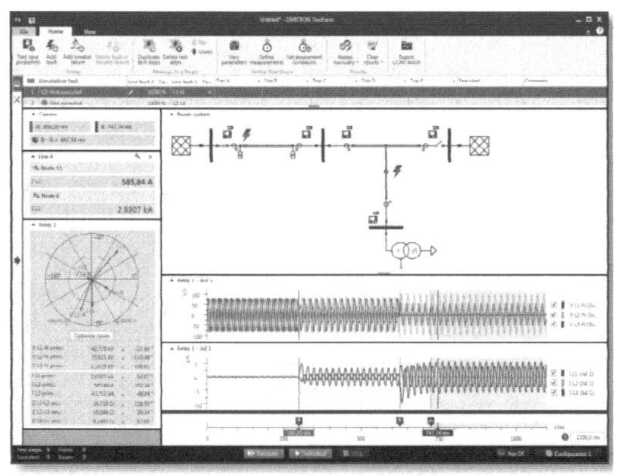

Figure 3. Test solution for system tests in the field

A state of the art solution fulfils all these requirements (Figure 3). With this solution the simulation is done within the software on the PC. Then the transient signals are transmitted to the test hardware. For distributed tests, the test hardware can also be time synchronized.

IV. Examples of System Test in the Field

To showcase the benefits of system tests, the following section will show some test that have been performed with such a system testing approach.

A. Distributed testing of a multi terminal line

In the southern part of Austria, the protection system of a multi-terminal double line had to be tested [4]. In this case a double line with a tied in station had been preexisting. Due to

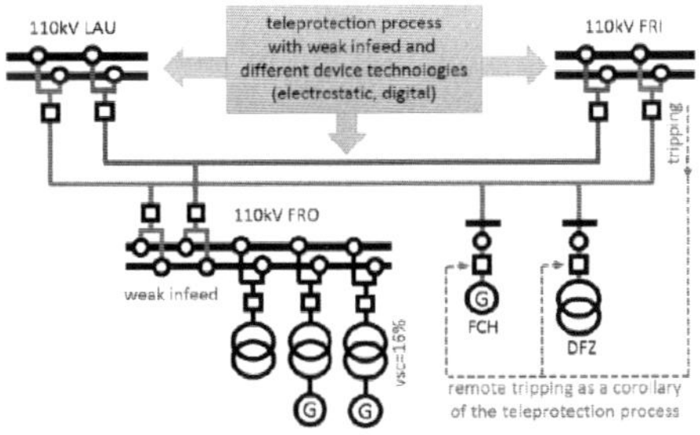

Figure 4. Tested teleprotection system

changed requirements in the tied-in station the topology had to

be extended to a three-terminal configuration. One of these lines also had two additional in-feeds. The protection system consisted of distance protection devices from different generations and manufacturers (microprocessor and solid state). The selectivity was ensured by a teleprotection system that included the three main stations. The two additional in-feeds were disconnected by a remote trip from one station.

For this test the test, the main application was run in one substation. The test sets at the remote ends were connected via an internal network as well as the internet. As a result of this test, two main problems have been found:

- One zone was set too long so that it overreached the intermediate station. This caused unselective operations for faults at the busbar.
- A logic print which was responsible for the time sequence of the teleprotection was found. This was only visible, when separate protection devices interacted with each other.

B. Testing of a busbar differential protection

The sheer size and complexity of a busbar protection system can make it very complicated to test conventionally [5]. In this case the busbar was modeled in the system testing tool. Then the measuring locations in the power system were assigned to outputs of the protection test sets. After this, multiple faults were simulated at various locations on the busbar and on different feeders (outside the protected area). With this test setup (see Figure 5), the whole busbar differential system could be tested at once.

The selectivity of the different protection zones could be tested with an iterative approach. The test software simulated the fault once and applied these quantities to the protection system. The faulty differential zone tripped first correctly. But as the test was still generating fault quantities, also the remaining busbar was tripped subsequently. In a second iteration of the test, however, the test system already knew the trip times of the first differential zones and could simulate the behavior of the circuit breakers accordingly. Therefore, only

Figure 5. Test setup for a busbar differential protection

the faulty differential zone tripped and the correct behavior of the protection system could be verified.

As all bays could be tested at the same time, the overall time effort for such system tests was much smaller [6]. Furthermore, the iterative testing feature allowed for an easy test of the zone detection logic. Overall the test was faster and more profound than a conventional parameter test.

V. Conclusions

Protection testing methods have been developed over the years, originating from electromechanical relays. These methods traditionally focused on testing threshold values and operating times. Today's digital microprocessor relays, however, offer much more complex protection functions and logics. But the testing methods that are typically applied today are still derived from the electromechanical relays.

The most frequent causes of protection system misoperations are incorrect settings/logic/design errors, relay failures/malfunctions and communication failures. Widening the scope from parameter tests to system tests, can detect many of these problems that cause misoperations of protection systems. A system test does not verify singular parameters. It rather tests whether a protection system properly works in the power system it was designed for.

Different methods of system testing have been used over the years. Mostly these were laboratory applications. The biggest benefits of system tests, however, can be achieved by using this approach in the field. There is an existing test solution that covers the simulation, test execution and the analysis of the results in one application [7]. Furthermore, it is suited for the use in the field because the simulation is easy to set up and adapt. It also allows for easy distributed testing by accessing remote test sets via a secure internet connection.

Uses in the field have shown that a proper system test can find more errors in the protection system, when they complement conventional parameter tests. The system testing approach also makes testing of complex logic functions easier and less time consuming. Therefore, system testing can improve the overall quality of the protection system while being more efficient.

References

[1] Protection System Misoperation Task Force, "Analysis of System Protection Misoperations", NERC, 2015

[2] Ch.Pritchard, D.Costello and K.Zimmerman, "Moving the Focus from Relay Element testing to protection System Testing,", PACW, 2015

[3] F.Steinhauser, "How network simulation improves protection testing and eases the application of end-to-end testing," OMICRON - Asia-Pacific Protection & Testing Conference, 2014

[4] Ch.Pritchard and T.Hensler, "Test and verification of a busbar protection using a simulation-based iterative closed-loop approach in the field,", OMICRON, 2014

[5] T.Hensler, „Application-oriented distance relay testing- New testing software opens up extend", OMICRON – 2015

[6] O.Skrbinjek and B.Cialla, "Simulation Based Testing and Measurements – Experiences," OMICRON - International Protection Testing Symposium, 2015

[7] J.R.G.Calderon, E.L.B.Gonzales, E.S.Arreola, M.R.Pena, E.Carvalheira, J.Coronel, "Simulation-Based Testing of the IEC 61850 Protection Logics at CFE Verónica Substation," PACW US, OMICRON – 2016

Design and Testing of Voltage Instrument Transformers for Railway Application

Lukáš Český, František Janíček, Filip Skudřík, Jozef Zuščak
Institute of Power and Applied Electrical Engineering,
Slovak University of Technology in Bratislava
Ilkovičova 3, 812 19 Bratislava, Slovakia
frantisek.janicek@stuba.sk

Abstract – **The paper deals with designation and testing of single-phase instrument transformer of voltage (VT) used in railway applications. In railway applications is VT stressed by overvoltage much more often than in another application. This is the reason, why we deal with special designation of VT. Manufactured VT was routine tested according to IEC 61869-3 in all three operation states. This article deals about type testing of VT according to additional standard IEC 62505-3-3.**

***Keywords*—voltage instrument transformer, railway application of VT, IEC 61869-3, IEC 62505-3-3.**

I. Introduction

Medium voltage instrument transformers are very important components for measuring of consumpted electric energy in railway applications. Instrument transformers are used for transformation of high values of current and voltage to standard values. Next functions are insulation of high voltage circuits from measuring devices and power supply of protection relays. Depending to transformed value we know current instrument transformers and voltage instrument transformers. In current transformers are standardized secondary currents 1 or 5 Amps. Voltage transformers in railways usually have rating 100 or 110 V for delta connection and 100 or 110 V for star in one-phase configuration.

This article describes design and testing of single pole voltage instrument transformer used in railway applications. In railway electrification network may easily generate ferro-resonance phenomena. With ferro-resonance is connected frequent overvoltage. To avoid the generation of ferro-resonance phenomena is necessary to design VT with right parameters. Requirements for VT used in railway applications are defined in standard IEC 62505-3-3.

II. Design of Voltage Instrument Transformer

Voltage instrument transformers consist of functional parts and insulation medium. Functional parts are primary winding, secondary winding and ferromagnetic core. In medium voltage instrument transformers is insulation medium usually epoxy resin. Epoxy resin has 17 times better insulation than the air. [1]

In the article is tested single pole one-phase voltage instrument transformer with following label values:
Manufacturer's name: ABB
Voltage transformer type: TJC 7.1
Year of manufacture: 2018
Serial number: 1VLT 5218018622
Mass: 53 *kg*
A-N: 27000 *V*
1a-1n: 100 V, 5 *VA*, 0,5
2a-2n: 100 V, 5 *VA*, 6P
Frequency: 50 *Hz*
Um: 36 *kV*
LI/SI/AC: 36 kV/80 *kV*/ 180 *kV*
Temperature: 40 *°C*/+65 *°C*
Fv: 1.9 for 8*h*
Insulation Class: E

All label values are according to IEC 61869-3. Label of VT is in Fig. 1. [2][5]

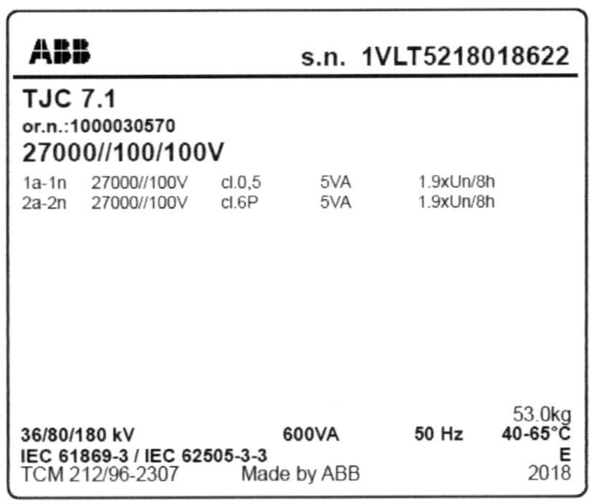

Figure 1: Label of TJC 7.1

Transformers in railway applications are connected as single phase. From label is obvious, that we have single pole VT with two secondary windings. Both windings are used for measuring purposes. In railway electrification network can be easily generated ferro-resonance phenomena (mainly in 25 kV, 50Hz). It is often connected with destruction of VT caused by overvoltage. By the correct design can VT avoid the generation of ferro-resonance phenomena. Standard IEC 62505-3-3 recommends high value of primary resistance R_{prim} – between 15 and 50 kΩ. The saturation point should not be reached. Scheme of connection is in Fig. 2.[6][7]

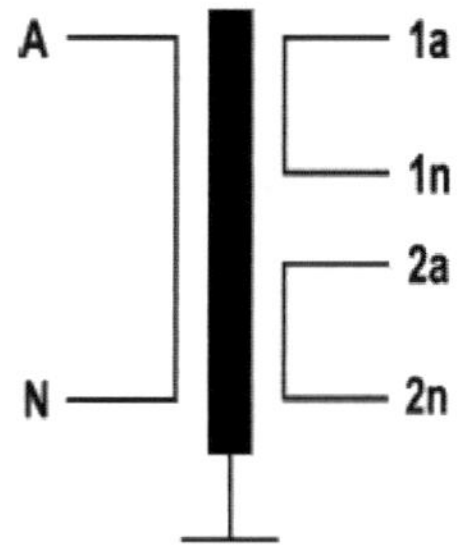

Figure2: Scheme of connection

A. Ferromagnetic C-core

Ferromagnetic C-core of voltage instrument transformer consists of sheets manufactured from ferromagnetic steel with thickness 0.35 mm. In our design is used defined C-core with 3 steps and constant cross section. Cross section of core is in Fig. 3.

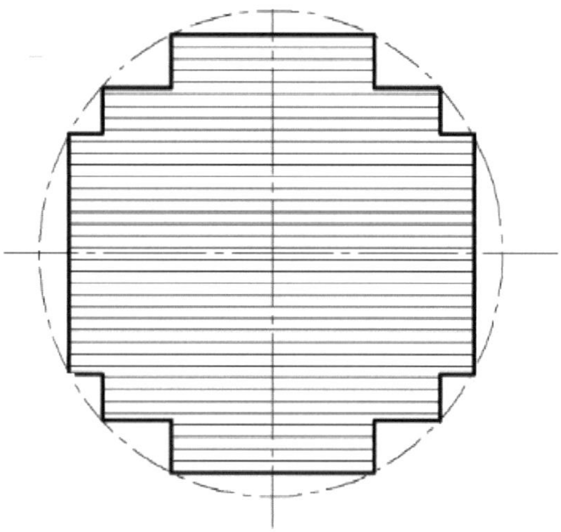

Figure 3: Cross section of core [4]

B. Primary coil

By design of primary coil is necessary to consider value of primary resistance. Primary resistance depends on middle length of primary conductor l_{prim} and its resistance per 1meter R. From equation 1(1) is clear, that for high value of R_{prim} is important middle length of conductor and higher diameter of wire.

$$R_{prim}=l_{prim} * R \qquad (1)$$

R_{prim} is resistance of primary coil [Ω],
l_{prim} is middle length of primary conductor *[m]*,
R is resistance of primary conductor per m [*Ω/m*].

Next important parameter for avoiding the ferro-resonance phenomena is value of rated electromagnetic induction Bn. Ferromagnetic core has maximal electromagnetic induction Material constant B_{max} is for transformer steel sheet usually between 1.7-1.8 T. Rated voltage factor is 1.9xU_n/8h. For elimination of ferro-resonance phenomena is necessary to avoid saturation. This is the reason, why real voltage factor is around 3.0xU_n/8h. Value of rated electromagnetic induction is given by equation (2):

$$B_n=\frac{B_{max}}{F_V} \qquad (2)$$

B_n is rated electromagnetic induction [*T*],
B_{max} is maximal electromagnetic induction [*T*],
F_v is real voltage factor [-].

In practice is trend to fill the coil for the highest possible voltage factor and highest possible primary resistance. Number of primary turns is given by equation (3):

$$N_1=\frac{U_1}{\frac{2\pi}{\sqrt{2}}*f*S*B_n} \qquad (3)$$

N_1 is number of primary turns [-],
U_1 is rated primary voltage [*V*],
f is rated frequency [*Hz*],
S is cross section of core [m^2],
B_n is rated induction [*T*].

Next important parameter of primary coil from dimension point of view is its wideness D. It is given by number of turns in one layer and diameter of wire d.

$$D=N*d\,, \qquad (4)$$

D is wideness of coil [*m*],
N is number of turns in one layer [-],
d is thickness of wire [*m*].

Thickness of coil H is given by number of layers, thickness of interlayer insulation and diameter of wire. From these parameters we can calculate dimensions of primary coil. Coil of voltage transformer is in Fig. 4.

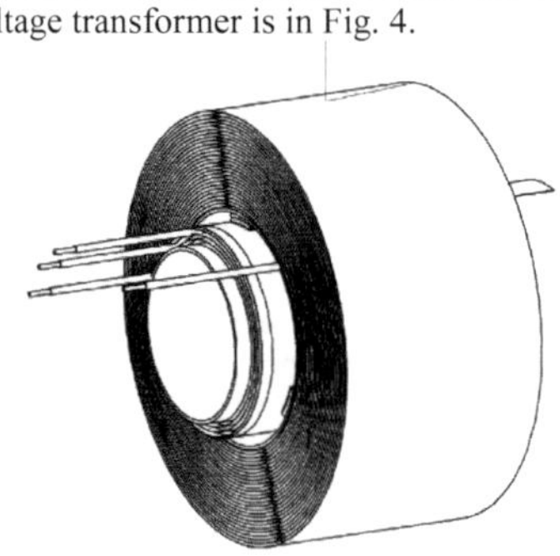

Figure 4: Coil of VT

C. Secondary coil

In secondary winding is important mainly number of secondary turns. It is given by equation. (5)[3]

$$N_2=\frac{U_2*N_1}{U_1}\,, \qquad (5)$$

N_2 is number of secondary turns [-].

III. Testing of Voltage Instrument Transformer

For verification of correct design is necessary to test transformer. According to standards has to each instrument transformer of voltage pass routine tests according to IEC 61869-1 and IEC 61869-3. The main test from railway applications point of view is ferro-resonance withstand test according to IEC 62505-3-3.

First routine test according to IEC 61869-3 is power-frequency voltage withstand test on primary terminals. The primary insulation of VT was subjected to the specified induced voltage test with an elevated frequency 120 Hz for 60 s. Test was provided by voltage 80 kV. Terminals 1n, 2n, N and frame were connected to earth. Configuration of test equipment is in fig. 5.

Figure 5: Induced AC voltage test and PD measurement

Partial discharges are reached while decreasing the voltage after power-frequency voltage withstand test on primary terminals. Test circuit for partial discharge measurement is in fig. 6. U represents high voltage supply, Z is filter, Z_{mi} is input impedance of measuring system, CC is connecting coaxial cable, C_a is test object, C_k is coupling capacitor LDIC 1000 pF, type CDHT 0704B11, CD is Coupling device LDIC, Type LDM-5/U, MI is measuring instrument LDIC, type LDS-6.

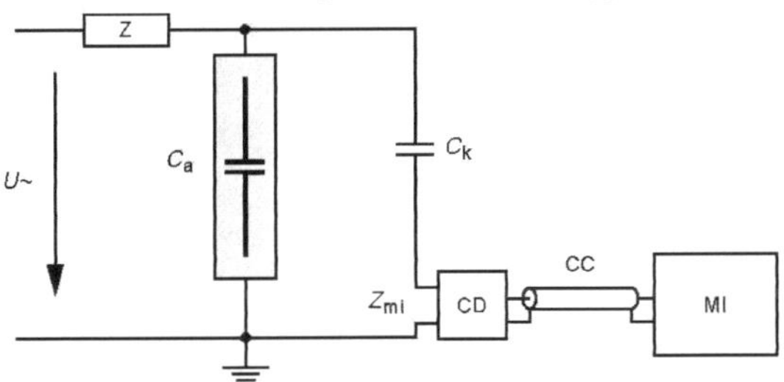

Figure 6: Test circuit for partial discharge measurement

After power-frequency voltage was applied the voltage decreased without interruption to $1.2xU_n$ and write down PD level q. After that voltage was decreased to $1.2xU_n/\sqrt{3}$ and write down the corresponding PD level q. The maximal PD level is 50 pC at $1.2xU_n$ and PD $\leq$ 20 pC at $1.2xU_n/\sqrt{3}$. Results of PD test are in table 1.

Table 1: Test results of PD measurement

PD test voltage	PD levels	Result
1.2xUn	q=2.7*pC*	satisfactory
1.2xUn/√3	q=2.3*pC*	satisfactory

Record of PD test by $1.2xU_n$ is in Fig.7.

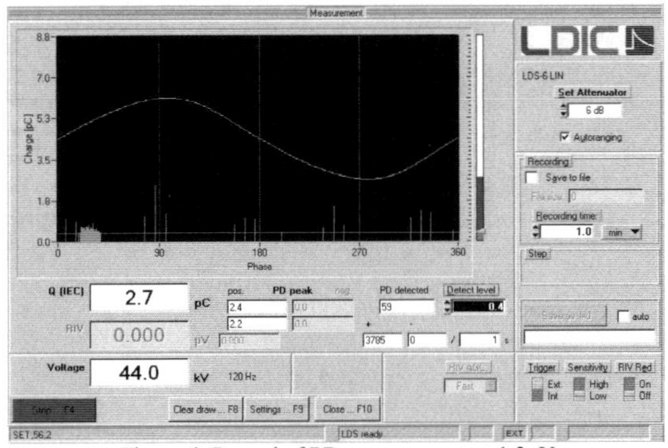

Figure 6: Record of PD measurement at $1.2xU_n$

Power-frequency voltage withstand test on secondary terminals and between sections is provided with test voltage 3 kV/50 Hz for 60 s. For our case it was applied on 1a-1n (2a-2n+N+PE+frame were connected to earth), 2a-2n (1a-1n+N+PE+frame were connected to earth) and N (1a-1n+2a-2n-PE-frame were connected to earth).

Test for accuracy is provided at 80 %, 1000 and 120 % of rated voltage U_n at rated frequency f with burdens between 25 % and 100 % at power factor 0.8. Criteria to pass the test are fulfil limits of ratio errors and phase displacement. Measured values are in table 2 -column e_i and d_i, limits are in column ratio error and phase displacement.

Table 2: Table of accuracy class test

TP.:	% Un	S / SN	Burden(VA)	COS ß	ei (%)	di (min)	Ratio error(+/- %)	Phase displ.(+ /-min.)
Winding-number: 1	Primary voltage: 27 000,00 V		Secondary-voltage: 100,00 V					
1	120,00	1 / 1	5,00	0,8	-0,338	-0,91	0,500	20,00
2	100,00	1 / 1	5,00	0,8	-0,337	-0,98	0,500	20,00
3	80,00	1 / 1	5,00	0,8	-0,336	-1,05	0,500	20,00
4	120,00	1 / 4	1,25	0,8	-0,281	-0,20	0,500	20,00
5	100,00	1 / 4	1,25	0,8	-0,280	-0,27	0,500	20,00
6	80,00	1 / 4	1,25	0,8	-0,278	-0,33	0,500	20,00
Winding-number: 2	Primary voltage: 27 000,00 V		Secondary-voltage: 100,00 V					
TP.:	% Un	S / SN	Burden(VA)	COS ß	ei (%)	di (min)	Ratio error(+/- %)	Phase displ.(+ /-min.)
1	190,00	1 / 1	5,00	0,8	-0,322	0,32	6,000	240,00
2	100,00	1 / 1	5,00	0,8	-0,321	-0,22	6,000	240,00
3	5,00	1 / 1	5,00	0,8	-0,398	0,41	6,000	240,00
4	190,00	1 / 4	1,25	0,8	-0,268	0,67	6,000	240,00
5	100,00	1 / 4	1,25	0,8	-0,267	0,13	6,000	240,00
6	5,00	1 / 4	1,25	0,8	-0,336	0,79	6,000	240,00

The last test procedure according to IEC 61869-3 is verification of terminal markings. In this procedure are verified nameplate and terminal markings.

From railway point of view is important mainly ferro-resonance test according to standard IEC 62505-3-3. It is the best verification of correct design. According to standard the best prevention against ferro-resonance phenomena is high value of primary resistance – between 15 and 50 kΩ. Transformer should not be saturated.

The ferro-resonance was provided as a temperature rise test at rated voltage, rated burden and at 1/3 of rated frequency. Ambient air conditions were temperature 22°C, air pressure 990 hPa and relative humidity 22 %. Test configuration is in Fig. 7.

Figure 7: Test configuration of temperature rise test

The worst case by temperature rise test was temperature rise test at 16 2/3 Hz. The temperature rise was calculated from difference of resistances at the start of test and in the end of test. Overheating Δv is given by equation (6).

$$\Delta \upsilon = \frac{R_2 - R_1}{R_1} \cdot (x + \upsilon_1) - (\upsilon_2 - \upsilon_1) \quad (6)$$

Δv is temperature rise [*K*],
x=234.5 is coefficient for cooper [-],
R_1 is the resistance at the start of test [Ω],
R_2 is the resistance in the end of test [Ω],
v_1 is ambient temperature at the start of test [*°C*],
v_2 is ambient temperature in the end of test [*°C*].

Values of overheating are around 5 K. By ambient temperature 65 °C and temperature class E – 115 °C VT passed type test. Value of primary resistance is 29.9 kΩ. It means that are fulfilled all recommendation according to standard. The results of temperature rise test by 16 2/3 Hz are in table 3.

Table 3: Resistance values and temperature rise of windings at 16 2/3 Hz

Winding	Resistance of winding before test R_1	Resistance of winding in the end of test R_2	Temp. rise Δv
Primary (A-N)	29.9 *k*Ω	30.65 *k*Ω	5.5 *K*
Secondary (1a-1n)	0.4747 Ω	0.4863 Ω	5.3 *K*
Secondary (2a-2n)	0.562 Ω	0.5750 Ω	4.9 *K*
Ambient temperature v	22 *°C*	23 *°C*	

IV. Conclusion

We studied single pole instrument voltage transformer for railway application from design point of view. Our goal was avoiding ferro-resonance phenomena. By calculation of VT we passed all requirements of standard IEC 62505-3-3. The resistance of primary winding was 29.9 kΩ. Real voltage factor was 3.0xUn for 8 hours. These values should eliminate ferro-resonance phenomena and protect VT from overvoltage. Next elimination would be possible with connecting of additive burden on secondary terminals. In our case was that transformer with rated burden 5 VA loaded only with 1 VA on secondary terminals. If there would be loaded rated burden, in VT should not occur ferro-resonance phenomena.

After calculation and manufacture of VT we tested transformer according to IEC 61869-1 and IEC 61869-3. We tested all routine tests - transformer passed power frequency withstand test on primary terminals. On primary was applied test voltage 80 kV for 60s by 120 Hz. Next test was partial discharges with result 2.7 pC by $1.2xU_n$ and 2.3 pC by $1.2xU_n/\sqrt{3}$. Transformer also passed test of accuracy, first secondary passed accuracy class 0.5 by rated burden 5VA, the second secondary passed accuracy class 6P by rated burden 5VA. Transformer was also tested with power-frequency withstand voltage test on secondary terminals and between sections and verification of terminal markings.

From railway application point of view were important mainly addition tests according to IEC 62505-3-3. According to the standard for indoor transformer is necessary ferro-resonance withstand test - clause 8.4. The ferro-resonance voltage, rated burden and at 1/3 of rated frequency. Criteria to pass the test were that the standard temperature rises shall not be exceeded. Overheating of VT by frequency of 16 2/3 Hz were 5.4°C, for first secondary winding it was 5.3°C and for second secondary it was 4.9°C. It means, that transformer had negligible temperature rise even by frequency 16 2/3 Hz. The test was fully passed.

Acknowledgment

Company ABB and Slovak University of Technology supported this research. We thank our colleagues from both organizations who provided insight and expertise that greatly assisted the research. This work was supported by the Slovak Research and Development Agency under the contract No. APVV-15-0326.

References

[1] J. Kopeček, M. Dvořák, "Přístrojové transformátory: měřící a jistící",1st edition Praha: Academia, 1966,492 p.

[2] IEC, IEC 61869-3: Instrument transformers – part 3: Adittional requirements for inductive voltage transformers, 1st edition, Geneva, 2011, 28 p.

[3] Ľ. Huttner, Ľ. Klug, "Elektrické stroje", Vydavateľstvo STU, 2011, Bratislava

[4] L. Český, F. Janíček, J. Kubica and F. Skudřík, "Overheating of primary and secondary coils of voltage instrument transformers", EPE 2017. Ostrava: Technical University of Ostrava, , 2017, p. 182-187

[5] IEC, IEC 62505-3-3:Railway applications – Part 3-3: Measurement, control and protection devices for specific use in a.c. traction systems-Single-phase voltage transformers, 1st edition, Geneva, 2009, 10 p.

[6] L. Český, F. Janíček, "Design of a Modern Medium Voltage Switchgear Utilizing the Latest Available Technologies in Medium Voltage Networks", Final thesis, Slovak University of Technology in Bratislava, 2016, p. 8

[7] F. Janíček, A. Arnold, J. Šedivý. I. Šulc, A. Cerman and P. Petrek, "Elektrické stanice", Vydavateľstvo STU, 2012, 189 p.

D: Power System Control, Liberalisation of Electricity Market

Electric Utility Costs Control by Means of Increasing Net Generation and Reducing Losses in the Distribution Grid

Nikita D. Mukhlynin
Ural Power Engineering Institute
Ural Federal University
Yekaterinburg, Russian Federation
mukhlynin@me.com

Andrey V. Pazderin
Ural Power Engineering Institute
Ural Federal University
Yekaterinburg, Russian Federation
a.v.pazderin@urfu.ru

Abstract — ***This paper is devoted to a new approach to reducing electric utility costs by means of controlling net generation and energy losses in the distribution grid, which are changing due to operation control devices. This approach allows to increase electric utility revenue using two criteria of grid optimization. The main criterion is increasing net generation taking into account static load characteristics in the grid nodes. Compared to reducing energy losses, it leads to higher benefits in short-term perspective until tariff revision. However, in long-term perspective the conventional criterion of energy losses is more suitable. The proposed optimization model takes into account both criteria flexibly with the possibility of changing their significance depending on current circumstances and calculation tasks. To solve the problem of optimization and state estimation, a new calculation flow model in this paper is used. The perspectives of using this model for distribution grids are based on its simplicity and the possibility of grid scheme representation as well as static load characteristics in single-phase and three-phase representations if necessary. To solve optimization task with discrete and continuous variables the algorithm including several steps based on the iterative Newton's method is used.***

Keywords—state estimation; optimization; distribution grid; flow model; constrains; smart grid; energy balances.

I. Introduction

The main trend in the development of distribution grids is aimed to realize two concepts, which are Smart Grid and Smart Metering. Development of IT technologies within reliability and high-speed information channels create a condition for introduction of new «clever» Smart information system, based on the analog-to-digital converters of telemetry and microprocessor meters (AMR) with two-way exchange of three-phase information transmission [1]. It creates new possibilities for using this information in the distribution grids with phase and amplitude imbalances. Besides, introducing distributed generation and flexible alternating current transmission systems (FACTS) in the grids leads to the complication of the operation modes in the distribution grids and change in the grid common principles of electric energy transmission and distribution, until recently focused on the transmission of electric energy in the one-sided feed mode [2]. In this regard, distribution grids with local closed-loop topologies with connection to the bus or single power supply section of feed substation, as well as global closed-loop topologies in the form of closed cellular polygons appear [3]. Thus, this leads to increase of controllability of local areas in distribution grids and brings them closer to their controlling capabilities and systems for collecting and transmitting measurement information to the higher voltage grids.

For distribution grid control, it is necessary to use a new calculation flow model. Widely spread classical calculation model for estimating operation mode parameters is mainly based on the classical equations where state vector consists of voltage magnitudes and angles at each grid node $X^T=\{U_i;\ \delta_i\}$ [4]. Such model takes into account modern development trends and features of distribution grids poorly, scilicet a modes and topologies diversity, a partial absence of circuit and equipment parameters and their large obstacle, a presence of short length electric lines and branches with zero resistance, and a large number of unobserved areas [5]. That is why the use of classical models in distribution grids is limited.

This paper is organized as follows: section II is devoted to the field of application of the power flow model in the optimization task for increasing electric utilities revenue. In section III a mathematical description of the new power flow model is given. Finally, in section IV a simple test case is presented. The conclusions are presented in section V.

II. Field of Application of State Estimation and Optimization Tasks Based on the Flow Model

The main companies in electric energy field, that want to develop distribution grid within incorporating automated devices are electric utilities. Reducing energy losses and increasing net generation are the ways to top up electric utilities revenue by means of voltage control in the local areas of distribution grid and Micro Grids with using controlling devices. These devices are the following: a transformer with on-load tap changing, active voltage conditioner, flexible reactive power sources and excitation systems of distributed generators etc. In addition, such approach allows to decrease transmission tariff by means of electric utility costs control. In a various time intervals a significance of mentioned above criteria may be different. In a long-term perspective within several years, the criterion of electric energy losses is more effective. But in the short-term perspectives within time of constant transmission tariff, increasing net generation according to static characteristic curves $P=f(U)$, $Q=f(U)$ in the grid nodes will be more suitable. As energy transmission tariff depends on common value of net generation in the denominator of equation being used for a new transmission tariff calculation [6]. It leads to decreasing an updated transmission tariff in the next period equal to one year, three and five years depending on cost justification method. That is why this criterion is applicable only for short time intervals.

The optimization model proposed in this paper, based on the state equation and calculation vector of the flow model takes into account both criteria flexibly with the possibility of changing their significance depending on current circumstances, time intervals and calculation tasks. These tasks are state calculation, state estimation and optimization. The paper presents examples that prove the effectiveness of work of the selected optimization criterion and show how simply calculation flow model allows to move from state estimation and state calculation tasks to solving an optimization problem in the local areas of a distribution grid using information from AMR systems.

III. Unified Power Flow Model

A. State Equations and Calculation Vector of Flow Model

The state equations of the proposed power flow model are the nodal and branch balances like in the energy flow problem – EFP (7). For active and reactive nodal power or energy in the described model, balance expressions, which take into account the state equations, can be written as follows:

$$P_i^{\text{calc}} = \sum_{j\in\omega_i} P_{ji} - \sum_{j\in\omega_i} \Delta P_{ji} - \sum_{j\in\gamma_i} P_{ih} - \Delta P_i \;, i = 1, 2 \ldots N_{\text{in}} \,;\, h = 1, 2 \ldots N_{\text{out}} \tag{1}$$

$$Q_i^{\text{calc}} = \sum_{j\in\omega_i} Q_{ji} - \sum_{j\in\omega_i} \Delta Q_{ji} - \sum_{j\in\gamma_i} Q_{ih} - \Delta Q_i \;, i = 1, 2 \ldots N_{\text{in}} \,;\, h = 1, 2 \ldots N_{\text{out}} \tag{2}$$

where P_{ji}, Q_{ji} are active and reactive flows which flow into the i node from all N_{in} nodes j; P_{ih}, Q_{ih} are active and reactive flows which go from the node i to node h for all N_{out}; γ_i is a set of nodes which accept power or energy.

These equations must be written for nodal loads or generation taking into account the corresponding flow signs. The ideology of forming equations (1) and (2) is used for recording an analytical expressions for all measurements in the state calculation and state estimation tasks and recording criteria of corresponding equal and non-equal constrains in the presented following optimization task. Calculation vector $\boldsymbol{X}^{\mathbf{T}}_{\mathbf{fm}}=\{P_{ij};\ Q_{ij};\ U_i\}$ of power flow model is extended as compared to the state vector of the classical state calculation model and contains the active and reactive flows in the conventional initial direction of all branches and the voltage magnitudes [8]. The following equations of losses are used for connecting the flows in a grid:

$$\Delta P_{ij}^{r} + j\Delta Q_{ij}^{x} = \frac{P_{ij}^2 + Q_{ij}^2}{U_i^2}\left(r_{ij} + jx_{ij}\right) + \left[\frac{\sigma_{P_{ij}}^2 + \sigma_{Q_{ij}}^2}{\overline{U}_i^2}\left(r_{ij} + jx_{ij}\right)\right], \tag{3}$$

$$\Delta P_i^{g} + j\Delta Q_i^{b} = U_i^2\left(g_i + jb_i\right) + \left[\sigma_{U_i}^2\left(g_i + jb_i\right)\right], \tag{4}$$

where U_i is a nodal voltage in the node i; $\overline{U}_i$ is an average voltage at the time interval T; σ^2_{Ui} is a voltage dispersion.

If the electrical mode on the time interval T is approximately constant, active (1) and reactive (2) balance equations are written in the power flows. If it does not, these balance equations are written in the energy flows as in the EFP. In the second case equations for power losses (3) and (4) should be supplemented with active and reactive power flows dispersions σ^2_{Pij}, σ^2_{Qij} respectively and average voltage should be used in the time interval T together with the voltage dispersion [7]. Using balance equations allows to change the calculation coordinate from power to energy W and vice versa easily.

$$\overline{P} = \frac{W_P}{T}; \quad \Leftrightarrow \quad \overline{Q} = \frac{W_Q}{T}. \tag{5}$$

Within electric utility costs control this key feature of the flow model is useful for solving problems in different time intervals.

In addition, for voltage calculation in the grid nodes in one calculation procedure [8] the balance equations should be supplemented with the corresponding application of the Ohm's law of the following expressions for all M grid branches:

$$U_j = t_{tr}\sqrt{\left(U_i - \Delta U_i(\boldsymbol{X}_{\mathbf{fm}})\right)^2 + \left(\delta U_i(\boldsymbol{X}_{\mathbf{fm}})\right)^2}\,,\ ij = 1, 2 \ldots M\,, \tag{6}$$

where U_j is nodal voltage at the end of all ij branches; $t_{tr}\leq 1$ is CT ratio for power transformer branches.

B. State Estimation and Power Mode Calculation tasks

State estimation is more general task as compared to state calculation. It allows to receive D operation mode parameters of the grid through the search of extremum of the following state estimation (SE) objective function with K equal constrains:

$$L_{\text{se}} = \sum_{i=1}^{D}\frac{1}{\left(\delta_{\text{mpe}_i} Z_i^{\text{meas}}\right)^2}\left(Z_i^{\text{calc}}(\boldsymbol{X}_{\mathbf{fm}}) - Z_i^{\text{meas}}\right)^2 + \sum_{i=1}^{K}\lambda_i\left(Y_i^{\text{calc}}(\boldsymbol{X}_{\mathbf{fm}}) - Y_i^{\text{fix}}\right), \tag{7}$$

where $\delta_{\text{mpe}i}$ is the maximum permissible error of each measurement unit, %; Z_i^{meas} is measured value; λ_i is the Lagrange factor; Y_i^{meas} and Y_i^{fix} are analytical expression and a value of equal constrain consequently.

In the SE task, equal constrains are the expressions for transient nods with zero injection. In the particular case, state calculation may be performed by means of recording all state equation (1) and (2) as equal constrains in the objective function without using weight least square (WLS) method. Thus, the modularity of the function for state estimation and state calculation is obtained, which allows to move from solving one problem to another more easily. In a common view, Eq. (7) is nonlinear. The task of searching the extremum of the objective function L_{se} can be solved by using iterative methods of the Newton and Gauss transformation.

C. Setting the Task of Optimization

Setting the task of optimization of utility costs includes forming an objective function which contains electric energy losses and net generation in the distribution grid. Such multicriterial setting is necessary for recording objective function with the use of the weighting factors k_l and k_t for both criteria consequently ($k_l+k_t=1$). These criteria should be written with the use of calculation vector $\boldsymbol{X}_{\mathbf{fm}}$ of the proposed unified mathematical model, which is based on power/energy balances in the nodes and branches of the distribution grid and does not contain voltage angles in the calculation vector. Each equation is used for forming analytic expression for criteria and for recording objective function and all G_{eq} equal or K_{neq} non-equal constrains in the optimization task:

$$L_{\text{ОПТ}} = k_l c_l \Bigg[\sum_{\substack{i=1 \\ i \neq j}}^{M} \frac{r_{ij}}{T}\left(\frac{W_{P_{ij}}}{\overline{U}_i^2} - \mathbf{0}\right)^2 + \sum_{\substack{i=1 \\ i \neq j}}^{M} \frac{r_{ij}}{T}\left(\frac{W_{Q_{ij}}}{\overline{U}_i^2} - \mathbf{0}\right)^2 +$$
$$+\sum_{i=1}^{N} g_i T\left(\overline{U}_i - \mathbf{0}\right)^2 \Bigg] - k_t c_t \sum_{i=1}^{N-1} W_{P_i} + \tag{8}$$
$$+\sum_{i=1}^{G_{\text{eq}}} \lambda_i \left(Y_i^{\text{calc}}(\boldsymbol{X}_{\text{fm}}) - Y_i^{\text{fix}}\right) + \sum_{i=1}^{K_{\text{neq}}} \psi_i \left(V_i^{\text{calc}}(\boldsymbol{X}_{\text{fm}}) - V_i^{\text{fix}}\right)^2,$$

where c_l are losses costs; c_t are energy net generation costs; V_i^{calc} and V_i^{fix} are analytical expression and a value of non-equal constrain consequently; $W_{P_{ij}}$, $Q_{P_{ij}}$ are active and reactive energy flows in the grid branches.

The equal constrains in the objective function (8) correspond to an active and reactive flow injections in the grid nods expressed thought the balance equations. Reactive power/energy balances at the nodes, where regulating voltage is possible, are exceptions. To solve this problem, the Lagrange factor is used. Non-equal constrains for nodal voltages increase are main constrains, because they directly affect grid losses and net generation. Accounting non-equal constrains is made by using the penalty function method. In the same style as the objective function (8), such constrains may be recorded in the following quadratic form:

$$\psi_i \left(U_i - U_i^{\text{fix}}\right)^2 = 0, \tag{9}$$

where U_i^{fix} is an upper limited of nodal voltages, driven by consumer's voltage requirements and level cap for insulation; Ψ_i is a coefficient for changing a form of penalty function curve.

There are many types of load models, which describe relation between active and reactive power and nodal voltage in the grid nodes. One of them is the static load model with polynomial characteristic [9]. This load model is more complicated but in the same time it is more suitable for calculating results. On the other hand, this form is similar to the one being used in the objective function (8). Static load characteristics should change the nodal nominal active P_{nom} and reactive Q_{nom} power according to the following static curves:

$$P_i = P_{\text{nom}_i}\left(\frac{a_{P_i}}{U^2_{\text{nom}_i}} U_i^2 \pm \frac{b_{P_i}}{U_{\text{nom}_i}} U_i + c_{P_i}\right) = P_{\text{nom}_i}\left(k_{P_i}\boxed{U_i} \pm f_{P_i}\right)^2, \tag{10}$$

$$Q_i = Q_{\text{nom}_i}\left(\frac{a_{Q_i}}{U^2_{\text{nom}_i}} U_i^2 \pm \frac{b_{Q_i}}{U_{\text{nom}_i}} U_i + c_{Q_i}\right) = Q_{\text{nom}_i}\left(k_{Q_i}\boxed{U_i} \pm f_{Q_i}\right)^2, \tag{11}$$

where a, b and c are polynomial coefficients of each i nodal loads; P_i, Q_i are phase or 3-phase active and reactive nominal powers; k and f are coefficients of quad of differences.

Depending on the form of grid elements representation, power/energy flows must be correct according to the losses equation (4). The form of objective function for state estimation and optimization is similar. It is a quadratic function, but in equation (8) measured values for flows and voltages are zero (Z_i^{meas}=0) and the weight coefficients $\boldsymbol{A}_{\text{op}}$ consist of resistant, costs and weight factors of choosen criteria.

- for losses: $A_{i,i} = k_l c_p r_{ij}$,
- for active nodal injections: $A_{i,i} = -k_t c_t P_{\text{nom}}$,
- for reactive nodal injections: $A_{i,i} = -k_t c_t Q_{\text{nom}}$,
- for penalty functions: $A_{i,i} = \psi_i$.

Thus, all components of objective function of optimization tasks are both of quadratic form (WLS method) with components expressed within calculation vector of flow model and also weighting coefficients, containing economic and technical parameters, like in the previous state estimation task. It allows to use one solving method based on well-known iterative method of Newton [10] with extra equal-constrains.

D. Solving State Estimation and Optimization Tasks

Quadratic form of objective function of both state estimation and optimization tasks allow to find its extremum strongly with using Gauss transformation in the matrix view. The transition from state estimation task (mode calculation) to optimization occurs through the object function replacement in the common matrix form of balance constrains.

State Estimation
$$\begin{bmatrix} \boldsymbol{H}^{\text{T}}\boldsymbol{A}_{\text{se}}\boldsymbol{H} & \boxed{\boldsymbol{R}^{\text{T}}} \\ \boxed{\boldsymbol{R}} & \mathbf{0} \end{bmatrix} \cdot \begin{bmatrix} \Delta \boldsymbol{X}_{\text{fm}} \\ \boxed{\boldsymbol{L}} \end{bmatrix} = \begin{bmatrix} \boldsymbol{H}^{\text{T}}\boldsymbol{A}_{\text{se}}\boldsymbol{Z}^{\text{calc}} \\ \boxed{\boldsymbol{Y}^{\text{fix}}} \end{bmatrix}$$
$$\Downarrow$$
Optimization
$$\begin{bmatrix} \boldsymbol{K}^{\text{T}}\boldsymbol{A}_{\text{op}}\boldsymbol{K} & \boxed{\boldsymbol{R}^{\text{T}}} \\ \boxed{\boldsymbol{R}} & \mathbf{0} \end{bmatrix} \cdot \begin{bmatrix} \Delta \boldsymbol{X}_{\text{fm}} \\ \boxed{\boldsymbol{L}} \end{bmatrix} = \begin{bmatrix} \boldsymbol{H}^{\text{T}}\boldsymbol{A}_{\text{se}}\boldsymbol{B} \\ \boxed{\boldsymbol{Y}^{\text{fix}}} \end{bmatrix}$$

} one calculation algorithm of power flow model

where $\boldsymbol{H}$ and $\boldsymbol{K}$ are the matrix coefficients of partial derivatives of the calculation vector $\boldsymbol{X}_{\text{fm}}$ (Jacobian); $\boldsymbol{A}_{\text{se}}$ is the matrix of weighting coefficients of state estimation task; $\boldsymbol{A}_{\text{op}}$ is the matrix of weighting coefficients of optimization task; $\boldsymbol{L}$ is the Lagrange factor; $\boldsymbol{R}$ is the matrix of partial coefficients at the Lagrangian; $\boldsymbol{B}$ is the matrix of residuals.

In this case, solving the tasks is carried out within unified iterative procedure and one calculation model. This mechanism provides an easy transition from state estimation and state calculation tasks to optimization and vice versa at various time interval without significant calculation model changing. In addition, using flow model allows solving the tasks within several iterations, due to better matrix $\boldsymbol{H}$ and $\boldsymbol{R}$ conditions [8].

E. One-Phase – Three-Phase Grid Representation

One of main complementary advantages of the power flow model is possibility of calculating operation mode parameters in three-phase representation for electrical grid areas, which are equipped with three-phase measurements. In other grid sections where there are no three-phase measurements, but there are single-line measurements or the degree of asymmetry of the mode parameters are insignificant, calculation problem in a single-line form is solved. This greatly facilitates the solution of the problems, which are described in this paper and creates additional benefits, especially for distribution grids.

Balance equations of flow model without angles in the calculation vector provide transition from one phase representation to three phase representation easily by means of record equations (1), (2) and (6) for A, B and C phases. However, introducing self r_{AA} and interaction Z_{AB} resistances complicate the expression (3) for losses. For example, for active power losses in the phase A change according to:

$$\Delta P_{\text{A}_{ij}} = \frac{\left[P^2_{\text{A}_{ij}} - Q^2_{\text{A}_{ij}}\right] r_{\text{AA}_{ij}}}{U^2_{\text{A}_i}} - \frac{\left[P_{\text{A}_{ij}} Q_{\text{B}_{ij}} - P^{\text{н}}_{\text{B}_{ij}} Q_{\text{A}_{ij}}\right] \operatorname{Im}(Z_{AB_{ij}} e^{j\delta_{\text{AB}_{ij}}})}{U_{\text{A}_i} U_{\text{B}_i}} -, \tag{12}$$
$$- \frac{\left[P_{\text{A}_{ij}} P_{\text{B}_{ij}} - Q_{\text{A}_{ij}} Q_{\text{B}_{ij}}\right] \operatorname{Re}(Z_{AB_{ij}} e^{j\delta_{\text{AB}_{ij}}})}{U_{\text{A}_i} U_{\text{B}_i}} - \Delta P_{\text{AC}_{ij}}$$

where P_A and Q_A are active and reactive power of phase A; ΔP_{AC} is interaction active power losses (record similarly).

The expressions for reactive power losses are recorded similarly. The expressions for the voltage drops (6) will be:

$$\Delta U_{ij} = \frac{P_{\text{A}_{ij}} r_{\text{AA}_{ij}} + Q_{\text{A}_{ij}} x_{\text{AA}_{ij}}}{U_{\text{A}}} + \frac{Q_{\text{B}_{ij}} \operatorname{Re}(Z_{AB_{ij}} e^{j\delta_{\text{AB}_{ij}}})}{U_{\text{B}}} +, \tag{13}$$
$$+ \frac{P_{\text{B}_{ij}} \operatorname{Re}(Z_{AB_{ij}} e^{j\delta_{\text{AB}_{ij}}})}{U_{\text{B}}} + \Delta U_{AC_{ij}}$$

$$\delta U_{ij} = -\frac{Q_{\text{A}_{ij}} r_{\text{AA}_{ij}} + P_{\text{A}_{ij}} x_{\text{AA}_{ij}}}{U_{\text{A}}} + \frac{P_{\text{B}_{ij}} \operatorname{Im}(Z_{AB_{ij}} e^{j\delta_{\text{AB}_{ij}}})}{U_{\text{B}}} -, \tag{14}$$
$$- \frac{Q_{\text{B}_{ij}} \operatorname{Re}(Z_{AB_{ij}} e^{j\delta_{\text{AB}_{ij}}})}{U_{\text{B}}} + \delta U_{AC_{ij}}$$

where ΔU_{AC} and δU_{AC} are the AC component of voltage drop.

The relationship between one phase A_{ij} and three-phase ij representation of various areas of distribution grid in one calculation model will be performed in the grid node and consist of the following power and voltage components:

$$\bar{U}_i = \frac{U_{A_i} + U_{B_i} + U_{C_i}}{\sqrt{3}} \quad (15)$$

$$P_{ij} = P_{A_{ij}} + P_{B_{ij}} + P_{C_{ij}} \quad (16)$$

$$Q_{ij} = Q_{A_{ij}} + Q_{B_{ij}} + Q_{C_{ij}} \quad (17)$$

The expressions (13) and (14) can be simplified under the assumption of insignificant asymmetry of the operation mode parameters in the distribution grid nodes [11]. If voltage in a phase differ more than 10 %, then the error is intent to exceed 5 %. If the voltages differ more than 15 %, then the iterative calculation procedure begins to diverge and a solution cannot be obtained. Such rules of constructing a hybrid calculation model allow representing in the three-phase as well as one-phase static load characteristics, measurements etc. in the state estimation and optimization tasks.

F. Three steps optimization algorithm

The tasks of distribution grid associated with voltage control by means of operating controlling devices are quite difficult due to using discrete and continuous parameters simultaneously as transformers with on-load tap and flexible reactive power sources correspondingly. In addition, depending on the characteristics, different in the switching resource and control principles, the control devices should be controlled at different time intervals. That is why it is necessary to use in the optimization task power and energy values. For clarification in Fig.1 three steps optimization algorithm is shown.

1. State Estimation ← current measurements

⇓ ← current estimated mode parametes (P, Q, U)

2. Optimization ← with using load forecast (energy W)

two – way ⇑ || ← control on-load tap of trasformer (tap changing)

|| ⇓ and form coefficient of penalty function curve

3. State Calculation ← prove the constrains in the grid scheme

Figure 1. Three steps optimization algorithm's diagrm.

Initial operation mode parameters obtained at the first step after solving the state estimation problem are used in the optimization problem for searching the optimal distribution of power/energy flows by means of using Newton's method. This information is needed to control flexible reactive power sources with continuous range. For solving optimization task with regard to discrete variables, it is necessary to use energy for forming control action in a long time intervals. Within step 2 the initial position of on-load tap of transformer will be set in the direction of decreasing object function of optimization. This process looks like the method of direct coordinatewise discrete optimization [12], but it uses a Newton's method inside iteration.

Under optimization process, all equal and non-equal constrains are taken into account. Nevertheless, due to solving optimization problem in the energy coordinate, it is necessary to verify none-equal constrains at a certain time interval by means of steady state calculation in power coordinates. In addition, this verification is needed to make a decision for changing on-load tap and correction of the form coefficients of penalty function curves (9) etc.

The control of the group devices with discrete and continuous characters should be performed non-synchronously at different time intervals, which are strongly fixed, and depends on commutation life of each device. For example, for on-load tap it is 8 hour and for flexible reactive power sources, it is less than 1 hour. All changes of grid mode due to using range limit of control devices should be taken into account in the process of rolling load forecast at various time intervals. Decision of control action forming is based on economic benefit calculation. Thus, all opportunities of the proposed power flow model are used in several algorithms within solving optimization task for increasing electric utilities benefits.

IV. Test Distribution Grid

The main advantages of the proposed algorithm for electric utility cost control are based on increasing net generation and shown in a simple test case in a Fig. 1. Test grid consists of three nodes, which are connected to each other at the radial topology. Branch, connecting nodes number 1 and 2, represents power transformer with on-load tap device. The load nodes 2 and 3 are represented with static load characteristics (SLC) for active and reactive power. In the node 3 flexible capacitor with possibility of changing the reactive power is installed. The value of the reactive power on the node number 3 depends on the voltage set in the automatic control system.

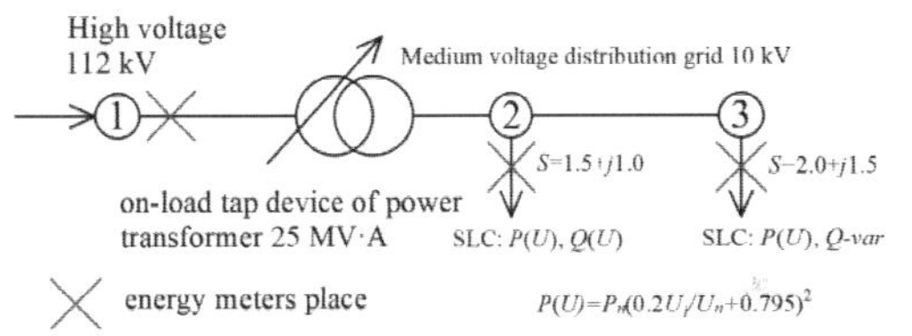

Figure 2. Three nodes test case.

First, the task of state estimation problem for estimation operating mode parameters is performed. At the initial mode configuration, the active power losses are 0.0298 *MW*. After the process of estimating mode parameters, it is necessary to do optimization of the distribution grid through regulating the reactive power source and on-load tap in cooperation. In the example, it is assumed that the loss costs are equal to 0.12 € per kW·h and the transmission tariff is 0.17 € per kW·h [13]. Depending on significance of the chosen criteria, various values of the objective function will be received. Table 1 shows comparison utility benefit under initial estimated mode of distribution grid and optimization results taking into account ratio between two main criteria. This ratio is 90 % and 10 %.

Table 1. Comparison of Calculation Results

Condition of Optimization => / **Criteria of Optimization**	***Initial steady-state mode (after state estimation)***	***Net generation 90 %, Losses 10 %***	***Losses 90 %, Net generation 10 %***
Utility cost for losses compensation, €	3.576	1.802	0.644
Utility benefit by means of net generation, €	595.000	613.158	593.421
The value of objective function of optimization (only benefits), €	591.424	**611.356**	592.777
On-load tap set	0	-1	3
Additional benefit, %	-	3.370	0.227

According to Table 1, total utility benefit is defined by not only voltage increasing in the power node supply number 1. Due to the opportunity of changing reactive power in node 3, control of the electric utility cost is performed by means of concerted actions with the use of both control devices simultaneously. As compared to decreasing power/energy losses, increasing net generation is more effective economically up to 5 %.

V. Conclusions

This paper introduces a method for increasing utility benefit by means of voltage control in the grid. Such approach allows decreasing power/energy losses in the distribution grid as well as increasing net generation at various time intervals. The last criterion is a new way of increasing utility benefit based on the voltage control according to static load characteristics at the grid nodes. Besides, depending on calculation tasks and time intervals, optimization may be performed with the use of both criteria with various weight coefficients. For linking all the parameters power-energy flow model was used. This model is more suitable for distribution grid. The proposed optimization algorithm and flow model have great perspectives for implementation in distributed automated control systems, which are installed in local areas of distribution grids.

Further work is aimed to extend the optimization algorithm using the possibility of representing the calculation grid model in one-phase and three-phase views.

Acknowledgment

The work was supported by Act 211 Government of the Russian Federation, contract № 02.A03.21.0006.

References

[1] S. Chren, B. Rossi and T. Pitner, "Smart grids deployments within EU projects: The role of smart meters," *2016 Smart Cities Symposium Prague (SCSP)*, Prague, 2016, pp. 1-5.

[2] Edward Jeroen Coster, "Distribution Grid Operation Including Distrib-uted Generation," Printed by Ipskamp drukkers: Enschede, 2010. P. 205.

[3] Nikander, Ari & Repo, Sami & Järventausta, Pertti. Utilizing the ring operation mode of medium voltage distribution feeders. 17th International Conference on Electricity Distribution. Barcelona, 12-15 May 2003, pp. 1-6.

[4] A. Abur, A. G. Expósito, "Power System State Estimation: Theory and Implementation," (Power Engineering (Willis)). s.l.:CRC Press.

[5] C. Arrigoni et al., "Smart Distribution Management System: Evolution of MV grids supervision & control systems," 2016 AEIT International Annual Conference (AEIT), Capri, 2016, pp. 1-6.

[6] M. E. Cebeci, O. B. Tor, C. Gencoglu and A. N. Guven, "A geographical zoning methodology for nodal transmission use-of-system tariffs," *2012 IEEE Power and Energy Society General Meeting*, San Diego, CA, 2012, pp. 1-8.

[7] A. V. Pazderin, E. A. Plesniaev and E. V. Mashalov, "An energy flow calculation method based on energy meters data," Electric Utility Deregulation and Restructuring and Power Technologies, 2008. DRPT 2008. Third International Conference on, Nanjuing, 2008, pp. 834-839.

[8] N. D. Mukhlynin and A. V. Pazderin, "Power-Energy State Calculation in Smart Grid on the Basis of Flow Model," *2018 International Conference on Smart Energy Systems and Technologies (SEST)*, Sevilla, 2018, pp. 1-5.

[9] T. Alexander, P. Andrey, M. Olga and C. Pavel, "Identification of static polynomial load model based on remote metering systems information," *2013 13th International Conference on Environment and Electrical Engineering (EEEIC)*, Wroclaw, 2013, pp. 213-216.

[10] S. Yeferev, A. Pazderin, "Combined newton's method for power flow calculation," *2013 6th IASTED Asian Conference on Power and Energy Systems (AsiaPES),* Phuket, 2013, pp. 104-107.

[11] D. Bellan and S. A. Pignari, "Circuit representation of voltage unbalance emission due to line asymmetry," *2015 IEEE Innovative Smart Grid Technologies - Asia (ISGT ASIA)*, Bangkok, 2015, pp. 1-5.

[12] N. D. Mukhlynin and M. Y. Komlev, "Method of direct coordinatewise discrete optimization in choosing the optimal normal open points in distribution grids," *2016 2nd International Conference on Industrial Engineering, Applications and Manufacturing (ICIEAM)*, Chelyabinsk, 2016, pp. 1-5.

[13] A. Singh, S. Saha, P. Das, S. Chatterjee, P. Chandrakar and S. Debbarma, "Power Tracing in Distribution Network in Deregulated Power Environment," *2018 International Conference on Computation of Power, Energy, Information and Communication (ICCPEIC)*, Chennai, 2018, pp. 442-448.

Comparative Evaluation of the Functioning and Development of the Energy Markets of Russia and Germany

Malinina Tatiana
Sant-Petersburg Polytechnical University
of Peter the Great
SPbPU of Peter the Great (СПбПУ Петра Великого)
Sant-Petersburg, Russia
e-mail: t_malinina@mail.ru

Novikova Olga
Sant-Petersburg Polytechnical University
of Peter the Great
SPbPU of Peter the Great (СПбПУ Петра Великого)
Sant-Petersburg, Russia
e-mail: novikova-olga1970@yandex.ru

Abstract — **Electricity markets are the main stage on the way to the formation of a sustainable and progressively developing national market. The specific organizational model of the energy market, pricing schemes for energy tariffs, structure of electric power systems can show the strengths and weaknesses of the country's economy.**

In order to determine the characteristics of the functioning of energy systems and the prospects of their development, the electricity markets of Germany and Russia are considered. The assessment of the state of the German energy market is based on an analysis of four large energy companies engaged in the production of energy on the basis of various types of resources and playing an important role in formation the German energy policy.

The analysis of the energy market of Russia was carried out both as a whole for the power system of Russia, and on the example of some large energy companies.

A comparative description of the structure and characteristics of the functioning of the energy markets of Russia and Germany was given, certain advantages and disadvantages of these markets are noted, and prospects for their development was considered.

On the basis of technical and economical indicators, an assessment of the efficiency of the energy market from the position of the producer and consumer of energy was made.

Keywords: energy companies, electricity and capacity market, efficiency indicators, electricity tariff

I. Introduction

In the modern world, electricity markets are the main stage on the way to the formation of a sustainable and progressively developing national market, the country's economy as a whole. Transformations taking place in the power industry are crucial for the development of countries and the well-being of their citizens.

The specific organizational model of the energy market, energy pricing schemes, the structure of electric power systems can show the strengths and weaknesses of a country's economy.

Features of the functioning of energy systems and the

prospects for their development were considered on the example of the countries of Germany and Russia.

II. Analysis of the Structure of Generation

The installed capacity of power plants of the Unified Energy System of Russia (UES of Russia) as of January 1, 2018 was 240 GW, which is 20% higher than in Germany. Electricity generation by UES of Russia power plants in 2017 amounted to 1,053.861 billion kWh, which is almost 80% more than in Germany.

The structure of capacities in Russia and Germany differ significantly. Thermal power plants (68%) dominate in Russia, of which heat and power plants with combined heat and power generation (CHP) account for about half. The share of hydroelectric and nuclear power plants is respectively 21% and 11%. Electricity generation from renewable energy sources (with the exception of large hydropower plants) in Russia is less than 1% [1,4].

The total installed capacity of power plants in Germany in 2017 reached 202 GW [3]. Renewable energy sources (RES) prevail in the structure of installed capacity, their share is more than 56%, and installed capacity based on traditional resources is 44%, however, electricity generation from RES has a smaller weight (37%). This is due to the smaller number of hours of use of installed power from RES.

The largest share of renewable energy falls on wind power plants (WPP) - 28%. In general, 78.6 billion kWh of electricity was produced on the basis of wind power plants in 2016, which accounted for 13.3% of the total electricity in Germany. This is due to the smaller number of hours of use of installed capacity (Tu) from RES

Currently, four large vertically integrated energy companies (EC) dominate the German energy market: RWE, E.ON, Energie Baden-Wurttemberg (EnBW), Vattenfall, which, along with electricity production, also engage in other activities: RWE (coal mining, production and electricity sales, trade and supply of consumers with natural gas), E.ON (production and sale of electricity), EnBW (production of electricity, owner of electrical networks), Vattenfall (production of energy and gas supply).

Analyzing the structure of the electricity generation of energy companies in Germany, we can conclude that a large share of production is made up of traditional energy sources, which are complemented by generation from renewable energy sources [5,6,7,8].

III. Structure of the Energy Market

Since the 90s of the 20th century, Germany, like other European countries, has been following the path of reforming the electricity sector in order to improve its work. The reform program included such measures as the division of energy companies by type of activity and the creation of competitive wholesale and retail electricity markets.

Restructuring of the energy market in Russia began a decade later than in Germany. On July 1, 2008, a transitional capacity market began to function, with which capacity trading mechanisms at free (unregulated) prices came into effect, and the liberalization of capacity trading began.

By 2011, the existing scheme of the wholesale and retail electricity and capacity market was basically formed in Russia. In contrast to Germany, in Russia, the capacity market functions at the same time with the electricity market [1,3].

The structure of the capacity market in Russia includes the following segments:

- competitive selection of power - KOM;

- free bilateral contracts of capacity - SDM;

- contracts of capacity submitting - DPM ("new" thermal generation capacity and agreements of purchase and sale capacity for new nuclear and hydroelectric power plants);

- regulated contracts - RD (the need for electricity and capacity of the population is provided at regulated prices);

- compelled generation market - VG (it includes CHP plants that have not passed on the KOM market, but cannot be excluded from the number of market participants due to the need to ensure the heat energy load).

Trade in capacity at unregulated (free) prices is carried out within the sectors of the KOM and SDM wholesale market. For segments of RD and VG of the capacity market, the price is determined by the Federal Tariff Service.

The purchase of capacity gives the wholesale market participant the right to demand that the generating equipment is ready for power buyers.

Germany did not take the path of development of the market for two goods. Reports by experts from the Federal Ministry of Economics and Energy of Germany showed that the market of a single commodity, electricity, is cheaper than the market for electricity and power. German power market models lead to an excess of power and, therefore, unnecessary additional costs compared to the electricity market [1,3].

In Russia, the capacity market was created in order to avoid a shortage, which was caused by the considerable deterioration of the existing equipment of power plants and power lines, as well as by the increase in electricity demand. Until 2008, there was a shortage of generating capacity in Russia, so when passing the annual peaks of power consumption, restrictions were imposed on consumers. However, due to the slowdown in economic development, reducing the need for electricity and the action of energy saving factors in Russia, there was a situation when the introduction of a capacity market, in particular such a segment as PDM (aimed at attracting investment in the construction of new capacities) led to a significant excess of capacity. For 2019, the excess capacity was 18 GW [9,13].

Given that the old model of the capacity market has lost its relevance in the context of stabilization of electricity consumption and fuel prices, in 2015 a new model of capacity market was accepted - the elastic demand model, which showed its adequacy to the current scenario conditions and the expediency of its application in the medium term [13].

The following market segments are present on the electricity market in Germany and Russia: a balancing market, a market of bilateral contracts, a day-ahead market. In addition to these segments, in Russia also has a regulated contracts market. The same principle of competitive pricing is used in the electricity market in Germany and in Russia - the method of marginal cost of electricity.

There is a competitive model in the wholesale electricity and capacity market in Russia, and there is no real competition in the retail market. Electricity is sold at regulated prices to the population and for all other consumers - at a residual principle, taking into account cross-subsidies. At the same time in Russia in the retail market the consumer (the population) cannot choose a supplier. Regional sales companies offer their services to industrial consumers, but the tariff does not depend on the type of generation, but mainly depends on the consumption mode.

In the retail electricity market in Germany, consumers can choose a supplier (including the type of generating facility) depending on the services provided and act as producers when there is a shortage of power in the network [1,3].

Another feature of the Russian energy market is a significant proportion of CHP (33%). In the combined generation of electricity and heat, the electric power of the CHP plant depends on the load of the heat consumers. This limits the changes in CHP electrical power for operation in the variable part of the load graph. For economic reasons, it is impractical to unload combined-cycle heat and power plants (TPP) with an efficiency of 50% or more, considering that the efficiency of the most efficient steam-powered TPPs does not exceed 40%. It is not profitable to transfer steam power TPPs from the base mode to the variable mode. As a result, a competitive market turns out to be incomplete, since taking into account the existing restrictions only TPPs can participate in competition (HPPs and NPPs do not participate in competition, because HPPs operate in the peak part of the load curve and NPPs - in the base part). Mainly competitive thermal power plants are powerful steam turbine power plants operating on supercritical steam parameters [12].

IV. THE EFFICIENCY OF THE ENERGY COMPANIES

It should be take into consideration that approaches to evaluating the effectiveness of a competitive market depending on the market participant - the producer, the consumer, and the state.

Producers of goods consider the maximum of their profits a criterion of the effectiveness of a competitive market. This criterion is not conducive to restraining price increases.

The consumer is interested in purchasing the necessary goods of the required quality at the lowest price. Competition, of course, significantly limits the rise in prices, but not always and not in the proper measure. The state needs economically reasonable, optimal prices.

For a comparative assessment of the efficiency of the electricity markets, we reviewed the main indicators of large generating companies in Russia and Germany. For Russia, two large energy companies (EC) were chosen - Mosenergo, and the territorial generating company TGK-1. Two companies were selected for Germany - E.ON and Energie Baden-Wurttemberg (EnBW).

Considering the different size of installed power of EC and various types of their activity, a comparative assessment of the efficiency of EC operation was carried out on the basis of specific indicators (Table 1) [5,6,10,11].

Table 1. Efficiency indicators of large energy companies in Russia and Germany

Indicators	E.ON	EnBW	Мос-энерго	ТГК-1
1. The number of hours of use of the installed capacity of power plants Tu, hour	3757	3887	4495	4246
2. Profit per employee of energy companie Pr/T_{EC}, euro/person	81395	102439	41600	17968
3. Specific number of personnel Kp, people/MW	1,74	1,5	0,62	0,9
4. Specific profit Pr/Wp, cent / kWh	3,7	4,0	0,55	0,35
5. Labor productivity Wp/T_{EC}, kWh/person	2,2	2,6	7,5	5,2

From the standpoint of the energy company, the financial effect of the enterprise is determined by the indicators of sales volume and profits. The profit indicators in relation to the number of personnel Pr/T_{EC} and the electricity sales volume Pr/Wp allow us to compare the staff work efficiency of various ECs.

To appreciate the labor productivity at power plants, the staff ratio Kp is often used, which is the ratio of the number of industrial personnel at power plants per unit of installed capacity, person/MW.

The staff ratio for German companies is higher than for Russian ones, and labor productivity is lower. The fact, this is due to that German companies along with the production of electricity are also engaged in the sale (E.ON) and transmission (EnBW) of electricity, so they have more staff.

In Germany, the profit per employee is 2-3 times higher than in Russia, which may be due to the higher tariff rate for consumers and the presence of other types of activity (transmission and sale of electricity). However, the volume of sold electricity per employee of Wp/T_{EC} in Russia is 2–3 times higher than in Germany, because in Russia only the number of personnel producing electricity and heat is taken into account, the number of EC personnel engaged in transmission and marketing is not taken into account. For these reasons, also specific number of personnel Kp in Russia is 2-3 times less than in Germany.

Thus, in terms of profit per employee, German ECs have an advantage, and in terms of labor productivity (Kp and Wp/T_{EC}), Russian EKs have an advantage.

V. EVALUATION OF THE EFFICIENCY OF THE ELECTRICITY MARKET FOR THE POPULATION

Analysis of the effectiveness of the operation of the electricity market from the position of the consumer can be carried out on the basis of electricity tariffs.

The absolute value of the tariff for household consumers (HC) and industrial consumers in Germany increased 2 times from 2000 to 2014, while in Russia over the same period, the tariff for HC increased 9 times. In addition, the tariff for the HC in Germany is approximately 2 times higher than the tariff for the industry. In Russia, the opposite trend is observed, when the tariff for HC is lower than for industrial consumers [10,11]. This is due to cross-subsidization in Russia, when industrial consumers cover part of the costs of household consumers.

A comparative assessment of tariffs in Russia and Germany, taking into account purchasing power parity, showed that the electricity tariff for industrial consumers in Russia is comparable to the tariff in Germany. However, the tariff for household consumers in Russia is 2 times lower than the tariff in Germany.

It should be noted also that in the structure of the electricity tariff in Germany, more than 50% are government price components (taxes and surcharge for the use of electricity from renewable sources). In Russia, the network component reach for more than 50%, it has increased by more than 3 times compared with the pre-reform period.

VI. Development of the energy market

Taking into account the strategic plans of energy companies and the state energy policy, it can be concluded that nuclear power is gradually excluded from the generation structure of German companies and the share of renewable energy significantly increases (up to 55-60% by 2035 and up to 80% by 2050 year) [2, 3]. The growth of renewable energy in the structure of electricity generation has already influenced the growth of tariffs. A further increase in the share of RES may lead to a subsequent increase in tariffs. However, the gradual perfecting of energy production based on renewable energy sources, as well as the perfection of energy storage systems can reduce the cost of electricity for German consumers.

In Russia, by 2035 it is expected to further increase electricity production based on nuclear energy and its growth in the structure of electricity generation, it is also envisaged to increase the share of renewable energy sources maximum to 20%, which is not yet confirmed by the actual commissioning rates of such facilities [2].

In 2019, a new capacity market development program was adopted (the CDM segment) to update the existing capacities of CHP plants without increasing the installed capacity of the main equipment. Expected consequences: an increase in electricity costs to compensation the investment costs, but the containment of the growth of electricity tariffs which is associated with the introduction of more economical equipment.

VII. Conclusions

1. The capacity structure and power generation of power plants in Russia and in Germany varies significantly. Russia is following the path of development of traditional energy, including nuclear, while Germany is actively developing renewable energy and is following the path of a moratorium on the development of nuclear energy.
2. The number of hours of use of installed capacity of power plants is lower in Germany than in Russia by an average of 15%, which is characterized by a low degree of capacity utilization.
3. The structure of the electricity and capacity market is different: in Russia, capacity is sold as a separate product, along with the electricity market, there is a capacity market that includes five segments, which greatly complicates the structure of the market.
4. The creation of a capacity market in Russia, on the one hand, played a positive role in the development of the country's energy sector, since it provided an inflow of investments, but on the other hand, this led to the presence of significant excess capacity due to a number of factors, including miscalculations in the macroeconomic forecasts of the country's economy.
5. Evaluation of the efficiency of the functioning of some energy companies in Russia and Germany showed the complexity of such an analysis due to differences in their capacity, structure and activities. For a more justified estimate, it is necessary to make such a comparison, taking into account the standard number of staff.
6. A comparative assessment of tariffs in Russia and Germany, taking into account purchasing power parity, showed that the electricity tariff for industrial consumers in Russia is comparable to the tariff in Germany. However, the tariff for domestic consumers in Russia is 2 times lower than the tariff in Germany, which is associated with cross-subsidies.

References

[1] System Operator of the UES, Report on the functioning of the UES of Russia in 2017, 2018 // [Electronic resource] / URL: https://soups.ru/index.php?id=pr2017.

[2] Project Energy Strategy of the Russian Federation for the period up to 2035 (revised from 02.02.2017) // [Electronic resource] / URL: https://minenergo.gov.ru.

[3] I. Savin, J. K. Hermann, "Evolution of the German electricity market," 2015 5. Annual report of the Ministry of Economy and Energy of Germany, 2017. // [Electronic resource] / URL: https://www.bmwi.de/Navigation/DE /Service/XML-Open-Data/xml-opendata.html.

[4] Annual report of the Ministry of Economics and Energy of Germany, 2017. // [Electronic resource] / URL: https://www.bmwi.de/Navigation/DE/Service/XML-Open-Data/xml-open-data.html .

[5] Annual report of the energy company E.ON, 2017 // [Electronic resource] / URL: https://www.eon.com/content/dam/eon/eoncom/investors/annualreport/EON_GB17_EN.pdf .

[6] EnBW 2017 Annual Production Report // [Electronic Resource] / URL: https://www.enbw.com/enbw_com/downloadcenter/annualreports/enbw-integrated-annual-report-2017.pdf .

[7] Annual report Vattenfall, 2017. // [Electronic resource] / URL:https://corporate.vattenfall.com/globalassets/corporate/investors/annual_reports/2018/vattenfall_annual_and_sustainability_report_2017_eng.pdf .

[8] Annual report of the energy company RWE, 2017 // [Electronic resource] / URL: https://www.group.rwe/en.

[9] Ernst & Young LLC, Overview of the Russian electric power industry, 2018 // [Electronic resource] / URL: https://www.ey.com/Publication/vwLUAssets/EY-ac-power-and-utilitiesregulations-newsletter-q2-2017/$FILE/EY-ac-power-and-utilitiesregulations-newsletter-q2-2017.pdf .

[10] PJSC Mosenergo, Annual Report, 2017 // [Electronic resource] / URL: http://www.mosenergo.ru/d/textpage/f9 /249/godovoj-otchetmosehnergo-2017.pdf .

[11] PJSC TGC-1, Annual Report, 2017 // [Electronic resource] / URL: http://www.tgc1.ru/ir/reports/ .

[12] V. I. Denisov, A. A. Dzyuba, "Economic reform and its consequences in the electric power industry. "Power plants", 2018, No. 7, p.2-5.

[13] F. Opadchy, A. Kataev, "Power Market. The change in the model during the transition from deficit to surplus" Energy Market, 2016, №7, p.34-41.

Problems of Ensuring the Balance Reliability of Perspective Development Schemes UES of Russia

Iurii Chukreyev
Institute of Socio-Economic and Energy Problems of the North, Komi Scientific Center, Ural Branch, Russian Academy of Sciences
Syktyvkar, Russia
chukreev@iespn.komisc.ru

Mikhail Chukreyev
Institute of Socio-Economic and Energy Problems of the North, Komi Scientific Center, Ural Branch, Russian Academy of Sciences
Syktyvkar, Russia
mchukreyev@gmail.com

Abstract —**The role of the task of ensuring balance reliability in the system of long-term planning of the development of the power industry of the country is revealed. The factors affecting the indicators of the balance reliability of electric power systems are given. The problems of the influence of their changes caused by the intellectualization of power systems on the standards of balance reliability and components of reserve capacity in the long-term planning of the UES of Russia are shown. The main factors affecting the different interpretations of the indicators of balance reliability and their normative values used in the domestic and foreign practice of energy planning are given. The problems and possibilities of their comparative analysis are presented.**

Keywords: UES of Russia; reliability criteria; power system; adequacy parameter; parallel algorithm

I. The Place of Balance Reliability in Modern Conditions of Long-Term Planning

The balance reliability (adequacy) in the task of prospective planning of the electric power industry is understood as the ability of the electric power system (EPS) to meet the aggregate demand for electric energy and power of consumers within specified limits on energy supply, taking into account planned and reasonably expected unplanned interruptions its elements, as well as operational limitations. In accordance with the definition, the tasks of balance reliability include only those whose solution is connected with the need to take into account system failures due to accidental damage to equipment and taking into account random deviations of loads from planned values [1]. It is important to understand that random states caused by unplanned interruptions in the tasks of long-term planning, in contrast to the tasks of operational management and short-term planning of regimes, can last several tens of days (emergency repair of equipment), and the depth of the possible power shortage can reach significant values in tens of GW, albeit with a low probability of their occurrence.

When planning the development of the industry, the so-called power balance is being developed for the UES of the country as a whole and its united EPS (UES) [2, 3]. Its form includes two positions (Fig. 1): the expenditure part is "demand" (maximum load, export, and standardized power reserve); incoming "coverage" (installed capacity, unused power during the period of

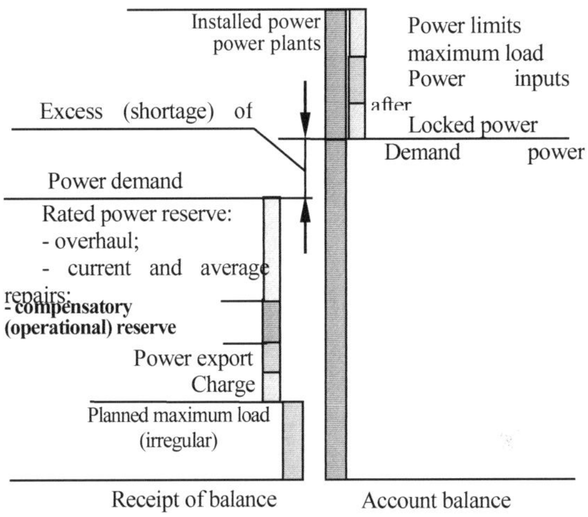

Figure 1 - Structure of the predicted power balance.

maximum passage, power inputs after maximum passage, under-utilization of power).

The question that concerns both consumers and suppliers of electrical energy is to justify the regulatory power reserve, which is one of the components of the expenditure side of the balance sheet. In the tasks of long-term planning of an EPS, it is called the full reserve, and is conventionally divided into three components: repair, strategic and compensation. If the first component is determined on the basis of the analysis of statistical information about the equipment operation, the second one is expert, the third one requires solving the problem of balance reliability.

Market relations in the power industry exacerbate the issues of justifying the compensation reserve of power. For NP "Market Council", representing the interests of consumers in order to reduce tariffs, it is important to reduce this component of the regulatory power reserve. In turn, JSC SO UES, representing energy supplying organizations, seeks to increase this component of the power balance in order to increase reliability in the future conditions of operation of the power system.

In the pre-reform period until 1991, the issue of ensuring balance reliability was addressed in the overall task of the long-term development planning of the UES of the country at the Energosetproekt design institute and its North-West branch. The reform of the industry in our country, begun in the mid-1990s, led to a decrease in interest in ensuring balance reliability when considering tasks related to the future planning of the UES of Russia. Many factors contributed to this: a sharp decrease in electricity consumption, a change in the form of ownership, a deterioration in the quality of managerial personnel, the closure of sectoral design institutes and much more.

In accordance with the Federal Law "On Electric Power Industry" [4], questions of justifying the means of ensuring reliability are entrusted to JSC SO UES. This company with the assistance of specialists from JSC "ENEGOSETPROEKT Institute" and ISE and EPS of Komi Scientific Center of Ural Branch of the Russian Academy of Sciences in 2011 initiated and completed in 2012 the development of new guidelines for designing the development of power systems*1. In them, as in the previous edition [5], there is a section on ensuring balance reliability. His recommendations on power reserve standards have been used in the development of the general layout for the placement of energy facilities for 15 years or more (developed by the Ministry of Energy of Russia) and the UES of Russia UES development plans and development programs for a seven-year period*2 (developed by JSC SO UES and Federal Grid Company UES ").

It should be noted that the new guidelines 2012 have not yet been approved by the Ministry of Energy of Russia. During the development of schemes and programs (DSP) of recent years, the systems developed by the centralized management of the industry in 2003 were used [5]. Modern conditions of the country's socio-economic development, giving the power systems intellectual content, especially in terms of information security, require the improvement of both the guidelines 2003 and the new 2012, not approved by the Ministry of Energy of Russia.

To some extent, the organization aimed at this at the beginning of 2017 by the Russian Ministry of Energy on the instructions of the Deputy Prime Minister of the Russian Federation A. Dvorkovich on the basis of the JSC "ENERGOSETPROEKT" Institute of a single center of competence (SCC). His intended activity is long-term planning of electric power facilities.

The energy industry planning structure proposed by the Ministry of Energy of Russia is becoming similar to that used earlier for the conditions of centralized management of the industry. At the same time, the new structure represented by the SCC should not lose any groundwork in terms of the information content of the balance reliability task received by SO UES and update the studies on the refinement of the new guidelines in terms of balance reliability. However, the year that has passed since the creation of the SCC did not change anything in resolving these issues.

II. Indicators of Balance Reliability and Influencing Factors.

It is known that an integral part in the justification of one of the components of the total power reserve - compensation (highlighted in Figure 1), has always been an assessment of indicators of balance reliability. The indicators characterizing the balance reliability of the EPS should be sufficiently sensitive to disturbances leading to its change. In domestic [1, 6, 7, etc.] and foreign publications [8, 9, etc.] this most fully answer:

*1 Guidelines for the design of energy systems / JSC "Institute" Energosetproekt ", 2012 (approved by NP" NTS UES ", section" Technical regulation in the power industry ").

*2 The rules for the development and approval of schemes and programs for the future development of the electric power industry, approved by the Government of the Russian Federation of October 17, 2009 No. 823.

- the probability of deficit-free operation of territorial zones of the EPS ($\rho = 1 - J_д$), where $J_д$ is the integral probability of power shortage;
- the average number of days of power shortage (loss of the anticipated load – LOLE [2]);
- the average number of hours per year of power shortages (loss of load hours – LOLH).
- When considering the design scheme of the UES of Russia, which includes territorial zones in the form of an UPS, the probabilistic indicator of balance reliability is used [6]. It is defined for the whole set of possible time intervals of the load change by the following expression:
- $$J_д = \sum_{i=1}^{T} Q_i \sum_{k=1}^{N} p_k(\overline{P}_{ik}^{н} - P_{ik}^{н}) = \sum_{i=1}^{T} \sum_{k=1}^{N} Q_i p_k(\overline{P}_{ik}^{н} - P_{ik}^{н})$$ (1)
- where $Q_i=1/T$ is the probability of deterministic steps of the load curve, usually equal 1/8760; T is number of stages of the graphics load (8760 hourly changes per year); $\overline{P}_{ik}^{н}$, $P_{ik}^{н}$ – respectively, required and provided available generating capacity and reserves capacity links load the k-th random state of the system; p_k – the probability of the k-th random state in which there is a shortage of capacity in the considered territorial area, i.e., when $\overline{P}_{ik}^{н} > P_{ik}^{н}$; N is the number of random states that are modeled on the i-th interval of the load change.
- The indicators of adequacy in the form $J_д$ is carried out by simulation of random States generating capacity. They are called emergency exits equipment. Modeling is carried out for each hour of the diurnal variation of the load (24 steps) for 365 days of the year or 250 working days. In foreign experience to assess the adequacy two probabilistic measures of adequacy – LOLE and LOLH are mainly used. The definition of the indicator LOLE associated with simulation of random states generating capacity only for peak maximum values [8] for all 365 days of the year. The definition of the indicator LOLH largely coincides with the definition of domestic rate $J_д$.
- It is known that indicators of the balance reliability of UES of Russia mainly depend on the following factors and random events [1, 6–11]:
- model of the design scheme of UES of Russia, selected territorial areas and connections between them;
- available capacity of some territorial areas;
- structure of generating capacities;
- scheduled repairs of equipment;
- reduction of generation capacity in rural areas due to accidental damage of units of power plants;
- regular and irregular highs loads of territorial zones and graphs of their changes in the context of day and year;
- random deviations and irregular load fluctuations;
- reserves bandwidth connections in normal and emergency modes between the selected in the model design scheme of UES of Russia territorial areas.

- It is obvious that all of the above mentioned factors were taken into account when assessing the indicators of adequacy in the process of preparing the reference document for the design of energy systems. While the content of tasks in modern conditions more adequately reflect the actual processes. That should naturally influence the quality of decisions taken.

III. Standards for Indicators and Means of Ensuring Balance Reliability.

The optimal distribution of the standard (full) power reserve and especially its compensation (operational) component across the territorial zones of the Unified Energy System of Russia depends largely on the accepted values of the standard indicators of balance reliability. In due time for the concentrated system*[3] the normative indicator of balance reliability was economically justified [10] $J_{Д}^{норм} = З_R/(y_0 T)$. The ratio links two contradictory indicators: unit costs to ensure reliability ($З_R$) and unit losses from unreliable power supply (y_0).

It should be noted that in North America and Western Europe there is no economic justification for standards of reliability. There, the regulations mainly reflect generally accepted indicators that suit everyone from consumers to suppliers. In the USA and Canada, the minimum value of the power reserve should correspond to the indicator of balance reliability not higher than the *LOLE* value = 0.1 days / year. The generally accepted standard standards for some European countries: France - *LOLH* = 3 h / year, UK - *LOLH* = 4 h / year, Ireland - *LOLH* = 8 h / year. It is also important that for several decades the spectrum of initial information and the adopted methods for obtaining indicators of balance reliability in these countries have not undergone significant changes.

In our country, the picture is somewhat different, the standard in the form $J_{Дi}^{норм} = 0{,}004$ [2] complied with certain rules for presenting source information and certain methodological principles for determining probabilistic indicators of balance reliability [6, 7, 11, 12]. The change in the information content of the task of balance reliability, both in terms of equipment accidents and power consumption characteristics, clearly requires their correction. Partially these questions are revealed in the next section.

IV. The Impact of the Intellectualization of Energy Systems on the Substantiation of the Components of the Regulatory Power Reserve

Consider the factors affecting the reliability (section 2), from the standpoint of changed conditions (information content of the task of assessing the balance sheet reliability indicators). The first item "the model of the UES of Russia design scheme", and the last "reserves of communication capacity between territorial zones" are related to each other. It is clear that it is almost impossible to create a model of the UES settlement scheme for the task of evaluating the indicators of balance reliability adequately reflecting the actual schemes. The most difficult in this task is the question of determining the reserves of bandwidth connections between the territorial zones selected in the model of the design scheme. Studies conducted by the authors together with specialists of JSC SO UES revealed that splitting the UES into smaller territorial zones leads to an increase in the power compensation reserve (up to 0.5%) [6].

The points "disposable capacities of territorial zones" and "structure of generating capacities" show no significant changes in the information content. Some problems of a methodical nature are caused by the presence of unjustified power surpluses, resulting, among other things, due to a sharp decrease in electricity consumption during the beginning of market transformations in the country and in the industry. To a large extent, the presence of excess power affects the rationale for the values of planned repairs of generating equipment.

For the following factors, "planned and emergency repairs of equipment", it should be noted that the launch of the capacity market in 2006, together with the above-mentioned presence of unjustified power surpluses, led to the loss of the normative role of such concepts as the duration and frequency of planned repairs, and promising five-year plans for repairs of the main equipment of power plants. But it should be noted that since 2010, experts of the JSC "SO UES" have been carrying out certain work in this direction. Information security makes it possible to process information about the operation of generating equipment and to obtain the average statistical indicators of its planned and emergency repairs. Unfortunately, this information is available only to a narrow circle of specialists of JSC SO UES. Given the upcoming legislative transfer of industry planning issues from this company to the SCC under the Russian Ministry of Energy, certain difficulties may arise in its use.

After processing the retrospective information on the operation of the generating equipment, there is a tendency to overestimate the norms of planned repair and, conversely, an understatement of the norms of emergency repair. This is connected both with the transfer by the generating companies of a number of emergency withdrawn units into the category of scheduled repairs, and with changes in the approaches to the conduct of medium and capital repairs in the time aspect. In market conditions, their conduct in the autumn-winter period is determined mainly by technical and financial capabilities, taking into account the operating balance situation in the power system, as well as the rules of the wholesale electricity (capacity) market. These conditions, due to the presence of sufficient power surpluses in the UES of the country (up to 30 GW) are not the limiters that were characteristic of the pre-reform period. This circumstance can explain the tendency of a significant increase in the values of power reserves for carrying out scheduled repairs represented in the power balance sheets, for example, in the work of the Scheme and Development Program of the UES of Russia for 2016–2022.

The next two points, "load maxima, graphs of their change, random deviations and irregular fluctuations", relate to the modeling of power consumption modes. The representation of the power consumption mode during the billing period (usually a year) and especially its random deviations from the average values caused by the temperature factor [6] has a significant impact on the indicators of balance reliability and, consequently, on the means of its provision. The possibility of obtaining sufficiently reliable information about the power consumption modes due to the introduction of operational-measuring complexes in the automated management system of the Unified Energy System of Russia allows to identify the correlation dependence of random load deviations between the UPS [13]. Studies have shown [14] that this factor has a significant effect on the indicators of balance reliability, which should naturally affect the normative indicators of balance reliability obtained in Section 3, obtained for the conditions of centralized control of the country's electric power industry [10, 11].

V. Conclusions

Modern conditions of the country's socio-economic development, giving the power systems intellectual content, especially in terms of information security, require improving

*[3] By concentrated is meant a system whose communication capacities do not affect the magnitude of the power deficit.

the sections relating to the prospective balance of power, not only in outdated Methodological Recommendations for the design of power systems (2003), but also in new 2012 not approved by the Ministry of Energy of Russia. The creation on the basis of the JSC "Institute" Energosetproekt "under the auspices of the Ministry of Energy of Russia of the Unified Center of Competences provided for the actualization of work in this direction. At the same time, it is necessary not to lose the performance of JSC SO UES in the period 2010-2016, which is responsible in accordance with the Law on Electric Power Industry for ensuring the reliability of the UES of Russia. However, it should be stated that after a year the expected results have not been received.

The conditions of intellectualization energy systems open up new possibilities in obtaining a more reliable information component of the task of balance reliability. Analysis of probabilistic characteristics of the operation of generating equipment and random deviations of the power consumption mode, obtained as a result of processing retrospective information, shows that their numerical values differ significantly from those used in the 2003 and 2012 power system design guidelines. In addition, these random characteristics are filled with new properties, by virtue of which they come into conflict with the existing ideas of taking them into account in the model-software complexes to substantiate balance reliability.

Important when justifying the means of redundancy in the management of the development of power systems have always been considered the issues of normative values of indicators of balance reliability used in justifying the values of compensation, and therefore the normative reserve of power. In this regard, the problems of comparing the standards of balance reliability applied in Russia and abroad are revealed. It is shown that they are mainly associated with different representations of the power consumption mode and their random fluctuations. For modern socio-economic conditions for the development of the country's electric power industry, under certain assumptions about the representation of the power consumption mode, the acceptability of the criterion used in the domestic practice of rationing indicators of balance reliability in the form of integral probability of deficit-free work ρ= 0.996 is shown.

References

[1] Yu.N. Rudenko, I.A. Ushakov "Reliability of energy systems" M.: Science, 1986. – 252 p.

[2] I.M. Vol'kenau, A.N. Zeiliger, L.D. Hobachev, "The economy of the formation of electric power systems." M.: Energoatomizdat, 1981. 320 p.

[3] Yu.Ya. Chukreyev, M.Yu. Chukreyev, "Status and problems of the regulatory definition of reserve power in modern conditions of planning and development of UES of Russia", Energy policy, 2017. Release 2. P. 39–50.

[4] Federal Law "On electric power industry" of 26.03.2003 №35-03.

[5] Guidelines for the design of energy systems. Approved by order of RF Ministry of energy June 30, 2003, № 281. M.: Ministry of energy Russian Federation, 2003.

[6] Yu.Ya. Chukreyev, M.Yu. Chukreyev, "Model indicators assessment of adequacy in the management of the development of electric power systems.", Syktyvkar: Komi SC UrO RAN, 2014. 207 p.

[7] A. Volkov, "Optimization of reliability electric power systems." M.: Scince, 1986. 116 p.

[8] R. Billinton, R.N. Allan, "Reliability Evaluation of Power Systems." Second Edition. New York and London, Plenum Press, 1996. 509 p.

[9] J. Enderny, "Modeling in the calculation of reliability in electric power systems", Per. from English, Under the editorship of Yu.N. Rudenko. M.: Energoatomizdat, 1983. 336 p.

[10] I.M. Markovich, "Modes of power systems.", M.: Energy, 1969. 351 p.

[11] Yu.Ya. Chukreyev, "Models of reliability of electric power systems.", Syktyvkar: Komi SC UrO RAN, 1995. 176 p.

[12] P.A. Malkin, "Shlifshtein V.A. On ensuring reliability in the power industry", Power plants, No. 6. 2010, pp. 2–7.

[13] Yu.Ya. Chukreyev, M.Yu. Chukreyev, V.S. Chuprov, "Characteristics of electricity consumption patterns in relation to the objectives of adequacy in managing the development of electric power systems", Methodical questions of research of reliability of large energy systems. Irkutsk: ISEM SB RAS, 2017, Issue 68 pp. 328–337.

[14] Yu.Ya. Chukreyev, M.Yu. Chukreyev, "The content of smart grid and its impact on ensuring an adequacy of", Methodical questions of research of reliability of large energy systems. Irkutsk: ISEM SB RAS, 2017. Issue 68 pp. 66–75.

Methodical and Information Aspects of The Task of Balance Reliability and Their Impact on Standards Power Reserve UES of Russia

Iurii Chukreyev
Institute of Socio-Economic and Energy Problems of the North, Komi Scientific Center, Ural Branch, Russian Academy of Sciences
Syktyvkar, Russia
chukreev@iespn.komisc.ru

Mikhail Chukreyev
Institute of Socio-Economic and Energy Problems of the North, Komi Scientific Center, Ural Branch, Russian Academy of Sciences
Syktyvkar, Russia
mchukreyev@gmail.com

Abstract — **The issues of the influence of the changed conditions for obtaining the likely-specific information on the operation of generating equipment and the modes of electrical consumption on the decisions taken to ensure the required level of redundancy of the territorial zones of the UES of Russia under the existing regulatory framework to indicators of balance reliability are considered. A comparison is made of the current conditions for presenting source information for assessing the indicators of the balance reliability of electric power systems with the conditions for developing guidelines for designing power systems in 2003 and 2012 that were not approved by the Russian Ministry of Energy.**

Keywords: reliability criteria; power system; adequacy parameter; power reserve; correlation of random load deviations

I. The Place of Advanced Planning in Modern Conditions

When planning the development of electric power systems (EPS), at the turn of the 80s of the last century, a fairly strict hierarchical system was used, within which a strategy was developed for the development of generating sources (including types, sizes and locations) and transmission lines for the future from 5 to 20 years. Within its framework, effective methods were developed, implemented in software development, allowing to solve the whole complex of tasks related to the management of the development of the UES of the former USSR power systems. The sharp decline in electricity consumption, changes in the form of ownership, deterioration in the quality of management personnel, the closure of industry institutions and much more led to a decline in interest in the long-term planning of the UES of Russia in the 21st century.

In accordance with the law "On Electric Power Industry" [1], JSC SO UES is responsible for development planning and ensuring reliability in the UES of Russia. Beginning in 2010, with the participation of specialists from various research institutes in the field of energy and institutes of the Russian Academy of Sciences, the following regulatory documents were developed: the draft Rules for the Technological Functioning of Electric Power Systems*1 and new guidelines for designing the development of power systems*2. They envisaged the development of major works on long-term planning in the power industry:

- the general layout of the placement of energy facilities for 15 years or more (developed by the Ministry of Energy of Russia);
- the program of development of the UES of Russia for a seven-year period*3 (developed by SO UES and FGC UES).

In addition to the above documents, the investment programs of the country's energy companies should also be included in the planning system.

Unfortunately, the documents listed above have weak links among themselves without an organizing beginning. In accordance with the request of the Deputy Prime Minister of the Russian Federation A. Dvorkovich, the Russian Ministry of Energy in 2017 created a single center of competence (SCC) for advanced planning in the electric power industry on the basis of the state JSC Energosetproekt Institute. At the same time, the powers regarding the issues of forward planning are fully concentrated in the Ministry of Energy of Russia. Thus, the planning structure (Fig. 1) is largely approaching that used in the centralized management of the industry. Its distinctive features are the creation of the SCC, the temporary synchronization of long-term planning documents, the strengthening of ties with the investment programs of subjects of the electric power industry. It should be noted that in the above-mentioned centralized planning system, the issues of long-term planning of the power industry of the country were also solved in the branches of the institute ENERGOSETPROEKT.

---

*1 Approved at the joint meeting of the Scientific Council of the Russian Academy of Sciences and the Scientific and Technical Board of non-commercial partnership «Scientific and Technical Council UES» May 16, 2011

*2 Guidelines for the design of energy systems / JSC "Energosetproekt", 2012 (approved by non-commercial partnership " Scientific and Technical Council UES", section "Technical regulation in the power industry" in July 2012, but so far not approved by the Ministry of Energy of the Russian Federation

*3 The rules for the development and approval of schemes and programs for the future development of the electric power industry, approved by the Government of the Russian Federation of October 17, 2009 No. 823.

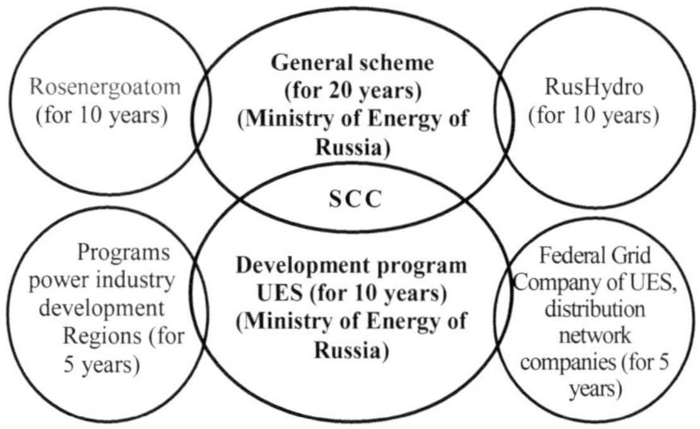

Figure 1 – Proposed forward planning structure.

II. The Task of Ensuring Balance Reliability in the System of Long-Term Planning of the Development of the Country's Power Industry.

The reliability of the power balance cover is ensured by maintaining the normative value of the reserve capacity, as applied to the selected territorial zones, which are usually used in this task by the combined EPS (UES). The question that concerns both consumers and suppliers of electrical energy is the correct substantiation of the components of the total power reserve, which cannot be achieved without solving the problem of balance reliability. The balance reliability (adequacy) in the task of long-term planning of the country's electric power industry is understood as the ability of EPS to meet the aggregate demand for electrical energy and the power of consumers within specified limits on energy supply, taking into account planned and reasonably expected unplanned interruptions in the operation of its elements, as well as operating restrictions.

In accordance with the definition, the tasks of balance reliability in managing the development of the UES of Russia include only those whose solution is related to the need to take into account system failures due to accidental equipment damage and to account for random deviations of loads from planned values. It is important to understand that random states caused by unplanned interruptions can last several tens of days, and the depth of a possible power shortage can reach tens of GW, albeit with a low probability of their occurrence.

When planning the development of the UES of the country as a whole and its united EPS, a power balance is developed [2]. Its form includes two positions:

- expenditure part "demand" - irregular maximum load, export and standardized power reserve;
- the incoming part "covering" - the installed capacity minus unused power, power inputs after passing through a maximum, under-utilizing power in a period or after passing an irregular maximum load.
- The expenditure part of the power balance includes the value of the normalized power reserve, called the full reserve. It is conventionally divided into three components: strategic, repair and compensatory power reserves. The greatest difficulty is in determining the last component of the reserve. This reserve is intended for compensation of unplanned conclusions in the repair of equipment and random deviations of loads from the planned values.
- The modern conditions of the country's socio-economic development, the giving of energy-based systems to intellectual content, especially in terms of information security, further aggravate this issue and require some improvement in the methodological principles of justifying the reserves of generating capacity. There are still problematic some issues that need to be improved both in 2003 and in the new 2012, not approved by the Russian Ministry of Energy.

III. Software Products Assess the Balance of Reliability and Their Modernization to Modern Conditions.

As mentioned earlier, the solution to the problem of assessing the balance reliability is necessary to substantiate the values of the power reserves and especially its compensation (operational) component. Their normalized values are used in the guidelines [3]. Justification of power reserves is impossible without developing a software for assessing the indicators of balance reliability (IBR). At the present stage, the most acceptable indicators of balance reliability from the standpoint of substantiating the means of ensuring reliability (power compensation reserve) are considered to be [5–8]:

- in the domestic practice of the probability of deficit-free operation of territorial zones $\rho = 1 - J_Д$ ($J_Д$ – integral probability of power shortage);
- in foreign - the average probability values of days of power shortage (Loss of Load Expectation, LOLE) in days / year [7] and hours of power shortage per year (Loss of Load Hours, LOLH).

It should be noted that the program complexes developed in the post-Soviet period, from the standpoint of their methodological content, are also quite acceptable for modern conditions. Currently, the Yantar (ISEM SB RAS) [4] and Orion-M (ICE and EPN of Komi SC UB RAS) [5] can be attributed to them. At the same time, the latter is more adapted to the changes that have taken place in the power industry of Russia in the recent period and, which is probably more important, it is more oriented towards solving the problem of justifying power reserves. This software was used in the development of 2003 guidelines [3] and not approved by the Ministry of Energy of Russia in 2012 guidelines.

The Orion-M software package is an upgrade of the Orion software package [6] widely used in design practice in the 1980s to analyze the balance reliability of promising options for the development of an EPS for 10–20 years (branch of the Energosetproject institute). In addition, the Orion, in the early 1990s, was used to monitor the reliability of the UES. B. USSR for the reporting and for the upcoming three years (OJSC "NIIPT", today non-commercial partnership "Scientific and Technical Council UES"). During the reform of the industry and the adoption of the above-mentioned Resolution of the Government of the Russian Federation of October 17, 2009, No. 823 from 2009 to 2013. In the "Orion-M" major changes were made [5], which to a certain extent influenced the decisions taken on the justification of the means of reservation in the UES of Russia.

IV. Analysis of the Justification of the Power Reserve on the Basis of Editions of 2003 Guidelines and of the not Approved of 2012 Guidelines

The authors of the article were involved in carrying out calculations to substantiate one of the component of the total power reserve - compensation in the designated guidelines. This was the basis for the study of a comparative nature in order to identify the features used in the calculations of the methodological principles for assessing the indicators of balance reliability [8]. The results of the distribution of the compensation reserve capacity, obtained in 2011-12. according to the Orion-M software package, they differ little from those calculated in accordance with the guidelines 2003. The other should not have been expected, since the probabilistic components of the accident rate of generating equipment and random deviations of the load from their average values did not change significantly. Confirmation of the above characterizes table 1. It shows the total power reserve values obtained 2011-2012 in accordance with guidelines 2003 and on the "Orion-M". For understandable temporary reasons, the results are given for the power balance of the past 2012 (SDP 2011-2017).

In rows 3 and 4 of table 1 shows the total power reserve, provided that the standards for scheduled repairs of generating equipment and strategic power reserve did not change compared to 2003 guidelines. Two scenarios for taking into account link capacity were considered (Table 1 - power line capacity) [5]. One without its optimization (line 4), the second with optimization for the conditions equality of unit costs in the development of one kW of generating capacity and reserve capacity of connections (line 3).

Table 1. The magnitude of the total reserve capacity for 2012

Power grid	Irregular maximum load, MW	Full power reserve in accordance with 2003 guidelines, MW /%	The calculated values of the total power reserve (MW /%) during the preparation of the 2012 guidelines		
			When optimizing bandwidth connections (ICEaEPN) [5]	*With existing bandwidth connections (ICEaEPN) [5]*	*2012guidelines (not approved Russian Ministry of Energy)*
UES of Russia without the East	147785	23738 13,06	24529 16,6	22974 15,55	30140 20,40
UES of Russia (Euro part)	116945	19881 17,0	20241 17,31	18980 16,23	23340 19,96
Ural	35005	6362 18,18	6944 19,84	6263 17,89	7010 20,0
Middle Volga	17316	2187 12,63	2213 12,78	1507 8,70	2860 16,5
South	14019	1988 14,18	2222 15,85	2049 14,62	2750 19,5
Northwest	14151	2982 21,07	2815 19,89	2933 20,73	2700 19,0
Centre	36454	6362 17,45	6046 16,59	6229 17,09	8020 22,0
Siberia	30840	3857 12,51	4291 13,91	3994 12,94	6800 22,0

In the edition of the guidelines 2012, the values of repair and strategic reserves of capacity were adjusted by the customers of the work of JSC SO UES and JSC Institute of ENERGOSETPROEKT (line 5 of table 1) [8]. According to the repair reserve, the main reasons for its change are the transition to the withdrawal of equipment for scheduled maintenance, as well as the opportunities that have opened up (a sharp decrease in electricity consumption) for conducting medium and capital repairs in the winter time period, including in December. For information in the former USSR, the reserves for capital repairs in almost all UPS were zero. The use of statistically processed retrospective information in the prevailing conditions led to a significant increase in the planned reserve, both for capital and for current repairs during their planning. Standards for strategic reserve in all integrated power systems have been increased from 1-2% to 3% [8].

V. The Impact of Information Support of Modern Intelligent Power Systems to Justify the Power Reserve.

This section presents the results of substantiation of the compensation reserve of capacity as applied to the forecast 2022 in accordance with the DSP of the UES of Russia for 2016-2022. Since the development of the non-approved Ministry of Energy of Russia 2012 guidelines, quite significant changes have occurred in the information on the task of assessing the IBR UES of Russia. This requires a certain analysis of their influence on the decisions made to substantiate the value of one of the components of the normalized power reserve - compensation (operational). The content will be understood as statistical information about the operation of generating equipment (unscheduled conclusions for repairs) and probabilistically certain information about load deviations from its hourly changes during the average daily schedule for December of the month [10, 11]. Their integrated characteristics are given in table 2. It compares the parameters as a percentage of the irregular maximum load (line 1, similar information for the 2012 guidelines is shown in line 1 of Table 1). The data on random parameters of the system, obtained as a result of statistical processing of retrospective information, and the corresponding information used in the preparation of 2012 guidelines from the reference literature [13], used in the 2003 guidelines, to be compared (lines 2-4 in Table 2) [3].

Table 2. Average statistical indicators of the generation of equipment in emergency and scheduled repairs as a percentage of the irregular maximum load

Options	UES name						
	UES of Russia without the East	*Ural*	*Middle Volga*	*South*	*Northwest*	*Centre*	*Siberia*
1. Irregular load max in accordance with the SDP UES of Russia for 2016-22.							
2022, MW	155860	37390	17096	16831	15151	39266	30126
2. Expected value of output in unscheduled repairs of power equipment							
2022	4,42	5,88	2,84	3,96	4,62	3,03	5,48
2012 guidelines	5,99	6,67	5,36	5,26	7,22	6,91	4,38
3. Random characteristics of the standard deviation of the load as a percentage of the irregular maximum							
2022	1,77	4,09	3,27	4,68	4,82	3,89	4,07
2012 guidelines	1,27	2,59	3,82	3,85	4,06	2,52	2,88
4. Power reserves for scheduled repairs of generating equipment							
2022	8,87	10,78	9,72	4,80	7,40	6,70	11,78
2012 guidelines	6,80	5,30	7,50	3,90	4,40	8,30	9,00
2003 guidelines	4,04	4,91	2,65	3,00	4,91	6,24	3,09

The results are shown in Table 2, clearly shows the significant changes in random operating parameters of generating equipment and deviations in the mode of electrical consumption that have occurred since the development of the guidelines in 2012 as a result of statistical processing of retrospective information [10, 11] (lines 2 and 3). A significant increase in the share of generating equipment that is in scheduled maintenance (line 4), as well as practically not used due to the presence of power surpluses, makes it necessary to carry out research on their accounting when modeling emergency equipment repairs in repair.

In table 3 presents the results of the influence of the noted changes in the information content of modern intelligent EPS on the value of the compensation power reserve. The first two positions table 3 shows the influence of the representation of power consumption modes on the means of ensuring reliability in the assumption of independence of random load deviations in the UES of Russia. At the same time, information on the accident rate of generating equipment for the purity of the experiment was taken from reference books, as in the 2012 guidelines. It would seem that in accordance with row 3 of the table 2, the effect of random load variances on reliability means should have been more significant. However, as can be seen from the table 3 in practice, this is not entirely confirmed.

Table 3. The magnitude of the compensation reserve capacity for 2022 (MW /%)

Options		UES of Russia without the East	UES name					
			Ural	*Middle Volga*	*South*	*Northwest*	*Centre*	*Siberia*
Information on random characteristics of output for repair of generating equipment in accordance with adopted in 2012 guidelines								
1.	s.d.- as in 2012 and 2003guidelines	14053 9,02	3584 9,59	628 3,67	1437 8,54	1879 12,4	4465 11,37	2060 6,84
2.	s.d.- based on retrospective information	15228 9,77	4034 10,79	628 3,67	1562 9,28	2004 13,23	4540 11,56	2460 8,17
Information about the random characteristics of the generating equipment and s.d. loads according to the retrospective data processing								
3.	s.d.- independent of UES	11920 7,65	3834 10,25	78 0,46	1194 7,09	1504 9,93	2855 7,27	2455 8,15
4.	s.d. - correlated in accordance with (1) and (2)	15610 10,02	4894 13,09	528 3,09	1544 9,17	1629 10,75	3990 10,16	3025 10,04

A more significant discrepancy in the results of substantiation of the compensation reserve of power is observed in the 3rd and 4th positions of the table 3. In them, for the same information on random states of generating equipment, two approaches to modeling random load deviations are considered. The first, as it was during the formation of guidelines 2003 and 2012 (line 3). The second (line 4) is when taking into account the correlation dependence between the UES of their random load deviations (see Section 2). It can be seen that accounting for the correlation leads to a significant increase in the share of the compensation power reserve for the UES of Russia as a whole from 7.65% to 10.02%.

Note that the compensation reserve of power for the UES of Russia as a whole (line 3) when using data on the state of generating equipment based on retrospective information processing decreased by 2.22% compared to its value when using the same information from reference literature (line 2) [13]. This is consistent with the first row of the table. 2, where the expected value (e.v.) of output for unscheduled repair of generating equipment for the UES of Russia is generally different for these calculations by more than one and a half points (4.42 and 5.99%). A more significant decrease in the power compensation reserve (2.22%) is easily explained by a decrease not only in m. on unscheduled equipment re-monitors, but also on dispersion (d) of the distribution density $f(P_r)$. For the UES of Russia and all of its constituent UES, these changes are shown in Figure 2. This picture shows the e.v. unscheduled conclusions of the generating equipment for repairs and the value of the necessary compensation reserve when using information on the accident rate of equipment obtained from the processing of historical data and in accordance with the reference literature. In all the UES, except for the Middle Volga, there is a strict dependence – e.v. always less compensation reserve. In the UES of the Middle Volga, the picture is somewhat different and this can be explained by the presence in this energy system of significant reserves of carrying capacity of connections with sufficiently powerful UES of the Center and the Urals (Figure 3).

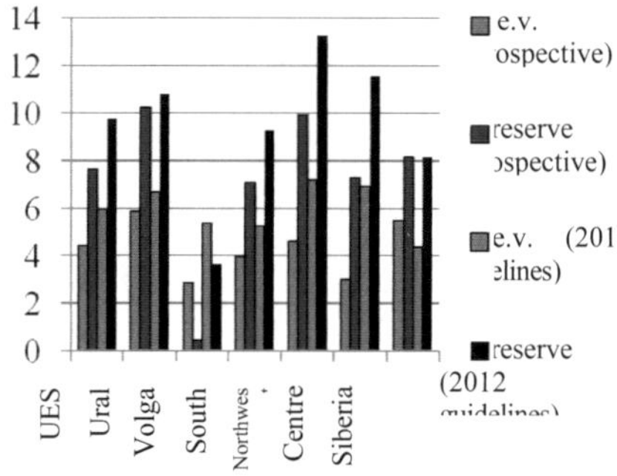

Figure 2. Interconnection of changes in random parameters of the power systems of the association and calculated parameters of the compensation reserve power.

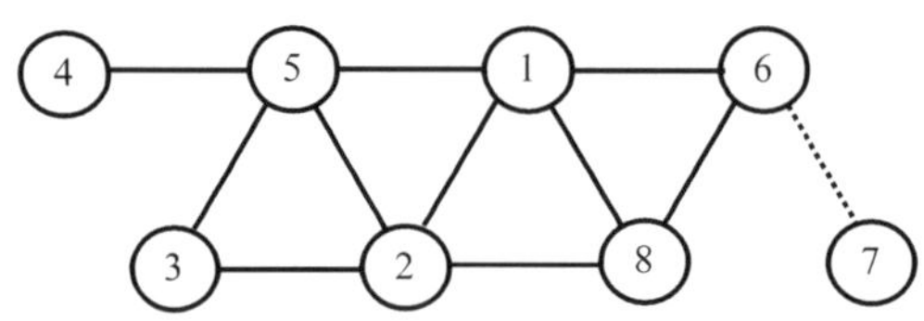

UES: 1 – Ural; 2 – Middle Volga; 3 – South; 4 – Northwest; 5 – Centre; 7 – Siberia; 7 – Far East; 8 – Kazakhstan.
Figure 3. Design model UES of Russia.

According to the results given in the last row of the table 3, one more important fact should be noted. Taking into account the correlation of random load deviations between the territorial zones in the form of the UES of Russia (Figure 3) sharply (5-7 times) reduces the likelihood of overloading of the transmission capacity of the links between the UES compared with the assumption of independence of these deviations (line 3). This again indicates that it is unnecessary to carry out calculations to substantiate the reserves of communications capacity, which should primarily be determined by ensuring the static and dynamic stability of power systems and transferring large volumes of power due to the difference in the load maxima in UES of Russia.

VI. Conclusions

Modern conditions of the country's socio-economic development, giving the power systems intellectual content, especially in terms of information security, require the improvement of not only outdated 2003 guidelines, but also the new ones in 2012, not approved by the Russian Ministry of Energy. The possibilities of obtaining, as a result of processing retrospective information, probabilistic characteristics about the operation of generating equipment and random load deviations come into conflict with similar information used in the preparation of the mentioned guidelines.

Taking into account changes in the standards of emergency and planned repairs of generating equipment and probabilistic load deviations of prospective power consumption modes in models for estimating indicators of balance reliability requires a rather serious software upgrade in terms of generating random load values and generating power by statistical modeling.

Comparative calculations showed a significant effect of changes in these parameters on the values of the normalized values of power reserves in the territorial zones in the form of a united UES of Russia. The creation of the Ministry of Energy of Russia on the instructions of A. Dvorkovich at the beginning of 2017 of a single center of competence at the JSC "ENERGOSETPROEKT" Institute provided for the actualization of work in this direction. However, it should be stated that after a year, the expected results were not received.

References

[1] Federal'nyj zakon Rossijskoj Federacii ot 26.03.2003 g. № 35-03«Ob jelektrojenergetike».

[2] I.M. Vol'kenau, A.N. Zejliger, L.D. Habachev, "EHkonomika formirovaniya ehlektroehnergeticheskih sistem." M.: EHnergoatomizdat, 1981. – 320 p.

[3] Metodicheskie rekomendacii po proektirovaniyu ehnergosistem. Utverzhdeny prikazom Minehnergo Rossii ot 30 iyunya 2003 g., № 281. – M.: Minehnergo Rossii, 2003.

[4] G.F. Kovalev, L.M. Lebedeva, "Nadezhnost' sistem ehlektroehnergetiki". Novosibirsk: Nauka, 2015. – 224 p.

[5] Yu.Ya. Chukreev, M.Yu. Chukreev, "Modeli ocenki pokazatelej balansovoj nadezhnosti pri upravlenii razvitiem ehlektroehnergeticheskih sistem.", Syktyvkar: Komi NC UrO RAN, 2014. – 207 p.

[6] Yu.Ya. Chukreev, "Modeli obespecheniya nadezhnosti ehlektroehnergeticheskih sistem.", Syktyvkar: Komi nauchnyj centr UrO RAN, 1995. – 176 p.

[7] R. Billinton, R.N. Allan, "Reliability Evaluation of Power Systems. Second Edition." New York and London, Plenum Press, 1996. – 509 p.

[8] Chukreev YU.YA., CHukreev M.YU. Sostoyanie i problemy opredeleniya normativnogo rezerva moshchnosti v sovremennyh usloviyah planirovaniya razvitiya EEHS Rossii // EHnergeticheskaya politika, 2017. Vypusk 2. – P. 39–50.

[9] Yu.Ya. Chukreev, "Sravnenie otechestvennyh i zarubezhnyh veroyatnostnyh pokazatelej balansovoj nadezhnosti ehlektroehnergeticheskih sistem", Izvestiya RAN. EHnergetika, 2012. № 6., P. 27–38.

[10] Yu.Ya. Chukreev, M.Yu. Chukreev, "Informacionnoe napolnenie intellektual'nyh ehnergosistem i ego vliyanie na obespechenie balansovoj nadezhnosti", Metodicheskie voprosy issledovaniya nadezhnosti bol'shih sistem ehnergetiki. – Irkutsk: ISEHM SO RAN, 2017. Vyp. 68. – P. 66–75.

[11] Yu.Ya. Chukreev, M.Yu. Chukreev, V.S. Chuprov, "Harakteristiki rezhima ehlektropotrebleniya primenitel'no k zadacham balansovoj nadezhnosti pri upravlenii razvitiem ehlektroehnergeticheskih sistem", Metodicheskie voprosy issledovaniya nadezhnosti bol'shih sistem ehnergetiki, Irkutsk: ISEHM SO RAN, 2017. Issue. 68. pp. 328–337.

[12] V.M. Verzhbickij, "Osnovy chislennyh metodov." M.: Vysshaya shkola, 2009. – 840 p.

[13] V.V. Ershevich, A.N. Zejliger, G.A. Illarionov, "Spravochnik po proektirovaniyu ehlektroehnergeticheskih sistem", Pod red. S.S. Rokotyana i I.M. SHapiro., M.: EHnergoatomizdat, 1985, pp. 352.

Evaluation of Various Programming Languages in Relation to Solving the Problem of Determining Indicators of Balance Reliability

Iurii Chukreyev
Syktyvkar Forest Institute
Syktyvkar, Russia
chukreev@iespn.komisc.ru

Mikhail Chukreyev
Syktyvkar Forest Institute
Syktyvkar, Russia
mchukreyev@gmail.com

Abstract — **Considers the use of various software tools to assess the possibility of obtaining indicators of balance reliability. A fairly new high-level, high-performance free programming language with dynamic typing Julia was used as a mathematical programming language. Traditional languages are represented by Fortran. The comparison of the speed of obtaining the result. The pros and cons of using each of the options are considered.**

Keywords: reliability criteria; power system; adequacy parameter; power reserve; software.

I. Introduction

The main part of the compilation of promising balances of power and electricity for the territorial zones (United EPS - UPS) Unified Electric Power System (UES) of Russia is the task of assessing their reliability indicators taking into account the possibilities of mutual assistance in backbone connections. Even in the presence of many subjects of market relations (generating, grid, power sales companies, etc.), the UES of Russia must operate as an integral and inseparable technological complex and meet the requirements of reliability and security at all levels of the time and territorial hierarchy of individual control zones and the entire power grid as a whole. In accordance with the law No. 35-FZ of March 26, 2003 "On Electric Power Industry", JSC "System Operator of the UES" is responsible for ensuring the reliability of the operation of the UES of Russia and the formation of promising schemes for the allocation of capacity reserves[1]. This became the basis for the creation of Technological rules of work. Based on this document, the planning of the development of an EPS should be carried out for a prospective period (up to 15 years) once every three years and for a period from 2 to 7 years annually. It also emphasizes the need to use mathematical models to evaluate the indicators of balance reliability (IBR) of EPS when managing their development. The balance sheet reliability of an EPS is understood to be the ability to meet the aggregate demand for electrical energy and the power of consumers within specified values and restrictions on energy supply, taking into account the planned and reasonably expected unplanned interruptions in the operation of its elements and operational limitations.

[1] Approved at the joint meeting of the Scientific Council of the Russian Academy of Sciences on the problems of reliability and safety of large energy systems and the Scientific and Technical Board of Non-commercial partnership "Scientific and Technical Council of the Unified Energy System" May 16, 2011

An important and necessary condition for the wide application of any mathematical models, including those aimed at solving the problem of ensuring the indicators of balance reliability when planning the development of the UES of Russia, is their respective content. Many factors (deterministic, random) [1, 2] affect the indicators of the balance reliability of the EPS and, consequently, the decisions taken to substantiate the means of their provision - power reserves of territorial zones and the maximum allowable power flows (throughput) of relations between them.

The modern development of computational tools promotes the introduction of automated systems for the preparation of information support and its connection with software systems for evaluating indicators of the balance reliability of the UES of Russia. This is one of the reasons for changing the existing software, aimed at assessing the indicators of balance reliability and justifying power reserves for completely different software, including the high-level, high-performance free programming language with dynamic typing discussed below.

An important role in the formation of information support for solving the problem of evaluating IBR is assigned to the development and specification of the UES of Russia design scheme. To include this task in the business process of SO UES OJSC, it is necessary, on the basis of the available retrospective information about the operation of generating and network equipment, the power consumption modes of various territorial zones of EPS unification, to prepare a storage system over a long period of time, which will allow further mathematical processing of this information to create a package the input data of the software complex.

The need to process a huge amount of information contained in the local database of Joint-Stock Company "System Operator of the Unified Energy System" and setting the topology of the calculation scheme implies the possibility of using many different tools capable of implementing the database and the calculation module within one software environment.

II. Models of Flow Distribution Analysis in the Problem of Estimating a Random System State.

Indicators of balance reliability in separate territorial zones of EPS depend not only on the physical condition of the generating and network equipment, but also on the adopted rules for the distribution of power shortages in long-term conditions caused by emergency exits of the main equipment. These rules differ in the formulation of the flow distribution problem for the conditions of the operational management and management of the development of an EPS. In the tasks of assessing the indicators of balance reliability with the current management, these rules are more developed and are determined by the relevant regulations. In managing the development of the industry, such regulations are not yet available.

There are many methods for calculating the flow distribution in EPS. They differ from each other in the speed and accuracy of obtaining results, as well as in the necessary computer RAM volume. Since the process of obtaining reliability indicators in the integration of EPS requires performing flow distribution calculations for each simulated random system state, the overall efficiency of the calculations largely depends on the algorithm used to analyze the flow distribution.

The methods for calculating the flow distribution based on the use of exact nonlinear equations describing the steady state EPS are quite complex and require a large amount of computer time to get the result. The use of such methods in calculations can somehow be justified for the analysis of the mode reliability of the operation of an EPS. Although here, there are problems in balancing the power system when a sufficiently unbalanced power condition arises on a random state, according to what criteria to distribute the power shortage arising in the system and many other issues.

Determining the indicators of the balance reliability of multi-zone EPS, and the UES of Russia in particular, involves conducting a huge series of flow distribution calculations for various combinations of generating power, load and link capacity due to the random nature of changes in their operational parameters. Therefore, the overall efficiency of calculations largely depends on the computational efficiency of the applied algorithm for estimating a random state. When solving the problem of assessing the indicators of balance reliability in managing the development of an EPS, there is a significant uncertainty in the initial information. This makes it possible to use either purely balance models or models in DC idealization for calculating the electric mode. Then the most acceptable methods for assessing the random state of the system are linear programming methods.

The task of flow distribution in models for evaluating indicators of balance reliability was called the task of power shortage distribution (PSD) [2, 3]. In the linear formulation, its solution is to minimize the power deficit:

$$\sum_{j=1}^{n} c_j^{\text{H}}(\overline{P}_j^{\text{H}} - P_j^{\text{H}}) \rightarrow min, \qquad (1)$$

under restrictions:

$$P_j^{\text{H}} - P_j^{\Gamma} \pm \sum_{l(i,j)\in V} P_{l(i,j)}^{L} = 0 \qquad j = 1, 2, .. ., n;$$

$$0 \leq P_j^{\text{H}} \leq \overline{P}_j^{\text{H}} \qquad j = 1, 2, \dots, n;$$

$$0 \leq P_j^{\Gamma} \leq \overline{P}_j^{\Gamma} \qquad j = 1, 2, \dots, n;$$

$$P_{l(i,j)}^{L}, \overline{P}_{l(i,j)}^{L}, \underline{P}_{l(i,j)}^{L} \qquad (i, j) = 1, 2, \dots, m,$$

where $P_j^{\text{H}}, \overline{P}_j^{\text{H}}, P_j^{\Gamma}, \overline{P}_j^{\Gamma}$ is, respectively, the covered demand and the power of demand for the load, the used and available generating capacities of the jth UES or territorial zone; $P_{l(i,j)}^{L}, \overline{P}_{l(i,j)}^{L}, \underline{P}_{l(i,j)}^{L}$ - power flow through the communication and its limiting values in the forward (from the *i-th* UES or territorial zone to the j-th) and reverse directions; c_j^{H} - coefficients reflecting the cost indicators of restricting consumers of the j-th UES or territorial zone, are assumed to be equal to one; V is the set of pairs of UES or territorial unification zone that have a systemic link between them; n (m) is the number of UES or territorial zones (strategic links) of the association.

The solution of this problem is possible by different methods. Consider two of them. The first is to use the appropriate package of one of the mathematical programming languages. This method is widespread in Western scientific school, because allows you to quickly conduct research without deepening in the mathematical apparatus. The dual simplex linear programming method [4] was adopted as the second method for solving this problem. Its application, firstly, makes it possible to drastically reduce the cost of machine time by using the optimal solution for the previous one in assessing the current state of the system and, secondly, it allows us to directly identify the worst zones and connections between them in EPS. The latter is used to solve optimization problems of reliability. The negative side of the linear formulation of the problem is the ambiguity in the distribution of the system power shortage in individual UPS or territorial zones. It should be especially emphasized that this ambiguity is precisely in the distribution of power deficits, but not in the probabilities of their occurrence across territorial zones. The elimination of ambiguity requires the second stage of solving the problem of minimizing the system power deficit (1). It is the solution of this stage for the conditions of market relations in the electric power industry that undergoes significant changes and the solution of the problem of determining the operational capacity reserve and the requirements for the carrying capacity of communications between the UPS and territorial zones depends on its formulation.

III. Information Support for Solving the Problem

To solve this problem, a fairly new high-level, high-performance free programming language with dynamic typing, created for mathematical calculations by Julia [5], was used as a mathematical programming language. The syntax of the language is similar to the syntax of other mathematical languages (for example, MATLAB and Octave), but it has some significant differences. Julia is written in C, C ++ and Scheme. The standard package includes a JIT compiler (technology to increase the performance of software systems that use bytecode by compiling bytecode into machine code or another format directly while the program is running) based on LLVM (Low Level Virtual Machine - universal analysis system, software transformation and optimization that implements a virtual machine. Can be used as an optimizing bytecode compiler into machine code for various architectures or for its interpretation and JIT compilation), thanks to which language authors, the application is completely written in the language, are not inferior in performance applications written in a statically compiled language like C or C ++. Most of the standard language library is written in it. The language also has built-in support for a large number of commands for distributed computing.

The language is dynamic, but uses JIT compilation. Due to this, high speed of work of applications written in a "pure" language is achieved without using low-level libraries and vector operations. One of the priorities in the development of the language is the support of distributed computing. There is a large number of standard constructions for code parallelization.

The undoubted advantage of Julia over other languages of this type is the absence of a barrier between high-level and assembler code. Julia in an interactive environment allows you to write code that works as fast as C code, gives you the opportunity to look at the LLVM representation of functions and their generated assembler code. This shows the main advantage of Julia - we can start with the prototype of the program, and finish with the optimization by SIMD instructions (single instruction, multiple data - the principle of computer calculations, allowing to ensure data-level parallelism) without leaving a single language. At the moment, the only drawback of the language is the lack of libraries. But he gradually solved. An additional advantage of this language is an active community, always ready to help.

The opponent was the code implementing the solution of the problem by the dual simplex method of linear programming written in Fortran.

IV. Comparative Results.

For the calculations, the UES of Russia scheme with the dimension of 56 nodes and 102 communications was chosen (Figure 1). It was developed in 2011-2012 on the instructions of JSC SO UES of STC UES JSC for solving the problems of ensuring balance reliability [6]. For this scheme, the maximum allowable power flows between the selected territorial zones were obtained. The initial information was presented by the report the scheme and program for the development UES of Russia for 2012-2019.

Conformity of numbers of knots:
1001; 1002; 1003; 1004; 1005 –Ural UES;
2001; 2002; 2003; 2004; 2005; 2006; 2007; 2008 – Middle Volga UES;
3001; 3002; 3003; 3004; 3005; 3006; 3007 – South UES;
4001; 4002; 4003; 4007; 4008; 4009 – Northwest UES;
5010; 5016; 5017; 5018; 5019; 5020; 5023; 5024; 5026; 5027; 5029; 5031 – Centre UES;
10001; 10002; 10003; 10004; 10005; 10006; 10007; 10008; 10009; 10010; 10011; 10012;
10013; 10014 – Siberia UES;
7000 – UES of Belorussia; 8000 – UES of Baltic;
9000 – UES of Kazakhstan; 6000 – UES of Ukraine

Figure 1. Model of the design scheme of balance reliability of the UES of Russia without the OES of the East (56 territorial zones and 102 system-forming communications)

To determine the accuracy of the results obtained in the process of solving a linear programming problem, a single discrete state of the system was used. It should be noted that for a more visual type of information received, the initial data were adjusted in the direction of increasing the deficit in the system. As a result of the calculation of 10 different system states, the total deficit for the analyzed calculation software coincided with an accuracy of up to 99.5% (Table 1). To determine the computational performance of these variants, a calculation of 10,000 states was launched. To obtain identical values of generating power and load, the same initial state was used, and the subsequent ones were formed by simple arithmetic operations similar to the software used. The time spent on the calculation was taken as the average of 10 calculations performed. The results are shown in Table 2. In which it is clear that using the Julia functional allows you to reduce the calculation time by up to 8 times. At the same time, it should be understood that this is achieved not by the speed of the language itself, but by the ability even of a researcher (not a professional programmer) to achieve close to ideal cleanliness of a program code. An additional advantage over the languages of the "old school" allows the selection of an optimal solver for a specific task (from the standpoint of the mathematical apparatus used). A number of solvers have implemented the ability to connect paralleling by means of the language itself, without the need to make appropriate edits to the code. It also adds performance compared to the same Fortran or C.

Table 1 - The total deficit in the system when calculated on different software tools

state	Total deficit in the system, MW	
	Julia	Fortan
1	9161,76	9207,81
2	9874,04	9913,69
3	10502,73	10566,12
4	12568,23	12657,83
5	13238,79	13345,56
6	14354,72	14477,78
7	15756,08	15899,17
8	16688,18	16493,07
9	17015,97	17135,92
10	17844,42	17988,33

Table 2 - Calculation time of 10,000 states on different software means

No. of calculation	Calculation time, sec	
	Julia	Fortan
1	0,9	7,3
2	1,2	9,1
3	1,1	8,4
4	1,1	7,2
5	0,8	9,1
6	1,2	8,7
7	1,0	7,9
8	1,0	8,2
9	1,1	8,1
10	0,9	7,6
average	1,03	8,16

V. Conclusions.

For the calculations, the The possibilities of obtaining, as a result of processing retrospective information, probabilistic characteristics about the operation of generating equipment and random load deviations come into conflict with similar information used in the preparation of the mentioned guidelines.

The possibility of using modern mathematical programming languages to solve individual components of the problem of determining the indicators of balance reliability is shown.

The use of programming languages with dynamic typing created for mathematical calculations, including Julia, allows, without loss of accuracy and speed of obtaining the result, to significantly reduce the time and complexity of writing software code. It requires less knowledge of programming languages and mathematical apparatus, through the use of ready-made packages, including the development of databases.

The above software comparison shows the enormous potential for using programming languages similar to Julia for research purposes.

REFERENCES

[1] Razrabotka metodiki i matematicheskoj modeli ocenki balansovoj nadezhnosti perspektivnyh variantov razvitiya EEHS Rossii s uchetom sovremennoj modeli rynka ehlektroehnergii i moshchnosti. EHtap № 1. Metodika i matematicheskaya model' ocenki balansovoj nadezhnosti perspektivnyh variantov razvitiya EEHS Rossii s uchetom sovremennoj modeli rynka ehlektroehnergii i moshchnosti / Otchet o nauchno-issledovatel'skoj rabote. ISEH i EHPS Komi NC UrO RAN. Syktyvkar, 2011, 120 p.

[2] Ju.Ja. Chukreev, "Modeli obespechenija nadezhnosti jelektrojenergeticheskih sistem.", Syktyvkar: Komi NC UrO RAN, 1995, 176 p.

[3] Yu.Ya. Chukreev, M.Yu. Chukreev, "Modeli ocenki pokazatelej balansovoj nadezhnosti pri upravlenii razvitiem ehlektroehnergeticheskih sistem.", Syktyvkar: Komi NC UrO RAN, 2014, 207 p.

[4] D.B. Yudin, E.G. Gol'shtejn, "Linejnoe programmirovanie.", M.: Mir, 1966, 276 p.

[5] The Julia Programming Language, avalaible at: https://julialang.org/

[6] Otchet o NIR «Ezhegodnaya aktualizaciya matematicheskoj modeli i bazy dannyh Edinoj ehnergeticheskoj sistemy Rossii i ezhegodnye issledovaniya perspektivnyh ehlektricheskih rezhimov Edinoj ehnergeticheskoj sistemy Rossii na perspektivu do 2020 goda». EHtap 1. Aktualizaciya raschetnoj modeli (ehkvivalentnoj skhemy) EEHS Rossii dlya provedeniya raschetov balansovoj nadezhnosti EEHS Rossii. Sankt-Peterburg: OAO «NTC EEHS (NIIPT) (po zadaniyu AO «SO EEHS»). 2012, 88 p.

Frequency Stability of Contineta1 Europe Synchronnous Area

Karel Máslo
Transmission System Analysis Dpt.
ČEPS, a.s.
Prague, Czech Republic
maslo@ceps.cz

Andrew Kasembe
Department of System Development
ČEPS, a.s.
Prague, Czech Republic
Kasembe@ceps.cz

Abstract — **The paper analyses influences of deterministic frequency deviations and load frequency control measurement errors on the frequency stability of the Continental Europe synchronous area. The results of the analysis can provide valuable insights into the dynamic behaviour of the system that can be incorporated into defence plans. Analysis and practical experiences from the power system operation show that these aspects (apart from traditional failures such as power modules outages or network separation into islands) can affect the frequency stability.**

Keywords: Frequency stability, Continental Europe (CE) synchronous area, Dynamic Stability Assessment (DSA), Transmission System Operator (TSO)

I. Introduction

A power system dynamic stability assessment (DSA) is an important part of a secure power system operation. According to the EU Regulation [1], each TSO shall monitor the dynamic stability of the transmission system and perform the dynamic stability studies at least once a year to identify the stability limits and possible stability problems in its transmission system. The dynamic stability includes rotor-angle, frequency and voltage stabilities. The first DSA Stakeholder Workshop [2] mentioned that in the CE synchronous area:

1. frequency stability assessment is covered by the 3 GW frequency control reserve provision,
2. beyond normal state (with frequency steady-state deviations greater than ±200 mHz) frequency stability is relevant for the defence plan (addressed in another network code) and
3. DSA is required only for rotor – angle and voltage stability.

Papers [3] and [4] were concerned with rotor – angle and voltage stability. This paper is focused on the frequency stability.

From the first two statements above it can be implied that the approach to the frequency stability is somewhat disjointed. This may be due to the fact that stability has a global nature and its examination requires specific knowledge of system dynamics (unlike angle and voltage stability, where simple principles can be applied – the equal area rule and nose curve).

On the other hand, the used models for the frequency stability assessment do not need to be based on a complicated and complex grid model, but a single-node model with a uniform mean frequency (see e.g. [5] or [6]) is usually sufficient. Alternatively a more complex multi-node model (if we consider interarea oscillations and different local frequencies) can be used (see e.g. [7]).

Frequency stability is usually investigated for loss of generation or load, possibly in combination with deterministic frequency deviations (caused by changes in trade schedules between different countries - see [8] for explanation).

A significant frequency drop of nearly 49.8 Hz in the Continental Europe synchronous area was registered on 10 January 2019 at 21:02. In addition to the deterministic deviations already mentioned, the second cause was a measurement error of cross-border exchanges between Germany and Austria (according to ENTSO-E press release [9]).

This paper analyses the influence of such failures on frequency stability using a dynamic model. Such analysis provides answers to questions: "What would happen if ..?". The MODES network simulator was used as a simulation tool.

II. Describtion of the frequency drop incident

In addition to the two reasons mentioned above, other publicly available sources presented additional causes of the frequency drop. For example according to RTE (French TSO) report [10] explained other factors that contributed to this drop in frequency were:

- changes of cross-border exchange schedules between European countries (control areas) and
- a permanent frequency deviation linked to the situation in Kosovo and in Serbia

The second issue is caused by insufficient power to control own power balance in the Serbia, Macedonia and Montenegro control block (and specifically Kosovo and Serbia) –see the press release [11].

More specific information about cross-border exchange changes was not available at the time of writing the article, but reference [12] presented important information on load shedding in France and also the frequency time course during the incident as depicted in the following figure:

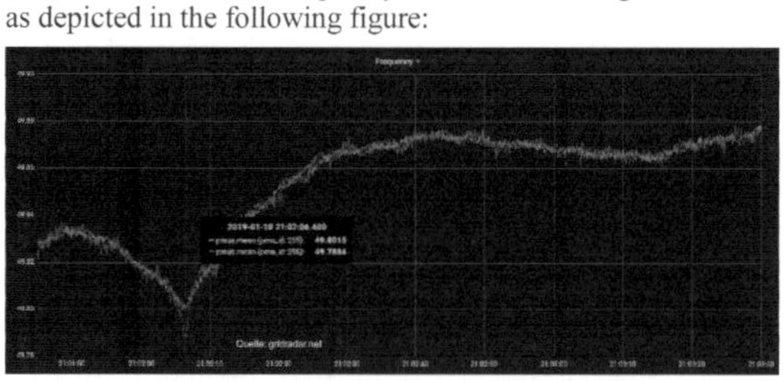

Figure 1. Frequency waveforms (source [12])

It can be seen from the figure that the frequency is not the same in all parts of the interconnection (CE synchronous area). The green frequency waveform is measured in Spain, while the orange one in Germany. The difference is caused by interarea oscillations mentioned in the introduction.

The fully automated French load shedding allows the disconnection of about 1500 MW of industrial loads (before normal system under frequency load shedding, which starts usually at 49 Hz). This information is also confirmed by the France's CRE energy regulator report [13].

Final publicly available information related to the drop in frequency to 49.8 Hz at 21:00 CET on 10 January [14] is a short but significant peak of power consumption caused by the switching-on of pump storage plants, including the 1.1 GW Goldisthal unit in eastern Germany.

III. DYNAMIC MODEL

The six control areas CE interconnection model described in [15] was used for the frequency stability simulation. This model enables to simulate all aspects mentioned in the preceding sections, namely:

1. deterministic frequency deviations (caused by changes in trade schedules between different countries or/and by changes of unit schedules for covering net load diagrams in particular control areas,
2. permanent frequency deviations caused by intentional non-compliance with the scheduled power balance,
3. measurement error of cross-border exchanges between control areas
4. under frequency load shedding

Moreover the model enables to simulate photovoltaics (PV) generation and therefore, it answers the question of what would happen if the event did not happen at night, but in the morning on a clear day with high PV generation. Figure 2 depicts control areas with high PV installed power (more than 2 GW).

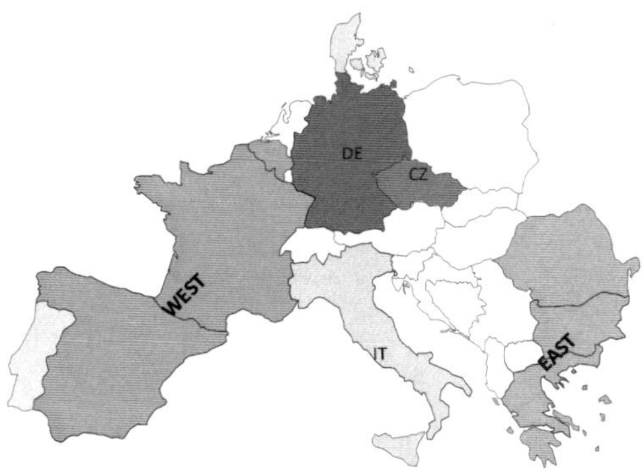

Figure 2. Geographical overview of modelled control areas

For simplification, Spain, France and Belgium were joined together into one control area named WEST and similarly Romania, Bulgaria and Greece were joined together into one control area named EAST. Italy, Germany and the Czech Republic were modelled as separate control areas.

The daily load diagram P_{LOAD}, estimated PV generation P_{PV} and net load diagram $P_{NetLOAD}= P_{LOAD} - P_{PV}$ were constructed for each control area in dependency on Coordinated Universal Time UTC. The Central European Time is equal UTC+1 and the East European Time (valid for the East control area) is equal UTC+2.

Following figures show time courses of the daily load diagrams, PV generation and net load diagrams of simulated areas. A spring day with clear sky during the morning is considered.

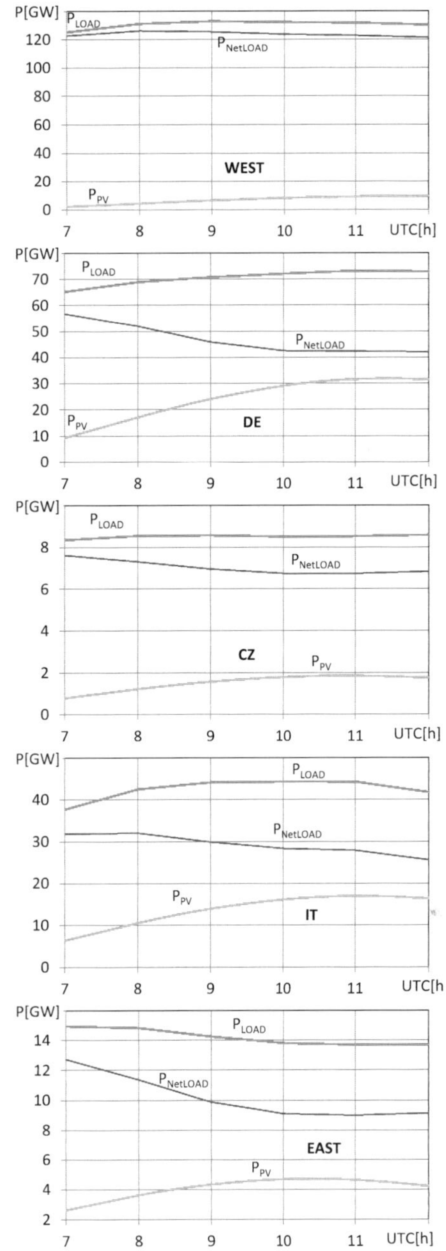

Figure 3. Daily load diagrams, PV generatin and net load diagrams

PV generation can be estimated as follows (T≡UTC):

$$P_{PV}=P_{PVp}*g \qquad g = \mathbf{A}e^{-0.5\left(\left|\frac{T-\mu}{\sigma}\right|\right)^3} \qquad (1)$$

An overview of used parameters is in TABLE 1.

TABLE I PARAMETERS FOR PV GENERATION MODELS

Area	P_{PVp}	A	μ	σ
WEST	10	0.893	11.799	3.276
EAST	5	0.914	10.39	3.249
IT	18	0.908	11.118	3.242
CZ	2	0.89	10.887	3.207
DE	35	0.882	11.405	3.206

The daily load diagram can be estimated as follows (T≡UTC):

$$P_{LOAD} = P_M[1 + \sum_{i=1}^{4}\left(A_i \sin\frac{iT\pi}{12} + \varphi_i\right)] \quad (2)$$

TABLE 2 PARAMETERS FOR DAILY LOAD DIAGRAMS APROXIMATION

Area	P_M	A_1	φ_1	A_2	φ_2	A_3	φ_3	A_4	φ_4
WEST	131	-.02	.91	-.08	2.7	.16	.15	.07	.58
EAST	14	.15	3.56	.15	4.2	.03	-.16	-.04	-2.2
IT	45	.31	3.34	.23	3.4	0	2.07	0	2.6
CZ	8.5	.07	2.61	.12	2.2	-.09	-1.0	.03	1.9
DE	70	.35	3.38	.30	3.5	-.08	0.92	.01	.65

The TABLE 3 shows nominal capacities P_n of conventional units, which were used to cover the net load diagram.

TABLE 3 NOMINAL CAPACITIES PN FOR CONTROL AREAS [GW]

Area	West	East	IT	CZ	DE
P_{nBASE}	47	1	0	3.5	15
P_{nTERC}	68	8	22	3.5	32
P_{nPRSEC}	24	5	19	3.5	25

Units with nominal capacities P_{nBASE} cover a base load, units with nominal capacities P_{nTERC} cover variable part of the net load diagram - they change operating points as a result of electricity trading depending on the trading interval (usually one hour). Units with nominal capacities P_{nPRSEC} are included in the load frequency control and they participate in the primary frequency control as well.

The overall primary control reserve (PCR) is 2932 MW, network power frequency characteristic of primary control is 14663 MW/Hz and the total kinetic energy stored in rotating mass (inertia) is about 1787 GJ.

The load frequency control (LFC) model is described in [16]. A standard turbine model is used for the prime mover modelling (detailed information about prime mover models were published at the previous symposium [17]).

The PV power modules dynamic model (published in [18]) contains tripping relay, which disconnect the module when the frequency exceeds f_{MAX} or falls below f_{MIN}. The capacities of PV with a risky frequency settings (between 49.5 and 50.5 Hz) are estimated (according to the dynamic study [19]) in TABLE 4.

TABLE 4 PV MODULES WITH RISKY FREQUENCY SETTINGS [GW]

Modules	P_{PV}	f_{MIN}	f_{MAX}
FPVR-5	540	47.5	50.5
FPVR+5	3200	49.5	50.5
BPVR+2	570	47.5	50.2
GPVR-5	2400	49.5	50.5
BPVR+3	1000	47.5	50.3
IPVR-3	2000	49.7	50.3
CPVR-5	800	49.5	50.5
CPVR-2	1000	49.8	50.2
DPVR-5	1000	49.5	50.5
DPVR+2	9000	47.5	50.2

Finally, the automatic under frequency load shedding is implemented in the model. It disconnects 5% of the load at 49.0 Hz in accordance to with standard S10 of ENTSO-E operational handbook Policy 5 [20]. A frequency self-regulation effect of load is neglected.

Now that the dynamic model is briefly described, we can proceed to the dynamic simulation.

IV. DYNAMIC SIMULATIONS

Three cases were simulated to demonstrate the influence of different aspects on frequency stability:

1. Base case - combination of deterministic frequency deviation caused by trade changes, intentional non-compliance with the scheduled power balance in Serbia (-100 MW) and the measurement error of the Germany cross-border exchange (-1000 MW),
2. Base case + a load shedding of 1500 MW in France,
3. Base case with disconnection of photovoltaics and
4. Base case with disconnection of photovoltaics + load shedding 1500 MW in France.

The deterministic frequency deviations were simulated by changing the scheduled power of P_{TERC} units (to cover load decrease in the market interval 8:00 - 9:00) and by control areas trade schedules between WEST area (-1200 MW import) and EAST and CZ areas (800 and 400 MW exports) at T=8:00. The significant peak of power consumption in eastern Germany was simulated by a unit outage in DE area (it caused the power deficiency 1000 MW).

The 1500 MW load shedding in France was simulated by automatics, which measure the frequency and, when this falls below the specified threshold, commands the switch-off of specified load volume. The threshold was derived from Figure 1 and set to 49.8 Hz (49.8 -50.2 Hz is a normal operating frequency range).

While in the previous two cases the photovoltaics switch-off by frequency protections was not considered, in the third and the fourth cases these protections on PV modules (according to TABLE 4) were activated. Figure 4 shows the results of two hours simulation (t=0 corresponds to time T=7:00).

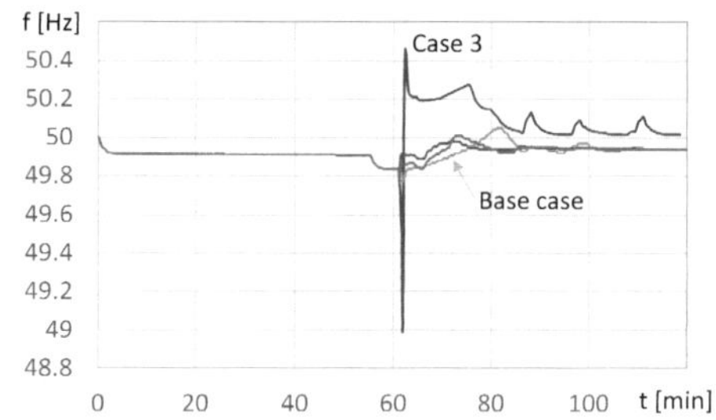

Figure 4. Frequency waveforms

The steady deviation about -90 mHz (caused by the power unbalance in Serbia and the measurement error in Germany) is seen during the first hour of the simulation. Changes of the scheduled power of P_{TERC} units starts in t=55 min. (these changes should start five minutes before the new trading interval and finish five minutes later according to operational rules). The unit outage (simulated at t=61 min.) caused the frequency decreasing under the 49.8 Hz limit. From that point onwards, the waveforms are different for each case, as shown in Figure 5.

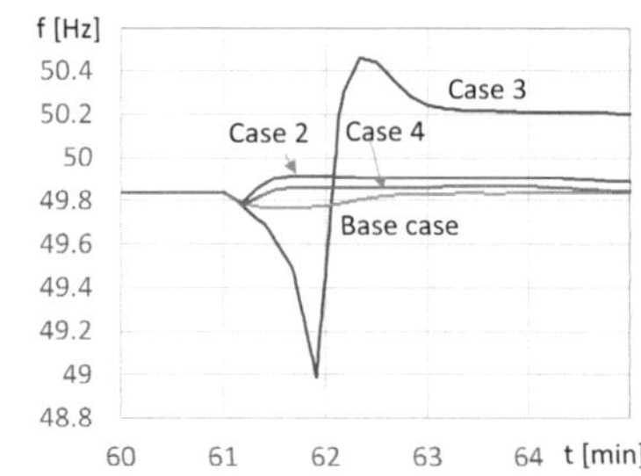

Figure 5. Frequency waveforms – time detail

In the base case, when frequency falls under 49.8 Hz for more than one minute the individual secondary controllers (LFC) are frozen (keeping the last signal of control) according to the standard S5 of ENTSO-E operational handbook Policy 5 [20]. This happened in t =62 min in the base case. Since then, the frequency is not regulated - the primary control reserve is already exhausted (it must be fully released at 49.8 Hz) and the secondary controllers are frozen. The power balance deficit is covered only by inertia. The frequency increases as the net load decreases from eight o'clock (see Figure 3). The individual secondary controllers were switched manually from the frozen mode to the normal operational mode successively from t = 75 min and they controlled power balance and frequency again.

A little bit different was the situation in the second case, where the load shedding of 1500 MW prevented a long-term frequency drop below 49.8 Hz and the secondary controllers stayed in the normal operation mode. This case corresponds to the actual incident as described in section II.

The frequency waveform in the third case (red line in Figure 4 and Figure 5) is fundamentally different from the previous cases. The frequency protections on PV modules were activated and no load shedding occurred before the frequency drop to 49 Hz. The frequency decreased and after reaching the band 49.8 - 49.5 Hz the PV modules with the total power of 4860 MW were switched off by under frequency protections.

When the frequency reached the limit value 49 Hz the under frequency load shedding relays disconnect of 16240 MW of load in all control areas. This caused an increase in frequency and between 50.2 and 50.5 Hz an additional 6561 MW of PV modules were switched off by over frequency protections, which had a positive effect and helped to stabilize the frequency.

Since the individual secondary controllers were frozen, the frequency grew similarly to the base case. To restore the frequency control, the secondary controllers were gradually switched to the so-called flat mode that controls only frequency and not exchange power (due to under frequency shedding the system is in emergency state without electricity trading). The flat mode restored the frequency to the nominal value of 50 Hz and the PV modules switched off by frequency protections were automatically reconnected to the network. These reconnections occurred at t=86 min, t=96 min and t=109 min. The reconnections and gradually increasing PV power resulted in small frequency peaks that were then regulated by the secondary controllers in the flat mode. The shedded load must be restored manually (it was not part of the simulation calculation).

The fourth case is similar to the second case with only one difference. Since frequency protections on the PV modules were activated, the PV module CPVR-2 was switched off due to a very narrow operational band. But the load shedding of 1500 MW prevented a further frequency drop to the 49.7 Hz limit, where other PV modules would be switched off. This is shown in Figure 6. The module was reconnected automatically when the frequency remained above the 49.8 Hz limit for 10 minutes.

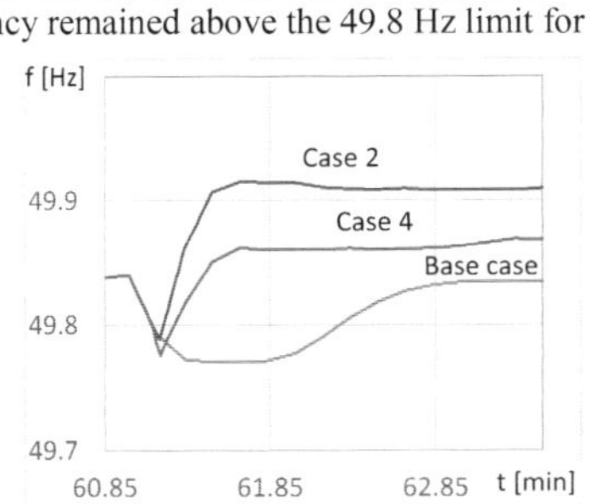

Figure 6. Frequency waveforms – time detail and frequency zoom

V. CONLUSIONS

The paper analyses a real incident - significant frequency drop of nearly - 200 mHz in the Continental Europe (CE) synchronous zone on 10 January 2019 at 21:02. The network simulator and dynamic model of the CE synchronous zone is used for analysis, so it is possible to simulate such incident at another time and a different situation and answers the question: "What would happen if it is a clear day in the morning with a relatively high generation in PV modules?".

Only public information ([9]-[14]) was used for the model preparation, because of an official ENTSO-E report was not available at the time of writing of the paper.

The paper describes all possible aspects that may affect frequency stability, which is traditionally analysed in terms of primary frequency control, the inertia (recently a very popular theme) and failure of the largest source in the system (or in an extreme case a separation of the system into islands).

However, the real incident shows that common phenomena such as deterministic deviations and even trivial measurement errors when they meet at one (inopportune) moment may endanger frequency stability and even security of supply.

The paper thus contains valuable knowledge for students and researchers who are interested in solving the problems of dynamic stability of the complex power system such as the synchronous interconnection of the continental Europe.

REFERENCES

[1] *Commission regulation (EU) 2017/1485 of 2 August 2017 establishing a guideline on electricity transmission system operation*

[2] Presentation during DSA Workshop on 23 May 2018, [on line] https://docstore.entsoe.eu/Documents/Events/2018/180523_DSA_SH_Workshop-Joint_presentation_with_conclusions.pdf

[3] K. Máslo, P. Toman, B. Bátora, J. Koudelka, "Methods for Dynamic Stability Assessment," Proc. 20th International scientific conference Electric Power Engineering (EPE 2019)

[4] K. Máslo, A. Kasembe, V. Barba, "Voltage collapse modelling in time domain simulation, " Proc. 13th International Scientific Conference Control of Power Systems (CPS 2018)

[5] Frequency Stability Evaluation Criteria for the Synchronous Zone of Continental Europe , ENTSO-E RG-CE System Protection & Dynamics Sub Group, March 2016, [on line] https://docstore.entsoe.eu/Documents/SOC%20documents/RGCE_SPD_frequency_stability_criteria_v10.pdf

[6] K. Maslo, A. Kasembe, S. Moroni, R. Pestana, "Frequency stability modelling of the future continental Europe power system," Proc. 52nd International Universities' Power Engineering Conference (UPEC 2017)

[7] K. Máslo, A. Kasembe, "Mitigation measures for photovoltaics retrofit", IEEE Transactions on Sustainable Energy, Vol. 9, No. 1 (Jan. 2018), pp. 333-339

[8] Deterministic frequency deviations – root causes and proposals for potential Solutions, EURELECTRIC & ENTSO-E Paper, Dec. 2011, [on line] https://www.entsoe.eu/fileadmin/user_upload/_library/publications/entsoe/120222_Deterministic_Frequency_Deviations_joint_ENTSOE_Eurelectric_Report__Final_.pdf

[9] Causes of frequency drop of 10 January still under investigation, ENTSO-E news, [on line] https://www.entsoe.eu/news/2019/01/16/frequency-drop-in-continental-europe/

[10] January 10th : RTE calls on interruptible industrialists, [on line] https://clients.rte-france.com/lang/an/visiteurs/services/actualites.jsp?id=9821&mode=detail

[11] Continuing frequency deviation in the Continental European Power System originating in Serbia/Kosovo: Political solution urgently needed in addition to technical, [on line] https://www.entsoe.eu/news/2018/03/06/press-release-continuing-frequency-deviation-in-the-continental-european-power-system-originating-in-serbia-kosovo-political-solution-urgently-needed-in-addition-to-technical/

[12] The 49.8 Hz incident on 10/01/2019 um 21h00 hours in the European electric grid, [on line] https://gridradar.net/Unterfrequenzereignis_Januar_2019.html

[13] French grid briefly curbed factory output to avert blackout, [on line] https://www.reuters.com/article/us-france-electricity/french-grid-briefly-curbed-factory-output-to-avert-blackout-regulator-idUSKCN1P60HS

[14] Data fault, demand spike caused 10 Jan grid issue – Tennet, [on line] https://www.montelnews.com/de/story/data-fault-demand-spike-caused-10-jan-grid-issue--tennet/970949

[15] K. Máslo, "Impact of Photovoltaics on Frequency Stability of Power System During Solar Eclipse", IEEE Transactions on power systems, Vol. 31, No. 3, Sept. 2016, pp. 3648-3655

[16] K. Máslo, A. Kasembe, "Long Term Dynamic of Power System with renewable energy sources and energy storage systems" Proc. International Conference on Renewable Energies and Power Quality (ICREPQ'19)

[17] K. Máslo, "Frequency stability in the context of the European network codes," Proc. 9th International Scientific Symposium on Electric Power Engineering, Sept. 2017, pp. 436-442

[18] M Pistora., K. Máslo, "Impact of embedded renewables on transmission and distribution network", Proc. 10th International Scientific Conference Control of Power Systems (CPS 2012)

[19] ENTSO-E Dynamic study, "Dispersed generation impact on continental Europe region security", 2014, [Online] https://www.entsoe.eu/publications/system-operations-reports/continental-europe/Pages/default.aspx

[20] ENTSO-E, "RGCE OH - Policy 5: Emergency Operations", 2015, [Online] https://docstore.entsoe.eu/Documents/Publications/SOC/Continental_Europe/oh/20150916_Policy_5_Approved_by_ENTSO-E_RG_CE_Plenary.pdf

The Flow-Based Capacity Calculation Approach and Comparison with Currently Used Method

Michal Grega, Tomáš Schnierer
Slovenská elektrizačná prenosová sústava, a.s.
Bratislava, Slovak republic
Michal.grega@sepsas.sk, tomas.schniere@sepsas.sk

***Abstract*—On 17 November 2016, the Agency for the Cooperation of Energy Regulators (ACER) decided to establish the new capacity calculation Core region. According to the network code capacity allocation (CACM) the Flow-Based approach shall be implemented within the Core region. The transmission system operators within this area, including SEPS, have launched the implementation. The Flow-Based approach represent a capacity calculation method in which energy exchanges between bidding zones are limited by constraints and available margins on critical network elements. Indeed, the more accurate calculation allows optimization of the transmission capacity and increases the level of grid security. This principle optimizes the calculation of capacity taking into account that power can flow via different paths in the highly meshed grid. In contrast to the current net transfer capacities (*NTC*) approach, Flow-Based method is much better because of considering physics laws instead of business laws. The individual market transactions are translated into the physical flows, while ensuring that the physical limits of the grid are respected. This fact explains the main advantage of the new approach. The purpose of this paper is to introduce Flow-Based capacity calculation method and compare this approach with the currently used *NTC*.**

Keywords: flow-based, capacity calculation, critical branch, available margin

I. Introduction

The European Commission regulation 2015/1222 has introduced the network code CACM [1] for calculating how much capacity is possible to use on cross-border lines with sufficient level of security. This is the first step towards an integrated European single market for electricity. The methodology recommends using Flow-Based (FB) approach as the most effective option for day-ahead (DA) and intraday capacity calculation. On the other hand FB is not sufficient for long-term capacity calculation where coordinated *NTC* should be used. The coordination among TSOs (transmission system operator) is essential because, as we know, the prediction of renewables is less accurate compared to conventional power plants, which means a great challenge for the TSOs. The operators are aware of deviation between commercial exchanges and physical flows and the main purpose of the Flow-Based mechanism is to reduce so called unscheduled flows. For example the commercial exchange between Austria and Germany also affects neighbouring countries in accordance with physical laws. In fact there is partial influence for example on Czech Republic and Poland because flow does not follow the direct path between Austria and Germany. However, this factor is of no concern to market participants because they just need information about maximal available transfer capacity between bidding zones and it is solely TSOs responsibility to provide most accurate information. TSOs are committed to providing all available capacity to the market while ensuring security grid. This poses a great challenge for TSOs in the next years because they are forced to provide as much capacity as possible. [2], [3].

II. Currently Used NTC Approach

The main aim of the currently used approach for the capacity calculation is to determine the maximum possible exchange between two areas for the particular market time unit (mostly for the hourly time frame), while ensuring the operational limits and basic security standards. In this case the maximum admissible capacity determined by each TSO is called *TTC* (Total Transmission Capacity). This value is calculated from reference scenarios starting from planed outages, configurations, load and generation pattern. Calculation is done by iteration process. Indeed, this means that generation in the area A is increased and the generation in the area B is decreased step by step (it can be done with the changing consumption as well). This process end until first limit is exceeded. In fact the *TTC* represents the maximum power exchange between two zones without affecting the system security. The *TTC* is the maximal theoretical value as depicted in the Fig. 1. The physical flow from A to B is increased until a security criteria is violated [4].

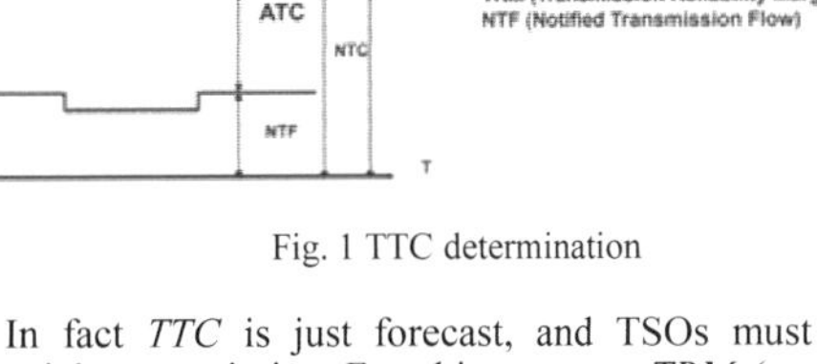

Fig. 1 TTC determination

In fact *TTC* is just forecast, and TSOs must consider potential uncertainties. For this purpose *TRM* (transmission reliability margin) is used and solely depends on TSOs experiences. The *TRM* mostly covers deviation between physical and market flows. After subtraction of the *TRM* from *TTC* we obtain *NTC* considering equation (1) below:

$$NTC = TTC - TRM \tag{1}$$

Taking into account *AAC* (already allocated capacity) from the previous market sessions TSOs shall subtract this capacity from *NTC* in order to calculate *ATC* (available transmission capacity) for the DA market:

$$ATC = NTC - AAC \tag{2}$$

III. FLOW-BASED APPROACH

Is it obvious that because of the aforementioned facts *NTC* values are not equal to the sum of the thermal capacities of lines, due to highly meshed grid it is necessary to consider the loop flows and impact on the neighboring countries. In order to mitigate this issue, the Flow-Based has been developed. The initial computation is strongly linked to the base case chosen in order to have best anticipation of the situation. The day-ahead capacity calculation process shall consist of three main stages:

- the creation of inputs for the capacity calculation purposes by the TSOs,
- the capacity calculation process by the CCC (capacity calculation coordinator),
- the capacity validation by TSOs in the coordination with the CCC. [5]

Each TSO shall provide a list of critical network elements (CNEs) with relevant contingencies, also known as CNECs based on the TSOs historical experiences. CNECs represent elements which could probably limit the exchanges between the zones. Indeed, the lack of capacity on particular CNEC means a price spread between two bidding zones. The impact of commercial transactions. i.e. change of Net Positions on flows through critical network element is considered by power transfer distribution factor (PTDF).

A. Mathematical description of the flow-based capacity calculation approach

The flow-based computation delivers three main classes of parameters which are needed for the creation of the flow-based domain: the power transfer distribution factor (PTDF), the generation and load shift key (GLSK) and the remaining available margin (RAM). The PTDF matrix represents the influence of a commercial exchange bidding zones on power flows on the considered combinations of CNEs and contingencies. Indeed, the goal of the PTDF is to capture the impact of the net position change on each CNE and contingency.

The effect on every CNE or CNEC is monitored and calculated as a percentage (e.g. if the net position change of 100 MW has an impact of 10 MW on a CNE, then the PTDF is 10 %, so we can consider this CNE to be relatively sensitive in according with article 15(1) of the CCM. The final list of CNECs after filtering step shall not include any network element with the contingency with a maximum PTDF below the 5 %. The CNEC filtering limit has been introduced due to the amount of CNECs in the whole Core region. The calculation of the PTDF matrix is performed on the basis of the common grid model (CGM) and the GLSK.

The GLSK defines how a change in net position is mapped to the active generator and load nodes in the CGM. In fact GLSK shows relation between the change in net position of the country and the change in output of every generating unit (or change of load) inside of this country. Every TSO should define a GLSK for its control area based on operational experiences. In general, the GSK includes power plants that are market orientated and flexible in changing the electrical power output (nuclear power plants are not suitable for this purpose). The GLSK values are given in percentage. For example value 0.01 for specific node means that 1 % of the change of the net position of the control area will be realized by this unit. In fact GLSK values are allocated to active units in the CGM [6].

The next important parameter called remaining available margin (RAM) could be calculated. RAM is the result of a simple subtraction as shown in Equation (3) below :

$$RAM = F_{\max} - FRM - F_{\mathrm{ref}} - FAV \tag{3}$$

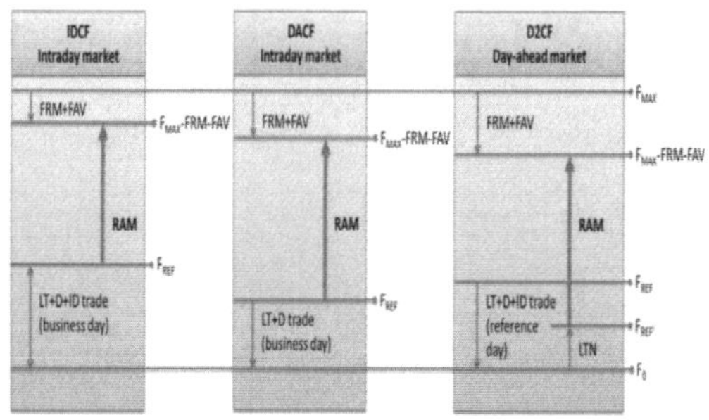

Fig. 2 The Flow-Based parameters overview

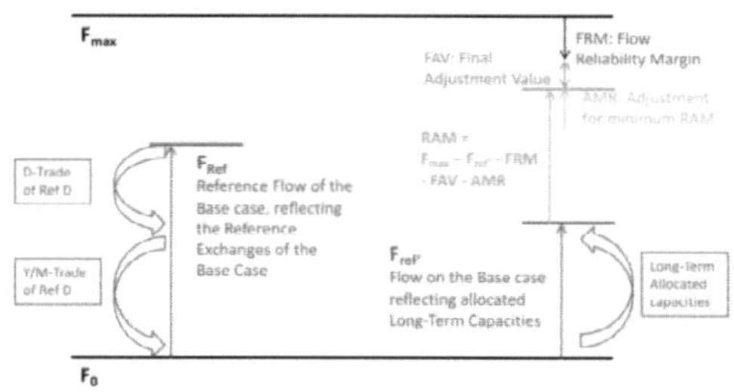

Fig. 3 Already allocated capacity

The essential parameter F_{MAX} describes the maximum admissible active power flow on a CNEC. F_{MAX} shall be calculated by the given formula:

$$F_{\max} = \sqrt{3} \cdot U \cdot I_{MAX} \cdot \cos\varphi \tag{4}$$

The reference flow F_{ref} is directly simulated from the *CGM* and reflects the loading of the critical branch induced by a reference program. F_{ref} should be positive in direct direction and negative in the opposite direction. The flow reliability margin (*FRM*) covers uncertainty (similar to the *TRM*) stemming from the fact that calculation process is based on forecast modes. The inputs are created two days before the delivery date of electricity. Therefore the deviation between the expected power flows and real time flows can occur. The FRM according to article 8(1) of the CCM shall cover the following forecast uncertainties:

- cross-zonal exchanges on bidding zone borders outside the CCR,
- generation pattern including specific wind and solar generation forecast,
- generation shift key,
- load forecast,
- topology forecast,
- unintentional flow deviation due to frequency containment process,
- flow-based capacity calculation assumptions including linearity and modelling of external (non-core) TSO's areas.

The *FRM* will reduce the remaining margin that can be provided to the market. The basic idea how to determinate the *FRM* is to compare the FB model with the physical flows in real time for the corresponding timestamp. The concept is depicted in the Fig. 4.

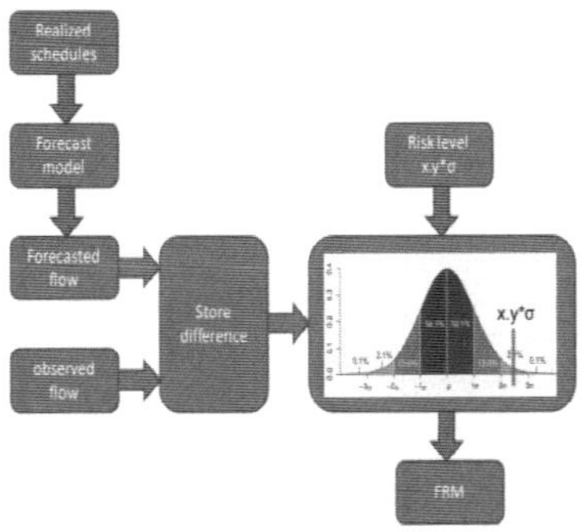

Fig. 4 The FRM determination

The difference between real time flow F_{real} and reference flow F_{ref} is calculated and results are stored for the further statistical analysis. The FRM is supposed to cover the 90th percentile on all CNECs. This means that the FRM will cover 90 % of the historical deviations. The FRM values will be updated every year in accordance of the previous observed period. CNECs which are not already used in existing flow-based initiatives, the FRM value shall be equal to 10 % of the F_{max}. Final adjustment value (FAV) serves to increase or decrease the capacity that can be allocated to the market for a specific CNEC. TSOs are allowed to adjust RAM during the validation step and it is the last chance for the TSOs to ensure security of supply (SoS) before the market session. In the first step of the validation TSOs shall assess whether they have sufficient RAs to avoid violations and if not, then TSOs shall use FAV in order to ensure operational security. The FAV shall have a positive value and may reduce the RAM only to the minimum degree that is needed. In generally the FRM shall ensure that the RAM remains non-negative [5].

B. *Long-term allocated capacity inclusion*

In the next step TSOs shall take into account the previously allocated capacity (long term). This step is called LTA inclusion. After each stage in calculating Flow-Based parameters TSOs shall verify if the already allocated capacity is included in the FB domain. For each CNEC a positive margin is expected in order to avoid pre-congestion, this means that already allocated capacity is not fully covered by FB domain. The crucial step for the LTA inclusion is to calculate flow for each CNEC in each combination of net positions resulting from the previously allocated capacity.

C. *Long-term nominations shift*

The long-term nominations (LTN) shift takes place during morning hours of day D-1. This means that *TSOs* shall provide shift from the LTA to the LTN as shown in the Fig. 5. This allows to calculate the flow for each CNEC F_{LTN}. This is done respecting the net positions reflecting the long-term nominations with the PTDF. The corresponding formula is described below:

$$\boldsymbol{F_{LTN} = PTDF \cdot NP_{LTN}} \qquad (5)$$

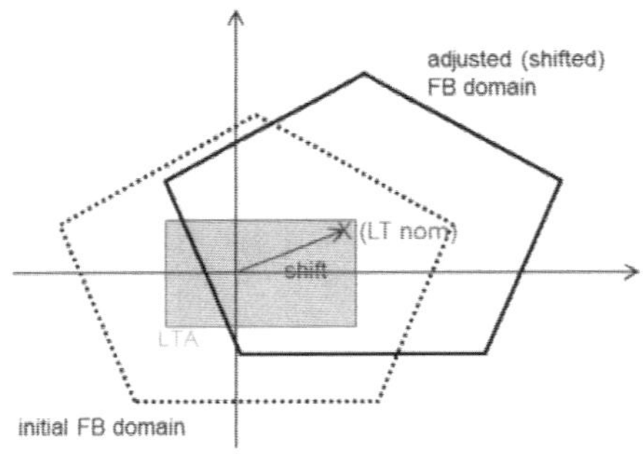

Fig. 5 The LTN shift

D. *Adjustment for minRAM*

The application of the minRAM adjustment is used when the RAM in situation without any commercial exchanges in the Core region (F_0) is lower than 20 %. The situation without any commercial exchanges means that all scheduled cross-border exchanges are set up to 0 MW. The F_0 is the flow per CNEC in the situation without any commercial exchanges. The aim of this step is to ensure a capacity of at least 20 % of the F_{max} on each CNEC as seen in the Fig 6. The adjustment of minimum RAM (AMR) aims to ensure that minRAM criterium is fulfilled. The non-zero values of the AMR indicate that on this specific CNEC the adaption is needed. As can be concluded from the tab. 1 below, the different minRAM factors have significant impact on the amount of the virtual capacity AMR. The minRAM with the 20 % value was proposed by TSOs, while minRAM equal to 70 % has been introduced by ACER. Moreover TSOs still have a possibility to introduce an action plan in order to achieve the minRAM equal to 70 % in 2026 at the latest. The action plan means that TSOs shall define the linear trajectory for calculation of minRAM for each year between 2020 and 2026. This action plan has been introduced in order to gradually increasing the minRAM year by year. According to this decision it is more than challenging for the TSOs to fulfill aforementioned requirements. The TSOs are worried that these plans are too ambitious and in some cases the endangering of the security of supply can occur. Comparison of the different minRAM volumes is shown in Tab. 1. Moreover, in this situation the significant grid reinforcements are needed and additionally the volume of the redispatch will be increased in the upcoming years in order to fulfill the minRAM. MinRAM is supposed to ensure that internal exchanges are not preferred and lead to more non-discriminatory way for the cross-border exchanges [5], [7].

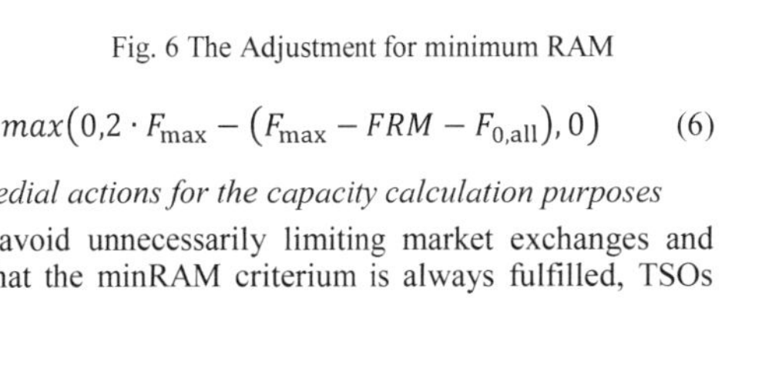

Fig. 6 The Adjustment for minimum RAM

$$AMR = max\left(0{,}2 \cdot F_{\max} - \left(F_{\max} - FRM - F_{0,\text{all}}\right), 0\right) \qquad (6)$$

E. *Remedial actions for the capacity calculation purposes*

To avoid unnecessarily limiting market exchanges and ensure that the minRAM criterium is always fulfilled, TSOs shall use RAs (remedial actions). These RAs will be used for the coordinated optimization of cross-border capacities. Each TSO may withhold some RAs which are needed to ensure operational security, this means that just a subset of RAs is used for the capacity calculation purposes. Each remedial action can have different properties, it can be either:

- Pre-fault (preventive) or post-fault (curative). Post-fault RAs allow short overload of the CNEC. Curative actions can be linked to a specific contingency. The preventive RAs (pre-fault) can be the opening of a bus-bar coupler, shifting of a tap position of a phase-shifting transformer and other topological measures. The tap positions are directly simulated in the CGM.
- Shared or non-shared. RAs can be available for the whole region or only specified for particular CNEC. In fact this means that if RA can heal a congestion just on a one specific CNE then it is considered like non-shared. If the RA of TSO A has impact on the grid of TSO B then it is shared [7].

Table 1. Initial Flow-Based parameters

Branch	F_{max}	*FRM*	F_{ref}	F_0	*RAM* F_{ref}	*RAM* F_0	**minRAM 20 %**	**minRAM 70 %**	*AMR* **20 %**	*AMR* **70%**	**RAM** F_0 **+** *AMR* **0,2**	**RAM** F_0 **+** *AMR* **0,7**
CNEC1	374	30	316	304	28	40	74,8	261,8	34,8	221,8	74,8	261,8
CNEC2	1734	140	1387	1361	207	233	346,8	1213,8	113,8	980,8	346,8	1213,8
CNEC3	1384	138	510	145	736	1101	276,8	968,8	0	0	1101	1101
CNEC4	311	25	80	4,9	206	281,1	62,2	217,7	0	0	281,1	281,1
CNEC5	1205	120	688	135	397	950	241	843,5	0	0	950	950
CNEC6	1384	138	529	254	717	992	276,8	968,8	0	0	992	992
CNEC7	1696	79	1039	858	578	759	339,2	1187,2	0	428,2	759	1187,2
CNEC8	1939	202	946	746	791	991	387,8	1357,3	0	366,3	0	1357,3
CNEC9	1912	180	795	610	937	1122	382,4	1338,4	0	216,4	0	1338,4
CNEC10	2182	269	1838	1431	75	482	436,4	1527,4	0	1045,4	0	1527,4

IV. FLOW-BASED VS ATC COMPARISON

Instead of supplying fixed commercial exchanges, the FB domain takes into account constraints (active CNECs) which reflect the physical limits of the grid. The FB domain is illustrated in the Fig. 7. Each CNEC is represented by a red line. The intersection of two CNECs represents the edge of the domain. Comparing both approaches, it is clear that the larger FB domain surrounds the ATC domain. This means increasing of the welfare because the more capacity is provided to the market. In the ATC approach TSO have to make choice how to split capacity among its borders (A to B and A to C) in contrast with the FB when the entire SoS domain is offered to the market. Given the security domain NTC/ATC constraints and the corresponding NTC/ATC domain are a choice made by the TSO. In contrast under the FB approach the domain is secure itself. The FB tries to be more transparent and clear process and additionally FB is expected to offer more trading opportunities with the same level of SoS [3].

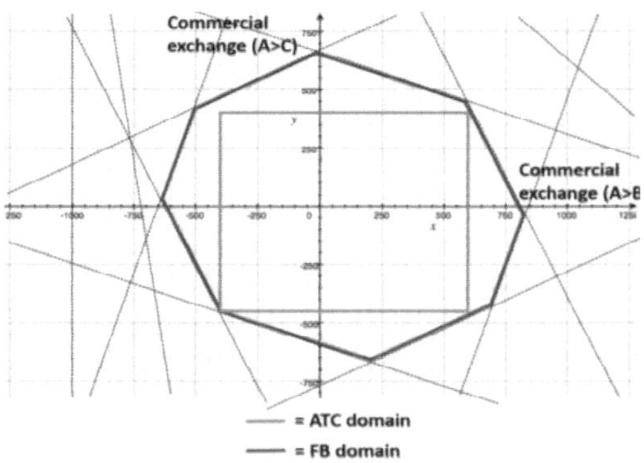

Fig.7 ATC vs FB domain comparison

V. CONCLUSION

The Flow-Based approach should lead to a more efficient use of the capacities compared to the *ATC* while taking into account the physical flows stemming from the market transactions. This means that TSOs are able to decrease volume of the reliability margin and provide more capacity to the market. The TSOs may better anticipate the cross border flows and ensure sufficient amount of the RAs if needed. However, it is necessary to mention that FB approach is more complex and complicated. The increasing of exchange volumes poses specific challenges on the cooperation among TSOs and more effort is needed and secondly the TSOs partially lose control of the capacity provided to the market because the FB requires more strict rules compared to the *ATC* where the volume of the provided capacity solely depends on the TSOs themselves. Apparently with flow-based it is even more important to know the exact rules and methods of the capacity calculation. Nevertheless, there has been no experience with FB in the capacity calculation during the whole process, including market participants, in the scope which it is planned for. Therefore, it is not certain how market participants will behave, since it is virtually impossible to predict. The fulfillment of described expectations will be known only after implementation of the procedure to the practice.

References

[1] Capacity Allocation and Congestion Management network code 1222/2015

[2] J. Salavec:. Congestion income distribution under Flow-Based method, VUT Brno 2016

[3] G. Plancke, K. De Vos, R.Belmans.: Efficient Use of Transmission Capacity for Cross-border Trading: Available Transfer Capacity versus Flow-based Approach, KU Leuven 2015

[4] Net transfer capacity assessment, Eléctrica de España presentation 2009

[5] Day-ahead capacity calculation methodology of the Core capacity calculation region

[6] CWE Enhanced Flow-Based MC feasibility report 2011

[7] CWE Market Coupling Flow-Based Forum, Amsterdam 2011

Modeling a Small Hydropower Plant

Pavol Fedor, Daniela Perduková
Dept. of Electrical Engineering and Mechatronics
Faculty of Electrical Eng. and Informatics TU Kosice
042 00 Kosice, Slovakia
pavol.fedor@tuke.sk, daniela.perdukova@tuke.sk

Richard Olexa
Dept. of Electrical Engineering and Mechatronics
Faculty of Electrical Eng. and Informatics TU Kosice
042 00 Kosice, Slovakia
richard.olexa@tuke.sk

Abstract — **Research, optimization and practical implementation of small hydroelectric power plants as a source of clean electricity presents one of the topical tasks in current energetics, which is virtually impossible to solve without powerful computer support due to the strongly nonlinear nature of such systems. The article presents a complete model of a small hydropower plant, modelling its individual subsystems and verifying its dynamic properties in basic operating modes. Also an influence of a major external fault, caused by a drop in the power grid frequency, was investigated. Full SHPP (Small-Hydropower Plant) model properties were verified using MATLAB software package. The obtained results confirmed correctness of the proposed model, which can be used for design of a SHPP control algorithm.**

Keywords: small hydropower plants, modelling and control of small hydropower plants

I. Introduction

Current trends in electricity generation are characterized by increasing criteria towards saving our environment. In the surrounding nature lot of accumulated amount of renewable energy occurs, esp. in the water resources (like lakes, rivers, artificial water reservoirs, etc.). This energy can be extracted from the water by creating suitable reservoirs of hydropower – either reservoirs utilizing water level or routes with continuous water flow) by using machines converting this energy to rotational mechanical energy (water wheels, turbines) and further converting the mechanical energy to the electrical one (in synchronous and asynchronous electrical machines). All this energy conversion must be controlled according to requirements of final consumer of the electric power. The described systems are called hydroelectric power plants and according to the IEEE Standard 1020-1988, these power plants are roughly described by the following attributes:

- the unit rating is between 100 V and 5 000 kW and the rated voltage from 480 V to 13.8 kV;
- they are fitted either by a synchronous or asynchronous generator;
- the total cost of the surroundings treatment should be minimal;
- they are designed for fully automated operation;
- they have a low-level tank only having a low level of water column (in Run-of-River Hydroelectric Plant).

Computer models of energy systems considerably simplify and speed up their design and also save costs at their realization. Several various models of hydropower generation were investigated by scientists where the existing models depend on supposed requirements involved in the study. Some of these models are simply analytical [1, 2] while others were developed from robust system models and show dynamic characteristics [3, 4]. Various models of hydro plant and techniques used to control the generation of power can be found in [5, 6]. The hydroelectric power plant model published in [7] allows design and investigation of the hydropower system in frequency domain. Moreover, there are existing models of linear and nonlinear hydro-turbine set with non-elastic and elastic water column effects [8, 9, 10, 11].

It is obvious that for detailed investigation of electromagnetic phenomena in the SHPP and interaction between electrical network and the hydraulic turbine, it is necessary to model in details electrical phenomena occurring in the electric generator [12, 13, 14]. An example for a detailed and high-quality model is presented in [15].

The article describes a model of a small hydroelectric power plant that is usable for designing its control, utilizing also artificial intelligence methods [16, 17, 18] or modern control methods used for investigation of the plant dynamic properties in chosen operating modes. The model is composed from particular models of SHPP physical subsystems that are suitable interconnected. Properties of the proposed model were verified by simulation in MATLAB/Simulink using real parameters of a micro hydro power power plant in Dobšiná (hydropower dam).

II. Overal Diagram of the Small Hydropower Plant Model

Model of a small hydropower plant (SHPP) is fairly complex non-linear dynamic system consisting of the subsystems shown in Fig. 1 ([20]).

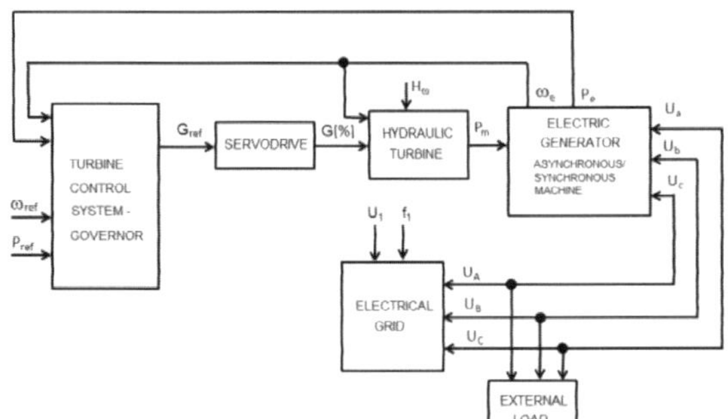

Figure 1 Block diagram of the micro-hydropower plant

The basic subsystem of the entire SHPP is the "*Hydraulic Turbine*" model in which the energy of water is converted into mechanical energy. The "*Servo Drive*" subsystem serves as an actuator to control this conversion by controlling the water amount flowing through the turbine or a turbine pressure, respectively. In principal, the turbine control has two basic functions:

1. turbine starting and running up to mechanical speed close to the grid frequency,
2. after being phased into the network (i.e., after synchronization) it controls the power supplied to the grid.

This is ensured by the turbine control subsystem, referred in the literature as a "*Governor*".

The conversion of the mechanical energy of a turbine into an electrical one is ensured by a subsystem labeled as "*Electricity Generator*", which can be either an asynchronous or a synchronous electrical machine. In SHPP this conversion can also be provided by static converters with a DC bus. When using a synchronous generator, it is also necessary to model its excitation circuit by a separate subsystem.

At output from the generator there is a three-phase electrical voltage, which can be used directly by own load, or it is possible to connect the SHPP via the transformer to an external electrical grid with external load. The most important grid parameters are the voltage U_l and its frequency f_l (or the angular frequency ω_l).

The complete SHPP model is obtained by modeling its individual subsystems and their interconnection according to Fig. 1. The accuracy of modeling of particular subsystems depends on purpose of using the complete model. Depending on the accuracy of the sub-system models, we obtain a more accurate or less accurate overall model.

III. Modeling Subsystems of Small Hydropower Plant

A. Hydraulic Turbine Model

All variables in the paper are normalized in respect to their maximum value, which corresponds to a value of 1 or 100% (e.g., at actuator opening).

The steady-state mechanical power of the water turbine is proportional to the height H of the water column above the turbine and the current water flow rate Q of the turbine according to the relationship

$$P_h = \rho g Q H \tag{1}$$

$$P_m = \eta_t P_h \tag{2}$$

where P_m is the mechanical power on the turbine shaft, P_h is the possible hydraulic power, ρ is the density of the water and η_t is the turbine efficiency.

The standard flow change is given by the relationship

$$\frac{dq}{dt} = (1 - h - h_l)/T_w \tag{3}$$

where T_w is the time constant of this change, $q = Q/Q_{max}$ is the standardized flowrate, $h = H/H_{max}$ is the standardized height of the water column, and h_l is the standardized loss of height which is equivalent to the losses in the pipelines.

The output mechanical power of the turbine depends on product of the current flow and the height of the water column in the turbine according to the relationship

$$P_m = A_t h(q - q_{nl}) - D_i G \Delta\omega \tag{4}$$

where A_t is the turbine gain, D_i is the turbine internal diameter (load damping constant), G is the servo drive position (in %), q_{nl} is the standard flow rate at unloaded turbine and $\Delta\omega$ is the shaft speed and power grid difference. The equations (1) – (4) correspond to the model of the hydraulic turbine subsystem (Fig. 2). Parameters for modeling are given in the Appendix.

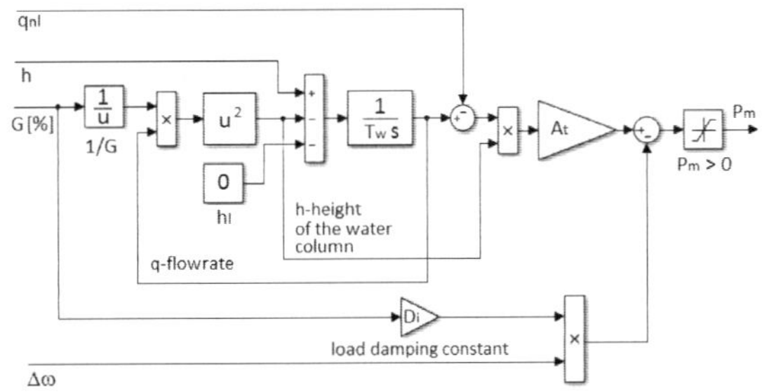

Figure 2 Block diagram of the subsystem titled "HYDRAULIC TURBINE"

The timer response of the turbine mechanical power P_m to a linear opening and closing of the valves by servo drive is shown in Fig. 3.

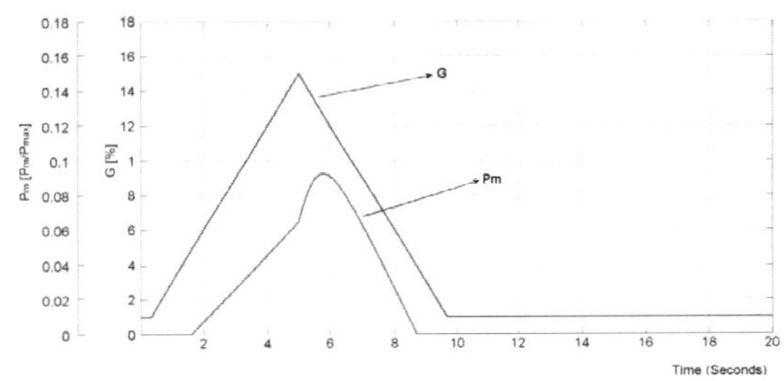

Figure 3 Time response of the mechanical power P_m on the hydraulic turbine shaft at opening and closing the valves by servo drive

B. Model of the Servo Drive

The servo drive subsystem serves as an actuator for controlling the hydropower flow to the hydraulic turbine. It is modeled by the second order dynamic system – Fig. 4, whose state variables represent the opening speed of blades of the inlet flap into the turbine and the position of opening of these flaps.

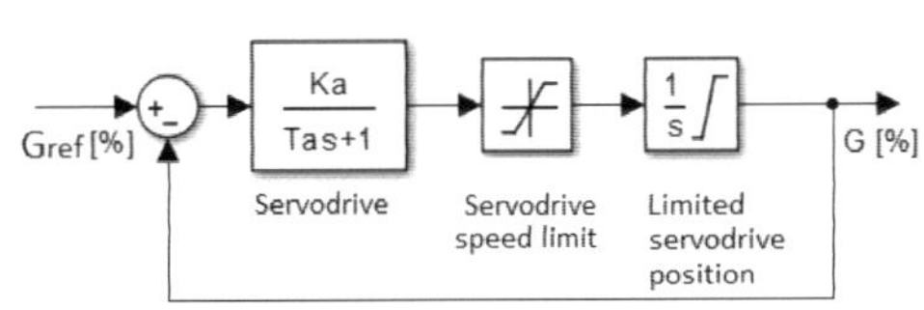

Figure 4 Block diagram of the subsystem named "SERVO DRIVE"

At higher levels of the water column above the turbine, strong pressure waves (so-called water hammer) are created during opening and closing the turbine inlet flaps, which could lead to turbine destruction. Therefore, for safety reasons, the opening speed of the flaps must be limited. Limiting the outlet position of the flaps in the model serves to set the real working area in which the turbine still responds to the opening of the servo drive. The rotor circuit gain K_a and rotor winding time constant T_a are servo drive parameters listed in the Appendix. The time course of opening and closing the actuator is shown in Fig. 5.

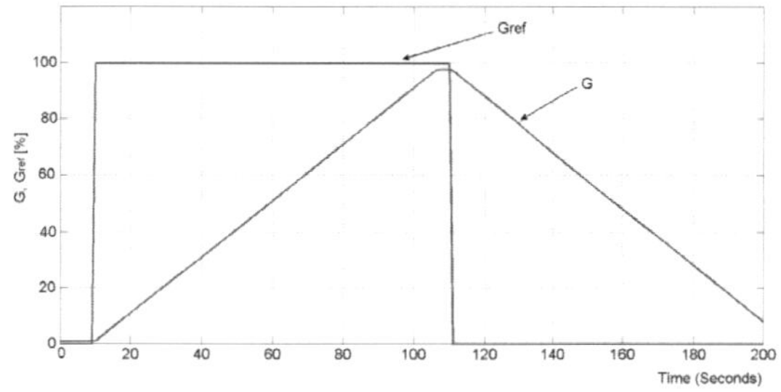

Figure 5 Opening and closing of the servo drive

C. Interconnecion of Models of Electric Generator and Hydro Turbine

An electric generator is an AC electric machine of synchronous or asynchronous type. Detailed models of these machines present non-linear dynamic systems of higher order in which internal electromagnetic dynamic phenomena lie within the millisecond range. In contrast to this, the hydrodynamic processes in the turbine are considerably slower and therefore, in most cases, it is meaningful to assemble an electromechanical model of the generator in terms of a torque equation for the common turbine and generator shaft. The hydraulic turbine and the generator are mechanically coupled through the common shaft for which the equation of equilibrium is valid

$$M_e + M_m - M_f = J_c d\omega_m/dt \quad (5)$$

$$M_f = K_t \omega_m \quad (6)$$

$$M_m = P_m/\omega_m \quad (7)$$

where M_e is the torque of the generator, M_m the torque of the turbine and M_f is the frictional torque proportional to the mechanical angular velocity ω_m and the constant K_t in the equation (6) represents the friction coefficient. The right side of the equation (5) represents the dynamic torque on the shaft, which accelerates the total moment of inertia on the shaft J_c. Here, P_e is the generated electrical power on the shaft, which in case of an asynchronous machine is proportional to product of its torque and the rotor angular velocity ω_m. The aforementioned assumption complies connection of models of the asynchronous electric generator and hydraulic turbine fixed on the common shaft (Fig. 6), where T_r is the rotor time constant of the asynchronous generator corresponding to the rotor resistance.

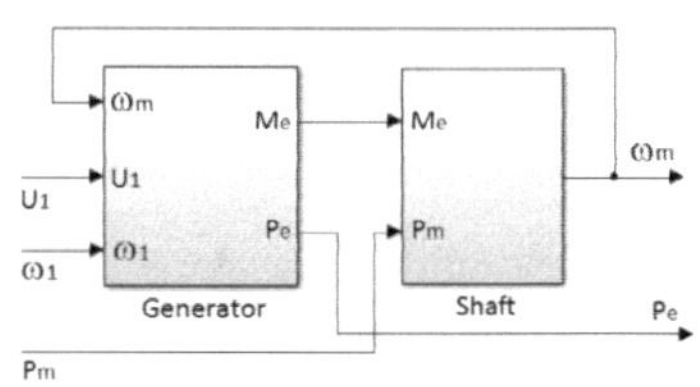

a)

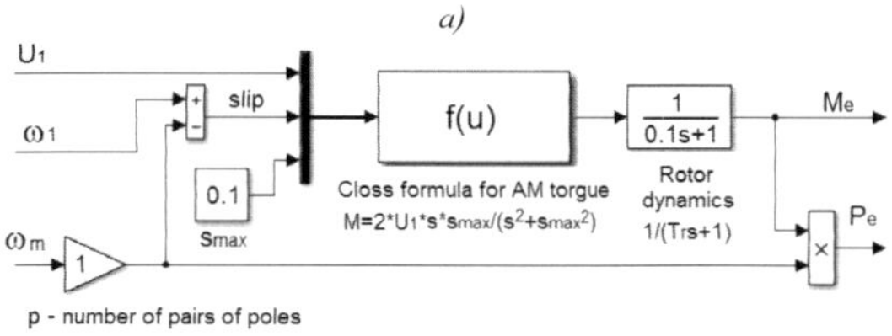

b)

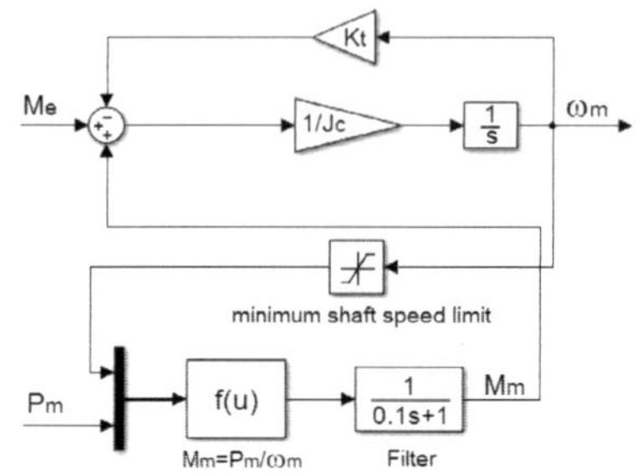

c)

Figure 6 Block diagram of the electromechanical subsystem "GENERATOR-LOAD":
a) Interconnection of the generator with the turbine on the common shaft
b) Asynchronous generator model
c) Model of the mechanical system on the shaft

Time course of mechanical quantities during turbine starting to the velocity close to grid frequency (the value $f_1 = 0.6$ corresponds to the frequency 50 Hz) are shown in Fig. 7.

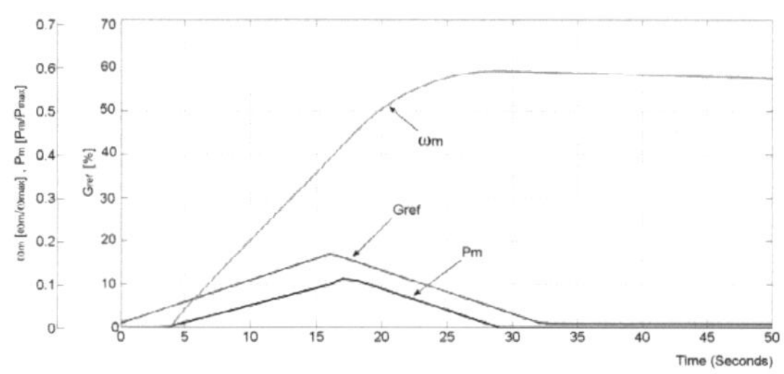

Figure 7 Turbine starting to the frequency of mains

D. Turbine Control Model – the Subsystem "GOVERNOR"

Turbine control system fulfils two tasks:

1. Turbine speed control within generator area of the electric generator.
2. Controlling the power supplied to the grid (if required).

A simple model of such control supposes using two standard PI controllers with the connection is shown in Fig. 8.

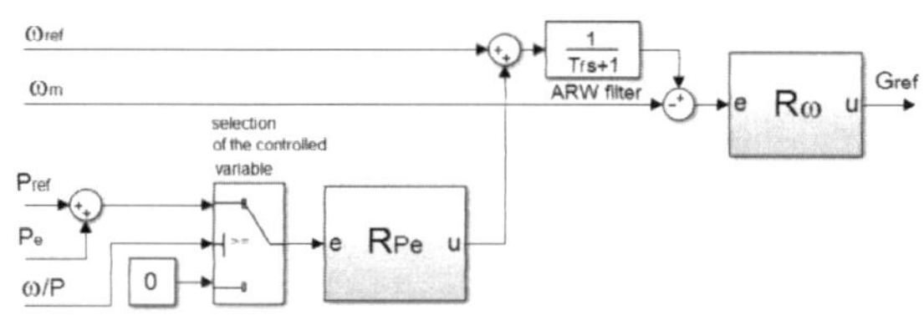

Figure 8 Block diagram of subsystem "GOVERNOR" for turbine control

In this control structure the controller R_ω of the PI type controls the angular velocity ω_m and the controller R_{Pe} controls the generator power. The logic signal ω/P provides switching between both controlling modes. The 1st order filter smooths the reference value ω_{ref} and serves to limit oscillations in the speed and power. Since the generated power is achieved by increasing turbine speed above the grid frequency, the output of the power controller represents the corresponding increment of the mechanical speed of the turbine and is added to the reference velocity ω_{ref}.

To obtain the PI controller parameters, a complete SHPP model was used according to Fig. 1. The algorithm of finding the proper values of the parameters of controllers was as follow: the whole range of possible values of the proportional gain K_P and the integral component K_I of the controllers was symmetrically divided with a defined density. For all combinations of the arguments the value of a criterion function was calculated. As the criterion the integral of quadratic deviation between the desired and actual angular velocity of the electric generator was used

$$J(K_P, K_I) = \int (\omega_{ref} - \omega_e)^2 dt \tag{8}$$

The optimal parameter pairs of PI controllers were selected in the instant when the criterion function (8) showed a minimum value. For finding optimal controller parameters a m - file in MATLAB was created. The specific controller parameters are given in the Appendix.

IV. Simulation of Basic Operating Modes of the SHPP

The basic operating modes of the hydro turbine are:

1. the start-up to the speed close to the grid frequency,
2. the power generator staging and
3. the electrical power control.

Fig. 9 shows an example of modeling these modes for the micro-hydropower plant in Dobšiná. The parameters are listed in the Appendix.

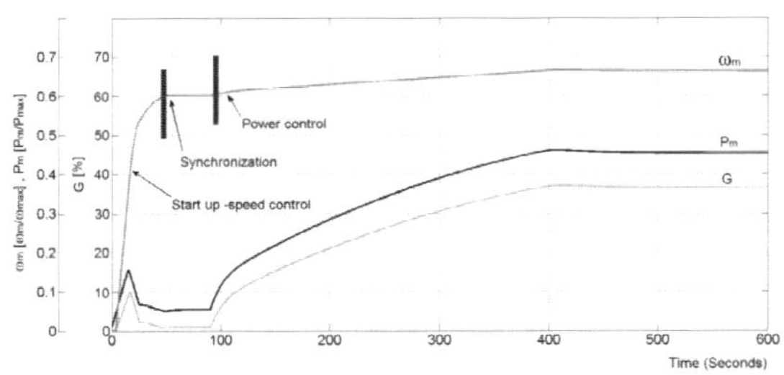

Figure 9 Modeling of turbine start-up, modeling, synchronization (phasing) and power control

At time $t = 0$, the turbine starts to run to the grid frequency without connecting the generator to the grid. After reaching the speed corresponding to the grid frequency, the generator is phased (synchronized), i.e. it is connected to the power grid. After the transient state has been stabilized at $t = 80$ s, the "governor" using the logic signal ω/P, switches the mode from speed control to power control. At time $t = 400$ s the turbine is in operation. The power is stabilized at a standard value of $P_m = 0.36$ which corresponds to the nominal turbine power.

Fig. 10 shows the turbine response to the grid frequency drop, which presents a basic type of external disturbance acting on the hydro generator.

One sees that at time $t = 600$ s, the frequency of the network dropped, and the control responded by reducing the speed of the hydro turbine and equalizing the power of the generator to the desired value.

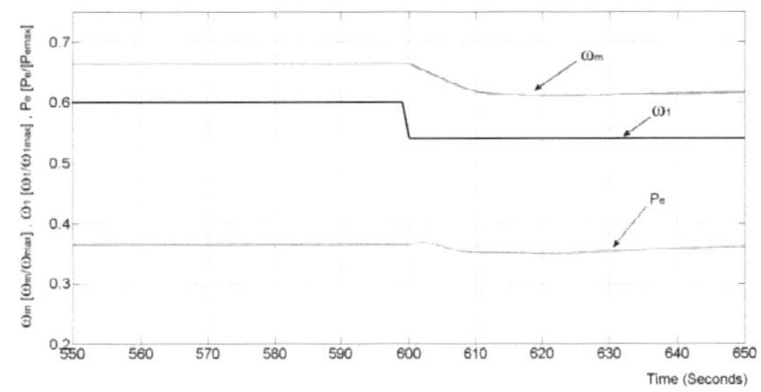

Figure 10 Modeling turbine start-up, synchronization and power control

Figure 11 shows the torque curve of the generator during synchronization, and Figure 12 displays the generator course when the grid frequency drops, which cause transient phenomena on the torque curve M_e of the generator.

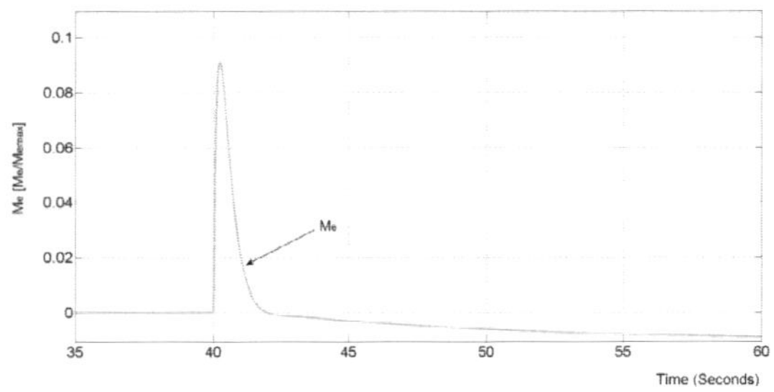

Figure 11 Generator torque after synchronization

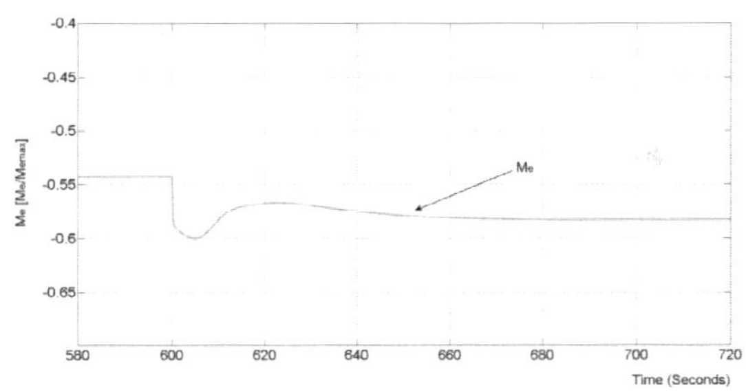

Figure 12 Generator torque at the grid frequency dropping

V. Conclusions

The paper shows the process of development a SHPP simulation model, which consists of three basic subsystems: a hydraulic turbine, an electric generator and a turbine controlling mechanism called a "governor" together with a servo drive. The resulting model was compiled from models of individual subsystems. The particular models were designed and verified separately in the MATLAB/Simulink program package.

The dynamic properties of the complete SHPP model have been verified by simulation of basic operating modes of the hydro turbine, such as its start-up to the speed close to the grid frequency, power generator synchronization and electrical power control. Also influence of an external disturbance, which is the power grid frequency drop, was investigated. The achieved results confirmed the correctness of the proposed model, which can be used to investigate its properties in different operating modes and especially for the design of various and optimal control structures using intelligent control methods.

Acknowledgment

The article was published with the support of the project No APVV-16-0206.

Appendix

Subsystem Governor

K_{PRe} = 0.15 K_{IRe} = 0.00064 $K_{PR\omega}$ = 6.5

$K_{IR\omega}$=0.068 T_f=10 [s]

Subsystem Servo Drive

K_a =3.3 [Ω^{-1}] T_a = 0.07 [s]

Subsystem Hydraulic Turbine

L = 1255 [m] penstock lenght

A = 1.54; [m2] penstock cross section

g = 9.81 [m/s2] h_base = 51.7 [m] q_base = 4.6 [m^3/s]

T_w = ((L/A)*(q_base/(h_base*g))) [s]

Di = 1.4 [m] M_m = 2.1 [MW]

A_t = 0.89 q_{nl} = 0.0434 [m3/s]

Subsystem Electric Generator

M_e = 2.1 [MVA] J_c = 7.99 [kgm^2] K_t=0.01 T_r = 0.1 [s]

p=1 s_{max} = 0.1

References

[1] C. Qijuan, X. Zhihuai, "Dynamic modeling of hydro turbine generating set," in IEEE International Conference on Systems, Man and Cybernetics, 2000, pp. 3427–3430. DOI: 10.1109/ICSMC.2000.886538.

[2] R.A. Naghizadeh, S. Jazebi and B. Vahidi, "Modelling Hydro Power Plants and Tuning Hydro Governors as an Educational Guideline," in International Review on Modelling and Simulations (I.RE.MO.S), 2012, Vol. 5, No. 4., pp.780-1790.

[3] D. Medved', M. Kolcun, Z. Conka, et al., "Simulation of Faults in the Power System and Their Elimination in the EMTP-ATP", Proc. of 7th International Scientific Symposium on Electrical Power Engineering (Elektroenergetika), Tech. Univ. Kosice, Stara Lesna, Slovakia, 2013, pp. 484-487.

[4] G. Singh, D.S. Chauhan, D. S., "Simulation and Modeling of Hydro Power Plant to Study Time Response during Different Gate States," in (IJAEST) International Journal of Advanced Engineering Sciences And Technologies, 2011,Vol. No. 10, Issue No. 1, pp. 042 – 047.

[5] IEEE Working Group., "Hydraulic turbine and turbine control models for system dynamic studies," in IEEE Trans on Power Syst., 1992, 7, pp.167–79.

[6] J. Tiwari, et al., "Modelling and Simulation of Hydro Power Plant using MATLAB & WatPro 3.0.," in Intelligent Systems and Applications, 2015, 08, pp. 1-8. DOI: 10.5815/ijisa.2015.08.01.

[7] Y.CH. Choo, K.M. MUTTAQUI, M. NEGNEVITSKY, "Modelling of Hydraulic Governor Turbine for Control Stabilisation," in ANZIAM Journal, 2007, 49, pp. C681-C698.

[8] O.P. Malik, G. S. Hope, G. Hancock, L. Zhaohui, Y. E. Luqing and W.E.I Shouping, "Frequency measurement for use with a microprocessor-based water turbine governor," in IEEE Trans Energy Conv, 1991, 6, pp.361–6.

[9] D.G. Ramey, J. W Skooglund, "Detailed hydro governor representation for system stability studies," in IEEE Trans on Power Apparatus and Systems, 1970, 89, pp.106–12.

[10] M.A. Bhaskar, "Non Linear Control of STATCOM," proc. of IEEE Internaltional Conference on Recent Trends in Information Telecommunication and Computing, India, 2010, pp. 190-195. DOI: 10.1109/ITC.2010.79.

[11] M. Sattou, "Simulation Model of Hydro Power Plant Using MATLAB/Simulink," in Journal of Engineering Research and Applications. 2014, Vol. 4, Issue 1, pp. 295-300.

[12] P. Hocko, M. Novak, M. Kolcun, Z. Čonka, et al., „Influence of photovoltaic power plants on the power system to the prediction for year 2020", proc. of 14th International Conference on Environment and Electrical Engineering (EEEIC), Krakow, Poland, 2014, pp. 116-121.

[13] J. Kurimský, R. Cimbala, I. Kolcunová, "Multi-scale decomposition for partial discharge analysis," in Przeglad Elektrotechniczny, 2008, Vol. 84, no. 9, pp. 169-173.

[14] I. Kolcunová, "Influence of electromagnetic shield on the high frequency electromagnetic field penetration through the building material," in Acta Physica Polonica A, 2017, Vol. 131, no. 4, pp. 1135-1137, ISSN 0587-4246.

[15] A. Acakpovi, E.B. Hagan, F.X. Fiftatin, "Review of Hydropower Plant Models," in Int. Journal of Computer Applications, 2014, Vol. 108, No. 18, pp. 33-38. DOI: 10.5120/19014-0541.

[16] P. Girovsky, J. Timko, J. Zilkova, et al., "Neural Estimators for Shaft Sensorless FOC Control of Induction Motor," Proc. of 14th International Power Electronics and Motion Control Conference (EPE-PEMC), Ohrid, Macedonia, 2010. DOI: 10.1109/EPEPEMC.2010.5606907.

[17] P. Fedor, D. Perduková, "Fuzzy Model Based Optimal Continuous Line Controller," Proc. of the 8th Int. Scientific Symposium on Electrical Power Engineering Elektroenergetika, Stará Lesná, 2015, pp. 404-407. ISBN 978-80-553-2187-5.

[18] D. Perduková, P. Fedor, "A Model-Based Fuzzy Control of an Induction Motor," in Advances in Electrical and Electronic Engineering, VSB-Technical University of Ostrava, 2014, Vol. 12, Iss. 5, pp. 427-434. ISSN 1336-1376.

[19] D. Perduková, P. Fedor, V. Fedák, S. Padmanaban, "Lyapunov Based Reference Model of Tension Control in a Continuous Strip Processing Line with Multi-Motor Drive," in Electronics, 2019, Vol. 8, Iss. 1. ISSN 2079-9292. Doi:10.3390/electronics8010060.

[20] A.M-H. German, P.M. Sa´ad, I.J. Dewi, Modeling and Controling Hydropower Plants, Springer-Verlag, London: Springer, 2013, ISBN 978-1-4471-6221-6.

Assessment of the Primary Load-Frequency Control Operation by Considering the Effect of Photovoltaic Power Plants

Zsolt Čonka, Pavol Hocko
Department of Electric Power Engineering
Technical university of Košice
Košice, Slovakia
zsolt.conka@tuke.sk, pavol.hocko@tuke.sk

Abstract — **In this paper, the influence of different volume of power installed in the Photovoltaic Power Plant on the support service activity is simulated on the power system mathematical models and it analyses the impacts on their quality and reliability.**

Keywords: component; formatting; style; styling; insert (key words)

I. Introduction

To ensure stable operation of the power system, Transmission system operator (TSO) regulates power on sources according to instantaneous consumption in a system that is partly predictable but enters the control as unknown. The current rapid development of renewable energy sources, and particularly of photovoltaic power plants (PPP), has caused the system to increase the proportion of installed power whose entire capacity must be redeemed and delivered to the grid. The problem of the increase in RES is the lack of electricity supply compared to conventional energy sources. PPPs energy supply is partly predictable (past year statistics from the same period, weather forecasts, etc.) but due to variable weather, the performance of PPP is changing considerably, causing a disruption to the power balance of electricity generation. This phenomenon has a detrimental effect on network management capabilities and increases the demand for support services and reserves held on conventional sources. In this paper, the influence of different volume of power installed in the PPP on the support service activity is simulated on the power system mathematical models and it analyses the impacts on their quality and reliability. In this paper, the influence of different volume of power installed in the PPP on the support service activity is simulated on the power system mathematical models and it analyses the impacts on their quality and reliability.

Support for the development of renewable energy sources brought an increased in the share of PPP in the total installed capacity, which currently (September 2018) accounts for about 4.0% of the total installed capacity of the SR. This value is about 660 MW of installed power in solar panels.

Specifically, the characteristic feature of PP is their daily production of electricity, which, when neglected by mechanical contamination of the panels, is directly proportional to the intensity of the solar radiation. On bright summer sunny days, when PPP output is maximal, the power-delivery curve has a characteristic course with an exponential increase from about 5:00 in the morning, with a peak at around 13 o'clock and an exponential drop to zero at around 22 o'clock (Fig. 1). The power output curve is smooth with no sharp changes. In case of variable weather conditions (cloud transition over the PPP area), the power delivered varies in the order of tens of percent of installed power. These weather conditions are applied in MODES as different sets of block parameter type parameters - power source for the PPP model. The analysis of the output power of the two PPP systems on clear and cloudy days is according to [49] in Fig. 1. This is the performance of the 24-hour performance on the two PPP systems installed on different places in Slovakia. The installed PPP performance was, according to [49], 550 MW for PPP1 and 150 MW for PPP2. The performance of the PPP has been chosen to represent the relative share of the PPP in the installed power equivalent to the actual share of the PPP in the power system of Slovakia. The source type parameters were set for the summer day for sunny and cloudy day.

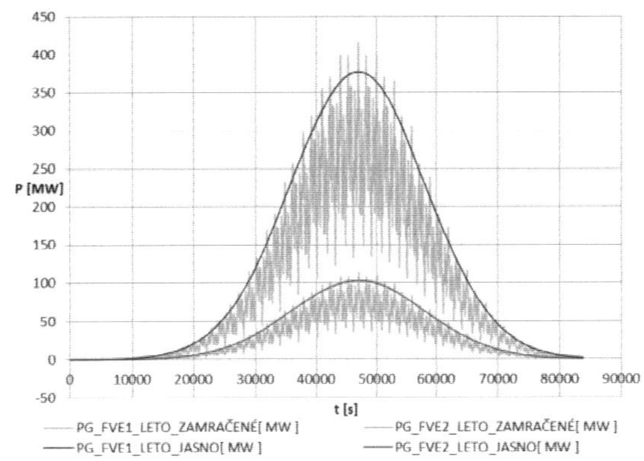

Figure 1 The generation during a clear and cloudy day [2]

Comparing the two power source models for PPP, there is a significant fluctuation in the delivered power from the PPP over time on a cloudy day. These changes are caused by the transition of the cloud, which changes the intensity of the solar radiation and consequently the delivered power from the PPP. The change in the delivered power causes a change in the frequency variation and the transmitted power in the system that activates the Primary frequency reserve PFR and thus changes the deviation of the cross-border profile. The high share of PPP in the power system thus imposes increased demands on support services and on the power reserves that need to be maintained for the proper functioning of these frequency restoration reserves.

The content of the following part of paper will examine the impact of RES (in this case, we will only talk about PPP for their characteristic dependence on electricity production and the high share of RES. Wind power plant can be neglected in this case for their minimum production share in Slovak power system).

II. Impact of RES on the dynamics of Power system of SlovakiA

The aim of the paper is to analyze the process of activation and activity of the PFR on the power system of Slovakia. For the purposes of modeling, the system was modified and supplemented by other lines and nodes in the surrounding TSOs to ensure the maximum possible value of the output data from the model based on available input data.

The system is monitored by PFR activity, which is activated during modeling of long-term dynamics (24h). The impact of PPP is modeled on several levels. The primary task is to compare the quality of PFR in case of favorable and unfavorable weather conditions. Subsequently, the share of production from PPP will be increased in an effort to model the impact of PPP on the system and PFR according to the forecasted development of renewable energy sources and to determine the anticipated boundary of the PPP share for reliable operation and quality of PFR in the power system.

A. Modificated model of Slovak power system

The PS SR system model was created in MODES program. All parameters of the main network elements as well as the dynamic parameters of the control elements use approximated data. From the available data, the selected parameters of the main transmission system elements were imported into the MODES program

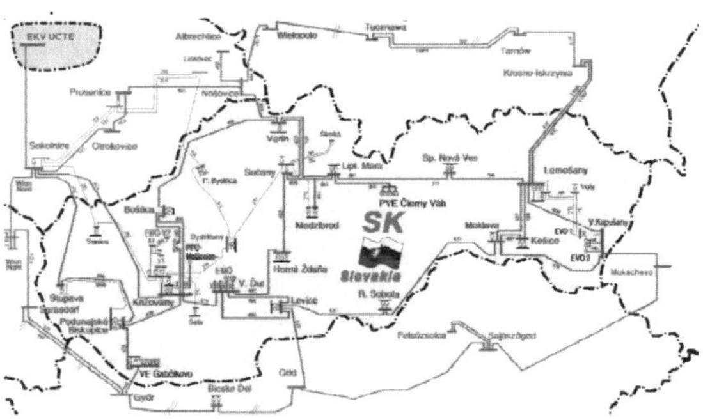

Figure 2. Modified model of Slovak power system

In order to achieve the most accurate dynamic results, the system model was supplemented with nodes and branches of neighboring countries (the so-called first loop) - modeled as separate ROs. Fig. 2 represents a single-pole scheme of a model network created in MODES.

Surrounding TSOs interconnected systems are modeled without resources with set balance value. The equivalent ENTSO-E node represents a part of the continental European interconnected system (SZKE). A ENTSO-E node operates a single block with an installed capacity of 241.366 GW representing conventional sources within the modeled interconnected area. The installed power and consumption of the equivalent system does not represent the entire continental European synchronous zone, only part of it as equivalent to the PPP involved in the modeled system. The block type parameters are set to represent a similar dynamic to the interconnected SZKE system (PFR, activation rate, etc.). The block is connected to the modeled SR system by an equivalent line at the 400 kV voltage level to the Kriz400 node.

B. Influence of Slovak PPP on P / f stability in system

The share of production from PPP on consumption in the Slovak republic power system during the day (24h) is shown in Fig. 3. The proportion of production in this case is set according to [5]. The DDZ is set at the sampling nodes for the average daily load during the summer day. The source of the PPP block (which represents the change in sunlight intensity during the day)

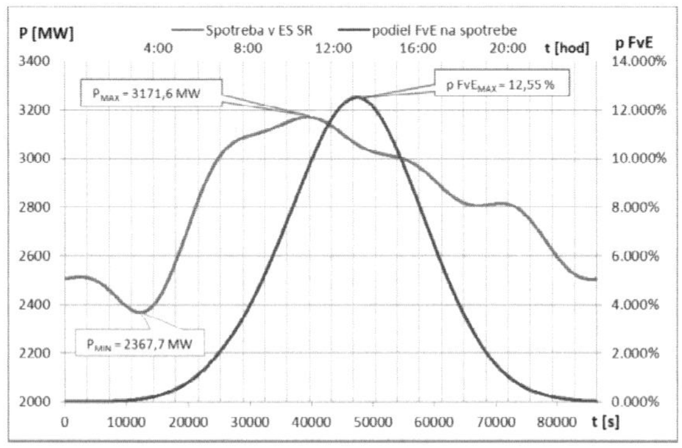

Figure 3. Share of production from PPP to consumption in SR PS

Consumption model during the day - DDZ was designed according to [5]. The difference between minimum and maximum consumption is ΔPS = 803.9 MW. The maximum share of PPP production in consumption reached 12.55% at 13:10. The outputs from the MODES program correspond to the real values in the PS of SR during the day when the minimum consumption was measured in the system (29 July 2017).

The effect of PPP on the system is more pronounced, resulting in a more significant system frequency deviation compared to active PRV in SRV assessment.

Thus, the time interval from 11:00 am to 3:00 pm was chosen, the average July day when, according to [5], system consumption is in its annual lows, production from PPP is at its highest level in the year, and the share of PPP in consumption is in the system at said time interval in the range of 10 to 12%. The course of the frequency variation during the conditions described above for the case of cloudy and clear day is shown in Fig. 4. The effect of cloudiness is particularly evident on frequency variations. In the case of a clear day when the performance of the PPP power supply is easily predictable, the frequency deviation is in the order of mHz (Fig. 4 shown in blue). The red curve represents the frequency deviation for the cloud day. The maximum frequency deviation value is Δf_{MAXobl} = 86.18 mHz. The black dashed line indicates the frequency deviation limit (± 70 mHz), when according to [1], the APS quality assessment is performed.

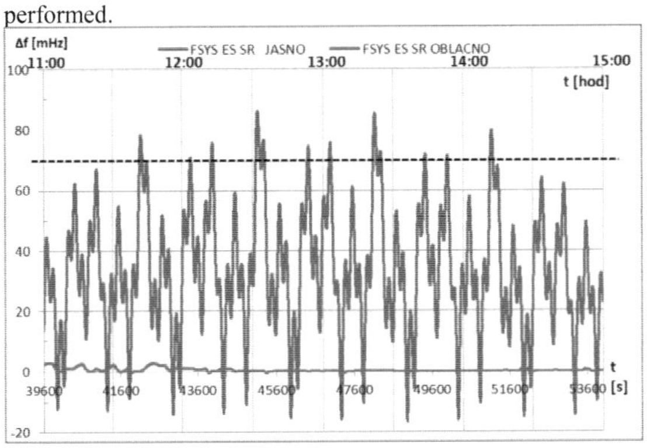

Figure 4. Frequency deviation course during cloudy and clear day

C. Primary Freguency Reserve Quality Assessment

The PFR Quality Assessment was conducted at the PS SR during the 24-hour dynamics modeling. Several scenarios have been modeled for this. The basic prerequisite for the evaluation of PFR quality on the blocks that are involved in this Support services (SS) is the transition of cloud through PPP, which causes fluctuation of the delivered power, which in relatively short time cycles (due to modeling of long-term dynamics of the day) disturbs the power balance in the system and causing deviation of the frequency and the agreed balance value [9]. These conditions place higher demands on the quality of block management and thus on the provision of PFR. Frequency deviation due to PPP under adverse weather conditions increases with the proportion of PPP production per consumption (Figure 3). Therefore, for the sake of clarity, only the time period of the day when the PPP impact is most significant will be monitored.

The RDP quality assessment was carried out on all RDP blocks. There are 21 blocks in total. An extract from the analysis of PFR program MODES of individual blocks providing PFR is in table Table 1.

Table 1. Analysis of PFR on resources in PS of SR

Blok	PRR+ [MW]	PRR- [MW]	N_{tMIN} [MW]	N_t [MW]	N_{tMAX} [MW]
TG31	20	-20	99	115	219
TG32	20	-20	99	115	219
TG41	20	-20	99	115	219
TG42	20	-20	99	115	219
1MKZ1	20	-20	99	169	219
1MKZ2	20	-20	99	169	219
2MKZ1	20	-20	99	169	219
2MKZ2	20	-20	99	169	219
TG1	4	-4	40	54	110
TG2	4	-4	40	54	110
TG5	4	-4	40	54	110
TG6	4	-4	40	54	110
TG21	5	-5	40	95	110
TG22	5	-5	40	95	110
TG23	5	-5	40	95	110
TG24	5	-5	40	95	110
ENOB1	5	-5	40	62	110
ENOB2	5	-5	40	62	110
ENOB3	5	-5	40	62	110
ENOB4	5	-5	40	62	110
OST	57	-57	100	401	600
Total	**273**	**-273**	**1372**	**2381**	**3672**

D. Quality assessment of RDP during normal operation of the EC SR

The analysis of system dynamics modeling on a cloudy day is based on the time interval from 11:00 to 15:00. At this time, the effect of the PPP is most pronounced and the frequency variation is greatest. The evaluation takes place for 15 minute intervals in which the frequency deviation is at least 0.07 Hz. The list of PFR quality evaluation is shown in Tab. 2 [10]. Analysis of RDP quality assessment was performed for model 24h cloud summer day on the system according to Figure 2. with PPP with total installed capacity = 660 MW.

Table. 2 Results of PFR Quality Assessment

Blok	B12	B13	B14	B15	P_{vyp}
TG31	219.182	125.864	119.975	159.380	25.000
TG32	54.779	125.837	119.957	159.317	25.000
TG41	54.779	125.837	119.957	159.317	25.000
TG42	54.779	125.837	119.957	159.317	25.000
1MKZ1	34.312	104.348	120.589	143.073	25.000
1MKZ2	34.313	104.364	120.606	143.047	25.000
2MKZ1	34.312	104.348	120.589	143.073	25.000
2MKZ2	34.313	104.364	120.606	143.047	25.000
TG1	20.698	47.896	57.718	68.278	10.000
TG2	20.698	47.896	57.718	68.278	10.000
TG5	20.698	47.896	57.718	68.278	10.000
TG6	20.698	47.896	57.718	68.278	10.000
TG21	14.282	37.059	45.669	63.298	12.500
TG22	14.282	37.059	45.669	63.298	12.500
TG23	14.282	37.059	45.669	63.298	12.500
TG24	14.282	37.059	45.669	63.298	12.500
ENOB1	17.225	21.177	35.082	47.420	12.500
ENOB2	17.225	21.177	35.082	47.420	12.500
ENOB3	17.225	21.177	35.082	47.420	12.500
ENOB4	17.225	21.177	35.082	47.420	12.500
OST	133.328	331.654	375.730	494.180	142.500

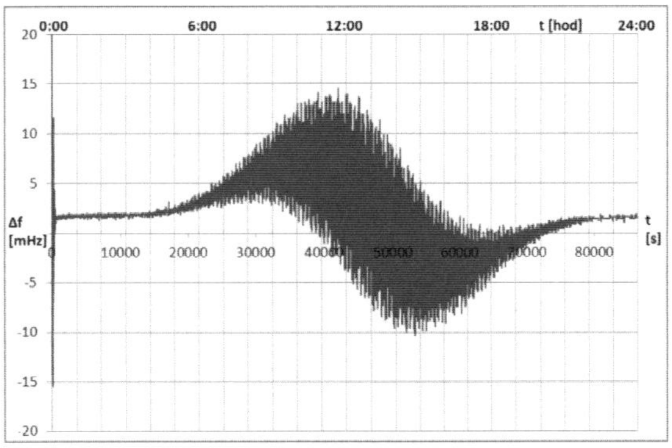

Figure 5. Deviation of system SVK PS frequency during cloudy day

PPPs operating in a system with installed capacity in the order of hundreds of MW do not have a significant impact on the stability of the system in the case of a cloudy day when the production prediction is difficult to achieve. The PFR is activated throughout the interconnected system and thus substantially compensates for the disruption of the power balance in the system due to the variable production of PPP [8]. The following chapters will be devoted to analyzing the impact of PPP installed in the surrounding TSOs on the quality provided by the Support Services. The total installed power of the PPP in the interconnected system moves in the order of tens of GW (see Table 3).

E. *Influence of Slovak PPP on P / f stability in system*

It follows from the definition of the principle of PFR activity that the primary regulation of active power in the whole interconnected system operates at the frequency stabilization. Therefore, the influence of PPP on surrounding TSOs is compared in this section.

The installed performance in PPP varies across EU countries. The EU as a whole has committed itself to raising the share of renewable energy to 20% of final electricity consumption by 2020, while the share of renewable energy in transport should reach 10% [12]. Most countries, including Slovakia, have the development of renewable energy sources in their national action plans. The installed capacities of individual countries whose TSOs belong to the ENTSO-E (TSO are part of the interconnected system) are shown in Tab. 3. The installed capacity of the blocks is according to [2].

According to [4], the total installed PPP power in the modeled system is set to PPPP = 18.8 GW. It is an PPP in the northern part of continental Europe (Poland, Germany, Belgium, Czech Republic, Slovakia, Hungary). PPP in Poland and Hungary have been neglected for small installed power. The total installed power in PPP is thus equivalent to the modeled system in terms of consumption, which is in the modeled system PODB = 190 GW[7]. The maximum PPP production share during a clear summer day of consumption thus constitutes approximately 7% of the PPP surrounding TSOs.

Table 3. List of PPPs operating in a modeled system

Country	Block name	Node name	S_N [MW]	Nt_{MIN} [MW]	Nt_{MAX} [MW]
Česká rep.	FV_CZ	FV_SZKE	1200	0	950
Germany	FV_DE	FV_SZKE	4000	0	4000
	FV_50H	FV_SZKE	2000	0	2000
	FV_TNT	FV_SZKE	8000	0	8000
Belgie	FV_BE	FV_SZKE	1000	0	600
France	FV_FR	FV_SZKE	2000	0	1600
Slovakia	FV_V	FV_Leme	115	0	144
	FV_S	FV_Medz	252	0	315
	FV_Z	FV_PodBi	144	0	180
Total			18711		17789

Where:

S_N - installed PPP power.

N_{tMIN} - Minimum Power of the Power Supply

N_{tMAX} - the power of the power supply (maximum power of the block)

The power output from the PPP in the surrounding TSOs of the interconnected system was in this case taken into account by adding equivalent blocks representing the production from the PPP in each country. PPPs are connected to an equivalent node representing SZKE. The dynamic parameters of the individual components of the blocks representing the PPP are according to [2]. The course of the balance on a clear day (24 h) is shown in Figure. 6. In this case, the balance has a minimum deviation. The performance of the power designated as SFR (secondary frequency reserve) represents the power activated by the equivalent block (UC1REG) representing SFR in the equivalent area of the SZKE. The SFR value was calculated according to [3] to 2% of the installed power in the modeled system. The total PFR in SZKE is approximately 3000 MW. TFR (therciary frequency reserve) is also involved. The automatics representing TFR changes the entered turbine power of the EKVIV block in case of SFR on the UC1REG control block. The course of the PS of SR and SZKE balance and the performance of the UC1REG block (SFR activated power) is shown in Figure. 6

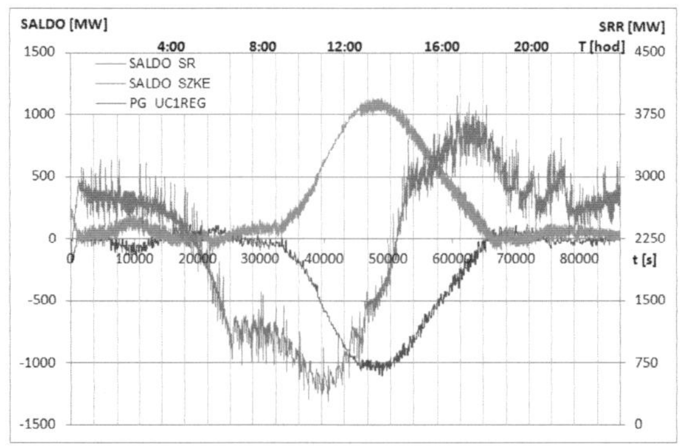

Figure 6. Performance of activated PFR in SZKE

The delivery of power from the PPP from the entire interconnected area does not have a more pronounced impact on the modeled system in terms of the rate of activation of the regulatory reserve on interconnected systems during the cloudless day [11]. For comparison, Figure. 7 shows the frequency and power deviation of the sources (including the production from the PPP) in the control area of the SR modeled system over a 24 h clear day. Since the frequency deviation has not reached a value of at least ± 0.07 Hz for the entire period under review, the quality of the PSS quality assessment is not being performed according to [1], [19].

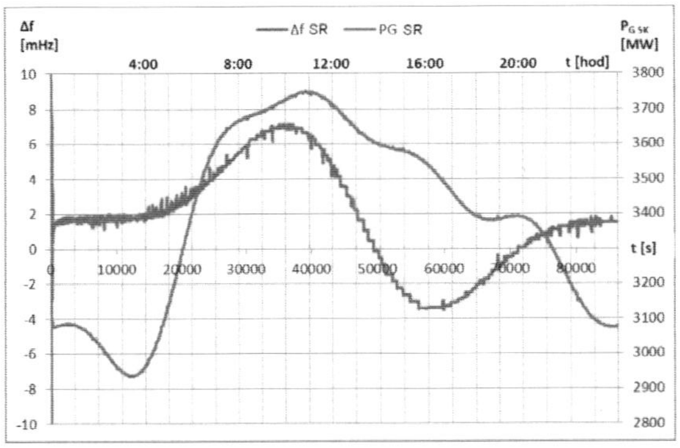

Figure 7. Course of deviation of frequency and power on blocks in PS of SVK

III. PFR Quality Assessment

The assessment of the performance of the primary power control activity taking into account the influence of the PPP in the surrounding TSO is described in this section of the article. The set parameters simulate cloud transition over the PPP and thus change the delivered power [12]. This phenomenon causes a disruption of the power balance in the interconnected system and consequently a frequency and balance deviation. The PFR responds to this change in the entire interconnected area. The P / f regulation is represented in the system by blocks in PS SVK and RO SZKE. Resources in these areas compensate for excess energy at a time when PPP production is the largest and thus reduces the power on blocks that are involved in it within the block control band [13]. The frequency deviation course between 10:00 and 16:00 is shown in Figure. 8. At this time interval, the frequency deviation exceeds 100 mHz. The PFR quality assessment is performed at 15 minute intervals when the frequency deviation exceeds ± 0.1 Hz (in the graph, the deviation limit is indicated by a horizontal dashed line). The course of the frequency deviation in the system is without the intervention of TFR and SFR for the purposes of PFR evaluation.

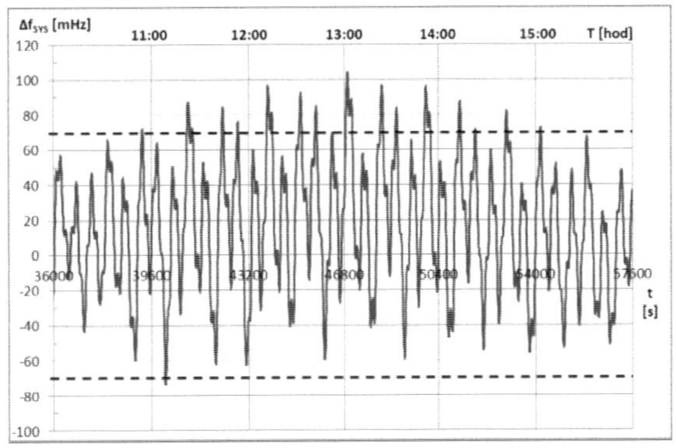

Figure 8. Frequency deviation in TSO SVK in the period from 10:00 to 16:00 – cloudy

The PFR quality assessment was carried out in a total of five business hours when the frequency deviation exceeded ± 0.07 Hz (see Figure 8). The evaluation was carried out at power plant units as a whole. The results of the PFR quality assessment are shown in Table 4. All blocks meet the quality criterion of [1] in all business hours monitored.

Table 4. Evaluation of Quality of RDP Resources in the Extended PS SR Model

Block	B11	B12	B13	B14	B15	P_{out} [MW]
TG31	29,607	25,907	28,958	29,016	25,906	25
1MK Z1	33,614	30,389	33,013	32,438	31,233	25
TG1	29,105	26,101	27,004	29,201	27,479	10
TG21	20,053	20,256	20,032	19,064	21,521	12.5
ENO B1	21,840	19,773	21,956	21,784	20,142	12.5
Other	166,749	151,730	165,301	160,689	155,804	142,5

Where:

b - Linear regression line directive evaluated at 15 minute intervals after G2 failure

|B| - the average of the straight line linear regression line direction values for a given business hour of the day.

P_{out} - 50% five times the PFR offered

Quality assessment was performed only in the case of a cloud day modeled throughout the interconnected system, where the impact of PPP is most pronounced. In other cases, the condition of minimum frequency variation according to [1] has not been fulfilled throughout the period under review. The PFR quality assessment is shown in Table 5.

Table 5. Evaluation of PFR quality in case of EMO failure

Blok	B11	B12	B13	B14	B15	B16	P_{VYP}	PRR+
TG31	29,656	31,068	28,556	34,279	28,728	26,412	25	10
TG1	29,334	30,050	27,352	28,077	28,109	27,182	10	4
TG21	16,443	18,294	17,734	18,841	17,681	17,355	12,5	5
ENO B1	20,471	20,158	19,380	20,325	19,214	19,839	12,5	5

The average of the absolute values of the directions calculated for each business hour of the day from 15 minute intervals when the absolute value of the frequency deviation exceeded 0.1 Hz that meet the quality criterion is shown in green.

Conclusion

The role of modeling dynamics was to analyze the impact of Photovoliatic power plants (PPP) on long-term, resp. medium-term dynamic behavior of the interconnected system, special resources working in TSO of Slovakia. For this purpose, the condition of quality assessment under [1] has been used. The quality of regulation of the power of the systems working in the system depending on the changes in the frequency in the system and the balance in TSO (evaluation of the quality of support services in the real system) is evaluated for various transients in the system. In the first step, the normal operation of the system is analyzed throughout the day with the consideration of Day-ahead load diagram. The aim was to compare the current day for different weather conditions. Above all, the comparison for clear day (the production prediction of the PPP is quite accurate and the dispatcher can provide sufficient SFR) and the cloud day when the change of generation from the PPP is relatively fast and difficult to predict. The latter case occurs under specific conditions in the system when passing a dense cloud over a territory with a high installed power density in PPP.

The principle of PFR quality assessment is to ensure the required performance value with respect to the frequency change in the interconnected system. In this case, all modeled blocks satisfy the condition defined in [1]. It should be noted here that any failure to comply with the condition does not necessarily imply an error in the power control process on the block or any part of it. As introduced in the individual dynamic simulations, the system's power control functioned correctly and at the required time interval according to the control principle.

Acknowledgment

This work was supported by the Scientific Grant Agency of the Ministry of Education of Slovak Republic and the Slovak Academy of Sciences by the projects VEGA No. 1/0372/18.

References

[1] SEPS a.s., Dokumenty, Technické podmienky, Technické podmienky prístupu a pripojenia, pravidlá prevádzkovania prenosovej sústavy, Dokument B – Technické podmienky prístupu a pripojenia do PS, online: http://sepsas.sk/seps/Dokumenty/TechnickePodmienky/TP-B-112016.pdf

[2] J. Kříž, K. Máslo, and A. Kasembe, "Impact of Distributed Generation on Grid Protection And Voltage Control," Paper No 0007 Page 1 / 4 CIRED Workshop - Lisbon 29-30 May 2012

[3] S1 – Strategy 1: Frequency and power control and performance, archive, operational handbook

[4] K. Máslo, and M. Pistora, "Dlouhodobá Dynamika Soustavy s Rozptýlenou Výrobou Včetně OZE," Tábor 9. a 10.11.2010 Konference ČK CIRED 2010 K. Máslo, M. Pistora - Sekce č. 4 / 4 © ČK CIRED 2010

[5] Annual report SEPS a.s. on-line: http://sepsas.SK/seps/DispSkladacka2012.asp?kod=559

[6] J. Zbojovský, and M. Pavlík, "Calculating the shielding effectiveness of materials based on the simulation of electromagnetic field," In: Elektroenergetika 2017. - Košice : TU, 2017 S. 369-372. - ISBN 978-80-553-3195-9

[7] B. Dolník, J. Kurimský, "Contribution to earth fault current compensation in middle voltage distribution networks," PRZEGLD ELEKTROTECHNICZNY (Electrical Review), ISSN 0033-2097, R. 87 NR 2/2011

[8] B. Dolník,"Contribution to Analysis of Daily Diagram of Supply Voltage in Low Voltage Network working days versus non-working days," In: Electric Power Engineering (EPE) 2015: May 20-22, 2015, Czech Republic. p.373-376. ISBN 978-1-4673-6787-5.

[9] J. Džmura, J. Petráš, J. Balogh, M. Bernát, "Modeling and Computer Simulation of Electrical Separation," 17th International Scientific Conference on Electric Power Engineering (EPE), MAY 16-18, 2016, Prague, Czech republic, p. 537-542, ISBN:978-1-5090-0908-4

[10] J. Džmura, J. Petráš, J. Balogh, J. Kurimský, R. Cimbala, I. Kolcunová, B. Dolník, M. Kolcun, "Separation of solid particles from flowing gases by AC high voltage," Journal of Electrostatics, Vol: 88, AUG 2017, P. 158-164, ISSN: 0304-388

[11] M. Pavlík, I. Kolcunová, L. Lisoň, "Measuring the shielding effectiveness and reflection of electromagnetic field of building material" 2015. In: Electric Power Engineering (EPE). - Ostrava VŠB-TU, 2015 P. 56-59. - ISBN 978-1-4673-6787-5

[12] J. Zbojovský, L. Kruželák, M. Pavlík, "Impedance Spectroscopy of Liquid Insulating Materials," In: Proceedings IEEE International Conference AND Workshop in Óbuda on Electrical and Power Engineering. - New York (USA) Institute of Electrical and Electronics Engineers s. 249-253. ISBN 978-1-7281-1153-7

[13] M. Kolcun, L. Beňa, J. Rusnak, "The solution of optimisation problems in the operation control of the electric power system using SOMA algorithm," PRZEGLAD ELEKTROTECHNICZNY, Volume: 84, Issue:9, Pages: 70-73, 2008, ISSN: 0033-2097

[14] M. Kolcun, K. Máslo, "Load-frequency control management in island operation," ELECTRIC POWER SYSTEMS RESEARCH, Volume: 114, Pages: 10-20 DOI: 10.1016/j.epsr.2014.03.030, 2014

[15] J. Rusnak, L. Bena, M. Kolcun, "Experiences with Examination of Dispersed Sources Impact on Distribution Power System," Electric Power Engineering 2010, proceedings of the 11th international scientific conference, 4.-6.5.2010, Brno, Czech Republic. 2010 P. 219-222, ISBN 978-80-214-4094-4

[16] S. Kušnír, L. Beňa, M. Kolcun, "The impact of FACTS devices to control the load flow," Electric Power Engineering 2010, proceedings of the 11th international scientific conference, 4.-6.5.2010, Brno, Czech Republic. 2010 P. 99-103. ISBN 978-80-214-4094-4

[17] M. Kolcun, D. Hlubeň, L. Beňa, N. Djagarov, Z. Grozdev, "Transformer use for active power flow control in the electric power system," In: Environment and Electrical Engineering, 2010 9th International Conference, Czech Republic, 16-19. May 2010, IEEE, 2010 P. 246-249. ISBN 978-1-4244-5374-0

[18] A. Hromadka, J. Jirickova, Z. Martinek, "Analysis of the unplanned Shutdowns and Estimation of Reliability in Temelin," Proceedings of the 2018 19th international scientific conference on electric power engineering (epe), IEEE, 2018, DOI: 10.1109/EPE.2018.839595

[19] J. Kurimsky, R. Cimbala, I. Kolcunova, "Multi-scale decomposition for partial discharge analysis," Prz. Elektrotechniczny, vol. 84, no. 9, pp. 191–195, 2008.

[20] P. Fedor, D. Perduková, "Fuzzy Model Based Optimal Continuous Line Controller. In: Proc. of the 8th Int. Scientific Symposium on Electrical Power Engineering ELEKTROENERGETIKA 2015. Stará Lesná, 2015, pp. 404-407. ISBN 978-80-553-2187-5

[21] D. Medved, M. Klesc, "Modeling of Electrical off-grid Network in the Simscape Power Systems," Proceedings of the 9th international scientific symposium on electrical power engineering, pp: 465-470, 2017, ISBN 978-80-553-3195-9

Stabilizing the Control of a Plant Material Drying Process in Off-Grid Power Systems

Natalia Shamarova
Konstantin Suslov
Dmitriy Gerasimov
Ilia Shushpanov
Irkutsk National Research Technical University
Irkutsk, Russia
souslov@istu.edu

Igor Altuhov
Irkutsk State Agrarian University
Irkutsk, Russia
altukhigor@yandex.ru

Pio Lombardi
Fraunhofer Institute IFF
Magdeburg, Germany
pio.lombardi@iff.fraunhofer.de

Przemyslaw Komarnicki
University of Applied Science Magdeburg-Stendal,
Magdeburg, Germany
przemyslaw.komarnicki@iff.fraunhofer.de

Abstract — **The study aims to develop an effective power supply control system to maintain the drying process parameters and stability in the off-grid power systems. The paper is focused on the process of wild plant material drying based on the electric heating technology. In order to ensure the effective control and high quality of drying, the electric heater operation stability should be provided with the supply voltage and current values set. This stability of the wild plant material drying process should be ensured by an appropriate control. The remoteness of the central power systems from the sites of plant material collection causes the need to use various power sources that are characterized by variability of parameters. The paper addresses the problem of stabilizing operating parameters of an off-grid power system to improve the efficiency and quality of drying wild plant material when using the energy storage control system. A net-zero test system is investigated for illustration.**

Keywords: electrical technologies, off-grid power supply system, storage devices

I. Introduction

The Russian population is mostly concentrated in a few macro areas, while about 60% of the Russian territory is poorly populated. These areas are generally located in the northern part of the Russian Federation and are far from the main national energy infrastructures (electric, gas and heat). In these Isolated Micro Grids (IMG) the energy (thermal and electric) is typically supplied through conventional generation systems such as diesel generators, gas turbines and steam turbines which operate either in a cogeneration process or together with boiler systems to generate electricity and the heat. The fuel (liquid and solid) for the power plants is mostly supplied by trucks and river ships [1]. The IMG in Russia are not business-attractive for the power generator companies since the density of power demand is relatively low [2]. As a consequence of all these aspects, the reliability of the electric power supply is low and the generation costs are high in the Russian IMG.

In this context, the use of Renewable Energy Sources (RES) such as wind, solar and the local biomass may contribute positively both towards reducing the energy generation costs and increasing the quality of power supply [3]. However, due to the stochastic nature of wind and solar power generators, their integration into the Russian IMG is limited by the low capacity flexibility of the conventional power plants that are used [4]. These generally operate as a heat-driven cogeneration process and do not have any backup system for the thermal power generation. The Energy Storage Systems (ESS), both electric and thermal, may increase the integration of the power generated by RES, [5]-[7] and also reduce the generation costs and improve the quality of the power supply [8]. Nevertheless the ESS, mostly the electric ones, are still expensive, therefore from the system point of view, it is essential to plan the optimal energy storage capacity [8],[9].

Remote areas do not have significant reserves of available electrical capacity near the areas where wild plants grow. Therefore, for the processing and drying purposes, it is advisable to use renewable energy sources capable to generate power directly at the wild-growing plant collection sites.

The process of drying wild plant raw material discussed in the paper involves the electric heating technology. In order to ensure efficient control and high quality of drying, the electric heater operation should be kept stable with preset values of the supply voltage and current. This stability of the drying process should be provided by an appropriate control system. The remoteness of central power systems from the collection sites generates the need to use various power sources, and they are characterized by variable parameters.

The unstable operating parameters and conditions of the central power system can also be caused by a variety of reasons. Monitoring of the load curves revealed relatively slow changes in operating parameters that are associated with variations in power flow. Fast variations in the values of voltage and current in the power supply system result from constant load fluctuations caused by the activity of consumers involved in real-time load control, while in alternative energy sources (wind and solar), such variations are caused by stochastic generation and sudden changes in operating parameters of power system in emergency situations.

Operation of the power system as part of a large-scale electric power system allows, to a certain extent, leveling the operation instability owing to the adaptation capabilities of the entire system. A more serious problem arises in the off-grid low-capacity power systems with high penetration of renewable energy sources. In this context, a rational solution for such systems is the use of an energy storage device.

II. Description of the Technological Process

There are various methods of drying plant material. Their comparative characteristics are presented in Table 1. In the studies [8], we determined that infrared radiation is the most optimal method of the known artificial methods of processing and drying. This method has proved to be effective in many parameters, including drying time, maintenance of temperature conditions, quality of obtained products, environmental friendliness.

Table 1. Comparative Characteristics of Various Drying Methods

Index	**Drying methods**				
	Infrared	***Convection***	***Conduction***	***Sublimation***	***Micro wave frequency***
Quality of obtained product (BAS losses)	10-20 %	40-60%	40-60%	5-10%	18-20 %
Drying time, hour	about 4	8-10	4,5-5	10-20	about 4
Restorability	85-95 %	60-70 %	60-70 %	85-95 %	85-95 %
Residual moisture reached, %	10 – 12	8,0	10-15%	8,0	8-12
Storageability	Above 1 year	0.3 – 0,5 year	0.3 – 0,5 year	Above 1 year	above 1 year
Energy consumption for evaporation, kWh/kg	0,9-1,2	1,8-3,0	1,5 - 1,7	2,7-3,0	2-5
Enviromental safety	safe	safe	safe	Unsafe (freon leackage)	unsafe (micorwave radiation)
Metal intensity	Low	Medium	High	High	High
Maintenance complexity	Low	Medium	Medium	High	High

An important element of this method is the source of infrared radiation. The research in [10] determined that pulsed ceramic emitters have a specific effect on the processed material. The specific effect of pulsed infrared radiation on the plant raw material is associated with the intensification of biochemical transformation processes. This circumstance places high demands on the quality of electricity.

A laboratory and production facility was developed to identify the general methods for pulsed infrared processing and drying.

The process of drying wild plant material is proposed to be implemented in a pulse-discontinuous mode with reduction in the operating time of the emitters in each subsequent cycle (Fig. 1).

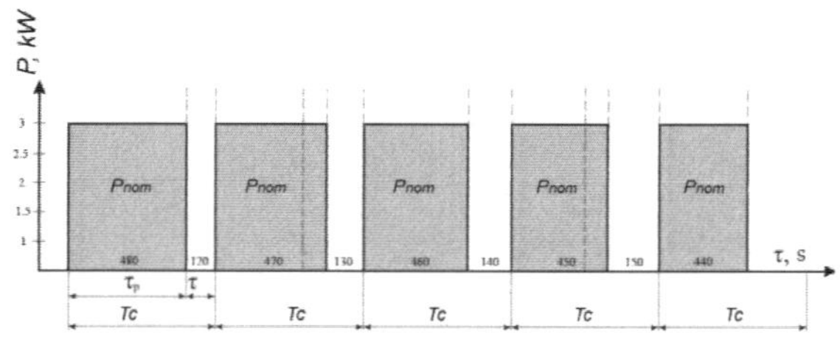

Figure 1 Pulse-discontinuous method of controlling infrared heating with a decrease in the operating time of infrared emitters in each subsequent cycle.

At the same time, the period of the T_c cycle, during which the infrared emitter is turned on and off, remains unchanged. Whereas at the beginning of the drying process, the amount of moisture in the material is maximum and it basically has a mechanical connection, during the period of intense heating, most of the moisture will be removed in the first cycles of the drying process with the minimum power consumed.

In the period of work τp, the process of controlling the surface temperature of a material according to the on-off principle implies its change along a saw-tooth curve near a predetermined value in the range from t_{min} to t_{max}, determined by the dead band of the regulator. Processing and drying of plant material is carried out in cycles of 10 minutes (600 s). In the first cycle, the operation of the emitters is τ_{p1} = 480 s, the period of a long pause τ_{pp1} = 120 s. The reduction in the period of τp in each cycle is 10 s. Accordingly, in the second cycle, τ_{p2} = 470 s, τ_{pp2} = 130 s, etc. The power density of the emitters under different power supply conditions was controlled with a single-channel software PID-controller. Before the drying plant starts to operate, the values of settings for the raw material heating temperature t_{max} and t_{min} are set in pulse-discontinuous mode. Scanning of temperature sensors is set taking into account the required measurement accuracy. The drying process lasts until the desired moisture is reached or according to a predetermined time. The temperature values of raw material and infrared emitters in various modes are archived on a PC. The total capacity of the production plant and all emitters used in the dryer is 35 kW.

III. Styling Analysis of the Power System Operating Conditions

The facilities to be considered as generators are: diesel generator, and solar panels. Consideration is also given to energy storage devices, since the renewable energy output is stochastic, their use is necessary to provide the required reliability of electricity supply to consumers of the off-grid systems. We consider an isolated (off-grid) system scheme presented in Fig.2. The experience gained in operating diesel generator reveals some serious problems when tuning the automatic control loops, namely:

1. Lack of a comprehensive approach, because power systems are considered separately from one another.

2. The problem of obtaining common algorithms for the power system control, which is related to the complexity of traditional mathematical tools.

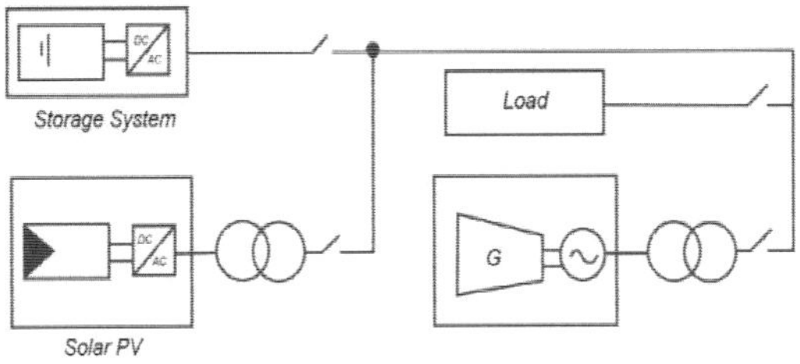

Figure 2. A scheme of an isolated (off-grid) system

The autonomous hybrid solar-diesel power plant is designed for fully automatic operation without human intervention. Maintenance functions include periodic external inspection, and control of basic operational characteristics.

The network inverter has a high efficiency and through it, power system receives power to cover the necessary load up to the maximum possible (rated). Storage inverters, in turn, charge storage systems and provide (or maintain) the required amount of electricity. The switching unit is connected to the network and storage inverters and automatically selects one of the operating conditions of the hybrid power plant.

This scheme ensures uninterrupted operation and automatically (in real time) determines the optimal condition, depending on solar radiation, charge of the storage systems and power requirements.

Figures 3-6 present monthly energy consumption for the existing scheme

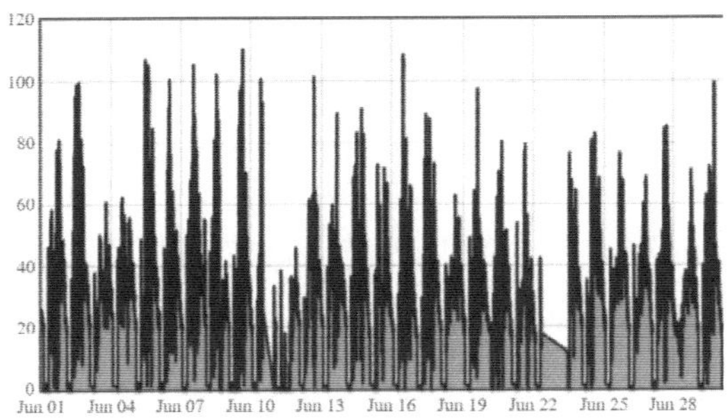

Figure 3. Electricity consumption, kW

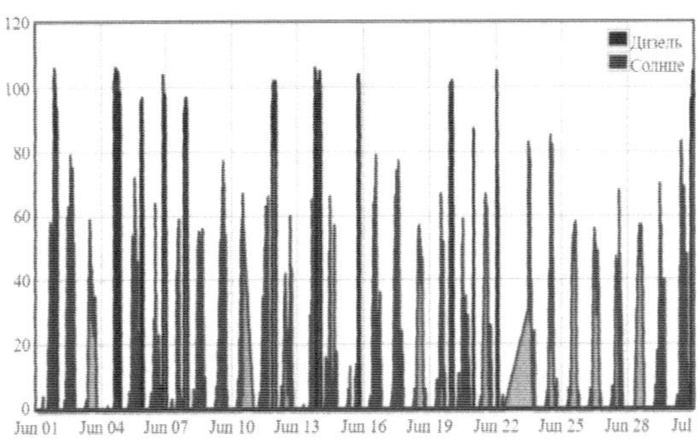

Figure 4 Storage device charging from solar plant and diesel generator, kW

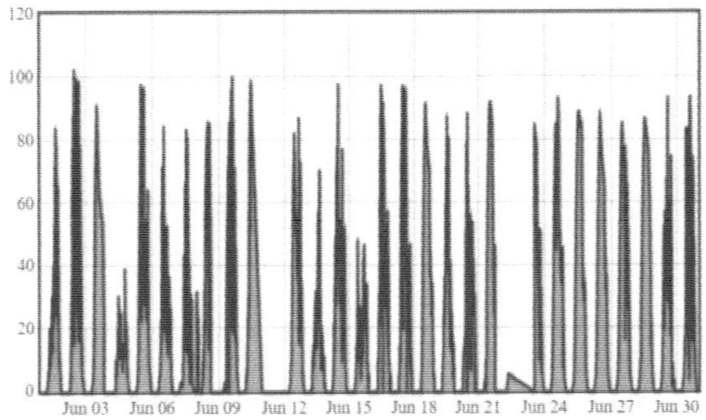

Figure 5. Electricity consumption from solar plants, kW

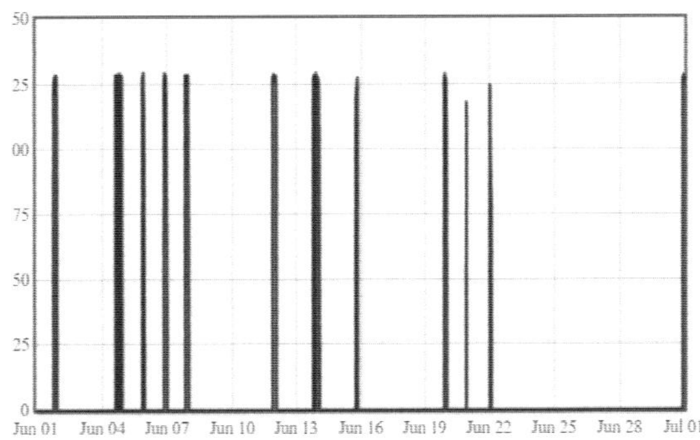

Figure 6. Electricity consumption from diesel generator, kW

IV. The Proposed Solution

An obvious way to improve the energy efficiency of such facilities of an off-grid system is to maximize the use of alternative and local energy resources. We proposed the use of wind turbines as an additional power source for the dryer of the plant raw material. An analysis of renewable energy sources, given the cyclical and seasonal nature of their operation, is indicative of the possibility of using the hybrid wind-solar plant.

The generation based on renewable energy sources is stochastic in nature, therefore, it is impossible to completely refuse from diesel generators when using it. These generators should be used as backup power sources.

Figure 7 demonstrates the proposed scheme of the diesel generator and wind-solar plant operation, considering the additional.

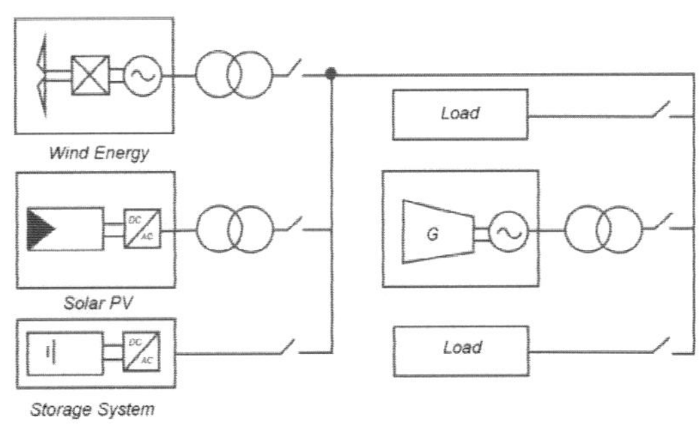

Figure 7. A scheme of an isolated (off-grid) system with eind energy systems

After the adoption of these plants, two block diagrams of joint operation of diesel generator and wind-solar plant (Fig. 8 and 9) have been developed.

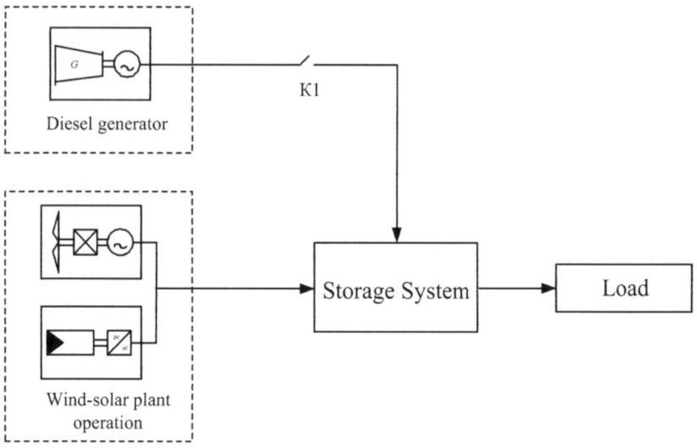

Figure.8 A bock diagram of the joint operation of a diesel generator and a wind-solar plant for an energy storage device.

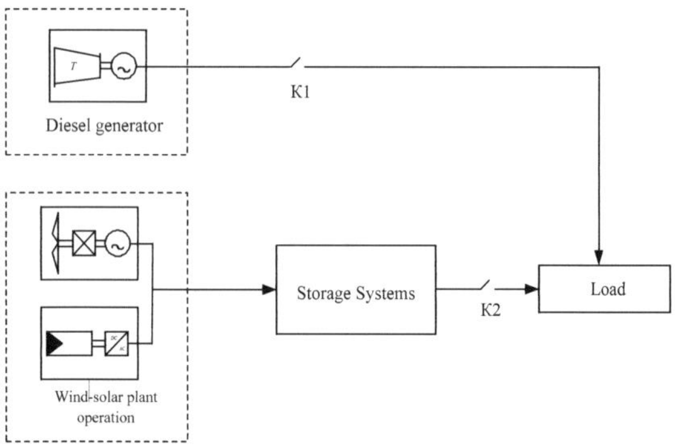

Figure 9. Separate operation of the diesel generator and wind-solar plant

In the first option (Fig.8), the wind-solar plant and the backup power source of the diesel generator are connected to and operate for energy storage devices, after which the load is supplied from them through inverters. In normal conditions, wind-solar plant is the main power source. A special controller (converter) is used to connect the wind generator to the network. In addition to the conversion, this unit will allow joint operation of solar panels and wind turbines. In emergency conditions, when the storage charge does not decrease below 20% of the rated value, the switch K1 is closed. Diesel generator is connected to the storage devices through rectifiers and delivers power current to the network and also ensures their charging. This happens until the charge reaches 80% of the rated value. After that, the switch K1 is opened and the proposed scheme continues to work in normal conditions.

In the second option (Fig. 9), in normal conditions, the wind-solar plant is also connected to the energy storage device, after which inverters are used to supply the load. In this case, in normal conditions, backup source, i.e. the diesel generator, does not work and the switch K1 is opened. In emergency conditions, if the storage device charge goes below 20% of the rated value, the switch K1 is closed and the switch K2 is opened. The power will be supplied to the equipment directly from the diesel generator. Operation in emergency conditions continues until the charge of the storage device reaches 80% of the rated value. In this case, the switch K1 is opened, and the switch K2 is closed. After that, the operation of this scheme continues in normal conditions.

Modeling of the presented options in the Matlab environment indicated that the first option is preferable for reliable and failure-free operation of the wind-solar-diesel system, given the technological features of the equipment and the location of the facility.

To determine the optimal capacities of solar panels and wind generator, we used a technique based on the following algorithm (Fig.10.):

• Set the climatic parameters of the region where the wind-solar plants are to be used (solar insolation by month, maximum and average wind speed by month);

• Set active power of the wind-solar plant and diesel generator, check their selection;

• Set active load of the electricity consumer;

• Set economic indices to analyze the effectiveness of adopting the alternative energy sources.

• Do calculations. As a result, we obtain generation-load diagrams which are used to analyze the effectiveness of the selected generation capacities.

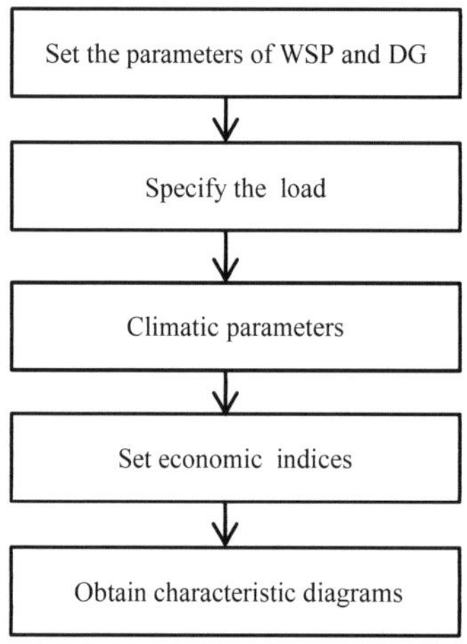

Figure 10. The calculation algorithm underlying the technique

Based on the calculation we select the capacities of solar panels and wind generators.

The cost-effectiveness of the proposed solutions is expressed in the reduced cost of fuel for diesel generators.

V. Conclusions

This paper is focused on the issues of uninterrupted power supply to the dryers of wild-growing plant material. Based on an analysis of climatic and geographical conditions, as well as solar and wind activity in the region, we have selected a set of generators necessary to ensure the process conditions. An analysis of renewable generation and consumer load curves has proved the economic viability of the hybrid wind- solar-diesel generator system.

References

[1] L. P. Konovalova. "Power Supply to Decentralized Consumers in Tomsk region on the Basis of Renewable Energy Sources": PhD Thesis , 2007, 202 p. [in Russian]

[2] I. Yu. Ivanova, S.P. Popov, A.N. Simonenko, T.F. Tuguzova. "Estimation of Markets for Renewable Energy Technologies in Russia's Eastern Regions", in Proc. Asian Energy Cooperation: Interstate infrastructure and energy markets. The 4th Int. Conf. AEC-2004, pp. 166-171.

[3] N. I. Voropai, K.V.Suslov, T.V. Sokolnikova, Z. A. Styczynski, P. Lombardi, "Development of Power Supply to Isolated Territories in Russia on the Bases of Microgrid Concept", Proceeding of 2012 IEEE Power & Energy Society General Meeting, 22-27 July 2012 San Diego, California USA

[4] R. Lasseter, et al., "The CERTS Microgrid Concept," Sacramento, CA, CEC Consultant Report P500-03-089F for the California Energy Commission, October, 2002

[5] M. Stadler, C. Marnay, A. Siddiqui, J. Lai, B. Coffey, H. Aki, "Effect of Heat and Electricity Storage and Reliability on Microgrid Viability:A Study of Commercial Buildings in California and New York States", Berkeley: LBNL, 2009, Available online on: http://eetd.lbl.gov/sites/all/files/publications/lbnl-1334e.pdf

[6] L.A.Wonga,V. K.Ramachandaramurthy, P.Taylora, J.B.Ekanayaked, S.L.Walker, S.Padmanaban "Review on the optimal placement, sizing and control of an energy storage system in the distribution network," Journal of Energy Storage vol. 21, , Pages 489-504 , 2019

[7] A. G. Olabi "Renewable energy and energy storage systems," Energy, vol.136, pp.1-6, 2017

[8] P. Lombardi, "Multi criteria Optimization of an autonomous virtual power plant with high degree of renewable energy sources", Proceeding of the 17th PSCC, 22-26 August 2011, Stockholm. ISBN 978-91-7501-257-5

[9] N. D. Hatziargyriou, A. Dimeas, A. G. Tsikalakis, J. Oyarzabal, J.A. Pecas Lopes, G. Kariniotakis, "Management of microgrids in market environment", 2009. Available online on: http://www.microgrids.eu/micro2000/presentations/39.pdf

[10] B.N.Dicevich, Ya.M.Ivan'o, A.A.Luzan, et al. Potencial'nye zapasy dikorastushchih resursov Irkutskoj oblasti [Potential reserves of wild resources of the Irkutsk region]. Irkutsk: Irkutskij GAU Publ., 2017. 156 p. (In Russian.).

How to Replace Electric Power System Inertia

Emil Dvorsky, Pavla Hejtmankova, Lenka Rakova
Department of Electrical Power Engineering and Environmental Engineering
University of West Bohemia
Pilsen, Czech Republic
dvorsky@kee.zcu.cz, hejtman@kee.zcu.cz, lencar@kee.zcu.

Abstract — **The Electric Power System (ES) must address two unique requirements - the need to maintain balance between generation and load, at the required time and with the required quality, and the need to adjust generation (or load) to manage power flows through individual transmission facilities. These requirements are performed mainly by synchronous generators (SG) which use for static transmission of the network the inertia of the power sources resulting from the direct electromechanical connections between the synchronous generators and the network. The Renewable Energy Sources (RES) utilization for power system supplying, notably inverter-connected electric generators that do not provide rotational inertia is effectively displacing conventional generators and their rotating machinery. This has implications for frequency dynamics and power system stability and operation. Second finding is that frequency dynamics are faster in power systems with low rotational inertia. This unwanted phenomenon preventing can be making by artificial inertia use with appliance and methods which are able to ensure ancillary services for power balance keeping in the electric networks. The use of rotary flywheels seems promising.**

Keywords: frequency regulation in power network; systems inertia; artificial inertia; distributed sources

I. Introduction

The basic requirements of ES are not new and had been respected at the beginning of electricity utilization, usually by vertically integrated utilities. With the electric power system development, it was necessary to impute the services to meet these requirements, now called "ancillary services". Power systems are dynamic systems with a high degree of complexity along several dimensions, i.e. time, space and grid hierarchy [1]. The system functions maintaining (power balance keeping among sources (generators – $\sum P_{G,i}$) and consumers - P_C (called Power System Control - PSC) is necessary along several time-scales (from milliseconds to yearly seasons) in great spatial distribution (from 100s-1000s km) with different voltage levels. The system control has hierarchical character and represents very complex system. This complex system keeps power balance on four levels – dynamic support (inertia) of synchronous generators (red square in Fig.1), primary, secondary and tertiary control - Fig.1. The power disturbance (in transmission systems with VHV, where active resistance of the power lines between sources and consumption may be neglected [2], is eliminated by the „Frequency Response"of individual sources that are connected to the system. Frequency response includes these time frames:

- Inertial response – dynamic response (up to a few seconds)
- Governor response ("Primary Response, Primary Control" – 1 s to 10s of seconds).
- AGC response - Automatic Governor Control, re-dispatch) ("Secondary Response, Secondary Control" – tens of seconds to tens of minutes).
- Tertiary control (power levels restoring of primary and secondary regulation – response after more than 15 min)

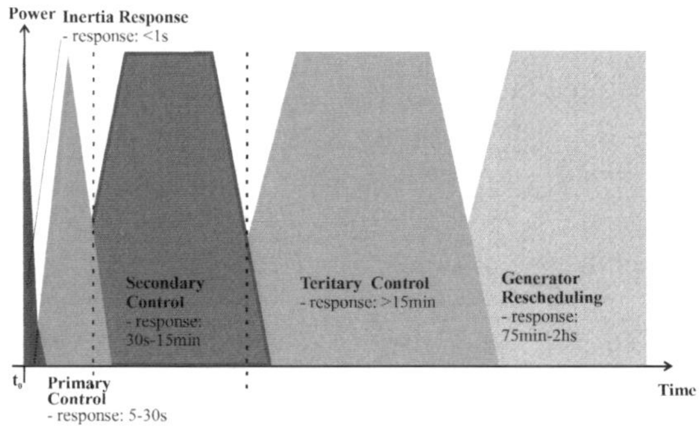

Figure 1. Grid inertial response in interconected power systems [2]

Due to the fact that tertiary level of the controlling system is focused towards optimization, for the stability of the power systems, there are the most important first three responses from the disturbance. For brief description and distinction between primary (decentralized) and secondary (centralized) level, (so-called LFC load frequency control) see Fig.2.

Primary frequency control is a decentralized function of the turbine governor and it is implemented by frequency correction of ES power reference value. Resulting static turbine characteristic (dependence of generator output P_G on frequency deviation Δf) is shown below the block labelled "$P_{control}$". Secondary control or Load Frequency Control (LFC) is a centralised automatic function to control the power generation in a "Control Area". Its main purpose is to keep interchange power flow P between consumption (P_c) and generation (P_G) at the scheduled value P_{Sched}, and to restore the frequency f in case of a frequency deviation originating from the control area.

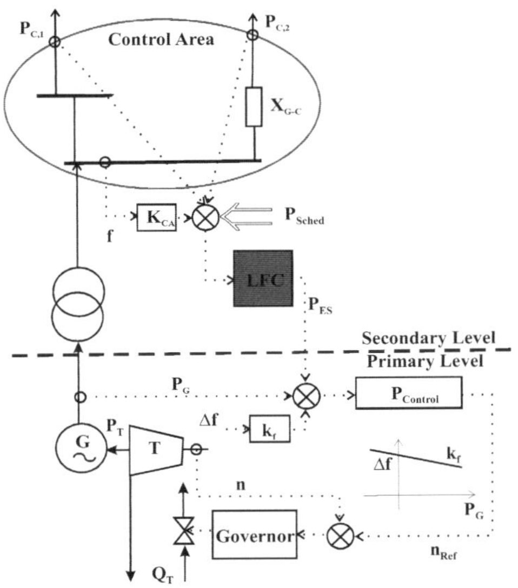

Figure 2. Grid inertial response in interconected power systems [3]

II. DYNAMIC BEHEVIOR OF GREAT POWER SYSTEMS

Under small signal conditions, the time derivative of the frequency deviation Δf from nominal value f_0 can be estimated as follows [1]:

$$\frac{d\Delta f}{dt}(t) = f_0 \frac{P_T(t) - P_G(t)}{2E_{kin}} = f_0 \frac{\Delta P(t)}{2E_{kin}} \tag{1}$$

where E_{kin} [MW.s] is the total kinetic energy stored at rated speed in the rotating masses of the considered synchronous area – control area, (mainly in the generators and turbine rotors). $P_T(t)$ [MW] is the mechanical power supplied by the generators and $P_G(t)$ is the electric power demand. The difference between them tends to accelerate (supply power to) and decelerate (draw power from) by rotating masses. The (1) is usually normalized in terms of per unit inertia constant H [MJ/MVA] in a given control of ES [3]:

$$\frac{d\Delta f}{dt}(t) = f_0 \frac{\Delta P^{pu}(t)}{2H_{ESs}} = f_0 \frac{\Delta P^{pu}(t)}{M_{ES}} \tag{2}$$

where:

$$H_{ES} = \frac{E_{ES}}{S_{ES}} = \frac{J(2\pi f_0)^2}{2S_{ES}} \tag{3}$$

J [kg.m^2] is the moment of inertia of the machine and H_{ES} [MW.s/MVA] and M_{ES} [s] are the system inertia and mechanical start time constants of the considered power system. Disturbances in this balance are compensated by the kinetic energy of the rotating variation in the system frequency f from its set-point value f_0 – Fig.3.

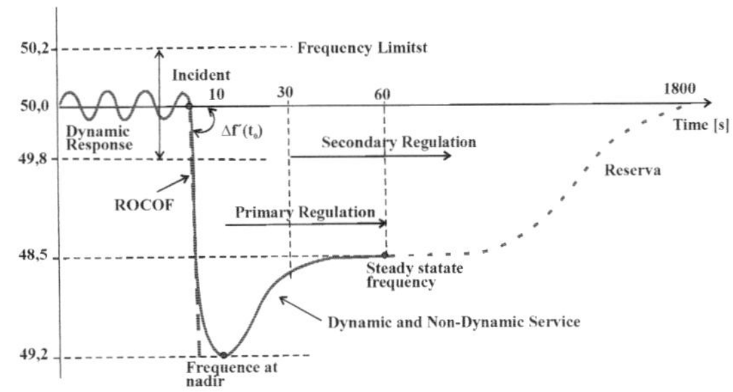

Figure 3. Grid response in huge power systems – Centrlised Electric Systems

Large disturbance, particularly trips of large generation plants, in t_0, cause unbalance that must be corrected by „Frequency Response". Early in the transient stage, the power imbalance is covered by inertial response of the remaining generators kinetic energy and turbine rotors, which causes speed and frequency to decrease with Rate of Change of Frequency (ROCOF) to nadir – Fig. 3. This rate is done by:

$$\frac{d\Delta f}{dt}(t_0) = f_0 \frac{\Delta P_0^{pu}}{E'_{ES}} \tag{4}$$

Where ΔP_0 [MW] is loss or surplus of power and E'_{ES} remaining kinetic energy.

Primary frequency control is provided within a few seconds after the occurrence of a frequency deviation. It provides power output proportional to the deviation (P control) with static frequency characteristic, stabilizing the system frequency but not restoring it to nominal frequency value f_0 but with deviation which is called Area Control Error (ACE). The process of the secondary control f and P is realized by sending the set-point power value P_{ES} from the secondary controller (LFC) to the control blocks – Fig. 2. The operation of the secondary control f and P should reset the nominal frequency and transmit power obviously to 15 min, so that ACE = 0, from the start point of the imbalance. In order we are able to restore governor and AGC response, there is necessary to input to the system not controlled power which is called tertiary frequency – Fig.1.

III. INERTIA RESPONSE IN ES WITH HIGH RES SHARES

The traditional assumption that grid inertia is sufficiently high with only small variations over time is not valid for power systems with high RES shares – where generators are connected to ES via inverters and doesn't have direct electromagnetic bond with system. To prevent this unwanted phenomenon can be made by artificial inertia use with appliance and methods which are able to ensure ancillary services. The artificial inertia can provide dynamic control support (DFCS) to ES and replace the classic inertia of rotating generators – Fig.4. In Fig.4 is presented LFC in Decentralised Energy Systems (DES) which obviously doesn't use secondary frequency regulation. Frequency regulation is there performed using automatic primary and manual tertiary controls. They must be able to maintain this response for at least 15 minutes to be enabling other generation assets to be brought on line. Typical property of DES is low inertia constant (green line in Fig.4). Storage technologies should be ideal suppliers of several ancillary services, including regulation, contingency reserves (spinning reserve, supplemental reserve, replacement reserve), and voltage support as well. The ES inertia increasing by the artificial inertia is red line in Fig. 4.

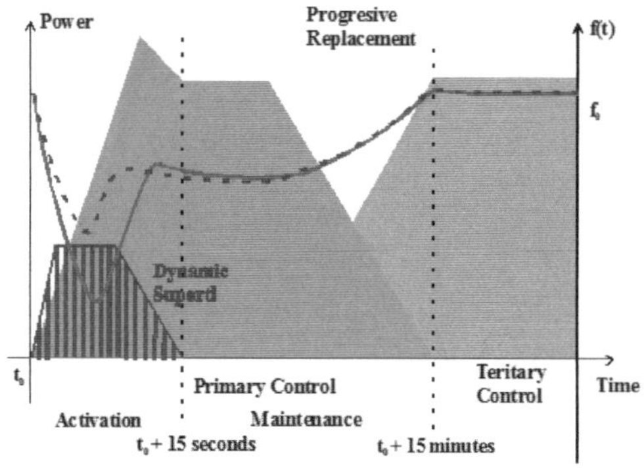

Figure 4. Grid response in Decentrlised Electric Systems [4]

IV. Dynamic Frequency Control Support

The proper control enables inverter-connected generation units to rapidly change their power output, via internal or external power sources, in order to emulate the Dynamic Frequency Response (DFR – red area in Fig.4) of conventional rotating generators to frequency variations [4]. It is Fast Frequency Response (FFR) that mitigates high ROCOF. A Fast Frequency Response is defined as a rapid active power injection (in 1-2 seconds or less), to arrest the frequency decline following a contingency event. FFR can be an important tool in mitigating high ROCOF, by very rapidly correcting the supply-demand imbalance following a contingency event. Challenges are the real-time tractability and the stringent operational constraints that need to be respected by any control scheme aiming to provide rotational inertia. One possibility to ensure DFC is Flywheels energy storage (FES) use.

The main objective of the paper is evaluating the possibilities of Flywheels Energy Systems (FES) units for DFR use. Therefore, there is necessary to evaluate especially performance FES characteristic. The possibility of using FES for primary and secondary frequency regulation has been demonstrated in the authors' published article [5].

A. FES design

The use of FES to store energy is not a new technology. Basic flywheels such as stone wheels were used to craft pottery thousands years ago. The amount of stored energy is limited by the tensile strength of the material. If the stresses in the flywheel exceed the tensile strength of the material the flywheel will break apart. The limiting stress in a thin rim is the tangential stress, which is given by (5):

$$\sigma_{max} = \rho\, r^2 \omega^2 \; \left[N/m^2\right] \qquad (5)$$

Where σ_{max} is the maximum tensile strength and ρ is the material density. The specific energy and energy density (mass and volume) for the thin rim flywheel can be expressed as:

$$e_{kin,m} = \frac{1}{2}\frac{\sigma_{max}}{\rho} \; [J/kg] \qquad (6)$$

$$e_{k,v} = \frac{1}{2}\sigma_{max} \; \left[J/m^3\right] \qquad (7)$$

Equation 6 shows that a high strength material with low density would be optimal for flywheels rotors. The factor in (8) and (9) is only valid for a flywheel with the shape of a thin rim.

$$e_{k,m} = K\frac{\sigma_{max}}{\rho} \; [J/kg] \qquad (8)$$

$$e_{k,v} = K\sigma_{max} \; \left[J/m^3\right] \qquad (9)$$

Where K is the so-called shape factor and can be described as a measure of how efficient the material of the flywheel is used. For a detailed derivation of the shape factor see. Fig. 5 shows the most common types of flywheel geometries.

Depending on the material used the optimal flywheel shape will differ. Isotropic materials, such as steel, have equal strength in both the tangential and radial direction. Anisotropic materials, like fibre-reinforced composites, have unidirectional tensile strength, i.e. are stronger in the longitudinal fibre direction. As shown in Figure 5, there are shapes better suited for these types of materials. A thin rim (hollow circular cylinder) flywheel is a very good candidate for composite material as this geometry exploits the high unidirectional strength of the fibres. Multiple sources state that a hollow circular cylinder is the optimal shape for a composite flywheel rotor. Maximum specific energy is calculated with (8) with K=0.5. For a flywheel rotor is a high strength material with low density. Composite materials have both high strength and low density and are ideal for flywheel rotors used for energy storage.

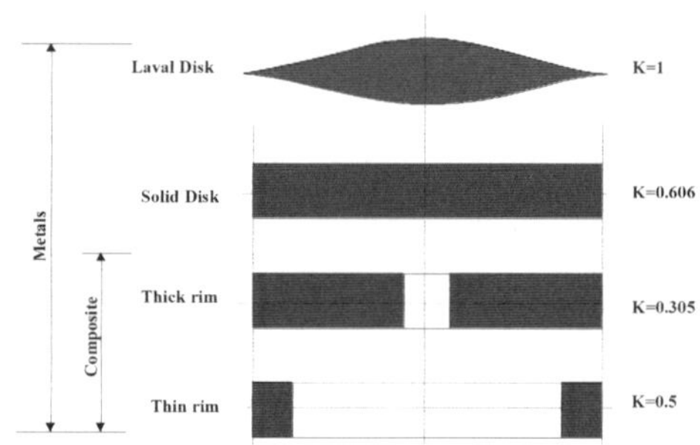

Figure 5. Types of flywheel geometries

B. FES for power systems

FES consists of a rotary flywheel and an induction machine that works both as a motor and a generator. During normal operation, electricity is transferred from the grid to the kinetic energy through the engine. If necessary, the stored kinetic energy is converted back to the electric generator – Fig. 6.

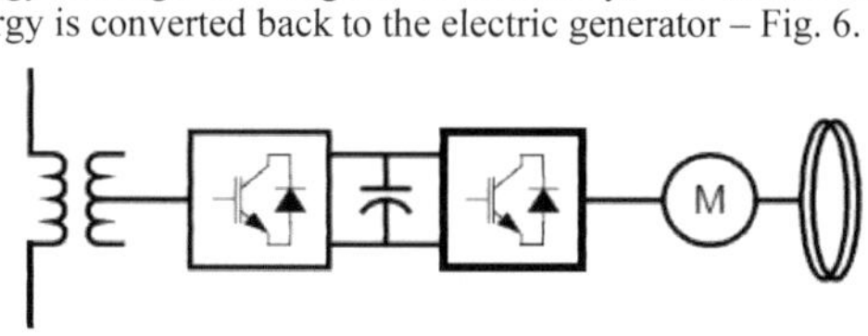

Figure 6. Principle of FES for power systems.

The instantaneous power injected by the torque depends only on the instantaneous angular speed and not on whether the angular speed increases, decreases, or remains constant while the torque is being applied. The flywheel can be charged or discharged during the operating period T at a constant speed, or at a rate increase or a decrease in speed.

The rate of speed change [6] is exponential and can be expressed for discharging in the given period by (10):

$$\omega(t) = \begin{cases} \omega_{max} & t \le T \\ \omega_{max}.e^{-\frac{K_0}{I}.t} & t \ge T \end{cases} \qquad (10)$$

T is period for which the flywheel works at ω_{max}. If this time is not available, it can be roughly determined using the rule of dependence between moment of inertia and flywheel power:

$$T_{max} = \frac{J\omega_{max}^2}{2P_{max}} \qquad (11)$$

The flywheel performance at exponential speed change can be deduced from (12) while respecting individual times:

$$P_{FES} = \begin{cases} P_{FES,max} & t \le T \\ \eta_{setr}\omega_{max}^2.e^{-\frac{K_0}{J}.t} & t \ge T \end{cases} \qquad (12)$$

The performance characteristics which were used for FES systems modeling are presented in [5].

The other possibility of FES performance characteristic use is data from FES producers. The characteristic of the most famous FES systems producer Bacon Power is presented in Fig. 7.

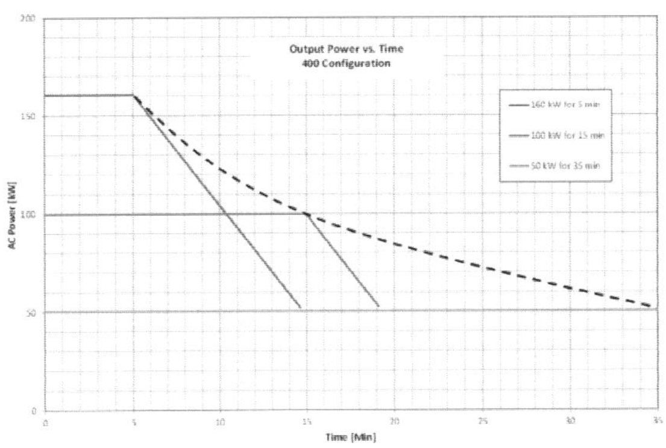

Figure 7. Performance charecteristics of Becon Power FES [7].

The FES lifetime depends on the number of charging (discharging) cycles; therefore there is a limit $P_{FES,min}$ (green line in fig.7) value of discharging [8].

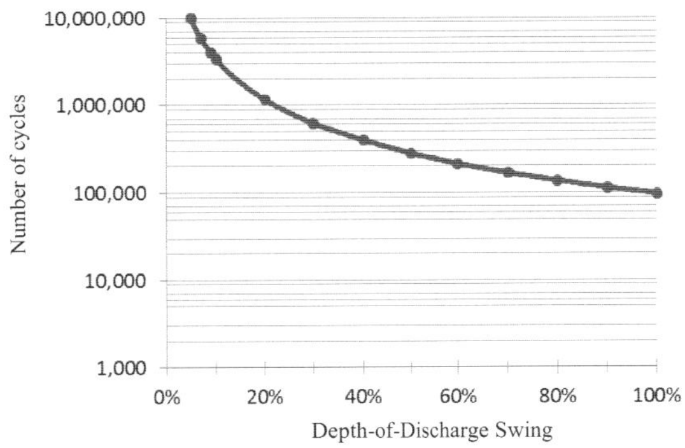

Figure 8. Performance charecteristics of Becon Power FES [7].

Then the power drop course FES can be replaced by a second degree polynomial:

$$P_{FES} = P_{FES,max} - a_1 t - a_2 t^2 \qquad (13)$$

In order the possibilities of FES for frequency control can be assessed, a model of DES was created [5]. The ES is composed by two sources, one is SG and second one is FES.

The analysis of the grid frequency control that can be secured by FES model of the island operation of the ES, showed possibilities FES to fulfill conditions of frequency control and they have the ability to increase ROFOC.

To know better answer on the question of ES inertia increasing, there is necessary to discuss FES dynamic behavior, which are equipped by invertors with IGBT transistors. The time constantans of these parts will be decisive for the FES utilization for the FES dynamic response. The time value guaranteed by the manufacturer:

- Response time (normal operations) - less then 1 second to full power,
- Ramp rate (normal) - up to 190 kW/second
- Response time (system event) - 100 milliseconds to full power.

V. Conclusion

The previous analysis of the grid frequency control that can be secured by FES which was presented on the model FES [5], confirmed the possibilities not only primary and secondary frequency regulation in decentralized systems but even for ancillary services. If the FES has possibility to fulfil conditions of DFCS, was discussed in this paper. FES can be used to create artificial inertia based on the analysis of the dynamic behaviour of the system after the energy instability and analysis of the dynamic behaviour of the FES controlled by IGB transistors and the available data. This fact will be demonstrated by subsequent calculations.

Acknowledgment

This paper was supported by the student research project SGS - 2018 – 023.

References

[1] A. Ulbig, T. Rinke, S. Chatzivasileiadis, G. Andersson: Predictive Control for Real-Time Frequency Regulation and Rotational Inertia Provision in Power Systems, 52nd IEEE Conference on Decision and Control December 10-13, 2013. Florence, Italy.

[2] E. Dvorsky, Pavla Hejtmankova, L. Rakova: Primary Voltage and Frequency Control by Means of two Photovoltaic Systems, EPE. Czech Republic, 2018.

[3] K. Máslo, Z Hruška: Control strategies for power system in island operation, Journal of Energy and Power Sources, ISSN: 2333-9144.

[4] G. Delille, B. François, G. Malarange: Dynamic Frequency Control Support: a Virtual Inertia Provided by Distributed Energy Storage, IEEE PES Innovative Smart Grid Technologies Conference Europe (ISGT Europe), 2010.

[5] E. Dvorsky, P. Hejtmankova, L. Rakova, Grid Frequency Regulation by Recycling Electrical Energy in Flywheels: Power Engineering 2017, Kosice, Technical University

[6] M. Ahrens, L. Kucera, R. Larsonneur, "Performance of a magnetically suspended flywheel energy storage device", IEEE Transactions on Control Systems Technology, Vol. 4, Issue 5, pp: 494 – 502, Sept. 1996.

[7] Beacon Power Flywheel Energy Storage Systems: www.beaconpower.com

[8] G. Delille, B. François and Gilles M. Schmutz: "Primary frequency control provided by battery", 10.1109/ISGTEUROPE.2010.5638887 Source: IEEE Xplore.

[9] B. Miller, Inertia in power system: We don't actually need that much, https://reneweconomy.com.au/inertia-power-system-dont-actually-need-much-65691/

[10] E. Spahic, D. Varma, G. Beck, G. Kuhn, V. Hild, Impact of Reduced System Inertia on Stable Power, CIGRE Paris Meeting, 2016

[11] R. Östergård. Flywheel energy storage a conceptual study, Teknisk- naturvetenskaplig fakultet UTH-enheten, 2011

[12] Union for the Coordination of Transmission of Electricity,Operation Handbook Appendix 1 - Load-Frequency Control and Performance. Paris: ENTSO-E/UCTE, www. entsoe.eu.

Short Circuit Calculations and Dynamic Stability Assessment within Dispatch Control Environment, Operational Requirements and SW Reality

Miloslava Chladová
EMS Department
ČEPS, a.s.
Prague, Czech Republic
chladova@ceps.cz

Karel Máslo
Transmission System Analysis Dpt.
ČEPS, a.s.
Prague, Czech Republic
maslo@ceps.cz

Vladimír Krištof
Gridman s.r.o.
Kosice, Slovakia
vladimir.kristof@gridman.sk

Abstract— **Real-time software tool used by ČEPS (Czech Transmission System Operator) dispatch control for monitoring and controlling operation security makes possible to simulate operating states of ČEPS at present in the dispatch control system. The paper presents parts of short circuit calculations (SCC) and dynamic stability assessment (DS) of observability area (OA) ČEPS. Limits of monitoring and control of operation security using supervision of substation short circuit resistibility, switching control and dynamic stability assessment are presented.**

Keywords- Transmission system, Czech Transmission System Operator (ČEPS), operation security, dispatch control, dispatch control system, EMS functions, Short circuit calculation, Stability, Transient stability

I. INTRODUCTION

The increasing complexity of transmission system operational conditions is presenting new challenges for operational planning as well as the operation of the Czech transmission system. Nowadays we encounter new situations that could pose a threat to operational security more and more often. Consequently, with the growing complexity also increases the importance of effective transmission system operational security monitoring and control processes. Operational security in this context means that the operating state of the system is resilient against sudden changes, in particular to short circuits or line outages and breaker switching. The system has to comply with grid protection and relay settings, preserve stable operation without overloading transmission lines, and return back to operational limits even in case of the occurrence of some of these events.

One of the topical functions of ČEPS's EMS (Energy Management System) control system is the supervision of operational security and recognition of treacherous system state during remote manipulations using switching scenarios in control system at times of lower security transmission system states [4]. An EMS analytical computational software tools "Short circuit calculation" and "Dynamic stability" perform the assessment.

II. REAL TIME MONITORING

A. Model extent and Data preparation

Observability Area (OA) within Control System ČEPS contains 400 kV and 220 kV grids, several parts of 110 kV grid, generation nodes and parts of neighboring TSO (Transmission System Operator) grids [3]. Adequate model of foreign TSO in real time is necessary particularly for simulations close borders with neighboring TSO grids, in periodical Power Flow (PF) calculations, during line outputs and N-1 criteria checking, for different types of grid Optimizations with remedial action searching and for breaker switching. OA model extent influences results accuracy directly during on-line analyses of transmission system operational security and interconnected parts of distribution systems side by side with transmission system. Of course, for short circuit calculations and Dynamic stability assessment the same model is used.

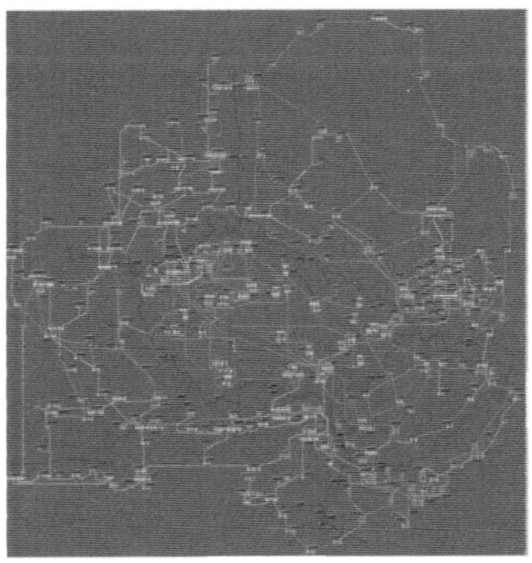

Fig. 1 Observability Area model extent

Contemporary and accurate in size dynamically measured data are the second inevitable condition of the correct transmission system model performance. FIX mode function was created for the purpose of serious mistake of measures and signals elimination [3]. FIX mode allows to fix input signals and like that to fix state of bay or whole substation. Operators use the FIX mode for hours when the substation is under reconstruction or when the test runs and when the control system gets false information. Monitoring of dispatch charts is temporarily divided into real power chart (FIX state) and chart with state of tests.

Transmission system topology is automatically checked. The islands are colored by different color in case of power system disintegration to islands. Operator is warned using alarms and hooter. The calculation contains each island with as minimum two nodes and one branch. Technical parameters of transmission system are necessary to continuously update. On time changes of the parameters at a change of operational state of power grid and hydrometeorological conditions can markedly change limits of transmission system components and results of EMS calculations [4]. Nowadays we use the automatic parameter transports from administrators of different types of devices periodically ones per day or ones per hour. Majority of parameters is modified very rarely. There are exceptions. One of the exceptions is dynamical overloading of lines between substations and wires inside substations. In case of considering of WAMS applications we will want to calculate average values of R, X, G a B in the lines.

Sets of technical or commercial values that describe limits of potential transmission system states are exploited for device parameters. These sets we use only partly, we want to use then within all types of EMS calculations.

B. Short circuit resistant check

SCC within substation goes each minute after State estimation completion. SCC serves and guards possible short circuit busbar resistance violations. It means short circuit node resistance violations within node oriented model (NOM) of transmission system. Three phases and one phases short circuits are calculated in every entered node of power system. If the short circuit node resistance violation happens in some node, control system puts an event on a record and the operator is monish of an alarm.

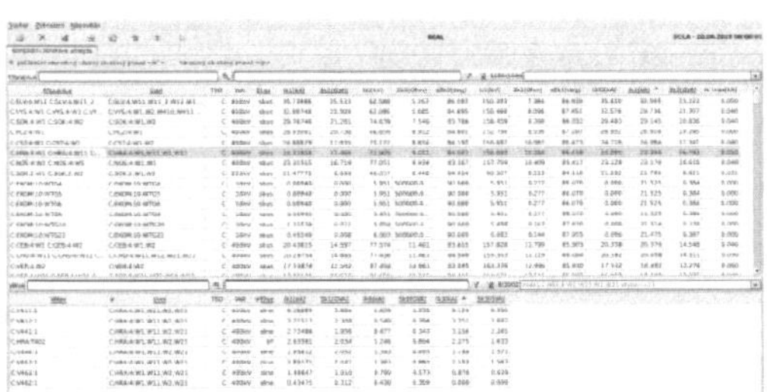

Fig. 2 Short circuit resistant check report with Substation Hradec selection

SCC calculation respects straight interconnection of two substations across busbar connection then the calculation is used in particular in 400 kV substations Hradec východ and Hradec západ. An extension to whole transmission system means extension of SCC for two phase grounded short circuit.

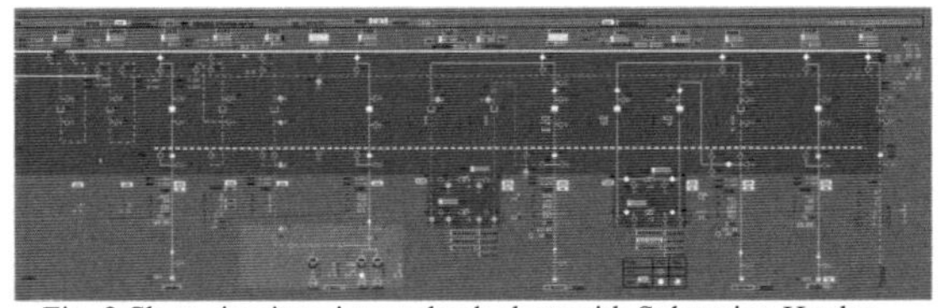

Fig. 3 Short circuit resistant check chart with Substation Hradec selection

C. Stability assessment (SA)

SA is divided to two parts. First part goes each minute after State estimation completion. The part of SA serves and guards possible active power impacts to nearest generators and distance protections penetration to different protection zones. Breaker switching or line outputs are initial evets. Limits of generator active power impact are presented as apparent power percentage.

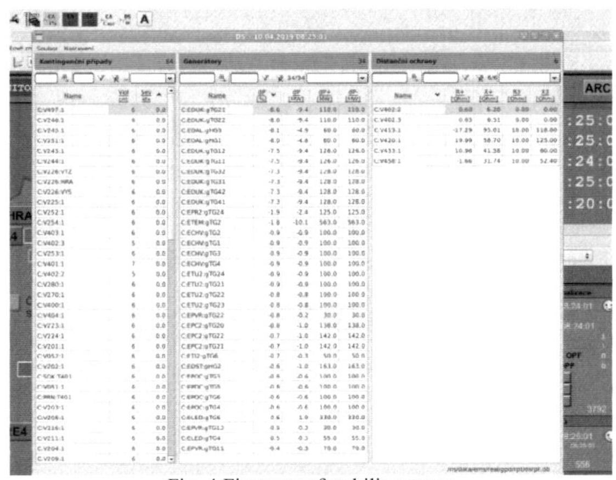

Fig. 4 First part of stability report

The calculation allows to find angle limits before breaker switching during remote control. Current impact within line is an input value for calculation of permissible angle, which can be compared with setting angle limit for switching and with settings of protection zones.

The second part goes each 5 minutes after State estimation completion and grid model export from control system in standard raw format. Five minutes is an interval when changes of topology inside OA happened with high probability. The part of SA presents dynamic angle stability. It calls dynamic stability assessment (DSA) and it serves and guards SCC impact to nearest generators. The part of DS presents transient angle stability and searches critical clearing time (CCT) of each entered generator [1]. CCT presents a time interval when the short circuit has to be switch off looking at it from generator and actual topology of OA.

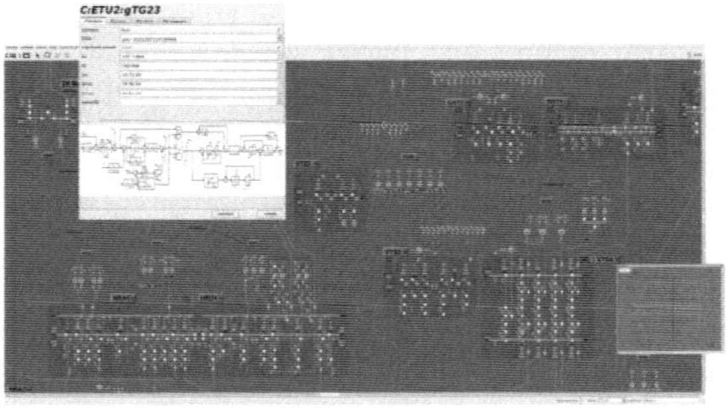

Fig. 5 Generator model

Angle stability estimation with CCT calculation runs out of the control system. Data between CCT and control system are transited using standard interface for supportive applications.

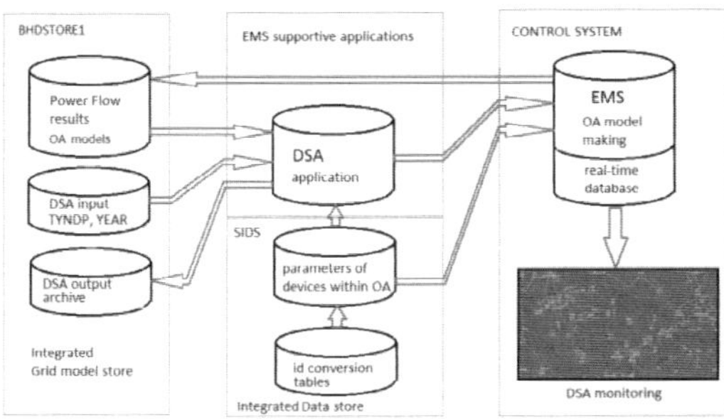

Fig. 6 DSA application interface

DSA application is doing calculations of transmission system states saved in Integrated grid model store. Initial grid models are saved in standard RAW format. Dynamic data and data from id conversion tables are stored within Integrated data store. The results of DSA application are transmitted to real-time database in SCADA/EMS control system. The data present short circuit location, generator/unit and calculated CCT. The calculation runs over historical snapshot from annual archive with 5 minutes granularity and turns back results to SCADA /EMS each 5 minutes.

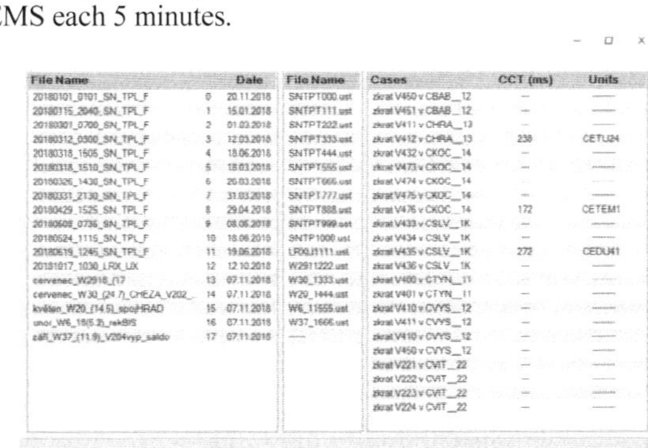

File Name		Date	File Name	Cases	CCT (ms)	Units
20180101_0101_SN_TPL_F	0	20.11.2018	SNTPT000.ust	zkrat V450 v CBAB__12	—	——
20180115_2040_SN_TPL_F	1	15.01.2018	SNTPT111.ust	zkrat V451 v CBAB__12	—	——
20180301_0700_SN_TPL_F	2	01.03.2018	SNTPT222.ust	zkrat V411 v CHRA__13	—	——
20180312_0300_SN_TPL_F	3	12.03.2018	SNTPT333.ust	zkrat V412 v CHRA__13	238	CETU24
20180318_1505_SN_TPL_F	4	18.06.2018	SNTPT444.ust	zkrat V432 v CKOC__14	—	——
20180318_1510_SN_TPL_F	5	18.03.2018	SNTPT555.ust	zkrat V473 v CKOC__14	—	——
20180326_1430_SN_TPL_F	6	26.03.2018	SNTPT666.ust	zkrat V474 v CKOC__14	—	——
20180331_2130_SN_TPL_F	7	31.03.2018	SNTPT777.ust	zkrat V475 v CKOC__14	—	——
20180429_1525_SN_TPL_F	8	29.04.2018	SNTPT888.ust	zkrat V476 v CKOC__14	172	CETEM1
20180608_0736_SN_TPL_F	9	08.06.2018	SNTPT999.ust	zkrat V433 v CSLV__1K	—	——
20180624_1115_SN_TPL_F	10	18.06.2018	SNTP1000.ust	zkrat V434 v CSLV__1K	—	——
20180619_1245_SN_TPL_F	11	19.06.2018	LRXUJ1111.ust	zkrat V435 v CSLV__1K	272	CEDU41
20181017_1030_LRX_UX	12	12.10.2018	W2911222.ust	zkrat V436 v CSLV__1K	—	——
cervenec_W2918_(17	13	07.11.2018	W30_1333.ust	zkrat V400 v CTYN__11	—	——
cervenec_W30_(24.7)_CHEZA_V202_	14	07.11.2018	W29_1444.ust	zkrat V401 v CTYN__11	—	——
květen_W20_(14.5)_spojHRAD	15	07.11.2018	W6_11555.ust	zkrat V410 v CVYS__12	—	——
unor_W6_18(5.2)_rekBIS	16	07.11.2018	W37_1666.ust	zkrat V411 v CVYS__12	—	——
září_W37_(11.9)_V204vyp_saldo	17	07.11.2018	——	zkrat V410 v CVYS__12	—	——
				zkrat V450 v CVYS__12	—	——
				zkrat V221 v CVIT__22	—	——
				zkrat V222 v CVIT__22	—	——
				zkrat V223 v CVIT__22	—	——
				zkrat V224 v CVIT__22	—	——

Fig. 7 DSA results table report

D. Security check during manipulations

Applications check operational states of transmission system during remote control after the choice of command request to breaker. The applications draw operator to problem and they can stop the intended manipulation in case of unsuitable or danger manipulation.

The applications check before manipulation with breakers:

- New island creation,
- Branches overloading,
- Overcrossing of short circuit resistance in a part of substation,
- A big decreasing of voltage or leaving voltage from entered limits,
- Active power impact to nearest generators.

EMS running during remote control with breakers uses Power Flow calculation, SCC and active power impact to nearest generators. All the applications come out from OA model acquired at latest State Estimation running. The EMS applications respect transmission system stability only partly and they covered:

- Voltage angle deviation between two connecting nodes from permissible value at synchro switching device,
- Active power impact impermissible deviation to nearest power plants calculation,
- Calculation of imagine impedance intersection to some from distance protection zones,
- Overcrossing of permissible current impact.

III. OFF-LINE CALCULATIONS IN SIMULATION ENVIRONMENT

Simulation environment id separated from live real time monitoring presentations using independent environment and database. Pictorial chart presentation is the same as a real time presentation. Input and calculated outputs values are possible to compare in tables. All EMS calculations are possible to run in simulation environment excluding DSA CCT calculation. The operator is able to use CCT calculation after export of adjusted initial OA model to Integrated Grid model store. Results of the simulation are present in report Fig. 7.

It is possible to make all calculations in simulation environment using latest real time transmission system state or using historical transmission system state. Both are states after State Estimation. The simulation environment is used for upcoming system state analysis or for historical system state analysis. The user can make changes to the model, to change grid connections, change active power inputs from generators or generally working point of generator or change loads in areas powered from load nodes. There are supportive applications like "transit" between two nodes or "international support" between two areas for power transmission simulations for a long distance. The user can change the selected configuration of the calculation.

Dispatchers, members of the operational planning department and analysts can calculate SCC and stability to verify potential states of the system, e.g. the system state following an intended manipulation in the grid in simulation environment.

User can insert the results of individual calculations in the simulation environment into a single line diagram and he can continue with other calculations or he can upload it to the power grid in the dispatch control system or export grid snapshot for further analysis.

The operators can also use the simulation environment for remote control simulation. Manipulation with the breaker is very important. Differently from cyclic on-line calculation there is a possibility to simulate switching-on with breaker after setting the connectors states. The switching-on of bus-bar breaker in substation Hradec is shown in Fig. 8. Results of stability assessment are presented in table reports only.

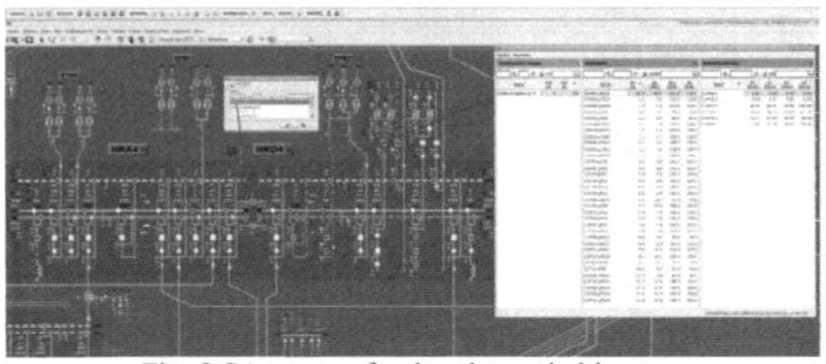

Fig. 8 SA report after breaker switching

Using dynamic stability CCT calculation requires export of grid model prepared for manipulation to RAW format. The export is possible from simulation environment, too. Results of CCT are automatically return back to the control system and they are presented to operators via report Fig.7.

IV. CONCLUSIONS – BENEFITS AND PROSPECTS

Development of grid calculations within Czech Transmission System Operator dispatch control system corresponds to permanently bigger and bigger attention to operational security during huge flow transmission, remote control switching and faults. That brings us to the limits of power system transmission abilities exploitation. These states of transmission system force us to verify stability of power system and use new functionalities and new devices.

Present calculations give at dispatcher disposal more powerful and more effective tools for solutions of emergency and exceptional situation in transmission grid especially for their preventions. At first it took implementation of on-line calculations like State Estimation and Contingency Analysis, application of ReConfiguration (topology optimizing), ReDispatch (power supply optimizing) and Security Constrained OPTimization (optimization of post contingency states) for unloading lines and checking of keeping operational limits of the devices. Today they are particularly short circuit calculations, active power impacts to nearest generators, remote control checking and dynamic stability checking. The operator can conduct these calculations during remote control and especially before each manipulation with breaker.

In the nearest future it will be necessary all of the above named calculations to be run on detailed breaker oriented OA model including surrounding transmission system Continental Europe node oriented model. The transmission power mass forces us to model dynamic behavior during remote switching within responsibility transmission grid especially line overloading or deviation from permissible limits of currents, active and reactive powers and voltages. The mere grid expansion won't be able to manage the challenges. It will be necessary to improve existing and to implement new functions, to use complex methods and minimize numbers of actions suggested to operators and to optimize components of calculations.

REFERENCES

[1] K. Máslo, P. Toman, B. Bátora, and J. Koudelka, "Methods for Dynamic Stability Assessment," Proc. 20th International scientific conference Electric Power Engineering (EPE 2019), May 2019

[2] K. Máslo, "Short circuit curret calculations," Proc. 13th International Scientific Conference Control of Power Systems (CPS 2018)

[3] B. Sadecký, M. Chladová, "Software assistance of ČEPS Power System Security Control," Conference EPE2011, Dlouhé Stráně, May 2011

[4] M. Chladová B. Sadecký, "Limitations of actual EMS functions in the dispatch control system ČEPS," Conference EE2011, Stará Lesná, 2011

Model-based fault diagnosis of wind turbines built on Takagi–Sugeno fuzzy observers

Dušan Krokavec, Anna Filasová
Department of Cybernetics and Artificial Intelligence
Technical University of Košice
Košice, Slovakia
dusan.krokavec@tuke.sk, anna.filasova@tuke.sk

***Abstract*—Unknown input Takagi–Sugeno fuzzy observer and H_∞ Takagi–Sugeno fuzzy observer techniques are utilized to generate fault residuals for fault diagnosis of wind turbines. To ensure the stability of the Takagi–Sugeno fuzzy observers error dynamics and to attenuate impact from disturbances, the linear matrix inequality approach is used to design the fuzzy observer parameters and disturbance decoupling. In consequence, simulating the designed fault detection implementation demonstrates main functionalities of the proposed approach.**

Keywords: fault diagnosis; fault residuals; unknown input fuzzy observer; linear matrix inequalities; wind turbine fuzzy model.

I. INTRODUCTION

Wind turbine technology is one of the appropriate ways for increased use of renewable energy. Thus, in order to improve a wind turbine behavior, profit and reliability, a reduction of faults in the parts of the wind turbine and fault tolerant control stay relevant objectives. A way to ensure this consists in introducing advanced fault detection, isolation, and accommodation schemes into the turbines [1]. The nonlinear nature of the system predetermines Takagi–Sugeno (TS) fuzzy inference reasoning [2] in fault tolerant control design (FTC) [3]. Using this approach, a continuous nonlinear system can be modeled as a multiple-model representation, corresponding to a number of regions of the state-space parameters. To achieve maximization of the wind power extraction, different approaches of model predictive control (MPC) based on the wind turbine TS fuzzy models are proposed [4], [5], substituting TS fuzzy models of reduced dimension. More accurate TS models can be find in design of FTC and fault monitoring systems [6], [7].

In this article, TS fuzzy unknown input observer based fault detection problems for wind turbines are addressed, while the unknown input TS fuzzy observer scheme is preferred to include robustness towards the uncertainty of the wind speed. Considering the wind turbine structured as several interconnected subsystems, a TS fuzzy representation is focused on nonlinear aerodynamic properties of the blades and wind speed to get a compromise between power and rotor speed relation with projection in the turbine operating range. This principle is subordinated to the synthesis algorithm for the residuals generation, which is based on TS fuzzy models.

The paper is organized as follows. Ensuing introduction in Sec. I, and continuing by description in Sec. II, the proposed unknown input observer and residual generator design are analyzed in Sec. III and Sec. IV. In response, Sec. V describes substantial details in wind turbine modeling and Sec. VI shows the applicability of the method using benchmark model example. Finally, Sec. VII gives some concluding remarks.

For sake of convenience, throughout this paper used notations reflect usual conventionality so that $\boldsymbol{x}^{\mathrm{T}}$, $\boldsymbol{X}^{\mathrm{T}}$ denotes transpose of the vector $\boldsymbol{x}$ and matrix $\boldsymbol{X}$, respectively, the notation $\boldsymbol{X} \succ 0$ means that $\boldsymbol{X} \in \mathbb{R}^{n\times n}$ is real symmetric positive definite, $\rho(\boldsymbol{X})$ reports the eigenvalue spectrum of $\boldsymbol{X} \in \mathbb{R}^{n\times n}$, $\boldsymbol{I}_n$ is the n-th order identity matrix, $\mathbb{R}^{n\times r}$ signifies the set of $n\times r$ real matrices.

II. TS CLASS OF NONLINEAR SYSTEMS

The systems under consideration fall in the class of nonlinear dynamic systems, which is represented as

$$\dot{\boldsymbol{q}}(t) = \sum_{i=1}^{s} \mathrm{h}_i(\boldsymbol{\theta}(t))\boldsymbol{A}_i\boldsymbol{q}(t) + \boldsymbol{B}\boldsymbol{u}(t) + \boldsymbol{F}\boldsymbol{f}(t) + \boldsymbol{E}\boldsymbol{d}(t) \tag{1}$$

$$\boldsymbol{y}(t) = \boldsymbol{C}\boldsymbol{q}(t) \tag{2}$$

where $\boldsymbol{q}(t) \in \mathbb{R}^n$, $\boldsymbol{u}(t) \in \mathbb{R}^r$, $\boldsymbol{y}(t) \in \mathbb{R}^m$, $\boldsymbol{f}(t) \in \mathbb{R}^v$, $\boldsymbol{d}(t) \in \mathbb{R}^p$ are vectors of the state, input and output variables as well as additive faults and disturbances, respectively, while $\boldsymbol{A}_i \in \mathbb{R}^{n\times n}$, $\boldsymbol{B} \in \mathbb{R}^{n\times r}$, $\boldsymbol{C} \in \mathbb{R}^{m\times n}$, $\boldsymbol{F} \in \mathbb{R}^{n\times v}$, $\boldsymbol{E} \in \mathbb{R}^{n\times p}$ are real matrices, It is assumed that $\boldsymbol{A}_i$ is constructed with respect to working sectors. Note, if no additive fault occurs then $\boldsymbol{f}(t) = \boldsymbol{0}\ \forall t \geq 0$.

It is considered that there exist normalized sector functions $\{\mathrm{w}_{lj}(\theta_j(t)), l = 1, 2, \ldots, h,\ j = 1, 2, \ldots, k_l\}$ such that

$$\mathrm{w}_{lk}(\boldsymbol{\theta}(t)) = 1 - \sum_{j=1}^{k_l-1} \mathrm{w}_{lj}(\theta_j(t)) \tag{3}$$

$$\boldsymbol{\theta}(t) = \begin{bmatrix} \theta_1(t) & \theta_2(t) & \cdots & \theta_q(t) \end{bmatrix} \tag{4}$$

is the vector of so-called premise variables. A premise variable represents any measurable variable (in a simple case it can be directly a state variable) and none of the premise variables does not depend on the input $\boldsymbol{u}(t)$.

If the number of the membership functions $\mathrm{h}_i(\boldsymbol{\theta}(t))$, $i = 1, 2, \ldots s$ is not a power of 2, the states of the system can be inferred in more appropriate form as follows

$$\dot{\boldsymbol{q}}(t) = \sum_{\phi=1}^{s_\phi}\sum_{\psi=1}^{s_\psi}\cdots\sum_{\omega=1}^{s_\omega} \mathrm{h}_{\phi,\psi,\ldots\omega}(\boldsymbol{\theta}(t))\boldsymbol{\Xi}_{\phi,\psi,\ldots\omega}(t) \tag{5}$$

$$\boldsymbol{\Xi}_{\phi,\psi,\ldots\omega}(t) = \boldsymbol{A}_{\phi,\psi,\ldots\omega}\boldsymbol{q}(t) + \boldsymbol{B}\boldsymbol{u}(t) + \boldsymbol{F}\boldsymbol{f}(t) + \boldsymbol{E}\boldsymbol{d}(t) \tag{6}$$

and

$$\mathrm{h}_{\phi,\psi,\ldots\omega}(\boldsymbol{\theta}(t)) = \mathrm{h}_\phi(\boldsymbol{\theta}(t))\mathrm{h}_\psi(\boldsymbol{\theta}(t))\cdots\mathrm{h}_\omega(\boldsymbol{\theta}(t)) \tag{7}$$

is the averaging weight representing the normalized grade of membership (membership function).

By definition, the membership functions have to satisfy the following convex sum properties

$$0 \leq \mathrm{h}_{\phi,\psi,\ldots\omega}(\boldsymbol{\theta}(t)) \leq 1 \tag{8}$$

$$\sum_{\phi=1}^{s_\phi}\sum_{\psi=1}^{s_\psi}\cdots\sum_{\omega=1}^{s_\omega} \mathrm{h}_{\phi,\psi,\ldots\omega}(\boldsymbol{\theta}(t)) = 1 \tag{9}$$

It is evident that the fuzzy model is achieved locally by fuzzy amalgamation of the linear subsystem models and at the boundaries between regions the linear interpolation is used.

III. Unknown Input TS Fuzzy Observer

Using the same set of membership functions, an unknown input observer can be designed for (1), (2) as

$$\dot{\boldsymbol{p}}(t)=\sum_{i=1}^{s}\mathrm{h}_i(\boldsymbol{\theta}(t))(\boldsymbol{P}_i\boldsymbol{p}(t)+\boldsymbol{Q}\boldsymbol{B}\boldsymbol{u}(t)+(\boldsymbol{J}_i+\boldsymbol{L}_i)\boldsymbol{y}(t)) \quad (10)$$

$$\boldsymbol{q}_e(t)=\boldsymbol{p}(t)+\boldsymbol{O}\boldsymbol{y}(t) \quad (11)$$

where $\boldsymbol{q}_e(t)\in I\!R^n$ is the state vector estimate, $\boldsymbol{p}(t)\in I\!R^n$ is the state vector of unknown input observer and $\boldsymbol{P}_i,\boldsymbol{Q}\in I\!R^{n\times n}$, $\boldsymbol{J}_i,\boldsymbol{L}_i,\boldsymbol{O}\in I\!R^{n\times m}$ are constant matrices.

The problem consists of designing the parameters such that the estimation error

$$\boldsymbol{e}(t)=\boldsymbol{q}(t)-\boldsymbol{q}_e(t) \quad (12)$$

converges asymptotically towards zero when $t\to\infty$ and the influence of unknown disturbance on the estimated state is minimized. But before going into the details it may be worth making some suitable general remarks about the nature of the problem.

Lemma 1: The equation describing dynamics of (12) in fault-free regime takes the autonomous form

$$\dot{\boldsymbol{e}}(t)=\sum_{i=1}^{s}\mathrm{h}_i(\boldsymbol{\theta}(t))\boldsymbol{P}_i\boldsymbol{e}(t) \quad (13)$$

if

$$\boldsymbol{P}_i=\boldsymbol{A}_i-\boldsymbol{O}\boldsymbol{C}\boldsymbol{A}_i-\boldsymbol{J}_i\boldsymbol{C} \quad (14)$$

$$\boldsymbol{Q}=\boldsymbol{I}_n-\boldsymbol{O}\boldsymbol{C} \quad (15)$$

$$\boldsymbol{L}_i=\boldsymbol{P}_i\boldsymbol{O} \quad (16)$$

$$(\boldsymbol{I}_n-\boldsymbol{O}\boldsymbol{C})\boldsymbol{E}=\boldsymbol{0} \quad (17)$$

while the necessary condition of a solution of (17) is prescribed by the relation

$$rank(\boldsymbol{S})=rank(\boldsymbol{E}),\quad \boldsymbol{S}=\boldsymbol{C}\boldsymbol{E} \quad (18)$$

Proof: Considering fault-free scene, then it follows from (1), (2) and (10), (11)

$$\dot{\boldsymbol{e}}(t)=\sum_{i=1}^{s}\mathrm{h}_i(\boldsymbol{\theta}(t))\dot{\boldsymbol{e}}_i(t) \quad (19)$$

$$\begin{aligned}\dot{\boldsymbol{e}}_i(t)&=\boldsymbol{A}_i\boldsymbol{q}(t)+\boldsymbol{B}\boldsymbol{u}(t)+\boldsymbol{E}\boldsymbol{d}(t)-\dot{\boldsymbol{p}}(t)-\boldsymbol{O}\dot{\boldsymbol{y}}(t)\\&=\boldsymbol{A}_i\boldsymbol{q}(t)+\boldsymbol{B}\boldsymbol{u}(t)+\boldsymbol{E}\boldsymbol{d}(t)-\boldsymbol{P}_i\boldsymbol{p}(t)\\&-\boldsymbol{Q}\boldsymbol{B}\boldsymbol{u}(t)-\boldsymbol{J}_i\boldsymbol{C}\boldsymbol{q}(t)-\boldsymbol{L}_i\boldsymbol{y}(t)\\&-\boldsymbol{O}\boldsymbol{C}(\boldsymbol{A}_i\boldsymbol{q}(t)+\boldsymbol{B}\boldsymbol{u}(t)+\boldsymbol{E}\boldsymbol{d}(t))\end{aligned} \quad (20)$$

Thus, it will be convenient to write using (11), (12),

$$\begin{aligned}\dot{\boldsymbol{e}}_i(t)&=\boldsymbol{A}_i\boldsymbol{q}(t)+\boldsymbol{B}\boldsymbol{u}(t)+\boldsymbol{E}\boldsymbol{d}(t)-\boldsymbol{J}_i\boldsymbol{C}\boldsymbol{q}(t)\\&-\boldsymbol{P}_i(\boldsymbol{q}_e(t)-\boldsymbol{O}\boldsymbol{y}(t))-\boldsymbol{Q}\boldsymbol{B}\boldsymbol{u}(t)-\boldsymbol{L}_i\boldsymbol{y}(t)\\&-\boldsymbol{O}\boldsymbol{C}(\boldsymbol{A}_i\boldsymbol{q}(t)+\boldsymbol{B}\boldsymbol{u}(t)+\boldsymbol{E}\boldsymbol{d}(t))\\&=\boldsymbol{A}_i\boldsymbol{q}(t)+\boldsymbol{B}\boldsymbol{u}(t)+\boldsymbol{E}\boldsymbol{d}(t)-\boldsymbol{J}_i\boldsymbol{C}\boldsymbol{q}(t)\\&-\boldsymbol{P}_i(\boldsymbol{q}(t)-\boldsymbol{e}(t)-\boldsymbol{O}\boldsymbol{y}(t))-\boldsymbol{Q}\boldsymbol{B}\boldsymbol{u}(t)\\&-\boldsymbol{O}\boldsymbol{C}(\boldsymbol{A}_i\boldsymbol{q}(t)+\boldsymbol{B}\boldsymbol{u}(t)+\boldsymbol{E}\boldsymbol{d}(t))-\boldsymbol{L}_i\boldsymbol{y}(t)\end{aligned} \quad (21)$$

and after manipulations of the above expression it becomes

$$\begin{aligned}\dot{\boldsymbol{e}}_i(t)&=\boldsymbol{P}_i\boldsymbol{e}(t)+(\boldsymbol{I}_n-\boldsymbol{Q}-\boldsymbol{O}\boldsymbol{C})\boldsymbol{B}\boldsymbol{u}(t)\\&+(\boldsymbol{A}_i-\boldsymbol{P}_i-\boldsymbol{J}_i\boldsymbol{C}-\boldsymbol{O}\boldsymbol{C}\boldsymbol{A}_i)\boldsymbol{q}(t)\\&+(\boldsymbol{P}_i\boldsymbol{O}-\boldsymbol{L}_i)\boldsymbol{y}(t)+(\boldsymbol{I}_n-\boldsymbol{O}\boldsymbol{C})\boldsymbol{E}\boldsymbol{d}(t)\end{aligned} \quad (22)$$

It is an immediate consequence that estimation error equation takes the form (13) if (14)-(17) are satisfied. This concludes the proof. ■

Lemma 2: Considering (17), for the matrix pseudoinverse principle in design formulation, then the parameter equation (14) can be easily translated into the notation

$$\boldsymbol{P}_i=\boldsymbol{A}_i^{\circ}-\boldsymbol{J}_i^{\circ}\boldsymbol{C}_i^{\circ},\quad i=1,2,\dots s, \quad (23)$$

where

$$\boldsymbol{\Upsilon}=\boldsymbol{I}_n-\boldsymbol{E}\boldsymbol{S}^{\ominus 1}\boldsymbol{C},\quad \boldsymbol{\Lambda}=\boldsymbol{I}_m-\boldsymbol{S}\boldsymbol{S}^{\ominus 1} \quad (24)$$

$$\boldsymbol{A}_i^{\circ}=\boldsymbol{\Upsilon}\boldsymbol{A}_i,\quad \boldsymbol{J}_i^{\circ}=\begin{bmatrix}\boldsymbol{J}_i & \boldsymbol{O}^{\circ}\end{bmatrix}\quad \boldsymbol{C}_i^{\circ}=\begin{bmatrix}\boldsymbol{C}\\ \boldsymbol{C}\boldsymbol{A}_i^{\circ}\end{bmatrix} \quad (25)$$

the pseudo-inverse of the matrix $\boldsymbol{S}$ is

$$\boldsymbol{S}^{\ominus 1}=(\boldsymbol{S}^{\mathrm{T}}\boldsymbol{S})^{-1}\boldsymbol{S}^{\mathrm{T}} \quad (26)$$

$\boldsymbol{O}^{\circ}$ is a design parameter matrix such that

$$\boldsymbol{O}=\boldsymbol{E}\boldsymbol{S}^{\ominus 1}+\boldsymbol{O}^{\circ}\boldsymbol{\Lambda} \quad (27)$$

and $\boldsymbol{\Lambda}$ is the orthogonal projector of $\boldsymbol{S}$, while the couples $(\boldsymbol{A}_i^{\circ},\boldsymbol{C}_i^{\circ})$ have to be observable for all i.

Proof: To obtain the observer attenuation towards $\boldsymbol{d}(t)$, equation (17), written more explicitly, are then given by the expression

$$\boldsymbol{O}\boldsymbol{C}\boldsymbol{E}=\boldsymbol{E} \quad (28)$$

Multiplying the right side of (28) by identity matrix of appropriate dimension means

$$\boldsymbol{O}\boldsymbol{C}\boldsymbol{E}=\boldsymbol{E}((\boldsymbol{C}\boldsymbol{E})^{\mathrm{T}}\boldsymbol{C}\boldsymbol{E})^{-1}(\boldsymbol{C}\boldsymbol{E})^{\mathrm{T}}\boldsymbol{C}\boldsymbol{E} \quad (29)$$

and with (18) it is an immediate consequence that the particular solution of (29) is

$$\boldsymbol{O}=\boldsymbol{E}\boldsymbol{S}^{\ominus 1} \quad (30)$$

where $\boldsymbol{S}^{\ominus 1}$ is defined in (26).

Since an implication of (28) and (25) is

$$\boldsymbol{C}\boldsymbol{E}-\boldsymbol{C}\boldsymbol{O}\boldsymbol{S}=\boldsymbol{S}-\boldsymbol{S}\boldsymbol{S}^{\ominus 1}\boldsymbol{S}=\boldsymbol{\Lambda}\boldsymbol{S}=\boldsymbol{0} \quad (31)$$

also every solution from the null space of the matrix $\boldsymbol{S}$ of the form

$$\boldsymbol{O}_s=\boldsymbol{O}^{\circ}\boldsymbol{\Lambda} \quad (32)$$

satisfies (28), where $\boldsymbol{O}^{\circ}\in I\!R^{n\times m}$ is an arbitrary matrix. To respect this fact, a general relationship for the parameter matrix $\boldsymbol{O}$ is (27). The important information, of course, is that $\boldsymbol{E}\boldsymbol{S}^{\ominus 1}\boldsymbol{S}\neq\boldsymbol{E}$ strictly.

Applying (26), the preceding form of the matrix $\boldsymbol{P}_i$ can be written more explicitly as

$$\begin{aligned}\boldsymbol{P}_i&=\boldsymbol{A}_i-\boldsymbol{O}\boldsymbol{C}\boldsymbol{A}_i-\boldsymbol{J}_i\boldsymbol{C}\\&=(\boldsymbol{I}_n-\boldsymbol{E}\boldsymbol{S}^{\ominus 1}\boldsymbol{C})\boldsymbol{A}_i\\&-\boldsymbol{O}^{\circ}(\boldsymbol{I}_m-\boldsymbol{S}\boldsymbol{S}^{\ominus 1})\boldsymbol{C}\boldsymbol{A}_i-\boldsymbol{J}_i\boldsymbol{C}\end{aligned} \quad (33)$$

Since relations (24) expressed in term of the matrix product with $\boldsymbol{C}$ are as follows

$$\begin{aligned}\boldsymbol{\Lambda}\boldsymbol{C}&=(\boldsymbol{I}_m-\boldsymbol{S}\boldsymbol{S}^{\ominus 1})\boldsymbol{C}\\&=\boldsymbol{C}-\boldsymbol{C}\boldsymbol{E}\boldsymbol{S}^{\ominus 1}\boldsymbol{C}\\&=\boldsymbol{C}(\boldsymbol{I}_n-\boldsymbol{E}\boldsymbol{W}^{\ominus 1}\boldsymbol{C})\\&=\boldsymbol{C}\boldsymbol{\Upsilon}\end{aligned} \quad (34)$$

the associated relation between $\boldsymbol{\Lambda}$ and $\boldsymbol{\Upsilon}$ comes from (34) which together with equation (33) implies that

$$\boldsymbol{P}_i=\boldsymbol{\Upsilon}\boldsymbol{A}_i-\boldsymbol{O}^{\circ}\boldsymbol{C}\boldsymbol{\Upsilon}\boldsymbol{A}_i-\boldsymbol{J}_i\boldsymbol{C} \quad (35)$$

Thus, using (25), then in matrix notation one obtains (23) from (35). This concludes the proof. ■

Equations (23) coincide with the standard equation for Luenberger observer design, except for the minor changes in notation, and it is evident that pole placement methods for multi inputs, multi outputs (MIMO) systems can be used in the synthesis if it is only an unknown input observer design task.

Considerably more challenging to explain fault residual filter design is treated in the following, focusing on linear matrix inequality principle prioritization to parameter solution for unknown input observers and fault residual filters.

Theorem 1: The unknown input TS fuzzy observer (10), (11) is asymptotically stable if there exist a positive definite symmetric matrix $\boldsymbol{V}\in I\!R^{n\times n}$, and matrices $\boldsymbol{W}_i,\boldsymbol{W}\in I\!R^{n\times m}$, $i=1,2,\dots,s$ such that for all i

$$\boldsymbol{V}=\boldsymbol{V}^{\mathrm{T}}>0 \quad (36)$$

$$\boldsymbol{V}\boldsymbol{A}_i^{\circ}+\boldsymbol{A}_i^{\circ\mathrm{T}}\boldsymbol{V}-\boldsymbol{W}_i^{\circ}\boldsymbol{C}_i^{\circ}-\boldsymbol{C}_i^{\circ\mathrm{T}}\boldsymbol{W}_i^{\circ\mathrm{T}}<0 \tag{37}$$

where

$$\boldsymbol{W}_i^{\circ}=\begin{bmatrix}\boldsymbol{W}_i & \boldsymbol{W}\end{bmatrix} \tag{38}$$

is structured matrix variable.

In the affirmative case, the matrices $\boldsymbol{J}_i^{\circ}$ are given by

$$\boldsymbol{J}_i^{\circ}=\boldsymbol{V}^{-1}\boldsymbol{W}_i^{\circ}=\begin{bmatrix}\boldsymbol{J}_i & \boldsymbol{O}^{\circ}\end{bmatrix} \tag{39}$$

the matrices $\boldsymbol{P}_i$ can be computed as

$$\boldsymbol{P}_i=\boldsymbol{A}_i^{\circ}-\boldsymbol{J}_i^{\circ}\boldsymbol{C}_i^{\circ} \tag{40}$$

the matrices $\boldsymbol{J}_i$ are separable from (39), the matrix $\boldsymbol{O}$ can be constructed using from (39) separable $\boldsymbol{O}^{\circ}$ as

$$\boldsymbol{O}=\boldsymbol{E}\boldsymbol{S}^{\ominus 1}+\boldsymbol{O}^{\circ}\boldsymbol{\Lambda} \tag{41}$$

and, subsequently, according (15), (16),

$$\boldsymbol{Q}=\boldsymbol{I}_n-\boldsymbol{O}\boldsymbol{C} \tag{42}$$

$$\boldsymbol{L}_i=\boldsymbol{P}_i\boldsymbol{O} \tag{43}$$

respectively, where the design parameters are given by relations (24)-(26).

Proof: Defining the positive definite Lyapunov function

$$v(\boldsymbol{e}(t))=\boldsymbol{e}^{\mathrm{T}}(t)\boldsymbol{V}\boldsymbol{e}(t)>0 \tag{44}$$

where $\boldsymbol{V}\in\mathbb{R}^{n\times n}$ is a positive definite symmetric matrix, then, evaluating the time derivative of (44), it has to be satisfied

$$\dot{v}(\boldsymbol{e}(t))=\dot{\boldsymbol{e}}^{\mathrm{T}}(t)\boldsymbol{V}\boldsymbol{e}(t)+\boldsymbol{e}^{\mathrm{T}}(t)\boldsymbol{V}\dot{\boldsymbol{e}}(t)<0 \tag{45}$$

Substitution of (13) in (45) gives

$$\begin{aligned}\dot{v}(\boldsymbol{e}(t)) &= \boldsymbol{e}^{\mathrm{T}}(t)\boldsymbol{V}\sum_{i=1}^{s}\mathrm{h}_i(\boldsymbol{\theta}(t))\boldsymbol{P}_i\boldsymbol{e}(t)+\\ &+\boldsymbol{e}^{\mathrm{T}}(t)\sum_{i=1}^{s}\mathrm{h}_i(\boldsymbol{\theta}(t))\boldsymbol{P}_i^{\mathrm{T}}\boldsymbol{V}\boldsymbol{e}(t)\end{aligned} \tag{46}$$

which means it has the following structure

$$\dot{v}(\boldsymbol{e}(t))=\sum_{i=1}^{s}\mathrm{h}_i(\boldsymbol{\theta}(t))\boldsymbol{e}^{\mathrm{T}}(t)(\boldsymbol{V}\boldsymbol{P}_i+\boldsymbol{P}_i^{\mathrm{T}}\boldsymbol{V})\boldsymbol{e}(t) \tag{47}$$

Then there exists a solution if, independently on the membership functions $\mathrm{h}_i(\boldsymbol{\theta}(t))$,

$$\boldsymbol{V}\boldsymbol{P}_i+\boldsymbol{P}_i^{\mathrm{T}}\boldsymbol{V}<0 \text{ for all } i \tag{48}$$

Inserting (23) then (48) gives

$$\boldsymbol{V}\boldsymbol{A}_i^{\circ}-\boldsymbol{V}\boldsymbol{J}_i^{\circ}\boldsymbol{C}_i^{\circ}+\boldsymbol{A}_i^{\circ\mathrm{T}}\boldsymbol{V}-\boldsymbol{C}_i^{\circ\mathrm{T}}\boldsymbol{J}_i^{\circ\mathrm{T}}\boldsymbol{V}<0 \tag{49}$$

and, denoting

$$\boldsymbol{V}\boldsymbol{J}_i^{\circ}=\boldsymbol{W}_i^{\circ} \tag{50}$$

then (49) implies (37). This concludes the proof. ■

IV. TS Fuzzy Fault Detection Filter

Working with additive faults, then (19) has the following structure

$$\dot{\boldsymbol{e}}(t)=\sum_{i=1}^{s}\mathrm{h}_i(\boldsymbol{\theta}(t))\dot{\boldsymbol{e}}_i(t)+\boldsymbol{Q}\boldsymbol{F}\boldsymbol{f}(t) \tag{51}$$

if (14)-(17) are satisfied. Without loss of generality, then on above given unknown input observer can be build the fault residual filter, defined as

$$\boldsymbol{r}(t)=\boldsymbol{R}\boldsymbol{C}\boldsymbol{e}(t) \tag{52}$$

where $\boldsymbol{R}\in\mathbb{R}^{m\times m}$ is the filter gain matrix, optimized for H_∞ norm of $\boldsymbol{G}(s)$, giving in the Laplace transform mapping

$$\hat{\boldsymbol{r}}(s)=\boldsymbol{G}(s)\hat{\boldsymbol{f}}(s) \tag{53}$$

Therefore, $\boldsymbol{r}(t)\in\mathbb{R}^m$ is the vector of fault residuals, approximately equal zero vector in fault-free regime.

Theorem 2: The fault residual filter built on observer with unknown input (10), (11) is asymptotically stable if there exists positive definite symmetric matrices $\boldsymbol{V},\boldsymbol{H}\in\mathbb{R}^{n\times n}$, matrices $\boldsymbol{W}_i,\boldsymbol{W}\in\mathbb{R}^{n\times m}$, $\boldsymbol{R}\in\mathbb{R}^{m\times m}$ and a positive scalar $\gamma\in\mathbb{R}$ such that for all i

$$\boldsymbol{V}=\boldsymbol{V}^{\mathrm{T}}>0,\quad \boldsymbol{H}=\boldsymbol{H}^{\mathrm{T}}>0,\quad \gamma>0 \tag{54}$$

$$\begin{bmatrix}\boldsymbol{V}\boldsymbol{A}_i^{\circ}+\boldsymbol{A}_i^{\circ\mathrm{T}}\boldsymbol{V}-\boldsymbol{W}_i^{\circ}\boldsymbol{C}_i^{\circ}-\boldsymbol{C}_i^{\circ\mathrm{T}}\boldsymbol{W}_i^{\circ\mathrm{T}}+\boldsymbol{H} & * & * \\ (\boldsymbol{V}\boldsymbol{\Upsilon}\boldsymbol{F}-\boldsymbol{W}\boldsymbol{C}\boldsymbol{\Upsilon}\boldsymbol{F})^{\mathrm{T}} & -\gamma\boldsymbol{I}_v & * \\ \boldsymbol{R}\boldsymbol{C} & \boldsymbol{0} & -\gamma\boldsymbol{I}_m\end{bmatrix}<0 \tag{55}$$

where the design parameters are given by (24)-(26).

When the above conditions are affirmative, the unknown input observer parameters are defined by (39)-(43) and the fault residual filter gain is defined directly by the resulting matrix variable $\boldsymbol{R}$.

Proof: Since a positive Lyapunov function

$$\begin{aligned}v(\boldsymbol{e}(i)) &= \boldsymbol{e}^{\mathrm{T}}(i)\boldsymbol{V}\boldsymbol{e}(i)+\\ &+\gamma^{-1}\int_0^t(\boldsymbol{r}^{\mathrm{T}}(\tau)\boldsymbol{r}(\tau)-\gamma^2\boldsymbol{f}^{\mathrm{T}}(\tau)\boldsymbol{f}(\tau))\mathrm{d}\tau>0\end{aligned} \tag{56}$$

is possible to be applied, then, evaluating the time derivative, the inequality (56) for stable error trajectories must apply

$$\begin{aligned}\dot{v}(\boldsymbol{e}(i)) &= \dot{\boldsymbol{e}}^{\mathrm{T}}(t)\boldsymbol{V}\boldsymbol{e}(t)+\boldsymbol{e}^{\mathrm{T}}(t)\boldsymbol{V}\dot{\boldsymbol{e}}(t)+\\ &+\gamma^{-1}\boldsymbol{r}^{\mathrm{T}}(t)\boldsymbol{r}(t)-\gamma\boldsymbol{f}^{\mathrm{T}}(t)\boldsymbol{f}(t)<0\end{aligned} \tag{57}$$

Therefore, substitution of (51) and (52) in (57) gives

$$\begin{aligned}\dot{v}(\boldsymbol{e}(t)) &= \boldsymbol{e}^{\mathrm{T}}(t)\boldsymbol{V}(\sum_{i=1}^{s}\mathrm{h}_i(\boldsymbol{\theta}(t))\boldsymbol{P}_i\boldsymbol{e}(t)+\boldsymbol{Q}\boldsymbol{F}\boldsymbol{f}(t))+\\ &+(\boldsymbol{f}^{\mathrm{T}}(t)\boldsymbol{F}^{\mathrm{T}}\boldsymbol{Q}^{\mathrm{T}}+\boldsymbol{e}^{\mathrm{T}}(t)\sum_{i=1}^{s}\mathrm{h}_i(\boldsymbol{\theta}(t))\boldsymbol{P}_i^{\mathrm{T}})\boldsymbol{V}\boldsymbol{e}(t)+\\ &+\gamma^{-1}\boldsymbol{e}^{\mathrm{T}}(t)\boldsymbol{C}^{\mathrm{T}}\boldsymbol{R}^{\mathrm{T}}\boldsymbol{R}\boldsymbol{C}\boldsymbol{r}(t)-\gamma\boldsymbol{f}^{\mathrm{T}}(t)\boldsymbol{f}(t)\end{aligned} \tag{58}$$

Forming new vector with the components $\boldsymbol{e}(t)$, $\boldsymbol{f}(t)$ as

$$\boldsymbol{e}_o^{\mathrm{T}}(t)=\begin{bmatrix}\boldsymbol{e}^{\mathrm{T}}(t) & \boldsymbol{f}^{\mathrm{T}}(t)\end{bmatrix} \tag{59}$$

the condition expressed in (58) can rewrite as

$$\sum_{i=1}^{s}\mathrm{h}_i(\boldsymbol{\theta}(t))\,\boldsymbol{e}_o^{\mathrm{T}}(t)\boldsymbol{V}_{oi}\boldsymbol{e}_o(t)<0 \tag{60}$$

where for all i

$$\boldsymbol{V}_{oi}=\begin{bmatrix}\boldsymbol{V}\boldsymbol{P}_i+\boldsymbol{P}_i^{\mathrm{T}}\boldsymbol{V}+\gamma^{-1}\boldsymbol{C}^{\mathrm{T}}\boldsymbol{R}^{\mathrm{T}}\boldsymbol{R}\boldsymbol{C} & * \\ (\boldsymbol{V}\boldsymbol{Q}\boldsymbol{F})^{\mathrm{T}} & -\gamma\boldsymbol{I}_v\end{bmatrix}<0 \tag{61}$$

Since, by definition of Schur complement, matrix inequality (61) implies that for all i

$$\begin{bmatrix}\boldsymbol{V}\boldsymbol{P}_i+\boldsymbol{P}_i^{\mathrm{T}}\boldsymbol{V} & * & * \\ (\boldsymbol{V}\boldsymbol{Q}\boldsymbol{F})^{\mathrm{T}} & -\gamma\boldsymbol{I}_v & * \\ \boldsymbol{R}\boldsymbol{C} & 0 & -\gamma\boldsymbol{I}_m\end{bmatrix}<0 \tag{62}$$

reflecting structure of (48) together with (49), (50), it can write

$$\begin{bmatrix}\boldsymbol{V}\boldsymbol{A}_i^{\circ}+\boldsymbol{A}_i^{\circ\mathrm{T}}\boldsymbol{V}-\boldsymbol{W}_i^{\circ}\boldsymbol{C}_i^{\circ}-\boldsymbol{C}_i^{\circ\mathrm{T}}\boldsymbol{W}_i^{\circ\mathrm{T}} & * & * \\ (\boldsymbol{V}\boldsymbol{Q}\boldsymbol{F})^{\mathrm{T}} & -\gamma\boldsymbol{I}_v & * \\ \boldsymbol{R}\boldsymbol{C} & \boldsymbol{0} & -\gamma\boldsymbol{I}_m\end{bmatrix}<0 \tag{63}$$

What it is need withal is an inclusion of (34), (41) and (42) which is adapted as follows

$$\begin{aligned}\boldsymbol{V}\boldsymbol{Q}\boldsymbol{F} &= \boldsymbol{V}(\boldsymbol{I}_n-\boldsymbol{O}\boldsymbol{C})\boldsymbol{F}\\ &= \boldsymbol{V}\boldsymbol{F}-\boldsymbol{V}\boldsymbol{E}\boldsymbol{S}^{\ominus 1}\boldsymbol{C}\boldsymbol{F}-\boldsymbol{V}\boldsymbol{O}^{\circ}\boldsymbol{\Lambda}\boldsymbol{C}\boldsymbol{F}\\ &= \boldsymbol{V}\boldsymbol{\Upsilon}\boldsymbol{F}-\boldsymbol{V}\boldsymbol{O}^{\circ}\boldsymbol{C}\boldsymbol{\Upsilon}\boldsymbol{F}\end{aligned} \tag{64}$$

Observing from (38) and (50) that

$$\boldsymbol{W}_i^{\circ}=\boldsymbol{V}\boldsymbol{J}_i^{\circ}=\begin{bmatrix}\boldsymbol{V}\boldsymbol{J}_i & \boldsymbol{V}\boldsymbol{O}^{\circ}\end{bmatrix}=\begin{bmatrix}\boldsymbol{W}_i & \boldsymbol{W}\end{bmatrix} \tag{65}$$

then, evidently,

$$\boldsymbol{V}\boldsymbol{Q}\boldsymbol{F}=\boldsymbol{V}\boldsymbol{\Upsilon}\boldsymbol{F}-\boldsymbol{W}\boldsymbol{C}\boldsymbol{\Upsilon}\boldsymbol{F} \tag{66}$$

Regularizing LMI using the Lyapunov-Krasovskii principle by a symmetric positive definite matrix $\boldsymbol{H}$ [8], then (63) implies (55). This concludes the proof. ■

V. TS Model of Wind Turbine

The overall model is given by interconnecting the models of the individual subsystems, considering a generic three blade horizontal wind turbine, driven by variable speeds, with a full converter coupling [9].

The aerodynamic power extracted from the wind is [11], [12]

$$\tau_r(t) = \frac{1}{2}\rho\pi R^3 C_q(\lambda(t),\beta(t))v^2(t) \tag{67}$$

where $v(t)$ is actual speed of wind [m/s], R is rotor radius [m], $C_q(*)$ is torque coefficient expressing what fraction of the power in the wind is extracted by wind turbine and ρ is air density [kg/m^3]. The torque coefficient is non-linearly depending on the tip speed ratio λ [rad] between the tangential speed of the tip of a blade and the actual speed of wind, parameterized by the blades pitch angle β [rad], where

$$\lambda(t) = \frac{\omega_r(t)R}{v(t)}, \quad C_q(\lambda(t),\beta(t)) = \frac{C_p(\lambda(t),\beta(t))}{\lambda(t)} \tag{68}$$

ω_r is rotor angular speed [rad/s], $C_q(\lambda(t),\beta(t))$ is the rotor torque coefficient. Consequently, the turbine speed is controlled to tracks its reference $\omega_{rt}(t)$, to maintain the tip-speed ratio at its optimal value λ_o, by relation implying from (68)

$$\omega_{rt}(t) = \lambda_m \frac{v(t)}{R} \tag{69}$$

and, substituting $\lambda(t)$ into (67), it can obtain

$$\tau_r(t) = \frac{1}{2\lambda^2(t)}\rho\pi R^5 C_q(\lambda(t),\beta(t))\omega_r^2(t) \tag{70}$$

Then the proposed two-mass drive train model is

$$\begin{aligned} J_r\dot{\omega}_r(t) &= \tau_r(t) - (B_{dt}+B_r)\,\omega_r(t) - \\ &-K_{dt}\theta_\Delta(t) + \tfrac{B_{dt}}{N_g}\omega_g(t) \end{aligned} \tag{71}$$

$$\begin{aligned} J_g\dot{\omega}_g(t) &= \tfrac{\mu_{dt}K_{dt}}{N_g}\theta_\Delta(t) + \tfrac{\mu_{dt}B_{dt}}{N_g}\omega_r(t) - \\ &-\left(\tfrac{\mu_{dt}B_{dt}}{N_g^2} + B_g\right)\omega_g(t) - \tau_g(t) \end{aligned} \tag{72}$$

$$\dot{\theta}_\Delta(t) = \omega_r(t) - \frac{1}{N_g}\omega_g(t) \tag{73}$$

where $\omega_r(t)$ is rotor angular speed [rad/s], $\omega_g(t)$ is the generator rotating speed [rad/s], $\theta_\Delta(t)$ is torsion angle [rad], $\tau_g(t)$ is aerodynamic torque [kg m^2/s^2], J_r is rotor inertia moment [kg m^2], N_g is gear ration, B_{dt} is torsion damping coefficient [Nms/rad], B_r is rotor external damping [Nms/rad], K_{dt} is torsion stiffness [Nm/rad], B_g is generator external damping [Nms/rad] and μ_{dt} is efficiency od drive train.

The hydraulic pitch system is modeled using second-order differential equation [10]

$$\ddot{\beta}(t) + 2\zeta\omega_n\dot{\beta}(t) + \omega_n^2\beta(t) = \omega_n^2\beta_r(t) \tag{74}$$

since, in order to limit the control effort, the reference speed signal is passed through the second order linear filter, while $\beta(t)$ is measured pitch angle [rad], $\beta_r(t)$ is reference pitch angle [rad], ω_n is pitch system filter natural frequency [rad/s], and ζ is damping factor.

Finally, the generator and converter dynamics are modeled by the first-order differential equation

$$\dot{\tau}_g(t) + \alpha\tau_g(t) = \alpha\tau_{gr}(t) \tag{75}$$

where $\tau_g(t)$ is the generator torque [Nm], $\tau_{gr}(t)$ is the generator torque reference [Nm] and α is the generator and converter model parameter [1/s].

Defining the state model variables

$$\begin{array}{lll} q_1(t)=\omega_r(t), & q_2(t)=\omega_g(t), & q_3(t)=\theta_\Delta(t) \\ q_4(t)=\dot{\beta}(t) & q_5(t)=\beta(t) & q_6(t)=\tau_g(t) \\ u_1(t)=\tau_{gr}(t) & u_2(t)=\beta_r(t) & \end{array} \tag{76}$$

then

$$\boldsymbol{A}_i(\boldsymbol{\theta}(t)) = \begin{bmatrix} a_{11}(\boldsymbol{\theta}(t)) & a_{12} & -\frac{K_{dt}}{J_r} & 0 & 0 & 0 \\ a_{21} & a_{22} & a_{23} & 0 & 0 & 0 \\ 1 & -\frac{1}{N_g} & 0 & 0 & 0 & 0 \\ 0 & 0 & 0 & -2\zeta\omega_n & -\omega_n^2 & 0 \\ 0 & 0 & 0 & 1 & 0 & 0 \\ 0 & 0 & 0 & 0 & 0 & -\alpha \end{bmatrix} \tag{77}$$

$$\boldsymbol{B}^{\mathrm{T}} = \begin{bmatrix} 0 & 0 & 0 & 0 & 0 & \alpha \\ 0 & 0 & 0 & \omega_n & 0 & 0 \end{bmatrix}, \ \boldsymbol{C} = \begin{bmatrix} 1 & 0 & 0 & 0 & 0 & 0 \\ 0 & 1 & 0 & 0 & 0 & 0 \\ 0 & 0 & 0 & 0 & 1 & 0 \\ 0 & 0 & 0 & 0 & 0 & 1 \end{bmatrix} \tag{78}$$

where

$$\boldsymbol{q}^{\mathrm{T}}(t) = \begin{bmatrix} q_1(t) & q_2(t) & q_3(t) & q_4(t) & q_5(t) & q_6(t) \end{bmatrix} \tag{79}$$

$$\boldsymbol{u}^{\mathrm{T}}(t) = \begin{bmatrix} u_1(t) & u_2(t) \end{bmatrix} \tag{80}$$

$$\boldsymbol{y}^{\mathrm{T}}(t) = \begin{bmatrix} y_1(t) & y_2(t) & y_3(t) & y_4(t) \end{bmatrix} \tag{81}$$

and

$$\begin{aligned} a_{12} &= \tfrac{B_{dt}}{J_r N_g}, & a_{21} &= \mu_{dt}\tfrac{B_{dt}}{J_g N_g} \\ a_{23} &= \mu_{dt}\tfrac{K_{dt}}{J_g N_g}, & a_{22} &= -\tfrac{\mu_{dt}B_{dt}}{J_g N_g^2} - \tfrac{B_g}{J_g} \end{aligned} \tag{82}$$

$$a_{11}(\boldsymbol{\theta}(t)) = \tfrac{\rho\pi R^5 C_q(\lambda(t),\beta(t))\omega_r(t)}{2J_r\lambda^2(t)} - \tfrac{B_{dt}+B_r}{J_r} \tag{83}$$

Respecting the structure of a single matrix non-linear element $a_{11}(\boldsymbol{\theta}(t)) = a_{11}(\omega_r(t),\lambda(t),\beta(t))$ the following three variables $\omega_r(t),\lambda(t),\beta(t)$ are selected as the premise variables.

Choosing the sector of $\omega_r(t)$ at the operating point as $\theta_1(t) = \omega_r(t) \in \langle\omega_{rn},\omega_{rm}\rangle$ then

$$\omega_r(t) = w_1^\phi(\theta_1(t)) + w_2^\phi(\theta_1(t)) \tag{84}$$

$$w_1^\phi(\theta_1(t)) = \frac{\omega_{rm} - \omega_r(t)}{\omega_{rm} - \omega_{rn}} \tag{85}$$

$$w_2^\phi(\theta_1(t)) = 1 - w_1^\phi(\theta_1(t)) \tag{86}$$

Choosing the sector of $\lambda(t)$ as $\theta_2(t) = \lambda(t) \in \langle\lambda_n,\lambda_s\rangle \cup \langle\lambda_s,\lambda_m\rangle$ then

$$\lambda(t) = w_1^\psi(\theta_2(t)) + w_2^\psi(\theta_2(t)) + w_3^\psi(\theta_2(t)) \tag{87}$$

$$\begin{aligned} w_1^\psi(\theta_2(t)) &= \tfrac{1}{2}\tfrac{\lambda_s-\lambda(t)}{\lambda_s-\lambda_n} + \\ &+ \tfrac{1}{2}\tfrac{\lambda_s-\lambda(t)}{\lambda_s-\lambda_n}\,\mathrm{sgn}(\lambda_s-\lambda(t)) \end{aligned} \tag{88}$$

$$\begin{aligned} w_3^\psi(\theta_2(t)) &= \tfrac{1}{2}\tfrac{\lambda(t)-\lambda_m}{\lambda_m-\lambda_s} + \\ &+ \tfrac{1}{2}\tfrac{\lambda(t)-\lambda_m}{\lambda_m-\lambda_s}\,\mathrm{sgn}(\lambda(t)-\lambda_m) \end{aligned} \tag{89}$$

$$w_2^\psi(\theta_2(t)) = 1 - w_1^\psi(\theta_2(t)) - w_3^\psi(\theta_2(t)) \tag{90}$$

and choosing the sector of $\beta(t)$ as $\theta_3(t) = \beta(t) \in \langle\beta_n,\beta_s\rangle \cup \langle\beta_s,\beta_m\rangle$ then

$$\beta(t) = w_1^\xi(\theta_3(t)) + w_2^\xi(\theta_3(t)) + w_3^\xi(\theta_3(t)) \tag{91}$$

$$\begin{aligned} w_1^\xi(\theta_3(t)) &= \tfrac{1}{2}\tfrac{\beta_s-\beta(t)}{\beta_s-\beta_n} + \\ &+ \tfrac{1}{2}\tfrac{\beta_s-\beta(t)}{\beta_s-\beta_n}\,\mathrm{sgn}(\beta_s-\beta(t)) \end{aligned} \tag{92}$$

$$\begin{aligned} w_3^\xi(\theta_3(t)) &= \tfrac{1}{2}\tfrac{\beta(t)-\beta_m}{\beta_m-\beta_s} + \\ &+ \tfrac{1}{2}\tfrac{\beta(t)-\beta_m}{\beta_m-\beta_s}\,\mathrm{sgn}(\beta(t)-\beta_m) \end{aligned} \tag{93}$$

$$w_2^\xi(\theta_3(t)) = 1 - w_1^\xi(\theta_3(t)) - w_3^\xi(\theta_3(t)) \tag{94}$$

Note, just with such simple structure of membership functions there exist 18 different values of elements in the position $\boldsymbol{A}(1,1)$, and consequently 18 different matrices $\boldsymbol{A}_i$.

It should be noted that the premise variables are measurable or calculable from measured variables.

VI. Illustrative Example

The parameters used in the benchmark model are [9]

$$\begin{array}{lll} \rho = 1.225\frac{\text{kg}}{\text{m}^3} & R = 57.5\,\text{m} & N_g = 95 \\ B_{dt} = 775.49\frac{\text{Nms}}{\text{rad}} & B_r = 7.11\frac{\text{Nms}}{\text{rad}} & B_g = 45.6\frac{\text{Nms}}{\text{rad}} \\ K_{dt} = 2.7\times 10^6\frac{\text{Nm}}{\text{rad}} & J_g = 390\,\text{kg}\,\text{m}^2 & J_r = 55\times 10^6\,\text{kg}\,\text{m}^2 \\ \mu_{dt} = 0.97 & \omega_n = 11.11\frac{\text{rad}}{\text{s}} & \zeta = 0.6 \\ \alpha = 50\frac{1}{\text{s}} & & \end{array}$$

which imply directly nonzero parameters of the matrix $\boldsymbol{B}$ and $\boldsymbol{C}$. The structure of the matrix $\boldsymbol{A}$ is

$$\boldsymbol{A}_{\phi,\psi,\omega}(\boldsymbol{\theta}(t)) = \begin{bmatrix} a(\boldsymbol{\theta}(t)) & a_{12} & a_{13} & 0 & 0 & 0 \\ a_{21} & a_{22} & a_{23} & 0 & 0 & 0 \\ 1 & a_{32} & 0 & 0 & 0 & 0 \\ 0 & 0 & 0 & a_{44} & a_{45} & 0 \\ 0 & 0 & 0 & 1 & 0 & 0 \\ 0 & 0 & 0 & 0 & 0 & a_{66} \end{bmatrix}$$

where

$$\begin{array}{lll} a_{12} = 1.484\times 10^{-2} & a_{13} = -49.0909 & a_{21} = 0.0203 \\ a_{23} = 7.0688\times 10^4 & a_{22} = -0.1171 & a_{32} = -0.0105 \\ a_{44} = -13.332 & a_{45} = -123.4321 & a_{66} = -50 \end{array}$$

Expressing the considered working sectors as

$$\begin{array}{lll} \beta_n = -0.05 & \beta_s = 0.05 & \beta_m = 0.15 \\ \lambda_n = 2.0 & \lambda_s = 2.4 & \lambda_m = 2.8 \\ \omega_{rn} = 1.1 & \omega_{rm} = 1.4 & C_{qs} = 0.25 \end{array}$$

the matrix element $\boldsymbol{A}(11)$ on the position $(1,1)$, indexed in the sequence (ϕ,ψ,ω) are as follows

$$\begin{array}{lll} a_{111} = -0.8749 & a_{112} = 0.8749 & a_{112} = 2.6247 \\ a_{121} = -1.0499 & a_{122} = 1.0499 & a_{123} = 3.1496 \\ a_{131} = -1.2249 & a_{132} = 2.9397 & a_{133} = 3.6746 \\ a_{211} = -1.1135 & a_{212} = 1.1135 & a_{212} = 3.3405 \\ a_{221} = -1.3362 & a_{222} = 1.3362 & a_{223} = 4.0086 \\ a_{231} = -1.5589 & a_{232} = 3.7414 & a_{233} = 4.6768 \end{array}$$

This statement was chosen so that it would not be necessary to increase the number of state variables.

The unknown input observer exploits the measured state variables $q_1(t)$, $q_5(t)$, $q_6(t)$ that is the observer measurement system and the input disturbance matrix are composed as

$$\boldsymbol{C}=\begin{bmatrix} 1 & 0 & 0 & 0 & 0 & 0 \\ 0 & 0 & 0 & 0 & 1 & 1 \end{bmatrix}, \quad \boldsymbol{E}^{\mathrm{T}}=10^{-1}\begin{bmatrix} 0.8 & 0.5 & 0.2 & 0.1 & 0.2 & 0.1 \end{bmatrix}$$

while, detecting actuator faults, $\boldsymbol{F} = \boldsymbol{B}$. Thus, the derived design parameters are

$$\boldsymbol{S} = \begin{bmatrix} 0.08 \\ 0.03 \end{bmatrix}, \quad \boldsymbol{S}^{\ominus 1} = \begin{bmatrix} 10.9589 & 4.1096 \end{bmatrix}$$

$$\boldsymbol{\Upsilon} = \begin{bmatrix} 0.1233 & 0 & 0 & 0 & -0.3288 & -0.3288 \\ -0.5479 & 1 & 0 & 0 & -0.2055 & -0.2055 \\ -0.2192 & 0 & 1 & 0 & -0.0822 & -0.0822 \\ -0.1096 & 0 & 0 & 1 & -0.0411 & -0.0411 \\ -0.2192 & 0 & 0 & 0 & 0.9178 & -0.0822 \\ -0.1096 & 0 & 0 & 0 & -0.0411 & 0.9589 \end{bmatrix}$$

Within the above given matrix parameters, by solving (54), (55) using SeDuMi package, the design conditions imply feasible solution, giving the common matrix variables for optimized γ,

$$\gamma = 11.1505, \quad \boldsymbol{R} = \begin{bmatrix} -0.0012 & -0.0002 \\ -0.0010 & -0.0038 \end{bmatrix}$$

$$\boldsymbol{V}=\begin{bmatrix} 68.0652 & -0.0048 & -8.7607 & -0.9163 & -6.4275 & -10.1365 \\ -0.0048 & 0.0102 & -0.0853 & -0.0007 & -0.0233 & -0.0051 \\ -8.7607 & -0.0853 & 68.2850 & -0.2458 & 4.6081 & 19.0811 \\ -0.9163 & -0.0007 & -0.2458 & 0.8871 & 0.9580 & 0.0059 \\ -6.4275 & -0.0233 & 4.6081 & 0.9580 & 64.8657 & 1.1119 \\ -10.1365 & -0.0051 & 19.0811 & 0.0059 & 1.1119 & 106.5696 \end{bmatrix}$$

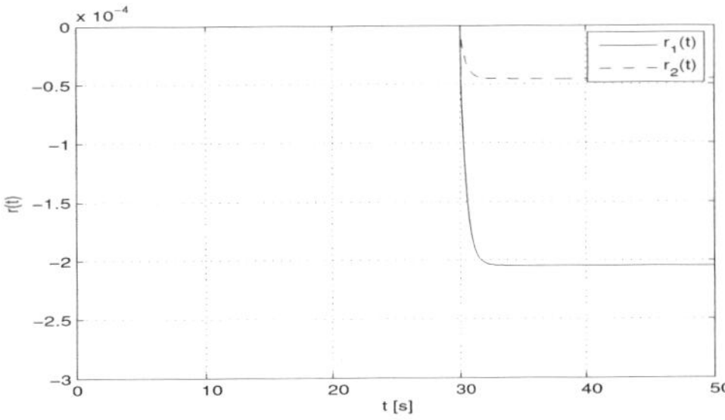

Figure 1. Residual generator time response

$$\boldsymbol{H}=\begin{bmatrix} 48.1801 & -0.0022 & -2.0616 & -0.0402 & 11.4225 & 0.7141 \\ -0.0022 & 0.0000 & -0.0003 & -0.0013 & 0.0020 & -0.0000 \\ -2.0616 & -0.0003 & 0.4558 & 0.1027 & 2.1509 & -0.0355 \\ -0.0402 & -0.0013 & 0.1027 & 1.4042 & -0.0191 & -0.0161 \\ 11.4225 & 0.0020 & 2.1509 & -0.0191 & 53.1137 & -0.1274 \\ 0.7141 & -0.0000 & -0.0355 & -0.0161 & -0.1274 & 59.9139 \end{bmatrix}$$

$$\boldsymbol{W}^{\mathrm{T}}=10^3\begin{bmatrix} 0.0008 & 0.0001 & -0.0020 & 0.0121 & -0.0001 & -6.7981 \\ -0.0152 & 0.0000 & 0.0181 & 0.0013 & 0.0002 & -0.5728 \end{bmatrix}$$

implying the common parameters of the TS fuzzy unknown input observers

$$\boldsymbol{O} = \begin{bmatrix} 0.8000 & -0.0000 \\ 0.5000 & 0.0001 \\ 0.2000 & 0.0000 \\ 0.1000 & -0.0004 \\ 0.2000 & 0.0000 \\ 0.0000 & 0.9999 \end{bmatrix}$$

$$\boldsymbol{Q} = \begin{bmatrix} 0.2 & 0 & 0 & 0 & -0.8 & 0.0000 \\ -0.5 & 1 & 0 & 0 & -0.5 & -0.0001 \\ -0.2 & 0 & 1 & 0 & -0.2 & -0.0000 \\ -0.1 & 0 & 0 & 1 & -0.1 & 0.0004 \\ -0.2 & 0 & 0 & 0 & 0.8 & -0.0000 \\ -0.0 & 0 & 0 & 0 & -0.0 & 0.0001 \end{bmatrix}$$

All other observer parameters differ for different indexes of linear sub-models. Due to the limited paper range, only values for $(\phi,\psi,\omega) = (1,1,1)$ are presented, where

$$\boldsymbol{P}_1 = \begin{bmatrix} -1.2895 & 0.0000 & -0.0487 & -0.0079 & -0.4215 & -4.3599 \\ -4.0533 & -0.1171 & 70.6885 & -0.0049 & -4.0779 & 17.7463 \\ 0.4075 & -0.0105 & 0.0001 & -0.0020 & -0.5942 & -0.8651 \\ 63.0343 & 0.0000 & 0.0001 & -13.3330 & -60.3987 & -6.8189 \\ -0.6735 & 0.0000 & 0.0001 & 0.9980 & -0.6752 & -0.7863 \\ -0.0928 & 0.0000 & -0.0049 & 0.0990 & -0.0062 & -1.1583 \end{bmatrix}$$

$$\boldsymbol{J}_1 = \begin{bmatrix} 0.4215 & 0.3986 \\ 4.0779 & -20.2158 \\ 0.5942 & -0.1237 \\ -63.0334 & 6.3021 \\ 0.6752 & -0.2034 \\ 0.0062 & 0.6583 \end{bmatrix}, \quad \boldsymbol{L}_1 = \begin{bmatrix} -2.0537 & -4.6245 \\ -9.6033 & 30.9991 \\ -0.1851 & -0.8955 \\ 14.9758 & -1.5589 \\ -0.6662 & -1.0442 \\ -0.0704 & -0.9538 \end{bmatrix}$$

while, obviously, $\boldsymbol{P}_1$ is Hurwitz. The parameter γ can be interpreted as an upper bound of the H_∞-norm, related to residual generator mapping.

To test the observer and filter performances, the system is stabilized in the nominal operating point [9]. To display residual simulations in the MATLAB and SIMULINK environment, the observer initial conditions are set as $\boldsymbol{p}(0) = \boldsymbol{0}$.

As the results Fig. 1 presents the TS fuzzy fault detection filter response, reflecting a step-like 50% gain loss of the second actuator at the time instant $t = 30\,$s. This examples illustrate the power that can be invoked through the prescribed H_∞ norm properties.

VII. Concluding Remarks

In the paper, a benchmark model is used for testing a fault detection scheme, applicable to wind turbines. Constructing the residual generator on Takagi-Sugeno fuzzy unknown input observer, the model is adapted for detection of pitch actuator faults. Since the fuzzy unknown input observer is designed for the TS model where the premise variables are measured state variables, the principle does not require redundancy in the sensor equipment.

Exploited fuzzy methodology has sufficient adaptability in correspondence to different operating points at which faults occurred, while primary quality of fault detection is covered by composed interconnection of non-trivial TS fuzzy rules. Simulation study indicates that the residual performance is satisfactory.

Extending of sector-nonlinearity modeling to polynomial cases, when the Taylor series are combined in TS models, exploiting membership-shape information as well as less conservative sufficient conditions for design based on TS fuzzy models are subject of future research in algorithmic field. From fault diagnosis point of view the possibility of isolating of single and multiple faults and residual directional properties and their quick responses are challenging subsequent research works in wind turbine diagnosis. It is worth to point out that the proposed method can be simply adjusted to general TS fuzzy systems with additive faults and unknown disturbances, which would find applications in wide industrial systems.

Acknowledgement

The work presented in this paper was supported by VEGA, the Grant Agency of Ministry of Education and Academy of Science of Slovak Republic, under Grant No. 1/0608/17. This support is very gratefully acknowledged.

References

[1] P.F. Odgaard and J. Stoustrup, "Fault tolerant control of wind turbines using unknown input observers," IFAC Proceedings Volumes, vol. 45, no. 20, pp. 313-318, 2012.

[2] T. Takagi and M. Sugeno, "Fuzzy identification of systems and its applications to modeling and control," IEEE Tran. Systems, Man and Cybernetics, vol. 15, no. 1, pp. 116-132, 1985.

[3] M. Sami and R.J. Patton, "Global wind turbine FTC via TS fuzzy modelling and control," IFAC Proceedings Volumes, vol. 45, no. 20, pp. 325-330, 2012.

[4] S. Bououden, M. Chadli, S. Filali, and A. El Hajjaji, "Fuzzy model based multivariable predictive control of a variable speed wind turbine. LMI approach," Renewable Energy, vol. 37, pp. 434-439, 2012.

[5] X. Feng and R. Patton, "Active fault tolerant control of a wind turbine via TS fuzzy model predictive control," Proc. 22nd Mediterranean Conf. Control and Automation MED 2014, Palermo, Italy, pp. 446-451, 2014.

[6] E. Kamal, A. Aitouche, R. Ghorbani, and M. Bayart, "Robust fuzzy fault-tolerant control of wind energy conversion systems subject to sensor faults," IEEE Tran. Sustainable Energy, vol. 3, no. 2, pp. 231-241, 2012.

[7] X. Liu, Z. Gao, and M.Z.Q. Chen, "Takagi–Sugeno fuzzy model based fault estimation and signal compensation with application to wind turbines," IEEE Tran. Industrial Electronics, vol. 64, no. 7, pp. 5678-5689, 2017.

[8] D. Krokavec and A. Filasová, "One approach to design the fuzzy fault detection filters for Takagi-Sugeno models," Advanced and Intelligent Computations in Diagnosis and Control, Cham: Springer, pp. 19-33, 2016.

[9] P.F. Odgaard, J. Stoustrup, and M. Kinnaert, "Fault-tolerant control of wind turbines. A benchmark model," IEEE Tran. Control Systems Technology, vol. 21, no. 4, pp. 1168-1182, 2013.

[10] H. Sanchez, T. Escobet, V. Puig, and P.F. Odgaard, "Fault diagnosis of an advanced wind turbine benchmark using interval-based ARRs and observers," IEEE Tran. Industrial Electronics, vol. 62, no. 6, pp. 3783-3793, 2015.

[11] K.E. Johnson, M.J. Pao, L.Y. Balas, and L.J. Fingersh, "Control of variable-speed wind turbines. Standard and adaptive techniques for maximizing energy capture," IEEE Control Systems Magazine, vol. 26, no. 3, pp. 71-81, 2006.

[12] J.F. Manwell, J.G. McGowan, and A.L. Rogers, Wind Energy Explained. Theory, Design and Application, New York: John Wiley & Sons, 2009.

The real test of black-start capacity and island operation in electricity system of Slovakia on June 20th, 2018

Stanislav Prieložný, Martin Jedinák, Peter Gamboš, Miroslav Kret, Michal Grega
Slovenská elektrizačná prenosová sústava, a.s.
Bratislava, Slovak Republic
stanislav.prielozny@sepsas.sk,
martin.jedinak@sepsas.sk, peter.gambos@sepsas.sk,
kret.miroslav@sepsas.sk, grega.michal@sepsas.sk,

Anton Beláň
Institute of Power and Applied Electrical Engineering,
Faculty of Electrical Engineering and Information Technology, Slovak University of Technology in Bratislava, Slovak Republic
anton.belan@stuba.sk

Abstract — **The defense and restoration plans development is a part of a transmission system operator's responsibilities and requires thorough calculations, simulations and publication of instructions. The plans focus on extensive failures that have a heavy impact on the system operation and could lead to a black-out. To restore the system after such fatal failure generation units that can start-up without outside help are essential. This ability is fairly easy to check. The units are certified regularly and represent black-start ancillary services for transmission system operator (TSO). However, the only way to check the practical applicability of the restoration plan as a whole is to test it in parallel with normal operation of the rest of the system. This process is called the real black start test and this paper deals with one such test in Slovak power system. Specifically a thermal power plant Nováky (ENO) started from a hydro power plant (HPP) Nosice and island operation with available loads followed by synchronization of the island to the rest of the power system that was done on June 20th, 2018.**

Keywords: black-out, restoration, black-start, test, voltage supply, self-consumption, frequency

I. Introduction

Despite the endeavor of transmission system operators (TSO) to design and operate the power system securely, i.e. at least by fulfillment of N-1 criterion, and a low probability of an occurrence of such serious failure as is a partial or a complete black-out, experience from real operation shows that this threat is real and quite possible [1].

Every year numerous disturbances within the contemporary power system are detected and recorded. The vast majority of them are handled without serious consequences by automatic designated systems or by human operators. Nevertheless, some of them can lead to a complete failure of a part or even the whole system. Changes in conditions related to energy market liberalization create high demands on technical ability of formerly stable system and consequently affect reliability of power generation, transmission and distribution and finally increase the likelihood of a critical event occurrence [2].

All black-out cases that happened so far were caused by an accumulation of several factors in combination with a staff failure [3][4].

The objective of power system restoration is to bring the system to the point in which as much as possible within the capacity of the remaining generation and transmission electricity is being supplied at normal frequency, voltage and security levels. This will be a moving target; it will depend upon the initial plan margins, but is often taken as all demand being supplied. In the event, restoration will be a combination of operator decisions and automatic control actions [5].

II. Black-start tests in Slovakia

SEPS as a TSO of the control area of Slovakia is according to the Dispatch rules for control of power system of Slovak republic [6] obligated to ensure a control of Slovakia power system during emergency situations which a necessity to secure a coordination and direct governance of defense and restoration plans of Slovak power system relates to. A verification of an ability to successfully restore the system is also done in accordance with SEPS's obligations resulting from Regional Group Continental Europe (RGCE) ENTSO-E rules.

A readiness for black-start of all entities involved in restoration plan is absolutely necessary for a successful recovery of the Slovakia's power system. One of the most important resulting tasks is a preparation of dispatchers and other operational staff for system recovery procedures. For this reason a regular black-start tests are essential because they provide a great basis for dispatchers' preparation for real events and consequently a preparation of trainings, adjustment of training simulators, adaptation of regulations and a verification of all kind of communication e.g. satellite communication and simultaneously recorded data can be used to verify a theoretical calculations of modeled situations. The restoration plan's philosophy in case of a black-out state is to, in the first step, restore the Slovak electricity system from neighboring systems (top-down strategy), if it is not possible for whatever reason, the second step is to recover it from internal sources (bottom-up strategy) according to standards in Policy 5 RGCE ENTSO-E [7].

The principle of the second step is to supply voltage to a system power plant from an electricity producer providing a black-start ancillary service. In case of a black-out state black-start ancillary provider has to be able to start-up his devices without any outside help and provide voltage for the system power plant for its self-consumption devices and sequential start-up, using transmission or distribution system operator's network elements. The supplied power is then used to initiate various pumps, fans, drives etc. Because they belong to a self-consumption of a system power plant, which is a large source (installed power greater than 100 MW), their input power is accordingly significant therefore, the source providing voltage has to be sufficient.

Simulations and calculations showed a need to test in reality another steps needed to be done in a situation when an extensive part of the system is without voltage. That means to practically try out to supply voltage from black-start ancillary service providing generation unit to the system power plant and consequently an island operation with a connection of a consumption. Besides two mentioned entities, smaller black-start source and system power plant, a consumer , who needs to disconnect from normal operation and stepwise connect his devices, which represents the island's load (as a substitution of commercial consumers) is also needed to take part in the test. The goal is to test a frequency, power and voltage control inside the island. Transmission and distribution system operators are not less important because they need to create a route in a state without voltage using their devices (lines and transformers) among all involved parties.

After evaluation of the test results from tests that were previously done [8], [9] [10], [11], [12], it was agreed to prepare complex test in the middle part of Slovakia for system power plant Nováky for June 20th, 2018.

The goal of the test was to check a starting ability of a hydro power plant Nosice (VE Nosice) with no voltage from outside, creation of route through 110 kV lines in order to provide voltage for self-consumption thermal power plant Nováky (ENO) recovering of loads of desulphurization, phasing of generator TG2 ENO into island and connection of loads in Fortischem company (carbide furnace). After a stabilization VE Nosice – ENO – Fortischem island should have been phased to Slovak´ electricity system in substation 400 kV Bošáca by busbar-breaker (KSP 400 kV). It was agreed that all of substations on the route would be on emergency supply of self-consumption (110 kV substations and 400 kV substation Bošáca) by diesel generators or batteries.

Slovenská elektrizačná prenosová sústava,a.s. (SEPS), Slovenské elektrárne,a.s. (SE), Stredoslovenská distribúcia, a.s. (SSD), Západoslovenská distribúcia, a.s. (ZSD), Fortischem, s.r.o. (Fortischem) and Syprin, s.r.o. (Syprin) were involved in this test.

Within preparation process for the test Study was prepared by company Syprin (technical support).

III. THE COURSE OF THE TEST

The test was performed on June 20th, 2018 using the route in Fig. 1.

Rms voltage, current, active and reactive power and frequency from fixed measuring devices installed in the individual SEPS or SSD, ZSD, Fortischem and SE substations were automatically recorded by their SCADA systems with 1 s interval. Additionally, phasor measurement units (PMU) and analyzers were installed in relevant measurements points. (Fig.2)

Test formally started at 8:16 h. with conference call between supervisors of all participating companies. Than continued by preparatory works that were finished at 9:32 h. Having finished preparatory works for the test a conference call among test supervisors of individual entities was created again in order to start test itself.

At 10:05 a loss of power supply to VE Nosice self-consumption on 0.4 kV level started followed by its automatic restoration using its diesel generators. This was successfully finished and after SEPS dispatcher's order at 10:11:06 h. TG2 VE Nosice generator was started to nominal speeds and voltage. After switching on of 110 kV V7776 line, voltage was provided to 110 kV P. Bystrica at 10:11:37 h. and the line 110 kV V7776 VE Nosice – P. Bystrica was switched on in substation 110 kV P. Bystrica. Than the test continued by creation of route to thermal power plant Nováky (ENO) when other lines 110 kV were switched-on - line 110 kV V7774 P. Bystrica – VE Ladce at 10:14:03 h., line V7773 VE Ladce – VE Dubnica at 10:16:12 h., V8759 VE Dubnica – Trenčín juh at 10:18:02 h., V8758 Trenčín juh – Bošáca at 10:19:32 h., V8740 Bošáca – Bánovce at 10:22:41 h., V8749 Bánovce – Bystričany at 10:24:20 h., V7631 Bystričany – NOT at 10:26:45 h., V7743 NOT – ENOA at 10:32:42 h., self-consumption transformer T10 ENO at 10:39:45 h. The route was created from 10:11h. – do 10:39 h. Then the voltage was provided on 6 kV busbar in ENO 1BBA and 1BBB, r25 and r22 and on busbar 0,4 kV r32 from 10:41h. – do 10:51 h. At 11:00 h. recovering of ENO unit 2 self-consumption started. Gradually loads of self-consumption unit 2 ENO and loads of desulphurization were restored as followed: oil start-up pump NOČ21 at 11:00:05h., cooling pump CHČ21 at 11:16:22 h., pump (ČVV21 a ČVV22) at 11:17:48h., electro feeder for feeding pump EN21 at 11:46:59h., condensation pump KČ23 at 11:49:26h., exhaust fan DV21 at 11:53:14 h., exhaust fan DV22 at 11:56:55h., water air pump and exhaust fan VV21+DV22 on power at 11:59:34h, water air pump and exhaust fan VV22+DV21 on power at 12:00:36h., cooling pump CHČ22 at 12:14:49h., circulating pump OBČ3 at 12:17:47 h., circulating pump OBČ5 at 12:21:35 h., blower OXD2 at 12:24:24 h., circulating pump OBČ6 at 12:28:22 h.. At 12:30h. was frequency in island by regulation TG2 VE Nosice adjusted from 49,6 Hz to 50,01Hz due to a danger of a breakdown of loads with frequency converters and other technological parts as well. Test continued by connection of other loads in self-consumption ENO - circulating pump OBC7 at 12:33:24 h., circulating pump OBČ1 at 12:36:08 h., cooling fans No. 1 and No. 3 at 12:40:03 and 12:40:48 h., coal crusher MO23 at 12:52:22h., recirculating exhaust fan (RECI) at 12:53h., coal feeder PP23 at 13:01 h. and condensation pump KČ22 at 13:02:31 h. Gradually all of operational thermal part´s parameters of thermal power plant Nováky were getting at requested values. On the basis of ENO staff request (high air temperature outside conditions) were switched-on other cooling fans No. 5. and No. 7. at 13:14:04 and 13:14:43h. with power 2x0,1MW.

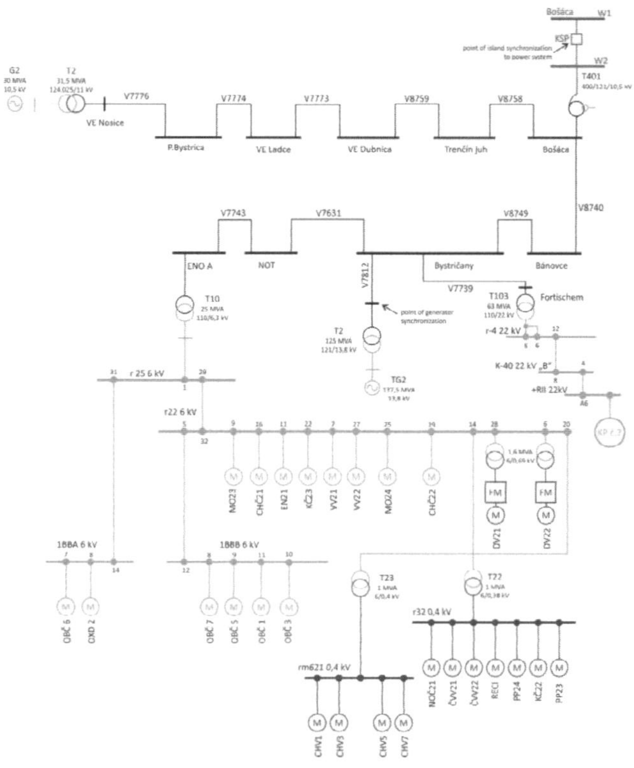

Fig. 1. Single-line diagram of the island operation during complete black-start test

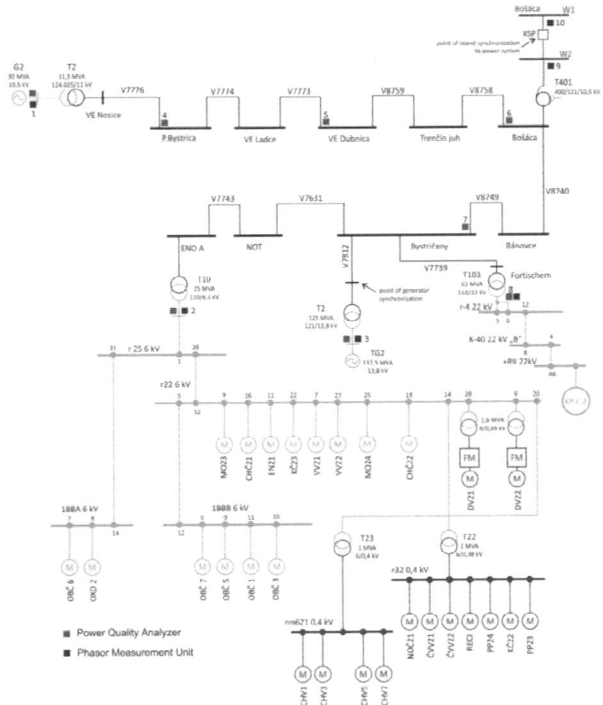

Fig.2. Location of phasor measurement units and power quality analyzers

After recovering of TG2 ENO self-consumption line 110 kV V7812 Bystričany – ENO was switched-on in substation 110 kV Bystričany at 13:40:44 h. At 13:47h. ENO staff announced preparedness for phasing unit 2 ENO to island with TG2 Nosice. Before phasing of unit ENO the island frequency was adjusted (decreased) to 49,84 Hz due to expected increasing of frequency in island after phasing unit ENO (expected increasing according to calculations from Study was cca +1,5 Hz). At 13:52:42 h. unit 2 ENO was successfully phased to island with TG2 Nosice. At 14:00h. on the basis on ENO staff request were switched-on other three cooling fans with power 3x0,1MW other loads of self-consumption ENO unit - crusher MO24 and coal feeder PP24 at 14:01:35 h.

The test continued by switching-on line 110 kV Bystričany – Fortischem V7739 into island at 14:17:01 h. and switching-on transformer T-103 in Fortischem at 14:23:44 h. Than the load in Fortischem was connected from 14:37:11 h. until 15:17 h. (final load connected in Fortischem was 10,2 MW).

After testing of regulation on units TG2 Nosice and TG2 ENO, where the whole consumption of island reached value 18MW, dispatcher SEPS started the process of island phasing to Slovak´s electricity system that was normal working within grid CE ENTSO-E. After switching-on of transformer T401 Bošáca 400/110 kV into free busbar 400 kV Bošáca and fulfilment of synchronization conditions the island was successfully phased with Slovak´s electricity system at 15:25:04h. Test was finished after VE Nosice, loads in Fortischem and loads of desulphurization were disconnected at 15:45h.

Relevant steps during test are described in table 1.

Table 1. Steps of the test

	Event No.	Event	time
Starting TG2 VE Nosice, creation of route	1	Starting of TG2 VE Nosice with no voltage from outside	10:11:06
	2	Providing voltage to substation P. Bystrica via 110 kV line V7776 (Nosice – P. Bystrica)	10:11:37
	3	Switching-on line 110 kV V7774	10:14:03
	4	Switching-on line 110 kV V7773	10:16:12
	5	Switching-on line 110 kV V8759	10:18:02
	6	Switching-on line 110 kV V8758	10:19:32
	7	Odbudenie generátora G2	10:21:25
	8	Switching-on line 110 kV V8740	10:22:41
	9	Switching-on line 110 kV V8749	10:24:20
	10	Switching-on line 110 kV V7631	10:26:45
	11	Switching-on line 110 kV V7743	10:32:42
	12	Switching-on transformer T10 ENO (odb. 15)	10:39:45
	13	Switching-on cable on r25	10:41:00
	14	Switching-on cable on r25 - r22	10:42:00
	15	Switchin-on cable on r25 - 1BBA	10:43:00
	16	Change of tap T10 (odb. 14)	10:44:36
	17	Switching-on cable on r22 - 1BBB	10:47:00
	18	Switching-on transformer T22 (odb. -1)	10:48:28
	19	Switching-on transformer T32	10:51:32
Connection of self-consumption loads TG2 ENO + desulphurization	20	Switching-on Oil start-up pump NOČ21	11:00:05
	21	Switching-on cooling pump CHČ21	11:16:22
	22	Switching-on pump ČVV21	11:17:48
	23	Switching-on pump ČVV22	11:17:48
	24	Increasing of generator´s excitation G2 Nosice	11:41:00
	25	Switching-on electro feeder for feeding pump EN21	11:46:59
	26	Switching-on Condensation pump KČ23	11:49:26
	27	Switching-on exhaust fan DV21 TR and frequency converter FM	11:53:14
	28	Switching-on exhaust fan DV22 TR and frequency converter FM	11:56:55
	29	Switching-on VV21 + DV22 on power	11:59:34
	30	Switching-on VV22 + DV21 on power	12:00:34
	31	Increasing of generator´s excitation G2 Nosice	12:14:02
	32	Switching-on cooling pump CHČ22	12:14:49
	33	Switching-on circulating pump OBČ3	12:17:47
	34	Switching-on circulating pump OBČ5	12:21:35
	35	Change of tap on transformer T10 (13)	12:22:39
	36	Switching-on blower OXD2	12:24:24
	37	Switching-on circulating pump OBČ6	12:28:22
	38	Increasing of generator´s excitation G2 Nosice	12:29:25
	39	Switching-on circulating pump OBČ7	12:33:24
	40	Switching-on circulating pump OBČ1	12:36:08
	41	Switching-on cooling fan No. 1	12:40:03
	42	Switching-on cooling fan No. 3	12:40:48
	43	Switching-on coal crusher MO23	12:52:22

	44	Switching-on recirculating exhaust fan RECI	12:53:00
	45	Switching-on coal feeder PP23	13:01:00
	46	Switching-on compressor KČ22	13:02:31
	47	Switching-on cooling fan No. 5	13:14:04
	48	Switching-on cooling fan No. 7	13:14:43
Phasing of TG2 ENO, connection loads of Fortischem, phasing island to ES SR	49	Switching-on line 110 kV Bystričany - ENO V7812	13:40:44
	50	Starting TG2 ENO, Switching-on transformer T2 ENO (odb. 9)	13:51:23
	51	Phasing TG2 ENO and increasing production on Pg = 5MW	13:52:42
	52	Switching-on coal crusher and coal feeder MO24 + PP24	14:01:35
	53	Switchin-on line 110 kV Bystričany - Fortischem V7739	14:17:01
	54	Switching-on transformer T-103 (odb. 16)	14:23:44
	55	Switching-on r-4 – K-40 + K-40 - +RII	14:30:00
	56	Switching-on carbide furnace KP No. 2	14:37:11
	57	Increasing of production TG2 on Pg = 13 MW	14:57:34 - 15:10:16
	58	Switching-on small compensation in Fortischem (Q = 4 MVAr)	15:06:15
	59	Increasing of power carbide furnace KP č.2 (P = 10.2 MW)	15:10:17 - 15:17:00
	60	Switching-on trasformer 400/110 kv T401 Bošáca from 110 kV (odb. -5)	15:19:07
	61	Island phased to ES SR	15:25:04

IV. TEST FINDINGS

In Fig.3 are frequency and voltage characteristics during island operation and in Fig.4 are characteristics of active power and reactive power at measurement point on VE Nosice TG2 terminal. In Fig.5 are characteristics of active and reactive power at measurement point T10 Nováky (measurement of self-consumption for ENO). In Fig.6 are characteristics of frequency, voltage, active and reactive power at measurement point Nováky TG2. In Fig.7 is active and reactive power characteristics at measurement point Fortischem.

The lowest value of voltage was reached after electro feeder for feeding pump EN21 in self-consumption in Nováky was connected to the island – 0,857(p.u.), (Fig.9).

Frequency was during the island operation within an interval from 49.279 Hz to 51.856 Hz (Fig.3), while the lowest values (49,276 Hz) was reached after load circulating pump OBČ6 in self-consumption in Nováky power plant was connected (Fig.10) and the highest value (51,856 Hz) of frequency was reached after phasing of TG2 Nováky to the island (Fig.11).

During frequency border values did not come to activation of frequency protections on VE Nosice (setting at 48 Hz and 52 Hz – time delay=1s) neither unit ENO (setting at 47, 5 Hz a 53 Hz, time delay=1s). After connection of electro feeder for feeding pump EN21 and reaching the lowest voltage value on self-consumption ENO during test did not come to activation of under-voltage protections setting on ENO self-consumption at 0,6 Un, resp. on exhaust fans with frequency converters at 0,76 Un.

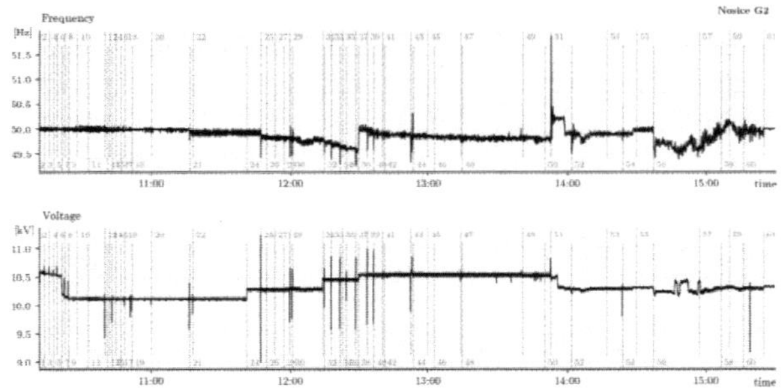

Fig.3 Detail of frequency and voltage characteristic at mneasurement point Nosice TG2 during test

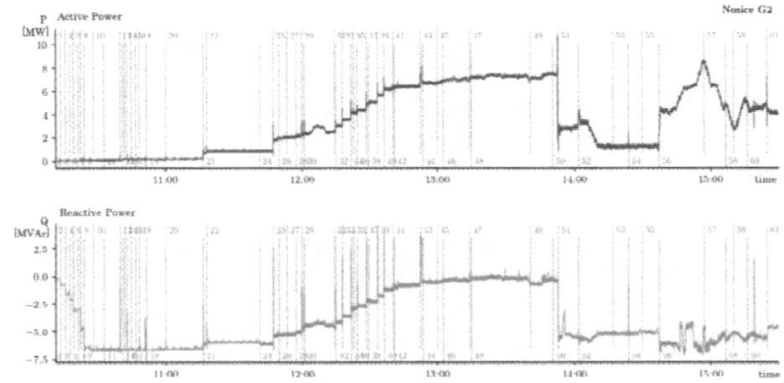

Fig.4 Details of active and reactive power characteristics at measurement point Nosice TG2 during test

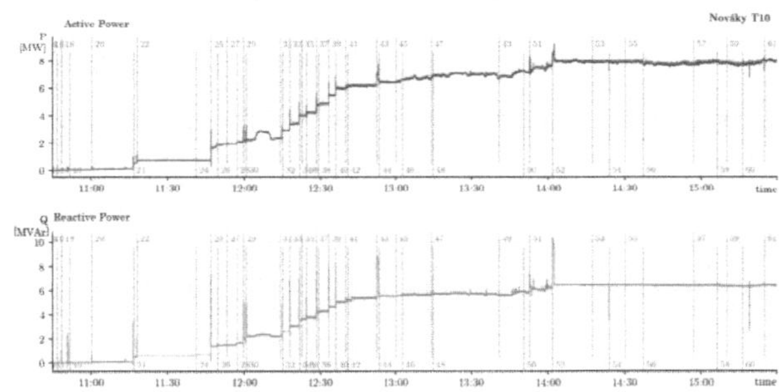

Fig.5 Details of active and reactive power characteristics at measurement point Nováky T10 during test

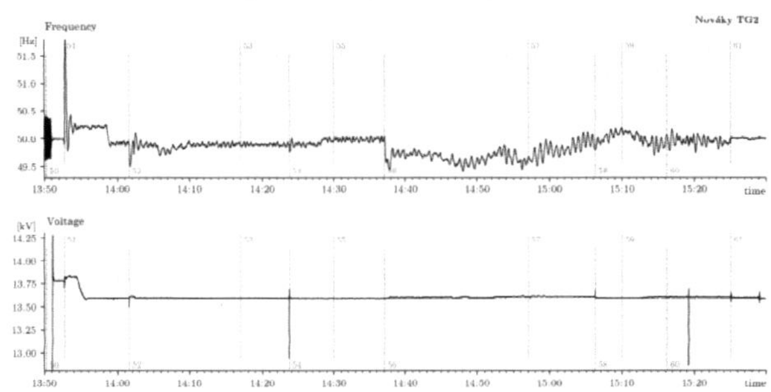

Fig.6 Details of frequency and voltage characteristics at measurement point Nováky TG2 during test

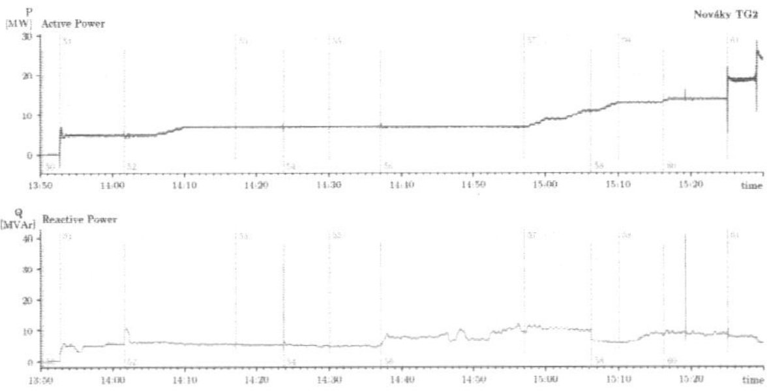

Fig.7 Details of active and reactive power at measurement point Nováky TG2 during test

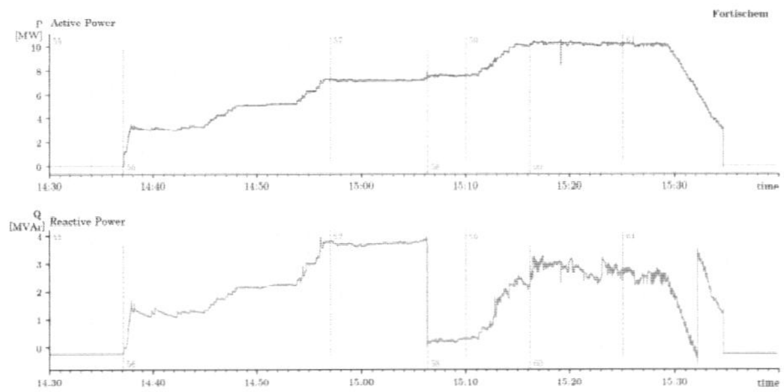

Fig.8 Details of active and reactive power at measurement point Fortischem during test

Phasing of island with normal synchronously working Slovak´electricity system.

In Fig. 12 are characteristics recorded by power quality analyzers in measurement points Bošáca island and Bošáca eletricity system.

The steps of synchronization were as follow:

- Operator command for synchronization - 15:23:27:040
- Start of synchronization - 15:23:44:910
- Order from synchrontakt to Bošáca breaker QM - 15:25:04:280
- Switching on breaker QM KSP 400 kV Bošáca - 15:25:04:340

At 15:25:04:340 h. the island was synchronized with the rest of the interconnected power system via 400 kV bus bar coupler in Bošáca substation.

The synchronization conditions settings for phasing via 400 kV Bošáca bus-bar coupler were: delta U = 10 %, delta f = 50 mHz, angle difference = 4°.

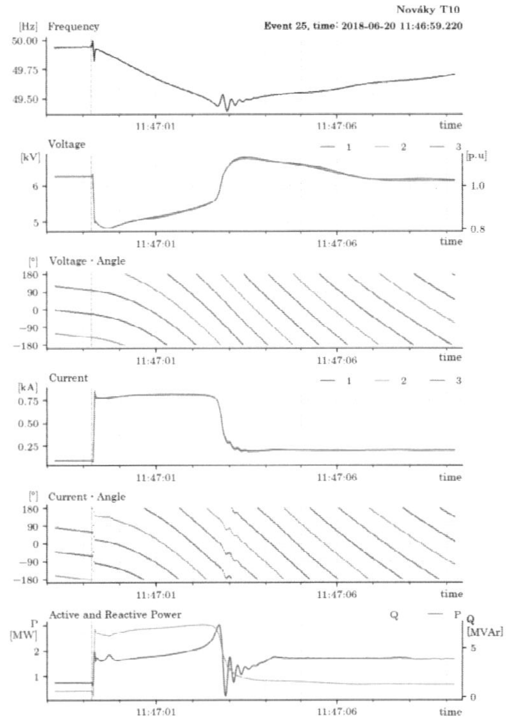

Fig.9. Details of characteristics after switching-on electro feeder for feeding pump EN21

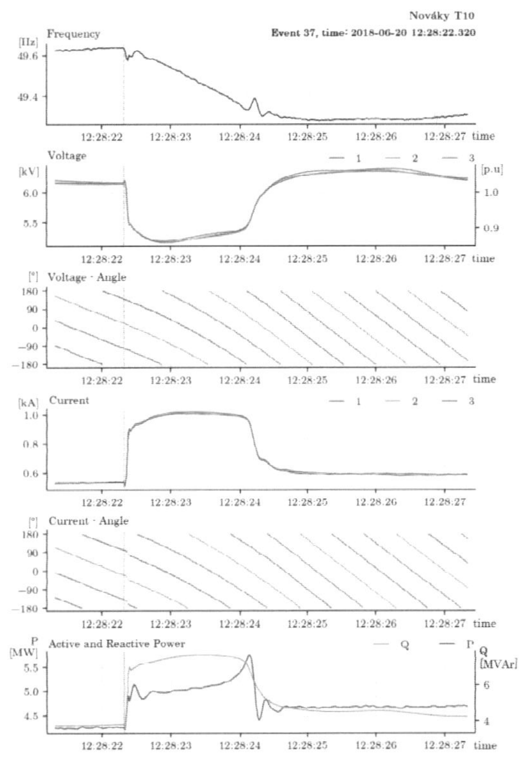

Fig 10 Details of characteristics after switching-on circulating pump OBČ6

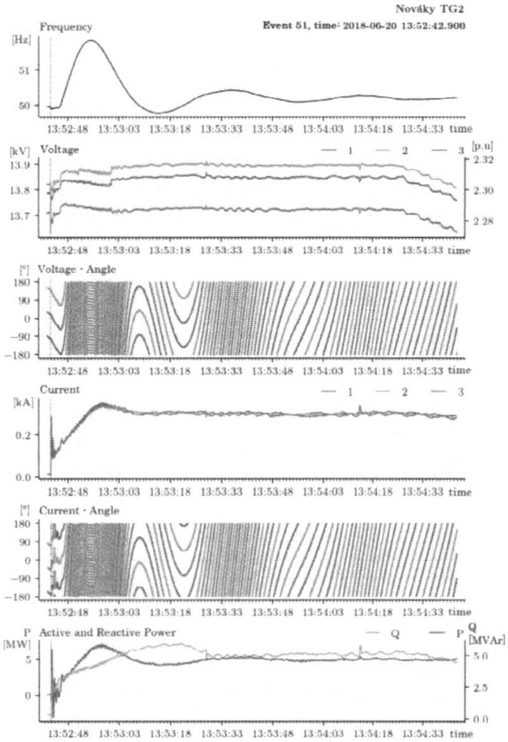

Fig. 11 Detail characteristics during state of phasing unit TG2 Nováky to the island

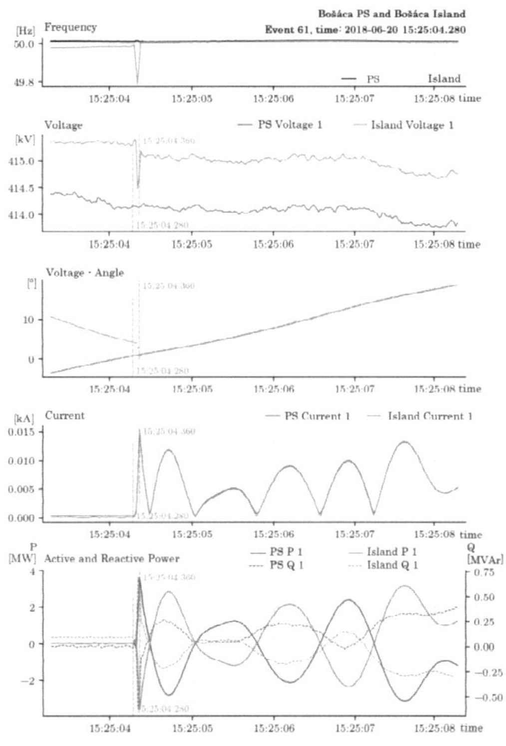

Fig.12 Details of characteristics during process of island´s phasing to electricity system by busbar-breaker in 400 kV Bošáca

V. Conclusion

Black-start test was done on June 20th, 2018 according to the schedule without significant problems and with a successful synchronization of the island with the rest of the Slovak system working in synchronously connected continental Europe, which verified the readiness and feasibility of restoration of Slovak power system within Slovakia in case of occurrence of black-out failure.

During the test as a most critical point was phasing of TG2 ENO to the island with TG2 VE Nosice when frequency reached value 51,856 Hz, however did not reach critical value activation of frequency protections on hydro turbine VE Nosice neither steam turbine on Nováky unit.

Although frequency increasing was expected within preparatory study, the real increase was higher than expected that can be a result of higher step change of basic loading of TG2 ENO after phasing (this value was within study only estimated) as well as influence of other time constants turbine itself. It would be useful to decrease step change of basic loading TG2 in ENO if technology allows it.

The biggest drops of frequency and voltage did not cause activation of voltage or frequency protections.

As showed experience during black-start tests is useful using analyzers with GPS synchronization.

References

[1] K. Máslo,"Řízení a stabilita elektrizační soustavy," 2003, AEM, p. 272, ISBN 978-80-260-44671-1,.

[2] J. Tuma, S. Rusek, Z. Martínek, I. Chemišinec and R. Goňo, "Spolehlivosť v elektroenergetice," CONTE spol s.r.o., ČVUT Praha, p. 291, ISBN 80-239-6483-6

[3] M. Eremia, M. Shahidehpour, "Handbook of Electrical Power System Dynamics", IEEE Press, 2013, p. 976, ISBN 978-1-118-49717-3.

[4] M. Kolcun and Ľ. Beňa, Využitie špecializovaných zariadení na reguláciu tokov výkonov v elektrizačných sústavách, EQUILIBRA, Košice, 2011, p.128, ISBN 978-80-553-0767-1

[5] U.G.Knight, Power Systems in Emergencies, From Contingency Planning to Crisis Management, John Wiley&sons LTD, 2001, p.393, ISBN 0-471-49016-4

[6] Dispach rules for control of Slovak power system, SEPS, Available at http://www.sepsas.sk.

[7] RG CE Operation Handbook, Policy 5: Emergency Operations Security, ENTSO-E, 2009. Available at: http://www.entsoe.eu.

[8] S. Prieložný, M. Jedinák, M. Kolcun, and F. Janíček, "The real test of „Black Start" units in Eastern Slovakia," In: Elektroenergetika 2015: 8th International Scientific Symposium on Electrical Power Engineering. Stará Lesná, Slovakia, September 16-18, 2015. - Košice: Technical University of Košice, pp.85-88, ISBN 978-80-553-2187-5.

[9] P. Gamboš, M. Jedinák, and S. Prieložný, "The real test of „Black Start" units in Eastern Slovakia," In: Blackout a Krajowy System Elektroenergetycny. Poznań 2016, Poland, June 1-3, 2016. Osrodek Wydawnictw Naukowych ICHB PAN, pp. 37-46, ISBN 978-83-7712-152-8.

[10] M. Grega, S. Prieložný, M. Jedinák, P. Gamboš, M. Kanálik and M. Kolcun. "The real test of Vojany power plant self-consumption restoration after a black-out," International symposium, Stará Lesná 2017, pp. 118-122, ISBN 978-80-553-3195-9.

[11] M. Grega, S. Prieložný, M. Jedinák, P. Gamboš, M. Kanálik and M. Kolcun.: The real test of black-start capacity and island operation in east part of Slovakia, paper, International symposium, Stará Lesná 2017, pp. 118-122., ISBN 978-80-553-3195-9

[12] M. Grega, S. Prieložný, M. Kret, M. Jedinák and Gamboš P.: The Slovak Power System Restoration after State of Blackout, 13th International Scientific Conference, 5-7 June, 2018, Tatranské Matliare, Control of Power Systems 2018, High Tatras, Slovak Republic.

Development a Partially Observable Markov Decision Processes-based Intelligent Assistant for Power Grids using Monte Carlo Tree Search

Nikita V. Tomin, Victor Kurbatsky
Dept. of Power Systems
Melentiev Energy Systems Institute SB RAS
Irkutsk, Russia
tomin.nv@gmail.com

Michael Negnevitsky
School of Engineering and ICT
University of Tasmania
Hobart, Australia
Michael.Negnevitsky@utas.edu.au

Abstract — **Autonomous control systems will make much "smarter" used automatic controls of modern power grids, as well as partially or completely replace the system operator, which may not be able sometimes to adequately respond to critical conditions due to psychological stress. Development of such systems can be solved by Monte-Carlo tree search algorithm that simulate ahead into the future, evaluate future states, and back-up those evaluations to the root of a search tree. We use the formalism of POMDPs (Partially Observable Markov Decision Processes) as the core of an intelligent assistant for power system control and dispatch. We demonstrate the feasibility of the approach to resolve the voltage and reactive power control in substation.**

Keywords: power system, Monte Carlo Tree Search, control, partially observable Markov decision processes, reinforcement learning

I. Introduction

Year by year control of normal and emergency conditions of up-to-date power systems becomes an increasingly complicated problem. With the increasing complexity, system operators are faced with a large amount of problems and information based on which they have to make decisions. To support such decisions, automatic assistants that provide operators with a list of suggested commands can be used. Either when the operator executes a command or when a unexpected disturbance occurs, a new list of recommended actions is presented. This recommendation process can be modelled as a sequential decision problem under uncertainty with an optimization criteria such as performance, availability, reliability, or security.

A promising solution is the use of Partially Observable Markov Decision Processes (POMDP), a well known stochastic method for sequential decisions, which allow for principled decision making under conditions of uncertain sensing. In this case, POMDPs suggest themselves as natural candidates to address observational uncertainty, and these have been recognized the natural solution in the development of intelligent agents (henceforth also referred to as POMDP-based intelligent assistants) [1], [2]. Such assistants must monitor and make decisions despite significant uncertainty in observations (so the true state of the world, for example a power grid, may not be known), and uncertainty in actions (outcome of agents decisions or actions may be uncertain). Agents may accomplish such monitoring and decision making fully autonomously or with adjustable autonomy.

Deep Reinforcement Learning recently emerged as one of the most competitive approaches for learning in sequential decision making problems with fully and partially observable environments, e.g., computer Go [3]. The development of deep reinforcement learning techniques in conjunction with online search allowed to create AI software designed to mimic a humans thinking and logical reasoning, which can present some similarities with power system control and dispatch problems. Recent developments in the field such as the AlphaGo system [4], Atari Deep Q-learning [5], show that high performance in the most challenging control domains can achieve with no human expert input and big historical datasets, when human knowledge and data may be too expensive, too unreliable or simply unavailable (especially given the commercial confidentiality of the power companies).

In this paper, we make an effort to develop an autonomous POMDP-based intelligent assistant, which could not only provide system operators with online guidance in the form of ordered recommendations, but and autonomously dispatch and control a power system without human. To this end, we use an advanced reinforcement learning algorithm - Monte Carlo Tree Search (MCTS), which is particularly well suited to be applied to POMDP. MCTS takes into consideration what might happen in the future by approximating what other entities/agents (components, devises or whole power system) might do via Monte Carlo simulations. The developed intelligent assistant will allow to deal with abnormal situations, unexpected events, or the occurrence of process transients under uncertainty conditions, which are occurred in power grids.

II. Monte Carlo Tree Search

MCTS is a policy-optimization algorithm for finite-horizon, finite-size Markov Decision Processes, based on random episode sampling structured by a decision tree. Monte Carlo methods work by approximating future rewards that can be achieved through random samplings [6]. The evaluation function of MCTS relies directly on the outcomes of simulations. Thus, the accuracy of this function increases by adding more simulations, so the optimal search tree is guaranteed to be found with infinite memory and computation.

MCTS relies on two key elements: (a) that the true value of an action can be approximated using simulations; and (b) that these values can be used to adjust the policy towards a best-first strategy. As result, MCTS will selectively try moves based on how good it thinks they are, thereby focusing its effort on moves that are most likely to happen.

The algorithm iteratively builds a tree until a condition is reached or satisfied, then the search is halted and the best performing action is executed. In the tree, each node represents a state, and directed links to child nodes represents actions leading to subsequent states (Fig. 1). Simulations in MCTS start from the root state (in our case, from the current time when an action for an electric device should be made) and are divided in two stages. When the state is added in the tree, a tree policy is used to select the actions. Otherwise, a default policy is used to roll out simulations to completion.

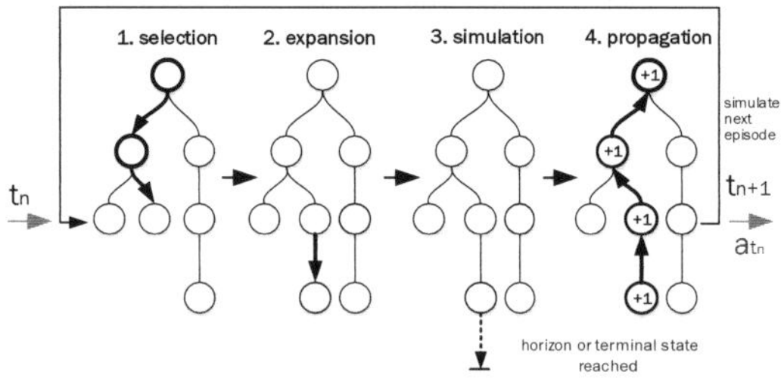

Figure 1. Steps of Monte Carlo tree search

MCTS does not expand all leaf nodes, as that would be very expensive. Instead, the selection process chooses nodes that strike a balance between being lucrative-having high estimated values-and being relatively unexplored-having low visit counts. A leaf node is selected by traversing down the tree from the root node, always choosing the child *i* with the highest upper confidence tree (UCT) score:

$$U_i = \frac{W_i}{N_i} + cP_i\sqrt{\frac{\ln N_p}{1+N_i}}\,, \qquad (1)$$

where W_i is the accumulated value of the *i*th child, N_i is the visit count for *i*th child, and N_p is the number of visit counts for the parent node. The parameter $c \geq 0$ controls the tradeoff between choosing lucrative nodes (low *c*) and exploring nodes with low visit counts (high *c*). It is often set empirically.

Thus, this selection mechanism works due to its emphasis on balancing both exploitation (first part of Eq. 1) and exploration (second part of Eq. 1). That is, every time a node is visited, the denominator of the exploration part increases resulting in decreasing its overall contribution. If, on the other hand, another child node of the same parent node is visited, the numerator increases, so the exploration values of unvisited children increase.

As result, the reward accumulated in the episode up the decision path taken by selection. Since the goal of MCTS is to optimize policy decisions from the root state, MCTS accumulates statistics from sampled episodes in each node. Usually, for MCTS algorithm is not use so-called action-value function $Q(s,a)$ to compute the policy π. Instead, π is derived from the visit count N_i.

$$\pi_i \propto N_i^{1/\tau}, \qquad (2)$$

After the initial iterations, moves with higher Q value will be visited more frequently. We used the visit count to calculate the policy because it is less prone to outliners. τ is a temperature controlling the level of exploration.

Thus, MCTS can be considered as a general technique for making decisions in perfect information scenarios. Therefore its use does not have to be restrained to games only and the diversification of MCTS in other research areas is already a reality. For example, in [7] several potential MCTS applications in electric power systems were proposed. This applications were categorized into two groups: game based (strategic wholesale market bidding, distribution-level market transaction, coordinated management of plug-in electric vehicles) and search based (online security assessment, system fault diagnosis, cascading outage prediction rolling optimization of renewable dispatch). In 2017, the cooperation of Google DeepMind with the large British national company National Grid UK also began to be used to apply the AlphaGo system to UK Power Grid management tasks [8].

III. POMDP-BASED INTELLIGENT ASSISTANT USING MCTS

A POMDP models a sequential decision problem, in which a system evolves over time and is controlled by an agent. At discrete time intervals the agent partially observes the state of the system and chooses an action [9]. In a POMDP there is a set of conditional observation probabilities O and a set of conditional probabilities $P(o|a,s), o \in O, a \in A, s \in S$, such that if the system makes a transition to state s with action a, it receives the observation label o with probability $P(o|a,s)$.

One major area of responsibility for POMDP-agents is to monitor the evolution of some process or state over time, including that of the human the agents are deployed to assist, and to make periodic decisions based on such monitoring [2], [10], [11]. Agents may accomplish such monitoring and decision making fully autonomously or with adjustable autonomy (dynamically reducing own autonomy for seeking human input).

MCTS is particularly well suited to be applied to POMDP, because the tree structure it relies on naturally contains the history of each state, i.e. all explored trajectories, without needing to increase the size of the state variables. That way, if each node of the tree contains either an observation *o* or an observation-action pair (*o*, *a*), then each node having a unique history contained in the tree, has enough information to sample possible hidden states for power grid.

In the power system control problem, an initial disturbance and/or control action would trigger a sequence of dependent events (outages, conditions changes etc.) in future time intervals. This in turn adds an even heavier computational burden to the system operator. To crack this computational intractability, if we assume that the procedure of future events in power grid is Markovian (or Partially Markovian) [7], then all the possible states can be formed as a tree structure and traversed using an MCTS, as shown in Figure 2. In the markovian tree structure, each node stands for a system state, and each branch represents control actions or disturbance. In this case the gain for MCTS-agent will consist in bringing in a power system in the normal steady-state or post-emergency conditions by means of the required control actions taking into account security, economic, market etc. constraints.

As shown in Figure 2, a complete control actions path begins from the root node (the initial state) to the leaf node (the terminal). The power system control and operation takes the forward tree traverse from the root until a leaf node is reached. Then, the security indicators of the states on this certain path are updated via a reversed tree traverse from a terminal to the root. The above process can be repeated for every control actions path starting from the current state. Finally, the control actions path, which lead to the better security index (for example, with minimum value) is selected as the most optimal policy for power system control.

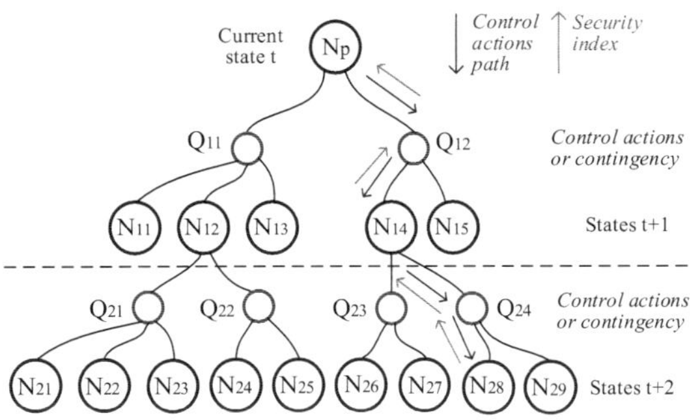

Figure 2. The MCTS for power system control and dispatch in a Markovian tree

IV. CASE STUDY

The most important methods for voltage and reactive power control in substation include N-Areas Method, which divided the control domain into N areas using a fixed upper and lower limit of voltage and reactive power or the intake power factor of the substation, is a more developed one of the voltage and reactive power control strategies [12], [13]. In this paper, an improved Seventeen Areas Method combined with reinforcement learning control strategy based on MCTS is used to resolve the voltage and reactive power control in substation (Fig. 3). The MCTS-agent is implemented in open- source language Julia (POMDPs library) [14]. The assistant is coupled to a test scheme simulator in Matlab/Simulink.

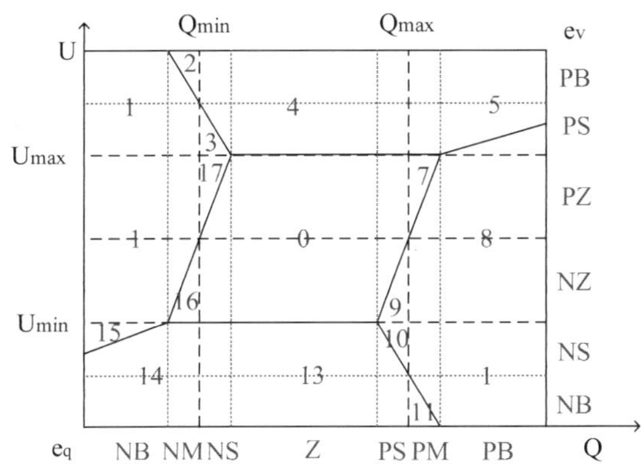

Figure 3. 17 areas sketch with fuzzy words after adjustment

By setting the system parameters we used the fuzzy control rules, when we have two inputs e_v, e_q, and two outputs nT, Q_c. Where evis the voltage deviation of transformer secondary side, eq is the reactive power deviation of system, nT is the action times of switching taps and Q_c is the operating amounts in capacitors. The test system mainly includes OLTC regulating transformer block, reactive power compensation block, a POMDP-based assistant controller block (Fig. 4).

The fuzzy words for control variables of taps nT can be DOWN, UC, UP. Where DOWN means tapping down or cutting down transformers ratio to boost the voltage, UP means tapping up or enlarging ratio to buck voltage, and UC means no action. The reactive power or power factor of the transformer input-side are demanded in qualified ranges. That is, reactive power cant be out of its upper and lower limits. Generally, several shunt capacitor banks are installed in a substation to adjust voltage and reactive power. When the system voltage and reactive power shift and its needed to invest (LEFT) or cut capacitor banks (RIGHT). Otherwise, do nothing (NONE). In practice the capacitor banks are operated one by one, not only one. So the fuzzy-words NB, NM, NS, Z, PS, PM, PB are chosen for both input variable e_q and output control variable Q_c (Fig. 3).

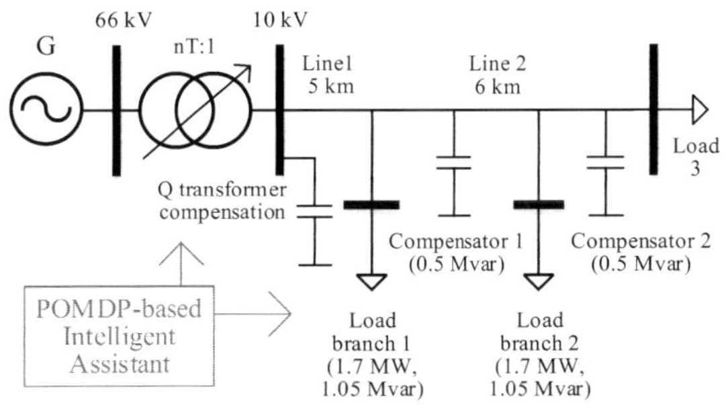

Figure 4. The substation structure test scheme

Every region may be controlled automatically or set manually. According to the real situation, the borders of voltage regulation should be relatively fixed and the borders of reactive power regulation are constrained by the voltage condition. When the voltage is higher, capacitance is not added if reactive power is not lack much or capacitors are cut a little more if the reactive power is not exceed the limits too much and when the voltage is lower, capacitance is added more or the capacitors isn't cut.

For training of MCTS-agent, we used a standard environment "Grid World", which is a 2D rectangular grid of size (N_y, N_x) with an agent starting off at one grid square and trying to move to another grid square located elsewhere. Such an environment is a natural one for applying reinforcement learning algorithms to discover optimal paths and policies for agents on the grid to get to their desired goal grid squares in the least number of moves. In this case, a 6x7 grid was created, where the rows are the e_v parameter and the columns are the e_q parameters. The discretization of the resulting grid cells determined the values of the input fuzzy variables, as shown in Figure 3. The MCTS-agent was given an action space in accordance with the output variables, where the movements "Down", "Up", "None" mean switching the OLTS taps, and the movements "Right" , "Up", "None" - regulation of the number of capacitor banks. In this case, the so-called sparse reward was used, when the reward is given only upon reaching the terminal state, and is absent in other states. In our case, the main goal of the agent was to reach area 0, where an optimal balance of voltage and reactive power is observed (Fig. 3).

To compare the effectiveness in finding the optimal policy, π^* various reinforcement learning methods were used: Q-learning, SARSA, dynamic programming, and MCTS. In addition, each of the methods required a different number of iterations to solve this problem (Table 1). As a result, all methods came up with a similar policy (Table 2), which is consistent with the control policy proposed in [13].

Table 1. The results of reinforcement learning for voltage and reactive power control problem

№	Method	Number of iterations
1	Dynamic programming	200
2	Q-learning	5000
3	SARSA	5000
4	MCTS	20000

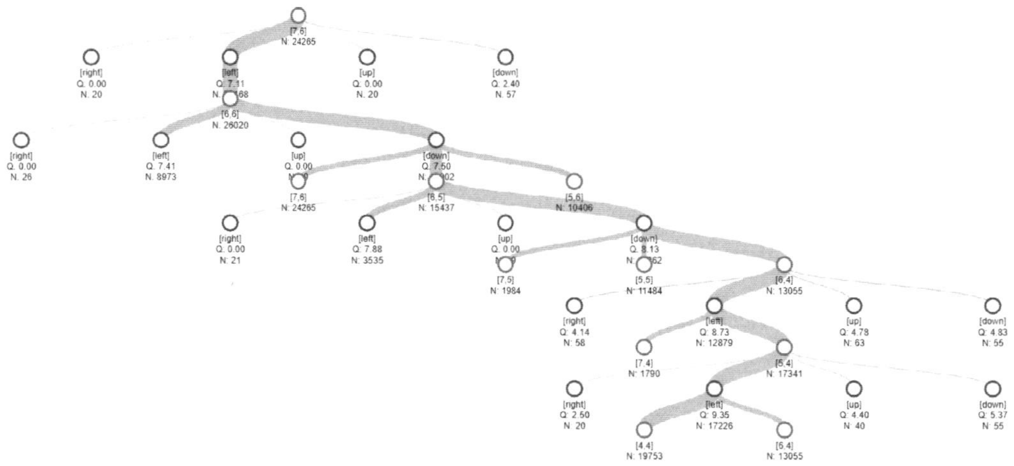

Figure. 5. Voltage simulated results at load 3 using traditional and POMPD-based assistant's regulation

Table 2. The optimal policy, π^* obtained using mcts method (where ↓= down, ↑= up, →= right, ←= left, 0 = none)

	NB	NM	NS	Z	PS	PM	PB
NB	↓	↓	↓	↓	↓	↓	↓
NS	→	→	↓	↓	↓	↓	↓
NZ	→	0	0	0	←	←	←
PZ	→	→	→	0	0	0	←
PS	↑	→	→	↑	↑	←	←
PB	↑	↑	↑	↑	↑	↑	↑

The study of the numerical values of the function $Q(s,a)$ for the policy, π^*, shows that in some cases the values of the resulting reward for some pairs of actions are close in meaning. For example, in the SARSA method for the state situation $e_v = PS$ ("small positive"), $e_q = NS$("small negative"), the "Right" action is performed with a reward of 7.00914, corresponding to the disconnection of the capacitor bank for stabilization voltage to nominal values. At the same time, the action "Up", corresponding to switching the OLTC, has a similar reward of6.96883.

To train MCTS-agent, 20.000 iterations were performed, as a result of which a tree was built, with the help of which one can visually track the sequence of control actions for different operational states at a substation. In Fig. 5 shows the obtained tree with a sequence of control actions (see the fat grey line) for the initial test system state characterized by the maximum voltage drop below U_{min} $(e_v = NB)$ and the highest possible reactive power above Q_{max} $(e_q = PB)$. At the first and final steps of actions, the MCTS-agent correctly chooses the policy power factor correction (the control action "Left", corresponding to investing capacitor banks) with the subsequent voltage increase at the second and third steps (the control action "Down" on the tap-changer). This corresponds to a real optimal control policy, noted in [13], and to policies obtained on the basis of dynamic programming , SARSA, and Q-learning. It is interesting to note that in certain nodes the Left and Down actions are similar in terms of the action-value function, $Q(s,a)$ and the visit count N_i which corresponds to the real situation when increased voltage and power factor correction.

Seeing Fig. 6, the voltage of the model using MCTS-agent control is less fluctuating and steadier than the model of without MCTS-agent control, which can get steadier system voltage and less transmission loss.

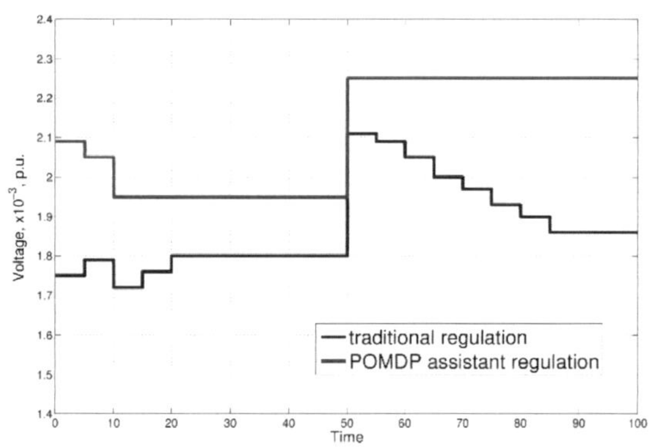

Figure 6. Voltage simulated results at load 3 using traditional and POMPD-based assistant's regulation

V. Conclusions and Future Work

This paper introduced a POMDP-based intelligent assistant concept using the advanced reinforcement learning techniques - MCTS, that allow to provide monitoring and decision making for power system control and dispatch with adjustable autonomy or fully autonomously (without human input - system operator). Last option can be realized using deep reinforcement learning ideology combining a deep neural network with a MCTS algorithm.

In many cases, the intelligent decision making systems, which includes POMDP-agents, leverage enormous amounts of human expertise and data. Our experience show that for electric power industry, especially for operative and emergency control, this human knowledge may be too expensive, too unreliable or simply unavailable. The recent developed AlphaGo Zero approach allow to bypass this step, creating algorithms that achieve superhuman performance in the most challenging domains with no human input.

We plan to develop an advanced form of a POMDP-based intelligent assistant is able to do this by using a novel form of reinforcement learning, in which POMDP-agent becomes its own teacher. The system starts off with a deep neural network that knows nothing about the power system control. It then plays games against itself, by combining this neural network with a MCTS algorithm. As it plays, the neural network is tuned and updated to predict control actions, as well as the eventual winner of the games (in our case, some optimal control action, which lead to normal operating condition of a power grid).

This updated neural network is then recombined with the search algorithm to create a new, stronger version of a POMDP-based intelligent assistant, and the process be- gins again. In each iteration, the performance of the system improves by a small amount, and the quality of the self- play games increases, leading to more and more accurate neural networks and ever stronger versions of a POMDP-based intelligent assistant.

Acknowledgment

This work was supported by the Russian Scientific Foundation (No. 19-49- 04108) under the project "Development of Innovative Technologies and Tools for Flexibility Assessment and Enhancement of Future Power Systems".

References

[1] P. R. Varakantham, R.Maheswaran, M.Tambe. "Implementation Tech- niques for Solving POMDPs in Personal Assistant Domains", in Proc. Programming Multiagent Systems: Third International Workshop, 2005, Utrecht, Netherlands, pp. 76-89.

[2] A. Reyes and M. T. Spaan, "Towards a POMDP-based Intelligent Assistant for Power Plants", in Proc. POMDP Practitioners Workshop ICAPS 2010, Toronto Canada, May 2010.

[3] Y. Li "Deep Reinforcement Learning", in arXiv, arXiv:1810.06339, 2018 [4]V.Mnih, K. Kavukcuoglu, D. Silver et al, "Playing Atari with

[4] Deep Reinforcement Learning," in arXiv, abs/1312.5602, url: http://arxiv.org/abs/1312.5602

[5] D. Silver, J. Schrittwieser, K. Simonyan, et al "Mastering the game of Go without human knowledge," in Nature, vol. 550, 2017, pp. 354359

[6] C. B. Browne et al., "A Survey of Monte Carlo Tree Search Methods," in IEEE Transactions on Computational Intelligence and AI in Games, vol. 4, no. 1, pp. 1-43, March 2012.

[7] F. Li and Y. Du, "From AlphaGo to Power System AI: What Engineers Can Learn from Solving the Most Complex Board Game," in IEEE Power and Energy Magazine, vol. 16, no. 2, pp. 76-84, March-April 2018.

[8] Googles AlphaGO Levels Board Games Power Grids, 2017. available from: https://www.wired.com/2017/05/googles-alphago-levels- board-games-power-grids/

[9] L. P. Kaelbling, M. L. Littman, A. R. Cassandra "Planning and acting in partially observable stochastic domains", in Artificial Intelligence, vol. 101, 1998, pp. 99134.

[10] Z. Feng and S. Zilberstein. "Region based incremental pruning for POMDPs", in Proc. Uncertainty in Artificial Intelligence, 2004.

[11] M. E. Pollack, et al. "Autominder: An intelligent cognitive orthotic sys- tem for people with memory impairment", in Robotics and Autonomous Systems, vol 44, 2003, pp. 273282.

[12] L. Sheng. "Strategy research of substation voltage reactive control basing on the ninth region plot", in Qinghai Electric Power, vol. 24, 2005, pp. 14

[13] X. Wu et al. "Fuzzy Control on Voltage/Reactive Power in Electric Power Substation", B. Cao, T.-F. Li, C.-Y. Zhang (Eds.): Fuzzy Info. and Eng., vol. 2, AISC 62, pp. 10831091.

[14] M. Egorov et al. "POMDPs.jl: A Framework for Sequential Decision Making under Uncertainty", in Journal of Machine Learning Research, vol. 18, no.26, 2017, pp. 15.

Evaluation of Energy Efficiency and RES Subsidies Metrics – Why is Important to do it?

Ing. Michaela Lachmanová
Dept. of Economics, Management and Humanities, Faculty of Electrical Engineering,
Czech Technical University,
Technická 2, 166 27 Praha, Czech Republic
Michaela.lachmanova@fel.cvut.cz

Abstract – **In December of 2018, process of adopting of new European regulatory energy framework was finished. New goals were set up and need to be reached. By 2030, the EU will have to cover 32 % of its energy consumption from renewable sources (RES) and meet the main energy efficiency target of 32.5 %. For fulfilling the targets it is necessary to properly set the combination of energy policy tools, which ensures the targeting in optimal way.**

This paper introduces the motivation why to concern about energy policy evaluation in the Czech Republic and presents actual situation in frame of European targets. The main objective of the paper is to briefly introduce development in renewables and savings field.

Keywords: strategy, energy consumption, legislative, energy efficeiency, renewables

I. Introduction

The principles of the European Union lie on energy policy, but a single complex energy policy cannot be presented this day. Currently, further integration is intended to provide consumers with secured energy supply at an affordable price, greater competitiveness and lower emissions. Over the next ten years, Member States will expect € 1 trillion of investment to diversify energy resources, connect infrastructures and replace equipment more efficiently and environmentally friendly [1].

These goals are embedded in a several documents, including the 2030 climate & energy framework, which sets out Key targets for 2030 (at least 40 % cuts in greenhouse gas emissions, at least 32 % share for renewable energy, at least 32.5 % improvement in energy efficiency) [1]. A high number of political, economic and financial instruments have been presented and also are being used to achieve these objectives.

These instruments (more widely, we are talking about all possible instruments to ensure reaching the targets – RES subsidies, ETC trading, but also financial instruments, like state-supported loans) require a huge amount of money to operate. For example, in the Czech Republic, it will be necessary to invest 400 - 600 bln CZK to fulfill our energy savings targets. But not only this investment is needed - also there are other direct and indirect costs associated with reaching these goals.

That's why it is absolute crucial to evaluate these instruments before, during and after the operation of the instrument. The evaluation of these instruments is very complex disciplines and the methodology varies on type of instrument (Energy efficiency policy instruments, RES subsidies, low-interests loan...). The evaluation includes not only financial evaluation (estimation of costs and benefits) but the question is much more complex. We should focus on all benefits and costs related to operation of these instruments, including reduction of emissions, reduction of fossil fuels consumption and higher energy savings.

This paper presents a very brief introduction to the situation, short summary of legislation and overview of actual situation in energy field in the Czech Republic. These data are crucial to understand, why the evaluation is the key tool to properly set a comprehensive portfolio of tools to achieve goals in an optimal way.

This paper is a very brief introduction, basic research to the field, and it is devoted as a basic for further research.

II. Legislative Overview

In December 2018, European Union finished adopting new legislative framework in energy field. By 2030, the EU will have to cover 32 % of its energy consumption from renewable sources and meet the main energy efficiency target of 32.5 %.

These new objectives are set out in the revised Energy Efficiency Directive and the revised Renewable Energy Directive. The Council also approved the so-called governance regulation, which establishes a framework for energy union governance and climate action. This step completes the final phase of the legislative procedure for negotiating the three acts, which are part of the clean energy package [2].

A. Energy efficiency directive

The revised Energy Efficiency Directive sets out a framework of measures designed primarily to ensure that the EU headline targets for 2020 and 2030 are met. Increasing energy efficiency will benefit the environment, reduce greenhouse gas emissions, improve energy security, reduce energy and household energy costs, help reduce energy poverty and contribute to growth and employment [6].

The main elements of the revised directive are:

- a major EU energy efficiency target of at least 32.5 % by 2030;
- the obligation to achieve annual energy savings of 0.8 % (0.24 % for Cyprus and Malta) from the annual final energy consumption of 2021 to 2030 and flexibility on how Member States fulfill this obligation;
- social provisions requiring Member States to take into account the need to alleviate energy poverty in designing policy measures to achieve energy savings.

B. Renewable Energy Directive

The revision of the Renewable Energy Directive will lead to a faster transition of Europe to clean energy through the use of a wide range of renewable energy sources such as wind, sun, water, tide, geothermal, biomass and biofuels. It sets the main target for 2030 when 32 % of EU energy consumption is to come from renewable sources [1].

Other key elements of the revised Renewable Energy Directive include:

- the introduction of renewable electricity production will be supported by market-based support schemes, simplified authorization procedures and consistent methods.
- the use of renewable energy sources in transport will be accelerated by the obligation for fuel suppliers to reach at least 14 % of the share of renewable energy in the transport sector by 2030, while phasing out the use of conventional high-risk biofuel biofuels.
- households wishing to produce their own energy from renewable sources, for example through solar roof panels, will be largely exempted from the energy consumption charges they have produced themselves.

C. Governance Regulation

One of the most important legislative act [3] sets out how Member States will work together with each other and with the European Commission to achieve the ambitious goals of the Energy Union, including in particular renewable energy and energy efficiency targets and the EU's long-term greenhouse gas emissions targets. It also sets out control mechanisms to help ensure that the objectives are met and that the range of proposed measures creates a coherent and coordinated approach.

Regulation 2018/1999 of the European Parliament and of the Council (Energy Union operation) introduces five dimensions of the Energy Union, which should ensure the interconnection of energy development aspects and ambitious climate protection objectives. The essential declared goal of the Energy Union is to ensure safe, sustainable, competitive and affordable energy for all types of consumers, together with support for research and innovation

Individual dimensions of the Energy Union, respectively. groups of criteria are in accordance with Regulation 2018/1999:

- Energy security,
- Internal energy market,
- Energy efficiency,
- Reducing carbon emissions,
- Research, development and competitiveness

Within the framework of the winter energy package – Regulation on the Administration of the Energy Union – Member States are obligated to elaborate the so-called National Climatic and Energy Plans. The States should finish the National Plan in 2019, while first National Plans should cover the period 2021-2030.

National plans should include a contribution to meeting the climate-energy targets for 2030. After the submitting of the draft National Plan, the iterative process will start, in which the European Council will assess the sufficiency of the European target. National plans should include the current status, objectives and policies to achieve these goals.

With all these plans, it is necessary to evaluate all tools needed to reach mentioned goals.

III. Situation in the Czech Republic

The direct impact (and the closest) is the obligation to submit the National Climatic and Energy Plan (NECP).

The Ministry of Industry and Trade has launched an external comment procedure on the Draft National Plan for Energy and Climate (also referred to as the National Climate and Energy Plan). After the handover, a discussion will take place with representatives of the European Commission and the plan will be finalized by the end of 2019.

The document contains objectives and main policies in all five dimensions of the Energy Union. Through this document, Member States are, inter alia, obliged to inform the European Commission of a national contribution in the field of greenhouse gas emissions, renewable energy, energy efficiency and electricity / transmission interconnectivity.

At the same time, the Ministry of Industry and Trade published the material on its website with the launch of the external comment procedure, thus launching a public consultation on this material. The proposal for the National Energy and Climate Plan will be finalized during 2019, also following the discussion with the European Commission.

This chapter is devoted to very brief overview of situation of energy field in the Czech Republic, based on NECP [4].

A. Total energy consumption

As we can see in Fig.1., final energy consumption is decreasing during last 20 years. The major problem is increasing energy consumption in last few years. It is due to the economic growth. National energy efficiency contribution until 2030 set maximum level of energy consumption on the basis of the real potential to reduce energy consumption (energy efficiency target at 32,5 %).

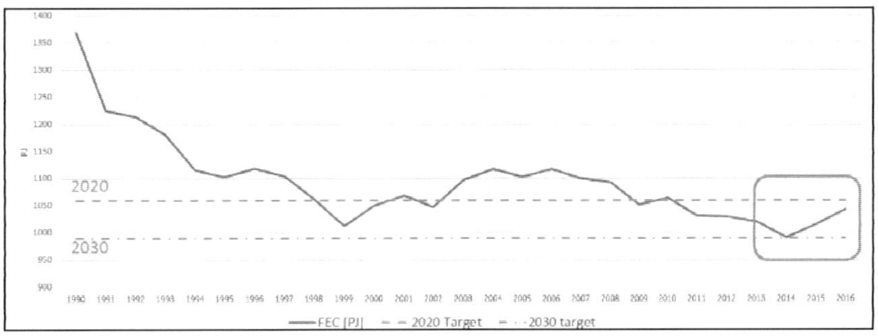

Figure 1. Final energy consumption development [4]

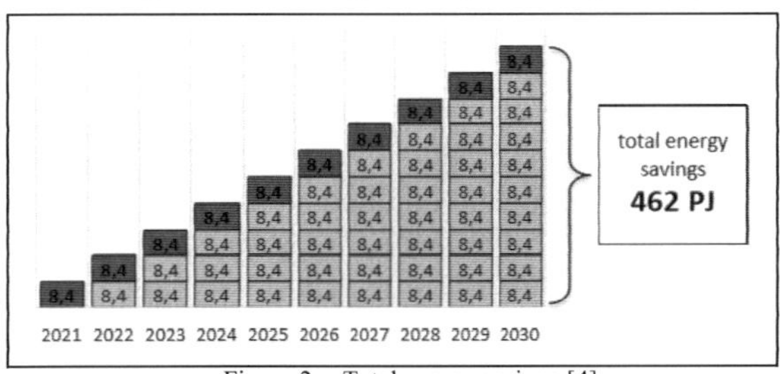

Figure 2. Total energy savings [4]

Energy savings obligation 2021-2030 set up cumulative end-use energy savings annual energy savings of 8,4 PJ –total of 462 PJ cumulative energy savings until 2030 (see Fig. 2.).

At first, we can observe tightening of mandatory annual savings (204 PJ (2014-2020) versus 462 PJ (2021-2030). Main reason why the Czech Republic voted against the adoption of the EED - implementation of Article 7 is binding at Member State level (compared with Article 3, which is "only" indicative).

At second, we can observe real energy savings potential for 2021-2030 Based on the realistic estimates of energy consumption trends and the potential to implement energy savings measures in the individual sectors in 2021-2030 [4].

At last we have to estimated cost to reach the energy savings obligation for 2021-2030 with the current alternative scheme is more than 600 billion CZK (23 billion EUR).

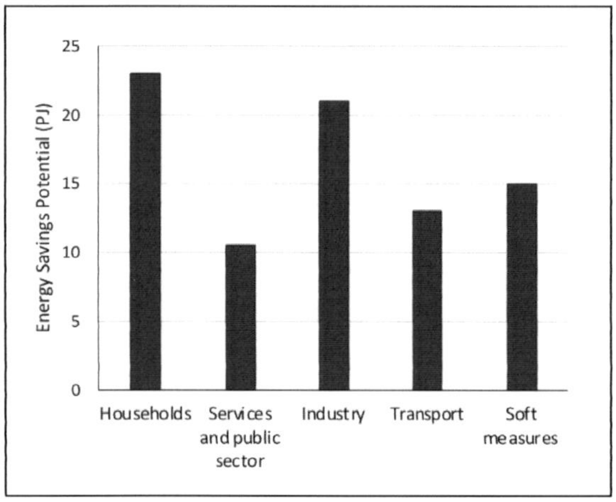

Figure 3. Energy savings potential [4]

B. Renewable sources

In NECP Czech Republic did not agree to increase the European ambition by more than 30 %, as we can observe from the following figures and tables, Czech Republic is already fulfilling its 2020 target of 13 % - proposed contribution of the Czech Republic for 2030 corresponds to 20-21 %.

We have number of sub-targets: mandatory RES share in transport: 14 % (+ 3.5 % advanced biofuel target) and indicative heating and cooling target at 1.1 % yearly growth. In order to fulfill the plan, the Ministry of Industry and Trade prepared an amendment to the law on the supported RES (currently following the comment procedure).

At first we should mention our 2020 target of RES share on gross energy consumption, which we had already fulfilled. (See Fig. 4). We have 2030 goal, which need to be reached with 68 PJ gap.

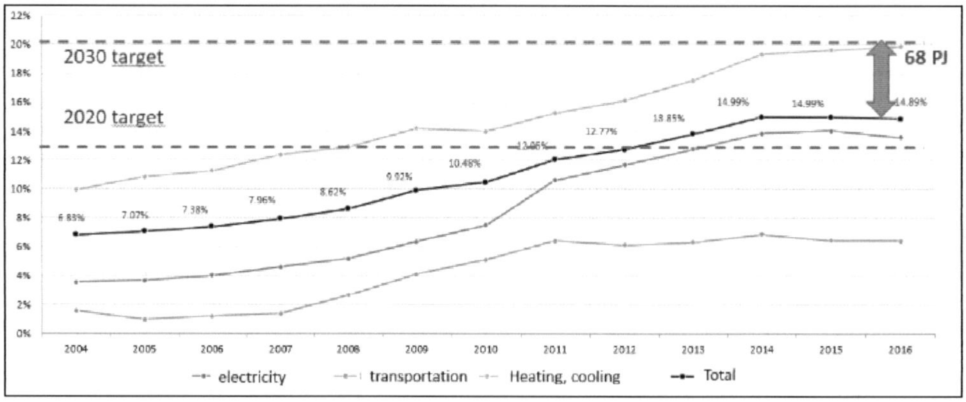

Figure 4. RES share on gross energy consumption. [4]

	Electricity	Transportation	Cooling and heating	Total final consump.
2010	7,52 %	5,12 %	14,02 %	**10,48 %**
2011	10,61 %	1,18 %*	15,29 %	**10,91 %**
2012	11,67 %	6,15 %	16,15 %	**12,77 %**
2013	12,78 %	6,34 %	17,56 %	**13,85 %**
2014	13,89 %	6,90 %	19,36 %	**15,00 %**
2015	14,07 %	6,45 %	19,66 %	**15,00 %**
2016	13,61 %	6,42 %	19,93 %	**14,91 %**

Table 1. RES share on gross energy consumption [4]

At second, we can see development of structure of RES production. In 2017: RES is 10,54 % on primary energy sources, decisive role of solid biomass: 65 %, incl. biogas and liquid biofuels: 88 % of RES contribution on PES.

At last we should comment development of electricity in the Czech Republic produced from RES. As we can see in Fig. 6, during last ten years huge amount of electricity from PVE was increased. So-called solar boom is one of the most significant problem in the Czech Republic caused by combination of high feed in tariff, low price of panels and unability of politics to properly react to new situation.

Unfortunately, this situation leads to misunderstanding role of RES in Czech energy mix and contributes to disreputed RES.

IV. Instruments and Evaluation

In mentioned goals, we should conclude, that there is significant gap to fulfill in next ten years and the major question is, how to reach them.

There is lot of instruments ensuring reaching the mentioned goals. Lot of them are already being operated on state level:

- Energy savings – Programme EFEKT (MET), OPPIK, New green savings…
- Renewable support – FIT, FIP for electricity production..

Lot of them are being operated on European level (Emission trading system (ETS)).

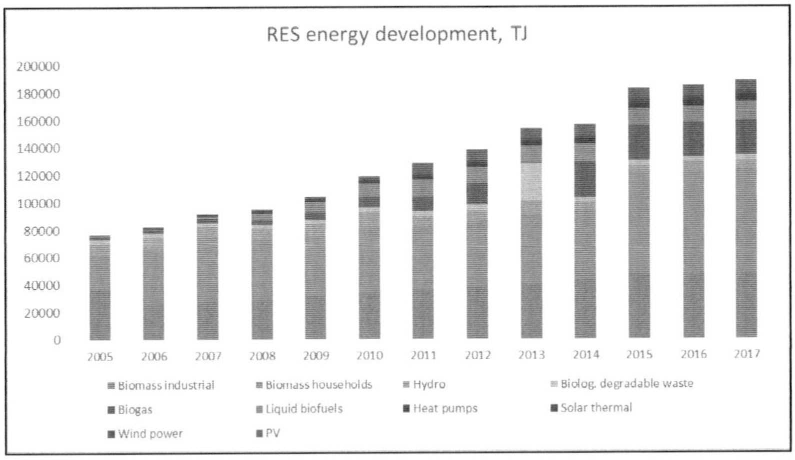

Figure 5. RES energy development [4], [5]

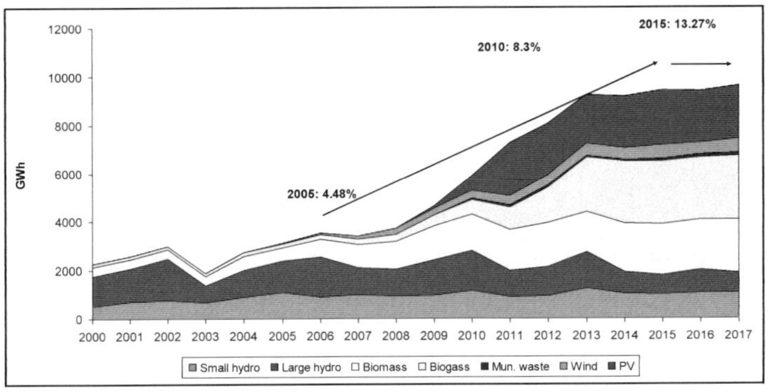

Figure 6. RES energy development [4], [5]

As we can see for example from solar boom in the Czech Republic, 2010-2012, it is crucial to evaluate costs and (not only financial) benefits of each programme.

The concept of the five dimensions of the Energy Union, as defined by the Regulation, is used as the basis for assessing the societal benefits of the energy policies. Each of the Energy Union dimensions monitors and evaluates the specific aspect of the energy sector, respectively. the energy sector as a whole in terms of meeting the Energy Union's own concept and in terms of meeting defined objectives (either for the given energy sector or for the energy sector as a whole) by 2030.

The EU is committed to the Paris Agreement at the 21st Conference of the Parties to the United Nations Framework Convention on Climate Change in 2015. These commitments are reflected in the Energy Union's concept and objectives, among other things: EU - "Ensure safe, sustainable, competitive and affordable energy for EU consumers, including households and businesses, and encourage research and innovation by attracting investment." A fully functioning and resilient Energy Union would turn the EU into a region that takes the lead in terms of innovation, investment, growth and social and economic development, which would be a good example of how closely the application of high ambition in mitigation is linked. climate change with measures to promote innovation, investment and growth.

The above general objective of further EU direction in the field of energy supply and energy consumption is then divided into five areas - Energy Union dimensions: Energy security, Internal energy market (and related economic aspects including energy poverty), Energy efficiency, Reducing carbon emissions and on Research, Development and Competitiveness. Each of the dimensions of the Energy Union then defacto is one of the sub-goals of further energy development.

Evaluation of the "social" effectiveness of a given development strategy then represents a comprehensive evaluation of all dimensions. This task is generally the task of multi-criteria evaluation, the individual dimensions of the Energy Union represent groups of criteria, within each group of criteria, several sub-evaluation criteria can be defined that are relevant to the particular objective - the Energy Union dimension (or the target values defined within each dimension).

A. Evaluation of energy policies

Evaluation is a process of making assessment of contribution of a project. Evaluation of a project in business enviroment is very often approach, but in policy enviroment is not so often.

But in recent years, it is crucial to evaluate instruments in energy state field. As we set, to reach the energy savings obligation for 2021-2030 with the current alternative scheme is more than 600 billion CZK (23 billion EUR) amount needed.

This huge amount of money need to be spent reasonably. But only few studies [7] is devoted to evaluation of energy policies. Research of these studies and evaluation of czech support system (mainly FIT and FIP) is a part of further research, following.

V. Conclusion

The key documents for the further development of the energy sector are the European Parliament and Council Regulation 2018/1999 of 11 December 2018 on Energy Union Management and Climate Action and the Czech Republic's National Energy and Climate Plan.

Regulation 2018/1999 establishes the necessary legal framework for the administration of the Energy Union and at the same time defines the basic quantitative targets and instruments for achieving them by 2030 in terms of decarbonisation (including RES) and energy efficiency. At the same time, the Regulation provides the basis for meeting the long-term objectives of the Energy Union in relation to the Paris Agreement concluded at the 21st Conference of the Parties to the United Nations Framework Convention on Climate Change 2015.

Regulation 2018/1999 introduces five dimensions of the Energy Union, which should ensure the interconnection of energy development aspects and ambitious climate protection objectives. The essential declared goal of the Energy Union is to ensure safe, sustainable, competitive and affordable energy for all types of consumers, together with support for research and innovation. The regulation recapitulates the current quantitative targets for reducing CO2 emissions - 40 % across the economy, 32 % RES in EU final energy consumption, improving energy efficiency by at least 32.5 % and at least 15 % electricity interconnections.

Regulation 2018/1999 sets up five dimenstions to develop the energy sector:

• Energy security,

• Internal energy market

• Energy efficiency,

• Reducing carbon emissions,

• Research, development and competitiveness

This paper was focused on brief introduction to the czech case, focusing on actual situation in energy field. These data shows, that evaluation of every programme is crucial tool to properly set the portfolio of energy policies to reach the goals in optimum way.

Acknowledgment

Research described in the paper was supervised by prof. Ing. Jaroslav Knápek, CSc., Department of Economy, Management and Humanities, FEE CTU in Prague and supported by the Student Grant Competition under grant No SGS16/154/OHK5/2T/13.

References

[1] Renewable energy. European comission [online]. [cit. 2019-3-20]. Available from: https://ec.europa.eu/energy/en/topics/renewable-energy

[2] Energetická účinnost, obnovitelné zdroje energie a správa energetické unie: Rada schválila 3 významné předpisy v oblasti čisté energie. *Rada Evropské unie* [online]. 4.12:2018 [cit. 2019-05-02]. Dostupné z: https://www.consilium.europa.eu/cs/press/press-releases/2018/12/04/energy-efficiency-renewables-governance-of-the-energy-union-council-signs-off-on-3-major-clean-energy-files/

[3] Návrh NAŘÍZENÍ EVROPSKÉHO PARLAMENTU A RADY o správě energetické unie, kterým se mění směrnice 94/22/ES, směrnice 98/70/ES, směrnice 2009/31/ES, nařízení (ES) č. 663/2009, nařízení (ES) č. 715/2009, směrnice 2009/73/ES, směrnice Rady 2009/119/ES, směrnice 2010/31/EU, směrnice 2012/27/EU, směrnice 2013/30/EU a směrnice Rady (EU) 2015/652 a zrušuje nařízení (EU) č. 525/2013 COM/2016/0759 final - 2016/0375 (COD). In: . Dostupné také z: https://eur-lex.europa.eu/legal-content/CS/TXT/?uri=CELEX %3A52016PC0759

[4] Návrh vnitrostátního plánu v oblasti energetiky a klimatu České republiky. *Ministerstvo průmyslu a obchodu* [online]. [cit. 2019-05-01]. Dostupné z: https://www.mpo.cz/cz/energetika/strategicke-a-koncepcni-dokumenty/navrh-vnitrostatniho-planu-v-oblasti-energetiky-a-klimatu-ceske-republiky--243377/

[5] Statistiky ERÚ. *Energetický regulační úřad* [online]. [cit. 2019-05-01]. Dostupné z: https://www.eru.cz/cs/

[6] EED Energy Efficiency Directive. Energetický regulační úřad [online]. [cit. 2019-05-01]. Dostupné z: https://ec.europa.eu/energy/en/topics/energy-efficiency/energy-efficiency-directive

[7] F. Polzin, F. Egli, B. Steffen, T. S. Schmidt, "How do policies mobilize private finance for renewable energy?—A systematic review with an investor perspective",Applied Energy,Volume 236,2019,Pages 1249-1268, ISSN 0306-2619, https://doi.org/10.1016/j.apenergy.2018.11.098. (http://www.sciencedirect.com/science/article/pii/S030626191831818X)

E: Smart Grids and Electromobility

A Universal, Modular and Bidirectional Active Balancing System for Lithium-Ion Cells

Andreas Ziegler
University of Applied Sciences Wuerzburg-Schweinfurt
Technology Transfer Centre for E-Mobility
Bad Neustadt an der Saale, Germany
andreas.ziegler@fhws.de

Prof. Dr. Bernhard Arndt
University of Applied Sciences Wuerzburg-Schweinfurt
Faculty of Electrical Engineering
Schweinfurt, Germany
bernhard.arndt@fhws.de

David Oeser
University of Applied Sciences Wuerzburg-Schweinfurt
Technology Transfer Centre for E-Mobility
Bad Neustadt an der Saale, Germany
david.oeser@fhws.de

Prof. Dr. Ansgar Ackva
University of Applied Sciences Wuerzburg-Schweinfurt
Technology Transfer Centre for E-Mobility
Bad Neustadt an der Saale, Germany
ansgar.ackva@fhws.de

Abstract — **This paper introduces a new, modular and universal active balancing system that is able to gain significantly more energy from aged and unbalanced batteries compared to other balancing concepts. One of the major issues when using battery packs is the possible unbalance within single battery cells. As this problem worsens with increasing age state of the art passive balancing methods reach their limits in terms of energy efficiency and applicability. After identifying the issues of standard balancing systems, requirements to overcome those problems are derived. To meet those requirements a new balancing hardware and the corresponding algorithm is implemented. The applicability and energy efficiency is proofed in the laboratory on a real hardware system and demonstrated by measurement results. Furthermore, a evaluation of a realistic application with a traction battery is executed.**

Keywords: Cell balancing; Active balancing; Battery; Battery management; Lithium-Ion; Algorithm

I. INTRODUCTION

Due to manufacturer-specific deviations and the unpredictable and unequal ageing of lithium-ion cells, there may be asymmetries in the State Of Charge (SOC) of a battery pack (1)–(6). Modern battery systems consist of many individual battery cells which are connected to each other via a series and parallel connection. Since the total usable capacity is determined by the weakest single cell in the series connection, unequal cell ageing leads to a limited usable energy of the entire battery. Fig. 1 illustrates the limitations when using battery packs with unbalanced cells. It is apparent that cell 1 in charge mode reaches the maximum cell voltage, typically defined as end-of-charge-voltage, first. As soon as one of the participating cells has reached this limit, the charging process must be aborted in order to avoid destruction.

However, this restriction results in all other cells not reaching the maximum cell voltage corresponding to unused capacity. A similar behaviour can also be observed in discharging mode. As soon as the first cell reaches the minimum cell voltage the discharging process is stopped even though the other cells still have unused capacity.

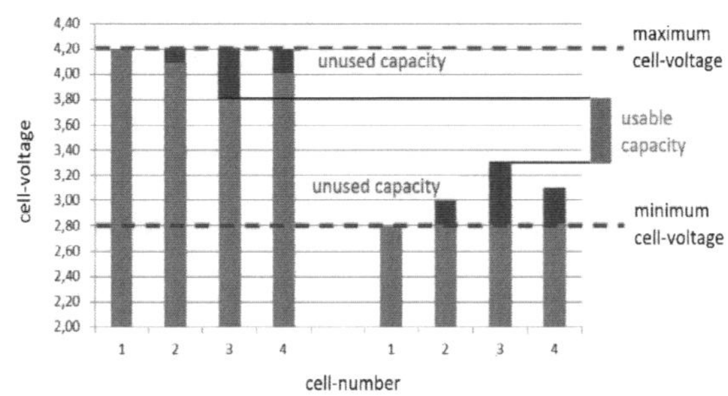

Figure 1. Usable capacity of an unbalanced battery.

A variety of different passive, as well as active balancing methods that are used to balance the described cell deviations, have already been developed (7)–(14). In most of the applications passive balancing systems are state of the art. Despite the known disadvantages, such as small balancing currents resulting in longer balancing time, they are often used due to lower costs (15).

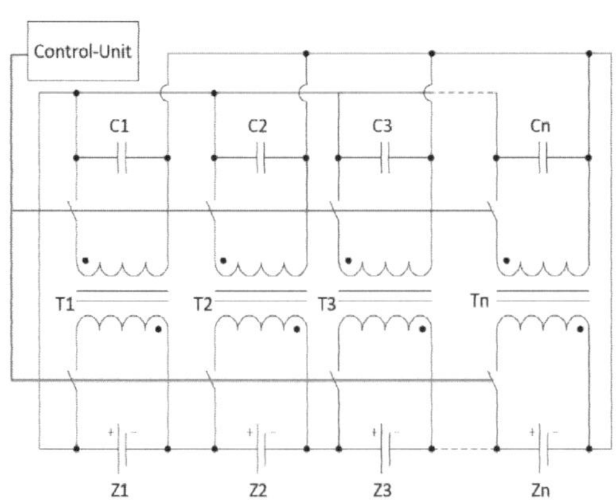

Figure 2. Multiple transformer circuit.

Many hardware concepts that are used in active balancing systems use a single and sometimes bidirectional DC/DC converter for each cell of a battery pack (16)-(17). Fig. 2 shows the basic concept of a multiple transformer circuit that is controlled by a central control unit. The primary side of the transformer is connected to the cell whereas the secondary side is connected to the whole battery pack. The maximum number of controllable cells in the proposed circuit depends on the specifications of the central control unit. To enable the operation with large battery packs multiple balancing controllers are necessary. When using this hardware implementation, it should be noted that energy can only be transmitted between cells of a single control unit, due to electrostatic potential problems. The exchange of information between the individual controllers is usually ensured using a serial communication interface.

II. Proposed Hardware Concept

The now introduced hardware concept is based on a redesign of the multiple transformer topology in such a way that energy transfer between cells of different control units is possible. The main ideas include:

- A power bus is used to transfer energy between two or more battery sections and thus to enable the balancing of cells connected to different control units.
- An isolated communication bus is used to overcome problems with different electrostatic potentials between different controllers.
- A modular concept ensures to set up a balancing system which is applicable to battery packs that contain variable arrangements of series and parallel connections of cells.

Beside the main ideas, additional components and properties must be taken into account and will be examined in more detail.

A. Power Bus

Each cell of the battery is connected to the primary side of a bidirectional flyback-converter whereas secondary side is connected to the newly introduced power bus. This circuitry enables the possibility to transfer energy between an theoretically unlimited number of converters.

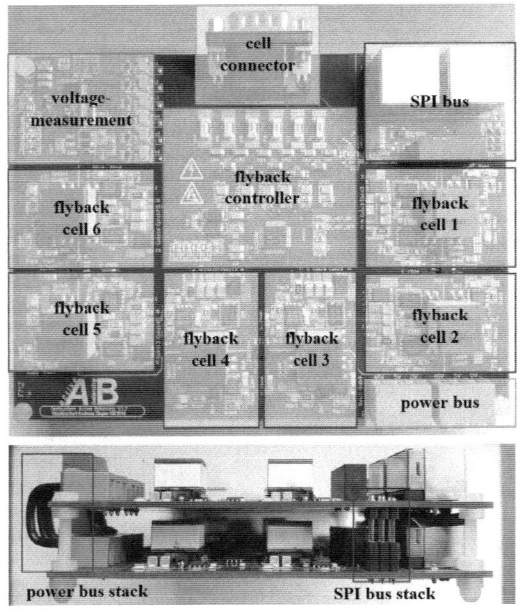

Figure 3. Modular balancing board.

B. Communication bus

To enable SPI communication using an isolated bus-system, multiple digital isolators have to be connected in parallel. In the parallel connection, each digital isolator wants to transfer its actual state to the bus system. This leads to signal overlays that make the actual signal of the controller just addressed unrecognisable. To avoid overlapping, all serial data outputs of the isolators are provided with diodes that allow the signal line to be pulled down for data transmission, but prevent pull it up. This allows the addressed controller to transmit its signal.

C. Modular concept

Due to individual structures, each battery type requires a specific balancing system. To guarantee re-usability a modular and universal concept was developed that can easily be extended to any type of battery. Fig. 3 shows an example of the proposed control unit and the possible extensibility.

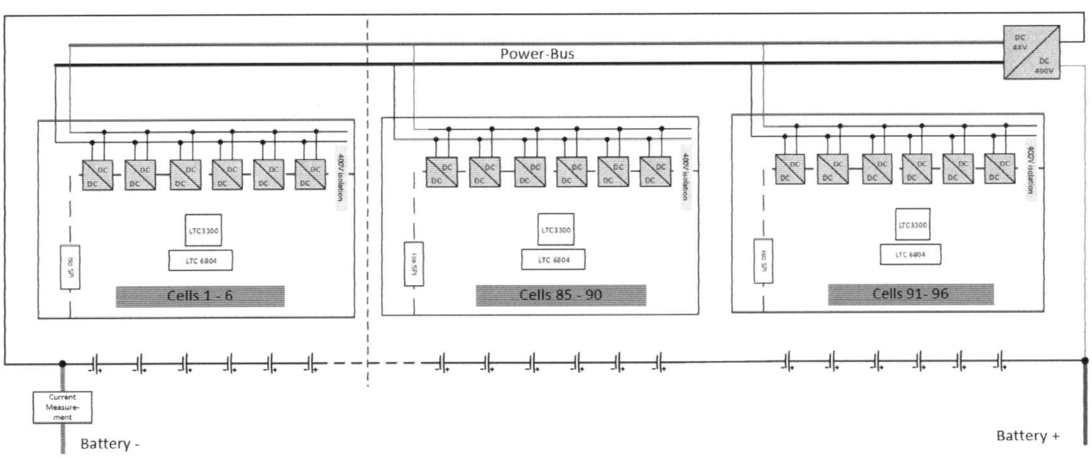

Figure 4. Overview of system topology.

The proposed concept allows an extension to a maximum of 16 control units, which enables balancing of up to 96 serially connected cells in a single battery pack.

D. *Current measurement*

In order to allow optimal balancing of the cells, a suitable and accurate current measurement must be provided. In the proposed implementation, current values with a resolution of 16 bits are sent via bus communication to the central control unit. In addition to the values of the currents, the measured cell voltages are used to determine the battery status. Most important states are for example "fully charged", "fully discharged" or an "up to date state of charge".

E. *Main specifications*

An example of the system topology consisting of n cells connected in series is shown in Fig. 4. The maximum possible number of connected cells is thereby only limited by the possible SPI addresses of the flyback-controller. Nevertheless, when using a second address range, up to 192 cells or, theoretically, an unlimited number of cells can be connected. Some of the major system specifications are variable, for example power bus voltage is adjustable between 12 V and 50.4 V and balancing currents are configurable from 0.5 A up to 4 A. The power consumption increases with the number of cells connected. In addition, the use of a higher number of cells also results in higher utilization and thus higher efficiency of the converter. A maximum overall system efficiency of 85.5% has been determined.

F. *Operational safety*

To ensure safe operation, three main protection components are provided in the hardware concept.

- Electrical isolation: The primary and secondary side of each converter is electrically isolated by a transformer to prevent short circuits between cells and power bus.
- Fuse on primary side: If a short circuit occurs on the primary side the current is only limited by the internal resistance and the characteristic impedance of the short circuit itself. To avoid damage, a fuse is provided that limits the current on the primary side.
- Fuse on secondary side: To avoid short circuit on secondary side a central fuse is required.

Due to these protection measures, a cell short circuit after a single fault is not possible. The resulting hardware concept including the protection components is shown in Fig. 5.

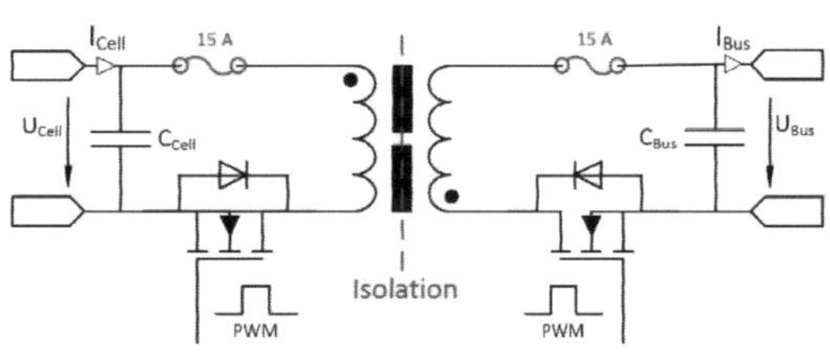

Figure 5. Operational safety steps of each flyback converter.

III. Control and Balancing Algorithm

To ensure a secure and proper cell balancing, the selected controller must have enough processing power to execute different commands, evaluate measurements, and perform calculations. Basically, as with already developed algorithms, there is the possibility to charge, discharge and stop the individual cells via commands and to carry out a fully automated cell balancing.

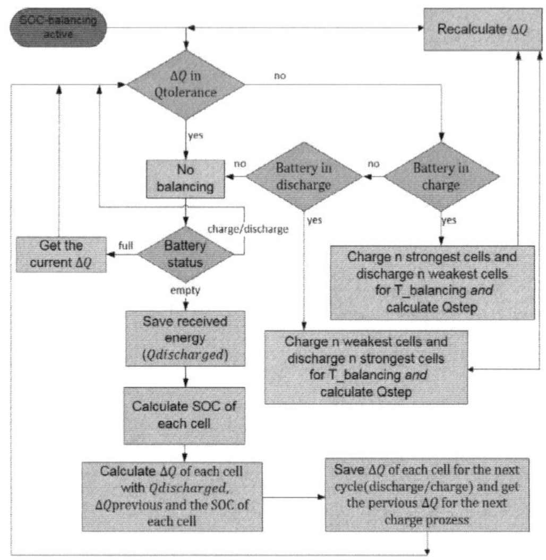

Figure 6. SOC-balancing algorithm.

A. *Balancing Manager Algorithm*

At the beginning, the balancing algorithm requires the measured values of cell voltages and load current of the battery as input parameters. With these values, battery status can be determined and a decision about the balancing algorithm to be used can be made. Once a decision has been made on the balancing method, the corresponding routine is executed.

B. *SOC-balancing-algorithm*

The principle working behaviour of the SOC balancing algorithm is highlighted in Fig. 6. In the beginning, ΔQ value of each cell is checked for deviations. If ΔQ of one of the cells is not within the predefined tolerance band, the actual battery state (charging, discharging, pause, SOC) is verified and the balancing algorithm is performed. During the balancing process, cell voltages and load current are measured, the battery state is calculated, the discharged energy amount is stored and the actual SOC as well as the new ΔQ values are calculated periodically within the time-step called $t_{balancing}$. The proposed algorithm is executed until the cell charge of all cells is equal.

C. *Top-balancing algorithm*

The top-balancing algorithm is performed during the constant voltage phase of the charging process of the battery pack. The measured cell voltages are first evaluated and sorted from highest to lowest. From the average value of all voltages, the deviation of each individual cell voltage can now be determined. If the voltage is too low the cell will be charged and if the voltage is too high it will be discharged. The process is restarted after the time $t_{balancing}$ until all cells lie in a predefined voltage range.

IV. Measurements and Results

A. *Laboratory Test on aged Batteries*

In order to evaluate and validate the hardware concept and the balancing algorithm, various tests and measurements were performed on the real hardware system. For the test setup six heavily aged batteries consisting of 12 cells connected in series with 5 parallel each were used. With the help of a High Power Battery Test System (HPS) from BaSyTec GmbH a defined and reproducible test cycle could be guaranteed. For a summary of test conditions, see Tab. 1.

Table 1. Test Conditions

Technical data of the test setup	Battery test system	
	Parameters	*Values*
Specification of the cycliser	Type	BaSyTec HPS
	Voltage range	0 – 60V
	Current range	+20A / -40A
	Current accuracy	±20mA
	Current rise time	<1ms
	Test batteries	
	Parameters	*Values*
Specification of the battery	Cell-type	Samsung 18650-26F
	Topology	6 packs, each with 60 cells (12s5p setup)
	Voltage range	35.5V (empty) – 50.1V (full)
	Nominal capacity	13Ah
	Test cycle	
	Procedure	*Explanation*
Specification of the cycle	Charging	CC-phase 3A, CV-phase starts at 50.1V
	End of Charge	Current <0.5A or one cell voltage >4.21V
	Discharging	Constant current 3A
	End of Discharge	Pack-voltage <35.5V or cell voltage <2.95V
	Pause	30min between charge and discharge

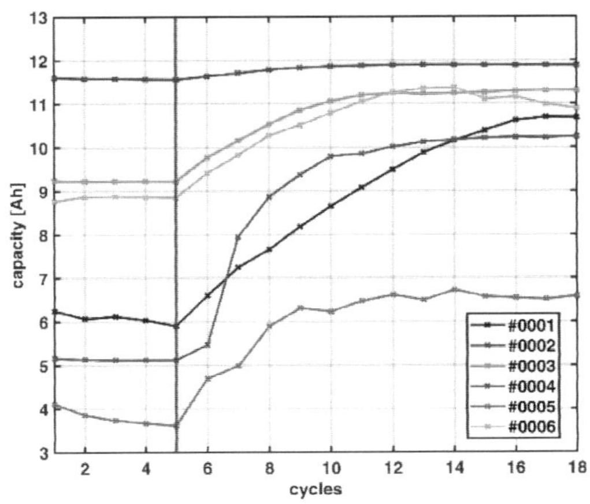

Figure 7. Capacity raise due to active balancing.

At the beginning of the test, five cycles were performed without active balancing to determine the exact capacity of every single battery. Then active balancing was activated and evaluated over 13 full cycles. The resulting discharge capacity as a function of the cycle count is shown in Fig. 7. It can be clearly seen that the capacity gain between the batteries varies a lot. Batteries aged only slightly, such as battery 2, benefit less from the performed balancing. Capacity can be barely increased. Batteries 3 and 6, near End Of Life (EOL), have a significantly larger increase in capacity. In batteries 1, 4 and 5, the capacity has been severely reduced due to the failure of several cell levels. Nevertheless, two of the three batteries can still be used normally with the aid of active balancing. To make a more precise statement about the possibilities of the presented algorithm, the resulting capacity for each battery is shown in Tab. 2.

Table 2. Capacity raise due to active balancing

Test results per battery	Values			
	Start Capacity [Ah]	*Cycles inbetween*	*End Capaxity [Ah]*	*Capacity increase [%]*
#0001	5.91	12	10.69	80.88
#0002	11.56	7	11.90	2.94
#0003	9.22	12	11.30	22.56
#0004	5.12	11	10.23	99.80
#0005	3.61	9	6.72	86.15
#0006	8.85	9	11.37	28.47

A preliminary analysis of the used batteries showed that the installed cells were not sufficiently matched. The batteries, therefore, had a partially stronger load on individual cell levels, which resulted in accelerated ageing. In order to demonstrate the operation and the benefits of the presented system, these batteries are very suitable.

B. Evaluation in real application

To estimate whether active balancing is reasonable for common applications, tests were carried out on a new traction battery of an BMW i3 electric vehicle. The battery pack was built up of 96 serially connected cells divided into 8 modules with 12 cells each. Via a suitable interface, information such as cell voltages, load current and cell temperature could be read periodically. The procedure for testing the battery in the laboratory environment will be briefly explained below.

1. A DC source was used to charge the battery pack up to a SOC of 100%.
2. A DC sink was used to discharge the battery to 0% SOC with 20 A constant current. As soon as the first cell reached the minimum cell voltage the load current was stopped and the 96 cell voltages were stored. Fig. 8 shows the resulting voltage of each cell in blue and in addition the separated 8 modules voltages in green.
3. With help of cell data sheet and measured voltages at turn off point, cell SOC and the corresponding energy was calculated independently of internal resistance to obtain a rough estimate of cell disbalance. This type of residual energy determination is legitimate in this case, as the new cells have an internal resistance of less than 1mΩ and discharge current was relatively low

There are two ways to balance the traction battery. Either all cells can be balanced with each other or only cells within a module can be balanced. The comparative results are shown in Tab. 3 for cell to cell balancing and in Tab. 4 for module balancing only. The unusable energy was calculated in relation to the weakest module. As shown, the energy transfer between all cells allows a gain in usable energy of 506 Wh. If Energy can only be distributed within a single module, the usable energy increases by 440Wh. From 505Wh of remaining energy in the modules of Tab. 3, 65Wh can not be used due to balance only inside the modules and so 440Wh can be achieved.

The results of the study of the traction battery show, that single cell balancing in and between the modules results in a gain in usable energy of up to 15% (increase from 440Wh to 506Wh) comparing to an module-intern cell balancing and thus is a relevant influence.

In case of a new battery, the difference is relatively small. However, with increasing age, cell to cell balancing can play a major role. For this, further tests must be carried out with aged traction batteries. For a more precise determination of the capacity differences and therefore cell disbalance, all cells must be measured individually.

Table 3. Remaining energy due to "balance all cells"

Test results per Cell	Values			
	Voltage [V]	*SOC [%]*	*Capacity [Ah]*	*Energy [Wh]*
1	2.899	2.35	1.411	4.092
2	2.942	2.86	1.717	5.050
3	2.983	3.35	2.008	5.989
4	2.858	1.87	1.121	3.203
5	2.801	1.19	0.716	2.006
6	3.007	3.63	2.178	6.549
7	2.957	3.04	1.823	5.391
8	3.020	3.78	2.270	6.856
9	2.963	3.11	1.866	5.528
10	2.953	2.99	1.785	5.300
11	2.971	3.20	1.922	5.711
12	2.984	3.36	2.015	6.012
...	...	...	...	...
Σ 1-12				61.685
Σ 1-96				506

Table 4. Remaining energy due to "balance only modules"

Test results per module	Values			
	SOC [%]	*Capacity [Ah]*	*Energy [Wh]*	*unusable Energy [Wh]*
1	2.89	1.727	61	6
2	3.27	1.964	70	15
3	3.43	2.060	74	19
4	3.00	1.797	64	9
5	2.79	1.676	59	4
6	3.00	1.799	64	9
7	2.61	1.566	55	0
8	2.75	1.650	58	3
ΣBattery			505	65

V. Conclusion

This paper proposes a novel hardware concept and an optimized balancing algorithm for active balancing of lithium-ion battery cells. With the help of a suitable test cycle, the potentials of active balancing are demonstrated with help of heavily aged batteries. In the specific example, huge capacity gains are achieved. Based on an application of a real traction battery of an electric vehicle it is additionally demonstrated that even with new batteries, a gain of up to 15% compared to other balancers may be possible. The results presented in this paper allow the conclusion that the use of active balancing shows a relevant influence and has an even clearer and significant influence as the battery ages.

As there will be more electric vehicles in the near future, the number of aged batteries will increase significantly. The use as a Second Life battery is therefore becoming increasingly interesting. Here, too, active balancing could maintain the usability of the battery over a longer period of time. For this purpose, more studies must be carried out in the area of Second Life batteries.

Acknowledgment

This work was funded by Preh GmbH and was carried out within a commissioned research project. All rights of the publication rest with the authors.

References

[1] T. Baumhofer, M. Bruhl, S. Rothgang, and D. U. Sauer, "Production" caused variation in capacity aging trend and correlation to initial cell performance," *Journal of Power Sources*, vol. 247, pp. 332–338, 2014.

[2] S. Paul, C. Diegelmann, H. Kabza, and W. Tillmetz, "Analysis of ageing inhomogeneities in lithium-ion battery systems," *Journal of Power Sources*, vol. 239, no. Supplement C, pp. 642 – 650, 2013. [Online]. Available: http://www.sciencedirect.com/science/article/pii/S037877531300116X

[3] S. F. Schuster, T. Bach, E. Fleder, J. Muller, M. Brand, G. Sextl," and A. Jossen, "Nonlinear aging characteristics of lithium-ion cells under different operational conditions," *Journal of Energy Storage*, vol. 1, no. Supplement C, pp. 44 – 53, 2015. [Online]. Available: http://www.sciencedirect.com/science/article/pii/S2352152X15000092S.

[4] F. Schuster, M. J. Brand, P. Berg, M. Gleissenberger, and A. Jossen, "Lithium-ion cell-to-cell variation during battery electric vehicle operation," *Journal of Power Sources*, vol. 297, pp. 242–251, 2015.

[5] M. Broussely, P. Biensan, F. Bonhomme, P. Blanchard, S. Herreyre, K. Nechev, and R. Staniewicz, "Main aging mechanisms in li ion batteries," *Journal of Power Sources*, vol. 146, no. 1, pp. 90 – 96, 2005, selected papers pressented at the 12th International Meeting on Lithium Batteries. [Online]. Available:http://www.sciencedirect.com/science/article/pii/S0378775305005082

[6] D. Oeser, A. Ziegler, and A. Ackva, "Single cell analysis of lithium-ion e-bike batteries aged under various conditions," *Journal of Power Sources*, vol. 397, pp. 25–31, 2018. [Online]. Available:https://www.sciencedirect.com/science/article/abs/pii/S0378775318307055

[7] M. Caspar, T. Eiler, and S. Hohmann, "Comparison of active battery balancing systems," in *Proc. IEEE Vehicle Power and Propulsion Conf. (VPPC)*, Oct. 2014, pp. 1–8.

[8] M. Caspar, V. Polinski, and S. Hohmann, "Structural comparison of battery balancing architectures with optimal control," in *Proc. IEEE Vehicle Power and Propulsion Conf. (VPPC)*, Oct. 2015, pp. 1–6.

[9] J. Cao, N. Schofield, and A. Emadi, "Battery balancing methods: A comprehensive review," in *Proc. IEEE Vehicle Power and Propulsion Conf*, Sep. 2008, pp. 1–6.

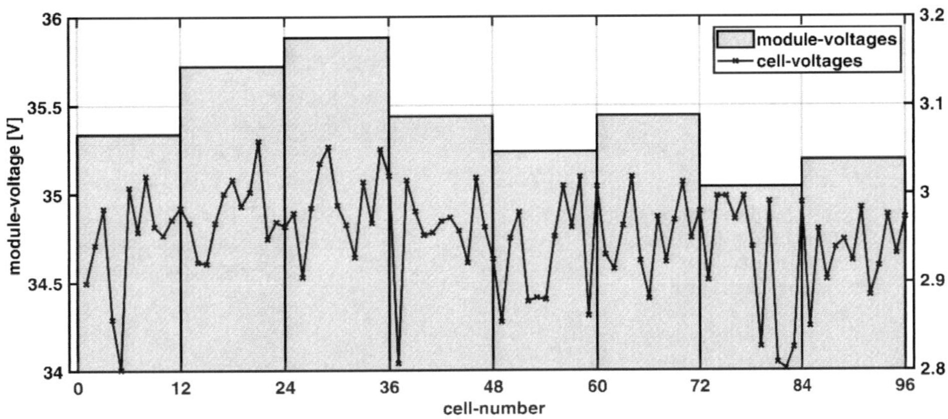

Figure 8. Voltages of a BMW i3 battery pack.

[10] F. Baronti, R. Roncella, and R. Saletti, "Performance comparison of active balancing techniques for lithium-ion batteries," *Journal of Power Sources*, vol. 267, pp. 603–609, 2014.

[11] M. Daowd, N. Omar, P. Van Den Bossche, and J. Van Mierlo, "A review of passive and active battery balancing based on matlab/simulink," *International Review of Electrical Engineering*, vol. 6, no. 7, pp. 2974– 2989, 2011.

[12] J. Gallardo-Lozano, A. Lateef, E. Romero-Cadaval, and M. I. Milanes-´ Montero, "Active battery balancing for battery packs," *Electrical, Control and Communication Engineering*, vol. 2, no. 1, pp. 40–46, 2013.

[13] J. Gallardo-Lozano, E. Romero-Cadaval, M. I. Milanes-Montero, and M. A. Guerrero-Martinez, "Battery equalization active methods," *Journal of Power Sources*, vol. 246, pp. 934–949, 2014.

[14] L. Valda and K. Kosturik, "Comparison of li-ion active cell balancing methods replacing passive cell balancer," in *Applied Electronics (AE), 2015 International Conference on*. IEEE, 2015, pp. 267–270.

[15] K. M. Lee, S. W. Lee, Y. G. Choi, and B. Kang, "Active balancing of liion battery cells using transformer as energy carrier," *IEEE Transactions on Industrial Electronics*, vol. 64, no. 2, pp. 1251–1257, Feb. 2017.

[16] M. M. U. Evzelman, K. Rehman, R. Hathaway, D. Zane, D. Costinett, and D. Maksimovic, "Active balancing system for electric vehicles with incorporated low-voltage bus," *IEEE Transactions on Power Electronics*, vol. 31, no. 11, pp. 7887–7895, 2016. [Online]. Available: http://ieeexplore.ieee.org/stamp/stamp.jsp?tp=&arnumber=7368921&tag=1

[17] Linear-Technology-Corporation, *Bi-Directional Cell Balancer Using the LTC3300-1 and the LTC6804-2, DEMO MANUAL DC2100A*, Linear Technology 1630 McCarthy Blvd. Milpitas, CA 95035, 2014. [Online]. Available: http://cds.linear.com/docs/en/demo-board-manual/dc2100afc.pdf

District Heating Sector in the Czech Republic

Status, Challenges and Perspectives

Tomáš Králík
Dept. of Economics, Management and Humanities
Czech Technical Univerzity in Prague, FEE
Prague, Czech Republic
tomas.kralik@fel.cvut.cz

Jiří Vecka
Associaton for District Heating
Prague, Czech Republic
vecka@tscr.cz

Jiří Vašíček
Dept. of Economics, Management and Humanities
Czech Technical Univerzity in Prague, FEE
Prague, Czech Republic
vasicek@fel.cvut.cz

Martin Beneš
Dept. of Economics, Management and Humanities
Czech Technical Univerzity in Prague, FEE
Prague, Czech Republic
benes@fel.cvut.cz

Abstract — **The paper provides an overview of the current state of the district heating sector in the Czech Republic with special focus on barriers for future development. The EU objectives, which the Czech Republic must gradually adopt in its concepts, constitute new and more specifically formulated challenges than in the current energy concept. The five dimensions of the Energy Union, that has been introduced by Regulation 2018/1999 and that will shape the future of energy sector, are briefly described. Last, but not least, the economy of heat production is discussed, with special focus on the influence of emission allowance price on final heat price.**

Keywords: centralised heat supply; heat price, emission allowances

I. CURRENT STATE OF HEAT SECTOR IN CZECH REPUBLIC

Currently, the Czech heating industry accounts for about a third of the total heat production of heating energy. At present (2016), the household sector accounts for the largest share of heat supplied from the centralized heating systems, almost 50 %. Another 28 % is taken by industry and the remaining 22 % by service sector.

Centralized heat generation in the Czech Republic is provided both by sources in the heating mode (producing only heat) and by sources in the cogeneration mode (producing heat and electricity). In 2017, 66 % of heat production came from Combined Heat and Power (hereinafter as "CHP").

In 2017, net heat production excluding technological heat consumption was 156.2 PJ. Heat supply to external entities amounted to 93.7 PJ. The remaining heat of 62.5 PJ consisted of heat supplied to consumers at the place of production and heat losses [1].

The volume of heat supply has been decreasing in the last few years, which is mainly due to the decreasing number of degree days (or the increase in average temperature in crucial months of the heating period in the Czech Republic) and the support for the implementation of energy-saving measures as shown in the following Figure 1.

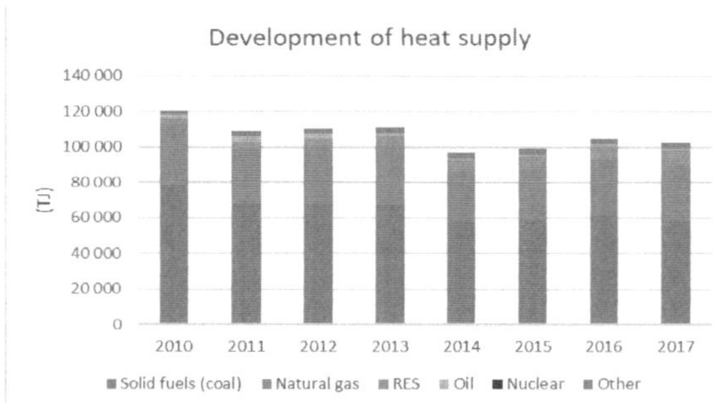

Figure 1: Development of heat supply from district heating [2]

The primary most used fuel in the Czech heating sector is coal with approximately 60 % of the fuel mix. However, compared to the fuel mix in 2010, this share decreased by about 5 %. The share of natural gas in heat supply remains more or less constant at around 30 % of the fuel mix. By contrast, the growing fuel consumption trend can be seen in the RES category (over 8 % of the fuel mix in 2017).

In absolute terms, the increasing share of biomass in heat supply (6 PJ in 2017 compared to approximately 2 PJ in 2010) is apparent from a more detailed view and the biogas share is doubled (0.5 PJ in 2017 compared to 0.25 PJ in 2010) compared to 2010 [3].

II. REGULATION FOR FUTURE DEVELOPMENT

The future development of the heat (not only) sector will be significantly influenced by Regulation 2018/1999 of the European Parliament and of the Council of 11 December 2018 on Energy Union Management and Climate Action and its follow-up The Czech Republic's national energy and climate plan.

Regulation 2018/1999 establishes the necessary legal framework for the administration of the Energy Union and at the same time defines the basic quantitative targets and instruments for achieving them by 2030 in terms of decarbonisation (including Renewable Energy Sources (hereinafter as "RES")) and energy efficiency. At the same time, the Regulation provides the basis for meeting the long-term objectives of the Energy Union concerning the Paris Agreement concluded at the 21st Conference of the Parties to the UNFCCC 2015.

Regulation 2018/1999 introduces five dimensions of the Energy Union, which should ensure the interconnection of energy development aspects and ambitious climate protection objectives. The essential declared goal of the Energy Union is to ensure safe, sustainable, competitive and affordable energy for all types of consumers, together with support for research and innovation.

The regulation recapitulates the current quantitative targets for reducing CO_2 emissions – 40 % across the economy, 32 % of RES in EU final energy consumption, improving energy efficiency by at least of 32.5 % and at least 15 % in electricity interconnections.

In assessing the societal benefits of the heating industry, this methodology is based on the logic of the five dimensions of the Energy Union, which consists of five different aspects of evaluation, and each dimension represents a group of criteria in the multi-criteria evaluation terminology, which includes sub-criteria assessing the individual aspects of a given Energy Union dimension.

Individual dimensions of the Energy Union, respectively groups of criteria, are declared by Regulation 2018/1999:

A. *Energy security*

In particular, in the context of this dimension it means diversification of energy sources; rate of import dependence on energy sources; energy storage; ability to respond to a temporary disruption of the energy supply and Demand Side Response (hereinafter as "DSR").

B. *Internal energy market*

It explicitly includes only electricity markets (and related services - support services, flexibility, non-frequency services) and pipeline gas and contains three primary objectives:

- the level of the interconnectivity of national electricity systems by 2030
- key infrastructure projects, including upgrading in the electricity and gas networks, which are needed to achieve the objectives and tasks of the Energy Union dimension
- national objectives related to other aspects of the internal electricity market, which are: increasing systemic flexibility, in particular through market pricing policies and measures; and market integration and coupling to increase the marketable capacity of existing interconnectors, smart grids, aggregation, demand-side responses, storage, distributed production, dispatching mechanisms, redispatching and production restrictions when needed and real-time price signals including a timeframe for achieving these objectives.

C. *Energy efficiency*

The energy efficiency criteria include:

- Primary energy / primary energy consumption
- Primary Energy Sources (hereinafter as "PES") consumption on Gross Value Added (hereinafter as "GVA")
- Final energy consumption

D. *Reduction of carbon emissions*

For this dimension, the Regulation introduces two fundamental objectives:

- Reduction of Greenhouse Gas (hereinafter as "GHG") emissions
- RES share in gross final energy consumption

For the Czech Republic, the basic objectives for these two areas arise:

- Reduction of CO_2 emissions in 2030: 44 $MtCO_2ekv$ compared to 2005 (−30 %)
- Share of RES in gross final consumption in 2030 up to 20.8 %

E. *Research, development and competitiveness*

The objectives in the area of the so-called research, competitiveness and innovation dimension are formulated only verbally (qualitatively) in the EU Regulation, and are not quantified in contrast to energy efficiency, RES share and emission reductions.

III. Problems and Barriers

A. *Asymmetric impacts of legislation*

As a sector, the heating industry is mostly affected by the asymmetric impacts of, in particular, environmental regulations, which reduce their competitiveness against smaller local heat sources or even individual heating.

Examples are the considerable investments made by the Industrial Emissions Directive, particularly in coal-fired power plants, with the requirements still not definitive. There is significant uncertainty in the long-term definition of BAT emission limits based on Best Available Techniques (hereinafter as "BAT") - implementation in the form of a BAT Reference Documents (hereinafter as "BREF") is still under preparation. The EU Directive defines only upper limits. Changes in these parameters can dramatically affect the competitiveness of the "big" heating industry and cause a fundamental asymmetric impact.

B. *Reduction of heat demand as a result of increasing energy efficiency*

The downward trend in heat demand from energy saving measures will be further strengthened as part of the implementation of the Energy Efficiency Directive and the use of funds from the sale of allowances. There is a significant potential for heat savings in the current panel and other residential buildings. According to the studies, it follows that panel buildings can be reconstructed to a passive standard. The reality is a general trend of reducing heat consumption for heating (but also industrial heat abstraction due to similar austerity measures and a trend towards the displacement of energy-intensive products). The need for heat supply is also reduced by the use of solar collectors.

The Czech Republic, as well as other EU member states, currently have binding energy efficiency targets for 2020 set by the 2009 climate and energy package. However, even after 2020, we can expect a further continuation of energy efficiency policy (energy savings), increasing the share of RES) and reducing greenhouse gas emissions. On 22 January 2014, the EU Commission presented a proposal for "climate and energy" targets by 2030. In addition to the goals for reducing greenhouse gas emissions and the binding share of RES, the proposal also foresees that the EU Member States will have to include energy efficiency in their energy strategies.

C. *Reduction of new investments in heating sector due to high risks for investors*

Restriction of new construction is combined with strict requirements on heat consumption. New consumptions are not enough to replace the decreasing demand of existing customers, even when there is no disconnection. A significant part of the new construction is an individual development, where centralized heating systems do not come into consideration in principle.

D. *Heat price regulation*

Tying regulation of heat prices makes it impossible to react flexibly within the pricing policy to the needs of individual customers, or to come up with competing offers of heat prices compared to natural gas offers (which are often priced for longer periods).

This is often due to inadequate ecological and other requirements, and restrictions applied selectively only to large sources – in many cases, the state interventions into the market environment in the heating industry exceed the rational framework in terms of efficiency (economic, energy and ecological). However, the applied requirements, for example for forced reconstructions, are not reflected in the current rules of regulation based on an annual assessment basis.

E. *Developments in the fuel market*

The current decade is characterized by significant changes in the energy commodity market. Due to the massive development of shale gas extraction in the US (and the increase in US consumption), black coal is being displaced, and its shipments to Europe are being released at substantially lower prices compared to the previous decade. This indirectly affects prices and other energy commodities in Europe. Other significant changes affecting the Czech Republic include the launch of the Nord Stream pipeline and the start of construction of the South Stream pipeline. Another factor affecting the natural gas market is the realization of new capacities at terminals for liquefied natural gas imported by ships. Another characteristic feature is the significant shift in contracts from long-term deliveries to spot trades. Gas suppliers can respond significantly to price changes on the market – they are so-called "on commodities" and are not tied to maintaining the price for all categories of suppliers – unlike heat suppliers, where non-discriminatory access is one of the principles of regulation of individual customer groups.

There are significant uncertainties in the area of natural gas prices arising from the unclear approach of EU countries to the development of shale gas extraction as well as from several uncertainties in the "classical" market. On the one hand, due to the end of US liquefied natural gas imports, significant supply capacities are being released, e.g. for Europe, but the development of demand for this commodity in countries such as China and Japan (for example, about nuclear program restraint) is currently not certain.

The question of further availability of domestic brown coal for heating plants is also unclear. This is primarily a political question of whether the so-called territorial limits will be revised based on Government Decree 444/1991 Coll. or whether the intended use of coal for the heating industry will be realized before the electricity production itself or, on the contrary, for the direct reduction of coal

F. *Emission allowance price*

The greenhouse gas emission allowance price has arisen almost five times during one year. Even at the beginning of September 2017, it was around five euros, while it currently oscillates around 25 euros as it is shown in Figure 2.

The heating plants receive part of the heat production allowances free of charge, but their amount is annually decreasing and for next year it reaches approximately one third of the amount of 2013. At present, heating plants have to buy most of the necessary allowances on the market. The obligation to purchase allowances covers equipment from 20 megawatts of installed thermal input using fossil fuels. In the heating industry according to Association for district heating of the Czech Republic [4], almost all of the heat production in coal-fired power plants and approximately 60 percent of heat generation in natural gas-fired power plants are affected by the purchase of allowances.

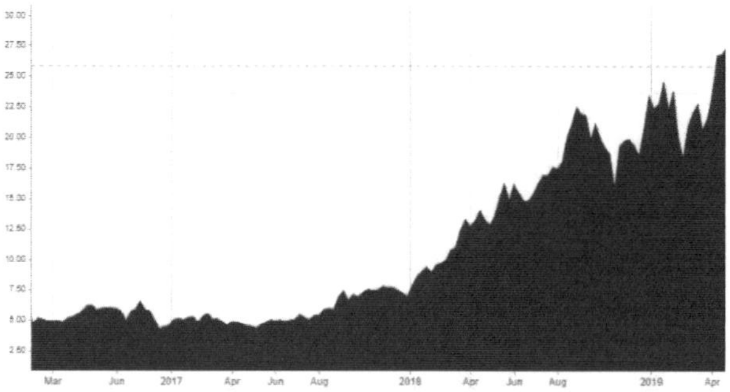

Figure 2: Emission allowance price [5]

To put this in to perspective – Czech heating sector will spend about four billion crowns on the purchase of heat production allowances in 2019 [6]. This amount of money will be spend unevenly according to CO_2 emission factors for individual heating source based by used fuels. The emission factors for the mostly used fuels are in the following Table 1.

Table 1: CO_2 emission factors

Fuel	Emission factor	
	t CO_2/MWh of fuel calorific value	***t CO_2/GJ of fuel calorific value***
Brown Coal	0.360	0.100
Hard Coal	0.330	0.092
Fuel Oil	0.260	0.072
Natural Gas	0.200	0.056

The above mentioned problems are considered as the crucial ones, however there are many other negatively influencing factors that has the potential to endanger centralized heating systems, such as:

- Higher efficiency of small boilers
- The absence of a long-term (realistically applied) state strategy in the heating sector
- The problem of oversized sources and networks in some district heating systems and outdated concepts of technical measurement and control solutions
- Electricity prices and low cogeneration support

- Learning curve effect on alternative heat generation technologies

IV. Economic impacts and Modelling

Price of heat is a very sensitive issue due to the fact, that roughly two thirds of the overall household energy bill is for the heat. The development of heat expenditures is shown in the following figure. The average annual household demand for energy is declining, mainly due to home insulation or window replacement. That is why the average energy bill is not changing even though the price for 1 GJ of heat has been rising during the past two decades - currently typical 4 person household consumes app. 25-33 GJ of heat per year, that is equal to 15 000 – 20 000 CZK. It is also interesting to investigate the net income of households in relationship to annual heat payment and compare it with following statistics:

- 30% of households, that buy heat from centralized heating systems, have net income below 20 000 CZK/month (785 EUR/month)
- next 20% of households have net income below 26 000 CZK/month (1020 EUR/month)

It is therefore clear that price of heat is also, in terms of the number of households, a crucial issue not only from technical or economic point of view, but it is also a social issue.

This open a question of energy poverty. Czech republic has currently this index below 3%, which is significantly lower than the EU average (8%). However this index can increase significantly with rising heat prices as poorer households (usually in rural areas) have no other means how to acquire heat from alternative sources.

This is the reason why we have followed suggested methodology in [7] and calculated and updated possible impacts under upcoming environmental legislation and with updated CO_2 prices. The calculation was done for two possible scenarios, for two price levels of emission allowances and for individual heating fuels and results are shown in Figure 4 and Figure 5. The scenarios are in [7] defined as:

- *Scenario 1 – Strictest implementation: Emission trading – no household rule, no derogation for electricity producers, free allocation according to the benchmark. Industrial Emissions Directive (hereinafter as "IED") – without any derogation for district heating, full application from 1 June 2020 (after expiration of IED Transitional national plan derogation)*
- *Scenario 2 – Pragmatic implementation: Emission trading – household rule, derogation for electricity producers. IED – with a derogation for district heating, full application from 1 January 2023*
- *CO_2 costs – Price of allowance: 2 price levels 20 EUR/t CO2 (current) and 45 EUR/tCO2 (estimated future price)*

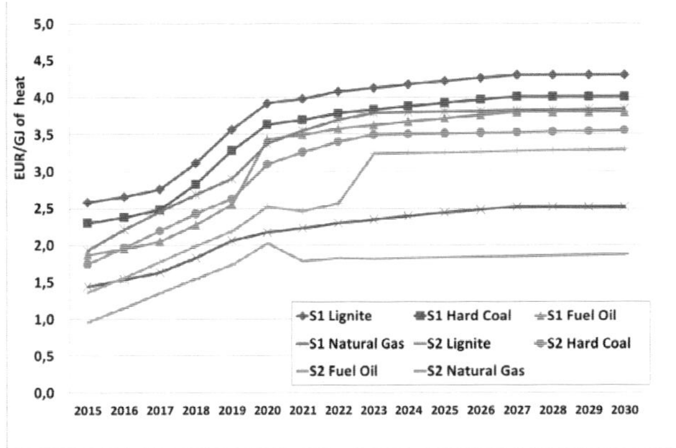

Figure 3: Impacts on heat prices in EUR/GJ – Comparison between strictest (S1) and pragmatic (S2) implementation of new environmental legislation, price of CO_2 20 EUR/tCO2

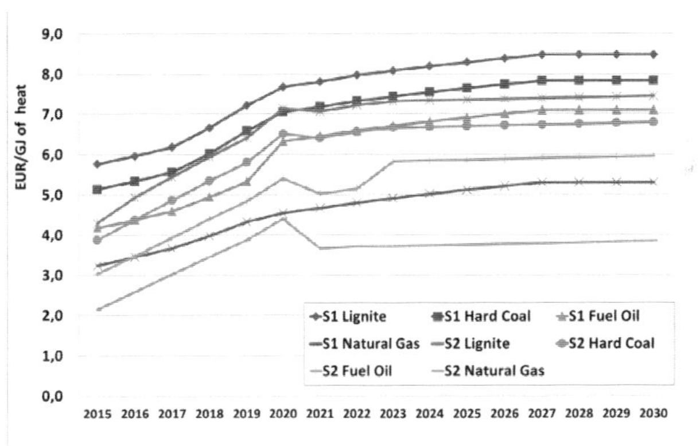

Figure 4: Impacts on heat prices in EUR/GJ – Comparison between strictest (S1) and pragmatic (S2) implementation of new environmental legislation, price of CO_2 45 EUR/tCO2

V. Conclusions and Disscussion

The paper analyses legislation induced impacts of relevant regulation tools (especially environmental dossiers) on the heating market participants in the Czech Republic with the focus on the district heating utilities and models series of possible development scenarios up to 2030.

Current average heat price in Czech Republic is app. 23 EUR/GJ from coal and 24 EUR/GJ from natural gas. It is obvious from obtained results, that the implementation of new regulation with the combination of higher prices of emission allowances has the potential to increase final heat price within the range of 1,5 to 5 EUR/GJ for heat from natural gas sources and even more for heat from coal and lignite (up to 4-8 EUR/GJ).

Such increase would be in direct contradiction with plans to keep the index of energy poverty low. It would also mean higher risk/probability of disconnection of current customers, what would lead to another price increase due to the increase of fixed cost for remaining customers.

It is obvious, that current situation in heating sector is not sustainable and there must be some changes, that allow centralized heating systems to maintain in operation under reasonable economic conditions. There are several ideas that can help with this situation, such as:

- Prohibition of disconnection from centralized heating systems
- Subsidy for lowering energy losses in heating systems (mainly in the grid)
- Carbon tax, that is imposed on all sources (not only big heating sources as it is currently)
- Lowering VAT on heat price (this is already approved in the Czech Republic)
- include all externalities – however it is problem how to properly calculate all externalities

The near future of centralized heating system will be influenced by the new legislation, that aims at lowering CO_2 emission and increasing energy efficiency. This can disadvantage and endanger systems that are mainly using coal and natural gas as currently cheapest heat sources. For the future existence of centralized heating systems, all main barriers and discriminatory access must be removed in existing relationships between all stakeholders. It is also necessary to set such support mechanisms so there is no discrimination – equal conditions for all participants.

It is also important to realize that in a number of locations, district heating systems may (and probably will) face a number of problems in the next 10 years, potentially leading to the risk of disintegration of these systems (loss of heat supply to customers). The main problem is, that in many location there is no cheap and available substitution.

Acknowledgment

This work was supported by the Technology Agency of the Czech Republic with grant number TK01010017.

References

[1] Czech Statistical Office, "Energy statistics - Public database", 2018, Available: https://vdb.czso.cz/vdbvo2/faces/en/index.jsf?page=vystup-objekt-parametry&pvo=PRU01-D&sp=A&skupId=1267&pvokc=&katalog=30835&z=T

[2] A. Bufka, J. Veverkova, D. Androvic, "Vývoj hrubé výroby elektřiny a tepla k prodeji v energetické bilanci ČR: výroba a dodávky v letech 2010-2017," Praha: MPO, 2019. Available: https://www.mpo.cz/assets/cz/energetika/statistika/elektrina-a-teplo/2019/1/Elektrina---teplo-2010-2017.pdf

[3] Energy Regulatory Office, "Yearly Report on the Operation of Heat Supply Systems in the CR", 2017, Available: http://www.eru.cz/documents/10540/4340114/Zprava_o_provozu_TS_2017_EN.pdf/7b1286f8-8b33-4c05-8200-7143b7f790a5

[4] Association for District Heating of the Czech Republic, „Energy in tables and graphs" 2018, Available: http://www.tscr.cz/?pg=0750&1559262944#

[5] CO2 EUROPEAN EMISSION ALLOWANCES. Market Insider [online]. [cit. 2019-05-28]. Available: https://markets.businessinsider.com/commodities/co2-emissionsrechte

[6] Onergetice, "Teplárenské sdružení: Cena tepla vzroste, hlavním důvodem jsou vyšší ceny povolenek", 2018, Available: https://oenergetice.cz/teplarenstvi/teplarenske-sdruzeni-cena-tepla-vzroste-hlavnim-duvodem-jsou-vyssi-ceny-povolenek/

[7] J. Vecka, "Model of medium-sized district heating development strategy based on new environmental legislation," 2016 2nd International Conference on Intelligent Green Building and Smart Grid (IGBSG), Prague, 2016, pp. 1-6. doi: 10.1109/IGBSG.2016.7539426G.

Model Selection for Wind-Driven Asynchronous Machines in Electric Power System State Estimation Problems

Glazunova Anna
Melentiev Energy Systems
Institute SB RAS
Irkutsk, Russia
glazunova@isem.irk.ru

Abstract — **The paper is concerned with two models of an asynchronous machine (a PQ model and an RX model) in the state estimation problems. When using the PQ model, the values of active power injection at a node is calculated as a sum of active power flows in the lines connected to this node. With the RX model, the value of injection is calculated through the voltage magnitude and generator slip. The study is focused on the problem of observability in electric power system state estimation depending on the asynchronous machine model selected.**

Keywords: asynchronous machine; electric power system; observability; state estimation.

I. INTRODUCTION

The efficiency of electric power system control is determined by the extent to which the data on the current state of the electric power system are valid and complete. These data are collected using the SCADA system (Supervisory Control and Data Acquisition) and PMUs (Phasor Measurement Units). In electric power system control, validity and completeness of data are provided by the state estimation procedure which calculates steady-state variables on the basis of a measurement snapshot (the measurements obtained at one and the same time point). Depending on the number and location of sensors, the measurements can be redundant, critical, or belong to a group of measurements with low redundancy.

Resistance of asynchronous machine depends on the load. This dependence can be taken into account when representing the generator by an RX model.

The problem of using the RX model in steady state calculations was examined in [1], [2]. In [1], there are two iteration processes in the steady-state problem. One of them is the calculation of the steady state and the other is the calculation of the induction generator slip. In terms of the steady state, the RX node behaves as a PQ node where active and reactive power equals zero, and, consequently, the additional iterations of the steady state calculation are not needed. Nevertheless, the calculation of slip requires additional 2 or 6 iterations. Based on the PQ model, the authors of [2] developed the models of WTGS (Wind Turbine Generating Systems), where active and reactive power depends on the magnitude of voltage, mechanical power, and equivalent circuit. They have a straight solution to obtain active power demand.

The EPS state estimation procedure is addressed in [3], [4]. The authors of [3] estimate the DFIG (Doubly Fed Induction Generator) using UKF (Unscented Kalman Filter). UKF provides highly accurate state estimates of the DFIG that is highly nonlinear. The authors of [4] use the RX model and introduce slip in the vector of state, thus increasing the dimensionality of the Jacobian matrix. However, one iteration process is necessary to obtain a steady state and slip. In [5], the authors use an extended state estimation method. According to this method, the models for different wind turbines are incorporated in the state estimation problem. In [6], DFIG is estimated using UKF, the High Gain Observer (HGO) and the Moving Horizon Estimation (MHE).

The paper is organized as follows. The second section presents models of an asynchronous machine. The third section explains the technique for state estimation. The fourth section explains an observability problem. The fifth section describes the modeling of wind speed and wind power generation. The sixth section is devoted to the online application of this technique. The last section presents the conclusions.

II. MODELS OF ASYNCHRONOUS MACHINE

There are two models of the asynchronous machine (a PQ model and an RX model) in the state estimation problems.

When using the PQ model, the values of active power injection at the node are calculated as a sum of the active power flows in the lines connected to this node:

$$P_i = \sum_{j=1}^{k} P_{ij} + U_i^2 G_i \,, \tag{1}$$

where k is the number of lines going from node i, G_i is conductance of shunt at node i, U_i is voltage magnitude, P_{ij} is active power flow calculated through the voltage magnitude and phase.

With the RX model, the value of injection is calculated through the voltage magnitude and generator slip by the formula [4]

$$P_W = -\frac{sR_r U^2}{s^2(X_s + X_r) + R_r^2} \,, \tag{2}$$

where s is generator slip, X_r, X_s are reactance of rotor and stator respectively, R_r is resistance of rotor.

III. STATE ESTIMATION

State estimation of power system implies the calculation of state variables, which is made on the basis of measurements obtained from the SCADA system and WAMS (Wide-Area Measurement Systems) [7].

The measurements look as follows:

$$\bar{y} = y_{true} + \xi_y \qquad \xi_y \in N(0, \sigma_y^2\ ) \,, \tag{3}$$

where σ_y^2 is measurement variance, y_{true} is true value, ξ_y is normal noise.

The vector of measurements looks as follows:

$$\bar{y} = (U_i, P_i, Q_i, P_{ij}, Q_{ij}, \delta_i), \quad (4)$$

where U_i – magnitudes of nodal voltages; P_i, Q_i – injections of active and reactive powers at nodes; P_{ij}, Q_{ij} – power flows in transformers and lines, δ_i – voltage phases at the nodes with PMUs.

The objective function in the state estimation with the PQ model of an asynchronous machine looks as follows:

$$J(U,\delta) = (\bar{y} - y(U,\delta))^T R_y^{-1} (\bar{y} - y(U,\delta)), \quad (5)$$

where $\bar{y}$ –vector of measurements, R_y – covariance matrix of measurement errors, $x = (U,\delta)$ – state vector.

In order to calculate the state vector, the derivative $\frac{\partial J(x)}{\partial x}$ is equated to zero, and the system of non-linear equations is solved:

$$H^T R^{-1} (\bar{y} - y(x)) = 0, \quad (6)$$

where $H = \frac{\partial y}{\partial x}$ – Jacobi matrix.

The objective function in the state estimation with the RX model of the asynchronous machine looks as follows:

$$\begin{aligned} J(U,\delta,s) &= (\bar{y} - y(U,\delta))^T R_y^{-1} (\bar{y} - y(U,\delta)) \\ &+ (\bar{P}_a - P_a(U,s))^T R_a^{-1} (\bar{P}_a - P_a(U,s)) \end{aligned}, \quad (7)$$

where $\bar{P}_a$ – a measurement of active injection of an asynchronous machine, R_a^{-1} – weighting coefficients of asynchronous machine active injection measurements.

In order to calculate the state vector, the systems of non-linear equations are solved:

$$H^T R^{-1} (\bar{y} - y(U,\delta)) = 0, \quad (8)$$

$$H_a^T R_a^{-1} \left(\bar{P}_a - P_a(U,s) \right) = 0. \quad (9)$$

Due to the nonlinear dependence $y(x)$, problems (6), (8), (9) are solved iteratively by Newton's method. The systems are linearized at each iteration and the systems of linear equations are solved, where the correction vector is calculated by the equation:

$$\Delta x^{(i)} = P^{(i)} (H^{T(i)} R_y^{-1} (\bar{y} - y(x^{(i)}))), \quad (10)$$

$$P^{(i)} = (H^{T(i)} R^{-1} H^{(i)})^{-1}. \quad (11)$$

Components of the state vector are calculated by the equation:

$$x^{(i+1)} = x^{(i)} + \Delta x^{(i)}, \quad (12)$$

where $x = (U,\delta)$ in the case of the PQ model and $x = (U,\delta,s)$ in the case of the RX model. If $\Delta x_i < d$, the iteration process converges, where d – the accuracy of the iterative process, and all state variables are calculated.

Then, using the obtained state vector we calculate all the other state variables.

IV. Observability

A node is observable if there are values of voltage magnitude and phase (the state vector) at this node. The state vector is measured or calculated by using other measurements.

With the PQ model, the flows are functions of the voltage magnitude and phase and, therefore, there are the derivatives of the injection with respect to phase and magnitude in the Jacobian matrix. In this case, if there are measurements of the injection at the wind farm, the values of the voltage phase and magnitude can be determined, and this node is observable. The voltage phase at node i is calculated by the formula that belongs to system (6):

$$\frac{\partial P_i}{\partial \delta_i} \Delta\delta_i = \bar{P}_i - P_i(U,\delta_i). \quad (13)$$

Observability of the node with respect to active power is available. With the RX model, there are derivatives with respect to slip and voltage magnitude in the Jacobian matrix. Consequently, the phase value in the absence of PMU at the node can only be determined by the formula in system (6) if there are measurements of power flows in the adjacent lines ($\bar{P}_{i-m}$):

$$\frac{\partial P_{i-m}}{\partial \delta_i} \Delta\delta_i = \bar{P}_{i-m} - P_{i-m}(U_i,\delta_i). \quad (14)$$

If the injection measurement is critical, then when the RX model is used this node becomes unobservable with respect to active power. In the case there is no redundancy of measurements at a node, to have observability it is necessary to carefully select the model of an asynchronous machine.

IV. Modeling of Wind Speed and Wind Power Generation

In this study, Weibull distribution is used to estimate the probability of wind speed. Fig. 1 shows wind speed at a=9.2, b=5, where a, b are constants used in Weibull distribution.

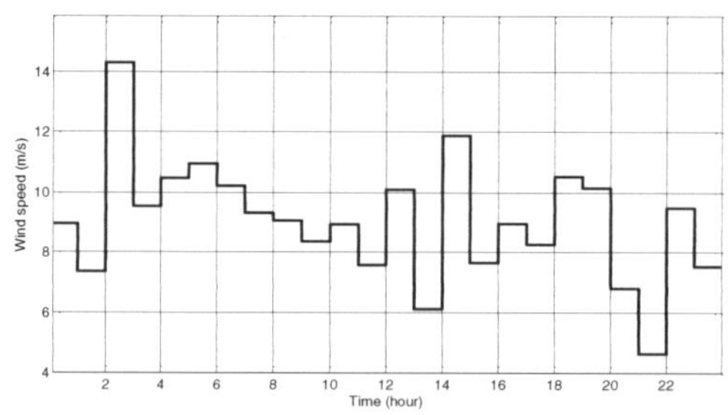

Figure 1. Wind speed

Fig. 2 shows that the wind turbine starts to work at wind speed $V_{cin} = 7 m/s$ and reaches rated power at wind speed $V_r = 12\ m/s$. Then, despite the increase in wind, the turbine power does not increase until $V_{cc} = 19\ m/s$ is reached.

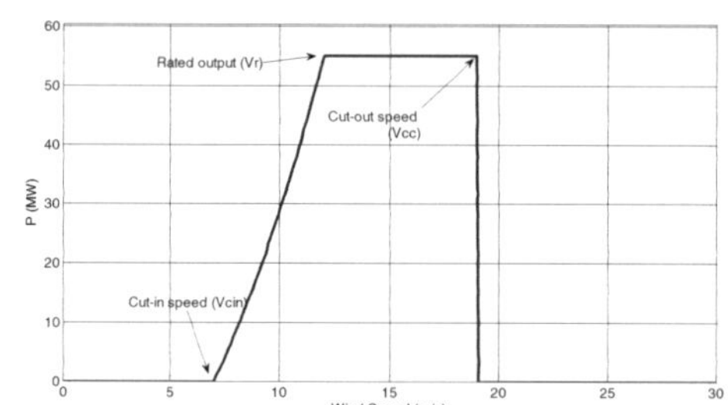

Figure 2. A typical curve of wind turbine

Using the known wind speed, we calculate the wind turbine output by the formula:

$$P_W = 1/2\rho A v^3 c_p \, , \tag{15}$$

where ρ – the density of air, v – the wind speed, A – the rotor area, c_p – the power coefficient. In this paper $\rho = 1.605 kg/m^2$ $A = 1124 m^2$.

Figure 3 demonstrates the wind turbine power.

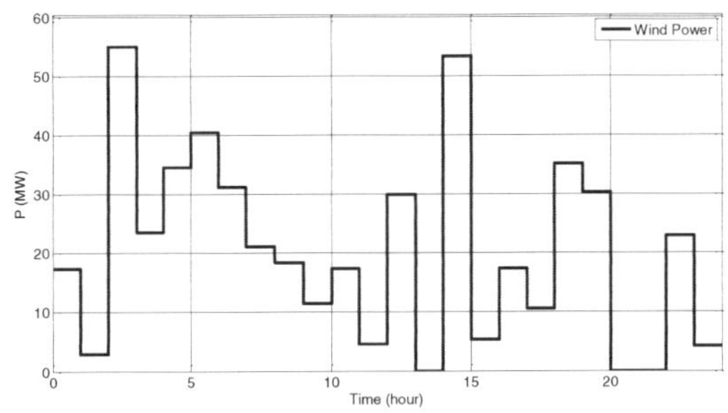

Figure 3. Active power of wind turbine

VII. CASE STUDY

A. Scheme and scenario

The calculations are performed using the IEEE 30-bus test system (Fig. 4) [8]. The wind farm (where an asynchronous machine is installed) is denoted by node 13. The balancing node is node 2. The accuracy of measurements P_{13}, P_{12-13} is the same. The calculations are performed according to the scenario: to perform the state estimation procedure for two cases using two archives:

1. Archive 1 contains 57 measurements including two measurements (P_{13}, P_{12-13}) that refer to node 13.

2. Archive 2 contains 56 measurements including one measurement (P_{13}) that refers to node 13.

B. Modeling of operating conditions

An archive of conditions is created on the basis of the mathematical model of the IEEE 30-bus test system under various operating conditions.

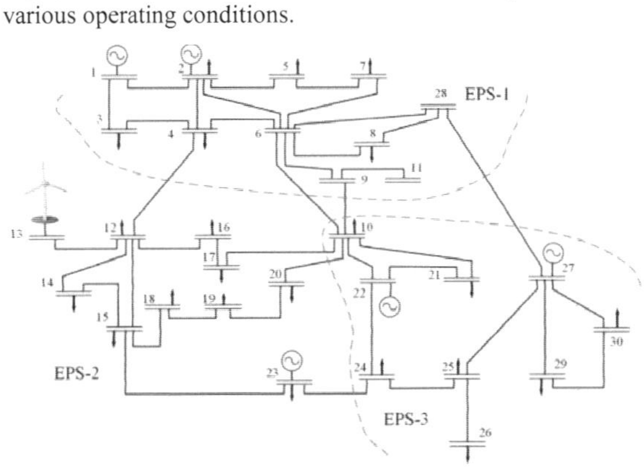

Figure 4. The IEEE 30-bus test system

Such a variety can be achieved by applying models of typical load profiles at chosen nodes. In this research, three types of typical load profiles are presented. Nodes 5 and 12 represent the load of three- and two-shift industrial facilities, respectively (Fig. 5, 6). At node 21, the load of a residential consumer (a city with a population of 200–250 thousand people) is modeled (Fig. 7). Wind generation is modeled at node 13, according to the plot presented in Fig. 3.

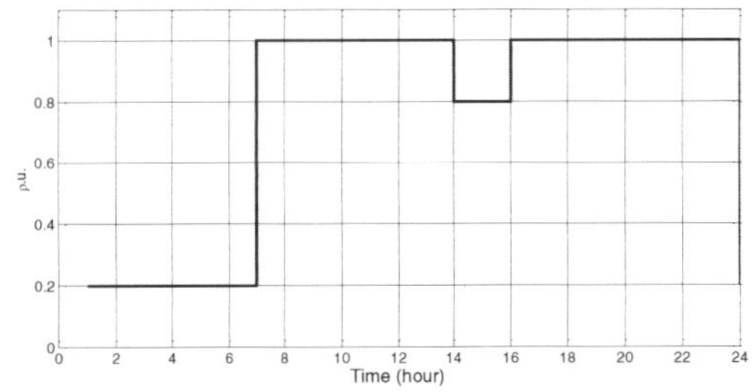

Figure 5. Load profile for a two-shift industrial facility

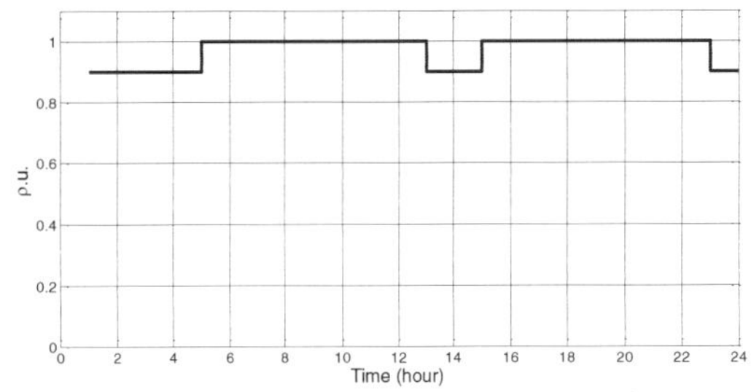

Figure 6. Load profile for a three-shift industrial facility

We obtain a load flow solution for each point of the load curves. These solutions are then used to create measurement snapshots. Chosen state variables of a particular solution are changed by adding a small random value according to the model of measurement

$$\bar{y} = y_{true} + x_{rand}\,\sigma_y \, , \tag{16}$$

where σ_y – measurement standard deviation, x_{rand} – random errors generated by a random number generator $x_{rand} \in N(0,1)$ y_{true}, – reference (the result of load flow solution).

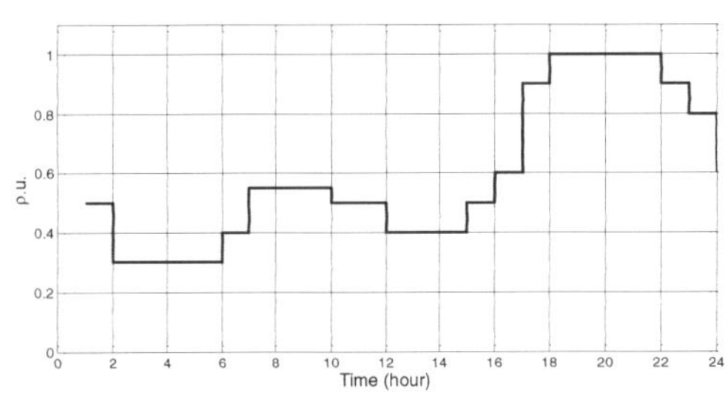

Figure 7. Load profile for a residential consumer (a city with a population of 200-250 thousand people)

Each solution can be used several times in a row to form the longevity of an unchanging state. In this research, each solution was used 10 times to form 240 snapshots of measurements.

C. *Results*

Case 1

Voltage phase at node 13 is calculated using the equation in system (6)

$$\frac{\partial P_{12-13}}{\partial \delta_{13}} \Delta\delta_{13} = \bar{P}_{12-13} - P_{12-13}(U_{13}, \delta_{13}, U_{12}, \delta_{12}). \quad (17)$$

The voltage phase δ_{13} is used to clarify the power flow P_{12-13}. Active power of generation from wind farm equals the power flow $P_{13} = P_{13-12}$.

Slip at node 13 is calculated using the equation in system (9)

$$\frac{\partial P_{13}}{\partial s_{13}} \Delta s_{13} = \bar{P}_{13} - P_{13}(U_{13}, s_{13}) . \quad (18)$$

The active power output at node 13 is calculated by (2), where $X_S = 0.2$, $X_r = 0.076$, $R_r = 0.007$, values of s and U depend on operating conditions.

Case 2.

In this case, it is possible to calculate either δ_{13} (voltage phase) or s_{13} (slip). In order to provide observability of node 13, we should have δ_{13}. That is why we use measurement P_{13} to calculate δ_{13}. Voltage phase at node 13 is calculated using the equation in system (6)

$$\frac{\partial P_{13}}{\partial s_{13}} \Delta\delta_{13} = \bar{P}_{13} - P_{13}(U_{13}, \delta_{13}). \quad (19)$$

Fig. 8 demonstrates the estimates and references of active power injection at node 13.

The estimation errors were calculated by the formulas:

$$ER_{PQ} = 1/N \sum_{i}^{N} (|P_{true} - P(U,\delta)| / P_{true}) * 100\% \ , \quad (20)$$

$$ER_{RX} = 1/N \sum_{i}^{N} (|P_{true} - P(U,s)| / P_{true}) * 100\% \ , \quad (21)$$

where ER_{PQ} – estimation error when the PQ model is used, ER_{RX} – estimation error when the RX model is used, P_{true}– reference, N – number of snapshots. For 240 considered snapshots of measurements ER_{PQ}=0.011%, ER_{RX}=0.01%. Based on the comparison of two results we can conclude that the quality of estimates obtained using different models of an asynchronous machine is the same.

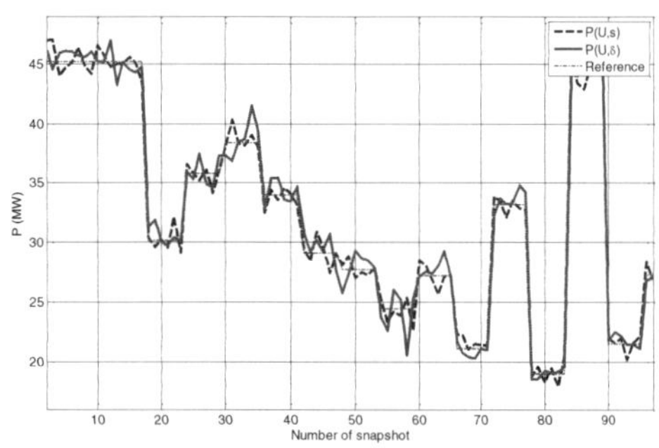

Figure 8. Estimates and references

Fig. 9 demonstrates the estimates of active power flow in line 12– 13. Fig. 8, 9 show, that $P_{13} = P_{13-12}$.

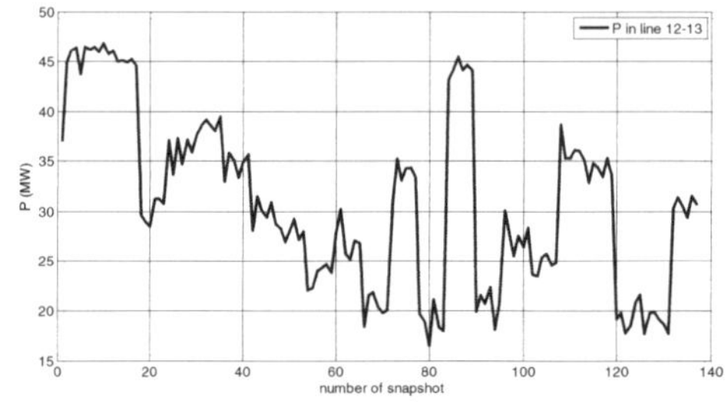

Figure 9. Estimates of active power flow in line 12–13

VIII. Conclusion

The paper considers the use of the PQ model and the RX model of induction generators installed at wind farms in the problem of state estimation. The voltage phase at the node under consideration can be determined using the PQ model. The use of the RX model enables the generator slip to be determined, which makes it possible to take into account the nonlinear load dependence of the generator resistance.

In the case the injection measurement at the node is critical it is necessary to select the PQ model of the asynchronous machine.

Acknowledgment

The research was carried out under State Assignment, Project III.17.4.2 (reg. № AAAA-A17-117030310438-1) of the Fundamental Research of Siberian Branch of the Russian Academy of Sciences.

Reference

[1] E. Andres, Feijdo and Jose Cidris, "Modeling of Wind Farms in the Load Flow Analysis", IEEE Transactions on Power Systems, vol. 15, no. 1, pp. 110–115, February 2000.

[2] P.Srihari1, G.Srinivasa Rao, "Load Flow Analysis of Distribution System Including Wind Turbine Generating System Models", International Journal Of Computational Engineering Research (ijceronline.com), vol. 2, no. 5, pp. 1521–1526, 2012.

[3] Yu Shenglong, Kianoush Emami, Tyrone Fernando, H.C. Iu Herbert, Kit Po Wong, "State Estimation of Doubly Fed Induction GeneratorWind Turbine in Complex Power Systems", IEEE Transactions on Power Systems, vol. 31, no. 6, pp. 4935–4944, 2016.

[4] S. Guo-qiang ,W. Zhi-nong, P. Bo, "Impact of wind generators on power system state estimation", International Conference on Sustainable Power Generation and Supply. Nanjing, China, 2009.

[5] Miranda-Blanco, B. Nieves, Díaz-Dorado, Eloy & Carrillo, Camilo & Cidrás J, «State estimation for wind farms including the wind turbine generator models», Renewable Energy, vol. 71, no. C, pp. 453-465, 2014.

[6] S. A. T. Ouambo1, A. T. Boum, A. M. Imano, "Parameters and States Estimation by Moving Horizon Estimation, High Gain Observer and Unscented Kalman Filter of a Doubly-Fed Induction Generator Driven", Journal of Engineering Science and Technology Review, vol. 11 (2) pp. 38- 47, 2018

[7] A.M. Glazunova, I.N.Kolosok, E.S.Korkina, "Monitoring of EPS operation by the state estimation methods," Innovative Smart Grid Technologies (ISGT Europe-2011*)*, Great Britain, Manchester, 2011.

[8] M. Shaaban, W. Li, Z. Yan, Y. Ni, F. F. Wu, "Calculation of total transfer capability incorporating the effect of reactive power", Electric power system research, no. 64, pp. 181-188, February 2003.

Model of Smart Greenhouse for Future Grid

Zoltán Kováč, Matej Sadloň, Ján Poničan, František Janíček
Faculty of Electrical Engineering and Information Technology
Slovak University of Technology in Bratislava
Bratislava, Slovak Republic
zoltan.kovac@stuba.sk, frantisek.janicek@stuba.sk, jan.ponican@stuba.sk, matej.sadlon@stuba.sk

Abstract — **Article describes how to use a novelty in the state of art with the highest possible added value. The respective novel multifunctional technical solution is suitable to be applied in greenhouses and to replace several currently used individual technical systems. It is capable of using the waste heat generated by LED light sources and thus to work with both thermal energy and artificial light. The most important feature of the device is an ability to create an active thermal barrier reducing winter thermal loss of the greenhouse. This leads to an all year greenhouse operability allowing for a close energy and horticulture sector cooperation. Active fluid cooling of main appliances (LEDs) increases sustainability and reliability of the whole system. The described model represents an energy efficient and sustainable electrical appliance. In addition to that, this article reveals its potential of becoming a smart electricity consumer, because of its ability to adjust its consumption according to the balancing needs. To sum up, the article is trying to find a way of achieving food production increase and a grid stability.**

Keywords: optimization of waste heat from LEDs; light and heat production; multifunctional energy system; continious greenhouse operation; cooling; active thermal barrier; demand and response; intermittent RES; grid stability support; smart grid;efficient appliance

I. Introduction

Robust continental power grid was not originally designed to handle a decentralized small-scale generation connected with renewable energy sources (RES). Key rule of power engineering is to keep the consumption and production in balance in order to avoid electricity quality issues. However, it is difficult to predict the RES generation with a sufficient amount of accuracy. There are two main barriers of a higher RES implementation. Firstly, the intermittency feature of RES energy generation, the barrier on grid operator´s side. Secondly, a long payback period of RES investment, the barrier on investor´s side. Inexpensive high capacity batteries would be revolutionizing. They would provide a sufficient energy storage and thus completely eliminate the intermittency issue. However, their current unprofitability and limited lifetime prevents their higher use.

A full RES potential utilization requires removing the aforementioned barriers. Profitable energy storage is nowadays still very hard to reach and therefore, it is important to find another solution. An option is a Demand & Response appliance (D&R), i.e. a device capable of a real-time consumption that could be adjusted according to balancing needs. [1] D&R devices are a perfect match for smart and micro grids, where RESs are a major electricity source supported by an energy accumulation system. Efficient microgrid balancing would reduce the storage capacity needs and thus would lead to a faster return on RES investment. Photovoltaic panels and wind turbines would be more attractive. To sum up, a smart D&R appliance attempts to eliminate both RES implementation barriers if properly located in a microgrid. [2]

II. Possible Interoperability of Energy and Horticulture Sectors

Modern greenhouses aim towards an efficiency maximization and thus need a suitable lighting, heating and cooling conditions provided by various technical systems. Greenhouses are usually located outside the residential area, where light intensity changes (smog) are irrelevant. Majority of plants do not require the light to be strictly of a particular intensity and thus can be increased or decreased for a short time as needed. This might be used as a power balancing tool that would be transferring a RES production peak into a higher lighting output and a RES production shortage into a light dimming in order to avoid producer´s balancing error. This would constitute a novel balancing service that could be utilized by RES electricity producers in order to help them in avoiding the balancing error payments in case they would be charged for it.

A combination of cold weather and low level of sunlight causes poor winter growing conditions in many temperate zone countries. Seasonal use neglects the greenhouse potential and causes an import need. Long-distance transportation has one major drawback, a negative impact on both environment and food quality. Moreover, a greenhouse effect causes that the implications of global warming are highlighted in greenhouses. The temperature rises even higher than outside. The annual increase is slowly affecting the food security. To address this issue a technical solution is needed. Greenhouse needs several novel characteristics: reliability, sustainability, high efficiency, uninterrupted availability, all-year operation and weather independence.

III. Replacing Existing Greenhouse Devices with One Multifunctional Device

Greenhouse farming represents a crucial part of food production. Global warming highlights a greenhouse cooling need. [3] In temperate zone countries, the air conditioning is used only seasonally and thus many farmers prefer a cheaper technical solution, i.e. sun protection shields and fans. These low-cost cooling solutions are usually insufficient and warmer weather can seriously damage the crop.

An efficient cooling process is always based on a thermal gravity principle requiring the space to be cooled from the top. However, in greenhouses it is important to keep in mind that the free transmission of sunshine must be still ensured. Therefore, it is necessary to minimize the number and dimensions of used devices, i.e. to minimize the shading. An efficient solution is to combine lighting and cooling devices into a one multifunctional device, showed in Fig. 1.

Figure 1. Replacing two standard devices with a multifunctional one.

The above-mentioned device uses an efficient LED light source. This narrowband light source provides a possibility to apply the optimal characteristics of artificial light on each single plant. Its variability makes it possible to customize the lighting solution to each stage of its growth. [4] Cold moderating fluid used for space cooling is also used to cool down the LEDs in order to increase their reliability and lifetime. To sum up, the device is capable of meeting all the basic needs of an organic life - the light, the air and the heat/cold, showed in Fig. 2. [5]

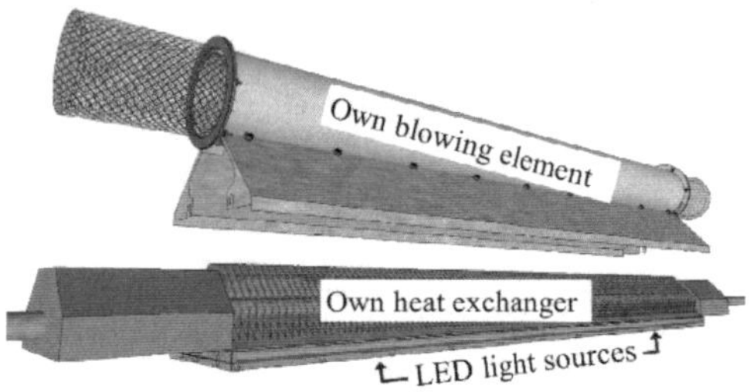

Figure 2. Elements of the device.

IV. Smart Greenhouse with Three Operating Modes

The device uses the active working fluid to cool down the LEDs and uses the waste energy in order to improve the greenhouse environment according to plant's needs. The heat power reaches 60-70 % of electricity consumption [6, 7]. These LED semiconductors loses constitute a significant energy value that should be utilized. Therefore, the key is a waste heat use optimization and an aim is an all-year greenhouse operation. In addition to the fact that the multifunctional device is capable of replacing two existing devices, a third additional feature is a necessity for a modern greenhouse. The waste heat blown from the top of the greenhouse may serve as a virtual insulation that allows for a proper heat distribution and minimizes the heat losses during the winter season.

Fig. 3 shows schemes of device's three operating modes (OMs). These are based on four valves that are opened and closed according to operational needs. The first OM reflects the function of the replaced LED luminaire. In this mode the LEDs are provided with an efficient active cooling. The second OM reflects the function of a conventional heating and cooling equipment that was replaced by the multifunctional device. After a heat or cold exchange, the working fluid (water) flows into a metal base with fixed LED light sources, where the LED generated waste heat is stored in the working fluid. In case of a cooling process, the waste heat is removed from the device and the space is cooled down by a cold water. In case of a heating process, the waste heat is used for water preheating in order to minimize the energy input needed for heating. The third OM needs a small integrated pump in order to provide a closed working fluid circulation and to utilize the generated waste heat from LEDs directly inside the greenhouse as the aforementioned virtual insulation.

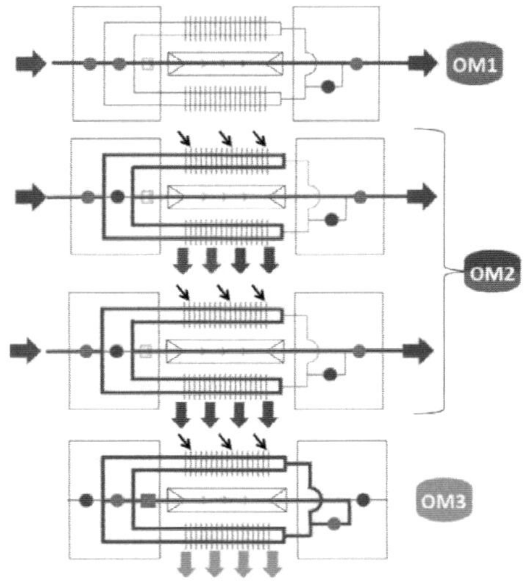

Figure 3. Three operating modes for optimizing waste heat.

These operational adjustments are achieved without human intervention via four controlled electrical valves [8]. A standard OM utilization is presented in Fig. 4.

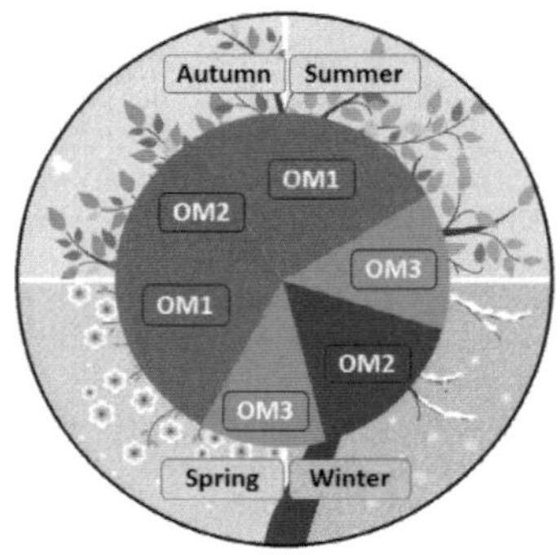

Figure 4. Ratio of annual operating modes.

V. Summer Operation of the Smart Greenhouse

Each greenhouse needs sufficient amount of water. Thus, it is always built next to a groundwater reservoir. Temperature of this water during the summer period is usually fluctuating around 12 °C. One of the operating modes allows the device to use this water as a greenhouse cooler in order to transfer the heat out of the greenhouse, i.e. to decrease the greenhouse temperature. Technology uses geothermal cold to achieve an ideal greenhouse temperature required for plants, showed in Fig. 5. This capability may remove the need of an air conditioner in greenhouse if used properly. Decreased investment and operational costs are thus achieved together with an optimal greenhouse temperature.

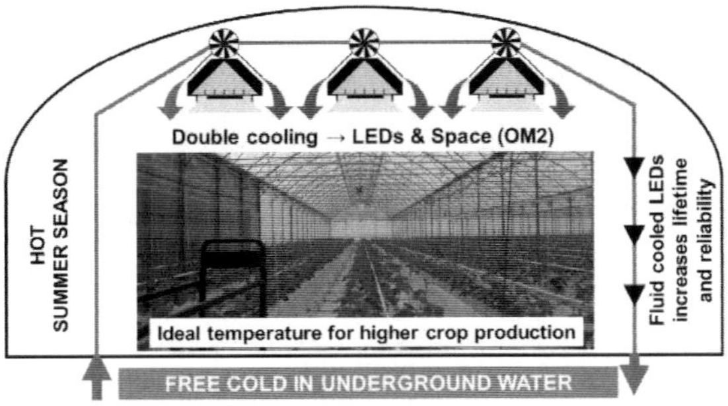

Figure 5. Cooling ability in hot summer period.

On the other hand in moderate parts of the summer season the LED waste heat produced during the night can be stored or used in order to heat the irrigation water and thus utilized in another form, showed in Fig. 6.

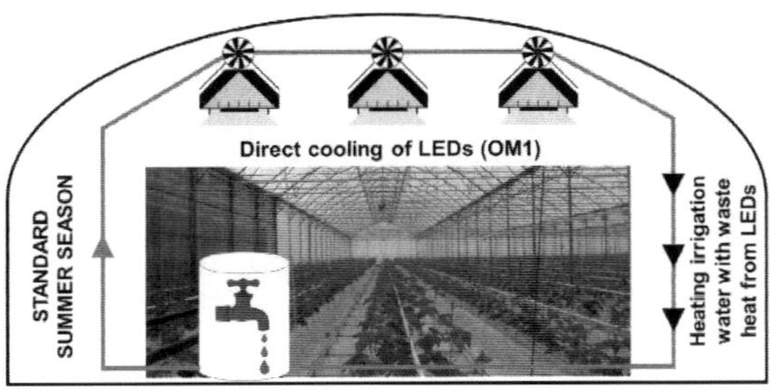

Figure 6. Efficient and sustainable lighting in standard summer season.

VI. Winter Operation of Smart Greenhouse

Minimal thermal insulation of glass and foil greenhouses limits their winter operation. This is reflected in price and quality of fruit and vegetables sold during the winter. The multifunctional device is based on a significant number of LED luminaires and uses their waste heat in order to provide a virtual insulation. Therefore, a greenhouse system of these devices may provide a homogeneous active thermal barrier decreasing the amount of heat losses in operating modes OM3 and OM2. The energy intensity of the winter operation is hence decreased.

Multifunctional device is capable of providing two different levels of the virtual insulation. First level utilizes solely the LED waste heat and is used under the OM3. The most suitable conditions for this operation is during the spring/ autumn to winter transition period, showed in Fig. 7.

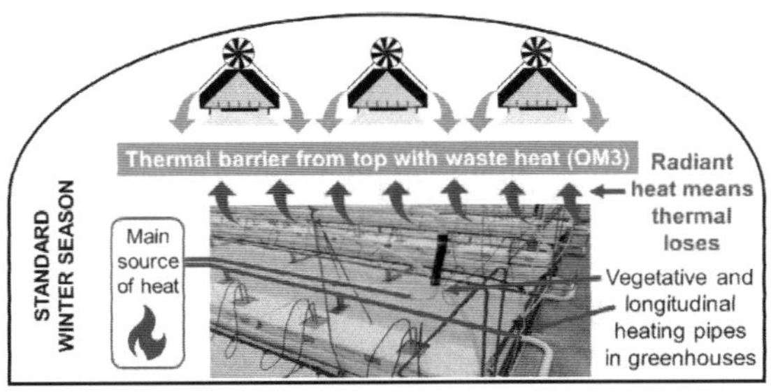

Figure 7. The first level of an active thermal barrier in moderate winter season.

Second level requires a higher temperature of the working fluid. Therefore, an additional water heating device must be used, showed in Fig. 8. However, the LED waste heat is used as a preheating energy source and thus decreases the operational costs of the main heat source. This regime is used under an OM2.

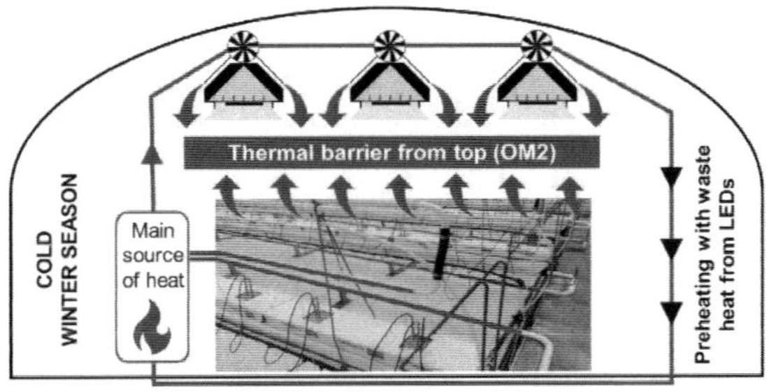

Figure 8. The second level of an active thermal barrier in cold winter season.

VII. High Performance Equipment for Smart Greenhouses

A fan integrated in the device may be activated anytime. It provides an air exchange that is relevant mostly in case of a dense planting. Air is dosed by joined ducts and can be transferred via the whole greenhouse as needed, showed in Fig. 9.

In addition to that, the water used as a working fluid can be automatically used as an LED optic cleaner in order to reach a sustainable lighting efficiency. This solution may be used also to increase the greenhouse humidity. This is usually required during the summer operation.

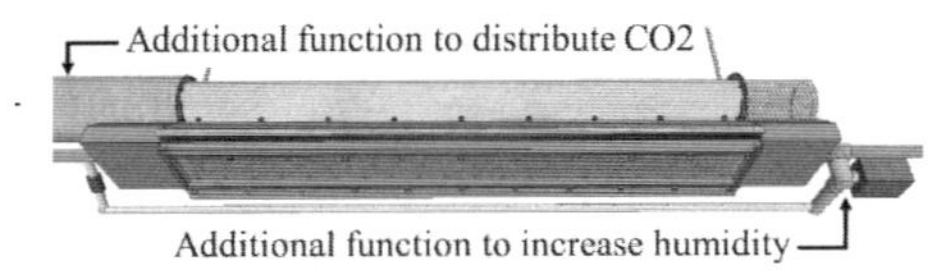

Figure 9. Additional functions of the multifunctional device.

To sum up, the multifunctionality represents the key feature that distinguishes the device from other currently available technologies [9]. Maximizing the efficiency of a technical solution is the easiest way of energy savings achieving. Multi-purpose technical solutions are far more efficient compared to single use seasonal technical solutions.

VIII. Smart Greenhouse as a Demand and Response Appliance in a Smart and Microgrids

A suitable operation of the multifunctional device makes the greenhouse operation easier in each single climate. The waste and geothermal energy utilization stands for significant operational costs savings. Moreover, the multifunctional characteristic of the device stands for significant investment savings. These facts constitute a competitive advantage for each single greenhouse. Some of the positive features of the device and their impacts are summarized in Table 1.

Table 1. Advantages for greenhouse connected with multifunctionality.

Feature	*Usage*	*Impact*
Air exchange	Air quality increase	Higher crop production
Combined light & heat production	Utilization of waste energy	Energy efficiency
Water cooled LEDs		Reliability & sustainability of LEDs operation
Optimal waste heat utilization	Active thermal barrier	Efficient winter operation

On the other hand, technology is still in the phase of an R&D and thus there are several challenges that needs to be tackled, summarized in Table 2.

Table 2. Challenges of the multifunctional system in greenhouse.

Feature	***Threat***	***Solution***
Total weight of the system	Statics of the greenhouse	Component dimensions
Purity of suction filters	Weak heat/cold exchange	Maintenance / air ducts connection
Uniformity of lighting, homogeneous thermal barrier, homogeneous cold distribution	Unoptimized lighting, thermal barrier and cooling	Dimensions of triangular base / shape of diffuser / properties of cover light optic

Another source of savings may be achieved by the farmer, if a system of these devices would be used as a demand and response appliance. As it was already mentioned, a real-time consumption adjustment could be offered to electricity market participants, who are responsible for the balancing error in order for them to avoid balancing electricity charges. Here, two optimal scenarios of this service may be identified. Firstly, a microgrid integration, where would this service allow for a simple balancing of the RES production peaks. And secondly, an ad hoc balancing service that could be offered widely on a short-term electricity market. [10]

IX. Provided and Renounced Capacity

A perfectly functioning D&R appliance should be able to adapt the power input anytime. The system of the devices installed in a greenhouse is capable of achieving this goal. Issues of an electricity surplus and shortage can be solved up to a certain extent by this system. From a long-term point of view, the LED light sources implemented in the system might be dimmed or brightened for short time periods even if it would deviate from the ideal lighting conditions for plants. Most of the cultivated plants used for the commercial food production would easily handle short time light intensity changes. Therefore, a part of the LED installed capacity could be turned on according to real-time market needs, i.e. a provided capacity. Conversely, if the request would constitute a consumption decrease need, the LEDs could be dimmed and thus provide a renounced capacity, showed in Fig. 10. [11]

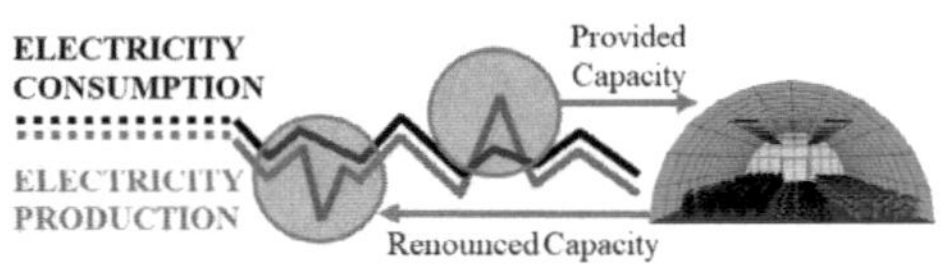

Figure 10. Principle of the greenhouse as a smart appliance.

A widely used lighting solution is used as starting point in defining the optimal installed capacity of the LEDs. In this scenario discharge lamps are replaced by LED light sources. This accounts for 60 % energy savings. The optimal specific power for LEDs is 44. 6 W/m^2, according by Table 3.

Table 3. Optimum LED power installed in greenhouse.

Existing greenhouse lighting system using SON-T 1000W [3]	Area with SON-T 1000W	40,709 m^2
	Total number of SON luminaires	4,536 pcs
	Grid spacing for SON luminaires	9.0 m^2/pc
	Specific power for SON-T	111.4 W/m^2
Discharge lamps replaced by LED light sources	Achievable energy savings SON-T to LED	60 %
	Optimal specific power for LED illumination	44.6 W/m^2

An area of the majority greenhouses commercially used for food production usually exceeds one hectare. Therefore, a reference area of one hectare is used as a unit. As shown in Table 4, the provided capacity is defined as 30 % of an optimal value. This number is selected due to an average radiation density safety limit related to commercially cultivated fruits and vegetable. The level of the renounced capacity is set to 70 % of the optimal value. This reflects the need of a minimal lighting conditions. To sum up, the total range of regulation is the same as the optimal installed capacity of the LEDs.

Table 4. Allocation of capacity per hectare of a greenhouse.

Design of multifunctional energy system with LEDs	Area of greenhouse	10,000 m^2
	Total optimal power of LEDs	446 kW
	Provided capacity (safety over-dimensioning of LED power)	+ 30%
		134 kW
	Renounced capacity	– 70%
		312 kW
	Total reserved capacity for LEDs	579 kW
	Range of regulation	446 kW

X. Main Barriers

A. Barriers on farmer's side

The interoperability may only be viable if it would be mutually beneficial. The renounced/provided capacity with a conveniently set price would surely be of an interest for the energy sector. However, on farmer's side, it would be connected with several issues.

1. The primary barrier represents farmer's distrust. Therefore, it would be crucial for the service to be viable to provide sufficient motivation resulting from financial benefits and minimal hassle. This would require an in-depth analysis assessing the impacts on particular cultivated crop and price optimization of such a service. Moreover, a complete automatization of the process would be a prerequisite. In this way a smart D&R greenhouse would not be only a balancing tool, but a profit-oriented market service for a greenhouse operator, too.

2. Increased power of LED lighting represents an increased investment into an installed system, i.e. higher installed capacity of LEDs, extra wiring, switchboard components, etc. Higher investment naturally prolongs the return of an investment (ROI). Thus, a ROI comparison of different installed capacities and their potential utilization must be performed in the installation designing stage in order to calculate profitability of various potential options and to choose the most suitable one. The increased installation costs of the farmer are described in the Fig. 11.

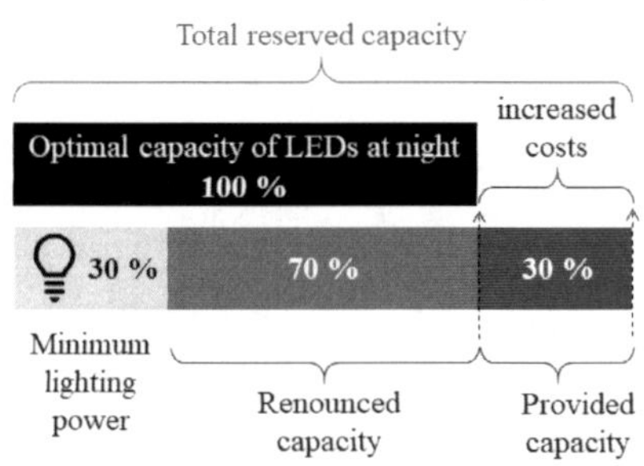

Figure 11. One of the barriers on the side of the farmer.

3. Both the renounced and provided capacity may only be provided on a short timeframe depending on the particular

characteristics of the cultivated plants. Therefore, the service would be more viable for those greenhouse operators who cultivate less demanding crop. It is important to state that a wrong service adjustment may cause serious financial loss connected with crop damage, crop reduction or even a damage of the multifunctional installation.

4. Each intervention into the greenhouse operation constitutes a financial loss for the farmer. Time required for multifunctional energy system with LED light and heat sources installation must be maximally limited. Future development must therefore solve this challenge by one of the two potential means. First option is to prepare a strict procedure for a quick installation that would take place during a time-limited shutdown of production connected with a crops change. Or the second option is to prepare a very simple installation procedure that could take place during the operation. However, this would require to prevent the impact on the cultivation process and on health of a growing environment

5. A sustainable farming requires plants to have a daily rest. According to a consultation with a specialist in the field of greenhouse agriculture, an average daily rest period represents 3 hours. However, this number may differentiate. In addition to that, light intensity, grid balancing conditions, weather and many other factors influencing the system may fluctuate throughout the year as well. This means that a prerequisite of the R&D appliance would be a sophisticated software based on the algorithms that would take all these details into an account in order to make it as simple for the farmer as possible.

The bottom line is that there are still several issues that need to be addressed before would the service be viable for the farmer. However, all of these have a potential solution.

B. Barrier in connection with photovoltaic installations

A summer season is very often reflected in critical sunny days. Naturally, using an artificial light in the greenhouse during such a day would have neither a positive, nor a negative effect. Unfortunately, these are the days when the photovoltaic sources usually generate a surplus electricity. On the other hand a multifunctional system provides cooling and air exchange, too. These features would be of a significant importance during these days. Electrical appliances related to this function, i.e. fans and pumps, can reach only up to 20 % of the optimal light capacity. In order to avoid a meaningless lighting and to fully utilize the surplus electricity, an energy storage would be required. Therefore, an integration of the greenhouse D&R appliance into a microgrid would be a perfect match. A microgrid requires an energy storage. And an optimized D&R service would lower the required battery capacity. [12]

XI. CONCLUSION

Devices in question connected via a pipeline grid create a novel system that can be adapted to any year season and to almost any type of the weather. This is achieved with a three-way active fluid circulation optimized within the above-mentioned operating modes. Seasonally optimized LED waste heat utilization represents a competitive advantage in the horticultural sector. Nowadays, market does not know this technology and thus it has an opportunity to make the cultivation far more efficient. Moreover, a utilization of the multifunctional system extends the viable greenhouse locations even to those regions, where the greenhouse operation would not be viable without the system. Plus, an all-year crop production in moderate geographical regions is far easier than before.

Waste heat stored in a working fluid is capable of providing a preheating for other technologies. If installed together with other devices and optimized properly, such system would increase the energy efficiency of solar technologies and heat pumps. This results in a better return on investment for these technologies, too. As already discussed above, a suitable microgrid installation combining various optimized electrical technologies seems like the best solution for a sustainable and energy efficient future. This in turn supports the RES development and thus decreases the carbon footprint. The basic integration of presented smart greenhouse into the future grid is described in Fig. 12. [13]

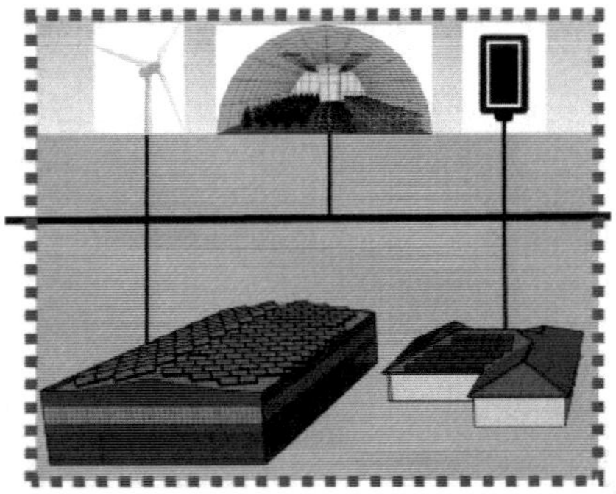

Figure 12. Smart and microgrid of the future.

To sum up, the respective novel system is a perfect solution for greenhouses. Their total global area represents approx. 450,000 hectares, according by Fig. 13. [15]. Most of the high-tech greenhouses do not have an artificial lighting, nor a cooling system. None of these has an active thermal barrier decreasing the heat loss. Thus, presented novel system is a perfect fit. It can maximize the usage time of greenhouse, decrease the technology investment costs needed for a high-end food production and to decrease the operational costs of a greenhouse. All of these aspects result in profit maximization related to high winter season market prices of fruit and vegetables. These assumptions are only highlighted with the D&R utilization of the system.

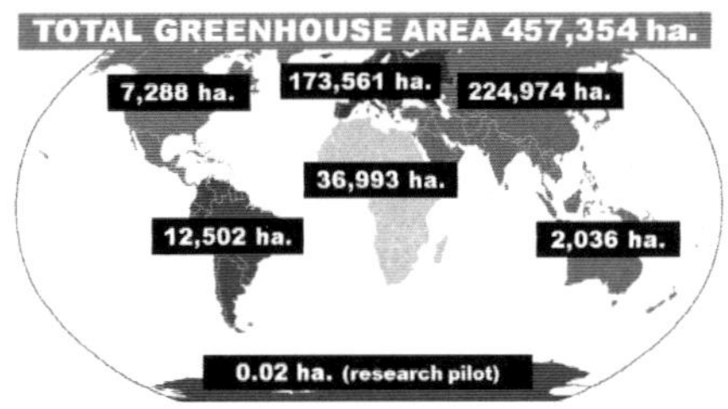

Figure 13. Global potential of greenhouses.

REFERENCES

[1] P. Siano, "Demand response and smart grids—A survey", Renewable and Sustainable Energy Reviews, vol. 30, pp. 461-478, 2014.

[2] A. Ipakchi, F. Albuyeh, "Grid of the future," IEEE Power and Energy Magazine, vol. 7, pp. 52-62, March-April 2009, ISSN 1558-4216.

[3] P.J.M. Van Beveren, J. Bontsema, G. Van Straten, E.J. Van Henten: "Minimal heating and cooling in a modern rose greenhouse," Applied Energy 137, pp. 97–109, 2015, ISSN 9066056215.

[4] D. Singh, Ch. Basu, M. Meinhardt-Wollweber, B. Roth, "LEDs for energy efficient greenhouse lighting," Renewable and Sustainable Energy Reviews, vol. 49, pp. 139-147, September 2015.

[5] Z. Kováč, F. Janíček: "Waste lighting heat generation as the key process to design the only technical device needed in interiors," 12th International Conference ELEKTRO 2018, May 2018, ISBN: 978-1-5386-4759-2.

[6] Y. Tang, L. Lin, S. Zhang, J. Zeng, K. Tang, G. Chen and W. Yuan, "Thermal management of high-power LEDs based on integrated heat sink with vapor chamber," Energy Conversion and Management, vol. 151, pp. 1-10, November 2017.

[7] Z. Kováč, F. Janíček, R. Dubnička: „Potential of usable waste heat generated from LED light sources in industry," International Science Conference Power Engineering 2018, June 2018, ISSN 2585-9684.

[8] Z. Kováč, "Multifunctional energy module and multifunctional energy system," Slovak utility model No. 8104 by application PUV50003-2017 and PCT International Patent Application No. PCT/IB2018/050234.

[9] A. Vadiee, V. Martin, "Energy management in horticultural applications through the closed greenhouse concept, state of the art", Renewable and Sustainable Energy Reviews, vol. 16, pp. 5087-5100, September 2012.

[10] M. Hussain, Y. Gao, "A review of demand response in an efficient smart grid environment", The Electricity Journal, vol. 31, pp. 55-63, 2018.

[11] G. Derakhshan, H. A. Shayanfar, A. Kazemi, "The optimization of demand response programs in smart grids", Energy Policy, vol. 94, pp. 295-306, 2016.

[12] F. Umbach, "Global energy security and the implications for the EU", Energy Policy, vol. 38, pp. 1229-1240, March 2010.

[13] A. O. Otuoze, M. W. Mustafa, R. M. Larik, "Smart grids security challenges: Classification by sources of threats", Journal of Electrical Systems and Information Technology, vol. 5, pp. 468-486, December 2018.

[14] F. Pallonetto, M. De Rosa, F. Milano, D. Finn, "Demand response algorithms for smart-grid ready residential buildings using machine learning models", Applied Energy, vol. 239, pp. 1265-1282, 2019.

[15] Hortidaily, "Nearly half of the world's greenhouse vegetable area located in Asia", Internet: www.hortidaily.com, 2017.

Measurement of Indoor Environment Parameters in a Classroom Equipped with Heat Recovery Unit

Jakub Jirinec, David Rot, Vojtech Senkyr
Department of Electrical Power Engineering and Environmental Engineering
University of West Bohemia
Pilsen, Czech Republic
jjirinec@kee.zcu.cz, rot@kee.zcu.cz, senkyrv@students.zcu.cz

Abstract — **The article deals with measurement of indoor environment parameters in a classroom equipped with heat recovery unit. The first part of the article is devoted to the own measuring system which is based on NI hardware (LabVIEW) in combinations with measuring sensors. Next part shows results of real time operation with emphasis on the observed CO_2 concentration. Finally, two situations were compared: installation with heat recovery unit (classroom EL319) and installation without heat recovery unit (classroom EL419).**

Keywords: air quality; CO_2 concentration; heat recovery unit; LabView; 2VV

I. Introduction

Building insulation reduces the energy consumption of the building but also prevents natural ventilation. This may cause in deterioration of an indoor air quality. To ensure the required air quality, the building should be ventilated by forced ventilation equipped with heat recovery units. The issue of classroom ventilation is mentioned in Decree 410/2005Sb., Government Regulation No. 217/2016Sb. and a Methodological Guideline for the Design of School Ventilation [1-2]. According to this regulations the main requirements are:

- 20 m^3/h per student and 50 m^3/h per teacher.
- Maximal concentration of CO_2 1500 ppm.
- Temperature difference between head (1.1 m) and ankles (0.1 m) maximally 3 °C.
- Maximal temperature 28 °C in the summer, minimal temperature 20 °C and maximal 28 °C in the winter.
- Maximal noise of the unit 45 dB (recommended 30-40 dB).
- Relative humidity between 30-65 %.
- Ventilation intensity 2.6 h^{-1}.
- Maximal air flow 0.2 m/s in high 1.8 m and 0.25 m from the wall.

To obtain information on how the heat recovery unit can affect the air quality in the building, a long-term measurement of indoor environment parameters in a classroom equipped with a 2VV Whisperair heat recovery unit has been performed. The main monitoring parameter was the evolution of CO_2 concentration and temperature in different parts of the room. The unit was operating in constant speed mode and in the CO_2 control mode. Thanks to the Modbus TCP / IP communication protocol, it was possible to use the sensors located directly in the unit and also to read the settings and the current operating status of the regenerative unit.

Figure 1. The classroom EL319 with Whisperair reperation unit

II. Measuring System EL319

Temperature, humidity and CO_2 concentration sensors were installed in the classroom (141.5 m^3) equipped with a 2VV Whisperair heat recovery unit to measure environmental parameters in the seating area of seated listeners. Furthermore, dust sensors and presence sensors were fitted. The classroom dimensions are 6.2 x 7 x 3.26 m. Pros of the decentralized heat recovery unit against the central heat recovery unit are:

- Easy and quick installation.
- No additional air pipeline installation required (easier fire protection).
- Suitable for existing buildings without the need for extensive construction work.

Cons of the decentralized heat recovery unit against the central heat recovery unit:

- Installing the entire recuperation unit directly in the class (relatively large dimensions).
- Condensate drainage in every room.

The unit was connected via an IME network analyzer Nemo 96HDLe. With this device, it is possible to monitor the actual power consumption of the heat recovery unit and to measure its feedback on the network. As a system controller, C-RIO from National Instruments, which was programmed with LabVIEW, was used. C-RIO provides communication with the heat recovery unit and network analyzer, storage of measured data and their visualization.

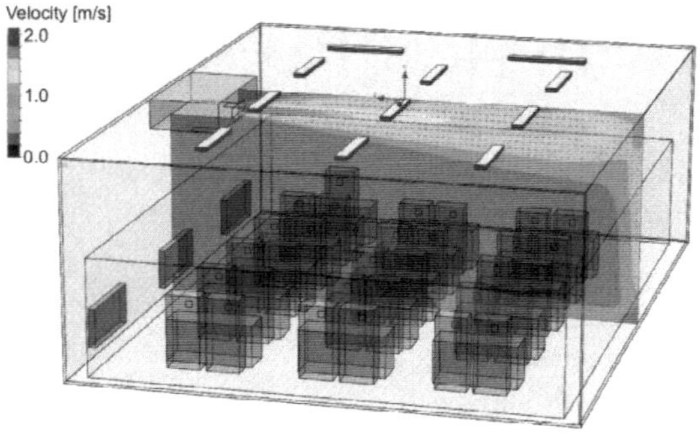

Figure 2. Air velocity in the classroom with heat recovery unit Whisperair [3]

The program itself is divided into several parts:

A. Classroom Air Quality Measurement

The C-RIO control unit enables connection of cards for measuring different types of signal from the measuring sensors. In this case, NI 9205 (0-10 V analog input) and NI 9035 (digital I / O) were used. All measuring sensors have been selected with a 0-10 V output. Temperature and humidity measurements in the classroom are provided by EATEC TTHIV sensors. The accuracy of relative humidity sensor is ±2% RH (20..80 % RH) + 2% FS and temperature sensor is ±0.3 °C (5..60 °C) + 1% FS. The sensors are installed on the classroom tables and placed on holders printed by 3D printer. The holders have been designed with regards to the appropriate airflow around the measuring sensors. On the other side of the holder, a CO_2 concentration sensor from 2VV (CI-CO2-M) was installed.

Temperature and humidity sensor (EATEC TTHDV) was installed to the outdoor air supply line. The same sensor was installed in a place where air is floating into the room and on the opposite side of the rooms. The accuracy of relative humidity sensor is ±2% RH (20..80 % RH) + 2% FS and temperature sensor is ±0.3 °C (5..60 °C). To measure the exhaust air parameters, the EATEC TTHIV sensor was fitted into the exhaust air intake area.

Dust sensors have been installed to the outdoor air supply line and to the place where air is floating into the room. This measurement can be used to determine the dust collection efficiency by using a supply air filter located in the recovery unit. The unit control system has a differential pressure measurement of the filter to measure filter clogging.

All the windows in the room were fitted with magnetic contacts to identify the open window. The magnetic contact was also mounted at the entrance door to the room. Thanks to the magnetic contacts it is possible to analyze the data with respect to the open window or door. A presence sensor was installed in the system for easy identification of people in the room. To determine the exact number of people, the classroom is equipped with an IP camera that stores images on a data repository every 30 minutes. The position of each sensor is shown in the Fig. 3.

B. Data Storage from the Heat Recovery Unit

The heat recovery unit enables communication via Modbus RS485 or Modbus TCP / IP. In our case, the unit was connected to C-RIO using Modbus TCP / IP. This connection allows reading of the values directly from the control unit of the heat recovery unit. Furthermore, it is possible to remotely control the unit and adjust its operation and settings. The individual positioning of the measuring sensors in the unit and their description is shown in Fig. 4. The unit has differential pressure measurements of the inlet and outlet air filters. As a result, filter clogging can be monitored and filters can be replaced only when it is necessary.

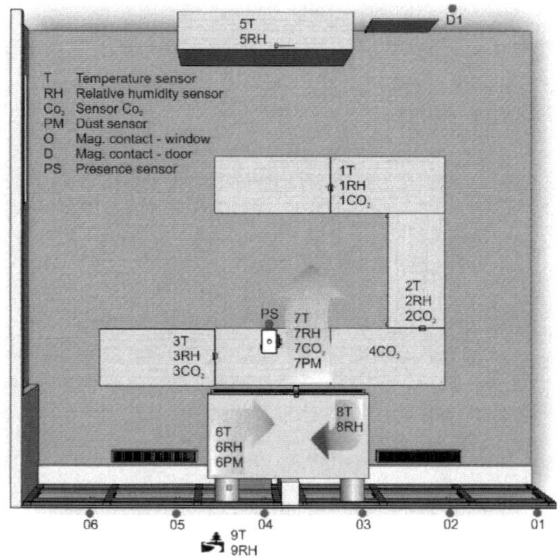

Figure 3. The position of sensors in room EL319

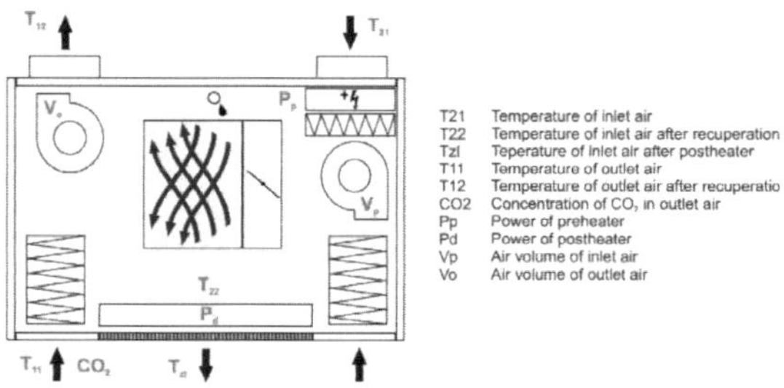

Figure 4. Measuring sensors integrated in recuperation unit

C. Measurement of Electrical Parameters

The heart recovery unit is connected to the grid by the Nemo 96 HDLe network analyzer from IME. The analyzer communicates with the C-RIO controller using the Modbus RTU communication protocol and serves to measure selected electrical parameters as shown in Fig. 5. For clarity, the measured quantities were divided into two tables. The first table lists the quantities measured in each phase, namely voltage, current, active and reactive power, total harmonic distortion of voltage and current. The second table lists the values for all three phases, namely active, reactive and apparent power, energy, frequency, and power factor.

	L1	L2	L3
U[V]	226.0	225.6	226.4
I[A]	0.295	0.000	0.000
P[W]	37.6	0.2	0.0
Q[VAr]	-10.80	0.25	0.00
THD U [%]	25.0	24.0	24.0
THD I [%]	145.0	0.0	0.0

P[W]	37.40
Q[VAr]	-10.55
S[VA]	66.45
Ec [Wh]	529.76
Ej [VArh]	0.00
cos f	0.560
kvadrant	2.0
f [Hz]	50.00

Figure 5. Electrical values measured by network analyzer

D. *Visualization and Data Storage*

For easy overview and control of the measuring system the web server for C-RIO was created. It is possible to connect to visualization from any computer in the network. The visualization itself is divided into five tabs with respect to the measured parameters. The first tab "Preview" serves for a quick visualization of the data in the schematic picture of the classroom and the regenerative unit. In the "RIO" tab, all values measured by analog and digital C-RIO inputs are displayed. For better clarity there is a graph, which is used to monitor the required quantities. The "Multimeter" tab is used to display data measured by the network analyzer. The data can also be displayed in a graph. The "Whisperair" tab is used to display data from the unit read via Modbus TCP / IP. The last "Saving" tab is used to enable or disable the measurement, data storage in the C-RIO memory and the flash drive. At the same time, the user can name the file as needed. The green LED indicator serves as a simple check of correct storage functionality.

III. Measurement of Heat Recovery Unit

A. *Dust concentration measurement*

The heat recovery unit draws in outdoor air through filters that clean the supply air and reduce the concentration of dust particles blown into the room. To verify the filtration capability, the heat recovery unit was equipped with dust sensors. Fig. 6 shows the concentration of dust particles during the day. It is clear from the graph that the inlet filters trap dust particles and clean the inlet air entering to the room. This may be particularly advantageous in areas with a high concentration of dust particles in outdoor air. It is necessary to keep the dust filters clean enough for proper functioning. To verify the clogging of the filter, the heat recovery unit measures the differential pressure before and after the filter. This makes it possible to indicate precisely the clogging of the filter and to alert the user to the need to replace the filter.

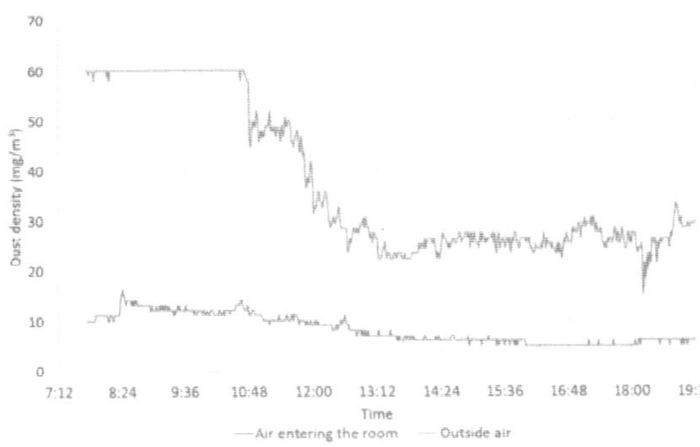

Figure 6. Concentration of dust particles during the day

B. *CO_2 concentration measurement*

The heat recovery units are primarily installed because of the high CO_2 concentration in the interior. Their main task is to ensure adequate air exchange in the room and to maintain the CO_2 concentration within the prescribed limits. For this reason, several measuring sensors were installed in the room. The recovery unit was set to automatically control of the fan speed according to the CO_2 concentration. Thus, the unit automatically regulates the amount of air supplied to the interior. The minimum air quantity is 300 m^3/h. Fig. 7 shows an increase in CO_2 concentration when people enter the room. Exceeding a certain concentration will automatically increase the amount of air entering the room and thus reduce the CO_2 concentration in the room. The blue wave in the graph shows the CO_2 concentration measured by the sensor placed on the bench directly by the seated persons. Reducing the air volume to zero is due to a schedule on which the heat recovery unit is turned off after 18:00.

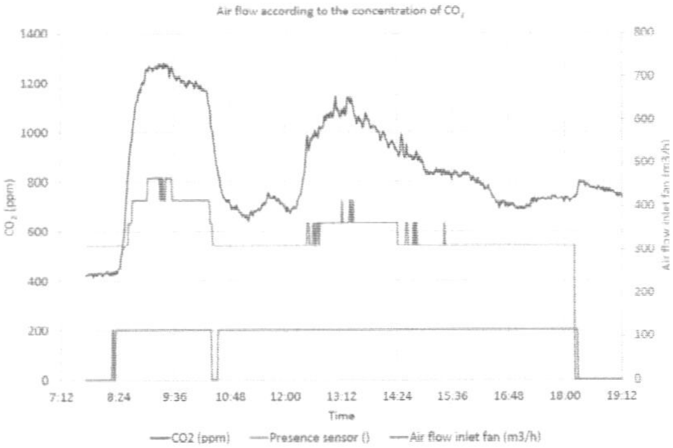

Figure 7. CO_2 concentration when people enter the room

IV. Measuring System EL419

To compare the air quality it the room with and without the heat recovery unit, an additional measuring system from Netatmo was installed. This system consists of a control unit and three measuring sensors (CO_2, temperature, noise and relative humidity) which are connected wirelessly. At first, the measuring system was operated in a classroom with a recovery unit (EL319). In this operation, the data measured by the Netatmo system and created measured system in LabVIEW were compared. The measurement results showed that the values of the systems are almost the same. After that, the Netatmo system has been moved to classroom without heat recovery unit (EL419), which serves the same purpose at floor upstairs. The position of sensors in this room (EL419) is shown in the Fig. 8.

The measurement takes place in both rooms from 17. 12. 2018 until today. The development of CO_2 concentration, relative humidity and number of people during this time period was monitored and compared with values from the classroom equipped by the heat recovery unit (EK319). It can be seen from the measured results that the heat recovery unit allows to maintain the CO_2 concentration within the prescribed limits. Three days were selected to show the development of the monitored parameters.

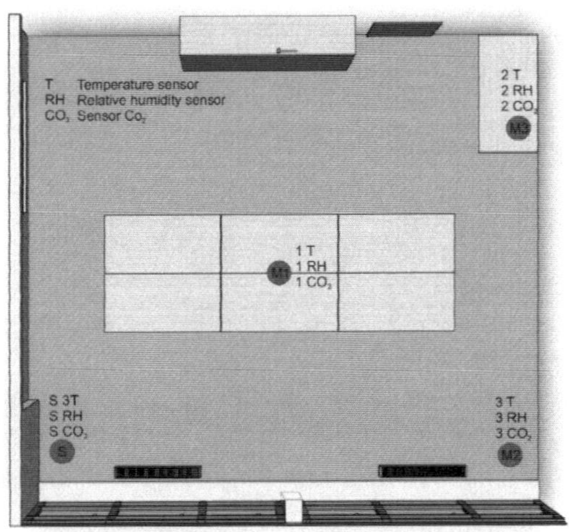

Figure 8. The position of sensors in room EL419

Three days were selected to show the development of the monitored parameters. Heat recovery unit was in the CO_2 control mode (ventilation according to the actual value of CO_2). Fig. 9 shows rapid increase of CO_2 concentration in EK319 caused by presence of 6 people. Heat recovery unit automatically increases air volume and keeps concentration at 850 ppm.

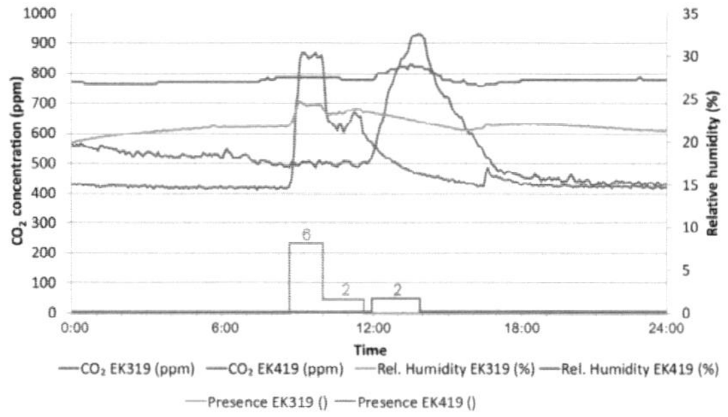

Figure 9. Comparison of CO_2 concentration and relative humidity in classrooms 9. 1. 2019

In the Fig. 10 is shown the same situation in the classroom without heat recovery unit. Presence of 6 people caused increase of CO_2 concentration to 4000 ppm. The subsequent reduction in concentration is probably caused by opening of the windows, this also shows decrease of temperature in the classroom.

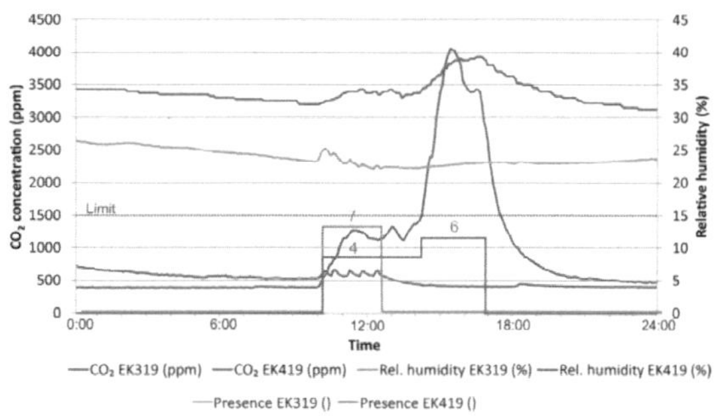

Figure 10. Comparison of CO_2 concentration and relative humidity in classrooms 18. 1. 2019

The last selected day shows how heat recovery unit can hold CO_2 concentration in prescribed limits in the classroom with 15 people (Fig. 11).

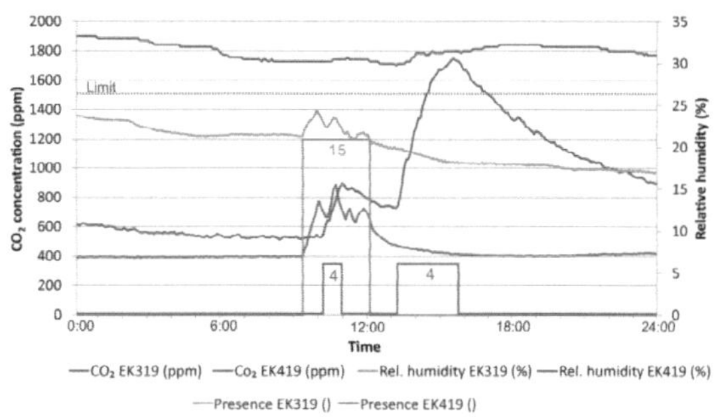

Figure 11. Comparison of CO_2 concentration and relative humidity in classrooms 8.2.2019

V. Conclusion

The system was operating over several months in various operating states. From the measured results it can be seen that the regenerative unit allows to keep the air quality within the required limits and reduces the dust particles in the inlet air thank to the filters. Heat recovery unit can be also used for cooling. During the night, cold air can cool down the classroom (thanks to opening the by-pass). Using CO_2 control mode can heat recovery unit be switched on only if necessary and it allows to keep the classroom colder, then in the case of standard ventilation by opening the windows. Improper adjustment of the regenerative unit can result in large operating costs. In order to achieve efficient operation, the control system of recuperation unit must be set up correctly.

Acknowledgment

This article was funded by the Ministry of Education within the project LQ1603 entitled "Research for SUSEN" and by the project SGS-2018-023.

References

[1] Vyhláška č. 410/2005 Sb. o hygienických požadavcích na prostory a provoz zařízení a provozoven pro výchovua vzdělávání dětí a mladistvých. In: *Sbírka zákonů*. Praha: Tiskárna Ministerstva vnitra, 2005, ročník 2005, částka 141.

[2] Nařízení vlády, kterým se mění nařízení vlády č. 272/2011 Sb., o ochraně zdraví před nepříznivými účinky hluku a vibrací, ve znění nařízení vlády č. 217/2016 Sb.

[3] J. Chlup, Distribuce vzduchu ve školní třídě s použitím větrací jednotky 2VV [online prezentace]., 2016 [cit. 2016-11-23]

[4] M. Székyová, Větrání a klimatizace. 1. české vyd. Bratislava: Jaga, 2006, 359 p., ISBN 80-807-6037-3

[5] J. Bašta, Regulace v technice prostředí staveb, Praha: České vysoké učení technické v Praze, 2014, ISBN 978-80-01-05455-0.

[6] F. C. Mcquiston, J. D. Parker and J. D. Spitler, Heating, ventilating, and air conditioning: analysis and design, 6th ed. Hoboken, N.J.: John Wiley, c2005. ISBN 978-0-471-47015-1.

[7] M. Jokl, Zdravé obytné a pracovní prostředí. Praha: Academia, 2002, ISBN 80-200-0928-0.

[8] P. Šafařík and M. Vestfálová, Termodynamika vlhkého vzduchu, Praha: České vysoké učení technické v Praze, 2016, ISBN 978-80-01-06020-9.

[9] Z. Mathauserová, Hygienické požadavky na vnitřní prostředí staveb, TZB-info, [Online] 25. 2 2013, https://vetrani.tzb-info.cz/vnitrni-prostredi/9595-hygienickepozadavky-na-vnitrni-prostredi-staveb#english_synopsis.

[10] 2VV, CI-CO2-M, Produktový katalog, [Online] http://www.2vv.cz/katalogproduktu/ci-co2-m.

[11] J. Votava, J. Kyncl and L. Straka, Optimization of local cogeneration stations,In 19th International Scientific Conference on Electric Power Engineering (EPE), 2018, pp. 1-6, IEE, DOI: 10.1109/EPE.2018.8396037, ISBN 978-1-5386-4612-0.

[12] LabVIEW User Manual, 2003, Austin Texas USA: National Instruments Corporation, 349 pages. Free download from: http://www.ni.com/pdf/manuals/320999e.pdf

Optimization of the Energy Consumption of a Building with PV

Jakub Jirinec, David Rot, Stanislav Jirinec
Department of Electrical Power Engineering and Environmental Engineering
University of West Bohemia
Pilsen, Czech Republic
jjirinec@kee.zcu.cz, rot@kee.zcu.cz, sjirinec@kee.zcu.cz

Abstract — **The article deals with optimization of the energy consumption of the building with photovoltaic system (PV). The first part briefly describes the building with implemented measuring and regulation system (the hardware platform, connection and real installation). Next part deals with possibilities of predictive control of the heating system in off-line and on-line mode. Finally the complete energy management of the building is described (air-to-water heat pump, PV, vehicle charging).**

Keywords: optimization; energy consuption; photovoltaic system (PV); measuring system; regulation system; programmable logic controller (PLC); heat pump

I. Introduction

At the present time, the commonly available control and regulation systems dealing with the control of technical equipment of building and photovoltaic systems (PV) usually work autonomously and their interaction is very small. The main task is to introduce a comprehensive control and measurement system for small and medium-sized residential or office buildings that would allow the automatic control of electricity consumption. Due to the practical verification of the proposed algorithms a multifunctional object, which is equipped with various technologies with a large measuring and regulation system, has been created to enable their validation in real operation. These are mainly: the predictive control of the heating system in on-line and the off-line mode, the control of the maximal consumption from the grid (quarter-hour peak monitoring), the speed control of the inverter's compressor depending on the production of the PV, efficient electric vehicle charging (in combination with the hybrid PV inverter) and the use of surplus electricity from PV to heat domestic water and heating water.

II. Description of the Building

The building which was selected for integration of control and measuring system is part of a larger complex in town of Domazlice (Fig. 1). The complex consists of three buildings (office building, warehouse and new building). The complex is connected to the distribution system via a common electricity meter and protected by a main circuit breaker 40A 3P/B. Due to the low value of the main circuit breaker, it is necessary to check the current drawn from the distribution grid in individual phases and in case of exceeding the maximum value disconnect selected types of load (heat pump, domestic hot water (DHW) heating, electric vehicle charging, etc.). Currently, two photovoltaic power plants are part of the complex. The PV 27 kWp is installed in the warehouse and delivers all the energy produced directly to the distribution network. This PV is not included in the control system. After the license expires, it would be possible to incorporate this PV into it and thus increase the independence on electricity supply from the grid. The second photovoltaic power plant is installed on the office building and operates in mode – excess electricity to the distribution network and is part of the created measurement and control system. The following year, a hybrid PV will be installed on the roof of the new building which can be fully controlled via the inverter communication interface. The inverter battery will be also recharged from the grid. As a result, it will be possible to accumulate the excess of electric energy produced from the existing PV into the batteries.

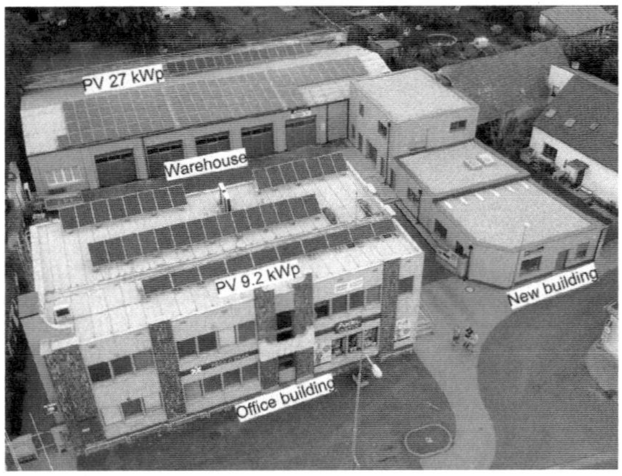

Figure 1. The complex of buildings

Due to the fact that the new building was built, it was possible to integrate into the building a relatively extensive system of measurement and regulation including its own design of the heating source and hydraulic connection. Heating of the new building is provided by the inverter split air-water heat pump Carrier (38AW 150H9 + 80AWH 150T9) with the output of 15 kW in combination with the underfloor heating system. The heat pump also serves to heat domestic hot water (DHW).

III. Measurement and Regulation System

The new building itself is divided into three independent sections of different purpose (apartment, offices). Each section is equipped with its own heating regulation system. For this purpose, the EVCO PLC (c-pro 3 NODE kilo) is used to create a program designed to control the heating system and to control roller shutters. Temperatures in individual locations are measured using Pt1000 temperature sensors. The hot water distributor for underfloor heating is equipped with actuators that switch heating circuits in individual rooms. Individual sections of the building operate autonomously.

As mentioned earlier, the main task is to create a measurement and control system that includes algorithms usable for efficient energy management within an object.

Considering its future deployment in other objects, it is also necessary to choose a suitable hardware platform, which would be a superior system that will communicate with individual meters and technical equipment of the building. Due to the high variability and price, a PLC (IO-750) from WAGO company was chosen, which has the necessary communication interfaces and measuring cards. The selected PLC allows to create user interface, data storage, remote management and cloud application connectivity. The communication scheme is shown in the Fig. 2.

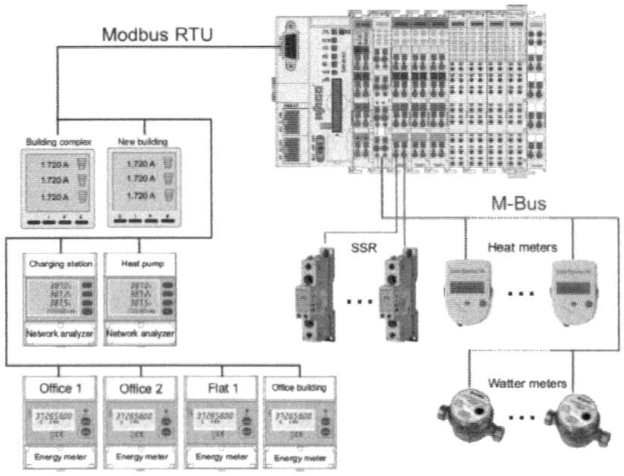

Figure 2. Communication scheme

The PLC WAGO IO-750 is used as a superior control system and communicates with individual EVCO PLCs by the Modbus TCP/IP protocol. The superior control system is used to control the processes in the technical room of the building (switching and controlling the power - compressor speed of the inverter heat pump), collecting data from individual measuring instruments and regulating excess to the distribution network from PV. The thermal energy supplied to the system from the heat pump and the heat energy supplied to each section of the building is measured using a calorimeter (Sontex Supercal 739). The heat pump is connected to the power supply via an industrial network analyzer (IME Nemo D4-Le). As a result, it is possible to evaluate the consumption and the impact on the network quite accurately, incl. determining the COP (Coefficient of Performance) of the heat pump. The power consumption of the building and other electrical parameters are measured using an industrial analyzer (IME NEMO96HDLe). The power consumption of each section is measured by energy meters (IME Conto D4-Pd). All devices used to measure electrical parameters communicate with the control system using the Modbus RTU protocol. The calorimeters and water meters communicate with the control system using the M-Bus protocol. The parts of power panel are shown in the Fig. 3.

Figure 3. Power panel

IV. Predictive control of the heating system in off-line / on-line mode

Current predictive control methods require powerful devices that are able to calculate the building's heat loss, download the necessary data from the Internet, and require detailed knowledge of the application and its modifications. This also brings higher investment costs. The aim of this chapter is to find management methods that are easy to apply to HVAC (heating, ventilating, air-conditioning) in general [5-7].

For heat recovery units, air-to-water heat pumps and other ventilation units, the efficiency of these devices is affected by the outside air temperature. The outside air temperature changes during the day, and thus the efficiency of these devices is changed. By operating the equipment at high efficiency, it is possible to save financial costs and thereby reduce their pay-back time. Standard heating systems respond only to the current outdoor temperature and do not allow to use this potential. The solution is a predictive control that takes into account the future development of the outside temperature and optimizes the operation of the whole plant (mainly regenerative units, air-water heat pumps and other ventilation systems). For simple devices (such as small refrigeration systems) that are difficult to connect to the Internet, the off-line prediction of outdoor temperature seems to be the right solution.

To determine the development of the outdoor temperature during the day, data from the weather station (Warioweather) located on the office building were used. In order to test and evaluate the proposed methods of regulation of HVAC, this building is equipped with an extensive measuring and control system into which the weather station will be integrated. The weather station stores every minute information about temperature, relative humidity, wind speed and direction, etc. The measured data are from January 2016 to March 2018. Average temperatures were determined over the course of the 30 minutes for each month. These courses are shown in Figure 4. The temperature during the day may vary greatly from region to region. In practice, it would be possible to import temperature courses for individual areas, for example from Wolfram Mathematica software. Here you can find average daily temperatures for any place on our planet (according to the nearest weather station).

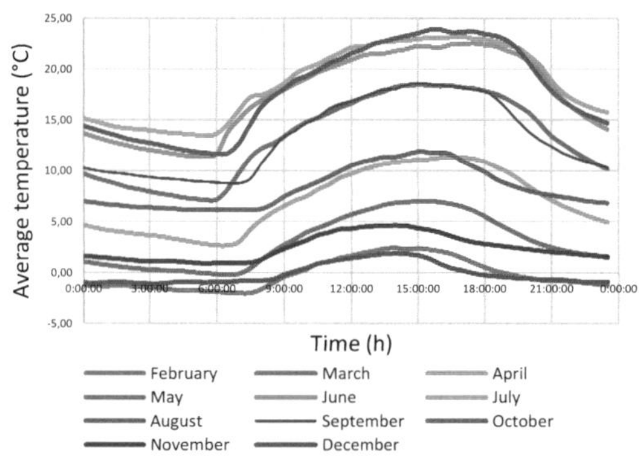

Figure 4. Average temperature according to the month

Figure 4 4 shows that the individual shapes of the average daily temperatures are very similar to each month. The average minimum temperature during the day is around 6 am, which is an area suitable for cooling. For example, during the summer months, a recuperation unit may be started in the area between the third and six o'clock in the morning to cool down the room. During the day, the recuperation can provide room ventilation according to hygiene standards, but because the initial low room temperature, there may be no need to air-condition the room (at least for a part of the day). The example with the indicated areas is shown in Fig. 5.

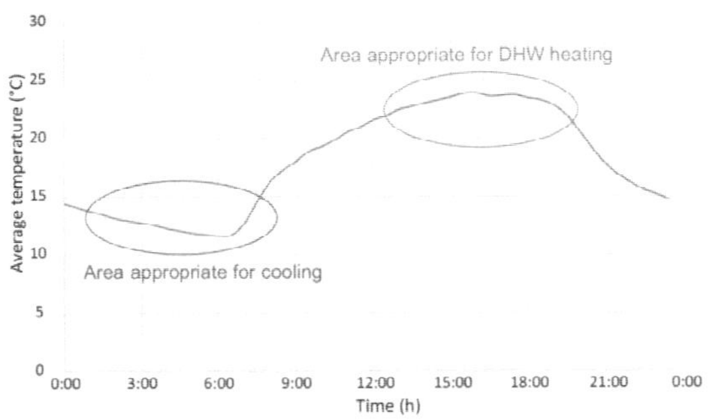

Figure 5. Average temperature during the day (in August) with marked areas suitable for cooling and for DHW heating

Conversely, the highest average temperature is around 3 pm. In the case of air-to-water heat pumps, their operation with an accumulation tank and underfloor heating (for the intended control system) is optimal. The heat pump can operate at the maximum output power (at high efficiency) in the heating area and accumulate the necessary heat in the storage tank for heating system, DHW tank and also in the concrete in the floor or other building structures. Due to the high inertia of the underfloor heating, the units can only be operated in the optimum area during transition periods. The size of the storage tank also plays an important role. As the temperature decreases, the heat loss of the building increases, and operation of the unit is also necessary outside the optimum area. By appropriately designing the heating and control system, the operation of the unit in the lowest efficiency area can be eliminated.

The most ideal way of predictive control is to use the on-line weather forecast for a given installation location. Weather forecasts can respond to the arrival of a warm or cold fronts and improve the daily forecast of temperature development. Controllers can be connected to a cloud application that provides weather forecast information for the controller. The disadvantage of this system is the necessity to connect the control system to the Internet, which may be problematic for a simple devices. Often, a complete change of the concept and replacement of used hardware is necessary.

Another option is to predict temperature development during the day off-line. It is obvious from average daily temperatures that the courses are very similar. The courses differ in the minimum and maximum temperatures and are shifted up or down the temperature axis by month. To predict the day temperature, it is necessary to take into account the month in which the system is located. When comparing all the days tested, it can be seen that the maximum time periods and the minimum measured temperatures during the day are almost identical. It would therefore be possible to use an off-line time management system, which would be particularly suitable for ventilation and heat recovery units, freezers and domestic water heating using a heat pump (air-water). Especially for systems that cannot be connected to an online forecast for some technical reason. As a result, it would be possible to streamline the operation of these devices to operate at full power only in the selected time zone, when they run optimally. Conversely, in a time zone where device efficiency is lower, they would only operate in restricted mode.

In the case of heat pumps (air-water) it would be possible to operate the DHW only in the time period with the highest average daily temperature. In the case of a well designed DHW tank volume, the heating would take place mainly during this time period and at higher efficiency.

For heat recovery units, it would be possible to use the lower outdoor temperatures in the morning to cool the air and walls in the room during the summer months. When the bypass is open (the air does not pass through the recuperator), cool outside air flows into the room and the warm indoor air is drained out. When the desired room cooling is achieved, the fans would be switched off, the system would measure the CO_2 concentration and, if the maximum allowed value is exceeded, the room would be ventilated by a standard recuperator (the warm outdoor air is cooled by cold room air). With this measurement it would be possible to reduce the room temperature and thus reduce the energy consumption needed for cooling (air conditioning).

V. Description of Energy Management

The heat pump will be primarily managed by an on-line weather forecast using a weather station (Warioweather) located on the roof of an office building, which uses internet forecasting and refines the forecast based on information from nearby weather stations. The control system will communicate with the weather station using the Modbus TCP/IP protocol. In the event of on-line communication failure, it will be reset to off-line mode. The current size of the generated electricity will also have a significant impact on the operation of the heat pump. In the case of charged batteries of the hybrid inverter (at the set limit) and excess electricity is generated, the heat pump power is controlled so that zero flow to the grid through the 4-quadrant meter is maintained. The excess electricity to the distribution network is shown in the Fig. 9.

A simplified diagram of the heating system is shown in Figure 6. The real installation is shown in the Figure 7 7. First, the heat pump will heat the domestic hot water in the storage tank. When the maximum set temperature is reached, the heating system (1000 l storage tank) is switched by a 3-way valve. As soon as the storage tanks are heated to the maximum temperature which can be supplied by the heat pump (in our case approx. 65 °C), the excess electricity will be used to heat the storage tanks electrically by means of SSR (Solid state relay). Thanks to the controllable mixing valves on the storage tanks, it will be possible to set the desired temperature to the heating system and the maximum domestic water temperature.

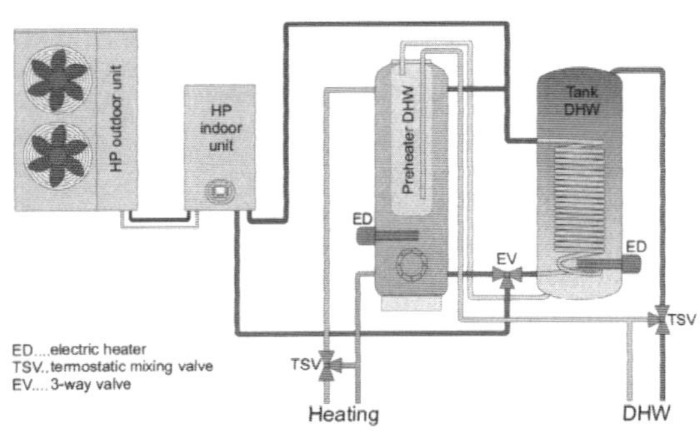

Figure 6. Schematic illustration of the hydraulic connection of the technical room

Figure 7. Real installation in the technical room

All electric equipment in the new building will be connected via a hybrid inverter to power the object even if the power supply fails. It will be possible to test the object disconnection from the grid and subsequent operation in island mode.

The consumption of the electric energy for complex of buildings is shown in the Fig. 8. The excess electricity to the distribution network is shown in the Fig. 9.

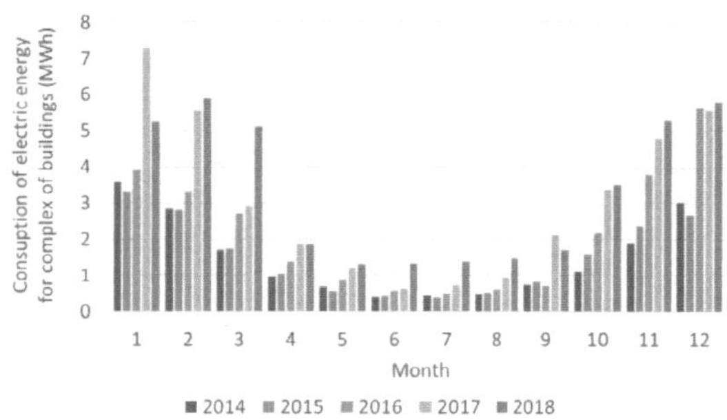

Figure 8. Consumption of electricity for complex of the buildings (2014-2018)

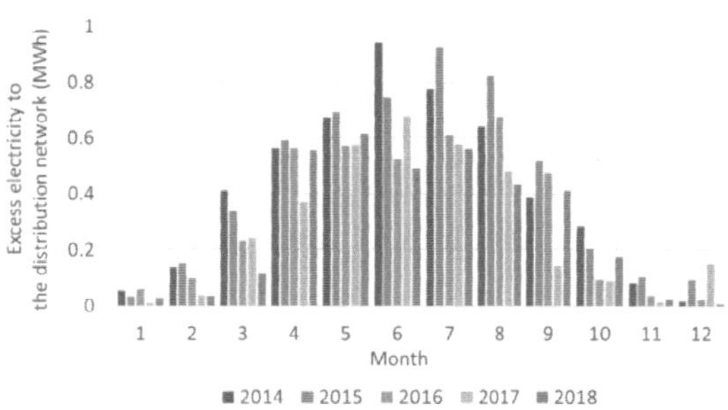

Figure 9. Excess electricity to the distribution network (2014-2018)

Charging of the electric vehicle will also enter to the regulation of control system. The building is planned with the installation of AC and DC electric car chargers. The system will be designed so that the electric car can be recharged as much as possible while maintaining the current size of the main circuit breaker in front of the energy meter of complex of the building. If you need to recharge quickly with the AC charger, all heating sources in the area will be temporarily disconnected and the hybrid inverter will deliver the maximum possible battery power to the electric vehicle. Further, a DC charger will be implemented to the hybrid inverter to allow more efficient recharging (loss elimination). The wiring diagram of the area is shown in Fig. 10.

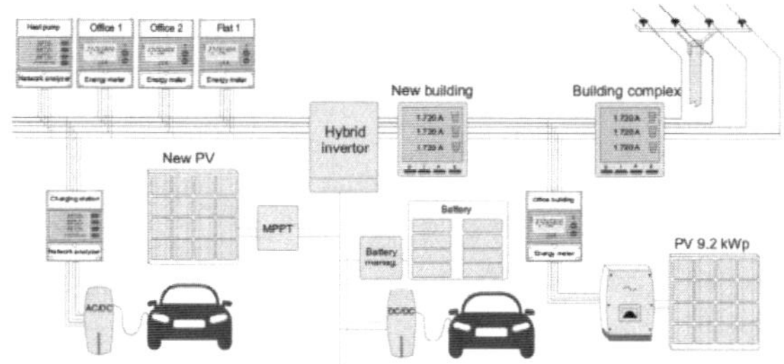

Figure 10. The schematic diagram of the area

VI. Conclusion

The article described proposed and installed measuring and regulation system which is able to combine control of heating system, use of excess electricity of the PV and charging of the electric vehicles. The article also shows several ideas how to control HVAC devices (in our case air-to-water heat pump) in combination with off-line and on-line prediction. This system of regulation allows to decrease energy consumption and to reduce operating costs of the building.

Acknowledgment

This article was funded by the Ministry of Education within the project LQ1603 entitled "Research for SUSEN" and by the project SGS-2018-023.

References

[1] M. Székyová, "Ventilation and air conditioning", Větrání a klimatizace. 1. czech ed. Bratislava: Jaga, 2006, 359 p., ISBN 80-807-6037-3

[2] I. Švarc, "Automatic control", Automatické řízení, Vyd. 2, Brno: Akademické nakladatelství CERM, 2011, ISBN 978-80-214-4398-3.

[3] J. Votava, J. Kyncl and L. Straka, Optimization of local cogeneration stations,In 19th International Scientific Conference on Electric Power Engineering (EPE), 2018, pp. 1-6, IEE, DOI: 10.1109/EPE.2018.8396037, ISBN 978-1-5386-4612-0.

[4] B. Alter, Environmental consulting fundamentals investigation and remediation, Online-Ausg, Boca Raton, FL: CRC Press, 2012, ISBN 9781439868911.

[5] Building automation: control devices and applications. Homewood, Illinois: American Technical Publishers, [2008]. ISBN 978-0826920003.

[6] H. Viot, A. Sempey, L. Mora, J.C. Batsale and J. Malvestio, Model predictive control of a thermally activated building system to improve energy management of an experimental building: Part II - Potential of predictive strategy. Energy and Buildings [online], Elsevier, 2018, (172), 385-396 [cit. 2018-06-16], ISSN 0378-7788, Available from: http://www.sciencedirect.com/science/article/pii/S0378778818311460

[7] H. Thieblemont, F. Haghighat, R. Ooka and A. Moreau, Predictive control strategies based on weather forecast in buildings with energy storage system, Energy and Buildings [online], Elsevier, 2017, (153), 485-500 [cit. 2018-06-16], ISSN 0378-7788, Available from http://www.sciencedirect.com/science/article/pii/S0378778816320333

Shared Battery Storage, a Key Solution for the Construction of Charging Stations in the Urban Area

Lukáš Dvořáček
Dep. Economic, Management and Humanities,
Faculty of Electrical Engineering
Czech Technical University
Prague
dvoral14@fel.cvut.cz

Abstract — **Following the EU 2030 targets, reduction of GHG emissions becomes ever more pressing. All stakeholders will have to seek solutions to reach the reduction of GHG emissions across sectors - not only in the electricity and heat generation, but also in the public and especially private transport sector. Mainly the field of new vehicles should change. The European Parliament specified that 40% of new cars are to be zero- or low-emission vehicles by 2030 targets.**

With the increase of electric vehicles (EVs), it will be important to find suitable points of charging with regard to transport infrastructure as well as to complete the existing system of energy network with superchargers. The existing literature and research have so far mainly focused on the location of the charging sites in the primary transport nodes, such as gas stations along the highways and roads outside cities and have rarely considered the fact that in the first stage, most EVs will be used as city vehicles. It is highly likely, that due to the parameters of current batteries and small distance, which the EVs are able to achieve, the EVs will initially be used for urban and suburban transport.

However, at the same time, many places, which have an excellent location in terms of transport infrastructure in the city, such as gas stations or large parking spaces, usually do not have a sufficient capacity of the electrical connection for EVs charging. Lower capacity of electrical connection can be the reason why it is not possible to create a charging site in the required place, or it is not possible to use sufficient number of charging stations with adequate power. One of the solutions to this problem may be an installation of battery storage at these points.

This paper evaluates the costs and benefits that the system with shared battery storage brings and compares the battery solution with the alternative of investing in increasing the capacity of the electrical connection in the required area. From technical and economic points of view, the model takes into account a number of economic aspects and the cheaper option is not always the better one. During the final choice, it is important to consider also other benefits for the owner of the charging site. For instance, elimination fines for exceeding the pre-determined value of the power absorbed can be imposed as well as the possibility to install more supercharger stations within the sites which will allow charging more EVs at the same time.

Keywords: Battery storage, fast-charger station, electric vehicles, electric taxis, charging site, consumption diagram, 2030 targets

I. INTRODUCTION

A key parameter, which needs to be taken into account when allocating the charging sites in existing energy infrastructure, is the fundamental difference between the charging an electric car and refueling a car with a combustion engine.

Home charging stations have created a new way users of electric vehicles are able to recharge their cars very comfortably. This option has to be considered before designing a network of recharging stations [1]. One of the most significant benefits of home charging of electric vehicles will be a lower price per kWh compared to a business fast-charging station [2], together with higher degree of comfort and less time spent charging in public charging sites [1], [2]. On the other hand, the disadvantage of home charging solution is the necessary upfront investment in the home charger set and surely, an investment in own parking space or a garage [3].

However, not all users (especially in urban areas) have their own garage or parking space. Most of the population in cities live in flats without a private garage or a private parking space [3]. These larger groups of users leave their cars in public parking places near apartments. They do not usually have the option to buy a home charging set and use it to charge electric cars.

The problem of charging electric vehicles outside home can be divided into two main parts. Firstly, charging a large number of electric cars during a long time with low energy power and secondly, charging electric vehicles with fast charging stations in the shortest time possible.

Technical design of charging stations, for instance, in large parking spaces near apartment buildings could resolve the issue of home recharging electric vehicles for many users living in respective urban area. In this large urban parking places it is expected that users park their cars and let them charge for a longer period, usually more than 6 hours. During this time and among a lot of electric vehicles recharging, it is possible to control that the network stability is not compromised and the energy capacity of connection in this area is fully used.

Equipping the charging site in urban areas with a battery could be a backup option for potentially increasing the number of charging stations. However, when simultaneous charging is expected, either a battery needs to be used, or an increase in charging station energy power needs to be applied. Another option is an increase of charging time without further interfering with the infrastructure of the superior network. A better place to build a charging point equipped with a large capacity battery is a city charging site with a large number of public fast charger stations or areas close to parking places for electric taxis. There we can expect high demand for fast recharge.

According to the efforts of all European cities to expand electric vehicles and electric taxi cars [4], the EU will be building new super-charging stations in the city centers especially in places reserved for taxi cars.

It can be assumed that the first wave of charging sites in the urban areas will be created by adding fast-charging stations to existing gas stations because existing gas stations usually have a perfect position in the city, in terms of traffic infrastructure. Also, it is expected that this upgrade of gas stations by new charging stations will mean lower investment than building new electrical charging site.

From an energy perspective, these areas can have a problem with the capacity of energy network, and it may not be possible to build a charging site due to the current stability of energy network.

However, in fast charging stations, where short charging time is a priority (as, for instance those for taxi vehicles), the method of controlling the delay of charging is not permitted. Also, it may not be possible to add another power line to the existing gas stations in urban areas, which in turn could stop this project of building enough volume of powerful charging stations. In such case, the shared battery storage is the only solution to obtain a charging site which satisfies energy consumption of electric vehicles without the need of additional infrastructure investment.

II. Issues of Charging Site Installation

This model assumes a location in an urban area that is connected to a 640 kW power transformer station. A charging site is not the only entity that is connected to the transformer station; another connection is a housing unit with 360 apartments. The model assumes that all flats are connected to a central heating system and thus do not have their own water heater and are not equipped with any other source of heat such as an electric boiler or electrical heating. This means that all apartments have a D02 or D01 tariff with a single-tariff electricity rate and flats have a circuit breaker 26A.

With this assumption, it is possible to extend the calculation by a type delivery diagram from 4 sections (TDD4). The delivery diagram characterizes yearly consumption of electricity at the customer and is also used to evaluate deviations. For calculation, the model uses an annual consumption diagram whose measured values also take into account the temperature development in the Czech Republic during the year [5].

The averages for each of the 24 hours of the day were calculated from the monthly measured values of the delivery diagram. To estimate the average value of the electric power, which the whole house unit requires each day, it is necessary to multiply these values by the difficulty factor, the number of apartments and ampere value of circuit breakers. The difficulty factor (1) takes into account the behavior of individual users and the possibility of switching on especially powerful appliances in individual apartments at the same time.

$$K_\infty + \frac{(1-K_\infty)}{\sqrt{n}} = K_n \quad (1)$$

Where K_n is the difficulty factor; K_∞ is the contemporaneity coefficient for non-finite number of flats and n is the number of flats [6].

It is important for the charging site that the difference between the reserve power for the housing unit and the maximum power level of the transformer station (TS) can be used to charge the battery.

It is important to consider the impact of battery charging on the consumption diagram. If the battery is charged by maximum power from the transformer station, it will reduce the charging time. The downside of the battery charging will be the undesired fluctuations in the consumption diagram. It will be particularly recognizable at intervals with low energy demand (Figure 1, 2).

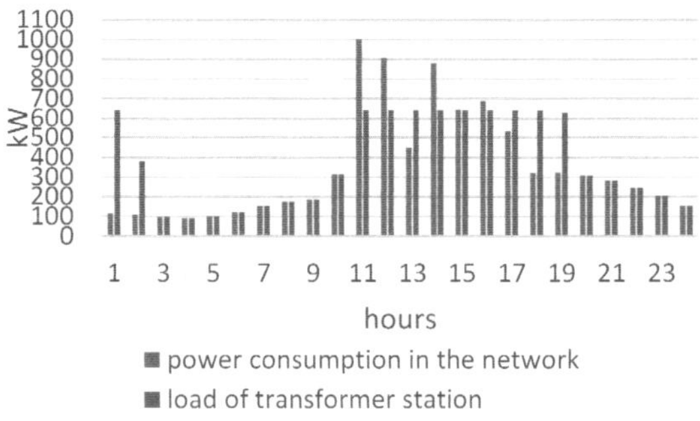

Figure 1. Utilization of transformer station power reserve from 100%

Figure 1 shows the short time needed to charge the battery and the power peak that the charging in the consumption diagram in May 2018 caused.

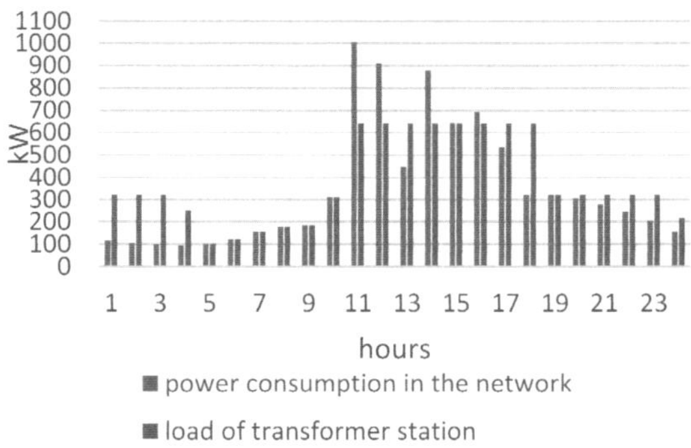

Figure 2. Utilization of transformer station power reserve from 50%

Figure 2 shows the consumption diagram when charging the battery with half the charging power of the transformer station than in Figure 1. This way of charging also involves undesired fluctuations in the consumption diagram in May 2018.

Therefore, the model evaluates controlled battery charging and charges the battery out of hours with high demand for energy to smooth out small variations in consumption. In the case of uncontrolled battery charging, the model achieves a faster battery charge but causes undesirable fluctuations in the consumption diagram. For this reason, the model consumes constant energy from the transformer station throughout the time interval with a low demand for electrical energy. The benefits of constant energy consumption are also smoothing of smaller fluctuations in the consumption diagram at time intervals beyond the main demand to a constant value. The stored energy from the intervals with less energy demand is used to smoothen the power peaks in the load diagram when the energy demand is higher. Very often these peaks are caused by charging a large number of EVs at the same time.

III. Model of a Charging Site

A. Technical part

The following model considers building a charging point with shared battery storage exclusively for electric taxis. The model uses the data of a study of charging frequency of electric taxis at a particular urban charging site in Singapore but with more frequent arrivals of electric taxis [7]. From the published values, 20 electric taxis were simultaneously charged during the most significant quarter-hour.

Another input condition of the model in [7] is the fixed working hours of taxi service. The first taxi cars leave at 6 am and return at 6 pm. During this time interval they can use the charging site to charge their cars. The charging site is not used after working hours.

During the working hours it is assumed that the electric taxis arrive to the charging site with almost empty batteries and after a 45-minute recharging, they depart with a fully charged batteries. To meet the 45-minute limit for almost full battery charge, considering the average battery capacity used in an electric car, it is necessary to use a 50 kW DC or higher power charging station for charging. It was found from the real measurement that a 50kW DC charger connected to a three-phase distribution system will charge a car battery to 10kWh in 12 minutes and 30kWh in another 24 minutes [8].

Table 1. Supply types and charging times. [8]

Supply Type / Charger Rating	*AC/DC*	*Rated power**	*Time to charge 10kWh*
3 phase, DC / Rapid	DC	50 kW	12 mins
3 phase, DC / Rapid	DC	120 kW	5 mins
3 phase, 32A per phase / Fast	AC	22 kW	27 mins
3 phase, 60A per phase / Rapid	AC	43 kW	14 mins

*Simplistically (ignoring Power Factor correction)

Therefore, the model assumes that in 45 minutes the charging station will consume 30 kWh of electric power. Considering the actual energy value that will be stored in a electric car battery after 45 minutes, it will be necessary to consider system-wide losses due to heating, transforming the AC voltage to DC and others. These losses range from 15% to 20% during charging [9].

However, for the model in question, the priority is the amount of the energy that has been released from the battery or the power network. The model knows from data collected by tracking the arrival of individual taxi cars to the charging point at 15-minute intervals, that the largest amount of consumed energy is 250 kWh (Figure 3).

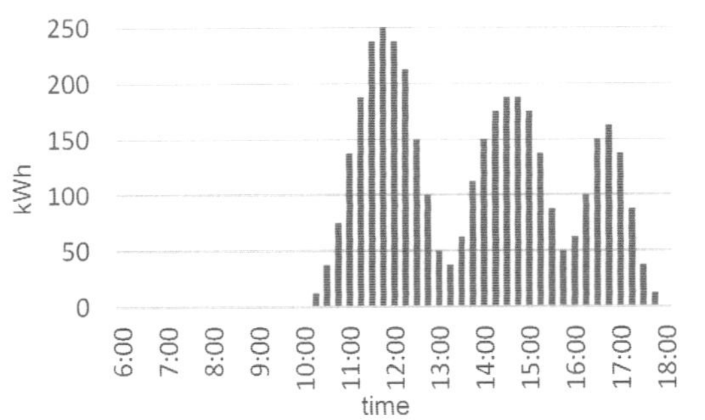

Figure 3. Energy consumption of EVs in the 15-minutes intervals.

An installation of a charging site in an urban area increases the demand for electric power especially in the working hours for taxi vehicles. The increase in consumption is mainly due to the charging of more electric vehicles at the same time. Due to the same start of working hours and similar mileage for each taxi, the charging time cannot be increased (Figure 4).

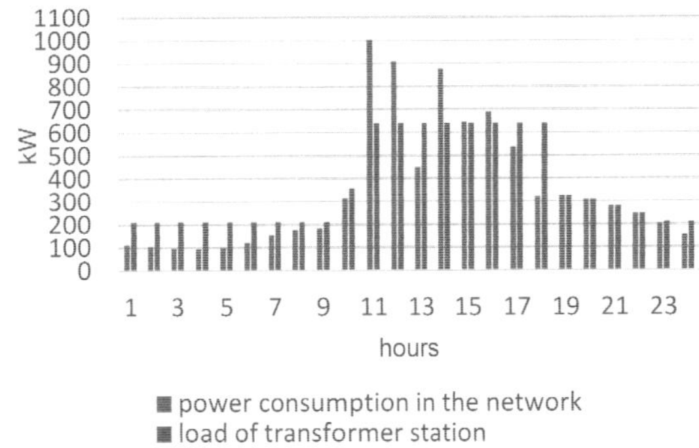

Figure 4. Consumption diagram for controlled battery charging.

A design of a respective charging site without the use of a battery would require a power boost not only for the local transformer station but also an increase of all power elements towards the power distribution network. In the urban area there are usually certain parts of the power grid, where increasing the necessary power elements in the network cannot be implemented. Limiting factors are generally power limits, current throughput, network stability or high investment to strengthen the entire power network.

Compared to increasing power limits of all elements of the power grid, the construction of a charging site with a large capacity battery has numerable benefits. One of the advantages is the possibility of battery installation in places where the power increase of the transformer station is not possible due to the stability of the whole network.

Furthermore, the possibility of installing a higher quantity of charging stations at a given location (charging site) and thus obtaining the possibility of charging a plurality of electric vehicles at a set time without jeopardizing the stability of the power network.

In terms of network stability, the battery located near the point of large energy consumption enables smoothing power peaks in the consumption diagram. It helps to stabilize the transmission network and, at the same time, creates a stable frequency in the power grid.

B. Model and results

The model has revealed a minimal energy capacity of battery required to charge electric taxis during ten year time period. For the calculation and testing, the model uses annual consumption diagrams which take into account the temperature development in the Czech Republic during each testing year from 2008 to 2018.

The model considered the most critical charging situation when all stationed electric taxi vehicles had to be charged. During these ten years there was not a situation, when the charging site including battery storage with 1 MW of energy power and energy connection with 640 kW transformer station did not have enough energy for charging, or, charging site operation jeopardized the stability of energy network.

An important point of the research was specified utilization of transformer station power reserve out of the primary charging hour of taxi vehicles and achieve there smoothing of power peaks in the consumption diagram. The minimum level of energy stored in the battery in each respective month was one of the necessary information for the specified utilization of transformer station power reserve (Figure 5).

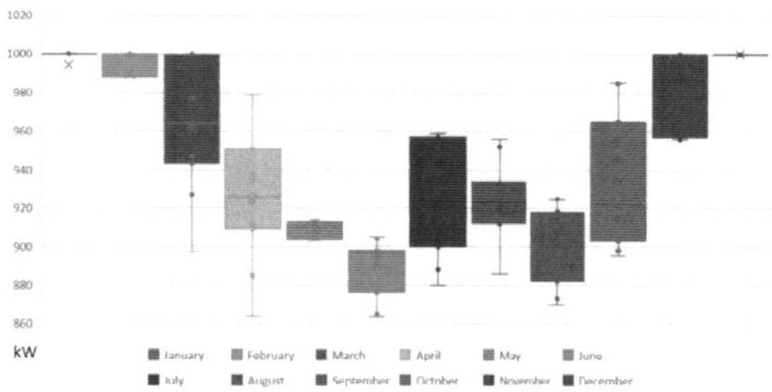

Figure 5. The minimum levels of energy stored in the battery.

A similar trend of minimal levels of energy stored in the battery can be observed in individual months in the year. In all ten years, the minimum charge level was from April to October; lower than in the other months. This means that the battery can store more energy power in the working hours and therefore the minimum charge level can be lower in these months. One of the ways of explaining the decrease in consumption diagrams in those months is that people spend more time out of apartments, and also they do not have to use lights so many hours as in the winter season.

The minimal level of energy stored in the battery each month in a particular year was calculated at the beginning of the following month. Consequently, the model predicted this minimum charge level throughout the testing period using stored values for individual months of each year. Method of EWMA - Exponentially Weighted Moving Average (2),(3) has been used to calculate minimal charge level in the following years. However, this method has not been able to sufficiently adapt to the changes and fluctuations of minimal charging levels from one year to another (Figure 6).

$$\alpha \sum_{k=0}^{n-1}(1-\alpha)^k \cdot x(n-k) + (1-\alpha)^n \cdot \hat{x}(0) = \hat{x}(n) \quad (2)$$

$$\frac{2}{N+1} = \alpha \quad (3)$$

Where $\hat{x}(n)$ is a vector of predicted minimal charging levels; α is coefficient used to weigh current and past predicted values; N is a coefficient, which defines the number of days in $\hat{x}(n)$; $\hat{x}(0)$ is initial battery charge.

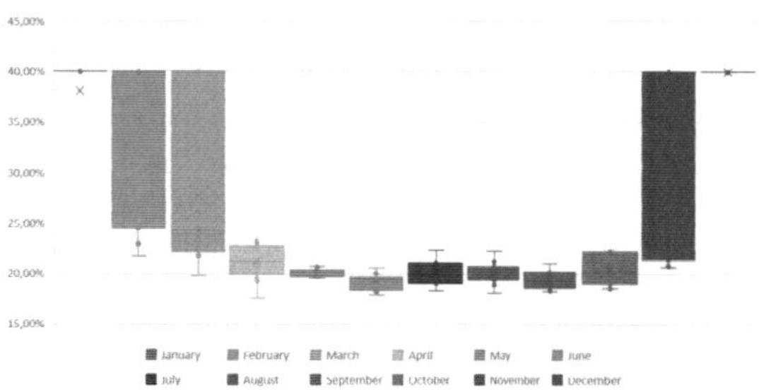

Figure 6. Optimal percentage of charging power of the TS

Even in the case of a simulation, where the average minimum levels of ten years have been used as input arguments for each month, the prediction of the method was not satisfactory. The result was a prediction of insufficient charging levels and thus a threat to the stability of charging electric taxi vehicles. Another situation would have arisen had we considered the possibility of shifting the charge of several electric cars to one of the following hours which would have enabled the battery recharge its capacity.

C. Economic part

Apart from technical aspects, economic factors need to be taken into account when constructing a charging point. One of the key issues will always be the place of construction. As already mentioned, from the perspective of transport infrastructure and accessibility, the already-built gas stations are ideal places to build a charging site. However, the gas stations may not have a sufficient capacity of energy connection. In particular, the requirements for the energy power of the local installed transformer station and the current throughput are usually Insufficient.

Provided that these technical requirements for increasing the power of the local transformer station and all the energy elements can be met towards the distribution network, it is also necessary to take into account all the costs related to strengthening the whole energy infrastructure.

Very often, the addition of a high-capacity battery charging station is a simpler and more cost-effective option which does not require additional investment to strengthen the surrounding infrastructure although the initial investment in battery storage that exceeds 80,000 EUR [10] is not negligible and may seem disadvantageous.

Economically challenging consideration will also be the one of charging site construction that will not be possible to complete in the required location without battery storage.

IV. Conclusion

The issue of construction of charging sites with the expected growth of electric cars will be necessary and up-to-date. There will be an increase in the demand for energy consumption at the points of implementation of these stations and will be necessary to take steps to ensure the stability of the entire power system.

One way to build a charging site at the required location and not jeopardize the stability of the entire power grid is to strengthen the local transformer station and all the energy elements towards the distribution network. Often, this method is usually not feasible from a technical or an economic point of view.

The second option is to use the battery storage, which is a backup source to reduce the energy fluctuations in the energy demand during the highest charging site load. The advantage of this solution is that it does not require another change in the energy network and helps to smoothen the fluctuation in the consumption diagram, as is also apparent from the model.

The implemented model presented a possible solution to the construction of the charging station by using battery storage. The model has verified that a one MW battery is sufficient due to the expected number of arriving electric vehicles to the charging site and the required power consumption in each hour without jeopardizing the stability of not only the local but the entire energy network. From the data collected by tracking the arrival of individual taxi cars to the charging point at 15-minute intervals, the model knew that the largest amount of consumed energy is 250 kWh which occurred between 11 and 12 o'clock. During this most significant quarter-hour when 20 electric taxis were simultaneously charged with 50kW DC chargers and 360 flats consumed 200 kW on average were both fully loaded 640kW transformer stations, and batteries reached their limits. Between 12 and 13 o'clock, the number of simultaneously charged electric taxis decreased, and one MW battery could be again recharged without a blackout or other jeopardizing the energy network stability.

However, the dispersion of these values for the same month from 2008 to 2018 and the prediction EWMA method has not produced satisfactory results. The EWMA method failed to respond quickly enough to changes in household consumption between each of the years. The reaction was particularly slow for February, March, and November when household consumption differences between years were the greatest.

For real implementation of the model, it is necessary to improve the prediction of the difference between the current household consumption and the expected consumption of electric vehicles primarily. With the use of the obtained data, it will be possible to better determine the required battery capacity and other costs necessary to construct the charging site.

In further research, it will be necessary to improve prediction skills of the model and use more sophisticated methods, which would use not only information from previous years, but could also respond to energy storage in the last months of the respective year.

Acknowledgment

Thanks to prof. Ing. Jaroslav Knápek, CSc. and prof. Ing. Oldřich Starý, CSc. for help and suggestions for this work. This work was supported by the Grant Agency of the CTU in Prague, grant No. SGS18/134/OHK5/2T/13.

References

[1] D. Lopez-Behar, M. Tran, T. Froese, J. R. Mayaud, O. E. Herrera, and W. Merida, "Charging infrastructure for electric vehicles in Multi-Unit Residential Buildings: Mapping feedbacks and policy recommendations," Energy Policy, vol. 126, no. May 2018, pp. 444–451, ISSN 0301-4215, 2019

[2] Cost of Charging an Electric Car", Pod Point'. [Online]. Available: https://pod-point.com/guides/driver/cost-of-charging-electric-car. [Accessed: 5-Mar.-2019].

[3] Z. Wei, Y. Li, Y. Zhang, and L. Cai, "Intelligent parking garage EV charging scheduling considering battery charging characteristic," IEEE Trans. Ind. Electron., vol. 65, no. 3, pp. 2806–2816, 2018, DOI: 10.1109/TIE.2017.2740834

[4] "Podnikatelé se mohou těšit na vyšší podporu elektromobility ze strany vlády: Podpora infrastruktury", Byznys&energie. [Online]. Available: https://www.byznys-energie.cz/autor/jan. [Accessed: 1-Mar.-2019].

[5] Normalizované TDD', OTE, a.s. [Online]. Available: https://www.ote-cr.cz/cs/statistika/typove-diagramy-dodavek-elektriny/normalizovane-tdd. [Accessed: 18-Mar-2019].

[6] ČSN 33 2130: Elektrické instalace nízkého napětí - Vnitřní elektrické rozvody, 3rd ed. 2015.

[7] H. Wang, R. L. Cheu, and E. Balal, "Operations of electric taxis to serve advance reservations by trip chaining: Sensitivity analysis on network size, customer demand and number of charging stations," Int. J. Transp. Sci. Technol., vol. 5, no. 2, pp. 47–59, 2016, DOI: 10.1016/j.ijtst.2016.09.002

[8] "SPIRIT Power change: EV Charging Knowledge Bank Understanding electric car charging". [Online]. Available: https://www.spiritenergy.co.uk/kb-ev-understanding-electric-car-charging. [Accessed: 1-Mar.-2019]

[9] E. Apostolaki-Iosifidou, P. Codani, and W. Kempton, "Measurement of power loss during electric vehicle charging and discharging," Energy, vol. 127, pp. 730–742, 2017, DOI: 10.1016/j.energy.2017.03.015

[10] "Česko má třetí velké bateriové úložiště s kapacitou 1,3 MWh: Smart City", IOT-network newa, 2018. [Online]. Available: https://cz.iot-nn.com/blog/2018/08/24/cesko-ma-treti-velke-bateriove-uloziste-s-kapacitou-13-mwh/. [Accessed: 18-Mar.-2019].

Economic Evaluation of Heating Systems for Energy Efficient Houses

A Case Study of Various Heating Systems of Family Houses in Conditions of Central Europe

Jiří Beranovský, Jaroslav Knápek
FEE, Department of Economics, Management and Humanities
CTU in Prague
Technická 2, 16627 Praha 6, Czech Republic
jiri.beranovsky@fel.cvut.cz, knapek@fel.cvut.cz

Abstract — **The case study works with a typical representative of low-energy family house (with its parameters close to the passive standard) on which different types of heating and hot water preparation are discussed. The paper discusses technologies for space heating and hot water preparation as follows: 1) electric direct space heating and boiler for hot water, 2) electric boiler for space heating, 3) heat pump for hot water preparation, 4) air/water heat pump for space heating and hot water preparation, 5) air/water heat pump for floor space heating, 6) condensing gas-fired boiler and floor heating, 7) air/air heat pump + heat pump for hot water preparation, 8) pellet boiler with heat accumulation. The economic evaluation of the options is performed using the basic criteria of economic efficiency, i.e., NPV, IRR, Ts (repayment period (simple)), Tsd (discounted time).**

The article analyses the advantages and disadvantages of individual technologies and their combinations both in economic terms and in terms of energy efficiency and comfort of use. At the same time, the paper presents estimates of changes in the consumption of electricity and other forms of energy for heating and hot water production from the point of view of the expected rate of construction of new and reconstruction of existing family houses.

Keywords: passive house; heating systems; heating costs

I. Introduction

With a certain degree of generalisation, it can be said that the proportion of new passive houses (houses with very low energy demand) in the Czech Republic has stabilised over the last two years (2017 and 2018) at approx. 6% of the total number of newly built family houses. In absolute figures, this translates to approximately more than 1000 new passive houses annually; the total number in the CR is estimated around 6000 [1]. According to the Czech Statistical Office, 19,156 family houses were completed in the CR in 2018, and 15,172 in 2017 [1, 2].

Compared to previous years, implemented European legislation has led to passive construction expanding into apartment and public buildings. However, it is still more of an exception in conventional development [1]. In addition to ordinary fear of the unknown and reluctance to critical thought among investors, the poor passive construction market is also the result of promotion of myths among some manufacturers and developers, who are satisfied with lower construction quality levels.

The most widespread myths say that passive buildings have "high investment costs that never return", that "thermally insulated buildings don't breathe" and that "you are not allowed to open the windows in a passive house". The last of the myths is sometimes taken to the absurd by stating that all the windows in a passive house are fixed (cannot open) and in order to prevent suffocation you have to install air-handling systems (with heat recovery), which will destroy all free ions that have a beneficial effect on the body and you will wither away, and if you don't, then you will suffocate the moment the power supply fails. The campaign "It is normal not to insulate" is a disgraceful example of this.

II. Economic Assessment Methodology

A. Investment costs

The analyses presented in this paper make use of budgets for various heating and hot water production technologies on the example of a specific executed family home in Prague-Vinoř [5], which can be used as a reference example of a family house in the passive standard. The calculations deal with Option 5 published in the study [4], which deals with economic assessment of building structures of passive houses. The whole of study [6] on economic assessment of heating and hot water production in energy-saving houses has been published as a book. Both books are intended for free download for the public as part of EKIS consultancy centres.

The table below shows an overview of investment costs of each heating option of the reference building analysed.

Table 1. Investment costs of each option

Investment costs of each option		CZK excl. VAT	CZK incl. 15% VAT
Opt. 0	0) el. convector heaters, el. water heater, window ventilation	71,000	81,650
Opt. 1	1) el. convector heaters, el. water heater, AHU with HR	206,000	236,900
Opt. 2	2) el. boiler + radiators, el. water heater, AHU with HR	259,000	297,850

Opt. 3	3) el. convector heaters, HP for DHW, AHU with HR	247,000	284,050
Opt. 4	4) air/water HP with radiators for CH and DHW, AHU with HR	416,500	478,975
Opt. 5	5) air/water HP with floor heating and DHW, AHU with HR	437,500	503,125
Opt. 6	6) ground source HP with floor heating and DHW, AHU with HR	482,500	554,875
Opt. 7	condensing gas-fired boiler, floor heating, AHU with HR	366,000	420,900
Opt. 8	8) MULTISPLIT+convector heaters, HP for DHW, AHU	349,000	401,350
Opt. 9	9) warm-air heating + el. boiler, HP for DHW, AHU with HR	284,500	327,175
Opt. 10	10) warm-air heating +air/air HP, HP for DHW, AHU with HR	342,600	393,990
Opt. 11	11) pellet stove+accum., warm-air heating, AHU with HR	351,500	404,225
Opt. 12	12) compact AHU with integrated HP, secondary convector heaters, AHU with HR	337,000	387,550

B. *Operating costs*

Besides costs of fuel consumption, the operating costs consist of costs of periodic maintenance and inspections, and servicing and reinvestment as necessary. Maintenance includes, for example, periodic replacement of filters; we assume CZK 180/year for the warm-air heating and CZK 120/year for the equal-pressure heating and compact units. The use of more user-friendly compact filters requires about 3.5 times the price of the ordinary ones. The assumed costs of inspections are CZK 300/year for electric boilers, CZK 1,400/year for gas boilers, CZK 600-1,800/year for heat pumps depending on quality and scope, and CZK 700/year for stove chimneys. Servicing costs can never be estimated precisely beforehand; they depend on product quality and extent of defects. Some suppliers, for example, cover minor servicing costs in the inspection price.

Costs of fuel consumption in the model are calculated from fuel prices, including prices of the fuel itself and constant (fixed) payments. For example, the fixed payments for electricity are always due, and they represent payments for the point of consumption, circuit breaker power, etc. To some extent, they can be understood as always required and, under some circumstances (especially when comparing options with electricity consumption at the same price rate and with the same circuit breaker rate), only payments for the electricity consumed can be compared.

The power invoice detail these items as payment per kWh, including, for example, power electricity, distributed quantity, system services, support to renewable sources of energy (SRSE). This approach is in order when comparing, for example, energy-saving measures with direct impact on electricity consumption but without an impact on the circuit breaker size or distribution rate. Since we are comparing entirely different concepts in this case, we employ a total cost methodology, meaning we compare total costs, including fixed payments. The majority of the options assume the D57d rate for electric heating, Option 0 has the convector heater rate D45d, and gas heating options also feature the distribution rate D02d. The prices are assumed including 15% VAT. The electricity prices are assumed from CZK 2.77 to 5.15 per kWh, the natural gas price is CZK 1.46/kWh and the price of pellets is CZK 1.02/kWh. The following prices are shown excl. VAT, which is either 15% or 21% depending on the service or product type.

Table 2. Approximate prices of annual operating costs for each option.

Operating costs (CZK/year) excl. cooling	**2017**
0) el. convector heaters, el. water heater, window ventilation	26,643.
1) el. convector heaters, el. water heater, AHU with HR	22,030
2) el. boiler + radiators, el. water heater, AHU with HR	22,183
3) el. convector heaters, HP for DHW, AHU with HR	18,246
4) air/water HP with radiators for CH and DHW, AHU with HR	14,439
5) air/water HP with floor heating and DHW, AHU with HR	14,363
6) ground source HP with floor heating and DHW, AHU with HR	13,953
condensing gas-fired boiler, floor heating, AHU with HR	23,899
8) MULTISPLIT+convector heaters, HP for DHW, AHU	15,079
9) warm-air heating + el. boiler, HP for DHW, AHU with HR	17,929
10) warm-air heating +air/air HP, HP for DHW, AHU with HR	15,283
11) pellet stove+accum., warm-air heating, AHU with HR	19,776
12) compact AHU with integrated HP, secondary convector heaters, AHU with HR	19,558

C. *Economic assessment*

The economic assessment of the options is made using basic economic effectiveness criteria, that is NPV (net present value), IRR (internal rate of return), Ts (repayment time (plain), and Tsd (repayment time (discounted).

The basic assessment is made using the criterion of present value of all expenditures for the assessment period, i.e., a period of 15 and 30 years with a 2% price growth and 3% discount rate (corresponding to realistic possibility for a household to appreciate free funds with longer-term zero-risk financial investments; building society account is the primary benchmark). The 15-year period is a time, within the bounds of permissible simplification, in which the heat pump or newer gas boilers or stoves are replaced once and the hot water heater twice.

The economic effectiveness calculation is enhanced with a calculation of the TCO parameter (total costs of ownership, i.e., for the whole building ownership or lifetime), which is made without escalating the prices (e.g., energy prices) and without reflecting the time value of money (i.e., without discounting).

Tables 1 and 2 show the input figures for the calculations; Table 3 shows the calculation results.

III. Methodology: Model Assessment Options

A. *Overall methodology approach*

To maintain continuity, the calculations work with the building from the quoted study [4], which has the following parameters:

The study house is a two-storey building. It has a saddle roof, but the boundary of the heated envelope is the ceiling structure for the unheated attic, which is why the roof structure is shown as transparent in the diagrams. The building has no basement. The plane view is a rectangle, its longer side facing in the SW–NE direction. The comparison calculations work with Option 5, which has the following construction and technical parameters. Walls: Lime and sand bricks 175 mm and PLS 300 mm of grey EPS. Floor: Foam glass 500 mm – a possible equivalent is insulation made of floor polystyrene 200 mm thick on a slab with foundation strips. Roof: Blown cellulose 600 mm. Windows and doors: Triple insulating glass Ug = 0.6 ($W/m^2.K$), frame U = 0.96 ($W/m^2.K$).

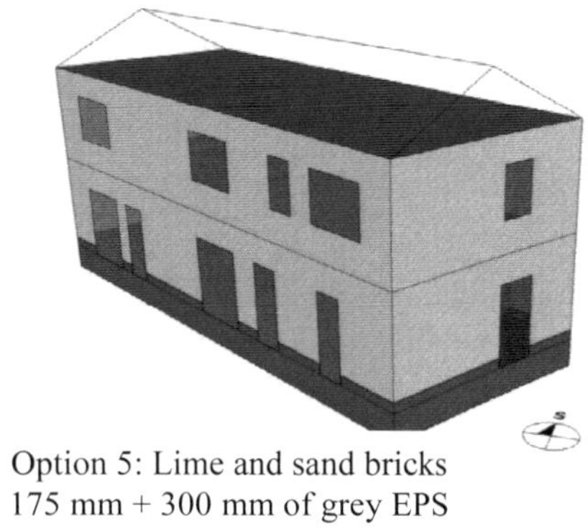

Figure 1. Building model.

To enable the heating system comparison, we assume an energy-saving building at the passive standard level. The heating and hot water production options represent various approaches and typical cases for a house with the design heat loss of 3.1 kW and a hear demand of 20 kWh/m^2.year). The heat recovery efficiency is assumed at 80%, and the reference power input is 0.19 $W/(m^2.hr)$.

The heating and hot water production options are selected and compiled based on economically effective, user-friendly and common solutions.

The study also takes into account the share of non-renewable primary energy, which, pursuant to MoIT Decree no. 78/2013 Coll., has to be lower than or equal to the reference value set for an identical building set to optimum prescribed parameters.

In terms of non-renewable primary energy (NPE), heating in ordinary family houses is ideally by way of heat pumps, natural gas or propane-butane, alternatively wooden pellets or firewood, or environmentally friendly combustion of brown or black coal, which are also low in terms of NPE. The pure use of electricity for heating of ordinary houses can be practically ruled out, since the compensation requires investment in large areas of photovoltaic panels or biomass heating.

The situation is different for energy-saving houses with very low energy intensity, which have a very low heat loss from the heating, frequently below 2 kW. Thanks to the low demand for energy inputs for heating, electricity is an option and the resulting value of NPE is not extreme. Optimisation may even help meet more stringent requirements of subsidy schemes.

All the options assume forced ventilation with heat recovery. To achieve the passive standard, sufficient efficiency of the heat recovery unit requires a maximum air exchange at an underpressure of 50 Pa: n50 ≤ 06 h-1 (so-called building envelope airtightness).

Unless expressly specified otherwise, the prices are assumed excl. VAT. The prices also exclude any additional equipment for compensating for the NPE criterion, since they are highly individual. Assumptions of the assessment options: Heating and hot water production

Option 0: We assume replacement of the convector heaters in 15-18 years, and the domestic hot water boiler in 10 years. The calculation thus assumes CZK 12,000 for the boiler in year 11 and CZK 54,000 for the convector heaters in year 16.

Option 1: Identical to Option 0. We assume replacement of the moving parts of the forced ventilation once every 15 years at approx. 30% of the total costs. The calculation includes CZK 47,000 in year 16.

Option 2: We assume replacement of the electric boiler every 15-18 years and the DHW boiler after 10 years. The calculation includes re-investment of CZK 17,000 and 47,000 for the electric boiler and the AHU parts in year 16 and CZK 12,000 for the DHW boiler in year 11.

Option 3: We assume CZK 54,000 for the convector heaters in year 16. We also assume replacement of the air/water heat pump for DHW at CZK 58,000 and re-investment in the moving parts of the AHU at CZK 47,000 in year 16.

Option 4: We assume replacement of the heat pump at CZK 185,000 and parts of the air-handling units at CZK 47,000 in year 16.

Option 5: Identical to Option 4.

Option 6: We assume replacement of the ground source heat pump at CZK 110,000 and parts of the air-handling units at CZK 47,000 in year 16. The distribution lines, boreholes and their equipment remain.

Option 7: We assume replacement of the gas boiler at CZK 85,000 and parts of the air-handling units at CZK 47,000 in year 16.

Option 8: We assume replacement, in year 16, of all the heat pump units, auxiliary convector heaters

and parts of the AHU, totalling CZK 110,000 + 46,000 + 58,000 + 47,000.

Option 9: We assume replacement if the DHW heat pump (CZK 58,000), the electric boiler (CZK 17,000) and the warm-air heating (CZK 85,000) in year 16.

Option 10: We assume re-investment, in year 16, of CZK 62,000 for the air/air heat pump for heating, CZK 58,000 for the DHW heat pump, CZK 85,000 for the warm-air heating, and CZK 17,000 for the auxiliary convector heaters.

Option 11: We assume re-investment, in year 16, of CZK 77,000 for the pellet stove, CZK 65,000 for the water storage tank and controls, and CZK 85,000 for the warm-air heating.

Option 12: We assume replacement of the compact unit at CZK 170,000 and convector heaters at CZK 54,000 in year 16.

B. Compensation to meet the non-renewable primary energy (NPE) criterion

Some of the designs have to be extended with equipment to reduce the consumption of non-renewable energy and improve the NPE criterion for the building to meet the building permit requirements and, as the case may be, NZÚ subsidy scheme requirements.

The cheapest auxiliary system is stoves or fireplaces fired by wood, pellets or wood briquettes, which differ in the fuel type, combustion method, air supply method, control and stoking and control comfort. Cheaper devices with manual stoking can be bought starting from CZK 12,000, but they typically get air for the burning from the interior and are not suitable for energy-saving (low-energy or passive) buildings, where a separate air intake directly to the combustion chamber is necessary. The separate air intake is typically achieved by a multi-layer chimney or a pipeline in the floor. The other option is less advantageous, since the air entering the combustion chamber is not pre-heated and the intake of external air cools down any adjacent pipelines.

Devices with automated stoking start at CZK 50,000 and may cost over CZK 100,000. A chimney can be had at approximately CZK 22-45 thousand . In any case, higher user comfort in the form of automated stoking requires a total investment of approximately CZK 100-120 thousand . With manual stoking, one can expect probably up to CZK 55-75 thousand . For passive and low-energy homes, the market offers wood-burning stoves with very low rated outputs of 4 kW or 5 kW, and some automated pellet stoves start from 2 kW.

A cheap but less comfortable option is bioethanol fireplaces, which can be bought for CZK 10-30 thousand . The advantage of bioethanol fireplaces is that they can be placed anywhere in the interior, and do not require a chimney. The disadvantage is the relatively expensive fuel, with 1 kWh of heat costing about CZK 16-80, depending on the fuel quality. Moreover, some of the cheaper bioethanol brands have a discernible odour and smell somewhat during operation. A recent option is a range of small-scale bioethanol "mini-heaters", costing CZK 750-900 per device. Bioethanol fireplaces are not eligible for the NZÚ subsidy scheme. These devices are only suitable for buildings with a ventilation system with heat recovery in order to vent off the CO_2.

A somewhat more costly auxiliary supply is photovoltaic panels (PVE), which will cost around CZK 150-300 thousand, excl. VAT. PVE are priced roughly approximately at CZK 45-50 thousand per kWp excl. batteries, and 90-100 thousand per kWp including batteries. The typical size of a PVE system is 2.5-3.5 kWp, sometimes up to 5 kWp. With the cheaper no-battery systems, it is necessary to arrange conditions for grid feed-in, and the normal feed-in tariff for power electricity is very low, approximately CZK 0.50/kWh. The more expensive battery systems make it possible to substitute for a part of own consumption with homemade electricity.

Subsidy under the NZÚ scheme can be drawn in the CR for PVE, but still the rate of return on investment is usually more than 10 years. For more information on options to connect PVE in a household in the CR, see, e.g., Doucha (2016)[10].

IV. Economic Assessment Results

The table summarises the assessment of options by total costs of ownership for the expected ownership period or TCO lifetime and the NPV criterion, including the discount rate. In the 30-year horizon, the best Options are 0, 1, 3, 4, 5, 6, 8 and 10, as shown in the table and chart below. The least economically attractive in the 30-year horizon are Options 2, 7, 9, 11 and 12. Options 1, 3, 8, are 10 more economically favourable after only 15 years. According to our expectations, however, some of them fall short of the NPE criterion. These options require investment of at least CZK 55-75 thousand for a full-fledged biomass heat source with manual stoking. For automated stoking, expected a sum of CZK 100-120 thousand . The extra for meeting the NPE criterion are even higher for PVE, but they allow drawing a subsidy under the NZÚ scheme.

Table 3. Results of economic assessment of options using the TCO and NVP criteria at 15 and 30 years.

Economic assessment	Opt. 0	Opt. 1	Opt. 2	Opt. 3	Opt. 4	Opt. 5	Opt. 6	Opt. 7	Opt. 8	Opt. 9	Opt. 10	Opt. 11	Opt. 12
Total costs over expected lifetime (15 years) = total costs of ownership (TCO) (CZK thousand)	646	717	826	705	810	831	870	866	749	803	759	870	822
TCO comparison against Option 0	0.0%	11.0%	27.9%	9.1%	25.4%	28.6%	34.7%	34.1%	15.9%	24.3%	17.5%	34.7%	27.2%
NPV of expenditures over assessment period (15 years) (CZK thousand)	625	700	805	680	791	812	852	841	729	775	738	844	797
NPV comparison against Option 0	0.0%	12.0%	28.8%	8.8%	26.6%	29.9%	36.3%	34.6%	16.6%	24.0%	18.1%	35.0%	27.5%
Total costs over expected lifetime (30 years) = total costs of	1,212.	1,199.	1,353.	1,125.	1,141.	1,159.	1,185.	1,311.	1,098.	1,277.	1,125.	1,337.	1,254.

ownership (TCO) (CZK thousand)													
TCO comparison against Option 0	0.0%	-1.1%	11.6%	-7.2%	-5.9%	-4.4%	-2.2%	8.2%	-9.4%	5.4%	-7.2%	10.3%	3.5%
NPV of expenditures over assessment period (30 years) (CZK thousand)	1,121	1,164	1,279	1,131	1,222	1,241	1,216	1,291	1,195	1,269	1,190	1,380	1,303
NPV comparison against Option 0	0.0%	3.8%	14.1%	0.9%	9.0%	10.7%	8.5%	15.2%	6.6%	13.2%	6.2%	23.1%	16.2%

As is evident, the TCO and the NPV criteria differ. For an investor, this means that the NPV criterion is more important to them if they can appreciate their money (by at least 3% annually in our case); failing that, TCO is sufficient.

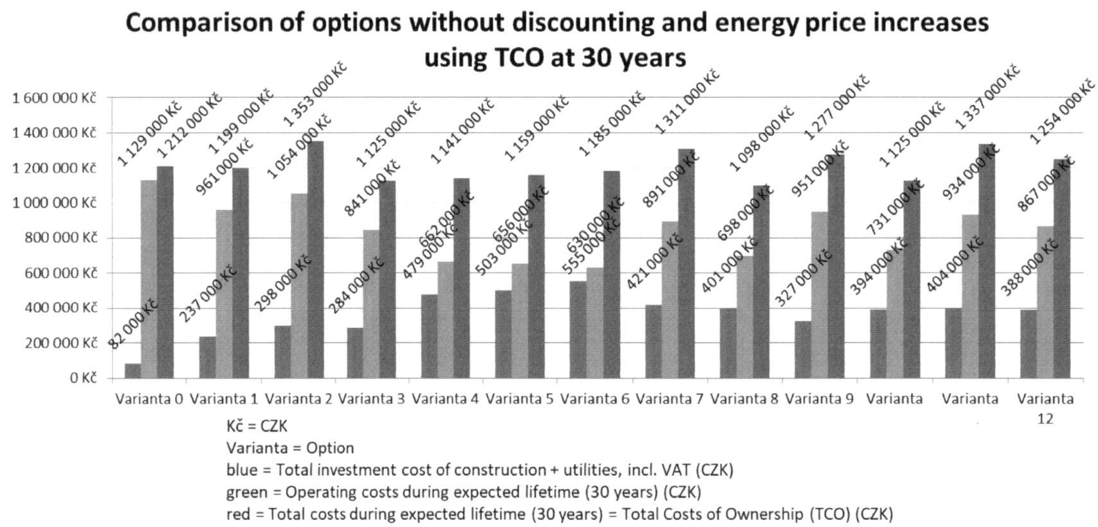

Figure 2. Comparison of options without discounting and energy price increases using TCO at 30 years.

The above comparison using TCO also shows investment and operating costs. It is evident from the chart that the options requiring lower investment tend to have higher operating costs and vice versa.

V. Summary of Options

Option 0 is a reference one, set deliberately as the cheapest solution, relatively practical and pragmatic for energy-saving buildings. Option 0 is admittedly the cheapest, but fails to meet the non-renewable primary energy (NPE) criterion. This is also the case of some of the other options that use mostly electricity for heating, if used in buildings with higher heat losses. Such buildings then fall short of the NPE criterion. For the purposes of building permit proceedings, these options have to have additional renewables, such as a photovoltaic power plant (PVE) or a wood-burning fireplace or stove, or at least a bioethanol fireplace, etc.

Options 1 and 2 meet the NPE criterion under certain circumstances. However, it does not meet NPE criterion of the NZÚ subsidy scheme, which is stricter. To meet that, the design has to be extended with, e.g., biomass heating or a PVE system, or have different domestic hot water production method along with improved building envelope parameters, etc.

Option 3 is at the threshold for meeting the NPE criterion. If the building thermal parameters enable heat demand for heating of approximately 14 kWh/(m^2.year), even such a building may comply with the parameters of NZÚ-B1 subsidy. However, it will not comply with the B2 parameters, and the design has to be modified again to do that.

Options 4-8 meet the building permit proceeding NPE criterion essentially without a problem, and even the B1 level subsidy from NZÚ can be attained; B2 if the construction is improved.

Option 9 is also at the threshold of compliance. If the building has very low energy intensity, it will meet the system requirements for non-renewable primary energy (NPE) for building authority execution approval. However, it will not meet the NPE criterion for the NZÚ subsidy scheme, and the system requires improvement with, e.g., a PVE system, air/air heat pump with a connection to the AHU equipment, and improved building envelope parameters, etc.

Options 10 and 11 meet the building permit proceeding NPE criterion, and even the B1 level subsidy from NZÚ can be attained; B2 if the construction is improved.

Option 12 is again somewhat of a decision-making threshold, since its acceptability depends on the share of heating and hot water production coverage by a heat pump in a compact air-handling unit. If the degree of utilisation of the heat pump is higher, the NPE requirements of the building permit proceeding are met, and a B1 level subsidy from the NZÚ can be attained.

Further concepts can be studied in the following interesting references. For example, the paper by Professor Kabele and Dr. Urban (2017) is interesting; it deals with various concepts of technical systems suitable for a typical single-family house [7]. Another resource for comparison of various family house standards for newly designed homes with near-zero energy consumption is Čejka, Antonín (2017) [8]. Last but not least, there is the case study of various designs for a family home and an apartment building in the near-zero standard by Antonín (2017)[9].

VI. Conclusions

Generally speaking, it can be concluded that among the economically attractive options (0, 1, 3, 4, 5, 6, 8 and 10), the best ones are those that make use of heat pumps (Options 4, 5, 6, 8 and 10), except Option 12, which belongs among the less economic. Options 0, 1 and 3, which are admittedly economically attractive for the time being, do not, or may not comply with the NPE criterion. They will have to include extra expenditures for compliance with the NPE criterion.

For the economically worse-performing options (2, 7, 9, 11 and 12), it can be concluded that the worst ones are those that use electric heating directly and those with little or not share of heat pumps (Options 2, 9 and potentially 12) or use fuels other than electricity. Those are Option 7 with gas and Option 11 with pellets. However, Option 11 can be improved in economic terms with a manual stoking boiler, whereas the options with direct electric heating can be expected to require extra investment in NPE.

Acknowledgement

This paper refers to the quoted publication [6]. The paper is made with the financial support of the State Programme for Support to Energy Savings 2017-2021, EFEKT 2 scheme for 2017. Project title: E-book: Efficient Heating in Family Homes, decision no. [122D 22200 7252], and own funds of EkoWATT z. s.

References

[1] ČTK release, "Podíl nových pasivních domů v ČR loni zůstal na šesti procentech". [online] Published 9.2.2018 | PRAHA (ČTK) Available at https://imaterialy.dumabyt.cz/rubriky/aktuality/prumysl-a-obchod/podil-novych-pasivnich-domu-v-cr-loni-zustal-na-sesti-procentech_46622.html

[2] ČTK. Podíl nových pasivních domů v ČR loni vzrostl na 6,5 procenta. [online] Ekolist editorial article. Published 9.2.2018. Available at https://ekolist.cz/cz/zpravodajstvi/zpravy/podil-novych-pasivnich-domu-v-cr-loni-vzrostl-na-6-5-procenta

[3] Redakce Tzb-info, "Jakým tématům se bude věnovat letošní veletrh FOR ARCH?" [online] Published: 8.9.2018: Available at https://www.tzb-info.cz/veletrh-for-arch/17856-jakym-tematum-se-bude-venovat-letosni-veletrh-for-arch

[4] J. Beranovský, J. Pokorný, "Je úsporný dům opravdu úsporný? Z čeho postavit úsporný dům", [online] EkoWATT z. s., Praha. ISBN: 978-80-87333-10-5. 2014. The paper was made with the financial support of the State Programme of Support to Energy Savings and Use of Renewable Sources of Energy for 2014 – the EFEKT scheme. Available at http://ekowatt.cz/cz/publikace/Je_usporny_dum_opravdu_usporny.

[5] M. Kozelková, J. Mádrová, "Rozpočty pro různé stavební varianty rodinného domu ve Vinoři", Company information of ORIGIS s. r. o., U Strouhy 264/13, 19600 Praha 9-Miškovice, tel.: 725 056 004, (2014) toll-free number: +420 800 884 884, m.kozelkova@origis.cz, j.madrova@origis.cz, http://www.origis.cz.

[6] J. Beranovský, M. Jindrák, V. Bejvlová, " Efektivní vytápění energeticky úsporných domů", [online] EkoWATT z. s., Praha. ISBN: 978-80-87333-14-3. 2017. The paper is made with the financial support of the State Programme for Support to Energy Savings 2017-2021, EFEKT 2 scheme for 2017. Available at http://ekowatt.cz/cz/publikace/.

[7] K. Kabele, M. Urban,. "Vliv koncepce technických systémů na energetické náročnosti rodinného domu", Univerzitní centrum energeticky efektivních budov, ČVUT v Praze. TZB-info information server. 2017 Available at http://stavba.tzb-info.cz/budovy-s-temer-nulovou-spotrebou-energie/16155-vliv-koncepce-technickych-systemu-na-energeticke-narocnosti-rodinneho-domu.

[8] M. Čejka, J. Antonín, "Budovy s téměř nulovou spotřebou – porovnání energetických standardů", [online] PORSENNA o.p.s., EnergySim. TZB-info information server. 2017 Available at http://stavba.tzb-info.cz/budovy-s-temer-nulovou-spotrebou-energie/15181-budovy-s-temer-nulovou-spotrebou-porovnani-energetickych-standardu.

[9] J. Antonín, M. Purkrtová, "Budovy s téměř nulovou spotřebou energie - Případové studie", [online] EnergySim. TZB-info information server. 2017 Available at http://stavba.tzb-info.cz/budovy-s-temer-nulovou-spotrebou-energie/15188-budovy-s-temer-nulovou-spotrebou-energie-pripadove-studie.

[10] P. Doucha, "Možnosti připojení domácí elektrárny v roce 2016". [online] Doucha Šikola advokáti s. r. o. TZB-info information server. 2016 Available at http://oze.tzb-info.cz/fotovoltaika/13918-moznosti-pripojenidomaci-elektrarny-v-roce-2016.

Infrared Monitoring of Industrial Uniterruptible Power Supply Batteries

Tadeusz Sikora, Tomáš Vantuch, Jan Fulneček, Lukáš Prokop, Stanislav Mišák
ENET Centre
VSB – Technical University of Ostrava
Ostrava, Czech Republic
tadeusz.sikora@vsb.cz

Abstract — **Various kinds of industries depend on uninterruptible power supplies (UPS) in a linear way, at least. With an increase in their complexity, the requirements for UPS gain on significance. UPS protect industrial technologies from being damaged by dangerous interruptions of an electric power supply. Advanced UPS monitoring performed by a battery management system (BMS) prevents the UPS from more serious damage due to its timely and accurate battery-fault detection mechanism based on voltage metering. While this advanced technique is widely used due to its high precision, it is also very expensive in the long-term, which often serves as a disadvantage for further applications of battery systems. Our research aims to develop low-cost UPS monitoring making use of infrared thermographic measurements. We assume that overheating of a battery implies its damaged state, which represents the leading factor behind our experiments. The paper brings a review of our current results, as well as shortly describes an extended automatized metering platform. We also review the options in design and development of a deep learning platform making use of the current software technologies and frameworks.**

Keywords: uninterruptible power supply battery; infrared diagnostics; deep learning; neural network

I. Introduction

A typical large scale uninterruptible power supply usually consists of a higher number of elementary batteries connected in series strings or series-parallel groups. Nowadays, most advanced uninterruptible power supplies (UPS) are equipped with a complex battery management system (BMS) in order to maintain their sustainable operability and prevent their degradation. The voltage of each battery is measured and the charging and discharging current of every battery in a string is controlled by balancers. This approach is necessary, especially for Li-Ion batteries which are sensitive to overcharge and deep discharge [1].

In industrial purposed UPS, the lead-acid batteries are most frequently used due to their lower price and higher endurance to overcharge and deep discharge processes [2]. This implies a long payback period for any BMS application and most of the lead-acid based UPS are operated without battery management. Their charging process is controlled only by the measured battery voltage.

Modern BMS evaluates the condition and charge level of the batteries according to their cells voltage level, dis/charging current and the individual cell temperature [3]. When a cell is damaged, overloaded or overcharged, it generates additional heat [4], which underlines the importance of temperature monitoring included as parts of a modern BMS. Temperature sensors must be placed directly on the cell surface, or inside the cell itself for proper function. In most applications, a simple thermistor is used for temperature monitoring. However, there are other types of sensors, developed especially for BMS purposes. Semiconductor CMOS sensors can provide high accuracy and linearity [4]. Flexible micro temperature sensors can be fitted into cells to provide high accuracy measurement with a short response time [5]. Temperature sensors may also be produced by a printed electronics technology [6], which can also positively affect their price. In special applications, an optical fiber can be used for the external and internal cell temperature monitoring [7]. This method is especially suitable in noisy environments.

Figure 1. Typical industrial large scale UPS

II. Typical Battery Failures

Contemporary secondary batteries have a relatively long lifetime in case of a proper care. There are eight typical battery failures in lead acid batteries. Further, we discuss different failure types in the context of heat generation.

Short-circuited cell due to failure of the separator between the positive and negative plates causes local cold strip around the cell bridged by short-circuit. This part of battery cell is no longer charging nor discharging and thus generates less or no heat. Short-circuited cell may be caused by production defects, shed plate material (see below) or too high charging current (high charging currents lead also to heavy wear of positive electrode).

A similar case is a short-circuited cell or cells due to a build-up of shed plate material below the plates. Such failure also generates no heat in damaged cells, or only less heat is generated only on the bottom part of a battery.

Sulfation after a long period of disuse in a low- or no-charge state causes more heat generation on affected cells. Gassing starts at lower voltages at sulfated cells, which is connected with a significant loss in capacity.

Corrosion or damage to the positive and negative terminals heats the vicinity of a terminal and also the terminal joint. Battery terminals are made of lead and thus show no big infrared radiation (temperature) rise due to low lead emissivity, battery case and terminal insulation have typically good emissivity (up to 0,96), so in IR range such failure will be more noticeable.

Broken internal connections as a result of corrosion in most cases cause whole string interruption and, therefore, no heat generation. Single serial string interrupts the whole system. Larger battery packs may be connected serial-parallel and then one branch is lost.

Broken plates due to corrosion and vibration cause a loss in the capacity of a cell. That cell then generates more heat during charging or discharging.

Damage to the battery case causes leakage of electrolyte. If that leakage is small and is in the beginning phase, the leaking electrolyte will cause a local cooler spot due to evaporating of water contained in electrolyte.

Low electrolyte (fluid) level may be caused by battery case damage mentioned above. But in most cases the fluid level is lower due to gassing – excessive electrolysis of water connected with wrong charging method or one of cell failures mentioned above. The lower electrolyte level, the more heat will be generated in remaining fluid.

III. Infrared Monitoring

Infrared monitoring becomes an option as industrial infrared imaging cameras have become cheaper in recent years. According to IR camera resolution, we get several temperature measurement points on a single cell of a battery, instead of one single spot at the contact measurement.

With a single sensor (IR Camera) we are able to monitor a whole battery pack. At once we can measure all 3 temperature related alarms for BMS: temperature difference in the battery within a string of cells, difference between battery and ambient temperature, and absolute battery temperature. For lead-acid batteries the temperature differences should be below 5 °C, absolute battery temperature should not exceed 45 °C [8]

Figure 3. shows the design of a measurement testbed for infrared monitoring of 40 NiCd battery cells. Batteries are installed in a matrix of 4x10 cells. Total nominal battery voltage is 48 V and capacity is 375 A.h (that is 18 kW.h).

Above the battery pack, we installed an industrial infrared camera Magda 640, which is designed for machine vision system. Power supply, communication and control are ensured by a single cable. Camera lens and distance to secondary battery pack is adjusted to get a whole picture of the monitored area.

Figure 2. Infrared camera Magda 640

Figure 3. Experimental design of a measurement testbed

Table 1. Camera parameters

Detector resolution	640x480
Spectral sensitivity	7.5 – 14 μm
Temperature range	High gain (-20 °C to +80 °C) Low gain (0 °C to +550 °C) Auto range selection
Detector type	Uncooled VOx (Vanadium oxide) microbolometer
Pixel pitch	17 μm
Sensitivity	Less than 50 mK @ 20 °C
Consumption	Less than 1.4 W
POE	PoE (nominal 48VDC, 802 3af)

Part of testbed is also monitoring all-in-one PC for data storing, processing and visualization.

IV. Thermal Image Processing

Relevant features from the thermal images must be extracted in order to detect the battery overheat. The thermal image is a color image of pixels whose values belong to RGB color space representing the temperature. Its exact level is based on a given scale defined by the camera adjustment (the scale may be located on the side of the thermal image). It may be seen that the scale consists of many color shades, and, unfortunately, the exact transform from the pixel color to the temperature level is not straightforward, because in most cases the relation between the pixel RGB value and its temperature is non-linear. This can be solved by obtaining an additional software provided by manufacturer with the capacity to estimate the temperature level. Because our motivation is to keep the costs of the final device minimal, we attempted to develop our own temperature level estimation, which, for the sake of simplicity, was based on the assumption that the color values of each pixel may be used as direct input features for our detection algorithm.

On the other hand, our previous experiment did not require exactly precise estimation of the temperature levels and the slightly incorrect estimations was still able to expose the differences among the battery cells, which was still considered as relevant. The relevancy was confirmed by several statistical experiments as well as with k-nearest neighbor algorithm.

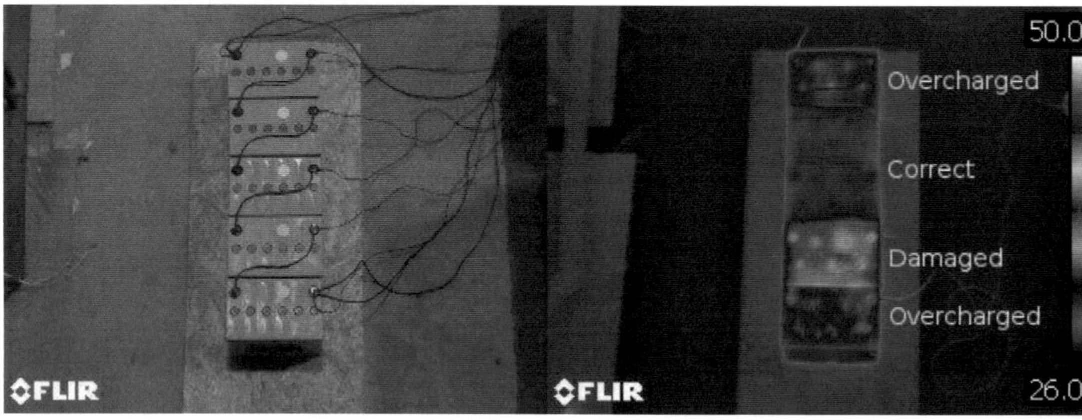

Figure 4. The standard image in RGB space (left); the thermal image with temperature scale on its side (right) [9]

In recent years, several types of feature extraction approaches have been developed and tested for thermal image processing in various fields of applications. The field of fault detection on electrical installation involved the maximum temperature difference between a hot spot and another point in its neighborhood [10]. Similarly, the combination of the spot temperature with the temperature difference of two similar components and the maximum admissible temperature of the component was used in [11, 12], the difference of the maximum and the minimum temperatures in the object, and the surrounding area, respectively, was designed. In [13], the authors used statistical features computed making use of the pixel intensities of the image regions. The considered features were the maximum, minimum, mean and the standard deviation of intensity level. The authors in [14] compared several features that can be extracted from thermal images, such as maximum region temperature, average region temperature, relative temperature difference between target and surrounding regions, or combinations of these features. The combination of those features was mostly considered the most relevant according to the presented results.

V. Image Processing with Deep Learning

Deep learning allows for computational models of multiple processing layers to learn and represent data with multiple levels of abstraction mimicking how the brain perceives multimodal information, thus implicitly captures intrinsic structures of the large-scale data. Deep learning is a rich family of methods, encompassing artificial neural networks, hierarchical probabilistic models, and a variety of unsupervised and supervised feature learning algorithms. The recent surge of interest in deep learning methods is due to the fact that they have been shown to outperform previous state-of the-art techniques in several tasks, especially complex data from different sources (e.g., visual, audio, medical, social, and sensor).

Among the most prominent factors that contributed to the huge boost of deep learning are the appearance of increase in complexity of the solved problems, along with the empowerment of parallel GPU computing, which enabled the transition from CPU-based to GPU-based training, thus allowing for significant acceleration in deep models' training. Additional factors may have played a lesser role as well, such as the alleviation of the vanishing gradient problem owing to the disengagement from saturating activation functions (such as hyperbolic tangent and the logistic function), the proposal of new regularization techniques (e.g. dropout, batch normalization, and data augmentation), and the appearance of powerful frameworks like *TensorFlow* [15], *theano* [16], and *mxnet* [17], which allow for faster prototyping. Deep learning has fueled great strides in a variety of computer vision problems, such as object detection (e.g., [18, 19]), motion tracking (e.g., [20, 21]), action recognition (e.g., [22, 23]), human pose estimation (e.g., [24, 25]), and semantic segmentation (e.g., [26, 27]).

VI. Conclusion

After primary design of infrared aided BMS (displayed in Figure 4.) made of laboratory components, an industrial BMS design was developed. In recent days we are testing its properties and limits.

Further work will focus on teaching infrared BMS machine vision system to identify basic failures and to differentiate its types.

Acknowledgement

This paper was supported by the following projects: LO1404: Sustainable development of ENET Centre; SP2019/159 and SP2019/28 Students Grant Competition and TACR TH02020191 and TACR TH777911, Czech Republic and the Project LTI17023 "Energy Research and Development Information Centre of the Czech Republic" funded by Ministry of Education, Youth and Sports of the Czech Republic, program INTER-EXCELLENCE, subprogram INTER-INFORM and project TN777901 National Centre of Energy.

References

[1] D. Andrea, Management Systems for Large Lithium-Ion Battery Packs; Artech House: Norwood, MA, USA, 2010.

[2] T.P. Crompton, Battery Reference Book; Newnes: Oxford, UK, 2000.

[3] J. Li, Advances in Battery Manufacturing, Service, and Management Systems; John Wiley & Sons: Hoboken, NJ, USA, 2016.

[4] A.A. Pesaran, M. Keyser, Thermal characteristics of selected EV and HEV batteries. In Proceedings of the Sixteenth Annual Battery Conference on Applications and Advances, Long Beach, CA, USA, 12 January 2001; pp. 219–225.

[5] J.B. Robinson, E. Engebretsen, D.P. Finegan, J. Darr, G. Hinds, P.R. Shearing, D.J. Brett, Detection of Internal Defects in Lithium-Ion Batteries Using Lock-in Thermography. ECS Electrochem. Lett. 2015, 4, A106–A109.

[6] J. Grosch, E. Teuber, M. Jank, V. Lorentz, M. März, L. Frey, Device optimization and application study of low cost printed temperature sensor for mobile and stationary battery based Energy Storage Systems. In Proceedings of the 2015 IEEE International Conference on Smart Energy Grid Engineering (SEGE), Oshawa, ON, Canada, 17–19 August 2015; pp. 1–7.

[7] S. Novais, M. Nascimento, L. Grande, M.F. Domingues, P. Antunes, N. Alberto, C. Leitão, R. Oliveira, S. Koch, G.T. Kim et al. Internal and external temperature monitoring of a Li-ion battery with Fiber Bragg grating sensors. Sensors 2016, 16, 1394.

[8] CEI/IEC/TR 62060:2001 Secondary cells and batteries – Monitoring of lead acid stationary batteries – User guide. IEC, Geneva, 2001

[9] T. Vantuch, J. Fulneček, M. Holuša, S. Mišák, J. Vaculík, An examination of thermal features' relevance in the task of battery-fault detection [online]. 2018 [cit. 2019-05-02]. DOI: 10.3390/app8020182. ISSN 2076-3417.

[10] Y.C. Chou, L. Yao, Automatic diagnostic system of electrical equipment using infrared thermography. In Proceedings of the SOCPAR'09 International Conference of Soft Computing and Pattern Recognition, Malacca, Malaysia, 4–7 December 2009; pp. 155–160.

[11] C.E. Prazzo, M. Antunes, A. Turra, J. Pereira, The use of fuzzy logic to classify heating severity of components applied in a thermography based maintenance. In Proceedings of the XVIII Congresso Brasileiro de Automática, Bonito, Brazil, 13–16 September 2010; Volume 12.

[12] J. Lu, C. Zhao, Z. Jiang, S. Guan, Y. Qian, D. Xia, Research of infrared thermal on-line detection technology of zero resistance insulator. In Proceedings of the Cross Strait Quad-Regional Radio Science and Wireless Technology Conference (CSQRWC), Harbin, China, 26–30 July 2011; Volume 2, pp. 1390–1393.

[13] M.M. Ahmed, A. Huda, N.A.M. Isa, Recursive construction of output-context fuzzy systems for the condition monitoring of electrical hotspots based on infrared thermography. Eng. Appl. Artif. Intell. 2015, 39, 120–131.

[14] M.S. Jadin, S. Taib, K.H. Ghazali, Feature extraction and classification for detecting the thermal faults in electrical installations. Measurement 2014, 57, 15–24.

[15] TensorFlow, Available online: https://www.tensorfow.org.

[16] B. Frederic, P. Lamblin, R. Pascanu et al., "Teano: new features and speed improvements," in Deep Learning and Unsupervised Feature Learning NIPS 2012 Workshop, 2012, http://deeplearning.net/sofware/theano/.

[17] Mxnet, Available online: http://mxnet.io.

[18] W. Ouyang, X. Zeng, X. Wang et al., "DeepID-Net: Object Detection with Deformable Part Based Convolutional Neural Networks," IEEE Transactions on Pattern Analysis and Machine Intelligence, vol. 39, no. 7, pp. 1320–1334, 2017.

[19] A. Diba, V. Sharma, A. Pazandeh, H. Pirsiavash, and L. V. Gool, "Weakly Supervised Cascaded Convolutional Networks," in Proceedings of the 2017 IEEE Conference on Computer Vision and Pattern Recognition (CVPR), pp. 5131–5139, Honolulu, HI, July 2017.

[20] N. Doulamis and A. Voulodimos, "FAST-MDL: Fast Adaptive Supervised Training of multi-layered deep learning models for consistent object tracking and classifcation," in Proceedings of the 2016 IEEE International Conference on Imaging Systems and Techniques, IST 2016, pp. 318–323, October 2016.

[21] N. Doulamis, "Adaptable deep learning structures for object labeling/tracking under dynamic visual environments," Multimedia Tools and Applications, pp. 1–39, 2017.

[22] L. Lin, K. Wang, W. Zuo, M. Wang, J. Luo, and L. Zhang, "A deep structured model with radius-margin bound for 3D human activity recognition," International Journal of Computer Vision, vol. 118, no. 2, pp. 256–273, 2016.

[23] S. Cao and R. Nevatia, "Exploring deep learning based solutions in fne grained activity recognition in the wild," in Proceedings of the 2016 23rd International Conference on Pattern Recognition (ICPR), pp. 384–389, Cancun, December 2016.

[24] A. Toshev and C. Szegedy, "DeepPose: Human pose estimation via deep neural networks," in Proceedings of the 27th IEEE Conference on Computer Vision and Pattern Recognition, CVPR 2014, pp. 1653–1660, USA, June 2014.

[25] X. Chen and A. L. Yuille, "Articulated pose estimation by a graphical model with image dependent pairwise relations," in Proceedings of the NIPS, 2014.

[26] H. Noh, S. Hong, and B. Han, "Learning deconvolution network for semantic segmentation," in Proceedings of the 15th IEEE International Conference on Computer Vision, ICCV 2015, pp. 1520–1528, Santiago, Chile, December 2015.

[27] J. Long, E. Shelhamer, and T. Darrell, "Fully convolutional networks for semantic segmentation," in Proceedings of the IEEE Conference on Computer Vision and Pattern Recognition (CVPR '15), pp. 3431–3440, IEEE, Boston, Mass, USA, June 2015.

Power Quality Analysis of Grid Connected Photovoltaic Inverter

Petr Bilik, Radek Martinek, Jan Vanus, Ojan M. Gorjani, Richard Velicka, Ludvik Koval
VSB Technical University of Ostrava
Ostrava, Czech Republic
petr.bilik@vsb.cz, radek.martinek@vsb.cz, jan.vanus@vsb.cz, ojan.majidzadeh.gorjani@vsb.cz, richard.velicka@vsb.cz, ludvik.koval@vsb.cz

***Abstract* – This paper describes the behaviour of photovoltaic power plant inverter firmware on flicker level and reactive power at Point of Common Coupling in low voltage distribution system. During routine measurement of PQ the photovoltaic power plant was suspicious as a source of flicker in the distribution system. During PQ measurement using the common PQ analyser the distribution system operator could not identified the reason of high flicker. Using a tailored power quality analyser authors identified pulsed active power delivery during low sun radiation. Much smooth active power delivery was identified after the firmware upgrade.**

I. Introduction

The flickermeter function is defined in international standard IEC 61000-4-15 [1]. Flicker is caused by time varying voltage drops on the impedance of the distribution system (DS). The flicker phenomenon is connected with DS short circuit impedance, steadiness of sources and steadiness of loads [2]. The flicker level depends on the level and frequency of voltage variations and it is one of several parameters describing Power Quality (PQ) [3], [4]. Steadiness of electrical power is problematic for the photovoltaic power plant (PVPP) and wind turbines because of instability of the primary energy sources: sunshine and wind [5]. The aforementioned power sources bring new issues by means of deteriorative PQ [6], [7], [8], [9]. There are some new approaches to evaluating the flicker level e.g. [10], [11] but for the purpose of this paper, the traditional approach [1] was used.

Short-time flicker (P_{st}), long-time flicker (P_{lt}) and active power were recorded over several days using an off-the-shelf power quality analyzer, see Fig. 1 and Fig.2. In all phases, the flicker and active power waveform were very similar, so the quantity waveforms for the L1 phase are shown only for clarity reasons. The 3 phases inverter rated power was 20kW.

II. Measurement Using The Common PQ Analyzer

It is obvious from the Fig.1 that the permitted P_{lt} value is repeatedly significantly exceeded in specific parts of the day, both in the morning and in the evening. All the measurements were always performed on the AC side of the inverter. The standard PQ analyzer recorded an active power waveform with an aggregation interval of 10 minutes as well as 200ms.

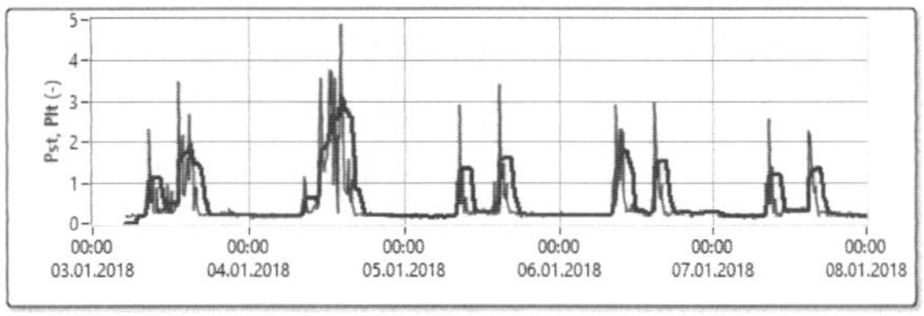

Figure 1. The P_{st}, P_{lt} waveform for a 5-day interval.

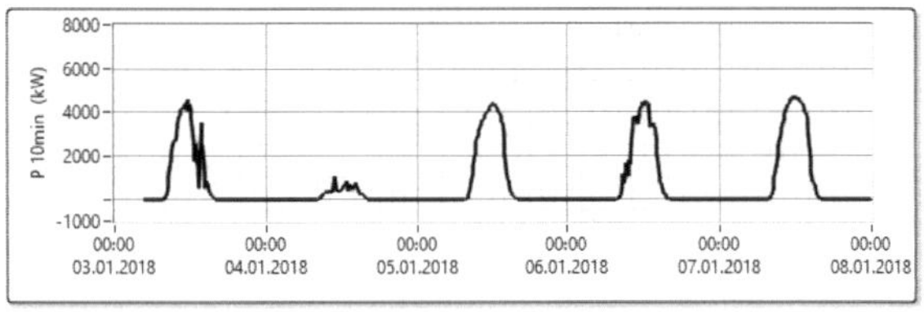

Figure 2. The inverter active power P_{L1} waveform for a 5-day interval, the aggregation interval of 10 min.

The waveform of the active power was almost identical for both aggregation intervals and, at the same time, the shape was typical of the PVPP without any signs of power fluctuation. On the inverter active power P_{L1} waveform with 10 minutes aggregation interval there are no suspicious artefacts. Thus, the cause of the high flicker value is not obvious from the active power waveforms.

III. Tailored PQ Analyzer

A PC-based PQ analyzer was used for the measurement of all the phenomena described in this paper. The authors used hardware of commercial PQ analyzer DEWE 561, which contains the following main blocks: a PC running a Windows operating system, voltage and current inputs signal conditioning and the data acquisition board with sampling rate up to 1MS/s. PQ analyzer DEWE 561 has direct voltage inputs and CA MN39 current clamps (two ranges 20/200 A) were used as a current sensor.

The common PQ analyzer functionality is defined in IEC 61000-4-30 [12]. Special analyzer firmware has been developed to detect the potential cause of a high flicker value due to very fast power changes. LabVIEW graphical programming language using the Electrical Power Toolkit library was used as a development environment to make an instrument prototype. In addition to the 10 min and 200 ms aggregation intervals, the firmware developed can record active power for an aggregation interval of 10 ms (synchronized half a period of 50 Hz fundamental frequency) and an instantaneous flicker for an aggregation interval of 20 ms.

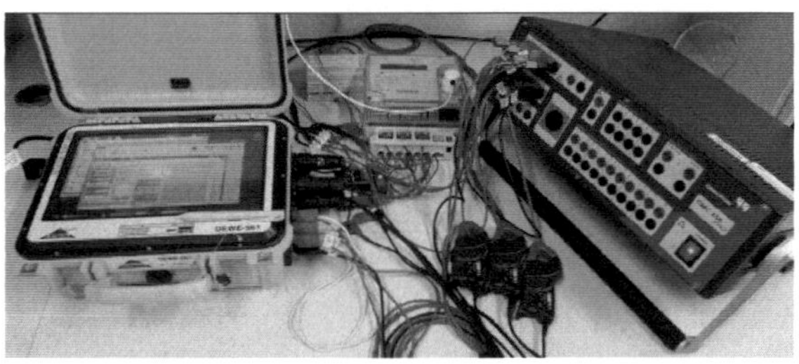

Figure 3. PQ analyzer DEWE561 and calibrator OMICRON 256+

The instantaneous voltage and current values, from which the value of the instantaneous active power is calculated, are the basis for the calculations in the firmware.

$$p(n) = u(n) \cdot i(n) \quad (1)$$

The firmware of the analyzer, once within the time window T_W, calculates the mean value of the active power (2).

$$P = \frac{1}{N}\sum_0^N p(n) = \frac{1}{N}\sum_0^N u(n) \cdot i(n) \quad (2)$$

In principle, the shortest time window $T_{W,}$ for which the active power mean value can be calculated, is equal to half of the period of the fundamental harmonic. From the mean active power, the energy value for T_w is calculated.

$$AP_{Tw} = Tw \cdot P = Tw \cdot \frac{1}{N}\sum_0^N p(n) \quad (3)$$

Reactive power is evaluated for each time window Tw as well as active power. The reactive component is the product of the voltage component with the current component perpendicular to the voltage. There are several methods to calculate reactive power:

- Calculation of instantaneous reactive power by product of current samples and phase shifted voltage samples.
- Calculation using the power triangle, the reactive power absolute value is calculated from P and S, but the result does not include the Q sign.
- Calculation of amplitude and phase spectrum of voltages and currents (FFT). This is a processor-intensive method, used mainly in precision performance analyzers.

Methods 1 and 2 do not produce correct results for non-sinusoidal waveforms (the error depends on the harmonic content). Methods 1, 2 and 3 are sensitive to fundamental harmonic frequency variation (but this can be eliminated by a suitable method).

The method most commonly used in electrometers or analyzers with a low-power processor is described further. The reactive power evaluation Q can be performed by moving the instantaneous voltage values by +90° in software phase, obtaining the value u_{90}, multiplying the shifted voltage u_{90} by the instantaneous current value, thereby obtaining an instantaneous reactive power value. Subsequently, the mean reactive power value is calculated.

$$Q = \frac{1}{N}\sum_0^N u(n) \cdot i(n) \cdot \sin(\varphi) = \frac{1}{N}\sum_0^N u_{90}(n) \cdot i(n) \quad (4)$$

To calculate the phase-shifted voltage, it is necessary to accurately evaluate the frequency of the network, which usually changes over time. If the network frequency is known, the period of time T can be calculated and from this data it is already known how many samples the voltage signal needs to be delayed to obtain a 90° phase shift. To calculate the exact frequency, the so-called "Zero-Crossing" method is often used, where the location on the timeline is determined, when the voltage changes polarity, or crosses the value of 0V. Reactive energy is calculated similarly to active energy by integrating power over time.

$$AQ_{Tw} = Tw \cdot Q = Tw \cdot \frac{1}{N}\sum_0^N u_{90}(n) \cdot i(n) \quad (5)$$

The flickermeter [1] provides two quantities: each 10 minutes, short time flicker P_{st}, each 2 hours, long time flicker P_{lt} [12], for regular use. According to the functional diagram of the IEC flickermeter [1], the instantaneous flicker P_{inst} is provided on the output of module 4 and P_{st} and P_{lt} are provided on the output of module 5. Module 5 provides statistical analysis of P_{inst}. Commercial flickermeters do not provide P_{inst}, as it is an internal quantity, so it is not used for PQ evaluation. The PQ analyzer firmware provides an instantaneous flicker for the purposes of flicker source identification and correlation of the flicker level with rapid changes of the active power.

Firmware measures developed and records following quantities:

- Voltage, current, active and reactive power with a 10 ms aggregation interval (each half-period of signals with a fundamental frequency of 50 Hz).
- Instantaneous flicker p_{inst} with 20 ms aggregation interval (each period of signals with a fundamental frequency of 50 Hz), flicker P_{st} (10 min) and P_{lt} (2 h).
- Oscilloscope waveforms of voltage and current can be recorded on demand with a sampling frequency of 9600 samples per second.
- Recording of all the above-mentioned quantities (except P_{st} and P_{lt}) is highly demanding in terms of hard drive space, but, on other hand, this way of data recording provides a very detailed view on the electrical quantities measured. The functionality of the firmware and the precision of the entire PQ analyzer were checked by an OMICRON CMC 256+ calibrator, see Fig.3.

IV. Measurement Results Using 10ms Time Window

On following figures Fig. 5 and Fig. 6 can be seen significantly different waveforms of active and reactive power for 10 ms aggregation interval in comparison with Fig. 2 where 10 min aggregation interval was used.

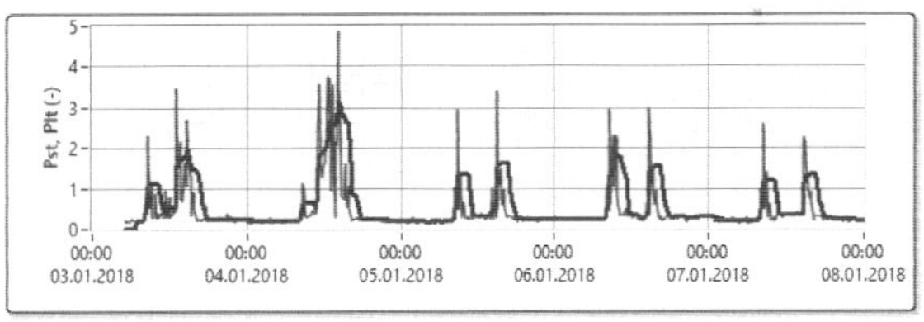

Figure 4. The P_{st}, P_{lt} waveform for a 5-day interval.

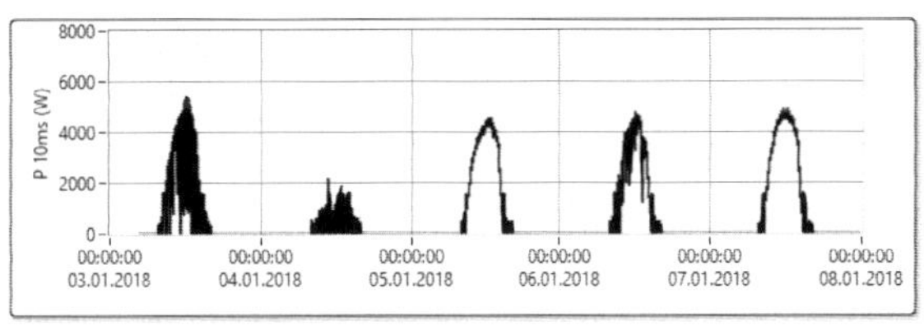

Figure 5. The inverter active power P_{L1} waveform for a 5-day interval, the aggregation interval of 10 ms.

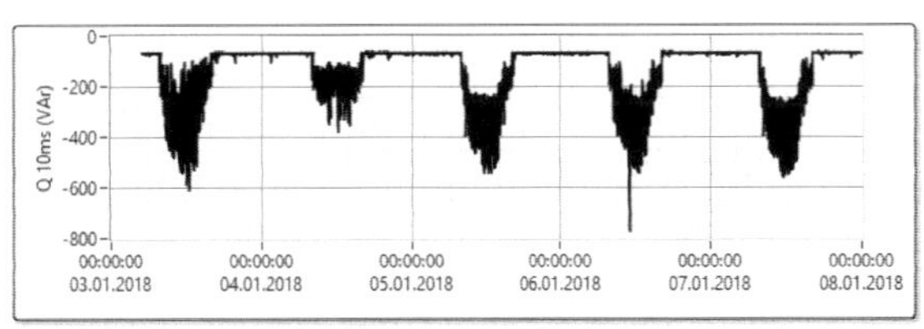

Figure 6. The inverter reactive power Q_{L1} waveform for a 5-day interval, the aggregation interval of 10 ms.

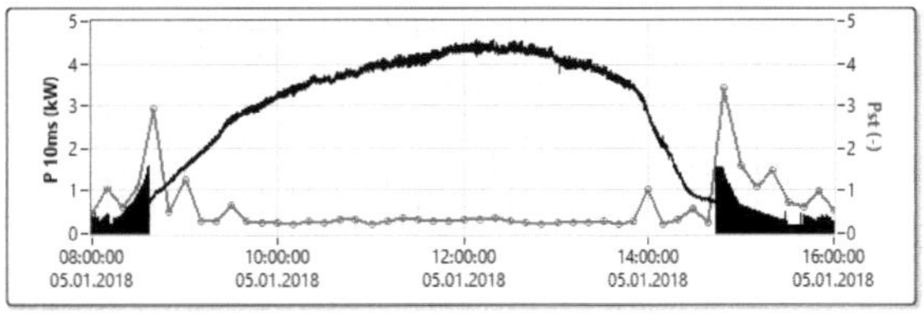

Figure 7. The inverter active power P_{L1} and P_{st} waveform for an 8-hour interval, the P_{L1} aggregation interval of 10 ms

Both active power and reactive power waveform for an 5-day interval recorded at an aggregation interval of 10 ms (see Fig. 5 and Fig. 6), detected a rapid power fluctuation in the morning and in the late afternoon, i.e. at times when the sunshine and PVPP active power was low. The details from Fig. 5 can be seen in Fig. 7 for a better understanding. Fig. 8 shows even smaller detail and shows the phenomenon of increased flicker even the amplitude of changes remains the same. The active power fluctuation has the same peak-peak amplitude, but different frequency. The flicker level depends on the voltage fluctuation amplitude and frequency.

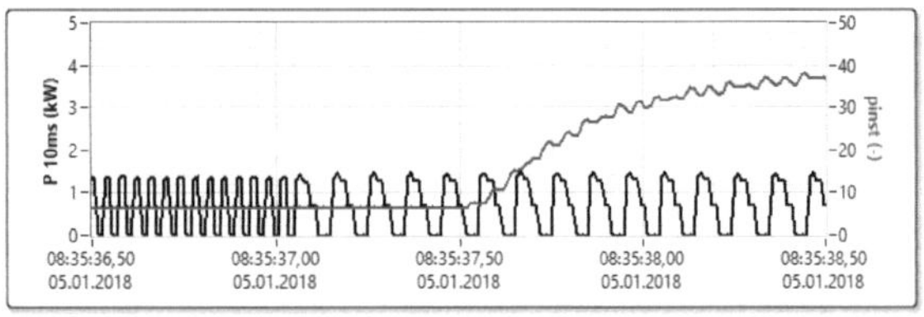

Figure 8. The inverter active power P and p_{inst} for a 2-sec interval, the aggregation interval of 10 ms

The active energy AP can be calculated by integration of active power over the time interval (3), by the same approach can be calculated reactive energy AQ (5). The average power factor PF is calculated from AP and AQ.

$$PF_{avg} = \frac{AP}{\sqrt{AP^2+AQ^2}} \qquad (6)$$

For the time interval (5 days) shown on Fig. 5 was the active energy AP_{L1}=82,704 kWh and reactive energy AQ_{L1}= -16,958 kWh. The power factor PF calculated by equation (6) was PF=0,979.

V. Inverter Behaviour

It is obvious that the high level of flicker is caused by rapid changes of the active power generated by the inverter (see Fig. 8). The frequency of the active power changes is too high to be caused by sunshine changes. There is exact relation between the level, the frequency of the active power changes and the level of the instantaneous flicker presented in Fig. 8, even the flicker is caused by the contribution of other loads in DS also, and the active power curve shows power from the PVPP inverter only. The improper behaviour of the PVPP inverter is this: at low power of the PV panels, instead of constant active power delivery, it deliver pulsed active power with a varying pulse frequency.

VI. Results After Inverter Upgrade

The firmware of inverter was upgraded to a improved version and then the measurement was repeated. The sunshine intensity and corresponding active power distribution during the days was significantly different in comparison with the first measurement, compare Fig. 5 and Fig. 10. From the measurement results (Fig. 10), it is obvious that even though the active power is higher in certain days than during first measurement (Fig. 5), the flicker level is much lower in

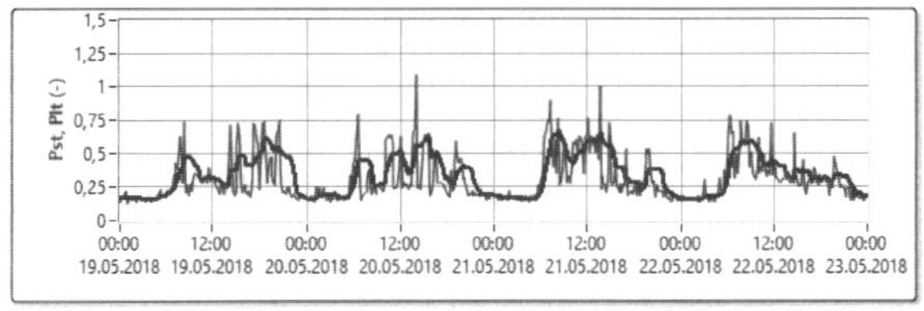

Figure 9. The P_{st}, P_{lt} waveform for a 4-day interval.

comparison with the situation before the inverter firmware upgrade. Figure 12 visualises the rapid power fluctuation in the early morning hours and in the evening again. In comparison with the active power fluctuation during low sunshine before the inverter firmware upgrade (Fig. 7), the behaviour of the inverter upgraded is much better: even the active power fluctuation during low sunshine unfortunately remains, the amplitude of fluctuation is much lower. As the consequence the flicker severity is much lower during low sunshine, compare Fig. 4 and Fig. 9, Fig. 7 and Fig. 12.

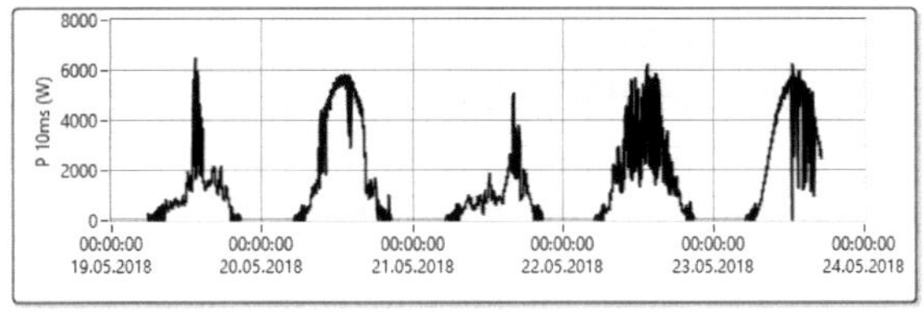

Figure 10. The inverter active power P_{L1} waveform for a 5-day interval, the aggregation interval of 10 ms.

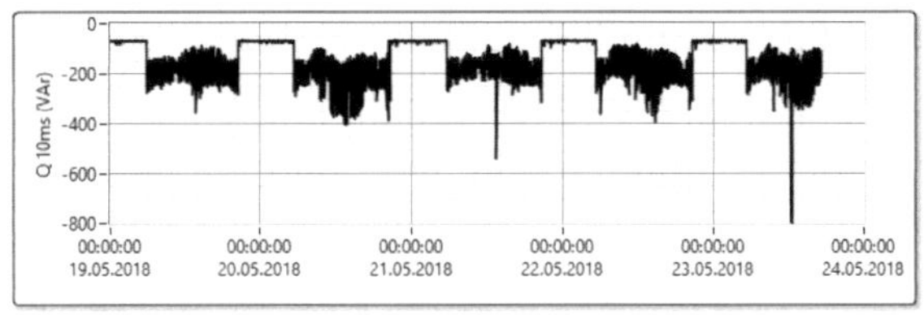

Figure 11. The inverter reactive power Q_{L1} waveform for a 5-day interval, the aggregation interval of 10 ms.

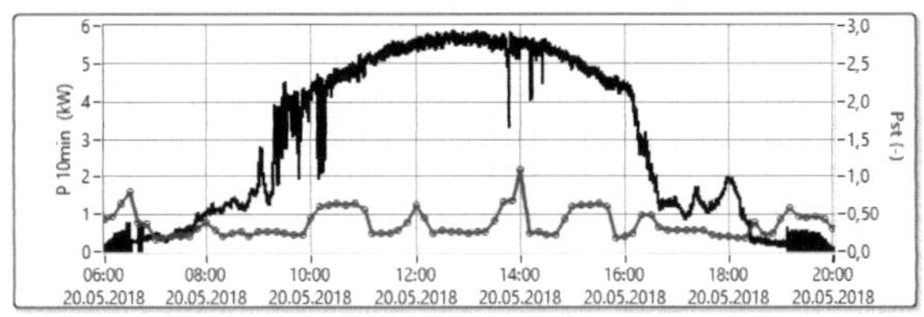

Figure 12. The inverter active power P_{L1} and P_{st} waveform for a 14-hour interval, the P_{L1} aggregation interval of 10 ms

For the time interval (5 days) shown on Fig. 10 is the active energy AP_{L1}=146,861 kWh and reactive energy AQ_{L1}= -16,076 kWh. The power factor PF calculated by equation (6) is PF=0,994.

VII. Conclusion

The measurement of the active power with an aggregation interval of 200 ms or even an aggregation interval of 10 minutes cannot detect rapid changes in the active power. Rapid changes are not noticeable as they are "averaged" by a long aggregation interval. Regular PQ analyzers detect high flicker, but without the possibility to conclusively identify the cause. A measurement conducted by the PQ analyzer developed with a short aggregation time of half period of the 50 Hz fundamental frequency (10ms) can record the rapid active power amplitude and sign changes. no international standard focused on the PQ requires a PQ analyzer to identify the causes of an improper VQ.

Acknowledgement

This work was supported by the Student Grant System of VSB Technical University of Ostrava, grant number SP2019/118.

References

[1] IEC 61000-4-15 ed2.0. Electromagnetic compatibility (EMC): Part 4-15: Testing and measurement techniques – Flickermeter – Functional design specifications. Geneva : International Electrotechnical Commission, 2010. 83 p. ISBN 978-2-88912-076-5.

[2] T. Spielböck and A. Beláň, "Design of a one-sided, impedance-based transmission line fault locator using line topology and source impedances," Electric Power Systems Research. vol. 161, 123-138. ISSN 0378-7796.

[3] M. Regula, A. Otcenasova and M. Roch, "SCADA system with power quality monitoring in Smart Grid model," Proceedings of IEEE 16th International Conference on Environment and Electrical Engineering (EEEIC), Florence, Italy, ISBN: 978-1-5090-2320-2

[4] M. Repak, A. Otcenasova and M. Regula, "Design of power quality analyzer," In: 2018 ELEKTRO [online]. IEEE, 2018, 2018, s. 1-6, DOI: 10.1109/EEEIC.2016.7555685. ISBN 978-1-5386-4759-2.

[5] M. Repak, A. Otcenasova, M. Regula and R. Bodnar, "Design of power converter for model of photovoltaic power plant," In: 2016 ELEKTRO [online]. IEEE, 2016, 2016, s. 359-363. DOI: 10.1109/ELEKTRO.2016.7512097. ISBN 978-1-4673-8698-2.

[6] V. Kus and T. Josefova,"Education strategy regarding the electromagnetic compatibility at low-frequency," Revue Roumaine des Sciences Techniques Serie Electrotechnique et Energetique, vol. 62, 48-52, ISSN: 00354066

[7] J. Dabrowski," Analysis of long-time efficiency of photovoltaic installation," PRZEGLĄD ELEKTROTECHNICZNY [online]. 2017, 1(2), 204-207 [cit. 2019-04-14]. DOI: 10.15199/48.2017.02.44. ISSN 0033-2097. Dostupné z: http://sigma-not.pl/publikacja-103822-2017-2.html

[8] K. Górecki and M. Szmajda, "The power quality in low-power solar off-grid system," Proceedings of 16th International Conference on Harmonics and Quality of Power, ICHQP, 244-248, ISBN: 978-146736487-4

[9] D. Chathurangi, U. Jayatunga, M. Rathnayake, A. Wickramasinghe, A. Agalgaonkar and S. Perera, "Potential power quality impacts on LV distribution networks with high penetration levels of solar PV," Proceedings of 18th International Conference on Harmonics and Quality of Power, ICHQP, Ljubljana, Slovenia, ISBN: 978-153860517-2

[10] W. Yao, Q. Tang, Z. Teng, Y. Gao and H. Wen, "Fast s-transform for time-varying voltage flicker analysis," IEEE Transactions on Instrumentation and Measurement, vol. 63, 72-79, ISSN: 00189456

[11] J. Drapela, R. Langella, J. Slezingr and A. Testa, "A Tunable Flickermeter to Account for Different Lamp Technologies," IEEE Transactions on Power Delivery [online]. 2017, 32(2), 872-880 [cit. 2019-04-14]. DOI: 10.1109/TPWRD.2016.2537699. ISSN 0885-8977.

[12] IEC 61000-4-30 Ed.2.0 Electromagnetic compatibility (EMC): Part 4-30: Testing and measurement techniques – Power quality measurement methods. Geneva : International Electrotechnical Commission, 2008. 134 p. ISBN 2-8318-1002-0.

Research of Hydrogen Fuel Cells Based on PEM Technology

Petr Moldrik
Department of Electric Power Engineering
VSB – Technical University of Ostrava
Ostrava, Czech republic
petr.moldrik@vsb.cz

Daniel Minarik, Artem Chesalkin
ENET Centre
VSB – Technical University of Ostrava
Ostrava, Czech republic
daniel.minarik@vsb.cz, artem.chesalkin@vsb.cz

Abstract — **This paper deals with the technology subject to current dynamic development - hydrogen fuel cells. There are several types of fuel cells, differentiated with respect to design, materials used, operation parameters and relevant method of use. The emphasis is laid on low-temperature hydrogen proton membrane fuel cells (PEMFC), which produce the electric power using pure hydrogen and gaseous oxygen. These are utilised in various stationary as well as mobile assemblies. The main part of this paper concerns the results of laboratory testing of the PEM module with fuel cells in order to optimise its operation. This paper provides description of its operation characteristics. The module subject to research is currently integrated within the hybrid energetic system to provide for accumulation of electric power produced by photovoltaic panels and it is capable to supply the generated electric power into a low-voltage local power grid via a coupling converter. The matter of accumulation of energy made from unstable renewable resources in order to improve their reliability has gained on crucial importance, while the number of units constructed (installed capacity), is rising. Hydrogen technologies seem a very progressive solution for energy accumulation purposes.**

Keywords: hydrogen; fuel cells; PEMFC; hybrid energetic system; operation characteristics

I. Introduction

Energetic systems with closed hydrogen cycle can be used to store electricity from renewable resources (solar and wind power plants). The hydrogen systems are able to eliminate fluctuations of the power output of these unstable resources. There are numerous hybrid power systems in the world, with on-grid or off-grid configurations. The laboratory hybrid energy system using hydrogen gas as the storage medium was also built at Technology Centre of Ostrava, VSB - Technical University of Ostrava, Czech republic [1].

II. Hydrogen Storage System with PEM Fuel Cells

Researched modules of hydrogen fuel cells are a part of the above mentioned hybrid energy system used for storage of solar energy. Its block diagram is shown on Fig. 1. This system, based on the hydrogen technology, ensures storage of electricity gained from the photovoltaic panels and its subsequent utilisation. The installed capacity of photovoltaic panels is 22.5 kWp. The storage of surplus electricity is ensured by hydrogen, which is produced by water electrolysis. The gasious hydrogen is stored in pressure vessels. Shortages

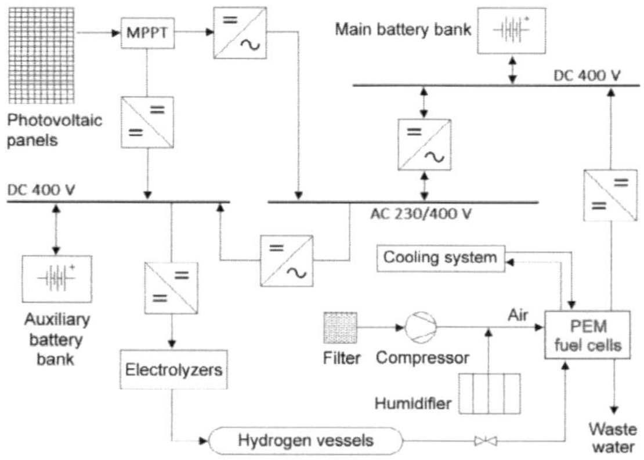

Figure 1. Block diagram of a hybrid energy system with researched hydrogen fuel cells

of electricity gained from photovoltaic panels are then covered by the electricity generated by modules of fuel cells. These modules consume stored hydrogen from vessels and oxygen from the air. The basic parameters of one module of fuel cells (type Nedstack FCS 8-XXL) are shown in Table 1.

Described hydrogen storage system is designed to operate four electrolyzers (i.e. hydrogen generators) with total input power of 17 kW with total hydrogen production of 4 m^3/h. The hybrid system also consists of lead traction batteries (main battery bank) with the total capacity of 930 Ah and a set of batteries type LiFePO (auxilliary battery bank) with the total capacity of 100 Ah. The accumulation with lead batteries enables storage of almost 500 kWh of power, with further 60 kWh that can be stored in LiFePO batteries [2]. These batteries accumulate energy for the photovoltaic panels and stabilize and optimize energy for electrolyzers. The linkage between DC and AC buses is ensure by means of solid-state power converters accompanied by the coupling transformers enable two-way transmission of power between buses. To secure correct function of hydrogen fuel cells and electrolyzers, other supporting operation technologies were created (gas mains, control and measuring armatures, cooling and humidification system).

Table 1. Basic Parameters of Measured Hydrogen Module [3]

Fuel cell count	64
Maximum power	8 kW_e
Maximum current	230 A
Operating voltage	61.9–34.8 V
Maximum H_2 consumption	102 Nl/min
Air humidification	80 % RH at 65 °C
Air consumption	488 Nl/min

Figure 2. Measured module of PEM fuel cells Nedstack FCS 8-XXL

Hydrogen fuel cells use the electrochemical reactions to transform chemical energy in the fuel (hydrogen) and oxidant to electricity, heat and water. This transformation is based on reversed principle of water electrolysis and occurs within catalytic reactions on both electrodes. The measured low-temperature module of hydrogen PEM (Proton Exchange Membrane) fuel cells, together with the control and measurement systems, forms a production unit with a rated electrical power of 8 kW. Fig. 2 shows its photo.

This module is made up of MEA (Membrane Electrode Assembly) structure that includes the end plates with positive and negative electrode (EP - End Plate), the plates with reaction gases distribution channels (FFP - Flow Field Plate / BP - Bipolar Plates), the catalyst and PEM membrane, where chemical reactions take place. The PEM membrane is formed by a 50–200 μm thick rigid film "Nafion" (perfluorinated polymer of sulphonic acid) covered with a layer of Platinum catalyser [4]. This membrane is a solid electrolyte that increases operational safety and the life span of fuel cells and reduces their corrosion. Cathodes and anodes created by paper with carbon fibres are porous and allow for transmission of reaction gases (hydrogen and oxidant). The flow field plates provide for minimising the contact of gas with the membrane assembly. These plates are made from graphite. The end plates with electrodes are made from steel.

Fig. 3 shows construction of one hydrogen PEM fuel cell with the following seven primary layers:

- PEM (Proton Exchange Membrane) - This membrane is only permeable to positive ions (H^+ protons).
- FP (Flow plates) - Fluid enters and exits the fuel cell through channels in these plates.
- GDL (Gas diffusion layers) - Fluid spreads through these layers.
- CL (Catalyst layers) - Fluid reacts in these layers according to the following electrochemical equations:

$$\text{Anode: } 2\,H_2 \Rightarrow 4\,H^+ + 4\,e^-$$

$$\text{Cathode: } O_2 + 4\,e^- + 4\,H^+ \Rightarrow 2\,H_2O$$

$$\text{Overall reaction: } 2\,H_2 + O_2 \Rightarrow 2\,H_2O$$

PEM fuel cell is provided with a simultaneous supply of the fuel gas (hydrogen in form of molecules H_2 on the anode side) and the oxidising agent (oxygen in form of molecules O_2 or air on the cathode side). The contact of H_2 hydrogen molecules with the platinum catalyser induces a reaction on the surface of proton membrane, whereas the hydrogen molecules decompose to individual atoms of H first, to be further degraded to H^+ protons and e^- electrons. The electrons are subject to the external electric load and they are received by the oxygen atoms on the cathode side to form the O^{2-} ion, which has been formed upon degradation of the O_2 molecules using platinum catalyser. A polymer membrane is able to conduct protons H^+ from anode to cathode. A single hydrogen fuel cell produces about 1.0 V at open-circuit and about 0.55 V at full power. The fuel is a hydrogen (with purity of 99.99 %) at a pressure of 0.3 bar. Hydrogen PEM fuel cells are sensitive to the sulphur compounds and the carbon monooxide in fuel. Oxygen is taken from humidified ambient air. Water produces by fuel cells is collected in the tank and is no longer used. The hydrogen modules are water-cooled [6].

III. Laboratory Testing of The Module of Fuel Cells

Our research of the experimental storage systems in the Laboratory of hydrogen technologies at VŠB-TU Ostrava focuses on investigating the realistic operating characteristics, properties and parameters of the above-mentioned low-temperature module of hydrogen fuel cells Nedstack FCS 8-XXL, which is the main constituent element of the storage system working on the principle of a closed hydrogen cycle.

A. *Basic operating characteristics*

Our measured load characteristics of the hydrogen module Nedstack FCS 8-XXL can be seen on Fig. 4. This V-A characteristic is significantly soft, that means the full load reduces its voltage to the half of open-circuit voltage (approx. 35–62 V). This undesirable condition (for the load) must be eliminated by a sufficiently broad range of the input voltage in the inverter or the binding DC/DC converter employed [7].

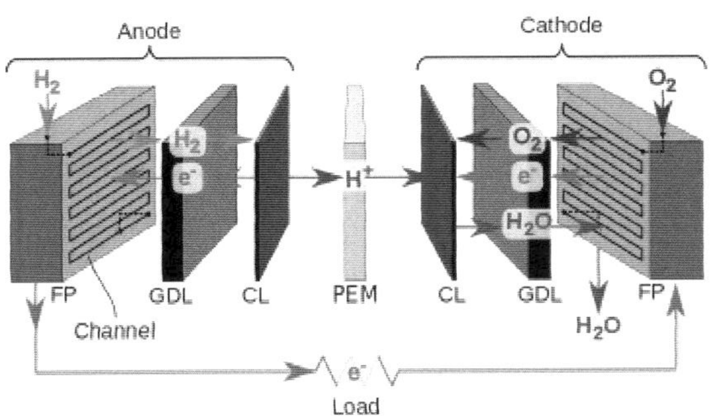

Figure 3. Construction of hydrogen PEM fuel cell [5]

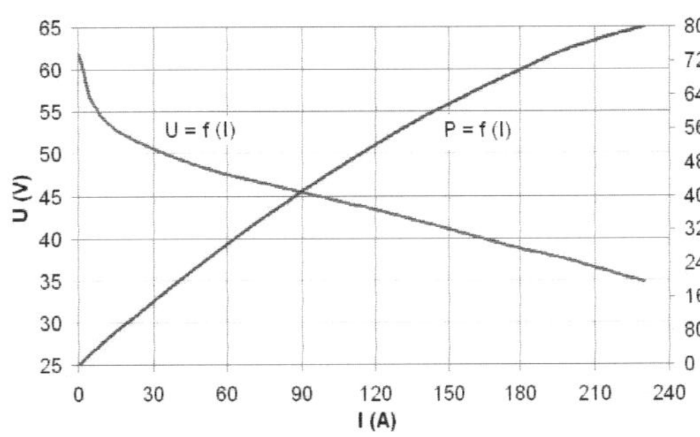

Figure 4. Load characteristics of researched module of fuel cells

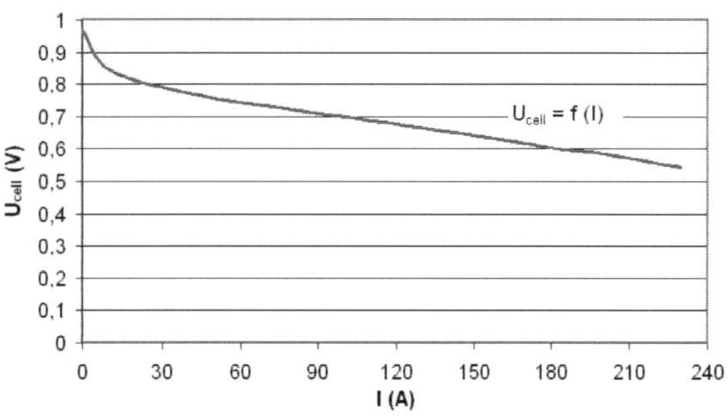

Figure 5. Voltage dependancy of one fuel cell on its current load

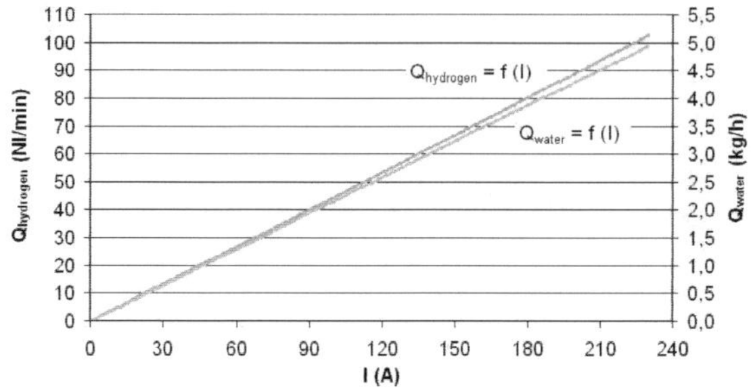

Figure 7. The dependency of the consumed hydrogen and the amount of produced water on the current load of the module

Due to the fact, that the measured hydrogen module consists of a total of 64 series-connected fuel cells, the voltage course shown on Fig. 4 is given by the sum of the individual voltages of these cells. A typical course of one PEM fuel cell can be seen on Fig. 5. The values of this voltage depend on the current load from ca 0.5 to 1 V (During maximum load, or during idle state).

A significant part of the energy contained in the fuel (hydrogen) and the oxidizer will be transformed into heat due to the electrochemical reactions taking place in the fuel cells during the running of hydrogen PEM [8]. The module Nedstack FCS 8-XXL is then basically a micro-cogeneration unit, which during full load (8 kW_e) produces a heat output of ca. 10 kW_t. Its dependancy on the current load is shown on Fig. 6. The operating temperature of the module can't be higher than 70 °C during maxmimum load. The water cooling system ensures that the temperature is maintained at the temperatue 65 °C. By way of heat exchangers, this output heat is used for the heating of DHW (Domestic hot water) at the whole research center in which this hydrogen storage system is running.

The next parameters, which we measured in testing the hydrogen module, was the amount of consumed fuel (hydrogen), oxidizer (air and oxygen) and the amount of produced water. The water is released at the cathode outlet in both as vapour and as liquid together with the used air. In the consumed gases we observed the dependency of the the consumption per minute (in NL/min) on the current load of the module (see Fig. 7 and Fig. 8). As can be seen from the measured values, the consumption of the oxidizer is in comparison with hydrogen consumption almost five-fold. In the case of water, we were interested in the amount that was produced per hour of operation of the module (in kg/h) during a specific load (see Fig. 7). These four dependancies are linear. Not only hydrogen, but even the oxidizer supplied to the tested

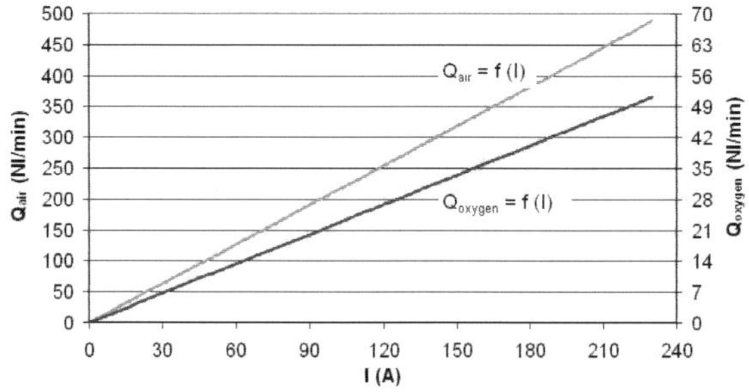

Figure 8. The dependency of the amount of consumed air and oxygen on the current load of the module

module has to comply with specific quality standards. Apart from filtration of dust particles, it is necessary to maintain the concentration of carbon monoxide and sulphur on a maximum (allowed) value of 25 ppm, or 0.01 ppm.

In considering the efficiency of the cooling system and its possible optimalization, it is necessary to know the temperature layout inside the module during its partial or full load. The course of the temperature along the entire length of the module, which means depending on the number of order of the individual fuel cells (1–64) is shown on Fig. 9. The measuring was performed in the ambient temperature 20 °C.

Fig. 10 shows the entire course of the temperature for various power loads of the module. For the presentation of these courses, the fuel elements with the numbers of order 1, 5, 15, 25, 35, 45, 55 a 64 were selected. The direction of this referencing is from the side of the electric outlet to the side of the input of media (cooling water, fuel, and oxidizer).

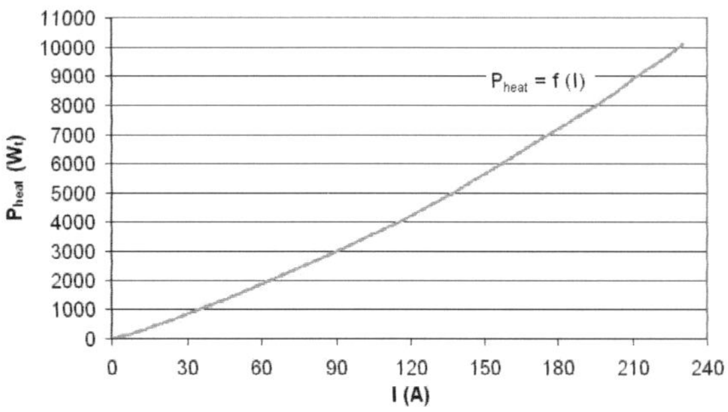

Figure 6. The dependency of the power of the researched module on its current load

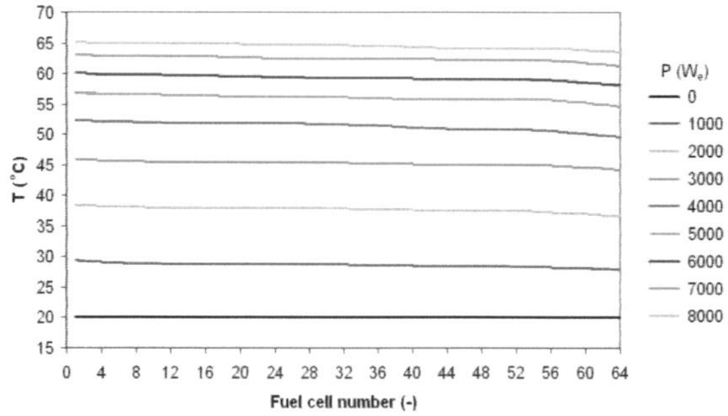

Figure 9. The temperature layout inside of the module during various power loads

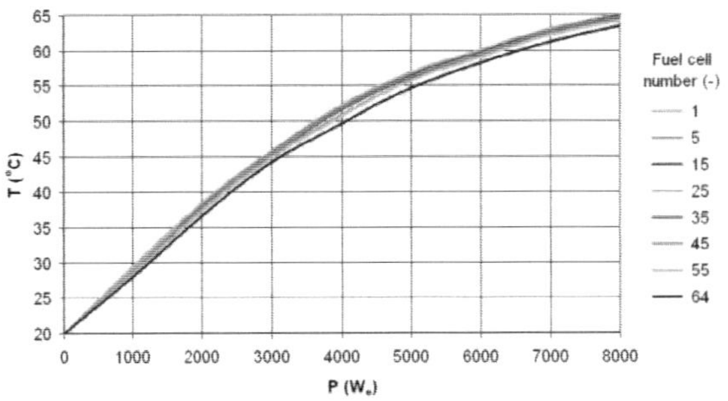

Figure 10. The temperature of the eight chosen fuel elements compared to the power load of the module

From both of these graphs, it is evident, that the fuel elements, which are located closer to the side of the module with media inputs, are better cooled. The temperature difference between both ends (sides) of the module is ca. 2 °C

The efficiency versus output power of measured hydrogen module is shown in Fig. 11. As you can see the efficiency at full power is approx. 40 %. Maximum efficiency is about 55 % and occurs at part load, approx. 600 W. The efficiency quickly declines for part loads less than 600 W, as the auxiliary loads begin to dominate the requirement for hydrogen consumption [9].

B. Dynamic operating characteristics

One of the important dynamic parameters of PEM fuel cells is an operating phenomenon called "purification". It is purification of the fuel part of the cell, that is, on the side of the anode, which has an impact on the size of the voltage output (see letter "P" on Fig. 12). This phenomenon happens in specific intervals depending on the load on the cell, or rather on the amount of consumed fuel (hydrogen). The necessity of the purification is due to the fact, that there is a build-up of dirt and condensation from the fuel on the cooler parts of the element. This causes clogging of the distribution channels and contamination of the structure of the catalyst (Pt layer), which decreases the reactive area of the cell. The purification also takes place, because the produced water (in the current of the oxidizer) gets through the membranes of the fuel cells and starts collecting in the fuel current [10].

Both named processes are the cause of the gradual unwanted power decrease of the fuel cells. The module's control system emits a signal to the so-called ventilation valve, which on its opening rids the anode area of the unreactive elements (dirt), which are (forced out with the fuel current) and the electric power of the cells is renewed. At the same time,

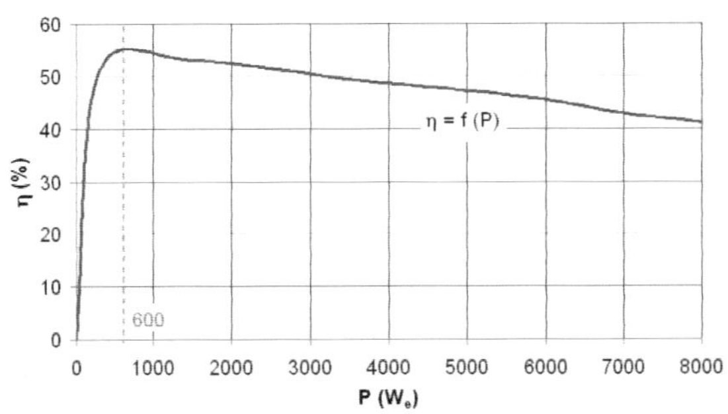

Figure 11. The dependency of the efficiency of the tested module on the electric power output

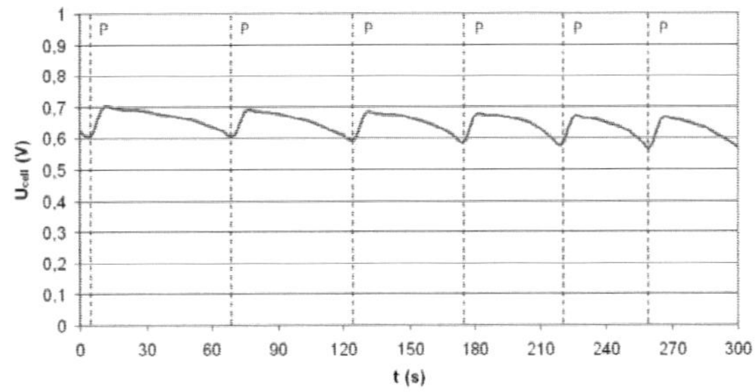

Figure 12. The time course of the output voltage of one fuel cell influenced by purification

a steep rise of voltage on elements takes place (see letter "P" on Fig. 12). This figure shows, apart from the changing number of purification, the gradual decrease of the ouput power, which corresponds to the load on the module of ca. 4000 W_e to ca 5500 W_e.

Fig. 13 shows the time plot of the output power of the measured module Nedstack FCS 8-XXL. The timeline corresponding to the output voltage of the module is then shown on Fig. 14. As can be seen from both figures, during ca. 100 minut the tested module was (in steps) moved from a no-load state to a state of maximum load (8000 W_e). The time interval, during which there was an impulse from the control system for an operating outage of the module is emphasized with color on both figures. The time "1" is the moment when the power of the module increased the first time (in single step) from value 6000 W_e to full power 8000 W_e. Due to the fact, that there wasn't an adequate increase in input fuel and oxidizer, the module was disconnected after the load. After this short outage, the module load returned in time "2" back to the value 6000 W_e.

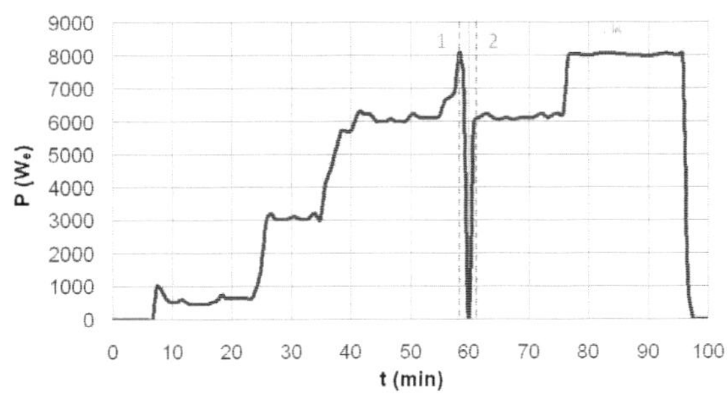

Figure 13. The time course of the output power of the researched module

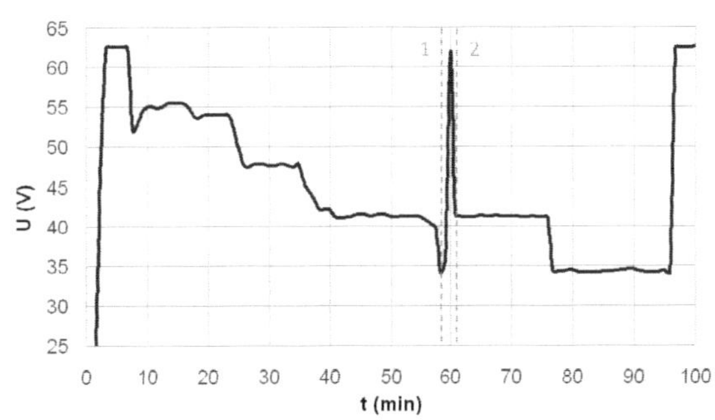

Figure 14. The time course of the voltage of the researched module

IV. Conclusion

Electric energy storage systems based on hydrogen technology are generally considered practical, especially for power plants, which use renewable energy sources, such as photovoltaics or wind power stations.

The performed measurements and the results, which are summarized in this paper, provide knowledge of the basic properties and parameters of the default and output variables of the module of hydrogen PEM fuel cells Nedstack FCS 8/XXL. This knowledge will be used in close future foremost for the modification of the control system of this module, leading to the optimization of its operation in collaboration with its associated technologies, with which it forms a hybrid storage system. The algorithm of the control system is crucial for the functioning of the module, as it has an impact on the dynamic running of the entire hybrid energy system. It is necessary to eliminate unwanted factors, such as operation outages to ensure problem-free operating.

The secondary objective of the research of the hydrogen modules with fuel cells will be the achievement of a higher efficiency of production of electricity from hydrogen. The results gained during the presented tested running of this hydrogen module will also be used for the creation of a precise numerical model.

Acknowledgment

This research was partially supported by the project LO1404: Sustainable Development of ENET Centre, by the project CZ.1.05/2.1:00/19.0389: Development of the Research Infrastructure of the ENET Centre, by the project TN01000007: National Centre for Energy and by the SGS grant from VSB-Technical University of Ostrava: SP2019/20.

References

[1] D. Minarik, B. Horak, P. Moldrik and Z. Slanina, "An Experimental Study of Laboratory Hybrid Power System with the Hydrogen Technologies," in Advances in Electrical and Electronic Engineering, vol. 12, 2014, pp. 518-528.

[2] M. Ney, Z. Hradilek, D. Minarik and P. Moldrik, "Technological centre for research of renewable resources and accumulation of electric power," 14th International Conference on Environment and Electrical Engineering, Ed. R. Tadeusiewicz, H. Schwarz, R. Gono, Z. Leonowicz, Krakow: IEEE, May 2014, pp. 65-68.

[3] Product Specifications of XXL stacks, [online], [2019-04-18] <http://www.nedstack.com/wpcontent/uploads/2017/05/nedstack product-specifications-of-xxl-stacks.pdf>.

[4] S. Srinivasan, "Fuel Cells From Fundamentals to Applications," Springer Science+Business Media, LLC, New York, USA, 2006.

[5] Layers and primary flows of PEMFC, [online], [2019-04-18] <https://build.openmodelica.org/Documentation/FCSys.html>.

[6] Y. Wang, K. S. Chen and S. Ch. Cho, "PEM Fuel Cells: Thermal and Water Management Fundamentals," Momentum Press, 2013.

[7] P. Moldrik and R. Chvalek, "PEM Fuel Cells - The Basic Characteristics," 10th International Conference on Environment and Electrical Engineering (EEEIC). Roma: IEEE Xplore, vol. 10, May 2011, pp. 566-569 .

[8] P. Kacor, D. Minarik and P.Moldrik, "CFD Analysis of Temperature Distribution of PEM Type Fuel Cell," 16th International Scientific Conference on Electric Power Engineering (EPE), Kouty nad Desnou: IEEE, 2015, pp. 507-12, ISBN 978-1-4673-6788-2

[9] P. Moldrik and R. Chvalek, " PEM Fuel Cells - The Effect of Fuel Parameters on Efficiency and Quality of Electric Power Supply," 10th International Conference on Environment and Electrical Engineering (EEEIC). Roma: IEEE Xplore, vol. 10, May 2011, pp. 570-573, DOI 10.1109/EEEIC.2011.5874674

[10] F. Barbir, "PEM Fuel Cells: Theory and Practice," Academic Press, 2012.

Small Renewable Energy Systems on Cars

Pavel Šimon
Institute of Aurel Stodola
FEIT, University of Žilina
Liptovský Mikuláš, Slovakia
simon@lm.uniza.sk

Michaela Holá
Institute of Aurel Stodola
FEIT, University of Žilina
Liptovský Mikuláš, Slovakia
hola@lm.uniza.sk

Abstract — **The article is dealing with preparing of experimental research. This research is meant to evaluate the possibility of using renewable energy sources with a vehicle moved by an internal combustion engine.**

Keywords: Photovoltaic; Internal Combustion Engine Vehicle; Energy Management in Car; Green House Gas

I. Introduction

The transport industry is causing a considerable amount of Green House Gas (GHG) emission. The transport itself has a portion of about 12 % and another fuel combustion in the energy industry even 42 %, see Figure 1 [1].

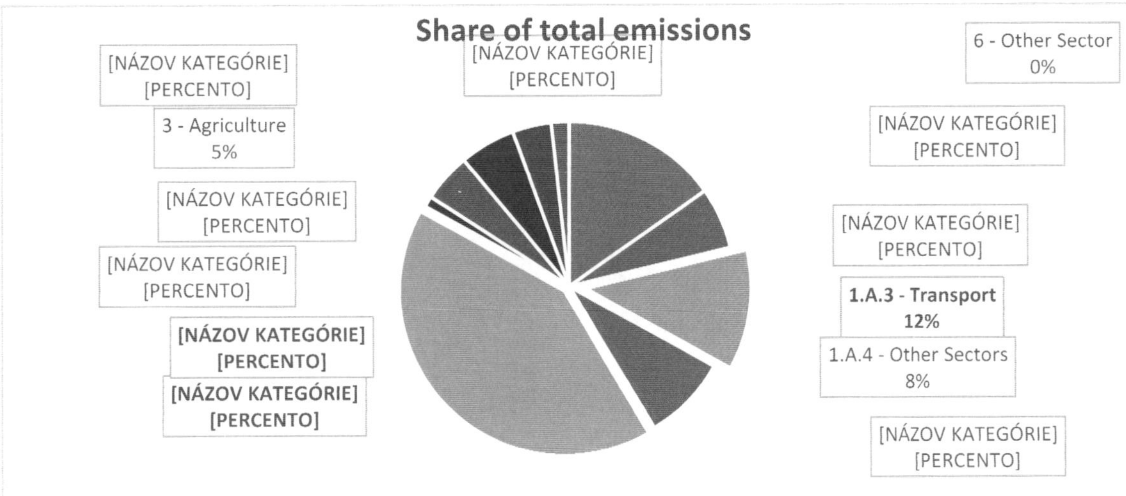

Figure 1. GHG portion of the industry in 2016 [1]

European Commission prepared A European strategy for low-emission mobility [2]. One answer for lowering GHG emission in transport is the electrification of transport. But this change requires time and money. And this change will require a lot of changes in energetics industry too.

However climatic changes are quicker than our lowering of people emission. Though during this process, we can bring a lot of small changes. One of them is a bigger portion of Renewable Energy Sources (RES) in any technology. And the conjunction of RES and vehicles is one of them.

Electronics in any car are more and more complex. With growing complexity is increasing demand for power to run those electronic even if a car is waiting for use. A huge portion of the car is parked in this time outside on open space. There is a lot of Sunshine, that can be used for powering some of those always online electronics. But it is a really good idea or even possible? We are preparing a test in real condition.

II. Is Renewable Energy Already Used in Vehicles?

The idea of using RES in vehicles is commonly used with electric cars. In this type is obvious to use top of the car for electric energy generation [3]. Example of using a car roof is the Fisker Karma [4]. Another car passenger car using top and side areas of the car is Sion form Sono Motors GmbH [5], German start-up. This car is heading to mass production in Sweden [6].

Using sunshine for movement is widely used in experimental car, where is even made world champions of the longest ride or top speed. One of the most famous challenges is The Bridgestone World Solar Challenge. Where solar powered cars are travelling about 3 000 km for the north to south of Australia [7].

Their design is unfortunately impractical for everyday use, see in Figure *2*. These cars are usually prototypes, often made by universities. Results of their effort are interesting, longest ride of the solar-powered car is hundreds of kilometres on one ride.

However, all examples of using sunshine for powering vehicles relates to an electric car. We are interested in using solar energy for using in standard Internal Combustion Engine (ICE) powered vehicle. Logically there couldn't be used directly for movements, but there is still place, where we can use solar energy for some purposes. It is fuelling onboard electronics, charging starting battery in a longer-term period of parking. Or even for some heating during the cold season and air-conditioning during hot sunny days. This leads us to prepare experiment which could help us to give us real data for research of how much solar energy could be used in ICE vehicles for everyday use.

Figure 2. Solar powered car in the challenge [7]

An ICE passenger car is in general object, in which is electricity used for starting the engine, to ignite sparks in gasoline engine and of course nowadays to run the car more effectively via many embedded computers and controllers. As additional use is electricity use for comfortable use of it. Inside the car is many electric energy-based devices, such as radio or bigger entertaining devices. Modern cars are using navigation systems, systems for semi or fully autonomous driving and many other systems. All of them are strongly dependant on electric energy which is stored in batteries and made during running of ICE.

One way how to prolong use of those electronic devices is to charge car battery from renewable energy, especially from photovoltaic modules. This solution will not only give some more energy for onboard electronics but would little bit cut fossil fuel consumption too. When devices used energy from another source than the car battery, then this battery will be charge less later from the engine alternator.

The main question still is, how to do it efficiently, not only based on energy efficiency but in investment efficiently too. But before thinking about efficiency, we need to design a working device. Thus, we start research on using photovoltaics device on the car roof.

III. What Can We Expect From Sunshine on a Car?

Evaluation of the amount of sunshine on standard fixed Photovoltaic (PV) systems is quite easy, there are plenty of tools for that purpose. European Commission JRC is running useful tool PV GIS as an example of free of charge one [8]. As a very good paid tool and service we can use pvPlanner from Solargis s.r.o. the company from Slovak republic [9], [10].

However, for our purposes there is a big disadvantage of those tools – they are meant to use in fixed systems on one location. They can be used on stationary solutions or on a tracked (moveable) system, but a place of systems is fixed. On the other hand, cars are objects with an unpredictable and changing location.

A. Preparing of Calculation of the Expected Result

Any cars are in two general states – parking on a fixed position and moving from these positions to another one. Predicting of insolation of moving a car is difficult or even impossible. But if the car has limited positions of parking, there is an opportunity to calculate possible insolation in each place. Evaluation of all possible places gives us the sum of irradiation. Combination of those evaluations and time spend on location would give us possible gained solar energy.

Unfortunately, even evaluation of fixed parking position is not easy as it sounds. Both tools (PV GIS and pvPlanner) are great tools for solar irradiation evaluation, however only for fixed systems with constant mounting. On other side parking of car is quite a stochastic phenomenon. Each parking of the car could be different in direction to South, position or shading of buildings. All these minor changes can significantly change the result of Sun shining.

As was written, there is difficulty to evaluate electricity of production during active use of the car. This leads us to prepare an experiment when we can confirm the theoretical evaluation of production with a parked car. Another part of the experiment will be the measurement of production during the car movement.

In the next chapter, we describe test equipment and then we will prepare an evaluation of two parking position where is the testing car usually parking.

IV. Design of PV Equipment

Design of photovoltaic system was done in April 2019 and was presented on the 40th conference of Unconventional sources of electrical energy (NZEE) [11]. The result of the design is shown in Figure *3*.

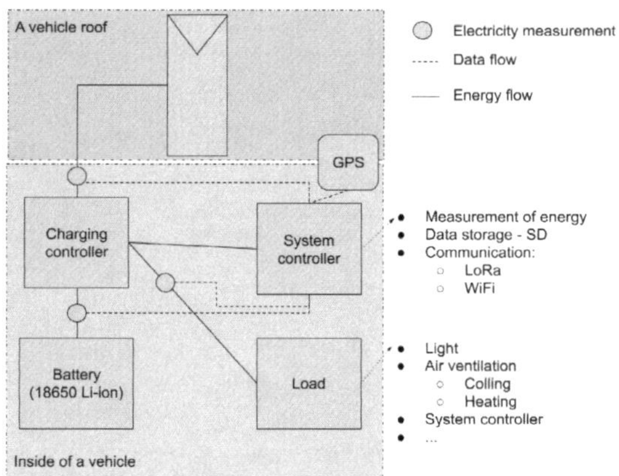

Figure 3. PV system on Car - block scheme [11]

During design process was 100 Wp flexible PV panel chosen as a main renewable energy source. The panel will be mounted on a roof of the testing car. The charging controller is based on an Arduino clone ESP32 module. It includes the MPTT (Maximum Power Point Tracking) algorithm for effective using sunshine and both a charging and discharging algorithm for Li-ion based storage. The load and the system controller are connected to the charging controller too.

The system controller is measuring an energy flow in three points for accurate evaluation. Measurement is done in three points: between the PV panel and the charging controller, energy flow from controller to storage and consumption of the load. All data are stored in the system controller. The system controller is equipped with SD card for off-line data storage. When Wi-Fi or LoRa communication port is in reach, then data are recorded to server. Both communication ways are used when is connection available, with Wi-Fi priority.

Design of car PV generator experiment includes a GPS receiver, which will generate the exact location of the car. This will help with a better understanding of the dynamics of PV production during car travel.

For storing of energy, we decided to use 78 pcs. of standard 18650 type Li-ion accumulators. They are in 3s26p (3 in serial, 26 in parallel) configuration. This allows us to use any standard 12 V type equipment as a load.

V. A Parking Places Evaluation

In preparation for an experiment, we evaluate two spots for the usual place of parking. There are two instances of evaluation in both spots. A first instance is a place without artificial shading, there is only a natural horizon taking into consideration. The second calculation is made involving surrounding buildings and other shading obstacles.

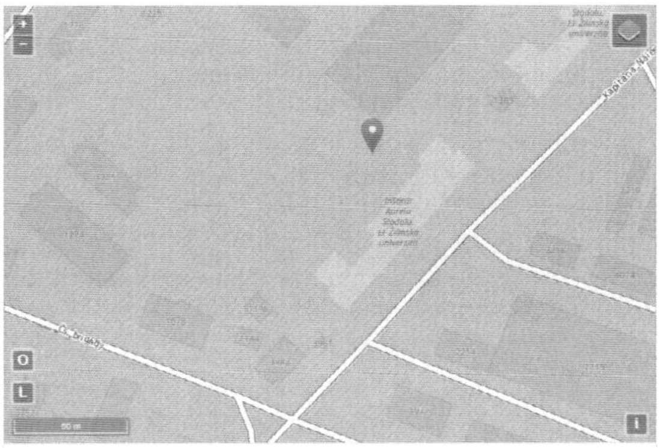

Figure 4. Parking place in front of the IAS in PV GIS map [8]

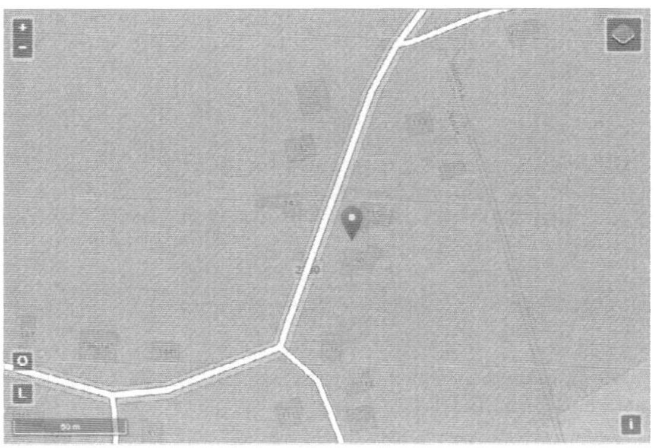

Figure 5. Home parking place in PV GIS map [8]

One place is a parking place in front of the Institute building, the second one is the usual point of home parking of the used car.

The evaluations were made in the newest version of JRC PV GIS tool [8]. All set up of the tool was in the default setting, except installed power and mounting slope. We evaluated 0,1 kWp PV production. The value was established as maximal possible to mount on the car roof. The mounting slope was set to 0° because is laying on a car roof in a horizontal position.

As a radiational database was used PVGIS-CMSAF [12]. In both places were measured shading angles based on PV GIS methods [8].

A. *Shading of Buildings Near Parking Places*

The shading of a building is highly dynamic fact. It varies during days and during the seasons. Evaluation of static places was published in many different publications. An example is research [13], where authors made Evaluation of predicted and operated parameters of photovoltaic power – station.

Figure 6. Shading of IAS building on PV GIS Google Satellite Map

Figure 7. Shading of a home on PV GIS Google Satellite Map

Figure *6* and Figure *7* are examples of building shading from PV GIS with Google Satellite Map. Even in sunny days of Summer is shading highly visible. However, these figures show only static shading during specific day and time.

B. *Evaluation of Two Places in Both Conditions*

Results of both places in both conditions are in the next tables. Where "Em" means average monthly electricity production from the given system [kWh], and "Hm" represent the average monthly sum of global irradiation per square meter received by the modules of the given system [kWh/m²].

Table 1. PV production on IAS spot

Month	Natural horizon		Buildings shading	
	Em	Hm	Em	Hm
January	2,010	26,0	1,440	19,1
February	3,360	41,2	2,440	30,9
March	6,900	82,9	5,600	68,2
April	10,500	128,0	8,660	107,0
May	11,500	143,0	10,300	129,0
June	11,600	148,0	11,400	144,0
July	12,200	158,0	11,400	147,0
August	11,000	141,0	8,940	116,0
September	7,510	95,2	6,060	77,7
October	4,740	59,8	3,410	44,2
November	2,490	32,3	1,690	22,6
December	1,660	22,2	1,200	16,3

Month	Natural horizon		Buildings shading	
	Em	Hm	Em	Hm
Year	***85,470***	***1 077,6***	***72,540***	***922,0***

Table *1* shows the result of possible electricity production in IAS parking spot place for both cases. The different result was achieved when we add buildings shading into consideration.

If we expect that car would be parking in the same place on each occasion, we can say that the building is lowering power gain from 85,47 kWh/year to 72,54 kWh/year.

Next evaluation was done with second place for parking – at the home. Data from the evaluation are shown in Table *2*. As expected, shading results are much smaller than with the natural horizon. This building is lower than the IAS building, but this parking is closer to the walls. This condition caused a bigger decrease in power generation with shade included. Power generation falls from 86,76 kWh/year to 59,48 kWh/year.

Table 2. PV production on a home parking spot

Month	Natural horizon		Buildings shading	
	Em	Hm	Em	Hm
January	2,000	25,8	1,470	19,3
February	3,340	40,9	2,210	28,0
March	6,960	83,4	4,300	53,3
April	10,600	130,0	6,920	86,2
May	11,900	147,0	8,920	112,0
June	11,800	150,0	8,950	115,0
July	12,400	160,0	9,060	118,0
August	11,200	144,0	7,440	97,3
September	7,620	96,2	4,540	59,3
October	4,750	59,8	2,740	35,9
November	2,530	32,6	1,710	22,6
December	1,660	22,1	1,220	16,5
Year	***86,760***	***1 091,8***	***59,480***	***763,4***

C. Result of Two Places to Harness the Solar Energy

Both places are influenced by building shading. Lower building with closer parking place is even more influencing power generation from sunshine. Combined results are in Table *3*.

Table 3. Comparison of all cases of PV generation

Place	Em	Difference	Hm	Difference
IAS clear	85,470	100,00%	1 077,6	100,00%
IAS shaded	72,540	**84,87%**	922,0	85,56%
B50 clear	86,760	100,00%	1 091,8	100,00%
B50 shaded	59,480	**68,56%**	763,4	69,92%

Proper selection of parking is significant for electricity production with a PV panel on a vehicle roof. Unfortunately, this is still only one factor, which needs to be considered.

D. Consideration of Parking Time

The car is used for everyday transportation. This means that it will not be parked in one place for the whole day. Evaluations from the previous chapter are thus only one step for a rating of PV panel used on the car. Another step is to include the time of parking in those places.

From common use is obvious, that the vehicle would stay in both places different time. Usually, it will stay about 8 to 9 hours in the IAS parking during working days and full day in home parking during weekends. But this involves only standard daily routines.

Calculation of common daily time influenced power generation is planned as one of the next steps in this research. We are preparing a model which will consider those daily routines.

VI. Evaluation of PV Production During the Ride

The previous chapter was dealing with evaluation of PV production while a vehicle is on a static position in parking. That evaluation was influenced by many random variables, such as the exact parking place. As was shown, evaluation is still possible, with some simplifications.

Evaluation of PV production of Solar panel on a vehicle roof brings much more dynamics into consideration. Those dynamics make it quite difficult. The streets, roads or highways are surrounded by many different objects. And the position of those objects could be in interference of sunshine to PV panels. Riding through that variable environment is making preliminary assessment of production highly dynamic process.

One main goal of this experimental research would be to collect data for evaluation future model of the Car PV systems. There is some possibility, some of them include even open-source tools [14], but main research of current state is on its way.

At that time, we were not able to make any significant evaluation of PV production during a ride. This evaluation is the next step in our work.

VII. Conclusion

The article is describing the first steps in experimental research which will require a much longer period for appropriate result to be taken. We consider it on as entrance to bigger project, which will allow us to do deeper research in field of small RES use.

Acknowledgement

This work was supported by Faculty of Electrical Engineering and Information Technology of the University of Žilina, Pavel ŠIMON s.r.o. and Auto-Moto servis, s.r.o. companies.

References

[1] European Environment Agency (EEA), "EEA greenhouse gas - data viewer," 2017. [Online]. Available: https://www.eea.europa.eu/data-and-maps/data/data-viewers/greenhouse-gases-viewer.

[2] European Commission, "A European strategy for low-emission mobility," 2016. [Online]. Available: https://ec.europa.eu/clima/policies/transport_en.

[3] M. Longo, F. Foiadelli and W. Yaïci, "Electric Vehicles Integrated with Renewable Energy Sources for Sustainable Mobility," IntechOpen Limited, Rijeka, 2018.

[4] wikipedia.org, "Fisker Karma," 2012. [Online]. Available: https://en.wikipedia.org/wiki/Fisker_Karma.

[5] Sono Motors GmbH, "Sion," 2018. [Online]. Available: https://sonomotors.com/en/sion/.

[6] Sono Motors GmbH, "Sono Motors to produce Sion in Sweden," 2019. [Online]. Available: https://sonomotors.com/en/press/press-releases/sono-motors-to-produce-sion-in-sweden/.

[7] South Australian Tourism Commission, "The Bridgestone World Solar Challenge," 2019. [Online]. Available: https://www.worldsolarchallenge.org/event-information.

[8] European Commission Joint Research Centre, "PHOTOVOLTAIC GEOGRAPHICAL INFORMATION SYSTEM," European Commission Joint Research Centre, 2019. [Online]. Available: http://re.jrc.ec.europa.eu/pvg_tools/en/tools.html.

[9] M. Šúri, T. Cebecauer and A. Skoczek, "SolarGIS: solar data and online applications for PV planning and performance assessment," in *26th European Photovoltaics Solar Energy Conference*, Hamburg, 2011.

[10] Solargis s.r.o., "pvPlanner," Solargis s.r.o., 2019. [Online]. Available: https://solargis.info/pvplanner/.

[11] P. Šimon and M. Holá, "Designing of using RES on personal vehicles," in *40. Nekonvenční zdroje elektrické energie*, Brno, unpublished, 2019.

[12] European Commission Joint Research Centre, "Overview of PVGIS data sources and calculation methods," 2017. [Online]. Available: http://re.jrc.ec.europa.eu/pvg_static/methods.html.

[13] J. Tomčejová, P. Haľko, E. Nováček and P. Tauš, "Evaluation of predicted and operated parameters of photovoltaic power - Station," *Acta Montanistica Slovaca,* vol. 17, no. 4, pp. 251-256, 2012.

The Use of Electromobiles to Align Daily Electrical Load Schedule of Power System

Inga Zicmane
Faculty of Power and Electrical Engineering
Riga Technical University
Riga, Latvia
Inga.Zicmane@rtu.lv

Kristina Berzina
Faculty of Power and Electrical Engineering
Riga Technical University
Riga, Latvia
Kristina.Berzina@rtu.lv

Anatolii Zamulko
National Technical University of Ukraine
Igor Sikorsky Kyiv Polytechnic Institute
37 Peremogy avenue,
Kyiv, Ukraine
zai_71@ukr.net

Yurii Veremiichuk
National Technical University of Ukraine
Igor Sikorsky Kyiv Polytechnic Institute
37 Peremogy avenue,
Kyiv, Ukraine
y.veremiichuk@kpi.ua

Abstract — **One of the most important trends in modern electric power industry development is the creation of an effective system for managing power grids load schedules. Analysis of power systems daily electrical loads schedules shows that their unevenness tends to increase, taking into account renewable sources growth in unregulated electricity production, jeopardizes the reliability of electrical energy supply and functioning of the electrical energy market.**

A significant contribution to addressing the alignment of electric power system load diagrams can make electric vehicles to become effective users (generators) - the regulators. Functional features of these devices are able to solve the problem as a failure of filling the night schedule of electrical loads and to provide the necessary balance of power generation throughout the day.

Statement of the problem is also devoted to mismatch factors considering energy consumption graphs isolated from the power consumer and its production by this energy source in renewable energy sources effectiveness evaluation.

The aim of this work is to develop proposals for evaluating the impact of systems V2G (Vehicle-to-grid) to smooth the electrical load curve in terms of unregulated renewable energy growth.

Keywords: electric energy consumer, energy storage, generation, smart grid; electric vehicle; charging system; charging station; vehicle-to-grid

I. Introduction

The non-uniformity of the electrical load during the day negatively affects the operation of the power system: operation of power generating units becomes more complex, specific fuel consumption increases, cost of electricity production increases. In order to contain and minimize these negative consequences, the power system, in an effort to economically interest electricity consumers in leveling the daily load schedule sets a differentiated tariff for consumers by the hour of the day: an increased charge during peak hours in the power system, a reduced charge. hours of nighttime electrical fall [1].

Prospective growth in power consumption requires an early search for ways to cover, the most important of which is the introduction of a new generation sources. At the same time, it is necessary to find an opportunity to streamline this growth, to make it manageable. Total power consumption is influenced by various social and economic factors, so its changes have a high degree of uncertainty. In addition to the probability of generation a shortage, there is always a probability that the power capacity will be underloaded or will have an inefficient mode of operation. Planning for generating capacity growth, taking into account the management of electrical load (demand side management), can be a buffer to mitigate the effects of sharp fluctuations in electricity consumption. When determining the degree of demand side management effectiveness, one should avoid its direct opposition to the buildup of generating capacity, means consider both of these activities as a reasonable and balanced complement to each other [2].

Today, to reduce daily loads unevenness, proposed to use electric vehicles as energy storage devices, based on the principle of pumped-storage hydroelectricity (PSH). This idea was called V2G (Vehicle-to-grid), as a concept of two-way use of electric vehicles and hybrids, which implies connecting the machine to the general grid for recharging the car and returning excess electricity back, which allows [3-7]:

- to sell electricity to power grid during hours when the machine is not in use, and to charge it during hours when electricity is cheaper;
- use electric vehicles as an uninterruptible power supply;
- use electric vehicles as a means to improve power distribution systems reliability.

V2G is conventionally considered as the delivery of power from a vehicle back to the power grid. However, vehicles can also provide useful services to the grid even without delivering power back to the grid. For example, by allowing the vehicle charging rate to be ramped up and down remotely through smart metering a vehicle can perform a grid ancillary service called regulation. Regulation is currently performed 24/7 by power plants in order to fine tune the balance between generation and load [8-11].

The increase in power consumption due to the massive use of electric vehicles can have an unexpectedly powerful impact on the electricity industry and require a significant change in the generating capacities structure and technical characteristics . The total equivalent capacity of the global motor vehicles has long exceeded the total capacity of the operating power plants, while the mass transition to electric vehicles will cause an electric power shortage. This determines the need to study the development of the electric power industry, taking into account the development of electric vehicles as a new electric power consumers class and development of concept for its effective and timely integration into the existing structure of electricity consumption.

II. The Problems Urgency for the Power System

The unevenness of the power systems daily electrical load schedule with its morning and evening peaks and a sharp decrease in load at night is an objective phenomenon, depending on the mode of power consumption of each consumer connected to the power system. With an increase in the share of large consumers, the unevenness decreases, with an increase in the share of enterprises in the services sector and in the burden of the housing and utilities sector it increases.

With the increase in the share of renewable energy sources in the power system balance and the probabilistic nature of manifestation during the year, the problem of energy consumption schedules and its production of renewable energy mismatch arises. The maximum values of the various renewable natural energy resources types potential indicators are observed in different periods of the year. Thus, the maximum wind speeds are characteristic for the autumn and spring periods. The annual schedule of water consumption of small rivers everywhere has a spring maximum due to floods and a significant decrease (up to dozens of times) during the winter months. The maximum values of the arrival of solar radiation are observed in the summer period. The mismatch of the maxima of consumption and energy production by a renewable source causes a significant under-utilization of its installed capacity. The neglect of this factor leads to an overestimation of the utilization rate of the installed capacity of a renewable energy source in assessing the economic efficiency of projects. This factor is aggravated by the prevalence of household consumers loads, which causes a pronounced winter maximum of the annual energy consumption graph. It should be noted that in the world a significant amount of research has been devoted to this topic [12,13] in which a number of solutions have been proposed. Among which stand out the use of electric vehicles as energy storage. In the majority of works, researchers considered electric vehicles in autonomous or small energy systems using renewable energy sources for power supply of electric vehicle charging stations on the territory [14-16].

In ukrainian energy sector , investments in the development of renewable energy sources are actively involved. In 2018, sources of renewable energy generated 2632.4 million kWh, which is 1.7% of Ukraine's balance. Also, the use of electric vehicles in our country has a positive tendency. As on April 1 2019 in Ukraine there are about 12119 electric cars, which averages 292 million kWh battery capacity.

The battery charge of electric vehicles users will always have a chance, but a common template for their use can be used in favor of the power system, such as transport habits, special energy tariffs. The owners of the network of charging stations can serve as aggregators of demand and involve electric car users to equalize demand.

To determine the possibilities of using electromobiles - as aggregators, which can take part in adjusting demand for power. Consider a typical daily schedule of electricity consumption in the UES of Ukraine and the generation of WP and PV “Fig. 1”. Taking into account the variability of renewable energy, we determine the areas in which electromobiles can be used in the charge -CHG and recharge -RCHG mode.

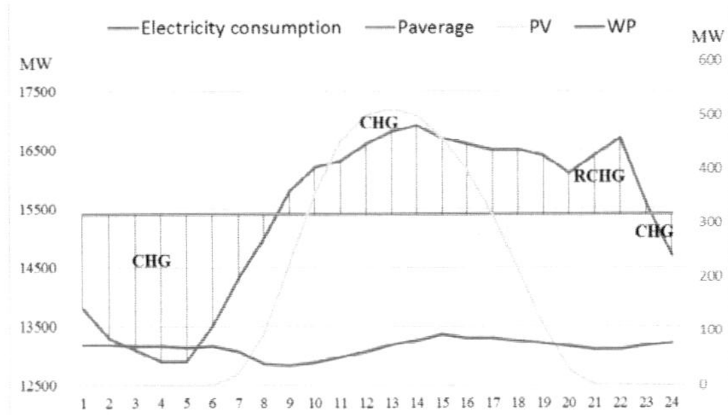

Figure 1. Graph of electricity consumption, generation WP and PV typical summer day 2018

The given graph shows one of many variants of possible interaction of the power system with electric vehicles as aggregators. Such work depends on many factors associated with the power structure structure and characteristics of the electric vehicle (EV), namely the installed power of the battery and their time of interaction with the network, represented by the formula:

$$Pev = (N, C_b, T) \quad (1)$$

N – number of cars, Cb – battery capacity, T – network interaction time (charge / discharge).

Therefore, we consider the methodology for assessing the potential power of electric vehicles in general generating capacities system for the UES of Ukraine.

III. Assessment Methodology for Electric Vehicles as Electric Energy Demand Cover

Adequacy assessment in the identified optimization models is provided on the basis of common approaches to the formalization of the UES of Ukraine generating capacity development and functioning processes and differs in the level of detail in the simulation, while the requirements for the provision of reserves are adopted in accordance with the provisions of the Code of the transmission system [17].

Simulation of electrical loads graphs (GEN) coverage for the most difficult conditions of the UES of Ukraine generation - days with maximum and minimum loads, working days and days of the flood, is carried out for the conditions of isolated power system operation.

Demand cover for electric power modeling is provided by introducing variables into the corresponding balance of supply-demand for electric power for each level of the GEN, when it is impossible to provide the specified balance, provided an opportunity to obtain the solution of the problem, and their significance determines the level of corresponding imbalances. If such variables take values in solving the problem, this indicates that it is impossible to ensure its implementation at the given limits for the installed power generation, demand management measures and accepted readiness rates. These variables are included in the optimization criterion with "penalty" values, means that their choice to cover the supply-demand balance will be the worst possible solution, and they will obtain more than zero values only in case of insufficient generating capacity to cover demand or with insufficient maneuverability of generation changes to compensation and fluctuations in wind and solar power plants generation.

Thus, in general, the supply-demand balances for the n-degree GEN for a characteristic day g, n=1÷N, g=1÷G, in t stage of the calculation period t=1÷T, in a simplified form, are formalized as follows:

$$\sum_{k=1}^{K} Y_{kngt} - \sum_{i=1}^{I} D_{ingt} - HM_{ngt} + HP_{ngt} = 0 \quad (2)$$

where: Y – power of k type of generation technology, which is used to cover demand at the appropriate level of the GEN and is determined taking into account own needs for electricity generation and availability factors that determine its "availability" for use, k=1÷K, D – demand for electric power of i-th consumer, taking into account losses on transportation and distribution of electric energy, i=1÷I, HP and HM are variables that provide an opportunity to get a solution in the event of the balance disturbance, due to the impossibility of ensuring that it is executed at the given limits for the installed generation power and accepted readiness rates

In more detail, the methodology for modeling the coverage of GEN using separate (typical) technologies for generating electricity and taking into account the aggregated use of electric vehicles on the V2G principle is realized using equations (3) - (11), using the following notation:

- Hng - the length of a degree GEN typical day hours;
- k- index of electrical installation type:
  - Con. – consumer,
  - NPP – nuclear power plant not maneuvering,
  - NPPM - nuclear power plant maneuvering ,
  - RPP – renewable power plants (WP, PV the other),
  - HPP – hydroelectric power plant,
  - TPP – thermal power plant is a power station in which heat energy is converted to electric power,
  - EVa – electric vehicle , which is used according to the V2G principle,
  - QS – installation with a quick start ,
  - PSH – pumped-storage hydroelectricity.
- Pkn - power plants producing electricity, MW;
- Zkn - power of electrical installations consuming electricity, MW;
- $\Delta P_{n\,RPP}^{UP}$, $\Delta P_{n\,RPP}^{DOWN}$ – magnitude of possible increase (UP) and decrease (DOWN) of the renewable power plants, MW;
- $\Delta P_{n\,Con}^{UP}$, $\Delta P_{n\,Con}^{DOWN}$ - magnitude of possible increase (UP) and decrease (DOWN) of the consumers load, MW;
- $\Delta Z_{n\,PSH}^{UP}$, $\Delta Z_{n\,PSH}^{DOWN}$ - magnitude of possible increase (UP) and decrease (DOWN) of power PSH, MW;
- P_n^{UP}- required capacity of loading reserve (increasing electric power generation) for balancing the power system and meeting the requirements for the required appropriate reserves, MW;
- P_n^{DOWN}- required capacity of the unloading reserve (reduction of electric power generation) for balancing the power system for balancing the power system and meeting the requirements for the required appropriate reserves, MW;
- $\Delta P_{n\,k}^{I\_UP}$, $\Delta P_{n\,k}^{I\_DOWN}$- value of power loading (UP) and discharge (DOWN) primary frequency control reserve and power, and to compensate for power fluctuations, particularly caused by workWP and PV, MW;
- $\Delta P_{n\,k}^{II\_UP}$, $\Delta P_{n\,k}^{II\_DOWN}$, $\Delta P_{n\,k}^{III\_UP}$, $\Delta P_{n\,k}^{III\_DOWN}$- value of power loading (UP) and discharge (DOWN) of the secondary reserve of frequency and power regulation, as well as the tertiary reserve, MW;
- $P_{n\,QS}^{UP}$- electrical power value of fast start load , MW;
- $\Delta P_{n\,PSH}^{UP}$, $\Delta P_{n\,PSH}^{DOWN}$- the value of power to boot (UP) and unloading (DOWN) PSH, MW;
- P_{nk}^{max} , P_{nk}^{min} , Z_{nk}^{max} , Z_{nk}^{min}- the value of the maximum and minimum power of generation or consumption of the corresponding electrical installation, MW;
- $P_{n\,HPP}$, $P_{n\,PSH}$ - power generated by HPP and PSH at the current level of the GEN, MW;
- $P_{n\,PSH}^{max}$ - the value of the maximum power that PSH can generate, MW;
- $P_{n\,EVa}$, $P_{n\,EVa}^{max}$- level of electric vehicles batteries charge at the beginning of the stage τ GEN and its maximum capacity, MWh;
- kzr_{EVa} - ratio of power generation to power consumption;

For each level of the GEN, the power balance is written as:

$$\sum_{n=1}^{N} P_{HPP} + \sum_{n=1}^{N} P_{RPP} + \sum_{n=1}^{N} P_{TPP} + \sum_{n=1}^{N} P_{NPPM} \sum_{n=1}^{N} + P_{PSH} + \sum_{n=1}^{N} P_{EVa} + \sum_{n=1}^{N} P_{NPP} - \sum_{n=1}^{N} Z_{EVa} - \sum_{n=1}^{N} Z_{PSH} - \sum_{i=1}^{I} D_{ni} = 0 \quad (3)$$

The maximum available power of electrical installations is superimposed on the upper limit. Their generalized formalization has the following form:

$$kg_{kg} * X_{kg} \geq P_{kg}^{max} \quad (4)$$

where: kg_{kg}– readiness rates, X_{kg} – installed power of k electrical installation. The output of the installed power of the electrical installation on the readiness factor determines the maximum available generation power. Readiness rates that are less than one provide for consideration of factors number that in the general case make it impossible to use all the installed capacity of the technologies indicated, in particular:

1. Technical and technological constraints.
2. Planned current and capital equipment repairs.
3. Conducting reconstruction works (upgrades).
4. Finding equipment in emergency repair.
5. Equipment conservation.
6. Environmental restrictions on usage possibility etc.

For each level of GEN the value of the required capacity of the load reserve (increasing the generation of electric power) for power system balancing (P_n^{UP}), as well as the necessary capacity of the unloading reserve (reduction of electric power generation) for power system balancing (P_n^{DOWN}) are determined according to expressions (5) та (6).

$$\Delta P_{n\,RPP}^{UP} + \Delta P_{n\,Con}^{UP} + P_n^{UP} \geq \sum_{k=\binom{TPP,EVa,}{NPPM}} \Delta P_{n\,k}^{I\,UP} + \sum_{k=\binom{TPP,EVa,}{NPPM}} \left(\Delta P_{n\,k}^{II\,UP} + \Delta P_{n\,k}^{III\,UP}\right) + P_{n\,QS}^{UP} + \Delta P_{n\,HPP}^{UP} + \Delta P_{n\,PSH}^{UP} + \Delta Z_{n\,PSH}^{DOWN} \quad (5)$$

$$\Delta P_{n\,RPP}^{UP} + \Delta P_{n\,Con}^{DOWN} + P_n^{DOWN} \geq \sum_{k=\binom{TPP,HPP,}{EVa}} \Delta P_{n\,k}^{I\_DOWN} + \sum_{k=\binom{TPP,EVa,}{HPP}} \left(\Delta P_{n\,k}^{II\_DOWN} + \Delta P_{n\,k}^{III\_DOWN}\right) + \Delta P_{n\,PSH}^{DOWN} + \Delta Z_{n\,PSH}^{UP} \quad (6)$$

The ability of individual power plants to participate in GEN coverage and the balancing of UES work are subject to certain limitations.

So the power of fast start technologies can be used only for secondary and tertiary regulation, which is formalized as follows:

$$P_{n\,QS}^{UP} \leq \Delta P_{nk}^{II\_UP} + \Delta P_{nk}^{III\_UP} \quad (7)$$

The opportunity to participate in the regulation of TPP, HPP, PSH is limited by other requirements.

For efficient operation of V2C technologies, it is necessary to ensure the sufficiency of the primary energy source for generating electricity and the free capacity of the drive for its consumption. This is formalized in the form of three such interrelation:

$$P_{n\,EVa}^{max} \leq \frac{B_{n\,EVa}}{H_{n\,k}} \quad (8)$$

$$Z_{n\,EVa}^{max} \leq \frac{B^{max}{}_{n\,EVa} - B_{n\,EVa}}{H_{n\,k}} \quad (9)$$

$$B_{n\,EVa} = B_{(n-1)\,EVa} + P_{n\,EVa} - Z_{n\,EVa} \quad (10)$$

In addition, the power of such electrical installations is subject to restrictions on their possible power in the modes of electricity production and consumption. This restriction is formalized as follows:

$$P_{n\,EVa} + kzr_{EVa} * Z_{n\,EVa} = P_{n\,EVa}^{max} \quad (11)$$

Demand for electricity is defined as the amount of power consumers at all GEN levels on typical days duration.

At same time, today's average power of thermal power plant used for regulation in Ukraine is 200 MW. This is actually the minimum required power in a system for electric vehicles power aggregators. According to preliminary estimates, the Ukrainian electric vehicles park has a rapid development. In this case, the power should be considered as a probabilistic indicator, which depends on many factors, including subjective nature.

The proposed approach to the assessment can be used to simulate the simulation at the level of both the power system of the country and the region to determine the possible impact of the V2G system (Vehicle-to-grid) on the alignment of the electric load schedules in the conditions of increasing unregulated power of renewable energy sources

IV. Conclusion

1. Management of electric energy production and consumption processes at the present stage requires the search for new organizational opportunities and technical implementations. The current opportunity is to use the power of electric vehicles in the mode of V2G by forming the organizational and technical methods of their aggregation.
2. Analysis of the forming possibilities for organizational and technical methods of electric vehicles aggregation shows that it is necessary to develop a modern tariff menu based on a combination of existing tariff systems forming methods (for example, two-zone tariff systems) with features of renewable energy sources usage.

References

[1] A. Mahnitko, K. Berzina, A. Zamulko, Y. Veremiichuk, V. Nakhodov, "Research of Electrical Power Consumers Reaction in Power Consumption Management System", the 9th International Scientific Symposium Elektroenergetika, 12–14 September 2017.

[2] Y. Veremiichuk, O. Yarmoliuk, I. Prytyskach, V. Opryshko, "Modeling energy hub operating modes with demand side management usage" EPE 2018 - Proceedings of the 2018 10th International Conference and Expositions on Electrical And Power Engineering (EPE 2018), Iaşi, Romania. October 18-19, 2018. Article number 8559639, pp. 972-976

[3] B. Slusarczyk "Conditions for ECV charging infrastructure development in Poland", Proceedings of the 9th International Scientific Symposium ELEKTROENERGETIKA 2017, 12.-14. 9. 2017, pp. 184-189.

[4] M. Endo, K. Tanaka "Evaluation of Storage Capacity of Electric Vehicles for Vehicle to Grid Considering Driver's Perspective" Published in IEEE International Conference on Environment and 2018 DOI:10.1109/eeeic.2018.8494218

[5] M. Ammous, M. Khater, and Mohammad AlMuhaini "Impact of Vehicle-to-Grid Technology on the Reliability of Distribution Systems" Conference: 2017 9th IEEE-GCC Conference and Exhibition (GCCCE). DOI: 10.1109/IEEEGCC.2017.8448072.

[6] Haneul Ko, Sangheon Pack "Mobility-Aware Vehicle-to-Grid Control Algorithm in Microgrids " IEEE Transactions on Intelligent Transportation Systems (Volume: 19 , Issue: 7 , July 2018, pp. 2165 – 2174

[7] Esra'a Alghsoon , Ahmad Harb and Mohammad Hamdan "Power Quality and Stability Impacts of Vehicle to Grid (V2G) Connection" 2017 8th International Renewable Energy Congress (IREC) 21-23 March 2017 DOI: 10.1109/IREC.2017.7925995

[8] K. Qian, Ch. Zhou, M. Allan, Yue Yuan, "Modeling of Load Demand Due to EV Battery Charging in Distribution Systems" IEEE TRANSACTIONS ON POWER SYSTEMS, VOL. 26, NO. 2, MAY 2011

[9] S. Jaiswal, M. S. Ballal, "Optimal Load Management of Plug-in Electric Vehicles with Demand Side Management in Vehicle to Grid Application" 2017 IEEE Transportation Electrification Conference (ITEC-India)

[10] H. Turker, S. Bacha " Optimal Minimization of Plug-in Electric Vehicle Charging Cost with Vehicle-to-Home and Vehicle-to-Grid concepts" DOI 10.1109/TVT.2018.2867428, IEEE Transactions on Vehicular Technology

[11] Tarek Hassan Mohamed, Heba Ahmed Hassan "Terminal Voltage and Power Regulation Using Vehicle-to-Grid (V2G) Schemes Connected to a Two-Area Power System" 2018 Twentieth International Middle East Power Systems Conference (MEPCON), Cairo University, Egypt

[12] Y. Veremiichuk, A. Zamulko, S. Zaichenko, A. Mahnitko, K. Berzina, I. Zicmane "Analysis of Electric Energy Supply Security Attached to Renewable Energy Sources Implementation" EPE 2018 - Proceedings of the 2018 10th International Conference and Expositions on Electrical And Power Engineering (EPE 2018), Iaşi, Romania. October 18-19, 2018.pp. 977-981

[13] M. Tyagunov, S. Sharapov, P. Shurkalov, "Hybrid energy complexes and their control algorithms" Vestnik MEI No. 4, M .: Publishing House MEI, 2013, p. 64-67.

[14] P. Shurkalov , M. Tyagunov, "Charging options electric vehicles from installations based on renewable energy sources". Vestnik MEI No. 5, M .: Publishing House MEI, 2013, p. 61-66.

[15] Harun Turker Ilhami Colak "Multiobjective optimization of Grid - Photovoltaic - Electric Vehicle Hybrid system in Smart Building with Vehicle-to-Grid (V2G) concept", 2018 7th International Conference on Renewable Energy Research and Applications (ICRERA), DOI: 10.1109/ICRERA.2018.8567002

[16] V. Khanaev, "Consumers-regulators: opportunities and prospects of application", Scientific and technical statements SPbGPU, 2008, №1, p. 59-63

[17] Report on conformity assessment (adequacy) of generating capacities. SE "NEC" Ukrenergo " P.175.

F: Works of PhD Students

The Influence of Ripple Control Signal on Protection Relay Operation

Zuzana Bukvisova, Kinan Wannous, Petr Toman, Jaroslava Orsagova, Pedro Henrique de Paula Alves, David Topolanek
Dept. Electrical Power Engineering, Faculty of Electrical Engineering and Communication
Brno University of Technology
Brno, Czech Republic
xbukvi00@stud.feec.vutbr.cz, xwanno00@vutbr.cz

Abstract —**A ripple control system is a form of load management used in many countries. It enables to remotely control devices using a high-frequency signal (usually between 100 – 1600 Hz) superimposed onto the fundamental frequency of the power system. Since the distance protection evaluates the fault distance using the fundamental voltages and currents, an error can occur if the analyzed signal contains a higher level of inter-harmonics. This paper describes the impact of inter-harmonic distortion on the fault location. The power system network is modeled utilizing the PSCAD software and the simulation outputs are used to test the distance protection. In the model, the ripple control source with frequencies of 183.3 Hz and 216.6 Hz is used. The performed tests show that the distance protection ability to locate the fault is not significantly influenced by the ripple control signal.**

Keywords: ripple control; interharmonics; PSCAD; power system; fault location.

I. Introduction

Load management is used to control and adjust the electrical load and create a balance between the power supply and the electrical load. The load management techniques allow to redistribute the power demand to off-peak periods and lower the operating costs. Usually, the prime objective is not reducing the electricity consumption but changing the usage pattern.

Controlling the load can be achieved by several ways. Special tariffs or pricing mechanisms can be used to influence the customers' behavior. Some methods focus on direct control of the load. The consumption pattern can be adjusted by remote connecting/disconnecting the devices by power utilities or time clocks can be installed by consumers.

A common load management method is ripple control. Its principle is based on superimposing a signal with frequency usually in range of 100 to 1600 Hz onto a fundamental frequency of the power system (e.g. 50 or 60 Hz). This signal is then transmitted via the grid to the receiver devices connected to controllable loads. This signal, however, causes additional harmonic or interharmonic distortion, which can affect the power system operation.

The harmonics impact on electromechanical and microprocessor relays was a matter of subject of several papers [1]-[12]. Some studies [1],[2] presented practical tests consisting of sending a current signal accompanied by harmonic distortion to the relays. It was found that the mixed-harmonics influence on the protective relays is minor, conversely, the influence of pure signals above the fundamental frequency showed a significant effect on protection relays operation.

Evaluation of the power quality impact on protective relays using simulation models was presented in [3],[4]. These studies were focused on the advantages and disadvantages of the relay algorithm. Problems associated with harmonic distortion caused by non-linear loads were also briefly discussed. A review related to various methodologies for harmonics detection and measurement was introduced in [5]. Some studies explored the research associated with quantifying the cost of power supply reliability and quality. Various power quality disturbances were investigated and possible methods of quantifying both the impact on protective relays and the costs were presented in [6][7]. In [8],[9], the harmonics impact on differential relays is analyzed using Electromagnetic Transients Program (EMTP) simulations.

Potential effect of the interharmonic distortion on the relay operation is discussed in [10]. The study [11] found that accuracy of digital relays that use time window in range from 1 to 3 cycles for evaluations could be influenced by the interharmonics presence. According to IEC 61000-2-1 standard, the interharmonics are defined as follows:

Interharmonics are frequencies which are not an integer of the fundamental. They appear as discrete frequencies. Generally, the impact of unwanted interharmonics in power system can be summarized with the following points:

- Saturation of current transformers,
- Interference with protection and controller signals,
- Thermal effects.

The impact of the interharmonics on the relay filters can be determined according to the window length and frequency response. The filter ability to remove unwanted components for 1/2 interharmonic needs 2 cycles, 1/3 interharmonic needs 3 cycles. A 1/3 interharmonic is defined as a multiple of 1/3 of the fundamental frequency with interharmonic, e.g. 56.7 Hz or 183.3 Hz. A 1/2 interharmonic is a multiple of 1/2 of the fundamental frequency with interharmonic, e.g. 25 Hz or 175 Hz.

II. Description of the Modelled Network

To test the interharmonics impact on the fault location, a series of simulations was performed. The studied power system (displayed in Figure 1.) was modelled using PSCAD software (version 4.6.0.0, Manitoba HVDC Research Centre, Winnipeg, MB, Canada). The model consists of a 110 kV source with a superimposed interharmonics voltage source, a 110/23 kV transformer and a transmission line with lengths of 40 km. To the end of the line, a 2 MW load is connected.

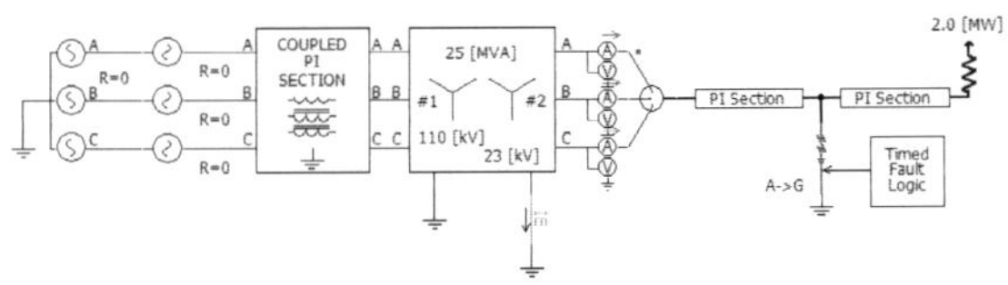

Figure 1. The PSCAD model

Parameters of the modelled system are listed in Table 1. The transmission line is divided into two parts, separated by functional blocks representing the single-phase-to-ground fault. Voltage and current data are measured at the beginning of the line. To record the data in a format commonly used by Intelligent Electronic Devices (IEDs), the RTP/COMTRADE Recorder component was implemented into the simulation.

Table 1. The model parameters

Source	
Voltage (phase-to-phase)	$U = 110$ kV
Frequency	$f = 50$ Hz
Positive-sequence impedance	$\overline{Z} = (1.981 + 19.809)\ \Omega$
Zero-sequence impedance	$\overline{Z}_0 = (1.348 + 13.484)\ \Omega$
Transformer 110/23 kV	
Connection	YN/yn
Power	$S = 25$ MVA
Positive-sequence leakage reactance	11 %
Eddy current losses	0.06 %
Copper losses	0.32 %
Transmission Line	
Length	$l = 25$ km
Positive-sequence resistance	0.259 Ω/km
Positive-sequence reactance	0.368 Ω/km
Positive-sequence capacitive reactance	685 kΩ/km
Ratio of zero-sequence resistance to positive-sequence resistance	1.5
Ratio of zero-sequence surge impedance to positive-sequence surge impedance	1.5

The ripple control signal was modelled using the superimposed ideal voltage source. The voltage level (phase-to-ground) and the frequency of this source was altered to simulate various scenarios. Two common ripple control frequencies were tested - 183.333 Hz and 216.666 Hz.

To determine the fault distance l_f, the voltage and current data were processed by the Fast Fourier Transform and then, the outputs were used to calculate the reactance to the fault X_{Lf}. For the single-phase-to-ground fault in the L1 phase, the equation (1) is applied [12]:

$$X_{Lf} = \frac{U_{L1E}}{I_{L1}} \frac{\sin\varphi_{L1} - \frac{I_E}{I_{L1}}\frac{R_E}{R_1}\sin\varphi_{L1E}}{1 - \left(\frac{X_E}{X_1} + \frac{R_E}{R_1}\right)\frac{I_E}{I_{L1}}\left(\cos\varphi_{L1} - \cos\varphi_{L1E}\right)^2 + \frac{X_E}{X_1}\cdot\frac{R_E}{R_1}\left(\frac{I_E}{I_{L1}}\right)^2}, \quad (1)$$

where U_{L1E} is the phase to ground voltage in the L1 phase, I_{L1} the short-circuit phase current in the L1 phase, I_E the short-circuit earth current, φ_{L1} the angle between U_{L1E} and I_{L1} and φ_E is the angle between U_{L1E} and I_E.

The real part of the residual compensation factor can be written as

$$k_{RE} = \frac{R_E}{R_L} = \frac{1}{3}\left(\frac{R_0}{R_1} - 1\right) \quad (2)$$

and the imaginary part as

$$k_{XE} = \frac{X_E}{X_1} = \frac{1}{3}\left(\frac{X_0}{X_1} - 1\right), \quad (3)$$

where $\frac{R_E}{R_L}$ and $\frac{X_E}{X_1}$ are the residual compensation factors set in the distance relay, R_1 and R_0 the positive- and zero-sequence line resistance and X_1 and X_0 the positive- and zero-sequence line reactance.

The fault distance can be computed using the ratio of the fault reactance X_{Lf} and the known value of the line reactance per kilometer X_k (Ω/km):

$$l_f = \frac{X_{Lf}}{X_k}. \quad (4)$$

In the modelled power system, single-phase-to-ground faults were simulated, and the voltage and current data measured at the beginning of the transmission line were used to calculate the reactance to the fault X_{Lf}. Values of the residual compensation factors and the line reactance X_k used for calculations are listed in Table 2. These values were estimated directly for the modelled transmission line.

Table 2. Values set in the distance relay

$\frac{R_E}{R_L} = 0.167$
$\frac{X_E}{X_1} = 0.417$
$X_k = 0.368$ Ω/km

The solution time step of the simulation was 100 µs, the duration of run 3 s. The simulated single-phase-to-ground faults emerged in time of 1.5 s.

III. Simulation Results

Figure 2. displays a situation, when a single-phase-to-ground fault occurred on the transmission line in a distance of 20 km from the terminal. Then, the voltages have been processed by the Fast Fourier Transform and the rms values of relevant harmonic components were plotted in Figure 3. and Figure 4. .

In Figure 3. , only fundamental frequency voltage source is assumed, so the harmonics are not present in the pre-fault and post-fault steady-state phase voltage. Figure 4. shows the fault voltage influenced by adding the ripple control signal source with voltage 5.716 kV (9 % of the fundamental voltage) and frequency 216.666 Hz. As seen in Figure 4. , not only the level of harmonics has changed, but also the fundamental frequency voltage significantly varied during the run of simulation. Therefore, the aim of this paper is to analyze the potential effect of these variations on the distance relay and the accuracy of the conventional fault locator algorithm.

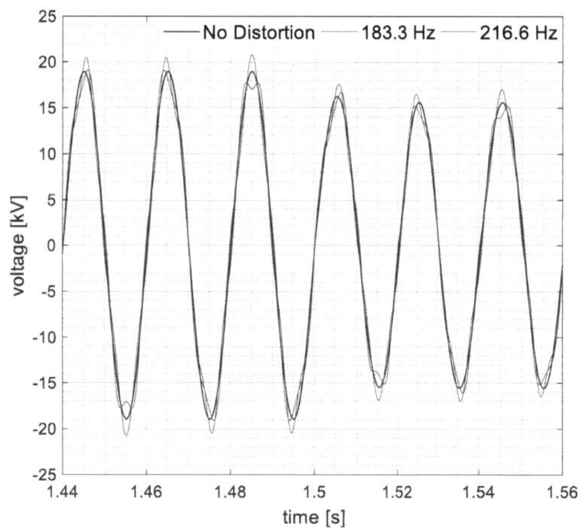

Figure 2. Comparison of voltages in the faulted phase with/without the ripple control signal (fault distance 20 km, fault resistance 1 Ω, ripple control voltage 5.716 kV).

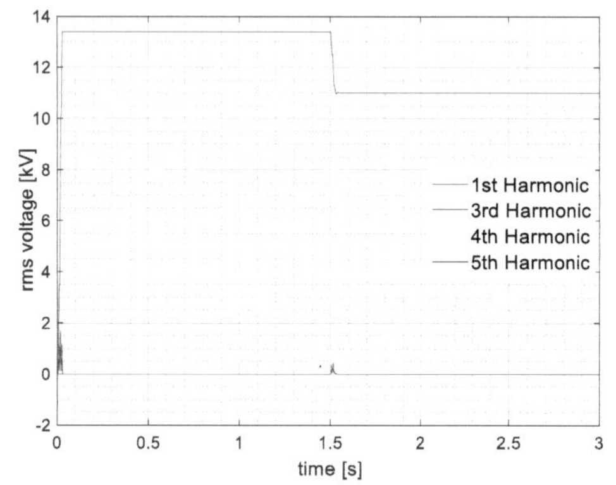

Figure 3. The rms values of the harmonic components of the faulted phase voltage without the ripple control influence (fault distance 20 km, fault resistance 1 Ω).

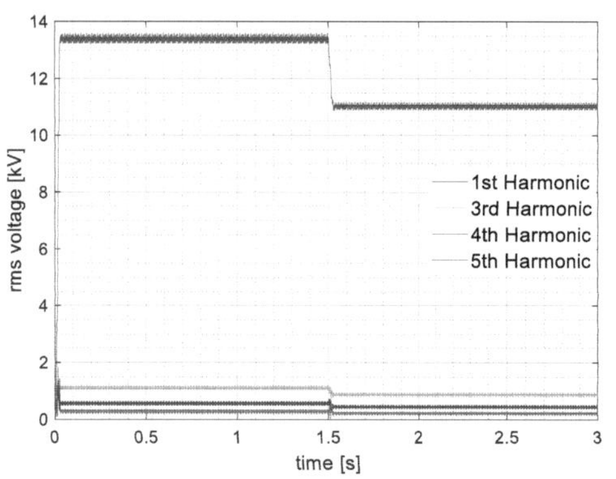

Figure 4. The rms values of the harmonic components of the faulted phase voltage considering the 216.666 Hz ripple control source with voltage level 5.716 kV (fault distance 20 km, fault resistance 1 Ω).

Table 3 summarizes results of the fault locator algorithm testing with varying fault distance. During all tests, the fault resistance was 1 Ω and the level of the ripple control signal (if present) was 9 % of the fundamental frequency voltage (i.e. 5.716 kV). The distance was evaluated in a time window of 20 periods. From the results, it is evident that the distance evaluated by the conventional fault locator algorithm defined by equation (1) is almost the same for all simulated scenarios, e.g. without/with interharmonic distortion.

The relative error was calculated according to equation (5) and the level of this error did not exceed 0.131 %.

$$\text{Rel. error} = \frac{|\text{Estimated distance} - \text{Actual distance}|}{\text{Actual distance}} \cdot 100 \quad (5)$$

Table 3. Results for various fault distances and the time window of 20 periods

Actual Distance (km)	No Interharmonics		With 183.3 Hz Interharmonic		With 216.6 Hz Interharmonic	
	Evaluated distance (km)	*Rel. error* (%)	*Evaluated distance* (km)	*Rel. error* (%)	*Evaluated distance* (km)	*Rel. error* (%)
5	5.006545	0.131	5.006537	0.131	5.006566	0.131
10	10.010162	0.102	10.010132	0.101	10.010201	0.102
15	15.014152	0.094	15.014096	0.094	15.014205	0.095
20	20.018629	0.093	20.018543	0.093	20.018697	0.093
25	25.023707	0.095	25.023898	0.094	25.023789	0.095

To evaluate the impact of the time window used for fault distance evaluation, another series of test was performed. The parameters of the simulation remained the same as for the previous test, but the time window was shortened to 3 periods. The results are in Table 4. The fault occurred in the distance of 10 km from the measuring terminal. To compute the fault distance, 3 periods (60 ms) of the signal in a short-circuit steady-state area are used. The short circuit is present in time from 1.5 s to 3 s. As seen in Table 4, in the scenario without interharmonic distortion, the distance computation results are the same for any 60 ms evaluation window of the analyzed signal. This does not apply for the signals distorted with interharmonics. Depending on the „position" of the 60 ms evaluation window, the relative error of the calculation slightly changes.

Table 4. Results for various fault distances and the time window of 3 periods

Time Range (s)	No Interharmonics		With 183.3 Hz Interharmonic		With 216.6 Hz Interharmonic	
	Evaluated distance (km)	*Rel. error* (%)	*Evaluated distance* (km)	*Rel. error* (%)	*Evaluated distance* (km)	*Rel. error* (%)
1.69 – 1.75	10.01016 2	0.10 2	10.00913 7	0.09 1	10.00988 8	0.09 9
1.94 – 2.00	10.01016 2	0.10 2	10.00845 8	0.08 5	10.01272 4	0.12 7
2.00 – 2.06	10.01016 2	0.10 2	10.00926 3	0.09 3	10.00876 9	0.08 8
2.30 – 2.36	10.01016 2	0.10 2	10.00903 6	0.09 0	10.01007 2	0.10 1
2.61 – 2.67	10.01016 2	0.10 2	10.01077 4	0.10 8	10.01102 9	0.11 0

One of the main factors that significantly influence the one-terminal algorithms used for the distance calculation is the fault resistance. As seen in Table 5, in a power system with a purely 50 Hz voltage source, the inaccuracy of the fault location estimation is the higher, the higher the fault resistance is. The same can be said about a system influenced by the ripple control signal. Moreover, the presence of a higher fault resistance causes slightly bigger differences between individual scenarios.

Table 5. Results for various fault resistances for the fault in the distance of 10 km

Fault Resistance (km)	No Interharmonics		With 183.3 Hz Interharmonic		With 216.6 Hz Interharmonic	
	Evaluated distance (km)	*Rel. error* (%)	*Evaluated distance* (km)	*Rel. error* (%)	*Evaluated distance* (km)	*Rel. error* (%)
0	9.998890	0.011	9.998902	0.011	9.998944	0.011
1	10.010162	0.102	10.010132	0.101	10.010201	0.102
5	10.055611	0.556	10.057629	0.576	10.057783	0.578
20	10.229981	2.299	10.229426	2.294	10.229705	2.297
50	10.586897	5.869	10.586267	5.862	10.586340	5.863

IV. EXPERIMENTAL ASSEMBLY

To test the interharmonic impact on a real distance protection, an experimental setup displayed in Figure 5. was assembled. This setup comprises a physical protection relay (7SA61x), Omicron® (CMC 256-6), manageable Ethernet switch (Hirschmann RS20), waveform recording instrument (Yokogawa DL850/DL850V ScopeCorder) and a PSCAD software.

Current and voltage data obtained from PSCAD simulations were stored in a COMTRADE file and the testing source Omicron CMC 256-6 was used to playback these records. Numerical filters of the digital protection 7SA61x perform Fourier analysis of the input data. The sampling frequency of the protection is 1 kHz. The aim of this experiment was to verify that even a real digital distance protection evaluates the fault distance correctly when a signal distorted with interharmonics is processed.

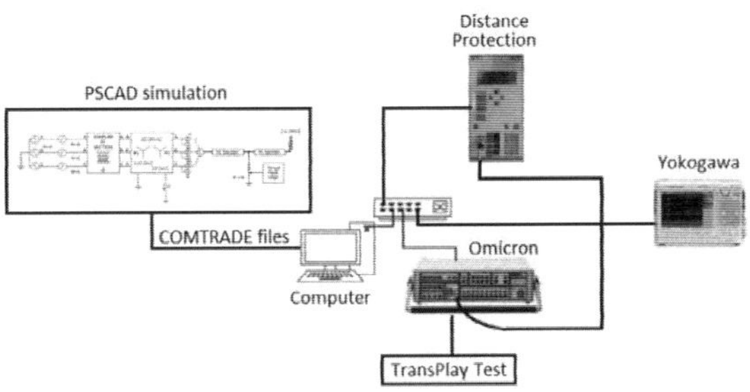

Figure 5. Laboratory setup for transmission line distance protection.

The results proved that the distance protection can eliminate the negative effect of the interharmonic distortion. Usually, the ripple control signal level is 1.6 to 2.5 % of the fundamental frequency voltage. The tested level of 9 % is therefore very high. However, even the interharmonic distortion of such level did not prove to have a significant effect on the distance protection operation.

V. CONCLUSION

This paper discusses an interharmonic impact on fault location and distance protection operation. The aim was to verify that the ripple control signal does not influence the fault distance calculation nor the protection relay operation. Computations based on the PSCAD simulation data showed only minor differences when comparing signals with and without the ripple control influence. Even the tested physical distance protection 7SA61x proved to be able to suppress the effect of interharmonics.

ACKNOWLEDGMENT

This research work has been carried out in the Centre for Research and Utilization of Renewable Energy (CVVOZE). Authors gratefully acknowledge financial support from the Ministry of Education, Youth and Sports of the Czech Republic under BUT specific research programme [project No. FEKT-S-17-4784].

REFERENCES

[1] L. Al-Musawi, A.Waye, W. Yu, N. Al-Mutawaly, "The effects of waveform distortion on powerprotection relays," In Proceedings of the International Protection Testing Symposium, Feldkirch, Austria, October 2015.

[2] G. Achleitner, L. Fickert, C. Obkircher, M. Sakulin, T.; IFEA, Ö.I.B. AG, "Earth fault distance protection," In Proceedings of the 20th International Conference and Exhibition on Electricity Distribution, Prague, Czech Republic, 8–11 June 2009; pp. 1–4

[3] T.W. Cease, S.A. Kunsman, A. Apostolov, J.R. Boyle, P. Carroll, D. Hart, G. Johnson, G. Kobet, M. Nagpal, K. Narendra, "Protective Relaying and Power Quality," IEEE PSRC Working Group Report; IEEE: Piscataway, NJ, USA, 2003; pp. 1–65.

[4] B. Kumar, "Design of Harmonic Filters for Renewable Energy Applications," Master's Thesis, Gotland University, Visby, Sweden, 2011.

[5] S.R. Durdhavale, D.D. Ahire, "A review of harmonics detection and measurement in power system," Int. J. Comput. Appl. 2016, 143, 42–45.

[6] P.S. Leao, G.C. Barroso, N.X. Melo, R.F. Sampaio, J.A. Barbosa, F.L. Antunes, "Numerical relay: Influencedby and accessing the power quality," In Book Power Quality, 1st ed. Eberhard, A., Ed. InTech: Rijeka, Croatia, 2011, pp. 213–236, ISBN 978-953-307-180-0.

[7] S. Elphick, P. Ciufo, V. Smith, S. Perera, "Summary of the economic impacts of power quality on consumers," In Proceedings of the Australasian Universities Power Engineering Conference (AUPEC), Wollongong, Australia, 27–30 September 2015, pp. 1–6.

[8] J.M. Ho, C.C. Liu, "The effects of harmonics on differential relay for a transformer. In IEE Conference Publication," Institution of Electrical Engineers: London, UK, 2001, pp. 2–34.

[9] P. Rafajdus, P. Bracinik, P. Hrabovcova, J. Saitz, J. Altus, M. Höger, J. Pyrhönen, "Examination of Instrument Transformers for Their Employment in New Fault Location Method," ADE: Bologna, Italy, 2012; pp. 4585–4595.

[10] F. Wang, M.H.J. Bollen, "Measurement of 182 Hz interharmonics and their impact on relay operation," In Proceedings of Ninth International Conference on Harmonics and Quality of Power (Cat. No. 00EX441), Orlando, FL, USA, 1-4 Oct. 2000, pp. 55-60.

[11] W. Jin, H. Liu, Z. Li, Z. Qiao, A. Ji, H. Wang, "Interharmonic measurement of three typical interharmonic sources and result analysis," Electric Power Automation Equipment, vol. 12, 2010, pp. 30-35.

[12] G. Ziegler, „Numerical Distance Protection," Munich, Siemens MCD 1999.

Modified Discounted Payback Period of Various Small-Scale Photovoltaic Installations

František Janíček, Ján Poničan, Matej Sadloň
Faculty of Electrical Engineering and Information Technology
Slovak University of Technology in Bratislava
Bratislava, Slovak Republic
frantisek.janicek@stuba.sk, jan.ponican@stuba.sk, matej.sadlon@stuba.sk

Abstract — **The article analyzes photovoltaic installation payback periods of various household customer categories under several defined scenarios. These include a real time consumption, an energy accumulation system use and a subscription to a virtual battery service provided by a specific electricity supplier. An indicator used for an installation optimization is a profitability index. Research shows that a household installation profitability largely depends on a consumption curve that thus constitutes a core element that should be considered when selecting the optimal PV installation. However, PV installation price, electricity price, state support schemes, energy accumulation systems and many other factors influence the payback period and hence are discussed in the article.**

Keywords: photovoltaic systems; payback period; net metering; renewable energy sources support; energy storage; profitability; optimization; standard consumption diagram

I. Introduction

EU energy policy framework Clean energy for all Europeans is aiming towards the clean energy transition via several regulatory instruments. The revised Renewable Energy Directive sets a new target of at least 32% share of energy from renewable sources in the Union's gross final consumption of energy in 2030. [1] This fact makes a pressure on Member States to enact more and more suitable support schemes in order to ensure the development of renewable energy sources (RES).

Renewable energy sources development potential of Slovakia lies in biomass and solar energy exploitation. [2] Installed capacity of both sources was heavily increased under support schemes encapsulated in the Act on RES Support almost a decade ago. Despite this fact, the room for further development in the exploitation of these sources in Slovakia is still huge.

However, not all technically feasible installations are also economically viable. Energy markets, especially the retail electricity market, is a typical example of a deformed nontransparent market without a demand response. Due to this, it appears to be much easier for an electrical installation producer to sell various power generating systems that might not be very efficient from an economical point of view.

This seems to be the case especially on the small-scale photovoltaic (PV) installations market. Final consumers considering an investment into a PV system have many options. There are various suppliers, various energy accumulation options, various types and rated powers of installations. In addition to that, these are not the only factors that need to be taken into consideration. The answer to a very frequently asked question, whether a PV installation is worth it or not, is a very complex one. Other facts, such as locality, load profiles, electricity prices and support schemes need to be considered.

Thus, the aim of the article is to identify and compare real payback periods of investments into a small-scale PV system for various final customer categories and for various scenarios. Market behavior of households and small businesses is hardly comparable, since their electricity consumption is usually located in different parts of the day. Small businesses consume the electricity mostly within working hours and their consumption peak is usually close to the production peak of the PV installation. Contrarily, households´ consumption peaks are located before or after the working hours. This has a significant impact on the amount of surplus electricity that cannot be consumed in a real time. Therefore, the article focuses only on household customers, the so-called prosumers. It appears that an in-depth assessment is much needed here, because payback periods stated by suppliers seem to be very misleading and do not reflect all the above-mentioned elements. Moreover, PV markets are very dynamic. Both energy accumulation appliances and PV panels are getting cheaper, electricity market participants come with novel services and the focus of a state support is changing.

II. Methodology

One of the simplest and still the most frequently used investment assessment methods is the payback period analysis. However, its major drawback is the fact that it does not take into consideration time value of money. Discounted payback period has overcome this error and reflects investor´s cost of capital. [3] On the other hand, none of these methods is suitable for all sorts of investment analyses, especially in case when investor is a household and not a profit-oriented entity. Thus, a full cost of capital calculation inclusion would be very misleading. Therefore, this article is based on a modified discounted payback period method.

Core of each payback calculation is a correct inflows and outflows definition. Small-scale PV installation is a specific investment, because its major part is made up of sunk costs, i.e. price of panels, wiring, project planning, inverter, switchgear and installation work performed by a certified undertaking. In comparison to these, its operating costs are insignificant and therefore negligible. Outflows represent electricity savings resulting from the consumption of self-produced electricity. Here, it is important to understand that, since a typical household consumption peaks are located before and after working hours and the PV installation produces electricity mostly during working hours, an efficiency maximization in principle requires the highest possible share of electricity used in real time and the lowest possible amount of surplus electricity. However, this principle might be altered by an energy accumulation appliance or a market service allowing the consumer to virtually store the surplus electricity. Thus, the produced electricity needs to be distinguished into electricity that can be used in a real time and a surplus electricity that is usually only donated to another market participant, usually the electricity supplier.

To make the calculations as precise as possible, catalogue data was used to set prices and technical details of the PV installations. Since the amount of a diffuse radiation is relatively high in our geographical conditions and polycrystalline panels are capable of its more efficient utilization, the article is based on a concrete PV panel designed by Jinko Solar, type JKM270P with specifications of 270 Wp, efficiency 16,5 % and a specific power of 165 Wp.m^{-2} achievable under standard STC test conditions, as a typical equipment reflecting a standard solution available on the market. [4]

When calculating electricity generation, irradiation, panel area, efficiency, performance factor and coefficient for losses need to be taken into consideration. Irradiation value depends on several factors. The major ones are locality and year season. This article only deals with Bratislava locality, assumes suitable PV panel location, no shading by surrounding obstacles and uses hourly irradiation data provided by PVGIS database in order to conclude as broadly applicable findings as possible. Average hourly irradiation values are used to define average hourly electricity generation values of PV installation.

As already mentioned above, household´s electricity savings depend on its ability to consume as much of the generated electricity as possible in the real time or its ability to store it physically or virtually in order to use it in periods when the PV installation does not generate enough electricity. Thus, another decisive input in defining the suitable PV installation is consumption. The article is based on values defined by standard consumption diagrams (SCDs) published by Západoslovenská distribučná a.s. (ZSD) and electrical connection details, i.e. number of phases and maximum reserved capacity of main circuit breaker. [5] Average daily consumption curve is crucial for inflows calculation, because it sets the amount of both surplus electricity and electricity that can be consumed in a real time. The surplus electricity is generated in those periods of the day, when the actual power of PV installation exceeds the power consumed by the household. Average hourly generation and consumption values are used to set profits (inflows). These are then compared to investment costs (outflows) in order to define the profitability index, which is used to select the optimal PV installation and its installed capacity. Examples of average hourly values of generation and consumption may be seen in Figure 1 and 2.

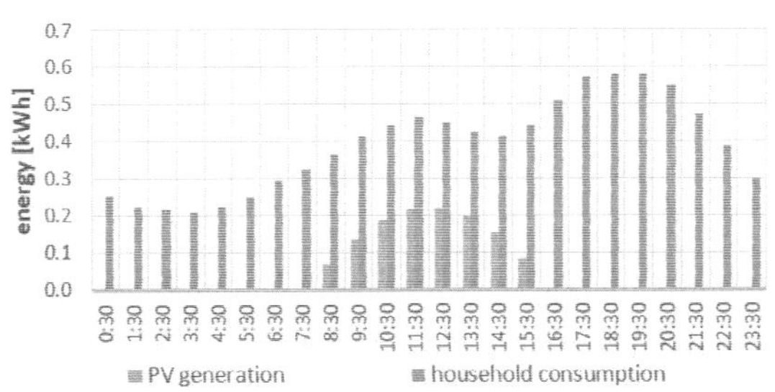

Figure 1. Example of an average hourly values of generation and consumption in January (SCD4, 3 MWh, 1,08 kWp PV installation).

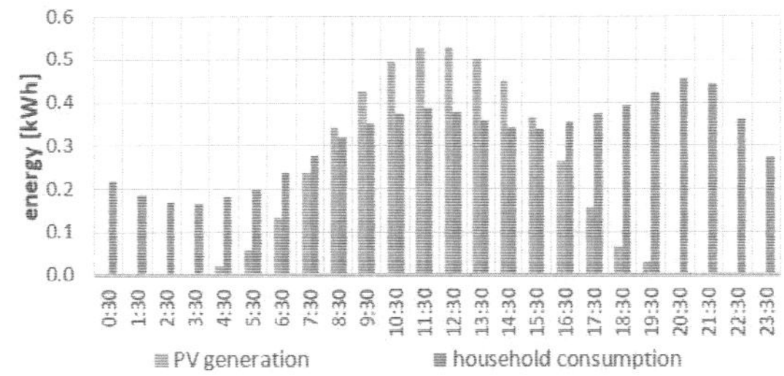

Figure 2. Example of an average hourly values of generation and consumption in June (SCD4, 3 MWh, 1,08 kWp PV installation).

However, an identification of optimal PV installation is far more complex and depends on other factors, too. This article focusses on comparison of optimal PV installations payback periods in three substantially different scenarios concerning households. First one can be defined as an online consumption, because it does not cover any form of accumulation solution, neither physical, nor virtual. Therefore, usable is only that portion of the electricity generation, which is consumed in the real time. Surplus electricity needs to be sent to the grid. But it is important to keep in mind that the electricity may only be sent to the grid in case it is sent to another market participant. In distribution grid operated by ZSD, households are allowed to dispose of their surplus electricity. Nevertheless, the electricity is usually only donated to the supplier. Second scenario discusses the payback period in case of energy accumulation system use. Originally lead-acid traction batteries were used for the surplus's storage. These are currently pushed out of the market by lithium, lithium-ion (Li-ion) or lithium iron phosphate (LiFePO4) batteries due to a longer lifetime. [6] The article is based on a concrete battery designed by LG, type Chem RESU 7 kWh with a life of approximately 6000 charging cycles, as a typical equipment reflecting a standard solution available on the market. [7] Physical accumulation use increases the usability of electricity generated by the PV installation, since the surplus electricity is stored in a battery and may be used in periods when power generation is insufficient. On the other hand, accumulation system investment is significant and therefore increases the outflows. And the last scenario deals with a novel service on Slovak electricity market "virtual battery" offered by Západoslovenská energetika, a.s. (ZSE). Service in question is basically a net metering allowing prosumers to virtually use the surplus electricity generated by their PV installation in periods when their PV generation is insufficient. However, this is in fact done only through invoicing, i.e. no physical energy storage is included. Naturally, a prerequisite of net metering is a smart meter measuring both generation and consumption of electricity. A settlement invoice states both surplus electricity generated by the PV installation, which is virtually stored in a grid, and grid electricity consumed in periods when the PV generation was insufficient. Surplus electricity, virtually stored in a grid, is deducted from the consumed grid electricity and an amount of grid electricity that was virtually the only consumed grid electricity is calculated. This means that the household is monthly invoiced for the whole amount of consumed electricity coming from the grid and only afterwards, at the annual settlement, its virtually stored electricity is deducted from the consumed grid electricity amount. Here it is important to understand that the electricity price has several components and savings related to this service cover only the cost of electricity supply, which stands for approximately 30 to 40 % of the electricity price, depending on customer´s tariff. In addition to that, the service in question is connected with a monthly fee of 2 € per month.

SCDs distinguish two major household customer categories. SCD4 represents single-tariff products specified for household customers without electric heating and SCD5 represents double-tariff products designed for household customers with storage heating and domestic hot water heating. The difference of the curve may be seen in Figure 3 and Figure 4. The article distinguishes these categories even further into 8 specific groups based on both the SCD and an average annual electricity consumption. SCD4 is therefore divided to households with annual electricity consumption of 3, 4, 5, 6 MWh and SCD5 is divided into households with annual electricity consumption of 4, 5, 6, 12 MWh.

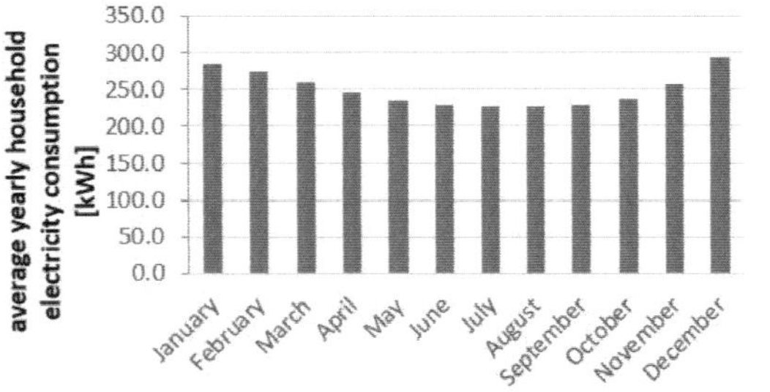

Figure 3. Household electricity consumption by months (SCD4, 3 MWh).

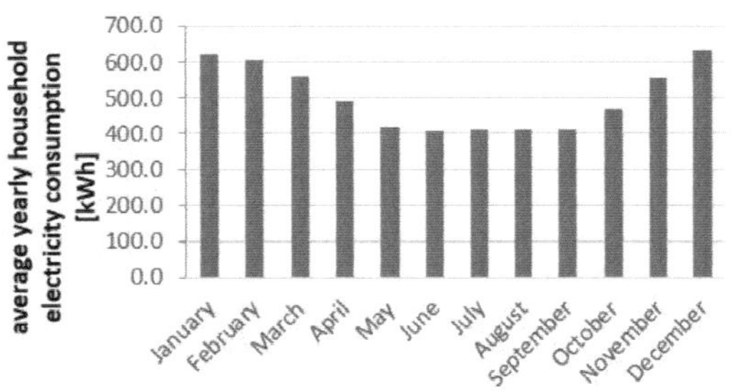

Figure 4. Household electricity consumption by months (SCD5, 6 MWh).

Finally, the article does not consider any forms of state support even though national RES support project "Zelená domácnostiam II." currently provides vouchers financially supporting small-scale PV-installations, i.e. installations with installed capacity of up to 10 kW. [8]

III. Findings

Fundamental economic indicator used for an investment assessment is the profitability index. It represents the ratio between inflows and outflows, in this case electricity savings price and PV installation investment. Therefore, it was essential to define it for each relevant PV installation. Moreover, in order to conduct a broad appraisal of all the relevant PV installation options concerning a particular defined household consumer category a line graph defining a relation between the profitability index and the installed capacity is the most relevant solution.

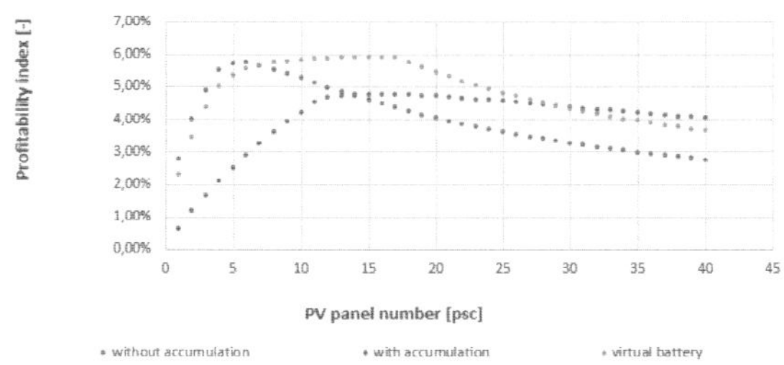

Figure 5. Example of profitability index fluctuation according to the PV panel number for individual scenarios (SCD5, 5 MWh).

By identifying the highest profitability index, the optimal PV installation was selected. Afterwards, the payback period was calculated. For the sake of clarity, the data is recorded in table 1 and 2. This approach seems to be the most appropriate due to the fact that the calculation model defines a particular PV installation with the most remunerative payback for each of the above described operational scenarios, i.e. online consumption, physical accumulation and virtual battery. Thus, this method allows to perform a clear economic comparison.

Table 1. PV installations optimized for household customer category SCD4 and their details.

PV instalation optimization		online consumption	physical accumulation	virtual battery
TDO4 household and yearly consumption ≈3000 kWh				
highest profitability index	[%]	4,91%	4,33%	4,98%
PV panels number	[pcs]	4	13	10
installed PV capacity	[kWp]	1,08	3,51	2,7
average PV installation price	[€]	3 300 €	12 000 €	5 500 €
payback period	[year]	19	21	19
average annual PV generation	[kWh]	1170	3810	2940
TDO4 household and yearly consumption ≈4000 kWh				
highest profitability index	[%]	5,64%	4,67%	5,65%
PV panels number	[pcs]	5	14	13
installed PV capacity	[kWp]	1,35	3,78	3,51
average PV installation price	[€]	3 700 €	12 400 €	6 400 €
payback period	[year]	17	20	17
average annual PV generation	[kWh]	1470	4410	3820
TDO4 household and yearly consumption ≈5000 kWh				
highest profitability index	[%]	6,25%	4,99%	6,21%
PV panels number	[pcs]	6	16	15
installed PV capacity	[kWp]	1,62	4,32	4,05
average PV installation price	[€]	4 200 €	12 900 €	7 000 €
payback period	[year]	15	19	16
average annual PV generation	[kWh]	1760	4700	4400
TDO4 household and yearly consumption ≈6000 kWh				
highest profitability index	[%]	6,78%	5,28%	6,71%
PV panels number	[pcs]	7	17	14
installed PV capacity	[kWp]	1,89	4,59	3,78
average PV installation price	[€]	4 500 €	13 200 €	6 800 €
payback period	[year]	14	18	14
average annual PV generation	[kWh]	2050	5000	4110

Table 2. PV installations optimized for household customer category SCD5 and their details.

PV instalation optimization		online consumption	physical accumulation	virtual battery
TDO5 household and yearly consumption ≈4000 kWh				
highest profitability index	[%]	5,16%	4,49%	5,38%
PV panels number	[pcs]	5	14	13
installed PV capacity	[kWp]	1,35	3,78	3,51
average PV installation price	[€]	3 700 €	12 400 €	6 400 €
payback period	[year]	18	21	18
average annual PV generation	[kWh]	1470	4410	3820
TDO5 household and yearly consumption ≈5000 kWh				
highest profitability index	[%]	5,73%	4,78%	5,91%
PV panels number	[pcs]	6	16	17
installed PV capacity	[kWp]	1,62	4,32	4,59
average PV installation price	[€]	4 200 €	12 900 €	7 600 €
payback period	[year]	17	20	16
average annual PV generation	[kWh]	1760	4700	5000
TDO5 household and yearly consumption ≈6000 kWh				
highest profitability index	[%]	6,22%	5,05%	6,36%
PV panels number	[pcs]	7	17	18
installed PV capacity	[kWp]	1,89	4,59	4,86
average PV installation price	[€]	4 500 €	13 200 €	7 900 €
payback period	[year]	15	19	15
average annual PV generation	[kWh]	2050	5000	5280
TDO5 household and yearly consumption ≈12000 kWh				
highest profitability index	[%]	8,34%	6,41%	8,42%
PV panels number	[pcs]	12	25	22
installed PV capacity	[kWp]	3,24	6,75	5,94
average PV installation price	[€]	6 200 €	15 300 €	9 000 €
payback period	[year]	12	15	12
average annual PV generation	[kWh]	3520	7340	6460

For example, in case of a household customer category SCD5 with an annual electricity consumption of 5 MWh, the highest profitability index among all the scenarios is 5, 91 % achieved with the virtual battery. This result concerns a 4, 59 kWp PV installation with 17 panels and an indicative investment cost of € 7600. Its expected annual electricity generation during the first year is 5000 kWh and the payback period is 16 years. Secondly, for the online consumption, the index is lower, but not significantly, with a number of 5, 73 %. This ratio represents a PV installation with 6 panels and an indicative cost of an investment of approximately € 4100. Despite the fact that the investment itself is significantly lower, the payback period is extended to 17 years. This is caused by a significantly smaller scale of an installation and thus a significantly smaller expected annual PV generation, i.e. only 1760 kWh in the first year. Finally, the highest profitability index in case of the last scenario, the physical accumulation stands only for 4, 78 %, i.e. 1, 13 % worse result than in case of the virtual battery. These numbers represent a 4, 32 kWp PV installation with 16 panels, an expected annual electricity generation of 4700 kWh in the first year and an indicative investment cost of € 12900. The heavily increased investment cost reflects the extension of the payback period up to 20 years.

The above described example confirms that for a majority of household customer categories, the virtual battery scenario is the most suitable one. However, profitability indexes of online consumption are very similar. This is caused by the fact that financial savings related to virtual battery scenario cover only a fraction of the virtually stored electricity price. Household´s savings related to virtual battery scenario may be distinguished into two categories. Firstly, savings resulting from the fact that a portion of consumed electricity is not bought from the supplier due to the online consumption of the PV generated electricity. And secondly, savings resulting from the settlement invoice, in which the electricity supply component of the electricity price related to the surplus electricity produced by the PV installation, which was virtually stored in a grid, is deducted from the electricity supply component of the electricity price paid for the grid electricity bought from the supplier in periods when the PV generation was insufficient. Simply put, with a virtual battery service, household still has to pay approximately 60 to 70 % of the electricity price of the surplus electricity, which was sent to the grid and virtually taken back. Yet, this is still justified by the fact that prosumer uses the grid to virtually store the electricity and thus creates expenditures for the distribution system operator. On the other hand, this service could potentially be reinforced by a state support consisting of a partial electricity price payment relief. In other words, prosumers could be relieved from some of the electricity price components in order to take into account their energy efficiency and RES implementation. Moreover, service itself is connected with a € 2/ month flat-rate charge. To sum up, a major portion of savings represents the real time consumption, because, in contrary to the surplus electricity stored in the virtual battery, it constitutes a cost cut reflecting 100 % of the electricity price. Thus, the prosumer is motivated to utilize as much of the PV generated electricity as possible in the real time. Nevertheless, the virtually stored electricity generally represents a profitable supplement to these savings.

On the contrary, in case of household customer categories SCD4 with an annual electricity consumption of 5 and 6 MWh, the profitability index is lower with the virtual battery service than in an online consumption scenario. This confirms the above-mentioned conclusion that savings are mainly made up of the online consumption. Therefore, an appropriate (relatively high) consumption values in periods of the PV generation, i.e. in working hours, constitutes high real time consumption savings and causes the profitability index of the online consumption to outrun the profitability index of the virtual battery. Calculations show that a SCD4 consumption curve causes the profitability index to be the highest in case of a virtual battery scenario only up to an average annual electricity consumption of 4, 3 MWh. For higher consumption values the highest profitability ratio is achieved with an online consumption scenario. Although, with a significantly lower installed capacity of a PV system and thus much lower profits after the investment return.

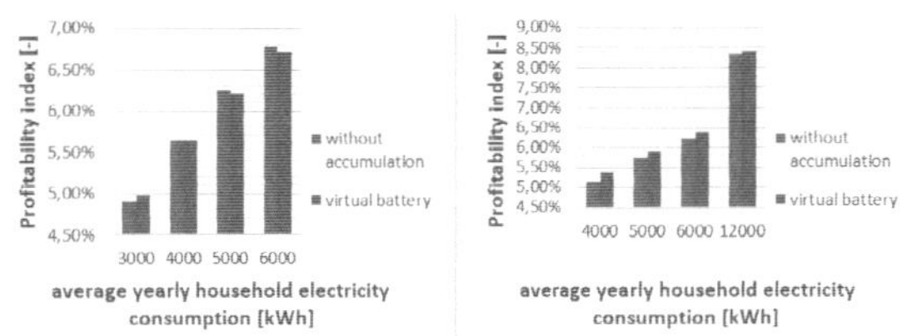

Figure 6. Profitability index comparison for individual household categories (left SCD4, right SCD5).

As already discussed, the results of the online consumption are usually very close behind the results of the virtual battery. On the other hand, the highest profitability indexes within the online consumption scenario were always achieved with very small PV installations. Graphs showed that after reaching the highest profitability ratio, a further increase of the installed capacity would mean a profitability index decrease. This is caused by the nature of the online consumption. The PV generated electricity may be used only in a real time and the daily household consumption curve causes that a majority of the PV generated electricity is only sent to the grid. This means that no electricity savings were generated, i.e. no profit was created. Moreover, the PV installation itself is relatively expensive and calculations showed that at a certain value, a further investment would be less profitable, i.e. the portion of generated electricity that could be used in a real time is decreasing and the cost of an investment is increasing.

Finally, in case of the physical accumulation, findings are straightforward. Numbers identified that a battery system is still a very expensive solution and extends the payback period due to a lower profitability index of such an investment. This may be seen on a complex profitability index appraisal showing that a battery accumulation used in a household is under current market conditions usually unprofitable and an increase in a battery capacity implies a profitability index decrease. Battery system price is fluctuating around € 800 / kWh of energy storage. Although, it is important to state that the price has a downward trend. Nevertheless, a physical accumulation system increases the investment costs more rapidly than it increases the electricity savings generated by the PV installation. In addition to that, the payback period of such an investment is generally longer than its battery life. Therefore, any investment into an accumulation system is currently irrational from an economic point of view.

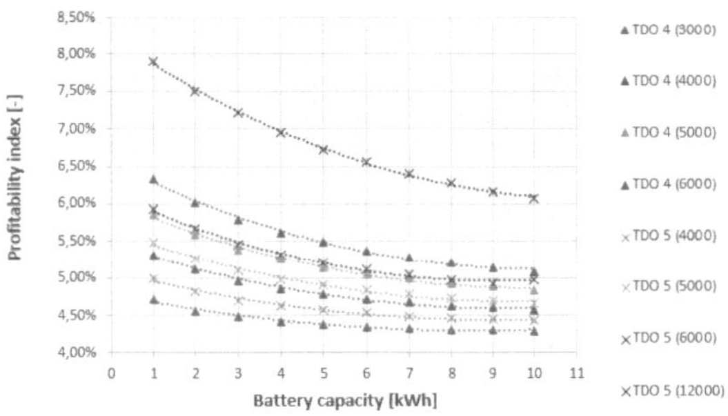

Figure 7. Graph showing a profitability index decrease caused by a battery capacity increase.

IV. Discussion

Virtual battery service is after all an effort of a private undertaking to implement a market solution proven abroad in order to convince household customers to invest into a PV installation. And, as findings showed, it does constitute a competitive advantage for ZSE and is an economically viable market solution for both contract parties, the supplier and the household. Nevertheless, on Slovak market, it is a novelty. This justifies the introductory conditions of the service allowing household customers to sign up for it only in a bundle with a brand-new PV installation from ZSE. This might have two economic reasons. Firstly, ZSE intends to perform a deep market impact analysis in order to consider a broader service offering with more convenient conditions. Secondly, ZSE solely intends to increase its share on the PV installation market by offering a unique selling point just in time when a state support scheme is being released.

Net metering is a RES support scheme originally introduced in the United States in order to combat non-dispatchable character of small-scale RES and to allow prosumers generating the electricity from RES to use it anytime, i.e. not only in the real time. [9] However, its nature is based only on a billing mechanism and hence it might be described by phrase virtual battery. The mechanism itself may have various versions. One of them is called net billing. Its major difference is that the surplus electricity sent to grid is not valued at the same price as the electricity consumed from the grid, whereas a standard net metering is the opposite. [10] Thus, in the net metering scheme prosumer can achieve a zero electricity bill provided that the annual amount of surplus RES generated electricity is equal or higher than the amount of consumed grid electricity. On the other hand, with net billing mechanism only a portion of the already paid grid electricity price would be returned, since the virtually stored electricity is priced at a lower value. The major advantage of the standard net metering is that it allows prosumers to use the electricity generated in summer during the winter season and the electricity generated in working hours during the evening. This reduces the need to generate electricity from standard large-scale installations. However, such a support comes with considerable costs for a support provider, i.e. the state, transmission or distribution system operator. This is caused by the fact that the market price of electricity is very dynamic, and the surplus electricity is usually injected to the grid in those daily periods when it is generally not needed and afterwards taken from the grid during the peak. These costs are subsequently reflected in the electricity price and thus shifted to all the customers. Therefore, a complex analysis comparing the impacts of previously used feed in tariffs and potentially used net metering under the Slovak conditions would be very relevant. Nevertheless, it might be assumed that net metering costs would be much lower.

As it may be seen from the above mentioned, it is clear that the ZSE service virtual battery is a specific version of net billing incorporated in a private sector. Now, a relevant question is whether the flat-rate charge of € 2/ month is set precisely, reflects all its costs and still provides a significant benefit for the prosumer. As it was discussed in findings, virtual battery under current conditions cannot be seen as a revolutionary support scheme, since the portion of the electricity price deducted from the settlement bill reflects only the electricity supply component and thus, may only be seen as a supplement helping to recover the PV investment costs in a faster way. Here it is crucial to keep in mind that it is not a role of a private undertaking to support the RES development. It is role of the state to adopt a suitable regulation providing sufficient incentives for the market to develop. To sum up, net metering even in such a reduced form is a very welcomed market solution for each household prosumer with a PV installation helping to utilize the surplus electricity generated by the PV installation.

In addition to that, as it was already mentioned in findings, prosumers subscribed to virtual battery service could be favoured by the regulator via partial electricity price relief. This should only concern the surplus electricity virtually stored in a grid and the component covering the system operation. Aforementioned is justified by the fact that prosumer contributes to the development of RES implementation, hence fulfilling the objective of respective price component. Such a measure would lead to approximately 10 % higher profit resulting from the virtual battery service. On the other hand, an intermittent grid injection of the surplus electricity brings certain risks to the distribution grid. A more difficult generation predictability goes hand in hand with balancing needs. Thus, it is up to a particular market participant, in this case supplier ZSE, to predict its electricity flows and minimize its deviations. This may be also the reason, why ZSE decided not to initially offer the respective service broadly, but only to a specific part of the market. Another aspect that should be considered is the fact that a virtual battery subscriber uses the distribution grid twice. Firstly, when the surplus electricity is injected to a grid. And secondly, when the virtually stored electricity is taken from the grid. This is naturally connected with double electricity losses incurred by the distribution system operator. Therefore, it may seem that there still is a significant room for further adjustment of the service.

Another interesting issue that should be discussed is what energy accumulation system price decrease is needed in order to make the battery system profitable for the prosumer. Modern battery accumulation systems have generally a life specified by the manufacturer of approximately 6,000 charging/discharging cycles. If a chemical battery degradation would be neglected and a one cycle per day operation would be assumed, meaning that the surplus electricity stored during the PV generation would be used by the household until the next day, a theoretical battery life of 15 years would be achieved. Thus, the longest possible payback period making the battery system theoretically profitable would be 15 years. Otherwise, the prosumer would need to make another investment reflecting an acquisition of a brand-new battery and possibly a replacement of some technical components, e.g. inverter, control electronics, etc. The required price decrease depends on details of the consumption, i.e. would be different for distinct household customer categories. For a household customer category SCD5 with an annual electricity consumption of 4 MWh a 7 kWh battery would be optimal. However, the price of accumulation system would have to be on a level of € 300/ kWh of storage capacity in order for a prosumer to have the investment back within 15 years. Naturally, this was calculated for the highest PV installation profitability index of 6, 31 %, i.e. installed PV capacity of 3, 24 kWp and a total investment of € 8260 covering both the PV installation and the accumulation system. In spite of the fact that the household accumulation system prices are decreasing from a long-term point of view, a € 500/ kWh decrease is very unlikely to occur in a relatively short timeframe. On the other hand, a major shortcoming of these calculations is neglecting an electricity price increase over time. This prognosis is highly probable and thus could potentially impact the profitability of household energy accumulation systems even more than the battery price reduction.

V. Conclusion

First of all, the aforementioned findings confirm the well-known premise stipulating, higher the household consumption, better the payback period in each analysed scenario. Secondly, numbers clearly show that the consumption curve has a crucial impact on the profitability index and thus, it is of a major importance to optimize the PV installation´s technical parameters according to all the above-mentioned factors, especially the consumption curve. Here the best practice would be to design the installation according to a real consumption data of the particular household. Nevertheless, the aim of the article was not to define a single specific PV installation as the most profitable one. The aim was to analyse what installation would be optimal for a standardized consumer category under common conditions. Thirdly, the accumulation system is still very expensive and thus unprofitable. Fourthly, profitability indexes are usually highest for the virtual battery scenario, but only slightly higher than in case of the online consumption. On the other hand, a closer investigation reveals drawbacks of the payment period method that does not assess the investment from a long-term point of view. It simply cannot address the situation after the return of the investment. Thus, in spite of the fact that the highest profitability indexes are very similar and that the payback periods are almost the same, the highest profitability index in case of the online consumption is always connected with a small installed capacity. Therefore, after the investment is returned, i.e. 12 to 19 years depending on the customer category, the profits related to virtual battery scenario will be generally much higher than in case of an online consumption. However, it is hard to predict, whether this service is going to be offered for more than two decades under at least as favourable conditions as today. Fifthly, under current market conditions, regardless of the assessed scenario, the prosumer is economically motivated to maximize the real time electricity consumption, i.e. to adjust the consumption according to the PV production. For this outcome to change a net metering introduction should be considered by the regulatory authority.

In addition to that, electricity price is very dynamic and thus may easily affect the PV installation payback period. The bottom line is, higher the electricity price, shorter the payback period. And, as was already stated above, an electricity price increase connected with the fossil fuel price increase might be expected in the upcoming decade. Another fact that should be taken into consideration is the state support project "Zelená domácnostiam II." currently providing vouchers for PV installations up to € 1500 and therefore constituting a significant factor influencing the return of an investment. For example, in case of a household customer category SCD5 with an annual electricity consumption of 4 MWh with optimal installations according to the above-mentioned findings, the payback period is 16 years for the virtual battery scenario, 17 years for the online consumption scenario and 20 years for the accumulation scenario. However, with a maximum subsidy eligible for the installed capacity of the particular PV installation, the payback period is reduced to 13, 14 and 17 years respectively. In case of SCD4 with an annual electricity consumption of 5 MWh, the payback periods of 16, 15 and 19 years respectively were reduced by the maximum subsidy eligible for concrete PV installation to 12, 13 and 17 years respectively. To sum up, the financial support helps the prosumer to recover the investment approximately 2 to 4 years faster.

Acknowledgment: This work was supported by the Slovak Research and Development Agency under the contract No. APVV-15-0326.

References

[1] European Commission. "Clean energy for all Europeans." Internet: https://ec.europa.eu/energy/en/topics/energy-strategy-and-energy-union/clean-energy-all-europeans, Mar. 20, 2019 [Mar. 25, 2019]

[2] F. Janíček, "Obnoviteľné zdroje energie 1: Technológie pre udržateľnú budúcnosť," Renesans. Pezinok, pp. 80

[3] F. Lefley, "The payback method of investment appraisal: A review and synthesis." International Journal of Production Economics, vol 44, Issue 3, pp. 207-22, Jul. 1996.

[4] Jinko Solar. "JKM270P-60 data sheet." Internet: https://jinkosolar.com/ftp/EN-JKM270P-60(4BB).pdf, Jan. 20, 2019 [Feb. 22, 2019]

[5] Západoslovenská distribučná, a. s. "Aktuálne Typové diagramy odberu." Internet: https://www.zsdis.sk/Uvod/Dodavatelia/Typove-diagramy-odberu/Hodnoty, Jan. 15, 2019 [Jan. 24, 2019]

[6] E. Martinez-Laserna, M. Berecibar, I. Gandiaga, L. Rodriguez-Martinez, I. Villarreal. " Realistic lifetime prediction approach for Li-ion batteries." Applied Energy, vol 162, pp. 839-852, Jan. 2016

[7] LG Chem. "LG Chem RESU 7H data sheet." Internet: http://www.windandsun.co.uk/media/803004/LG-Chem-RESU-LV-HV-Datasheet_Utility.pdf, Jan. 25, 2019 [Feb. 25, 2019]

[8] Slovenská inovačná a energetická agentúra. "Zelená domácnostiam – Podpora využívania obnoviteľných zdrojov energie v domácnostiach." Internet: Slovenská inovačná a energetická agentúra, Mar. 25, 2019 [Feb. 26, 2019]

[9] J. Stoutenborough, M. Beverlin, "Encouraging Pollution-Free Energy: The Diffusion of State Net Metering Policies." Social Science Quarterly, vol 89, Issue 5, pp. 1230-1251, Dec. 2008.

[10] R. Dufo-Lopez , J. L. Bernal-Agustín, "A comparative assessment of net metering and net billing policies. Study cases for Spain." Energy, vol 84, pp. 684-694, Feb. 2015.

Determination of Overhead Power Lines Ampacity Based on CIGRE 207 Brochure

Jakub Urbanský
Department of Electric Power Engineering
FEI TU of Košice
Košice, Slovakia
Jakub.Urbansky@tuke.sk

Michal Špes
Department of Electric Power Engineering
FEI TU of Košice
Košice, Slovakia
Michal.Spes@tuke.sk

Ľubomír Beňa
Department of Electric Power Engineering
FEI TU of Košice
Košice, Slovakia
Lubomir.Bena@tuke.sk

Daniel Pál
Department of Electric Power Engineering
FEI TU of Košice
Košice, Slovakia
Daniel.Pal@tuke.sk

Abstract — **With increasing consumption of electricity, the need for its production naturally rises as well. Thus, it is important to know if it's still possible to use existing overhead power lines or if it's necessary to replace or extend them. Ampacity, or current-carrying capacity, is defined as maximum amount of current that can flow through power line without damaging the conductor. It depends primarily on the electrical and mechanical properties of the conductor, their ability to spread generated heat and the ambient weather conditions. Design of power lines as any electrical device is subject to the valid standards. In general, the environmental conditions are not constant over time. They depend on the climatic conditions of the country, area, that are different climatic nature. Each manufacturer is providing the maximum allowable current load of the conductors so that its maximum operating temperature does not exceed the maximum safe value. This article is dealing with confirmation of maximal allowable current that is resulting maximal operating temperature provided by manufacturer of the conductor. For that purpose, designing and assembling a measuring apparatus to determine ampacity in overhead power lines were realized. The measurements and calculations based on standard CIGRE brochure 207, were made. The measured results were compared with calculation. It was proven that standard CIGRE brochure 207 has sufficient accuracy for the use in practice.**

Keywords: ampacity; overhead power line; current-carrying capacity; ACSR conductor

I. Introduction

Power system consists of electric power components, which are used for generation, transformation and electricity transmission. For transmission of electricity, we distinguish the transmission and distribution power lines. A transmission lines are representing at ultra-high voltage system, which provide a transmission of power from sources to the three local areas of distribution system. Also cross-border power lines belong to the transmission system of power lines [1] [2].

Current capacity of every power lines depends on their constructions and used types of the conductor. Manufacturer of the conductors specified maximal allowable temperature for the operation of the power line. With maximal allowable temperature is related a current that can flow through the conductor and leads to achieve maximum operating temperature [3]. A maximal current value that flow through the conductors by not excited allowable temperature is marked as capacity of power lines or ampacity of power lines [3].

Capacity of the power lines depends on the electrical and mechanical features, thermal features of the conductor insulation and ambient conditions [3].

Every construction of electricity facilities is subject to the standard. This standard described methods for construction these facilities and to determination of maximum allowable conditions, that are non-destructive for their operation.

During a preparing of project to construction of power lines it is necessary to determine a capacity of power lines. For determining a capacity of power lines exist a standard EN 50341 where are defined ambient conditions, that have main influence on the value of maximal current.

II. Defining Research Problems

1) Current capacity

Current capacity of the power lines means a current value that can be transmitted through a power line without destroying conductors. For this state deals steady state equation, which depends on determinants that represents the produced and consumed heat (1) [5] [6].

$$P_Z + P_S = P_k + P_r \quad (1)$$

Where:

P_Z is heating the conductor due to current flow,

P_S is heating the conductor due to solar radiation,

P_k is cooling the conductor due to forced convection,

P_r is cooling the conductor due to natural convection.

By substitution equation above, we obtained a following expression for determining current capacity [5] [6]:

$$I = \sqrt{\frac{P_r + P_k - P_S}{R_{ac}}} \quad (2)$$

Where:

R_{ac} *is* the ac resistance of the conductor at 20 °C in Ω.

For the design of power lines is currently applicable standard EN 50341 where are defined ambient conditions for calculation of maximum allowable current value. This standard recommended maximum operating temperature 70°C of conductor [7].

Ambient conditions for calculation of maximum allowable current value of conductor according to the standard are:

- The ambient temperature is 35 °C,
- Wind speed is 0.5 m.s^{-1} at 45° angle of impact,
- Solar irradiance is 1000 W.m^{-2},
- Absorption coefficient is 0.5,
- Emissivity coefficient of 0.5 [7].

It is necessary to say that conditions given by a standard are the worst case of ambient conditions, which is rarely found.

For determining current capacity standard CIGRE Technical Brochure 207 describes equations for calculation based on the climatic conditions and temperature of conductor. [8]

Based on this hypothesis, we compare temperature of conductor during a heating by nominal current value and calculated temperature for examined conductor.

2) Construction of ultra high voltage power lines

In practical terms, for the line of 400 kV voltage level are used trunked conductors where one phase consists of three conductors each and electrically connected at a distance, thereby enhanced radius of the conductor of one phase [9].

As conductors of transmission lines are used aluminum cables with steel core. Their advantage is greater mechanical strength, which allows its use for large distance. Among their other advantages include greater flexibility, more uniform structure. When wires material mistake can degrade the whole wire, but with ACSR ropes, tearing of one wire not damage the whole conductor [9].

ACSR rope used in high voltage grid was used to investigate due to limited conditions in the laboratory.

A rope 24-AL1/4-ST1A was chosen for research with following parameters:

Table 1 Parameters of 24-AL1/4-ST1A

Conductor type		24-AL1/4-ST1A
Cross section (mm^2)	AL	23,64
	ST	3,94
	Overall	27,58
Number of wires	AL	6
	ST	1
Diameter of wires (mm)	AL	2,24
	ST	2,24
Rope diameter (mm)	Inner section	2,24
	Overall	6,72

Table 2 Electromechanical specifications of 24-AL1/4-ST1A

Nominal weight (kg.km^{-1})	95,46
Nominal strength (kN)	8,87
Maximum DC resistance at 20 °C (Ω.km^{-1})	1,1823
Final modulus of elasticity (MPa)	79 000
Coefficient of thermal expansion (10^{-5} K^{-1})	1,86
Current load capacity (A)	125,9

III. The Design and Assembling of Measuring Apparatus to Determine Ampacity in Overhead Power Lines

Basic hypothesis for this article is verification of proposed mathematical model. It was determined based on empirical equations of current-carrying capacity of conductors under the influence of ambient conditions according to CIGRE standard. Main element for the investigation was ACSR conductor troughs which was flowing electric current.

This electric current is causing conductor warming caused by Joule losses neglecting ambient conditions. Based on theoretical knowledge, two main elements have influence at conductor warming, ambient temperature (which also contributes in Joule losses) and intensity of solar radiation. In contrast, when conductor is being cooled the main influence has temperature drop caused by forced convection and temperature drop forced by natural convection.

Another essential element for measuring apparatus is transformer. With the transformer it is possible to continuously regulate the value of the electric current until desired value is reached (current load capacity of examined conductor).

Temperature probes are part of the measurement. They are measuring temperature of conductor surface.

Based on theoretical analysis for measuring apparatus, laboratory measuring site was compiled.

Block diagram is shown in Figure 1. At the input of measuring site is three phase power supply SST 250/20 with voltage regulation 0 – 250V and electric current regulation 0 – 20A.

Power supply SST250/20 is exciting large current transformer (TR) in the range 0 – 1000A.

Between SST250/20 and large current transformer was connected device WPQT1H. WPQT1H is primary intended for testing the electrical protection. In measuring apparatus its protection functions are used for protection against TR damage. On the secondary side of TR is connected examined ACSR rope with temperature probes.

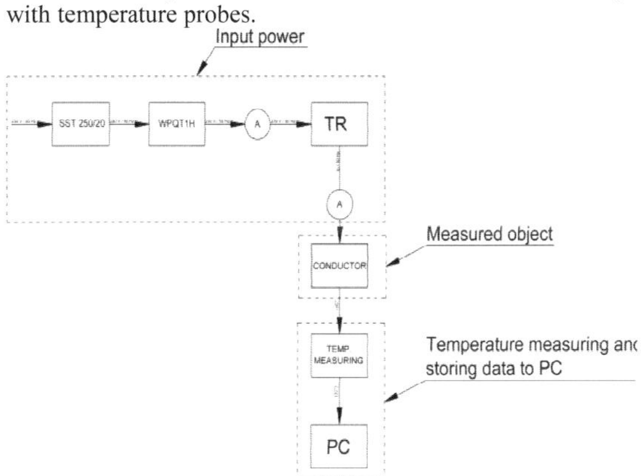

Figure 1 Block diagram of measuring apparatus

Based on measured temperature of the conductor, when the maximal allowable current is flowing through, the maximal allowable conductor temperature can be determined.

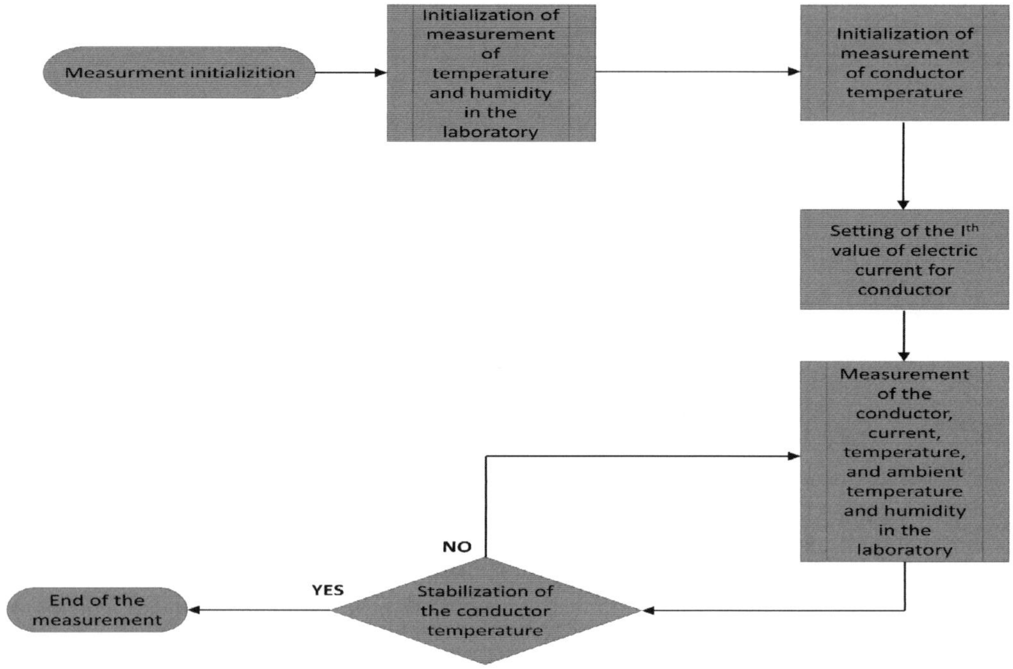

Figure 2 Flow chart of measurement

Measurement method in form of flow chart is shown in Figure 2.

After measurement initialization of ampacity is concurrently started measurement of temperature and humidity in laboratory. Subsequently with using of voltage source is current set to nominal value. To end the measurement, temperature of conductor has to be in steady state, i.e. it has not upward or downward trend.

IV. Measuring of ambient temperature influence on ampacity of 24-AL1/4-ST1A conductor

To determine the influence of ambient temperature on ampacity of conductors was based on proposed experimental measuring apparatus and measuring flow chart (Figure 2) carried out experiment.

Due to technical and security restrictions in laboratory were experiment carried out on ACSR ropes used in high voltage level.

From the point of view of ensuring credibility of measurements were individual measurements repeated 10 times.

The measurement procedure was based on a flowchart shown in Figure 2. Measured temperature was arithmetically averaged.

The ambient conditions were as follows:

- Ambient temperature $T_a = 27$°C,
- Wind speed $v = 0$ m.s^{-1}, i.e. windlessness / wind speed $v = 3$ m.s^{-1},
- Intensity of solar radiation $I_s = 0$ W.m^{-2}.

In the process of verification of the mathematical model, measurements were made on the basis of the proposed experimental set-up and technological procedure for the measurement of overhead power lines.

In Figure 3 is shown time course of measured temperature of 24-AL1/4-ST1A conductor in a step load of nominal current. As shown in the Figure 3, temperature of conductor was measured in advance, i.e. before switching on the measuring circuit.

The temperature measurement time for unloaded ACSR rope was 10 minutes. The measuring circuit was turned on in 10th minute. The set current value was almost nominal, equal to 125A.

There was a sharp rise in driver temperature between 00:10:00 and 00:20:00. From a conductor temperature of 25-27 ° C at 10 repeat measurements, the temperature increased from 73 to 85 °C. For averaged conductor temperature, this meant a temperature rise from 26.2 ° C to 80.5 ° C.

A dampened increase in driver temperature was recorded from 00:20:00. In the case of individual measurements, the temperature rises of the conductor at 00:20:00 for individual measurements ranged from 77 °C to 85 °C until 00:50:00 from 77 °C to 90 °C. For average temperatures, an increase from 80.5 ° C to 85 ° C was recorded.

At Figure 3 is possible to observe, that in case of sub-measurements 1 – 10, was temperature increase different, however characteristic of partial temperature increases were almost identical.

This state was caused by the inconsistency of the surrounding conditions. In the laboratory was recorded 1.3 °C difference in temperature.

The temperature of the conductor is also influenced by its surface treatment.

In the manufacture process of these ropes they are treated against oxidation in air to maintain their mechanical properties.

To verify the proposed mathematical model for the determination of the static and dynamic part, a calculation was made for determining the temperature of the conductor in the dynamic and static parts. From a solution point of view, the determined steady-state conductor temperature was then compared to the average conductor temperature obtained on the basis of the repeat measurement.

The following ambient conditions were used to determine the static temperature:

- Ambient temperature T_a = 27°C,
- Room temperature was between 25 – 27 °C, however in calculation is necessary taken in to account higher one.
- Intensity of solar radiation I_S = 0 W.m^{-2},
- Wind speed, windlessness v = 0,2 m.s^{-1}.
    - Beaufort's wind force scale defines windlessness at a speed interval 0,0 until 0,2 m.s^{-1}. Given that it was not possible to achieve 0.0 m.s^{-1} in the laboratory; higher wind value is considered.
- Nominal current I = 125 A.

By gradual adjustment of the conductor temperature by the iteration method, the determined conductor temperature was T_S = 74.6 ° C. The factors involved in heating and cooling the conductor was equal to:

- Warming due to current flow P_J = 22,88 W.m^{-1},
- Warming due to solar irradiation P_S = 0 W.m^{-1},
- Cooling due to forced convection P_C = 19,01 W.m^{-1},
- Cooling due to natural convection P_r = 3,88 W.m^{-1},
- Difference ΔP is equal to -0,01 W.m^{-1}.

The dynamic value of the conductor temperature is the conductor temperature that changes over time after the conductor current change or when the ambient conditions suddenly change.

To calculate the conductor temperature at a jump load by the nominal current value, it starts from the static conductor temperature obtained by the calculation in the previous subchapter. It is equal to θ_m = 74.6 °C. This temperature represents the steady-state conductor temperature after a step change in current or ambient conditions. The overall time course of temperature after a step change in the current that flowed through the investigator and the comparison is shown in Table 3.

Table 3 Comparison of calculated and measured values for 24-AL1/4-ST1A conductor

Measured average temperature of conductor (°C)	Calculated temperature of conductor (°C)	Difference (°C)	Difference (%)
26,2	35,5	9,3	26,13
67,9	42,4	25,5	60,04
80,5	48,1	32,4	67,19
84,3	52,9	31,4	59,50
85,8	56,7	29,1	51,27
85,8	59,9	25,9	43,24
85,7	62,5	23,2	37,09
86	64,7	21,3	32,99
85,2	66,4	18,8	28,25
85,2	67,9	17,3	25,51
85,2	69,1	16,1	23,34
85,2	70,1	15,1	21,61
85,2	70,9	14,3	20,22
85,2	71,5	13,7	19,11
85,2	72,1	13,1	18,21
85,2	72,5	12,7	17,47
85,2	72,9	12,3	16,88
85,2	73,2	12,0	16,40
85,2	73,4	11,8	16,00
85,2	73,7	11,5	15,68
85,2	73,8	11,4	15,41
85,2	74,0	11,2	15,20
85,2	74,1	11,1	15,02
85,2	74,2	11,0	14,88
85,2	74,2	11,0	14,76
85,2	74,3	10,9	14,66
85,2	74,4	10,8	14,58
85,2	74,4	10,8	14,51
85,2	74,4	10,8	14,46
85,2	74,5	10,7	14,41
85,2	74,5	10,7	14,38
85,2	74,5	10,7	14,35
85,2	74,5	10,7	14,32
85,2	74,6	10,6	14,21

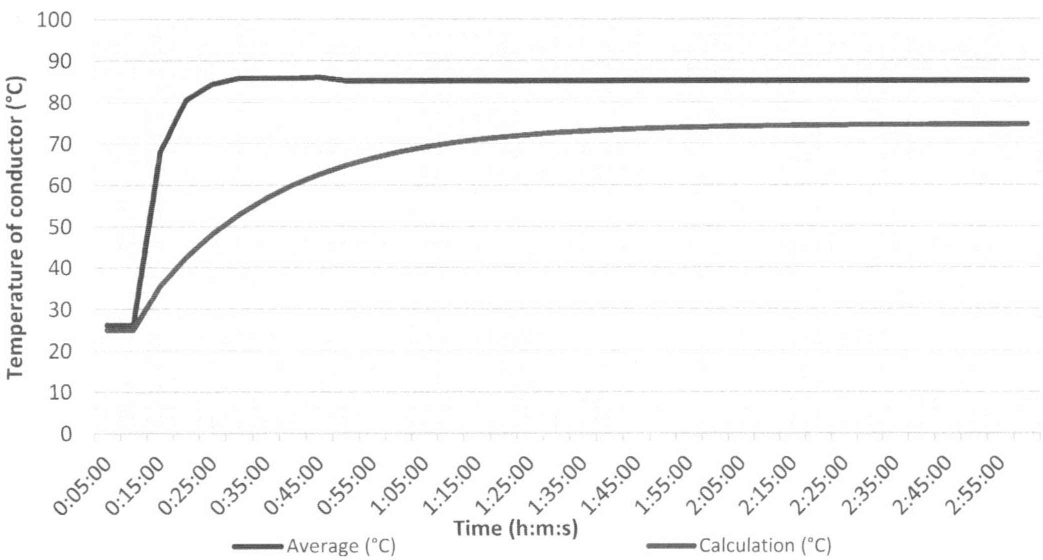

Figure 2 Graphical comparison for 24-AL1/4-ST1A conductor

As can be seen from the Table 3, when comparing the average conductor temperature obtained from 10 measurements and the calculated conductor temperature, there is a significant difference between these temperatures.

The biggest difference was in the first points of measurement, when the difference was up to 67.19%.

With increasing time of measurement and calculation, the difference between the measured temperature and the calculated temperature decreased to 14.21%.

The CIGRE Technical Brochure 207 standard defines a temperature difference up to 20% for smaller wind speeds of 0.5 $m.s^{-1}$.

Measurement uncertainty also has a significant impact. When comparing the driver's static temperature values, this difference is up to 20% by standard.

Due to the low wind speed, the convection is mixed, which is not taken into account by the conductor temperature calculation standard.

The graphical representation of the comparison is shown in the Figure 4.

V. Conclusion

Power lines are one of the most important part of power system. Temperature of the conductor is function of the current value, ambient conditions, type of used conductor and their properties.

Ambient conditions have most influence on the actual value of current capacity. If it is possible to accurately determine the ambient conditions in real time, current capacity can be determined under these terms and adapted on operation of power system or power lines. Increasing current capacity of existing power lines is one of the way how to operate power system in the short term. On the other side, is necessary to build, expand with new power lines which is however time-consuming and economically demanding.

For the static part, i.e. steady-state conductor temperature at rated electric current was found to be 14.21% difference for 24-AL1/4-ST1A. In point of view of such a high difference in measurement and calculation temperatures, the CIGRE 207 standard defines a situation where the air flow rate is less than 0.5 $m.s^{-1}$. In this case difference is up to 20%. In our case, considering the combined measurement uncertainty, this result is excessively accurate.

In the dynamic part, i.e. after switching the measuring circuit, when the nominal current flowed through the conductor, the difference was significantly higher at 67.19% for the 24-AL1 / 4-ST1A rope.

These differences in temperature are beyond the explanation in CIGRE 207. The problem with these parts of the results and its verification is that the CIGRE 207 methodology does not specify a sufficiently accurate calculation option for a state where the wind speed is below 0.5 $m.s^{-1}$ in the dynamic part of the load.

The missing parameter is the effect of mixed convection.

VI. Acknowledgment

This work was supported by the Scientific Grant Agency of the Ministry of Education of Slovak Republic and the Slovak Academy of Sciences by the projects VEGA No. 1/0372/18.

References

[1] M. Kolcun, V. Griger: *Controlling the operation of the power system (Riadenie prevádzky elektrizačnej sústavy)*, Mercury - Smékal, Košice, 2003, 288 pages, ISBN 80-89061-57-5.

[2] *Guide for selection of weather parameters for bare overhead conductor rating*, CIGRE standard, 2006.

[3] A. Bracale, "Probabilistic index for increasing hourly transmission line rating," in *Int. Journal of Emerging Electric Power Systems*. 2007, pp. 119.

[4] T. Košický, Ľ. Beňa, M. Kolcun, "Analysis of utilisation energy storage systems for frequency regulation," in *Acta Electrotechnica et Informatica*. Vol. 14, Nr. 3 (2014), p. 36-42. – ISSN 1335-8243

[5] R. Jakubčak, Ľ. Beňa, M. Kmec, "Possibilities of Using Facts Devices In Power System," in *Acta Electrotechnica et Informatica*. Vol. 13, No. 3 (2013), p. 8-11. – ISSN 1335-8243

[6] Slovenský hydrometeorologický ústav (2019, Feb. 8.), *Klimatické podmienky na Slovensku* [Online]. Available: http://www.shmu.sk/sk/?page=1064

[7] A. Klenovcanova, T. Brestovic: "Possibilities of utilization of photovoltaic cells for electricity production in Kosice area," *in Acta Mechanica Slovaca*, 2007, ELFA, s.r.o., 11, 4-D, 511-516, ISSN: 1335-2393

[8] *The Thermal Behaviour of Overhead Conductors*, CIGRE Technical Brochure 207, August 2002.

[9] M. Špes: "Powerline ampacity system," in *Proc. SCYR 2016*. Košice : TU, 2016 pages 202-205. ISBN 978-80-553-2566-8

[10] M. Špes, Ľ. Beňa, M. Mikita, M. Márton, H. Wachta, "Possibilities of increasing transmission capacity overhead lines, " in *Acta Electrotechnica et Informatica*. Vol. 16, No. 3 (2016), p. 20-25. – ISSN 1335-8243

[11] L. Varga, P. Leščinsky, Ľ. Beňa: "Mechanical calculation of ovearhead power line conductor under combined load," in *Acta Electrotechnica et Informatica*. Vol. 1, no. 1 (2001), p. 28-31. – ISSN 1335-8243

[12] *Elektrická venkovní vedení s napětím nad AC 1 kV – Část 1: Obecné požadavky – Společné specifikace,* EN 50341-1 ED.2 (333300).

[13] D. Solus, L. Ovseník, J. Turán, "Optical correlator in vertical traffic signs inventory system". in *2015 IEEE 13th International Scientific Conference on Informatics*, INFORMATICS 2015 - Proceedings, 247-251. doi:10.1109/Informatics.2015.7377841

[14] D. Solus, L. Ovseník, J. Turán, "Signal processing - object detection methods with usage of optical correlator". in *2016 26th International Conference Radioelektronika, RADIOELEKTRONIKA 2016*, 315-318. doi:10.1109/RADIOELEK.2016.7477434

[15] Š. Fecko, J. Žiaran, L. Varga, *Power lines – Overhead power lines*, SVŠT Bratislava, 1990.

[16] *Sag-tension Calculation Methods for Overhead Lines*, CIGRE Technical Brochure 324, June 2007.

[17] *Alternating Current (AC) Resistance of Helically Stranded Conductors*, CIGRE Technical Brochure 345, April 2008.

[18] *IEEE standard for calculating the current temperature of bare overhead conductors*, IEEE Power Engineering Society, 2007.

Analysis of Batteries Impact on a Distribution Network with Distributed Generation

David Motyka, Martina Kajanova, Peter Bracinik, Marian Tomasov, Marek Roch
Department of Power Systems and Electric Drives
Faculty of Electrical Engineering and Informatics Engineering, University of Zilina
Zilina, Slovak Republic
david.motyka@fel.uniza.sk

Abstract — **Distributed networks are getting more and more penetrated with distributed generation. The largest shares of distributed generation have photovoltaic and wind power plants. In Slovakia, these are mainly photovoltaic power plants, whose development is subsidized by the European Union. These sources have a significant impact on the distribution network. Because of the adverse impact of the sources, the individual distribution companies defined the conditions for connecting the distributed generations to their networks. The main issues of distribution companies with distributed generation are the change of power flow, voltage rise at the point of connection, instability of resources. Various references have pointed out that the benefits of distributed generation can only be achieved by choosing the right size of installed power and connecting a source to a suitable distribution network location. The issue of placement and dimensioning of distributed generation is continually investigated through research, with different approaches and optimization of objective functions. This work focuses on simulating battery systems, photovoltaics, and load in a given node of the distribution network. The simulation is done on the 34 Node Test Feeder. The network model is adapted to the elements used in the Slovak power system. Several batteries and photovoltaic sources are implemented in the selected network. The created network model is simulated using software GridLab-D, which uses real-time weather data. Weather data are the real data collected from several meteorological stations. These files provide more realistic simulation model results when using photovoltaics. The issue of distributed generation could be partially suppressed by batteries installed in distribution networks. Therefore, this issue is addressed and the impact of batteries on distribution networks with distributed generations is pointed out in the paper.**

Keywords: distributed network; power flov; GridLab-D; batteries impact; distributed generation

I. Introduction

Smart grid brings the new way for communicating grid events and information between utilities and their customers. It has also enabled better utilization of demand response programs and distributed generation (DG) in more effective and efficient way [1]. In the smart grid concept, power produced by DG, such as solar and wind power, can be integrated and coordinated efficiently with the main utility grid. DG can help support the increased demand for electricity, and may improve power generation efficiency and network reliability [2]. However, high penetration of DG with intermittent character also brings challenges for power system, such as reverse power flow, voltage unbalance, changes in network short circuit level and consequently changes in the setting of protective relay systems [3]-[4].

The main aim of Wang and Nehrin study [5], one of the first DG studies, was to find the optimal DG placement but not to optimize size. In this paper, the authors proposed various analytical methods for optimizing radial networks. Other references that have also used analytical approaches are more described in [6] - [8]. Acharya et al. [6] proposed an analytical expression to calculate the optimal size to identify the best location from a loss perspective. Gozel and Hocaoglu [7], using the Jakobi matrix, suggested a faster calculation of the DG's optimal location, using the same methodology as Acharya. The resulting optimization was for radial networks only. Similarly, Hung et al. [8] expanded the work presented by Acharya and colleagues. Conventional optimization techniques are mostly used to optimize the amount of installed power. Keane and O'Malley [9] used linear programming to determine the optimal allocation for all relevant technical constraints. The objective of the proposed method is to maximize the installed DG power.

Energy storage systems can be a solution at LV distribution network to store excessive electric energy generated by DG and then supply power back to the network during low generation times. Storage systems can also help the system to overcome the problems with intermittency of renewable energy [10].

In Germany, solar power can be produced for 0.12 €/kWh. In contrast, distribution system utilities currently charge an average of 0.25 Euro for electricity for households. Solar storage systems can increase on-site consumption by up to 70 %. According to EuPD research survey, almost 90 % of photovoltaic power plant operators are already thinking about adding additional storage systems [11]. Small size Lithium ion batteries are available in the range 1–10 kWh. Energy storage system integrated with PV can maximize consumption of renewable energy by using energy stored during off-peak times [12]. The batteries can also support the system when transfers into islanding mode. Battery storage are used in [13] to improve the security of supply for grid connected PV, curtail overvoltage problem, peak shaving and to improve transient stability.

This work investigates the high penetration level of DG and impact of batteries utilization on the distribution network.

II. Simulation of 34 Node Test Feeder

Distributed generation, such as photovoltaic power plants, wind turbines, fuel cells, energy storage technologies, and batteries, is a quick solution to the growing demand for energy. This section is dedicated to describing and modifying a selected IEEE 34 Node Test Feeder according to the characteristics of the distribution systems used in Slovakia, where the impact of distributed resources will be investigated. 34 Node Test Feeder is based on the network in the United States of America, Arizona. It is characterized by long and slightly loaded transmission lines. The frequency was changed to 50 Hz and the rated voltage to 22 kV, because the original IEEE 34 Node Test Feeder model operates at 60 Hz with a nominal voltage of 24.9 kV. Single phase and three-phase unbalanced current loads with constant impedance are present in individual network nodes [14].

A. Modifying of the test feeder

The configuration of power lines, cross-sections of conductors as well as the voltage level have been adjusted with regard to commonly used elements in distribution networks in the central Slovakia [15]. Furthermore, the power load was changed along the network model (Tab. 1) so that the permissible load current of the conductors is not exceeded. The total network load is 1 MV·A. The loads are connected to the distribution network at a voltage level of 22 kV. Loads represent power flows through distribution transformers to LV voltage levels. All loads are constant, symmetrical, connected to the line-to-line voltage in delta connection (D-PQ).

Table 1. Demand in the model of the test feeder

Node	Constant demand for individual phases					
	A	A	B	B	C	C
	P (W)	Q (V·Ar)	P (W)	Q (V·Ar)	P (W)	Q (V·Ar)
802	9166.7	4833.3	9166.7	4833.3	9166.7	4833.3
806	9166.7	4833.3	9166.7	4833.3	9166.7	4833.3
808	2666.7	1333.3	2666.7	1333.3	2666.7	1333.3
810	2666.7	1333.3	2666.7	1333.3	2666.7	1333.3
812	0.0	0.0	0.0	0.0	0.0	0.0
814	0.0	0.0	0.0	0.0	0.0	0.0
850	0.0	0.0	0.0	0.0	0.0	0.0
816	833.3	333.3	833.3	333.3	833.3	333.3
818	5666.7	2833.3	5666.7	2833.3	5666.7	2833.3
820	5666.7	2833.3	5666.7	2833.3	5666.7	2833.3
822	22500	11666.7	22500	11666.7	22500	11666.7
824	833.3	333.3	833.3	333.3	833.3	333.3
826	6666.7	3333.3	6666.7	3333.3	6666.7	3333.3
828	1833.3	833.3	1833.3	833.3	1833.3	833.3
830	1166.7	500	1166.7	500	1166.7	500
854	666.7	333.3	666.7	333.3	666.7	333.3
856	666.7	333.3	666.7	333.3	666.7	333.3
852	0.0	0.0	0.0	0.0	0.0	0.0
832	2500	1166.7	2500	1166.7	2500	1166.7
890	150000	75000	150000	75000	150000	75000
858	2500	1166.7	2500	1166.7	2500	1166.7
864	333.3	166.7	333.3	166.7	333.3	166.7
834	5333.3	2833.3	5333.3	2833.3	5333.3	2833.3
842	1500	833.3	1500	833.3	1500	833.3
844	9000	4666.7	9000	4666.7	9000	4666.7
846	11333.3	5666.7	11333.3	5666.7	11333.3	5666.7
848	3833.3	1833.3	3833.3	1833.3	3833.3	1833.3
860	24333.3	12166.7	24333.3	12166.7	24333.3	12166.7
836	6666.7	3333.3	6666.7	3333.3	6666.7	3333.3
862	4666.7	2333.3	4666.7	2333.3	4666.7	2333.3
838	4666.7	2333.3	4666.7	2333.3	4666.7	2333.3
840	6666.7	3333.3	6666.7	3333.3	6666.7	3333.3

The size of the network was taken from the original model of a standardized network, i.e. the network's vastness is 93.91 km. The basic electrical properties of the used conductor are shown in Tab. 2.

Table 2. Types and parameters of the used conductors

Type	GMR_{50Hz} (m)	*ØAlFe* (mm)	R_{20} ($\Omega\cdot km^{-1}$)	S_{AL} (mm^2)
AlFe 70/11-1	0.0026	11.70	0.4402	66.27
AlFe 110/22	0.0033	14.96	0.2663	110.34
AlFe 150/25	0.0038	16.96	0.2479	143.40

The distribution network model is supplied from a 110 kV upstream system via a transformer whose parameter is S_n = 2.5 MV·A, R = 1%, X = 8%. This data is unchanged and taken from the original IEEE 34 Node Test Feeder.

B. Simulation model in GridLab-D

GridLab-D is a open-source software based on utilization of complex elements that provides unprecedented capabilities to model the power flow of smart grids and different concepts in the power system. It was created by the United States Department of Energy, the Pacific Northwest National Laboratory and the academic community. GridLab-D includes an advanced power system, temperature, control and economy modeling techniques so that fully functional smart grid models can be created. These models include and address power flows, power models, market models, distribution network models, and so on, and therefore GridLab-D is an useful tool to explore in detail the interaction of smart grid components.

The advantage of the GridLAB-D program is the possibility of implementing a metrology module in the form of *.tmy2* (typical meteorological year) file, which was also used. GridLab-D provides models for several different types of distributed power sources, including diesel generators, solar systems and wind power plants. Power sources are modeled as parametric load with negative values. GridLab-D allows to implement your own meteorological data. The user is able to create the weather data in the *.csv* file. *.csv* files are in predefined form. Using a CSV reader, GridLab-D implements the created data into a network model.

A modified test feeder model for GridLab-D was created in Notepad ++.

III. Simulation Model Analysis

28 photovoltaic (PV) sources were added to the created 34 IEEE Node Test Feeder (Fig. 2). GridLab-D, as metioned above, offers implementation of own meteorological data. Using meteorological data, power generation at the PV sources were simulated. These meteorological data, together with PV resources, have been implemented into the network model for the entire calendar year. The meteorological data were selected according to the approximate parallel of Slovakia in the USA, these were data for city Minot in North Dakota, "ND Minot.tmy2". Based on the current requirements for connection of renewable sources to the network, the installed power of individual PV power plants was determined (Tab. 3). The DG connection requirements are determined by the maximum voltage change at the connection point to ± 2% of the original voltage value. After the implementation of the PV sources, the flow of power through the reference node 800 was compared. In cases of PV production and non-compliance with the technical requirement for constraint of the power flow from PV sources into the distribution system [16], there is a flow of power into the upstream 110 kV (Fig. 1) system during some hours of the day (negative values).

Table 3. Installed capacity of the PV sources

Node	Power *P* (kWp)	Node	Power *P* (kWp)
802	320.625	832	32.066
806	346.275	858	25.650
808	320.625	864	19.238
810	282.150	834	6.413
816	23.085	842	12.825
818	23.085	844	15.390
820	23.085	846	12.825
822	23.085	848	15.390
824	38.475	890	89.775
826	38.475	860	32.063
828	23.085	836	32.063
830	25.650	840	12.825
854	25.650	862	19.238
856	19.238	838	19.238

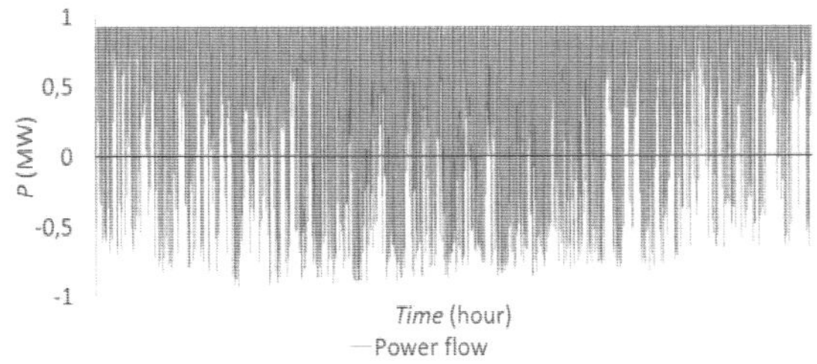

Figure 1. Power flow in node 800 with PV sources

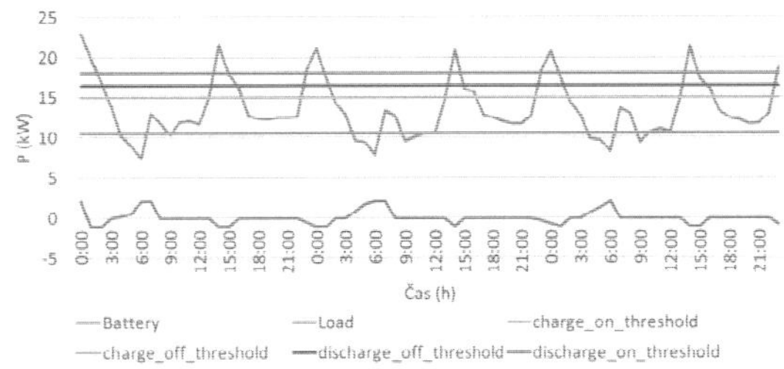

Figure 2. Battery operation during three days

Changing the flow of power can cause problems in the control of distribution systems. Obviously, the distribution system operator therefore tries to protect distribution system by the technical requirements for DG connection. This issue could be solved to a certain extent by suitable allocation of batteries in some nodes of network.

A. Implementation of batteries

Six batteries were added into selected nodes of the network (Fig. 2). The capacity of the individual batteries was determined by the size of the demand in the given node (Tab. 4). The batteries are LI_ION type and have set an initial capacity to 50 % of the total capacity. The battery mode is defined by four values. Charging of batteries starts when the demand falls below 50 % (charge_on_treshold). Charging of battery stops when demand is at 55 % of the nominal load (charge_off_treshold). The start and end of the discharge is set to demand at 70 % of nominal power (discharge_on_treshold, discharge_off_treshold).

Table 4. Installed capacity of the batteries

Node	Capacity battery (kW)
806	27.5
820	17
826	20
830	3.5
842	4.5
838	14

To simulate the battery object, it was necessary to change the static nature of the load to dynamic one that is changing over time. The nominal load on each node remained the same but was multiplied by the normalized day demand NDD that GridLab-D had read for the entire simulated year from the .csv file. NDD values changed every hour and represented household consumption, including the heating system and DHW heating. Subsequently, simulations of the network were performed. For all batteries, a maximum discharge power of 2 kW and a charging power of slightly less than 1.8 kW have been determined. These charging and discharging limits were set indicatively. Fig. 3 shows that if the demand rises above 70 % of the nominal load, the required power is supplied also from the battery (negative values of the blue line). If the load falls below 55 % of the nominal load, the battery starts to charge (storage of the energy – positive values of blue line).

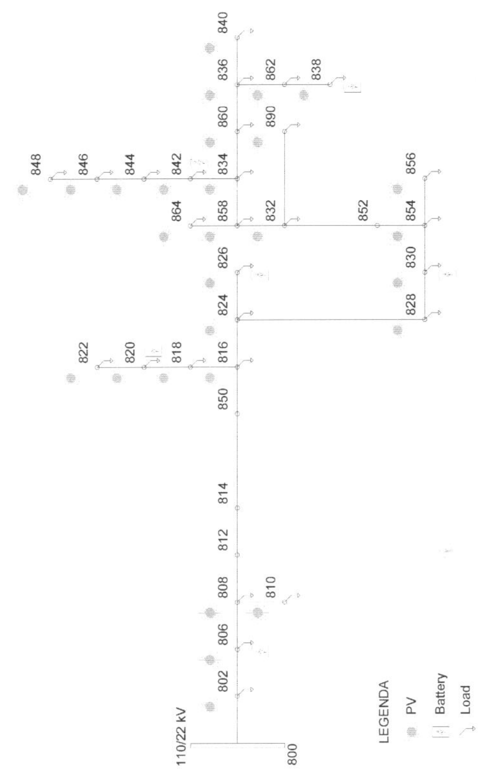

Figure 3. Allocation of the PV sources and batteries in the model network

When analyzing the data we found out that during some days there was no charging or discharging of the batteries. Such days were for more than the half of the simulated year. Fig. 4 shows the days during which the battery has not been used. The reason for the non-utilization of the battery was that the demand at node 806 did not exceed 70 % of the nominal load value (set value of the discharge_on_treshold parameter), so the battery was not discharged. Although the demand has fallen below 50% of the nominal value (charge_on_treshold) during the reporting period, the battery has no longer been able to store any additional energy since it was already fully charged from the previous day.

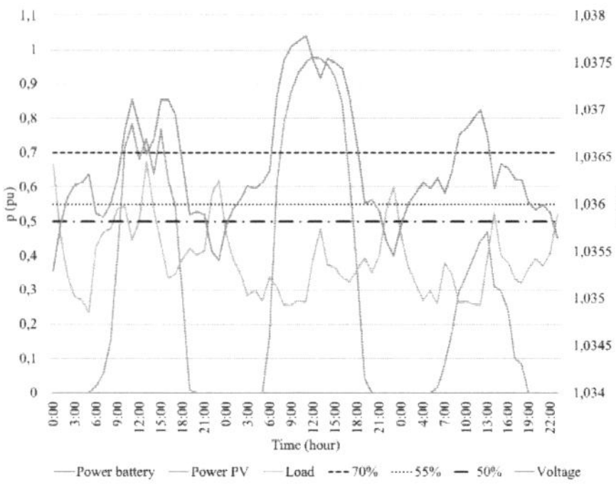

Figure 4. Curves of selected parameters in node 806

However, there were also days when energy was charged and discharged in node 806 (Fig. 5). When the demand is increased above 70 % of the nominal value (dotted line), the battery has started to release power to the grid (negative values). When the demand dropped below 70 %, the battery terminated the power supply to the grid. When the demand falls below 50 % (dashdotted line), the battery has started to charge (positive values) and when the demand is increased again above 55 % (dashed line), the battery charging has been completed. Adequately to these changes and changes in power generation in PV, the voltage at the node has also changed.

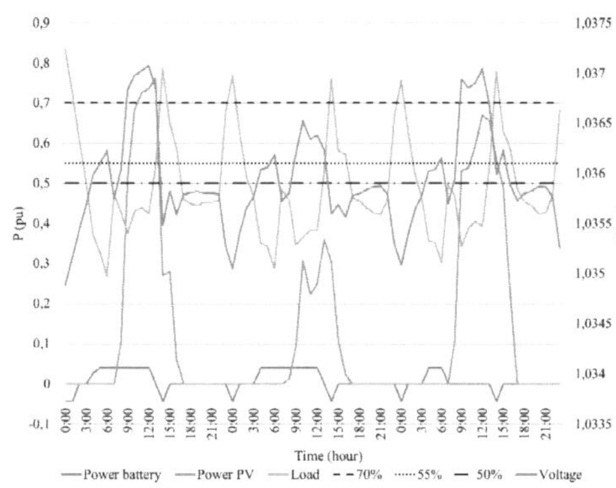

Figure 5. Battery charging and dischargin in node 806

As mentioned above, DG influence can significantly affect distribution system control, but the impact may be smaller in case of their cooperation with batteries. Therefore, it is important to set not only the operation parameters of individual batteries (fig. 4 and fig. 5), but also their location within distribution networks. Last but not least, it is also necessary to choose the right battery capacity to provide the necessary control functions together with economic benefits.

IV. CONCLUSION

The simulation shows that even if the maximum voltage change at the connection point is maintained for the renewable energy sources, the power flow can change. The next simulation scenario was complemented with batteries. Battery storage model functionality has been provided by changing the static load to dynamic one. The inverter that charges the battery monitors the demand at the node and charges or discharges the batteries according to the specified actual value of demand. If there was a static load, the battery object would not work. Choosing nodes for batteries was experimental. From the results of the simulation can be said that the battery storage had a positive effect on the distribution network. More detailed analysis of battery storage, economic deployment, proper placement in the distribution network model, capacity size and finding the right charging and discharging threshold values will be addressed in the next work.

ACKNOWLEDGMENT

This work was supported by the Slovak Research and Development Agency under the contract No. APVV-15-0464.

REFERENCES

[1] F. Rahimi, and A. Ipakchi, "Demand response as a market resource under the smart grid paradigm," IEEE Trans. on Smart Grid., 2010; 1(1):82–88.

[2] I. Waseem, "Impacts of distributed generation on the residential distribution network operation,". MSc thesis. Virginia Polytechnic Institute and State University. Virginia; 2008.

[3] Hadi, Azah, and Hussain, Marjan, Impact of distributed generations on power system protection performance. International Journal of the Physical Sciences, 2011; 6(16):3999-4007.

[4] C. Barbier, A. Maloyd, and G. Putrus, "Embedded Controller for LV Network with Ddistributed Generation." UK: Department of Trade and Industry; 2007.

[5] C. Wang, and MH. Nehrir, "Analytical approaches for optimal placement of distributed generation sources in power systems." IEEE Trans Power Syst, 2004;19:2068–76.

[6] N. Acharya, P. Mahat, and N. Mithulananthan, "An analytical approach for DG allocation in primary distribution network." Int J Electr Power Energy Syst,2006;28:669–78.

[7] T. Gozel, and MH. Hocaoglu, "An analytical method for the sizing and siting of distributed generators in radial systems." Electr Power Syst Res 2009;79:912–8.

[8] Do. Hung, N. Mithulananthan, RC. Bansal, "Analytical expressions for DG allocation in primary distribution networks." IEEE Trans Energy Convers 2010;25:814–20.

[9] A. Keane, M. O'Malley, "Optimal allocation of embedded generation on distribution networks." IEEE Trans Power Syst 2005;20:1640–6

[10] N. K. Bangash, M. E. A. Farrag, A. H. Osman, "Impact of energy storage systems on the management of fault current at LV network with high penetration of distributed generation." International Journal of Smart Grid and Clean Energy, Glasgow 2018, 978-1-5386-2910-9

[11] S. Rentzing, "When will solar batteries become economical." [Online]. Available: http://www.solarenergystorage.org/en/wann-werden-solarakkus-wirtschaftlich-2/, April 2013

[12] P. Denholm, E. Ela, B. Kirby, M. Milligan, "The role of energy storage with renewable electricity generation." Research report. National Renewable Energy Laboratory, 2010.

[13] M. Bahadornejad, N. Nair, "Solar PV, battery storage and low voltage distribution network." A discussion on the benefits of distributed battery storage. Power systems group, University of Auckland, October 2013

[14] Modeling and Simulation in XENDEE, IEEE 34 Node Test Feeder, [Online] : https://www.xendee.com/IEEE/Xendee_ASU_IEEE_34_BUS.pdf., December 2018

[15] Katalóg prvkov a funkčných celkov, [Online]: https://online.sse.sk/pls/kpfc/kpfc_web.kpfc_main, December 2018

[16] Technické podmienky prevádzkovateľa distribučnej sústavy. Stredoslovenská Distribučná. [Online]: https://www.ssd.sk/buxus/docs/dokumenty/o_nas/legislativa/Technicke%20podmienk y%20platne%20od%2001072015.pdf, February 2019.

Analysis of a Power Looses and Local Factors affecting the Performance of the Photovoltaic Power Plant

Marek Novak, Pavol Belany, Marek Siranec, Marek Roch
Department of Power Systems and Electric Drives
Faculty of Electrical Engineering and Information Technology, University of Zilina
Zilina, Slovakia
marek.novak@fel.uniza.sk

Abstract—**This paper discusses about a looses of energy in a photovoltaic power plant. There are also defined a local factors affecting a performance of the photovoltaic power plant. The local factors selected for this analysis are: partial shading of panels due to the device located immediately before the photovoltaic panels, pollution of photovoltaic panels and mechanical damage of photovoltaic panels. These factors are explained and then are described the possibilities of eliminating these factors. Also, the measurement of the influence of the local factors on the performance of the photovoltaic panels and their negative impact on the performance of the photovoltaic panels is presented.**

Keywords: pollution; partial shading; photovoltaic panels damage,

I. Introduction

With the growing number of photovoltaic power plants, public interest in providing the required efficiency is also increasing. Also, no PV plant operator wants to have any losses that can be avoided or removed.

Low efficiency is one of the fundamental problem of photovoltaic power plants. In a research of the increasing a photovoltaic panels efficiency were invested a lot of money but the present photovoltaic panels do not reach efficiency more than about 27 % with common monocrystalline panels [1]. Theoretical tandem cell efficiencies up to 45 % [2]. Therefore, any increase in efficiency, which is only 1%, means a significant shift in this area. There are many ways to increase this parameter. One way is to choose such a material of photovoltaic cells, which is better able to use the whole solar spectrum [3]. Another possibility is to focus on the fact, that the best efficiency of conversion of solar energy to electrical is achieved by photovoltaic panels when the direction of the rays is perpendicular to the surface of the photovoltaic panel. Due to the unfavorable influence of warming on the efficiency of the photovoltaic panels, their cooling seems to be another effective way [2]. Also, the elimination of a partial shading and an impact of a pollution of the photovoltaic panels should be avoided [4].

II. Local Factors that affect Photovoltaic Panels Performance

The electricity generation in photovoltaic power plants also brings several problems. A solar irradiance falls on the panels only for a limited time. The angle of a solar irradiance is constantly changing. Also the photovoltaic panel temperature affects its performance. However, this study focuses on the influence of another local factors on photovoltaic panels. One of these factors is partial shading of the photovoltaic panel (PV panel). It some cases, the location of photovoltaic panels on a roofs of a buildings is not ideal and a PV panels are permanently shaded by other equipment. For example, a retrofitted air condition on the roof of a building where a photovoltaic power plant has already been installed is presented in the figure 1. Another local factor is a pollution of the PV panels. The level of the pollution differs from location and length of operation. During the operation of the photovoltaic power plant, a layer of impurities of different composition and thickness are deposited on the PV panels. This will cause a partial decrease in the PV plant performance. Another significant factor is a mechanical damage of the PV panels. Also, on the photovoltaic panels are often micro cracks, which can be caused by a storm or a hail. Such damage has a significant impact on the photovoltaic panel performance.

Figure 1. Photo of a partially shaded SV60-255 photovoltaic panel.

III. Possibilities of Reducing the Impact of the Local Factors

The local factors mentioned in this paper are the most neglected influences that reduce the efficiency of photovoltaic systems. Depending on the PV panels type, age and location of the photovoltaic system, it is necessary to check the photovoltaic system and evaluate the solution procedure based on the measurement results. The possible ways are mentioned below.

A. Cleaning of the Photovoltaic Panels

The neglect of cleaning the PV panels will be interpreted by their lower performance and by shortening the lifetime of the PV panels. A partial rain and wind cleaning of the photovoltaic panels will not completely remove all impurities that are deposited on the surface of the photovoltaic panels. Therefore, there is a situation of permanent pollution of the PV panels, which leads to a decrease in the efficiency of the conversion of a solar energy to an electric energy. Neither a tilting of the PV panels will ensure their adequate cleaning with the rain and wind. In a cities, dust, grease and various chemical compounds from emissions are deposited on the PV power panels much more than in countryside (Figure 2.). In the meadows, photovoltaic panels are polluted by the dust and a frequent pollution is caused by a bird feces. Therefore, a study was carried out to determine the effectiveness of the surface cleaning of photovoltaic panels in 2017 by Milk the Sun GmbH and MeteoControl GmbH. The testing was carried out on a sample of seventeen large PV plants throughout Germany. All the selected PV panels have not been cleaned ever since their initial start-up in 2007 to 2013. The tests were subjected on three types of the PV panels, namely polycrystalline, monocrystalline and thin-walled. Each PV plant was cleaned with an area of 30 kW and its output was compared to the power of the rest of the PV plant over a period of seven months. The photovoltaic panels were cleaned only with demineralized water using a rotating brush without the use of a chemical cleaners. The results of the study confirmed that the cleaning of the photovoltaic panels has an impact on the increase in the PV plant power. Immediately after cleaning, the power of most PV power plants increased. At some PV power plants, the increase was up to 5%. In the case of heavily polluted PV power plants found, for example, in industrial areas, the power can be increased by 10% after cleaning. In countries with higher dust levels, the benefits of the PV panel cleaning are of course even more pronounced. The frequency of cleaning depends on the type of the PV panels, a place of the PV power plant placement, the inclination of the PV panels. Under normal conditions, however, it is advisable to clean the panels twice a year [5].

Figure 2. Pollution of the photovoltaic panel

B. Elimination of partial shading of a Photovoltaic Panels

The shading of a part of the photovoltaic panels negatively affects the amount of produced electricity. If only one photovoltaic cell is shaded, all other photovoltaic cells connected to it in series supply only the current supplied by the shaded cell. As a result, the total output power produced by the photovoltaic panel will decrease, thus reducing the efficiency of the conversion of solar energy to electricity [6]. Including a part of the photovoltaic panel will also cause this part not to be treated as a source but as an appliance. This is subsequently represented by increasing the temperature of the shaded cell. In extreme cases, the rising temperature can damage the aforementioned part of the panel, which will cause the entire photovoltaic panel to be damaged. Therefore, in the design and construction of a photovoltaic power plant, it is important to prevent any partial shading the PV panels by trees, surrounding buildings, power lines, etc. Also, a correct distances between individual PV panels is a significant one. These distances are usually determined for the worst case scenario, which is the lowest position of the Sun above the horizon. For our territory (Slovakia), there is the smallest angle below which Sun rays hit the surface of the Earth in the winter period of about 17 ° [6].

IV. Measurement of the Influence of Local Factors on Photovoltaic Panels

The aim of the measurement was to verify the influence of the local factors on the PV panels and to find how much they affects the PV panels performance. The measurement was focused on polycrystalline and amorphous photovoltaic panels. The first case was influence of surface pollution of photovoltaic panels on their performance. Measurements were also performed on a broken and partially shaded photovoltaic panel. For the measurement was used volt-ampere method. Each measured PV panel was disconnected from its string. Then was connected voltmeter, ammeter and variable resistive load. Also, the temperature of the photovoltaic panels and the Solar irradiance were measured in all cases. For all the PV panels were made photos by infrared camera to find out how their individual parts are heated in specific cases. The measurement was carried out at the University of Žilina, on a Research Center of the University of Žilina.

A. Measurement of the Impact of a partial shading of the Photovoltaic Panel

This measurement was performed on a photovoltaic panel which was partially shaded by air condition (Figure 1.). The purpose of this measurement was to find out how the partial shading of the photovoltaic panel affect on its performance and heating. The location of the air condition in front of the photovoltaic panels was very close, causing a large shadow. In the Figure 3 and 4. there are the images of the SV60-255 PV panel made by infrared camera, while the panel is loaded with resistors. In the figure 3, there is shown a temperature of the PV panel measured in the middle of it. A big increase of the PV panel temperature was measured on shaded part of the panel shown in the figure 4.

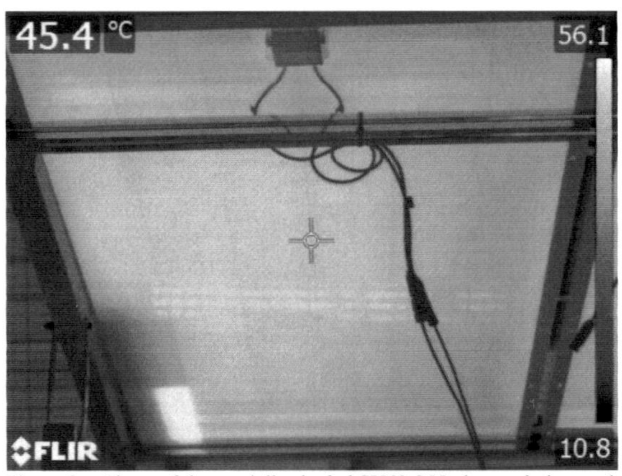

Figure 3. Photo of the partially shaded SV60-255 photovoltaic panel from the back site made by infrared camera.

Figure 4. Photo of the partially shaded cells of the SV60-255 photovoltaic panel

The measurement was performed for a specific case when only four cells were shaded. All of the shaded cells were only from one of three bypassed parts of the panel. The SV60-255 PV panel consists of 60 cells and each 20 cells are bypassed by diode. This resulting in decreasing the PV panel performance by one third. The results are interpreted by figure 5.

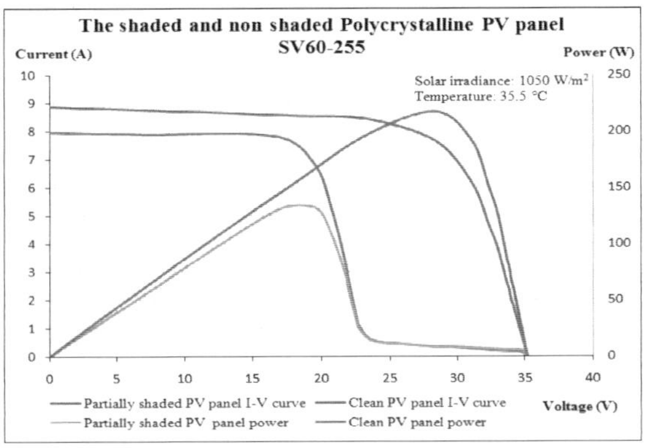

Figure 5. I-V and P-V curves of the partially shaded SV60-255 photovoltaic panel.

B. *Measurement of a polluted and clean photovoltaic panels*

This measurement was aimed at the change in photovoltaic panel performance depending on its pollution. Panels were polluted by Saharan dust from the first quarter of the 2018 year. First, the polluted PV panels were measured by volt-ampere method, then the same panels were cleaned and the measurement was repeated. The measurement was done on polycrystalline panels and two types of amorphous photovoltaic panels. The results are presented below.

1) *Measurement on the SV60-255 PV panel*

By the measurement represented in figure 6, it was found that after cleaning the photovoltaic panels, the power at the maximum power point (MPP) of the PV panels increased by 4.3 % The current increase at the short-circuit current point was about 5.6 %.

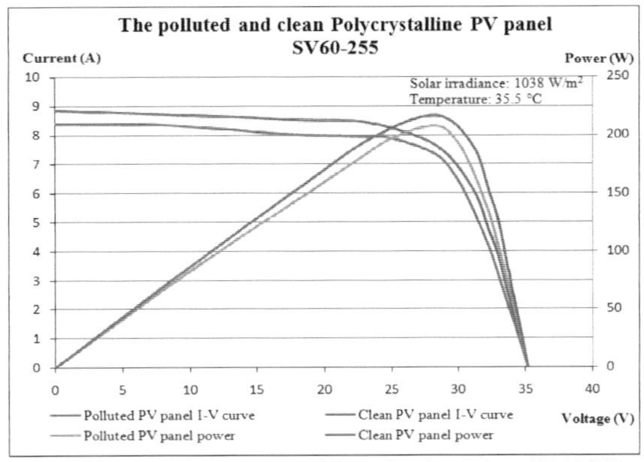

Figure 6. I-V and P-V curves of the polluted and clean SV60-255 photovoltaic panel.

2) *Measurement on the FS-277 PV panel*

The results from the measurement of the FS-277 amorphous PV panels are quite similar to the measurement of polycrystalline panel, but the impact was twice as low as that of the polycrystalline panels. The power of the PV panels in MPP was increased after cleaning by 2.3 %. The current increase at the short-circuit current point was 2.7 % in this case.

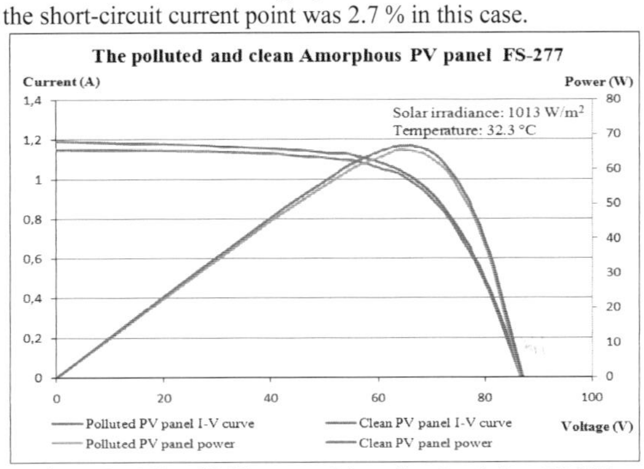

Figure 7. I-V and P-V curves of the polluted and clean FS-277 photovoltaic panel.

3) *Measurement on the PowerMax 3.5-125 PV panel*

The biggest impact of the dust was found on the PowerMax 3.5-125 amorphous PV panels. The PV panel power increase at MPP after cleaning was 11.2 % but the current increase at the short-circuit current point was only 0.6 %.

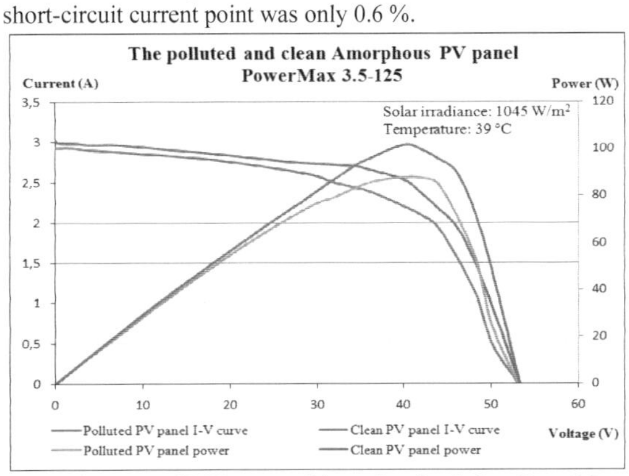

Figure 8. I-V and P-V curves of the polluted and clean PowerMax 3.5-125 photovoltaic panel.

C. Measurement of a cracked photovoltaic panel

The last case was measurement made on a cracked amorphous photovoltaic panel FS-277 shown in the figure 9. The measurement was focused on the impact of the PV panel crack on its performance. Measured results were compared with measurement on the same type of this PV panel which was in good condition. As shown in the figure 9, the crack extends almost along the entire PV panel.

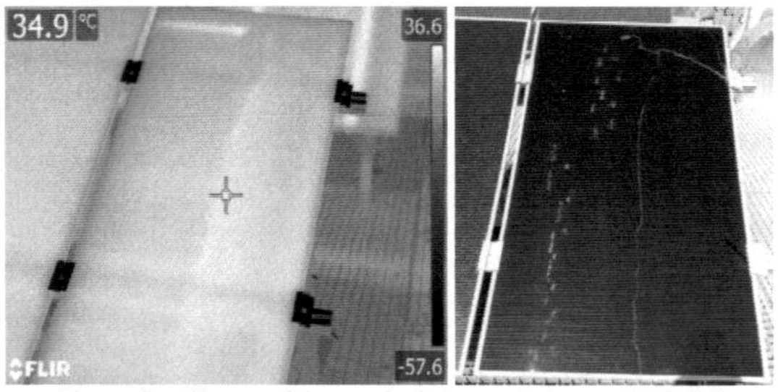

Figure 9. Photos of the cracked photovoltaic panel FS-277.

It has been found that the power of mentioned cracked photovoltaic panel FS-277 is up to 47.2 % less than the performance of the same type of the new PV panel. The limited operation of this cracked panel is also documented by infrared camera, where is visible lightning strip on the PV panel. This strip represents higher temperature along the crack of the PV panel. The performance of the cracked PV panel is also represented by figure 10, where this significant difference between cracked and good PV panel is plotted.

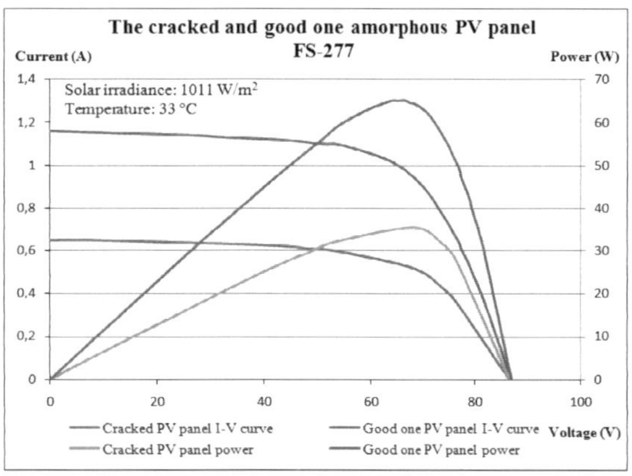

Figure 10. I-V and P-V curves of the FS-277 photovoltaic panel.

CONCLUSION

This study was done to analyzing selected local factors that affect the performance of the photovoltaic power plants. The measurement was aimed at the influence of the panel's partial shading, pollution of the PV panels and PV panel damage on the performance.

In the case of a partially-shaded SV60-255 polycrystalline panel, its performance was about 1/3 less than power of the same type of the photovoltaic panel which was not shaded. This results from the fact, that the shadow only hit the cells bridged by one of three bypass diodes of the PV panel. The shaded part of the PV panel had considerably higher temperature which in long-term will have an impact of the PV panel life time and efficiency. In view of this fact, the elimination of the partial shading of the panels is an important task in the design, construction, but also during the operation of the photovoltaic power plant.

The difference between power of the polluted and clean PV panels varied from 4.3 % to 11.2 % in maximum power point. The biggest difference of PV panels power was found on PowerMax 3.5-125 amorphous PV panels. However, unlike the other two measured types of PV panels, this amorphous type had a minimum short-circuit difference between measurement of the polluted and clean PV panel. As mentioned in the figures, the temperature and solar irradiance were the same for all the measurements of one case. Also the short-circuit current coefficient of the PowerMax 3.5-125 PV panel from a datasheet is 0 milliamperes per degree of Celsius. This differences in the PV panels performance were similar to the study mentioned above, but in this study, PV panels have been cleaned up several times during PV panels life time.

The performance of the crack amorphous panel was up to 47% less than that of the same undamaged panel. In addition, the damaged photovoltaic panel permanently limits the performance of a PV string. For this reason, when damaging the photovoltaic panel, it is necessary to replace mentioned PV panel with a new undamaged one.

The whole photovoltaic power plant must be also solved according to reliability in future. Firstly, it is necessary to know the complete data about failures of the photovoltaic power plant. After that, these data can be used for creation analysis of reliability. This analysis can determine key components and the failures, which are crucial for the total reliability of the power plant. Finally, the failure impact of the key components must be minimized by correct setup of maintenance and regular replacements of the components for the better sustainability and for performance prediction of a PV power plant [7].

ACKNOWLEDGMENT

This work was supported by Slovak Research and Development Agency under the contract No. APVV-16-0505: The short-term prediction of photovoltaic energy production for needs of power supply of Intelligent buildings – PREDICON.

REFERENCES

[1] J. Ayoub, R. Rammal, A. Assi, I. Assi, "Two techniques used to improve the efficiency of existing PV panels: Thermal management & PERC technology", 2017 29th International Conference on Microelectronics (ICM), Lebanon, Beirut, IEEE Press, Jan. 2017, 978-1-5386-4049-4

[2] S. Essig, et al., "Boosting the efficiency of III-V/Si tandem solar cells", 2016 IEEE 43rd Photovoltaic Specialists Conference (PVSC), Portland, USA, Jun 2016, ISBN 978-1-5090-2724-8

[3] A. K. Dutta, "Prospects of nanotechnology for high-efficiency solar cells", 2012 7th International Conference on Electrical and Computer Engineering, Dhaka, Bangladesh, March 2012, , ISBN 978-1-4673-1436-7

[4] D. Pera, J. A. Silva, S. Costa and J. M. Serra, "Investigating the impact of solar cells partial shading on photovoltaic modules by thermography," 2017 IEEE 44th Photovoltaic Specialist Conference (PVSC), Washington, DC, 2017, pp. 1979-1983. 978-1-5090-5605-7

[5] J. Dorda, "Vyplatí sa investovať do čistenia solárnych panelov? Prvé výsledky nemeckej štúdie to potvrdzujú!", 2017, [Online]. Available: http://www.solarnenovinky.sk/?spravy/2017112001/aktualne-vyplati-sa-investovat-do-cistenia-solarnych-panelov-prve-vysledky-nemeckej-studie-to-potvrdzuju

[6] P. Mastný. "Obnovitelné zdroje energií", 2011, (1st edition), Prague, Czech, ISBN: ISBN 978-80-01-04937-2 [Paper book]. Available:https://www.mpoefekt.cz/upload/7799f3fd595eeee1fa 668 75530f33e8a/cvut-2-oze.pdf

[7] A. Hromadka, , J. Jirickova, Z. Martinek. "Analysis of the unplanned shutdowns and estimation of reliability in Temelin". 19th International Scientific Conference on Electric Power Engineering (EPE), Brno, Czech, IEEE, Jun 2018. pp. 551-556. ISBN: 978-1-5386-4612-0 ISSN: 2376-562

Methodological Approaches to the Choice of Energy Storage and Optimization of Their Parameters to Improve the Electric Power Quality in Various Types of Electric Power Systems

Kamil Bakhteev, Alexander Fedotov, Nataliya Chernova, Rinat Misbakhov
Engineering center, Computer modeling, and engineering in the field of power engineering and power engineering
Kazan State Power Engineering University
Kazan, Russia
kam1609@mail.ru, zerdex@bk.ru, fed.ai@mail.ru

Abstract — **The relevance of the study is the need for the implementation of energy saving policy on remote oil production areas, where there is a possibility of utilization of associated gas as a fuel in a distributed generation (DG). The use of energy storage systems (ESS) complete with a natural-gas-fueled reciprocating-engine power plant equipped with a synchronous generator (SG) allows you to prevent load rise and frequency dip of the alternating current under disturbing loads, as well as to maintain the required level of residual voltage during short circuits in the grid.**

Keywords: mathematical modeling, electrical grids, synchronous machine, supercapacitor, battery, power quality, voltage dip

I. Introduction

The modern architecture of the Russian electric power industry is a combination of centralized power distribution with power distribution to consumers from local power systems with distributed generation (DG) with low-power power plants using gas, local fuel or renewable sources. Small power plants are installed directly at consumers and can work in autonomous mode or be connected to the electricity distribution grid in the voltage range of 6 - 35 kV [1]. The speed and ease of installation and maintenance, as well as their lower cost compared to connecting to the public grid, all contribute to the large-scale development of the DG.

The flagship of the DG in the Russian Federation is the oil and gas industry, which is explained both by considerations of economic feasibility (availability of own energy resources, remoteness of many fields from large generation sources and grids), and requirements of environmental legislation obligating to increase the degree of utilization of associated petroleum gas. Oil refineries create their own generation, which has the opportunity to reduce the cost of purchasing energy resources, at the same time solving the problem of waste disposal, as well as enterprises of the petrochemical and chemical industries that use energy that is available in production to generate electricity. In general, the presence of objects of their own generation has a positive effect on the electric power quality, since their owners have full control over the output power and the voltage value of the DG. For remote areas where delivery of liquid fuel to power plants is necessary, the use of gas engine generator (GEG) working on associated petroleum gas is hampered by their high sensitivity to stepwise load changes and short-term power outages (STPO), accompanied by voltage dips (VD). DG facilities are sensitive to any even the shortest (millisecond-long) VD due to the continuous nature of the production cycle, therefore the problem of ensuring reliable power supply becomes even more acute for them, which requires finding new effective technical solutions. However, there are negative consequences of the DG. So, due to the wrong choice of the type or power of gas engine generator, the specific fuel consumption, the minimum process unit utilization, load rises and drops may be unreasonably overestimated.

Supplementing the GEG with ESS (batteries, supercapacitors) allows them to function at various disturbing influences on the electric power supply systems and to obtain the most significant effect in monetary terms, both by eliminating imported liquid fuels and by reducing negative impacts on consumers from VD. This article discusses aspects of the use of ESS for limiting the depth of VD, which consist both in their use to power SG exciter during short circuits, and forcing short-term feeding of the short circuit site from ESS to increase the residual voltage level. It is necessary to consider both the cost indicators of the components of the electrotechnical complex consisting of a natural-gas-fueled reciprocating-engine power plant equipped and ESS and the current costs of their operation. Expanding the scope of possible application of ESS. Their use may be appropriate for power facilities on diesel fuel, gasoline, fuel oil, where its delivery is associated with high costs.

II. Application of Electric Energy Storages for Limiting the Depth of Voltage Dips

Interest in the use of ESS in various areas of the electric power system (in grids, substations, generation, individual consumers) has increased. Strengthening the development trend of DG facilities along with the massive industrial development of lithium-ion electric battery, which is a kind of building blocks of today's most common ESS, and a sharp decline in their cost are the main factors for the development and use of battery energy storage system (BESS).

The purpose of this study is to select the ESS for solving two problems at once: preventing short-term VD in supply networks with the voltage of 6, 10 kV when working in parallel with the power system, and also to limit the frequency dip during autonomous operation of the GEG. To do this, we analyzed the technical characteristics of existing types of energy storages (ES), their advantages, and disadvantages [2-13]. Based on the analysis, it was concluded that the most promising for these purposes are lithium-based batteries, namely Li-Ion, LTO, LFP: compared to other types of batteries, they have the highest specific energy capacity (120-200) Wh·h / kg and very low internal resistance of the elements. For a long time, due to their high cost, the use of lithium-based batteries was considered economically inefficient. At present, as a result of price competition among leading battery manufacturers, their prices have dropped sharply, and by 2030, according to the BNEF forecast, a further price reduction is expected to be 3.7 times [14]. The main objective of this study is to determine the criteria for the optimal choice of an electric battery to ensure proper quality of electricity in the power supply system, where the "SG-ES" system is used when exposed to the STPO, using the SG capabilities in terms of excitation forcing. Since it is important to take into account not only the specific energy consumption of ES but also the time of its full discharge, to prevent short-term VD, in cases where a short-term power requirement arises and there is a possibility of fast charging, supercapacitors (SC) are the ideal choice. SC can be charged and discharged the almost unlimited number of times, their self-discharge significantly exceeds the indicator of other types of electrochemical ES (the system resource can exceed 100,000 discharge charge cycles). The combined use of BESS and supercapacitor energy storage system (SESS) as a hybrid energy storage system (HESS) can level the shortcomings of both elements. In addition, it is possible to several times increase the short-term capacity of the ES in the emergency mode of the grid, as well as cover short-term peak loads without reducing the overall life of the ES.

Confirm the validity of the findings will help develop the simulation model of the electrical complex using the ES and the simulation results presented in Figures 1,2. In this work, the model of the electrical complex is being developed for a typical industrial enterprise with a total power consumption of 1 MW and higher. Figure 1 shows the schematic diagram of the electrical complex. Figure 1 shows the circuit diagram of the electrical complex.

In Russia, an experimental installation of low power was assembled, developed by the Research and development center at the federal grid company of unified energy system [13]. By analogy with the concept shown in Figure 1, a simulation model was developed in Matlab-Simulink [15]. To create a HESS, a large number of batteries connected in parallel will be required, to provide industrial enterprises with a total power consumption of more than 1 MW, a simulation model is presented in Figure 2.

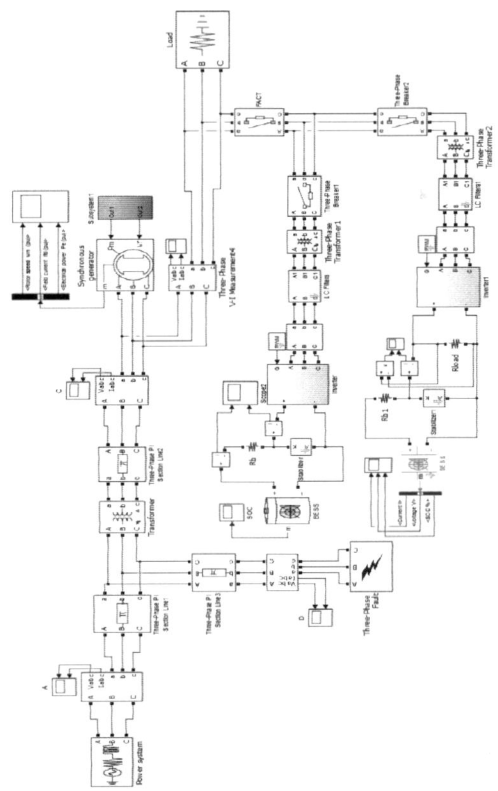

Figure 1. Circuit diagram of the connection of HESS, SG, working in parallel with the power system

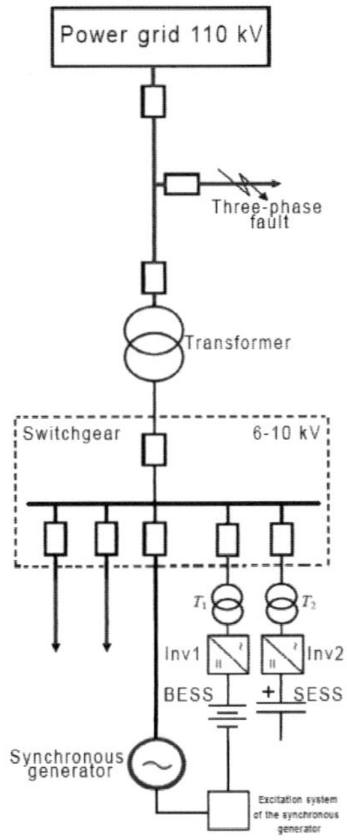

Figure 2. The use of HESS to limit VD

The main characteristics of BESS, assembled from 147 LT-LFP 700P batteries: one 700 A·h capacity, total voltage - 470 V. Assumptions adopted in the HESS model [15]:

- The internal resistance is considered constant during charge and discharge cycles and does not change with the current amplitude.

- The model parameters are derived from the discharge characteristics, are considered the same for charging.

- Self-discharge of the battery is not provided. It can be imagined by adding a large resistance parallel to the terminals of the SB.

- Temperature does not affect the behavior of the model;

- The electric battery has no memory effect.

- The electric battery capacity does not change with the current amplitude (Pekert effect).

Limitations of the model [14]:

- The minimum battery voltage is 0 V, and the maximum battery voltage is 2·E0.

- The minimum battery capacity is 0 A/h, and the maximum cannot be more than 100%.

SESS is assembled of 78 SC with a capacity of 500 F, the voltage of 16 V. Total voltage of SESS is 526 V. SESS operates on a load of 1 MW through a stabilizer, which limits the voltage to 470 V. All types of ES are connected to the grid through power electronics devices - current or voltage converters. The simulation results are presented in Figure 3.

Figure 3. The use of HESS to limit VD

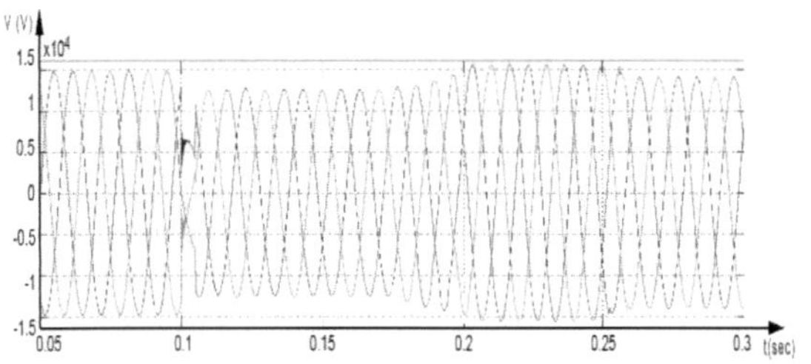

The level of residual voltage at VD was 40%. After 10 ms, SESS was turned on, which ensured almost instantaneous recovery of the voltage to the level of residual voltage of 93%. After 100 ms, the BESS is connected after VD. For a short time they work in parallel, then the SESS is switched off, and work continues on BESS.

In addition to the use of short-term supply the fault location from the ES to increase the level of residual voltage, another option may be the use of ES to limit the depth of the VD in terms of their use for the supply of excitation system of the synchronous generators at short-circuiting [18-24]. Excitation forcing (i.e., a rapid increase) of SG is the simplest and the cheapest way to increase the voltage level in the grid. However, the static excitation valve systems used under the influence of the STPO become, in essence, self-excitation systems with external faults, since they are powered directly from the sections of the buses on which SG work. Therefore, when VD has not achieved the expected result of the force, since the reduced rectified voltage on the exciter. Switching exciters to ES allows you to fully use the capabilities of SG to increase the level of residual voltage.

To identify the regularity of influence of the field-forcing ratio (with independent power supply from ES and supply of the exciter from its own buses), a series of experiments with SG 8 MW and length of an overhead power line, on which three-phase fault occurs, were conducted. The system parameters are as follows: the resistive resistance of the power source is 0.04 ohms, the inductive resistance is 0.016 Gn. Figure 10 shows a comparison of the graphs of the Critical length of a line (CLL) for SG.

Figure 4. Change in residual voltage level

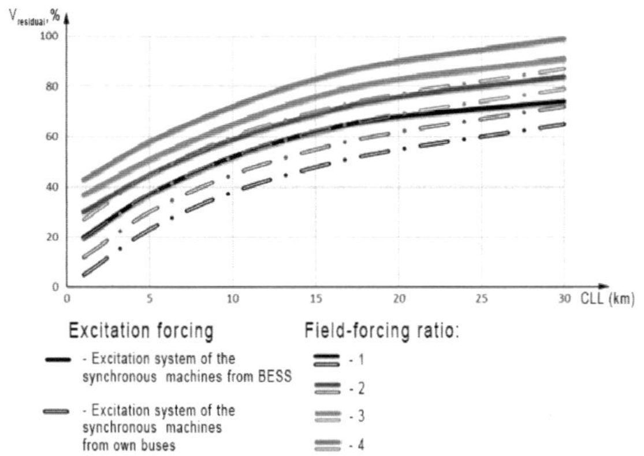

Comparing the graphs for SG with a system powered by its own buses and an independent system of excitation from ES, shows that under the same initial conditions at the level of a given minimum allowable residual voltage of 80% of the nominal (taken as an example, depending on the generator load can be more , and less). The CLL for SG with an independent excitation system from ES is less than for SG with an exciter power supply system from its own buses for 18 kilometers. If we compare the effectiveness of the excitation multiplicity of the SG excitation with an independent excitation system from ES, then a four-fold excitation forcing provides a two-fold reduction in the CLL of the power lines.

III. Imitation Model of Working Gas Engine Generator with Current Sources

Among the types of power units, the GEG is distinguished by simplicity, the reliability of the construction and the highest electrical efficiency. The electrical efficiency of modern GEG when working on Russian natural gas is approximately 41–44% [16]. Their wide distribution is associated with the relative ease of transportation and maintenance, as well as with low specific investment. At the same time, the operation of power plants on liquid fuel is associated with the need for constant delivery and seasonal storage, as well as the need to work in non-permanent modes of following a load of consumers, which increases fuel consumption and contributes to accelerated development of the technical resource of the power plant. Under extreme climatic conditions, the procedure for fuel delivery is a technically complex and costly operation, with the implementation of which the share of the transport component in the cost of fuel can reach 70% [17]. The widespread use of GEG is currently limited by their sensitivity to significant load rise and load droppings, which lead to the output of frequency values beyond the limits of normatively established values and to the corresponding work of emergency control automation.

The solution of the problems described and the general increase in the efficiency of the functioning of closed power systems is possible through the sharing of the GEG with the ESS, which allows the GEG to be transferred to the basic working mode with a constant load. Changes in load are fulfilled with ES. Research of the effectiveness of the ES in case of disturbances from the load side was carried out using the example of a typical stand-alone power system scheme, Figure 5.

Figure 5. Circuit diagram of the connection of GEG, SG working stand-alone

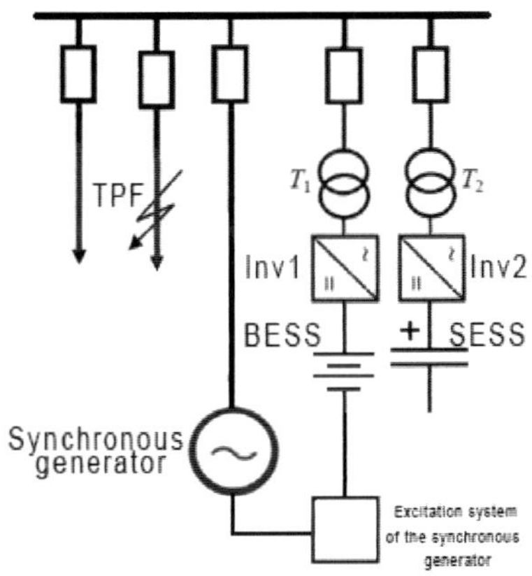

The corresponding simulation model is presented in Figure 6. Where the ES are presented in a simplified form as controlled sources of alternating current.

Figure 6. The simulation model of the GEG in parallel with the sources of alternating current

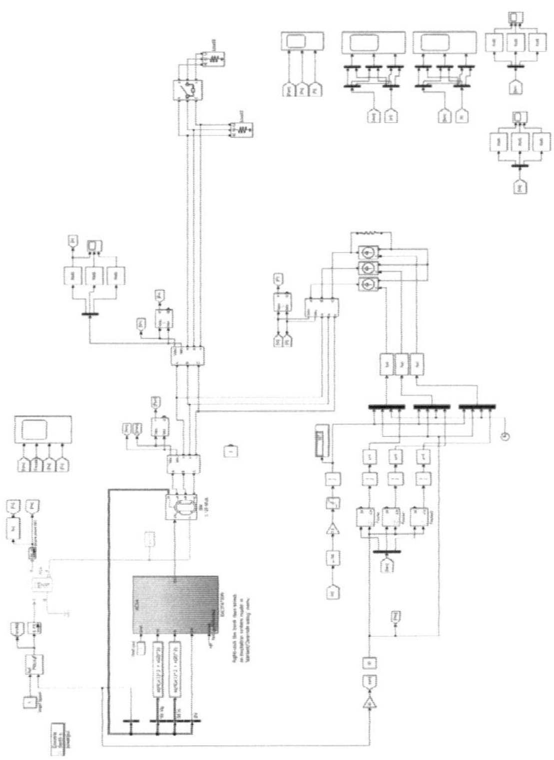

The combination of ESS with an autonomous generator implies the solution of a number of issues related to the choice and optimization of the parameters of the created electrotechnical complex.

Figure 7. The results of the simulation work of the GEG without ESS

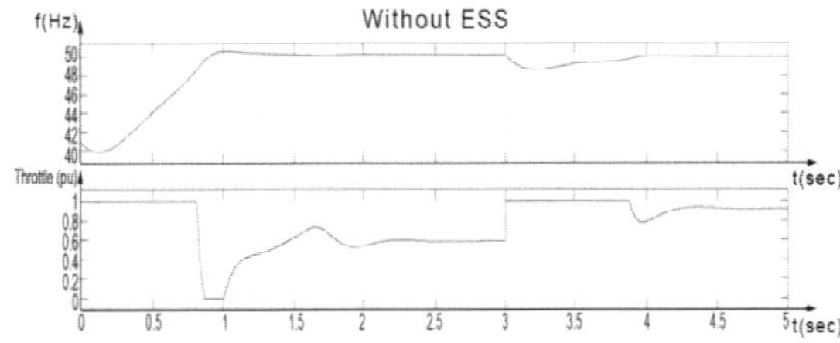

As can be seen from the oscillogram shown in Figure 7, at a time point of 3 seconds a load rise of 300 kW occurs, the frequency drops to 1.57 Hz and, as a result, the number of motor revolutions decreases, the performance of machines and mechanisms decreases, which leads to marriage and undersupply products. According to GOST 13109-97 in normal mode, a frequency dip (± 0.2 ÷ 0.4 Hz) is permissible. The use of ESS allows limiting the frequency dip, Figure 8.

Figure 8. The results of the simulation work of the GEG with ESS

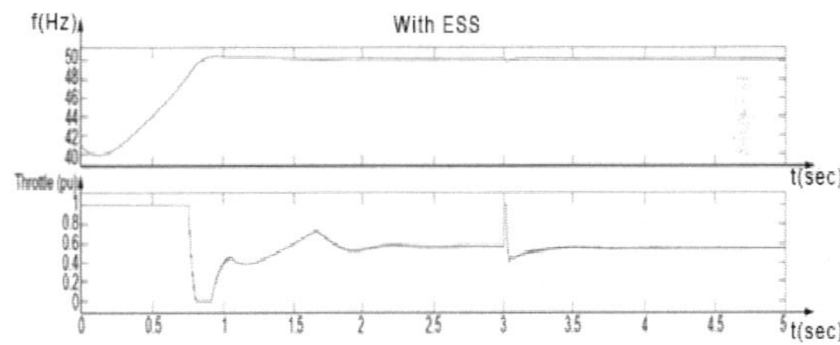

Analysis of the results shows that when using ESS with the same load rise, it is possible to limit the voltage drop to 0.18 Hz, which, according to GOST, is allowed in normal mode. Figure 9 shows how the load rise on a GEG without ESS is fulfilled and in figure 10 a GEG with ESS with the same load rise of 300 kW. Analysis of changes in the power of the SG and the load without the use of ESS shown on Figure 9.

Figure 9. Analysis of changes in the power of the SG and the load without the use of ES

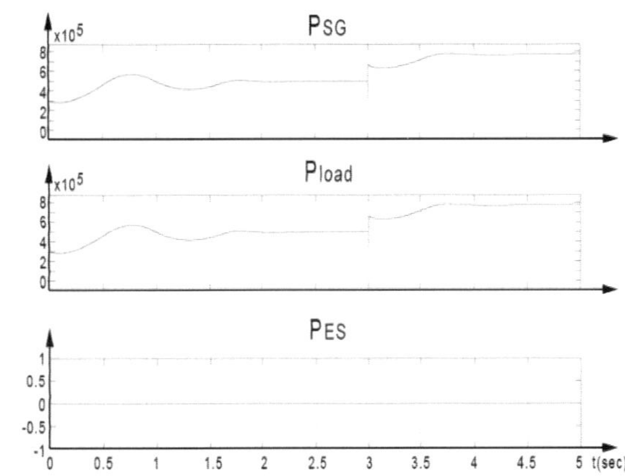

Figure 10. Analysis of changes in the power of the SG and the load using the ES

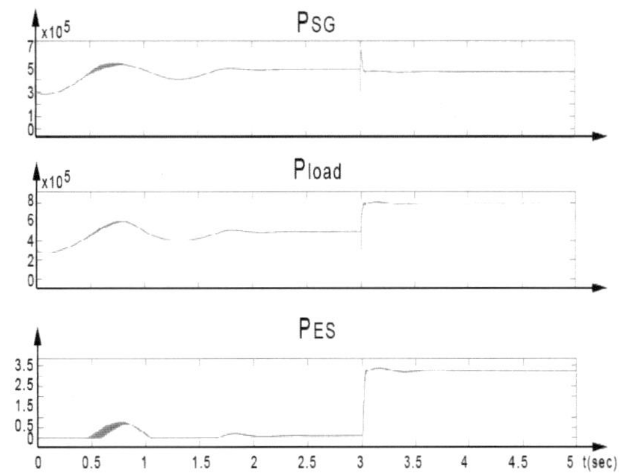

When the frequency drops, the EMF decreases generator (because the speed of the exciter decreases) and the voltage in the grid decreases too.

IV. Conclusion

The use of HESS allows limiting the depth of VD with various duration, affecting industrial enterprises with a total power consumption 1 MW and higher. At present, it seems possible and promising to use ESS based on Li-ion batteries to solve the problem of fuel economy at distributed generation facilities, which are based on gas engine and gas turbine generators.

There is an effect of frequency stabilization in an autonomous electrical grid with disturbances from the load in the case of using the ESS as a regulator, which is included in the complex with the GEG, working on the associated petroleum gas. This testifies to the economic efficiency and feasibility of the use of ES in the composition of the GEG in isolated power systems and other remote and hard-to-reach areas of Russia, reducing the volume of imported liquid fuels. The introduction of ESS will contribute to the differentiation of power supply schemes and the overall increase in energy efficiency.

The depth of the VD directly depends on the power of the power supply system and the remoteness of the three-phase fault point from the buses of the 110 kV distribution substation. The higher the power of the system and the deeper the damage, the less influence the residual voltage of consumers have on the presence of SG working at a voltage of 6-10 kV.

The field-forcing ratio of SG with an independent excitation system from ESS, then a four-fold forcing ensures a two-fold reduction in the critical length of the line (CLL). When planning the power supply system, it is possible to recommend the placement of several SG (with higher than normal field-forcing ratio) with a uniform load on the main step-down substation switchboard sections, from which the responsible consumers are powered. This will significantly reduce the depth of payload at close damage in a 110 kV distribution grid.

Acknowledgment

Research is made with financial support of the Ministry of Education and Science of the Russian Federation within implementation of the federal special program «Research and development in the priority directions of scientific and technological complex of Russia for 2014-2020», the agreement on granting a subsidy № 075-02-2018-190-1 stage, unique identifier of applied scientific research (project) RFMEFI57418X0188All research articles should have a funding acknowledgment statement included.

References

[1] N.I. Voropai, "Distributed generation in power systems," International Scientific and Practical Conference "Small Energy-2005". (http://www.combienergy.ru/stat/983-Raspredelennaya-generaciya-v-elektroenergeticheskih-sistemah)

[2] A.I. Fedotov, K.R. Bakhteev, "Improving the quality of power supply to industrial consumers through the use of electricity storage devices with voltage dips // Dynamics of nonlinear discrete electrical and electronic systems: Proceedings of the XII All-Russian Scientific and Technical Conference. Cheboksary, 2017, pp.308–311.

[3] A. Kuruvita, "Optimization of energy storage in electrical networks. Ph.D. thesis," Moscow, 2003, 129 p.

[4] "Which energy storage device is the most energy-intensive," (https://altenergiya.ru/accumulator/kakoj-nakopitel-energii-samyj-energoemkij.html)

[5] Y. Yoo, M.-S. Kim, J.-K. Kim, Y. Kimb, W. Kim, "Fast-response supercapacitors with graphitic ordered mesoporous carbons and carbon nanotubes for AC line filtering," Journal of Materials Chemistry, 2016, vol. 14.

[6] K.R. Bakhteev, "Creation of a big power hybrid electric energy storage for prevention short-term interruption of power supply to industrial consumers" / Proceedings of higher educational institutions, Energy problems, 2018. vol. 20. Issue 3-4. pp. 36-44., doi.org/10.30724/1998-9903-2018-20-3-4-36-44

[7] O. Abramova, "Supercapacitor and Battery Comparison", April 2016 (https://best-energy.com.ua/support/battery/bu-209)

[8] Lithium-ion battery manufacturers. Official site of Litotech LLC (http://www.liotech.ru/liamanufacturers)

[9] B. Olek, M. Wierzbowski, "Local Energy Balancing and Ancillary Services in Low-Voltage Networks With Distributed Generation, Energy Storage, and Active Loads" Industrial Electronics, IEEE Transactions on Year: 2015, Volume: 62, Issue: 4 Pages: 2499 – 2508.

(https://ieeexplore.ieee.org/document/6975183.)

[10] K.R. Bakhteev, Prevention of voltage dip in industrial enterprises with the use of electrical power storage devices / In the collection: Electric Power Industry through the Eyes of Youth – 2017, Materials of the VIII International Scientific and Technical Conference, 2017, pp. 419-420.

[11] R. Hari, S. Sobhit, S. Anjan, "Internal Hybrid System of Li-Ion Battery and Supercapacitor," Journal of Basic and Applied Engineering Research, Online ISSN: 2350-0255, vol. 2, Number 11, April-June, 2015, pp. 937-940.

[12] M. C. M Laura, F. M. R. Diego, A. H. Johann, "Behavioral Analysis of a Lithium-Ion Battery, a Nickel Metal Hydride Battery and a Supercapacitor in an Isolated Photovoltaic Solar System," Energy and Power Engineering, ISSN Online: 1947-3818 2017, vol. 9, pp. 293-308.

[13] R.N. Berdnikov. V.E. Fortov. E.E. Son. K.K. Denshchikov. A.Z. Zhuk. N.L. Novikov. Yu.G. Shakaryan., "Hybrid electricity storage for the UNEG on the basis of batteries and supercapacitors," Energy of the Unified Network, 2013, Issue 1.

[14] Brian Eckhouse. The Battery Boom Will Draw $620 Billion in Investment by 2040.

(https://www.bloomberg.com/news/articles/2018-11-06/the-battery-boom-will-draw-1-2-trillion-in-investment-by-2040)

[15] MATLAB-Simulink (https://www.mathworks.com/)

[16] D. N. Cheremnykh, E. V. Tashlykova, M. G. Razepina, "Gas piston installations as an alternative method of generating electricity" // Young Scientist, 2014, Issue 21, pp. 245-247.

[17] V.Yu. Kononenko, V.P. Bilashenko, "Effects of the use of energy storage devices in isolated power systems of Russia" / Arctic: Ecology and Economy Issue. 2 (14), 2014

[18] A.Ya.Shklyarskiy, "Reducing the magnitude and duration of voltage dips in the electrical networks of oil producing

enterprises. Electronic scientific and practical journal," Modern scientific research and innovation", Issue 5 (37) May 2014, (http://web.snauka.ru/issues/2014/05/34992)

[19] A. Fedotov, K. Bakhteev, A. Leonov Influence of the small distributed generation on the level of residual voltage at short circuits / Proceedings of higher educational institutions, Energy problems, 2016, Issue. 3-4, pp. 45-49.

[20] Seymour J., "The Seven Types of Power Problems. White Paper 18.Revision 1," APC by Schneider Electric, 2018, (https://www.apc.com/salestools/VAVR-5WKLPK/VAVR-5WKLPK_R1_EN.pdf)

[21] A. Fedotov, K. Bakhteev, Influence excitation forcing of the synchronous machines on the level of residual voltage in case of short-time violations of electrical power supply / Proceedings of higher educational institutions, Energy problems, 2016, Issue. 7-8, pp. 64-71.

[22] A. Fedotov, E. Fedotov, K. Bahteev, "Application of local Fourier transform to mathematical simulation of synchronous machines with valve excitation systems" / Latvian Journal of Physics and Technical Sciences, 2017, vol. 54. Issue 1, pp. 31-40.

[23] K. Bakhteev, A. Fedotov and R. Misbakhov, "The Improving quality of power supply to industrial consumers using high-power energy storage," 2018 IEEE 59th International Scientific Conference on Power and Electrical Engineering of Riga Technical University (RTUCON), Riga, Latvia, 2018, pp. 1-5. doi: 10.1109/RTUCON.2018.8659834

[24] A.E. Veselov, A.S. Karpov, V.V. Yaroshevich, G.P. Fastiy, E.A. Tokareva, "Development of a new structure for electricity distribution networks of 6 and 10 kV industrial power supply systems, "Vestnik of Kolskogo center of the Russian Academy of Sciences, vol.4, 2015, October 2018.

The Impact of the Retrofitted Lighting System to the Electrical Power Grid

Pavol Belany, Marek Novak, Marek Siranec, Marek Roch
Department of Power Systems and Electric Drives
Faculty of Electrical Engineering and Information Technology, University of Zilina
Zilina, Slovakia
pavol.belany@fel.uniza.sk

Abstract — **This article deals with the impact of buildings retrofit on the electrical grid. By retrofitting, the original devices are replaced to newer ones. However, this replacement brings problems in the area of the influence of these devices on the electrical grid. These devices very often contain semiconductor components. By their operation are generated harmonic, which decrease the power quality. The main aim is comparing the impact of the original system with the new system after applying the retrofit concept of the selected system in buildings. The lighting system was chosen as the reference system. This system originally contained low-pressure mercury lamps. These lamps have been replaced by new energy-saving LED lighting. Lighting retrofitted system is managed by two types of control: switching by presence and dimming according to the daylight level. The dimming via triac causes the greatest deterioration in power quality. The last part of the article was a proposal for a countermeasure to mitigate or eliminate this impact. This proposal can only be done after an analysis of the measurement of the power quality each lighting systems.**

Keywords: power quality; retrofitted lighting system; Matlab; simulation; power quality measurement; LED lights

I. Introduction

On one hand, in present, a number of electronic devices use an electronic converter and ballast. These devices have not positive impact on power quality. These devices are, for example, compact fluorescent lighting (CFL), LED lighting, etc. [1].

There are many studies in the world that deal with the effects of non-harmonic currents. These studies deal with the possibility of identifying high current amplitudes sources. Low-amplitude current sources such as lighting have not been researched yet. The main result of these studies is a description of these sources and a description of how to improve power quality [1], [2].

It is a fact that harmonics have an effect on electrical equipment and can cause malfunctions etc. The impact on the measuring system is not sufficiently precisely confirmed, so further research is needed. This will be verified by using a programmable power source. This consistency has been chosen to such an extent that it can generate accurate harmonics values [1], [3].

The base of these analyses is the measurement of the lighting systems and according to these values, it is possible to make the next steps. After determination of priority non-harmonic sources, it is possible to elaborate countermeasures, which this adverse effect reduced or eliminated. The power quality measurement was made using the ENA 330A Quality Analyzer on two independent retrofitted LED systems (switching and dimming) [1].

On the other hand, another point of view of LED lights is the impact on reliability. The solution of the total reliability is important for better sustainability of the system [3], [4].

All electric devices are tightly connected with the bathtub curve. Classical lights, which is based on resistance or fluorescent phenomenon, have quite often a problem with lower lifetime. It means that a lifetime of LED lights is much longer than another type of lights [4].

The second area of the bathtub curve represents time with a constant intensity of failures. The general effort is to lengthen this area as long as possible. Thanks to LED lights, this effort is met. LED lights to have the longest lifetime from all known type of lights i.e. one of the biggest advantages of LED lights. Other advantages of LED lights can be higher light power and lower self-consumption than other ones [3], [4], [5].

This article contains an analysis of lighting before and after a retrofit. Based on the measured values of individual varieties of the light system, the simulation in Matlab Simulink was made to improve the power quality.

II. The Harmonic in Grid

The harmonic current has been in the grid for many years. The original sources of harmonics were mercury rectifiers used in electrical traction. Currently, the number of devices that produce harmonics is increasing. In the past, only certain areas of communication and security technologies have influenced harmonics. Today, this problem is becoming more general because it affects customers who also source decreasing of power quality. Harmonic frequencies are integer multiples of the fundamental harmonic supply voltage. This means that the third harmonic with 150 Hz, the fifth harmonic with 250 Hz, etc. It is for basic harmonic with 50 Hz [1].

A. Analysis of non-sinusoidal waveform

Each non-harmonic waveform can be divided into a sum of several sine parts. These waveforms have given amplitude, frequency and phase. These parameters can be obtained by Fast Fourier analysis. The basic advantage of using only sinusoidal waveforms is the ability to use conventional methods to calculate the steady state [1], [2].

If we analyze the grid in which the deformed waveforms of voltage and current occur, we perform the Fast Fourier decomposition at first in order to obtain other parts. These parts will be only sinusoidal with frequencies that are integer multiples of the fundamental frequency. Next step is to use the conventional calculation methods for each harmonic [1], [6].

B. *Adverse effects of harmonics*

Harmonic parts cause adverse effects on equipment and the grid. These effects can be divided into long and short term.

Short-term effects are associated with malfunctions or lower quality levels during operation of the equipment. Large harmonics amplitudes can cause malfunctions in the switching and protection devices [1].

The long-term effects are essentially thermal. Further losses, overheating cause faster ageing, and they can even cause damage to capacitors and rotating machines. They start to work after 10 minutes of interference [2].

LED lighting retrofit problems:

- Harmonic

The LED system may cause harmonic distortion. Harmonics can cause malfunctions on equipment or in power grid, causing increased maintenance and downtime. In some cases, the harmonic voltage may go unnoticed until the device fails [1], [2].

- Power factor

In some cases, depending on the installed type of LED lights, they may cause power factor problems. The low power factor can cause device degradation, power loss, voltage drops, and maintenance fees [1].

- Inrush current

The very high efficient LED lighting can cause high switching current and overvoltage. Damage and equipment failure and operational deficiencies are just some of the results [2], [6].

- Voltage unbalance

The primary source of voltage unbalance is often defined as a single-phase source. Since the LED light systems have low power and are therefore often operated as a single-phase load. Voltage unbalance can cause individual phase wire overload, loss of equipment and damage to equipment [7].

III. Lighting Retrofit

The lighting retrofit is one of the stages in the process of modernizing buildings. This process is located at the beginning because lighting has a significant impact on other systems, for example, it affects heating, cooling, and energy quality [3], [5].

Lighting consumes about 35% of the electricity used in commercial buildings and it has affects other systems through electrical requirements and heat generation. Modernization of lighting systems with new efficient devices and their accessories and controls can reduce energy consumption and improve the visual environment [5], [8], [9].

In order to ensure that retrofitting leads to improved system efficiency, the following key guidelines need to be introduced [3]:

- Create a light design that provides enough light to perform tasks in space.
- Use daylight, if possible, but avoid direct sunlight. Installing a control system to automatically change the light intensity depending on daylight.
- Use of high-efficiency light sources.
- Automatic switches or dimmers for lighting control [3].

Retrofitted lighting system allows the use of more advanced control capabilities. The problem arises by switching and dimming because of using semiconductor by the operation and then overall harmonics components will be rise [5].

Automatic control can be divided into:

- On / Off switching

It is the easiest way to reduce energy consumption. Luminaires have the option of manual switching, but this function is often not used. To increase the savings, the system switches to automatic mode. Timers or motion detectors collect data for this switch. The problem is this case is in switching because SSR relay is used. [3], [6].

- Dimming

Dimming regulators are usually used to adjust the level of illumination for human needs. For collect data for setting the light level, the light sensors were used. Dimming can be carried out in either a stepped or continuous method. There are new advanced dimming technologies that use individually addressable ballasts or wireless technologies. For dimming, the triac control is used, This technology of dimming will have a big effect on overall current [3], [6].

The practical implementation of the additional lighting installation was carried out in the common areas of the office building. This common space was chosen to implement both of the above control systems [8].

A lighting retrofit is to be understood as an innovation of lighting in a building. This innovation must keep the basic requirements of the standard. The total investment cost must be lower than saving. It is appropriate to use parts of the original system and reduce the initial costs [8], [10].

IV. Measurement of Retrofitted Lighting System

The power quality measurement was made using the ENA 330A. The connection of the analyzer into the circuit was realized in a switchboard. The connection is shown in the block diagram in figure 1.

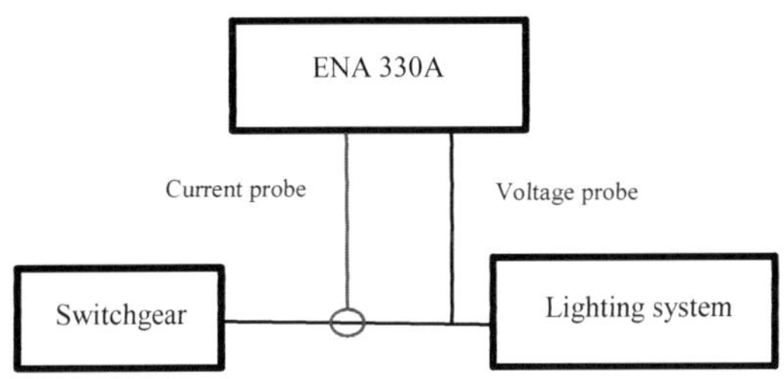

Figure 1. Block diagram of measurement

The measurement was realised on for each system (switching, dimming) to compare individual systems with regard to power quality. From the measured data was made an analysis for each system. This analysis is given in the following section.

A. *Measurement of the original system*

The original system consisted of low-pressure mercury lamps. These luminaires provided enough light but for their operation, they needed additional equipment, which increases the energy consumption. The benefit of this system was continuous operation and then some problems with power quality have not occurred. In the next figure 2 is graphical dependence of voltage and current.

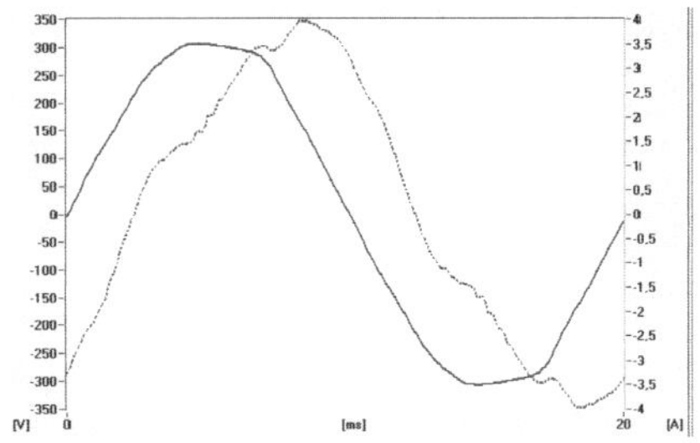

Figure 2. Graphical dependence voltage and current

In figure 2 is shown that the waveforms of the voltage and the current are almost sinusoidal. The current is phased shifted due to the compensation by choke.

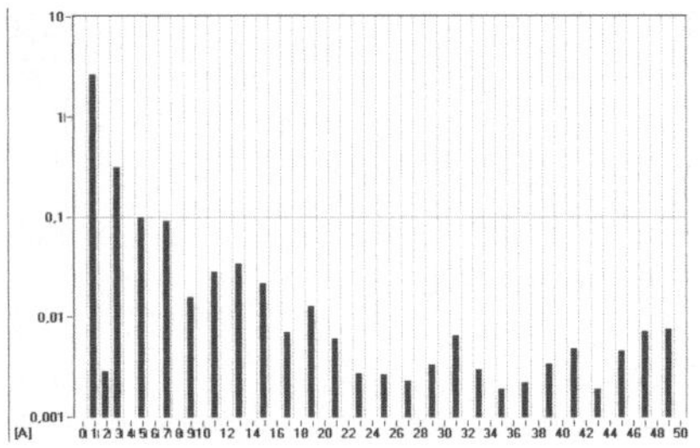

Figure 3. Harmonic spectrum of current for original system

In figure 3 is the FFT analysis of current. The harmonics components decrease according to the amplitude law. The harmonic components are made by the operation of the light source.

Table 1. Power quality parameters of the original system

P[W]	384.1
Q [var]	458.7
S [VA]	597.9
PF [-]	0.6414
U_{rms} [V]	223.4
I_{rms} [A]	2.67
THD_I [%]	12.99
THD_U [%]	2.949

Table 1 shows an overview of the power quality parameters of the original system. The worst parameter is PF and has value 0.6414 and it does not reach the prescribed values of 0.95-1. The other parameters meet the EN 50160 standard.

B. Measurement of the LED 1 system

LED 1 system consists of 18 LED tubes with a total power 180W. This system used switching control because it is not possible to dimming this source. In this case, problems with electric power quality are caused by inrush current and electronics converter which generate harmonics by their operation.

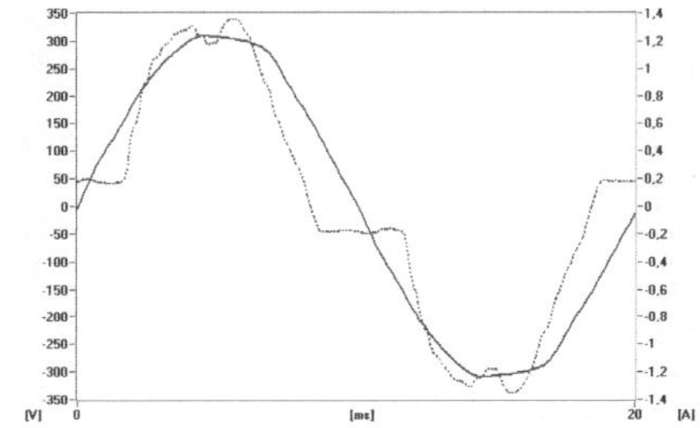

Figure 4. Graphical dependence voltage and current

In figure 4 is shown the harmonic distortion of the current. This distortion is due to the above-mentioned factors, which are made by operation of the LED 1 system.

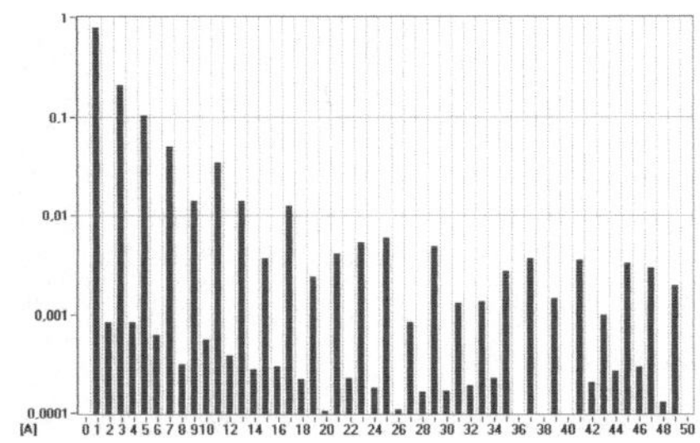

Figure 5. Harmonic spectrum of current for LED 1 system

Figure 5 shows the harmonic spectrum of the current. The harmonics components are higher than in the original system. This state is made using a semiconductor converter on the input of individual LED tubes.

Table 2. Harmonic spectrum (voltage and current) of the LED 1 system

h[-]	U [V]	I [A]
0	0	0
1	225.2	0.805
2	0.1	0.002
3	3.1	0.214
4	0.1	0.001
5	4.8	0.119
6	0.1	0.001
7	3.5	0.053
8	0.1	0.001
9	1.9	0.027
10	0.1	0.001
11	1.8	0.045
12	0.1	0.001
13	1.6	0.017
14	0.1	0.001
15	0.8	0.015
16	0	0.001
17	0.8	0.012
18	0	0.001
19	0.5	0.005
20	0	0

In table 2 are the harmonics components of the voltage and the current to 20 harmonics order. Higher harmonics orders than 20 are negligible. The highest values have odd harmonics. This state comes out from the theory of power quality because these tubes have on input a rectifier or electronics converter. The current harmonics are higher than voltage harmonics and the higher harmonic have order 3 and 5.

Table 3. Power quality parameters of LED 1 system

P[W]	176.2
Q [var]	-64.15

S [VA]	187.6
PF [-]	0.9397
U_{rms} [V]	224.1
I_{rms} [A]	0.8368
THD_I [%]	30.26
THD_U [%]	2.526

Table 3 shows the basic measured power quality parameters. Compared to the original system, there is decreasing apparent and active power. Reactive power has a negative value, which means that the light system delivers capacitive power to the grid. This is probably caused using capacitors on the input of the rectifier. The PF is higher than in the original system (0.9397), THDI value is 30.26%. This value is caused by the combination of switching and operation of the light system.

C. Measurement of LED 2 system

This second system works with the use of dimmable LED tubes. Overall operation is realized differently than with the LED 1 system. With the use of dimming tubes, other light parameters can be achieved.

Triac control is used for dimming. This control uses the triac firing control to change the total RMS value of the input current. The disadvantage of this method of control is the use of a thyristor which generates harmonics and the resulting current is more affected.

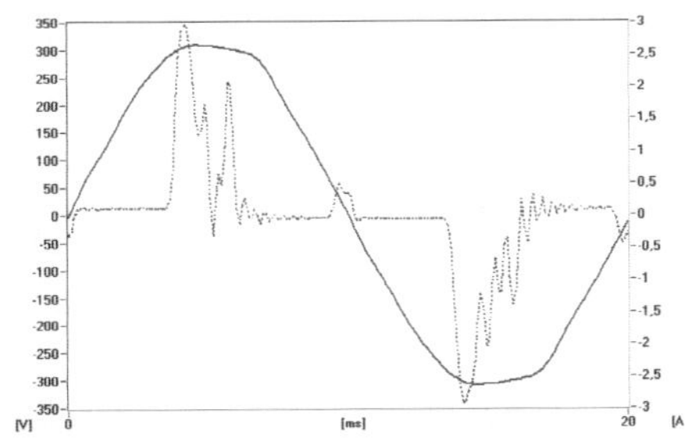

Figure 6. Graphical dependence voltage and current

Figure 6 shows a significantly distorted total current. A peak value of the current is about twice bigger than the current in the previous case. The current contains the glitches that are caused by the control during measurement. But control cannot be influenced, changed or turn off due to the character and configuration of the light system. These glitches are directly proportional to the influence of daylight because this system regulates the resulting illumination intensity to a constant level with the utilization of daylight. These effects will be even more pronounced by the fast change of daylight.

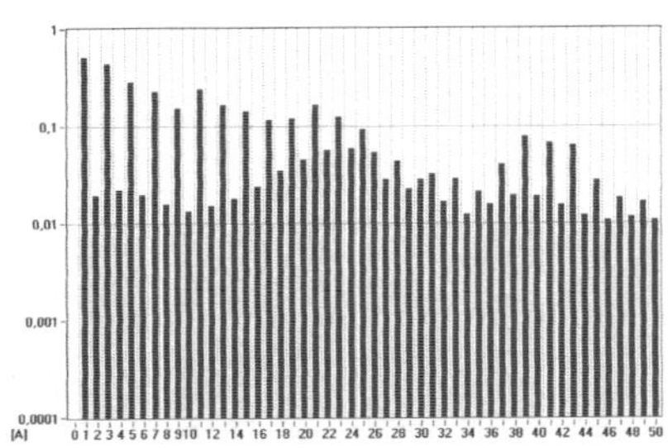

Figure 7. Harmonic spectrum of current for LED 2 system

Figure 7 shows FFT analyses of the measured current. Individual harmonics reach high values even though they tend to fall. In this case, the amplitude law does not apply because h + 1 harmonic has a higher value than h harmonic. This state is due to triac regulation with combination with the LED tube input circuits.

In table 4 there is a list of the harmonic components up to the 20th harmonic order with their amplitudes. The harmonic components of the voltage are negligible with comparison to the basic harmonic because they only achieve minimal values. Current harmonics achieve high values with comparison the basic harmonics.

Table 4. Harmonic spectrum (voltage and current) of LED 2 system

h[-]	U [V]	I [A]
0	0	0
1	225.4	0.561
2	0.1	0.03
3	2.5	0.447
4	0.1	0.024
5	5.8	0.296
6	0.1	0.02
7	3.9	0.308
8	0.1	0.025
9	2.2	0.209
10	0.1	0.03
11	1.7	0.262
12	0.1	0.034
13	1.2	0.216
14	0.1	0.038
15	0.7	0.192
16	0	0.042
17	0.4	0.167
18	0	0.044
19	0.2	0.157
20	0	0.042

List of basic power quality parameters is in table 5. It is possible to determine the state of art of the system. The high reactive negative power is caused by the LED tube connection (the input circuit is a rectifier with a capacitive filter). The most significant value in comparison to others is THDI as it is up to 149.7% in this case. This high value is due to the triac control with combination with the light system input circuits.

Table 5. Power quality parameters of the original system

P[W]	119.7
Q [var]	-186.1
S [VA]	221.3
PF [-]	0.5411
U_{rms} [V]	223.5
I_{rms} [A]	0.9909
THD_I [%]	149.7
THD_U [%]	2.282

Correct applied retrofit can achieve savings in electricity consumption. The disadvantage of a retrofit is that lighting system has an impact on the power quality in the grid. This impact is not negligible because of the massive expansion of such systems. It is necessary to implement countermeasures that mitigate or eliminate this negative effect.

V. DESIGN OF FILTRATION

After the massive expansion of retrofitted systems, there will be a problem with power quality. These systems can lead to a superposition of harmonics and the resulting current will have a major impact on the device, which is connected to the grid or has an impact on the measurement accuracy of the smart meter.

The design of the filtering device was carried out in the Matlab Simulink and simple grid and load simulation were created. The input data for each simulation is obtained from the analyses given above.

The load is formed by a programmable current source, the power supply is formed by a single sine wave source. As a filter was used harmonic filters for individual harmonics, which are dominated in the grid. The model is illustrated in figure 8.

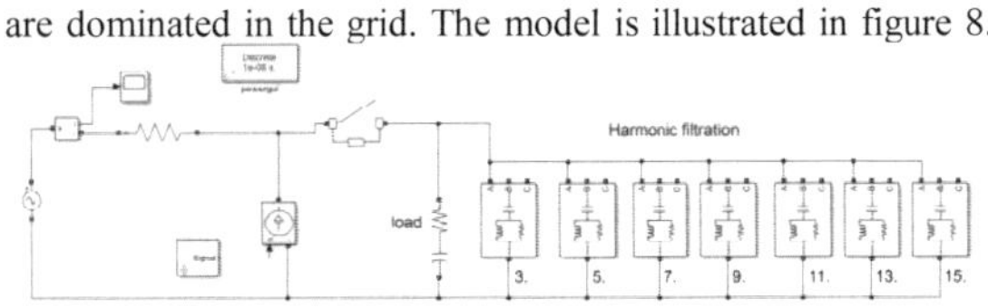

Figure 8. Model in Simulink

The following table 6 compares the values before and after filtration for each system. After filtration, the resulting power quality parameters were improved. The problem will be with the initial cost of filtering equipment.

Table 6. Power quality values before and after filtration

	THDI[%]		THDU [%]		PF[-]	
	Before	After	Before	After	Before	After
Original	12.99	-	2.949	-	0.6414	-
LED1	30.26	18.15	2.526	2.314	0.9397	0.9581
LED2	149.70	58.73	2.282	2.132	0.5411	0.7614

After filtration, there is decreasing THDI, THDU, and PF. Improvement of THDI has a direct impact on THDU. By improving this parameter, the overall power quality will be improved and losses due to power quality will be reduced.

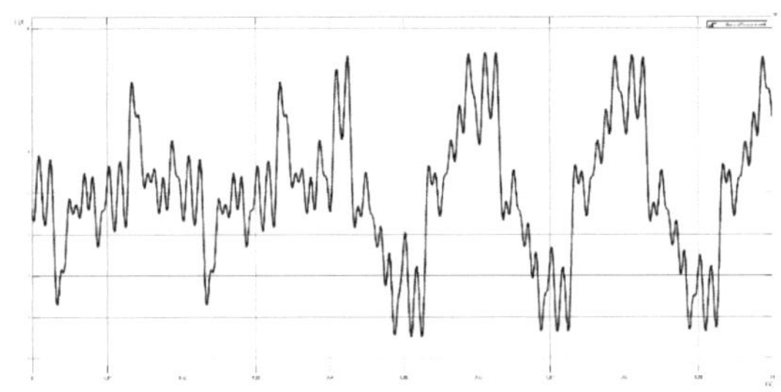

Figure 9. Current curve before and after filtration connection

In figure 9 is a current curve before and after the filtration connection. Current after filtration is less distorted, but it still has the harmonic components, which cannot be filtered.

VI. CONCLUSIONS

This paper describes the effect of the retrofitted light system on power quality. The impact of retrofit lighting on quality is a much-discussed topic today and analysis of individual systems is needed.

The main aim of this article was to measure and then analyse individual retrofitted lighting systems. The first system is based on lighting switching and the second one is based on dimming and maintaining a constant level of illumination in space.

The conclusion of the article describes the possibilities of improving this adverse effect. Analysis and simulation were done in Matlab Simulink. The application of filtration equipment has reduced total THDI. The THDI is a direct indicator of the power quality in the grid and its reduction will have also affect the resulting THDU and on overall improvement the power quality in the grid.

It is also necessary to think about the consequences that it will cause implementing a retrofit light system. On the one hand, there will be a reduction in electricity consumption. On the other hand, there will be a problem with the degradation of power quality. In the future, it will be necessary to install filtration devices to mitigate this impact. Complete eliminating it will not be possible because it is technically very complex and very expensive.

By determining the final configuration of a retrofitted system or filtering device, it is necessary to evaluate the individual technical characteristics of each system and find a consensus between price and desired parameters.

ACKNOWLEDGMENT

This work was supported by the Slovak Research and Development Agency under the contract No. APVV-15-0464

REFERENCES

[1] R. C. Dugan, M. F. McGranaghan, S. Santoso, H. W. Beaty, " Electrical Power Systems Quality," 3th edition, The McGraw-Hill Companies, Inc., NY, USA , ISBN:978-0-07-176155-0

[2] B. Singh, A. Chandra, K. Haddad, "Power Quality Problems and Mitigation Techniques," Wiley, 2015, ISBN: 9781118922057

[3] D. L. Dilaura, K. W. Houser, R. G. Mistrick, G. R. Steffy, "The Lighting Handbook: Reference and Application," Illuminating Engineering; 10 edition, pp. 1087, July 2011, ISBN: 978-0879952419

[4] A. Hromádka, J. Jiřičková, Z. Martínek, "Analysis of the unplanned shutdowns and estimation of reliability in Temelin," In Proceedings of the 2018 19th International Scientific Conference on Electric Power Engineering (EPE). Piscataway: IEEE, 2018. s. 551-556. ISBN: 978-1-5386-4612-0 , ISSN: 2376-562

[5] M. Zalesinska, S. Szwedek and A. Pawlak, "Evaluation of Lighting Parameters at the Workplace with the use Replacements for Incandescent Lamps," 2018 VII. Lighting Conference of the Visegrad Countries (Lumen V4), Trebic, 2018, pp. 1-5.

[6] R. P. B Silva, R. Quadros, F. P. Santilio,L. C. P. Silva, B. C. Carvalh, E. A. S. Silva, "A New Electronic Device for Analysing the Impact of Non-Linear Loads in the Electrical Power Measurement for Billing Purposes", IEEE Latin America Transactions, 2018, pp. 88-95, ISSN: 1548-0992

[7] EN 50160, Voltage characteristics of electricity supplied by public electricity networks

[8] R. Ganslandt, H. Hofmann, "Handbook of Lighting Design," ERCO Leuchten GmbH, pp. 289, 1992

[9] L. B. Oliveira, R. A. Salles, A. P. Fragoso, M. Z. Fortes and G. M. Tavares, "Lighting retrofit using LED technology — Efficiency analysis and environmental impacts," 2018 Simposio Brasileiro de Sistemas Eletricos (SBSE), Niteroi, 2018, pp. 1-5Z.

[10] Wu, K. Zhao and X. Xia, "Lighting retrofit and maintenance models with decay and adaptive control," in IET Control Theory & Applications, vol. 12, no. 5, pp. 593-600, 2018

Mathematical Model to Design Experiment Arrangement of Induction Furnace with Cold Crucible

Jan Hrbek
Nuclear Fuel Cycle Department
Research Centre Rez
Husinec-Rez, Czech Republic
jan.hrbek@cvrez.cz

Wladimir Ebel
Institute of Electrotechnology
Leibniz Universität Hannover
Hannover, Germany
ebel@etp.uni-hannover.de

Abstract — **This article deals with a mathematical model of induction furnace with a cold crucible for melting of electrically non-conductive materials. The main purpose of the research is creating a mathematical model enabling fast and correct determination of geometrical and electrical parameters for an arrangement of an induction furnace with the cold crucible for various experiments. The mathematical model was created using ANSYS APDL and it includes the electromagnetic problem. The two kinds of verification considering several electrical parameters were carried out and the calculated results were compared with the measured ones.**

Keywords: mathematical modeling; electromagnetic field; induction furnace with cold crucible; non-conductive materials

I. Introduction

The induction furnace with the cold crucible (IFCC) is a technology offering many possibilities of using for melting different materials. It is useful for melting the electrically conductive materials and non-conductive materials (oxides, glass, etc.) [1]. Using the induction melting in the cold crucible of the electrically non-conductive materials is possible for various purposes, several noteworthy examples - crystal growth, remelting of powder to ingot, synthesis of materials, vitrification of radioactive waste, investigation of material properties at high temperatures and modeling of severe nuclear accidents. In the Cold Crucible Laboratory at Research Centre Rez, the induction melting in the cold crucible is used mainly for the investigation of properties and behavior of the corium under the condition of a severe nuclear accident. Corium means a molten mixture consisting of materials that locate in the reactor vessel, the main components are UO_2 and ZrO_2 with liquidus temperature above 2500°C.

In Fig. 1, a basic geometrical arrangement of the induction furnace with the cold crucible is shown. The principle of induction furnace with cold crucible is the same as in the case of conventional induction crucible furnace. The difference between induction crucible furnace and induction furnace with cold crucible lies in the construction of crucible itself. In the cold crucible case, the bottom and the segments are made of copper and they are intensively water-cooled. It results in a fact that the temperature in the inner surface of cold crucible lies around 100 °C. Therefore the material is not melted there, it

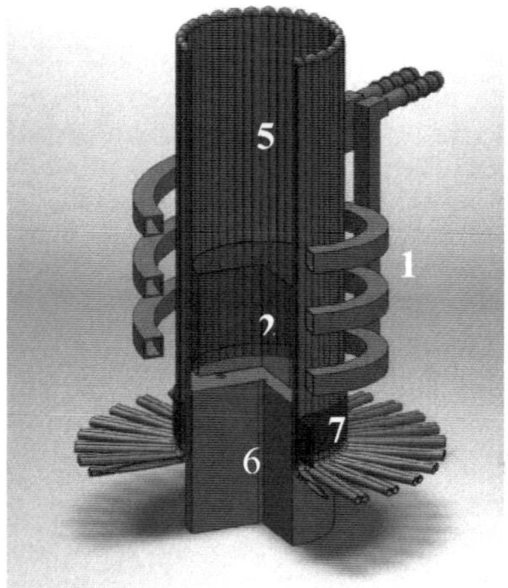

Figure 1. Induction furnace with the cold crucible. 1 – inductor, 2 – melt, 5 – cold crucible segments, 6 – cold crucible bottom, 7 - band.

forms a protective layer that prevents the contact between the melt and the crucible wall. This protective layer is called a skull [2]. It comes to the situation that the material is melted in itself, so the high purity of the product can be achieved. In the case of melting electrically non-conductive materials, the so-called starting phase of the melting process needs to be done. One of the possibility to start the melting process is to add electrically conductive material (typically Zr, Al, U) to the mixture before melting. The exothermic reaction starts after heating the additional material and it leads to the melting of the electrically non-conductive material around. Thus the electrical conductivity of the originally non-conductive material is increased because its temperature is higher and the material can be melted itself by the electromagnetic field [3].

As mentioned before, the described technology of melting is used for simulation of severe nuclear accidents in the Cold Crucible Laboratory at Research Centre Rez. It is necessary to decrease possible risks given by working with radioactive materials at high temperature. Also, it is necessary to minimize the amount of radioactive waste from the experiment. From these reasons, each experiment needs to be prepared precisely. It led to a necessity to create a tool for the design of the experiment arrangement from the induction melting viewpoint. The most important parameters are the Joule losses in the melt, voltage at the inductor, output power, and inductor input impedance.

II. Mathematical Model

The model of the electromagnetic task was created using the programming language APDL in the ANSYS software [3].

As obvious from the view in Fig. 1, it is not an acceptable simplification to consider the geometry of the mathematical model for the electromagnetic field as two-dimensional [4], [5]. The band is neglected in the model.

The geometrical arrangement is shown in Fig. 2 and Fig. 3. In the pictures, indexed volumes and boundaries have the following meaning: Ω1 - turn of the inductor, Ω2 - melt, Ω4 - air around the cold crucible and inductor, Ω5 - cold crucible segment, Ω6 - cold crucible bottom, Γ1 - boundary at z-axis, Γ2 - boundary around volume Ω4 simulating the infinity.

The equation (1) based on the $\boldsymbol{A}$-φ formulation describes the distribution of the electromagnetic field in the model volume [6].

$$\Delta\underline{\boldsymbol{A}} - j\gamma\omega\mu\underline{\boldsymbol{A}} = -\mu\underline{\boldsymbol{J}}_s, \qquad (1)$$

The equation presented above is applicable to the time-varying harmonic electromagnetic field.

In the formulation, $\boldsymbol{A}$ represents magnetic vector potential, φ denotes electric scalar potential and $\boldsymbol{J}_s$ is current density delivered to the system from external source. Other parameters used in the equation (1) are electrical conductivity γ, magnetic permeability μ and angular frequency ω.

To decrease computing time, it is necessary to reduce the degrees of freedom (DOFs). To save DOFs, the geometry is not considered as in Fig. 1 but the smallest symmetrical part is used for the model as shown in Fig. 2 and the Fig. 3 with corresponding boundary condition taking into account the symmetry at z-axis and the asymmetry at boundaries lying in the xz plane. On both melt and bottom sides lying in the xz plane is electric scalar potential $\varphi = 0$. On one inductor side in the xz plane is electric scalar potential $\varphi = 0$ as well and on the opposite one is applied electric current I.

Table 1. Description of dimensions shown in Fig. 2 and Fig. 3

Label	Description
r_1	Inner radius of the inductor
r_2	Radius of the melt
r_4	Radius of the external boundary
r_5	Outer radius of the cold crucible wall
r_6	Radius of the cold crucible bottom
d_1	Diameter of the inductor conductor
h_1	Distance between inductor turns
h_2	Height of the melt
h_5	Height of the cold crucible wall
h_6	Height of the bottom
N	Number of inductor turns

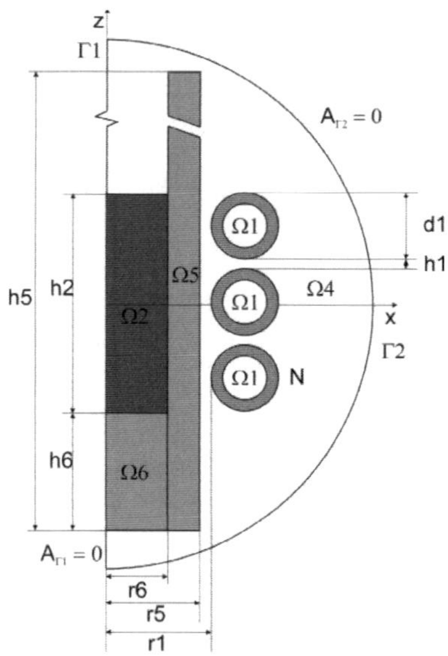

Figure 2. The geometry of the model in the xz plane with dimensions explained in the Tab.1, indexed areas and boundary conditions. Ω1, Ω5, Ω6 - copper, Ω2 - melt, Ω4 - air

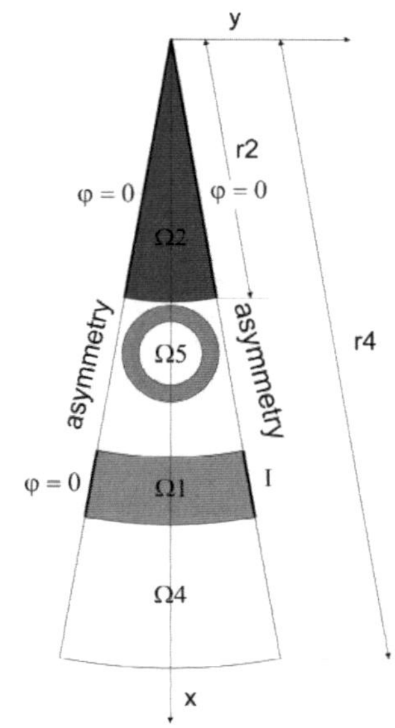

Figure 3. The geometry of the model in the xy plane with dimensions explained in the Tab.1, indexed areas and boundary conditions.

III. Illustrative Example

The illustrative example was calculated to prove the usability of the created mathematical model to design an experiment arrangement in a reasonable range of accuracy. The material parameters and required values of inductor input impedance, the voltage at inductor and the Joule losses in the melt were used as input data for the calculation. The geometry of the experiment arrangement was obtained as a result and following it, the device was manufactured and its properties were tested to verify the correctness of calculations.

Required values of quantities are denoted in Tab. 2 and used material properties in Tab. 5. The size of the cold crucible depends on the amount of the melt, material properties of the melt, cooling system capacity and others. For the considered experiment purposes, it takes to prepare cold crucible with 18 sections and inner diameter 116.0 mm.

The geometrical dimensions of the model are listed in the Tab.3. It is obvious from the Tab. 3 that the wanted dimensions from the calculation are the inner radius of the inductor r_1, the distance between inductor turns h_1 and the number of turns N to reach needed values shown in Tab. 2. There is assumed thin non-conductive layer between the cold crucible bottom and the melt with height equals to 1 mm to simulate the skull layer on the bottom.

A. Determination of Copper Electrical Resistivity

To receive the most accurate results as possible, the physical properties of used materials had to be precisely determined as well. Therefore, the resistivity of the copper used for the manufacturing of cold crucible and inductor was measured.

The measurement was carried out with one sample. For the measurement, several values of electric current were used and the temperature of the sample was measured during the test.

The schematic view at the copper sample is shown in Fig. 4. Electric current entering to the sample through surface A and leaving the sample through surface B causes a voltage drop between contact point CP1 and contact point CP2. The Ohm's method for measurement of small values of resistance was used. To have a statistical file of data, measurement with each value of electric current was repeated ten times.

The results are listed in Tab. 4 and the data represent the mean values obtained from ten measurements for each level of the electrical current.

Table 2. Required values of key quantities

Quantity	Value	Description
P_2 (kW)	50 or more	Joule losses in the melt
L (μH)	from 1.9 to 3.3	Inductance of the inductor
U_{ind} (V)	6 600	Voltage at the inductor

Table 3. Geometrical dimensions used in the calculation

Labe l	Dimension (mm)	Description
r_2	57.0	Radius of the melt
r_4	1000.0	Radius of the external boundary
r_5	68.0	Outer radius of the cold crucible wall
r_6	58.0	Radius of the cold crucible bottom
d_1	10.0	Diameter of the inductor conductor
h_2	$N \cdot d_1+(N-1) \cdot h_1$	Height of the melt
h_5	400.0	Height of the cold crucible wall
h_6	100.0	Height of the bottom

Table 4. Results of copper resistance measurement

I (A)	U (mV)	R (μΩ)
2.57	2.33	907.11±6.60
5.11	4.61	901.28±10.08
10.14	9.18	904.99±12.91

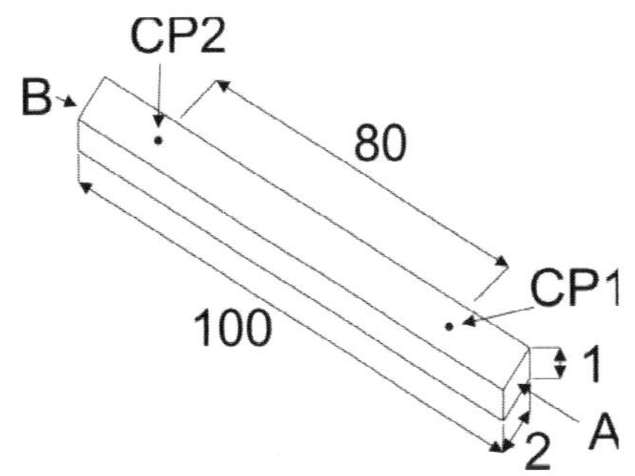

Figure 4. Schematic view at the copper sample with dimensions in millimeters.

The final resistivity of the copper sample equals to 2.27±0.03E-8 Ω·m at room temperature 20 °C.

B. Numerical Calculation

As mentioned above, the wanted dimensions are the inner radius of the inductor r_1, the distance between inductor turns h_1 and the number of turns N to reach needed values of key quantities. The height of the melt is changing according to inductor height – the height of both the melt and the inductor is supposed to be the same.

The geometry of the model created in ANSYS APDL is shown in Fig. 5 and it is depicted in detail in Fig. 6.

In the calculation, the limits of searched dimensions with their steps are listed in Tab. 6. The limits were chosen with respect to practical aspects from the electrical and mechanical viewpoint.

Following the calculation, the ideal dimensions should be r_1 = 85.0 mm, N = 4 turns and h_1 = 12.0 mm (it corresponds to h_2 = 76.0 mm). For this configuration, the induction system parameters obtained from the calculation are shown in Tab. 8.

The distribution of power in the melt for the configuration defined above is shown in Fig. 7. The lower power losses in the area near to the bottom is caused by the effect of the cold crucible bottom [7].

The presented results provided background information for the manufacturing of new inductor and tuning of the generator.

Table 5. Used material properties and parameters

Quantity	Value	Description
ρ_2 (Ω·m)	0.001	Resistivity of the melt
ρ_{Cu} (Ω·m)	2.27E-8	Resistivity of the copper parts
μ_r (-)	1	Relative permeability of all parts
f (MHz)	1.7	Frequency of the inductor current

Table 6. The limits of searched dimensions with their steps

Dimensio n	Minimum value	Maximum value	Step
r_1 (mm)	80.0	90.0	1.0
h_1 (mm)	10.0	20.0	1.0
N (-)	1	4	1

C. Verification of Calculated Results

The two different kinds of verification of the results were realized on the manufactured inductor with cold crucible according to the design based on the results from the numerical calculation:

- Measurement of electrical parameters of the induction system consisting of the inductor and cold crucible using precision RLC meter.
- Calorimetric measurement of power losses in the inductor and cold crucible during idle mode (empty cold crucible without the melt) at various inductor current levels.

Generally, material properties of the melt depend on the temperature and purity of the material. The properties are not well-known for many mixtures at high temperatures. Moreover, the electrical resistivity may be several times higher depending only on the purity or temperature of the material. Therefore, an unacceptable error caused by incorrect material properties used in the calculation may occur at verification including the melt. This fact is the main reason to verify calculations using idle mode.

The view on the inductor with cold crucible is shown in Fig. 8. Dimensions of all parts correspond to the values used in the calculation.

The precision RLC meter IET Labs 7600 Plus was used to measure input impedance, resistance and inductance of the inductor with the cold crucible at the resonant frequency the same as in the case of the mathematical model. The resonant frequency equals to 1.7 MHz. The results are shown in Tab. 7.

As the second type of verification, the idle mode was carried out. During the test, the power losses in all parts of the induction system were calorimetrically measured. The electrical characteristics of the high-frequency generator were recorded as well. The measurement consisted in setting the level of the inductor current, stabilization of all parameters and after three minutes, the inductor current was increased. This procedure was repeated with different levels of the inductor current.

The results obtained from idle mode can be compared with calculated data and measured using RLC meter ones in the Tab. 8. The comparison of calculated and measured values of power losses in the cold crucible with the bottom depending on the inductor current is shown in Fig. 9. The power losses in the inductor dependence on the inductor current are shown in Fig. 10.

Table 7. The comparison of calculated and measured electrical parameters of the inductor with cold crucible

Quantity	Measured values		Calculated values
	RLC meter	Idle mode	
R (mΩ)	141.91±29.93	178.87±88.78	114.50
X (Ω)	24.03±1.20	24.08±2.71	22.95
L (μH)	2.25±0.11	2.25±0.25	2.15
Z (Ω)	24.03±1.20	24.09±2.71	22.96

Table 8. Results of key quantities obtained from the calculation

Quantity	Value	Description
P_2 (kW)	52.88	Joule losses in the melt
L (μH)	1.98	Inductance of the inductor
U_{ind} (V)	6599.80	Voltage at the inductor

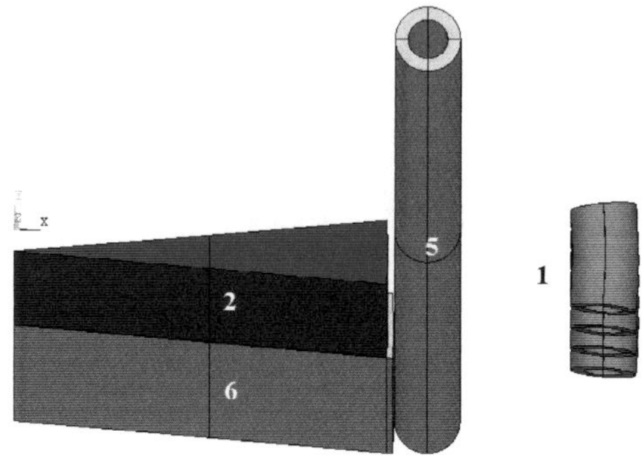

Figure 5. The geometry of the model created in ANSYS APDL. 1 – inductor, 2 – melt, 5 – cold crucible wall segment, 6 – cold crucible bottom.

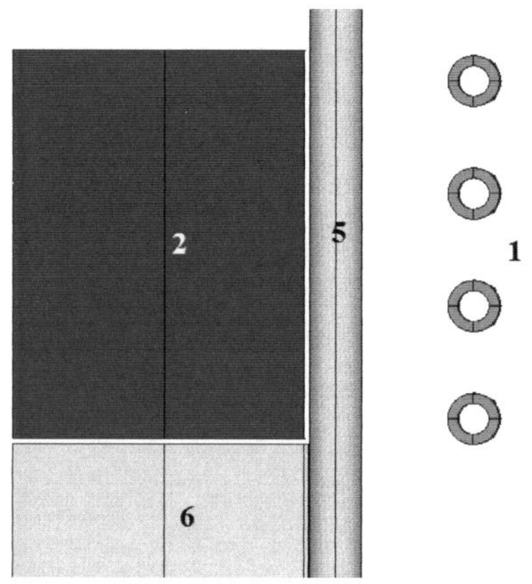

Figure 6. The detail view of the geometry of the model created in ANSYS APDL in the melt area. 1 – inductor, 2 – melt, 5 – cold crucible wall segment, 6 – cold crucible bottom.

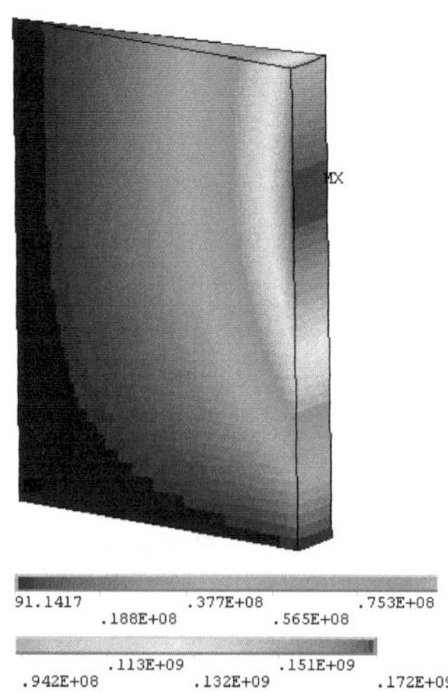

Figure 7. The distribution of power generated in the melt obtained using calculation for the inductor current equal to 220.28 A.

Figure 8. The inductor with cold crucible arrangement during both types of verifications.

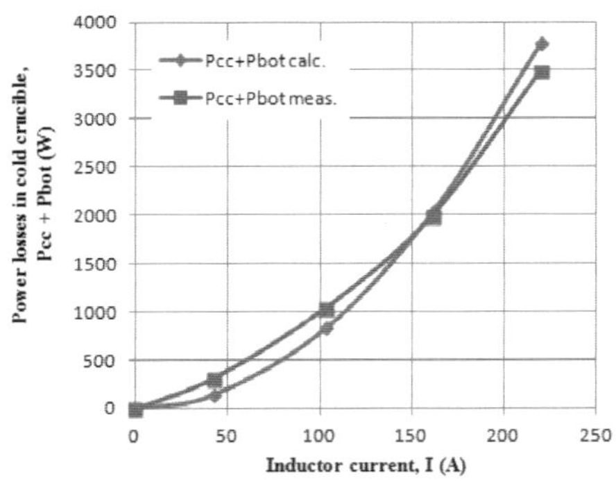

Figure 9. The power losses in the cold crucible and the bottom $P_{cc}+P_{bot}$ depending on inductor current I. The calculated results compared with the data measured during the idle mode test.

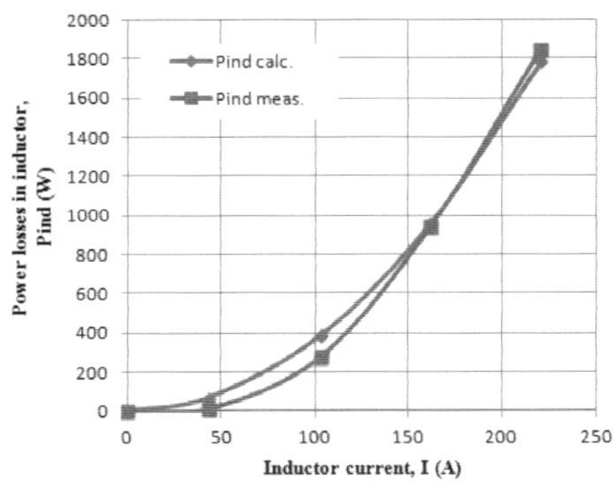

Figure 10. The power losses in the inductor P_{ind} depending on inductor current I. The calculated results compared with the data measured during the idle mode test.

IV. CONCLUSION

The mathematical model created using ANSYS APDL is suitable for the design experiment arrangement of an induction furnace with the cold crucible. Nevertheless, the problem could arise with modeling of the large induction system, for instance, the cold crucible with an inner radius more than 85.0 mm and height of cold crucible wall more than 300.0 mm. Such large dimensions of the cold crucible lead to an increase in the number of discretization mesh elements of the model and a significant increase in the computing time. For these cases, the model should be optimized.

The results of the illustrative example show good accordance between calculation and two kinds of verification. The electrical parameters are in the range of uncertainty. The presented uncertainty is A-type uncertainty.

In the case of power losses in the inductor, the measurements well correspond with the calculated results and the differences do not exceed about 6.0 % for the inductor current higher than 140 A. The higher difference between measured and calculated power losses values in the case with lower inductor current could be caused by inaccuracy of the calorimetric measurement.

In the case of power losses in the cold crucible and the bottom, the difference between calculations and the measured results does not exceed about 9.0 % for the inductor current higher than 140 A.

ACKNOWLEDGMENT

The presented work was financially supported by the Ministry of Education, Youth and Sport Czech Republic Project LQ1603 (Research for SUSEN). This work has been realized within the SUSEN Project (established in the framework of the European Regional Development Fund (ERDF) in project CZ.1.05/2.1.00/03.0108).

REFERENCES

[1] A. Aliferov, E. Baake, J. Barglik, M. Forzan, A. Jakovics, S. Lupi, B. Nacke, S.Pavlovs. *Theoretical background and aspects of electrotechnologies: Physical principles and realization.* St.Petersburg: Publishing house of ETU, 2013, p 247-250..

[2] B. Nacke, E. Baake, S. Lupi, F. Dughiero, M. Forzan, J. Barglik, A. Smalcerz, D. Dolega, A. Jakovics, S.Pavlovs, A. Aliferov, L. Goreva. *Induction heating technologies.* St.Petersburg: Publishing house of ETU, 2012, p 130-150.

[3] P. Kohne. *Theory Reference for the Mechanical APDL and Mechanical Applications.* 2009, p.185-267.

[4] D. Rot, S. Jirinec, J. Kozeny and I. Poznyak. "Electrical efficiency of induction furnace with cold crucible via different segments width." In *Proceedings* of the 16th International Scientific Conference on Electric Power Engineering (EPE), 2015,pp 17-20.

[5] D. Rot, S. Jirinec, J. Kozeny and I. Poznyak. "Electrical efficiency of induction furnace with cold crucible with varied number of rectangle segments." In *Proceedings* of the 8th International Scientific Symposium on Electrical Power Engineering, ELEKTROENERGETIKA 2015, pp. 269-272.

[6] E. Baake, J. Barglik, D. Dolega, M. Forzan, A. Jakovics, S. Lupi, B. Nacke, S. Pavlovs, Yu. Pleshivtseva. Power supplies. Mathematical simulation and optimization. St.Petersburg: Publishing house of ETU, 2013, pp. 128-145.

[7] V. Rudnev, D. Loveless and R. Cook. Handbook of induction heating. Boca Raton, FL: CRC Press, Taylor & Francis Group, 2017, p. 78-85.

Evaluation of Power Line Parameter Estimation Accuracy Based on Data from Airborne LiDAR

Marek Siranec, Marek Hoger, Alena Otcenasova
Department of Power Systems and Electric Drives
Faculty of Electrical Engineering and Information Technology, University of Zilina
Zilina, Slovakia
marek.siranec@fel.uniza.sk

Abstract — **LiDAR system mounted on an aircraft or a helicopter provides quick and efficient data collection about the state of overhead power line and its corridor. Using the appropriate software, the digital model of a power line and its surroundings can be processed and obtained data can provide information about the exact position of each conductor, the geometry of the pylons, terrain profile and positions and shapes of all ground objects including trees, medium vegetation, buildings and other potentially interesting objects. From this data, it is possible to extract information about the state of the scanned power line not only for operational conditions at the time of scanning, but it is also possible to estimate the state of the power line during extreme climatic conditions. This article deals with the analysis of the achievable accuracy of methods which are used for the reconstruction of the conductor geometry from the original point clouds and for the estimation of conductor geometry during the extreme climatic conditions.**

Keywords: conductor sag; conductor temperature; LiDAR; power line inspection; state equation;

I. Introduction

Transmission and distribution of electricity is currently realized mostly by overhead power lines. In contrast to cable lines, overhead power lines are exposed to many adverse climatic conditions during their lifetime, such as strong wind, icing and high temperatures.

At the same time, mechanical properties of used conductors are changed by aging which leads to increasing of conductor sag in individual spans. Aluminium conductors are during their lifetime subjects to a permanent elongation process known as creep. Creep leads to increasing of conductor sag and the height of the conductor above the terrain or above objects under the power line can decrease under limits requested by technical standards. Sag of conductors is therefore a very important factor that must be checked regularly. The size of conductor sag is primarily dependent on the conductor temperature – conductors of overhead power lines increase their sag with increasing temperature (or additional icing load) and vice versa [1], [2].

Overhead power lines are also threatened by growing vegetation under the power lines or close to them. Therefore, it is very important to keep the power line corridor free of high vegetation and other potentially dangerous objects. To ensure safe and reliable operation of a power line, its corridor must be regularly inspected, and the waxing vegetation must be removed [1].

According to the above-mentioned reasons, it is necessary to make regular inspections of power lines. The main problem of power line corridor inspections is the fact, that overhead power lines are typically several tenths of kilometers long and are often crossing difficult terrain conditions, so their inspection and monitoring is technically difficult and time consuming. Nowadays, most of the inspection procedures are based on visual inspection and assessing the state of power line by a technician who performs this control. Transmission system operators are often using helicopters to monitor the most exposed sections of transmission lines. From the helicopter, the power line is inspected visually and using an infrared camera. Distribution lines (110 kV and lower voltage levels) and their corridors are mostly inspected simply by a technician walking along the power line. Moreover, neither of these two methods can provide a comprehensive picture of real state of the power line and its corridor [1].

In contrast to these traditional inspection methods, airborne LiDAR scanners provide a more advantageous method of power lines inspection and monitoring. These scanners, thanks to their high reflectivity and easy definition of the shape of conductors and power lines, offer a much faster and more specific point of view about actual state of the power line and its corridor. The output from LiDAR scanning is a large number of points (cloud of points) that has to be processed to a digital model of scanned area [1].

Basically, there are three possible options how to realize an airborne LiDAR scanning of a power line. First possibility is to use a drone (UAV), second possibility is to use a helicopter and the third option is using a fixed wing aircraft as the LiDAR carrier. The choice of the LiDAR carrier significantly affects the resulting cost (via different cost per flight hour) and accuracy of the resulting point cloud. In general, using a fixed wing aircraft as LiDAR carrier reduces the scanning time and price compared to other options, however the higher speed and above ground altitude of flight results to lower point cloud density and worse point cloud accuracy, compared to a scan using helicopters or drones.

Processing of scanned data represents a considerable part of total costs in comparison with classical methods of overhead power lines state inspection that were described above. However, on the other hand, the transmission or distribution system operator gets a much more comprehensive view of the power line and its corridor state. Processed data also allow to predict corridor state over some period based on the growth rate of vegetation. It is also possible to predict, with a certain inaccuracy, the lifetime of conductors and their increasing sag by the influence of metallurgical creep [1], [2], [3].

This article deals with the analysis of the achievable accuracy of methods which are used for the reconstruction of the conductor geometry from the original point clouds and for the estimation of conductor geometry during the extreme climatic conditions.

II. Advanced Diagnostics Of Overhead Power Lines Based On LiDAR Scanning

Using the appropriate software, the digital model of scanned area can by processed and obtained data can provide information about power line and its corridor such as:

- exact position of each conductor, its sag and mechanical tension,
- height and position of towers and possibly their inclination,
- state of power line corridor and the distance of conductors from vegetation and other objects,
- permanent elongation of conductors due to metallurgical creep.

Software TerraScan was used to process the scanned data.

A. *Automated detection of minimal height of conductor above the terrain*

Commercial LiDAR data processing software usually contains tools for automatic identification of conductors and determination of their minimal height above the terrain. However, this height corresponds to conductor temperature only at the time of scanning. In accordance to theoretical assumptions, changes of outside temperature during the day and during the season, as well as changes in current loading cause a change of conductor sag which leads to change of its minimal height above the ground. The relationship among the sag, mechanical tension and temperature of conductor is defined in mounting tables that are derived from maximal permissible stress of conductor under the critical conditions. Critical conditions according to the technical standards represent a situation at the temperature of -5 °C with additional icing burden. In order to create a mounting table, the conductor state equation must be solved, where index "0" indicates the initial state and index "1" indicates the state during installation. The state equation is defined as [1], [2], [3]:

$$\sigma_{H1}^3+\sigma_{H1}^2\cdot\left[\frac{\gamma^2\cdot E}{24}\cdot\left(\frac{a\cdot z_0}{\sigma_{H0}}\right)^2+\alpha\cdot E\cdot(\vartheta_1-\vartheta_0)-\sigma_{H0}\right]=\frac{\gamma^2\cdot E}{24}\cdot(a\cdot z_1)^2 \qquad (1)$$

where σ_{H1} [MPa] is the mechanical tension of conductor in the state "1", σ_{H0} [MPa] is the mechanical tension of conductor in the state "0", υ_1 [°C] is the temperature of conductor in the state "1", υ_0 [°C] is the temperature of conductor in the state "0", E [MPa] is the conductor modulus of elasticity, α [°C^{-1}] is coefficient of thermal expansion, γ [N.m^{-3}] is the nominal weight of conductor, a [m] is the tower span, z_1 [-] is the overloading factor of conductor in state "1" and z_0 [-] is the overloading factor of conductor in state "0".

Subsequently, the resulting maximal sag of the conductor f_m [m] is possible to calculate using the equation:

$$f_m=\frac{a^2\cdot\gamma\cdot z}{8\cdot\sigma_H} \qquad (2)$$

Overhead power lines in the Slovak Republic are designed for temperatures in range of -30 ÷ +40 °C. Since aerial scanning of overhead power lines is mainly performed under the good weather conditions, it is necessary to recalculate the position of conductors to the most critical state – i.e. when the sag of conductor is the largest. Typically, the largest conductor sag occurs either at temperature of +40 °C or at temperature of -5 °C with additional icing burden. In order to recalculate mechanical tensions and sags of conductor to the critical climatic conditions correctly, it is necessary to know the mechanical properties of scanned conductor and it is also necessary to determine the temperature and actual mechanical tension of the conductor during the scanning.

Subsequently, it is possible to plot a new curve representing the conductor at new, recalculated temperature, e.g. +40 °C, then detect the conductor sag, its minimal height above the terrain as well as check the clearance of power line corridor.

The value of actual mechanical tension can be calculated according to equation (2) from the measured value of conductor sag. Alternatively, we can calculate the actual mechanical tension based on equation (3), if the catenary constant c [m] for the conductor is known.

$$\sigma_{H1}=c\cdot\gamma \qquad (3)$$

The value of catenary constant for a given conductor can be determined directly from the digital model using specialized software tools or can be calculated manually based on the conductor sag in a given point of the catenary.

B. *Metallurgical creep of overhead power line conductors*

Conductors of overhead power lines are during their lifetime exposed to different thermal and mechanical stresses. It is known, that due to the aging of conductors and thus due to the changes in molecular structure of their material, the permanent elongation of conductors occurs, resulting in increasing their sags. This phenomenon is called metallurgical creep. The characteristic of such permanent conductor elongation as a function of time at constant temperature and mechanical tension is shown in Fig. 1 [1], [3], [4].

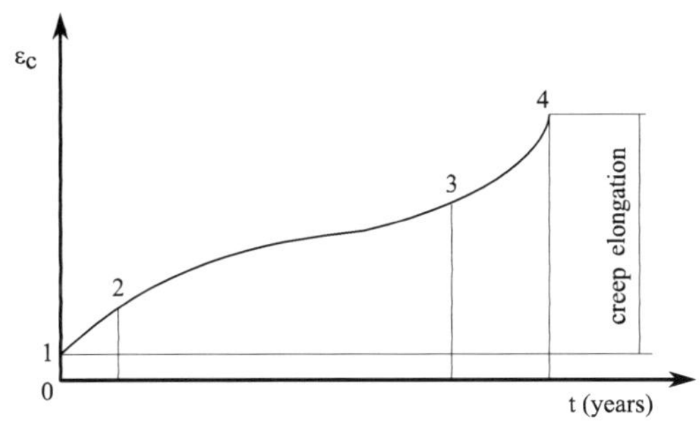

Figure 1. Permanent elongation of power line conductor [1]

Section 0-1 represents the immediate deformation after the short-term mechanical loading. It occurs immediately after conductor installation. Then, during a short period after installation (section 1-2), the conductor increases its length significantly. This is called the primary creep. In section 2-3, this deformation slows down – secondary creep occurs. The section of the characteristic between points 3-4 represents tertiary creep, where the deformation accelerates and ends with a refraction in point 4 [1], [3].

When power lines that have been in operation for a long time (several years) are scanned, it is very likely, that conductors will be elongated due to metallurgical creep. This means, that their sag at the same temperature will be bigger compared to the post-installation conditions. As the magnitude of permissible mechanical tension varies with ageing of the conductor material, mounting tables of the same power line will be different for the time of installation and today. It is quite difficult to determine the lifetime of individual conductors. For example, after a ten-year period, the conductor stressed by icing will be in different condition in comparison with conductor that is unstressed by icing.

In commercial programs, by using real scanned data, it is possible to create a vectorized representation of conductors with respect to the operational conditions at the time of scanning and then, based on theoretical calculations, recalculate their position at the time of installation. From the resulting difference of their sags, it is (at least theoretically) possible to determine the constant, that represents the conductor elongation during some time period due to metallurgical creep. If the calculated relative elongation reaches values from interval 3-4 in Fig. 1, it is a signal for power line operator to replace the conductor. Using this way, it is also possible to determine the approximate remaining lifetime of the conductor in terms of its mechanical properties. The big advantage of this approach is, that sags of conductors in the time of scanning, at the time of installation and also in the future can be easily compared and plotted for each span of the power line.

III. Analysis Of Factors Influencing Accuracy Of Conductor Geometry Recalculation

In order to correctly recalculate and analyse the individual parameters of the overhead power line conductors as described in previous chapter, it is necessary to accurately determine all the input variables of the state equation. All analyses were performed for an ACSR conductor 240/39, whose mechanical parameters are summarized in table 1.

Table 1. Mechanical parameters of selected ACSR 240/39 conductor

ACSR 240/39				
E **[MPa]**	γ **[N.m^{-3}]**	α **[°C^{-1}]**	σ_{H0} **[MPa]**	υ_0 **[°C]**
73 861	0.03427	$18.991\cdot10^{-6}$	23,45	-5

The state equation of selected ACSR conductor 240/39 was calculated in MATLAB for a symmetrical tower span of 250 m. Subsequently, analysis of impacts of input parameters inaccuracies on the accuracy of recalculated conductor sag and mechanical tension were performed. The analysed parameters were:

- the impact of conductor temperature determination error on recalculated sag and mechanical tension,
- the impact of inaccuracy of basic mechanical parameters of the conductor, specifically the conductor modulus of elasticity on the recalculated sag and mechanical tension,
- the impact of conductor initial mechanical tension determination error on the recalculated sag and mechanical tension of the conductor.

The sensitivity analysis of input parameters was performed in such a way, that one particular variable was changed in certain range while the other variables were at constant values.

A. The impact of conductor temperature determination error on the recalculated sag and mechanical tension of overhead power line conductor

Temperature is a very important factor affecting the design and operation of overhead power lines. Temperature of conductors influences their sag and mechanical tension. High temperatures can also decrease their mechanical strength and shorten the lifetime. The conductor temperature is influenced by various factors: the temperature of surrounding air, the wind speed, solar irradiation, the angle of solar rays and also by the loading current [5], [6].

As mentioned in chapter II, it is very important to know the exact temperature of conductors during the scanning, because based on this temperature, the mechanical tension and sag of conductor are recalculated. An error in temperature determination (measurement or estimation errors) could possible contribute to incorrect calculations of conductor sags and mechanical tensions.

Currently, the methods applied to the conductor temperature measurement can be classified as direct and undirect methods. In direct methods, the conductor temperature is measured directly or by measuring a particular conductor parameter, which is temperature dependent such as sag, mechanical tension, conductor resistance, the conductor distance above the terrain etc. In undirect methods of conductor temperature monitoring, the temperature of a conductor is estimated using a specific mathematical model, that uses the measured values of meteorological parameters and the value of loading current as input parameters [7].

Measurement of weather conditions may contain some error. As reported in **Chyba! Nenašiel sa žiaden zdroj odkazov.**, current methods based on determining the conductor temperature from ambient weather data and loading current produce a standard error of ±2 °C with a confidence interval of 68.2 %.

This chapter deals with the analysis of the influence of conductor temperature determination error during the scanning on its resulting sag and mechanical tension. For every value of temperature from mounting table, i. e. -30, -20, -10, -5, 0, 10, 20, 30 a 40 °C, an error value of ±2 °C was added. For all these temperatures, the resulting mechanical tension and sag for an ACSR 240/39 conductor and symmetrical span of 250 m were calculated. The initial mechanical tension, conductor temperature and mechanical constants are described in table 1.

Subsequently, the differences in sags and mechanical tensions caused by incorrect temperature determination were analysed. The results are shown in tables 2 and 3, where f is the conductor sag at the selected temperature from mounting table (-30 ÷ +40 °C), $f_{\upsilon+2°C}$ is conductor sag at corresponding temperature from mounting table increased by +2 °C, $f_{\upsilon-2°C}$ is conductor sag at corresponding temperature from mounting table reduced by -2 °C, $\Delta_{f\,\upsilon+2°C}$ is difference between sag value f and sag value $f_{\upsilon+2°C}$, $\Delta_{f\,\upsilon-2°C}$ is the difference between sag value f and sag value $f_{\upsilon-2\,°C}$, σ_{H1} is mechanical tension of the conductor at the selected temperature from mounting table (-30 ÷ +40 °C), $\sigma_{H1\,\upsilon+2°C}$ is mechanical tension of the conductor at corresponding temperature from mounting table increased by +2 °C, $\sigma_{H1\,\upsilon-2°C}$ is mechanical tension of the conductor at corresponding temperature from mounting table reduced by -2 °C, $\Delta_{\sigma H1\,\upsilon+2°C}$ is difference between the value of mechanical tension σ_{H1} and the value of mechanical tension $\sigma_{H1\,\upsilon+2°C}$, $\Delta_{\sigma H1\,\upsilon-2°C}$ is the difference between the value of mechanical tension σ_{H1} and the value of mechanical tension $\sigma_{H1\ \upsilon-2°C}$.

Table 2. The impact of conductor temperature determination error on the recalculated sag

υ [°C]	-30	-20	-10	-5	-5+icing
f [m]	10.9341	11.1297	11.3221	11.4172	11.9687
$f_{\upsilon+2°C}$ [m]	10.9735	11.1684	11.3603	11.4551	-
$f_{\upsilon-2°C}$ [m]	10.8946	11.0908	11.2839	11.3793	-
$\Delta_{f\ \upsilon+2°C}$ [m]	-0.0394	-0.0387	-0.0382	-0.0379	-
$\Delta_{f\ \upsilon-2°C}$ [m]	0.0395	0.0389	0.0382	0.0379	-
υ [°C]	0	10	20	30	40
f [m]	11.5116	11.6983	11.8823	12.0637	12.2426
$f_{\upsilon+2°C}$ [m]	11.5492	11.7354	11.9188	12.0997	12.2781
$f_{\upsilon-2°C}$ [m]	11.4740	11.6612	11.8457	12.0276	12.2070
$\Delta_{f\ \upsilon+2°C}$ [m]	-0.0376	-0.0371	-0.0365	-0.0360	-0.0355
$\Delta_{f\ \upsilon-2°C}$ [m]	0.0376	0.0371	0.0366	0.0361	0.0356

Table 3. The impact of conductor temperature determination error on the recalculated mechanical tension

υ [°C]	-30	-20	-10	-5	-5+icing
σ_{H1} [MPa]	24.4862	24.0559	23.6470	23.4500	93.4263
$\sigma_{H1\ \upsilon+2°C}$ [MPa]	24.3983	23.9725	23.5676	23.3725	-
$\sigma_{H1\ \upsilon-2°C}$ [MPa]	24.5749	24.1402	23.7272	23.5282	-
$\Delta_{\sigma H1\ \upsilon+2°C}$ [MPa]	0.0879	0.0834	0.0794	0.0775	-
$\Delta_{\sigma H1\ \upsilon-2°C}$ [MPa]	-0.0887	-0.0843	-0.0802	-0.0782	-
υ [°C]	0	10	20	30	40
σ_{H1} [MPa]	23.2577	22.8865	22.5321	22.1934	21.8691
$\sigma_{H1\ \upsilon+2°C}$ [MPa]	23.1821	22.8143	22.4632	22.1274	21.8059
$\sigma_{H1\ \upsilon-2°C}$ [MPa]	23.3341	22.9594	22.6017	22.2599	21.9328
$\Delta_{\sigma H1\ \upsilon+2°C}$ [MPa]	0.0756	0.0722	0.0689	0.0660	0.0632
$\Delta_{\sigma H1\ \upsilon-2°C}$ [MPa]	-0.0764	-0.0729	-0.0696	-0.0665	-0.0637

From the comparison of the results presented in tables 2 and 3, it can be seen, that the error of ±2 °C in determining the conductor temperature will cause a difference in resulting sag of conductor approx. ±3.5 – 4 cm and the difference in resulting mechanical tension of conductor approx. ±63.2 – 88.7 Pa. It can by also seen, that these differences are approximately the same for all analysed temperatures. The results in this article are presented only for one value of span (250 m). For other span values, the error of determining the resulting conductor sag and mechanical tension is approximately the same. Based on the results achieved in this chapter, it can be stated, that the error of determining the conductor temperature by ±2 °C does not have a significant impact on the correct determination of resulting sag and mechanical tension of power line conductor. Despite the achieved results, great attention has to be devoted to correctly determine the conductor temperature, because error in temperature determination more than ±2 °C could cause bigger differences in determination of resulting sag and mechanical tension.

B. The impact of conductor elasticity modulus determination error on the recalculated sag and mechanical tension of overhead power line conductor

Another analysed factor was the impact of errors in conductor modulus of elasticity determination. The variations in elasticity modulus of conductor should also be considered due to variations in production technology and variations in conductor loadings during the normal operation. Variations in value of elasticity modulus can lead to different values of the resulting sag and mechanical tension of the power line conductor. Therefore, it is necessary to determine the impact of the error in elasticity modulus determination on these output parameters [4].

Currently, the latest and most sophisticated methods of the resulting sag calculation are based on using the variable value of elasticity modulus for a given type of conductor. As mentioned in article [4], the biggest difference between the maximal and minimal value of elasticity modulus for an ACSR conductor does not exceed ±20 %. Article [4] also states, that if the mean value of elasticity modulus will be chosen instead of variable value of elasticity modulus, the error in determining the modulus of elasticity does not exceed ±10 %.

Therefore, based on this knowledge, the tabular value of elasticity modulus for the conductor ACSR 240/39 was changed by error of ±10 %. Since the tabular value of elasticity modulus for ACSR 240/39 is 73 861 MPa, the error of ±10 % from this value is ±7386.1 MPa. Subsequently, new mounting tables were calculated for 3 values of elasticity modulus ($E_{-10\%}$, E, $E_{+10\%}$) for temperatures in range of -30°C ÷ +40 °C. Changes in the mechanical tensions and sags of conductors were analysed from calculated mounting tables in the case of such determination of the elasticity modulus. The results are shown in tables 4 and 5, where f is the conductor sag at tabular value of elasticity modulus, $f_{E+10\%}$ is the conductor sag at the value of elasticity modulus increased by the error of +10 %, $f_{E-10\%}$ is conductor sag at the value of elasticity modulus reduced by the error of -10 %, $\Delta_{f\ E+10\%}$ is the difference between the sag value f and sag value $f_{E+10\%}$, $\Delta_{f\ E-10\%}$ is the difference between sag value f and sag value $f_{E-10\%}$, σ_{H1} is mechanical tension of conductor at tabular value of elasticity modulus, $\sigma_{H1\ E+10\%}$ is the mechanical tension of conductor at the value of elasticity modulus increased by the error of +10 %, $\sigma_{H1\ E-10\%}$ is mechanical tension at the value of elasticity modulus reduced by the error of -10 %, $\Delta_{\sigma H1\ E+10\%}$ is the difference between the value of mechanical tension σ_{H1} and the value of mechanical tension $\sigma_{H1\ E+10\%}$, $\Delta_{\sigma H1\ E-10\%}$ is the difference between the value of mechanical tension σ_{H1} and the value of mechanical tension $\sigma_{H1\ E-10\%}$.

Table 4. The impact of conductor elasticity modulus determination error on the recalculated sag

υ [°C]	-30	-20	-10	-5	-5+icing
f [m]	10.9341	11.1297	11.3221	11.4172	11.9687
$f_{E+10\%}$ [m]	10.9328	11.1289	11.3219	11.4172	-
$f_{E-10\%}$ [m]	10.9357	11.1306	11.3224	11.4172	-
$\Delta_{f\ E+10\%}$ [m]	0.0013	0.0008	0.0002	0.0000	-
$\Delta_{f\ E-10\%}$ [m]	-0.0016	-0.0009	-0.0003	0.0000	-
υ [°C]	0	10	20	30	40
f [m]	11.5116	11.6983	11.8823	12.0637	12.2426
$f_{E+10\%}$ [m]	11.5119	11.6990	11.8834	12.0652	12.2444
$f_{E-10\%}$ [m]	11.5114	11.6975	11.8810	12.0619	12.2404
$\Delta_{f\ E+10\%}$ [m]	-0.0003	-0.0007	-0.0011	-0.0015	-0.0018
$\Delta_{f\ E-10\%}$ [m]	0.0002	0.0008	0.0013	0.0018	0.0022

Table 5. The impact of conductor elasticity modulus determination error on the recalculated mechanical tension

υ [°C]	**-30**	**-20**	**-10**	**-5**	**-5+icing**
σ_{H1} **[MPa]**	24.4862	24.0559	23.6470	23.4500	93.4263
$\sigma_{H1\ E+10\%}$ **[MPa]**	24.4891	24.0576	23.6475	23.4500	-
$\sigma_{H1\ E-10\%}$ **[MPa]**	24.4826	24.0539	23.6464	23.4500	-
$\delta_{\sigma H1\ E+10\%}$ **[MPa]**	-0.0029	-0.0017	-0.0005	0.0000	-
$\delta_{\sigma H1\ E-10\%}$ **[MPa]**	0.0036	0.0020	0.0006	0.0000	-
υ [°C]	**0**	**10**	**20**	**30**	**40**
σ_{H1} **[MPa]**	23.2577	22.8865	22.5321	22.1934	21.8691
$\sigma_{H1\ E+10\%}$ **[MPa]**	23.2572	22.8852	22.5301	22.1907	21.8658
$\sigma_{H1\ E-10\%}$ **[MPa]**	23.2583	22.8881	22.5347	22.1966	21.8730
$\delta_{\sigma H1\ E+10\%}$ **[MPa]**	0.0005	0.0013	0.0020	0.0027	0.0033
$\delta_{\sigma H1\ E-10\%}$ **[MPa]**	-0.0006	-0.0016	-0.0026	-0.0032	-0.0039

From the comparison of the results presented in tables 4 and 5, it can be seen, that the biggest absolute difference in the determination of resulting sag is at the level of 0.22 cm and the biggest absolute difference in the determination of resulting mechanical tension is at the level of 3.9 Pa. From these results, it is clear, that the error of ±10 % in determining the elasticity modulus has a negligible impact on correct determination of resulting conductor sags and mechanical tensions over the entire range of analysed temperatures. Analyses were performed only for one value of span (250 m). For lower span values have the results similar character, for span values bigger than 300 m, the effect of the error in elasticity modulus determination has virtually no impact on determining the resulting sag and mechanical tension of power line conductor.

C. Impact of point cloud accuracy on the initial tension estimation

As mentioned in chapter II, the initial tension can be estimated from the power line geometry according to equation (*2*) or (*3*). However, the accuracy of such estimation is dependent on the overall accuracy of the point cloud representing the power line corridor.

Regarding the characteristics of different LiDAR carriers form chapter I, we decided to perform an accuracy validation of a point cloud from a power line scan realized using a fixed wing aircraft, because this is the worst-case scenario from the accuracy point of view.

The evaluated scan was performed using a Piper Seneca III aircraft equipped with a Trimble Harrier 68i LiDAR scanner. The flight speed during the scan was about 210 km·h^{-1} and the flight altitude was approximately 650 m about ground level [1]. The target area was re-flown multiple times, so the resulting point cloud density was about 20-25 pt·m^{-2}. The basic accuracy level specified for this LiDAR scanner is ±15 cm horizontally and ±25 cm vertically.

The target area is crossed by multiple powerlines of different type, including one 400 kV and several 110 kV and 22 kV powerlines.

To get some insight about the absolute accuracy of the scan, we need to check, how well the acquired point cloud represents an object with known dimensions. For our purpose, a very suitable object with a known geometry is a bundled conductor. We know, that the 400 kV line in the target area has bundled conductors with three ACSR conductors per bundle, with a fixed distance of 40 cm. Figure 2. shows a detailed section-view of three bundled conductors from a short segment of this 400 kV powerline.

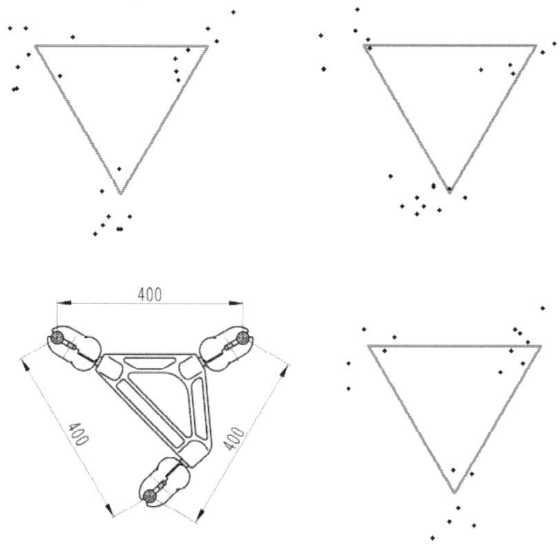

Figure 2. Point cloud representation of three different bundeled conductors for a short segment of a 400 kV power line

The green triangles are representing the 40 cm distance holders used on this power line, black dots are representing points acquired by the LiDAR scanner. From this picture, we can see the dispersion of points representing the individual conductors in the bundle. The diameter off all point groups representing a conductor is less than 20 cm. With respect to the fact that the distance between conductors must be 40 cm (as represented by the green triangles), the real position of a conductor is always somewhere close to the middle of each point group, thus the resulting accuracy must be ±10 cm or better in both axis (significantly better than the ±25 cm vertical accuracy declared in the datasheet of the scanner).

However, to cover the worst-case scenario, we decided to perform the analysis of the impact of point cloud accuracy on the tension estimation for a point cloud inaccuracy of ±25 cm (according to the worst-case vertical accuracy specified for Harrier 68i). According to equation (*2*), the mechanical tension in the conductor can be calculated based on the measured value of conductor sag as follows:

$$\sigma_H = \frac{a^2 \cdot \gamma \cdot z}{8 \cdot f_m}. \tag{4}$$

Assuming all other variables to be constant, we have calculated the tension in a conductor for different values of conductor sag f (according to the values of f_m from table 2). The original values of conductor sag f were modified by ±0.25 m, respecting the maximal inaccuracy of the scanner, obtaining values for $f_{+\Delta f}$, $f_{-\Delta f}$. Using equation (4), the tension σ_H was calculated for the original value of conductor sag and for estimated values of tension $\sigma_{H+\Delta f}$, $\sigma_{H-\Delta f}$ for the modified values of conductor sag. Comparing these estimated values to the original value of σ_H, we can calculate the relative error of these estimates $\Delta\sigma_{H+\Delta f}$, $\Delta\sigma_{H-\Delta f}$. The accuracy of the tension estimation for different values of conductor sag is presented in table 6.

Table 6. Impact of conductor sag measurement errors on tension estimation accuracy

υ [°C]	-30	-20	-10	0	10	20	30	40
f [m]	10.93	11.13	11.32	11.51	11.70	11.88	12.06	12.24
$f_{+\Delta f}$ [m]	11.18	11.38	11.57	11.76	11.95	12.13	12.31	12.49
$f_{-\Delta f}$ [m]	10.68	10.88	11.07	11.26	11.45	11.63	11.81	11.99
σ_H [MPa]	24.49	24.06	23.65	23.26	22.89	22.53	22.19	21.87
$\sigma_{H+\Delta f}$ [MPa]	23.94	23.53	23.14	22.76	22.41	22.07	21.74	21.43
$\sigma_{H-\Delta f}$ [MPa]	25.06	24.61	24.18	23.77	23.39	23.02	22.66	22.32
$\Delta\sigma_{H+\Delta f}$ [%]	-2.2	-2.2	-2.2	-2.1	-2.1	-2.1	-2.0	-2.0
$\Delta\sigma_{H-\Delta f}$ [%]	2.3	2.3	2.3	2.2	2.2	2.1	2.1	2.1

From these results, we can see that the maximal error in tension estimation is approximately 2.3 % (for the value of conductor sag approximately 11 m). As we can see, with the increasing values of conductor sag, the relative error decreases, for lower sag values, the relative error will increase. For very small values of conductor sag, the error of the estimated tension value will significantly increase.

Now, when an estimation of the conductor tension was made, we can investigate the influence of the tension estimation accuracy on the accuracy of conductor geometry recalculation to different operation conditions.

The evaluation was made based on the value of conductor sag from table 2 (f=10.93 m) for the temperature of -30 °C. At this temperature, the conductor sag is minimal, thus the accuracy of sag measurement is most affected by the accuracy of LiDAR scanner. With respect to the scanner accuracy of ±0.25 m, two initial tension estimations σ_{H0min} and σ_{H0max} have been calculated using equation (*4*). These estimated values have been used to solve equation (*1*) and calculate conductor tension and the corresponding conductor sag according to equation (*2*) at different temperatures. Results are compared to the original values of sag (f) and tension (σ_H) from table 2.

Table 7. Impact of initial tension estimation errors on the accuracy of conductor sag estimation at different operating temperatures

υ [°C]	-30	-20	-10	0	10	20	30	40
f [m]	10.93	11.13	11.32	11.51	11.70	11.88	12.06	12.24
f_{min} [m]	10.68	10.88	11.08	11.27	11.46	11.65	11.84	12.02
f_{fmax} [m]	11.18	11.38	11.56	11.75	11.93	12.11	12.29	12.47
σ_H [MPa]	24.49	24.06	23.65	23.26	22.89	22.53	22.19	21.87
σ_{Hmin} [MPa]	23.94	23.54	23.15	22.79	22.44	22.10	21.78	21.47
σ_{Hmax} [MPa]	25.06	24.60	24.16	23.75	23.36	22.98	22.62	22.28

As we can see, the maximal error in estimation of conductor sag is close to the inaccuracy of the LiDAR scanner, thus during the recalculation process, the inaccuracy is not significantly amplified. The maximal error of tension estimation is approximately 0.5 MPa. From these results, we can conclude that the accuracy of the original point cloud (±0.25 m) is sufficient for recalculation of the conductor geometry to different operational conditions with reasonable accuracy.

IV. Conclusion

From the results presented in chapter III, it seems that accuracy of most currently used methodologies for conductor temperature estimation is satisfactory for the purpose of conductor geometry recalculation. The impact of temperature inaccuracy in the range of ±2 °C on the accuracy of conductor sag recalculation is very small (the maximal error in conductor sag estimation was less than 0.1 m).

Similar results were obtained during examination of the impact of elasticity modulus inaccuracy. Changing the value of elasticity modulus in the range of ±10 % of nominal value for a given conductor caused only neglectable errors in estimation of conductor sag at different operation temperatures.

From the last analysis, it seems that the impact of the original point cloud accuracy on the accuracy of the recalculation of the conductor geometry to different temperatures is similar to the accuracy of the original point cloud. In this case, the maximal error was approximately 0.25 m.

These partial results presented in this paper are suggesting, that the recalculation of conductor geometry to different operation conditions is possible with a reasonable accuracy, enabling more comprehensive evaluation of the power line safety, more effective planning of maintenance and hypothetically also estimation of the remaining conductor lifetime via the calculation of conductor elongation.

However, it must be noted, that in this paper, the influence of these three basic input variables has been evaluated "per partes", considering inaccuracies only in one input variable at the time. In fact, all three examined variables will be affected by a certain error at the same time and the resulting inaccuracy of the recalculated conductor sag won't by just a linear combination of errors caused by all individual input variables.

Therefore, additional research is necessary to determine the effects of different combinations of errors in conductor temperature, elasticity modulus and initial tension estimation on the accuracy of the conductor geometry recalculation.

Acknowledgement

This work was supported by the project: *"Broker center of air transport for transfer of technology and knowledge into transport and transport infrastructure, ITMS 26220220156".*

References

[1] A. Otcenasova, M. Hoger and J. Altus, "Possible use of airborne LiDAR for monitoring of power lines in Slovak Republic," *Proceedings of the 2014 15th International Scientific Conference on Electric Power Engineering (EPE)*, Brno, 2014, pp. 477-481.

[2] A. Hatibovic, "Derivation of Equations for Conductor and Sag Curves of an Overhead Line Based on a Given Catenary Constant", Periodica Polytechnica Electrical Engineering and Computer Science, vol. 58, no. 1, pp. 23-27, 2014.

[3] M. Michael, P. Stephan and J. Stefan, "Elongation of overhead line conductors under combined mechanical and thermal Stress," *2008 International Conference on Condition Monitoring and Diagnosis*, Beijing, 2008, pp. 671-674.

[4] A. Polevoy, "Impact of Data Errors on Sag Calculation Accuracy for Overhead Transmission Line," in *IEEE Transactions on Power Delivery*, vol. 29, no. 5, pp. 2040-2045, Oct. 2014.

[5] T. O. Seppa, "Factors influencing the accuracy of high temperature sag calculations," in *IEEE Transactions on Power Delivery*, vol. 9, no. 2, pp. 1079-1089, April 1994.

[6] E. Albizu, A. J. Fernandez, J. M. Bengoechea, "Influence of the conductor temperature error on the overhead line ampacity monitoring systems," in *IET Generation, Transmission & Distribution*, vol. 5, no. 4, pp. 440-447, April 2011.

[7] A. Pavlinić, V. Komen, "Direct monitoring methods of overhead line conductor temperature," *2017 Engineering Review*, 2017, pp. 134-146.

[8] C. Rakpenthai and S. Uatrongjit, "Power System State and Transmission Line Conductor Temperature Estimation," in *IEEE Transactions on Power Systems*, vol. 32, no. 3, pp. 1818-1827, May 2017.

The Impact of the Placement of a Renewable Energy Resources on Power Losses in the Electricity Distribution System

Daniel Pál
Department of Electric Power Engineering
FEI TU of Košice
Košice, Slovak Republic
daniel.pal@tuke.sk

Jakub Urbanský
Department of Electric Power Engineering
FEI TU of Košice
Košice, Slovak Republic
jakub.urbansky@tuke.sk

Ľubomír Beňa
Department of Electric Power Engineering
FEI TU of Košice
Košice, Slovak Republic
lubomir.bena@tuke.sk

Michal Špes
Department of Electric Power Engineering
FEI TU of Košice
Košice, Slovak Republic
michal.spes@tuke.sk

Abstract — **This article discusses the possibility of reducing power losses using renewable energy resources that arise in the operation of distribution systems. Renewable energy resources have become increasingly widespread in recent years. Small photovoltaic panels or wind turbines are being used to generate electricity in households, large power plants such as water, geothermal or solar power plants can replace currently used fossil power plants and can therefore produce electricity without harmful emissions. In the past, centralized production was typical, where the electricity produced in power stations was transmitted via transmission and distribution lines to the point of consumption. Today, electricity is still dominated by centralized power plants, but with renewable energy resources, it is starting to change to decentralized, when electricity is produced in several smaller power stations close to consumption. Households can also produce electricity with small wind turbines or with photovoltaic cells, which can become self-sufficient, respectively, may reduce the take-off from the distribution system, thereby reducing the amount of transmitted electricity through transmission lines. The paper will analyse the impact of the placement of new energy sources in the distribution system on the resulting power losses.**

Keywords: power losses, renewable energy resources

I. Introduction

Renewable energy sources (RES) are in recent years used more and more often. Traditional power plants such as fossil fuel power plants are known environmental burden, they are polluting environment and causes greenhouse effect. On the contrary RES represents so called "green" energy sources. These green sources do not produce any harmful emission during their operation. According to the Paris agreement it is necessary that the global temperature does not increase by more than 2 °C [1]. This goal can be only achieved with more increased usage of RES in near future.

European Union have set a target for year 2020 that 20 % of final energy consumption will be covered by energy produced from RES. The target for year 2030 was increased to 27 % [2].

RES can be divided into categories shown at Fig. 1:

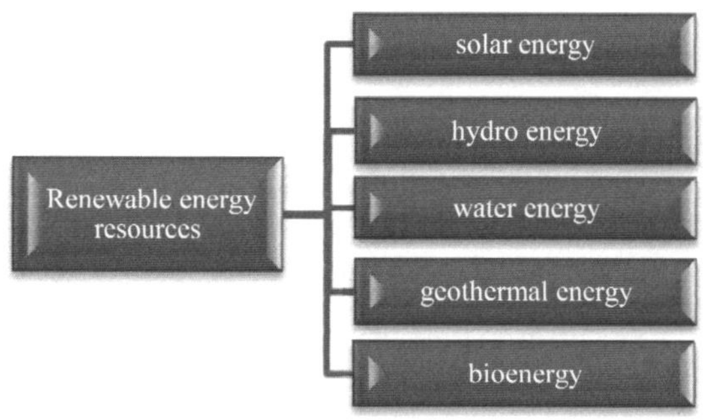

Figure 1 Renewable energy resources [3]

One of the oldest used RES is kinetic energy of water. As an example it is possible to mention watermills, which were used to convert kinetic energy of water in to the mechanical. Hydro power started to be used for electricity generation purpose at the end of the 19th century [4]. However, on the other hand, there are newer sources with higher annual global growth rates such as photovoltaics or wind. The Fig. 2 shows the average annual global growth rates of various energy sources.

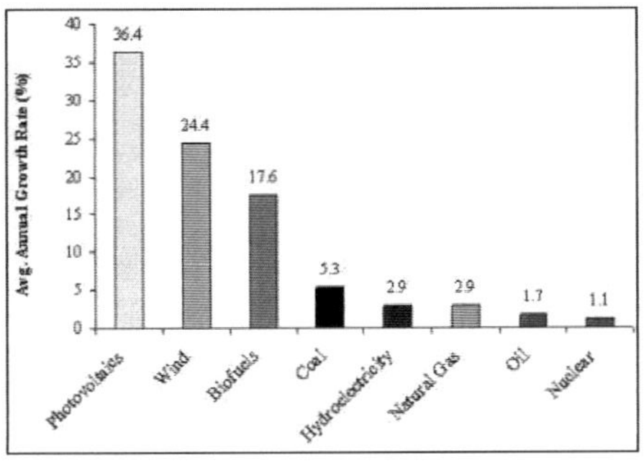

Figure 2 Average annual global growth rates of various energy sources [5]

II. STEADY-STATE MODEL

In order to investigate the impact of the placement of a renewable energy resources on power losses in the electricity distribution system it is necessary to solve steady-state. Steady-state is a condition where the line parameters does not change. At the beginning there are defined parameters, which are applied until the end of calculation.

The Fig. 3 shows the single-line network diagram that was used in the calculations.

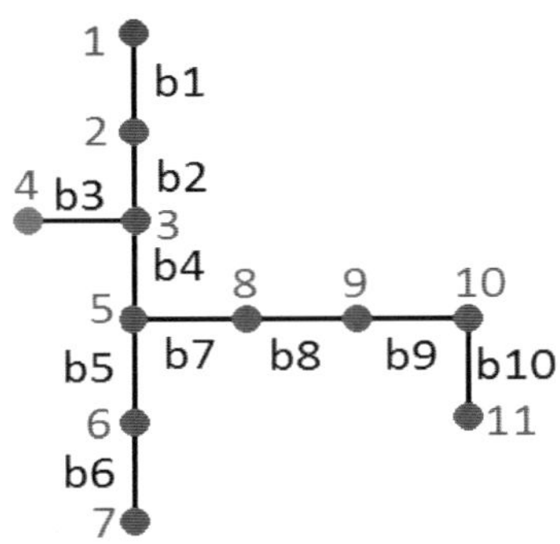

Figure 3 Single-line network diagram

The parameters of the line and the node are shown in the Table 1 and Table 2. It can be seen that examined network has 1 node where electricity is injected. Other nodes represents solely consumption. At node 1, the voltage is 22.8 kV, which characterizes the outgoing voltage of the transformer.

Table 1 Branch parameters

Branch	R (Ω)	X (Ω)
1	0.8100	0.8910
2	1.2600	1.3860
3	2.4150	1.2390
4	0.8100	0.8910
5	1.8036	1.2420
6	2.1042	1.4490
7	1.5000	1.7250
8	0.4830	0.2478
9	2.0010	1.0266
10	4.2780	2.1948

Table 2 Active and reactive power at nodes

	Production		Consumption	
Node	P (MW)	Q (MVAr)	P (MW)	Q (MVAr)
1	0	0	0	0
2	1	0	0	0
3	0	0	0	0
4	0	0	0.362	0.119
5	0	0	0	0
6	0	0	0.762	0.238
7	0	0	0.762	0.238
8	0	0	0.762	0.238
9	0	0	0	0
10	0	0	1.086	0.357
11	0	0	0.724	0.238

Newton's iteration method was used to solve steady-state. It is an iteration method which starts with initial value V_0. Calculation runs until the difference between two iteration steps is smaller than a predefined limit ε. The stopping conditions for the iteration is when: $|V_i - V_{i-1}| < \varepsilon$. Absolute value is necessary because final value can be negative.

The Newton method tends to converge very quickly, however, the speed of convergence depends strongly on the chosen initial value V_0 .

The following equations were used in the calculations:

$$\overset{*}{S_i} = \overset{*}{U_i} * \sum_{j=1}^{n} \dot{V_j} * \dot{Y}_{i,j} \tag{1}$$

Where:

S_i – apparent power (VAr)

$$\overset{*}{S_i} = P_i - j * Q_i \ \text{(VAr)} \tag{2}$$

V_i – voltage in the i-node (V)

V_j – voltage in the j-node (V)

$$\dot{V_i} = V_{ai} - j * V_{ri} \ \text{(V)} \tag{3}$$

$$\dot{V_j} = V_{aj} - j * V_{rj} \ \text{(V)} \tag{4}$$

$Y_{i,j}$ – admittance matrix (Ω)

$$\dot{Y}_{i,j} = g_{ij} - j * b_{ij} \ (\Omega) \tag{5}$$

From (1), (2), (3), (4) and (5) it is possible to state following formula:

$$P_i - j * Q_i = (V_{ai} - j * V_{ri}) * \sum_{j=1}^{n} (V_{aj} - j * V_{rj}) * (g_{ij} - j * b_{ij}) \tag{6}$$

ΔP, ΔQ is possible to calculate with equation 7:

$$\begin{bmatrix} \Delta P \\ \Delta Q \end{bmatrix} = \begin{bmatrix} \frac{\partial P}{\partial V_a} & \frac{\partial P}{\partial V_r} \\ \frac{\partial Q}{\partial V_a} & \frac{\partial Q}{\partial V_a} \end{bmatrix} * \begin{bmatrix} \Delta V_a \\ \Delta V_r \end{bmatrix} \quad (7)$$

Algorithm for solving steady-state with the Newton iteration method [6]:

1. Input data – parameter lines, active and reactive power at nodes, …
2. Initial approach – V_a^0, V_r^0
3. Calculation ΔP, ΔQ
4. ΔP, ΔQ is calculated until the difference is smaller than predefined limit ε.
   a. When the difference is smaller than a predefined limit ε – next step.
   b. When the difference is bigger than a predefined limit ε - it is necessary to calculate the partial derivative of active and reactive power. Using the equation 7 calculate ΔV_a, ΔV_r and correct the initial approaches of voltage. Then continue from the steps 3.
5. Results.

Total active and reactive power losses were calculated with following equations:

$$\Delta P_\Sigma = [V_{a\Sigma}]^T * [G] * [V_{a\Sigma}] + [V_{r\Sigma}]^T * [G] * [V_{r\Sigma}] \text{ (MW)} \quad (8)$$

$$\Delta Q_\Sigma = [V_{a\Sigma}]^T * [B] * [V_{a\Sigma}] + [V_{r\Sigma}]^T * [B] * [V_{r\Sigma}] \text{ (MVAr)} \quad (9)$$

With application of equation 8 and equation 9 were calculated voltages, total active and reactive power losses in steady-state. These values are included in the Table 3.

Table 3 Steady-state

	Voltage (kV)		
node	**Va**	**Vr**	**|(Va + j*Vr)|**
1	22.8	0	22.8
2	22.6082	-0.0851	22.608
3	22.2539	-0.2785	22.256
4	22.2078	-0.2852	22.210
5	22.0443	-0.3929	22.048
6	21.8905	-0.4371	21.895
7	21.8006	-0.4629	21.806
8	21.7952	-0.5341	21.802
9	21.7474	-0.5404	21.754
10	21.5497	-0.5662	21.557
11	21.3798	-0.5882	21.388
Total active power losses (MW)	0.1766		
Total reactive power losses (MVAr)	0.1753		

III. PLACEMENT OF A RENEWABLE RESOURCE INTO THE SYSTEM

With correct placement of new RES in to the power grid it is possible to reduce power losses and improve its parameters. One of these parameters is a voltage in the nodes.

The distribution generation (DG) technologies and their typical module size with RES are the following:

- Small hydro power plant: 25 kW – 1 MW,
- Large hydro power plant: 1 MW – 100 MW,
- Wind power plant: 200 W – 3 MW,
- Solar power plant: 20 W – 100 kW,
- Biomass: 100 kW – 20 MW,
- Geothermal power plant: 5 MW – 100 MW,
- Ocean power plant: 100 kW – 5 MW [7].

The new RES has a same power as existing source, which generate electricity in the grid, so the active power is 1 MW. Table 4. show changes in the voltages at the nodes depending on where is new RES placed. Fig. 4 and Fig. 5 shows the power losses and the voltages changes with new RES.

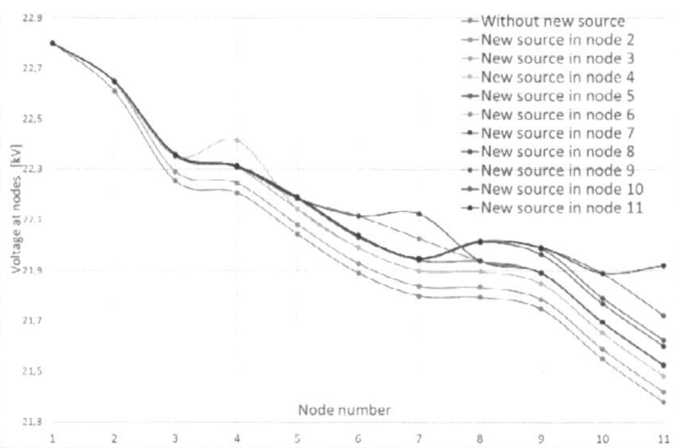

Figure 4 Voltage profile at nodes, when a new source is connected

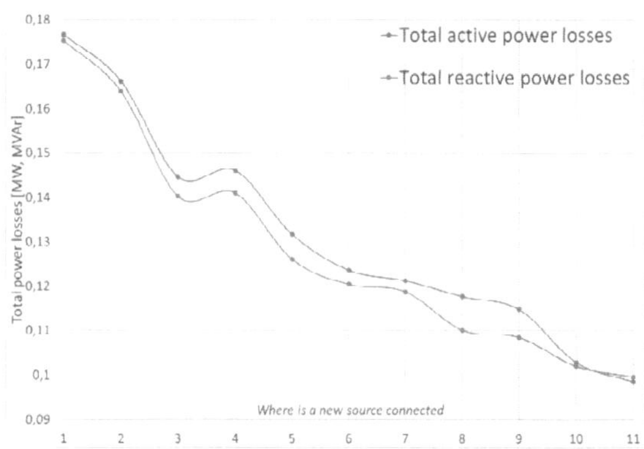

Figure 5 Depending of total power losses to which node is the new renewable source connected

Table 4 Node voltage with new RES

	RES at node:									
	2		3		4		5		6	
node	**Ua (kV)**	**Ur (kV)**	**Ua (kV)**	**Ur (kV)**	**Ua (kV)**	**Ur (kV)**	**Ua (kV)**	**Ur (kV)**	**Ua (kV)**	**Ur (kV)**
1	22.8	0	22.8	0	22.8	0	22.8	0	22.8	0
2	22.6446	-0.0460	22.6463	-0.0460	22.6462	-0.0460	22.6473	-0.0460	22.6478	-0.0459
3	22.2912	-0.2397	22.3513	-0.1786	22.3511	-0.1787	22.3539	-0.1786	22.3552	-0.1783
4	22.2452	-0.2465	22.3055	-0.1855	22.4136	-0.1310	22.3081	-0.1855	22.3094	-0.1852
5	22.0822	-0.3543	22.1432	-0.2935	22.1430	-0.2936	22.1833	-0.2539	22.1851	-0.2535
6	21.9287	-0.3987	21.9903	-0.3382	21.9901	-0.3382	22.0308	-0.2988	22.1154	-0.2433
7	21.8390	-0.4246	21.9009	-0.3642	21.9007	-0.3643	21.9416	-0.3250	22.0267	-0.2696
8	21.8337	-0.4957	21.8958	-0.4352	21.8956	-0.4353	21.9367	-0.3958	21.9385	-0.3954
9	21.7861	-0.5020	21.8484	-0.4416	21.8482	-0.4417	21.8893	-0.4023	21.8912	-0.4019
10	21.5888	-0.5281	21.6517	-0.4682	21.6515	-0.4683	21.6931	-0.4292	21.6949	-0.4288
11	21.4192	-0.5504	21.4827	-0.4909	21.4825	-0.4910	21.5244	-0.4522	21.5263	-0.4518

	(Continued) RES at node:									
	7		8		9		10		11	
node	**Ua (kV)**	**Ur (kV)**	**Ua (kV)**	**Ur (kV)**	**Ua (kV)**	**Ur (kV)**	**Ua (kV)**	**Ur (kV)**	**Ua (kV)**	**Ur (kV)**
1	22.8	0	22.8	0	22.8	0	22.8	0	22.8	0
2	22.6479	-0.0459	22.6484	-0.0460	22.6485	-0.0460	22.6492	-0.0457	22.6495	-0.0456
3	22.3556	-0.1783	22.3568	-0.1787	22.3572	-0.1785	22.3589	-0.1779	22.3596	-0.1777
4	22.3098	-0.1851	22.3109	-0.1855	22.3114	-0.1854	22.3131	-0.1848	22.3137	-0.1845
5	22.1857	-0.2534	22.1873	-0.2540	22.1879	-0.2538	22.1903	-0.2529	22.1912	-0.2526
6	22.1163	-0.2432	22.0348	-0.2988	22.0354	-0.2986	22.0378	-0.2978	22.0387	-0.2975
7	22.1237	-0.2051	21.9456	-0.3250	21.9462	-0.3248	21.9487	-0.3240	21.9496	-0.3237
8	21.9391	-0.3953	22.0113	-0.3190	22.0122	-0.3186	22.0159	-0.3173	22.0172	-0.3168
9	21.8917	-0.4018	21.9641	-0.3256	21.9872	-0.3143	21.9912	-0.3130	21.9926	-0.3125
10	21.6955	-0.4287	21.7686	-0.3531	21.7919	-0.3419	21.8888	-0.2952	21.8908	-0.2947
11	21.5268	-0.4516	21.6006	-0.3766	21.6241	-0.3654	21.7219	-0.3190	21.9216	-0.2210

Fig. 5 shown that, loss can be reduced depending on where we place the new RES. The best result is achieved when new RES is placed in to the farthest node. In this case it is node 11. Another explanation of the new RES placement is because in the node 10 and 11 is highest consumption. This means that is not necessary to transport produced electricity far away from new RES, thus power loss caused by transport is lower.

The other parameter is voltage in the nodes. It can be seen, that voltage in the nodes has increased in this case.

IV. CONCLUSION

This article was dealing with total active and reactive loss depending on placement of RES in power grid. In the nearest future, it is expected massive installation of new RES in order to reduce the greenhouse effect. Another reason is that fossil fuel reserves are limited, so it is necessary to replace them with sustainable one.

Other aspect such as quality of electricity, flicker effect, harmonics was not addressed in this article.

It is possible to expand this topic more in more details in future articles. For example, it is possible to take in the account ambient conditions such as sunlight a wind speed. As photovoltaic panels do not produce electricity through whole day and wind speed is not constant as well.

ACKNOWLEDGMENT

This work was supported by the Scientific Grant Agency of the Ministry of Education of Slovak Republic and the Slovak Academy of Sciences by the projects VEGA No. 1/0372/18.

REFERENCES

[1] The Paris Agreement [online]. Available at: <https://unfccc.int/process-and-meetings/the-paris-agreement/the-paris-agreement>. Accessed on April 7,2019.

[2] Renewable energy, Mowing towards a low carbon economy [online]. Available at: <https://ec.europa.eu/energy/en/topics/renewable-energy>. Accessed on April 7, 2019.

[3] O. Ellabban, H. Abu-Rub, F. Blaabjerg, "Renewable energy resources: Current status, future prospects and their enabling technology". Elsevier, 2014. Volume 39, pp. 748-764. Available at: <https://www.sciencedirect.com/science/article/pii/S1364032114005656>

[4] Volker Quaschning, "Understanding renewable energy systems", 2005, ISBN 1-84407-128-6.

[5] The coming energy revolution [online]. Available at: <http://www.peopleandtheplanet.com/index.html@lid=26272&section=36&topic=44.html>. Accessed on 10. April, 2019.

[6] M. Kolcun, Ľ. Beňa, A. Mészáros, „Optimalizácia prevádzky elektrizačnej sústavy", Technická univerzita v Košiciach 2009, ISBN 978-80-553-0323-9

[7] L. I. Dulă, M. Abrudean, D. Bică, „Optimal Location of a Distributed Generator for Power Losses Improvement" in Procedia Technology. Elsevier Ltd. 2016, Volume 22, pp. 734-739. Available at: <https://www.sciencedirect.com/science/article/pii/S2212017316000335>

Verification of Synchronous Generator Time Constants Given by Manufacturers Using the Short-Circuit Current Calculation

Anastázia Margitová, Martin Kanálik, Michal Kolcun
Department of Electrical Power Engineering
Technical University of Košice
Košice, Slovakia
anastazia.margitova@tuke.sk, martin.kanalik@tuke.sk, michal.kolcun@tuke.sk

Abstract — **Synchronous generators are an essential part of the power system, they perform a number of important tasks related to the production of electrical energy, voltage control and so on. The article deals with the verification of synchronous generator time constants given by manufacturers using by short-circuit currents calculation. The procedure of mathematical dynamic modeling of electromagnetic properties of the synchronous generator was used for this verification. Results obtained by short-circuit current calculation by this approach are compared to results obtained by short-circuit current calculation method using generator time constants.**

Keywords: synchronous generator modeling; short-circuit current calculation; synchrnous generator time constants

I. Introduction

Synchronous generators form the principal source of electric energy in power systems. The commercial birth of the synchronous generator can be dated to 1891, when the first alternate transmission of electricity from Lauffen to Frankfurt (Germany, length 175 km, 20 kV line) was realized. Many large loads are driven by synchronous motors. Synchronous compensators are used to compensate the reactive power and voltage control. These devices operate on the same principle and are collectively referred to as synchronous machines [1].

The basic operating state of each power system (and also of a synchronous generator) is its steady state (electrical and mechanical) of operating system quantities. Any change in system operating parameters leads to disturbing a steady state (equilibrium), resulting in a transient phenomenon. One of the most common causes of electromagnetic transient phenomena in the power system is short circuits. Short-circuit current generation is a steady state disturbance resulting in a corresponding voltage and current change. After the short circuit, the total network impedance decreases and the current increases, resulting in reduced voltage near the short circuit. The ideal short circuit is characterized by zero (or negligibly small) short-circuit impedance. The short-circuit current caused by the ideal short circuit is always greater than the short-circuit current caused by a non-ideal short circuit (created through an impedance that cannot be neglected) if the same conditions are considered. Therefore, the ideal short circuit is considered to determine maximum short-circuit currents. If all three phases are affected by the short circuit, it is referred to as a symmetrical short circuit. All other short circuits are listed as unsymmetrical [2], [3], [4], [5].

II. Mathematical Model of a Synchronous Generator for the Short-Circuit Current Calculation

The basic mathematical model of the synchronous generator (turbogenerator as well as hydrogenerator) is based on its design and consists of equations for stator voltages and rotor voltages. For turbogenerators (modeled with two amortisseurs in the q-axis), after applying dq0 transformation, a system of first order linear differential equations is obtained [1], [2], [6]-[9]:

$$\left[\frac{\mathrm{d}i}{\mathrm{d}t}\right] = -[N]^{-1}[M][i] + [N]^{-1}[u] = [A][i] + [B], \qquad (1)$$

$$[i] = \begin{bmatrix} i_{\mathrm{d}} \\ i_{\mathrm{q}} \\ i_{\mathrm{fd}} \\ i_{\mathrm{1d}} \\ i_{\mathrm{1q}} \\ i_{\mathrm{2q}} \end{bmatrix} \begin{matrix} \text{(d-component of the stator current)} \\ \text{(q-component of the stator current)} \\ \text{(rotor field current)} \\ \text{(rotor amortisseur 1 current in the d-axis)} \\ \text{(rotor amortisseur 1 current in the q-axis)} \\ \text{(rotor amortisseur 2 current in the q-axis)} \end{matrix},$$

$$[u] = \begin{bmatrix} u_{\mathrm{d}} \\ u_{\mathrm{q}} \\ u_{\mathrm{fd}} \\ u_{\mathrm{1d}} = 0 \\ u_{\mathrm{1q}} = 0 \\ u_{\mathrm{2q}} = 0 \end{bmatrix} \begin{matrix} \text{(d-component of the stator voltage)} \\ \text{(q-component of the stator voltage)} \\ \text{(rotor field voltage)} \\ \text{(rotor amortisseur 1 voltage in the d-axis)} \\ \text{(rotor amortisseur 1 voltage in the q-axis)} \\ \text{(rotor amortisseur 2 voltage in the q-axis)} \end{matrix},$$

$$[M] = \begin{bmatrix} -R_{\mathrm{a}} & \omega L_{\mathrm{q}} & 0 & 0 & -\omega L_{\mathrm{aq}} & -\omega L_{\mathrm{aq}} \\ -\omega L_{\mathrm{d}} & -R_{\mathrm{a}} & \omega L_{\mathrm{ad}} & \omega L_{\mathrm{ad}} & 0 & 0 \\ 0 & 0 & R_{\mathrm{fd}} & 0 & 0 & 0 \\ 0 & 0 & 0 & R_{\mathrm{1d}} & 0 & 0 \\ 0 & 0 & 0 & 0 & R_{\mathrm{1q}} & 0 \\ 0 & 0 & 0 & 0 & 0 & R_{\mathrm{2q}} \end{bmatrix},$$

$$[N] = \begin{bmatrix} -L_{\mathrm{d}} & 0 & L_{\mathrm{ad}} & L_{\mathrm{ad}} & 0 & 0 \\ 0 & -L_{\mathrm{q}} & 0 & 0 & L_{\mathrm{aq}} & L_{\mathrm{aq}} \\ -L_{\mathrm{ad}} & 0 & L_{\mathrm{ffd}} & L_{\mathrm{ad}} & 0 & 0 \\ -L_{\mathrm{ad}} & 0 & L_{\mathrm{ad}} & L_{\mathrm{11d}} & 0 & 0 \\ 0 & -L_{\mathrm{aq}} & 0 & 0 & L_{\mathrm{11q}} & L_{\mathrm{aq}} \\ 0 & -L_{\mathrm{aq}} & 0 & 0 & L_{\mathrm{aq}} & L_{\mathrm{22q}} \end{bmatrix},$$

where

$[i]$ is the matrix of output unknown quantities, it is formed by the matrix of stator currents and rotor currents in amperes or per units,

$\left[\frac{di}{dt}\right]$ is the matrix of time derivations of unknown rotor and stator currents in amperes or per units,

$[u]$ is the matrix of input known quantities with constant coefficients, it contains the matrix of stator voltages and rotor voltages in volts or per units,

$[M]$ is the matrix with constant coefficients, it contains parameters of the synchronous generator (resistances and inductive reactances) in ohms or per units bound to unknown currents,

$[N]$ is the matrix with constant coefficients, it contains parameters of the synchronous generator (inductances) in henrys or per units bound to time derivations of unknown currents,

R_a is the (stator) armature resistance per phase in ohms or per units,

L_d is the (stator) synchronous inductance in the d-axis in henries or per units,

L_q is the (stator) synchronous inductance in the q-axis in henries or per units,

L_{ad} is the mutual inductance bound to rotor circuits in the d-axis in henries or per units,

L_{aq} is the mutual inductance bound to rotor circuits in the q-axis in henries or per units,

ω is the rotor angular velocity in radians per second or per units.

R_{fd}, R_{1d}, R_{1q}, R_{2q} are the (rotor) resistances of the field winding, amortisseur winding 1 in the d-axis, amortisseur winding 1 in the q-axis and amortisseur winding 2 in the q-axis in henries or per units. L_{ffd}, L_{11d}, L_{11q}, L_{22q} are the (rotor) inductances of the field winding, amortisseur winding 1 in the d-axis, amortisseur winding 1 in the q-axis and amortisseur winding 2 in the q-axis in ohms or per units.

The solution of the system (1) consists of determining the initial conditions (for example generator no-load operation): unknown currents (the matrix $[i]$) at time 0. The initial conditions are defined by the steady state solution of the synchronous generator: simplification of the system (1). During steady state, time derivations do not occur in the equations (1) for describing the synchronous machine. Also, all currents flowing in amortisseur windings are equal to 0. They are closed circuits in which currents are induced only in the case of a transient event. The zero component is skipped and the angular velocity is not changed (is equal to the nominal) [2].

The control and regulation of the synchronous generator voltage is realized by changing the excitation current, respectively the excitation voltage of the synchronous generator. As a result, the excitation voltage (which is determined from the steady state of the synchronous generator [6]) is required as the single input to the calculation (whether the steady state or transient phenomena). This can be achieved by expressing the stator voltages u_d and u_q in the system (1) via an external load ($R_{ext} + j\omega L_{ext}$) connected to synchronous generator terminals. In other words, the stator voltages u_d and u_q can be moved from the input matrix $[u]$ to the matrices $[M]$ and $[N]$ of synchronous generator parameters. The ideal three-phase short circuit at synchronous generator terminals can be realized by setting the external load (R_{ext} and L_{ext}) to zero. Some elements of matrices $[u]$, $[M]$, $[N]$ change to:

$[u](1,1)=0$, $[u](2,1)=0$,

$[N](1,1)=-(L_d+L_{ext})$, $[N](2,2)=-(L_q+L_{ext})$,

$[M](1,1)=-(R_a+R_{ext})$, $[M](1,2)=\omega(L_q+L_{ext})$,

$[M](2,1)=-\omega(L_d+L_{ext})$, $[M](2,2)=-(R_a+R_{ext})$.

Applying the inverse dq0 transformation determines time courses of phase currents i_{ka}, i_{kb}, i_{kc} (index k in the case of the ideal three-phase short circuit) in amperes or per units [2]:

$$\begin{pmatrix} i_{ka} \\ i_{kb} \\ i_{kc} \end{pmatrix} = \begin{pmatrix} \cos\theta & -\sin\theta & 1 \\ \cos(\theta-120°) & -\sin(\theta-120°) & 1 \\ \cos(\theta+120°) & -\sin(\theta+120°) & 1 \end{pmatrix} \begin{pmatrix} i_d \\ i_q \\ i_0 \end{pmatrix}, \quad (2)$$

where i_0 is the zero component of the stator current. Zero component i_0 is neglected in the case of symmetric systems, it is $i_0 = 0$.

III. Time Constants and Reactances of a Synchronous Generator

Following a disturbance, currents are induced in the synchronous machine stator and rotor circuits. Some of these currents decay more rapidly than others. Accordingly, parameters of the synchronous machine are divided into three types (reactances and time constants) [7]:

- subtransient (characterized by two commas above the quantity symbol),
- transient (characterized by one comma above the quantity symbol),
- synchronous (without comma above the quantity symbol).

The duration of dynamic events in the synchronous generator is described by time constants. Time constants are based on basic resistances and inductances of the synchronous generator. Basic time constants (in seconds or per units) in the d-axis and the q-axis are [10]:

T'_{do} is the transient open-circuit time constant in the d-axis, it expresses the rate of decay or buildup of the field current when the stator is open circuited and there is zero resistance of the field winding,

T''_{do} is the subtransient open-circuit time constant in the d-axis,

T'_d is the transient short-circuit time constant in the d-axis, it expresses the rate of decay of the transient component of the stator current in the d-axis under an ideal three-phase short circuit at synchronous generator terminals,

T''_d is the subtransient short-circuit time constant in the d-axis, it expresses the rate of decay of the subtransient component of the stator current in the d-axis under an ideal three-phase short circuit at synchronous generator terminals,

T_a is the armature (stator) time constant, it expresses the rate of decay of the DC component of the short-circuit current under an ideal three-phase short circuit at synchronous generator terminals.

Other important data belonging to the synchronous generator are its inductances, respectively inductive reactances (in henries, respectively ohms or per units) [10]:

X_l is the leakage reactance, it expresses the reactance due to flux setup by armature windings, but not crossing the air gap. It can be divided into end-winding leakage and slot leakage,

X''_{d} is the subtransient reactance in the d-axis, it expresses the magnitude of the subtransient component of the stator current in the d-axis under an ideal three-phase short circuit at synchronous generator terminals,

X'_{d} is the transient reactance in the d-axis, it expresses the magnitude of the transient component of the stator current in the d-axis under an ideal three-phase short circuit at synchronous generator terminals. It is the reactance after the currents in amortisseurs windings have decayed, but before the transient event ending in the field winding,

X_{d} is the synchronous reactance in the d-axis, it is the reactance related to the steady-state stator current in the d-axis after or before the ideal three-phase short circuit at synchronous generator terminals,

X_2 is the negative sequence reactance, it is the reactance encountered by a voltage of reverse phase sequence applied to the stator, with the machine running. Negative sequence flux revolves opposite to the rotor and is at twice the system frequency,

X_0 is the zero sequence reactance, it is the reactance related to the zero sequence current,

X_{p} it is the Potier reactance, it is the reactance used for calculation of field current when open circuit and zero power factor curves are available.

The time constants T'_{qo}, T''_{qo}, T'_{q}, T''_{q} and the reactances X''_{q}, X'_{q}, X_{q} in the q-axis have the same meaning as the time constants in the d-axis, but they are related to currents in the q-axis. Typically, manufacturers list the above time constants and reactants in data sheets. Other parameters (for example, rotor reactances and inductances in Chapter II) must be calculated according to the equations given in [11]. There are also interconnections between the manufacturers' time constants and reactances mentioned above. Time constants T'_{d}, T''_{d}, T_{a} can be determined from the equations:

$$T'_{\mathrm{d}} = T'_{\mathrm{do}} \frac{X'_{\mathrm{d}}}{X_{\mathrm{d}}}, \tag{3}$$

$$T''_{\mathrm{d}} = T''_{\mathrm{do}} \frac{X''_{\mathrm{d}}}{X'_{\mathrm{d}}}, \tag{4}$$

$$T_{\mathrm{a}} = \frac{1}{\omega R_{\mathrm{a}}}\left(\frac{2X''_{\mathrm{d}}X''_{\mathrm{q}}}{X''_{\mathrm{d}}+X''_{\mathrm{q}}}\right) = \frac{X_2}{\omega R_{\mathrm{a}}}. \tag{5}$$

IV. Simplified Model of a Synchronous Generator for the Short-Circuit Current Calculation

The simplified model of the synchronous generator for the calculation of short-circuit currents is based on the subtransient, transient and steady-state parameters of the synchronous generator (time constants and reactances mentioned in Chapter III). Only d-axis parameters are considered, the model is suitable for round-rotor generators where $X''_{\mathrm{d}} \cong X''_{\mathrm{q}}$. Using this theory, the ideal three-phase short circuit at synchronous generator terminals is modeled. It is considered that at the moment of the short circuit, the synchronous generator operates in the no-load state [4], [5].

For the RMS value of the subtransient I''_{k}, transient I'_{k} and the steady-state I_{k} component of the short-circuit current in amperes or per units it is possible to write (U_{n} is the nominal voltage of the synchronous generator in volts or per units) [4], [5]:

$$I''_{\mathrm{k}} = \frac{U_{\mathrm{n}}}{\sqrt{3}X''_{\mathrm{d}}}, \tag{6}$$

$$I'_{\mathrm{k}} = \frac{U_{\mathrm{n}}}{\sqrt{3}X'_{\mathrm{d}}}, \tag{7}$$

$$I_{\mathrm{k}} = \frac{U_{\mathrm{n}}}{\sqrt{3}X_{\mathrm{d}}}. \tag{8}$$

Time courses of the current flowing in each phase i_{ka}, i_{kb} and i_{kc} in the case of the ideal three-phase short circuit at generator terminals can be expressed as the sum of the AC short-circuit component $i_{\mathrm{k\,AC}}$ and the DC short-circuit component $i_{\mathrm{k\,DC}}$. The instantaneous values of stator phase currents $i_{\mathrm{ka}}(t)$, $i_{\mathrm{kb}}(t)$, $i_{\mathrm{kc}}(t)$ at time t can be determined from (φ_{k} is the short-circuit angle between voltage and current at generator terminals during a short circuit, α is the voltage angle in degrees or radians) [4], [5]:

$$i_{\mathrm{ka}}(t) = i_{\mathrm{ka\,AC}}(t) + i_{\mathrm{ka\,DC}}(t), \tag{9}$$

$$i_{\mathrm{ka\,AC}}(t) = \sqrt{2}\left[\left(I''_{\mathrm{k}} - I'_{\mathrm{k}}\right)\mathrm{e}^{-\frac{t}{T''_{\mathrm{d}}}} + \left(I'_{\mathrm{k}} - I_{\mathrm{k}}\right)\mathrm{e}^{-\frac{t}{T'_{\mathrm{d}}}} + I_{\mathrm{k}}\right]\sin\left(\omega t + \alpha + \varphi_{\mathrm{k}}\right),$$

$$i_{\mathrm{ka\,DC}}(t) = \sqrt{2}I''_{\mathrm{k}}\sin\left(\alpha + \varphi_{\mathrm{k}}\right)\mathrm{e}^{-\frac{t}{T_{\mathrm{a}}}},$$

$$i_{\mathrm{kb}}(t) = i_{\mathrm{kb\,AC}}(t) + i_{\mathrm{kb\,DC}}(t), \tag{10}$$

$$i_{\mathrm{kb\,AC}}(t) = \sqrt{2}\left[\left(I''_{\mathrm{k}} - I'_{\mathrm{k}}\right)\mathrm{e}^{-\frac{t}{T''_{\mathrm{d}}}} + \left(I'_{\mathrm{k}} - I_{\mathrm{k}}\right)\mathrm{e}^{-\frac{t}{T'_{\mathrm{d}}}} + I_{\mathrm{k}}\right]\sin\left(\omega t + \alpha + \varphi_{\mathrm{k}} - 120°\right),$$

$$i_{\mathrm{kb\,DC}}(t) = \sqrt{2}I''_{\mathrm{k}}\sin\left(\alpha + \varphi_{\mathrm{k}} - 120°\right)\mathrm{e}^{-\frac{t}{T_{\mathrm{a}}}},$$

$$i_{\mathrm{kc}}(t) = i_{\mathrm{kc\,AC}}(t) + i_{\mathrm{kc\,DC}}(t), \tag{11}$$

$$i_{\mathrm{kc\,AC}}(t) = \sqrt{2}\left[\left(I''_{\mathrm{k}} - I'_{\mathrm{k}}\right)\mathrm{e}^{-\frac{t}{T''_{\mathrm{d}}}} + \left(I'_{\mathrm{k}} - I_{\mathrm{k}}\right)\mathrm{e}^{-\frac{t}{T'_{\mathrm{d}}}} + I_{\mathrm{k}}\right]\sin\left(\omega t + \alpha + \varphi_{\mathrm{k}} + 120°\right),$$

$$i_{\mathrm{kc\,DC}}(t) = \sqrt{2}I''_{\mathrm{k}}\sin\left(\alpha + \varphi_{\mathrm{k}} + 120°\right)\mathrm{e}^{-\frac{t}{T_{\mathrm{a}}}}.$$

V. Verification of Synchronous Generator Time Constants By the Short Circuit Current Calculation

In this chapter, a three-phase ideal short circuit at the terminals of different turbogenerators (G1, G2, G3) is simulated and compared by two methods (Chapter II and IV). At the moment of the short circuit, synchronous generators operate in the no-load state. A principal single-line diagram of an ideal three-phase short circuit at synchronous generator terminals is shown in Fig. 1. Manufacturers' generator parameters (nominal data, reactances and time constants [12]) are shown in Table 1, the calculated generator parameters (resistances and inductances of field and amortisseurs windings according to [11]) are shown in Table 2. A comparison of manufacturers' and calculated (according to equations (3), (4), (5)) generator (G1, G2, G3) time constants T''_{d}, T'_{d}, T_{a} is given in Table 3. All generators are modeled with two amortisseurs in the q-axis. The only input is the field voltage, which is constant during the calculation, so the voltage regulation is not considered. The model is analyzed in physical units.

Table 1. Manufacturers' parameters of modeled synchronous generators

Manufacturers' parameters		**G1**	**G2**	**G3**
Rated voltage	U_{nG} (kV)	13.8	18	24
Rated power	S_{nG} (MVA)	25	192	384
Rated frequency	f_{nG} (Hz)	60	60	60
Rated power factor	$\cos\varphi_{nG}$ (–)	0.8	0.85	0.85
Synchronous reactance in d-axis	X_d (Ω)	9.522	2.648	2.697
Transient reactance in d-axis	X'_d (Ω)	1.767	0.547	0.486
Subtransient reactance in d-axis	X''_d (Ω)	0.914	0.42	0.39
Synchronous reactance in q-axis	X_q (Ω)	9.293	2.612	2.667
Transient reactance in q-axis	X'_q (Ω)	5.447	1.549	1.577
Subtransient reactance in q-axis	X''_q (Ω)	0.914	0.419	0.383
Leakage reactance	X_l (Ω)	1.021	0.344	0.29
Armature resistance	R_a (Ω)	0.0107	0.0027	0.0021
Transient open-circuit time constant in d-axis	T'_{do} (s)	4.75	5.9	5.21
Subtransient open-circuit time constant in d-axis	T''_{do} (s)	0.059	0.033	0.042
Transient short-circuit time constant in d-axis	T'_d (s)	0.882	0.829	0.159
Subtransient short-circuit time constant in d-axis	T''_d (s)	0.035	0.023	0.035
Transient open-circuit time constant in q-axis	T'_{qo} (s)	1.5	0.535	1.5
Subtransient open-circuit time constant in q-axis	T''_{qo} (s)	0.21	0.078	0.042
Transient short-circuit time constant in q-axis	T'_q (s)	–	0.415	0.581
Subtransient short-circuit time constant in q-axis	T''_q (s)	0.035	0.023	0.035
Armature time constant	T_a (s)	0.177	0.254	0.45

Table 2. Calculated parameters of modeled synchronous generators

Calculated parameters		**G1**	**G2**	**G3**
Field winding resistance	R_{fd} (Ω)	0.0052	0.0011	0.0013
Amortisseur winding 1 resistance in d-axis	R_{1d} (Ω)	0.0294	0.026	0.0254
Amortisseur winding 1 resistance in q-axis	R_{1q} (Ω)	0.0315	0.024	0.0092
Amortisseur winding 2 resistance in q-axis	R_{2q} (Ω)	0.0546	0.0437	0.0876
Mutual d-axis inductance	L_{ad} (H)	0.0225	0.0061	0.0064
Mutual q-axis inductance	L_{aq} (H)	0.0219	0.006	0.0063
Field winding inductance	L_{ffd} (H)	0.0247	0.0067	0.007
Amortisseur winding 1 resistance in d-axis	L_{11d} (H)	0.0223	0.0064	0.0069
Amortisseur winding 1 resistance in q-axis	L_{11q} (H)	0.0472	0.0128	0.0137
Amortisseur winding 2 resistance in q-axis	L_{22q} (H)	0.0217	0.0062	0.0066

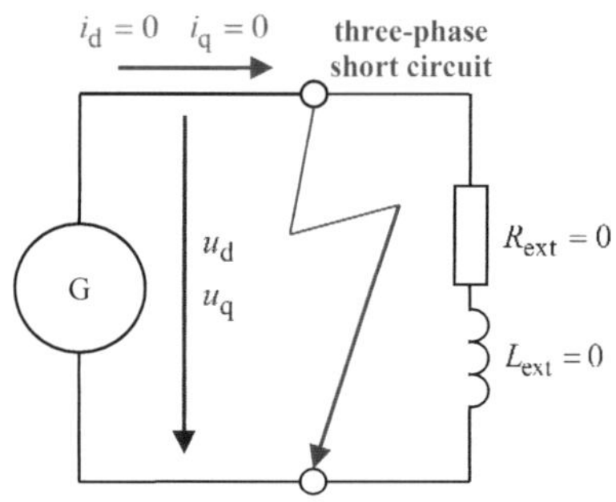

Figure 1. Principal single-line diagram of an ideal three-phase short circuit at synchronous generator terminals

Table 3. Comparison of manufacturers' and calculated time constants of modeled synchronous generators

Time constant (s)		**G1**	**G2**	**G3**
Subtransient short-circuit time constant in d-axis	Manufacturers'	0.035	0.023	0.035
	Calculated	0.031	0.025	0.034
Transient short-circuit time constant in d-axis	Manufacturers'	0.882	0.829	0.159
	Calculated	0.881	1.219	0.939
Armature time constant	Manufacturers'	0.177	0.254	0.45
	Calculated	0.227	0.412	0.488

If the short circuit occurs when the phase voltage passes the maximum positive value, then the short-circuit current in the corresponding phase is symmetrical and the DC component is zero. The peak short-circuit current has a positive value. Otherwise, if the short circuit occurs when the phase voltage passes the maximum negative value, then the short-circuit current in the corresponding phase is symmetrical and the DC component is zero. The peak short-circuit current has a negative value. If the voltage reaches zero from a negative half-period and the short-circuit occurs, the unsymmetrical short-circuit current with the maximum positive DC component is generated. If the voltage reaches zero from a positive half-period and the short-circuit occurs, the unsymmetrical short-circuit current with the maximum negative DC component is generated [1].

In Fig. 2 are shown time courses of the generator G1 (18 kV and 192 MVA) current and voltage after the ideal three-phase short circuit at its terminals (determined by analytical model consisting of differential equations, Chapter II). At the moment of the short circuit, the phase a voltage was zero (from the negative half-period) and the generator G1 was operating without load. In the phase a, the maximum positive DC and peak short-circuit current of 22.8 kA were generated.

The representation of generator G2 phase currents, field and amortisseur currents in the d-axis and the q-axis is shown in Fig. 3 (determined by analytical model consisting of differential equations, Chapter II). After the short circuit, there was a strong armature (stator) reaction. It means that the stator winding (magnetic flux caused by a stator short-circuit current flowing) was acting against the main magnetic flux (caused by the field current and rotation of the rotor). In order to reduce the armature reaction, the excitation current value rapidly increased as well as the amortisseur current values. At the end of the transient phenomena (ideal three-phase short circuit at generator G2 terminals), the field current was the same as before the short circuit, and the amortisseur currents were equal to zero.

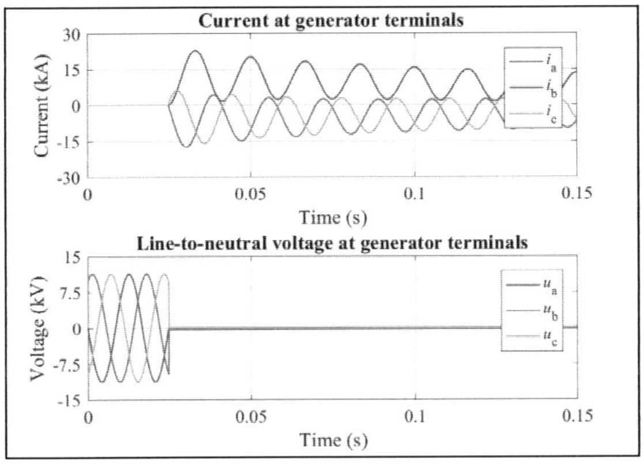

Figure 2. Time courses of the generator G1 current and voltage after the ideal three-phase short circuit at its terminals

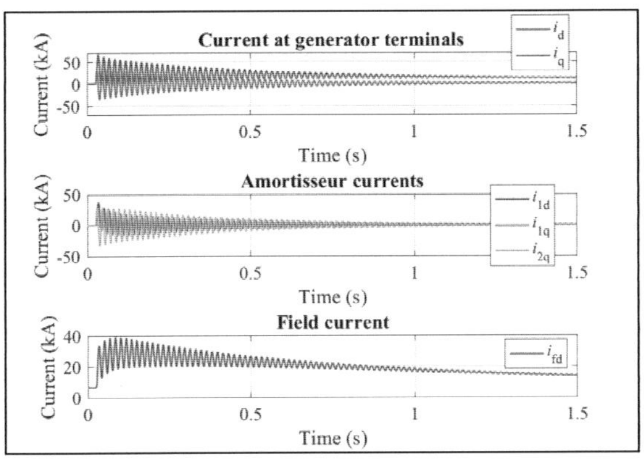

Figure 3. Time courses of the generator G2 stator, field and amortisseur currents after the ideal three-phase short circuit at its terminals

In Fig. 4 to Fig. 6 are shown time courses of the generator G1, G2 and G3 stator current in phase a after the ideal three-phase short circuit determined by the analytical method (system of differential equation) and the simplified model (with data sheet or calculated time constants). In the case of generators G1 and G2, the maximum positive DC component was generated (voltage passed through zero), in the case of the generator G3, the DC component was zero (voltage passed through maximum). The upper graph compares the analytical method (differential equations) with a simplified model with calculated time constants (Table 3). The lower graph compares the analytical method (differential equations) with a simplified model with manufacturers' time constants (Table 2).

Comparison of manufacturers' and calculated generator G1 time constants implies that the calculated armature time constant $T_{\mathrm{a}} = 0.227$ s differs significantly from the data sheet value $T_{\mathrm{a}} = 0.177$ s. This difference is also seen in Fig. 4 where the DC component of the time course determined by the simplified method using the data sheet armature time constant $T_{\mathrm{a}} = 0.177$ s differs from the DC component of the time course determined by the method of differential equations. According to the manufacturers' value of the armature time constant $T_{\mathrm{a}} = 0.177$ s, the DC component disappears sooner than with considering the calculated armature time constant $T_{\mathrm{a}} = 0.227$ s.

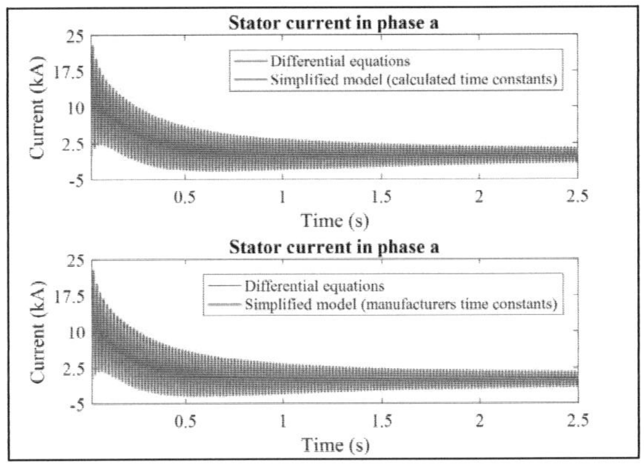

Figure 4. Time courses of the generator G1 stator current in phase a after the ideal three-phase short circuit at its terminals

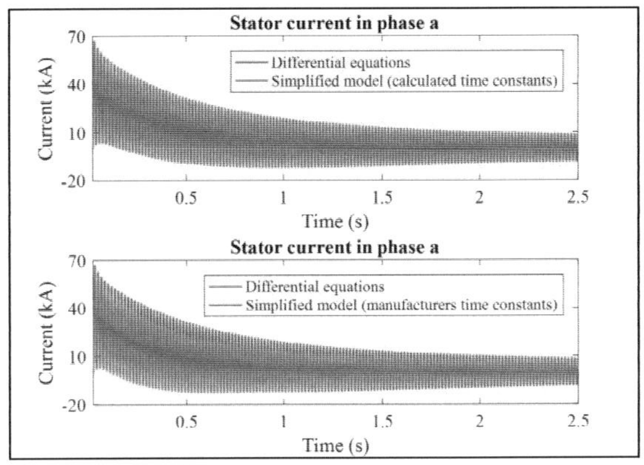

Figure 5. Time courses of the generator G2 stator current in phase a after the ideal three-phase short circuit at its terminals

In the case of the generator G2, in addition to the armature time constant ($T_{\mathrm{a}} = 0.254$ s and calculated $T_{\mathrm{a}} = 0.412$ s), the data sheet $T'_{\mathrm{d}} = 0.829$ s and computed $T'_{\mathrm{d}} = 1.219$ s transient short-circuit time constants in the d-axis differed. Fig. 5 shows that when using the data sheet value of the transient short-circuit time constant in the d-axis, the transient short-circuit current component disappeared faster than when applying the calculated transient short-circuit time constant in the d-axis. In the case of the generator G3, the difference between the data sheet $T'_{\mathrm{d}} = 0.159$ s and calculated $T'_{\mathrm{d}} = 0.939$ s transient short-circuit time constant was even more pronounced compared to the generator G2 (Fig. 6).

In Table 4 and Fig. 7 is shown the (maximal) deviation of the simplified synchronous generator model (manufacturers' or calculated time constants) from the analytical synchronous generator model (differential equations). The results show that when considering the calculated time constants (simplified synchronous generator model), the deviation is considerably smaller for all analyzed generators (G1, G2, G3). This is also visible in Fig. 4 to Fig. 6, where the upper time courses of currents are almost identical. The times at which the greatest deviation was reached also point to the difference in data sheet and calculated DC and transient short-circuit time constants. These times also point to the difference between the DC and transient short-circuit current component calculated by the simplified method versus the current time courses determined by the analytical method.

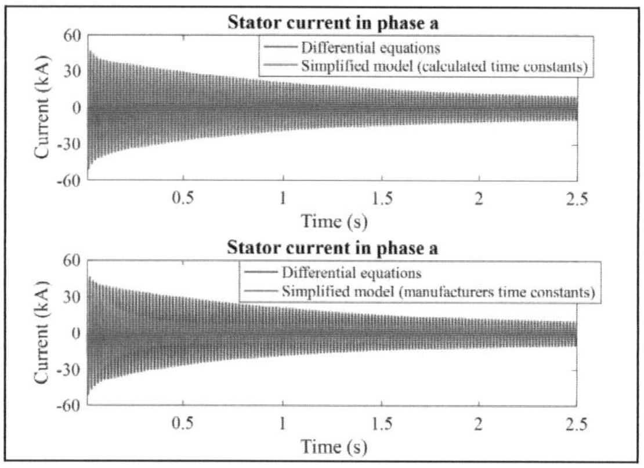

Figure 6. Time courses of the generator G3 stator current in phase a after the ideal three-phase short circuit at its terminals

Table 4. Maximum deviation of the simplified synchronous generator model (manufacturers' or calculated time constants) from the analytical synchronous generator model (differential equations)

Maximum deviation		G1	G2	G3
Manufacturers' time constants	Value (%)	17.26	25.8	68.53
	Time (s)	0.28	0.53	0.56
Calculated time constants	Value (%)	7.98	2.05	3.57
	Time (s)	1	1.36	0.02

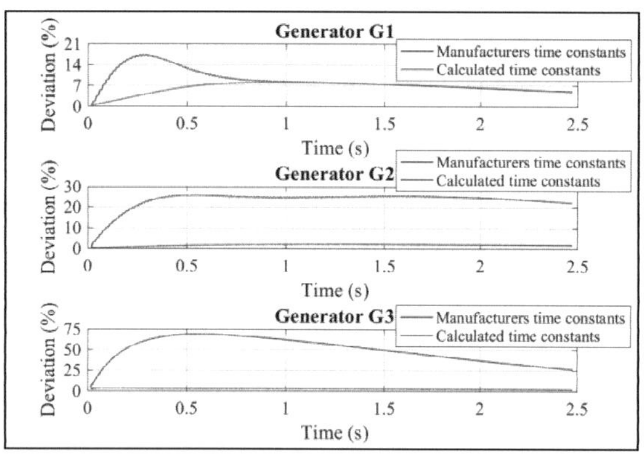

Figure 7. Deviation of the simplified synchronous generator model (manufacturers' or calculated time constants) from the analytical synchronous generator model (differential equations)

VI. Conclusion

This paper deals with the verification of manufacturers' time constants of three different synchronous generators using the short-circuit current calculation by two different methods.

The first of these methods, the complex analytical model of the synchronous generator, consists of a system of differential equations. The parameters entering the calculation, in addition to the nominal data (voltage, power, etc.), are resistances, self and mutual reactances, respectively inductances of each generator circuit (stator and rotor). The resistances and reactances of stator circuits are determined by manufacturers, the resistances and reactances of rotor circuits are calculated from the steady-state, transient and subtransient reactances and also from open-circuit time constants of the synchronous generator.

In the case of the second simplified method, only the synchronous generator nominal voltage, the steady-state, transient and subtransient reactance, the transient and subtransient short-circuit time constant and armature time constant enter the calculation. However, only the d-axis parameters are considered in this model.

It is presented in this paper that some manufacturers' values of synchronous generator time constants (given in [12]) do not correspond to calculated time constants. This difference caused a significant deviation of the analytical model from the simplified synchronous generator model (with manufacturers' time constants). When comparing the analytical model with the simplified model with calculated time constants, there was a relatively large match of current time courses. If it is considered that the synchronous generator model assembled in the form of differential equations is more accurate, even the simplified theory (with calculated time constants) yields satisfactory results. However, this only applies to turbogenerators, because there are no large differences between the parameters in d-axis and q-axis. It follows that it is necessary to consider the choice of parameters of synchronous generators as well as the methods of calculation according to the specified objectives.

Acknowledgment

This work was supported by the Ministry of Education, Science, Research and Sport of the Slovak Republic and the Slovak Academy of Sciences under the contract No. VEGA 1/0372/18.

References

[1] A. Margitová, Modeling of Transient Behavior of Synchronous Machines (Modelovanie prechodných javov synchrónnych strojov). Technical university of Košice, 2017, diploma thesis.
[2] M. Kanálik, A. Margitová, and M. Kolcun, "Modeling of Synchronous Machines Including Voltage Regulation," 14th International Scientific Conference FORECASTING IN ELECTRIC POWER ENGINEERING PE 2018, in press.
[3] M. Kolcun et al., Power System Analysis (Analýza elektrizačnej sústavy). Technical university of Košice, 2005.
[4] L. Reiss, Electrical Power Engineering Theory (Teoretická elektroenergetika II). Bratislava: Alfa, 1979.
[5] M. Kanálik, " Influence of Synchronous Generator Parameters on Arc burning during the short-circuit switching off (Vplyv parametrov synchrónneho generátora na dobu horenia oblúka pri vypínaní skratu)," in Elektroenergetika, vol. 7, 2014, pp. 23-27.
[6] M. Kanálik and A. Margitová, "Three-Phase Short-Circuit Currents at the Terminals of a Synchronous Generator Determined According to the Standard STN EN 60909-0 and Using the Mathematical Model of the Synchronous Generator (Trojfázové skratové prúdy na svorkách synchrónneho generátora určené podľa normy STN EN 60909-0 a pomocou matematického modelu synchrónneho generátora)," in Elektroenergetika, vol. 11, 2018, pp. 35-44.
[7] P. Kundur, Power System Stability and Control. McGraw-Hill, 1994.
[8] R. H. Park, "Two-Reaction Theory of Synchronous Machines – Generalized Method of Analysis – Part I," in AIEE Trans., vol. 48, 1929, pp. 716-727.
[9] A. Barakat, S. Tnani, G. Champenois, and E. Mouni, "Analysis of Synchronous Machine Modeling for Simulation and Industrial Applications," in Elsevier, vol. 18, 2010, pp. 1382-1396.
[10] J. C. Das, Power System Analysis. Marcel Dekker, 2002.
[11] A. Margitová, (Výskum dynamických javov v elektrizačnej sústave Slovenskej republiky). Technical university of Košice, 2019, dissertation thesis essay.
[12] P. M. Anderson and A. A. Fouad, Power System Control and Stability. The Iowa State University Press, 1977.

XLPE Ageing Under Non-Standard Voltage Stresses

Ondrej Sefl
Department of Electrical Power Engineering
Faculty of Electrical Engineering, CTU in Prague
Prague, Czech Republic
seflondr @fel.cvut.cz

Radek Prochazka
Department of Electrical Power Engineering
Faculty of Electrical Engineering, CTU in Prague
Prague, Czech Republic
xprocha3@fel.cvut.cz

Abstract — **This paper is focused on the effect of high-frequency oscillations on the ageing rate of XLPE insulation material. Thin XLPE sheets (~200 microns) were aged under two distinct electric stress regimes – pure 50-Hz sinusoidal and 50-Hz sinusoidal with superposed high-frequency transients. These stresses were applied at several different levels, with each pair at the same RMS value of electric field strength. The gathered data were analysed via Weibull statistics, and a comparison between both regimes was made. Finally, an ageing model was applied to the data, and its results were discussed.**

Keywords: XLPE, electrical ageing, high-frequency oscillations, non-sinusoidal voltage

I. Introduction

Insulation degradation (or ageing) is an unwanted phenomenon that affects practically all electric systems. Degradation gradually worsens insulation properties and ultimately leads to a state of unusability of the system, either via a breakdown or via reaching critically low values of properties. After such a state is reached, the insulation system (in most cases also the afflicted machine) needs to be replaced. The size and cost of an insulation system are proportional primarily to its rated voltage. Therefore, medium and high voltage insulation systems need to be designed in a way that minimizes the rate of their degradation.

Among other factors, the electric field also plays a vital role in the insulation systems ageing. The influence of electric field on the ageing rate of materials has been being researched for several decades, yet no universal theory that would explain all the phenomena in different materials has been presented. Numerous scientists presented their theories of polymer electric field ageing, along with ageing models that estimate the average time to failure of polymeric insulation for a given set of conditions. Some models have already been disproved, such as the threshold model based solely on partial discharge activity, some are merely empirical and provide no insight into the mechanism of ageing, such as inverse power law model, while some are based on solid physical background and utilize real parameters of the material, such as Crine's ageing model [1].

Lately, a significant number of power electronic systems, such as photovoltaics, off-shore wind farms, or frequency-controlled drives, has been incorporated into the electrical grid. These systems create higher-frequency oscillations that pollute not only the particular system but its neighbours as well. Some articles focused on this topic have already been published. The consensus is that the frequency of the higher-frequency oscillation pulses is in the range of units of kilohertz [2]. These pulses should, in theory, increase the stress on insulation systems, and hence increase their ageing rate. Individual research groups have already delved into this problematic, but their focus was primarily aimed at lower order harmonic oscillations (up to 40th). An ageing model that considers the influence of these lower order harmonics has also been published by Giovanni Mazzanti [3], but, as shown in [4], it is improper for voltages distorted by frequencies over a kilohertz and will require some editing before becoming suitable for specific cases, such as this one.

II. Theoretical Background

Certain polymers, such as PP, PVC, PE, and XLPE, are excellent insulation materials in insulation systems due to their high dielectric strength, low permittivity and dissipation factor, and also due to the low cost and ease of their manufacturing. They are particularly useful in cable insulation systems, where their manufacturing process, extrusion, allows for simple coupling with the conductive core.

Polymers have a unique semicrystalline structure, which among other things determines their behaviour when subjected to electric fields. Individual polymer chains are spread out in the whole bulk – in some places, they are organized tightly next to each other in periodic structures called crystallites (crystal phase), whereas in other places they occur in an unordered state with no prevalent orientation (amorphous phase). In the amorphous phase, polymer chains are scarcer, and therefore more free volume exists between the individual chains. In these free volumes (also called submicrovoids), the omnipresent free electrons are accelerated unobstructedly by the electric field, possibly gaining enough energy to disrupt intermolecular (Van der Waals) bonds. After such a scission occurs, the afflicted molecules start to move to each other, eventually rearranging and enlarging the previous free volume. The new, larger free volume enables free electrons to be accelerated even more. The process repeats, ultimately causing some electrons to gain enough energy to disrupt intramolecular bonds, causing a practically instantaneous breakdown of the material [1].

Based on these theoretical assumptions, Jean-Pierre Crine et al. published an aging mode that describes a potential barrier between unaged local state (bound molecules) and aged local state (disrupted intermolecular bond). The height of the barrier is equal to the Gibbs free energy required for the (disruption) reaction to occur. When no electric field is applied, both unaged and aged state are equiprobable, meaning that, from a macroscopic point of view, for each disrupted bond another one is created and vice versa. The balance is thus kept, and the material does not age electrically. However, when an electric field is applied, free electrons are being accelerated inside free volume of the material, and the scission probability is increased at the expense of reverse reaction (bonding). In terms of the potential barrier, energy proportional to the mean free path of electrons and the electric field strength is added in favour of the unaged state, while also lowering the energy of the aged state. In other words, it is much simpler for the reaction to occur from the unaged to the aged state than vice versa. Hence, the reaction probabilities are not equal anymore, and the resulting disbalance sets off the degradation process. The individual probabilities can be derived from the Arrhenius equation by including the energy change caused by the electric field. The ageing model then estimated the expected time to breakdown as the inverse of the difference in probabilities [1]:

$$t \approx \frac{h}{2kT} exp\left(\frac{\Delta G_0}{kT}\right) csch\left(\frac{e\lambda E}{kT}\right), \qquad (1)$$

where t is the expected time to breakdown, h and k are the Planck and Boltzmann constants, respectively, T is the thermodynamic temperature, ΔG_0 is the initial Gibbs energy required for the reaction, e is the elemental charge, λ is the submicrocavity size, and E is the applied electric field strength. Both λ and ΔG_0 can be determined from the slope and intercept of a fitted line in $E = f(t)$ linear-log plot for strong fields. The model also works with a so-called critical field level E_c, below which is the ageing process significantly slowed down. In plots, this level is characterized as a transition between a line with slope given by (1) and a horizontal line.

III. EXPERIMENTAL SETUP

A. Samples

Approximately 200-µm thick films were extruded from PE granules and underwent crosslinking in peroxide vapours. These films were rather large (15 by 15 cm) and their thickness was not constant. Thus, they were cut into smaller pieces (2.7 by 2.7 cm) and sorted according to their thickness. For each measurement, all samples had the same thickness to minimize the variation of the applied electric field strength.

B. Electrode System

An electrode system capable of ageing up to ten samples was designed for the experiment. The system consisted of a grounded copper plate (15 by 7 cm) on which the samples would be placed and fixed. On top of them, brass electrodes with additional weights were placed. The weights ensured a close contact between the electrode, sample, and copper plate. Each electrode had a specific position in a poly(methyl methacrylate) (PMMA) matrix and was connected to a voltage source by individual conduit. The copper plate, samples, and electrodes were placed in a PMMA vessel, as they had to be immersed in insulating natural oil. This vessel was placed in an even larger airtight box to prevent possible ingress of air vapours. The space around the vessel was filled with silica gel to minimize the humidity of the inside air.

The main idea of this design was to allow for parallel ageing of several samples. Whenever one of them broke down, it would be possible to remove the electrode above the afflicted sample and carry on with the measurement. This idea determined the spacing between every two electrodes (and the size of samples), as it was necessary to suppress any surface discharges. As per the previous experience with the employed insulating oil, the minimum pitch was chosen to be 2.6 cm. Two similar electrode systems were constructed, one for the use in standard sinusoidal ageing regime, and the other one for the non-standard, distorted ageing regime.

C. Measurement & Recording

A calibrated voltage probe with broad frequency range connected to an oscilloscope was employed for voltage setup. The voltage was calculated from the required electric field strength and thickness of the tested set of samples. Voltage would remain constant throughout the whole measurement; therefore, it was sufficient to determine and adjust the settings of the waveform generator and amplifier only at the beginning. After the setup had been finished, the probe was connected to a laboratory multimeter, from which the value of the AC voltage was read by a PC via a GPIB link. The voltage was read every ten seconds, and its value was recorded in a spreadsheet. Whenever a sample had broken down, a sudden voltage drop was recorded due to the disconnection of the voltage source. Hence, the time to the breakdown of a sample could be easily determined from the spreadsheet. Afterward, the electrode above the afflicted sample was removed, and the measurement was continued.

Block scheme of the whole circuit is shown in Fig. 1. A circuit breaker that deenergizes the amplifier is also included. The breaker is activated whenever a large enough current flows through a sensing resistor in the "Current Sensing" block.

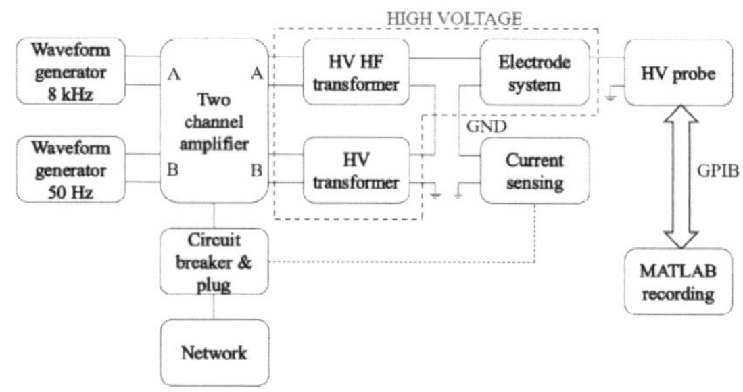

Figure 1. Block scheme of the circuit employed in the experiment

D. Parameters

A short-term breakdown test of the samples was carried out before the start of the experiment. Its results were evaluated via Weibull statistics and the fitted Weibull distribution along with plotted 90% confidence intervals can be seen in Fig. 2. The dielectric strength of the employed samples was determined as the scale parameter of the fitted Weibull distribution, α = 123.6 kV/mm. The initial voltage level for ageing tests was set to correspond with the applied electric field strength of 70 kV/mm (RMS value). Per previous experience, this would cause the samples to break down due to the accelerated ageing in a manner of weeks.

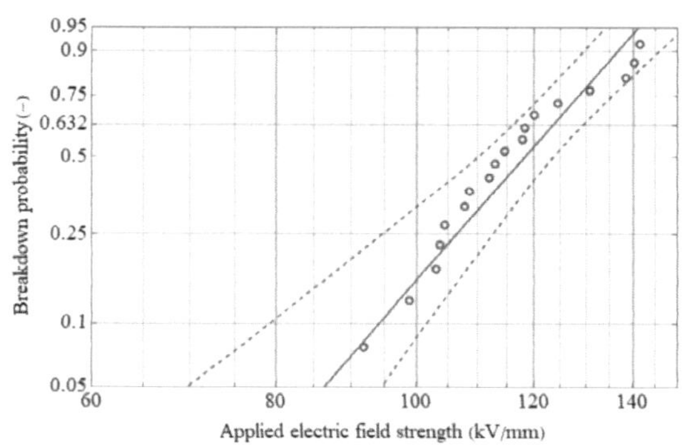

Figure 2. Short-term breakdown test results

The setup of voltage for sinusoidal, standard ageing was simple, requiring only the adjustment of the RMS value of 50 Hz waveform to the previously calculated value. However, in case of the distorted, non-sinusoidal ageing, several additional parameters had to be set specifically, so the resulting waveform would represent voltage distortions found in real systems. These parameters were fundamental frequency, repetition frequency, peak-to-peak voltage of the superposed waveform, and the number of cycles. Respectively, they were chosen as 8 kHz, 700 Hz, 10 % of the 50 Hz waveform's peak-to-peak voltage, and one cycle. The magnitude of the superposed voltage was set in unison with the base voltage so that the overall RMS value would be the same as in the case of standard ageing. The utilized waveform for non-standard ageing and a detail view of the superposed voltage are shown in Fig. 3 and 4, respectively.

Figure 3. The voltage waveform of the non-standard ageing regime

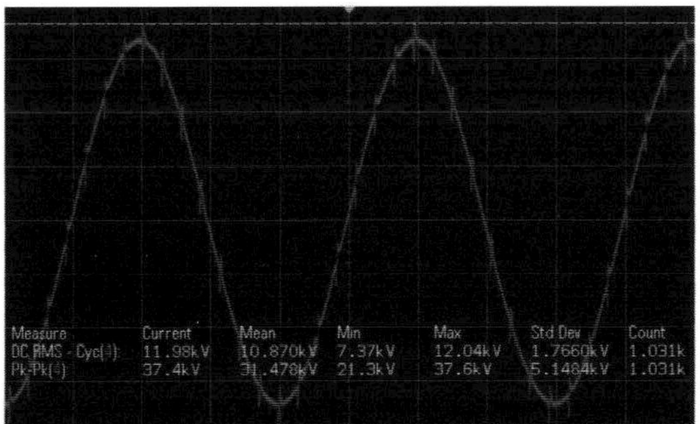

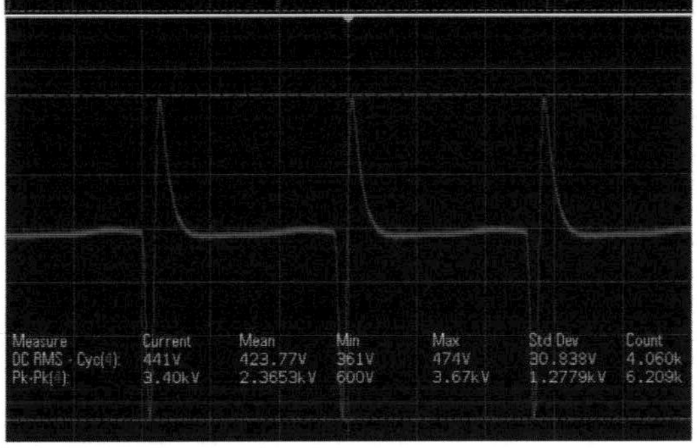

Figure 4. Detail view of the higher-frequency oscillations

IV. RESULTS

Due to their extended duration, the standard 50 Hz ageing tests were conducted only at two different electric field strengths: 70 kV/mm and 65 kV/mm. Their results are shown in Fig. 5. Since the non-standard ageing regime degraded the samples faster (as was expected), it was possible to conduct another ageing test at an electric field strength of 60 kV/mm, in addition to the previously mentioned ones. The results of all three tests can be seen in Fig. 6.

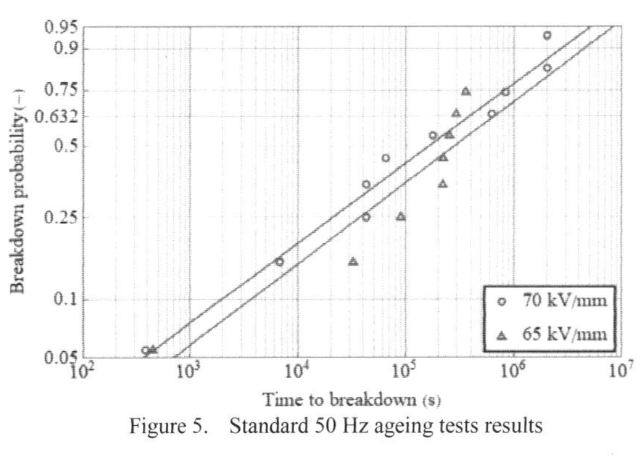

Figure 5. Standard 50 Hz ageing tests results

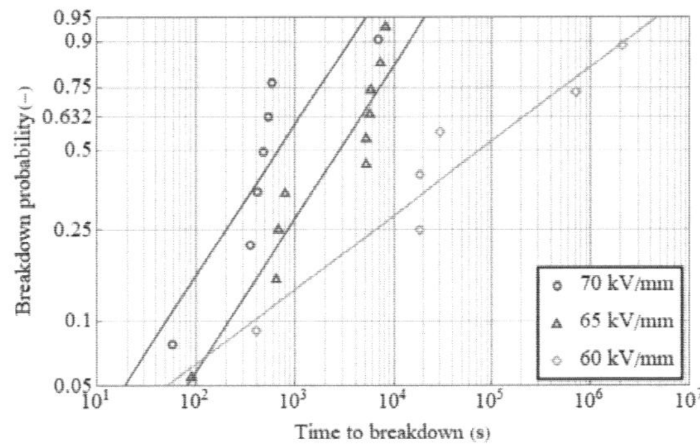

Figure 6. Non-standard ageing tests results

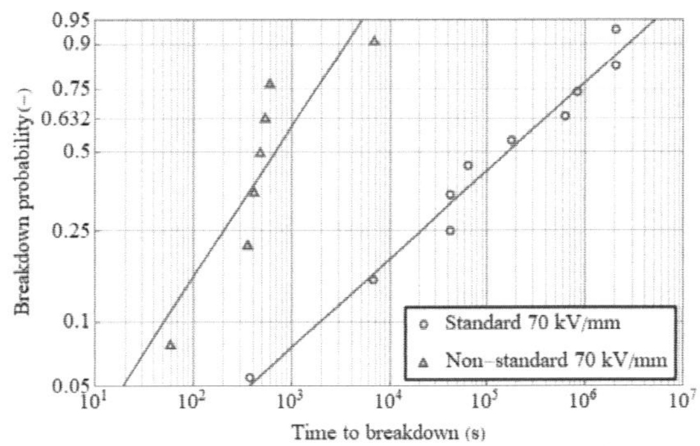

Figure 7. Ageing tests results for electric field strength 70 kV/mm

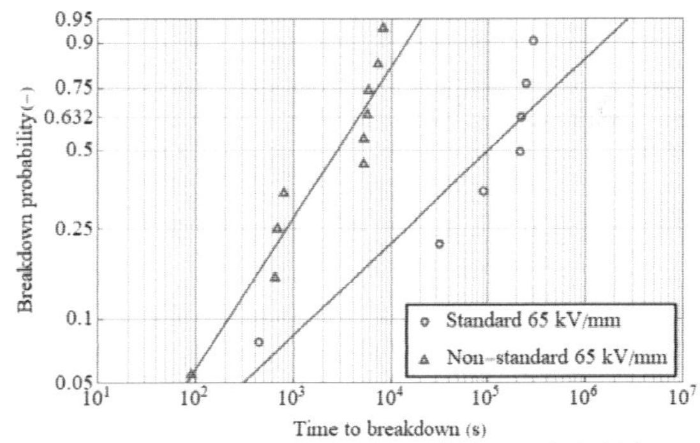

Figure 8. Ageing tests results for electric field strength 65 kV/mm

To better illustrate the impact of ageing regime on degradation rate, the results of tests for the same electric field strengths are shown in Fig. 7 and 8 (70 kV/mm and 65 kV/mm, respectively).

For the sake of clarity, parameters of each fitted distribution are listed in Table I (ST - standard ageing regime, NS - non-standard ageing regime). Apart from the scale and shape parameters of fitted Weibull distributions, correlation coefficients are also shown. These coefficients determine whether the calculated parameters fit the results well; in other words, sufficient correlation implies that the employed method is correct and that the individual samples are interchangeable without affecting the results. A check mark is shown next to it if the correlation is higher than the required value by IEC 62539:2007 standard.

TABLE I. ESTIMATED PARAMETERS OF INDIVIDUAL TESTS

Ageing type	α	β	Correlation	Good fit?
ST 70 kV/mm	389 000 s	0.43	0.988	✓
ST 65 kV/mm	657 000 s	0.43	0.926	✓
NS 70 kV/mm	1 130 s	0.73	0.881	×
NS 65 kV/mm	4 600 s	0.74	0.948	✓
NS 60 kV/mm	212 000 s	0.36	0.959	✓
Short-term	124 kV/mm	8.28	0.962	✓

V. APPLICATION OF CRINE'S MODEL

Fig. 9 shows the data from Table I fitted with Crine's ageing model. An additional point determined from the short-term breakdown tests was given for the standard ageing regime. In this case, the time to breakdown was estimated from voltage regulation speed as 10-2 s.

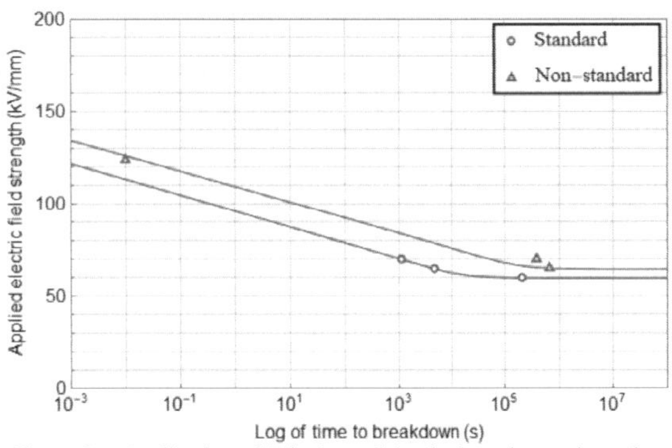

Figure 9. Application of Crine's model to both ageing regimes data

Estimated parameters of fits from Fig. 9 can be found in Table II. These parameters include critical field strength E_c at which the ageing slows down significantly; this phenomenon can be seen in Fig. 9 as a smooth transition of each line to the horizontal position. Barrier height parameter (Gibbs energy) is listed in Joules to enable a quick comparison with results in related literature.

VI. DISCUSSION

As can be seen from the results, the fundamental assumption that the superposed higher-frequency oscillations degrade insulation faster has been verified for XLPE film samples.

TABLE II. CRINE'S MODEL ESTIMATED PARAMETERS

Ageing type	ΔG_0 (10^{-19} J)	λ_{max}(nm)	E_c (kV/mm)
Standard	1.60	6.60	64.6
Non-standard	1.50	6.43	59.9

The differences between both ageing regimes for the same applied electric field strengths are substantial - for both 70 and 65 kV/mm they are of two orders of magnitude.

Application of Crine's ageing model also yielded particularly significant results. The barrier heights and maximum submicrocavity sizes are almost similar for both types of ageing and are in good agreement with estimations from previous works [1]. Ideally, both parameters would be the same since they represent material constants. However, this would require the implementation of a waveform shape factor into the model. Taking into account the number of results gathered so far, such an addition is not yet feasible. In this research, the effect of increased ageing rate is best illustrated on the value of a critical electric field strength Ec. The difference between these critical fields is not marginal (~5 kV/mm) and, as can be seen in Fig. 9, clearly determines the shift in ageing rates for longer degradation times.

It is worth noting that in the case of standard 65 kV/mm ageing, two samples have not broken down in reasonable time and hence are not shown. Nevertheless, the breakdown probability was still evaluated as if all ten samples were present and therefore the two samples still had at least an indirect effect on the results.Non-standard 70 kV/mm ageing results did not meet the correlation criterion given by the respective standard. Presumably, this has been caused by the short breakdown times, between which the effect of repetitive voltage ramp when reenergizing the circuit came into play. The applied electric field strength level might thus be the turning point between the feasibility of a parallel and series ageing method for the given waveform and samples.

Next, the experimental results were evaluated via Weibull statistics. A significant enhancing effect was evident for the non-standard stress. The viability of the used method and homogeneity of the samples was verified via correlation criterion from IEC 62539:2007 standard.

VII. CONCLUSION

XLPE film samples underwent long-term parallel breakdown tests that were designed to determine the impact of superposed higher-frequency oscillations on samples' ageing rate. A sinusoidal 50 Hz waveform (standard) and a 50 Hz waveform with superposed 8 kHz oscillations (non-standard) of the same RMS value were employed for each type of breakdown test. For the standard waveform, two stress levels were examined, and a dielectric strength test was conducted. In the case of non-standard waveform, results of tests from three different stress levels were gathered. Finally, Crine's model was applied to the processed data and the samples' ageing parameters were estimated for each type of ageing regime. The estimated parameters were much alike in both cases and showed good agreement with those in other publications for the same material. The decrease of critical electric field strength was evident in case of the nonstandard ageing and further solidified the conclusions from the statistical evaluation.

ACKNOWLEDGMENT

This work was supported by the Grant Agency of the Czech Technical University in Prague, grant No. SGS18/133/OHK3/2T/13.

REFERENCES

[1] J. L. Parpal, J. P. Crine and Chinh Dang, "Electrical aging of extruded dielectric cables. A physical model," in IEEE Transactions on Dielectrics and Electrical Insulation, vol. 4, no. 2, pp. 197-209, April 1997.

[2] F. Endrejat and P. Pillay, "Resonance Overvoltages in Medium-Voltage Multilevel Drive Systems," in IEEE Transactions on Industry Applications, vol. 45, no. 4, pp. 1199-1209, July-aug. 2009.

[3] G. Mazzanti, "Reliability evaluation of insulation subjected to harmonic voltages within the limits set by international standards," in IEEE Transactions on Dielectrics and Electrical Insulation, vol. 21, no. 5, pp. 2037-2046, Oct. 2014.

[4] O. Sefl, M. Knenicky and R. Prochazka, "Life-Time of Oil Filled Insulation Paper under Nonstandard Voltage Stresses," 2018 International Conference on Diagnostics in Electrical Engineering (Diagnostika), Pilsen, 2018, pp. 1-4

The Impact of the Smart Grid on the Distribution

Michal Ivančák, Juraj Kurimský
Department of Electric Power Engineering
Technical University of Košice, FEI
Košice, Slovak Republic
michal.ivancak@tuke.sk, juraj.kurimsky@tuke.sk

Abstract — **Intelligent networks or smart grid are the most frequently mentioned theme in this time. We can say that research in this area is expected. First part of this paper describes modelling electrical network through software like Neplan that replaces mathematical computation, facilitates and speeds up this process. The data are used from real distribution system from the average village in Slovakia. In this paper is simulated effect of PV panels and e-car charger on the small part of the distribution system.**

Keywords: smart grid, modelling of network, e-car charging, RES

I. Introduction

Smart grids are a topic that is often referred to in these years, not only in the scientific circles but also in common users of electricity. The constant development of new technologies as well as the increased informatization brings many new questions about how electricity distribution will work in the future. Significant facts that are already visible today are higher electricity consumption per capita, more and more electronics within the household as well as the arrival and increasing number of electric cars not only in the EU but also in Slovakia.

The impact of smart grids can be divided into 3 parts - the impact on the transmission system, the impact on the distribution system and, last but not least, the impact on the customer. As part of the Smart Grid's impact on the transmission system, the effects on system and support service volumes, voltage stability, short-circuit ratios, static and dynamic stability are being analyzed. In the last part are analyzed mainly impacts on electric energy quality for end customers.

Several methods can be used to achieve a fairly accurate analysis. One of them is the use of software programs where, thanks to the adaptive graphical interface and data from real measurements, we can model an existing power network and test its performance in critical conditions on such a network. At the same time, such software has a number of mathematical models to calculate short-circuit conditions, network optimization, and load calculations.

Another option is to test conditions in a smart grid lab that no longer only works with mathematical models. Depending on the size and capabilities of the laboratory, it is then possible to do such testing and to determine the behavior of the network model again under different conditions. Such a form of smart grid designation is also called microgrid.

II. Smart Grid Systems

A. The main features of smart grid

One of the main concepts closely related to SG is Distributed Production (DG). Distributed production is or even decentralized production is the production of electrical energy from a large number of small energy sources. Most of the electricity is currently produced in large central power plants, whether coal-fired power plants, nuclear power plants, or hydroelectric power plants. Consequently, the energy produced in such power plants is transmitted over long-distance power lines to the point of consumption. Most of these power plants are built on remote locations due to the economic, security or geological conditions. DG is significantly reducing the losses that arise from the transmission and distribution of electricity. The dispersion of these sources, low maintenance, small dirt areas and high efficiency are the main features of DG. By building such microgrids, ultimately, it can reduce emissions, reduce public threats. The most commonly used sources of this type are photovoltaic panels, wind turbines and cogeneration units **Chyba! Nenašiel sa žiaden zdroj odkazov.**, [2].

The most commonly used sources are photovoltaic panels. Due to their size they are placed on the roofs of family houses but also on the roof of industrial buildings. Building on fields is no longer supported by legislation. Polycrystalline panels are most commonly used in various sizes. The potential of a photovoltaic power plant is most often connected to a grid or as an island operation, sometimes with battery accumulation. An inverter is also an integral part of such a system. Such a system for family houses envisages a return on investment of about 6-8 years [5],[6].

Wind turbines in the Slovak Republic do not have very suitable conditions. For their use an average wind speed above 5 m/s is required. Therefore, their use, but lesser performance, is usually combined with PV panels.

Cogeneration units – i.e. combined production of electricity and heat by burning natural gas are mainly used in industrial buildings or in larger residential complexes. For renewable energy sources, RES predicts wind and sun prediction. Therefore, the MG is looking for suitable combinations for reliable power supply [7].

B. Legislative changes in the energy sector

As in the last European Directive 2014/94/EU of the European Parliament and of the Council of 22 October 2014 on the deployment of alternative fuels infrastructure, Directive 2009/28/EC of the European Parliament and of the Council of 23 April 2009 on the promotion of the use of energy from renewable sources and Directive 2010/31/EU of the European

Parliament and of the Council of 19 May 2010 on the energy performance of buildings. There the idea is suggested, that new buildings need to take into the account:

- Cogeneration
- Heat pumps
- Block or district heating or cooling, particularly where it is partially or entirely supplied from renewable energy sources
- Decentralised energy supply systems on energy from RES

Article 9 of the Directive is focused on Nearly zero-energy buildings committed to build only houses with nearly zero energy by 31. December 2020. To fulfil this criterion is needed to install to all buildings photovoltaic power plant, and there are suggestions to build on all new houses electro mobile charger [3]**Chyba! Nenašiel sa žiaden zdroj odkazov.**.

III. Modeling in simlation software

Electricity modeling is a quite often used tool for theoretical verification as a practice preparation. Optionally, simulations to test critical network situations whose parameters are based on measurements on existing devices.

In the following lines there is a simulated behavior of the electric network, in the village in eastern Slovakia. It is a part of this village, namely a single street that is powered from one transformer branch. In this model, photovoltaic panels were designed for each house and an electric vehicle charger. The following table shows the number of elements used in the model.

Table 1. Elements used in project

Line	63
Nodes	51
Loads	37
Charging for Electric Vehicles	37
Photovoltaic	37
Transformer	1

A. Loads

The loads represented in the simulations are family houses in the village. One street from this village was tested to verify stability. The values of these loads were used from the data provided directly from the distribution service provider. The values in the table below are in the state before the PV panels are connected.

Table 2. Network parameters

P Loss	Q Loss	P Imp	Q Imp	P Gen	Q Gen	P Load	Q Load
kW	**kVar**	**kW**	**kVar**	**kW**	**kVar**	**kW**	**kVar**
4.03	9.42	139.8	54.05	139.8	54.05	135.8	44.63

B. PV panels

All-Black CS6K-300M monocrystalline panels were used as renewable electricity sources for each sampling. Eleven pieces were used to create an optimal 3.3 kW output. The value of this performance was determined because of the amount of the subsidy for renewable electricity sources in Slovakia. In the software was used a dispersion generator with a preset type photovoltaic, which simulates the production of the PV panel and the inverter simultaneously.

The parameters of the panel used can be seen in table no. 3, under this text.

Table 3. PV panel - parametres

Nominal Max. Power (Pmax)	300 W
Opt. Operating Voltage (Vmp)	32.5 V
Opt. Operating Current (Imp)	9.24 A
Open Circuit Voltage (Voc)	39.7 V
Short Circuit Current (Isc)	9.83 A
Module Efficiency	18.33%
Operating Temperature	-40°C ~ +85°C
Max. System Voltage	1000 V
Module Fire Performance	TYPE 1 (UL 1703) or CLASS C (IEC 61730)
Max. Series Fuse Rating	15 A
Application Classification	Class A
Power Tolerance	0 ~ + 5 W

C. Electric car charger

The model also includes an electric car charger for every household. This is due to the changing European legislation mentioned in Chapter 2. The same 11 kW charger has been designed for each house.

There are some basic charging scenarios/profiles when charging an electric car. All these scenarios are based on people's working hours:

Profile #1: People going to work with a possibility to charge their car at work.

Profile #2: People going to work with no possibility to charge their car at work.

Profile #3: People going to work with a possibility to charge their car at work but with a longer ride

Profile #4: People staying at home.

Profile #5: People working on a night shift.

It is important to note that in the future, when an electric vehicle is connected to the network, it will be considered as a battery that will assist the distribution network in its operation.

IV. Network analysis

In this model, several analyzes have been made to assess the impact of renewable resources on this small system. Based on these analyzes, it is then possible to evaluate the design itself and the network functionality itself.

Three scenarios were created as part of the network analysis.

Scenario 1: Turning PV panels on step by step

Scenario 2: Starting electric cars chargers without PV panels

Scenario 3: Starting e-cars chargers in full PV panel operation.

For these scenarios was made Table 4, 5, 6 and Figure 1. to Figure 6. After every table there is a comparison of losses (active and reactive) according to number of connected devices and also comparison of generated active and reactive power.

In table 4. is scenario 1 where was PV panels turning on step by step. In Figure 1. we can see that losses are going down because of energy that PV panels made. In Figure 2. we can see that the energy supplied to the network decreases as the number of connected PV panels increases. Based on these findings, we can say that by increasing renewable energy sources, the energy intensity of a given part of the village is reduced and the network losses are reduced. Based on these findings, we can say that by increasing renewable energy sources, the energy intensity of a given part of the village is reduced and also the network losses are reduced. However, in the event of a load decrease, there could be conditions where there would be a surplus of energy in the network and the electricity flow would change.

Table 4. Simulated values during scenario 1

PV	P Loss	Q Loss	P Imp	Q Imp	P Gen	Q Gen	P Load	Q Load
(-)	kW	kVar	kW	kVar	kW	kVar	kW	kVar
0	3.25	7.78	124.2	54.1	124.22	54.05	130.87	43.02
2	2.97	7.19	117.3	55.6	117.34	55.63	130.87	43.01
6	2.67	6.34	103.8	59.1	103.84	59.12	130.87	43.01
10	2.48	5.68	90.45	62.8	90.45	62.8	130.87	43.02
14	2.22	5.05	77	66.5	77	66.5	130.87	43.02
18	1.95	4.49	63.52	70.3	63.52	70.28	130.87	43.02
22	1.68	4.03	50.06	74.2	50.06	74.16	130.87	43.02
26	1.51	3.75	36.68	78.2	36.68	78.22	130.87	43.02
32	1.49	3.71	23.47	82.5	23.47	82.52	130.87	43.02
37	1.52	3.8	10.29	87	10.29	86.95	130.87	43.02

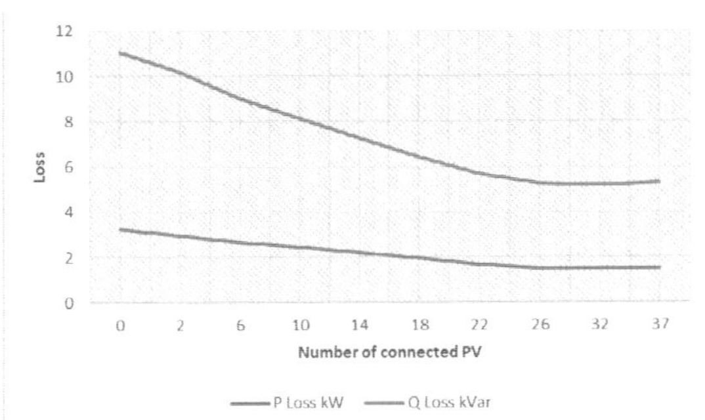

Figure 1. Losses of active power and reactive power in the network when running scenario 1

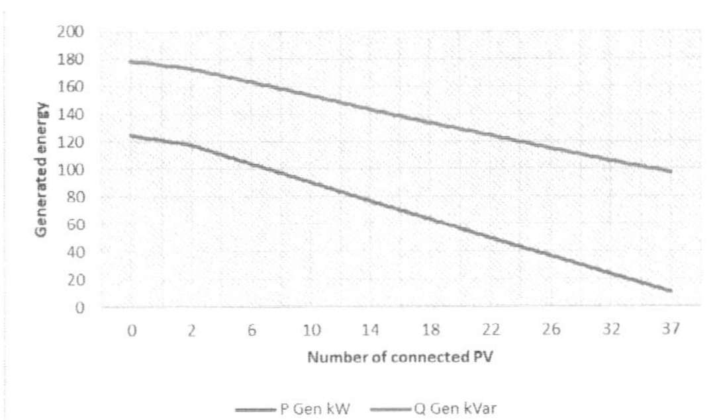

Figure 2. Generated energy in the network when running scenario 1

In the scenario two, chargers for electric vehicles were switched on while the PV panels were turned off. When the 17th charger was switched on, there was an increased load on the network and overloading the lines. In Figure 3. we can see that the losses started to grow sharply. Even after connecting all the devices, the network remained running even in extreme traffic.

In Figure 4. , the energy supplied to the network began to rise more sharply as the network became more congested. The chargers were switched on gradually from the location to the transformer closest to the outermost charger. In the case of the opposite direction of switching, network overload could occur earlier, especially at the edge of the network.

Table 5. Simulated values during scenario 2

	P Loss	Q Loss	P Imp	Q Imp	P Gen	Q Gen	P Load	Q Load
	kW	kVar	kW	kVar	kW	kVar	kW	kVar
0	4.03	9.42	139.81	54.05	140	54.05	135.79	44.63
4	5.08	14.21	184.86	73.3	185	73.3	179.79	59.09
8	7.36	21.41	231.15	94.97	231	94.97	223.79	73.56
12	11.17	31.45	278.95	119.5	279	119.47	267.79	88.02
16	14.1	41.88	325.89	144.4	326	144.36	311.79	102.48
17	15.18	45.07	337.97	151.2	338	151.17	322.79	106.1
21	21.09	60.55	387.88	181.1	388	181.11	366.79	120.57
25	35.49	93.22	468.28	235.5	468	235.48	432.79	142.26
30	47.67	120.43	524.46	277.2	524	277.16	476.79	156.73
34	66.02	161.26	597.8	336.1	598	336.06	531.79	174.81
38	75.57	181.4	629.36	363.4	629	363.44	553.79	182.04

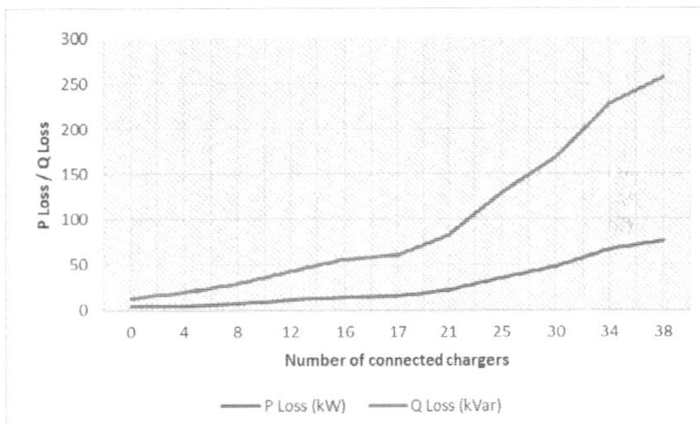

Figure 3. Losses of active power and reactive power in the network when running scenario 2

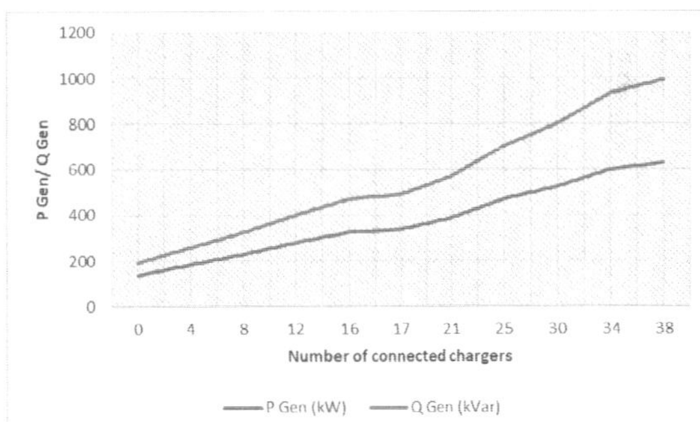

Figure 4. Generated energy in the network when running scenario 2

In scenario 3, the chargers switched on again step by step, but with the PV panels in fully operation. Network overload occurred only when the 19th charger was turned on. When comparing graphs or tables, it is possible to observe reduced performances that have been beneficial to the operation of the network and partial reduction in energy load.

PV panels installed on the home of individual offsets have a positive impact on the operation of the distribution system. First and foremost, they reduce household consumption, while reducing losses as energy is directly consumed at the site of production. The use of PV panels, thanks to subsidies, will be increasingly a solution to save household electricity.

Table 6. Simulated values during scenario 3

	P Loss	Q Loss	P Imp	Q Imp	P Gen	Q Gen	P Load	Q Load
	(kW)	(kVar)	kW	kVar	(kW)	(kVar)	kW	kVar
0	1.62	4	15.31	88.76	15.3	88.76	135.79	44.63
4	2.02	5.81	59.71	105	59.7	105.04	179.79	59.09
8	2.89	9.17	104.58	122.9	105	122.86	223.79	73.56
12	4.42	14.32	150.1	142.5	150	142.47	267.79	88.02
16	6.92	21.68	196.61	164.3	197	164.3	311.79	102.49
19	9.43	28.72	232.11	182.2	232	182.18	344.79	113.33
20	10.46	31.44	244.14	188.5	244	188.52	355.79	116.95
24	15.41	43.77	289.79	216.4	290	216.4	399.79	131.41
28	22.18	59.97	340.57	247.1	341	247.07	443.79	145.88
32	31.72	81.05	394.1	282.6	394	282.61	487.79	160.34
36	43.77	107.04	450.16	323.1	450	323.06	531.79	174.8
38	50.98	122.25	479.37	345.5	479	345.5	553.79	182.04

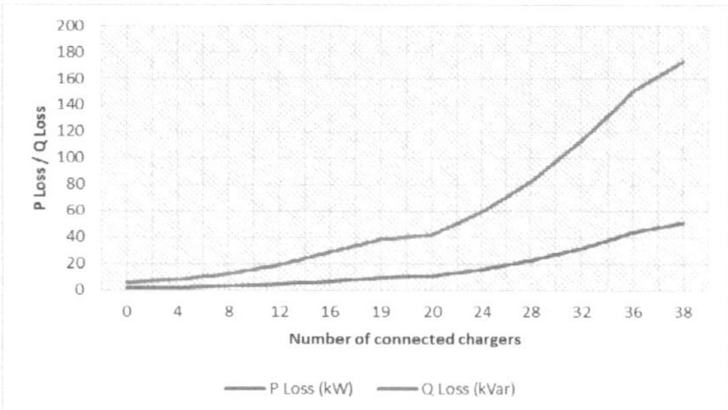

Figure 5. Losses of active power and reactive power in the network when running scenario 2

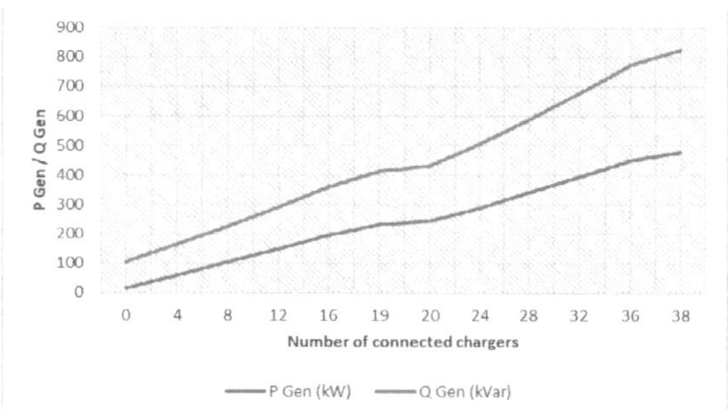

Figure 6. Generated energy in the network when running scenario 3

The average household household reaches a maximum output of 3-5 kW. In the case of electric underfloor heating it can be even more. Photovoltaics on the roofs of these homes would be higher than the current household consumption when simulated during a sunny day. Therefore, it would be ideal to add a battery pack to each PV power plant. By accumulating, the energy produced on a family home could be consumed to a greater extent, thereby reducing household consumption.

V. CONCLUSION

Smart grids will have a significant impact on the operation of distribution systems not only in Slovakia. The effect of PV panels and chargers on e-cars was simulated on this paper. It should be noted that there is little likelihood that all chargers will run simultaneously. If a maximum of 15 out of 38 chargers were running at the same time, we can say that the simulated network would do it if it was a day (PV panels go) or even if it was night (PV panels won't go).

This simulation is only approximate as the software offers only limited possibilities to simulate network behavior. Values are static rather than dynamic. If we wanted to be more precise, it would be necessary to simulate the change of load during the day. It would also be necessary to simulate the course of solar radiation during the day in different conditions and different seasons. E-car charging itself could then be divided into different parts of the day based on charging scenarios.

Finally, smart grids play an important role now. Currently, we can only perceive it through smart metering, but several pilot projects are already running not only abroad but also in our country. Even this paper can be a small grain of sand in the desert of smart grids.

ACKNOWLEDGMENT

This work was supported by the Scientific Grant Agency of the Ministry of Education of Slovak Republic and the Slovak Academy of Sciences by the projects VEGA No. 1/0372/18.

This work was supported by the Slovak Research and Development Agency under the contract No. APVV-17-0372.

REFERENCES

[1] P. Hocko, M. Novak, M. Kolcun, and Z. Conka, "Influence of photovoltaic power plants on the power system to the prediction for year 2020," presented at the 2014 14th International Conference on Environment and Electrical Engineering, EEEIC 2014 - Conference Proceedings, 2014, pp. 116–121..

[2] M. Shamshiri, C. K. Gan, and Chee Wei Tan, "A review of recent development in smart grid and micro-grid laboratories," in 2012 IEEE International Power Engineering and Optimization Conference Melaka, Malaysia, 2012, pp. 367–372. doi:10.1109/PEOCO.2012.6230891

[3] "Directive 2014/94/EU of the European Parliament and of the Council of 22 October 2014 on the deployment of alternative fuels infrastructure." [Online]. Available: https://eur-lex.europa.eu/legal-content/EN/TXT/?uri=celex%3A32014L0094. [Accessed: 11-Jun-2019].

[4] "Directive 2010/31/ of the European Parliament and of the Council of 19 May 2010 on the energy performance of buildings." [Online]. Available: https://eur-lex.europa.eu/legal-content/EN/TXT/HTML/?uri=CELEX:32010L0031&from=SK. [Accessed: 11-Jun-2019].

[5] S. Borlase, Ed., Smart Grids: Infrastructure, Technology, and Solutions, 1 edition. Boca Raton, FL: CRC Press, 2012. 607 p. ISBN-13: 978-1439829059.

[6] J. A. Momoh, Smart Grid: Fundamentals of Design and Analysis, 1 edition. Hoboken, N.J: Wiley-IEEE Press, 2012. 232 p. ISBN-13: 978-0470889398.

[7] F. P. Sioshansi, Smart Grid: Integrating Renewable, Distributed & Efficient Energy. Academic Press, 2011. 510 p. ISBN 978-0-12-386452-4

High-Frequency Resonance Overvoltages in Distribution Network with Large Photovoltaic Power Plant

Martin Knenicky, Radek Prochazka
Faculty of Electrical Engineering
Czech Technical University in Prague
Prague, Czech Republic
knenimar@fel.cvut.cz, xprocha3@fel.cvut.cz

***Abstract*—The wide usage of power electronics and distributed resources in modern power systems bring new challenges for power quality and stability. Switching of power electronic components produces a wide spectrum of voltage and current interferences which are superimposed on power frequency. High-frequency components are usually monitored and filtered up to 40th harmonic under many technical standards. However significant harmonics and impulses with fundamental frequencies up to tens of kilohertz can arise in power systems as a consequence of mutual interaction between passive parameters (resistance, inductance, and capacitance) of the distribution network and switching of power electronic devices. These components on voltage deal repetitive significant overvoltages which excessively stress insulation systems of medium and high voltage devices. The theoretical study presented here is based on numerical simulation to prove the development and propagation of the high-frequency harmonic distortions produced by a large photovoltaic power plant. The modeled power plant consists of some smaller power units with filters and own power transformers which operate in parallel into medium voltage distribution level. The medium voltage distribution system contains self-consumption, cable and overhead power lines and connection to the high voltage transmission system via power transformer. All modeled components are substituted by extended equivalent circuits respecting the propagation of high-frequency components of currents and voltages. Sensitivity analyses of some parameters of the distribution network to the formation of high-frequency distortions were carried out and discussed.**

Keywords—Photovoltaic power plant, high-frequency modeling, power system simulation, distribution network, medium voltage, harmonic analysis, high-frequency overvoltages

I. INTRODUCTION

Originally, electric power systems were designed for the central generation of energy with large power sources and subsequent transmission and distribution to local customers. Nowadays many distributed resources are installed directly into distribution networks, and in many cases, their generated power is larger than consumed power in the same part of the electrical system. These mostly renewable resources (photovoltaic and wind power plants) and components for power control (e.g., Flexible AC Transmission Systems) are based on solid state devices. The wide usage of power electronics in modern distribution networks brings new challenges for power quality and stability.

Despite the installations of high-frequency filters, switching of solid state devices produces harmonic distortions of voltages and currents in distribution systems. Most standards, like EN 50160 or IEC 61000-3 series, are focused only on distortions up to 40th order of the fundamental harmonic. However, higher harmonics with significant amplitudes can arise in power systems, and they are not usually monitored. Some studies introduced in the last years are focused on simulation and confirmation of this effect for the medium voltage systems [1]-[4]. It was found that some combinations of the capacity of medium voltage cables and inductance of power transformer windings can create significant resonances of the impedance of medium voltage power system. These resonances are close to emitted harmonics which come from the switching of power electronic devices and resonance distortions of power voltage and current can occur. Moreover, individual distortions caused by different devices usually interact between themselves in real complex large systems. These phenomena can lead to the formation of significant high-frequency resonance overvoltages which cause faster degradation of insulation systems and thereby increasing the risk of failure [3], [5].

This study focuses on the numerical model of part of the distribution system containing the large photovoltaic power plant. The proposed model respects the high-frequency behavior of individual power components. The aim of numerical simulation is to confirm the presence of high-frequency distortions in the low voltage system and their transmission to the medium voltage system via a power transformer.

Several recently published papers deal with the origin and propagation of high-frequency distortions in photovoltaic power plants [6]-[8]. It was shown that photovoltaic power plants produce significant harmonics in the frequency range of 3 – 18 kHz. These distortions are usually caused by interactions between switching of

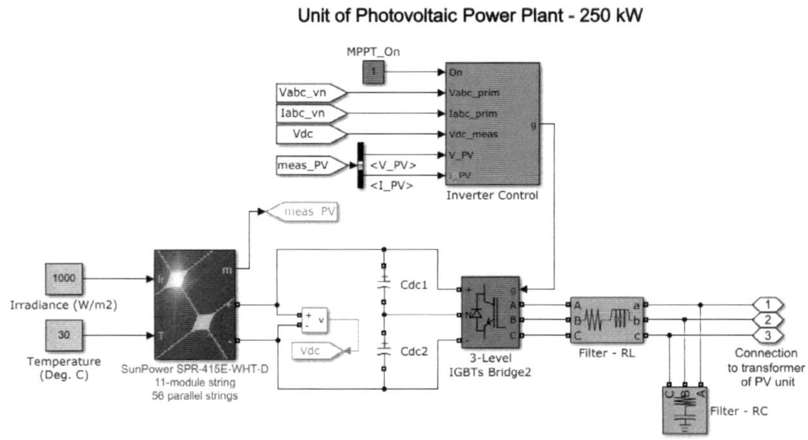

Fig. 1. The model of the unit 250 kW of the photovoltaic power plant used in the simulation of resonance overvoltages in a distribution network

voltage source converters and parameters of power transformer windings, filters, and shunt capacitors. The impedance of the upstream network also has an influence on the resulting disturbances.

From the previous findings, it is evident that the occurrence of high-frequency distortions in distribution network strongly depends on the structure of the electrical system and many other factors, such as switching frequency of power electronic devices, operating conditions, weather conditions, etc. Therefore, the calculation or prediction of this phenomenon is very complicated and specific for each electrical power system.

II. Design of Model of a Distribution Network with a Large Photovoltaic Power Plant

The whole model was constructed in Matlab Simulink software. A photovoltaic power plant consisting of four 250 kW units (see Fig. 1) is considered as the source of distortions. Units of the photovoltaic power plant were assembled with recommendations stated in appropriate guidelines for power system simulation of photovoltaic systems [9]. Stable values of solar power (1000 W/m^2) and temperature of photovoltaic cells (30 °C) were applied in all simulations. Each unit uses controlled 3-level IGBT converter and sinusoidal filter with an inductance of 300 μH and a capacitive reactive power of 25 kVAr. These values were systematically chosen for sufficient filtering of produced power according to power quality standards (see chapter below). The switching frequency of IGBTs was set to 3 kHz (60th harmonic order).

Each photovoltaic power plant unit is connected to the distribution network by power transformer 250 kVA 400 V/22 kV (D1/Y). The real three-phase transformer was modeled as three single-phase units, whose equivalent circuits respected high-frequency behavior [10], see Fig. 2.

In comparison with the common models of the transformer, the capacitance between windings C_p and capacitance between windings and ground (transformer tank) C_g must be considered. Such a high-frequency transformer model can be used for frequency range up to a few tens of kilohertz. Values of resistances and inductances in equivalent circuit come from the real distribution transformer. Capacitances were estimated according to the procedure published in [11]. The summary of all used parameters is presented in TABLE I.

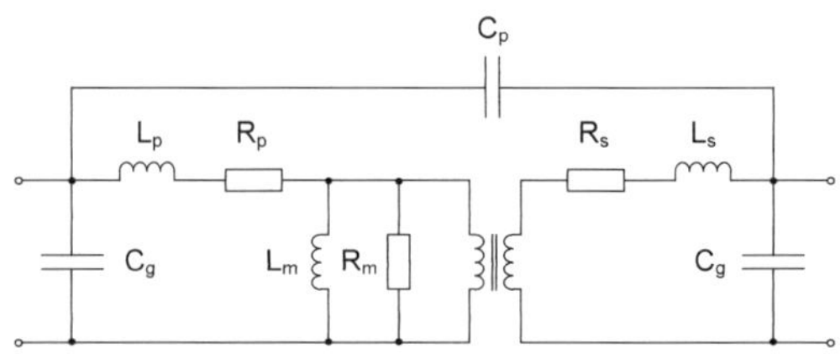

Fig. 2. The equivalent circuit of one-phase transformer which was used in the model of the photovoltaic power plant

TABLE I. Parameters of Used Transformers for Units of the Photovoltaic Power Plant

Transformer 250 kVA 400 V/ 22 kV (D1/Y)		
Primary winding resistance (MV) R_1	46	Ω
Primary winding self-inductance (MV) L_1	270	mH
Secondary winding resistance (LV) R_2	0.015	Ω
Secondary winding self-inductance (LV) L_2	0.090	mH
Mutual inductance L_m	75	H
Core equivalent resistance R_m	1885	kΩ
Ground capacitance C_g	10	pF
Capacitance between windings C_p	200	pF

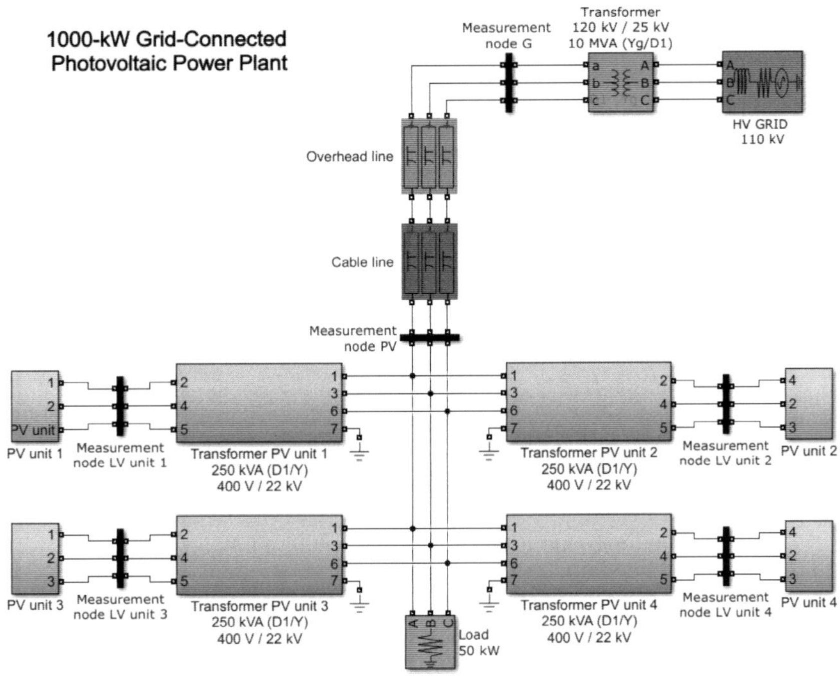

Fig. 3. The part of the distribution network with the 1000 kW photovoltaic power plant used for the simulation of resonance overvoltages

TABLE II. PARAMETERS OF THE FEEDER OF THE PHOTOVOLTAIC POWER PLANT

Cable resistance	0.32	Ω/km
Cable inductance	0.40	mH/km
Cable capacitance	0.31	μF km
Overhead line resistance	0.24	Ω/km
Overhead line inductance	1.1	mH/km
Overhead line capacitance	4.5	nF/km
Transformer 10 MVA 120 kV/25 kV:		
Both winding resistances	0.003	pu
Both winding self-inductances	0.008	pu
Mutual inductance	500	pu
Core equivalent resistance	500	pu
Upstream grid 110 kV: X/R ratio	7	-

Power plant units and 50 kW local consumption are connected to the common node in the distribution network (see Fig. 3). The whole power plant is then connected to 110 kV high voltage grid by short cable line, long overhead line and standard power transformer 10 MVA 120 kV/25 kV (Yg/D1). Parameters of lines and transformer are presented in TABLE II. Both power lines were modeled like π equivalent circuit.

III. HIGH-FREQUENCY VOLTAGE DISTORTIONS IN DISTRIBUTION NETWORK

The Total Harmonic Distortion factor (THD_V) was chosen to compare voltage distortions in the distribution network:

$$THD_{Vn} = \frac{\sqrt{\sum_{h=2}^{n} V_h^2}}{V_1} \cdot 100\ (\%)\,, \qquad (1)$$

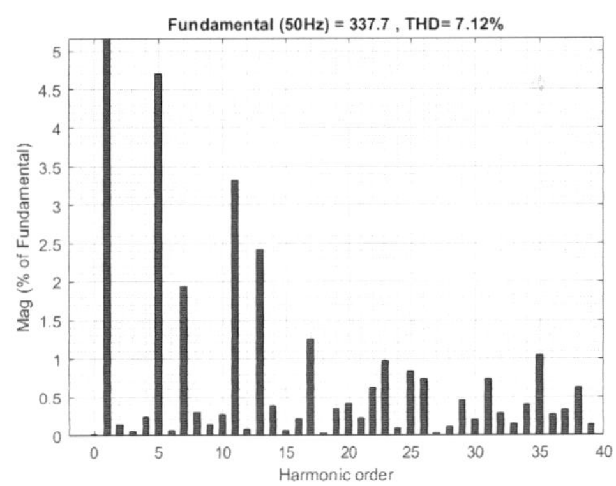

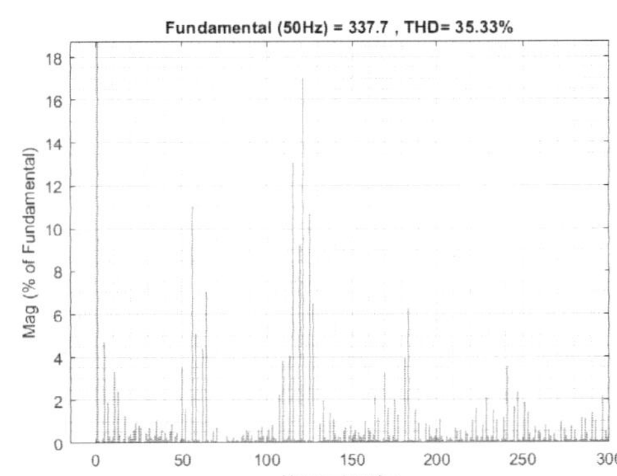

Fig. 4. Harmonic distortions of the voltage in node LV; parameters of the network: 100 m cable line, 30 km overhead line, 100 MVA short-circuit power of 110 kV grid on the top: considered harmonics up to 40th order; down: considered all harmonics

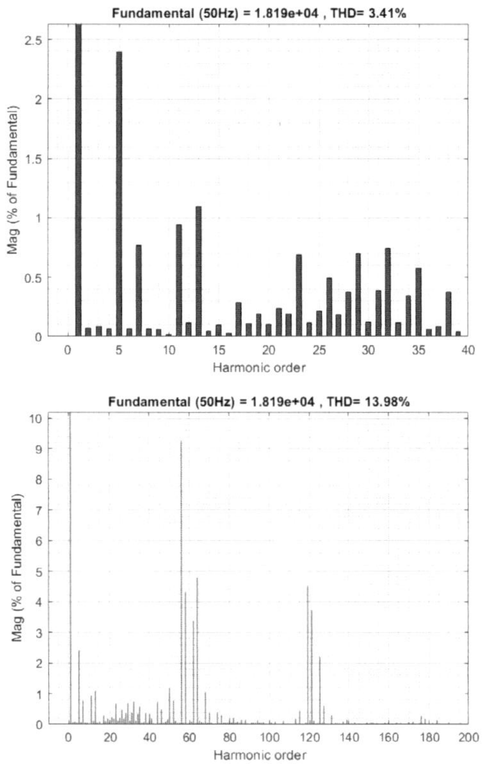

Fig. 5. Harmonic distortions of the voltage in node PV; parameters of the network: 100 m cable line, 30 km overhead line, 100 MVA short-circuit power of 110 kV grid; on the top: considered harmonics up to 40th order; down: considered all harmonics

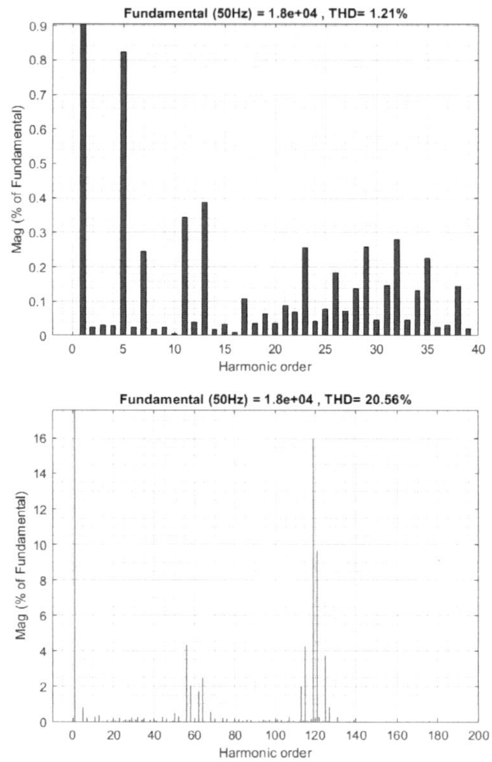

Fig. 6. Harmonic distortions of the voltage in node G; parameters of the network: 100 m cable line, 30 km overhead line, 100 MVA short-circuit power of 110 kV grid; on the top: considered harmonics up to 40th order; down: considered all harmonics

where V_h is the voltage of h-th order of harmonics and V_1 is the voltage of first harmonics (for frequency 50 Hz). According to standards EN 50160 or IEC 61000-3 series, the values of THD_{Vn} for $n = 40$ were determined and compared with the values calculated for all harmonics up to Nyquist frequency.

The performed numerical simulations are focused on influences of the length of power lines and size of short-circuit power of the upstream grid on the occurrence of overvoltage resonances. Calculations of THD from voltage waveforms were carried out in nodes LV, PV, G (see Fig. 3). LV node is located between low voltage level sinusoidal filter and power transformer. PV and G nodes are located on medium voltage level on the side of the photovoltaic power plant and high voltage grid, respectively.

The results show that THD_{V40} in node LV is slightly below 8 % for all simulated cases which are in line with the requirements of standards. Values for individual harmonics are also under the limits given by standards. However, the determined value of THD_{Vall} was around 35 %. This value can be realistic in practice due to the very different quality of installed converters and filters. The example of a content of voltage harmonics in node LV is shown in Fig. 4. Dominant harmonics are around switching frequency their multiples. Magnitudes of these significant harmonics are very high (over 10 % of first harmonic).

These high-order harmonics propagate through the power transformer to medium voltage power system where are attenuated or amplified depending on impedances of the distribution network. The example of such harmonic distortions for nodes PV and G are shown in Fig 5 and Fig. 6, respectively. System parameters of stated examples were set 100 m long cable line, subsequent 30 km long overhead line and 100 MVA short-circuit power of 110 kV grid. Some harmonics may be attenuated, and contrary some may be amplified over the same feeder line as is shown from the figures. In this case, harmonics around 60th are attenuated and around 120th are amplified on the contrarily. The resonance frequency of the distribution system is close to 120th harmonic for these network parameters.

Dependencies of THD on short-circuit power of 110 kV grid and lengths of cable and overhead lines for nodes PV and G are shown in Fig. 7. THD considering harmonics up to 40th harmonics according to the power quality standards is low as the standards required (under 8 %) for almost simulation setting. Only 10 MVA short-circuit power of the 110 kV grid is very low to required attenuation of common harmonics. On the opposite side, THD considering all harmonics is over 8 % limit in many settings of network parameters. Short-circuit powers over 100 MVA of 110 kV grid attenuate voltage distortions under the standard limit in node G. However, if the power lines are long enough, it does not

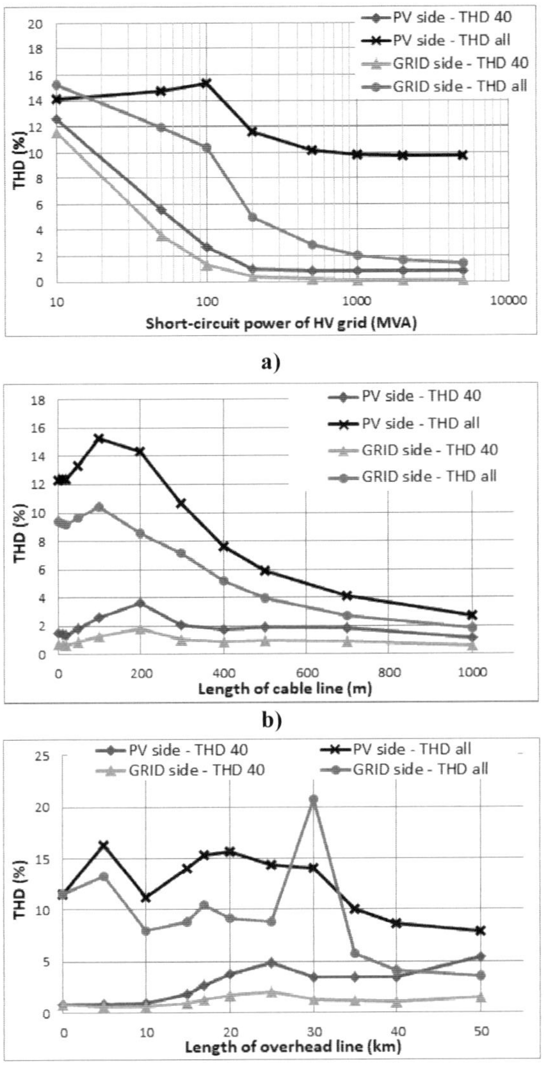

Fig. 7. Dependence of THD on: a) short-circuit power of 110 kV grid (100 m cable line, 30 km overhead line); b) length of the cable line (100 MVA short-circuit power of 110 kV grid, 30 km overhead line); c) length of the overhead line (100 MVA short-circuit power of 110 kV grid, 100 m cable line)

significantly influence THD considering all harmonics in node PV. The worst case is 100 MVA short-circuit power. From dependency THD on a cable line, the worst length is 100 m where THD considering all harmonics is over 10 % in both nodes PV and G. Further, the size of THD decreases with increasing length and capacity of the cable line. Lengths of the overhead line up to 35 km cause THD considering all harmonics above 8 % limit in both nodes PV and G. Longer overhead lines cause harmonic distortions under the limit. The worst case is a length of 30 km overhead lines.

Parameters of the distribution network, causing its resonance frequency close to the frequency of emitted harmonics with high amplitudes from low voltage level, are 100 m cable line, 30 km overhead line and 100 MVA short-circuit power of the upstream 110 kV grid in this model.

IV. Conclusion

Simulation results of the modeled distribution network with the large photovoltaic power plant show possible presence of high-order harmonics with inconsiderable magnitudes in the electrical system with power electronic devices. Actual power quality standards, like EN 50160 or IEC 61000-3 series, do not consider the occurrence of these high-order harmonics. Continuous disturbances may cause problems with medium voltage insulation system, a correct function of sensitive devices installed in electrical networks or faster degradation of overvoltage protections. This effect occurs only in systems with a resonance frequency close to switching frequencies of solid state devices and their multiples. These risk frequencies tend to be in the range of 3 – 15 kHz. However, these high-frequency distortions are difficult to measure on site or predict during the design phase of electrical systems. Nevertheless, the simulations performed in this study can help to detect weak points of distribution network where the insulation system should be inspected, protected or strengthened.

Acknowledgment

This work was supported by the Grant Agency of the Czech Technical University in Prague, grant No. SGS18/133/OHK3/2T/13.

References

[1] K. Temma, F. Ishiguro, N. Toki, I. Iyoda and J. J. Paserba, "Clarification and measurements of high frequency harmonic resonance by a voltage sourced converter," in *IEEE Transactions on Power Delivery*, vol. 20, no. 1, pp. 450-457, Jan. 2005.

[2] K. Lee, D. Carnovale, D. Young, D. Ouellette and J. Zhou, "System Harmonic Interaction Between DC and AC Adjustable Speed Drives and Cost Effective Mitigation," in *IEEE Transactions on Industry Applications*, vol. 52, no. 5, pp. 3939-3948, Sept.-Oct. 2016.

[3] L. Paulsson, B. Ekehov, S. Halen, T. Larsson, L. Palmqvist, A. Edris, D. Kidd, A. J. F. Keri and B. Mehraban, "High-frequency impacts in a converter-based back-to-back tie; the Eagle Pass installation," in *IEEE Transactions on Power Delivery*, vol. 18, no. 4, pp. 1410-1415, Oct. 2003.

[4] S. K. Rönnberg, M. H. J. Bollen, H. Amaris, G. W. Chang, I. Y. H. Gu, Ł. H. Kocewiak, J. Meyer, M. Olofsson, P. F. Ribeiro and J. Desmet, "On waveform distortion in the frequency range of 2 kHz–150 kHz - Review and research challenges," in *Electric Power Systems Research*, vol. 150, pp. 1-10, September 2017.

[5] T. Bengtsson, F. Dijkhuizen, Li Ming, F. Sahlén, L. Liljestrand, D. Bormann, R. Papazyan and M. Dahlgren, "Repetitive fast voltage stresses-causes and effects," in *IEEE Electrical Insulation Magazine*, vol. 25, no. 4, pp. 26-39, July-Aug. 2009.

[6] T. Joshi, G. Heydt and R. Ayyanar, "An Analysis Approach for the Assessment of Pulse-width Modulated Carrier Frequency Spectral Components in Power Distribution Systems with Photovoltaic Resources," in *Electric Power Components and Systems*, vol. 43, no. 18, pp. 2094-2104, September 2015.

[7] L. Jessen and F. W. Fuchs, "Investigation of renewable energy generation and load impact on the grid impedance at different points of connection in public low voltage grids to support grid integration of renewable energies," *2016 18th European*

Conference on Power Electronics and Applications (EPE'16 ECCE Europe), Karlsruhe, 2016, pp. 1-10.

[8] O. S. Nduka and B. C. Pal, "Harmonic Domain Modeling of PV System for the Assessment of Grid Integration Impact," in *IEEE Transactions on Sustainable Energy*, vol. 8, no. 3, pp. 1154-1165, July 2017.

[9] A. Yazdani, A. R. Di Fazio, H. Ghoddami, M. Russo, M. Kazerani, J. Jatskevich, Kai Strunz, S. Leva and J. A. Martinez, "Modeling Guidelines and a Benchmark for Power System Simulation Studies of Three-Phase Single-Stage Photovoltaic Systems," in *IEEE Transactions on Power Delivery*, vol. 26, no. 2, pp. 1247-1264, April 2011.

[10] A. S. Alfuhaid, "Frequency characteristics of single-phase two-winding transformers using distributed-parameter modeling," in *IEEE Transactions on Power Delivery*, vol. 16, no. 4, pp. 637-642, Oct. 2001.

[11] L. Dalessandro, F. da Silveira Cavalcante and J. W. Kolar, "Self-Capacitance of High-Voltage Transformers," in *IEEE Transactions on Power Electronics*, vol. 22, no. 5, pp. 2081-2092, Sept. 2007.

Investigation of State-of-the Art Solar Cell Using Colour Filters

Balázs Hörömpöli*,**
* Óbuda University, Doctoral School for Applied Informatics and Applied Mathematics, Bécsi út 96/B
** Power Systems Department at the Kandó Kálmán Faculty of Electrical Engineering at the Óbuda University Bécsi út 96/B
1034 Budapest, Hungary
horompoli.balazs@kvk.uni-obuda.hu

Dr. Ervin Rácz**
** Power Systems Department at the Kandó Kálmán Faculty of Electrical Engineering at the Óbuda University 1034 Budapest, Hungary
racz.ervin@kvk.uni-obuda.hu

Abstract — ***In our article, we analysed the electric parameters of the DSSC demo cell depending on the irradiation with different wavelength-intensity spectra. The solar cell was made by the Solaronix company in Switzerland. We put the solar cell in a Pasan Flasher professional flash light tester and measured the electrical characteristics of the solar cell – such as electrical current-voltage, electrical power-voltage characteristics –, for various irradiations. We could change the degree of irradiation with the Pasan Flasher control systems, while different wavelength-intensity spectra were generated by color filters. In this way, we analysed different electrical characteristics of a demo cell - exactly one Solaronix PICTO 1010DEMO green solar cell - on different irradiations and spectra. Our experiments were based on an article entitled "Statistical Analysis of Power Measurements made on Mono- and Polycrystalline Solar Cells" which we presented at the 2018 SAMI Conference. In that case we investigated the effect of color filters on a DSSC solar cell, the main question was whether there is some similarity between the effect of filters on different types of solar cells.***

Keywords: Solar cells; DSSC cell; pasan flasherreflektive layer, colour filter

I. Introduction (Heading 1)

One of the most pressing issues of the 21st century is the growing demand for electric energy. Therefore, engineers build, develop and upgrade several power plants operating on different principles. These power plants can be categorized in many ways, such as renewable and non-renewable. Of the non-renewable energy sources, the solar cell is the only one which is capable of converting solar energy directly into electricity, meanwhile not emitting pollutants and other harmful materials.[1],[2],3]

Nowadays, conventional silicon-based semiconductor solar cells are manufactured and used for the production of an industrial amount of electricity. There are three main types of these, which depend on the type have different electrical parameters, efficiency, power etc. performance these. amorphous, polycrystalline, monocrystalline. [1-4]

The so-called third generation Dye Sensitized Solar Cells (DSSC) has been developed during the development of solar cells. It's structure is significantly different from Silicon (SI) based solar cells. The basis for these solar cells is the Gratzel cell which was published in 1991 by Grätzel and O'Regan. Dye Sensitized Solar Cells, directly converting solar energy into electrical energy, but the absorption of photons and the electric energy transmission happens in two different layers.

Our research groups are engaged in the research and development of electrical parameters of these modern 3rd generation experimental solar panels. In our current article, we present various tests and results of a green prototype demo cell (Solaronix PICTO 1010DEMO green solar cell) which sold by a Swiss company the SOLARONIX. [1], [7-8]

II. The Operation of the Dye Sensitized Solar Cells

A conventional Dye Sensitized Solar Cells (DSSC) is structured as shown in Figure 1

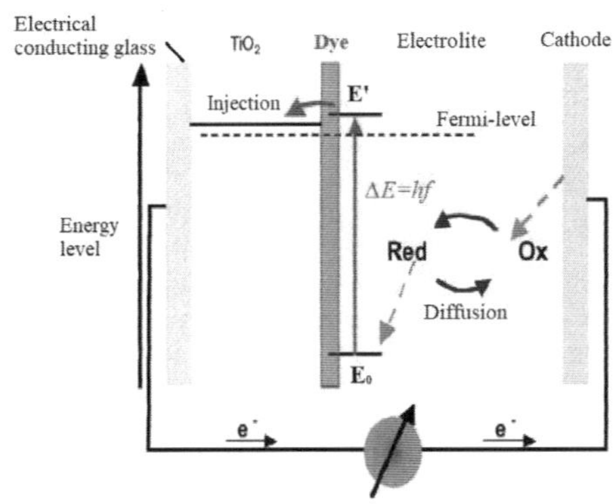

Figure 1. Dye Sensitized Solar Cell elvi felépítése [9]

The light energy is absorbed by the photosensitive dye layer Dye layer), and the resulting electrons, (charge carriers) are transferred to a semiconductor TiO2 layer. The (N-type) TiO^2 layer leads electrons to an electrode. This collecting electrode is one of the contacts on the electrical circuit. The dye layer acts as a potential barrier between the electrolyte layer and the TiO2 layer, and the missing electrons are replaced by the electrolyte solution, which can be - for example - iodine ions (I^-, I_3^-). The electrolyte replaces the electron deficiency through the cathode from the circuit. In most cases, the two electrodes are made of a transparent conductive material. The magnitude of the voltage generated by incident light is the same as the potential difference between the electron solution semiconductor oxide and the Fermi level of electrons injected into the chemical redox potential of the electrolyte solution. [4-7]

III. Used tools, equipment ,The process of measurement

Our experiments are based on the previously mentioned Solaronix (company) made PICTO 1010DEMO green type DSSC solar cell. The solar cell was placed in a professional Pasan Flasher measuring device. The conceptual structure of the equipment and the description of the process of measurement are described in detail in our previous article, which is cited as [8]. We will further describe the measurements based on what is described there. [8], [10-12]

A. The Dye Sensitized Solar Cell

The manufacturer of the PICTO 1010DEMO green solar panel has provided minimum of information for us, which contained in [6] reference. The demo cell is made of a green, photosensitive dye layer. By measuring, we found that despite the fact that the solar cell is transparent, it shows orientation-dependence in terms of illumination. So, we can select a front panel and a backsheet, according to the direction of the light. This is also important to note, because a test tube with a previously tested different color-sensitive dye layer did not show this type of direction dependency. Accordingly this was investigated from one direction of illumination. [8], [10-11]

B. The color filter

For our experiments, we wanted to produce variants of a specific homogeneous light spectrum of a calibrated light source. For this we have used color filters that are provided by the Alternative Energy Resources Center at the University of Óbuda. Unfortunately, we didn't get any manufacturer parameters of the filter, so this was screened the spectral filtering of these filters by placing them in a professional HQI Halogen Sunsimulator as described in [2] our article. Four differently colored filters were used for the experiment: blue, green, orange, and purple. The filters were placed at the same distance from the light source in the Sun simulator and examined using a Spectrophotometer. The results are described in detail in Chapter IV, Section A. [12]

C. The Pasan Flasher professional measurement devices

These were made the experiments in one of the laboratory of Institut für Energie- und Antriebstechnik at the Hochschule Ulm, in Ulm, in Germany with the Pasan Flasher flashing system. As described in [8], This was performed experiments at several different irradiation values. However, only the data obtained from the 1000W / m^2 irradiation value were used for the evaluation. This was performed each measurement series twice, and a measurement sequence contained three measurements of 1000 W / m^2, therefore, on the basis of the six measurements, we were able to decide on the accuracy of the values obtained. There was no significant difference between the results of the six measurements in the experiment, and the characteristics were consistent. [10],[11]

D. The process of measurement

During the experiment, the solar cell was placed in the dark chamber of the Pasan Flasher professional measuring device. We set up the measurement system then, measured the electrical parameters and characteristics of the solar cell in standard state (without filters). After that, different color filters were placed in front of the solar panel. These color filters were used to filter different parts of the light spectrum. The measurement inside the dark chamber is shown in the Figure 2 and Figure 3. [10-12]

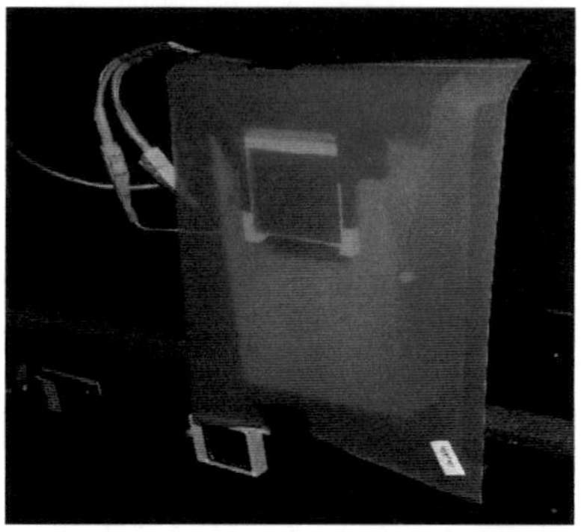

Figure 2. The Dye Sensitized Solar Cell whit the magenta color filter in the dark room of the Pasan Flasher measure system

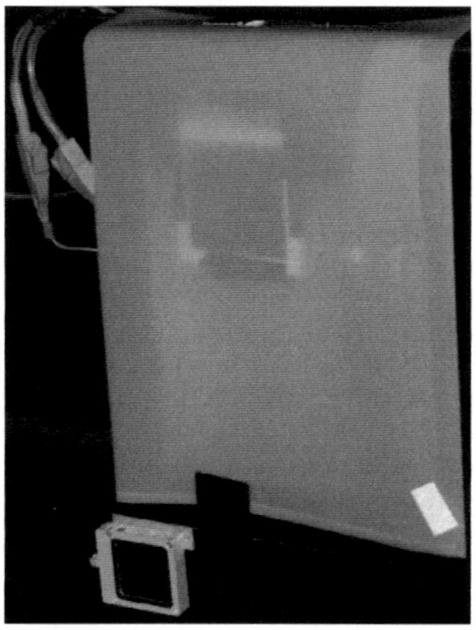

Figure 3. The Dye Sensitized Solar Cell whit the green color filter in the dark room of the Pasan Flasher measure system

In this part of our experiment, we wanted to investigate if our filtering of different parts of the light spectrum of the incoming light on the solar panel – with a green light-sensitive dye layer – would cause a significant change in the main electrical parameters of the solar cell. In this publication, this was examine the electric power-voltage characteristics of the Dye Sensitized Solar Cell. Therefore, the electric power-voltage curves formed from the measurement data we examined together. That is, we compared the electric power-voltage curve which this was measured in the standard (filterless) state with the curves which we got when we used different filters. The results obtained will be described later. [10-12]

Another main direction of our experiment was to test the effect of the reflective background which this was placed on the back of the solar cell on the electric power-voltage characteristic of the DSSC cell. To this end, the measurements were repeated in the state of standard (filterless), as well as when these were used the color filters, but we put reflective background on the back of the solar panel. The results thus obtained were compared with the data without the reflective layer. We examined the state of the standard (filterless) and the states of the filters (separately for each filter) separately.

IV. Experimental Results

A. Effect of the color filters on the light spectrum

As mentioned in the Section III / B, the effects of filters on the light spectrum were measured using a spectrophotometer. The spectra obtained are shown in the following pictures. It can be clearly seen that the different filters have just filtered a particular portion of the light spectrum significantly.

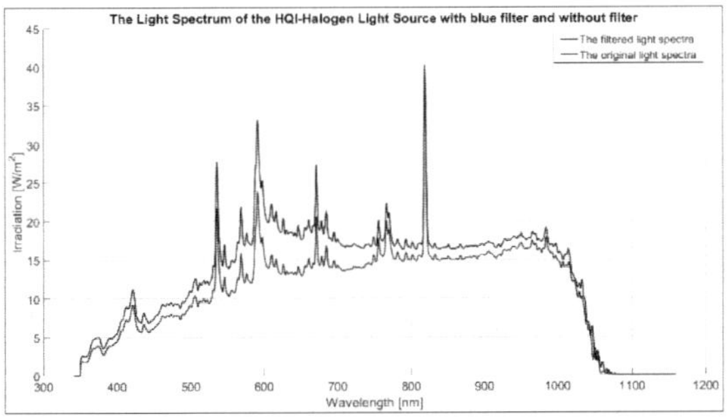

Figure 4. The light spectrum of the HQI-Halogen light source with blue filter and without filter

The Figure 4 clearly shows that the blue color filter filters the spectral range of 600 nm-780 nm (nanometers) significantly, but shows minimal filtering across the entire spectrum of light. In contrast, the green filter (Figure 5.) exhibits significant filtration at a spectral range of 350 nm to 850 nm. That is, it filtered the most of the visible light range and it cut off almost completely the spectrum between 350 nm and 450 nm

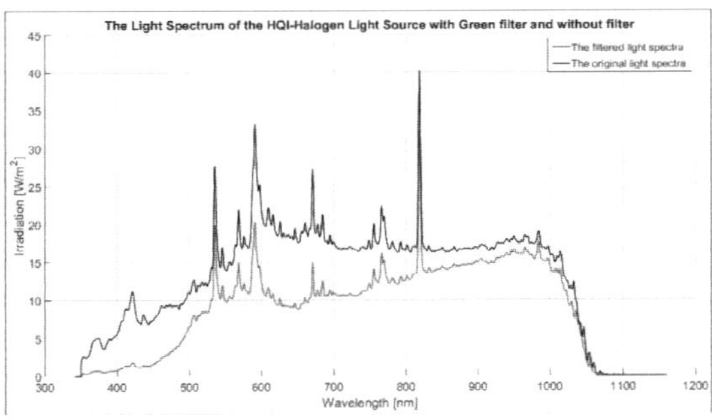

Figure 5. The light spectrum of the HQI-Halogen light source with green filter and without filter

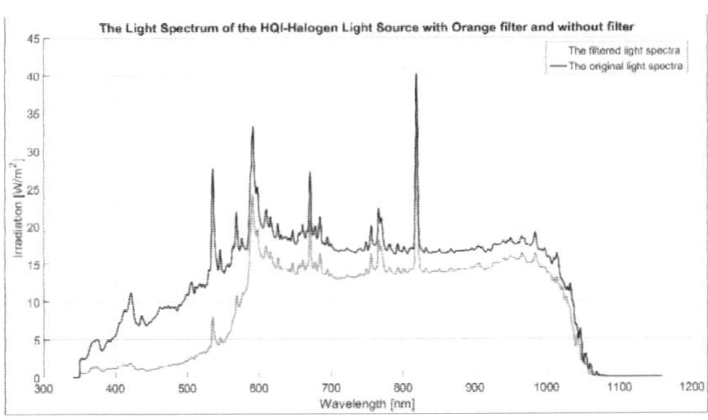

Figure 6. The light spectrum of the HQI-Halogen light source with orange filter and without filter

Filtration of the orange (Figure 6.) filter also extends over the entire spectral range, but at the same time, It is almost completely filtered the spectrum of 350 nm - 550 nm, like the green filter

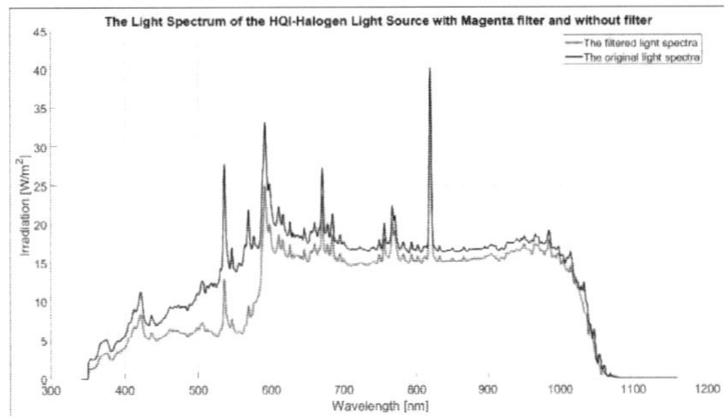

Figure 7. The light spectrum of the HQI-Halogen light source with blue magenta and without filter

The magenta filter (Figure 7.) filters the range of the spectral more significantly -which we examined - between 450 nm and 600 nm. But this filtering is less powerfull ont he full spectra than the green or orange filter screening.

After the measurements, we calculated the numerical integral of the irradiation values of the spectra by the standard (filterless) and the filtered irradiation values. We subtracted from the numerical integral of the irradiation of the non-filtered spectrum one by one the numerical integral of the spectra of the obtained by the filters. As a result, we have been able to set the approximate order between the effects of the four filters. The Table 1. represent the results of the subtraction of the numerical integrals obtained from the data measured in the HQI-Halogen sun simulator. We can see that the most spectral filtering was achieved with the green filter and then with the orange, the purple. The smallest light filtering was of the blue color filter.

Table 1. Irradiations

	Type of filter				
	Standard	**Blue**	**Green**	**Orange**	**Magenta**
I	21008.23 [W/m^2]	17601.2 [W/m^2]	14265.3 [W/m^2]	14698.8 [W/m^2]	17240.9 [W/m^2]
ΔI	0 [W/m^2]	3406.99 [W/m^2]	6742.88 [W/m^2]	6309.35 [W/m^2]	3767.24 [W/m^2]

(I:Numerical integral of the light Irradiation [W/m^2]; ΔI: Difference of Numerical integrals of the light Irradiation [W/m^2])

V. Generated Electric Power-Voltage Caracterictics Without the Reflexive Background

Figure 8 shows the power voltage characteristics of the modern solar cell which we used. The electric power is marked on the vertical axis, and the electric voltage is marked on the horizontal axis. The power voltage curves are marked as follows:

- The black curve represents the standard (without the filter) results
- The blue curve represents the blue filter results
- The magenta curve represents the magenta filter results
- The orange curve represents the orange filter results
- The green curve represents the green filter results

The result was what we expected. That is, the highest power (based on the curves) characterizes the state without the filter, the maximum value of which was 0.01991 W. Out of the electrical power values measured with color filters, the highest power value was obtained when this was used the blue color filter, this value was 0.01727 W. The next maximum electrical power value we got when used the magenta color filter this value was 0.01602 W. The maximum power value when we used the orange color filter has just been slightly higher than the lowest value. When we used the orange filter the maximum electric power value was 0.01096 W. The lowest power maximum was obtained when we used the green filter and Then the maximum power of the solar cell was 0.00951 W. The obtained are as expected based on the sequence set up for the screening of the filter spectrum of the filters.

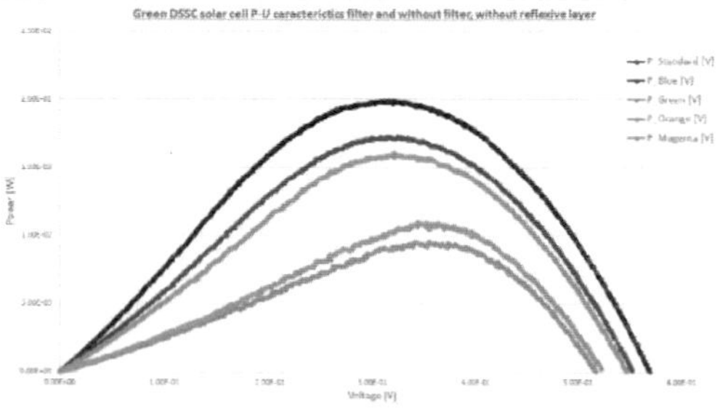

Figure 8. The solar cell electric power-voltage characteristics with filter and without filter, without reflexive layer

VI. GENERATED ELECTRIC POWER-VOLTAGE CARACTERICTICS WITH THE REFLECTIVE BACKGROUND

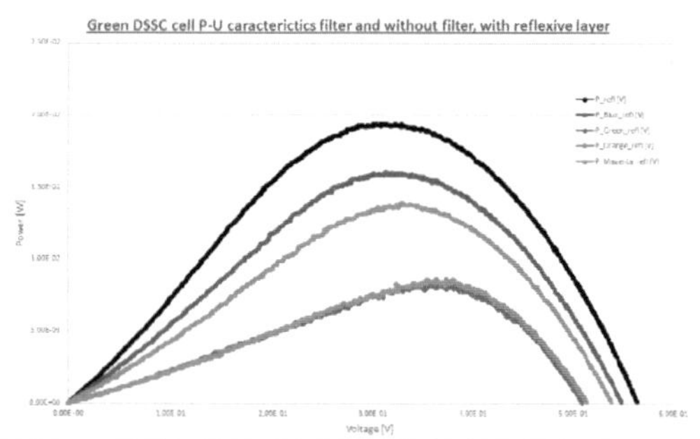

Figure 9. The solar cell electric power-voltage characteristics with filter and without filter, with reflexive layer

Figure 9 shows the characteristics of the electrical power-voltage of the solar cell when we used the reflection background with and without filters. In the Figure, the colors of the curves are exactly the same as before. That is, the Black curve is the standard power-voltage characteristic of the solar cell, but here we already have used the reflective background. The blue curve represents the power voltage curve obtained with the blue filter, but now we were using reflective background, and so on. The electric power-voltage curves, and thus the order of the electric power peaks, are identical to the order of the maximum curves obtained without the original reflection layer. Maximum power was for reflection background, without filter 0.01940 W with the blue filter 0.01606 W with the magenta filter 0.01390 W with the orange filter 0.00860 W, with the green filter 0.00819 W. The measured values were as expected, but there is a clear difference between the course of each curve. Table 2 summarizes the measured power peaks.

Table 2. Maximum of the electric power

	Type of filter				
	Standard	**Blue**	**Green**	**Orange**	**Magenta**
P_{max} [W]	0.01991	0.01727	0.00951	0.01096	0.01602
P_{max_refl} [W]	0.01940	0.01606	0.00819	0.00860	0.01390

(P_{max}: Maximum electric powers
P_{max_refl}: Maximum electric powers when we used reflexive background)

Therefore, in the case of a reflective background, the distance between the curves obtained using each color filter have changed compared to when we had not used a reflection background. This is illustrated the Figures 10 to 14. On these Figures, the curves obtained with the smooth (without reflection background) and the reflective background are shown separately for each filter and according to the standard (filterless) state.

Figure 10 clearly shows the effect of the reflection background. The electric power-voltage curve shifted only minimally compared to the curve obtained without the smooth reflective backing.

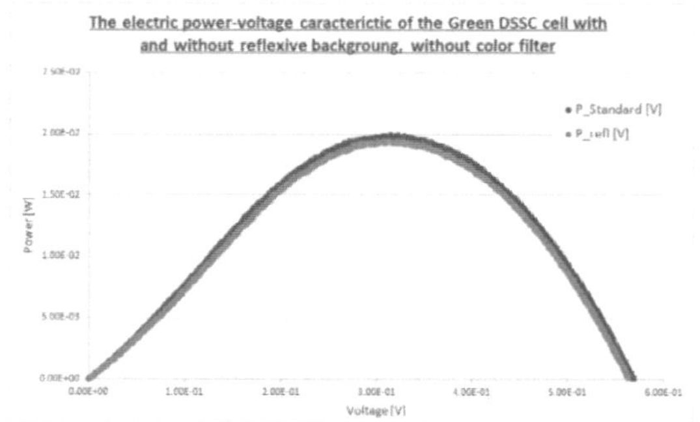

Figure 10. The electric power-voltage characteristics of the Green DSSC cell with and without reflexive background without color filter

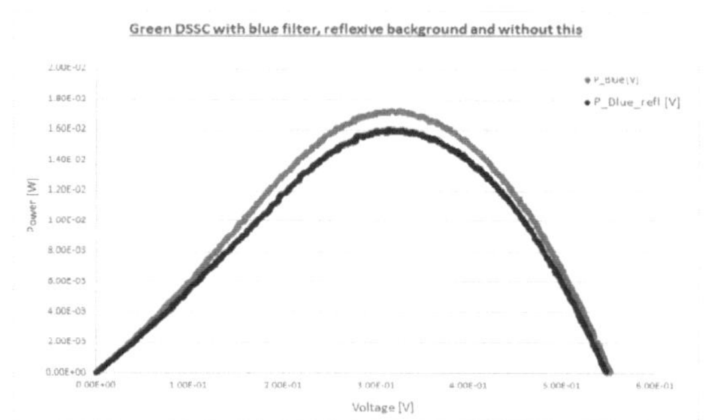

Figure 11. The electric power-voltage characteristics of the Green DSSC cell with and without reflective backgroung, with blue color filter

A similar, but somewhat greater, discrepancy can be observed when using the blue filter. The measured power voltage characteristics are shown in Figure 11. The difference between the two curves is the most significant point around the working point (the maximum power point).

Figure 12 shows the power-voltage curves obtained with the magenta filter as a function of the presence of the reflective backing. The shift between the curves (with and without reflexive background) had shown earlier is also present here. When we used the magenta filter, the difference between the curves obtained with and without the reflective backing increased further. The difference between the curves in this case is not constant for the whole characteristic. The biggest difference is on the range around the working point.

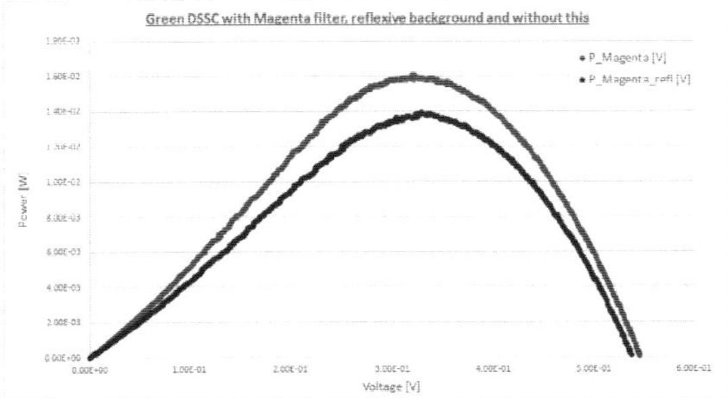

Figure 12. The electric power-voltage characteristics of the Green DSSC cell with and without reflective backgroung, with magenta color filter

Figure 13 shows the power voltage characteristics obtained when we had used a orange filter. The change of shifting of the curves can also be observed here. The shift of the two curves relative to each other was the greatest, when we had used this filter. The distance between the curves for the whole characteristic still not constant. The most significant difference is around the maximum performance point. Interestingly, if we look at the light filtering order (Table 1.), then this filter does not have the greatest effect of the light spectrum filtering.

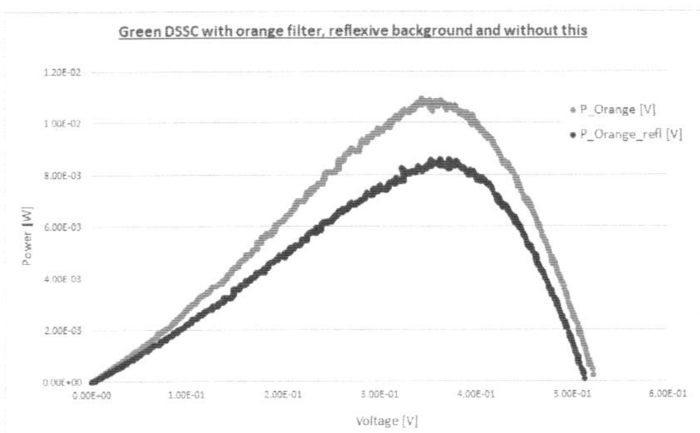

Figure 13. The electric power-voltage characteristics of the Orange DSSC cell with and without reflective backgroung, with orange color filter

Figure 14 shows a graphical representation of the power-voltage curves obtained for green filters, depending on the presence of the reflexive background. In the same way as before, the two characteristics have shifted here as well. However, the result was surprising, because the biggest difference is not when we have used the green filter. The shift of the curves is significant but not as large as that of the orange filter. Despite the fact that the light spectrum tested is best filtered by the green filter.

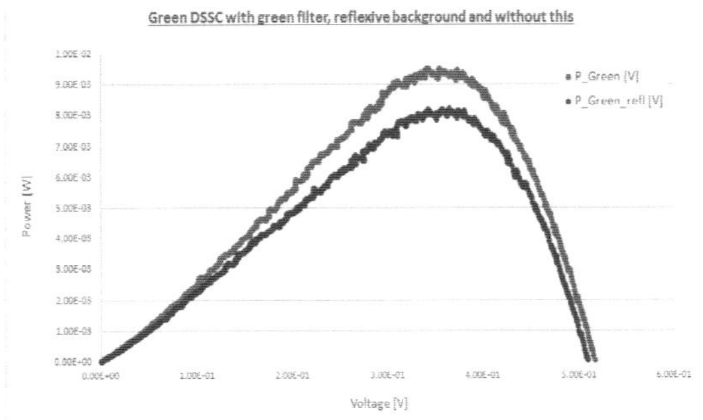

Figure 14. The electric power-voltage characteristics of the Green DSSC cell with and without reflective backgroung, with green color filter

VII. SUMMARY

In our article, we have tested a modern 3rd generation solar panel in a professional Pasan Flasher light measure device. The Dye Sensitized Solar Cell was a Solaronix-based demo cell with a green photosensitive dye layer. The solar cell was placed in the dark chamber of the device, the labeled side have faced to the light, and its electrical parameters and characteristics were measured. Subsequently, various color filters were placed between the light source and the solar cell, and this was examined the effect of the filters on the electrical power-voltage characteristics and maximum power of the solar cell. Finally, with a reflective backsheet on the back of the solar cell, the measurement sequence was repeated with and without color filters.

Before to the start of the measurements, we determined the filtered range and filtering size of the color filters which we have gotten from the Alternative Energy Resources Knowledge Center. A Sun simulator was used for the measurements.

Our experiments were successful because we were able to make measurable reproducible measurements. The data was processed using Microsoft Excel and MatLab from the .xlsx and .csv formats. We have used Microsoft Excel and MatLab software to data process. The format of data was .xlsx and .csv format.

The results of the different color filters to the light spectrum filtering are shown in Figures 4-7. These results are summarized in Table 1. We have calculated the numerical integral of irradiation value of the light spectra with and without filter. We subtracted from the numerical integral of the irradiation of the non-filtered spectrum one by one the numerical integral of the spectra of the obtained by the filters. Table 1. lists these values. Based on this, we have determined a filtering order. It was based on that the filters how much filtering the light which we examined.

Figures 8 and 9 are illustrating the electrical power-voltage characteristics of the solar cell with and without the color filters. Figure 8 shows the results of the solar cell without reflexive background. Figure 9 shows the power voltage curves obtained using the reflexive background. The results are evident, that the reflection background caused a different change in the power-voltage curve of the solar cell for each filter. The values of the power maxima are shown separately in Table 2.

For transparency and comparability, see, in the case of a reflection background, the distance between the curves obtained using each color filter have changed compared to when we have not used a reflective background. This is illustrated on the Figures 10 to 14. On these Figures, the curves obtained with the smooth (without reflective background) and the reflective background are shown separately for each filter and according to the standard (filterless) state. The Figures show that the power-voltage curves slipped relative to one another. The degree of slip between the curves differs with every filter and is not constant for the entire characteristics. The size of the shift is the highest around the work points (maximum power points) in every case. On the basis of the characteristics, it can be stated that the discrepancies observed in the power-voltage curve experienced with the application of the reflective background during the present measurement do not follow the light-filtering sequence for the color filters defined previously (Table 1).

VIII. Conclusion

The experiment was a success based on the results, but many more questions were raised. The aim of our study was to determine the extent to which the screening of a spectral range influences the electrical parameters (power-voltage characteristics) of the DSSC cell produced with the green color photosensitive dyes we investigated. Is there a spectral range which if filtered, would not cause a significant deterioration in the electric power of the solar cell? Or, on the contrary. Is there a spectral range that causes the power of the solar cell to drop drastically?

The color filters used for the measurement did not only filter one spectral range. Also, the filters did not filter to the same extent the light spectrum. Nevertheless, we have been able to set up a filtering sequence among them. It was based on the filtering of a Sun simulator under the same conditions. This was a good starting point. The solar cell power-voltage characteristics followed the sequence which this was specified before. In the future, we would like to continue this series of measurements using filters which have identical light transmission parameters but limited to different spectral domains.

Another main direction of our experiment was to test the effect of the reflective background which this was placed on the back of the solar cell on the electric power-voltage characteristic of the DSSC cell. To this end, the measurements were repeated in the state of standard (filterless), as well as when we used the color filters, but this was put reflexive background on the back of the solar panel. In this case, we also got usable results and the measurement results followed the previous system. At the same time, the reason for further studies be that we investigated the effect of the reflexive background on each filter. At the same time, the reason for further studies, that we had investigated the effect of the reflexive background on each filter separately. The electrical power-voltage curves of the solar cell changed differently for each filter with and without reflexive background. These differences did not follow the order of filtering to the light spectrum for color filters. Questions arising from them: what can be the reason? What may affect? What direction can be changed? We cannot answer these, but we are planning to continue our research in the near future, with a different color-sensitive dye layer made DSSC cells and high-quality color filters.

IX. ACKNOWLEDGMENT

Authors gratefully say thanks for the support of the Óbuda University, Doctoral School of Applied Informatics and the Alternative Energy Sources Knowledge Center.

Hungarian Authors say many thanks for Hochschule Ulm, Institut für Energie und Antriebstechnik, Ulm, Germany for using their experimental apparatuses, and for the great help and assistance provided during the experiments and for the long-term friendly cooperation

Thank you for the help of Erik René Neuchel, member of Hochschule Ulm, Institut für Energie und Antriebstechnik, Ulm, Germany, who introduced us to the reliable operation of Pasan-Flasher measurement device.

Authors are also appreciative for the ERASMUS program and the Vice Rector responsible for Science and International Connections at the Óbuda University to support our students' visit to Ulm, Germany.

Gratefully say thanks for the help of Kristóf Tóth, who helped our research with outstanding professional knowledge.

References

[1] B. Hörömpöli, E. Rácz, B. Kis „ Analysis of electric operation of a dye sensitized solar cell at different incoming light spectra" 2019 IEEE 17th World Symposium on Applied Machine Intelligence and Informatics (SAMI), Kosice

[2] M. Kolcun, Z. Conka, et. al."Improvement of Transmission Capacity by FACTS devices in Central East Europe power system" In: IFAC PAPERSONLINE, Volume: 49, Issue: 27, Pages: 376-381, ISSN: 2405-8963, DOI: 10.1016/j.ifacol.2016.10.756

[3] Z. Conka, M. Kolcun, G. Morva, "Utilizing of Phase Shift Transformer for increasing of Total Transfer Capacity " In: ACTA POLYTECHNICA HUNGARICA, Volume: 13, Issue: 5, Pages: 27-37, 2016, ISSN:1785-8860, DOI: 10.12700/APH.13.5.2016.5.2

[4] B. O´Regan, M. Grätzel: "A low cost, high efficiency solar cell based on dye-sensitized colloidial TiO2 films", Nature 353, pp. 737-740, 24 October 1991., DOI: https://doi.org/10.1038/353737a0

[5] H. J. Snaith, "Estimating the Maximum Attainable E-ciency in DyeSensitized Solar Cells". Adv. Funct. Mater. 20, 13-19 (2010

[6] B. Mátravölgyi, T. Hergert, A. Thurner, B. Varga, N. Sangiorgi, R. Bendoni, L. Zani, G. Reginato, M. Calamante, A. Sinicropi, A. Sanson, F. Faigl, A. Mordini, "Synthesis and Investigation of Solar - Cell Photosensitizers Having a Fluorazone Backbone" Eur. J. Org. Chem, 2017, p 1843.

[7] J. Burschka, N. Pellet, Soo-Jin Moon, R. Humphry-Baker, Peng Gao, M. K. Nazeeruddin, M. Grätzel," Sequential deposition as a route to high-performance perovskite-sensitized solar cells." Nature 499, 316-320 (2013)

[8] http://shop.solaronix.com/picto-1010demo-1.html.(access time: 2019.04.29)

[9] X. Gao, X. Li, X. Gan: Enhancing the Light Harvesting Capacity of the Photoanode Films in Dye-Sensitized Solar Cells (2013): Figure 1. b: Working principle of a typical DSC employing an iodide/triiodide-based redox electrolyte and N719 as a sensitizer. http://www.intechopen.com/books/solar-cells-research-and-application-perspectives/enhancing-the-light-harvesting-capacity-of-the-photoanode-films-in-dye-sensitized-solar-cells

[10] E. Racz, B. Horompoli, A. Varga, E. R. Neuchel, „Experimental investigation based on analyzation of electric currentvoltage characteristics on Pasan-flasher flashlight irradiated photovoltaic cells aged by various processes" Elektroenergetika, pp. 718 September 2017.

[11] E. R. Neuchel, "Measurement of the characteristics of PV modules", Dezentrale Energiesysteme, Laborversuch, Institut für Energie- und Antriebstechnik, Hochschule Ulm, Ulm, Germany, July 2016, pp.I.-1.

[12] B. Hörömpöli, E. Rácz „Statistical Analysis of Power Measurements made on Mono- and Polycrystalline Solar Cells" 2018 IEEE 16th World Symposium on Applied Machine Intelligence and Informatics (SAMI), Kosice DOI: 10.1109/SAMI.2018.8324003

Power Loss in Smart Grids

Maksym Oliinyk
Department of Electric Power Engineering; Faculty of Electrical Engineering and Informatics
Technical University of Košice
Košice, Slovak Republic
maksym.oliinyk@tuke.sk

Jaroslav Džmura
Department of Electric Power Engineering; Faculty of Electrical Engineering and Informatics
Technical University of Košice
Košice, Slovak Republic
jaroslav.dzmura@tuke.sk

Abstract — **SMART GRID is a system of transmission of electricity from the producer to the consumer, monitoring and distributing the flow of electricity to achieve the highest energy efficiency. The basis of the new generation network is the use in the general power system of numerous various distributed energy sources, which include low-power stations and installations that operate, including renewable energy sources. Using distributed energy sources makes it possible to reduce power losses during power transmission. In today's market conditions, in financial calculations between the energy system and consumers, it is important to analyze the energy balance, which includes the technological consumption of electricity for its transmission. Therefore, the issues of determining electricity losses are very relevant. However, with introducing information-measuring systems, the information security of calculations increases, so it became possible to take into account additional factors while calculating losses, determining the structure of losses in the entire transmission and distribution system of electricity, and identifying the place of increased losses. Gathered information makes it possible to take measures for the rational reduction of losses. Energy storage systems are rapidly developing a class of high-tech devices that open new opportunities in the energy power industry. They make electricity stored and portable, eliminating the need for strict simultaneity of its generation and consumption - the fundamental constraint of power balance, which was a key factor to make a modern architecture of energy systems operating around the world. Electric power storages are fundamentally new element of their architecture, complementing the objects of generation, transmission, and consumption and opening wide opportunities for the effective realization of the potential of new types of distributed generation. The new active consumers will be using already commissioned and new energy capacities and operate energy systems in general.**

Keywords: Smart Grid, Electricity storage systems, renewable energy sources

I. Conversion Of Classic Electrical Networks to SMART GRID

At present, the Smart Grid concept is being introduced into power supply systems. To create a power supply system according to the concept, Smart Grid requires a large number of technologies. They provide consumers with improved power supply reliability and high quality power [1].

To create a smart network, technologies from the following areas are mainly used:

1. Modern measurement systems
2. Communication systems
3. Control systems
4. Automation systems
5. Renewable energy sources and energy storage systems

Renewable energy sources transform centralized systems into a distributed system. Intelligent power management systems are a promising direction for energy development. This system comprising traditional energy sources of renewable energy sources. Optimization of the combined work of various energy sources, increasing energy efficiency and reliability of the system is an important issue in the modern energetic industry. [2]

One of the important elements of smart grids is electricity storage systems. Electricity storage systems (ESS) are developing a class of high-tech devices that open new opportunities for developing the electric power industry. They are sufficient technologies that create all the conditions for smart energy management [1].

Traditional cars will become a thing of the past as the use of traditional energy sources is reduced. So, according to the Deloitte forecast, in 2030 annual sales of electric vehicles, including plug-in hybrids (PHEV) [3], will reach 21 million units (20% of global sales of light vehicles). This is quite different from the latest BNEF forecast (30 million electric cars by 2030) and even slightly less than in the baseline scenario of the International Energy Agency, which differs in a rather conservative approach to forecasting the development of new energy technologies (21.5 million electric vehicles in 2030 d) [4]. According to a report by analysts J.P. Morgan by 2030 the share of electric vehicles will reach approximately 39% of the total number of cars [5].

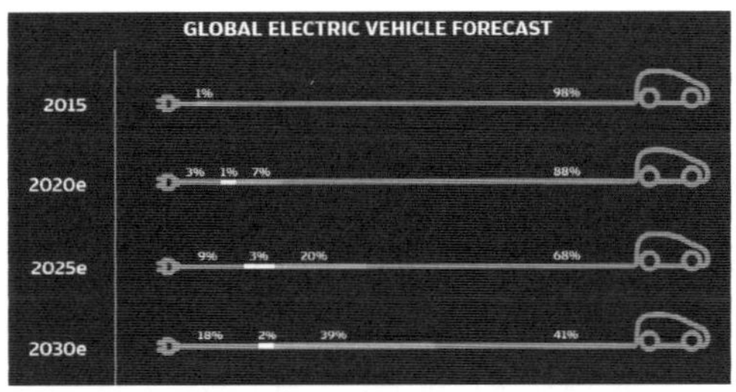

Figure 1. Forecast number of electric cars to 2030

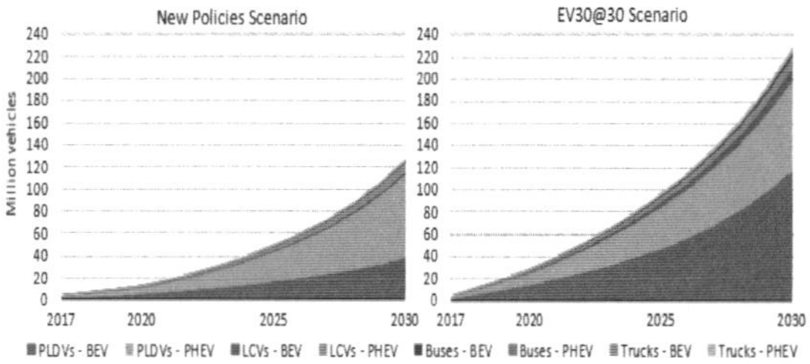

Figure 2. Forecast number of electric cars to 2030

II. Legislative Changes in the Energy Sector

As in the last European Directive 2014/94/EU of the European Parliament and of the Council of 22 October 2014 on the deployment of alternative fuels infrastructure [6], Directive 2009/28/EC of the European Parliament and of the Council of 23 April 2009 on the promotion of the use of energy from renewable sources [7] and Directive 2010/31/EU of the European Parliament and of the Council of 19 May 2010 on the energy performance of buildings. [8] In article 6 about new buildings suggesting that all new building need to take into account:

- Cogeneration;
- Heat pumps;
- Block or district heating or cooling;
- Decentralised energy supply systems on energy from RES.

December 2020. To fulfil this criterion is needed to install to all buildings photovoltaic power plant, and there are suggestions to build on all new houses electro mobile charger [6].

III. Methods for Reducing Power Loss During Transmission

Measures to reduce energy losses can be divided into four groups[9]:

1. **Measures to improve the management of electrical network modes.** These activities are aimed at the ability to optimize the generation and consumption of electricity during peak periods due to: control of electrical equipment of consumers; the use of distributed generation and alternative sources of electricity at the consumer (batteries, solar batteries and other renewable sources), as well as increasing the capacity of transmission lines [10];
2. **Measures to automate the management of electrical network modes.** This direction is to increase the number of active elements, allowing to change the topological parameters of the network [11].
3. **Electrical network reconstruction activities.** The main source of reducing losses in this direction is to reduce the number of system failures and the time it takes to restore normal operation, by creating a self-diagnostic system. Which would track the current state of the system, and as a result of the accident, independently reconfigure the network [12].
4. **Measures to improve electricity metering, methods and means of information transfer.** In this direction, the focus is on creating communication systems, which will lead to improved consumption forecasting. Based on this, it will be possible to optimize the operation of the electrical system [13].

This article discusses the use of energy storage systems to regulate the load curve.

IV. Creating a Grid Model

To analyze the effect of accumulation systems, a village electric grid model was created. This model includes the elements shown in Table 1. For calculations in the article were used used matlab and Excel.

Table. 1. Elements of the model

Elements in the model	Number of elements	Installed power, kW
Line	91	-
Loads	58	-
battery	4	48
Photovoltaic	58	191,4
Transformer	2	-

A. Loads

The load model is the average home consumption per year. Each place consumption within the model represents a real home. The load is expressed in the annual load schedule Figure 3.

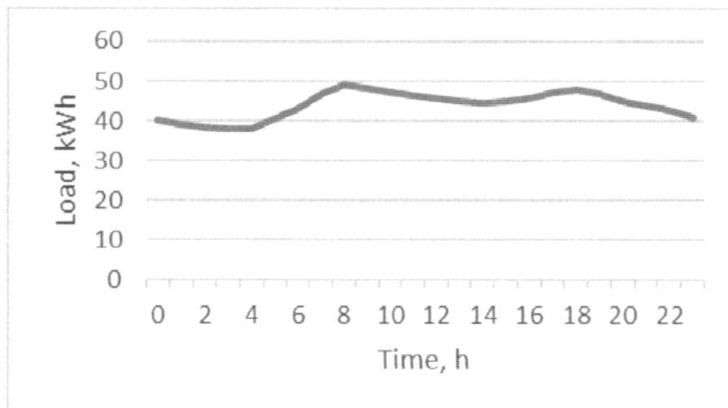

Figure 3. Annual load schedule

B. Renewable energy sources

To create a model of solar panels, real data were used. The All-Black CS6K-300M model from Canadian solar with a maximum output of 3.3 kW was chosen for the simulation. The parameters of the solar panel are presented in Table 2[14].

Table 2. Parameters of photovoltaic [14]

Nominal Max. Power (Pmax	300 W
Opt. Operating Voltage (Vmp)	32,5 V
Opt. Operating Current (Imp)	9,24 A
Open Circuit Voltage (Voc)	39,7 V
Short Circuit Current (Isc)	9,83 A
Module Efficiency	18,33%
Operating Temperature	-40°C ~ +85°C
Max. System Voltage	1000 V
Module Fire Performance	TYPE 1 (UL 1703) or CLASS C (IEC 61730)
Max. Series Fuse Rating	15 A
Application Classification	Class A
Power Tolerance	0 ~ + 5 W

And data on solar radiation were obtained by measuring our department. For calculations, we selected the average power per hour for the whole calendar year.

C. Electric car charger

To create a model of an electric car, an analysis was made of possible charging stations. Charging is made according to the nominal values. The rated power of the selected charging station is 3.7 kW. [15] Charging time depends on the degree of battery discharge during the day. To study the influence of the state of the batteries of electric vehicles, it was analyzed the possible discharge of an electric vehicle by an ordinary user under different usage scenarios. According to the results of this analysis, a battery discharge scenario was made in the design model; it is presented in Figure 4.

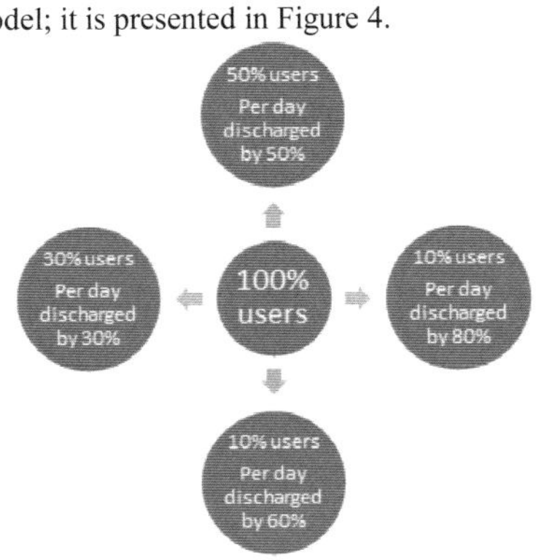

Figure 4. The distribution of discharge on the total number

As a prototype electric car was used Nissan Leaf 24kWh. [16] Fully charging this vehicle with the selected charging station takes about 7 hours.

V. SIMULATION RESULTS

As part of the simulation, three options for charging electric cars were analyzed.

- 19:00-1:00
- 22:00-4:00
- 17:00-23:00

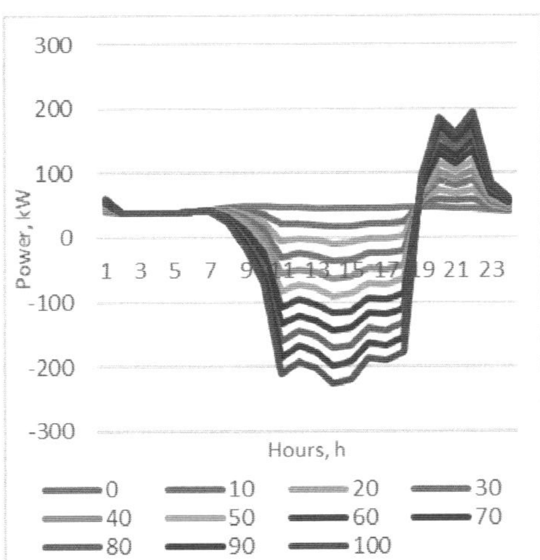

Figure 5. Schedule of power consumption by the network when charging electric vehicles in the period 19:00-1:00. Curves indicate the percentage of use of smart elements in the network.

As we see from the graphs, with an increase in the number of smart elements, losses in the network begin to grow [13]. In the daytime, this is due to the fact that solar panels produce a large amount of electricity that is transmitted over a low voltage network of 0.4 kV. In the evening, the increase in losses is caused by the fact that the load on the network increases significantly, and consequently also the losses. As we can see on the graph of the losses of the third network operation mode, the optimal solution is to charge electric vehicles at the time of producing power on solar panels. But as a rule, people are at work during the daytime and they cannot put their electric car on charge.

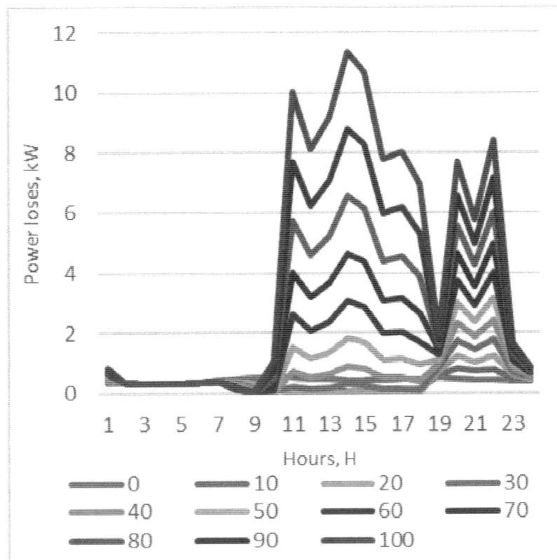

Figure 6. The graph of the loss of active power when charging electric vehicles in the period 19:00-1:00. Curves indicate the percentage of use of smart elements in the network.

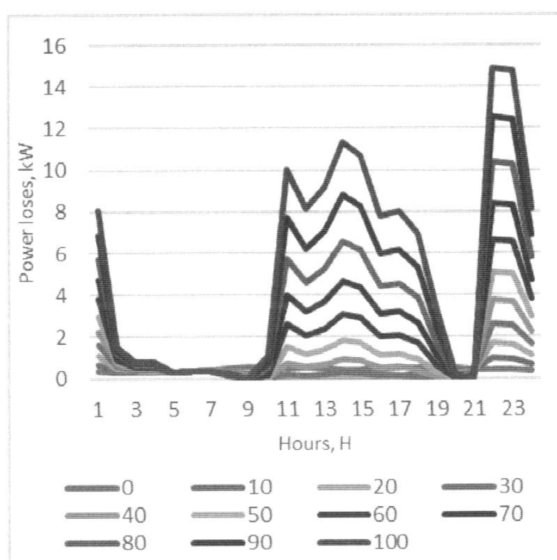

Figure 7. The graph of the loss of active power when charging electric vehicles in the period 22:00-4:00. Curves indicate the percentage of use of smart elements in the network.

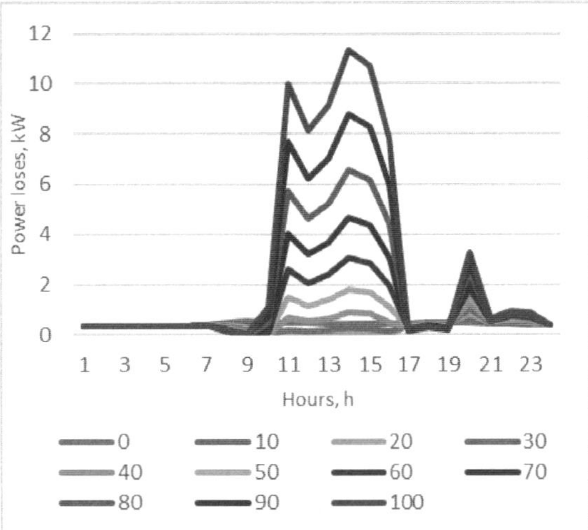

Figure 8. The graph of the loss of active power when charging electric vehicles in the period 17:00-23:00. Curves indicate the percentage of use of smart elements in the network.

To reduce power losses in the network, an increased power energy storage system was installed.

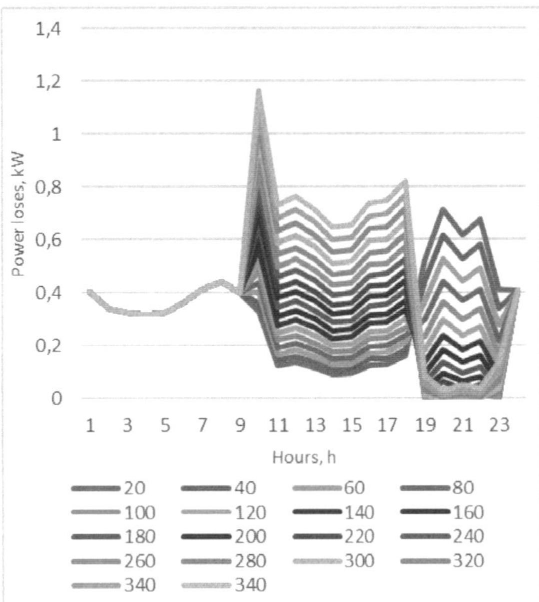

Figure 9. Graph of changes in losses in the electrical network when using different powers of accumulation systems (range 20-360 kW). Using 10% of smart elements.

Figure 9 shows the change in losses due to changes in battery capacity. The calculation is made using only 10% of smart elements and a charging period of 19:00-1:00. Analyzed the results of this calculation, we can conclude that high-capacity batteries can also cause increased losses. However, if you install a charge and discharge control system, you can maintain an optimal level of losses in the network [17].

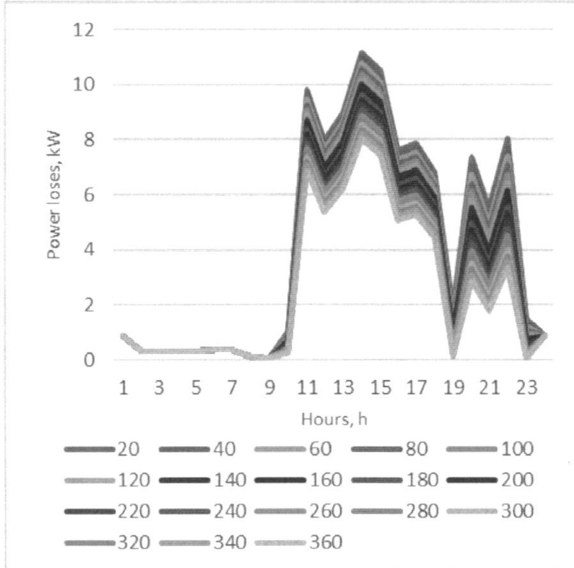

Figure 10. Graph of changes in losses in the electrical network when using different powers of accumulation systems (range 20-360 kW). Using 10% of smart elements.

Figure 10 shows the change in active power loss depending on the change in battery capacity from 20 to 360 kW. The analysis was carried out for the most difficult mode, that is, 100% of smart elements were installed in the network. Losses have decreased by approximately 25% during the generation of power by solar panels. And by almost 50% during the purchase of electric cars with the maximum capacity of the energy storage system. It is worth noting that when installing a control system, it is possible to adjust the losses in the electrical network with sufficiently high accuracy. That is, to increase or decrease the amount of energy that the battery consumes or vice versa gives into the network [18].

VI. Conclusion

As part of the analysis of the use of accumulation systems to reduce the loss of active power was established:

- Accumulation systems need to develop and install a power control system
- Electric energy storage systems are flexible enough to use.
- Accumulation systems can reduce power losses, voltage deviations, reduce the current load in the lines.
- It is quite easy to increase power.

They also have a number of drawbacks, for example, they are:

- Expensive
- Have a short lifespan
- Require careful adjustment.

Based on forecasts for the future, the number of renewable energy sources will increase, the number of electric vehicles will increase, and consumption of ordinary consumers will also increase[19]. Current electrical networks will not be able to fully perform their tasks. They will have to be updated, and the installation and use of energy storage systems is one of the ways to solve future problems[20] [21].

Acknowledgment

This work was supported by the Scientific Grant Agency of the Ministry of Education of Slovak Republic and the Slovak Academy of Sciences by the projects VEGA No. 1/0372/18

References

[1] M. Shamshiri, C. K. Gan, a C. W. Tan, "A review of recent development in smart grid and micro-grid laboratories", v 2012 IEEE International Power Engineering and Optimization Conference Melaka, Malaysia, 2012, s. 367–372.

[2] SAMI BS. An Intelligent Power Management Investigation for Stand-Alone Hybrid System Using Short-Time Energy Storage. International Journal of Power Electronics and Drive Systems. 2017 ;8(1):pp. 367.

[3] Deloitte, New market. New entrants. New challenges. Battery Electric Venicles 2018, Available on internet: https://www2.deloitte.com/content/dam/Deloitte/uk/Documents/manufacturing/deloitte-uk-battery-electric-vehicles.pdf

[4] IEA, Global EV Outlook 2018, May 2018, Available on internet: https://webstore.iea.org/global-ev-outlook-2018

[5] J.P. Morgan Driving into 2025: The Future of Electric Vehicles, October 2018 Available on internet: https://www.jpmorgan.com/global/research/electric-vehicles

[6] Directive 2014/94/EU of the European Parliament and of the Council Available on internet: https://eur-lex.europa.eu/legal content/EN/TXT/?uri=celex%3A32014L0094

[7] Directive 2010/31/eu of the european parliament and of the council Available on internet: https://eur-lex.europa.eu/legalcontent/EN/TXT/HTML/?uri=CELEX:32010L0031&from=SK

[8] Directive 2009/28/EC of the European Parliament and of the Council Available on internet: https://eur-lex.europa.eu/legalcontent/EN/ALL/?uri=celex%3A32009L0028

[9] N. Nikmehr a S. N. Ravadanegh, "Optimal Power Dispatch of MultiMicrogrids at Future Smart Distribution Grids", IEEE Transactions on Smart Grid, roč. 6, č. 4, s. 1648–1657, júl. 2015.

[10] M. Pavlík, I. Kolcunová, B. Dolník, J. Kurimský, A. Mészáros, D. Medveď, M. Kolcun and J. Zbojovský, "Measuring of Dependence of Shielding Effectiveness of Wet Materials on The Frequency of Electromagnetic Field in the High Frequency Range," Acta Electrotechnica et Informatica. Vol. 13, No. 3 (2013), s. 12-16. - ISSN 1335-8243.

[11] Z. Čonka, M. Kolcun, " Impact of TCSC on the Transient Stability" in: Acta Electrotechnica et Informatica. Roč. 13, č. 2 (2013), s. 50-54. - ISSN 1335-8243 Available on internet: www.versita.com/aei. (2013)

[12] J. Petras, J. Kurimsky, J. Balogh, R. Cimbala, J. Dzmura, B. Dolnik, I. Kolcunova, „Thermally Stimulated Acoustic Energy Shift in Transformer Oil ", In: ACTA ACUSTICA UNITED WITH ACUSTICA, Volume: 102 Issue: 1 Pages: 16-22, ISSN: 1861-9959

[13] I. Kolcunova, M. Pavlik, L. Lison, „Assessment of long thermal ageing on the oil-paper insulation "In: ADVANCES IN ELECTRICAL AND ELECTRONIC ENGINEERING Volume: 14 Issue: 5 Pages: 506511, DOI: 10.15598/aeee. v14i5.1836

[14] Datasheet-All-black_CS6K-MS_EN Available on internet: https://www.canadiansolar.com/upload/b46f34f7016c1aab/aeda548a156 10168.pdf

[15] Catalog 2019 Electric vehicle charging solutions schneider-electric Available on internet: https://download.schneider-electric.com/files?p_enDocType=Catalog&p_File_Name=COM-POWER-VE-CA3-EN+%28web%29.pdf&p_Doc_Ref=COM-POWER-VE-CA3-EN

[16] Nissan Leaf manual Available on internet: http://mynissanleaf.ru/docs/manual-nissan-leaf-2013-en.pdf

[17] J. Bačík, D. Perduková, P. Fedor , "Design of fuzzy controller for hexacopter position control", Advances in Intelligent Systems and Computing. Volume 347. Springer-International Publishing Switzerland 2015, pp. 193- 202. ISSN 2194-5357

[18] Z. Martinek, A. Hromadka, J. Hammerbauer, "Reliability characteristics of power plants", In: Advances in electrical and electronic engineering, Vol. 15, Issue:1, Pg. 37-45, 2017, DOI: 10.15598/aeee.v15i1.2043

[19] S. Borlase, Smart Grids: Infrastructure, Technology and Solutions, 577, (CRC, 2013) ISBN: 978-1-4398-2905-9

[20] J. Momoh., "Smart Grid: Fundamentals of Design and Analysis", p. 216, (IEEE P., 2012) ISBN:978-0-470-88939-8

[21] J. Zbojovský, L. Kruželák, M. Pavlík, "Impedance Spectroscopy of Liquid Insulating Materials", In: Proceedings IEEE International Conference AND Workshop in Óbuda on Electrical and Power Engineering. - New York (USA) : Institute of Electrical and Electronics Engineers pp. 249-253 ISBN 978-1-7281-1153-7, 2018.

Prediction of CO2 concentration in smart homes using IBM SPSS software tool

Ojan M. Gorjani, Jan Vanus, Jan Kubicek, Petr Bilik
Department of Cybernetics and Biomedical Engineering, Faculty of Electrical Engineering and Computer Science
VSB–Technical University of Ostrava
Ostrava, Czech republic
ojan.majidzadeh.gorjani@vsb.cz, jan.vanus@vsb.cz, jan.kubicek@vsb.cz, petr.bilik@vsb.cz

Abstract — **The growth of measured data from intelligent buildings in the recent years provides new opportunities for improvements of quality of the indoor environment. This paper examines the possibilities of increasing the accuracy of CO2 predictions in the smart homes using IBM SPSS software tools. The accuracy of this implementation was evaluated at daily and monthly intervals. Radial basis function (Feedforward neural network) was used in order to predict CO2 concentration levels from the measured relative humidity, indoor and outdoor temperatures. Daily interval resulted in the most accurate prediction results.**

Keywords: Smart Homes; monitoring; prediction; Artificial Neural Network (ANN), Radial Basis Function (RBF), SPSS (Statistical Package for the Social Sciences) IBM

I. Introduction

The term “Smart home” is commonly referred to the homes that contain lights, heating, blinds, appliances which can be controlled and monitored. These components of smart homes often can communicate with each other through technologies such as KNX and BACnet [1]. There are many contributions in the field of smart homes. These researches are typically aimed at automation, communication, control, activity recognition, visualization, monitoring and improving the health and wellbeing of the users [2-12]. As an example Coelho et al. created a solution for collection of data from multiple heterogeneous sensors that were providing different types of information at a variety of locations in a smart home [13]. These types of examples often result in big data. The Oxford dictionary defines big data as “Extremely large data sets that may be analyzed computationally to reveal patterns, trends, and associations, especially relating to human behavior and interactions” [14].

One of the possibilities that come with big data processing is a predictive analysis which provides predictions about the future or otherwise unknown events. The predictive analysis comes with a wide range of applications such as insurance risk calculations [15], social networking, healthcare, mobility, finance, marketing, etc. The predictive analysis can be performed with many different statistical techniques such as predictive modelling, machine learning data mining. Predictive modelling has a wide range of applications such as business, weather forecasting, fraud detection, Bayesian spam filters, advertising and marketing, etc. Predictive models are based on variables that are most likely to influence the outcome. These variables are also known as predictors [15]. There are many types of predictive models, such as neural networks and decision trees. Ahmad et al. compared different methods of predictive modelling for solar thermal energy systems such as random forest, extra trees, and regression trees [16].

The IBM offers large number of services in terms of predictive analysis such as Watson analytics and SPSS. Additionally, these services can be integrated with IBM Cloud (formerly known as Bluemix) with other IBM tools and services, including Artificial Intelligence (AI) and the Internet of Things (IoT) [17]. One of the possibilities with IBM services is the Watson platform, which includes services such as Watson Analytics, Watson Machine Learning, Watson Assistance etc. Watson Analytics is a powerful tool which takes advantages of machine learning and AI technologies combined with natural language querying that is also intuitive and easy to use. The only shortcoming of this platform is a lack of real-time data streaming [18].

There is a wide range of IBM Watson Analytics applications, for example, Nagwanshi et al. used IBM Watson Analytics processes for distinguishing features of human footprint images [19]. The IBM SPSS Statistics comes as a software package which is mainly used for statistical analysis. It is considered as one of the world’s leading statistical software which offers advanced statistical analysis, a vast library of machine-learning algorithms, text analysis, open-source extensibility, integration with big data and seamless deployment into applications. SPSS can be driven externally by a Python or a VB.NET program using supplied "plug-ins" which provides endless opportunities for custom solutions [20]. There are many examples of project and researches implemented within IBM platform [21-26], Murtaza investigated an effective strategy to train IBM Watson question answering system for purpose of Big Data Analytics [25]. Cervenka defined various ways to apply cognitive systems for unstructured data analysis and management for further use in marketing analytics with IBM Analytics Tools for analyzing unstructured data found on social media [26].

In this paper, the authors focused on designing a methodology for processing data obtained from sensors through KNX and BACnet technologies which measured non-electrical quantities in a smart home environment. The obtained result may be used for monitoring the presence of people in smart home environment. The paper examines the possibilities of increasing the accuracy in CO2 predictions in Smart Home using IBM SPSS Modeler software tools. To predict CO2 levels from the measured indoor and outdoor temperatures and relative humidity, Radial Basis Function method (feedforward neural network) was applied. The accuracy of CO2 predictions from the processed data was compared and evaluated at daily, weekly and monthly intervals for the spring and autumn periods.

II. Proposed Methods for CO2 Concentration Prediction

The IBM SPSS Modeler is based on an object-oriented user interface that provides a simplified and easy-to-use way for performing data analysis. This software package comes with variety of built-in algorithms, such as prediction, classification, segmentation and association detection. Implement data analysis with IBM SPSS Modeler should be performed as the following three basic steps:

1. Importing the data into IBM SPSS Modeler
2. Performing series of analysis on the imported data
3. Evaluation and exporting the data

This sequence is often called "Data stream", because the data is flowing from the source to each analysis node and then to the output. These streams are created by drawing diagrams of relevant data operations. IBM SPSS Modeler's Node Palette displays the available data and modeling tools that can be implemented by the user. The node palette items are divided into few main categories, such as source (importing data from an external sources), record ops (for operations on data records, such as selecting, merging, and appending), field ops (for operations such as filtering, deriving new data fields, and determining the measurement level for given fields), graphs, modeling (neural networks, decision trees, clustering, etc.), output, export, etc. [27].

The IBM SPSS Modeler offers multiple types of neural networks for predictive analysis. For this implementation the radial basis function was selected as main predictive model. The RBF network is a feed-forward network which requires supervised learning. This network consists of only one hidden layer. There are three layers in RBF network: input layer, RBF layer (hidden layer) and output layer. Overall, comparing to the other neural network models the RBF holds the advantage of faster training time. IBM SPSS algorithm guide describes mathematical models of these layers [28-30] as following:

Input layer: $J_0 = P\ units,\ a_{0:1}, \dots, a_{0:J_0}$ with $a_{0:j} = X_j$

RBF layer: $j_1\ units, units, a_{1:1}, \dots a_{1:j_1}$; with $a_{1:j} = \emptyset_j(X)$

$$\emptyset_j(X) = \frac{e^{\left(-\sum_{p=1}^{P}\frac{1}{2\sigma_{jp}^2}(x_p-\mu_{jp})^2\right)}}{\sum_{j=1}^{J_1} e^{\left(-\sum_{p=1}^{P}\frac{1}{2\sigma_{jp}^2}(x_p-\mu_{jp})^2\right)}} \quad (1)$$

Output layer: $j_2 = R\ units,\ a_{I:1}, \dots a_{I:j_2}$ with $a_{I:r} = \omega_{I:r} + \sum_{j=1}^{j1} \omega_{rj}\emptyset_j(X)$

Where:

$X^{(m)}$: Input vector

I: Number of layers (for RBM = 2)

Ji: Number of units in layer i

$\emptyset_j(X^{(m)})$: jth unit for input $X^{(m)}, j = 1, \dots, j_1$.

μj: Center of $\emptyset_j$

σj: Width of $\emptyset_j$

$a_{i:j}^{m}$: Unit j of layer i

ω_{rj}: weight connecting rth output unit and jth hidden unit of RBF layer

The RBF training starts with determining the basis function by clustering methods and then it determines the weights given to the basis function. The SPSS measures the accuracy of neural networks by calculating the percentage of the records for which the predicted value matches the observed value. For the continues values, the accuracy is calculated by 1 minus the average of the absolute values of the predicted values minus the observed values over the maximum predicted value minus the minimum predicted value (the following formula) [27].

$$\text{Accuracy} = \frac{1}{n}\sum_{m=1}^{M}\left(1 - \frac{|y_r^{(m)} - \hat{y}_r^{(m)}|}{max_m(y_r^{(m)}) - min_m(y_r^{(m)})}\right) \quad (2)$$

In order to carry out comparison of the obtained results, the following parameters are considered:

Mean Absolute Error (MAE) represents estimator measuring of difference between two continuous variables. The MAE is given by the following expression

$$MAE = \frac{1}{n}\sum_{i=1}^{n}|y_i - \hat{y}_i| \quad (3)$$

Mean squared error (MSE) represents estimator measuring the average of the error squares between two signals. The MSE represents a risk function which corresponds with the expected value of the squared or quadratic error loss. The MSE is given by the following expression:

$$MSE(x_1, \hat{x}_2) = \frac{1}{n}\sum_{i=1}^{n}\left(x(i) - \hat{x}(i)\right)^2 \quad (4)$$

III. Implemetation of CO2 Concentration Prediction

As it was described earlier, the main goal is to predict the CO2 concentration based on data collected by outdoor temperature, indoor humidity and indoor temperature. The sample data were collected from a smart home that was equipped with the indoor temperature sensor Ti (°C), (QPA 2062 range: 0 to 50°C/-35 to 35°C, accuracy ± 1K) and relative humidity rH (%) sensor, (QPA 2062 range: 0 and 100%, accuracy ± 5%), outdoor temperature T0 (°C) sensor (AP 257/22). The procedure of this implementation had been divided into few steps as following

1. Pre-processing the data
2. Developing a data stream using IBM SPSS Modeler
3. Testing the implementation using different intervals

A. Pre- processing

Data Normalization using min-max method (often known as feature scaling) was used as a pre-processing method. The normalized values were calculated using the formula (5).

$$Normalized\ value = \frac{current\ value - \min value}{\max value - \min value} \quad (5)$$

B. Developing a data stream using IBM SPSS Modeler

The figure 1 shows the data stream developed in IBM SPSS Modeler. In order to define default targets (CO2 concentration) and default inputs or predictors (humidity, indoor temperature, and outdoor temperature), a data type selection operator was added from the field operator category. Commonly K-fold, V-fold, N-fold and Partitioning validation methods are used in IBM SPSS Modeler, the User manual of this software tool recommends using the partitioning method for large datasets due to the faster processing time. Partitioning method uses randomly divided data sets. These data sets are usually divided into three section of training, testing and validation. The partition node from field operator category was added to the stream in order to divide the data into three parts (division ratio: 40% training, 30% testing, 30% validation). The training partition is used for the training of the neural network. The testing partition is used for selecting the most suitable model and prevention of overfitting. The validation partition is dedicated to validation of the developed model, using only predictors and comparing the prediction with the reference signal. In other words, the validation partition is used to determine how well the model truly performs. [26].

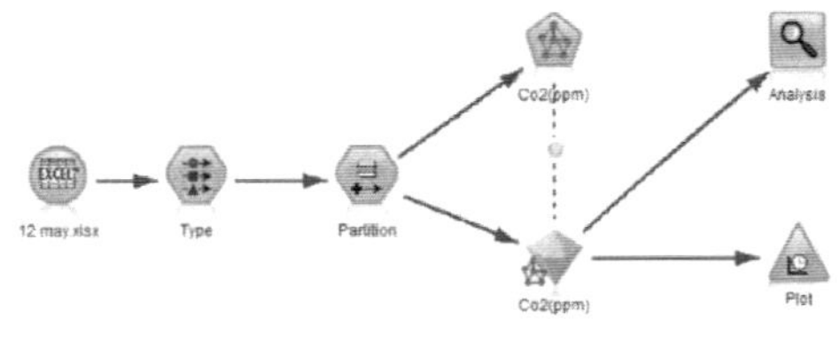

Figure 1. The Developed stream using IBM SPSS Modler

The neural network node was configured to using the RBF model with multiple implementations for various numbers of neurons and intervals. The resulting model from the trained neural network is represented by a nugget gem. Additional nodes were connected to this nugget gem for further interpretation of the results. Time plot node was connected to the nugget gem in order to display the time plot of reference versus predicted values. Additionally, an analysis node was added to the data stream. This node displays details such as linear correlation, mean absolute error, etc.

C. Testing various implementations using different intervals

For evaluation of this implementation, several different intervals from spring and fall of 2018 were randomly selected. These intervals were based on the rate of one sample per minute. In the first stage two intervals with length of one month were examined, first the month of May and then the month of November. The validation of the interval month May have implied that the model with 300 neurons shows the highest accuracy of prediction (76.6%) and relevantly small error (MAE=0.006, MSE=9.75E-5). By re-performing the experiment with the November as internal, an insignificant improvement on the accuracy of all models could be observed. Unlike the month May, the model with 400 neurons held the best overall results in aspects of accuracy (80.6%), linear correlation (0.898), MAE (0.011) and MSE (1.553*10-3).

For further evaluations, the interval lengths were reduced to one week. This reduction should reduce the overfitting and create better generalization. In the both of the week long experiments, the models with 400 neurons showed the finest accuracy (6th to 13th of May 2018: 98.2% - 6th to 13th of November 2018: 96.2%), and the lowest error (6th to 13th of May 2018: MSE:3.14*10-6 MAE:0.001 – 6th to 13th of November 2018: MSE:4.12*10-5, MAE:0.003). Beside improvement in accuracy, a consistency in the obtain results was observed. For further examination of the relationship between interval lengths and accuracy, additional day long intervals (15th of May and 15th of November) were examined. The obtained results indicated a noticeable improvement in terms of accuracies, which further supports the theory of better generalizing with the reduction of interval lengths. The results from interval of 15th of May indicated that the model with 100 neurons provides an impressive 99.9% of accuracy. This result was closely followed by the model with 400 neurons (99.8%). The obtained results from the interval of November 15 (fig. 2) showed that the model with 400 neurons presents the highest accuracy (99.7%), (linear correlation: 0.895, MAE: 0.003 and MSE: 6.45*10-5).

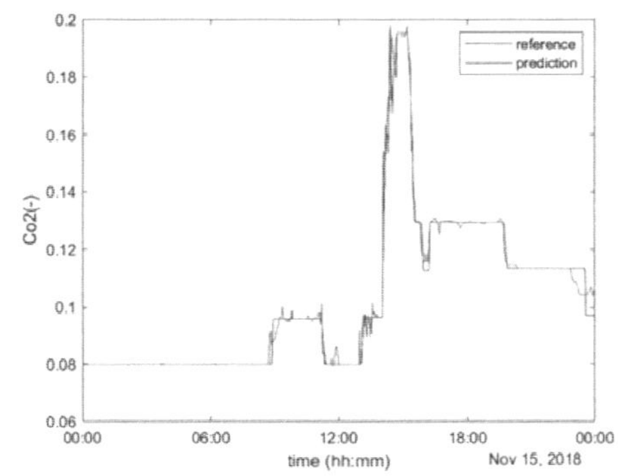

Figure 2. Example of prediction results (The model with 50 neurons trained and validated using data interval of 15th of November 2018

The average of accuracy, linear correlation MAE and MSE in each experiment is represented in table 1. It is apparent that with increase in lengths of the intervals, the accuracy of predictions decreases. It can also be observed that the experiment with 15th of May as period holds the highest average accuracy (99.5%), and relevantly low error values. Additionally, the experiment with interval length of a week in May shows slightly less average accuracy and linear correlation (0.967) but it has overall lower error values. The table 2 demonstrates the relationship between the number of neurons and the average results. The overall trend of this table, points toward improvement of accuracy with increasing number of neurons. This increase may be insignificant after a certain point. As an example it can be observed that the models with 200, 250, 300, 350 and 400 share similar average accuracy, linear correlation, MAE, MSE.

Table 1. Type Styles

Interval	***Accuracy (%)***	***Linear correlation***	***MAE***	***MSE***
May	63.1	0.78	7.44*10^{-03}	1.51*10^{-04}
November	72.2	0.848	1.30*10-^{02}	1.45*10^{-03}
May 6 to 13	94.7	0.967	1.33*10^{-03}	7.52*10^{-06}
November 6 to 13	89.9	0.939	5.00*10^{-03}	1.17*10^{-04}
May 15	99.5	0.995	1.78*10^{-03}	2.44*10^{-05}
November 15	92.0	0.912	3.22*10^{-03}	9.67*10^{-05}

Table 2. Type Styles

Number of neurons	***Accuracy (%)***	***Linear correlation***	***MAE***	***MSE***
10	59.8	0.75	1.15*10-02	4.99*10-04
50	86.1	0.886	6.33*10-03	3.06*10-04
100	79.8	0.919	5.00*10-03	2.69*10-04
150	88.6	0.926	4.5*10-03	2.87*10-04
200	90.0	0.934	4.17*10-03	2.79*10-04
250	90.4	0.936	4.00*10-03	2.78*10-04

Number of neurons	*Accuracy (%)*	*Linear correlation*	*MAE*	*MSE*
300	90.5	0.936	4.00*10-03	2.78*10-04
350	90.5	0.936	4.00*10-03	2.78*10-04
400	91.5	0.935	4.16*10-03	2.97*10-04

The figure 3 clearly demonstrates the relation between the experiments interval length, the number of neurons and accuracy. By observing all of the obtained results, it was clear that the model with 100 neurons and trained with the data from 15th of May, holds the highest accuracy (99.8%), linear correlation (0.996) and relevantly small error values. The model with 400 neurons shows the best overall average accuracy (table 2) and in the case of the experiments with the interval of 15th May, the accuracy difference between this model and the most accurate model (model with 100 neurons) is negligible. Therefore, the model with 400 neurons and training interval of May 15th was selected as the best overall result.

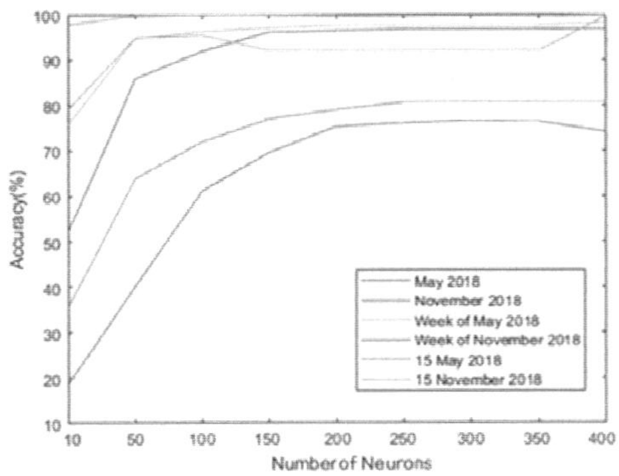

Figure 3. The accuracy of the implementaion for each interval

IV. Using the Template

The authors of the paper focused on designing a methodology that specifies a procedure for processing data measured by sensors in a smart home environment. The paper explores the possibilities of improving accuracy of CO2 predictions in smart home environment using IBM SPSS software tools. The radial basis function method (feedforward neural network) was applied to predict CO2 levels from the measured indoor and outdoor temperatures and relative humidity. The prediction results evaluated at daily, weekly and monthly intervals for the spring and autumn periods. The most accurate results were provided by experiments with the daily intervals (accuracy up to 99%), while the monthly intervals resulted in the least accurate results (up to 80%). Therefore, reduction in training interval, results in better generalizations. In future works, the performance and the limitations of the predictive model in environment with external sources of CO2 and humidity (such as kitchen) must be examined. Overall, the result of this implementation proved to be capable of predicting the CO2 concentration values.

Acknowledgment

This work was supported by the Student Grant System of VSB Technical University of Ostrava, grant number SP2019/118. This work was supported by the European Regional Development Fund in the Research Centre of Advanced Mechatronic Systems project, project number CZ.02.1.01/0.0/0.0/16_019/0000867 within the Operational Programme Research, Development and Education.

References

[1] J.Vanus, J.Kubicek, O. M. Gorjani and J.Koziorek."Using the IBM SPSS SW Tool with Wavelet Transformation for CO2 Prediction within IoT in Smart Home Care," ,'Sensors (Basel, Switzerland), 19(6) doi:10.3390/s19061407.

[2] J.S. Beaudin, S.S. Intille and M.E Morris. To Track or Not to Track: User Reactions to Concepts in Longitudinal Health Monitoring. Journal of Medical Internet Research [online]. 2006, 8(4), e29- [cit. 2018-04-04]. DOI: 10.2196/jmir.8.4.e29.

[3] M.J. Booysen, Machine-to-Machine (M2M) Communications in Vehicular Networks. KSII Transactions on Internet and Information Systems [online]. 2012, - [cit. 2017-11-07]. DOI: 10.3837/tiis.2012.02.005.

[4] D. Basu, G. Moretti, G. S. Gupta and S. Marsland. Wireless sensor network based smart home: Sensor selection, deployment and monitoring. In: 2013 IEEE Sensors Applications Symposium Proceedings [online]. IEEE, 2013, 2013, s. 49-54 [cit. 2017-11-07]. DOI: 10.1109/SAS.2013.6493555.

[5] S. Fleck, and W. Strasser. Smart Camera Based Monitoring System and Its Application to Assisted Living. Proceedings of the IEEE [online]. 2008, 96(10), 1698-1714 [cit. 2017-11-07]. DOI: 10.1109/JPROC.2008.928765.

[6] B.Xu, J.Zheng, Q.Wang "Analysis and Design of Real-Time Micro-Environment Parameter Monitoring System Based on Internet of Things," 2016 IEEE International Conference on Internet of Things (iThings) and IEEE Green Computing and Communications (GreenCom) and IEEE Cyber, Physical and Social Computing (CPSCom) and IEEE Smart Data (SmartData).

[7] Q. Min, Y. F. Ding, T. Xiao et al., "Research of Visualization Monitoring Technology Based on Internet of Things in Discrete Manufacturing Process," 2015 2nd International Symposium on Dependable Computing and Internet of Things (Dcit), pp. 128-133, 2015

[8] Y. Wang, J. Song, X. Liu et al., "Plantation Monitoring System Based on Internet of Things" 2013 IEEE International Conference on Green Computing and Communications and IEEE Internet of Things and IEEE Cyber, Physical and Social Computing pp. 366-369.

[9] Y. E. Windarto, and D. Eridani, "Door And Light Control Prototype Using Intel Galileo Based Internet of Things," 2017 4th International Conference on Information Technology, Computer, and Electrical Engineering (Icitacee), pp. 176-180, 2017.

[10] F. AlFaris, A. Juaidi, and F. Manzano-Agugliaro, "Intelligent homes' technologies to optimize the energy performance for the net zero energy home," (in English), Energy and Buildings, Review vol. 153, pp. 262-274, Oct 2017.

[11] M. Alirezaie et al., "An Ontology-based Context-aware System for Smart Homes: E-care@home," (in English), Sensors, Article vol. 17, no. 7, p. 23, Jul 2017, Art. no. 1586.

[12] M. Bassoli, V. Bianchi, and I. De Munari, "A Plug and Play IoT Wi-Fi Smart Home System for Human Monitoring," (in English), Electronics, Article vol. 7, no. 9, p. 13, Sep 2018, Art. no. 200.

[13] C. Coelho, D. Coelho, M. Wolf et al., "An IoT Smart Home Architecture for Long-Term Care of People with Special Needs," 2015 Ieee 2nd World Forum on Internet of Things (Wf-Iot), pp. 626-627, 2015.

[14] Oxford dictionaries, "Definition of big data in English". Available online: https://en.oxforddictionaries.com/definition/big_data (accessed on 25/11/2018)

[15] M. Rouse. "Predictive modeling,". Available online: https://searchenterpriseai.techtarget.com/definition/predictive-modeling (accessed on 25/11/2018).

[16] M. W. Ahmad, J. Reynolds, and Y. Rezgui, "Predictive modelling for solar thermal energy systems: A comparison of support vector regression, random forest, extra trees and

regression trees," Journal of Cleaner Production, vol. 203, pp. 810-821, Dec, 2018.

[17] "IBM SPSS software,". Available online: https://www.ibm.com/analytics/spss-statistics-software (accessed on 25/12/2018).

[18] "IBM Watson,". Available online: https://www.ibm.com/watson/. (accessed on 25/11/2018).

[19] K. K. Nagwanshi, and S. Dubey, "Statistical Feature Analysis of Human Footprint for Personal Identification Using BigML and IBM Watson Analytics," Arabian Journal for Science and Engineering, vol. 43, no. 6, pp. 2703-2712, Jun, 2018

[20] T. Perumal, S. K. Datta, C. Bonnet et al., "IoT Device Management Framework for Smart Home Scenarios," 2015 Ieee 4th Global Conference on Consumer Electronics (Gcce), pp. 54-55, 2015.

[21] O. Arnold, L. Kirsch, A. Schulz, and Ieee, "An Interactive Concierge for Independent Living," (in English), 2014 Ieee 3rd Global Conference on Consumer Electronics (Gcce), Proceedings Paper pp. 59-62, 2014.

[22] J. R. Carvalko and Ieee, "LAW AND POLICY IN AN ERA OF CYBORG-ASSISTED-LIFE THE IMPLICATIONS OF INTERFACING IN-THE-BODY TECHNOLOGIES TO THE OUTER WORLD," in 2013 Ieee International Symposium on Technology and Society(IEEE International Symposium on Technology and Society, New York: Ieee, 2013, pp. 204-215.

[23] Y. Chen, E. Argentinis, and G. Weber, "IBM Watson: How Cognitive Computing Can Be Applied to Big Data Challenges in Life Sciences Research," (in English), Clinical Therapeutics, Review vol. 38, no. 4, pp. 688-701, Apr 2016.

[24] E. Kolker and V. Ozdemir, "How Healthcare Can Refocus on Its Super-Customers (Patients, n=1) and Customers (Doctors and Nurses) by Leveraging Lessons from Amazon, Uber, and Watson," (in English), Omics-a Journal of Integrative Biology, Review vol. 20, no. 6, pp. 329-333, Jun 2016.

[25] S. S. Murtaza, P. Lak, A. Bener, and A. Pischdotchian, "How to Effectively Train IBM Watson: Classroom Experience," in 2016 49th Hawaii International Conference on System Sciences, T. X. Bui and R. H. Sprague, Eds. (Proceedings of the Annual Hawaii International Conference on System Sciences, Los Alamitos: Ieee Computer Soc, 2016, pp. 1663-1670.

[26] P. Cervenka, I. Hlavaty, A. Miklosik, and J. Lipianska, "Using cognitive systems in marketing analysis," (in English), Economic Annals-Xxi, Article vol. 160, no. 7-8, pp. 56-61, Oct 2016.

[27] IBM SPSS Modeler 18 user guide Guide. Available online: ftp://public.dhe.ibm.com/software/analytics/spss/documentation/modeler/18.0/en/ModelerUsersGuide.pdf (accessed on 25/11/2018).

[28] IBM SPSS Modeler 16 Algorithms Guide. Available online: ftp://public.dhe.ibm.com/software/analytics/spss/documentation/modeler/16.0/en/AlgorithmsGuide.pdf (accessed on 25/12/2018).

[29] IBM SPSS Modeler 17 Algorithms Guide. Available online: ftp://public.dhe.ibm.com/software/analytics/spss/documentation/modeler/17.0/en/AlgorithmsGuide.pdf (accessed on 25/12/2018)H. Goto, Y. Hasegawa, and M. Tanaka, "Efficient Scheduling Focusing on the Duality of MPL Representatives," Proc. IEEE Symp. Computational Intelligence in Scheduling (SCIS 07), IEEE Press, Dec. 2007, pp. 57-64, doi:10.1109/SCIS.2007.357670.

[30] IBM SPSS Modeler 18 Algorithms Guide. Available online: ftp://public.dhe.ibm.com/software/analytics/spss/documentation/modeler/18.0/en/AlgorithmsGuide.pdf (accessed on 25/12/2018

The Visibility of Pedestrians in the Traffic Area, Using Different Light Source Technologies of Car Headlights

Richard Baleja, Ondřej Dolejší, Petr Běčák
Faculty of Electrical Engineering and Computer Science
VSB-TU Ostrava
Ostrava, Czech Republic
richard.baleja.st@vsb.cz, ondrej.dolejsi.st@vsb.cz, petr.becak.st@vsb.cz

Karel Sokanský, Tomáš Novák
Faculty of Electrical Engineering and Computer Science
VSB-TU Ostrava
Ostrava, Czech Republic
karel.sokansky@vsb.cz, tomas.novak1@vsb.cz

Abstract—**The aim of this publication is to compare the individual light source technologies of car headlights illuminating a traffic area. The measurement was carried out on a sample section of a road, which took place on the safe driving polygon in Ostrava. The measurement was carried out for two model situations of a public lighting: car headlights in a low beam mode and a combination of a public lighting together with car headlights in a low beam mode. Vehicles of the same type with headlights with halogen, xenon and LED sources were used for the measurements. The illuminating of the road corresponded to the real conditions that can occur on the roads in a normal operation. Luminance of the space analysis and vertical illumination measurements were performed on the model road. The aim of this work is to assess the visibility of road users in particular real traffic situations.**

Keywords: car headlights; traffic area; correlated color temperature; low beam mode; luminance analyzer

I. Introduction

In November 2018, the Comparative Measurement of Night Traffic Area took place at the LIBROS safe driving polygon in Ostrava. The aim of the measurement was to compare the illuminance distribution on the model road section [1], which was illuminated by different types of the car headlights. Three Škoda Octavia vehicles with halogen, xenon and LED light sources in the front headlights were selected for the measurements. The measurement took place on a sample section of the polygon road, which was illuminated by a one-sided lighting system. Metal-halogen vapor of the PHILIPS brand with a power of 70 W and a correlated color temperature of 4000 K were installed in the luminaires of the public lighting [2].

Table 1. Measured sample situations in night traffic area

Headlights	Situation in the night traffic area
Halogen	Low beam mode without the public lighting
	Low beam mode with the public lighting
Xenon	Low beam mode without the public lighting
	Low beam mode with the public lighting
LED	Low beam mode without the public lighting
	Low beam mode with the public lighting

II. The Measurement of the Road Illuminance

As a model section was chosen a road with two lanes with a total width of 7.2 meters. The total length of the model section of the road was 90 meters with a luminaire spacing of about 30 meters [3]. The measured road section is shown in Figure 1.

Figure 1. Model road section at the safe driving polygon in Ostrava.

The luminance distribution on the sample road section was measured using a luminance analyzer according to the relevant standards. The luminance analyzer was placed at a height of 1.5 meters above the road at a distance of 60 meters from the measured sample section [4]. The measurements were taken when public lighting was turned on and also when it was turned off to detect light from distant ambient light sources in order to detect potential influence of an adaptation luminance. In the figures below, there are model road luminance maps when public lighting is on (Figure 2), and when public lighting is turned off (Figure 3). The measured average luminance of the road illuminated by public lighting is 0.29 cd/m^2 and the ambient contributions on the section are of hundredths of cd/m^2.

Figure 3. Luminance map of the situation with the public lighting turned on.

Figure 2. Luminance map of the situation with the public lighting turned off.

III. The Measurement of the Traffic Area

The measurement of the traffic area was carried out on a straight road under different meteorological conditions. Using smoke grease, the fog situation was simulated, and a wet road was simulated using a fire truck. It would be very extensive to describe the results of all combinations of measurements, therefore there is described only the measurement of a straight dry section in this article. The measurements were carried out for three types of car headlights with technology (LED, xenon and halogen) in a low beam mode for two standard distances of pedestrians from the vehicle. The distance between the pedestrian and the vehicle was determined by the vehicle's braking distance for a speed of 50 km/h to 35 meters and for a speed of 90 km/h at a distance of 83 meters [5]. Measurements were made by using two luminance analyzers. One luminance analyzer was placed in the vehicle (checking the luminance from the driver's view from the car), the other was located outside the vehicle at the same height. The height of the two analyzers corresponded to the height of the driver's eyes, who is sitting in the car. The luminance analyzers have the same luminance parameters as ISO, aperture, and shutter length settings in order to assess the impact of the car's windscreen on the perception of the environment. The car windscreen impact is not a part of this publication and it will be described in detail in the further visibility analyzes under the terms of safety in night traffic area.

Figure 4 shows the measured dummies as well as the Škoda Octavia cars, on the left a vehicle with LED headlights, in the middle a vehicle with halogen headlights and on the right a vehicle with xenon headlights. The straight dry road model was measured when the public lighting was switched on and switched off for a distance of the dummy of 35 and 83 meters while car has the low beam headlight mode. On the luminance maps, the red figurant is always placed on the left and the blue one is placed on the right side. The first series of luminance maps is made for a distance of 83 meters from the vehicle, the second series for a distance of 35 meters from the vehicle. The luminance maps have always been used to assess the background luminance as the average luminance of the entire image and the luminance of the dummies. Subsequently, the contrast between the dummy and the background was calculated from the measured values, and then it is possible to determine unambiguously whether it is possible to see the figurant in the traffic area well in advance. In the following chapter, examples of calculated contrasts for figurants at a distance of 35 meters with public lighting on and off are always

Figure 4. Measured cars Škoda Octavia and dummies representing pedestrians in the traffic.

given.

IV. The Assessment of Measurement of Straight Dry Road Module

When assessing the calculated contrasts between the observed dummies and a background, it was found that when the entire luminance image is used to determine the average background luminance, the differences are minimal between the calculated contrasts in the situation with the public lighting on and off in case of the red dummy, which is located on the right side, even if the luminance differences of the dummy itself are doubled. This finding applies to all headlights and to both measured distances. This is due to the fact that when the public lighting is switched on, the luminance of the dummy increases, as well as the average luminance of the whole image. Therefore, there is an increasing of both luminances in approximately the same ratio. In the case of the blue dummy placed on the right side, the situation is different. The differences in the calculated contrasts are significantly higher in the case of the road illuminated by the public lighting. The significant difference is caused by the placement of the dummy, the blue dummy on the right is located on the public lighting side, the red dummy on the left is located on the opposite side, so its average luminance is approximately twice lower than the luminance of the dummy placed under the public lighting.

However, if we do not assess the contrast to the average luminance of the whole image, but only to the near surroundings of the dummy, the situation will be slightly different. It is important to note that in the case of a dummy placed on the left, its background is formed by an empty asphalt surface, and therefore the background luminance is very low at about 0.03 cd/m^2. The situation is similar in the case of the public lighting switched on and off. In the case of the dummy placed on the public lighting side on the right, its background is formed by the building facade. In the case when the public lighting is off, the average luminance is practically the same as in the case of a pedestrian placed on the left, but when the public lighting is turned on, the background luminance increases approximately ten times. Therefore, when assessing the result, it was found that the dummy placed on the left was more visible when the public lighting was turned on. The calculated contrast for the right-positioned dummy was approximately the same when the public lighting was turned on and off. In the case of the dummies located at a bigger distance from the vehicle, the dummies cannot be illuminated by the low beam mode of headlights, and therefore the only source of vertical illumination of dummies is the public lighting. However, it is important to note that the position of the dummy plays a significant role in the case of the vertical component of the illumination. The difference is when the dummy is in front of or behind the public lighting post from the perspective of an incoming car. In Figure 5 we can see the most critical position of pedestrians in the traffic area, in front of the public lighting post from the perspective of an incoming car, when the lowest vertical illuminance values are generated towards the pedestrian.

Figure 5. Model situation of a combination of the public lighting and low beam mode of headlights.

From these findings it can be stated that the public lighting helps to see the pedestrians and obstacles in the traffic area, but it must be designed and implemented according to the relevant standards and regulations with the emphasis on generating the vertical component of the illumination from the direction of the incoming drivers. Pedestrian visibility is also significantly influenced by their background. High luminances on the facades of surrounding buildings can negatively affect its visibility. The figures below show the individual traffic situations in the measurement, and the sample luminance and contrasts assessment for a distance of 35 meters.

The formula for calculating the contrast between the dummy and background luminance:

$$K = |L_a - L_b|/L_b \qquad (1)$$

K – contrast (-)

L_a – pedestrian luminance (cd/m^2)

L_b – background luminance (cd/m^2)

A. A Road Illuminated by LED Headlights

In the case of a road illuminated by the low beam LED headlights there is high-quality illumination of the departure zones in the periphery area that applies both to the illuminance and a light color, while providing luminous flux up to 40 meters in front of the driver.

Figure 6. The public lighting switched on, dummies - 83 m.

Figure 7. The public lighting switched off, dummies - 83 m.

Figure 8. The public lighting switched on, dummies - 35 m.

Figure 9. The public lighting switched off, dummies - 35 m.

Table 2. Measured illuminances and calculated contrasts for the situation, when the public lighting is on and the dummies are at a distance of 35 m

The average luminance of the left dummy (cd/m^2)	The average luminance of the right dummy (cd/m^2)	The average luminance of the background (cd/m^2)	The contrast of the left dummy/ background	The contrast of the right dummy/ background
0,086	0,202	0,553	0,845	0,635

Table 3. Measured illuminances and calculated contrasts for the situation, when the public lighting is off and the dummies are at a distance 35 m

The average luminance of the left dummy (cd/m^2)	The average luminance of the right dummy (cd/m^2)	The average luminance of the background (cd/m^2)	The contrast of the left dummy/ background	The contrast of the right dummy/ background
0,048	0,096	0,567	0,915	0,830

B. Straight Dry Road Model Illuminated by the Xenon Headlights

The xenon headlights have excellent light beam trimming that reduces the dazzling of oncoming vehicles. However, there may be a situation (headlights adjustment) where the low beam headlights illuminate the area in front of the driver only within 30 meters. High Tc of the xenon headlights are suitable for peripheral vision areas with low adaptation luminance.

Figure 10. The public lighting switched on, dummies - 83 m.

Figure 12 The public lighting switched off, dummies - 83 m.

Figure 11. The public lighting switched on, dummies - 35 m.

Figure 13. The public lighting switched off, dummies - 35 m.

Table 4. Measured illuminances and calculated contrasts for the situation, when the public lighting is on and the dummies are at a distance of 35 m

The average luminance of the left dummy (cd/m^2)	The average luminance of the right dummy (cd/m^2)	The average luminance of the background (cd/m^2)	The contrast of the left dummy/ background	The contrast of the right dummy/ background
0,074	0,141	0,372	0,800	0,621

Table 5. Measured illuminances and calculated contrasts for the situation, when the public lighting is off and the dummies are at a distance of 35 m

The average luminance of the left dummy (cd/m^2)	The average luminance of the right dummy (cd/m^2)	The average luminance of the background (cd/m^2)	The contrast of the left dummy/ background	The contrast of the right dummy/ background
0,037	0,036	0,354	0,896	0,898

C. Straight Dry Road Model Illuminated by Halogen Headlights

The classic halogen headlights is capable of illuminating up to an area of about 35 metrs in front of the driver. However, it is important to note that in the right and left boarding zones we only move in the area of 20 - 25 meters.

Figure 14. The public lighting switched on, dummies - 83 m.

Figure 15. The public lighting switched off, dummies - 83 m.

Figure 16. The public lighting switched on, dummies - 35 m.

Figure 17. The public lighting switched off, dummies - 35 m.

Table 6. Measured illuminances and calculated contrasts for the situation, when the public lighting is on and the dummies are at a distance of 35 m

The average luminance of the left dummy (cd/m^2)	**The average luminance of the right dummy (cd/m^2)**	**The average luminance of the background (cd/m^2)**	**The contrast of the left dummy/ background**	**The contrast of the right dummy/ background**
0,080	0,167	0,316	0,748	0,472

Table 7. Measured illuminances and calculated contrasts for the situation, when the public lighting is off and the dummies are at a distance of 35 m

The average luminance of the left dummy (cd/m^2)	**The average luminance of the right dummy (cd/m^2)**	**The average luminance of the background (cd/m^2)**	**The contrast of the left dummy/ background**	**The contrast of the right dummy/ background**
0,043	0,057	0,284	0,850	0,801

V. Conclusion

In the case when the road is illuminated by a combination of low beam headlights (xenon, halogen, LED) and by the public lighting, it is apparent from the above-mentioned results that by this combining of the headlights of the vehicle in the low beam mode and the public lighting system, the visibility of the obstacles is increased in the whole traffic area. It was also found by measurement that the visibility of the obstacle in the right-hand lane in the direction of incoming cars with the headlights in low beam mode without public lighting is up to 40 meters from the vehicle. This is an alarming finding, as we are able to stop the vehicle within this distance on wet roads only at speeds of up to 55 km/h.

Based on the assessment of luminance maps, it can be concluded that the public lighting increases the visibility of pedestrians in the traffic area. Nevertheless, it is important to note that only the public lighting, which is designed according to the relevant standards with the emphasis on achieving the necessary vertical illumination in the entire section between two light points (luminaire spacing), has a positive impact on pedestrian noticing in the traffic area. It is important to keep the required even distribution when designing public lighting and do not dazzle the users of the traffic area. Of course, compliance with normative requirements does not necessarily mean that a pedestrian will be seen by the driver in all circumstances. In practice, there may occur a situation where a pedestrian is located directly below the luminaire or in front of the luminaire from the perspective of an incoming vehicle, see the Figure 5, and thus the pedestrian will not be sufficiently illuminated by the incoming driver. Thus, the driver's response to the obstacle may be delayed, or the driver may fail to notice the pedestrian completely. The type of headlights also have a significant impact on the perception of obstacles in the traffic area, see the assessments in the previous chapter. Firstly, the headlights should be adjusted properly in order to illuminate the road in the length needed for early detection of obstacles and to avoid dazzling other road users.

Acknowledgments

This article was developed within the framework of the research grant VI2VS / 571 "Analysis of visibility of road users for the purpose of increasing their safety at dusk and at night" provided by the Ministry of the Interior and supported by the Traffic Police Service of the Police Presidium of the Czech Republic.

References

[1] ČSN 73 6110, "Design of local roads," Czech standardization institute, Prague, 2006.

[2] ČSN CEN/TR 13201-1, "Road Lighting: Part 1: Guidance for selecting lighting classes," Czech standardization institute, Prague, 2016.

[3] EN 13201-3, "Road Lighting: Part 3: Calculation," Czech standardization institute, Prague, 2016.

[4] EN 13201-2, "Road Lighting: Part 2: Requirements," Czech standardization institute, Prague, 2016.

[5] Road traffic regulations, "Road traffic regulations - duties of persons enjoying diplomatic privileges and immunities in the Czech Republic," Act No. 361/2000 concerning road traffic and amending certain acts (Road Traffic Act), 2000.

Smart Crash Barrier Based on Off-Grid Technology

Vojtech Blazek, Jan Fulnecek, Michal Petruzela, Lukas Prokop, Stanislav Misak
Department of Electrical Power Engineering, Faculty of Electrical Engineering and Computer Science
VŠB - Technical University of Ostrava
Ostrava, Czech Republic
vojtech.blazek@vsb.cz, jan.fulnecek@vsb.cz, michal.petruzela@vsb.cz

David Seidl
Department of Computer Science, Faculty of Electrical Engineering and Computer Science
VŠB - Technical University of Ostrava
Ostrava, Czech Republic
david.seidl@vsb.cz

Abstract — **The paper presents an idea to use road warning signalisation system to increase road safety for which an independent power source to operate the warning system had to be identified. We tested two energy harvesting methods of harnessing energy of airflow created by a passing vehicle. The methods applied piezoelectric element with a wind flap and small vertical axis wind turbine. To find the proper shapes of wind flap and wind turbine, we used software ANSYS Workbench 18. An electric circuit was developed to efficiently utilise energy harvesting. Finally, we shifted from energy harvesting to an off-grid based system as there was a very low amount of energy in this airflow. The paper described the design of the off-grid system based on practical experience, while keeping in mind its modularity and future upgrades.**

Keywords: Off-Grid, Road Warning System, Signaling, Dangerous road sections, Micro Grid, Smart Grids

I. INTRODUCTION

Road safety has always been important and nowadays along with an increase in traffic density, it is of an increasing concern on the part of authorities and road users. Efforts are exerted to decrease the number of traffic accidents or at least their severity.

According to statistics, the accident ratio in the Czech Republic is decreasing [1]. However, there are still many, especially on road sections with limited sight distance. Based on the experience from Czech roads, we introduce an idea to create a system that improves road safety and is energy self-sufficient. The idea to improve road safety lies in the application of signal LEDs situated on road sections with limited sight distance. These LEDs mounted on a crash barrier are meant to inform drivers about an oncoming vehicle.

First, we investigated the idea of energy harvesting as an energy source for the system [2][3]. There are many methods of energy harvesting – piezoelectric, electromagnetic, thermoelectric, etc.

First, we decided to test piezoelectric, which is one of the most common energy harvesting methods. In [4], there is an example of piezoelectric energy harvesting application for a railway monitoring system energy supply. Piezoelement has a few downsides, for example, to work efficiently a frequency of oscillations needs to be as close as possible to the rated mechanical resonation frequency. Piezoelement efficiency may be in the order of units of % [5], so the amount of produced energy was insufficient for our purpose.

Another idea was the usage of a small vertical axis wind turbine connected with dynamo instead of piezolement with wind flap. Unlike the wind flap, there is no need for a rectifier. But on the other hand, there is also a disadvantage: the smaller the dynamo is, the less efficient it is. Research in [6] and [7] inspired us to use ANSYS Workbench to design the shape of wind turbine.

Both energy harvesting systems were designed with the idea of mounting wind flap or wind turbine on the crash barrier. But the final concept is photovoltaic-based system where only signal LEDs are mounted on a crash barrier and the rest of the system is located on a construction next to this crash barrier.

II. CONCEPTS

A. System Based on Piezoelement as a Power Source

We constructed several prototypes of piezoelement with a wind flap. The purpose of these prototypes was to find out if they produce enough energy to supply signal LED. First of all, we tested and evaluated these prototypes in the laboratory. After this step, selected prototypes were tested on a real crash barrier in VSB-TUO campus. Figure 1 shows a wind flap prototype in the shape of a cup.

Figure 1. Prototype with piezoelement

Simultaneously we analysed a wind flap in ANSYS Workbench 18 with a goal to optimize its shape. Kinetic energy of a wind flow created by passing by vehicle is captured by this wind flap. This energy moves the wind flap and causes bending of piezoelement and thus generates electric energy.

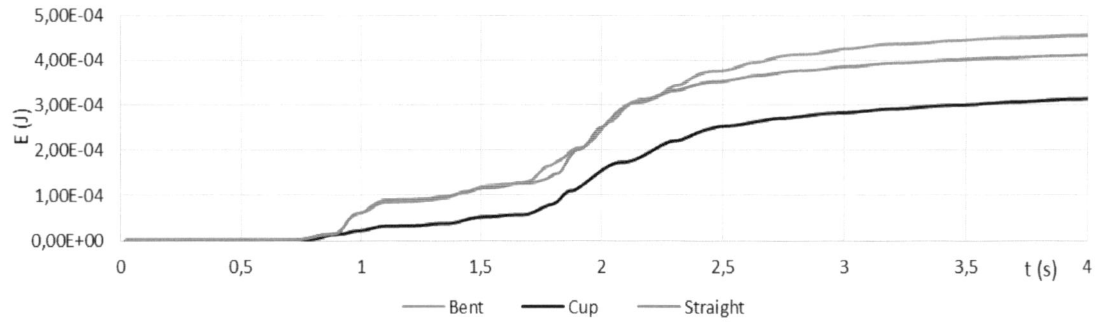

Figure 2. Energy generated over time by different wind flaps

The final shape of the wind flap was designed according to the results from analysis. Figure 3 shows the analysed shapes of wind flap. Analysis was done having used finite elements method in software ANSYS Workbench 18.0. The used method was two-way fluid structure interaction, this method utilises two modules - CFX for fluid flow and transient structural for solid bodies. Generated energy is calculated according to the following equation:

$$E = \sum_{t=1} T_t \cdot (\varphi_t - \varphi_{t-1}) \quad (1)$$

where t is time step, φ_t and φ_{t-1} is rotation of assembly in radians while affected by torque T in Nm. Figure 2 shows the total amount of energy generated over time. According to the results, it is obvious that the bent shape generated more energy than others. However, it was not sufficient for our purposes.

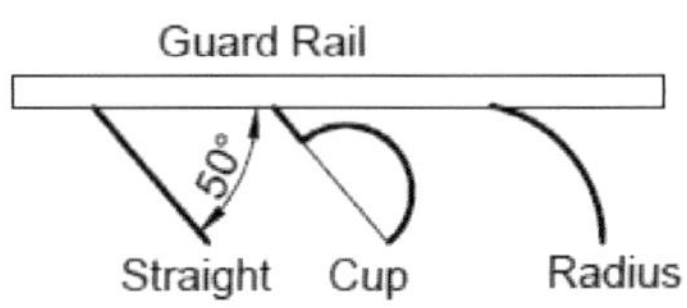

Figure 3. Shapes of tested wind flaps

Electric part of the prototype was based on operational amplifier AD820R. This amplifier has one important feature and that is its standby power consumption around hundredths of mA. This fact allows to supply energy to the operational amplifier from the piezoelement. Other components of this circuit are a full wave rectifier and two electrolyte capacitors.

Mechanical movement of the wind flap is transformed into bending moment in piezoelement, which generates electric energy. This AC voltage is rectified and used to charge capacitors. The amount of accumulated energy is estimated from capacitor voltage. When capacitor voltage reaches the threshold of comparator set by voltage divider, the transistor switches into an on state and accumulated energy is available on the output connector, where the LED is connected. We performed numerous tests, analyses and evaluations of energy required for signal LED and self-consumption of the system, and we compared it to possibilities of piezoelement as a power source. And we decided to replace the system power source.

B. System Based on Small Wind Turbine as a Power Source

We replaced wind flap by wind turbine as a power source of our system. For this purpose, several prototypes of Savonius type wind turbine were designed. We decided for a drag-type wind turbine even though in steady state, it is less efficient than a lift-type. But a drag-type wind turbine starts up faster than lift type and its nominal rotating speed is slower. Figure 4 shows one of those prototypes.

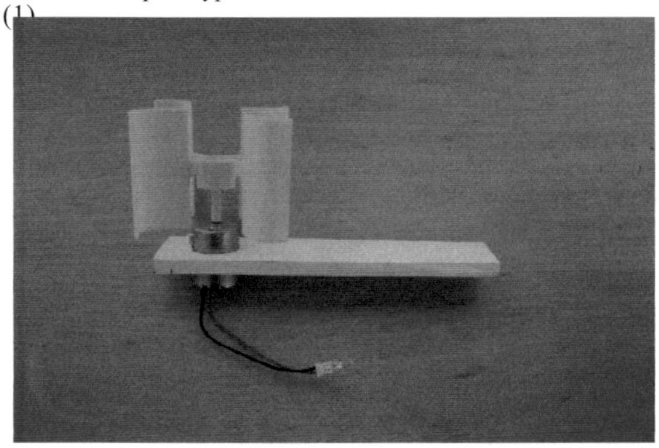

Figure 4. Prototype with wind turbine

The operation of wind turbine is the same as wind flap: energy of wind flow, created by a passing car, is harnessed by wind turbine. Wind turbine blades, attached to the dynamo rotor axis, are pushed by oncoming wind. This causes revolution of rotor and thus electric energy generation. Dynamo output voltage is DC so there is no need for a rectifier and LED can be connected directly to the dynamo output.

During the turbine development, shape of wind turbine blades was utilised by software ANSYS workbench 18. In analysis, we set the speed of wind to 5 m/s and turbines were examined and compared in steady state. Figures 6-9 show the analysed shapes of wind turbine. Figure 5 shows the mechanical power generated by different wind turbines in steady state.

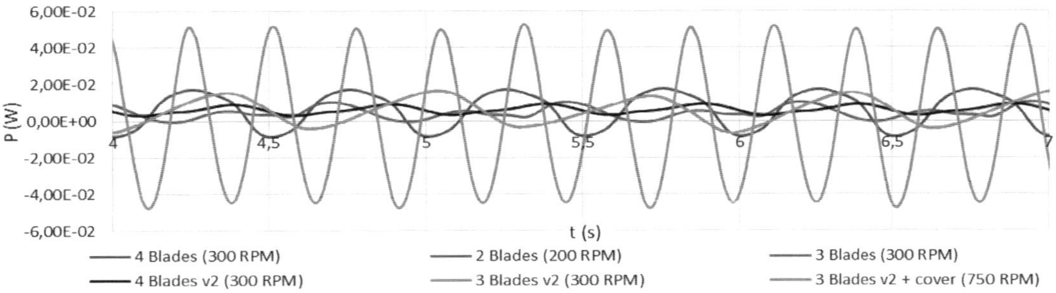

Figure 5. Power generated by analysed wind turbines (continuous flow)

Figure 6. 3D model of analysed turbine – Savonious 4 blades

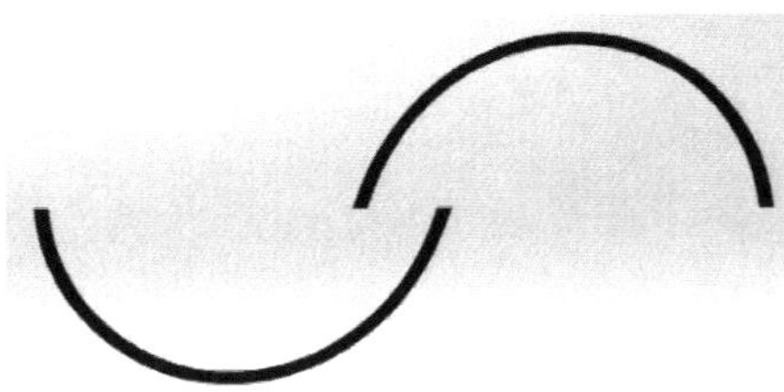

Figure 7. Savonious wind turbine 2 blades

Figure 8. Savonius wind turbine 3 blade a Savonius wind turbine 4 blades

Figure 9. Savonius wind turbine 3 blades version 2, and Savonius wind turbine 3 blades version 2 with a cover

According to the results shown Fig. 5, we chose Savonius type wind turbine with 3 blades and cover. In the field test, this turbine did not have satisfactory results either. Therefore, we decided to abandon the concepts based on the energy harvesting from an air flow, created by a passing car. The reason for abandoning these concepts is simple: energy in wind flow is quite small and with combination of very low efficiency, these energy harvesting methods give us insufficient amounts of electric energy to work with.

Finally, we had to redirect our research to a different concept. In this last concept, power consumption was covered by photovoltaic panel and batteries.

C. System Based on Photovoltaic Panel as a Source

Last concept is based on an Off-grid technology. We already had experience in the design and operation of this platform type with control system, such as Active demand system management (ADSM) [8],[9],[10],[11].

Figure 10 shows the prototype of a system that has been designed while keeping in mind its high modularity and applicability on Czech roadways.

Figure 10. Laboratory testing of Off-grid based system

Figure 11 shows the principle of the system. The system works in the following manner – a laser rangefinder is responsible for vehicle detection. When the rangefinder detects a vehicle, it sends a signal to the microcontroller which manages switching of addressable RGB LED strip mounted on a crash barrier.

The whole system is powered by lead acid batteries with total capacity of 12 Ah. The batteries are recharged by a photovoltaic solar regulator. One photovoltaic panel 125 Wp is connected to this regulator and is mounted on a structure next to the crash barrier. Figure 12 shows a 3D model of the whole system – crash barrier with LED strip, structure that carries the photovoltaic panel and also a switchboard which also contains solar regulator, batteries and microcontroller.

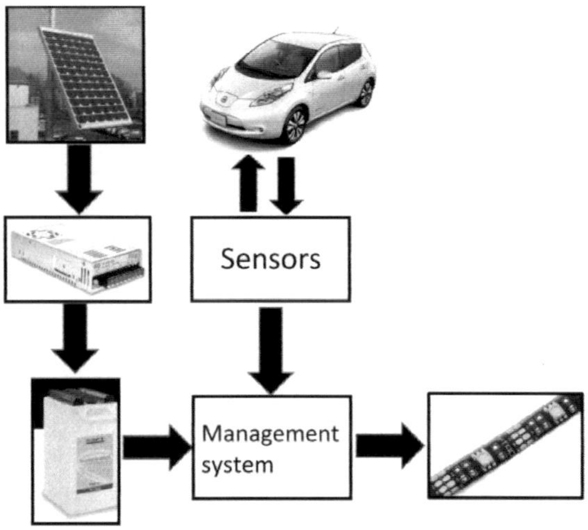

Figure 11. Block diagram of system

We chose a laser sensor LIDAR (Light detection and ranging) as a device for vehicle detection. LIDAR is placed on a crash barrier and measures the distance in the following manner – it illuminates a vehicle with laser pulse and measures how much time it takes for the pulse to return to the sensor. This method ensures good accuracy of distance measurement. Utilised LIDAR is able to measure distances in the range from 5 m to 40 m and sends information to the microcontroller for the evaluation. This microcontroller is Arduino UNO clone that is fully compatible with original Arduino. This microcontroller is based on ATmega328 with in-circuit system programming, 16 MHz crystal, 6 analog inputs, 16 digital input/outputs and USB connectivity.

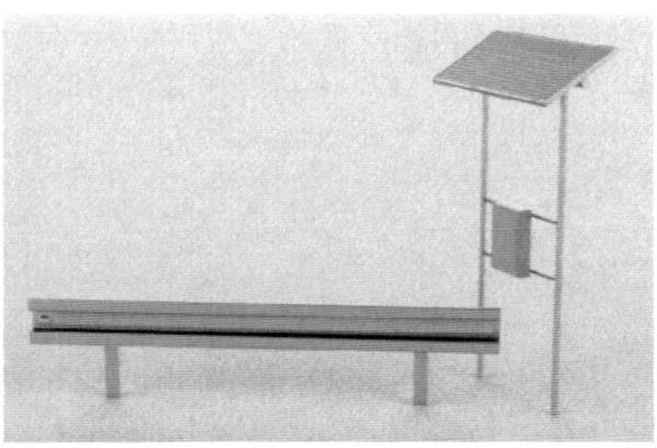

Figure 12. 3D model of the whole system

Control program for Arduino was written in C++ programming language in Espressif IoT Development Framework with standard libraries. There is an array of 1800 structures and one structure is responsible for state of one RGB LED. Then in the main loop, these arrays are transferred to LED strip via the serial communication and rewrites state of all LEDs. The next step in program is to delete array's last column and replace it with previous one, this repeats until program reaches the first column. This way, we achieved a moving text or pictures. Animation is played when laser sensor detects closing in object three times in a row.

Heating element in a form of resistor is another system component that is connected to power supply by a bimetal thermostat. Heating element is located inside the switchboard and its purpose is to protect thermally sensitive parts of the system from cold weather in winter. Figure 13 shows a circuit diagram of the system.

Most recent version of the system is powered by a valve regulated lead acid battery (VRLA). This battery is composed of 4 cells that are connected in series. Voltage of each cell is 6 V and has capacity 12 Ah. Total capacity of battery is 250 Wh of usable energy, this value includes reserve, to avoid battery deep discharge and shortening its lifespan.

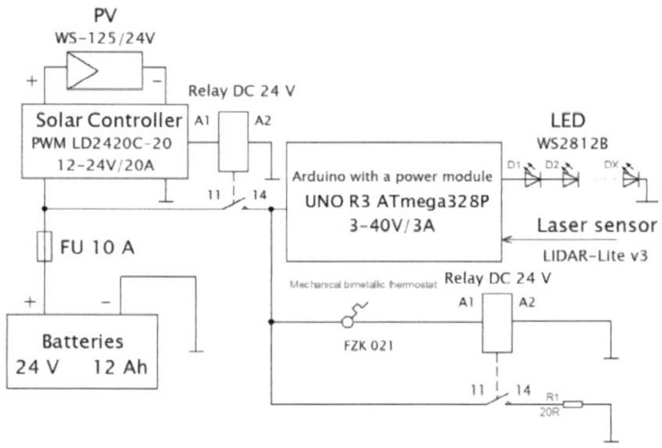

Figure 13. Electric schematic of system

Rangefinder, LED strip and microcontroller together have 17 W consumption in total. Heating element itself has 28.8 W consumption. It is 45,8 W in total, considering battery capacity it gives us 8 hours and 40 minutes of nonstop operation. Power consumption in real application is lower. Because heating element turns on only when internal temperature drops below -5°C. LED strip – there is a programmed number of shining red LEDs and only when a vehicle is detected. The system was tested for half a year in the VSB-TUO campus, and next it was relocated to a municipality called Suche Lazce, next to the public roadway for further testing. Figure 14 shows a photo of the system located in Suche Lazce.

Figure 14. System located in Suche Lazce

III. CONCLUSION

The first two concepts were based on the idea of energy harvesting of wind flow generated by a passing vehicle and exploitation of this energy to supply the whole system. We conducted many analyses and tests resulting in the conclusion that there was not enough energy for reliable operation of the system. This led us to the decision to utilise off-grid technology.

While developing the off-grid system, we had in mind its high modularity, low price and accessibility of parts. As a result, we can easily modify the system by adding more batteries, connecting different photovoltaic panels or RGB LED strips. The system, based on this technology, is currently in service.

The next version of the system that is currently under development will have an extended signal ability. This will be achieved by connecting more LED strips, and allow us to implement signalization in the form of text or symbols. For example, text and symbol combination that warns drivers about the possibility of ice or any other danger on the road. Also, it is planned to increase energy capacity by adding more batteries. The last planned thing is incorporation of internet of things (IoT) into the system. IoT should be used to process information and control traffic with the aim to further improve traffic fluency.

ACKNOWLEDMENT

This paper was supported by the following projects: LO1404: Sustainable development of ENET Centre; SP2019/159 and SP2019/28 Students Grant Competition and TACR TH02020191 and TACR TH777911, Czech Republic and the Project LTI17023 "Energy Research and Development Information Centre of the Czech Republic" funded by Ministry of Education, Youth and Sports of the Czech Republic, program INTER-EXCELLENCE, subprogram INTER-INFORM and project TN01000007 National Centre for Energy.

REFERENCES

[1] T. Kuzel and J. Zly, "Dopravní nehodovost na území moravskoslezského kraje za rok 2018," [online]. In: . 2019 [cit. 2019-03-01]. Online: https://www.policie.cz/clanek/dopravni-nehodovost-v-roce-2018.aspx

[2] C. Wei and X. Jing, "A comprehensive review on vibration energy harvesting: Modelling and realization," Renew. Sustain. Energy Rev., vol. 74, no. December 2016, pp. 1–18, 2017, DOI: 10.1016/j.rser.2017.01.073.

[3] A. Khaligh, Peng Zeng and Cong Zheng, "Kinetic Energy Harvesting Using Piezoelectric and Electromagnetic Technologies - State of the Art," IEEE Trans. Ind. Electron., vol. 57, no. 3, pp. 850-860, March 2010, DOI: 10.1109/TIE.2009.2024652.

[4] M. Gao, Y. Li, J. Lu, Y. Wang, P. Wang, & L. Wang, "Condition monitoring of urban rail transit by local energy harvesting," International Journal of Distributed Sensor Networks, vol 14, no 11, November 2018, DOI: 10.1177/1550147718814469.

[5] Z. Yang, A. Erturk, J. Zu, "On the efficiency of piezoelectric energy harvesters," Extreme Mechanics Letters, vol. 15, pp. 26-37, September 2017, ISSN: 2352-4316

[6] P. Kacor, S. Misak, and L. Prokop, "Optimization and redesign of vertical axis wind turbine for generator of independent source of energy", *Annals of DAAAM for 2010 & Proceedings of the 21st International DAAAM Symposium*, 20-23rd October 2010, Zadar, Croatia, ISSN 1726-9679.

[7] P. Kacor, S. Misak and L. Prokop, "Modification of construction design of vertical axis wind turbine," Annals of DAAAM for 2011 & Proceedings of the 22nd International DAAAM Symposium, Volume 22, No. 1, ISSN 1726-9679.

[8] T. Vantuch, S. Misak, T. Jezowicz, T. Burianek and V. Snasel, "The Power Quality Forecasting Model for Off-Grid System Supported by Multiobjective Optimization" IEEE Transactions on Industrial Electronics, vol. 64, no. 12, pp. 9507-9516, 2017, DOI: 10.1109/TIE.2017.2711540.

[9] T. Vantuch, S. Misak and J. Stuchly, "Power quality prediction designed as binary classification in AC coupling Off-Grid system," IEEE 16th International Conference on Environment and Electrical Engineering (EEEIC), Florence, pp. 1-6, September 2016, DOI: 10.1109/EEEIC.2016.7555551

[10] S. Misak, J. Stuchly, J. Vramba, T. Vantuch and D. Seidl, "A novel approach to adaptive active relay protection system in single phase AC coupling Off-Grid systems" Electric Power Systems Research, vol. 131, pp. 159-167, February 2016.

[11] T. Burianek, J. Stuchly and S. Misak, "Solar power production forecasting based on recurrent neural network," Advances in Intelligent Systems and Computing, vol. 427, pp. 195-204, January 2016, ISBN 978-3-319-29503-9

Optimisation of Earthing System Design through Touch Voltage Minimisation Using Ansys Maxwell

Vaclav Vycital, Michal Ptacek, David Topolanek, Petr Toman
Faculty of Electrical Engineering and Communication
Brno University of Technology
Brno, Czech Republic
vycital@feec.vutbr.cz, ptacekm@feec.vutbr.cz, topolanek@feec.vutbr.cz, toman@feec.vutbr.cz

Abstract — **Earthing system design is usually assessed based on the permissible touch voltage curve of EN 50522. The touch voltage is driven by the spatial placement of the earthing system electrodes, i.e. rods, strips, rings etc. Basically, if one wants to optimize the design, appropriate spatial placement of these electrodes has to be found. As this optimization could be quite challenging, it is usually based on the designing engineer experience and the final solution might not always be the optimal one. Thus, in this paper**
a software based (Sequential Non-Linear Programming) optimization technique is used to search for the 'optimum' solution of earthing system design by Ansys Maxwell. A simple double ring earthing system model was placed in horizontally stratified soil model with two layers and through the optimisation the burial depths of both rings were found. As the optimisation objective function the actual touch voltage was used. It was found that in most cases the Ansys Maxwell optimiser found the optimal design, however in a few cases more variations of earthing system were found with equal value of objective function meaning that possible local and global minimums can be expected.

Keywords: earthing; earthing system; optimisation; Ansys Maxwell

I. Introduction

Earthing system (ES) designs should be compliant with international standards EN 50522:2010 [1] and IEC 61936-1:2010 [2], or its American counterpart IEEE Std. 80 [3]. Nevertheless which approach is adopted, generally the design still has to fulfil some common requirements. Basically, following requirements can be distinguished: safe and reliable operations of the ES requirements, the personal safety requirements etc. That might be

- Anti-corrosion treatment.
- Ability to dissipate the fault current to earth in case of fault occurrence.
- The earth potential rise (EPR) of the ES should not exceed certain limits.

Also, the step voltages (SV) and touch voltages (TV) should be below the permissible step and touch voltages limits. From human safety point of view the ES design is assessed based on the total EPR of the design, the TV and the SV. The approaches in case of standards EN 50522 and IEEE 80 are both based on the assessment of ES EPR, TV and SV. In case of IEEE Std. 80 the term TV is replaced by a mesh voltage. Even though both mean similar things, TV will be further used throughout this paper. However, smaller or greater differences might be found if safety assessment of both standards is compared. Both standards require the engineer to conduct the following steps:

1. The engineer needs to propose the initial ES design. The shape is usually based on his empirical experience with ES designing. In some cases a commonly used types of designs for specific applications are used. Together with the soil resistivity measurement data the engineer is able to determine the resistance/impedance to earth. In case of European Standard EN 50522 basic simplified formulas (1-3) are introduced in its Annex J. In case of IEEE 80 kind of different formulas based on the work of Schwartz [4] (4), Sverak [5] and Nieman [6] are adopted. However, these formulas provide still only simplified solutions of the actual spatial electromagnetic problem [7, 8] and provide only approximate solution if soil model with more layers is used.

$$R_{ring} = \frac{\rho}{\pi^2 D} \ln \frac{2\pi D}{d} \quad (1)$$

$$R_{rod} = \frac{\rho}{2\pi L} \ln \frac{4L}{d} \quad (2)$$

$$R_{strip} = \frac{\rho}{\pi L} \ln \frac{2L}{d} \quad (3)$$

$$R_{grid} = \frac{\rho}{\pi L_c} \left[\ln \frac{2L_c}{L_r} + \frac{k_1 \cdot L_c}{\sqrt{A}} - k_2 + 1 \right] \quad (4)$$

2. After determination of ES resistance the fault current as the source of hazardous voltages has to be determined. This might be achieved with the help of method of equivalent voltage source method of EN 60909 [9] and symmetrical components method.
3. As a first approach, both standards allow the engineer to assess the safety limit of the ES design based on the total EPR. If the EPR is lower than the permissible touch/mesh voltage the ES is assessed as safe.

However, further assessment is different in case of both standards.

4.1. In case of IEEE Std. 80, the mesh voltage (i.e. the maximum TV within grounding grid) as well as SV has to be smaller than the permissible TV and SV. The necessity of evaluating the ES safety with mesh voltage in other words means that the designing engineer is required to determine the maximum TV. Due to the complexity of the actual electromagnetic field problem [10], the determination of this TV might not be easy, especially if soil model different from uniform is considered. IEEE Std. 80 proposes some formulas to determine the mesh voltage in case of horizontally stratified soil model with only two layers.

4.2. In case of EN 50522 a simplification is allowed that the maximum TV is taken as one half, or even one quarter of the entire EPR and this value should be less than the permissible touch voltage. If neither of these two reduced EPR conditions are met, the designing engineer needs to determine the actual TV within the earthing system (i.e. the equivalent of mesh voltage). Here, the European Standard does not provide the engineer with any formula that might be used to determine the actual touch voltages. Also, the EN 50522 expects that the SV does not need to be assessed if permissible TV criterion has been met.

As the permissible touch voltage limit both standards adopted different TV limit curves. The EN 50522 is based on the work of Biegelmeier [11] whereas the IEEE 80 is based on the work of Dalziel [12]. Both limit curves give similar results as was presented in [13].

If TV of proposed design is above the permissible touch voltage, the designing engineer has to modify the design so, that the conditions of 4.1 and 4.2 will be met. This might be managed by changing the ES design dimension. In case of EN 50522 approach the designing engineer has to solve kind of optimisation problem with an objective function of minimising the EPR. The approach of IEEE Std. 80 might be more difficult as the true max TV (mesh voltage) has to be minimised. The problem of minimising EPR/TV can be addressed by:

a) Changing the ES spatial layout.
b) Increasing the ES dimensions.
c) Burying the ES in greater depth.
d) Adding earthing rods.
e) Connection of ES to the global earthing system.

The abovementioned changes a) and b) might be done either empirically by the designing engineer (e.g. it is commonly known fact that increasing the ES area/ring diameter etc. would lead to lower ES resistance and thus also lower EPR), or by some more sophisticated optimisation method. Thus, many researches have devoted their attention to the problem of optimisation of ES designs. Many different approaches have been presented throughout publications [14-22]. Optimisation methods with different objective functions were proposed, i.e. the method of using compression ratio to find the spacing of the horizonal earthing grid with lowest EPR [14], methods to find design with shortest ES horizontal and vertical conductors [15, 16], method of reducing costs associated with excavation work [15, 17, 18], or method of adding special chemicals [19, 20] in the vicinity of the ES – low resistivity material – to decrease the earthing resistance of ES. Also, a method of optimising the distance to some other external earthing grid was proposed by paper [21].

Basically, to optimise the ES design some kind of objective function is necessary and also some kind of step direction functions as well. The objective function quantifies if the desired optimum has been reached, whereas the step direction function determines in which direction the design has to be changed to reach the optimum. Different objective functions proposed by the researches [14-22] were introduced in the previous paragraph. In case of step direction function the researches used either simple parametric/sensitivity analysis or some kind of more complex approach based on non-linear programming methods - genetic algorithm optimization, particle swarm optimization, hybrid particle swarm genetic algorithm optimization etc.

Because conducting detailed sensitivity analysis might be tedious and time demanding, using of more sophisticated methods might be desirable (i.e. like using of mentioned non-linear programming optimisation methods). Also, it is allowed by the standards [1, 3] to conducted computer aided designing of the ES. With the help of earthing oriented software like CDGES [23], SKM GroundMat, Ansys Maxwell [24] the ES design might be treated more accurately. The approximation assumptions of equations (1-4) and also mesh voltage equations (4.1) of IEEE Std. 80 might be determined more accurately because of using appropriate soil model, actual ES design layout and also the potential profiles on the earth surface are determined based on the solution of actual electromagnetic field (EMF) problem in its spatial nature.

Thus, in this paper the use of Sequential Non-Linear Programming (SNLP) optimiser in Ansys Maxwell is demonstrated on a simple ES design optimisation. The true TV 1 m apart from the ES was adopted as the optimisation objective function and the SNLP optimiser was used to determine the necessary design changes so that the design was optimised (step direction function).

II. Ansys Maxwell Optimetrics Component

Ansys Maxwell [24] is a software for solving general EMF problems. The range of application is wide, however in this paper the software was used for optimisation of ES design. The solution of EMF is conducted by the discretisation of the model into mesh of small finite elements – Finite Element Method (FEM). Basically, the engineer has to define/import the desired spatial model, define excitation, boundary and meshing condition and after performing the analysis the desired electromagnetic quantities can be determined from postprocessing of EMF solution data [25]. This software has among other things also Optimetrics Component. This module enables the engineer to perform parametric sweep analysis as well as optimisation analysis. The Optimetrics component [24] features with six different optimisation methods:

1) *Quasi Newton (QN)*
2) *Patern Search (PS)*
3) *Sequential Non Linear Programming (SNLP)*
4) *Sequential Mixed Integer Non Linear Programming (SMINLP)*
5) *Genetic Algorithm (GA)*
6) *Matlab optimiser (MO)*

In case of this paper only the SNLP optimiser was used. Basically, this optimizer creates a response surface by using Taylor series approximation of finite element analysis (FEA). By this approximation the optimizer can find location of improving points and determine next step direction. The SNLP optimizer is supposed to be more accurate and reliable because it is not constrained by the problem of stucking in local minima like other optimisers. That is also caused by the possibility to overcome the local minima by taking larger steps within optimisation variable constraints limits.

III. Modelled Earthing System Design

A simple double ring ES subject to the optimisation was placed in a soil model with two horizontally stratified layers, see Fig. 1. The ES model consisted of smaller inner ring with diameter D_1 and burial depth h_1 and a bigger outer ring with diameter D_2 and burial depth h_2. Both rings were connected by a strip and the strip was brought up above the ground in the centre of the ES, where it was assumed as accessible for touch. The ES was modelled as being made of FeZn strip with cross section 30x4 mm^2. The surface soil resistivity was denoted with ρ_1 = 500 Ωm, the bottom soil resistivity with ρ_2 = 100 Ωm and the transition depth of both layers as H = 2 m. The soil was modelled as a semi sphere with radius of 200 m. The ES was excited by a current of 30 A.

IV. Ansys Maxwell ES Optimisation Results

The ES of section III. was modelled in Ansys Maxwell and it was optimised by the SNLP optimiser integrated in the Optimetrics component. In the optimisation process only burial depth of both rings was searched. The other ES dimension and soil model parameters remained the same. It might be possible to optimise also the diameters of both rings, however for the purpose of this paper these remained constant as D_2 = 4 m, D_1 = 2 and 3.5 m (two different initial values but excluded from optimisation), and the soil model with more resistive surface layer was used as described in section III. The optimisation always started with some initial burial depths h_{1i}, h_{2i} and through the optimisation the optimised depths h_{1o}, h_{2o} were found.
The maximum and minimum burial depth optimisation constraints were set as h_{min} = 0.3 m and h_{max} = 1.5 m. The objective function of the optimisation was the determined TV 1 m apart from the centre of the ES (Fig. 1) denoted as TV_{1m}. For the optimised design also TVs at the distance of 2 and 3 m were determined from the solution data as TV_{2m} and TV_{3m}.
The optimisation results data were further supplemented by the optimised design EPR, the time t it took the optimiser to perform total number of ItN iterations (simulations) in optimisation analysis and the number of iteration where the optimum design was found ItO (out of total iterations ItN). As a user defined stopping condition apart from finding the optimum also the maximum number of iterations for the analysis was set to 300. The results of the optimisation together with the reported stopping condition are in Tab. 2. The reported stopping condition is stated in the last column of Table 1 as one of these options:

- S – for solved, stopped by finding minimum by Ansys Maxwell optimiser.
- M – for stopped by maximum number of iterations, the best suiting result was selected.
- F – for simulation failed. Again, the best suiting result was selected, however the optimisation ended prematurely.

The optimisation was performed for two values of inner ring diameter and primarily for 4 different combinations (extreme values) of initial burial depth values as stated in Tab. 1

In case of inner ring diameter D_1 = 2 m, all four initial conditions led to the optimum design with the inner ring close to the surface and the outer ring close to the maximum depth constraint as approx. h_1 = 0.3 m and h_2 = 1.5 m. However, in case of inner ring diameter D_1 = 3.5 m, two different optimum designs with same value of objective function (TV_{1m}) of about 180 V were found. With another 5 different initial starting conditions selected, the optimisation analysis always ended in either one of these optimal designs with approx. h_1 = 0.4 m, h_2 = 1.5 m or h_1 = 1.5 m, h_2 = 0.3 m. In case that the diameter D_1 would be as fixed value 3.5 m or 2 m, these both ES designs assessed on the TV_{1m} objective function are equal. In case when other objectives would be incorporated (i.e. like assessment of TV_{2m}, EPR etc.) the design with outer ring close to the surface has better performance as the TV_{2m} is only about 190 V compared to 260 V. It is also interesting to note, that the complete switch of burial depths of both rings does not have significant influence on design EPR with only approximate change of 30 V.

Table 1. Results of the optimisation

Depth combination no.	h_{1i} (m)	h_{2i} (m)
1	0.3	0.3
2	0.3	1.5
3	1.5	0.3
4	1.5	1.5

From the results it is further worth pointing out, that the analysis not always ended with successful finding of optimum design. In some cases, the optimiser failed during the analysis and stopped the optimisation prematurely. In other cases, the maximum number of iterations had been reached and so possible not optimal design could have been found. In about 30 % of all performed analysis the 300 iteration limit was not enough and thus even more iterations might be necessary. However, based on the results it can be expected that even though the maximum number of iterations had been reached, the solution close to the optimum was still found and listed in the optimisation analysis results.

Other quite unfavourable thing is the necessary time needed to perform the optimisation analysis where the average time of the optimisation was about 7 hours. The analyses were performed on a common low-end consumer personal computer with i3 8100 processor (3.6 GHz), 8 GB RAM memory and with integrated GPU.

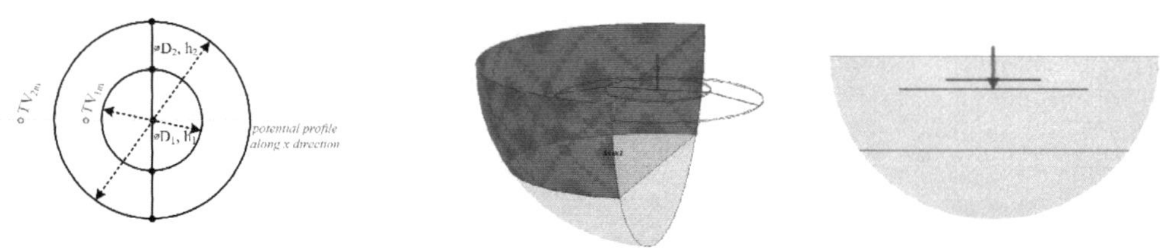

Figure 1. ES model, top view (left), spatial view (middle), side view (right)

Table 2. Results of the optimisation

No.	D_1	D_2	h_{1i}	h_{2i}	h_{1o}	h_{2o}	TV_{1m}	TV_{2m}	TV_{3m}	EPR	t	ItO	ItN	Status
-	m	m	m	m	m	m	V	V	V	V	min	-	-	-
1	2	4	0.5	0.7	0.31	1.5	164	448	604	876	191	16	66	S
2	2	4	1.4	1.4	0.30	1.40	172	457	625	907	89	12	28	F
3	2	4	1.5	0.3	0.3	1.34	166	452	628	924	629	165	212	S
4	2	4	0.3	1.5	0.3	1.5	166	447	604	876	108	1	42	S
5	3.5	4	0.3	0.3	1.5	0.36	180	187	446	764	793	272	300	M
6	3.5	4	0.7	0.7	1.48	0.37	180	183	441	762	802	187	290	S
7	3.5	4	0.9	0.9	1.5	0.4	180	189	437	754	833	268	300	M
8	3.5	4	1.1	1.1	0.46	1.5	182	259	479	784	394	61	135	F
10	3.5	4	0.3	1.5	0.4	1.44	182	254	492	805	127	11	47	F
11	3.5	4	1.5	0.3	1.5	0.38	180	190	446	764	194	74	74	F
12	3.5	4	0.9	0.7	0.37	1.45	181	248	488	800	796	275	300	M
13	3.5	4	0.7	0.9	1.5	0.42	180	192	437	755	756	211	300	M

Slight changes in the resulting TV_{1m} for same design (also TV_{2m}, TV_{3m} and EPR) might be caused by the randomness of created finite element mesh that in case of set analysis percent error of 1 % can be expected. The effect of this solution noise can be partly filtered out by decreasing this analysis percent error, however the solution time needed will increase rapidly.

V. Conclusion

This paper presents the results of the optimisation of ES by Ansys Maxwell. It was found that the Ansys Maxwell Optimetrics component can be used to find optimum design of ES. However, it seems that the TV_{1m} objective function might not be simple convex function and thus multiple location of local minima can be found. To determine the position of the global minimum methods of nonconvex optimisation should be adopted.

Another beneficial finding is that by changing the burial depth of both rings two opposite designs were found, however one with much better TV_{2m}. Thus, using Ansys Maxwell software or some other equivalent software, it is possible to optimise the design with help of some multiobjective optimisation method and thus get better design within desired constraints what might be kind of tricky if the optimisation is performed empirically or by try and fail method.

Acknowledgment

This research work has been carried out in the Centre for Research and Utilization of Renewable Energy (CVVOZE). Authors gratefully acknowledge financial support from the Ministry of Education, Youth and Sports of the Czech Republic under BUT specific research programme (project No. FEKT-5-17-4784).

References

[1] *Earthing of Power Installation Exceeding 1 kv a.c.*, CENELEC Standard EN 50522:2010, 2010.

[2] *Power Installations exceeding 1 kV a.c. – Part 1: Common rules*, International Electrotechnical Commission IEC Standard 61936-1:2010, ed.2, 2010.

[3] *IEEE Guide for Safety in AC Substation Grounding*, IEEE Standard 80, 2015.

[4] S. J. Schwarz, "Analytical expression for resistance of grounding systems," *AIEE Transaction on Power Apparatus and Systems*, vol. 73, no. 13, pp. 1011-1016, 1954.

[5] J. G. Sverak, "Sizing of ground conductors against fusing," *IEEE Transaction on Power Apparatus and Systems*, vol. PAS-100, no. 1, pp. 51-59, 1981.

[6] J. Nieman, "Unstellung von Hochspannungs-Erdungsalagen Aufden Betrieb Mit Starr Geerdetem Sternpunkt," Electrotechnische Zeitschrift, vol. 73, no. 10, pp. 333-337, 1952.

[7] V. Vycital, D. Topolanek, P. Toman and M. Ptacek, "Sensitivity Analysis of Earthing System Impedance for Single and Multilayered Soil," *CIRED 2017, The 24th International Conference and Exhibition on Electricity Distribution*, Glasgow, UK, pp. 1-5, 2017.

[8] V. Vycital, M. Ptacek, D. Topolanek and P. Toman, " Earthing System Resistance Calculation Using Analytical Approach with Mutual Coefficients," *Proceedings of 18th Internation Scientific Conference on Electric Power Engineering (EPE)*, Ostrava, Czech Republic, pp. 67-72, 2017.

[9] *Short-Circuit Currents in Three Phase a.c. Systems: Part 0: Calculation of Currents*, International Electrotechnical Commision, IEC 60909-0:2016.

[10] F. Dawalibi, Transmision Line Grouding, Electric Power Research Institute, EPRI Research Report 1494-1, vol. 1, Montreal, Canada, 1982.

[11] G. Biegelmeier and W.R. Lee, "New Considerations on the Threshold of Ventricular Fibrillation for a.c. Shocks at 50-60 Hz," *IEEE Proc*, vol. 127, no. 2, pp. 103-110, 1980.

[12] C.F. Dalziel and W.R. Lee, "Reevaluation of Lethal Electric Currents," *IEEE Transaction on Industry and General Applications*, vol. IGA-4, no. 5, pp. 467-476, 1968.

[13] C.-H. Lee, AP. Meliopoulos, "A Comparision of IEC 479-1 and IEEE Std. 80 on Grounding Safety Criteria," *Proc. Natl. Sci. Counc. Repub. China*, vol. 23, no. 5, pp. 612-621, 1999.

[14] H-S. Lee, J-H. Kim, F. P. Dawalibi, and J. Ma, "Efficient Ground Grid Designs in Layered Soils," *IEEE Transaction on Power Delivery*, vol. 13, no. 3, pp. 745-751, 1998.

[15] A. Taher, A. Said, T. Eliyan, and A. Hafez, "Optimum Design of Substation Grounding Grid Based on Balancing Parameters using Genetic Algorithm," *(MEPCON) - Twentieth International Middle East Power Systems Conference*, Egypt, pp. 352-360, 2018.

[16] C. L. B. Silva, D. N. Oliveira, T. G. Pires, J. W. L. Nerys, W. P. Calixto, and A. J. Alves, "Optimization of Grounding Grid´s Multidesign Geometry," *IEEE (EEEIC) 16th International Conference on Environment and Electrical Engineering*, Italy, pp. 1-6, 2016.

[17] B. Alik, M. Teguar, and A. Mekhaldi, "Minimization of Grounding System Cost Using PSO, GAO, and HPSGAO Techniques," *IEEE Transaction on Power Delivery*, vol. 30, no. 6, pp. 2561-2569, 2015.

[18] J.-W. Perng, Y.Ch. Kuo, and S.-P. Lu, "Grounding System Cost Analysis Using Optimization Algorithms," *Energies*, vol. 11, no. 3484, pp. 1-19, 2018.

[19] A.A. Al-Arainy, Y. Khan, M.I. Quereshi, N. H. Malik, and F.R. Pazheri, "Optimized Pit Configuration for Efficient Groudning of Power System in High Resistivity Soils using Low Resistivity Materials," *Fourth International Conference on Modeling, Simulation and Applied Optimization*, Malaysia, pp. 1-5, 2011.

[20] A. Boualegue, and F. Ghodbane, "A new Parametric Approach for Groudning Analysis in Multilayered Soils with Finite Volumes," *First International Conference on Renewable Energies and Vehicular Technology*, Tunisia, pp. 274-278, 2012.

[21] W. Pan, H. Sun, S. Chai, and J. Zhou, "Research of Optimal Distance of External Grounding Grid," *IEEE (POWERCON)*

International Conference on Power System Technology, Australia, pp. 1-4, 2016.

[22] A. J. Datta, R. Taylor, and G. Ledwich, "Earth Grid Safety Criteria Determination with Standards IEEE-80 and IEC-60479 and Optimization of Installation Depth," *(AUPEC) Australian Universities Power Engineering Conference*, Australia, pp. 1-5, 2015.

[23] F. Dawalibi and F. Donoso, "Integrated Analysis Software for Groudning, EMF, and EMI," *IEEE Computer Applications in Power*, vol. 6, pp. 19-24, 1993.

[24] Ansys Maxwell package, EMF simulating software, version 18.

[25] V. Vycital and P. Toman, "Modelling of Electrical Installations Earthing Systems in Ansoft Maxwell" *Proceedings of 24th Conference STUDENT EEICT 2018*, Brno, pp. 487-491, 2018.

Analysis of daily load diagrams for active and reactive energy

Daniel Janik, Branislav Batora, Petr Toman
Brno University of Technology
Brno, Czech Republic
janikd@feec.vutbr.cz, batora@feec.vutbr.cz, toman@feec.vutbr.cz

Abstract - **The article compares and evaluates daytime active and reactive power consumption, exploiting data supplied by electricity quality meters. The purpose of the research is to determine and evaluate the amount of reactive energy taken by different buildings. The measurement took place in residential and commercial buildings over several weeks during various parts of the year. The analyzed structures included single-family homes, a school, an office building, a manufacturing plant, and a shop. The paper presents the daily load diagrams related to the active and reactive energies, with discrimination between workdays and weekends. The actual result consists in not only demonstrating the reactive energy consumption compared to active energy use in different types of customers but also analyzing possible reasons for reactive energy utilization.**

Keywords—active energy, reactive energy, power consumption

I. Introduction

At present, the consumption of active and reactive energy is a widely discussed problem, especially in view of the rising consumption rates. The character of energy use and the amounts of power consumed during the day vary greatly depending on the customer; thus, for instance, the requirements of a regular single-family house differ from those of a shop or an office or school building, which follow a clearly defined schedule. In such commercial and public facilities, the weekday and weekend consumption volumes differ markedly. Other types of houses or premises are characterized by the fact that some appliances draw the energy continuously, regardless of the day of the week. Residential buildings then provide a classic example of random consumption, with minimal correlation to associated factors and events; the actual consumption usually depends solely on the residents' preferences.

Controlled consumption also participates in the daily electricity consumption time pattern. Directly controlled appliances, such as electric heating or domestic water heaters, can enter the pattern simultaneously with multiple other facilities; depending on the collective remote control used, these appliances normally respond well to delayed start.

The reactive part of consumption depends on the appliances connected to the object. Any analysis of consumption and connected appliances based only on active and reactive consumption is a very complex task; advantageously, however, some effects can be observed from the daily diagrams of the consumption of individual objects.

The data to be analyzed were provided by the electricity distributor E.ON Distribuce, a.s.

II. Power factor in the distribution network

The power factor is a parameter that specifies information about the efficiency of a transmission system in use; the factor assumes values of <0; 1>, can be inductive or capacitive, and indicates how much apparent energy is effectively convertible to active power. Unlike active energy, reactive power only serves to create an electric or magnetic field for capacitors and inductors: it does not perform work. Typical devices that require reactive power include transformers and motors; presently, however, switching power supplies that have a rectifier and a capacitor at the input are frequently assigned to this group too. Most of the current consumer electronics connected to the network are equipped with these resources. Desirably, the power factor should approach 1 in order for the transmission system not to be unnecessarily exploited by the transmission of inefficient energy. When working with a real network, it is necessary to distinguish between $\cos\varphi$ and the power factor (λ). $\cos\varphi$ can be employed only in harmonic conditions within the network and at linear load, where the angle φ can be unambiguously defined [1][2].

In an electricity distribution system, the distributor defines the customer requirements and the $\cos\varphi$ values that the customer must comply with. By default, $\cos\varphi$ should be of an inductive character within the range of <0.95; 1>. This applies especially to high and very high voltages; at the low voltage level, $\cos\varphi$ has not yet been evaluated, and thus it ought to be focused on to minimize losses in the power system and to optimize the operation. In Slovakia power factor monitoring has recently been introduced in low level voltage networks within public lighting, and values outside the required range have been penalized [1][3].

Compensation is utilized to suppress the negative impact of the reactive energy consumed; the procedure embodies a measure to reduce not only the line current load and resistance losses in the power line but also the apparent power output. Devices requiring a reactive power component to operate will cause a phase shift between the mains voltage and the current. Compensation is aimed to produce an equally large phase shift, albeit one of an opposite character. The most common types of low voltage level network compensation rest in the *controlled* and *fixed* approaches. The former includes a battery of capacitors with a control unit that evaluates the current amount of reactive energy consumed and connects accordingly a combination of, for example, suitably large capacitors; the latter then finds use in devices with constant reactive power consumption for individual load compensation. The principle of compensation is presented in the diagram in Fig. 1, where U_1, U_2 is line and load voltage (V), R is line resistance (Ω), X_L is line reactance (Ω), I_k is line current (A), I is load current (A), Z is load impedance (Ω), I_C is compensation current (A), X_C is compensation reactance (Ω). Without compensation, the network – between the power generation and the reactive energy consumers - is loaded with both active and reactive components. In the case of compensation, the line is relieved by the most reactive power, and the apparent power decreases [2][5].

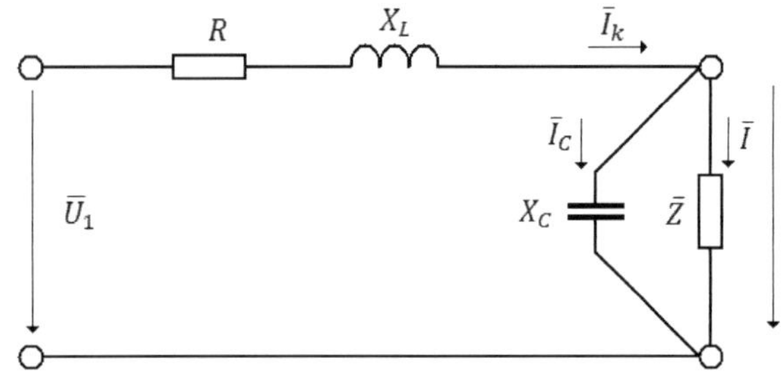

Figure 1. The basic scheme of reactive power compensation [4].

III. Power factor analysis

The analysis was centered on data obtained by E-ON quality meters at several sampling points, including single-family houses, office rooms, schools, and public lighting. The data (active and reactive power, *cos φ*) had been measured for different periods of time and at various instances over a year. In order to compare the individual data items, we had to yield the data average from identical times measured during a day, as expressed in Equations 1 and 2. To demonstrate the differences between a weekday and a weekend, corresponding averages were introduced; on weekdays, the average consumption is usually higher, even in single-family homes, depending on the type of traffic (multi-shift or continuous). We have

$$P = \frac{1}{n}\sum_{i=1}^{n} P_i, \tag{1}$$

where P is average value of active power (W),
P_i is one of measured value of active power (W),
n is number of averaged values (-),

$$Q = \frac{1}{n}\sum_{i=1}^{n} Q_i, \tag{2}$$

where Q is average value of reactive power (var),
Q_i is one of measured value of reactive power (var),
n is number of averaged values (-).

As regards the power factor (Fig. 2) value in the six monitored sampling points, more than 50 % of cases remains within the desired interval <0.95; 1> (inductive character). Most of the values outside the interval relate to single-family houses with low active power consumption, usually below 1 kW. In this context, we should note that such a house may contain a variety of small appliances and consumer electronics where the compensation and reactive power consumption are left unresolved by the manufacturers. At the next sampling point, an office building operating during the week and the weekend, the power factor is only around the value of 0.9 (capacity character).

Special supply points consist in public lighting; this paper discusses two of these spots. They are characterized by the same type of appliances and constant operating consumption, possibly at several levels if some degree of control is applied. Details on the *cos φ* are presented in Fig. 2, where it becomes evident in one case that the public lighting does not deviate from the cos φ interval <0.95; 1> of an inductive character,

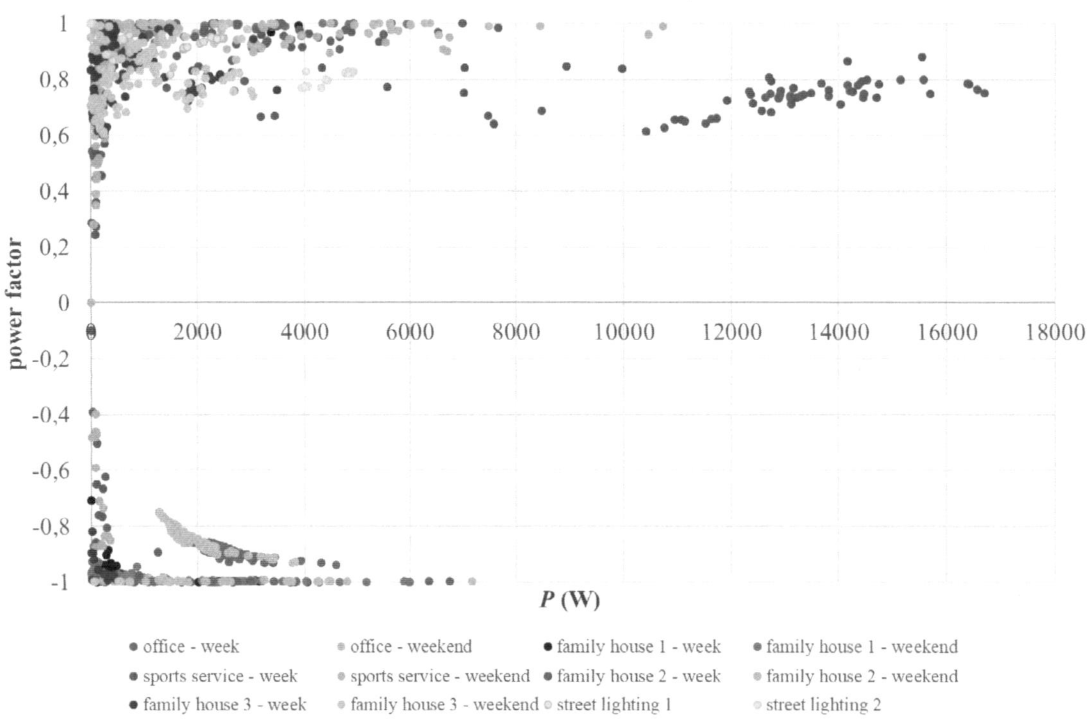

Figure 2. The relationship between the power factor and the active power consumed.

while in the other case the *cos φ* reaches up to about 0.65. The size of *cos φ* in this latter case can be markedly influenced by the age of the luminaires, especially in old luminaires with a choke and a compensating capacitor which has lost its original capabilities over the years and does not fulfill the compensation function, wholly or in part. In new LED luminaires, the electronic ballast provides the correct *cos φ* value [6].

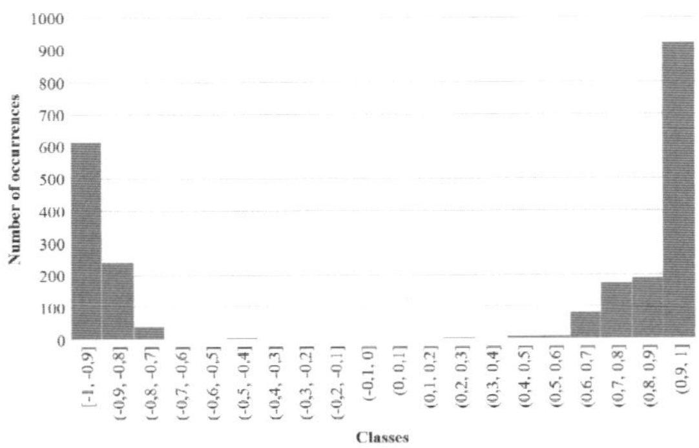

Figure 3. A power factor histogram.

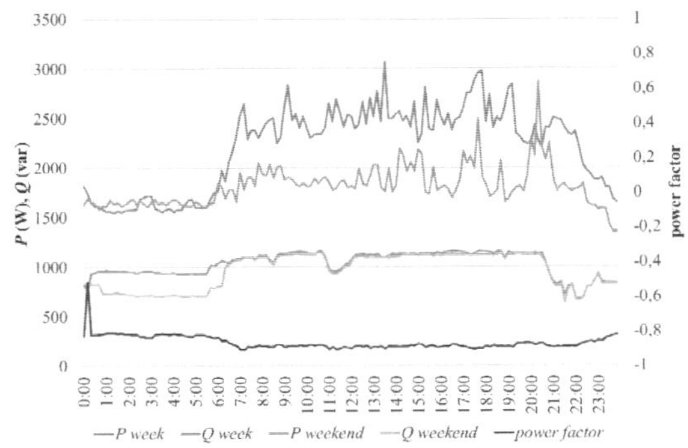

Figure 4. A daily load diagram: the office rooms.

IV. DAILY LOAD DIAGRAM OF THE MEASURED LOCATIONS

The analysis focuses on data measured over several weeks at various supply points, including office rooms, a primary school, a sports shop with service, several single-family houses, and street lights as very specific supply points. The measured data were averaged separately in relation to all weekdays on one hand and weekends and holidays on the other. The individual consumption points are very diverse, stemming from the nature of the consumption given by the composition of the appliances and their operating times.

Fig. 4 shows that the investigated office building is in use during the week and the weekend, and the appliances have a strong capacitance according to the *cos φ* due to the number of switching power devices. The *cos φ* range equals <-0.97; -0.8>. The subscription at the sampling point applies to weekdays and weekends, but the consumption is lower, meaning that only some office rooms or other spaces are likely to be in operation if there is nonstop traffic; in this context, we can mention a gatehouse as a suitable example.

Another sampling point, which exhibits very good parameters in terms of the reactive power and power factor, is shown in Fig. 5. The relevant diagram represents the operating mode of a typical school building; here, we can identify the operational consumption (with people in the building to participate in classes or other forms of instruction) and its constant counterpart. The power factor is in the ideal range of <0.95; 1> (inductive).

Figure 5. A daily load diagram: the elementary school.

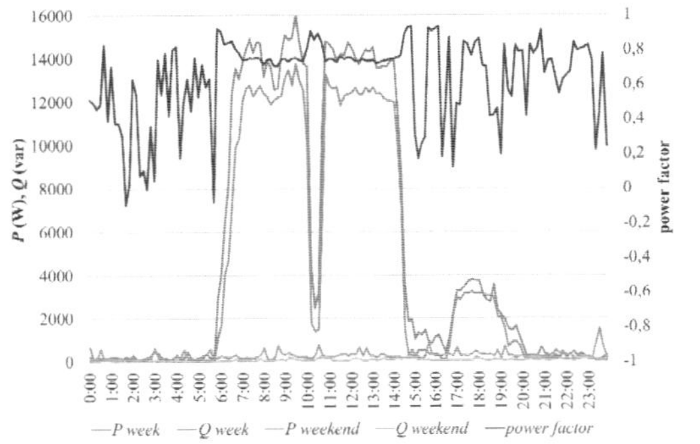

Figure 6. A daily load diagram: the sports service.

The third monitored supply point is a sports equipment and repair shop. The offtake point has almost no permanent consumption, drawing energy only during the operating hours and on workdays; the daily load diagram in Fig. 6 visualizes the operating time, including the break between 10:00 and 10:30. The reactive energy consumption is approximately 85 % of the active energy consumed. The power factor pattern during the largest draw remains below 0.8, fluctuating significantly during the lower draw, which, however, does not constitute a major problem due to the amount of energy consumed.

The single-family homes embody a type of consumption point with very few recurring events. All consumption depends on several factors, such as the habits of the residents; season; outdoor temperature; or just day of the week (Fig. 7). In a working week, the residents generally start their days less than two hours earlier compared to weekends. Fig. 8 shows that the day of the week is irrelevant in this respect; in both diagrams, we can only comment that the weekend consumption is significantly higher, especially in the morning. The reactive energy consumption is proportional to the amount of the active energy consumed. The power factor is also highly individual and depends solely on the connected appliances. In the first house examined, the power factor varies considerably but always has an inductive character; in the other building, the power factor is also variable over time but remains predominantly capacitive. Regarding the sustained consumption observable at night, namely, when nothing happens in the house, it is minimal in each case, oscillating around 100 watts on average [7][8].

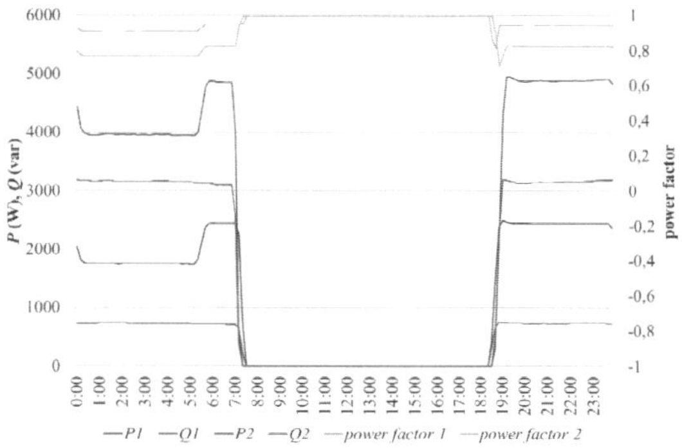

Figure 7. A daily load diagram: single-family house 1.

The last, and very special, type of supply point is street lighting. From the perspective of timing during the day, this is the opposite of all the above-mentioned consumption points. The consumption reaches the highest levels during the night, whereas in the day the lighting is switched off, and the consumption equals zero if the consumption of the switching control circuit is not considered. In Fig. 9, an energy saving mode is set for both of the examined street lighting consumption points; some luminaires are switched off at low traffic - in these cases approximately from 0:00 to 5:00. However, this noticeable decline is only reflected in the active power consumed. It can be seen that the first of the draw-off points in all the cases fulfills the *cos φ* condition in the interval <0.95; 1> of an inductive character; in the second case, the cos φ is slightly worse. As the lighting system is public, there is no difference between weekday and weekend traffic; Fig. 9 thus shows the daily courses of the offtake points and the average for working and non-working days.

V. Conclusion

At the sampling locations, the power factor reached the highest values usually during the workday and often also at weekend daytimes, frequently exhibiting a highly inductive nature. The lowest values were observed at night times. Even though the offtake points (as street lighting) provide the greatest power during the night, they were sufficiently compensated in the cases examined. For the other offtake points, the nighttime consumption was so small that even at poorer power factor values the absolute reactive power value was negligible. Individual compensations for appliances significantly improve the situation in households and smaller offtake points, but it is necessary to check the technical condition of the compensation because it may not be as effective in older appliances.

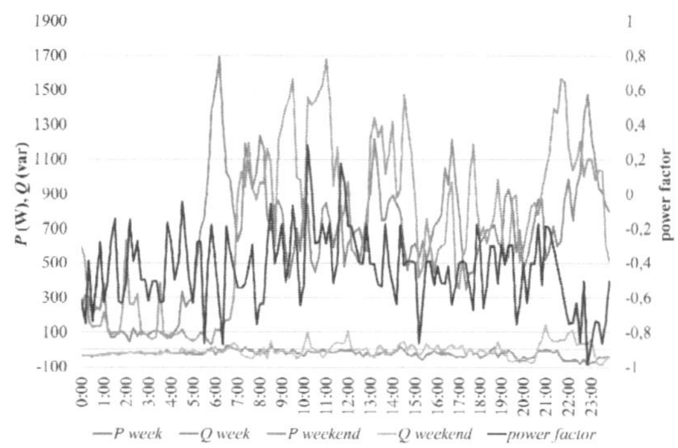

Figure 8. A daily load diagram: single-family house 2.

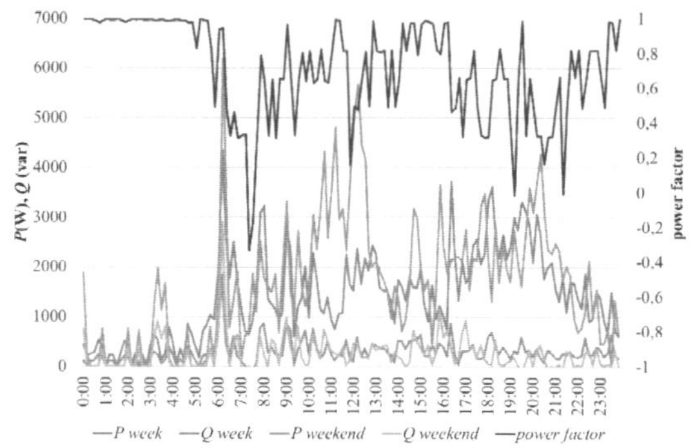

Figure 9. A daily load diagram: the street lighting.

ACKNOWLEDGEMENT

The research was carried out at the Centre for Research and Utilization of Renewable Energy (CVVOZE). The authors gratefully acknowledge financial support from the Ministry of Education, Youth and Sports of the Czech Republic under BUT specific research programme (project No. FEKT-S-17-4784).

REFERENCES

[1] J. Popek, "Návrh koncepce kompenzace jalového výkonu v průmyslové síti," (Design of reactive power compensation concept in industrial network, in Czech). Brno: Brno University of Technology, 2012. 55 p.

[2] M. Synek, "Analýza účinníku v distribučních sítích," (Power factor analysis in distribution networks). Praha: Czech Technical University in Prague, 2014. 86 p.

[3] P. Janiga, D. Gasparovsky, L. Lipnicky and D. Minovski," Reactive Power in Street Lighting Netwirk and Relation with Smart Metering in Slovakia," In Proceedings 19th International Scientific Conference on Electric Pover Engineering ISBN 978 - 1 - 5386 - 4611 – 3. p 463 - 466

[4] V. Blazek and P. Sklada," Distribuce elektrické energie," (Electrical Energy distribution, in Czech). Brno : Brno University of Technology, 138 p.

[5] M. Pour," Pravidla provozování lokální distribuční soustavy: Příloha 4 Pravidla pro paralelní provoz výroben a akumulačních zařízení se sítí provozovatele lokální distribuční soustavy," (Rules of operation of the local distribution system: Annex 4 Rules for parallel operation of production facilities and storage facilities with the local distribution system operator's network). ERU (Energy Regulatory Office) , 2018.

[6] M. Srinivasan, S.U. Praba and R. Suganthi," Harmonic and Power Factor Analysis with Reactor and Capacitor in Adjustable Speed Drive," *Advanced Materials Research* [online]. Trans Tech Publications, 2013, pp.317-322, ISSN: 1662-8985.

[7] J. Haluza and J. Macháček," Spotřeba elektrické energie domácností, predikce a potenciální úspory pomocí BACS," (Household electricity consumption, prediction, and potential BACS savings). Tzb info [online]. 2012. Available from: https://elektro.tzb-info.cz/8570-spotreba-elektricke-energie-domacnosti-predikce-a-potencialni-uspory-pomoci-bacs.

[8] Roční zpráva o provozu ES ČR (Annual report on the operation of the electricity grid in the Czech Republic). ERU (Energy Regulatory Office) [online]. 2017. Available from: http://www.eru.cz/documents/10540/462820/Rocni_zprava_provoz_ES_2017.pdf/521bff99-fdcf-4c86-8922-3a346af0bb88

An Estimation of Appliance Impact on Frequency and Power Factor Parameters in Off-Grid Environment

Vojtech Blazek, Jan Vysocky, Michal Petruzela, Stanislav Misak, Lukas Prokop
ENET Centre
VSB – Technical University of Ostrava
Ostrava, Czech Republic
vojtech.blazek@vsb.cz, jan.vysocky@vsb.cz, michal.petruzela@vsb.cz , stanislav.misak@vsb.cz, lukas.prokop@vsb.cz

Abstract — **Nowadays, for many people living in a remote area, powering their homes or small businesses by use of a small island system with renewable energy sources is becoming an interesting option. Today, these systems can be driven by algorithms based on intelligent data analysis methods, real-time optimizations, and predictive models. Systems, which use these algorithms, are called the Smart Grids. The smart-grid technology makes the electricity generation and consumption more effective efficient. It has the potential to reduce greenhouse-gasses emissions and to improve environmental quality. For the correct design of the control algorithm, we need the appropriate data of the behavior of each device connected to the electrical system. Our paper presents an analysis of the various appliances impact on the off-grid system. This analysis shows how the usual appliances affect frequency (FREQ) and power factor (PF) in the off-grid system.**

Keywords: Off-grid electrical system, Frequency, Power factor, Power quality, usual appliances

I. Introduction

Our vision is to improve off-grid house systems into fully autonomous energy platform with energy supply from the renewable sources (RESs) and with an application of an energy storage system. Now the platform is being developed. The stochastic character of the RESs which are part of system causes low short-circuit power. Short circuit power decreases the stability of power quality (PQ) parameters like the off-grid-system frequency (FREQ) [1], [2]. Part of the development is the measurement of the appliance behavior in this system. Our objective is to find out which appliances have the impact on FREQ and power factor (PF) in the Off-grid system. This knowledge is important to be able to avoid poor PQ parameters. Poor PQ parameters may affect the operation of the other devices connected to the Off-Grid house [3].

The first concept of Off-Grid house system with management algorithms was proposed and designed in a study by Mišák [4] [5]. This concept was titled as Active Demand Side Management (ADSM). In new version ADSM was suggested power quality forecasting module by Vantuch et al. [5]. This experiment is first of the many measurements to improve forecasting model.

II. EXPERIMENT DESCRIPTION

The off-grid system under normal conditions is powered by two 2-kWp photovoltaic panels and NiCd batteries KPL 375 FERAK (see figure 1). The capacity of each battery is 375 Ah and the nominal voltage is 1.2 V. A total of 40 batteries are connected in series. Batteries are connected to DC bus. Photovoltaic panels and hybrid inverter Conext XW+ 8545 (see in figure 2) are also connected to this DC bus. This hybrid inverter transforms direct current to the alternating one for the appliances connected to Off-grid.

Figure 1. NiCd KPL 375

Figure 2. Hybrid inverter

Hybrid inverter has nominal power 6.8 kW. The grid's AC part and local appliances are connected to the hybrid inverter via a switchboard. The switchboard is controlled by a control system which utilizes an exact switching schedule. A power analyzer measures on the AC output of the hybrid inverter.

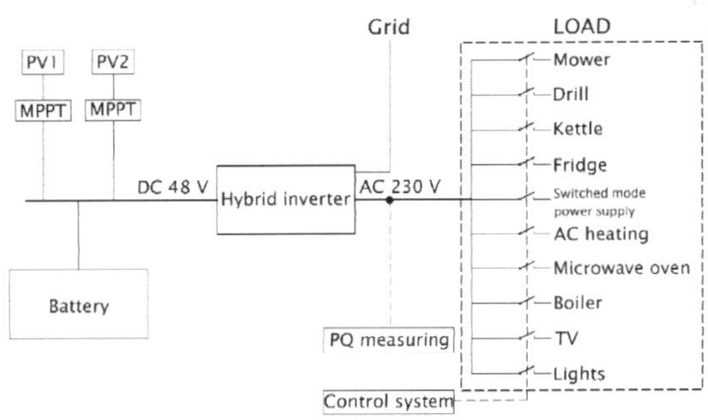

Figure 3. Off-Grid scheme

Fig. 3 shows the electrical scheme of our off-grid system located in campus VŠB – Technical University of Ostrava.

Table 1. Usual appliances

Appliance	Load (W)		PF		Characteristic			Power supply	
	avg	min	max	avg	Inductive	Capacitive	Resistive	Continuous	Switch
Mower	537.60	532.10	549.27	0.52	True	False	False	True	False
Drill	157.10	149.50	167.00	0.49	True	False	False	True	False
Kettle	619.10	617.00	628.30	1.00	False	False	True	True	True
Fridge	207.60	195.50	219.50	0.72	True	False	False	True	False
Switched mode power supply	410.00	409.70	420.20	0.78	False	True	False	True	True
AC heating	880.00	852.50	910.00	0.91	True	False	False	True	False
Microwave	203.00	76.80	1348.30	0.84	True	False	False	False	False
Boiler	307.00	305.80	346.50	0.99	False	False	True	True	False
TV	44.00	42.80	50.50	0.60	False	True	False	True	True
Lights	156.00	152.50	165.10	0.84	False	True	False	True	True

In the experiment, we used ten various usual appliances (see table 1) to create 120 combinations of three. There were always three appliances turned on because only one running appliance is uncommon in a house. Every combination was switched on at a defined time. Each combination was measured for 12 minutes. We omitted the first and last minute as a transient phenomenon. Appliances were supplied by batteries without connection to the external grid and RESs. During measure intervals, we measured PF and FREQ. When the measured interval had ended, batteries were charged from the external grid. Battery charging time was 18 minutes long to ensure that the desired state of charge (SoC) was always reached.

FREQ and PF values were measured by the power analyzer KMB SMC 144 with a one-minute timescale. For further analysis, we stored minimum, average and maximum measured values. We measured all combinations at two different battery SoCs (fully and half charged).

The hybrid inverter with changing load parameters could have a problem with FREQ because Off-grid have low short-circuit power. That is why we made the measurements for SoC 50% and SoC 100% to verify that the inverter itself will not have a problem on the AC side.

III. TESTING AND RESULTS

A. Power factor

Fig. 4 and fig. 5 show PF values of each combination measured with the half and fully charged batteries.

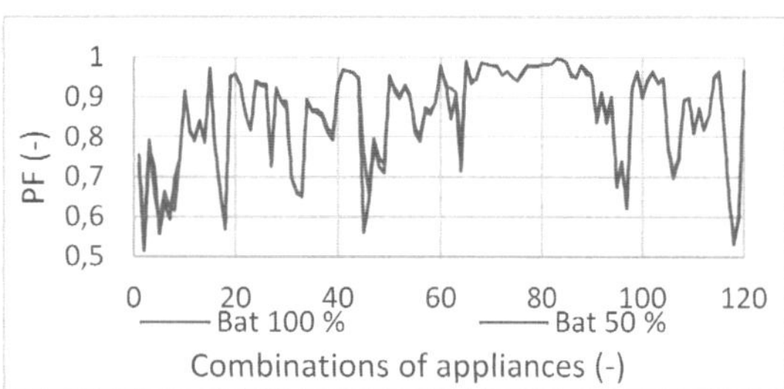

Figure 4. Measured PF values

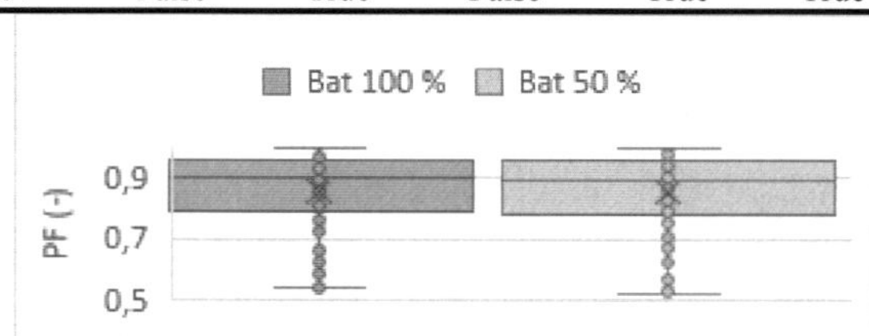

Figure 5. Measured PF values

Fig. 6 and fig. 7 show the data series of average PF values for SoC 50% and 100%. These data points were calculated for the individual measurement periods (each data point is calculated as the average value of one measurement-period values). Since the system was loaded by three different appliances during each measurement period, the average PF value is calculated for one given period and then it is assigned to the data series of all three appliances that were connected to the electrical system in the given period. Each appliance was switched on in 36 periods of total 120 periods. Table 2 shows the arithmetic average of PF values of each combination for individual appliances.

Table 2. Arithmetic average of power factor for each appliance during interval measurement

Appliances	Power factor (-)	
	SoC 100%	SoC 50%
Kettle	0.935	0.935
Switched mode power supply	0.914	0.913
Boiler	0.890	0.887
AC heating	0.873	0.875
Lights	0.870	0.869
TV	0.858	0.853
Microwave oven	0.857	0.854
Fridge	0.837	0.829
Drill	0.820	0.808
Mower	0.726	0.725

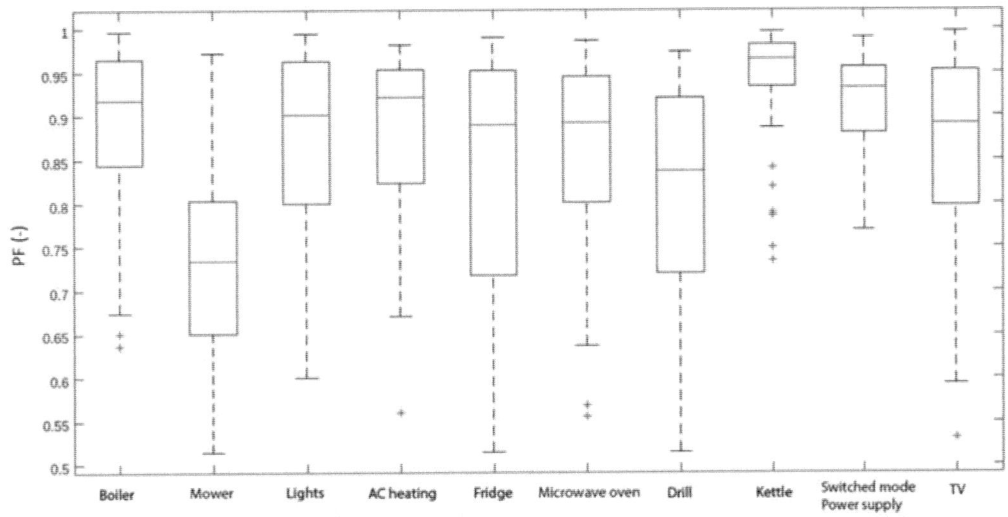

Figure 6. Box plot for all used appliances – impact of appliance operation on PF, SoC 50%

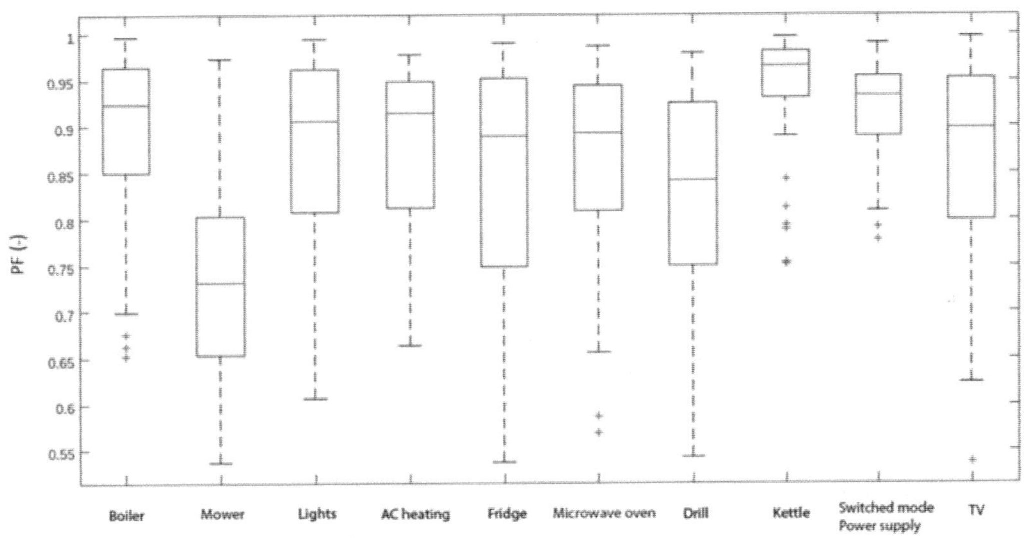

Figure 7. Box plot for all used appliances – impact of appliance operation on PF, SoC 100%

B. Frequency

Fig. 8 and fig. 9 show FREQ values of all combinations measured with the half and fully charged batteries. Measured FREQ values range from 50.0071 Hz to 50.0073 Hz for SoC 100% and from 50.0022 Hz to 50.0072 Hz for SoC 50%. Fig. 6 shows that the average value of the frequency fluctuates more at SoC 50% than at SoC 100%. Measured FREQ values are in accordance with European Standard EN 50160: 2010 + A1: 2015.

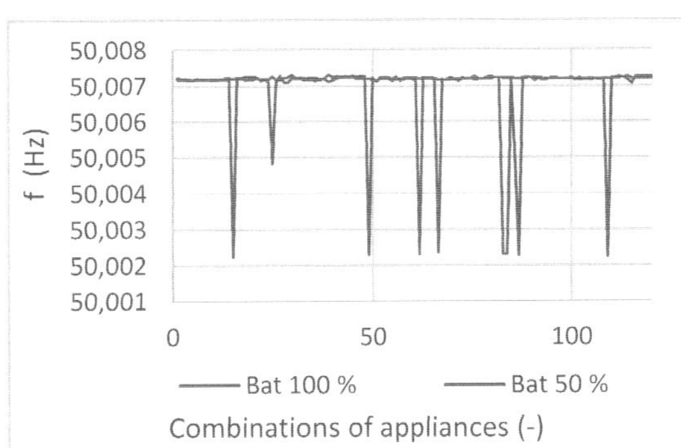

Figure 8. Measured FREQ values

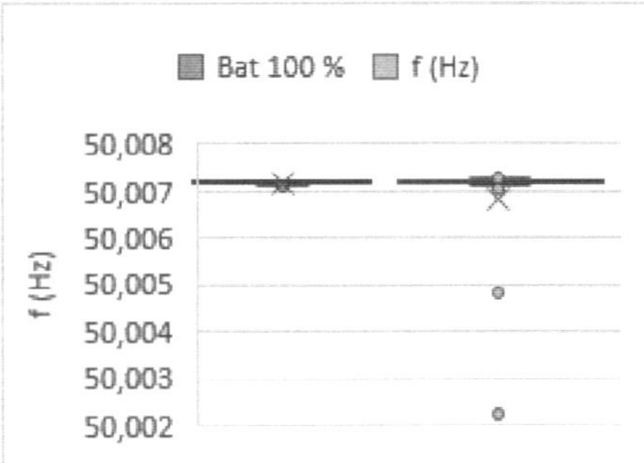

Figure 9. FREQ values during measured interval

Fig. 10 and fig. 11 show the data series of average FREQ values for at SoC 50% and 100%, the data are presented for individual appliances. These data series were calculated analogously like the data series presented in fig. 4 and fig. 5.

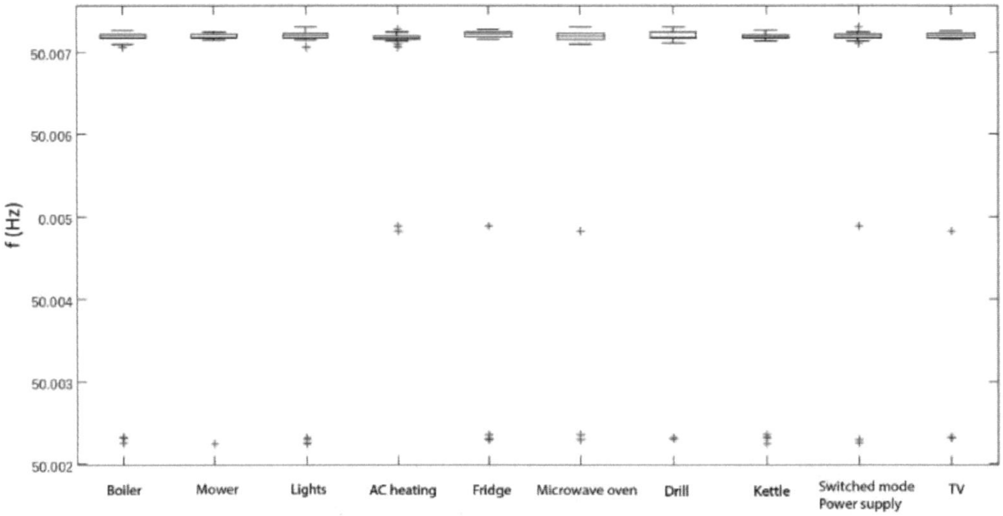

Figure 10. Box plot for all used appliances – impact of appliance operation on FREQ, SoC 50%

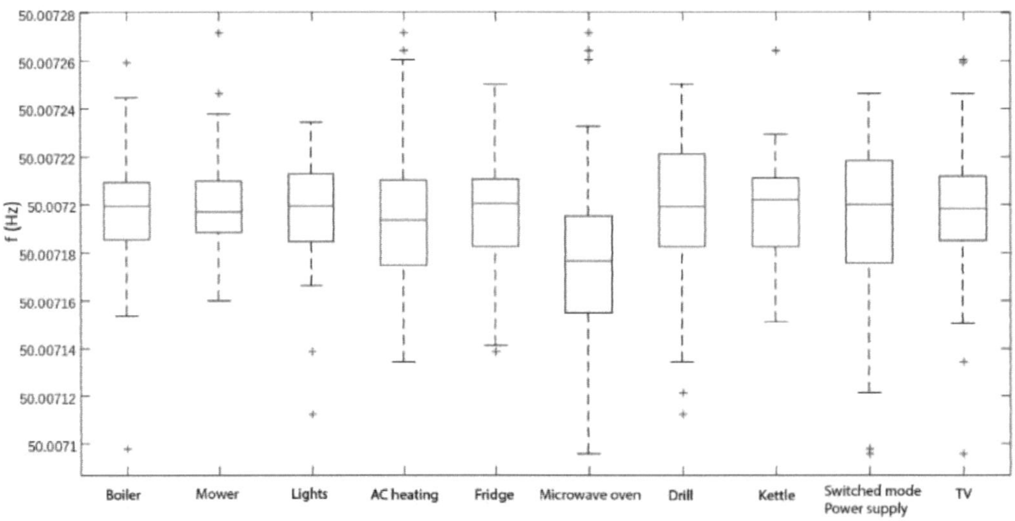

Figure 11. Box plot for all used appliances – impact of appliance operation on FREQ, SoC 100%

Table 3 and fig 12,13,14 shows arithmetic average of FREQ for each appliance during measurement interval for SoC 100% and SoC 50%. During measurement at SoC 100%, each appliance had the same average FREQ values. At SoC, 50% FREQ values during measurement had insignificant differences.

Table 3. Arithmetic average of frequency for each appliance during interval measurement

Appliances	Frequency (Hz)	
	SoC 100%	SoC 50%
Kettle	50.0072	50.0067
Switched mode power supply	50.0072	50.0069
Boiler	50.0072	50.0067
AC heating	50.0072	50.0070
Lights	50.0072	50.0067
TV	50.0072	50.0069
Microwave oven	50.0072	50.0069
Fridge	50.0072	50.0067
Drill	50.0072	50.0069
Mower	50.0072	50.0071

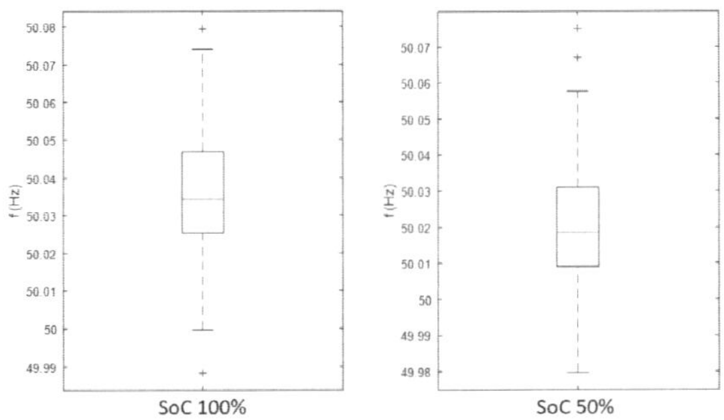

Figure 12. Box plots of measured average FREQ during measurement interval

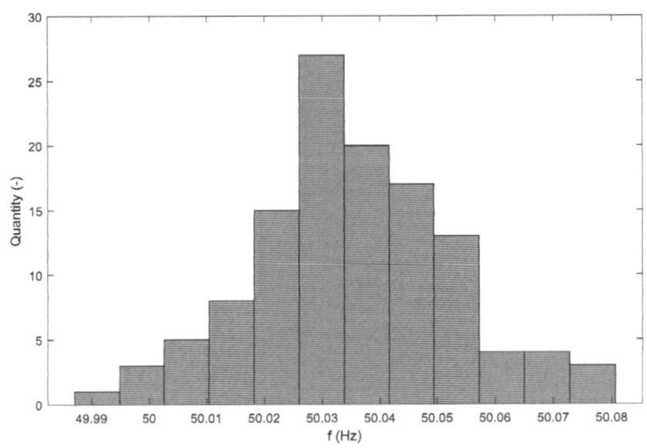

Figure 13. Quantity FREQ with SoC 100%

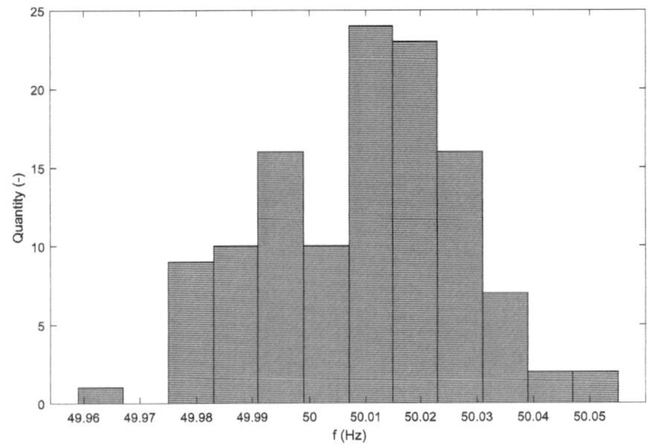

Figure 14. Quantity FREQ values with SoC 50%

IV. Conclusion

The composition of turned-on appliances combination has minimal impact on FREQ and PF and observed changes in these quantities' values were statistically insignificant. During the measurement at both SoCs 100% and 50%, the lowest PF value was measured when the electric mower was turned on, whilst the highest PF value was measured when the kettle was turned on. This is in accordance with a generally considered assumption that the kettle is a resistive load. The comparison of the measured PF values indicates that the difference of the impact of the various-appliances operation on this quantity value for the system at SoC 50% or 100% is small and statistically insignificant. The impact on FREQ value is also small and statistically insignificant. Measured FREQ values are in accordance with European Standard EN 50160: 2010 + A1: 2015.

The longer-time measurement within a shorter timescale will be our future work. This measurement should give us more precise values for further processing. Another task is to increase the number of used appliances number and perform new combinations. This will increase load. Our hybrid inverter has a nominal power of 6.8 kW and the power input of the largest load which was turned on during the measurement was 1.8 kW. So the inverter during experiment was operating in the low load area.

Acknowledgment

This paper was supported by the following projects: LO1404: Sustainable development of ENET Centre; SP2019/159 and SP2019/28 Students Grant Competition and TACR TH02020191 and TACR TH777911, Czech Republic and the Project LTI17023 "Energy Research and Development Information Centre of the Czech Republic" funded by Ministry of Education, Youth and Sports of the Czech Republic, program INTER-EXCELLENCE, subprogram INTER-INFORM and project TN01000007 National Centre for Energy.

References

[1] D. Eckbert and P. Larsson-Edefors, "Interconnect-driven short-circuit power modeling," In: Digital Systems Design. Euromicro Symposium, pp. 414-421., 2001

[2] M. Saradarzadeh, S. Farhangi, J.-L. Schanen, P.-O. Jeanin and D. Frey, "Combination of power flow controller and short-circuit limiter in distribution electrical network using a cascaded h-bridge distribution static synchronous series compensator". In: IET Generation, Transmission & Distribution, vol. 6, no. 11, pp. 1121-1131, IET Digital Library, 2012

[3] S. Bhattacharyya, J. Myrzik, and W. Kling, "Consequences of poor power quality – an overview." In: Universities Power Engineering, UPEC 2007 42nd international conference on, pp. 651-656., 2007

[4] S. Mišák, J. Stuchlý, J. Platoš and P. Krömer, "A heuristic approach to active demand side man-agement in off-grid systems operated in a smart-grid environment," In: Energy and buildings, vol. 96, pp. 274-284, 2015, ISSN: 0378-7788

[5] S. Mišák, J. Stuchlý, J. Vramba, L. Prokop and M. Uher, "Power Quality Analysis in Off-Grid Power Platform," In: Electrical and Electronic Engineering, vol. 12, no. 3, pp. 177-184, 2014

[6] T. Vantuch, S. Mišák, T. Ježowicz, T. Buriánek and V. Snášel, "The power quality forecasting model for off-grid system supported by multiobjective optimization." In: IEEE Transactions on Industrial Electronics, vol. 64, no. 12, pp. 9507-9516, 2017, DOI: 10.1109/TIE.2017.2711540

Review of Radial Distribution Network Reconfiguration Algorithms Based on Mathematical Optimization Theory

Jan Vysocky, Vojtech Blazek, Michal Petruzela, Stanislav Misak, Lukas Prokop
ENET Centre
VSB – Technical University of Ostrava
Ostrava, Czech Republic
jan.vysocky@vsb.cz, vojtech.blazek@vsb.cz, michal.petruzela@vsb.cz , stanislav.misak@vsb.cz, lukas.prokop@vsb.cz

Abstract — **The radial distribution network topology optimization problem (the reconfiguration problem) can be described as a problem of the determination which network feeder is to supply the electrical energy to individual consumption nodes; in other words, it can be described as a problem of the determination what the state of individual network switches should be (if opened or closed) to achieve the optimal network operation state. This optimal network operation state is mathematically described as a state where the value of the so-called object function is minimal or maximal (depending on the formula of this function). This type of problem is solved in many papers and the individual algorithms finding the optimal problem solution described in these papers differ from one another in various aspects. For example, these algorithms differ in their internal structure, computational complexity, or quality of the solution found. According to their internal structure, all reconfiguration algorithms are divided into four groups. This paper details one algorithm group and it is the group of algorithms based on mathematical optimization theory. This paper also describes the Optimal power flow (OPF) problem, a typical problem solved by the reconfiguration algorithms during the reconfiguration process, and six OPF-based reconfiguration algorithms that together provide a good understanding of the modern reconfiguration algorithms operation. The aim of this review paper is to explain to its readers how OPF-based reconfiguration algorithms find the optimal topology of a radial distribution network, and what properties these algorithms have.**

Keywords: distribution network reconfiguration; mathematical optimization algorithms, optimal power flow

I. Introduction

The distribution system (DS) is one part of the power generation, transmission and consumption system. The task of the DS is to distribute the electrical energy received from the transmission system to individual consumption nodes. Electrical energy consumers (whether they are households, service suppliers or industrial consumers) today want the quality of all the parameters of the power supplied to be high, and want an uninterrupted power supply under all external conditions. Institutions which supervise and regulate activities related to the of DS operators (DSO) business, as well as institutions which create legislation affecting this business (the individual state governments are the legislators), also want the services provided by these companies to be of high quality. Experience shows that the reliability of the entire power delivery system is most affected by the DS operation reliability since most of the power delivery reduction events occur in the distribution networks (DN). In addition, the value of the DN operation parameters is reflected in the value of some power quality parameters. Primarily, this is reflected in the magnitude of voltage fluctuations in the individual network nodes.

In order for DSOs to be able to deliver power permanently to meet the high expectations of electrical energy consumers, DSOs have to continually upgrade their DN and increase their operational capabilities. As part of these DN upgrades, new measuring and control devices are installed in the DN, and the DN's basic power infrastructure is expanded to include communication networks so that the DN devices can be controlled and read remotely. All these new features transform the old DNs system into a new system, the system of the so-called Smart Grids.

The DN topology change is one of the abilities provided by the modern smart information and communication technology devices implemented in the DNs: the DN topology change is the process of disconnecting some network nodes from the first feeder, and connecting them to another feeder. This paper describes algorithms which enable DSOs to determine the steps of an optimal reconfiguration process, that is, to determine the DN switch to be opened and the DN switch to be closed. This review paper focuses on reconfiguration algorithms which seek optimal topology of the radial DN and which are based on mathematical optimization theory. Power grid operation optimization is currently one of the main research directions in electrical power engineering, so various documents which describe this research issue in an understandable and compact form are very appreciated. Also, the authors of this review paper tried to write the paper which would bring compact information about algorithms searching for optimal radial topology, and which would be understandable for readers who are not experienced in solving optimization problems. This review paper also describes the Optimal power flow problem which is the core of many optimization algorithms.

The paper is structured as follows. Section II. details the objectives of DN reconfiguration processes, Section III. classifies reconfiguration algorithms, Section IV. describes the algorithms based on the mathematical optimization theory and Section V. presents six examples of reconfiguration algorithms.

II. DN Reconfiguration Process Goals

As mentioned above, governments, different regulatory authorities and electrical energy consumers motivate DSOs to ensure that the DN operation parameters are as high as possible. Specifically, this means that DSOs take various actions to achieve the highest values of electrical power quality parameters, the highest values of parameters describing the electrical energy supply reliability, and also the lowest values of individual DN parts' active power losses. Active power losses are generated, inter alia, during electrical power transmission through power lines. In common DNs powered by AC voltage sources (such as synchronous generators), the active power loss in the electrical conductor j equals

$$\Delta P_j = R_j \cdot I_j^2 \quad (\mathrm{W;\Omega,A}) \, , \tag{1}$$

where R_j is the electrical AC resistance of the electrical conductor j and I_j is the RMS value of the electrical current flowing through the conductor j. Generally, therefore, the active power loss of some part of the electrical power line is directly proportional to a) the electrical resistance of this electrical power line part and b) the square of the magnitude of the electrical current flowing through this electrical power line part. Reducing the DN active power losses can thus be achieved by reducing the electrical resistance of the individual DN power lines (it means increasing the cross-section of the individual power lines' conductors or increasing the number of parallel conductors installed) or reducing the current flowing through the individual power lines. DSOs achieve a long-term reduction in the current flowing through the individual DN parts by increasing the voltage level of the DN part. The apparent power transmitted by the power line equals

$$S_j = \sqrt{3} \cdot U_j \cdot I_j^* \quad (\mathrm{VA;V,A}) \, , \tag{2}$$

where U_j is the electrical voltage of the power line's end node j, and I_j is the electrical current flowing out of the power line in the node j. When transmitting the constant value apparent power, an increase in the voltage magnitude leads to a reduction in the current magnitude, and so it also leads to a reduction in the active power losses generated by the transmission of this apparent power. In practice, this method of active power losses reduction can only be applied when the DN physical infrastructure is under reconstruction, because the DN operation at a higher voltage level requires the use of power devices designed to be operated with such high voltage (higher voltage levels require installation of devices equipped with larger insulation distances, or devices equipped with higher-dielectric-strength insulations). The second way to achieve a active power loss reduction is to optimize the power flows in each DN part. For the DN power-flows optimization, various optimization actions are taken here; however, in a radial-topology DN, change of its topology, i.e. the DN reconfiguration, is the most effective action [1].

However, a DS's total active power loss reduction is not the only achievement gained using the DN topology optimization by DSOs. By connecting the individual DN power lines to make the optimum topology, the voltage drops over the individual DN lines are reduced (thus, the network voltage profile will get better, i.e. there will be lower deviations of the individual DN nodes' voltage values from their nominal value), voltage fluctuations in the individual DN nodes are reduced, and the individual DN power lines' transmission capacity reserves are increased.

In the previous paragraphs we described the targets of the DN reconfiguration processes which are performed continuously in the DN when there is a normal operating state in the DN (i.e., when in DN, there is no fault limiting the DN in its ability to deliver the electrical energy to individual customers). Some DSOs reconfigure their DN only if a fault occurs in the DN (or if another event which restricts a proper electrical energy distribution to individual DN nodes occurs). The purpose of these DN reconfigurations is the same as in DN-normal-operation-state reconfiguration systems, that is, to close and open DN switches to transform a DN to an "optimal topology", but this optimal topology is defined here differently than the optimal topology in the DN-normal-operation-state reconfiguration systems. Here, the optimal topology is the topology which will enable to restore the electrical energy delivery to the highest possible number of consumption nodes, and which will enable restoration of this electrical energy delivery to be achieved as quickly as possible when the reconfiguration system is applied at the DN. In these reconfiguration systems, the optimal reconfiguration process can be defined as the one that minimizes the total value of the electrical energy which was non-delivered to individual consumption nodes during that one restoration process.

III. Types of DN Reconfiguration Algorithms

The paper's previous section outlined reasons why DSOs change the DN topology to the optimal one and this optimal topology was defined here. This section describes those algorithms which provide DSOs with the optimum topology for each situation in their DN. There are many reconfiguration algorithms and all these algorithms can be categorized into several groups according to their internal structure:

a) algorithms based on mathematical optimization theory

b) switches exchange algorithms

c) algorithms seeking the optimal power flow pattern

d) algorithms based on artificial intelligence

In Section IV, algorithms based on mathematical optimization theory are detailed. A description of switches exchange algorithms was presented in [2], a description of algorithms seeking the optimal power flow pattern was presented in [3] and a description of algorithms based on artificial intelligence was presented in [4].

IV. Algorithms Based on Mathematical Optimization Theory

These algorithms describe the DN reconfiguration problem using mathematical models. The algorithms then solve the DN reconfiguration problem using various mathematical optimization methods. The individual mathematical optimization methods are described in the following subsections.

A. Linear programming

Linear Programming (LP) is a mathematical optimization method solving the problem of finding the minimum or maximum of a linear function of one or more variables, where the set of possible solutions is described by a system of linear inequalities. The task of linear programming is to solve the optimization problem

$$\min_{x \in M}(\mathbf{c}^T \mathbf{x}) \, , \tag{3}$$

where $c^T x$ denotes the scalar product of vectors c and x. The set of possible solutions M is described by a system of linear equations

$$\mathbf{Ax} = \mathbf{b}, \quad x \geq 0 \, , \tag{4}$$

where A is a matrix of the dimension $m \times n$ (m is the number of rows of matrix A and thus also the number of equations describing the analyzed system, n is the number of columns of matrix A and thus also the number of system variables), b is the m-dimensional vector, and c and x are n-dimensional vectors [5].

The radial DN reconfiguration problem was first solved using LP in [6]. The algorithm presented in [6] was derived from an LP method called Branch and bound (B&B). The B&B method is usually used to solve problems from discrete optimization and combinatorial optimization. The B&B method systematically analyzes all potential problem solutions, and large subsets of inappropriate potential solutions are removed from the area under investigation at once. During the cuts, the estimation of the upper and lower values is used and these estimations are refined in the individual iterations of the algorithm, so the algorithm gets closer to the solved problem's optimal result in each iteration [5]. The reconfiguration algorithm in [6] solves the problem as follows. In the first step, the algorithm closes all tie switches. So, the individual DN feeders are connected to the closed loops now. Then the algorithm starts opening one power-line switch after the other and does so until the DN topology is radial again. Each time a switch is opened, the algorithm calculates the current value of total active power losses generated in the DN, and if these are now higher compared to the active power losses value calculated at the previous algorithm's step, the algorithm again closes the last open switch and opens a different one. When the algorithm is no longer able to reduce the current value of total active power losses by opening any DN switch, the network topology created by the last successful switch opening is the optimal radial topology of the DN. The reconfiguration problem's optimal solution found by this algorithm does not depend on the initial topology of the radial DN. The weaknesses of this algorithm are as follows: phases of node voltage phasors and network operation limits are ignored; the algorithm uses DC calculation.

B. *Nonlinear programming*

Nonlinear Programming (NLP) is a mathematical optimization method solving the problem of finding the minimum or maximum of a nonlinear function of one or more variables, where the set of possible solutions is described by a system of inequalities. Special types of NLP are convex programming and quadratic programming. The task of nonlinear programming is to solve the optimization problem

$$\min_{x \in M}\left(f(x)\right), \tag{5}$$

where $f(x)$ is any nonlinear function. The set of possible solutions M is described by a system of inequalities

$$g_i(x) \le 0, \quad i = 1, \dots, m, \tag{6}$$

where $g_i(x)$ are real functions, m is the total number of functions describing the analyzed system. There are many algorithms for solving NLP problems, e.g. the Greatest gradient method, the Conjugate gradient method or the Penalty method [5].

The radial DN reconfiguration problem was first solved using NLP in [7]. The radial DN reconfiguration problem solved in [7] is described as a problem of nonlinear integer programming with quadratic objective function and binary state variables. The reconfiguration algorithm searches for the optimal radial topology in a way where one network switch is closed and another one is opened in each algorithm iteration. In each iteration, the algorithm calculates the change in the value of total active power losses generated in the DN. If this value increases, the algorithm tries to transfer consumption nodes from one feeder to another in the opposite direction compared to the direction of consumption nodes transfers used in the algorithm's earlier steps. If the algorithm cannot find a topology giving a lower value of DN active power losses, then the topology obtained in the last successfully completed iteration is the optimal topology of that radial DN.

C. *Integer programming*

Integer Programming (IP) is a mathematical optimization method solving the problem of finding the minimum or maximum of a function of one or more variables, where the set of possible solutions is described by a system of inequalities, and state variables are restricted to be integers. The task of integer programming is to solve the optimization problem

$$\min_{x \in M}\left(f(x)\right), \tag{7}$$

where the individual elements of the vector x can be only integers. The most common type of integer programming problem is the linear integer programming problem

$$\min_{x \in M}(\mathbf{c}^T\mathbf{x}), \tag{8}$$

where $c^T x$ denotes the scalar product of vectors c and x. The set of possible solutions M is described here by the system of equations

$$\mathbf{Ax} = \mathbf{b}, \quad x \ge 0, \quad x_i \in Z, \quad i \in C, \tag{9}$$

where A is a matrix of the dimension $m \times n$ (m is the number of rows of matrix A and thus also the number of equations describing the analyzed system, n is the number of columns of matrix A and thus also the number of system variables), b is the m-dimensional vector, and c and x are n-dimensional vectors. $C \subseteq \{1, \dots, n\}$ is a set of indexes of variables to be integers. There are many algorithms for solving IP problems, such as the Cutting-plane method or the Branch & bound method. If some decision variables of the solved problem are not an integer the problem is called a mixed-integer programming problem [5].

V. Reconfiguration Algorithms Based on the Optimum Power Flow Problem

Most reconfiguration algorithms are based on the so-called Optimal power flow (OPF) problem. The OPF problem is a mathematical problem seeking the power grid's state where the power grid's power flows are optimal given the defined optimization objectives, and where the OPF problem results are within the limits of the operation parameters. The OPF problem is a non-linear problem, so there is no guarantee that the algorithm solving a standard OPF problem finds a globally optimal solution of this problem. To be able to solve this problem and be sure that the solution found is the global optimum, the OPF problem is reformulated into another problem that can be solved by standard integer programming methods. The most effective way is different for various optimization problems (the selection of the most effective method is influenced by the extent of the optimized problem, the object function's formula, the power grid's operation constraints, etc.).

In [8], a heuristic algorithm is presented which solves the problem of nonconvexity of AC OPF problem using second-order cone programming relaxation. The algorithm does not solve any mixed-integer optimization problem and optimizes "the successive branch reduction approach". In [8], the mathematical proof is presented which says that the proposed algorithm always finds the optimal solution for radial DNs with two feeders. Reconfiguration tests of four more-complex test DNs are presented in the paper and these tests show that the presented algorithm is able to find the optimal topology even for large complex radial DNs. The main advantages of this algorithm are the efficiency (the computing time is linearly proportional to the total number of switches which need to be opened), the accuracy of the optimal solution found and the versatility (in the algorithm, there are no parameters and initialization which need to be tuned for different DNs). There is also a simplified algorithm proposed in this paper, whose time of searching for optimum configuration is constant for variously-complex DNs, but the solution found by this simplified algorithm is still close to the optimal solution.

In [9], the authors explore how the distributed power generation (DG) hosting capacity of active distribution systems can be increased by means of DN reconfiguration. In the paper, both static reconfiguration (DN reconfiguration at the planning stage) and dynamic reconfiguration (DN reconfiguration using remotely controlled switches) are examined. In the paper, a reconfiguration algorithm based on the mixed-integer, nonlinear, multi-period OPF problem is proposed. This algorithm aims to maximize the DG hosting capacity under thermal and voltage constraints. This algorithm has some aspects that are missing in other reconfiguration algorithms: this algorithm considers a different optimization goal, the DG, the inherent time-varying (renewable) generation patterns and the load behavior which are aggregated into multi-time periods, and the Active network management schemes. The algorithm's effectiveness was tested by a simulation of the reconfiguration of the widely used 34-bus 12.66-kV DN which was modified here to consider eight potential sites for the DG.

Paper [10] presents a reconfiguration algorithm which solves the problem of finding the optimal DN topology using Benders decomposition. The reconfiguration problem is divided here into two stages: the first stage is the Master problem and the second stage the Slave problem. The master problem is formulated as a mixed-integer NLP problem and determines the DN radial topology. Constraints related to power flows flowing through DN power lines are included in this stage. This fact accelerates the system's convergence process since the master problem solution is close to the final solution. The slave problem is formulated as an NLP problem. This stage is used to define the feasibility of the master problem solution by means of an OPF, and provides information with which to formulate the linear Benders cuts which transfer information from the slave to the master problem. The last iteration of the slave subproblem defines the optimal voltage profile and the optimal power flows in individual DN power lines. If the DN data per phase is available, this algorithm can be executed per phase to solve unbalanced DN reconfiguration. The effectiveness and robustness of the algorithm are shown by means of three test cases.

Paper [11] presents an algorithm which reconfigures radial DN with respect to both the usual network operation constraints (maximum value of power line currents, maximum and minimum value of node voltages, ...) and the Fault current constraints (FCC). The algorithm is designed for DSOs which face higher fault levels than the DN protection equipment can handle. The algorithm can use a set of remedial actions (the bus splitting, the power line opening or the transformer disconnection), which can reduce expected fault currents within the limits imposed by protection specifications. In the first step, the algorithm solves the classical OPF problem considering only traditional operation constraints. Its solution is used as a basis for fault analysis. If there are violations of FCCs, the algorithm solves the FCC-OPF problem (it is an enhanced OPF problem considering both traditional operation constraints and FCCs). If the algorithm does not find an FCC-OPF problem solution which would satisfy all FCCs, then the algorithm selects the most effective remedial action which would help the algorithm meet all those FCCs. In the paper, the algorithm was tested on a medium-scale network and the results demonstrate the algorithm's efficiency.

Paper [12] presents a reconfiguration method based on a modified rolling horizon-like (RH) algorithm which decomposes the multi-period optimum power flow (MP-OPF) problem into a desired number of sub-problems of decreasing size while relaxing some binary variables associated with the dynamic DN reconfiguration. The RH algorithm treats binary variables as fixed, binary, or relaxed (i.e. continuous). In the case of MP-OPF composed of three periods, the RH algorithm treats binary variables as follows: the first sub-problem instance solves the MP-OPF problem with the status variables modeled as binary in the first period and as continuous in all subsequent periods, while the last sub-problem instance solves the MP-OPF problem including only the constraints of the last period in which all status variables are modeled as binary. The primary goal of this reconfiguration method is minimizing DG energy curtailment. This reconfiguration method was developed for planning purposes but is also suitable for day-ahead or intraday optimization. The tests of the method's optimization capabilities showed the method's great effectiveness in terms of DG curtailment savings.

Paper [13] presents a reconfiguration algorithm for the Active distribution network (ADN). The algorithm is based on the branch flow OPF problem which is reformulated by the second-order cone convex relaxation to the form of a mixed integer second-order cone (MISOCP) problem. Since the algorithm uses the convex relaxation of the original OPF problem, any globally optimal solution (if it is feasible for the OPF) is also the optimal solution of the OPF problem. The MISOCP problem is then solved by Branch & Cut algorithm to find global optimality. The solved problem takes into account the European context of unbundling where the DSO-owned generation is impossible except for system security. In the problem, the upward dispatching of generation is totally eliminated. In the algorithm, exact models of discrete-nature flexibilities like the on-load tap changer and the reconfiguration are directly incorporated in the relaxed operation planning model through reformulation techniques. The short-term economic aspects of flexibility operation are also incorporated in the algorithm. The algorithm's goal is the DS operation costs minimization. The algorithm is tested on a modified test DN, and the results show a high reduction in DS operation costs.

VI. Conclusion

Today, major changes are taking place in the electrical power sector (the advent of distributed renewable energy sources and electric cars; there are changes in the electricity market), and these changes are making electrical DS control increasingly difficult. In order for DSOs to be able to operate their DS in a reliable and economical manner, they must use various modern tools. These tools include software searching for optimal values of DS parameters. This paper presented algorithms which search for optimal DN topology (the reconfiguration algorithms) and detailed algorithms based on mathematical optimization theory.

Algorithms based on the mathematical optimization theory describe DN reconfiguration problem with mathematical models and solve the problem using various mathematical optimization methods (such as linear programming, nonlinear programming, etc.). The standard problem solved by reconfiguration algorithms is the Optimal power flow problem, which enables to find optimal values of various parameters of the electric network considering various operation constraints. Because the OPF problem is nonlinear, there are various modifications to the basic form of the OPF problem presented in many research papers. The presented reformulations of the standard OPF problem transform the OPF problem into a new problem which can be solved by mathematical optimization methods and guarantees the finding of the system's global optimum. In this paper, six reconfiguration algorithms differing in the optimization goals and the method of reforming OPF problem were presented. All papers cited in this review present both reconfiguration algorithms and operational capability tests of these reconfiguration algorithms. These tests were performed on standard radial DNs test models. The results of these tests showed that algorithms based on mathematical optimization theory are able to find the system's global optimum quickly and reliably.

ACKNOWLEDGMENT

This paper was supported by the following projects: LO1404: Sustainable development of ENET Centre; SP2019/159 and SP2019/28 Students Grant Competition and TACR TH02020191 and TACR TH777911, Czech Republic and the Project LTI17023 "Energy Research and Development Information Centre of the Czech Republic" funded by Ministry of Education, Youth and Sports of the Czech Republic, program INTER-EXCELLENCE, subprogram INTER-INFORM and project TN01000007 National Centre for Energy.

REFERENCES

[1] Y. Hu, N. Hua, Ch. Wang, J. Gong and X. Li, "Research on distribution network reconfiguration," *2010 International Conference on Computer, Mechatronics, Control and Electronic Engineering*, Changchun, 2010, pp. 176-180. doi: 10.1109/CMCE.2010.5610464

[2] S. Civanlar, J. J. Grainger, H. Yin et al ., "Distribution Feeder Reconfiguration for Loss Reduction" *IEEE Trans on Power Delivery*, 1988, pp. 1217-1223. doi: 10.1109/61.193906

[3] D. Shirmohammadi and H. W. Hong, "Reconfiguration of electric distribution networks for resistive line losses reduction," in *IEEE Transactions on Power Delivery*, vol. 4, no. 2, pp. 1492-1498, April 1989. doi: 10.1109/61.25637

[4] R. Qiu, X. Lv, S. Chen, "A Survey on Artificial Intelligence Algorithm for Distribution Network Reconfiguration" *Robotic Welding, Intelligence and Automation. Lecture Notes in Electrical Engineering*, vol 88. Springer, Berlin, Heidelberg, 2011

[5] J. W. Chinneck, "Practical Optimization: a gentle introduction", Dated of the cited version: June 23, 2015. Available online at: www.sce.carleton.ca/faculty/chinneck/po.html

[6] A. Merlin, H. Back, "Search for a minimal-loss operating spanning tree configuration in an urban power distribution system" *Proceedings of 5th Power System Computation Conference*, Cambridge, UK, 1975, pp. 1-18.

[7] J. Fan, L. Zhang and J. D. McDonald, "Distribution network reconfiguration: single loop optimization," *IEEE Transactions on Power Systems*, vol. 11, no. 3, pp. 1643-1647, Aug. 1996. doi: 10.1109/59.535709

[8] Q. Peng, Y. Tang and S. H. Low, "Feeder Reconfiguration in Distribution Networks Based on Convex Relaxation of OPF," *IEEE Transactions on Power Systems*, vol. 30, no. 4, pp. 1793-1804, July 2015. doi: 10.1109/TPWRS.2014.2356513

[9] F. Capitanescu, L. F. Ochoa, H. Margossian and N. D. Hatziargyriou, "Assessing the Potential of Network Reconfiguration to Improve Distributed Generation Hosting Capacity in Active Distribution Systems," *IEEE Transactions on Power Systems*, vol. 30, no. 1, pp. 346-356, Jan. 2015. doi: 10.1109/TPWRS.2014.2320895

[10] H. M. Khodr, J. Martinez-Crespo, M. A. Matos and J. Pereira, "Distribution Systems Reconfiguration Based on OPF Using Benders Decomposition," *IEEE Transactions on Power Delivery*, vol. 24, no. 4, pp. 2166-2176, Oct. 2009. doi: 10.1109/TPWRD.2009.2027510

[11] P. N. Vovos, H. Song, K. Cho and T. Kim, "A network reconfiguration algorithm for the reduction of expected fault currents," *2013 IEEE Power & Energy Society General Meeting*, Vancouver, BC, 2013, pp. 1-5. doi: 10.1109/PESMG.2013.6673000

[12] F. Capitanescu, "A relax and reduce sequential decomposition rolling horizon algorithm to value dynamic network reconfiguration in smart distribution grid," *2017 IEEE PES Innovative Smart Grid Technologies Conference Europe (ISGT-Europe)*, Torino, 2017, pp. 1-6. doi: 10.1109/ISGTEurope.2017.8260094

[13] B. Swaminathan, V. Debusschere and R. Caire, "Short-term active distribution network operation with convex formulations of power flow," *2017 IEEE Manchester PowerTech*, Manchester, 2017, pp. 1-6. doi: 10.1109/PTC.2017.7981084

The Influence of FACTS Devices on Power System

Vladimír Kohan, Zsolt Čonka
Department of Electric Power Engineering
Technical university of Košice
Košice, Slovakia
vladimir.kohan.2@student.tuke.sk, zsolt.conka@tuke.sk

Abstract — **The present article is aimed to issue of FACTS systems and their utilization for improvement of transient stability of power systems. This article is concerned with simulations of SVC, STATCOM, TCSC and UPFC systems with major objective of a practical representation of the influence these devices in NEPLAN software. This scheme examined transient phenomena (dynamic stability) adjusted to 3-phase short-circuit and subsequently observed power system oscillation using flexible devices with and without the use of it. The final article chapter focuses on the recommendation of the practice of FACTS systems, which are used in the usual practice, and what is more they help to manage the power flow, stabilize voltage and to have a positive impact on the transient phenomena.**

Keywords: FACTS, SVC, TCSC, STATCOM, UPFC, controller, dynamic stability, transient phenomenon

I. Introduction

This chapter focuses on implementing FACTS devices and is compared to different types of controllers. The modeling of the elements in the NEPLAN program shows the regulator's response to voltage and active and reactive power at the short circuit.

The operation of each FACTS system is different, and therefore, when it is installed, it will be determined in advance what type of controller is to be dealt with in order to provide the most optimal solution to the individual problems in a given line or node, whether global or local. In almost any case, any version that is suitably set up for a given network greatly improves the quality of either transmission or transients and the stability of electricity.

The application of FACTS systems is mainly limited by the investment costs due to power semiconductor devices and the overall complexity of the technology structure. However, due to their high reliability and efficiency, the return on these devices exceeds the costs associated with their purchase and installation.

Using the NEPLAN computational program, a simple diagram (Figure *1.*) was constructed to observe transient events with the implementation of FACTS systems and to compare their response and ability to regulate voltages and power flows with the best possible ability to maintain system stability. In the given scheme, disturbances are defined, where the line "LINE23" is disconnected at 0.2 s from both nodes and in the node "BUS3" at a time of 0.1 s a three-phase short-circuit with a duration of 0.1 s is set (thus up to 0.2 s).

In the following subchapters, the individual effects of the controllers will be graphically presented, starting with a system where the generator is not using flexible systems (only the wake-up controller is used) and then compared with individual FACTS devices (SVC, TCSC, STATCOM with UPFC - in this order). Each of the waveforms shows the time on the x axis (from 0s to 5s) and the y-axis shows the bus voltages and active and reactive power on the generator in proportional units. All elements of the FACTS system will connect to a node called "BUS4".

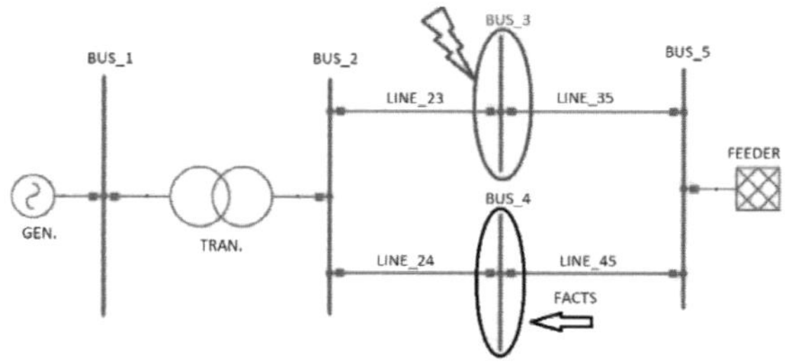

Figure 1. Fundamental scheme

The scheme consists of a generator, a transformer, a series-parallel line and a feeder. All of these elements are enough to keep track of dynamic influences, stability, and voltage fluctuations in a variety of unwanted phenomena in the system. In our case, disconnecting the line and applying a three-phase short circuit in the node can be observed in the oscillations of the voltage and power, thereby evaluating why it is necessary and advantageous to use the control FACTS devices.

II. Simplified model of simulation of dynamic phenomena without and with using exciter in NEPLAN programme

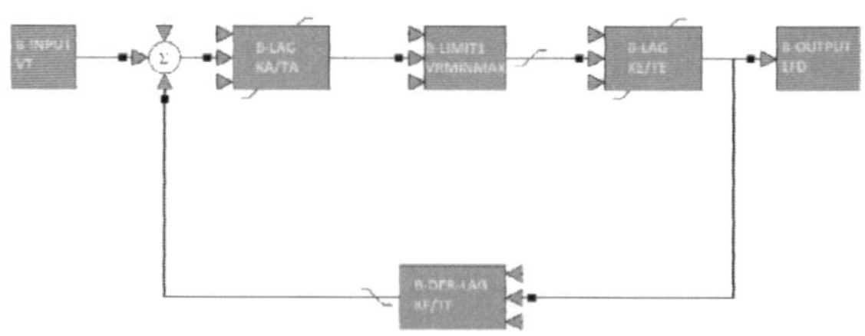

Figure 2. Controller exciter

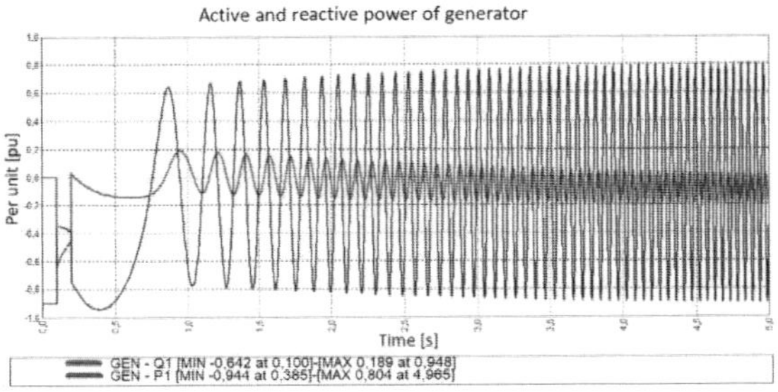

Figure 3. Without exciter P, Q

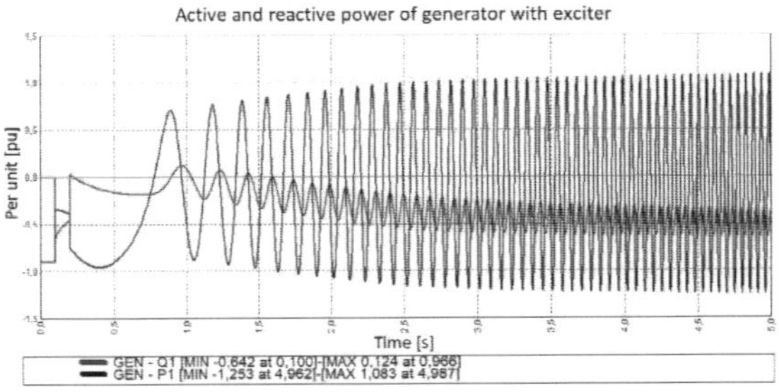

Figure 4. With Exciter P, Q

Figure *3*. and Figure *4*. shows the active (blue wave) and reactive (red wave) waveforms of the generator without and using the exciter controller. From the waveforms it can be seen that with the set failure, the generator has grown to the point that it has not been able to stabilize itself to a constant power value. The generator has dropped out of synchronism while the amplitude of the oscillation increases with increasing time - generator shutdown is required (power supply stabilization). The exciter response can also be noted by the fact that the reactive power oscillation has stabilized at a lower value (without amplitude change) compared to the non-exciter generator.

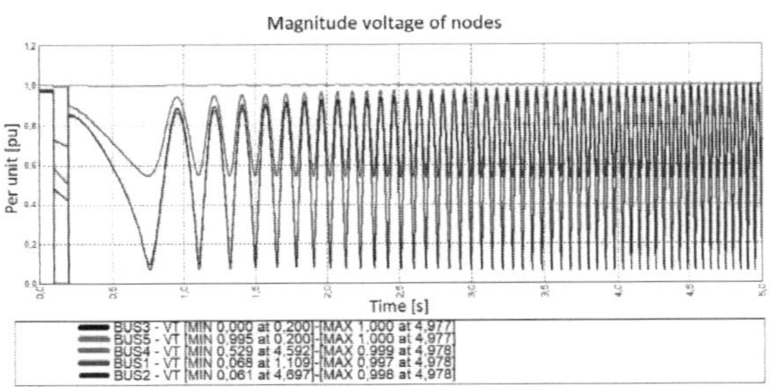

Figure 5.Voltage without exciter

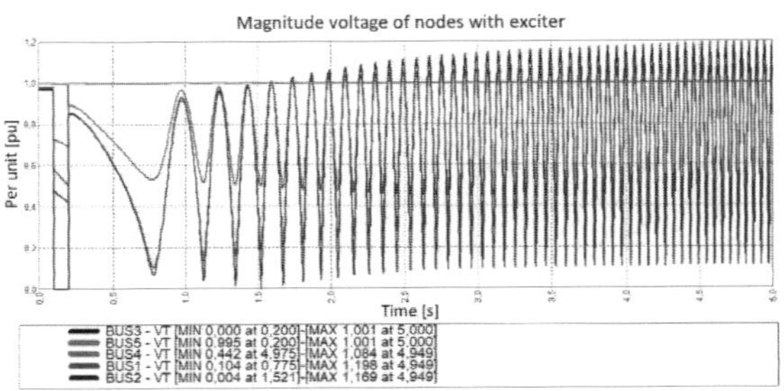

Figure 6. Voltage with exciter

Disconnection of the line "LINE23" and the formation of a three-phase short-circuit on the bus "BUS3" due to large currents and interaction of the entire system oscillated the voltage as shown in Figure *5*. and Figure *6*. As a result of the exciter controller, the voltage fluctuations at the "BUS1" generator outlet were shifted (similar to reactive power) above 1 p.j. and similarly they could not stabilize. The voltages "BUS5" have almost constant value throughout the monitored time. This is due to the fact that "FEEDER" is considered to be a so-called. a hard network - a network capable of delivering any amount of power without much voltage change (used to calculate and simulate waveforms for NEPLAN). Thus, we can conclude that the driver cannot stabilize and compensate for the serious failure and is not sufficient to stabilize and regulate the dynamic events in the system.

III. Dynamic Model of Svc Controller

The main purpose of the SVC regulator is to increase the transmission capacity of the power system. This can be achieved if voltage is provided (by stabilizing it) and by increasing system stability boundaries. In order to stabilize the voltage at the receiving end of the transmission line and to contribute to the improvement of transient stability, the SVC essentially functions as a voltage regulator. While reactive power changes to reduce and quickly damp oscillations during voltage fluctuations, and especially after major bus failures.

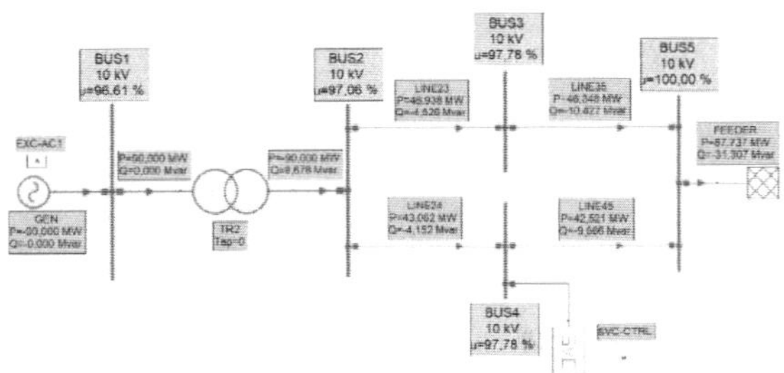

Figure 7. Model with SVC

The FACTS system technology, called SVC, is placed in the model in the "BUS4" node, with each FACTS element having its own controller, which controls the individual parameters for power transmission based on inputs and outputs. It is a modern trend that SVC controllers have also been widely used to control offtake in industry or household to improve a power factor.

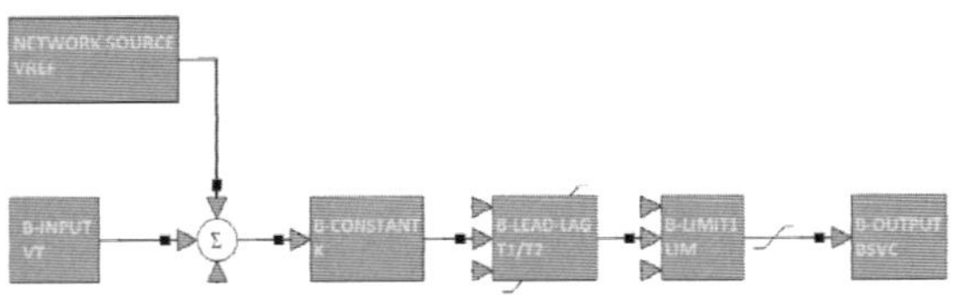

Figure 8. Controller of SVC system

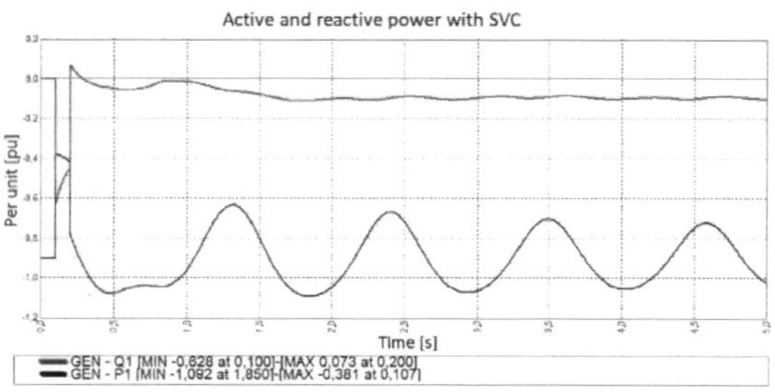

Figure 9. Active and reactive power with using of SVC system

In Figure *9.* we can observe the response and influence of the controller and SVC system on generator performance compared to its use or using other FACTS systems (listed in other subchapters). The reactive power the SVC was able to stabilize to an almost constant value while the active power settled at low oscillations, which no longer set a constant value with the controller.

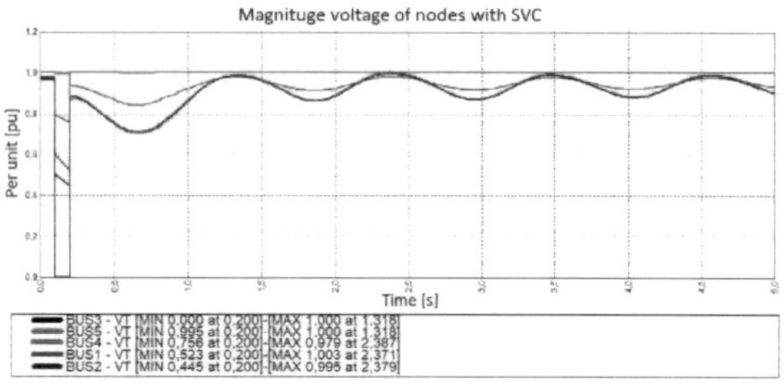

Figure 10. Magnitude voltage with using of SVC system

The SVC system greatly influenced the voltage fluctuations in all system nodes except the "BUS3" fault and the "BUS5" hard network node. Voltage fluctuations were managed by SVC to much lower values, which the network can handle much better, thereby delaying system disruption, disconnecting power supplies, or preventing generator outage from synchronism.

IV. Dynamic Model of Tcsc Controller

For studies of dynamic stability and oscillation events, a TCSC device may be represented by a variable reactance that is modeled as a variable reactance at a base frequency due to failure and during frequency variation - the frequency remains constant.

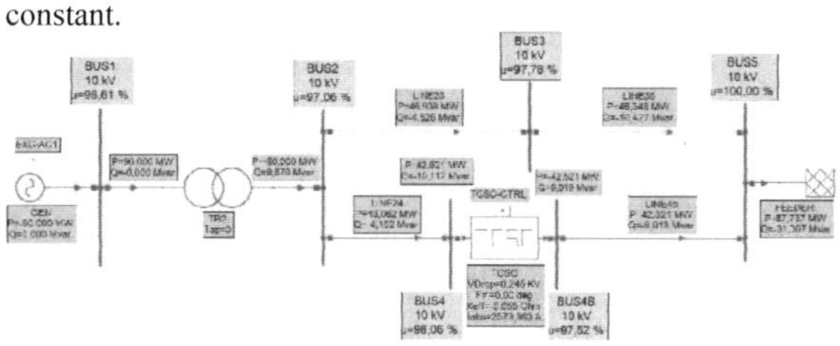

Figure 11. Model with TCSC device

Each of the FACTS devices is designed for the purpose of controlling or controlling the parameters in the system. Thus, the regulator may have entered different input values (voltage, current, power, etc.), which it monitors during operation and thereby controls the output value by means of the members included in the diagram, which may also differ from other regulators (susceptance - change of admittance, inductance - impedance change and others).

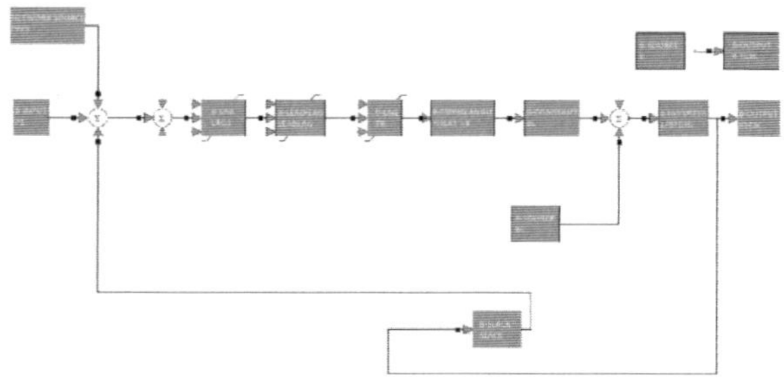

Figure 12. Controller of TCSC system

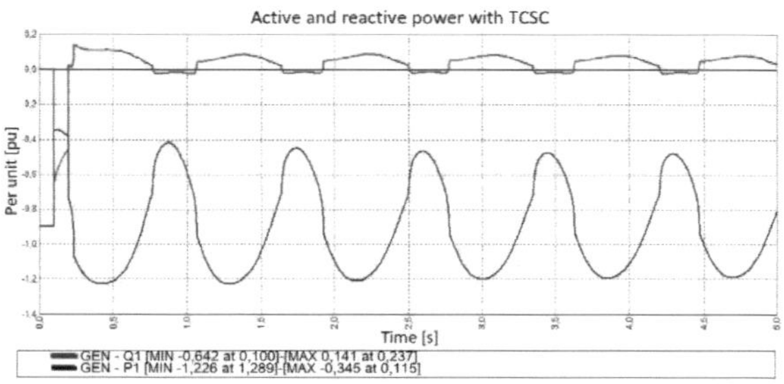

Figure 13. Active and reactive power with using of TCSC

On the curves of active and reactive power of the generator of Figure *13.* we can also observe a marked improvement in performance oscillation compared to a non-FACTS-based scheme. Compared to the SVC system, the module thus failed to stabilize TCSC reactive power without oscillations. From the graph (red waveform) we can see when the regulator has turned the controller on and off to achieve the best possible power transmission operation by the power system.

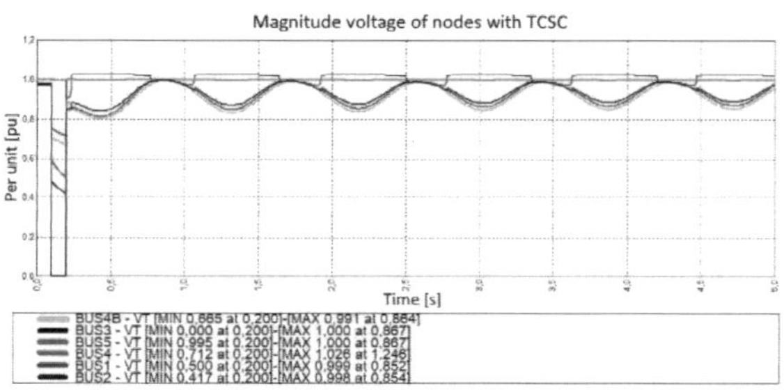

Figure 14. Magnitude voltage with using of TCSC system

Switching on the TSCS module also affected the voltage that the controller was able to settle for minimal fluctuation to ensure its subsequent amplification amplitude as was the case without using a FACTS device. When switching the semiconductor elements in the TCSC, the voltage variation in the "BUS4" node was changed step by step when the oscillations and voltage oscillations were changed due to the impedance change (see Figure *14*).

V. Dynamic Model of STATCOM Controller

Dynamic models of STATCOM devices are based on dynamic SVC systems, the difference being that each of them is constrained by a different value. STATCOM is restricted by current (between IC max and IL max) while SVC is limited by susceptance (between BC and BL).

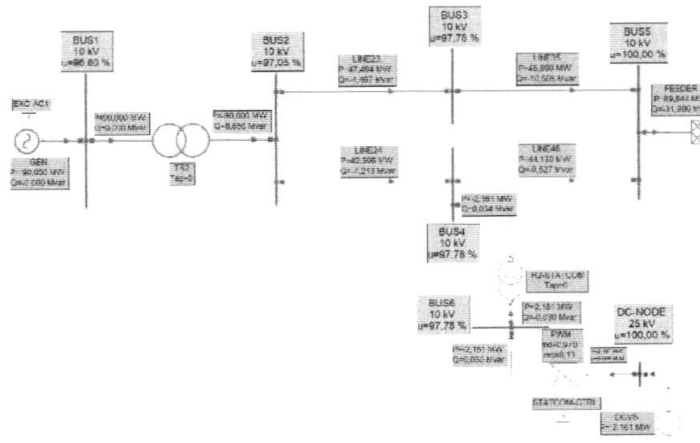

Figure 15. Model with STATCOM device

Each FACTS system has its own controller structure and is, in a way, unique in design and control of given parameters in a system or compensated area. The same technology of a given type of FACTS system can have many designs with a different controller control principle. It means different concurrency of input and output information - another mathematical model of the controller diagram. While the output of them is always connected and it is connected to the switching and switching control pulses of semiconductor components.

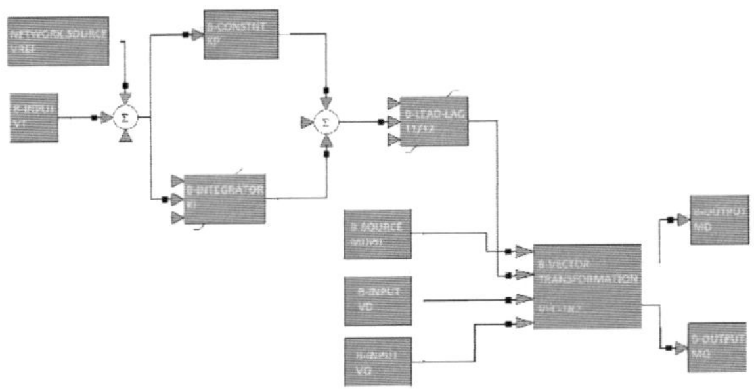

Figure 16. Controller of STATCOM system

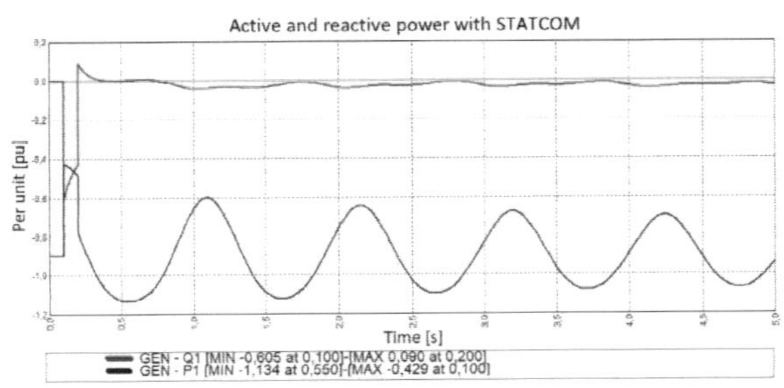

Figure 17. Active and reactive power with using of STATCOM system

As with previous simulations of FACTS controllers used to improve system stability and dynamics, STATCOM technology is able to similarly regulate and reduce voltage and active and reactive power oscillations even when using other equipment designs. From the graphs shown in Figure *17*. and Figure *18*. we can observe the influence of the STATCOM controller system, which has stabilized the reactive power to almost constant value, and the active power has substantially reduced the amplitude of the oscillations and limited their rapid step changes as opposed to without the FACTS controllers.

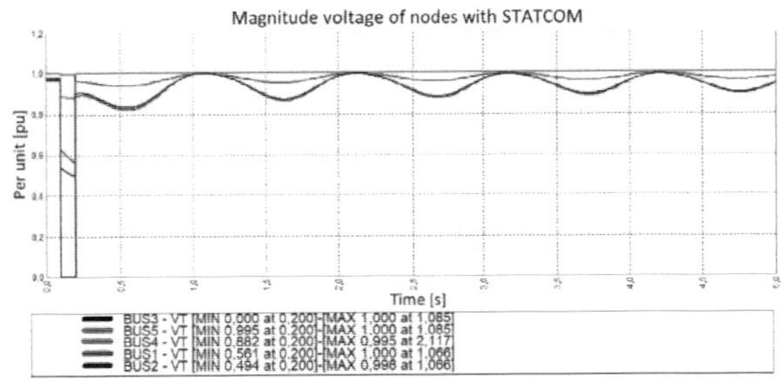

Figure 18. Magnitude voltage with using of STATCOM system

VI. Dynamic Model of UPFC Controller

The UPFC is a combination of serial and parallel controlled compensation. The UPFC system may appear to be interconnected two known as devices, namely STATCOM and SSSC interconnected via VSC converters with a DC link. This device can independently control both active and reactive power transmitted by the transmission line.

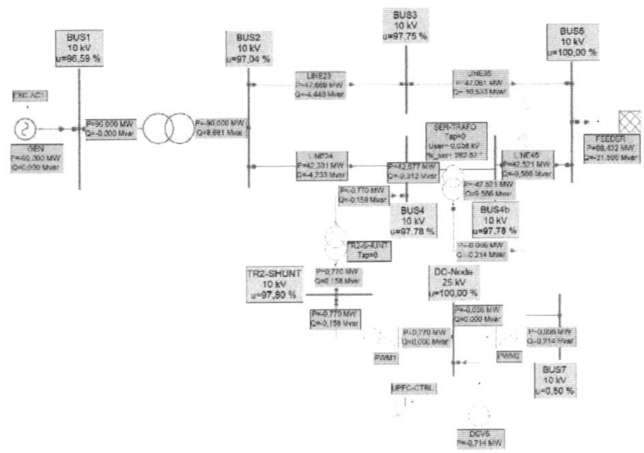

Figure 19. Model with UPFC device

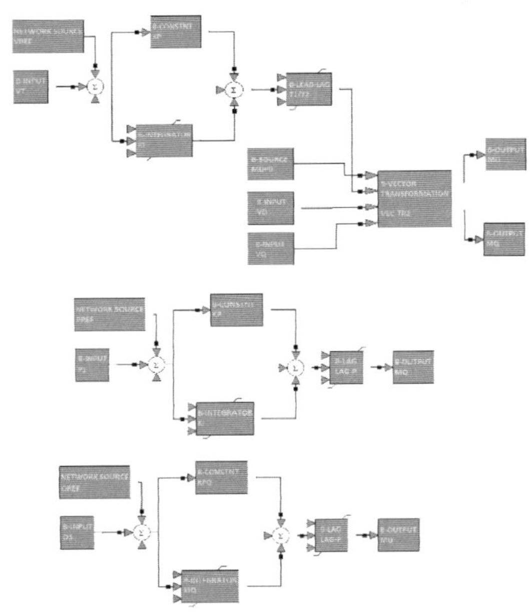

Figure 20. Controller of UPFC system

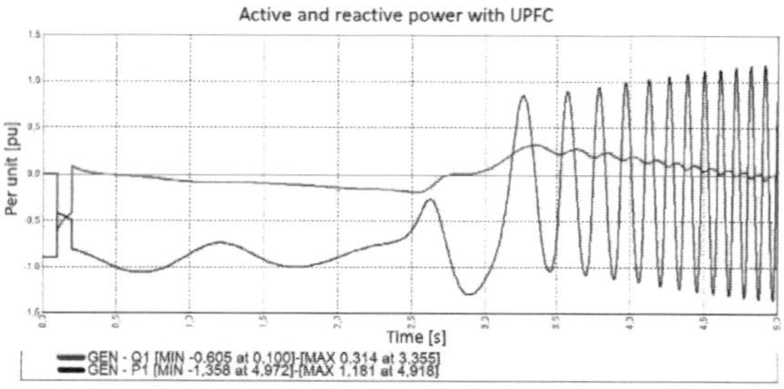

Figure 21. Active and reactive power with using of UPFC system

In Figure *21*. and Figure *22*. we can see the response of the UPFC system controller, which failed to compensate for the failure, stabilize or mitigate the effects of oscillation of power, voltage, and other system parameters. By approximately 2.9 seconds when the first greater active power oscillation was generated, the controller was unable to stabilize this oscillation. As a result, the amplitude of power and voltage increased with increasing time and would oscillate until the local area in the network or the entire system was broken down. In this case, we can say that the synchronous generator has dropped out of synchronism and the UPFC controller has failed to improve the power transmission in the system.

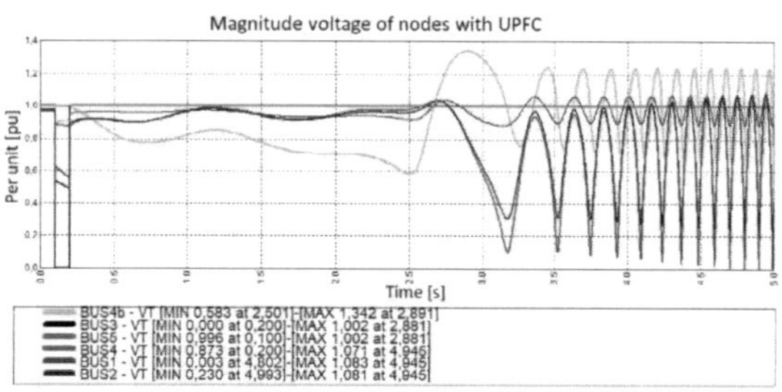

Figure 22. Magnitude voltage with using of UPFC system

CONCLUSION

From the individual developments and research on the impact of FACTS devices in this model, we can evaluate that in the vast majority of these devices are able to greatly ensure the improvement and reliability of the power system. Each device has its own controller and its design and location is suitable for various applications that need to be solved for their installation. This is an analysis for the purpose of applying the device to the system, and which of the electricity quality indicators need to be improved or optimized at a given location. It can be a power factor improvement, reactive power compensation, node voltage stability, increased transmission capacities, and especially the fastest possible dynamic events in various failures, resulting in rapid assistance for synchronous generators and preventing their outage from synchronism. Through FACTS systems, we can manage power flows, ensure continuous production, transmission and supply of electricity, and thus strengthen the whole system as a whole, thereby avoiding blackouts or even a major problem such as the aforementioned "Blackout" systems. Due to the rapidly evolving demand for electricity, it can be stated that FACTS systems will be largely implemented in different locations, whether transmission or distribution systems. However, before they are put into practice, an analysis is needed to address what equipment is best suited to the issue and to which part of the network to use the selected system to make it as reliable as possible to optimize electricity transmission.

ACKNOWLEDGEMENT

This work was supported by the Scientific Grant Agency of the Ministry of Education of Slovak Republic and the Slovak Academy of Sciences by the projects VEGA No. 1/0372/18.

REFERENCES

[1] M. Eremia , Liu, CH. CH., Edris, A. A., "Advanced Solutions in Power Systems"HVDC, FACTS, and Artificial Intelligence, 1. ed. New Jersey: Wiley-IEEE Press, 2016. 271-717 p. ISBN 978-1-119-03569-5.

[2] R.M. Mathur, R.K. Varma, ,"Thyristor-Based FACTS Controllers for Electrical Transmission System" New York: Wiley-IEEE Press, 2002. 1-493 p. ISBN 978-0-471-20643 1.

[3] N.G. Hingorani, L. Gyugyi," Understanding FACTS; Conceps and Technology of Flexible AC Transmission Systems", New York: IEEE Press, 2000. 1-425 p.ISBN 0-7803-3455-8.

[4] M. Goňo, M. Kyncl, R. Goňo, I. Kłosok-Bazan,"Experience with the production of electricity from biogas at sewage treatment plant in the Czech Republic", Przeglad Elektrotechniczny, Volume 89, Issue 11, 2013, Pages 12-15

[5] M. Eremia, M. Shahidehpour, "Handbook of Electrical Power System Dynamics; Modeling, Stability, and Control", New Jersey: IEEE Press, 2013. 291-724 p. ISBN 978-1-118-49717-3.

[6] T. Gönen, "Electrical Power Transmission System Engineering; Analysis and Design. 3. ed. Boca Raton: CRC Press, 2014. 140-176 p. ISBN 978-1-4822-3222-6.

[7] K. R. Padiyar, "HVDC Power Transmission System. 2. "ed. Kent: New Academic Science Ltd., 2011. 155-170 p. ISBN 978-1-906574-77-2.

[8] J. Kurimsky, R. Cimbala, I. Kolcunova, "Multi-scale decomposition for partial discharge analysis," Prz. Elektrotechniczny, vol. 84, no. 9, pp. 191–195, 2008.

[9] M. Mikita, M. Kolcun, M Špes M. Vojtek, M. Ivančák, " Impact of electrical power load time management at sizing and cost of hybrid renewable power system" (2017) *Polish Journal of Management Studies*, 15, pp. 154-162. DOI: 10.17512/pjms.2017.15.1.15

[10] J. Rusnak, L. Bena, M. Kolcun,"Experiences with Examination of Dispersed Sources Impact on Distribution Power System" 11th International Scientific Conference Electric Power Engineering 2010, Pages: 219-222

[11] S. Kusnír, L. Bena, M. Kolcun, "The impact of FACTS devices to control the load flow" In: Electric Power Engineering 2010 : proceedings of the 11th international scientific conference : 4.-6.5.2010, Brno, Czech Republic. 2010 P. 99-103. ISBN 978-80-214-4094-4

[12] J. Zhu," Optimization of Power System Operation. 2. ed. New Jersey: Wiley-IEEE Press, 2015. 13-576 p. ISBN: 978-1-118-85415-0

[13] R. Gono, S. Rusek, M. Kratky, M.Slivka,"Component reliability parameters of distribution network", Proceedings of the 8th International Scientific Symposium on Electrical Power Engineering, ELEKTROENERGETIKA 2015, pp. 364-367.

[14] P.Fedor, D. Perduková,"Fuzzy Model Based Optimal Continuous Line Controller" In: Proc. of the 8th Int. Scientific Symposium on Electrical Power Engineering ELEKTROENERGETIKA 2015. Stará Lesná, 2015, pp. 404-407. ISBN 978-80-553-2187-5

[15] M. Kolcun, K. Rusek"Analysis of prices for electricity at the Polish power exchange [Analiza cen energii elektrycznej na towarowej giełdzie energii] (2018) Polish Journal of Management Studies, 17, pp. 155-164. DOI: 10.17512/pjms.2018.17.1

[16] D. Medved, M. Klesc,"Modeling of Electrical off-grid Network in the Simscape Power Systems", Proceedings of the 9th international scientific symposium on electrical power engineering, pp: 465-470, 2017, ISBN 978-80-553-3195-9 Available from: http://toc.proceedings.com/35784webtoc.pdf